Provided as a reference for the "Viewing Society in Global Perspective" maps that appear throughout the text

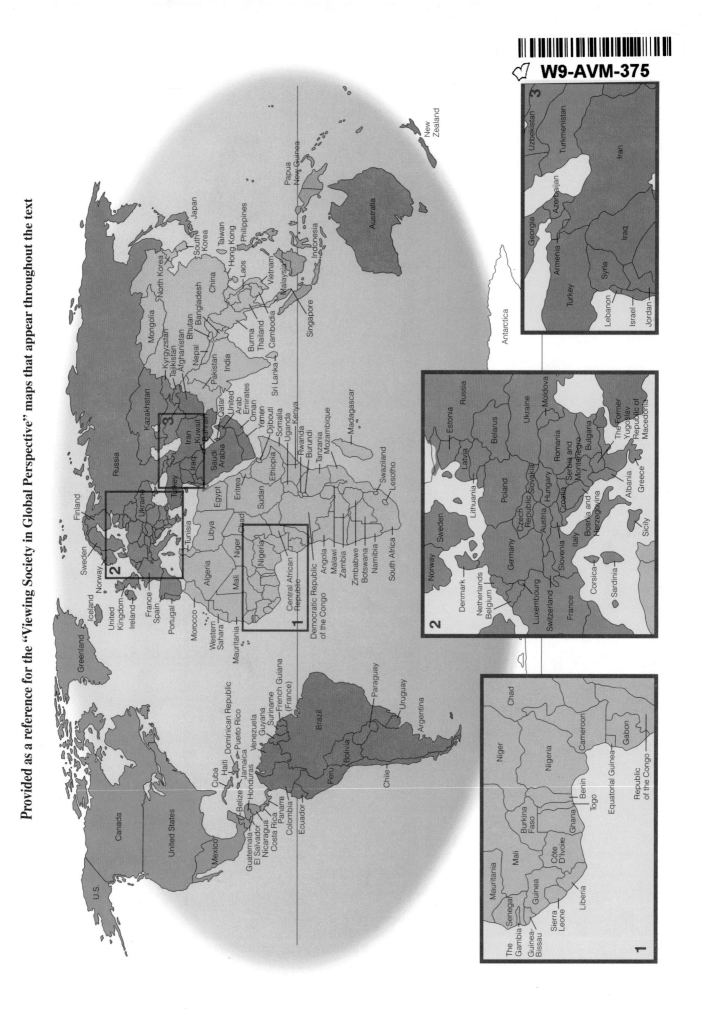

Sociology

Understanding a Diverse Society

FOURTH EDITION

Margaret L. Andersen
University of Delaware

Howard F. Taylor
Princeton University

THOMSON

WADSWORTH

Australia • Canada • Mexico • Singapore • Spain • United Kingdom • United States

THOMSON

WADSWORTH

Sociology Editor: *Robert Jucha*
Development Editor: *Julie Sakaue*
Assistant Editor: *Elise Smith*
Editorial Assistant: *Christina Cha*
Technology Project Manager: *Dee Dee Zobian*
Marketing Manager: *Wendy Gordon*
Marketing Communications Manager: *Linda Yip*
Project Manager, Editorial Production: *Cheri Palmer*
Creative Director: *Rob Hugel*
Print Buyer: *Karen Hunt*

Permissions Editor: *Sarah Harkrader*
Production Service: *Dusty Friedman, The Book Company*
Text Designer: *Lisa Buckley*
Photo Researcher: *Myrna Engler*
Copy Editor: *Yonie Overton*
Illustrator: *Impact Publications*
Cover Designer: *Yvo Riezebos*
Cover Printer: *Phoenix Color Corp*
Compositor: *Thompson Type*
Printer: *R.R. Donnelley/Willard*

This cover and title painting (Passionate Fantasy) and the details of paintings used in the interior are from the work of Hessam Abrishami, whose colorful art reflects the diversity of society. Hessam's passion for art began in his homeland of Iran when he was in the eighth grade. Later, as a teenager, he won two drawing contests, which encouraged him to pursue his passion and to obtain a master's degree in fine arts. His work is exhibited in Asia, Europe, and the United States.

Dedication

To Richard and Pat,
with love

For more information about our products, contact us at:
Thomson Learning Academic Resource Center
1-800-423-0563
For permission to use material from this text or product, submit a request online at **http://www.thomsonrights.com.**
Any additional questions about permissions can be submitted by email to **thomsonrights@thomson.com.**

Library of Congress Control Number: 2004096993

Student Edition: ISBN 0-534-61716-6

Instructor's Edition: ISBN 0-495-00060-4

International Student Edition: ISBN 0-495-00649-1
(Not for sale in the United States)

ISBN 0-495-00489-8

ISBN 0-495-00490-1

Thomson Higher Education
10 Davis Drive
Belmont, CA 94002-3098
USA

Asia (including India)
Thomson Learning
5 Shenton Way
#01-01 UIC Building
Singapore 068808

Australia/New Zealand
Thomson Learning Australia
102 Dodds Street
Southbank, Victoria 3006
Australia

Canada
Thomson Nelson
1120 Birchmount Road
Toronto, Ontario M1K 5G4
Canada

UK/Europe/Middle East/Africa
Thomson Learning
High Holborn House
50–51 Bedford Row
London WC1R 4LR
United Kingdom

Latin America
Thomson Learning
Seneca, 53
Colonia Polanco
11560 Mexico
D.F. Mexico

Spain (including Portugal)
Thomson Paraninfo
Calle Magallanes, 25
28015 Madrid, Spain

Brief Contents

Contents

Chapter 4
Socialization 81

Chapter 5
Social Interaction and Social Structure 111

Chapter 12
Gender 301

Chapter 13
Sexuality 333

Chapter 17
Religion 447

Chapter 18
Economy and Work 477

Chapter 22

Collective Behavior and Social Movements 589

Chapter 23

Social Change in Global Perspective 617

Boxes

A SOCIOLOGICAL EYE ON THE MEDIA

KEY SOCIOLOGICAL CONCEPTS 20

Maps

MAPPING AMERICA'S DIVERSITY

VIEWING SOCIETY IN GLOBAL PERSPECTIVE

Preface

Sociology: Understanding a Diverse Society, Fourth Edition, introduces students to the basic concepts and theories of sociology. We believe that the sociological perspective is useful to people in any walk of life. It is thus our intent to excite students about a sociological perspective, despite knowing that most students taking an introductory course will not become sociology majors. Even so, both of us have met countless people who, upon learning that we are sociologists, volunteer "Oh, sociology was one of my favorite subjects!" That is the appreciation of sociology we hope to inspire in the students who read this book.

Major Themes

Sociology: Understanding a Diverse Society focuses on several major themes:

Diversity

The study of diversity is central to our book. Unlike other introductory texts that add diversity onto a pre-existing approach to sociology, we see diversity as part of the fabric of society. As the title of our book suggests, diversity is central to how society is organized; how inequality is manifested in society, thus shaping the experiences of different groups; and how diversity is shaping (and is shaped by) contemporary social changes. Both of us are known for our scholarship on various aspects of diversity (especially race, gender, and class). Thus, our attention to diversity pervades the book and also is reflected in some of the book's special features (see especially the box, Understanding Diversity).

We define *diversity* to include the differences in experience created by social factors such as race, ethnicity, class, gender, age, religion, sexual orientation, and region of residence, to name only some, but we also see diversity as contributing to the rich texture of society through the diverse cultures and identities of different groups in society (see our working definition of diversity on page 11 of Chapter 1). We stress the positive aspects of a diverse society, as well as its problems. We do not think of diversity as just the study of victims, although clearly systems of disadvantage are part of society. But diversity also is reflected in the social and cultural values and contributions of different groups. We think this is just as important for students to recognize and understand. We are pleased that our thorough integration of diversity has led reviewers to comment that our book provides the most comprehensive coverage of diversity of any book on the market.

Current Theory and Research

According to those who have used and reviewed the book, our presentation of *current theory and research* is one of the strongest features of this text. We use the most current research throughout the book to show students the value of a sociological education. The new edition continues this strength, with updated research throughout and new areas of research also included. We know however that introductory students can get bogged down in the details of research if it is not crisply and succinctly presented. We think that, compared to other introductory books, our book has more depth in its coverage of research, while still being accessible and engaging to undergraduate students. The addition of MicroCase® Online research exercises in every chapter—exercises we have designed—strengthens the research focus of this edition and allows students to do their own "hands-on" investigations of national and international data on a variety of topics.

We want students to understand not just the concepts and procedures of research methods, but also how sociologists do their work and how the questions they ask are linked to the methods of inquiry that they use. Thus the feature, Doing Sociological Research, showcases different and specific research studies. Each box opens by identifying the question asked by the researcher, then describes the method of research, presents the findings and conclusions, and briefly discusses the implications. These boxes also show students the diverse ways that sociologists conduct research and thus feature the rich and varied context of the discipline.

We also want students to understand the contributions that different theoretical traditions offer. Thus, we present various theoretical perspectives in a balanced way. All chapters explore how alternative theoretical frameworks reveal different aspects of the particular subjects. We help students see this with the tables we include in every chapter; these tables concisely compare different theoretical viewpoints, showing how each illuminates certain questions and principles. We think this helps students understand an important point: Starting from a different set of assumptions can change how you interpret different social phenomena.

Critical Thinking and Debunking

Critical thinking is a term widely used but often vaguely defined. We use it to mean the process by which students learn to apply sociological concepts to observable events in society. Thus, throughout the book, we ask students to use sociological concepts to analyze and interpret the world they inhabit. This is reflected in the Thinking Sociologically feature that is present in every chapter.

We know that contemporary students are much influenced by the *mass media,* so we have added a feature to this edition that asks students to think more critically about the media. Consistent with our research focus, A Sociological Eye on the Media is grounded in actual sociological research on subjects that students find inherently interesting, such as eating disorders, racial stereotypes, and media monopolies. This feature examines how sociological research challenges some of the ideas and images portrayed in the media. This not only fosters students' critical thinking skills, but also shows them how research can debunk these ideas and images.

Our focus on critical thinking also is reflected in the critical questions in the Doing Sociological Research boxes. These Questions to Consider challenge students to think more about the research being presented.

Critical thinking also is linked to the debunking theme because it helps students see how a sociological perspective differs from common sense or taken-for-granted understandings about society and social issues. We use the theme of *debunking* in the way first developed by Peter Berger—to look behind the facades of

everyday life, challenging the ready-made assumptions that permeate commonsense thinking. We use the approach of debunking to help students understand how society is constructed and sustained. Throughout the book, we show how sociology can help them see society differently with a firm grounding in empirical research. The debunking theme is highlighted in the Debunking Society's Myths feature found in every chapter, but it also is embedded in how we present sociological ideas and research.

Social Action/Social Policy

We want students to understand the utility of the sociological perspective. We broadly define social policy to include how sociological research and theory can be applied to social issues. This can include formal policies (such as federal and state legislation) but also can include practices that show how social action makes a difference in people's lives. We want students to understand the social forces that shape people's lives, but we also want them to understand that society changes as the result of what people do. Our feature, Taking on Social Issues reflects this idea. This box asks students to apply what they are learning in sociology to analyze current social issues (such as gay marriage, national security and civil liberties, or global epidemics).

The Taking on Social Issues boxes include a component called "Taking Action," directing students to our companion website. Here we provide links to organizations working on specific policy issues, as well as critical thinking questions that ask students to interpret different sociological features of these different action groups. This feature serves several purposes: It helps students identify groups that are organized around different policy and social change issues; it enriches critical thinking by asking students to analyze the structure, goals, strategies, and constituents of different action groups; and it connects sociological research to the activities of organized groups, thus letting students see how sociological knowledge can be applied. This feature has the added benefit of directing students to organizations where they can potentially pursue internships, service learning projects, or possibly careers. Together, Taking on Social Issues and Taking Action show students the different ways that sociological research and theory can be applied in "real" life.

The box feature Sociology in Practice also shows students how sociologists have used their knowledge in concrete, applied ways. This helps students see not only how sociology matters in social policy and social action, but also how people can make a difference in society—even when the social forces of society are strong and sometimes seem immutable to students once they understand the sociological perspective. In other words, we want students to understand the power of social forces but not become immobilized by this understanding.

Social Change

The sociological perspective helps students see society as characterized both by constant change and social stability. The box Forces of Social Change allows students to reflect on how certain social realities have evolved by including a historical dimension. By bringing a historical perspective to bear on sociological thinking, students can better see some of the social forces that operate in the present. The social change theme is especially significant in light of the events of September 11, 2001. What effect have these events had on society? This is a question we address in various discussions throughout the book. We also include a box in Chapter 23 on the devastating *tsunami* waves that struck several Southeast Asian and East African countries in December 2004. The social changes caused by such a disaster are likely to be extensive.

Global Perspective

Diversity includes the growing global character of society. The United States is increasingly being changed by *globalization*. We use a global perspective to examine how global changes are affecting all parts of life within the United States as well as other parts of the world. This means more than including cross-cultural comparisons. It means, for example, examining such phenomena as migration and immigration, the formation of world cities, the increasing cultural diversity found within the United States, and the impact of a global economy on work within this country. The global perspective within this book is found in the research and examples cited throughout, as well as in various chapters that focus directly on the influence of globalization on particular topics, such as work, culture, and crime. The map feature Viewing Society in Global Perspective found in virtually every chapter also brings a global perspective to the subject matter under study.

New to the Fourth Edition

One of the reasons that sociology is so fascinating is that its subject matter constantly evolves as society itself changes. This makes writing new editions of our book particularly exciting. Although the core ideas of sociology remain important, new topics, concepts, and research findings emerge quickly. We have tried to capture this vitality in the new edition. Therefore, the Fourth Edition includes new material on subjects such as media monopolies (Chapter 3), ethnocentrism (Chapter 3), corporate wrongdoing (Chapters 7 and 8), gay marriage (Chapter 15), educational reform (Chapter 16), and war and terrorism (Chapters 22 and 23), to name a few. We also have increased our coverage of the media in this edition.

There are also several new features and themes that we have added in this edition. Every chapter now includes a MicroCase Online research exercise. We developed these ourselves so that students can practice doing actual sociological research on their own using this free easy-to-use online feature. For each MicroCase exercise, we have developed a research question, provided instructions on analyzing the data they find, and developed questions to help students interpret and explain their findings.

We also have enhanced the map program in the book. The U.S. maps now include many county-level maps to provide a closer, more detailed examination of the topics at hand. In addition, we have expanded our captions to link the maps more directly to the text material and to give students more guidance in interpreting the maps from a sociological perspective. Other major revisions in the organization and content of the book follow.

The Organization of the Book

Sociology: Understanding a Diverse Society is organized in five major parts: "Introducing the Sociological Perspective" (Chapters 1 and 2); "Individuals in Society" (Chapters 3 through 8); "Social Inequalities" (Chapters 9 through 14); "Social Institutions" (Chapters 15 through 20); and "Social Change" (Chapters 21 through 23). **Part I** introduces students to the unique perspective of sociology, differentiating it from other ways of studying society, such as the individualistic framework students tend to assume. This section includes a chapter on the methods of sociological research.

In **Part II** ("Individuals in Society"), students learn some of the core concepts of sociology. It begins with the study of culture (Chapter 3) and socialization (Chapter 4), followed by chapters on social interaction and social structure (Chapter 5), then groups and organizations (Chapter 6). We include separate chapters on deviance (Chapter 7) and crime and criminal justice (Chapter 8) because we think deviance is not just about crime and because some instructors want more information on criminal justice topics.

Part III ("Social Inequalities") focuses on the different sources of social stratification. This section includes separate chapters on social class and social stratification, global stratification, race and ethnicity, gender, sexuality, and age and aging. In this edition, we have located the chapter on sexuality in this section rather than in Part II. Sexuality is increasingly being recognized in sociological research as not just a matter of interpersonal relationships, but also as one of the sources of inequality in society. The study of sexuality from a sociological perspective is also relatively new—and growing—and we also wanted to give this topic the rich sociological focus it deserves.

Part IV ("Social Institutions") of the book examines the different social institutions that compose society. These chapters explore the structure of social institutions and the different experiences groups have in them. We also examine how different theoretical perspectives within sociology help us interpret the structure of social institutions. The institutions we include are families, education, religion, the economy and work, politics and government, and health care.

Finally, Part V ("Social Change") focuses on the theme of social change, with three chapters included. In this section, students examine changes that come from long-term processes, such as population change, but they also see how the mobilization of collective action, such as in social movements, produces change. The different topics included here are population, urbanization, and environment, collective behavior and social movements, and a chapter focusing entirely on social change in global perspective. Detailed in the next section are the specific revisions and themes found in each of the individual chapters of the book.

Chapter-by-Chapter Changes

Part I. Introducing the Sociological Perspective

Chapter 1 ("Developing a Sociological Perspective") introduces students to the basics of the sociological perspective. The chapter briefly reviews the development of sociology as a discipline, with a focus on the classical frameworks of sociological theory. Incorporated are the contributions of people sometimes excluded from the history of sociological thought, such as W. E. B. DuBois and Jane Addams. New to this chapter is a section on why a sociological perspective is helpful to students even if they are not sociology majors. We also illustrate Mills's concepts of "troubles and issues" by exploring research on beauty and appearance—a subject we think students find particularly interesting. In this edition, we also have added more material on Erving Goffman and the dramaturgical model.

Chapter 2 ("Doing Sociological Research") introduces students to the basics of sociological research methods. We now introduce the chapter with Mitchell Duneier's intriguing study, "Sidewalk," in which he ex-

amines the social organization of street vendors. There is now a stronger focus in this chapter on qualitative research, thereby further balancing the presentation of both quantitative and qualitative research methods. We also have included a new diagram of how sociological research proceeds (see Figure 2.1, a drawing designed to capture the basic elements of research) and have expanded the discussion of inductive and deductive reasoning.

Part II. Individuals in Society

Chapter 3 ("Culture") now includes new material on media monopolies and the enormous influence that the media have on contemporary culture. One way we capture this is through the new discussion of media blackouts—a point included to help students see the enormous influence of the media in their lives. We also have included more on gays and lesbians in the media. Because of the significance of world events, we have included more material on ethnocentrism, including discussion of cultural differences between the United States and nations such as Iraq and Iran—subjects we think students need to know more about. In addition, we have included a new section on symbolic interaction and its interpretation of the sociology of culture.

Chapter 4 ("Socialization") points out the significance of socialization, including attention to the influences of the media, family, sports, and other agents of socialization. We have added new material on eating disorders and gender and self-esteem. There is now more on functionalism and conflict theory. We have reorganized the chapter so that the section on theory follows the section on agents of socialization. We think this improves the flow of the chapter. Also included is new material on social identity and on transitions over the life course.

Chapter 5 ("Social Interaction and Social Structure") has been reorganized to begin with a micro-sociological perspective and move into a macrosociological perspective. This introduces students to the nuances in the study of social interaction, moving on to the complex structure of society and social institutions. New to this chapter are updated discussions of social construction, ethnomethodology, nonverbal communication, and interpersonal attraction, exchange theory, game theory, and interaction in cyberspace.

In **Chapter 6** ("Groups and Organizations"), we study social groups and formal organizations, using sociology to understand the complex processes of group influence, organizational dynamics, and the bureaucratization of society. This chapter also includes material on gender, ethnicity, social class in organizations, the "McDonaldization" of society, and consumerism. There are new discussions of social networks and the "six degrees of separation" controversy, the behavior of guards and Iraqi prisoners at Abu Ghraib Prison, and a new section on leadership in groups and organizations.

Chapter 7 ("Deviance") includes a look at the sociological theories of deviance. The chapter has been somewhat reorganized by putting psychological explanations of deviance in the first section in order to better contrast them with sociological theories of deviance and by giving better coverage of the medicalization of deviance. There is a new figure (Figure 7.2) on Merton's structural strain theory to complement the text coverage.

Chapter 8 ("Crime and Criminal Justice") examines types of crimes; courts and the criminal justice system; the role of race, class and gender in crime; and international aspects of crime. New to this chapter is more material on corporate crime, including the various corporate scandals, such as Enron, that have made international news. There is also new material on the question of decriminalizing drugs, on international terrorism, and consistent with our focus on the media, on how the media influence public perceptions of crime.

Part III. Social Inequalities Chapter 9 ("Social Class and Social Stratification") provides an overview of basic concepts central to the study of social stratification, as well as current research on class inequality, the impact of welfare reform, the Black middle class, and the rising levels of debt and insecurity among the middle class. This revised chapter includes the very insightful "double diamond" model of class structure provided by Robert Perrucci and Earl Wysong. There is also new material on how the media perpetuate class stereotypes. Data on class inequality, including rising rates of poverty, have been updated throughout.

Chapter 10 ("Global Stratification") follows with a particular emphasis on understanding the significance of global stratification—the inequality that has developed among, as well as within, different nations throughout the world. Throughout this text, we see globalization as a process transforming many societies, including the United States. Here we examine global events and processes, including the growth of world cities, transnational migration, world poverty, and the use of sweatshop labor—both abroad and in the United States.

Chapter 11 ("Race and Ethnicity") is a comprehensive review of the significance of race and ethnicity in society. Although these concepts are integrated throughout the book, because of our focus on diversity, they also require particular attention to learn how race and ethnicity differentiate the experiences of diverse groups in society. Included here are topics such as multiracial classification in the U.S. census and an expanded treatment of racism, including the distinctions made among different forms of racism (traditional, aversive, laissez-

faire, color-blind, and institutional). There are updates on defining "race" and an extended section on affirmative action.

Chapter 12 ("Gender") focuses on gender as a central concept in sociology—one that is closely linked to systems of stratification in society. There is more material in this chapter on balancing work and family, as well as an expanded discussion of homophobia and its connection to gender socialization.

Chapter 13 ("Sexuality"), now moved from the section on individuals to the section on inequality, has been rewritten to emphasize sexuality as a matter of inequality. The chapter also has been substantially reorganized for better flow. The emphasis in this chapter is on the social construction of sexuality and the inequality that results from sexual orientation. Throughout, the connection between sex and social structure is emphasized, with new research and theory on topics such as sexual identity, reproductive issues, and queer theory.

Chapter 14 ("Age and Aging") examines age as a dimension of stratification, including age stereotyping, age and social structure, aging and diversity, age prejudice and discrimination, and influences over the life course. This chapter has been reorganized to emphasize the social construction of age. The chapter has an expanded discussion of death and dying, including more on euthanasia. There is new research included on Social Security, including explaining to students how this system works. There is also new research on retirement, especially since retirement earnings are influenced by race and gender. The Fourth Edition includes more cross-cultural discussion of aging, new research on aging and mental health, and a new box on privatizing Social Security.

Part IV. Social Institutions Chapter 15 ("Families") includes new material on fatherhood and more material on the current same-sex marriage debate, including new research on gay and lesbian marriage. There is a new and expanded discussion of various forms of care work—an expanding area of sociological research. Cohabitation has been given its own section, to separate it from the discussion of singles. Finally, there is a new box on "hooking up"—a common form of social interaction among young college students.

Chapter 16 ("Education") examines the institution of education. The primary focus of this chapter is on educational inequality, with a thorough discussion of the effects of social class on mobility, a global view of education and mobility, tracking, and the shortchanging of girls and women by the educational system. Also included is new research on how stereotypes affect academic performance. The section on school reform has been significantly revised to give attention to research on high-stakes testing, school voucher programs, and charter schools. Finally, new material has been added on home schooling and the *No Child Left Behind Act*.

Chapter 17 ("Religion") reviews such topics as the social correlates of religion, the changing significance of religion in society, and diverse patterns of religiosity. There also is a discussion of religious cults. Because religion is a topic of increasing concern and attention, this chapter includes more material on religion as a source of current world conflicts, including a new section on religious extremism and research on the consequences of intertwining state politics and religion. We have drawn new examples from such issues as the controversy over ordaining gay priests, and there is a fuller discussion of the global surge in religious fundamentalism, including Christianity. At the reviewers' suggestion we have clarified that Protestantism and Catholicism are both forms of Christianity and we have added new material on Asian Americans and religion.

Chapter 18 ("Economy and Work") compares different economic systems and analyzes contemporary patterns in the social organization of work. A theme in the chapter is the impact of economic restructuring, including global restructuring, on work in the United States. Data on work have been updated throughout, including new material on declining job benefits and the impact of a "24/7" economy.

Chapter 19 ("Government and Politics") is framed by a discussion of power and authority as well as the structure of government institutions. The chapter has been updated to include the results of the 2004 election. We have added a discussion of the iron law of oligarchy, monarchy, and dictatorships. We illustrate these concepts with cross-cultural examples, such as Iraq. We also include new material on the Iraq war, including important data on race, social class, and the casualty rate in the war on Iraq. And, there is a new box on politics and the media.

Chapter 20 ("Health Care") details the social organization of health care. This chapter has been completely reorganized and is framed by a new chapter opener focusing on the current crisis in health care coverage. There is new material on the international AIDS crises and a much stronger focus on current issues in health care, such as corporate control of health care, prescription drug costs, and the patients' bill of rights. We have added a section on illnesses that will be of special interest to students, including eating disorders, AIDS, smoking, and obesity. There is added material on specialization and bureaucratization in the health care industry and, finally, new research on disability.

Part V. Social Change **Chapter 21** ("Population, Urbanization, and Environment") connects the study of demography to the phenomenon of growing urbanization in society, including information on minorities in suburbia, and increasing environmental problems. In addition, it reviews the subjects of population growth, pollution, and environmental racism.

Chapter 22 ("Collective Behavior and Social Movements") focuses on how people mobilize to effect social change. Since 9/11, collective behavior and social movements have become an even more significant area of study in sociology. Thus, this chapter will be of even more interest to students, especially because they have likely observed increased social movement activity in society. There is extensive discussion in this chapter of terrorism as a transnational social movement, a concept we introduced in the last edition. We also have added material on Smelser's structural strain theory of social movements.

Chapter 23 ("Social Change in Global Perspective") looks at the broad dynamics of social change, emphasizing the broad patterns of change associated with modernization, globalization, and technological development. The discussions of theories of social change have been streamlined somewhat. This chapter now includes a section on revolution (originally found in the chapter on politics and government), and it includes an expanded section on terrorism and war. There is new attention to the media as a source of change. The chapter includes a new Forces of Social Change box discussing potential social changes for societies struck by the tsunamis in 2004.

Features and Pedagogical Aids

The special features of this book flow from its major themes: diversity, current research and theory, social action and social policy, social change, and a global perspective. The features are designed to help students develop critical thinking skills so that they can apply abstract concepts to observed experiences in their everyday lives and learn how to interpret different theoretical paradigms and approaches to sociological research questions. The following features accomplish these goals in various ways.

Fostering Critical Thinking Skills

The feature Thinking Sociologically takes concepts from each chapter and asks students to think about them in relationship to something they can easily observe in an exercise or class discussion. The feature Debunking

Society's Myths takes certain common assumptions that are taken for granted and shows students how the sociological perspective would inform such assumptions and beliefs. In addition, many other box features, such as Sociology in Practice—include questions designed to foster critical thinking skills.

Unparalleled Integration of Web-Based Resources

Instructors will find that several technology-based teaching enhancements are integrated throughout the text, making this book the best conceived in using the tools of the Internet for teaching and learning introductory sociology. At the end of each chapter, for instance, is a list of stable Web-based resources that students and faculty can use to explore data and information pertinent to the chapter topic. We have deliberately selected sites we know are stable and provide the latest information or resources on a given subject.

The new MicroCase feature developed in this section will be especially valuable for faculty who want to expose students to actual sociological research. Following the Key Terms at the end of each chapter, we have developed a MicroCase Online exercise that explores some dimensions of the chapter's subject. These research exercises are simple to do and we think will help students discover the excitement of doing sociological research.

Each chapter includes a feature, Taking on Social Issues, which is designed to help students explore current public issues and demonstrate how the sociological perspective can provide different viewpoints about such issues. The Taking Action component that we have added to this box integrates Web-based instruction by giving instructors and students the option of linking to the website of an organization that works specifically on the topic under discussion. Critical thinking questions about each organization allow students to use sociological analysis to consider the structure and goals of each organization. This feature also shows students how sociology can be used to make a difference in the world and has the added benefit of pointing them to organizations where they might pursue careers.

An additional tool for using the Internet in sociology research is InfoTrac® College Edition, a powerful online library providing access to the articles contained in thousands of periodicals and journals. The InfoTrac College Edition tool is integrated into the Doing Sociological Research feature to allow students to further explore research on the topics presented in each of the boxes.

An Extensive and Content-Rich Map Program

Our book includes two map features in almost every chapter: One map feature is called "Mapping America's Diversity" and the other, "Viewing Society in Global Perspective." These maps have multiple instructional applications, not the least of which is instructing students about world and national geography. The maps have been designed primarily to present country, state, or county data—with an expanded emphasis in this edition on county-based data—depending on the question being explored and the availability of map data. These maps help students visualize regional differentiation in such areas as access to technology (see Map 10.2 on the global digital divide) and the distribution of racial and ethnic minorities (Map 1.1) and of the aged population (Map 14.1) in different U.S. counties. We also have amplified the map captions in this edition to provide more detailed information and questions to help students use a sociological perspective to analyze the data they see in the maps. For example, Map 18.2 shows the percent of disabled persons by county in the United States. The caption asks several questions, such as how regional differences in the percent disabled might affect the needs of the labor force in those parts of the country. To take another example, Map 14.1 in the chapter on aging shows the percent of the U.S. population that is age 65 or older by county and state. From this it is seen that older persons are more concentrated in some areas than in others, thus having implications for the kinds of social services needed in different regions.

High-Interest Theme Boxes

We use six high-interest box themes to embellish our focus on diversity and sociological research throughout the text. Understanding Diversity explores the approach to diversity taken throughout the book. In most cases, these boxes provide personal narratives or other information designed to teach students about the experiences of different groups in society. Because many are written as first-person narratives, they can invoke student empathy toward groups other than those to which they belong—something we think is critical to teaching about diversity. We hope to show students the connections between different race, class, and other social groups that they otherwise may find difficult to grasp.

Doing Sociological Research boxes are intended to show students the diversity of research questions that form the basis of sociological knowledge and, equally important, how the question a researcher asks influences the method used to investigate the question. We see this as an important part of sociological research—that how one investigates a question is determined as much by the nature of the question as by allegiance to a particular methodological strategy. Some questions require a more qualitative approach, others a more quantitative approach. In developing these boxes, we ask, What is the central question sociologists are asking? How did they explore this question using sociological research methods? What did they find? and What are the implications of this research?

We deliberately selected questions that show the full and diverse range of sociological theories and research methods, as well as the diversity of sociologists themselves. These boxes also include Questions to Consider at the end of each box to ask students to think further about the implications and applications of the research. And we have tied the InfoTrac College Edition terms more closely to the questions to assist those students who are interested in following up on the questions posed.

Our box feature Sociology in Practice is designed to show the application of the sociological perspective in different contexts. Thus we show examples where sociologists have testified before Congress, advised presidents, changed an organization, or made a difference in society in some way that students can identify. Related to this box is our feature Taking on Social Issues, which enables students to use the sociological perspective they are acquiring to explore contemporary social issues and to see how some groups have organized to respond to these issues.

The feature Forces of Social Change highlights some of the major changes currently affecting society, such as the influence of participation on the socialization of women (Chapter 4), the influence of consumerism on religion (Chapter 17), and the phenomenon of "hooking up" as a new form of social relations on college campuses (Chapter 15), to name a few.

Finally, new to this edition is a feature entitled A Sociological Eye on the Media. This box is intended to help students think more critically about the influence of the mass media in many areas of society. Because of the enormous power of the media, we think this is increasingly important in educating students about sociology. Each box raises a question about the media's influences that is relevant to the chapter topic and then informs students, based on contemporary research, about how sociologists have investigated this topic. We think students will find all these boxes readable and interesting, all the while encouraging them to foster a more critical eye on the media.

In-Text Learning Aids

In addition to the features just described, there is an entire set of learning aids within each chapter that promote student mastery of the sociological concepts.

Chapter Outlines A concise chapter outline at the beginning of each chapter provides students with an overview of the major topics to be covered.

Theory Tables Each chapter includes a table that summarizes different theoretical perspectives, comparing and contrasting how these theories illuminate different aspects of a variety of subjects.

Researching Society with MicroCase Online At the end of each chapter, we have included new exercises that allow students to experience computerized statistical data analysis for free using MicroCase Online. Detailed instructions are provided for each exercise, plus questions to help students analyze the data they find. Students can easily access MicroCase Online via the book's Companion Website.

SociologyNow Icons These icons appear at the beginning of the chapter and in the Chapter Summary to remind students to take advantage of SociologyNow, a helpful study and review tool that instructors may select to bundle for free with new copies of the text.

Chapter Summary in Question-and-Answer Format Questions and answers that highlight major points in each chapter provide a chapter summary and a quick review of major concepts and themes covered in the chapter.

Key Terms Major terms and concepts are in bold type when first introduced in the chapter. A list of the key terms is found at the end of each chapter with page references to their definitions to help make students' studying more effective. Definitions of the key terms also may be found in the Glossary.

Suggested Readings and Web Resources An annotated list of suggested readings and Web resources is included at the end of each chapter as a source for further study.

A Glossary All key terms from each chapter, and a complete **Bibliography** for the entire text is found at the back of the book.

Supplements

Sociology: Understanding a Diverse Society, Fourth Edition, is accompanied by a wide array of supplements prepared to create the best learning environment inside as well as outside the classroom for both the instructor and the student. All the continuing supplements for *Sociology: Understanding a Diverse Society,* Fourth Edition, have been thoroughly revised and updated, and several are new to this edition. We invite you to take full advantage of the teaching and learning tools available to you.

Supplements for the Instructor

Instructor's Resource Manual This supplement offers the instructor brief chapter outlines, student learning objectives, extensive chapter lecture outlines, lecture/discussion suggestions, student activities, chapter worksheets, suggested resources, questions for discussion, Internet activities, InfoTrac College Edition exercises, and creative lecture and teaching suggestions. Also included is a list of additional print, video, and online resources, including a table of contents for the CNN® Today Sociology Video Series and concise user guides for both InfoTrac College Edition and WebTutor™.

Test Bank This test bank consists of 75–100 multiple-choice questions and 20–30 true/false questions for each chapter of the text, all with answer explanations and page references to the text. Each multiple-choice item has the question type (fact, concept, or concept application) indicated. Also included are 10–15 short-answer and 5–10 essay questions for each chapter.

ExamView Computerized Testing for Macintosh and Windows Create, deliver, and customize printed and online tests and study guides in minutes with this easy-to-use assessment and tutorial system. ExamView includes a Quick Test Wizard and an Online Test Wizard to guide instructors step by step through the process of creating tests. The test appears on screen exactly as it will print or display online. Using ExamView's complete word processing capabilities, instructors can enter an unlimited number of new questions or edit questions included with ExamView.

Classroom Presentation Tools for the Instructor

JoinIn™ on TurningPoint® Transform your lecture into an interactive student experience with JoinIn. Combined with your choice of keypad systems, JoinIn turns your Microsoft® PowerPoint® application into audience response software. With a click on a handheld device, students can respond to multiple-choice questions, short polls, interactive exercises, and peer-review questions. You also can take attendance, check student comprehension of concepts, collect student demographics to better assess student needs, and even administer quizzes. In addition, there are interactive, text-specific slide sets that you can modify and merge with any of your own PowerPoint lecture slides. This tool is available to qualified adopters at **http://turningpoint .thomsonlearningconnections.com.**

MultiMedia Manager Instructor Resource CD: A 2005 MicrosoftPowerPoint Link Tool With this one-stop digital library and presentation tool, instructors can assemble, edit, and present custom lectures with ease. The MultiMedia Manager contains figures, tables, graphs, and maps from this text, pre-assembled PowerPoint lecture slides, video clips from CNN and DALLAS TeleLearning, ShowCase presentational software, tips for teaching, the instructor's manual, and more.

Videos Adopters of *Sociology: Understanding a Diverse Society,* Fourth Edition, have several different video options available with the text. Please consult with your Thomson Learning sales representative to determine if you are a qualified adopter for a particular video.

Wadsworth's Lecture Launchers for Introductory Sociology and Sociology: Core Concepts Exclusive offerings created jointly by Wadsworth/ Thomson Learning and DALLAS TeleLearning, these products present video highlights taken from the *Exploring Society: An Introduction to Sociology* telecourse (formerly the *Sociological Imagination*). The Lecture Launcher product offers 3–6-minute video segments and the Core Concepts product offers 15–20-minute video segments. The selections have been chosen to enhance and enliven class lectures and initiate discussion of key topics covered in any introductory sociology text. The videos cover topics such as the sociological imagination, stratification, race and ethnic relations, and social change. Brief descriptions accompany each product, along with suggested discussion questions to help effectively incorporate the material into the classroom. They are available on VHS or DVD.

CNN Today Sociology Video Series, Volumes I–VIII Illustrate the relevance of sociology to everyday life with this exclusive series of videos for the introduction to sociology course. Jointly created by Wadsworth and CNN, each video consists of approximately 45 minutes of footage originally broadcast on CNN and specifically selected to illustrate important sociological concepts.

Wadsworth Sociology Video Library Bring sociological concepts to life with videos from Wadsworth's Sociology Video Library, which includes thought-provoking offerings from Films for Humanities, as well as other excellent educational video sources. This extensive collection illustrates important sociological concepts covered in many sociology courses.

Supplements for the Student

SociologyNow This online tool provides students with a customized study plan based on a diagnostic "pretest" they take after reading each chapter. The study plan provides interactive exercises, videos, and other resources to help students master the material. After the study plan has been reviewed, students can take a "posttest" to monitor their progress in mastering the chapter concepts. Instructors may bundle this product for their students with each new copy of the text for free!

Study Guide This student study tool contains both brief and detailed chapter outlines, brief chapter summaries, a list of key terms and key people with page references to the text, questions to guide student reading, Internet and InfoTrac College Edition exercises, and practice tests consisting of 20–25 multiple-choice questions, 10–15 true-false questions, 3–5 fill-in-the-blank and short-answer questions, and 3–5 essay questions. All multiple-choice, true/false, and fill-in-the-blank and short-answer questions include answer explanations and page references to the text.

Practice Tests Designed to help students test their knowledge of chapter concepts, this booklet contains 50–60 multiple-choice questions, 10–20 true/false questions, and 4–6 short-answer questions, all with answer explanations and page references, for each chapter of the text.

Wadsworth's Sociology Online Resources and Writing Companion, *First Edition* This valuable guide shows students how they can use Wadsworth's exclusive online resources—InfoTrac College Edition, the Opposing Viewpoints Resource Center (OVRC), and MicroCase Online—to assist them in their study of sociology and to build essential research and writing skills. Part One provides informative user guides that introduce each of these powerful research tools. Part Two contains directed exercises designed to develop research and critical thinking proficiency for each of the core topics in sociology. Part Three provides an overview of some of the research and writing tools available online, such as *InfoWrite* and the *OVRC Research Guide,* and shows students how they can effectively integrate their research findings into class assignments.

Understanding Society: An Introductory Reader Edited by Margaret Andersen, Kim Logio, and Howard Taylor, this reader complements the *Sociology: Understanding a Diverse Society,* Fourth Edition, text. It includes articles with a variety of styles and perspectives, with a balance of the classic and contemporary. The editors selected readings that students would find accessible yet intriguing and that maximize the instructional value of each selection by prefacing each with an introduction and following each with discussion questions. The articles center on the following five themes: classical sociological theory, contemporary research, diversity, globalization, and the application of the sociological perspective.

Current Perspectives: Readings From InfoTrac College Edition: Introductory Sociology Research Updates, First Edition Hand-selected by subject-area experts to correlate with each major topic area of an introductory course in sociology, the articles in this new reader are drawn from InfoTrac College Edition's vast database of full-length, peer-reviewed articles from more than 5000 top academic journals, newsletters, and periodicals. Ideal to supplement your main textbook, each reading raises a significant issue that can enhance students' understanding of the field, and each is accompanied by questions that can be assigned as homework or used to promote critical thinking and group discussion.

Internet-Based Supplements

InfoTrac College Edition with InfoMarks Available as a free option with newly purchased texts, InfoTrac College Edition gives instructors and students four months of free access to an extensive online database of reliable, full-length articles (not just abstracts) from thousands of scholarly and popular publications going back as much as 22 years. Among the journals available 24/7 are *American Journal of Sociology, Social Forces, Social Research,* and *Sociology.* InfoTrac College Edition now also comes with InfoMark, a tool that allows you to save your search parameters, as well as save links to specific articles. (Available to North American college and university students only; journals are subject to change.)

WebTutor Advantage on WebCT and Blackboard This web-based software for students and instructors takes a course beyond the classroom to an anywhere, anytime environment. Students gain access to a full array of study tools, including chapter outlines, chapter-specific quizzing material, interactive games and maps, and videos. With WebTutor Advantage, instructors can provide virtual office hours, post syllabi, track student progress with the quizzing material, and even customize the content to suit their needs.

Virtual Society: The Wadsworth Sociology Resource Center at http://sociology.wadsworth .com Combine this text with Virtual Society's exciting range of Web resources, and you will have truly integrated technology into your learning system. Virtual Society provides instructors and students with a wealth of FREE information and resources, such as Sociology in Action, Census 2000: A Student Guide for Sociology, Research Online, a Sociology Timeline, a Spanish glossary of key sociological terms and concepts, and more.

Companion Website for Sociology: Understanding a Diverse Society, *Fourth Edition, at http://sociology.wadsworth.com/andersen_ taylor4e/* The book's companion site includes chapter-specific resources for instructors and students. For instructors, the site offers a password-protected instructor's manual, PowerPoint presentation slides, and more. For students, there is a multitude of text-specific study aids, including the following: tutorial practice quizzes that can be scored and emailed to the instructor, Web links, InfoTrac College Edition exercises, flash cards, MicroCase Online data exercises, CNN Video exercises, crossword puzzles, Virtual Explorations.

And much more!

Acknowledgments

We relied on the comments of many reviewers to improve the book, and we thank them for the time they gave in developing very thoughtful commentaries on the different chapters. Thanks to

Deborah Carr, Rutgers University
Vanessa Eslinger-Brown, Strayer University
David Gauss, San Diego State University

Michael Granata, Community College Southern Nevada
Joanna Claire Grey, Pikes Peak Community College
Jianjun Ji, University of Wisconsin–Eau Claire
Virginia S. Mulle, University of Alaska Southeast
Lesley Williams Reid, Georgia State University
Richard Sweeney, Modesto Junior College
S. Rowan Wolf, Portland Community College

We also would like to thank the many reviewers who have provided valuable commentary as we worked on the first three editions: Jan Abushakrah, Portland Community College; Peter Adler, University of Colorado; Susan Albee, University of Northern Iowa; Angelo Alonzo, The Ohio State University; Elena Bastida, University of Texas–Pan American; Jane Bock, University of Wisconsin at Green Bay; Rebecca Brooks, Kent State University; Valerie Brow, Cuyahoga Community College East; Phil Brown, Brown University; Russell Buenteo, University of South Florida; Jeffery Burr, State University of New York at Buffalo; Cynthia C. Calhoun, State Technical Institute at Memphis; Kathy Dennick-Brecht, Robert Morris College; Marlese Durr, Wright State University; John Ehle, Northern Virginia Community College; Charles F. Emmons, Gettysburg College; Jess Enns, University of Nebraska–Kerry; Kevin Everett, Radford University; Grant Farr, Portland State University; Kenneth Fidel, DePaul University; Michael Goslin, Tallahassee Community College; Andrea R. Greenberg, University of Alaska, Fairbanks; Joanna Grey, Pikes Peak Community College; Shelly K. Habel, Georgetown University; Richard Halpin, Jefferson Community College (New York); C. Allen Haney, University of Houston; Dean Harper, University of Rochester; Gary Hodge, Collin County Community College; Michael Hodge, Morehouse College; Matt Huffman, George Washington University; Shauntey James, University of Dayton; Wanda Kaluza, Camden Community College; Mary E. Kelly, Central Missouri State; Alice Able Kemp, University of New Orleans; Edward Kick, University of Utah; Keith Kirkpatrick, Victoria College (Texas); Margie L. Kiter, University of Delaware; Hadley Klug, University of Wisconsin at Whitewater; Russell L. Long, Del Mar College; Dale A. Lund, University of Utah; Koorps Mahamoudi, Northern Arizona University; Susan Mann, University of New Orleans; Patrick Mcguire, University of Toledo; Beth Mintz, University of Vermont; Charles Norman, Indiana State University; Tracy Orr, University of Illinois—Champaign; Thomas A. Petee, Auburn University; Carol Ray, San Jose State University; David Redburn, Furman University; Chad Richardson, University of Texas—Pan America; Paulina X. Ruff, St. Cloud

University; Tahmoores Sarraf, University of California at Davis; Glenn Sims, Glendale Community College; Michael Smith, St. Philip's College, Alamo Community College District; Larry Stern, Collin County Community College; Brett Stockdill, California Polytechnical State University at Pomona; Kathleen Tierney; University of Delaware; Rance Thomas, Lewis and Clark Community College; Gale A. Thoen, University of Minnesota; Christopher K. Vanderpool, Michigan State University; Glenna Van Metre, Wichita State University; N. Ree Wells, Missouri Southern State College; Mark Winton, University of Central Florida; J. R. Woodward, Montana State University; John F. Zipp, University of Akron.

We also thank the following people, each of whom provided critical support in different, but important ways: Blanche Anderson, Vicky Baynes, Cindy Gibson, Linda Keen, and Judy Watson. We especially thank Bethany Brown for her research assistance.

We are fortunate to be working with a publishing team with great enthusiasm for this project. We thank all of the people at Wadsworth who have worked with us on this and other projects, but especially we thank Bob Jucha and Julie Sakaue for their careful and expert editorial efforts. In addition, we have had strong support from many others at Wadsworth and we appreciate their support for our work. They include Julie Aguilar, Eve Howard, Cheri Palmer, Christina Cha, Wendy Gordon, and Dee Dee Zobian. We especially thank Dusty Friedman of The Book Company for her extraordinary attention to detail; we appreciate enormously her talent and perseverance. Our copyeditor, Yonie Overton, also has a great eye for detail and we thank her for her careful reading of the manuscript. And we thank Myrna Engler for her work as the photographic editor. Finally, our special thanks also go to Richard Morris Rosenfeld and Patricia Epps Taylor for their ongoing support of this project.

About the Authors

Margaret L. Andersen is Professor of Sociology and Women's Studies at the University of Delaware. She received her Ph.D. from the University of Massachusetts, Amherst, and her B.A. from Georgia State University. She is the author of ***Thinking About Women: Sociological Perspectives on Sex and Gender*** and the best-selling Wadsworth text ***Race, Class, and Gender: An Anthology*** (with Patricia Hill Collins). She is the co-editor of the forthcoming book, ***Race and Ethnicity in Society: The Changing Landscape*** (with Elizabeth Higginbotham). She is chair of the National Advisory Board for the Stanford University Center for Comparative Studies of Race and Ethnicity, the past president of the Eastern Sociological Society, and former editor of *Gender & Society.* She has won two teaching awards at the University of Delaware, including the Outstanding Teaching Award from the College of Arts and Sciences and the University Excellence in Teaching Award. In 2004 she won the SWS Feminist Lecturer Award, an award given annually to someone whose work has benefited women. She lives on the Chesapeake Bay in Maryland with her husband Richard Rosenfeld.

Howard F. Taylor was raised in Cleveland, Ohio. He graduated Phi Beta Kappa from Hiram College and has a Ph.D. in sociology from Yale University. He has taught at the Illinois Institute of Technology, Syracuse University, and Princeton University, where he is presently Professor of Sociology and formerly Director of the African American Studies Program. He has published over fifty articles in sociology, education, social psychology, and race relations. His books include ***The IQ Game*** (Rutgers University Press), a critique of hereditarian accounts of intelligence; ***Balance in Small Groups*** (Van Nostrand Reinhold); and the forthcoming ***Race, Class, and the Bell Curve in America.*** He has appeared widely before college, radio, and TV audiences, including ABC's ***Nightline.*** He is past president of the Eastern Sociological Society and a member of the American Sociological Association and the Sociological Research Association, an honorary society for distinguished research. He is a winner of the DuBois-Johnson-Frazier Award, given by the American Sociological Association for distinguished research in race and ethnic relations, and the President's Award for Distinguished Teaching at Princeton University. He lives in Pennington, New Jersey, with his wife, a corporate lawyer.

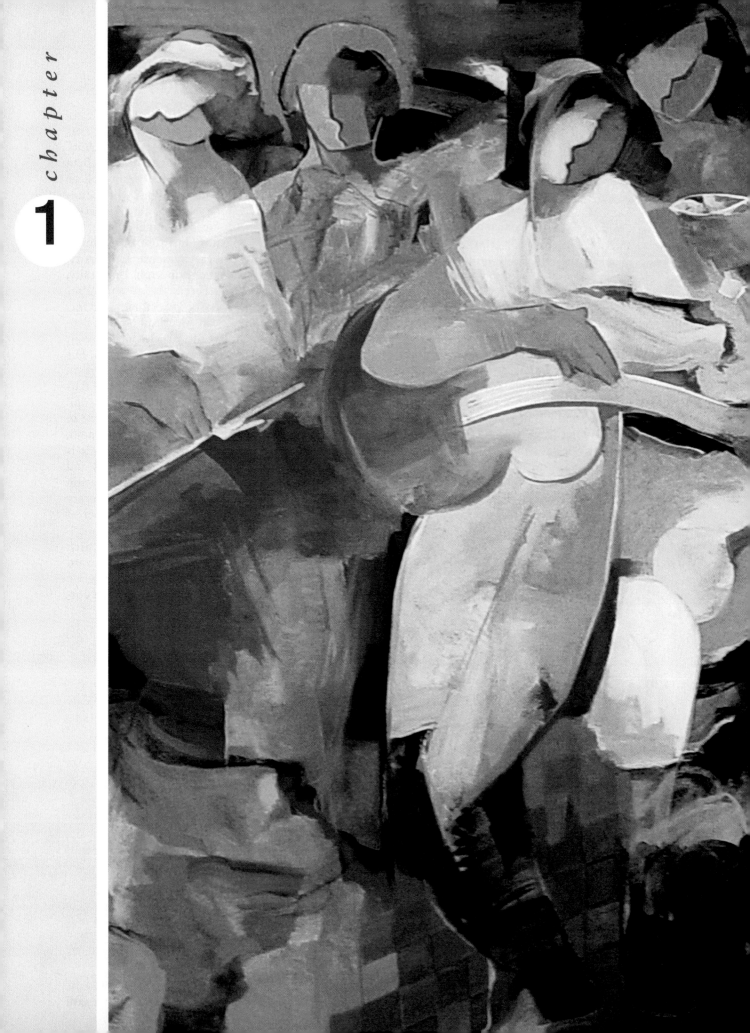

Developing a Sociological Perspective

Imagine that you had been switched with another infant at birth. How different would your life be? What if your accidental family was very poor or very rich? How might this have affected the schools you attended, the health care you received, the possibilities for your future career? If you had been raised in a different religion, how would this have affected your beliefs, values, and attitudes? Taking a greater leap, what if you had been born another sex or a different race? What would you be like now?

We are talking about changing the basic facts of your life—your family, social class, education, religion, sex, and race. Each has major consequences for who you are and how you will fare in life. These factors play a major part in writing your life script. Social location (meaning one's place in society) establishes the limits and possibilities of a life.

Consider this:

• Men who father children before marriage leave school earlier, have lower earnings, and are more likely to live in poverty than men who do not have children before marriage (Nock 1998).

• Black Americans who kill Whites are much more likely to face the death penalty than Blacks who kill other Black people (Paternoster 2003).

• Men who work in jobs traditionally defined as "women's work" behave in ways that emphasize their masculinity; doing so brings rewards since they tend to be promoted faster than similarly qualified women in the same occupations with the same education (Williams 1995).

• The physical transformations that female impersonators must make to deliver a convincing performance are more easily mastered than the behavioral changes they have to make (Tewksbury 1994).

These conclusions, drawn from current sociological research, describe some of the consequences of particular social locations in society. Although people may take their place in society for granted, social position has a profound effect on their lives. The power of sociology is that it teaches how society influences people's lives, and it helps to explain the consequences of different social arrangements.

Sociology also has the power to help us understand the influence of major changes on people. Currently, rapidly developing technologies, increasing globalization, a more diverse population in the United States, and changes in women's roles are affecting everyone in society, although in different ways. How are these changes affecting your life? Perhaps you rely on a cell phone to keep in touch with friends, or maybe your community is witnessing an increase in immigrants from other places, or maybe you see women and men trying hard to balance the needs of both work and family life. All of these are issues that guide sociological questions. Sociology explains some of the causes and consequences of these changes.

Although society is always changing, it is also remarkably stable. People generally follow established patterns of human behavior, and you can generally anticipate how people will behave in certain situations. You can even anticipate how different social conditions will affect different groups of people in society. This is what sociologists find so interesting: *Society is marked by both change and stability.* Societies continually evolve, creating the need for people to adapt to change while still following generally established patterns of behavior. ••

What Is Sociology?

Sociology is the study of human behavior in society. Sociologists are interested in the study of people and have learned a fundamental lesson: All human behavior occurs in a social context. That context includes culture, groups, social interactions, and social institutions—all of which shape what people do and think. Sociology is a scientific way of thinking about society and its influence on human groups. Observation, reasoning, and logical analysis are sociologists' tools. Sociology is inspired by the fascination people have for the thoughts and actions of other people, but it goes far beyond casual observations about people and events. It builds upon observations that are *objective* and *accurate* to create analyses that are reliable and can be validated by others.

Every day, the media in their various forms (television, film, video, and print) bombard people with social commentary. Whether it is Oprah Winfrey, Jerry Springer, or someone else, media commentators and celebrities provide endless opinion about behavior in society. Sociology is different. Sociologists may study the same subjects that the media examine, such as domestic violence, religious cults, or race relations, and many sociologists make media appearances, but sociologists use specific research techniques and well-tested theories to examine and explain social issues.

Much of what one hears in the media and elsewhere about society, though it may be delivered with perfect earnestness, is misstated and sometimes completely wrong. Sociology provides the tools for testing whether the things said and believed about society are true. The lessons and practices of sociology can be applied to media pronouncements and other pieces of social lore to find objective understanding of social issues that are often clouded by people's opinions.

Some of the students reading this book will become sociology majors; many will not. Either way, sociology can be useful in your life because it helps you understand the context of human behavior. It thus reveals the underlying basis for many of the social issues one might face in a business organization, watching or reading about current events, or even just seeing new dimensions to the images that surround us in popular culture. As we will see throughout this book, sociology gives you a greater awareness of what makes up society and thus helps you become an informed and active citizen who can interpret and understand what you see around you. In this chapter, we introduce the sociological perspective and some of the essential concepts and theories that guide sociological thinking.

Question:

What do the following people have in common?

Dan Akroyd (actor; comedian)

Debra Winger (actress)

Saul Bellow (novelist; Nobel Prize recipient)

Joe Theissman (NFL quarterback)

(Former) Senator Robert Torricelli

Congresswoman Maxine Waters

Senator Barbara Mikulski (from Maryland)

Regis Philbin (TV personality)

Rev. Jesse Jackson

Ruth Westheimer (the "sex doctor")

Robin Williams (comedian; actor)

Ahmad Rashad (sportscaster)

Rev. Martin Luther King, Jr.

Ronald Reagan

Answer:

They were all sociology majors!

Source: Compiled by Peter Dreier, Occidental College. Full list available on the home page of the American Sociological Society: **www.asanet.org**.

Sociology: A Unique Perspective

The subject matter of sociology is everywhere. The routines of everyday life, how people fare, the rapid changes taking place in society are all topics of sociological study. Psychologists, anthropologists, political scientists, economists, and others also study social behavior and social change. Along with sociologists, these disciplines make up the *social sciences*. The difference between sociology and other disciplines is not in the topics that each studies, but in the perspective each discipline brings to the subject.

Psychology analyzes individual behavior. Sociologists share the interest of psychologists in individuals, but the unit of analysis for psychology is the individual; for sociologists, it is society—the

Lara Jo Regan/Liaison/Getty Images

Sociology is the study of human behavior in society, including the significance of diversity in society.

SOCIOLOGY IN PRACTICE

Careers in Sociology

When students first encounter sociology, they often ask, "What can I do with a degree in sociology?" Some will become professional sociologists, entering careers as professors, policy advisors, or researchers. Most students see sociologists in their roles as educators and that is an important part of the work sociologists do. But sociologists are also frequently called upon as experts for the U.S. Congress, by the media, by state and federal agencies, and by private organizations to consult on a variety of subjects. Sociologists have given briefings to congressional staff on family–work connections to help pass the Family and Medical Leave Act. Several sociologists have been direct advisors to U.S. presidents.

Even without an advanced degree in sociology (such as a Master's degree or a Ph.D.), an undergraduate education in sociology can lead to a variety of careers. The skills you acquire from your sociological education are useful for jobs in business, health care, criminal justice, government agencies, various nonprofit organizations, and other job

venues. For example, the research skills one gains through sociology can be important in analyzing business data or organizing information for a food bank or homeless shelter. Students in sociology also gain experience working with and understanding those with different cultural and social backgrounds; this is an important and valued skill that employers seek. Also, the ability to dissect the different causes of a social problem can be an asset for jobs in various social service organizations. Some sociologists have worked in their communities to deliver more effective social services. Some are employed in business organizations and social services, where they use their sociological training to address issues such as poverty, crime and delinquency, population studies, substance abuse, violence against women, family social services, immigration policy, and any number of other important issues. Sociologists also work in the offices of U.S. representatives and senators doing background research on the various issues addressed in the political process.

These are just a few examples of how sociology can prepare you for various careers. For more information about careers in sociology, see the booklet *Careers in Sociology* available through the American Sociological Association. You can find it online at either www.asanet.org/student/career/homepage.html or http://sociology.wadsworth.com

Critical Thinking Exercise

1. Read a national newspaper over a period of one week and identify any experts who use a sociological perspective in their commentary. What does this suggest to you as a possible career in sociology? What are some of the different subjects about which sociologists provide expert information?

2. Identify some of the students from your college who have finished degrees in sociology. What different ways have they used their sociological knowledge?

•••

whole configuration of group life. In a society that cherishes individualism and individual rights, a tendency exists to develop psychological explanations of all behavior. People tend to think that behavior always stems from personality differences or different motivations. From a sociological point of view, psychological explanations are not wrong, just incomplete. Sociologists explain that people's behavior arises not only from motives and attitudes internal to the person, but also from the social context in which people live. One person's behavior might be attributed to personality—a psychological explanation—but when it becomes apparent that there are consistent patterns of thought and behavior across the whole society, a larger perspective than individual analysis is needed. Sociological inquiry extends to the largest social unit of all, society itself.

Anthropology is the study of human cultures. Anthropologists see culture as the basis for society, studying how people live in different cultures and how cultures evolve. Sociologists study culture too, seeing it as part of other social systems that together compose society. Generally, sociologists are more likely to study a society of which they are a part. Anthropologists are more likely to study faraway and remote cultures, although some anthropologists now also study various cultural dimensions of U.S. society and many sociologists do research in different countries. The primary difference in the two disciplines is not where sociologists and anthropologists do their work, but the central emphasis that anthropologists place on culture.

Economics and political science are the other pillars of the social sciences. *Political science* is the study of politics, including political behavior, political philosophy, and the organization of governments and political parties. *Economics* scrutinizes the production, distribution, and consumption of goods and services. Political scientists and economists study particular social institutions that shape political and economic behavior, respectively. Sociologists are interested in all **social institutions,** defined as established and organized systems of social behavior with a recognized purpose. In a nutshell, sociologists are interested in how all social forces, including cultural, economic, and political forces, affect human behavior.

Many students enter the field of sociology because they want to help people and think they might become social workers. Sociology and social work are closely allied, but they are not the same. *Social work* is an applied field that draws from the social sciences to serve people in need. Social work draws heavily from studies of families and communities, but it typically addresses problems of individuals. Studying sociology can prepare you for a career in social work because it helps social workers better understand the plight of various individuals, but sociologists do not see individual solutions as adequate for addressing society's problems. Sociologists are likely to argue for societal-level changes to address problems people face.

To understand the differences among the social sciences, consider how practitioners in each of these fields might study the family. Psychologists would be interested in how individuals form personalities within families. Anthropologists would be curious about the diverse family structures that develop in different cultures. Political scientists might scrutinize how a policy decision affects families differently or how voting behaviors and political opinions are passed on within families. Economists expend great effort interpreting family consumer patterns and assessing changes in the economy that affect employment, the mainstay of family economics. Social workers would be most concerned with delivering social services to families in need. Sociologists, however, would be interested in how families are shaped by society and how changes in society are affected by changes in family structure.

The boundaries between the social sciences are not rigid. Each is capable of making significant contributions to our knowledge of people, and some of the best work in each field draws on work in others. Many sociologists engage in interdisciplinary research, using the perspectives of multiple disciplines. Sociology is also linked to several new academic fields that were previously ignored or misperceived by scholars in the traditional disciplines. African American studies, Chicano and Latino studies, Asian American studies, Native American studies, women's studies, lesbian and gay studies, and Jewish studies draw from sociological ideas and have enriched the research and theory that sociologists use to understand the lives of diverse people.

Sociology in Practice

Sociology is not merely fascinating, it is also useful. One basic application of sociology is its teaching mission, that is, using sociological research and theory to educate people about the diverse groups that make up society. This is no small task. The media emit an endless stream of conventional and sometimes prejudicial explanations of social issues. Learning to think like a sociologist can teach one to challenge unexamined assumptions about society, values, and people.

Sociologists have used their sociological perspective for many purposes. When the publication of *The Bell Curve* created great public controversy over race and intelligence, Howard F. Taylor, an expert on race and intelligence testing (and coauthor of this textbook), appeared on ABC TV's *Nightline* in a debate with author Charles Murray. Margaret L. Andersen (the other author of this textbook) has used her knowledge about gender, race, and feminism to work with social scientists in Mexico City who are developing social policies to challenge gender and race stereotyping in Mexico. Sociologists also testify before Congress on topics as diverse as teenage pregnancy, acquired immune deficiency syndrome (AIDS) funding, and street violence, and sometimes provide expert testimony in criminal or civil legal cases. They may consult with organizations trying to develop fair employment practices, or work with community action groups trying to solve local problems of housing, safety, and education.

The practical applications of sociology are as diverse as the people who become sociologists. What sociologists hold in common is a commitment to rigorous study as the path to understanding society and making a difference in people's lives. You will see some of the different ways that sociologists have put their knowledge to use in the feature of this book called "Sociology in Practice."

The Sociological Imagination

Think back to the vignette that opened this chapter. We asked you to imagine growing up under completely different circumstances to make you feel the stirring of the **sociological imagination,** which is the ability to see the societal patterns that influence individual and group life. The beginnings of the sociological imagination can be as simple as the pleasures of watching people or wondering how society influences people's lives. Many students begin their study of sociology because "they are interested in people." Sociologists convert this curiosity into the systematic study of how society influences different people's experience within it.

C. Wright Mills (1916–1962) was one of the first to write about the sociological imagination. In his classic book *The Sociological Imagination* (1959), Mills wrote that the task of sociology was to understand the relationship between individuals and the society in which they lived. Sociology should be used, Mills argued, to reveal how the context of society shapes our lives. He described this as understanding the intersection between biography and history. Mills thought that to understand the experience of a given person or group of people, one had to have knowledge of the social and historical context in which people lived.

To visualize the junction of biography and history, think about the time and effort that many people put into their appearance. Although, on the one hand, you can think about this merely as personal grooming or an individual attempt to "look good," there are very social origins for this behavior. An individual may stand in front of the mirror, perhaps thinking about how society is present in that reflection. But we also look in the mirror to see how others see us; therefore, this is a very social act. Furthermore, if we are trying to achieve a particular look, that look has likely been established by the social forces that tell us to achieve a particular ideal—ideals produced by particular industries that profit enormously from the products and services that people buy. Some industries suggest that you should be thinner or curvier, your pants should be baggy or straight, your breasts should be minimized or maximized; either way you need more products. Maybe you should even be totally "made over!" Many people then go to great lengths—sometimes extreme lengths that are hazardous to physical and mental health—to try to achieve a constantly changing beauty ideal, one that is probably not even attainable (i.e., flawless skin, hair never out of place, perfectly proportioned body parts).

The point is that the "ideal" is produced by social factors that extend far beyond people's immediate concerns with their appearance. These ideals are produced in particular social and historical contexts. People may come up with all kinds of personal strategies for achieving a presumed ideal—they may buy more products, try to lose more weight, get a Botox treatment, or even become extremely depressed and anxious if they perceive their efforts as failing. These personal behaviors are individual issues, but they have social origins. The sociological imagination allows you to see the social structure behind individual lives. Sociologists care about individuals, but they direct their attention to the social and historical context that shapes the experiences of individuals and groups.

A fundamental concept for organizing the sociological imagination is the distinction Mills made between troubles and issues. **Troubles** are privately felt problems that spring from events or feelings in one individual's life. **Issues** affect large numbers of people and originate in the institutional arrangements and history of a society (Mills 1959). This distinction is the crux of the difference between individual experience and **social structure,** defined as the organized pattern of social relationships and social institutions that together constitute society. Issues shape the context where troubles arise and sociologists employ the sociological imagination to understand how issues are shaped by social structures.

Mills used the example of unemployment to explain the meaning of troubles versus issues. When a person becomes unemployed, that person has a personal trouble. In addition to financial problems, the person may feel a loss of identity, become depressed, lose touch with former work associates, or have to uproot a family and move. The problem of unemployment, however, is

Spencer Platt/Getty Images

Personal troubles are felt by individuals who are experiencing problems; social issues arise when large numbers of people experience problems that are rooted in the social structure of society. Violence against women is another example.

deeper than the experience of one person. Unemployment is rooted in the structure of society; this is what interests sociologists. What causes unemployment? Who is most likely to become unemployed at different times? How does unemployment affect an entire community when a large plant shuts down or all older workers when corporations downsize and lay people off? Sociologists know that unemployment causes personal troubles, but understanding the cause and effect of unemployment is more than understanding one person's experience. It requires understanding the social structural conditions that influence people's lives.

The specific task of sociology, according to Mills, is to comprehend the whole of human society—its personal and public dimensions, historical and contemporary, and its influence on the lives of human beings. Mills had an important insight: People often feel that the situation they are in is larger than their experience alone because they feel the influence of society in their lives, even when the specific forces shaping their lives are not always readily apparent.

Consider this: Most likely you remember what you were doing on September 11, 2001, when you first heard that terrorists had flown planes into the World Trade Center in New York City. Obviously, this affected people's personal lives, but its impact—and its causes—go beyond the personal troubles it produced. Events like this influence our lives in profound ways, even though we may not know how at the time. Now we can use the sociological perspective to explain many dimensions of this event and its aftermath, including, for example, the significance of cultural symbols, how people respond to unexpected events, and how religious, ethnic, and national identities are created in society. Of course, the social forces that influence people's lives are not always so drastic and also include the ordinary events of everyday life. One of the contributions that sociology makes is revealing the forces at work that shape society and people within it.

Revealing Everyday Life

June Jordan, the noted African American poet and essayist, says, "If you understand the sociological perspective, you can never be bored" (1981: 100). Why would Jordan think that sociology relieves boredom? Waiting in an airport terminal for an overdue flight can be boring, but to the sociologist, even this situation can be brimming with observable phenomena. Sociologists might observe the rituals people engage in when they meet and greet each other. They might observe how workers behave in different jobs, perhaps inspiring thoughts about how work conditions influence workers' attitudes about work and about each other. The keen sociological observer might also detect differences in who is employed in what jobs and may perceive the influence of gender and race on employment patterns at the airport.

Arlie Hochschild, a contemporary sociologist, has observed how flight attendants interact with passengers and with other attendants. She concluded that flight at-

TAKING ON SOCIAL ISSUES ●●●

Sexual Abuse

Sexual abuse is a good example of the distinction C. Wright Mills makes between personal troubles and public issues. To the person experiencing the abuse (and, perhaps, to the abuser), this is a personal trouble. Sociologists would add that sexual abuse is a public issue. It is more common than one

would like to believe and is rooted in some of the social structures of society, such as the power of men over women, conflict within the family, the use of children as sexual objects, and so forth. How does sexual abuse illustrate Mills' two concepts of personal troubles and public issues?

Taking Action

Go to the Taking Action Exercise on the companion website—at http://sociology.wadsworth.com/andersen_taylor4E/—to learn more about an organization that addresses this topic.

●●●

tendants are trained to produce certain feelings in passengers, such as being a guest in someone's home, not hurtling through space at 400 miles per hour. Hochschild also noted that flight attendants are evaluated by supervisors according to how well they succeed in producing the desired emotional state. Based on her research, she wrote a fascinating book on how service industries expect employees to "manage" the emotions of others (Hochschild 1983). The result is the commercialization of human feeling. Perhaps Hochschild got the initial idea for her study while sitting in an airplane stuck on the tarmac, trying not to be bored!

Revelations about everyday life are part of the appeal of sociology, but sociology is much more than simply thinking about things you see around you. Sociology uses rigorous methods of research to examine its assumptions and conclusions. To see the difference between social musing and sociological research, consider the common explanation of inequality in the United States. Many people believe that poor people do not get ahead because they do not want to work hard enough. A sociologist would not assume this to be true, but rather use research to find out whether people less well off have less of a work ethic than those who are better off. In fact, sociologists have found that attitudes toward work are a poor explanation of the likelihood that someone will be poor or on welfare (Santiago 1995; Wagner 1994). Forced to look to other factors to explain why there are poor people in society, sociologists have studied educational and job opportunities and asked whether the labor market and the educational system give some people better opportunities than others. They do. By finding societal explanations of why there is inequality, sociologists are led beyond commonsense explanations. Sociologists usually discover that social factors, not just individual states of mind, are the most compelling explanations of behavior.

A big difference between sociology and common sense is that *sociology is an empirical discipline*. The **empirical** approach to knowledge requires that conclusions be based on careful systematic observations and not on previous assumptions. For empirical observations to be useful to other observers, they must be gathered and recorded rigorously. Although the specific methods that sociologists use to examine different problems vary (see Chapter 2), the empirical basis of sociology is what distinguishes it from mere opinion or other forms of social commentary.

Debunking in Sociology

The power of sociological thinking is that it helps us see everyday life in new ways. Sociologists question actions and ideas that most people take for granted. Peter Berger (1963) calls this process **debunking.** De-

bunking refers to looking behind the facades of everyday life—what Berger calls the "unmasking tendency" of sociology (1963: 38). In other words, sociologists look at the behind-the-scenes patterns and processes that shape the behavior they observe in the social world.

The example of schooling shows how the sociological perspective debunks assumptions about education long taken for granted. A commonsense perspective sees education as helping people learn and get ahead. A sociological perspective on education, however, reveals that more than learning takes place in schools; other social processes are at work. Social cliques are formed where some students are "insiders" and others are excluded. Young schoolchildren acquire not just formal knowledge, but also the expectations of society and their place within it. Relative to boys, girls are often short-changed by the school system and receive less attention and encouragement (American Association of University Women 1998; Sadker and Sadker 1994). Race and class conflicts are often played out in schools (Lewis 2003). Poor children seldom have the same resources in schools as middle-class or elite children. They are often assumed to be incapable of doing schoolwork and are treated accordingly. The somber reality is that schools may stifle the opportunities of some children instead of launching all children toward success.

Debunking is sometimes easier to do when looking at a culture or society different from one's own. Consider how behaviors that are unquestioned in one society may seem positively bizarre to an outsider. For a thousand years in China, the elite classes often bound the feet of young girls to keep their feet from growing bigger. This practice allegedly derived from one of the emperor's concubines, and bound feet became a sign of delicacy and vulnerability. A woman with large feet (defined as more than four inches long) was thought to bring shame to her husband's household. Footbinding was supported by the belief that men were highly aroused by small feet, even though men never saw the naked foot. If they had, they might have been repulsed because a woman's foot was U-shaped and often decayed and covered with dead skin (Blake 1994).

Outside the social, cultural, and historical context in which it was practiced, footbinding seems bizarre, even dangerous. Chinese women were actually crippled by this practice, making them unable to move about freely and therefore more dependent on men (Chang 1991). Many behaviors that ordinarily go unquestioned can be debunked by an alert, imaginative, and careful observer.

Suppose that you were to use the debunking process to examine practices that typically go unquestioned in U.S. culture. Strange as the practice of Chinese footbinding may seem to you, how might someone from another culture view the practice of implanting silicon in

Cultural practices that seem bizarre to outsiders may be taken for granted or defined as appropriate by insiders.

women's breasts to enhance their size? Or piercing the tongue, nose, or ears and filling the hole with jewelry? Or mailing catalogs full of glossy pictures of women in lacy underwear to millions of consumers every month? These practices of contemporary U.S. culture are taken for granted by many. Until these cultural processes are debunked and unraveled in a social context—as if seen for the first time—they might seem normal. With sociology, one can step away from cultural familiarity to glimpse the social structures behind them.

> ### THINKING *SOCIOLOGICALLY*
> School violence is now widely presented in the public media. How is school violence generally depicted in commonsense explanations? What questions might sociologists ask to *debunk* these everyday explanations of school violence?

Establishing Critical Distance

Debunking requires critical distance—that is being able to detach from the situation at hand and view things with a critical mind. In other words, not taking things for granted. Some people live in circumstances that make this easier for them. *Marginal people* are those who share the dominant culture to some extent but are blocked from full participation because of their social status. One need not be a marginal person to be a sociologist, but marginality has often provided the critical distance necessary to inspire a thriving sociological imagination. For example, many of the leaders of the civil rights movement in the American South during the 1950s and 1960s studied sociology. Martin Luther King Jr., Jesse Jackson, and Ella Baker (one of the major leaders of the student civil rights movement in the American South) were sociology majors who understood that problems such as racial inequality have their origins in the structure of society, not just in the minds of individuals (Carson 1981). Most likely, their experiences as African American women and men helped them acquire a sociological imagination and defy the time-honored beliefs in U.S. society that supported racial segregation. In this way, they (and others) benefited from the critical distance their lives provided as they analyzed how the structure of society produced racism.

The role of critical distance in developing a sociological imagination is well explained by the early sociologist **Georg Simmel** (1858–1918). Simmel was especially interested in the role of *strangers* in social groups. Strangers have a position both inside and outside social groups; they are part of a group without necessarily sharing the group's assumptions and points of view. Because of this, the stranger can sometimes see the social structure of a group more readily than people who are thoroughly imbued with the group's world view. Simmel suggests that the sociological perspective requires a combination of nearness and distance. One must have enough critical distance to avoid being taken in by the group's definition of the situation, but be near enough to understand the group's experience.

Sociologists are not typically strangers to the society they study, nor do they have to be marginal to acquire the sociological imagination. One can acquire critical distance through willingness to question the forces that shape social behavior. Often, sociologists become interested in things because of their own experiences. The biographies of sociologists are rich with examples of how their personal lives informed the questions they asked. Among sociologists are former ministers and nuns now studying the sociology of religion, women who have encountered sexism who now study the significance of gender in society, rock-and-roll fans studying music in popular culture, and sons and daughters of immigrants now analyzing race and ethnic relations (see the box "Understanding Diversity: Becoming a Sociologist" on page 12).

Discovering Unsettling Facts

In studying sociology, it is critical to work with an open mind, even when observing the most disquieting facts. Revealing troubling facts about society is the price of seeing society for what it is. Consider the following:

- Despite the idea that Asian Americans are seen as a "model minority," poverty among Asian American families is higher than that among White American families and has increased in recent years; among certain Asian American groups, namely Laotians and Cambodians, poverty strikes two-thirds of all families (Proctor and Dalaker 2003; Lee 1994).

- Women's income (on average) has generally risen, while men's has generally fallen. Still, women with a college degree earn less on average than men with only some college experience (DeNavas-Walt et al. 2003; Mishel et al. 2003).

- Infant mortality in the United States, especially among African Americans and Hispanics, exceeds that of many impoverished nations (U.S. Census Bureau 2004).

- The United States has the highest rate of imprisonment of all nations in the world (Mauer 1999; The Sentencing Project 2000; Wolmsley 2000).

These facts provide unsettling evidence of persistent problems in the United States, *problems that are embedded in society, not just in individual behavior.* Sociologists try to reveal the social factors that shape society and affect the chances of success for different groups. Doing so often challenges the widely held view that all people have a fair chance to succeed if they just try hard enough. The sociological perspective reveals instead that many social factors affect people's lives and opportunities. Some never get the chance to go to college; others are likely never to go to jail. These divisions persist partly because of people's location within society.

As we mentioned earlier, sociologists do not study only the disquieting side of society. Sometimes, sociologists ask questions for no reason except that they are curious. Sociologists may study questions that affect everyday life, such as how families adapt to changing societal conditions (Lempert and DeVault 2000), how street people live (Duneier 1999), how children of immigrants fare (Portes and Rumbaut 2001), or how racism has changed in recent years (Zuberi 2001; Bonilla-Silva 2000). There are also many insightful and intriguing studies of unusual groups, including dog owners (Sanders 1999), cyberspace users (Turkle 1995; Kendall 2000), transvestites (Bullough 1993), and tattoo collectors (Irwin 2001). The subject matter of sociology is vast. Some work illuminates odd corners of society; other work addresses urgent problems of society that may affect the lives and futures of millions.

The Significance of Diversity

The analysis of diversity is one of the central themes of sociology. Differences among groups are significant in any society, but they are particularly compelling in a society as diverse as the United States. With so much cultural diversity, people will share some experiences, but not all. Experiences not held in common can in-

Understanding diversity is important in a society comprising so many different groups, each with unique, but interconnected, experiences.

© Richard Lord/The Image Works

MAP 1.1 Minorities in the United States

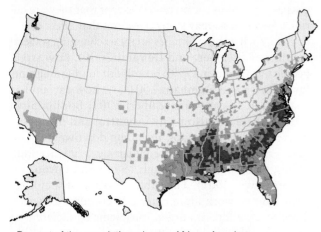

Percent of the population who are African American

| | 0–5.6 | | 17.3–32.3 | | 51.6–86.5 |
| | 5.7–17.1 | | 32.5–51.2 | | |

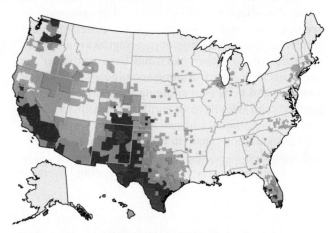

Percent of the population who are Hispanic

| | 0.1–5.9 | | 17.4–34.7 | | 72.2–99.7 |
| | 6.0–17.3 | | 35.1–67.6 | | |

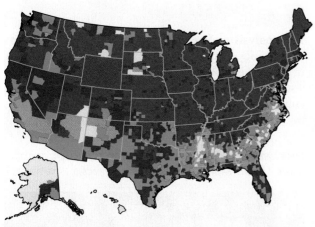

Percent of the population who are White

| | 4.5–42.9 | | 63.3–78.3 | | 90.7–99.7 |
| | 43.2–63.1 | | 78.4–90.6 | | |

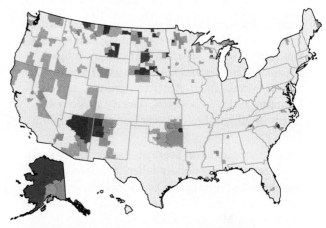

Percent of the population who are American Indian
or Alaskan Natives

| | 0–3.0 | | 10.7–24.3 | | 59.7–94.2 |
| | 3.1–10.3 | | 28.3–55.8 | | |

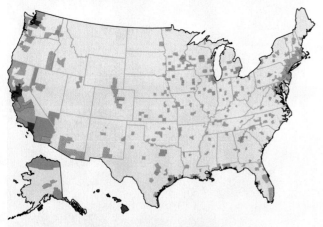

Percent of the population who are Asian American

| | 0–1.1 | | 3.6–8.9 | | 24.6–46.0 |
| | 1.2–3.5 | | 9.4–20.4 | | |

These maps show the percentage of persons from different racial-ethnic groups in the different regions of the United States. Many factors influence this regional distribution, including historical patterns of slavery for African Americans, urban migration as different groups (including immigrants) move to cities in search of work, the forced relocation of American Indians, and contemporary immigration patterns, among others. Looking at your region of the country, what factors influence the proportion of racial-ethnic groups? How does this influence the social issues that affect your region? Where do you see the largest concentration of different racial-ethnic groups? What social changes does this produce?

Source: U.S. Census Bureau. 2000. *American FactFinder.* Website: www .census.gov.

clude some of the most important influences on social development, such as language, religion, and the traditions of family and community. Understanding diversity means recognizing this diversity and making it central to sociological analyses.

Diversity: A Source of Change

Perhaps the most basic lesson of sociology is that people are shaped by the social context around them. Today, the United States is comprised of people from all nations and races. In 1900, one in eight Americans was not White; today, racial and ethnic minority groups, including African Americans, Latinos, American Indians, and Asian Americans, are one-quarter of the U.S. population and that proportion is growing (see Figure 1.1; Grieco and Cassidy 2001). These broad categories themselves are internally diverse, including those with long-term roots in the United States, as well as Cuban Americans, Salvadorans, Cape Verdeans, Filipinos, and many others.

Defining Diversity

In this book, *we use the term **diversity** to refer to the variety of group experiences that result from the social structure of society. Diversity is a broad concept that includes studying group differences in society's opportunities, the shaping of social institutions by different social factors, the formation of group and individual identity, and the process of social change.* Diversity includes the study of different cultural orientations, although diversity is not exclusively about culture. Understanding diversity is critical to understanding society because the fundamental patterns of social change and social structure are increasingly influenced by diverse group experiences.

> **THINKING *SOCIOLOGICALLY***
> What are some of the sources of *diversity* on your campus? What questions might a sociologist pose about the significance of diversity in this environment?

Race, class, and gender are fundamental to understanding diversity because they have been so critical to shaping social institutions in the United States, but are not the only sources of diversity. Age, nationality, sexual orientation, religion and region of residence, among other social factors, also differentiate the experience of diverse groups in the United States. Because the world is now interconnected through global communication and a global economy, the study of diversity also encompasses a global perspective. It is an understanding of the international connections existing across national borders and their impact on life throughout the world.

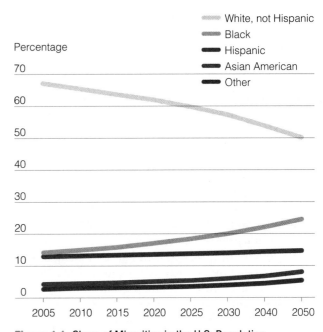

Figure 1.1 Share of Minorities in the U.S. Population
It is a social fact that in 2050 the U.S. population is projected to be half non-Hispanic whites. How do you think this will affect your life? What will the population percentage be for different racial-ethnic groups when you are forty years old? How does this illustrate how social facts influence the lives of individual people and diverse groups in society?

Source: U.S. Census Bureau. 2004. *Statistical Abstract of the United States 2003.* Washington, DC: U.S. Department of Commerce, p. 18.

The study of diversity within society is also important because different groups have widely different experiences. For example, if you are rich, White, well-insured, and enter a hospital on the referral of your family doctor, you are likely to have a very different

"*Actually, Lou, I think it was more than just my being in the right place at the right time. I think it was my being the right race, the right religion, the right sex, the right socioeconomic group, having the right accent, the right clothes, going to the right schools . . .*"

experience than if you enter the hospital poor, Latino, or uninsured. Diversity also lies at the heart of many difficult social conflicts, and using sociology to analyze these issues is one of the compelling values of a sociological perspective. Diversity is not just about the study of problems. The complex weave of society is a source of great cultural richness and human achievement. Different social groups often use their social position to generate social change and sociologists have provided studies of how they do so, such as how poor women develop survival strategies to support their children (Edin and Lein 1997).

When studying diversity, it is crucial to remember that the various sources of diversity exert their influence not independently, but interactively. For instance, in the context of race, class, *and* gender, being a man is not the same for all men. Class and race may affect some men's lives more than their gender, but gender is also significant in men's experiences. Class, race, and gender form overlapping categories of experience for all groups. Thus, an Asian American woman's experience is shaped by her gender, race, and social class simultaneously. She is confronted with the expectations attached to being a woman *and* an Asian American *and* having the resources of those in her social class of middle class, working class, rich, or poor. Moreover, her identity as an Asian American woman will be different from that of an Asian American man. Under-

standing diversity means understanding how these social influences blend.

Society in Global Perspective

No society can be understood apart from its global context. The social and economic system of any one society is increasingly intertwined with those of other nations around the world. Coupled with the possibility of travel and rapid telecommunication, a global perspective is necessary to understand change both in the United States and throughout the world.

To understand globalization, you must look beyond the boundaries of your own society to see how patterns in any given society are continually shaped by the connections between societies. Sociologists and other social scientists have typically used a cross-cultural or comparative perspective to study different societies. Comparing and contrasting societies across different cultures is valuable. It reveals patterns in society that you might otherwise take for granted, and it enriches your appreciation of the diverse patterns of culture that mark human society and human history. A global perspective, however, is more than comparing different cultures: A global perspective shows how events in one place may be linked to events on the other side of the globe.

UNDERSTANDING DIVERSITY

Becoming a Sociologist

Individual biographies often have a great influence on the subjects sociologists choose to study. The authors of this book are no exception. Margaret L. Andersen, a White woman, studies the sociology of race and women's studies; Howard F. Taylor, an African American man, studies race, social psychology, and especially, race and intelligence testing. Here, each of them writes about the influence of their early experiences on becoming a sociologist.

Margaret L. Andersen • As I was growing up in the 1950s and 1960s, my family moved from California to Georgia, then to Massachusetts, and then back to Georgia. Moving as we did from urban to small-town environments and in and out of regions of the country that were very different in their racial character, I probably could not help becoming fascinated by the sociology of race. Oakland, California,

where I was born, was highly diverse; my neighborhood was mostly White and Asian American. When I moved to a small town in Georgia in the 1950s, I was ten years old, but I was shocked by the racial norms I encountered. I had always loved riding in the back of the bus—our major mode of transportation in Oakland—and could not understand why this was no longer allowed. Labeled an outsider by my peers, because I was not southern, I painfully learned what it meant to feel excluded just because of "where you are from."

When I moved again to suburban Boston in the 1960s, I was defined by Bostonians as a southerner and ridiculed because of it. Nicknamed "Dixie," I was teased for how I talked. Unlike in the South, where, despite strict racial segregation, Black people were part of White people's daily lives, Black people in Boston were even less visible. In my

high school of 2500 or so students, Black students were rare. To me, the school seemed little different from the strictly segregated schools I had attended in Georgia. My family soon returned to Georgia, where I was an outsider again; when I later returned to Massachusetts for graduate school in the 1970s, I worried about how a "southerner" would be accepted in this "Yankee" environment. Because I had acquired a southern accent, I think many of my teachers thought I was not as smart as the students from other places.

These early lessons, which I may have been unaware of at the time, must have kindled my interest in the sociology of race relations. As I explored sociology, I wondered how the concepts and theories of race relations applied to women's lives. So much of what I had experienced growing up as a woman in this society was completely unex-

For instance, return to the example of unemployment that Mills used to distinguish troubles and issues. One man may lose his job in Peoria, Illinois, while a woman in Los Angeles may employ a Latina domestic worker to take care of her child while she pursues a career. On the one hand, these are individual experiences for all three people, but they are linked in a pattern of globalization that shapes each person's life. The Latina domestic may have a family whom she has left in a different nation so that she can afford to support them. The corporation for which the Los Angeles woman works may have invested in a new plant overseas that employs cheap labor, resulting in the unemployment of the man in Peoria. The man in Peoria may have seen immigrant workers moving into his community, and one of his children may have made a friend at school who speaks a language other than English. These processes continue to shape many of the subjects examined in this book, such as work, family, education, and politics. Without a global perspective, you cannot fully understand the experience of any of the three people just mentioned or how society is shaped by the global context.

In this book, we introduce a global perspective to understand those aspects of U.S. society that are most influenced by worldwide developments. The U.S. economy is one obvious example that extends far beyond the nation's borders. Processes such as technological development, the distribution of natural and economic

A. Ramey/PhotoEdit Inc.

Globalization brings diverse cultures together, but it is also a process of the penetration of western markets throughout the world.

resources, unemployment, and poverty simply cannot be understood anymore without analyzing the global basis of contemporary society. Something as seemingly simple as what toys children play with can be fully un-

amined in what I studied in school. As the women's movement developed in the 1970s, I found sociology to be the framework that helped me understand the significance of gender and race in people's lives. To this day, I write and teach about race and gender, using sociology to help students understand the significance of both in society.

Howard F. Taylor • I grew up in Cleveland, Ohio, the son of African American professional parents. My mother, Murtis Taylor, was a social worker and the founder and then president of a social work agency. She is well known for her contributions to the city of Cleveland and was an early "superwoman," working days and nights, cooking, caring for her two sons, and being active in many professional and civic activities. I think this gave me an early appreciation for the roles of women and the place of gender in society, though I surely would not have articulated it as such at the time.

My father was a businessman in an all-Black life insurance company. He was also a "closet scientist," always doing experiments and talking about scientific studies. He encouraged my brother and me to engage in science, so we were always experimenting with scientific studies in the basement of our house. In the summers, I worked for my mother as a camp counselor and in other jobs in the social service agency where she worked. Early on, I contemplated becoming a social worker, but I was also excited by science. As a young child, I acquired my father's love of science and my mother's interest in society. In college, the one field that would gratify both sides of me, science and social work, was sociology. I wanted to study human interaction, but I also wanted to be a scientist, so the appeal of sociology was clear.

At the same time, growing up African American meant that I faced the consequences of race everyday. It was

always there, and like other young African American children, I spent much of my childhood confronting racism and prejudice. When I discovered sociology, in addition to bridging the scientific and humanistic parts of my interests, I found a field that provided a framework for studying race and ethnic relations. The merging of two ways of thinking, coupled with the analysis of race that sociology has long provided, made sociology so fascinating to me.

Today, my research on race and intelligence testing and Black leadership networks seems rooted in these early experiences. I do quantitative research in sociology and see sociology as a science that reveals the workings of race, class, and gender in society.

Further Resources: For a collection of essays where sociologists examine various biographical influences on the work they now do, see: Barry Glassner and Rosanna Hertz. 2003. *Our Studies, Ourselves: Sociologists' Lives and Work.* New York: Oxford University Press. ●●●

derstood only by analyzing the international economic system in which these toys are produced and distributed. Throughout this book, we will use a global perspective to understand some of the developments that shape contemporary life in the United States.

The Development of Sociology

Like the subjects it studies, sociology is a social product. It is an intellectual framework deeply influenced by the social and historical conditions in which it is produced. Sociology first emerged in western Europe during the eighteenth and nineteenth centuries. In this period, the political and economic systems of Europe were rapidly changing. Monarchy, the rule of society by kings and queens, was disappearing. Religion was eroding as the system of authority, giving way to law and scientific authority. At the same time, capitalism grew. Contact between different societies increased, and worldwide economic markets developed. Traditional ways of the past were giving way to a new social order. The time was ripe for a new understanding.

The Influence of the Enlightenment

The **Enlightenment** in eighteenth- and nineteenth-century Europe had an enormous influence on the development of modern sociology. Also known as the Age of Reason, the Enlightenment was characterized by faith in the ability of human reason to solve society's problems. Intellectuals believed that there were natural laws and processes in society to be discovered and used for the general good. This belief was linked to the development of modern science, which was gradually supplanting traditional and religious explanations for natural phenomena with theories confirmed by scientific experiments.

Several strands of thought emerged during the Enlightenment that influenced the development of sociology. **Positivism** is a system of thought in which accurate observation and description is considered the highest form of knowledge, as opposed to religious dogma or poetic inspiration. The modern scientific method, which guides sociological research, grew out of positivism. **Humanitarianism** is the principle that human reason can direct social change for the betterment of society. The thread of social reform that has always been a part of sociology can be traced to the humanitarian movement. The ideals of the Enlightenment are positivism, humanitarianism, and faith in human reason; they continue to permeate sociological thought (Nisbet 1970).

The Development of Sociology in Europe

The earliest sociologists promoted a vision of sociology grounded in careful observation. **Auguste Comte** (1798–1857), who coined the term *sociology,* believed that just as science had discovered the laws of nature, sociology could discover the laws of human social behavior. He is often identified as the founder of modern sociology.

Comte thought that understanding the laws of behavior would enable people to solve society's problems. In this sense, he asserted positivism as the basis for sociological research. He declared that scientific knowledge developed in stages, astronomy being first, followed by physics, chemistry, biology, and then sociology. Each was progressively more complex, with sociology, according to Comte, being the most highly evolved because it involved the study of society, which is a complex and changing system. Comte thought that sociology would become the "Queen of the Sciences."

Alexis de Tocqueville (1805–1859), a French citizen, traveled to the United States as an observer beginning in 1831. The book that followed from his experience, *Democracy in America,* is an insightful analysis of U.S. society and its democratic culture and structure. De Tocqueville thought that democratic and egalitarian values in the United States not only influenced American social institutions for the better but also transformed personal relationships. Less admiringly, he felt that in the United States the tyranny of kings had been replaced by the "tyranny of the majority." He was referring to the ability of a majority to impose its will on everyone else in a democracy. De Tocqueville also felt that despite the emphasis on individualism in American culture, Americans had little independence of mind and that individualism made people self-centered and anxious about their social position.

Another early sociologist was **Harriet Martineau** (1802–1876). One of eight children of a successful British surgeon and deaf from her teen years, Martineau embarked on a long tour of the United States in 1834. Like de Tocqueville, Martineau was fascinated by the newly emerging culture in America and she carefully analyzed the social customs she observed. Her subsequent book, *Society in America* (1837), was overlooked for many years, probably because she was a woman, but it is now recognized as a classic. Her other book, *How to Observe Manners and Morals* (1838), discusses how to observe behavior when one is a participant in the situation being studied—a classic work on the research method called participant observation (see Chapter 2).

Martineau was a strong feminist and abolitionist whose travels in the United States caused considerable controversy. She thought the subservience of women in the United States was comparable to slavery, and she compared women in England and the United States to servants. She thought slavery would tear the United States apart, although she respected the independence in American culture. Her abolitionist and feminist views finally forced her to restrict her travels to the northern United States because of threats to her life, but she continued to insist on the right of women to speak their minds (Hoecker-Drysdale 1992; Rossi 1973).

As one of the earliest observers of American culture, Harriet Martineau used the powers of social observation to record and analyze the social structure of American society. Long ignored for her contributions to sociology, she is now seen as one of the founders of early sociological thought.

Classical Sociological Theory

Of all the contributors to the development of sociology, the giants of the European tradition were Emile Durkheim, Karl Marx, and Max Weber. The works of these scholars form a classical tradition in sociology that continues to inform sociological thought.

Emile Durkheim established the significance of society as something larger than the sum of its parts. Social facts stem from society but have profound influence on the lives of people within society.

Emile Durkheim During the early academic career of the Frenchman **Emile Durkheim** (1858–1917), France was in the throes of the notorious Dreyfus Affair. Alfred Dreyfus was a Jewish captain in the French army who was scapegoated by the Army and the government in an attempt to cover up a spy scandal. Amid much public controversy, Dreyfus was convicted of treason in 1894, subjected to the humiliating ritual of having his badges of rank stripped from his uniform, and sent to Devil's Island. With the ritual demotion, the com-motion surrounding the affair quieted for a while. For years, France was bitterly split over the case, which rested on little more than clumsily forged documents and anti-Semitism (hatred of Jews).

Durkheim, himself Jewish, was fascinated by how the public degradation of Dreyfus in 1894 seemed to unify a large segment of the divided French public. Durkheim later wrote that public rituals have a special purpose in society, creating *social solidarity;* that is, social bonds link the members of a group. Some of Durkheim's most significant works explore the question of what forces hold society together.

According to Durkheim, people in society are glued together by belief systems (Durkheim 1947/1912). The rituals of religion and other institutions are public ceremonies that symbolize and reinforce the sense of belonging felt by insiders of a society or group. Drawing on this insight, Durkheim made a creative leap that helps explain social deviance. Deviant behavior is defined as actions that others perceive as violating the customary ways of doing things (see Chapter 7). But deviance has a purpose in society, Durkheim argued. By labeling some people as deviant, the society confirms a sense of normalcy in others, thus social solidarity results from condemning some behavior as deviant. Durkheim thought that deviance, like public rituals, sustains moral cohesion in society.

Durkheim viewed society as an entity larger than the sum of its parts. He described this as society *sui generis* (meaning, "a thing in itself"). Accordingly, society should be studied as something more than the sum of the individuals who compose it. Society is external to individuals, yet its existence is internalized in people's minds because people believe what society expects them to believe. Durkheim conceived of society as an integrated whole, each part contributing to the overall stability of the system. His work is the basis for *functionalism,* an important theoretical perspective that we will examine later in this chapter.

Durkheim's major contribution was the discovery of the *social* basis of human behavior. Durkheim conceptualized **social facts** as those social patterns that are *external* to individuals. Thus, it is a social fact that some groups are more likely to commit suicide than others. It is also a social fact that people with limited opportunities for work are more likely to commit crimes than others (though they are also more likely to be caught!). Social facts are not to be explained by biology or psychology, but are the subject matter of sociology. Durkheim held that social facts have their existence outside individuals, but pose constraints on individual behavior.

He proposed that social systems could be known through the discovery and analysis of social facts; this, he wrote, is the central task of the sociologist (Bellah 1973; Coser 1977; Durkheim 1950/1938, 1951/1897).

Karl Marx It is hard to imagine another scholar who has had as much influence on intellectual history as **Karl Marx** (1818–1883). Along with his collaborator, Friedrich Engels, Marx changed intellectual history, and world history, too.

Marx's work was devoted to explaining how capitalism shaped society. **Capitalism** is an economic system based on the pursuit of profit and the sanctity of private property. Marx used a class analysis to explain capitalism, describing capitalism as a system of relationships among different classes.

Karl Marx analyzed capitalism as an economic system with enormous implications for how society is organized, in particular, how inequality between groups stems from the economic organization of society.

Under capitalism, the capitalist class owns *the means of production,* the system by which goods are produced and distributed. To say that the capitalist class owns the means of production does not just mean that this class owns property; it also owns the system by which wealth is accumulated. The factories and machinery by which goods are produced; the ships, railroads, and airlines by which goods are distributed; the banks and financial institutions by which profits are managed; and the communications systems by which ideas supporting capitalist values are disseminated all comprise this system. In Marx's view, profit, the goal of capitalist endeavors, is produced by exploiting the working class. Workers sell their labor for wages, while capitalists make certain that wages are worth less than the value of what workers produce. The difference is profit for the capitalist class. According to Marx, the capitalist system is inherently unfair because it rests on workers getting less than they give.

Marx thought that the economic organization of society was the most important influence on people's lives—both on how they behave and how they think. He argued that the beliefs of the working people tend to support the interests of the capitalist system, not the interests of the workers themselves. Why? Because the capitalist class controls both the production of goods and the production of ideas. It owns the publishing companies, endows the universities where knowledge is produced, and controls the images seen in the media.

We will return to Marx's theories in Chapter 10. For now, the important point is that Marx considered society to be fundamentally shaped by economic forces. Laws, family structures, schools, and other institutions all develop to suit economic needs under capitalism.

Like other early sociologists, Marx took social structure as his subject instead of individual action. It was the *system* of capitalism that dictated people's behavior. Marx saw social change arising from the conflict between the capitalist and working classes.

Marx's ideas are often misperceived by U.S. students because communist revolutionaries throughout the world have claimed Marx as their guiding spirit. It would be naive to reject his ideas solely on political grounds. Much that Marx predicted has not occurred. For instance, he claimed that a worldwide revolution of workers was inevitable, and this has not happened. Still, he left an important body of sociological thought springing from his insight that society is systematic and structural and that class fundamentally shapes social behavior.

Max Weber The German sociologist **Max Weber** (1864–1920; pronounced "Vay-ber") was greatly influenced by Marx's work, but whereas Marx saw economics as the basic organizing element of society, Weber theorized that society had three basic dimensions: political, economic, and cultural. To Weber, a complete sociological analysis must recognize the interplay between economic, political, and cultural institutions (Parsons 1947). Weber is credited with developing a *multidimensional* analysis of society that goes beyond Marx's one-dimensional focus on economics.

Max Weber used a multidimensional approach to analyzing society, interpreting the economic, cultural, and political organization of society as together shaping social institutions and social change.

Weber also theorized extensively about the relationship of sociology to social and political values. He did not believe that a value-free sociology could exist, because values would always influence what sociologists studied. Weber thought sociologists should acknowledge the influence of values so that ingrained beliefs might not interfere with objectivity. He professed that sociologists should teach students the uncomfortable truth about the world. Teachers, he thought, should not use their positions to promote their political opinions, rather they have a responsibility to examine all opinions, including unpopular ones, and to use the tools of rigorous sociological inquiry to understand why people believe and behave as they do.

An important concept in Weber's sociology is **verstehen** (meaning, "understanding"; pronounced, "vershtay-en"). Verstehen refers to understanding social behavior from the point of view of those engaged in it. Weber believed that to understand social behavior one had to understand the meaning that a behavior held for social actors. He did not believe sociologists had to be born into a group to understand it. In other words, he did not believe "it takes one to know one," but he did think sociologists had to develop some *subjective* understanding of how other people experience their world. One of Weber's major contributions was the definition of **social action** as a behavior to which people give meaning (Gerth and Mills 1946; Parsons 1951b; Weber 1962). Social behavior is more than just action. People do things in a context and use their interpretive abilities to understand and give meaning to their action. Weber's work emphasized this interpretive dimension of social life.

Weber's work is far-reaching. He was interested in the rise of bureaucracy in the modern world, the nature of power and authority, the operation of class systems, the foundation of the work ethic in capitalist society, and the role of religion in supporting other social institutions. Weber's theories continue to have a profound impact on sociology.

The Development of American Sociology

American sociology was built on the earlier work of Europeans, but unique features of U.S. culture contribute to its distinctive flavor. Pragmatism is the belief in practicality and has led many early American sociologists to believe they could alleviate some of the consequences of social problems if they understood their causes.

Sociology and Social Change Early sociologists in both Europe and the United States conceived of society as an organism, a system of interrelated functions and parts that work together to create the whole. This perspective is called the **organic metaphor.** Sociologists saw society as constantly evolving, like an organism, but they debated whether humans could shape the evolution of society.

Some sociologists were influenced by the work of British scholar **Charles Darwin** (1809–1882), who had revolutionized biology when he identified the process by which evolution creates new species. **Social Darwinism** was the application of Darwinian thought to society. According to the Social Darwinists, the "survival of the fittest" is the driving force of social evolution. In Britain, **Herbert Spencer** (1820–1903) conceived of society as an organism that evolved from simple to complex in a process of adaptation to the environment. Spencer believed that society was best left alone to fol-

low its natural evolutionary course. Because he had faith that evolution always took a benign course toward perfection, he advocated a *laissez-faire* (that is, hands-off) approach to social change. Social Darwinism was a conservative mode of thought; it assumed that the current arrangements in society were natural and inevitable. Although Social Darwinism departed from some of sociology's general emphasis on social reform, it had an important influence on the development of sociology. It characterized society as a system that was stable, yet constantly evolving.

The American sociologist **William Graham Sumner** (1840–1910) was an ultraconservative thinker who claimed that survival of the fittest justified inequality in society. The rich and powerful, he felt, were those who were most successful in the competition for scarce resources. He called millionaires the "bloom of a competitive society" (Hofstadter 1944: 58). Utterly convinced that society followed an evolutionary path, Sumner thought that tampering with the social system would bring certain disaster.

Most other early sociologists in the United States took a more hands-on approach. **Lester Frank Ward** (1831–1914) thought sociology could be used to engineer social change. He called this *social telesis,* which is the idea that human intervention in the natural evolution of society would advance the interests of society. Ward and the reform-minded sociologists of the late nineteenth and early twentieth century were optimistic about the ability of sociology to address society's problems.

Early American sociologists identified industrialization and urbanization as the causes of social problems and wanted to use sociology to improve these conditions. This tradition has left an important legacy for contemporary sociologists: **Applied sociology** is the use of sociological research and theory to solve human problems.

The Chicago School The emphasis on application was especially evident at the University of Chicago, where a style of sociological thinking known as the *Chicago School* developed. Sociologists in the Chicago School were interested in how society shaped the mind and identity of a person, and they used social settings as human laboratories where sociologists could do scientific studies intended to address human needs.

Charles Horton Cooley (1864–1929), though at the University of Michigan most of his career, was one of the foremost theorists of the Chicago School. Cooley theorized that individual identity developed through people's understanding of how they are perceived by others. This was further elaborated in the work of **George Herbert Mead** (1863–1931), who like Cooley was interested in specifying the processes that linked the individual to society. Cooley and Mead saw the individual and society as interdependent. Individuals developed through the relationships they established with others. Society, the ever-changing web of social relationships, existed because it was imagined in the minds of individuals. We return to the work of Cooley and Mead in Chapter 4 when we discuss socialization, the process by which individuals are shaped to fit within society.

The focus on the individual espoused by the Chicago School reflects the importance of individualism in U.S. culture. The concern for the individual also stemmed from perceptions about modern society, which some feared isolated people and diminished community feeling.

Sociologists in Chicago worked close to home, using the city they lived in as a living laboratory. Race riots, immigrant ghettoes, poverty, and delinquency were all features of "modern" life in prewar Chicago, seeding the work of the Chicago theorists. The five-volume study of Polish immigrants in Chicago produced by **W. I. Thomas** (1863–1947) and **Florian Znaniecki** (1882–1958; pronounced, "Znan-yets-ki") remains a classic study of ethnicity, immigration, neighborhood segregation, and urban life (Thomas 1918, 1958/1919). W. I. Thomas promoted the use of personal documents to reveal how people interpreted their experiences. There are legends of Thomas sifting through the garbage of Polish immigrants to retrieve personal letters, which he used to study their social relationships and to learn how they felt about their lives in the New World. Thomas also thought that the influence of society was so strong that even if confronted with evidence to the contrary, people would behave according to what they thought was true. Thomas's famous dictum (developed in collaboration with Dorothy Swaine Thomas, sociologist and wife of W. I. Thomas) was, *"If men define situations as real, they are real in their consequences"* (Thomas and Thomas 1928).

Robert Park (1864–1944) was another founder of sociology from the University of Chicago. Originally a journalist who worked in several midwestern cities, Park was an experienced witness to urban problems and interested in how other races interacted with Whites. At the turn of the century, he was a regular correspondent and adviser to the African American leader Booker T. Washington. Fascinated by the sociological design of cities, he noted that cities were typically sets of concentric circles. The very rich and the very poor lived in the middle, ringed by slums and low-income neighborhoods. Progressing to the outer rings, the residents became increasingly affluent (Coser 1977; Park and Burgess 1921). Although this particular geometric shape of cities has changed with the growth of suburbs and the concentration of poverty in inner cities, Park would still be intrigued by how boundaries are defined and maintained in urban neighborhoods. You might notice this yourself in cities. A single street crossing might de-

lineate a Vietnamese neighborhood from an Italian one, a White affluent neighborhood from a barrio. The social design of cities continues to be a major subject of sociological research.

Many of the early sociologists of the Chicago School were women whose work is only now being rediscovered. Jane Addams, Marion Talbot, Edith Abbott, and others managed to produce excellent work at the University of Chicago, although sexist practices of that time excluded them from the careers only available to men. George Herbert Mead was among the men who publicly acknowledged their admiration for the work of these women, yet women were frequently marginalized in the profession, and their contributions have been largely neglected (Deegan 1990).

Jane Addams (1860–1935), close friend of and collaborator with W. E. B. Du Bois, was one of the most renowned sociologists of her day and a leader in the settlement house movement. Hull House, where Addams worked, provided community services and did systematic research designed to improve the lives of slum dwellers, immigrants, and other dispossessed groups. Although the University of Chicago was progressive in admitting women when it opened in 1892, it barred

Jane Addams, the only sociologist to win the Nobel Peace Prize, used her sociological skills to try to improve people's lives.

© CORBIS

hiring women as full professors. Addams, the only practicing sociologist ever to win a Nobel Peace Prize (in 1931), never had a regular teaching job. Instead, she used her skills as a research sociologist to develop community projects that assisted people in need (Deegan 1988).

Other women, including Edith Abbott and Sophinista Breckenridge, worked closely with some of the men at the University of Chicago, completing numerous studies on education, housing, working women, urban children, and other subjects. Many women, excluded from the universities, made their contribution in the newborn profession of social work instead of as academic sociologists. Over time, theoretical sociology and social work became highly gender-segregated, with men in universities doing abstract, theoretical work that was better paid and women doing applied work for less pay.

The Segregated Academy Two prominent sociologists (one Black and one White) have written, "As long as sociology has been in existence in the United

States, despite the segregation of American intellectual life, there has been a powerful tradition of the study of the Black community by Blacks" (Blackwell and Janowitz 1974: xxi). African American thinkers and others whose work has not been properly recognized are now being credited for their contribution to the history and evolution of the discipline (Barber et al. 1998), despite formidable obstacles to their success.

W. E. B. Du Bois (1868–1963; pronounced "due boys") was a prominent Black scholar, a co-founder in 1909 of the NAACP (National Association for the Advancement of Colored People), a prolific writer, and one of America's best minds. He studied for a time in Germany, hearing several lectures by Max Weber. His Ph.D. from Harvard was the first awarded by Harvard to a Black person in any field. While he was an Assistant Professor of Sociology at the University of Pennsylvania (1896–1898), he was assigned to a research project investigating neighborhood conditions in the Black community in Philadelphia. Du Bois was required to live in the settlement he studied, a situation he found to be harsh and unsatisfactory. His book from this study, *The Philadelphia Negro,* in collaboration with Isabel Eaton, a White sociologist from Columbia University, is one of the first empirical community studies. Du Bois was censured for denying the inferiority of African Americans, harassed for his leftist pronouncements by the U.S. government (which rescinded his passport in 1959), and indicted as an alleged "foreign agent." Optimistic in his youth, he grew bitter over racial issues in America by the end of his life. In 1961, at the age of ninety-three, he emigrated to Ghana, where he died in 1963 (Lewis 1993).

Du Bois's vision of sociology differed in some ways from that of the White European–American mainstream. He was deeply troubled by the racial divisiveness in society, writing in a classic essay published in 1901 that "the problem of the twentieth century is the problem of the color line" (Du Bois 1901: 354). Like many of his women colleagues, he envisioned a community-based, activist profession committed to social justice (Deegan 1988). At the same time, like Weber, he believed in the importance of a scientific ap-

An insightful observer of race and culture, W. E. B. Du Bois was one of the first sociologists to use community studies as the basis for sociological work. His work, long excluded from the "great works" of sociological theory, is now seen as a brilliant and lasting analysis of the significance of race in the United States.

proach to sociological questions. He thought that sociology should be used to study the pressing issues of the time. Although he believed convictions always directed one's studies, he also thought sociology should use the most rigorous tools of sociological research to ensure accuracy.

Sociology was not immune from the patterns of exclusion found in other areas of society. **Oliver Cromwell Cox (1901–1974)** is another African American theorist and researcher whose work has been largely neglected. Cox analyzed the racial prejudice, discrimination, and segregation of American society. He was especially interested in the origins of capitalism and used a comparative and global perspective in this work (Hunter and Abraham 1987). However, his minority status and his Marxist orientation led many sociologists to shun him. When **E. Franklin Frazier (1894–1962)** became president of the American Sociological Society in 1948 (now the American Sociological Association, or ASA), he was the first Black person elected. Since then, only two other Black people have been elected, William Julius Wilson in 1990 and Troy Duster in 2004.

The sociological profession is more diverse than it has ever been. Professional caucuses such as the Association for Black Sociologists, Sociologists for Women in Society, and the American Sociological Association's section on Latino Sociology and section on Asia and Asian America work to promote the inclusion of women and minorities in the discipline. These organizations have made the discipline more attentive to the experiences of these groups and have advocated social policies to make society more racially just. In doing so, they carry on the work of sociologists such as Du Bois, Addams, and others who were denied the same opportunities.

Theoretical Frameworks in Sociology

The founders of sociology have established theoretical traditions that ask basic questions about society and inform sociological research. For many students, the idea of theory seems dry because it connotes something that is only hypothetical and divorced from "real life." To the contrary, sociological theory is a tool that sociologists use to interpret real life. Sociologists use theory to organize their empirical observations, to produce logically related statements about observed behavior, and to relate observed social facts to the broad questions sociologists ask: How are individuals related to society? How is social order maintained? Why is there inequality in society? How does social change occur? Sociologists work within theoretical traditions to answer these questions. The major theoretical frameworks make different assumptions about society and

KEY SOCIOLOGICAL CONCEPTS

As you build your sociological imagination, you must learn certain key concepts to begin understanding how sociologists view human behavior. *Social structure, social institutions, social change,* and *social interaction* are not the only sociological concepts, but they are fundamental to grasping the sociological imagination.

Social Structure

Earlier, we defined **social structure** as the organized pattern of social relationships and social institutions that together constitute society. This is a critical, though abstract, concept in sociology. Social structure is not a "thing," but it refers to the fact that social forces, not always visible to the human eye, guide and shape human behavior. It is as if there were a network of forces embracing people and moving them in one direction or another. Acknowledging that social structure exists does not mean humans have no choice in how they behave, only that those choices are largely conditioned by one's location in society. We examine this concept further in Chapter 5.

Social Institutions

In this book, you will also learn about the significance of **social institutions,** defined as established and organized systems of social behavior with a particular and recognized purpose. The family, religion, marriage, government, and the economy are examples of major social institutions. More than individual behavior, social institutions are abstractions, but they guide human behavior nonetheless. C. Wright Mills understood that social institutions confront individuals at birth as given systems of behavior and that they transcend individual experience. Understanding the character of social institutions is one product of the sociological imagination.

Social Change

As you can tell, sociologists are also interested in the process of **social change.** As much as sociologists see society as producing certain outcomes, they do not see society as fixed or immutable, nor do they see human beings as passive recipients of social expectations. In other words, sociologists do not take a determinist approach to human behavior as if all is fixed and without any possibility for change, or even surprise. Society exists because people interact with one another in socially meaningful ways. Society is, after all, an abstraction, real only in that it reflects the total pattern of people's social interaction.

Social Interaction

Sociologists see **social interaction** as behavior between two or more people that is given meaning. Through social interaction, people react and change, depending on the actions and reactions of others. Society results from social interaction; thus, people are active agents in what society becomes. Because society changes as new forms of human behavior emerge, change is always in the works. As you read this book, you will see that these key concepts of social structure, social institutions, social change, and social interaction are central to the sociological imagination. ••●

---

thus have different answers to these perennial sociological questions.

Within sociology, the different theoretical perspectives can be grouped into macrosociological and microsociological approaches. In the realm of *macrosociology* are theories that strive to understand society as a whole. Durkheim, Marx, and Weber were macrosociological theorists. Theoretical frameworks that center on face-to-face social interaction are known as *microsociology.* Some of the microsociological work that has been derived from the Chicago School is research that studies individuals and group processes in society. Although sociologists draw from diverse theoretical perspectives to understand society, three broad traditions form the major theoretical perspectives that they use: functionalism, conflict theory, and symbolic interaction.

Functionalism

Functionalism originates in the work of Durkheim, who analyzed how society remains relatively stable.

Functionalism interprets each part of society as contributing to the stability of the whole. As Durkheim suggested, society is more than the sum of its component parts. The different parts are primarily the institutions of society, each organized to fill different needs, and each with particular consequences for the form of society. The major institutions of society include: families, education, religion, the economy, government, and health care (examined in Part IV of this book). Together, these institutions compose society.

From a functionalist perspective, each part of society is dependent on other parts, and each institution has consequences for the whole of society. The family, for example, has multiple functions. At its most basic level, the family has a reproductive role. Within the family, infants receive protection and sustenance. As they grow older, children learn the patterns and expectations of their culture. Across generations, the family integrates people into society, giving people a sense of continuity with the past and future. These aspects of family contribute to the stability and prosperity of society. The same is true for other institutions. The economy supplies a framework for the production and distribution of goods and services. Religion propagates a set of beliefs, which has proved to be one of the greatest sources of stability within societies.

The functionalist framework emphasizes consensus and order in society, focusing on social stability and shared public values. From a functionalist perspective, *disorganization* in the system leads to change because societal components must adjust to achieve stability. This is a key part of functionalist theory. When one part of society is not working (or *dysfunctional*), it affects all the other parts and creates social problems. Change may be for better or worse. Changes for the worse stem from instability in the social system, such as a breakdown in shared values or a social institution no longer meeting people's needs (Eitzen and Baca Zinn 2004).

Functionalism was the dominant tradition in sociology for many years, and one of its major figures was **Talcott Parsons** (1902–1979). In Parsons's view, all parts of a social system are interrelated. Different parts of society have four basic functions: (1) adaptation to the environment, (2) goal attainment, (3) integrating members into harmonious units, and (4) maintaining basic cultural patterns. This creates a system of interrelated parts with the parts being these behavioral goals. To Parsons, society worked like a machine and was always moving toward social stability. If one part malfunctioned, the other parts would quickly adjust to return the system to equilibrium (Parsons 1968/1937), as when a disaster strikes a community, and people organize to help one another.

Functionalism was further developed by **Robert Merton** (1910–2003). Merton elaborated an important point in functionalist theory: Social practices have consequences for society that are not always immediately apparent, and they are not necessarily the same as the stated purpose of a given practice. Durkheim found rituals to produce group solidarity, a function that may be remote from the apparent purpose of a ritual, such as the celebration of religious beliefs. Thus, solidarity is one function of ritual (Merton 1968). Merton suggested that human behavior has both manifest and latent functions. **Manifest functions** are the stated and open goals of social behavior. **Latent functions** are the unintended consequences of behavior (Merton 1968). For example, reforming social welfare programs may have the manifest function of reducing federal budget expenditures, but the policy may also have the latent function of increasing homelessness.

THINKING SOCIOLOGICALLY

What are the *manifest functions* of grades in college? What are the *latent functions*?

Conflict Theory

Conflict theory emphasizes the role of coercion and **power,** which is the ability of a person or group to exercise influence and control over others, in producing social order. Functionalism emphasizes cohesion within society, but conflict theory emphasizes strife and friction. Derived from the work of Karl Marx, conflict theory pictures society as fragmented into groups that compete for social and economic resources. Social order is maintained by domination, not consensus, with power in the hands of those with the greatest political, economic, and social resources. According to conflict theorists, when consensus exists it is because people are united around common interests, often in opposition to other groups (Dahrendorf 1959; Mills 1956).

In the conflict perspective, inequality is unfair but exists because those in control of a disproportionate share of society's resources actively defend their advantages. Coercion and social control, not shared values and conformity, bind people to society. Groups and individuals struggle over control of societal resources, trying to advance their own interests. Those with the most resources exercise power over the others; inequality is the result. Conflict theory gives great attention to class, race, and gender in society because these are seen as the grounds of the most pertinent and enduring struggles in society. One of the greatest contributions to sociology from conflict theory is its emphasis on class, race, and gender inequality and their influence on all dimensions of social life.

Conflict theorists see inequality as inherently unfair, unlike functionalists who find inequality benefiting society. The dominance of the most advantaged group even extends to the point of shaping the beliefs of others, by controlling public information and influencing institutions such as education and religion, where beliefs and ideas are produced. From the conflict perspective, power struggles between conflicting groups are also the source of social change.

Again, families provide an example of how conflict theorists analyze social institutions. Whereas functionalists see families as contributing to the stability of society, conflict theorists would be more likely to see families as reflecting systems of power in society. Thus, within families, gender roles are shaped by power relationships between men and women in society at large, resulting in the fact that men tend to have more power in families than women. But as economic and political change occurs in society, the power balance within families also changes—for example, as women become more financially independent. Conflict theorists would also interpret families in terms of their relationship to other systems of inequality. Family stability, for example, is influenced by poverty. From a conflict perspective, all families are situated within larger systems of power and inequality—systems that affect family life.

Conflict theory has been criticized for neglecting the importance of shared values and public consensus in society while overemphasizing inequality and social control. To critics, this overemphasizes the role of conflict in society while understating the importance of shared values and social cohesion. Like functionalist theory, conflict theory finds the origins of social behavior in the structure of society, but it differs from functionalism in emphasizing the importance of power.

Symbolic Interaction

The third major framework of sociological theory is **symbolic interaction theory.** Symbolic interaction considers immediate social interaction to be the place where "society" exists. Because of its emphasis on face-to-face contact, symbolic interaction theory is a form of microsociology, whereas functionalism and conflict theory are more macrosociological. Derived from the work of the Chicago School, symbolic interaction theory analyzes society by addressing the subjective meanings that people impose on objects, events, and behaviors. According to symbolic interactionists, people behave based on what they *believe,* not just on what is objectively true. Thus, society is considered to be *socially constructed* through human interpretation (Berger and Luckmann 1967; Blumer 1969; Shibutani 1961). Symbolic interactionists see meaning as constantly modified through social interaction. People interpret one another's behavior, and these interpretations form social bonds.

A study of African American mothers and sickle cell anemia illustrates symbolic interaction theory. Sickle cell anemia is a disease resulting from a defective gene that causes impaired circulation of the blood. The disease is painful, debilitating, and can result in early death. Why would a mother who was told she carries the sickle cell trait decide nevertheless to have children, knowing her child would be susceptible to this disease if the father also carries the sickle cell trait? The sociologist who studied this found that such mothers construct a definition of the situation that shapes their reproductive behavior. As low-income African American women, they mistrust the medical system and thus define the warnings as unreliable. Moreover, especially among low-income women who are devalued by society, motherhood can confer high status (Hill 1994). In the end, the mothers' definition of the situation is more influential in determining their behavior than the objective presence of a potentially dangerous condition.

Symbolic interaction theory interprets social order as being constantly negotiated and created through the interpretations people give to their behavior. Symbolic interactionists seek not simply facts but "*social constructions,*" which are meanings attached to things, whether they are concrete symbols (such as a tattoo) or nonverbal behaviors. To a symbolic interactionist, society is highly subjective; it exists in the imaginations of people even though its effects are real. For example, thinking that the Holocaust never happened, despite all historical evidence to the contrary, is likely to produce anti-Semitic (that is, anti-Jewish) behavior.

Symbolic interaction theory is also found in the work of sociologist **Erving Goffman** (1959, 1963a). Goffman worked in a **dramaturgical model** of society—a perspective that sees society like a stage (that is, a drama) wherein social actors are "on stage," projecting and portraying social roles to others (that is, their audience). The dramaturgical model of society stems from theories of symbolic interaction. For Goffman, people present themselves to others by giving off impressions that are consistent with the definition of a given situation. Goffman conceptualized society—and its stability—as coming from people's performances in social roles. People act as if society is real and, therefore, it is! We examine Goffman's work further in Chapters 3 and 5.

Symbolic interaction theory is criticized by some for its emphasis on subjectivity. It is regarded as understating power relations and inequality in society, except as they appear in immediate social interaction. Functionalists and conflict theorists, however, consider society to have an objective reality. It takes each theoretical perspective to develop a complete understanding of how society works.

Functionalism, conflict theory, and symbolic interaction theory are not the only theoretical frameworks in sociology, but they provide the most prominent general explanations of society. Functionalism highlights the order and cohesion that characterizes society. Conflict theory emphasizes inequality and power in society. Symbolic interaction emphasizes the meanings that humans give to their behavior. Most sociologists working on current studies examine their data from a variety of theoretical approaches. Together, these frameworks provide a rich, comprehensive perspective on society, individuals within society, and social change (see Table 1.1).

Diverse Theoretical Perspectives

In addition to the three frameworks just discussed, other theories are used within sociology. Contemporary sociological theory has been greatly influenced by the development of *feminist theory,* which analyzes the status of women in society with the purpose of using that knowledge to better women's lives. Feminist theory has created vital new knowledge about women and trans-

Table 1.1
Three Sociological Frameworks

Basic Questions:	Functionalism	Conflict Theory	Symbolic Interaction
What is the relationship of individuals to society?	Individuals occupy fixed social roles.	Individuals are subordinated to society.	Individuals and society are interdependent.
Why is there inequality?	Inequality is inevitable and functional for society.	Inequality results from a struggle over scarce resources.	Inequality is demonstrated through the importance of symbols.
How is social order possible?	Social order stems from consensus on public values.	Social order is maintained through power and coercion.	Social order is sustained through social interaction and adherence to social norms.
What is the source of social change?	Society seeks equilibrium when there is social disorganization.	Change comes through the mobilization of people struggling for resources.	Change evolves from an ever-evolving set of social relationships and the creation of new meaning systems.
Major Criticisms:	This is a conservative view of society that underplays power differences among and between groups.	The theory understates the degree of cohesion and stability in society.	There is little analysis of inequality and it overstates the subjective basis of society.

formed what is understood about men. Feminist scholars have shown that traditional theoretical perspectives have distorted women's experiences and thus given an inadequate picture of society. When Karl Marx investigated social class, for example, labor was analyzed exclusively in the context of wage earners; traditional Marxist thinkers have therefore been ill-equipped to analyze the social class of housewives. Feminist scholarship in sociology, by focusing on the experiences of women, provides new ways of seeing the world and contributes to a more complete view of society.

Exchange theory argues that the behavior of individuals is determined by the rewards or punishments they receive in day-to-day interaction with others (Homans 1961, 1974). Exchange theory has been especially useful in analyzing subjects such as social networks. *Rational choice theory* is used by political scientists and economists as well as sociologists. Rational choice theory posits that the choices human beings make are guided by reason. Society is seen as the sum of these individual decisions and actions (Smelser 1992b).

Many contemporary theorists are increasingly influenced by **postmodernism.** Postmodernism is based on the idea that society is not an objective thing, but is found in the words and images—or *discourses*—that people use to represent behavior and ideas. Postmodernist theory challenges the idea that anything, including society, is objectively real. Postmodernists believe that images and texts reveal how people think and act.

Such studies typically involve detailed analyses of images, words, film, music, and other forms of popular culture. Because society is now saturated by images from the mass media, postmodernist analyses illuminate much about society. Postmodernist thinkers see contemporary life as involving multiple experiences and interpretations, but they avoid categorizing human experience into broad and abstract concepts such as institutions or society (Rosenau 1992).

Some sociologists rely primarily on one theoretical perspective and choose the topics they study accordingly. Others draw from multiple theoretical perspectives, guided by the perspective that best explains a question at hand. Someone studying the sociology of families, for example, might draw from postmodernism to examine the significance of cultural imagery in producing popular definitions of family life. Using conflict theory, you might study how corporate collapse affects a family's retirement options. Feminist theory might reveal how power within the family is shaped by gender. Relying on a single theory can produce an incomplete explanation, so sometimes several theoretical models are needed. Whatever theoretical framework one uses, theory is evaluated in terms of its ability to explain observed social facts. The sociological imagination is not a single-minded way of looking at the world. It is the ability to observe social behavior and interpret that behavior in light of societal influences.

Chapter Summary

What is sociology?

Sociology is the study of human behavior in society, and the *sociological imagination* is the ability to see societal patterns that influence individual and group life. Sociology is an *empirical* discipline, relying on careful observations as the basis for its knowledge.

What does debunking mean?

The *debunking* tendency in sociology refers to the ability to look behind things taken for granted, looking instead to the origins of social behavior.

Why is diversity important to the study of sociology?

One central insight of sociology is its analysis of social diversity and inequality. Understanding *diversity* is critical to sociology because it shapes most social and cultural institutions and is necessary to analyze these institutions.

What influenced the development of sociology as a field of study?

Sociology emerged in western Europe during the Enlightenment and was influenced by the values of *critical reason, humanitarianism,* and *positivism. Auguste Comte,* one of the earliest sociologists, emphasized sociology as a positivist discipline. *Alexis de Tocqueville* and *Harriet Martineau* developed early and insightful analyses of the American future.

What are some of the basic insights of classical sociological theory?

Emile Durkheim is credited with conceptualizing society as a social system and with identifying *social facts* as patterns of behavior that can be explained as external to individuals. *Karl Marx* showed how capitalism shaped the development of society; he theorized economic systems as basic to the structure of society. *Max Weber* thought that society had to be explained through cultural, political, and economic factors. Weber also wrote that sociologists should use *verstehen,* which is the ability to see things from the point of view of others to understand social behavior.

What is distinct about the development of sociology in the United States?

Most American sociologists believed in using sociological research and theory to solve social problems. The Chicago School is characterized by its concern with the relationship of the individual to society and the use of society as a human laboratory. Numerous scholars are identified with this school, including *Charles Horton Cooley, George Herbert Mead, W. I. Thomas, Florian Znaniecki, Robert Park,* and *Jane Addams.*

What roles have women and racial minorities had in the history of sociology?

Many women and African American sociologists have made important contributions to the development of the discipline even though their work has gone largely unrecognized until recently. *W. E. B. Du Bois* used his sociological training to develop extensive studies of race in American society.

What are the major frameworks of sociological theory?

There are three major theoretical frameworks in sociology: functionalism, conflict theory, and symbolic interaction theory. *Functionalism* emphasizes the stability and integration in society. *Conflict theory* sees society as organized around the unequal distribution of resources, held together through power and coercion. *Symbolic interaction theory* emphasizes the role of individuals in giving meaning to social behavior, thereby creating society. Different theoretical perspectives are used by sociologists, depending on the questions they are trying to answer. All theories should be evaluated in terms of their ability to explain observed social facts.

Key Terms

applied sociology 17	positivism 14
capitalism 16	postmodernism 23
conflict theory 21	power 21
debunking 7	social action 17
diversity 11	Social Darwinism 17
dramaturgical model 21	social facts 15
	social institution 4
empirical 7	social structure 5
Enlightenment 14	sociological imagination 5
functionalism 20	
humanitarianism 14	sociology 2
issues 5	symbolic interaction theory 22
latent functions 21	
manifest functions 21	troubles 5
organic metaphor 17	verstehen 16

Researching Society with MicroCase Online

You can see the results of actual research by using the Wadsworth MicroCase® Online feature available to you. This feature allows you to look at some of the results from national surveys, census data, and other data sources. You can explore this easy-to-use feature on your own, but try this example. Suppose you want to know:

What percentage of the population says their views are conservative and does this change over time?

To answer this question, go to http://sociology.wadsworth .com/andersen_taylor4e/, select MicroCase Online from the left navigation bar, and follow the directions there to analyze the following data.

Data file: U.S. Trends

Analysis Task: Historical Trends

Variable One: Pol.View

Questions

Once you have your results, answer the following questions:

1. What did you find?
2. Has conservatism increased or decreased over the last thirty years or so?
3. Do you think there is any connection between the growth or decline of conservatism and political events in the society?
4. What does this tell you about the sociological imagination—that is, the influence of social and historical context on people's lives and viewpoints?

The Companion Website for Sociology: Understanding a Diverse Society, *Fourth Edition*

http://sociology.wadsworth.com/andersen_taylor4e/

Supplement your review of this chapter by going to the companion website to take one of the Tutorial Quizzes; use the flash cards to master key terms; and check out the many other study aids you'll find there. You'll also find special features such as GSS Data and Census 2000 information, data and resources at your fingertips to help you with that special project or do some research on your own.

Suggested Readings and Web Resources

Becker, Howard. 1986. *Writing for Social Scientists.* Chicago, IL: University of Chicago Press.
This engaging book is not only an excellent discussion of how to improve one's writing, but it is also a good guide to sociological thinking. Premised on the idea that poor writing produces errors in thinking, Becker's book will help students at all levels improve their sociological perspective and their ability to communicate it.

Berger, Peter. 1963. *Invitation to Sociology.* New York: Doubleday.
Berger's classic book provides an introduction to how sociologists conceptualize society and individuals within it. It is a good way to engage students in sociological thinking.

Du Bois, W. E. B. 1994/1903. *The Souls of Black Folk.* New York: Dover Publications.
This classic book explores African American life and consciousness in a racially unequal society. It showcases the sociological mind of one of the great thinkers of the twentieth century.

Johnson, Allan G. 1997. *The Forest and the Trees: Sociology as Life, Practice, Promise.* Philadelphia, PA: Temple University Press.
This brief introduction to sociology provides an excellent overview of the basic concepts of the field and the application of sociology to interpreting contemporary social issues. The book also grounds its discussion in an awareness of the diversity among different groups in the United States.

Lengermann, Patricia Madoo, and Jill Niebrugge-Brantley. 1998. *The Women Founders: Sociology and Social Theory, 1830–1930, A Text with Readings.* New York: McGraw-Hill.
This book documents the many contributions of women sociologists. Overlooked for years, these scholars have had productive careers and added to sociological knowledge and practice.

Mills, C. Wright. 1959. *The Sociological Imagination.* New York: Oxford University Press.
This book is a classic statement of the meaning and significance of the sociological perspective.

American Sociological Association
www.asanet.org
The American Sociological Association is a national organization of sociologists. Its home page provides information on annual meetings, resources available for sociological research, information on careers in sociology, special reports on current initiatives in social policy, and information on how students can get involved in this sociological organization, including the Minority Fellowship Program.

Society for the Study of Social Problems
funnelweb.utcc.utk.edu/~SSSP
This national professional group sponsors an annual meeting and publishes one of the major journals in sociology, Social Problems.

Society for Applied Sociology
www.appliedsoc.org
This is an international organization of those applying sociology in a variety of settings.

Regional Sociological Associations
There are several regional sociological associations, all of which welcome student participation:

Eastern Sociological Society
www.essnet.org

MidSouth Sociological Society
www.oakron.edu/hefe/mssapage.html

Midwest Sociological Society
www.drake.edu/MSS/home.html

North Central Sociological Association
www2.hanover.edu/ncsa/

Pacific Sociological Society
www.csus.edu/psa/

Southwestern Sociological Association
www.sfasu.edu

Southern Sociological Association
www.MsState.edu/org/SSS/sss.html

● ● ●

Doing Sociological Research

Sociological research is the tool sociologists use to answer questions. Sociologists do research in a variety of ways, all of which involve rigorous observation and careful analysis. The method that you use depends on the kind of question you are asking.

Suppose you wanted to know how homeless people lived. What is life like for them? How dangerous is it? Where are the homeless to be found? Do they interact and associate with each other? Do they work at all, and if so, at what? Are they organized? Do they feel rejected by society? How do other people on the street treat them? Do they really sleep on park benches at night? What if it is too cold and the homeless shelters are all filled up?

Sociologist Mitch Duneier (1999) in his study "Sidewalk" wanted to know all these things, plus more. So he decided to study a group of homeless people by living with them. And that is exactly what he did: He lived with them on their park benches and in doorways on New York City's lower East Side. He spent

four years with them. He interacted with them. He worked with them—a group consisting largely of African American men, who sold books and magazines on the street. Duneier himself is White: He tells how becoming accepted into this society of African American men was itself an interesting process. Contrary to popular belief, he discovered that these men comprised a rather well organized "mini-society," with a social structure, rules, norms, and a culture. He discovered many fascinating and largely unknown elements of this "sidewalk" society.

Duneier was engaged in what is called *participant observation*—a sociological research technique in which the researcher actually becomes both participant in as well as observer of that which she or he studies.

There are other kinds of research that sociologists do, as well. Some approaches are somewhat more structured and focused than participant observation, such as survey research. These different approaches to socio-

Sociology⊛Now™
Reviewing is as easy as ❶❷❸.

Use SociologyNow to help you make the grade on your next exam. When you are finished reading this chapter, go to the chapter review for instructions on how to make SociologyNow work for you.

logical research reflect some of the different research methods sociologists use to answer the questions they ask. Some questions require more direct involvement and participation with the individuals one is studying, as does participant observation; other methods involve more statistical analysis. Either way, the chosen research method must be appropriate to the sociological question being asked.

In this chapter, we will examine the research methods that sociologists use to answer the different kinds of questions they ask. As you will see, research is fundamental to the sociological enterprise—whether it is the analysis of surveys or interviews, the observation of people in unusual settings, scrutiny of media documents, examination of historical records, or any other method sociologists use. Research is an engaging and demanding process. It requires careful observation and the ability to think logically about the things that spark your sociological curiosity. Sociologists ask many different questions, and their techniques for answering questions are limited only by the horizon of the sociological imagination. •••

The Research Process

When sociologists do research, they engage in a process of discovery. Like scientists who do research in a laboratory, sociologists organize their research questions and procedures systematically. Their laboratory is the social world. Through research, they organize their observations and interpret them.

Sociology and the Scientific Method

Sociological research derives from the **scientific method,** originally defined and elaborated by the British philosopher **Sir Francis Bacon** (1561–1626). The scientific method involves several steps in a research process, including observation, hypothesis testing, analysis of data, and generalization. Since its beginnings, sociology has attempted to adhere to the scientific method.

The research process involves several operations that can be performed on the computer, such as entering data in numerical or text form and writing the research paper.

© Brian Strickland/Zuma Press

To the degree that it has succeeded, sociology is a science; yet, there is also an art to developing sociological knowledge. The process is difficult to capture as a simple series of steps or procedures, as sociological research does not necessarily proceed in a sequence of rigid steps.

For more than a hundred years, a lively debate has been ongoing about whether sociology is a "true" science. Many argue that it is a science, albeit a "little science"—it employs the scientific method, but not to the same degree as some other sciences, such as physics or chemistry (Mazur 1968). Sociology aspires to be both scientific and humanistic, whereas purists contend that it can be only one or the other. The crux of the scientific method is observation and the testing and building of theory. Science is **empirical,** as we noted in Chapter 1, meaning it is based on careful and systematic observation, not just conjecture or belief. Sociological studies may be based on surveys, observations, and many other forms of analysis, but always they depend on an empirical foundation. Although some sociological studies are highly quantitative and statistically sophisticated, others are qualitatively based—that is, based on more interpretive observations, not statistical analysis. Both quantitative and qualitative studies are empirical. Both are important in sociology.

Sociological knowledge is not the same as philosophy or personal belief. Philosophy, theology, and personal experience can deliver insights into human behavior, but at the heart of the scientific method is the notion that a theory must be testable. This requirement distinguishes science from purely humanistic pursuits such as theology and literature. Imagination and fascination with humanity may drive the creation of theories in theology and literature, as much as in the social sciences.

One wellspring of sociological insight is **deductive reasoning,** wherein specific inferences are derived from general principles, such as sociological theory. Here is an example of deductive reasoning: one might reason deductively that because Catholic doctrine forbids abortion, Catholics answering a survey would therefore be less likely than other religious groups to support abortion rights. **Inductive reasoning**—another source of sociological insight—reverses this logic, by arriving at general conclusions from specific observations (see Figure 2.1). Observing that most of the demonstrators protesting abortion in front of a family planning clinic are evangelical Christians, one might infer that strongly held religious beliefs are important in determining human behavior. This is an example of inductive reasoning. This in turn might stimulate a research project on religious behavior.

Developing a Research Question

Sociologists use various methods to explore questions, but a lot of sociological research is an organized practice that can be described roughly in a series of steps. However it needs to be stressed that by no means does all sociological research rigidly follow the steps outlined in Figure 2.1. In fact, a large amount of sociological research is largely inductive, even at the start, rather than deductive.

The first step in sociological research is to develop a research question. One source of questions is past theory or past research. For any number of reasons, the sociologist might disagree with a research finding and decide to carry out further research or develop a detailed criticism of the prior research or theory. Someone else's work may suggest an idea, seem incomplete, or appear to be seriously in error. Criticism of past research has been an important inspiration for many investigations, and the research of other workers can often be refined. The best research frequently inspires other sociologists to study a question further.

Developing a sociological research question typically involves reviewing the existing literature on the subject, such as past studies and research reports. (The sociologist often uses the term "literature" to mean past studies and past research reports.) A subject interesting enough to spur research has probably attracted the attention of other researchers, and their work can help define a research question or guide a researcher toward an original path. Digital technology has vastly simplified the task of reviewing literature. Researchers who once had to burrow through paper indexes and card catalogs to find material relevant to their studies can now scan much larger swaths of literature in far less time, using online databases. The catalogs of most major libraries in the world are accessible via the Internet, as are specialized indexes, discussion groups, and other research tools developed to assist sociological researchers. Increasingly, many journals that report new sociological research are now available online and can

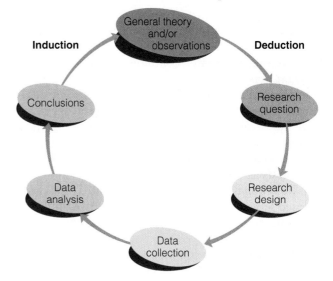

FIGURE 2.1 The Research Process

often be read electronically in full text, by means of JSTOR or other computerized databases that allow full-text viewing on-screen. Although some books are now available online, most research is still done with paper books.

Sometimes a review of past research may cause you to wonder if the same result would be found again if the study were repeated, perhaps examining a different group. Research that is repeated exactly, but on a different group of people or in a different time or place, is called a **replication study**. As an example, suppose earlier research found that women managers have fewer opportunities for promotion than men. You might want to know if this still holds several years later. You would then replicate the original study, probably using a different group of women and men managers, but asking the same questions that were presented earlier.

Sociological research questions can also come from casual observation of human behavior. Perhaps you have observed the seating patterns in your college dining hall and wondered why people sit with the same group day after day. Does the answer point to similarity on the basis of race, gender, age, or perhaps political views (Tatum 1997)? Answering this question would be an example of induction: Going from a specific (your observation of seating patterns at lunch) to a generalization (a theory about the effects of race and gender).

Social policy or practical goals are frequently the source of research questions. As politicians have debated welfare policy in recent years, they have often indulged in careless and inaccurate pronouncements about welfare recipients. For example, they may claim that welfare encourages women to have more children for higher benefits. Is this true? Sociological research shows that women do not have more children to receive benefits nor do most become dependent for a long time. (Butler 1996; Mink 1999).

Sociological research is often developed to test an existing theory; in this respect, sociological theory becomes the source of the research question. A theory can stand only as long as research continues to support it. This is what makes sociology a scientific endeavor. Some sociological theories have withstood years of challenges. Other theories are in vogue only a short time before the weight of additional research causes them to be rejected. Commonly, a new theory is substituted for the one rejected. The new theory then undergoes the same tests as the old one.

We saw in the previous chapter that sociologists have devised several theoretical frameworks to interpret what they observe. One's theoretical orientation can channel what questions are asked and how they are answered. For example, suppose you want to understand why some women are more likely to be poor than men. You may have concluded already that conflict theory offers the best approach to your question. You might then do a study that examines, among other things, differences in the work done by men and women, how much they earn, how much they spend on themselves, and how much they spend on their children. If you want an extremely precise answer, you might separate the data for Latinos, Whites, Asians, Native Americans, and African Americans because poverty is known to vary among different racial groups. When done, you might ask what additional insights are given by also considering symbolic interaction theory or functional theory.

Creating a Research Design

Once you have developed a research question, you must decide how to proceed with your investigation. A **research design** is the overall logic and strategy underlying the research. Research design consists of choosing the observational technique best suited to a particular research question. Sociologists engaged in research may distribute questionnaires, interview people, or make direct observations in a social setting. They might analyze cultural artifacts, such as magazines, newspapers, television shows, or other media. Some researchers use historical records; others base their work on the analysis of social policy. All these techniques for gathering information are forms of sociological observation.

Suppose that you want to study the career goals of student athletes. In reviewing past studies, you might find an article that discusses how athletics is related to academic achievement (Schacht 1996; Messner 1996). You may also have read an article in your student newspaper reporting that the rate of graduation for women college athletes is much higher than the rate for men athletes and you wondered if women athletes are better students than men athletes. In other words, are athletic participation, academic achievement, and gender interrelated and, if so, how?

Your research design lays out a plan for investigating these questions. It includes details of your study. Which athletes will you focus on? How will you study them? To begin, you must get sound data on the graduation rates of the groups you are studying to verify that your assumption of better rates among women athletes is true. Perhaps you think the differences between men and women are not so great when the men and women play the same sports, or perhaps the differences depend on other factors, such as how the students were recruited and what kind of financial support they get. Do graduation rates vary for women and men depending on the kind of college they attend? If you want to answer these questions, you have to build a comparison of college types into your research design and study different colleges. The details of your research design flow from the specific questions you ask.

The research design often involves the decision to do research that, for the most part, is either qualitative, quantitative, or perhaps some combination of both. **Qualitative research** is somewhat less structured, yet focused on a question being asked. Qualitative research does not make extensive use of statistical methods and allows for more interpretation and nuance in what people say and do; thus, it can provide a very in-depth look at a particular social behavior. **Quantitative research** uses statistical methods. Both forms of research are useful, and both are used extensively in sociology. Quantitative studies provide data from which one can calculate average incomes, the percent employed, and so forth. Qualitative research also yields data, but data that cannot be easily reduced to numbers. It is simply called *qualitative data*. Thus, if you wanted to study the measurable factors that produce differences in group earnings, you would likely use a quantitative approach. But if you were more interested in how employees perceive their work environment, you would likely use a more qualitative research design. Both approaches yield data, but qualitative data are not easily reduced to numbers (Babbie 2004).

Some research designs involve the testing of hypotheses. A **hypothesis** (pronounced "hy-POTH-isis") is a prediction, a tentative assumption that one intends to test. If you have a research design that calls for the investigation of a very specific hunch, you might formulate one hypothesis or, as is true in many sociological studies, a series of hypotheses ("hypothesis" is singular; "hypotheses" is plural). Hypotheses are often formulated as if–then statements. For example:

> Hypothesis: If a person's parents are racially prejudiced, then that person will, on average, be more prejudiced than a person whose parents are relatively free of prejudice.

This is a hypothesis, not a demonstrated fact or finding. Having phrased the hypothesis, the sociologist must then determine if it is true or false. If the hypothesis is shown to be correct, that tends to support (though not necessarily prove) the reasoning behind it. To test the preceding hypothesis, one might take a large sample of people and determine their level of prejudice via interviews or some other mechanism. One would then determine the level of prejudice of their parents. According to the hypothesis, one would expect to find an association between relatively prejudiced children and prejudiced parents, and between nonprejudiced children and nonprejudiced parents. If the association is found, the hypothesis is supported. If no such association is found, then the hypothesis would be rejected.

Not all sociological research follows the model of hypothesis testing. *Exploratory research* is more open-ended, or based on broader observations of human behavior. Any research design, whether exploratory or not, includes a plan for how **data** will be gathered and also analyzed. Quantitative data are numerical and can be analyzed using statistical techniques. Although not numerical in nature, qualitative data can be logically interpreted, classified, and analyzed. Sociologists often try to convert their observations into a quantitative form, using techniques that are discussed later in the chapter in "Statistics in Sociology." Quantifying the responses to questionnaires or interviews, for example, helps categorize the data, makes it possible to analyze qualitative responses statistically, and removes an element of subjectivity from the interpretation of research results. The point is that data can be qualitative or quantitative; either way, they are still data. (Note: "Data" is the plural form; one says "are still data," not "is still data.") Research data are the heart of sociological research, and the research design specifies what kinds of data an investigator plans to use.

A **variable** is a characteristic that can have more than one value or score. A variable can be relatively straightforward, such as age or income; or a variable may be more abstract, such as social class or degree of prejudice. In much sociological research, variables are analyzed to understand how they influence each other. With proper measurement techniques and a good research design, the specific relationships between different variables can be discerned.

Sociologists frequently design research to test the influence of one variable on another. Most of the time, they are trying to establish cause and effect. The **independent variable** is the one that the researcher wants to test as the presumed cause of something else. The **dependent variable** is the one upon which there is a presumed effect. That is, if X is the independent variable, then X leads to Y, the dependent variable. In the previous example of the hypothesis, the amount of prejudice of the parent is the independent variable, and the amount of prejudice of the child (as an adult) is the dependent variable. The researcher here is arguing that

FIGURE 2.2 The Analysis of Variables
Sociological research often seeks to find out whether some independent variable (X) affects an intervening variable (Z), which in turn affects a dependent variable (Y).

the higher the parent's prejudice, then the higher the child's prejudice. In some sociological research, **intervening variables** are also studied: that is, variables that fall between the independent and the dependent variables (see Figure 2.2). In the hypothesis example, the researcher might reason that higher parental prejudice (X) results in the parent negatively stereotyping minorities (Z, the intervening variable), which in turn makes the child more prejudiced (Y). In this respect, it is the intervening variable Z that *connects* the independent and dependent variables.

There may be other variables that affect those being studied. Sociological variables are often sensitive to a variety of influences, and trying to control for the influence of factors other than the ones being studied is a challenge to the creativity of the researcher.

Sociological research proceeds via the study of concepts. A **concept** is any abstract characteristic or attribute that can be potentially measured. Prejudice is a concept, as are some of the terms introduced in Chapter 1, such as social class, power, and social institution. These are not things that can be seen directly, although they are key concepts in the field of sociology. When sociologists want to study concepts, they must develop measures for "seeing" them.

THINKING *SOCIOLOGICALLY*

If you wanted to conduct research that examined the relationship between student alcohol use and family background, what *indicators* would you use to measure these two variables? How might you *design* your study?

Variables are sometimes used to show more abstract concepts that cannot be directly measured, but can only be indirectly measured—such as the concept of social class. In such cases the sociologist uses **indicators**—variables that point to or reflect the abstract concept. An example is shown in Map 2.1 on the United Nation's "human development index." Here, the human development index is an *indicator* used to show different levels of well-being around the world. The index involves several items, including life expectancy, educational attainment, and standard of living, all combined together to show the levels of well-being in different countries.

MAP 2.1 The Human Development Index

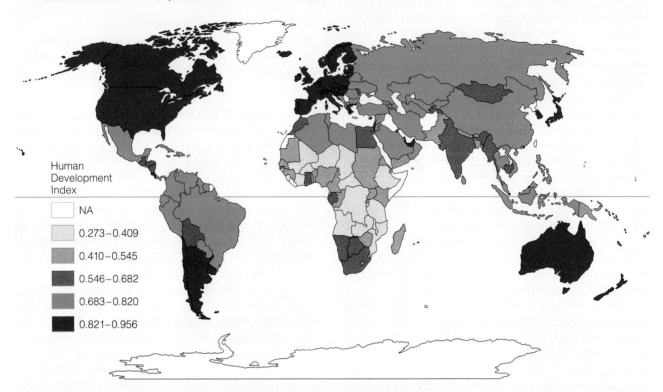

Human
Development
Index

□	NA
	0.273–0.409
	0.410–0.545
	0.546–0.682
	0.683–0.820
	0.821–0.956

The Human Development Index is an indicator developed by the United Nations used to show the differing levels of well-being in nations around the world. The index is calculated using a number of measures, including life expectancy, educational attainment, and standard of living. Are these reasonable indicators of well-being? What else might you use?

Data: Human Development Report 2003. Website: www.undp.org

The **validity** of a measurement (indicator) is the degree to which it accurately measures or reflects a concept. One step in research design is deciding whether to develop new procedures or to borrow from previous research. Other researchers may have investigated the same topic and designed questionnaires or other research instruments that measure the concept one is working on. In this case, a researcher must make a decision about the validity of earlier research. Is it accurate? Is it up-to-date? To ensure the validity of their findings, researchers usually measure more than one indicator for a particular concept. If two or more chosen indicators of a concept give similar results, the measurements likely are giving an accurate—that is, valid—depiction of the concept.

Sociologists also must be concerned with the **reliability** of their research results. A measurement is reliable if a repeat of the measurement gives the same result. If a person is given a test daily, and every day the test gives different results, then the reliability of the test

is poor. One way to ensure that sociological measurements are reliable is to use measures that have proved to be consistent in past studies. Another technique is to have a variety of people gather the data, in order to make sure that the results are not skewed by the tester's appearance, personality, and so forth. The researcher must be sensitive to all factors that affect the reliability of the study.

Sometimes sociologists want to gather data that would almost certainly be unreliable if the subjects knew they were being studied. Knowing that they are being studied may cause people to change their behavior, a phenomenon in research called the *Hawthorne Effect*—an effect first discovered while observing work groups at a Western Electric plant in Hawthorne, Illinois. An example of this effect is a professor who wants to measure student attentiveness by observing how many notes are taken during class. Students who know they are being scrutinized will magically become more diligent.

Gathering Data

After the research design comes data collection. During this stage, the researcher interviews people, observes behaviors, or collects facts that throw light on the research question. When sociologists gather original material, the product is known as *primary data.* Examples include the answers to questionnaires or notes made while observing group behavior. Often, sociologists rely on *secondary data,* which have already been gathered and organized by some other party. A large collection of data is known as a *data set* or *database* and would include national opinion polls, census data, national crime statistics, or data from an earlier study made available by the original researcher. Secondary data may also come from official sources, such as university records, city or county records, national health statistics, or historical records. Historical archives are a source of secondary data that can give insight into social change or give historical perspective on contemporary issues. Such historical data could be used for qualitative analysis, rather then quantitative analysis.

Since 1972, the National Opinion Research Center (NORC) at the University of Chicago has annually conducted the General Social Survey (GSS), an important source of secondary data frequently used by sociologists. A sample of about 1500 Americans is surveyed each year, producing a gold mine of data for secondary analysis. Questions are asked on a large range of topics. Basic demographic data are gathered, such as occupation, religion, and hours worked, as well as opinions on topics such as abortion, marijuana use, consensual homosexual sex, welfare spending, gun control, and capital punishment.

Another major source of raw material for sociologists has been census data generated by the U.S. Census Bureau. Census data yield detailed information on standards of living, income, educational attainment, language spoken in the home, birth and death rates, and many other specific items, broken down by race and ethnic group, gender, region, age, and many other variables. The census is a giant database from which a great deal of information about the United States can be extracted. Sociologists have used it extensively in their research.

An ongoing issue affecting the use of census data is the *census undercount problem.* Certain groups are more likely to be undercounted than others: transients and homeless people, criminals, both legal and illegal immigrants, those who fear, dislike, or distrust the government, and those who are unable to fill out a questionnaire because of illiteracy, mental illness, or other causes. It is known that members of minority groups are undercounted, particularly African Americans and Hispanics. Minorities tend to be less trusting of government bureaucracies such as the U.S. Census Bureau,

© Rob Crandall/The Image Works

Taking the census involves people completing what is called a survey. Here, a person is completing the Census Bureau form.

resulting in levels of compliance that are lower than the general population. Sociologists need to estimate the extent of the undercount in census data and take it into account in their research.

Because many federal, state, and local government programs distribute money based, in part, on census data, the undercount problem has potentially serious repercussions. Cities and special-interest groups have been particularly active in addressing the inaccuracies in the U.S. Census. If the census undercounts a city, then it will be shortchanged when public funds are distributed.

Analyzing the Data

After data have been collected, whether through direct observation (primary analysis) or the use of existing data (secondary analysis), they must be analyzed. **Data analysis** is the process by which sociologists organize collected data to discover the patterns and uniformities that the data reveal. The analysis may be statistical, or it may be qualitative. Survey data require statistical treatment, whereas a study based on an analysis of rap music lyrics, for example, would likely have a large qualitative component. Conclusions and generalizations can be made only when the data analysis is completed.

If data have been gathered through a survey or interviews, they will likely have to be converted into numerical form. Coding is the process by which data are categorized and put into a form suitable for either quantitative or qualitative analysis. Careful coding is critical because inaccuracies would pollute the data set.

MAP 2.2 Regional Attitudes Toward Abortion

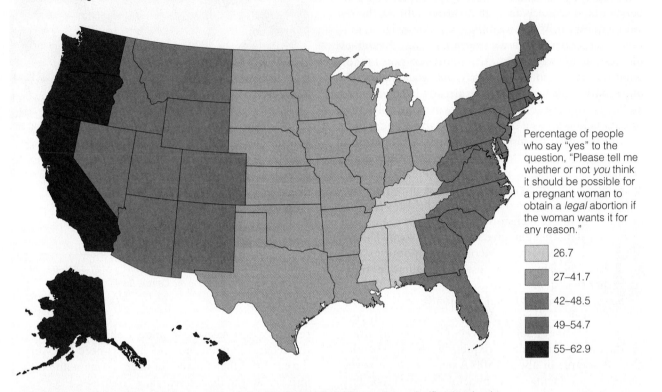

Percentage of people who say "yes" to the question, "Please tell me whether or not *you* think it should be possible for a pregnant woman to obtain a *legal* abortion if the woman wants it for any reason."

	26.7
	27–41.7
	42–48.5
	49–54.7
	55–62.9

Attitudes on various public opinions can vary by region. The General Social Survey shows significant regional differences in attitudes toward legal abortion. What regional differences does this survey (the General Social Survey) show? In general, what overall region is least in favor of legal abortion? In general, what region is most in favor? What would you study to explain such differences? Can you locate your own state/county? How did the people in it feel about abortion?

Data: General Social Survey, 1994

Data analysis is labor intensive, but it is also an exciting phase of research. Here is where research discoveries are made. Sometimes while pursuing one question, a researcher will stumble across an unexpected finding, referred to by researchers as *serendipity.* A serendipitous finding is something that emerges from one's study that was not anticipated, perhaps the discovery of an association between two variables that one was not looking for or some pattern of behavior that was outside the scope of one's research design. Such findings can be minor sidelines to one's major conclusions or, in some cases, lead to major new discoveries. They are a part of the excitement of doing sociological research. (A classic example of serendipity was the discovery, in the Hawthorne studies, that it was not the variables being studied (lighting, the pay system, length of rest periods, and so on) that affected the productivity of people in work groups; is was the fact of their being studied that increased their productivity.)

Computers greatly facilitate the process of analyzing data. Computational operations that were once done by hand, perhaps over weeks, can now be performed by computer in seconds. Especially valuable is the ability to perform complex operations on large data sets, those containing many thousands or even millions of data points. Once, large data sets were stored on cards with holes punched in them. Data sorted by date might then be laboriously resorted by region. Nowadays, computers can instantly resort data by any number of variables. Once data are entered in a spreadsheet or statistical program, they can be stored, retrieved, sorted, compared, and analyzed with ease. Perhaps most useful of all, they can be shared because digital data are easily reproduced, transported, and merged with other data. Many data are now instantly accessible on the Internet, and researchers continue to create public electronic archives for their own data and that of colleagues working in similar areas.

Reaching Conclusions and Reporting Results

After the data are gathered and analyzed, the background reading is done, and the files are organized, the researcher arrives at the final stage in research: developing conclusions, relating findings to sociological theory and to past research, and reporting the findings in the professional literature.

An important question researchers will ask at this stage is whether their findings are generalizable. **Generalization** is the ability to draw conclusions from specific data and apply them to a broader population (see Figure 2.1). One asks: Do my results pertain only to those people who were studied, or also to the world beyond? Assuming that the results have wide application, one

can then ask if the findings refine or refute existing theories. In addition to generalization, you can also ask if the research has direct application to social issues. Using the earlier example of measuring prejudice, if you found that increased contact between members of different ethnic groups lessened group prejudice, what programs for reducing prejudice might your study suggest?

As you make conclusions from research, you typically relate your conclusions to the theoretical questions you asked from the start, and you are also likely to think about new questions that your research inspires. This would also be an example of inductive reasoning. You might also think about the implications of your study, either for social policy or for further understanding of the issue you have addressed (see the box "Sociology in Practice: What Causes Violence?"). An im-

SOCIOLOGY IN PRACTICE

What Causes Violence?

A bomb in Olympic Park, drive-by shootings, nightly news of murders in the cities, children carrying guns to school: These incidents all raise fears that the United States is an increasingly violent society. What causes the violence that we witness in society? The American Sociological Association (ASA), the national professional organization for sociologists, has addressed this question and identified violence as an important area for new sociological research. Using funds from the Spivack Program in Applied Social Research and Social Policy, the ASA sponsored a national workshop on the social causes of violence, asking these experts to develop new research questions that would shed new light on the causes of violence. With this new research, sociologists hope to identify social policies that would help reduce the violence in society.

What questions did the experts think were most important to address?

- How can violence within and among different racial and ethnic groups be

This person is a victim of criminal violence. The many causes of violence are of concern to the sociologist.

© Chris Brown/CORBIS

explained? Too often statistical data on violence look only at two categories—White and Black. Other groups are vulnerable to violence, and we need to know more about them.

- How and why are poverty and inequality linked to violence? Sociologists already know much about this question, but are societies with more rigid social class boundaries—or those moving in that direction—most likely to experience increased violence?

- Why is violence prevalent among young people? Young people seem increasingly vulnerable to violence, yet this area has not been studied much.

- What are the characteristics of different communities that foster or discourage violence? Sociologists know that social context (a community, a group, a society) is an important factor in shaping behavior, but little research has been done looking at communities, not just violent of-

fenders, as an explanation of violence. Similarly, how do social institutions, especially families, schools, and workplaces, encourage or discourage violent behavior?

- What is the effect of the media on the promulgation of violence? This is a much debated question, especially because violence is increasingly used as a form of entertainment. Research evidence shows that violence in the media does cause at least some violence, especially among children.

- How do different people, groups, and institutions respond to violence in effective or counterproductive ways? Answering this question might identify constructive ways to reduce violence.

Critical Thinking Exercise

1. If you were an expert on this ASA panel, what questions would you pursue?

2. What would your answers show federal, state, and local groups if they want to direct their resources to reduce violence?

Source: Levine, Felice J., and Katherine J. Rosich. 1996. *Social Causes of Violence: Crafting a Science Agenda.* Washington, DC: American Sociological Association.

•••

portant part of your research, however, is sharing your results with others.

Researchers report their results in several ways. They often discuss their research with others, either students in a classroom, colleagues at a regional or national conference, or perhaps the media. Results are typically also reported in writing, perhaps via publication in a professional sociology journal, such as the *American Sociological Review*, the *American Journal of Sociology, Social Problems*, or *Gender & Society*. Research may also be published as a book, in which case it is called a *research monograph*. Many sociologists also write books for more general audiences, sometimes even producing two books based on the same research. One book is more technical and methodologically detailed and another version interprets the results in a more accessible style for a general readership. Another alternative is to publish in the form of a *research report* for limited distribution, say, to interested colleagues, certain members of the U.S. Congress, or to some federal, state, or local agency.

The Tools of Sociological Research

Sociologists use several tools to gather data. Among the most important are survey research, participant observation, controlled experiments, content analysis, historical research, and evaluation research. These are not the only sociological research methods. Other methods are used less often, and innovative scholars can even invent entirely new ones, but the research tools we are about to examine are the ones that have consistently proved themselves most useful.

The Survey: Polls, Questionnaires, and Interviews

Whether in the form of a questionnaire, interview, telephone poll, or email surveys are among the most commonly used tools of sociological research. Typically, a survey questionnaire will solicit data about the respondent, such as income, occupation, educational level, age, race, and gender, coupled with additional questions that throw light upon a particular research subject. In *closed-ended* questions, people must reply from a list of possible answers, like a multiple-choice test. For *open-ended questions,* the respondent is allowed to elaborate on her or his answer. Closed-ended questions are generally analyzed quantitatively, and open-ended questions are generally analyzed qualitatively. Thus, a survey can involve both qualitative as well as quantitative research.

Questionnaires are typically distributed to large groups of people. A fundamental requirement of successful questionnaires is an efficient mechanism for getting the questionnaires returned. The *return rate* is the percentage of questionnaires returned of all those distributed. A low return rate introduces possible bias because the small number of people may not represent the whole population. The researcher must be clever about following up after the original distribution and inducing recipients to complete and return the questionnaire.

Like questionnaires, interviews provide a structured way to ask people questions. They may be conducted face to face, by phone, or by email. Interview questions may be open-ended, closed-ended, or a combination of the two.

As a research tool, surveys offer characteristic advantages that are linked to their structured formats. The orderly presentation of questions and answers makes it possible to systematically compare respondents, to ask specific questions about many topics, and then to perform sophisticated analyses in search of patterns and relationships among variables. The disadvantages of surveys arise from their rigidity (see Table 2.1). Whereas data analysis may be greatly simplified by questions that require only prescribed responses ("a, b, c, d, or none of the above"), such responses may not accurately capture the opinions of the respondent. Furthermore, written surveys, and even those conducted by phone, fail to capture nuances in behavior and comportment that would be obvious and informative to a direct observer. Such nuances in behavior can

These individuals are interviewing people by phone in a survey and at the same time entering the responses into the computer for analysis.

© Jeff Greenberg/Photo Researchers

Table 2.1
A Comparison of Six Research Techniques

Technique (Tool)	Qualitative Analysis or Quantitative Analysis	Advantages	Disadvantages
The survey (polls, questionnaires, interviews)	Usually quantitative	Permits the study of a large number of variables; results can be generalized to a larger population if sampling is accurate	Difficult to focus in great depth on a few variables; difficult to measure subtle nuances in people's attitudes
Participant observation	Usually qualitative	Studies actual behavior in its home setting; affords great depth of inquiry	Is very time-consuming; it is difficult to generalize beyond the research setting
Controlled experiment	Usually quantitative	Focuses on only two or three variables; able to study cause and effect	Difficult or impossible to measure large number of variables; may have an artificial quality
Content analysis	Can be either qualitative or quantitative	A way of measuring culture	Limited by studying only cultural products or artifacts (music, TV programs, stories, other), rather than people's actual attitudes
Historical research	Usually qualitative	Saves time and expense in data collection; takes differences over time into account	Data often reflect biases of the original researcher and reflect cultural norms that were in effect when the data were collected
Evaluation research	Can be either qualitative or quantitative	Evaluates the actual outcomes of a program or strategy; often has direct policy application	Limited in the number of variables that can be measured; maintaining objectivity is problematic if research is done or commissioned by administrators of the program being evaluated

often be captured by participant observation, to which we now turn.

Participant Observation

A unique and interesting way for sociologists to collect data and study society is to become part of the group they are studying. This is the method of **participant observation,** and it represents a type of qualitative research in sociology. Two roles are played at the same time: subjective *participant* and objective *observer*. Usually, the group is aware that the sociologist is studying them, but not always. The group might be a youth gang, religious sect, table dancers, a class of first-year medical students, or denizens of a corner bar. Participant observation is sometimes called **field research,** a term derived from anthropology. Like anthropologists, participant observers typically go to the places where research subjects are found and, to some degree, adopt their ways.

Participant observation combines subjective knowledge gained through personal involvement and objective knowledge acquired by disciplined recording of what one has seen. The subjective component supplies a dimension of information that is completely lacking

The men in this bar, as shown by participant observation studies, reveal status differences among themselves such as "regulars" versus "winos."

in survey data and similar techniques. The potency it adds to interpretations of human interaction has proved to be well worth the burden of managing the rich and sometimes subjective raw material.

Street Corner Society (1943), a classic work by sociologist William Foote Whyte, documents one of the first qualitative participant observation studies ever done. Whyte studied the Cornerville Gang, a group of Italian men whose territory was a street corner in Boston in the late 1930s and early 1940s. Although not Italian, Whyte learned to speak the language, lived with an Italian family, and then infiltrated the gang by befriending the gang's leader, whose pseudonym was Doc. Doc was the **informant** for Whyte, a person with whom the participant observer works closely to learn about the group. For the duration of the study, Doc was the only gang member who knew that Whyte was doing research on the group.

Most social scientists of the 1940s thought gangs were socially disorganized, random deviant groups, but Whyte's study showed otherwise. He found that the Cornerville Gang, and by implication other urban street-corner gangs, was a highly organized mini-society with its own social hierarchy, morals, and practices, and its own punishments for deviating from the norms or rules of the gang.

DEBUNKING *SOCIETY'S MYTHS*

Myth: People who are just hanging out together and relaxing do not care much about social differences between them.

Sociological research: Even casual groups, such as the street-corner gang in William Foote Whyte's "Street Corner Society" research, or the street people in Duneier's "Sidewalk" study, have organized social hierarchies. That is, they make distinctions within the group that give some higher status than others.

The power of participant observation studies was shown dramatically in another study by a Black sociologist, Judith Rollins, who studied Black domestic workers (Rollins 1985). Rollins examined the relationships between Black domestic workers and their White employers by posing as a maid and hiring herself out to employers in the Boston, Massachusetts, area. She discovered that Black women domestics were treated by their employers as virtually invisible and inferior. White employers, women as well as men, often openly discussed the domestic worker with other Whites, right in front of the domestic worker, as if the Black woman herself were not even present. One woman employer even turned down the heat in the house when she left even though Rollins was expected to stay there. Through participant observation, Rollins was able to observe directly and intimately how social class status combines with race and gender to shape the world of Black domestic workers.

Participant observation has a few built-in weaknesses as a research technique. It is very time-consuming, since the researcher must cull meaningful data and comparisons from vast amounts of field notes and possibly even tape recordings. The studies usually focus on fairly small groups, and care must be taken not to generalize too widely from such a group. These limitations aside, participant observation has been the source of some of the most arresting and valuable studies in sociology (see for example the box "Doing Sociological Research: 'Sidewalk,' A Participant Observation Study").

Controlled Experiments

Controlled experiments are highly focused ways of collecting data and are especially useful for determining a pattern of cause and effect. To conduct a controlled experiment, two groups are created. An *experimental group* is exposed to the causal factor one is examining, and the *control group* is not. All other conditions must be equal for both groups. In a controlled experiment, external influences are either eliminated or *equalized*, that is, held constant. This is necessary to establish cause and effect.

Suppose you wanted to study whether violent television programming causes aggressive behavior in children. You could conduct a controlled experiment to investigate this question. The behavior of children would be the dependent variable (the presumed effect); the independent variable (the presumed cause) would be exposure to violent programming. To investigate your question, you would expose an experimental group of children under controlled conditions to a movie con-

Playing a violent videogame often causes the player, especially if a man, to be somewhat more aggressive afterward.

© AP Wide World Photo/Rodney White

taining lots of violence like martial arts, or gunfighting. The control group would watch a second movie that is free of violence. Aggressiveness in the children is measured twice: a *pretest* measurement made before the movies are shown, and a *posttest* measurement made after the movies. You would take pretest and posttest measures on both the control and the experimental groups.

In one such series of studies (Taylor et al. 2003; Bushman 1998; Berkowitz 1974, 1984) an experimental group of children saw a film clip showing boxers fighting. A control group of children watched a neutral film clip. For the posttest, aggressiveness was measured by determining the willingness of each child to deliver a painful electric shock to another child, and the amount, in volts, of the electric shock. Aggressiveness was also measured on both groups, before seeing the movie (the pretest), to establish a baseline measure of aggressive-

ness. It was found that the children in the experimental group delivered more electric shocks and more voltage per shock than the control group children.

In another example of a controlled experiment, researchers wanted to know if viewing pornography makes men more violent toward women. The pornography industry has argued against a link between violence against women and pornography, but many feminist organizations argue the opposite. Who is right?

An insightful series of experiments was conducted to answer this question (Mullin and Linz 1995; Donnerstein and Berkowitz 1981). Men were first assigned randomly to three separate groups:

- The control group saw a nonsexual film.
- One experimental group saw a nonviolent erotic film showing consensual sex between a man and a woman.

DOING SOCIOLOGICAL RESEARCH

"Sidewalk," A Participant Observation Study

Participant observation, which consists for the most part of *field research,* has become an important cornerstone of sociological knowledge. One such study is the recent detailed field research of Mitchell Duneier, *Sidewalk* (1999), which is filled with hundreds of excellent photographs as well. Researcher Duneier spent the better part of four years literally living as a homeless person on the streets of New York City's Greenwich Village. He interacted on a daily basis with the mainly Black, mainly poor street vendors there who sold ("hawked") magazines and used books on the street. He also interacted with other persons engaged in similar activities, such as providing (mostly unwanted) windshield washes to cars stopped at traffic lights. He got to know many of these vendors and others intimately as he lived with them, often on park benches and homeless shelters, and learned in great detail the language and ways of the vendor street society.

Most people would probably think that these street vendors are primarily homeless and simply represent a deviant and poor group existing on the fringes of society. Duneier found otherwise: These street vendors constitute a highly socially organized subsociety

that is closely interdependent with outer society and exists with it in a mutually beneficial relationship. The street vendors benefit from the "outside," and those from the outside society who come into contact with them also benefit. Street vendor society has its own culture of language, norms, habits, practices, "ways," and even songs, and its own stratified social structure.

Hakim Hasan, a successful vendor on the street, was greatly admired by the other vendors and occupied a high-status position of leadership in street vender society. A number of the older and knowledgeable vendors (called "old heads" in this culture) act as mentors to younger ones, teaching them the ins and outs of vendoring, as well as how to survive on the street.

The study asks such questions as: How do the sidewalk vendors manage to live in the face of exclusion, stereotyping, and stigmatization by others on the basis of both race and class? How do their acts intersect with a city's mechanisms to regulate its public spaces?

Finally, Duneier's study contains a useful appendix on the methodology of participant observation, which is filled with good advice. He was acutely aware of his own role as participant observer, and especially sensitive to how he

would be perceived by the Black and largely poor community of vendors. Duneier is White, Jewish, and of upper-middle class origins. He had to cope with a variety of perceptions that the vendors initially had of him: a naïve White man; a Jew who was there only to make money "off of" them while writing an expose on street people; or a journalist or writer who was simply trying to "state the truth about what was going on." Eventually, Duneier was accepted as somewhat of a fixture on the street. They called him "professor." He sold magazines and used books himself.

Questions to Consider

1. Do you think you might want to study sidewalk culture yourself? If so, how long do you think it would take you? *Keywords: sidewalk culture*

2. How much about street vendors do you feel you already know? *Keywords: street vendors*

We have included InfoTrac College Edition keywords at the end of each question to make it easier for you to find more to read on these topics. Go to www.infotrac-college.com, an online library, to begin your search.

Source: Duneier, Mitchell. 1999. *Sidewalk.* New York: Farrar, Straus and Giroux. •••

• Another experimental group saw a film that combined sex and violence, showing a man tying up a woman, tearing her clothes off, slapping her, and then raping her.

The researchers measured the effects of the movies on the men by testing the willingness of the men in each group to deliver an electric shock to a woman who posed as a subject for the experiment, but was secretly working with the researchers. A questionnaire given after the experiment confirmed that almost all men did not know that the woman was in alliance with the researchers.

There were striking results. The men who watched the nonsexual film or the purely erotic film had little willingness to shock the woman secretly collaborating in the experiment. The men who saw the sexually violent film were willing to deliver considerably more electric shocks and shocks of higher intensity. The researchers concluded that sexually violent pornography causes men to behave with more aggression and hostility toward women; purely erotic film does not. Given that all the men were randomly assigned across the three groups beforehand, you cannot conclude that the men in the group seeing sexual violence were more inclined beforehand to behave in a hostile way toward women.

Among its advantages, a controlled experiment can establish causation, and it can zero in on a single independent variable. On the downside, controlled experiments can be artificial. They are for the most part done in a contrived laboratory setting (unless it is a field experiment), and they tend to eliminate many real-life effects. Analysis of controlled experiments requires judgments about how much the artificial setting has affected the results (see Table 2.1, p. 37).

Content Analysis

Researchers can learn a vast amount about a society by analyzing cultural artifacts such as magazine stories and magazine ads, TV commercials, novels, biographies, soap operas, movies, popular music, rap music, graffiti, and all other kinds of cultural elements. **Content analysis** is a way of using cultural artifacts to measure what people write, say, see, and hear. The researcher studies not people but the communications or documents they produce as a way of creating a picture of the society. Content analysis involves primarily qualitative, but also quantitative, analysis. Content analysis is frequently used to measure cultural change and to study different aspects of culture such as manners, morals, slang, customs, and so on.

Children's books have been the subject of many content analyses, in acknowledgment of their impact on the development of youngsters. Three sociologists compared images of Black Americans in children's books from the 1930s to the present (Pescosolido et al. 1997). The researchers focused on four things: (1) how frequently the images appeared, (2) how they had changed over time, (3) the portrayal of interaction between African American adults, and (4) whether multicultural portrayals were consistent across different kinds of children's books. The authors obtained four important findings. First, they found a declining representation of African Americans from the 1930s through the 1950s, with practically no representation from 1950 through 1964. Beginning in 1964, an increase in representation lasted until the mid-1970s when the appearance of African American characters leveled off. Second, they found that the images of African Americans did change significantly over time. In the 1960s—a period of much racial protest—African Americans were mostly portrayed in "safe," distant images. Third, they found few portrayals of Black adults in intimate, egalitarian, or interracial relationships. Finally, they found that award-winning books were more likely to represent African Americans.

Content analysis has the advantage of being an *unobtrusive* (or "nonreactive") measure. It has little effect on the people being studied, because the cultural artifact has already been produced, and the individuals producing it are not studied directly. Thus, they cannot "react" to it. Content analysis is limited in what it can study, given that it is based only on communication, either visual, oral, or written. It cannot reveal what people think about these images or whether they affect people's behavior. Other methods are required to answer those questions.

Historical Research

Historical research examines sociological themes over time. It is commonly done in historical archives, such as official records, church records, town archives, private diaries, or oral histories. Historical research is usually qualitative, but quantitative approaches are also used—for example analysis of trends.

Oral histories have been especially illuminating, most dramatically in revealing the unknown histories of groups that have been ignored or misrepresented in other historical accounts. For example, when developing an account of the spirituality of Native American Indians, one would be misguided to rely solely on the records left by Christian missionaries or U.S. Army officials. These records would give a useful picture of how Whites perceived Native American Indian religion, but they would be a very poor source for discovering how Native American Indians understood their own spirituality. In a similar vein, the writings of a slave owner can deliver fascinating insights into slavery, but a slave owner's diary will certainly present a different picture of slavery as a social institution than will the written or oral histories of former slaves themselves.

Sociologists must be discriminating in how much credence they give to historical materials about different groups, as well as by different groups, and they must be creative in discovering the material that brings light to a darkened past. Handled properly, comparative and historical research is rich with the ability to capture long-term social changes and is the perfect tool for sociologists who want to ground their studies in historical perspectives.

Evaluation Research

Many sociologists use their research to make social policy recommendations. On the opposite side of the coin, research is often used to evaluate the effectiveness of social policy that has already been implemented. **Evaluation research** assesses the effect of policies and programs on people in society. If the research is intended to produce policy recommendations, then it is called **policy research.** The research can be either qualitative or quantitative.

Suppose you want to know if an educational program is improving student performance. You could design a study that measured the academic performance of two groups of students, one that participates in the program and one that does not. If the academic performance of students in the program is better, and if the groups are alike in other ways, then you would conclude that the program was effective.

Closely allied with evaluation research is **market research.** In market research, the aim is to evaluate the sales potential of some product or service. Perhaps a company wants an estimate of the number of people in a population who are likely to purchase a new toothpaste. The company could hire a researcher who would survey a number of people to find out how they rate various types and brands of toothpaste. Or, the researcher might invite a group of people to participate in a sales pitch and then study whether these persons increase their ratings after the sales pitch compared to their ratings before the sales pitch. A still further stage in the market research could determine whether people's attitudes (what they think) predict their behavior (what they do). The researchers would then see if a connection could be made between what they thought about the toothpaste (the attitude) and whether or not they purchased it later (the behavior).

Prediction, Sampling, and Statistical Analysis

Not all sociological research is based on scientific precision, but all sociologists learn to understand some basic scientific concepts and techniques. Central to scientific work is the ability to predict certain outcomes.

Likewise, knowing how to draw a scientific sample and organize one's data into meaningful statements is critical to sociological studies. In this section, we review the basic concepts of prediction and probability in sociology, the importance of sampling, and some elementary uses of statistics in sociological research.

Prediction and Probability

The essence of science is prediction and explanation. If you understand the principles of a physical system, you can predict how it will behave. You can predict the outcome of chemical reactions and the time that the moon will rise, but can you ever predict what humans will think or say or do under specified conditions? Are human beings too unique and too mercurial to conform to scientific predictability? The answer to both questions is, roughly, yes. Human social, behavioral, and attitudinal characteristics can be measured and, within a margin of error, can be predicted. This notion is a cornerstone of sociology; however, there are qualifications. Humans are not mechanical beings, and there is a special quality to the predictions sociologists make about people.

Sociology analyzes, explains, and predicts human social behavior in terms of trends and probabilities. A **probability** is the likelihood that a specific behavior or event will occur. You cannot predict that a specific individual will die within the next five years, yet you can predict how *likely* it is that a person will die in the next five years knowing only his or her age, and you can predict how many people in a large population will die in the next five years to within a fine margin of error. Age is an excellent predictor of the chances, or odds, of dying: The older a person is, the greater the chances of dying within the next five years. Actuarial tables that compare risks and premiums are commissioned by insurance companies that use probability statistics and a wide range of data to tabulate life expectancies.

Many sociological conclusions are stated in terms of probability. Here is an example:

> If you are a Black male, aged sixteen to twenty-nine, you are six times more likely to be arrested for a given offense than a White male of the same age, from the same neighborhood and the same socioeconomic background, for the same offense, and with the same prior arrest record (D'Allessio and Stolzenberg 2003; Zuberi 2001).

These statistics about arrests reflect a general pattern. However, it is not possible to say with certainty that a specific person will adhere to the pattern. This difference between pattern and prediction can be seen in research on violence. Data show that nearly all men who murder their wives have previously assaulted them (Browne and Williams 1993). Yet, sociologists cannot predict which of the men who assault their wives will go on to murder them, and which will not. Sociologists can predict patterns of violence, but they cannot predict specific incidents of violence (Best 1999).

Sampling

Sociologists typically study groups, but often the groups they want to study are so large or so dispersed that research on the whole group is impossible. To construct a picture of the entire group, they take data from a subset of the group and extrapolate to get a picture of the whole. A **sample** is any subset of a population. A **population** is a relatively large collection of people (or other units) that a researcher studies and about which generalizations are made. Suppose a sociologist wants to study the students at your school. All the students together constitute the population being studied. A survey could be done that reached every student, but conducting a detailed interview with every student would be highly impractical.

A smiling Harry S. Truman, who was elected president of the United States in 1948, holds up the infamous morning newspaper headline, based on an erroneous Gallup Poll result, stating incorrectly that his opponent (Thomas E. Dewey) has won the election.

How is it possible to draw accurate conclusions about a whole population by studying only part of it? The secret lies in making sure that the sample is representative of the population as a whole. The sample should have the same mix of people as the larger population, and in the same proportions. Obtaining what is called a scientific **random sample** approximates this. With a random sample, everyone in the population is given an equal chance of being selected from the population to be in the sample. The sample is said to be *biased* if it is not representative of the entire population, namely, if not everyone in the population is given an equal chance of being in the sample.

A biased sample can lead to grossly inaccurate conclusions. A famous example is the Gallup Poll done in 1948 before the November presidential election. After surveying U.S. voters, Gallup confidently predicted that Thomas E. Dewey, the Republican candidate and governor of New York, would win the election over Harry S. Truman, the Democratic candidate. Faith in the Gallup Poll was strong enough that one major metropolitan newspaper, the *Chicago Tribune*, went to press late on the night of the election with a banner headline wrongly proclaiming Dewey the winner! Despite the polling data, Harry Truman took office as the thirty-third president of the United States.

How could the Gallup researchers have made such an embarrassing error? The answer is that they inadvertently used a biased sample. In the three prior presidential elections, the Gallup Poll organization had used a *quota sample*. A set number of people (the quota) were polled in a selection of key counties across the na-

tion and had successfully predicted the winners of the past elections. The same sample was used again in 1948, but in the 1930s and 1940s, people had begun to migrate in great numbers from rural farms and towns to the cities. The United States had become less rural and more urban, as well as more ethnically diverse. Because the Gallup organization did not revise their quota sample to reflect the change in the U.S. population, rural people were *overrepresented* in the poll and urban people were *underrepresented*. Urban people proved more likely to vote Democratic than rural people. Gallup missed the changing trend, and the photo of a smiling president-elect Truman holding aloft the headline "Dewey Wins!" has become an icon reminding researchers of the consequences of botched data gathering. This underscores how sampling and bias are more than methodical terms: They can affect outcomes as important as electing a president (Dershowitz 2001).

Statistics in Sociology

Certainly not all sociological research is quantitatively based, but quantitative research is an important part of sociology, and all sociologists must have at least basic quantitative skills to interpret the research in their field.

The fundamental statistical tools of research are the percentage, rate, mean, and median. A **percentage** is the same as parts per hundred. To say that 22 percent of U.S. children are poor tells you that for every one hundred children randomly selected from the whole population, approximately twenty-two will be poor. A rate is the same as parts per some number, such as per 10,000 or 100,000. The homicide rate in the year 2003 was 5.7 killed per 100,000 (Federal Bureau of Investigation 2003). That means that for every 100,000 in the population, 5.7 were murdered. A rate is meaningless without knowing the numerical base on which it is calculated—it is always the number per some other number.

A **mean** is the same as an average. For example, adding a list of fifteen numbers and dividing by fifteen gives the mean for the list. The **median** is often confused with the mean but is different. The median is the midpoint in a series of values arranged in numerical order. In a list of fifteen numbers arrayed in numerical order, the eighth number (the middle number) is the median. If there were sixteen numbers, the median would be the number halfway between the eighth and ninth numbers. In some cases, the median is a better measure than the mean because the mean can be skewed by extremes at either end. Another often-used measure is the **mode**, which is simply the value or score that appears most frequently in a set of data.

To illustrate the difference between mean and median using national income distribution, suppose that you have a group of ten people. Two make $10,000 per year, seven make $40,000 per year, and one makes $1 million per year. If you calculate the mean (the "average"), it comes out to $130,000. The median is $40,000, which is a figure that more accurately suggests the income profile of the group. That single million-a-year earner dramatically distorts, or skews, the picture of the group's income. If we want information about how the group in general lives, we are wiser to use the median income figure as a rough guide. Note that here the mode is the same as the median: $40,000.

Sociologists frequently choose to relate different variables to each other. **Correlation** is a widely used technique for analyzing the patterns of association, or correlation, between pairs of variables like income and education. We might begin with a questionnaire that asked for annual earnings (Y) and level of education (X). Correlation analysis delivers two types of information: the *direction* of the relationship between X and Y, and the *strength* of that relationship. The direction of a relationship is positive (that is, a positive correla-

tion exists) if X is low when Y is low *and* if X is high when Y is high. But there is also a correlation if Y is low when X is high (or vice versa); this is a negative, or inverse, correlation, but it is a correlation nonetheless. The strength of a correlation is simply how closely or tightly the variables are correlated or associated, regardless of the direction of correlation. For example, if all those persons who were low in education (X) were also low in earnings (Y), then the correlation is strong and positive, but if most (but not all) of those low in education were low in earnings, then the correlation would still be positive, but less strong.

When interpreting correlations, one must realize that a correlation does not necessarily imply cause and effect. A correlation is simply an association, one whose cause must be explained by other means than simple correlation analysis. A *spurious correlation* exists when there is no meaningful causal connection between apparently associated effects. A frequently quoted example is the finding that the number of women reporting physical assaults by their husbands increases on Super Bowl Sunday. You might conclude, wrongly, that watching the Super Bowl causes husbands to be violent. In fact, this is a spurious correlation. The real correlation is between wife beating and alcohol consumption. Men watching the Super Bowl are prone to drink more than usual. The increased consumption of alcohol correlates meaningfully with the rise in wife beating: The more men drink, on average, the more likely they are to become violent with their spouses.

Another widely used method for analyzing sociological data is **cross-tabulation,** in which data are broken down into subsets for comparison. Table 2.2 reports on a Gallup Poll that asked people, "Do you feel that the laws covering the sale of firearms should be made more strict, less strict, or kept as they are now?" The answers are broken down by gender. The table cross-tabulates the answers, "more strict," "less strict," "kept as they are," and "no opinion," with the gender of the respondent, male or female. The table is thus a cross-tabulation of opinion on gun control by gender.

Table 2.2
A Cross-Tabulation of Opinion on Gun Control by Gender

Question asked: "Do you feel that the laws covering the sale of firearms should be made more strict, less strict, or kept as they are now?"

	More Strict	Less Strict	Kept as They Are	No Opinion	Totals
Women	72%	2%	24%	2%	100%
Men	52%	8%	39%	1%	100%

Source: The Gallup Poll, 2000. www.gallup.com/poll/indicators/indGuns.asp

THINKING SOCIOLOGICALLY

Suppose that you asked a *random sample* of students on your campus if they approved or disapproved of pre-marital sex. What questions would you ask, and what would you expect to find? Now suppose you extended your sample to include a campus very different from your own (perhaps one with more commuting adult students or a religiously affiliated college). What *hypotheses* would you suggest?

Did men and women answer the question differently? The cross-tabulation shows that they did. A large majority of women (72 percent) thought the laws should be more strict, 2 percent said less strict, and 24 percent thought they should be kept as they are. In contrast, fewer men (52 percent) said the laws should be more strict, only 8 percent said they should be less strict, and 39 percent thought they should be kept as they are. Thus, in general, women are more likely than men to think that gun control is not as strict as it should be. Survey answers might differ not only by gender, but also by race, class, region, or an infinite variety of other variables. By cross-tabulating, the researcher is able to discover such relationships among the variables.

The Use and Misuse of Statistics

Statistical information is notoriously easy to misinterpret, willfully or accidentally. Unfortunately, statistics put forward by interest groups are often of little real value because they are incomplete, out of context, or misinterpreted. Worse than misinterpretation are statistics that are falsified. Fudging is the term for falsifying statistics, and it includes any tampering with data. Fudging is not only a misuse of statistics, it is intellectual vandalism, and in some cases it is even a criminal offense (Best 2001; Zuberi 2001).

Examples of some statistical mistakes include the following (as well as the box "A Sociological Eye on the Media: Research and the Media" on page 46):

• *Citing a correlation as a cause.* A correlation reveals an association between things, nothing more. Correlations do not necessarily indicate that one thing causes the other. A warning often used by sociological researchers is, *Correlation is not proof of causation.*

• *Overgeneralizing.* Statistical findings are limited by the extent to which the sample group reflects or represents the population from which the sample was obtained. Generalizing beyond the population is a misuse of statistics. Studying only men, and then generalizing conclusions to both men and women, would be an example of overgeneralizing.

• *Interpreting probability as certainty.* Probability is a statement about chance or likelihood only. For ex-

ample, finding that women are more likely than men to favor strict gun control only means that women have a higher probability (a greater chance) of favoring strict gun control than men. It does not mean that all women favor strict gun control and all men do not favor it (see Table 2.2).

• *Building in bias.* In a famous market research advertising campaign, public taste tests were offered between two soft drinks. A wily journalist verified that, in at least one site, the brand sold by the sponsor of the test was a few degrees colder than its competitor when it was given to testers, which helped its scores. Statistics derived from this test were bunk. In a similar but less loathsome vein, bias can be built into a questionnaire by little more than careless wording.

• *Faking data.* Perhaps one of the worst misuses of statistics is making up, or faking, data. A famous instance of this occurred in a study of separated identical twins (Burt 1966). The researcher wished to show that, despite their separation, the twins remained similar in certain traits such as measured intelligence, thus suggesting that their (identical) genes caused their striking similarity in intelligence. It was later shown that the data were fabricated (Kamin 1974; Hearnshaw 1979; Taylor 1980, 2002; Mackintosh 1995).

• *Using data selectively.* Sometimes a survey includes many questions, but the researcher reports on only a few of the answers. Doing so makes it easy to misstate the findings. Often, researchers will not report findings that show no association between variables, but these can be just as telling as associations that do exist. For example, researchers on gender differences typically report the differences they find between men and women but seldom publish their findings when the results for men and women are identical. This tends to exaggerate the differences between women and men, and it falsely confirms certain social stereotypes about gender differences.

Is Sociology Value-Free?

The topics dealt with by sociology are often controversial. People have strong opinions about social questions, and they may have deeply felt commitments. In some cases, the settings for sociological work are highly politicized. Imagine spending time in an urban precinct house to do research on police brutality, or doing research on acquired immune deficiency syndrome (AIDS) and sex education in a conservative public school system. Under these conditions, can sociology be scientifically objective? How do researchers balance their own political and moral commitments against the need to be objective and open-minded? Sociological knowledge has an intimate connection to political values and social views. Often, the very purpose of sociological research is to gather data as a step in creating social policy. Can sociology be value-free? Should it be value-free? These are important questions without simple answers.

Political Commitments and Sociological Research

Sometimes a political component is a valuable instigator to research. For many years, sociological research was *androcentric;* that is, it centered on the experiences of men. Without necessarily being aware of it, sociologists, almost all of whom were men, often studied only males in surveys or looked at their subjects through the perspectives and experiences of men. The result was that much research either excluded women altogether or did not represent their experience accurately. Feminist scholars have asked if the concepts and theories generated from men's experiences are entirely valid when applied to women's lives. Based on these criticisms, new questions have been asked, new data generated, and new theories and concepts developed that have transformed sociological thought (Andersen 2003). In this case, political motivations improved the scientific grounding of sociology by removing an artificial constraint on its breadth. Likewise, Asian American, African American, Native American Indians , and Latino studies, as well as gay and lesbian studies—groups that have been previously ignored—have raised new questions for sociology and have generated insights that extend the frontier of sociological knowledge.

This does not mean that research should necessarily be driven by activism or politics. The result would be work that reproduced researchers' opinions, not tested their theories. The goal is not to confirm researchers' political values, but to generate new knowledge. As theorist Max Weber pointed out, this may require questioning the status quo. Sociology cannot be perfectly value-free, but researchers have an obligation to make their research as objective and value-free as possible. Sociologists should not deny the existence of their own values and biases; they should make them known. They should also ask many questions, be open-minded, and listen to the evidence.

Research is produced in specific social, historical, political, and economic contexts. The context can influence research findings. For example, can scientists at a major chemical company objectively assess the impact of chemical pollution on the environment? They can if they have the freedom and independence to pursue all possible research questions. However, if the company does not allow publication of results unless they are favorable to the company's interests, and if it suppresses incriminating research, then the organization is unlikely to produce objective research. To preserve objectivity, investigators must be free to ask questions without political constraint.

Claiming to be value-free when you are not is probably more damaging to your research credibility than making your commitments clear. Sociologists should take care not to hide their assumptions and should try to make apparent their possible sources of bias, including identifying sponsors of research, should they exist (such as government agencies and corporate sponsors). However, they are not always conscious of the assumptions they make. Those who produced *androcentric* research studies, were probably unaware that they were doing so. Only with the development of feminism did scholars question the assumptions they were making about women. In the spirit of objectivity, scholars should try to be aware of how their experiences, values, and commitments shape the research they do.

The Insider–Outsider Debate

Assuming that you can overcome your personal values in favor of objectivity, another question emerges: Is it possible to understand a person or group whose experience is greatly different from your own (Merton 1972)? Do you have to be an insider to understand the people you are studying? Does one, for instance, have to be Asian American to understand what it is like to be Asian American in the United States? Does one have to be gay to understand gay issues?

An interesting controversy in sociology is the insider–outsider debate, which raises the issue of whether the researcher should or should not be a member of the group, society, or culture that the researcher is studying.

© Jeremy Hartley/Panos Pictures

No one will ever have another's experience. If only insiders could understand the experiences of a group, sociologists would never be able to empathize with others or learn from their experiences. Insiders can have a distorted understanding of their own experience, the sort of distortion that is overcome by stepping away from assumptions long taken for granted.

Max Weber argued that sociologists could understand others through *verstehen*, a form of understanding achievable when researchers imagine themselves in the place of their subject (see Chapter 1). Verstehen,

A SOCIOLOGICAL EYE ON THE MEDIA
Research and the Media

On any given day, if you watch the news, read a newspaper, or search the Web, you are likely to learn about various new research studies purporting some new finding. How do you know if the research results reported in the media are accurate?

Most people are not likely to check the details of the study or have the research skills to verify the study's claims. But one benefit of learning the basic concepts and tools of sociological research is to be able to critically assess the research frequently reported in the media. Based on your knowledge of sociological research, what questions might you ask to assess research claims in the media? Consider these:

1. **What are the major variables in the study? Are the researchers claiming a causal connection between two or more variables? If so, is this a causal relationship or is it a spurious correlation (see page 43)?**

   A spurious correlation has no direct causal connection between two variables, even if there appears to be an association between them. For example, recent media reports claim that researchers have found a causal connection between teen sex and parental smoking. The press reported that one way parents can reduce the chances of their children becoming sexually active at an early age is to quit smoking (O'Neil 2002). The researcher who conducted this study actually claimed there was no direct link between parental smoking and teen sex, although she did find a correlation between parents' risky behaviors—smoking, heavy drinking, and not using seat belts—and children's sexual activity. She argued that parents who engage in unsafe activities provide a model for their children's own risky behavior (Wilder and Watt 2002).

   Correlation is not cause. Seeing parental behavior as a model for what children do is hardly the same thing as seeing parents' smoking as the cause of early sexual activity!

2. **How have researchers defined the major topics of their study? How were the variables in the study measured?**

   For example, if someone claims that 10 percent of all people are gay, how is "being gay" defined? Does it mean having had one such experience over one's entire lifetime or does it mean actually having a gay identity? The difference matters because one definition will likely inflate the number reported. Sometimes you must look up the original study to learn how things are defined or how they are measured, yet if you read press reports carefully, you can often see this without further work. And, with the widespread availability of electronic information, you can often find the full text of original articles online.

3. **What is the sample of people on which the research was based? Is it scientifically generated? Is it a random sample or is it biased by the way research subjects were selected?**

   Again, you might have to go to the original source of the study to learn this, but often the sample will be reported in the press (even if in nonscientific language). For example, a study widely reported in the media in 2002 had headlines ex-

claiming, "Study Links Working Mothers to Slower Learning" (Lewin 2002). But if you read even the news report closely, you will learn that this study included only white, non-Hispanic families. Black and Hispanic children were dropped from some of the published results because there were too few cases in the sample to make meaningful statistical comparisons (Brooks-Gunn et al. 2002). A later study by the same research team found that there were no significant effects of mother's employment on children's intellectual development among African-American or Hispanic children (Waldfogel et al. 2002). However, this finding generated no media reports.

4. **Has there been false generalization either in the study itself or in the way it is reported in the media?**

   Often a study reported in the media has many more limited claims in the scientific version than what is reported in the media. Using the example just given about the connection between maternal employment and children's learning, it would be a big mistake to generalize from the study's results to all children and families.

5. **Can the study be replicated?**

   Unless there is full disclosure of the research methodology (that is, how the study was conducted), this will not be possible. But you can ask yourself how the study was conducted, whether the procedures used were reasonable and logical and whether the researchers made good decisions in constructing their research question

like empathy, is a way of sharing in the experiences of other people and an essential part of doing good research, just as empathy is an essential part of human relationships.

The insider–outsider debate frequently occurs over questions about studies of race. Can White sociologists do meaningful research on the experience of oppressed groups? Latinos, African Americans, Asian Americans, and Native Americans as research subjects may be suspicious of Whites, making them reserved in interviews, or they may fail to absorb Whites fully into their social interactions. Some have concluded that minority

and research design. Although you will likely have to read the original study to know this, you can always ask yourself how you would proceed to investigate the conclusions reached in a given research report.

6. **Who sponsored the study and do they have a vested interest in the study's results?**

Find out if a group or organization with a particular interest in the outcome sponsors the research. If so, did they use objective methods of research or did they bias the research results by developing the study in a way that would produce a desired outcome? Research sponsored by interested parties does not necessarily negate research findings, but it can raise questions about the researchers' objectivity and the standards of inquiry they used. For example, would you give as much validity to a study of environmental pollution that was funded and secretly conducted by a chemical company as compared with a study on the same topic conducted by independent scientists who openly report their research methods and results?

7. **Who benefits from the study's conclusions?**

Although this question does not necessarily challenge the study's findings, it can help you think about whom the findings are likely to help. This, in turn, can help you think through some of the assumptions made by the researcher and the logic of what the researchers have concluded.

8. **What assumptions did the researchers have to make to ask the question they did? If they**

were to start from a different set of assumptions, would they have asked different questions or come up with different results?

For example, if you started from the assumption that poverty is not the individual's fault but is the result of how society is structured, would you study the values of the poor or perhaps the values of policy makers? Especially when research studies explore matters where social values influence people's opinions, it is important to identify the assumptions made by certain questions.

9. **Do these questions mean you should never believe anything you hear in the media?**

Of course not. Thinking critically about research does not mean being negative or cynical about everything you hear or read. Being overly cynical will make you just as ignorant as if you never read or thought about anything at all. The point is not to reject all media claims as out of hand, but instead to be able to evaluate good versus bad research. All research has limitations. Knowing the limitations of a given study can teach you just as much about the subject of study as do the particular conclusions that are reached. The point is to know how to interpret what you read and to do so with a critical eye. Just asking the questions discussed here will make you a better informed citizen. Learning the basic tools of research, even if you never conduct research yourself or pursue a career where you would use such skills, can make you a better in-

formed citizen and prevent you from being duped by claims that are neither scientifically nor sociologically valid.

10. **What sort of considerations should be taken into account to accurately assess the credibility of data you get off of the internet?**

There are four criteria for distinguishing useful from useless information online. Keep these criteria in mind as you engage in your next online search:

- **Credibility of source.** Where does the information come from? Does it come from an academic institution (having the .edu abbreviation on the source's URL), or a commercial source (the .com abbreviation), or some other source? You should be aware of the origin of the information and make some judgment of its reliability.

- **Accountability.** Is there an author or sponsor identified on the Web page, with an email link? Is there a link on the page back to its "home"?

- **Timeliness.** Find out when the Web page was last updated. Online information tends to become obsolete quickly.

- **Scope and coverage.** How well does the topic appear to be covered when you compare the information to a known paper source such as a book chapter or professional journal article with which you are already familiar? Try to distinguish sources with complete coverage from email spam. •••

researchers are better positioned to generate new insights and hypotheses, especially ones that challenge the traditional frameworks on which scholarship about minorities has been based.

Research Ethics

The importance of ethics in human endeavor has been brought to the forefront by recent scandals in the business world, such as Enron Corporation's wiping out its employees retirement "nest egg" accounts, or domestic-life guru Martha Stewart's lying to government officials about her stock sales.

Sociological research often raises ethical questions. Ethical considerations of one sort or another exist with any type of research. In a survey, the person being questioned is often not told the purpose of the survey or who is funding the study. Is it ethical to conceal this type of information? In controlled experiments, deception is often employed. Many experiments depend on respondents giving natural, unconsidered responses to staged situations. Researchers often reveal the true purpose of an experiment after it is completed in a session called a *debriefing*. The deception is therefore temporary. Does that lessen the potential ethical violation?

Participant observation research, if done without the knowledge of the people being studied, is also a form of deception. Whyte's *Street Corner Society* study is an example. A participant observer may choose to work covertly if it appears that the known presence of a researcher would affect the behavior of the group, tainting the observer's findings. None of the Cornerville Gang, except the leader, Doc, knew they were being studied. Years later, however, the gang members found out about the study, read the book, and became angry at Whyte for "fooling" them and trading on their relationship to get famous. The discontent of the Cornerville Gang confirms that an ethical issue exists. Do researchers have the right to hide their intentions? Did Whyte have the right to decide that the discontent of the Cornerville Gang (which was not unexpected) was less important than the needs of the study? Did gaining the collaboration of the gang leader, Doc, bring with it the right to study the rest of the group secretly?

When Judith Rollins did her research on Black women domestic workers, she posed as a maid and did not tell her employers that she was a Ph.D. student writing her dissertation. Had she done so, her study would have been ruined because surely the subjects would have acted differently toward her. Was her research unethical? Most sociologists would say no, as long as no harm came to her subjects. The definition of harm, however, can be debated. Some of Judith Rollins's sub-

jects might have been embarrassed if they had recognized themselves in her study, especially if they felt that others might recognize them, too. Does that count as mental harm? Most sociologists would say no, or that any harm caused was extremely minor. The professional code of ethics clearly states that if a research subject is at risk of physical, mental, or legal harm, then the subject must be informed of the rights and responsibilities of both researcher and subject. Sociologists also take measures not to identify their subjects by using pseudonyms or no names at all.

THINKING SOCIOLOGICALLY

Based on your knowledge of Judith Rollins's research, would you say it is *value-free*? Drawing on the concept of *verstehen* from Max Weber, would you say one has to be a Black person to understand Rollins's observations of Black domestic workers and the employers?

One of the clearest ethical violations in the history of science has come to be known as the *Tuskegee Syphilis Study*. The study was conducted at the Tuskegee Institute in Macon County, Alabama, a historically Black college. For this study, begun in 1932 by the government's United States Health Service, a sample of about 400 Black males who were infected with the sexually transmitted disease syphilis, were allowed to go untreated for forty years. Another 200 Black males who had not contracted syphilis were used as a control group. The purpose of the study was to examine the effects of "untreated syphilis in the male negro." Untreated syphilis causes blindness, mental retardation, and death, and this is how many of the untreated men fared over the forty-year period. During this period, in the early 1950s, penicillin was discovered as an effective treatment for infectious diseases, including syphilis, and was widely available. Nonetheless, the scientists conducting the study decided not to give penicillin to the infected men in the study, on the grounds that it would "interfere" with the study of the physical and mental harm caused by untreated syphilis. The U.S. government itself authorized the study to be continued until the early 1970s. By the mid-1970s pressure from the public and the press caused the federal government to terminate the study. But by then it was too late to save the approximately 100 men who had already died of the ravages of untreated syphilis, to say nothing of many more who continued to live but suffered with major mental and physical damage. The ethical violations of this research are now widely recognized and admitted.

What if a court were to demand that Judith Rollins reveal the identities of her study subjects? The law is not yet clear on whether sociologists must reveal the identities of their subjects or hand over their data to investigating authorities. Journalists have long relied on the constitutional right of freedom of the press to protect their sources. Sociologists and other scholars are claiming the same rights to protect their research subjects. The courts have not decisively acknowledged the rights of either. The American Sociological Association has taken the position that guarantees of privacy and confidentiality to research subjects are essential to open scientific inquiry and, therefore, must be protected under the law. At this point in the development of the law, judges make decisions on a case-by-case basis as to whether scholars and journalists can be forced to testify against people from whom they have acquired information. Such cases show that questions about research ethics has two sides—protecting the rights of individuals, and guaranteeing the possibility of open scientific inquiry, particularly when the subject is controversial or sensitive.

Sociology ⊛ Now™
Reviewing is as easy as ❶❷❸.

1. *Before you do your final review, take the SociologyNow diagnostic quiz to help you identify the areas on which you should concentrate. You will find information on SociologyNow and instructions on how to access all of its great resources on the foldout at the beginning of the text.*

2. *As you review, take advantage of SociologyNow's study videos and interactive Map the Stats exercises to help you master the chapter topics.*

3. *When you are finished with your review, take SociologyNow's posttest to confirm you are ready to move on to the next chapter.*

Chapter Summary

What is sociological research?

Sociological research is used by sociologists to answer questions. The research method one uses depends upon the question asked.

How is sociological research scientific?

Sociological research is derived from the *scientific method*, which means that it relies on empirical observation and data and that it often involves the testing of *hypotheses*. The research process often, though not always, involves several steps: developing a research question, designing the research, collecting the data, analyzing the data, and developing conclusions. Different research designs are appropriate to different questions, but sociologists have to be concerned about the *validity*, *reliability*, and *generalization* of their studies.

What are validity, reliability, and generalization?

Validity refers to whether something accurately measures what it is supposed to measure, namely, the concept being studied. *Reliability* is whether the same results would be found by another researcher using the same measurement. *Generalization* means being able to make claims that extend beyond the specific observations of a given research project. Applying the results from a *sample* to a broader *population* is an example of generalization.

What is the difference between qualitative research and quantitative research?

Qualitative research is research that is relatively unstructured, does not rely heavily upon statistics, and is closely focused on a question being asked. *Quantitative research* is research that uses statistical methods. Both kinds of research are used in sociology.

What are the techniques of sociological research?

The most common techniques of sociological research are *surveys, participant observation, controlled experiments, content analysis, historical research,* and *evaluation research,* including *market research.* Each tool has its own strengths and weaknesses. Some techniques are primarily qualitative, some are primarily quantitative, and some involve both methods (see Table 2.1).

What are some of the statistical concepts in sociology?

Through research, sociologists are able to make statements of *probability*, or likelihood. They often do so by means of a *sample*, which is a subset of units (such as people) used from a *population*. Sociologists use *percentages* and *rates*. The *mean* is the same as an average. The *median* represents the midpoint in an array of values or scores. The *mode* is the most common value or score. *Correlation* and *cross-tabulation* are statistical procedures that allow sociologists to see how two or more different variables are associated. There have been instances of misuse of statistics in sociology, and these have resulted in incorrect conclusions.

Can sociology be value-free?

Although no research in any field can always be value-free, sociological research nonetheless strives for objectivity while recognizing that the values, and biases, of the researcher may have some influence on the work. Furthermore, value considerations as well as political agendas can often result in corrected and refocused research with new insights.

What are the ethical considerations in research?

Ethical considerations include whether one should collect data without letting research subjects know they are being studied or observed. Other ethical dilemmas include whether researchers have the right to hold their data in confidence from public officials such as the courts or police, without naming their research subjects.

Key Terms

concept 31

content analysis 40

controlled
     experiment 38

correlation 43

cross–tabulation 43

data 31

data analysis 33

deductive reasoning 28

dependent variable 31

empirical 28

evaluation research 41

field research 37

generalization 35

hypothesis 30

independent
     variable 31

indicator 31

inductive reasoning 29

informant 38

intervening variable 31

market research 41

mean 43

median 43

mode 43

participant observation
     37

percentage 42

policy research 41

population 42

probability 41

qualitative research 30

quantitative research 30

random sample 42

rate 43

reliability 32

replication study 29

research design 30

sample 42

scientific method 28

validity 32

variable 31

Reseaching Society: Doing Sociological Research with MicroCase Online

You can see the results of actual research by using the Wadsworth MicroCase® Online feature available to you. This feature allows you to look at some of the results from national surveys, census data, and other data sources. You can explore this easy-to-use feature on your own, but try this example. Suppose you want to know:

How do different states compare in the percentage of births to mothers under age 20?

To answer this question, go to http://sociology.wadsworth.com/andersen_taylor4e/, select MicroCase Online from the left navigation bar, and follow the directions there to analyze the following data.

Data file: STATES

Task: Mapping

Variable 1: MA AGE 20-PERCENTAGE OF BIRTHS TO MOTHERS UNDER 20 YEARS OLD

View: Map

Questions:

Once you have your results, answer the following questions:

Which region of the country has the highest rate of births to mothers under age 20?

Now select the Back button, and select the Rank Table view:

1. Which state has the highest rate of births to mothers under age 20?

2. What is the percentage for this state?

3. Which state has the lowest rate of births to mothers under age 20?

4. What is the percentage for this state?

5. What is the ranking of your own home state?

The Companion Website for Sociology: Understanding a Diverse Society, Fourth Edition

http://sociology.wadsworth.com/andersen_taylor4e/

Supplement your review of this chapter by going to the companion website to take one of the Tutorial Quizzes, use the flash cards to master key terms, and check out the many other study aids you'll find there. You'll also find special features such as GSS Data and Census 2000 information, data and resources at your fingertips to help you with that special project or do some research on your own.

Suggested Readings and Web Resources

Babbie, Earl. 2004. *The Practice of Social Research*, 10th ed. Belmont, CA: Wadsworth.
  One of the most widely used texts on research methods and techniques, this book is both comprehensive and readable. It gives a complete overview of a wide range of research techniques and procedures.

Best, Joel. 2004. *Damned Lies and Statistics: How Numbers Confuse Public Issues*, 2nd ed. Berkeley, CA: University of California Press.
  A useful and readable compendium of how political, administrative, social, and other orientations can result in serious misuse and misinterpretation of quantitative and statistical data and results.

Duneier, Mitchell. 1999. *Sidewalk*. New York: Farrar, Strauss, and Giroux.
  Among the broadest and most thorough examples of participant observation research done in sociology, and containing extensive photography, this work is a study of homeless men on New York's lower East Side who support themselves as street vendors, how they stratify themselves, and how their group is interdependent with—not separate from—society.

Gubrium, Jaber F., and James A. Holstein. 1997. *The New Language of Qualitative Methodology*. New York: Oxford University Press.

This book is an interesting account of the techniques, approaches, and foundations of qualitative and field research.

Miller, J. Mitchell and Richard Tewksbury (eds). 2001. *Extreme Methods: Innovative Approaches to Social Science Research*. Boston, MA: Allyn and Bacon. *This interesting book covers a variety of techniques and approaches in qualitative field research. Covered are such unusual topics as participant observation research in prisons, in women's motorcycle gangs, and among table dancers, and discussions of the ethics of deception.*

Reinharz, Shulamit. 1992. *Feminist Methods in Social Research*. New York: Oxford University Press. *Reinharz's book reveals how a variety of feminist approaches has resulted in new research questions as well as new findings. Includes a discussion of the past invisibility of women as research subjects.*

The Psychology of Cyberspace
www.rider.edu/users/suler/psycyber/psycyber.html
This site includes an outline book about social interaction and cyberspace with links to other sites exploring the implications of cyberspace for interpersonal interaction.

Center for Social Infomatics
www-slis.lib.indiana.edu/ssi/
The Center for Social Informatics at Indiana University does research on the influence of technology and computerization on society; the website includes information on publications, conferences, and other resources.

• • •

Culture

In one contemporary society known for its advanced culture, many women would not consider going to work in the morning without encasing their legs in nylon sleeves. The nylon covers them from waist to toe and is believed to make their legs more attractive. If, as often happens, the delicate nylon filaments tear, the women are embarrassed. Many women put on their leg sleeves during a morning ritual that includes smearing oily pigments on their faces to achieve striking, unnatural hues, and dusting their faces with powder to eliminate the skin's natural shininess. The hair on the scalp is slathered with gels and liquids and then modeled into decorative forms. The hair may be augmented with knots; strips of cloth; and small items of hardware made of plastic, wood, or metal. Beads, metals, and glittering objects are often inserted into holes punched into the women's earlobes. Plastic may be glued onto the fingernails and then adorned with bright colors, paint, and sparkles. All this effort is expended in the belief that it makes women more attractive to others.

Most men in this culture begin their day by scraping their faces with sharp-bladed tools to remove hair. The scraping tools of the wealthiest men may be embellished with silver and brass, but most men use plain metal or plastic tools that are sold in local markets. After removing hair from their faces, the men douse their skin with healing spirits that are usually scented with fragrances thought to be sexually provocative. Some men modify this ritual by arranging their facial hair in elaborate designs, even packing the hair above their lips with wax and modeling it into dramatic shapes. Men usually slather the hair on their scalps with many of the same substances used by women, but they generally do not augment their hair designs with the ornamental objects commonly used by women. Individuals incur social penalties if they do not engage in these rituals. Men who appear in public without removing the day's growth of facial hair are considered unkempt. Women who

Sociology ⊛ Now™
Reviewing is as easy as ❶❷❸.

Use SociologyNow to help you make the grade on your next exam. When you are finished reading this chapter, go to the chapter review for instructions on how to make SociologyNow work for you.

do not wear nylon sleeves on their legs may be considered ugly or unfeminine, and they may jeopardize their chances of getting the highest-paying and most prestigious jobs. Elaborate markets have developed to sell these products. People in this culture exchange millions of the tokens they are paid for work to buy these bodily decorations.[1]

From outside the culture, these practices seem strange, yet few within the culture think the rituals are anything but perfectly ordinary. They sometimes wish they did not have to go through the elaborate morning ritual, and many forgo the elaborate grooming and costuming on weekends. Most of the time, people neither analyze these rituals nor spend time thinking about their meaning.

You have surely guessed that the practices described here are taken from U.S. culture. When viewed from the outside, cultural habits that seem perfectly normal take on a strange aspect. The Tchikrin people are a remote culture of the central Brazilian rain forest who paint their bodies in elaborate designs. Painted bodies communicate meanings to others about the relationship of the person to his or her body, society, and the spiritual world. The designs and colors symbolize the balance the Tchikrin believe exists between biological powers and the integration of people into the social group. The Tchikrin associate hair with sexual powers, and lovers get a special thrill from using their teeth to pluck an eyebrow or eyelash from their partner's face (Turner 1969). To the Tchikrin people, these practices are no more exotic than the morning rituals adhered to in the United States.

To analyze culture and measure its significance in society, we must separate ourselves from judgments such as "strange" or "normal." We must see a culture as it is seen by insiders, but we cannot be completely taken in by that view. One might say we should know culture as insiders but understand it as outsiders. •••

Defining Culture

Culture is the complex system of meaning and behavior that defines the way of life for a given group or society. It includes beliefs, values, knowledge, art, morals, laws, customs, habits, language, and dress. Culture includes ways of thinking as well as patterns of behavior. Observing culture involves studying what people think, how they interact, and the objects they make and use. As stated by two sociologists, "Culture appears to be 'built into' all social relations, constituting the underlying assumptions and expectations on which social interaction depends" (Wuthnow and Witten 1988: 50).

In any society, culture defines what is perceived as beautiful and ugly, right and wrong, good and bad. Culture helps hold society together. It gives people a sense of belonging, instructs them in how to behave, and tells them what to think in particular situations. Culture gives meaning to society.

Culture is both material and nonmaterial. **Material culture** consists of the objects created in a given society, which are its buildings, art, tools, toys, publications, and other tangible objects. In the popular mind, material artifacts constitute culture because they can be collected in museums or archives and analyzed for what they represent. These objects are significant because of the meaning they are given. A temple, for example, is not merely a building or a place of worship. Its form and presentation signify the religious culture of the faithful. **Nonmaterial culture** includes the norms, laws, customs, ideas, and beliefs of a group of people. Nonmaterial culture is less tangible than material culture, but it has a strong presence in social behavior. Examples of nonmaterial culture are numerous and found in the patterns of everyday life. In some cultures, people eat with silverware; in others, with chopsticks; and in some, with their fingers. Such are the practices of nonmaterial culture, but note that the eating utensils are part of material culture.

Characteristics of Culture

Across societies, sociologists note certain features of culture. These include:

1. *Culture is shared.* Culture has significance because people hold it in common. Culture is not idiosyncratic; it is collectively experienced and agreed upon. The shared nature of culture makes human society possible, but may be difficult to see in complex societies where groups have different traditions and perspectives. In the United States, different racial and ethnic groups have unique histories, languages, and beliefs. Even within these groups, there are diverse cultural traditions. Latinos, for example, comprise many groups with distinct origins and cultures yet still share the Spanish language and some values and traditions. The different groups constituting Latino culture also share a culture that is shaped by their common experiences as minorities in the United States. Similarly, African Americans have created a rich and distinct culture that is the result of their unique experience within the United States. What identifies African American culture are the practices and traditions that have evolved

---

[1]These illustrations are adapted from a classic article by Horace Miner. 1956. "Body Ritual Among the Nacirema." *American Anthropologist* 58: 503–507. "Nacirema" is "American" spelled backward.

Cultural traditions represent the different values and beliefs of diverse groups. Here, Hannukah (upper left), Christmas (upper right), Ramadan (lower left), and Kwanza (lower right) are celebrations of diverse cultures.

© Francene Keery/Stock Boston

© Bob Krist/CORBIS

© Annie Griffiths Belt / CORBIS

© Lawrence Migdale/Photo Researchers

from both the American experience and African traditions. Placed in another country, such as one of the African nations, African Americans would likely recognize elements of their culture, but they would also feel culturally distinct as Americans.

Within the United States, culture also varies by age, region, gender, ethnicity, religion, class, and other social factors. A person growing up in the South is more likely to develop different tastes, modes of speech, and cultural interests than a person raised in the West. Despite these cultural diversities, certain symbols, language patterns, and belief systems, are distinctively American and form a common culture.

2. *Culture is learned.* Cultural beliefs and practices are usually so well learned that they seem perfectly natural, but they are learned nonetheless. How do people come to prefer some foods to others? How is musical taste acquired? Culture may be taught through direct

instruction, such as a parent teaching a child how to use silverware or teachers instructing children in songs, myths, and other traditions in school.

Culture is also learned indirectly through observation and imitation. Think of how a person learns what it means to be a man or a woman. Although the "proper" roles for men and women may never be explicitly taught, one learns what is expected from observing others. A person becomes a member of a culture through both formal and informal transmission of culture. Until the culture is learned, the person will feel like an outsider. The process of learning culture is referred to by sociologists as *socialization* (see Chapter 4).

3. *Culture is taken for granted.* Because culture is learned, members of a given society seldom question the culture they belong to unless they begin to question the usual cultural expectations. People engage unthinkingly in hundreds of specifically cultural practices every day. Culture makes these practices seem "normal." If you suddenly stopped participating in your culture and questioned each belief and every behavior, you would soon find yourself feeling detached and perhaps a little disoriented. You would also become increasingly ineffective at functioning within your group. Little wonder that tourists, even when they are well informed, stand out so much in a foreign culture because they typically approach the society from their own cultural orientation.

Think how you might feel if you were a Native American student in a predominantly White classroom. Many, though not all, Native American people are raised to be quiet and not outspoken. If students in a classroom are expected to assert themselves and state what is on their minds, a Native American student who is not assertive may feel awkward. The professor may not be aware of these cultural differences and may penalize students who are quiet, perhaps resulting in a lower grade for the student from a different cultural background. Culture unifies people, but lack of communication across cultures can have negative consequences.

4. *Culture is symbolic.* The significance of culture lies in the meaning people give to **symbols** for things or behavior. The meaning is not inherent in the symbol but is bestowed by the cultural significance. The U.S. flag, for example, is literally a piece of cloth. Its cultural significance derives not from the cloth of which it is made, but from its meaning as a symbol of freedom and democracy, as was witnessed by the widespread flying of the flag after the terrorist attacks on the United States on September 11, 2001. But, symbols mean different things in different contexts. Insurgents in Iraq see the U.S. flag as a symbol of American im-

Cultural values can clash when groups have strongly held, but clashing, value systems. A good example is the conflict that developed over a State Supreme Court judge in Alabama placing the Ten Commandments in the lobby of the county courthouse. When he refused to remove them because they violated the constitutional separation of church and state, he was forced to resign.

perialism. Likewise, a cross on a church altar has a meaning different from a cross burning in someone's front yard. This tells you something important about culture: Cultural meaning is created in specific social contexts.

Symbols are powerful expressions of human life. Protests over the flying of the Confederate flag provide

DOING SOCIOLOGICAL RESEARCH

Tattoos: Status Risk or Status Symbol?

Not so long ago tattoos were considered a mark of social outcasts. They were associated with gang members, sailors, and juvenile delinquents. But now tattoos are in vogue and a symbol of who's trendy and hip. How did this once stigmatized activity associated with the working class happen to become a statement of middle-class fashion?

This is what sociologists Katherine Irwin wanted to know when she first noticed the increase in tattooing among the middle class. Irwin first encountered the culture of tattooing when she accompanied a friend getting a tattoo in a shop she calls Blue Mosque. She

started hanging out in the shop, eventually married the shop's owner, and began a four-year study using participant observation in the shop, along with interviews of people getting their first tattoos. At one point, Irwin also interviewed some of the parents of tattooees and potential tattooees.

She found that middle-class tattoo patrons were initially fearful that their desire for a tattoo would associate them with low-status groups. They reconciled this by adopting attitudes that associated tattooing with middle-class values and norms. Thus, they defined tattooing as symbolic of independence, liberation, and freedom

from social constraints. Many of the women defined tattooing as symbolizing toughness and strength, which are values they thought rejected more conventional ideals of femininity. Some saw tattoos as a way to increase their attachment to alternative social groups or to gain entrée into "fringe" social worlds. Although tattoos held different cultural meanings to different groups, people getting tattooed used various techniques (what Irwin calls "legitimation techniques") to counter the negative stereotypes associated with tattooing. Irwin concludes that people try to align their behavior with legitimate cultural val-a good example. Those who object to the flag being displayed on public buildings see the Confederate flag as a symbol of racism and the legacy of slavery. Those who defend the flying of the flag see it as representing Southern heritage, a symbol of group pride and regional loyalty. Similarly, the use of Native American mascots to name and represent sports teams is symbolic of the exploitation of Native Americans. (Think of the Washington Redskins, the Cleveland Indians, the "tomahawk chop," and various college mascots). Native American activists and their supporters see these mascots as derogatory and extremely insulting, representing gross caricatures of Native American traditions. The protests that have developed over symbols such as the Confederate flag and sports mascots are indicative of the enormous influence of cultural symbols.

One interesting thing about culture is the extent to which symbolic attachments guide human behavior. For example, people stand when the national anthem is sung and may feel emotional by displays of the cross or the Star of David. Under some conditions, people organize mass movements to protest what they see as the defamation of important symbols, such as the burning of a flag or the burning of a cross. The significance of the symbolic value of culture can hardly be overestimated. Learning a culture means not just engaging in particular behaviors but also learning their symbolic meanings within the culture.

5. *Culture varies across time and place.* Physical and social environments vary from one society to another, and because people are creative in adapting culture to the challenges they face, culture is not fixed from one place to another. In the United States, there is a strong cultural belief in scientific solutions to human problems. Consequently, many think that problems of food supply and disease can be addressed by scientific breakthroughs, such as gene splicing to create genetically engineered food or stem cell research to take human genetic material and use it to alter the course of a disease. Those from strict religious cultures might think these scientific practices trespass on divine territory. In some cultures, science may even be seen as creating more problems than it solves (Harding 1998). Contemporary debates about the genetic engineering of food or stem cell research are examples of the culture conflicts that can develop in response to changing societal conditions.

Because culture varies from one setting to another, the meaning systems that develop within a culture must be seen in their cultural context. **Cultural relativism** is the idea that something can be understood and judged only in relationship to the cultural context in which it appears. This does not make every cultural practice morally acceptable, but it suggests that without knowing the cultural context, understanding why people behave as they do is impossible.

Understanding cultural relativism gives insight into some controversies, such as the international debate about the practice of clitoridectomy—a form of genital mutilation. In a clitoridectomy (sometimes called female circumcision), all or part of a young woman's clitoris is removed, usually not by medical personnel, often in very unsanitary conditions, and without any pain killers. Sometimes, the lips of the vagina may also be sewn together. Human rights and feminist organizations have documented this practice in some countries on the African continent, in some Middle Eastern nations, and in some parts of Southeast Asia; estimates are that around

ues and norms, even when that behavior seemingly falls outside of prevailing standards.

Questions to Consider

1. Do you think of tattoos (or body piercing) as fashionable or as *deviant*? What do you think influences your judgment about this and how might your judgment be different were you in a different culture, age group, or historical moment? *Keywords: body art, fads, tattoo artist*

2. Are there fashion adornments that you associate with different social classes? What are they and what kinds of judgments (positive and negative) do people make about them? Where do these judgments come from and why are they associated with social class? *Keywords: status symbol, tattoo collecting*

We have included InfoTrac College Edition keywords at the end of each question to make it easier for you to find more to read on these topics.

Cultural norms establish what people think of as desirable and beautiful. Tattooing, once associated with the working class, is now a form of "collecting" for many.

AP/Wide World Photos

Go to **www.infotrac-college.com**, an online library, to begin your search.

Source: Katherine Irwin. 2001. "Legitimating the First Tattoo: Moral Passage through Informal Interaction." *Symbolic Interaction* 24 (March): 49–73. ●●●

two million girls per year, worldwide, are at risk. This practice is most frequent in cultures where women's virginity is highly prized and where marriage dowries depend on some accepted proof of virginity (Toubia and Izett 1998; Amnesty International 2004).

From the point of view of Western cultures, clitoridectomy is genital mutilation and an example of violence against women. Many feminist organizations have called for international intervention to eliminate the practice of clitoridectomy (Morgan and Steinem 1980; Toubia and Izett 1998), but Western feminists have also debated whether their disgust at this practice should not be balanced by a reluctance to impose Western cultural values on other societies. Should cultures have the right of self-determination, or should cultural practices that maim people be treated as violations of human rights? This controversy is unresolved. The point is to see that understanding a cultural practice requires knowing the cultural values on which it is based. This does not make the practice right, but it illuminates why it occurs, thereby better informing efforts for change.

Just as culture varies from place to place, it also varies over time. As people encounter new situations, the culture that emerges is a mix of the past and present. Second-generation immigrants to the United States, raised in the traditions of their culture of origin, typically grow up with both the traditional cultural expectations of their parents' homeland and the cultural expectations of a new society. Adapting to the new society can create conflict between generations, especially if the older generation is intent on passing along cultural traditions. The children may be more influenced by their peers and may choose to dress, speak, and behave in ways that are characteristic of their new society but unacceptable to their parents (Portes and Rumbaut 2001).

To sum up: Culture is concrete in that we can observe the cultural objects and practices that define human experience. Culture is abstract in that it is a way of thinking, feeling, believing, and behaving. It links the past and the present because culture constitutes the knowledge that makes people part of human groups. It is culture that gives shape to human experience.

Humans and Animals: Is There a Difference?

It is cultural patterns that make humans so interesting. Animal species do develop what we might call culture. Chimpanzees, for example, learn behavior through observing and imitating others—a point proved by observing the different eating practices among chimpanzees in the same species, but raised in different groups (Whiten et al. 1999). Others have observed elephants picking up the bones of dead elephants and fondling them, perhaps evidence of grieving behavior (Meredith 2003). Many animal species have also developed systems of communication that they use for various purposes, including attracting mates and alerting others to danger. Dolphins, for example, are now known to have a complex auditory language. Some animals use tools, develop language, and form social structures that are organized, relatively stable, and exhibit persistent patterns of behavior. Is this evidence of culture?

Like humans, animals adapt to their environment, and many species return to traditional nesting sites each year. If culture is what makes us human, does it distinguish humans from animals? The distinction between humans and animals is not as clear as was once believed.

British ethologist Jane Goodall lived for many years among the chimpanzees in the Gombe Stream Chimpanzee Reserve in Tanzania, Africa. She learned to recognize each chimp and was able to simulate the expressions and gestures that they used to communicate (Goodall 1990). She observed chimpanzees stripping the leaves from a stick, licking the end to make it sticky, and then using the tool to catch termites in their nests. She argued that tool use among the chimpanzees was purposeful and learned behavior, not just instinctual. Apparently then, humans are not unique just because they make and use tools. Toolmaking among humans, however, is far more complex. Think of the difference between using a twig to catch termites and manufacturing an automobile. The difference between humans and animals as toolmakers is not so much an absolute distinction as a matter of degree.

Some animals also use elaborate systems of communication for a variety of purposes, such as attracting mates and alerting others to danger. Do animals develop language-based cultures, as humans do? The most dramatic examples of language capability among animals have come from studies of nonhuman primates. Gorillas and chimpanzees do not have the vocal apparatus necessary to make the sounds of human speech, but under the tutelage of Beatrice and Allen Gardiner (1969), a female chimpanzee named Washoe acquired a vocabulary of 160 standard signs in Ameslan, the American Sign Language system (McGrew 1992; Wallman 1992). By the time Washoe was two years old, she had independently put the words she knew into simple sentences.

One of the most advanced cases of language learning is that of Koko and Michael, a female and male gorilla who learned more than 500 Ameslan signs and used 1000 signs at least once. Koko used complex sentences, saying such things to her teacher as "Hurry, go drink," while pointing to a vending machine, or "Want apple eat want" (Patterson 1978: 454). Koko showed further social development when she adopted a small kitten that she named All Ball. She played affectionately with the cat and at one point signed to her teacher, "Koko love visit Ball" (Vessels 1985: 110). When her friend Michael died in 2000, Koko showed many signs of grieving. You can see Koko using sign language, and learn more about her, at her website (**www.koko.org**).

Despite evidence of social skills among animals, no animal group has developed the elaborate, symbol-based culture characteristic of human societies. Even primates, after several years of intense instruction for several hours every day, have not acquired the level of linguistic skill routinely expected of a three-year-old human. Scientists generally conclude that animals lack the intelligence required to develop the elaborate symbol-based cultures common in human societies.

Studying animal groups reminds us of the interplay between biology and culture. Human biology sets limits and provides certain capacities for human life and the development of culture. Similarly, the environment in which humans live establishes the possibilities and limitations for human society. Nutrition, for instance, is greatly influenced by environment, thereby affecting human body height and weight. Not everyone can drive a golf ball like Tiger Woods, slam-dunk like Shaquille O'Neal, or lob a tennis ball like Venus or Serena Williams, but with daily training and conditioning, people can enhance their physical abilities. Biological limits exist, but cultural factors have an enormous influence on the development of human life.

The Elements of Culture

As we have seen, culture consists of both material objects and abstract thoughts and behavior. Several elements of culture are of particular interest to sociologists: language, norms, beliefs, and values.

Language

Language is a set of symbols and rules that, put together in a meaningful way, provides a complex communication system (Cole 1988). The formation of culture among humans is made possible by language. Learning the language of a culture is essential to becoming part of a society.

Becoming a part of any social group such as a friendship circle, fraternity or sorority, or a professional organization involves learning the language the group uses. Otherwise, you cannot participate fully in its culture. You have to learn the common language to become a member of any social group. Lawyers have their own vocabulary and their own way of constructing sentences of intricate prose commonly known as "legalese." If you are not adept in the language of the law, you are likely to be at a disadvantage in any legal proceeding. Likewise, becoming a sociologist requires learning the words and concepts that sociologists use to communicate their ideas.

Language systems are fluid and dynamic; language evolves in response to social change. Think of how the introduction of computers and other electronic systems of communication have affected the English language. People now talk about needing "downtime" and providing "input." Not that long ago, had you said you were going to "IM" your friends, no one would have known what you were talking about. Even now, those without exposure to advanced electronic technologies might not understand that "IM" refers to the instant messaging that is increasingly common among email users. IM has also introduced its own language—"BFN" (bye for now"), "TTYL" ("talk to you later"), and "GTG" ("got to go")—all new forms of language shared among those in the IM culture.

Language reflects the values of a culture. Thus, Internet language is short and succinct, reflective of the speedup in an increasingly fast-paced, highly technological society. Other cultural changes are also reflected in ordinary language. And, it is no accident that in a society based on the economic system of capitalism, terms such as "buying into" something, "banking on it," and getting to the "bottom line" have entered the daily vocabulary.

Does Language Shape Culture? Edward Sapir (writing in the 1920s) and his student Benjamin Whorf (writing in the 1950s) thought that language was central in determining social thought. Their theory, the **Sapir–Whorf hypothesis**, states that language determines other aspects of culture because language provides the categories through which social reality is understood. In other words, Sapir and Whorf thought that language determines what people think because language forces people to perceive the world in certain terms (Sapir 1921; Whorf 1956). It is not that you perceive something first and then think of how to express it, but that language itself determines what you think and perceive.

If the Sapir–Whorf hypothesis is correct, then people who speak different languages have different perceptions of reality. Whorf used the example of the construction of time to illustrate cultural differences in how language shapes perceptions of reality. He noted that the Hopi Indians conceptualize time as a slowly turning cylinder, whereas English-speaking people conceive of time as running forward in one direction at a uniform pace. European languages place great importance on verb tense, and things are located unambiguously in the past, present, or future.

Sapir and Whorf did not think that language singlehandedly dictates the perception of reality, but it undoubtedly has a strong influence on culture. Scholars now see two-way causality between language and culture. Asking whether language determines culture or vice versa is like asking which came first, the chicken or the egg. Language and culture are inextricable. Each shapes the other, and to understand either, we must know something of both (Aitchison 1997; Hill and Mannheim 1992).

Consider again the example of time. Contemporary Americans think of the week as divided into two parts: weekdays and weekends. The words *weekday* and *weekend* reflect the way Americans think about time. When does a week end? Language that defines the weekend encourages people to think about the weekend in specific ways. It is a time for rest, play, and chores. In this sense, language shapes thoughts about the passage of time (looking forward to the weekend, preparing for the work week), but the language itself (the very concept of the weekend) stems from patterns in the culture, specifically, the work patterns of advanced capitalism. Language and culture shape each other.

The work ethic in American culture also shapes our language since the rhythm of life in the United States is the rhythm of the workplace. The phrase "9 to 5" defines an entire lifestyle. The capitalist work ethic makes it morally offensive to merely "pass the time;" instead, time is to be managed for maximum productivity. Concepts of time in preindustrial, agricultural societies involve time and calendars that follow a seasonal rhythm. The year proceeds according to the change of seasons, not arbitrary units of time such as weeks and months.

Social Inequality in Language The significance of language in culture is particularly apparent in how patterns of race, gender, and class inequality are reflected in language. The language of any culture reflects and reinforces attitudes that are characteristic of the culture. Language can also reproduce these inequalities through the stereotypes and assumptions that may be built into what people say (Moore 1992). As shown in the box "Understanding Diversity: The Social Meaning of Language," names for different groups of people reflect assumptions about those groups. What someone is called can impose an identity on that person. This is why the names for various racial and ethnic groups have been so heavily debated. Even conventions such as whether to capitalize the word *Black* in Black American

---

UNDERSTANDING DIVERSITY
The Social Meaning of Language

Language reflects the assumptions of a culture. This can be seen and exemplified in several ways:

- *Language affects people's perception of reality.*

  Example: Researchers have found that men tend to think that women are not included when terms such as "man" are used to refer to all people. Studies find that when college students look at job descriptions written in masculine pronouns, they assume that women are not qualified for the job (Hyde 1984; Switzer 1990).

- *Language reflects the social and political status of different groups in society.*

  Example: A term such as "woman doctor" suggests that the gender of the doctor is something exceptional and noteworthy. The term "working woman" (used to refer to women who are employed) also suggests that women who do not work for wages are not working. Ask yourself what the term "working man" connotes and how this differs from "working woman."

- *Groups may advocate changing language referring to them as a way of asserting a positive group identity.*

  Example: Some advocates for the "disabled" challenge the term "handicapped," arguing that it stigmatizes people who may have many abilities, even if they are physically distinctive. Also, though someone may have one disabling condition, they may be perfectly able in other regards.

- *The implications of language emerge from specific historical and cultural contexts.*

  Example: The naming of so-called races comes from the social and historical processes that define different groups as inferior or superior. Racial labels do not come just from physical, national, or cultural differences. The term "Caucasian," for example, was coined in the seventeenth century when racist thinkers developed alleged scientific classification systems to rank different societal groups. Alfred Blumenthal used the label Caucasian to refer to people from the Caucuses of Russia, who he thought were more beautiful and intelligent than any group in the world.

- *Language can distort actual group experience.*

  Example: Terms used to describe different racial and ethnic groups homogenize experiences that may be unique. Thus, the terms "Hispanic" and "Latino" lump together Mexican Americans, island Puerto Ricans, U.S.-born Puerto Ricans, as well as people from Honduras, Panama, El Salvador, and other Central and South American countries. "Hispanic" and "Latino" point to the shared experience of those from Latin cultures, but like the terms *Native American* and *American Indian,* they obscure the experiences of unique groups, such as the Sioux, Nanticoke, Cherokee, Yavapai, or Navajo.

- *Language shapes people's perceptions of groups and events in society.*

  Example: Native American victories during the nineteenth century are

can reflect assumptions about the status of different groups. You may have noticed that in this book, the authors capitalize the word *Black* whenever it is used as a proper noun. Likewise, the word *White* is capitalized when used to refer to the specific group of White Americans. This convention is relatively new and still debated: Do you treat the color designator as an adjective (as if only color marks the group's experience), or do you recognize that these labels describe particular group experiences and, therefore, treat the name like a proper noun (as you would American Indian, Latino, and Jewish American)?

Language reflects the social value placed on different groups, which is why it is so demeaning when derogatory terms are used to describe a group or group member. Throughout the period of Jim Crow segregation in the American South, Black men, regardless of their age, were routinely referred to as "boy" by Whites. Calling a grown man "boy" is an insult; it diminishes his status by defining him as childlike. Referring to a woman as a "girl" has the same effect. If girls become young women when they reach puberty, why are young women well into their twenties routinely referred to as

"girls?" As with calling a man "boy," this diminishes women's status. Who are you likely to consider a serious intellectual or leader, a girl or a woman? Many women, and not just feminists, believe that being called a "girl" is belittling.

Note, however, that terms such as "girl" and "boy" are pejorative only in the context of dominant and subordinate group relationships. Whereas African American men would be insulted if called "boy" by a White man, it is not necessarily insulting when African American men refer to each other that way. Similarly, "girl," when used between those of similar status, is not seen as derogatory (as when African American women refer to each other as "girl" in informal conversation). When used by someone in a position of dominance, such as when a man calls his secretary a "girl," it is demeaning. Likewise, terms like "dyke," "fag," and "queer" are terms lesbians and gay men sometimes use without offense in referring to each other, even though the same terms are offensive to lesbians and gays when used about them by others. By reclaiming these terms as positive within their own culture, lesbians and gays build cohesiveness and solidarity (Due

---

typically described as "massacres"; comparable victories by White settlers are described in heroic terms. The statement that Columbus "discovered" America implies that Native American societies did not exist before Columbus "found" the Americas. Indians are frequently referred to as "savages," whereas White colonists are called "settlers" (Moore 1992). Likewise, the word tribe to describe nations of African or Native American people suggests primitive, backward societies. Even the term "American" is problematic. It is used synonymously with "United States" when the Americas include Canada, Mexico, Central America, and South America. To say that the culture of the United States is "American" culture suggests that the United States encompasses the other Americas or that the other cultures of the Americas are not worth notice.

• **Terms used to define different groups change over time and can originate in movements to assert a positive identity.**

Example: In the 1960s, "Black American" replaced the term 'Negro' because the civil rights and Black Power movements inspired Black pride and the importance of self-naming (Smith 1992). Earlier, "Negro" and "colored" were used to define African Americans. Currently, it is popular to refer to all so-called racial groups as "people of color." This phrase was derived from the phrase "women of color," created by feminist African American, Latina[2], Asian American, and Native American women to emphasize their common experiences. Some people find the use of "color" in this label offensive since it harkens back to the phrase "colored people," a phrase generally seen as paternalistic and racist because it was a label used by dominant groups to refer to African Americans prior to the civil rights movement. The phrase "women of color" now has a more positive meaning than the earlier

term "colored women" because it is meant to recognize common experiences, not just label people because of their presumed skin color.

In this book, we have tried to be sensitive to the language used to describe different groups. We recognize that the language we use is fraught with cultural and political assumptions and that what seems acceptable now may be offensive later. Perhaps the best way to solve this problem is for different groups to learn as much as they can about one another, becoming more aware of the meaning and nuances of naming and language and more conscious of the racial assumptions embedded in the language. Greater sensitivity to the language used in describing different group experiences is an important step in promoting better intergroup relationships.

Further Resources: Ellen J. Goldner and Safiya Henderson-Holmes. 2001. *Race and (E)Racing Language: Living with the Color of Our Words.* Syracuse, NY: Syracuse University Press; Mary Crawford. 1995. *Talking Difference: On Gender and Language.* Thousand Oaks, CA: Sage Publications.

[2]*Latina* is the feminine form in Spanish and refers to women; Latino, to men.

••••

1995). These examples show that power relationships between groups supply the social context for the connotations of language.

As Sapir and Whorf argued, language constructs a sense of truth. Thus, language can either reproduce social stereotypes or be used to change how people think about different social groups—and themselves. This is why racism and sexism in language is such an important issue.

Norms

Norms are the specific cultural expectations for how to behave in a given situation. A society without norms would be in chaos; with established norms, people know how to act, and social interactions are consistent, predictable, and learnable. Norms govern every situation. Sometimes they are *implicit*; that is, they need not be stated for people to understand what they are. For example, when waiting in line, an implicit norm is that you should not barge in front of those ahead of you. Implicit norms may not be formal rules, but violation of these norms may produce a harsh response. Implicit norms may be learned through specific instruction or by observation of the culture. They are part of a society's or group's customs. Norms are *explicit* when the rules governing behavior are written down or formally communicated, and specific sanctions are imposed for violating explicit norms.

In the early years of sociology, **William Graham Sumner** (1906), whose work was described in Chapter 1, identified two types of norms: folkways and mores. **Folkways** are the general standards of behavior adhered to by a group. You might think of folkways as the ordinary customs of different group cultures. Men wearing pants and not skirts is an example of a cultural folkway. Other examples are the ways that people greet each other, decorate their homes, and prepare their food. Folkways may be loosely defined and loosely adhered to, but they nevertheless structure group customs and implicitly govern much social behavior.

Mores (pronounced "more-ays") are stricter norms than folkways. Mores control moral and ethical behavior such as the injunctions, legal and religious, against killing others and committing adultery. Mores are often upheld through rules or **laws,** the written set of guidelines that define right and wrong in society. Laws are formalized mores, and violating mores can bring serious repercussions.

When any social norm is violated, sanctions are typically meted out against the violator. **Social sanctions** are mechanisms of social control that enforce norms. The seriousness of a sanction depends on how strictly the norm is held. The strictest norms in society are **taboos**—those behaviors that bring the most serious sanctions. Dressing in an unusual way that violates the folkways of dress may bring ridicule, but is usually not

seriously punished. In some cultures the rules of dress are strictly interpreted, such as among Islamic fundamentalists requiring women who appear in public to have their bodies cloaked and faces veiled. It would be considered a taboo for a woman in this culture to appear in public without being veiled. The sanctions for doing so can be as severe as whipping, branding, banishment, even death.

Negative sanctions may be mild or severe, ranging from subtle mechanisms of control, such as ridicule, to overt forms of punishment, such as imprisonment, physical coercion, and death. Enforcement of the segregationist practices in the American South offers a historical example of how norms are enforced. Segregation was controlled through strict social norms that governed everyday forms of behavior. In one horrendous incident in 1955, Emmett Till, a fourteen-year-old Black youth from Chicago who was visiting relatives in Mississippi, was shot, beaten to death, mutilated, and then dumped in the Tallahatchie River after he whistled at a White woman and called her "baby" in a local store. Segregationist norms were so strong at the time that the courts acquitted the White men accused of his murder (Carson et al. 1987). The teen's violation of the social norms was more strictly punished than the act of murder.

Sanctions can also include rewards, not only punishments. When children learn social norms, correct behavior may elicit positive sanctions and the behavior is reinforced through praise, approval, or an explicit reward. Early on, parents might praise children for learning to put on their own clothes; later, children might get an allowance if they keep their rooms clean. Bad behavior earns negative sanctions, such as getting grounded.

One way to study social norms is to observe what happens when they are violated. Once you become aware of how social situations are controlled by norms, you can see how easy it is to disrupt situations where adherence to the norms produces social order. **Ethnomethodology** is a technique for studying human interaction by deliberately disrupting social norms and observing how individuals respond. The idea is that the disruption of social norms helps one discover the normal social order (Garfinkel 1967). In a famous series of ethnomethodological experiments, college students were asked to pretend they were boarders in their own homes for a period of fifteen minutes to one hour. They did not tell their families what they were doing. The students were instructed to be polite, circumspect, and impersonal; to use terms of formal address; and to speak only when spoken to. After the experiment, two of the participating students reported that their families treated the experiment as a joke. Another family thought the daughter was being extra nice because she wanted something. One family believed that the student was hiding some serious problem. In all the other cases, parents reacted with shock, bewilderment, and anger. Parents accused students of being mean, impolite, and inconsiderate, and demanded explanations for their behavior (Garfinkel 1967). Through this experiment, the student researchers were able to see that even the informal norms governing behavior in one's home are carefully structured. By violating the norms of the household, the norms were revealed.

> **THINKING SOCIOLOGICALLY**
>
> Identify a *norm* that you commonly observe. Construct an experiment in which you, perhaps with the assistance of others, violate the norm. Record how others react and note the sanctions engaged through this norm violation exercise. **NOTE: Be careful not to do anything that puts you in danger or causes serious problems for others.**

Ethnomethodological research teaches that society proceeds on an "as if" basis. That is, society exists because people behave *as if* there were no other way to do so. Most of the time, people go along with what is expected of them. Culture is enforced through the social sanctions applied to those who violate social norms. Usually, specific sanctions are unnecessary because people have learned the normative expectations.

Beliefs

As important as social norms are the beliefs of people in society. **Beliefs** are shared ideas people hold collectively within a given culture, and these beliefs are also the basis for many of the culture's norms and values. An example in the United States is the widespread belief in God. Some beliefs are so strongly held that people find it difficult to cope with ideas that contradict their beliefs. Someone who devoutly believes in God may find atheism intolerable. Those who believe in reincarnation may seem irrational to those who think life ends at death. Similarly, those who believe in magic may seem merely superstitious to those with a more scientific and rational view of the world.

Beliefs orient people to the world by providing answers to otherwise imponderable questions about the meaning of life. Beliefs provide a meaning system around which culture is organized (see Chapter 17). Whether belief stems from religion, myth, folklore, or science, it shapes what people consider to be possible and true. Although a given belief may be unprovable or even logically impossible, it nonetheless guides people through their lives.

Sociologists study belief in a variety of ways, and each theoretical orientation provides different insights into the significance of beliefs for human society. Functionalists see beliefs as a functional component of society in that they integrate people into social groups.

Conflict theorists interpret beliefs as potentially competing world views. Those with more power are able to impose their beliefs on others. Symbolic interaction theorists are interested in how beliefs are constructed and maintained through the social interaction people have with each other.

Values

Deeply intertwined with beliefs are the values of a culture. **Values** are the abstract standards in a society or group that define the ideal principles of what is desirable and morally correct. Thus, values determine what is considered right and wrong, beautiful and ugly, good and bad. Although values are abstract, they provide a general outline for behavior. Freedom, for example, is a value held to be important in U.S. culture, as is equality.

Values can provide rules for behavior, but can also be the source of conflict. For example, the political conflict over abortion is a conflict over values. Pro-choice groups value reproductive freedom and women's right to control their bodies. Those opposed to abortion value the life of the fetus over the woman's right to choose to terminate her pregnancy. Abortion law has attempted to strike a balance among the values behind three competing rights: the right to privacy, the right of the state to protect maternal health, and the right of the state to protect developing life. You can see this value conflict played out in the language the different groups use to promote their value position.

Society's norms are a reflection of underlying values. An example of the impact that values can have on people's behavior comes from an American Indian society known as the Kwakiutl (pronounced "kwok-eeudul"), a group from the coastal region of southern Alaska, Washington state, and British Columbia. The Kwakiutl developed a practice known as *potlatch*, in which wealthy chiefs would periodically pile up their possessions and give them away to their followers and rivals (Benedict 1934; Harris 1974). The object of potlatch was to give away or destroy more of one's goods than did one's rivals. The potlatch reflected Kwakiutl values of reciprocity, the full use of food and goods, and the social status of the wealthiest chiefs in Kwakiutl society. (By the way, chiefs did not lose their status by giving away their goods because the goods were eventually returned in the course of other potlatches. At times, they would even burn large piles of goods, knowing that others would soon replace their accumulated wealth through other potlatches.)

Contrast this practice with the patterns of consumption in the United States. Imagine the chief executive officers (CEOs) of major corporations regularly gathering up their wealth and giving it away to their workers and rival CEOs. In the contemporary United States, *conspicuous consumption* (consuming for the sake of displaying one's wealth) celebrates values opposite those of the potlatch. High-status people demonstrate their position by accumulating more material possessions than those around them (Veblen 1899). Conspicuous consumption encourages consumerism throughout the society so that people feel as though they always have to have more. Values and norms together guide the behavior of people in society. It is necessary to understand the values and norms operating in a situation to understand why people behave as they do.

Cultural Diversity

A society is rarely culturally uniform. As societies develop and become more complex, different cultural traditions appear. The more complex the society, the more likely its culture will be internally varied and diverse. The United States, for example, hosts enormous cultural diversity stemming from religious, ethnic, and racial differences, as well as regional, age, gender, and class differences. Currently, 11 percent of people living in the United States are foreign-born. In a single year, immigrants from more than 100 countries come to the United States. Whereas earlier immigrants were predominantly from Europe, now Central/South America and Asia are the greatest sources of new immigrants. Now, 18 percent of young people (under age five) speak a language other than English at home, the largest increase being among Black, non-Hispanics (see also Map 3.1; U.S. Census Bureau 2004). Cultural diversity is clearly a characteristic of contemporary American society.

The richness of American culture stems from the many traditions that immigrants have brought with them to this society, as well as from the new cultural forms that emerged through their experience within the United States. Jazz has roots in the musical traditions of the slave community and African cultures, but is one of the few musical forms to originate in the United States. Jazz is an *indigenous* art form, one originating in a particular region or culture. Since the birth of jazz, cultural greats such as Ella Fitzgerald, Count Basie, Duke Ellington, Billie Holiday, and numerous others have not only enriched the jazz tradition, but have also influenced other forms of music, including rock and roll.

With such great variety, how can the United States be called one culture? The culture of the United States, including its language, arts, food customs, religious practices, and dress can be seen as the sum of the diverse cultures that constitute its society. Altogether, these behaviors and beliefs constitute American culture, but cultural diversity is a major feature of this culture as a whole.

DEBUNKING **SOCIETY'S MYTHS**

Myth: Americans share a single culture derived from Western European traditions.

Sociological perspective: The United States is an increasingly diverse society, and many of its dominant cultural traditions have been derived from African Americans, Latinos, and other groups whose experiences are often neglected (Takaki 2002).

Dominant Culture

Two concepts from sociology help in understanding the complexity of culture in a given society: the *dominant culture* and *subcultures*. The **dominant culture** is the culture of the most powerful group in society. It is the cultural form that receives the most support from major institutions and constitutes the major belief system. Although the dominant culture is not the only culture in a society, it is commonly believed to be "the" culture of a society, despite the other cultures present. Social institutions in the society perpetuate the dominant culture and give it a degree of legitimacy that is not shared by other cultures. Often, the dominant culture is the standard by which other cultures in the society are judged.

The Harlem Renaissance produced some of the most distinctive and accomplished artists, including some of the greatest jazz musicians ever. In this photo, taken by Art Kane in Harlem in 1958, the musicians are gathered on 126th street. For a list of the artists' names in this photo, see the last page of this book.

MAPPING AMERICA'S DIVERSITY

MAP 3.1 Language as Evidence of Cultural Diversity

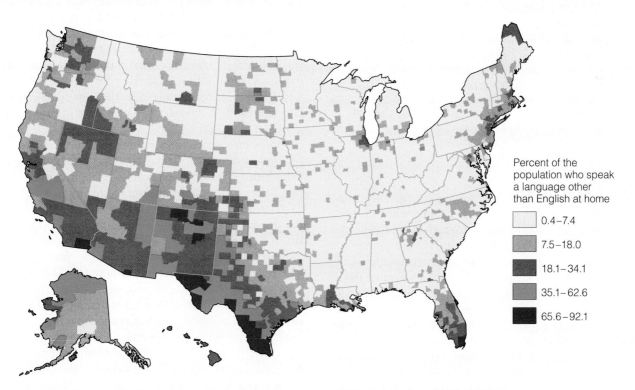

Percent of the population who speak a language other than English at home

	0.4–7.4
	7.5–18.0
	18.1–34.1
	35.1–62.6
	65.6–92.1

With increased immigration and greater diversity in the U.S. population, evidence of cultural diversity can be seen in many homes—language being one type of evidence. This map shows the regional differences in the percent of the population over age 5 who speak a language other than English at home. For the U.S. as a whole, 17.9 percent of the population—almost one-fifth— fits into this category. Eight percent of the population say they speak English less than very well. What implications does this have for the regions most affected? How might it influence relations between different generations within households?

Source: U.S. Census Bureau. 2004. *American FactFinder.* **www.census.gov**

A dominant culture need not be the culture of the majority of people; it is simply the culture of the group in society that has enough power to define the cultural framework. As an example, think of a college or university that has a strong system of fraternities and sororities. On such a campus, the number of students belonging to fraternities and sororities is probably a small percentage of the total student body, but this cultural system originated by the Greeks may dominate campus life. In a society as complex as the United States it is hard to isolate a single dominant culture, although the widely acknowledged middle-class values, habits, and economic resources constitute the dominant culture. This culture is strongly influenced by such instruments of culture as television, the fashion industry, and Anglo-European traditions, and includes such diverse elements as fast food, Christmas shopping, and professional sports.

The dominant culture of the middle class has been the subject of political debate, as the nation responds to increasing diversity within the population. Groups sensitive to the interests of minority cultures have criticized the dominant culture for being based so exclusively on White, male, European cultural traditions. One political debate is about the movements in many states to establish English as the official language. Through public referenda and legislative action, several states have adopted laws designating English as the official state language. Proponents of this concept insist that English is already the de facto official language of America, and mastering English is essential for educational and economic success. To opponents, this movement reflects an anti-immigrant backlash growing from hostility toward groups perceived as outsiders (Portes 2002). Given the cultural diversity within the United States, this debate is likely to continue.

Subcultures

Subcultures are the cultures of groups whose values and norms of behavior differ from those of the dominant culture. Members of subcultures tend to interact frequently with one another and share a common world view. They may be identifiable by their appearance (style of clothing or adornments) or perhaps by language, dialect, or other cultural markers. You can view subcultures along a continuum of how well integrated into the dominant culture they are. Subcultures typically share some elements of the dominant culture and coexist within it, although some subcultures may be quite separated from the dominant culture—either because they are unwilling or unable to assimilate into the dominant culture—that is, share its values, norms, and beliefs (Dowd and Dowd 2003).

Rap and hip hop music first emerged as a subculture where young African Americans developed their own style of dress and music to articulate their resistance to the dominant White culture. Now, rap and hip hop have been incorporated into mainstream youth culture. Indeed, they are now global phenomena, as cultural industries have turned hip hop and rap into a profitable commodity. Even so, rap still expresses some revolt against dominant values. In Italy, rap is associated with the political left-wing; in other national contexts, rap expresses an oppositional identity for Black and White youth who feel marginalized by the dominant culture (Haines 1999; Wright 2000; Kaplan 2000; Briggs and Cobley 1999; Rumbo 2001).

Sometimes subcultures are imposed, as when groups are excluded from participation in the dominant group. Prohibiting a group from expressing their culture may drive the culture underground, resulting in a *culture of resistance*, one that is a specific challenge to the dominant culture. In this case, cultural exclusion produces cultural solidarity. The history of Asian Americans in California provides an example. Immigration of Chinese and Japanese laborers to the West Coast near the turn of the nineteenth century generated racist fears about the "Yellow Peril" threatening Western civilization. In California, the San Francisco school board, with the support of the governor and the president of the United States, Theodore Roosevelt, directed "all Chinese, Japanese, and Korean children to attend the 'Oriental School'" (Takaki 1989: 201). Forced into their own schools, excluded from mainstream trades, neighborhoods, and professional associations, Asian Americans developed their own neighborhoods, associations, and social clubs, fostering a high degree of solidarity among themselves. Ironically, they were then accused by Whites of being "clannish."

Contemporary rock and roll can also be seen as the result of a flourishing subculture that originated when Black musicians were earlier excluded from the White-dominated music industry in the 1940s and 1950s. The musical form known as *race music* was produced on independent Black labels and sold primarily to Black audiences. Studios such as Stax Records in Memphis, Tennessee, produced a vast repertoire of music and

TAKING ON SOCIAL ISSUES ●●●

Bilingual Education

Culture may seem like a relatively neutral concept, but in recent years, it has become the basis for numerous social conflicts and policy debates. The movement to make English the official language of the United States and related debates about bilingual education, such as the 1998 referendum in California to ban it, signify the significance of culture in the creation of social policy. What are the arguments for and against bilingual education, and how is this policy debate linked to increasing diversity in the United States?

Taking Action

Go to the Taking Action Exercise on the companion website—at **http://sociology.wadsworth.com/andersen_taylor/4e**—to learn more about an organization that addresses this topic. ●●●

supported a large number of Black artists. As race music gathered a following, White artists and producers copied it, recording and marketing it through White-owned companies, giving little or none of the profits to the original artists and songwriters. Classic Black musicians, such as Big Mama Thornton, Sam Cooke, and countless others seldom received more than a pittance for their work, while White singers such as Pat Boone, Elvis Presley, and many others made fortunes recording songs originally written and performed by Black artists. Yet, race music thrived as a subcultural expression of Black experience, giving birth to rock and roll.

Subcultures also develop when new groups enter a society. Puerto Rican immigration to the U.S. mainland, for example, has generated distinct Puerto Rican subcultures within many urban areas. Although Puerto Ricans partake of the dominant culture, their unique heritage is part of their subculture, some of which has now entered the dominant culture. Salsa music, now heard on mainstream radio stations, was created in the late 1960s by Puerto Rican musicians to reflect the experience of barrio people. It mixes the musical traditions of other Latin music, including rumba, mambo, and cha-cha (Sanchez 1999). As with other subcultures, the boundaries between the dominant culture and the subculture are permeable, thus resulting in cultural change as new cultures intermix.

Countercultures

Countercultures are subcultures created as a reaction against the values of the dominant culture. Members of the counterculture reject the dominant cultural values, often for political or moral reasons, and develop cultural practices that explicitly defy the norms and values of the dominant group. Nonconformity to the dominant culture is often the hallmark of a counterculture.

Women's communities are examples of countercultures created to resist the dominant culture. Women's communities are based on woman-centered and feminist social networks, relationships, and cultural activities. Typically they are not geographically separate from the mainstream society; however, specific places, such as feminist bookstores, community activist groups, and women's shelters, promote a feeling of belonging that some women may find absent in the dominant culture. For those within women's communities, this solidarity is a relief from the male domination and harassment that they find in the dominant society (Taylor and Rupp 1993).

The contemporary militia movement is a counterculture that directly challenges the dominant political system. Not only have militia members withdrawn from society into isolated compounds that they have created in rural areas, but they have also armed them-

© Jeffrey Greenburg/PhotoEdit

Subcultures, *such as hip hop, develop their own styles—including dress, language, and other cultural markers.*

selves for what they perceive as a countercultural revolution against the authority of the U.S. government. Like other countercultures, militia groups have unique modes of dress, a commonly shared worldview, and a distinctive lifestyle (Ferber 1998; Stern 1996).

Ethnocentrism

Ethnocentrism is the habit of seeing things only from the point of view of one's own group. To judge one culture by the standards of another culture is ethnocentric. A White American who believes the Latino culture is deficient because it is different from Anglo culture is engaging in ethnocentric thinking. An ethnocentric perspective prevents a person from understanding the world as it is experienced by others, and it can lead to narrow-minded conclusions about the worth of diverse cultures. Any group can be ethnocentric, including members of racial minority groups, if they believe their way of living is superior to others. Any group that sees the world only from its own point of view is engaging in ethnocentrism.

Ethnocentrism creates a strong sense of group solidarity and group superiority, but it also discourages intercultural or intergroup understanding. A good example is seen in *nationalism,* the sense of identity that arises when one group exalts its own culture over all other groups and organizes politically and socially around this principle. Nationalist groups tend to be highly exclusionary, rejecting those who do not share their cultural experience and judging all other cultures to be inferior. Nationalist movements tend to use extreme ethnocentrism as the basis for nation-building. Taken to its limits, ethnocentrism can lead to overt political conflict, war, terrorism, even *genocide* (the mass killing of people based on their membership in a particular group). You might wonder how someone could believe so strongly in the righteousness of their religious faith that they would murder people to express their commitment. Understanding ethnocentrism does not excuse such behavior, but it can explain how it occurs. Think of al Qaeda and its extremist belief that terrorism is a justified tactic of *jihad,* which is a religious struggle to defend Islamic faith. Nationalism and the deep ethnic conflicts associated with it have produced many of the wars of contemporary world affairs. You can see some evidence of the problems that ethnocentrism creates in Figure 3.1, which shows how cultural values in the Islamic world clash with those of the West (see Figure 3.1).

Ethnocentrism can be extreme, as in the preceding examples, but it can also be subtle, as in the example of social groups who think their way of life is better than any other group. Is there such a ranking among groups in your community? Fraternities and sororities often build group rituals around such claims; youth groups see their way of life as superior to adults; urbanites may think their cultural habits are more sophisticated than those labeled "country hicks."

A concept antithetical to ethnocentrism is **multiculturalism,** referring to modes of thinking that view society through the plural experiences of its diverse membership. Instead of limiting one's perspective to a single group, multiculturalism values the perspective of multiple groups, recognizing the rich and diverse cultural patterns now part of U.S. society. Proponents of multicultural education, including those in African American and Latino studies, gay and lesbian studies, and women's studies, have argued that studying culture, history, and society only from the perspective of any one group distorts the diverse experiences of everyone else and may misrepresent the favored group as well. Multicultural education is intended to remedy ethnocentric thinking.

The Globalization of Culture

The infusion of Western culture throughout the world seems to be accelerating, as the commercialized culture of the United States is marketed on a worldwide basis. One can go to distant places in the world and see familiar elements of U.S. culture, whether it be McDonald's in Hong Kong, The Gap in South Africa, or Disney products in western Europe. From films to fast food, the United States dominates international mass culture, largely through the influence of capitalist markets, as conflict theorists would argue. The diffusion of a single culture throughout the world is referred to as **global culture.** Despite the enormous diversity of cultures worldwide, fashions, foods, entertainment, and other cultural values are increasingly dominated by U.S. markets, thereby creating a more homogenous global and capitalist culture. One result is that tourists can step off the Star Ferry in Hong Kong and, instead of eating Chinese food, have a familiar Big Mac, just as they can in most major cities of the world.

To sociologists, one of the intriguing questions about the globalization of culture is whether economic development supplants the traditional values of diverse world cultures. Sociologists have found that economic development results in a shift from absolutist norms and values to more rational, tolerant, trusting, and participatory norms and values. However, global economic change is not the sole influence on national cultures. The broad cultural heritage of diverse societies, often reflected in religious and political orientation, such as Roman Catholic, Protestant, Confucian, Orthodox, or communist, also has an effect on the enduring value systems of societies. Educational and religious institutions, along with the mass media, transmit these values. Sociologists conclude that *both* economic changes

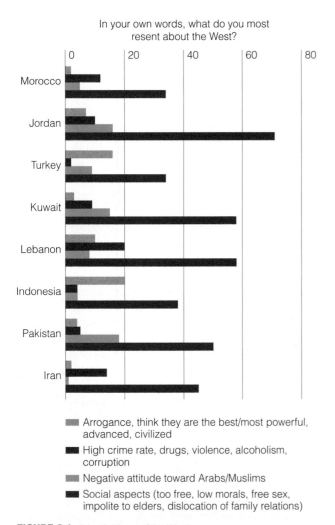

In your own words, what do you most resent about the West?

- ▨ Arrogance, think they are the best/most powerful, advanced, civilized
- ■ High crime rate, drugs, violence, alcoholism, corruption
- ▨ Negative attitude toward Arabs/Muslims
- ■ Social aspects (too free, low morals, free sex, impolite to elders, dislocation of family relations)

FIGURE 3.1 *Islamic Views of the West*

Data: Burkholder, Richard. 2003. "Iraq and the West: How Wide Is the Morality Gap?" *The Gallup Poll,* Princeton, NJ. **www.gallup.com**

and traditional cultural values shape the national culture of different societies (Inglehart and Baker 2000).

The increasing grip of global culture nonetheless has produced a more commercially based culture. Some argue that a number of the conflicts in modern history stem from a cultural struggle between the values of a consumer-based, capitalist western culture and the traditional values of local communities. Benjamin Barber (1995) expresses this as the struggle between "McWorld" and "Jihad," which he interprets as the tension between global commerce and parochial ethnic or tribal values. As some people resist the influence of market-driven values, movements to reclaim or maintain ethnic and cultural identity can intensify. Thus, you can witness a proliferation of identity-based movements among various racial and ethnic groups both in the United States and worldwide.

Popular Culture

Some aspects of culture pervade the whole society, such as common language, general patterns of dress, and dominant value systems. **Popular culture** comprises the beliefs, practices, and objects that are part of everyday traditions. This includes mass-produced culture, such as popular music and films, mass-marketed books and magazines, large-circulation newspapers, and other parts of the culture that are shared by the general populace (Gans 1999). Popular culture is distinct from *elite culture,* that which is shared by only a select few, but highly valued, such as yacht racing or opera. Unlike elite culture (sometimes referred to as "high culture"), popular culture is mass-produced and mass-consumed. Popular culture has enormous significance in the formation of public attitudes and values, and it plays a significant role in shaping the patterns of consumption in contemporary society.

The distinction between elite and popular culture is relatively new. The industrialization of modern life in late nineteenth-century America fragmented public life, creating greater cultural divisions among people and establishing a more distinct hierarchy within cultural institutions. Even such simple things as hierarchical seating arrangements in theaters or the development of separate theaters for the "truly refined" reflect changes in the socioeconomic arrangements of society. During the early twentieth century, as a larger middle class and upper middle-class emerged, the distinction between elite and mass culture widened. Whereas in the early nineteenth century Shakespeare and other classic writers were read by a wide spectrum of the reading public, now they are considered more refined (Levine 1984).

This trend has continued, but with some blurring of the distinctions. Although elites may derive their culture from expensive theater shows, opera performances, or private libraries, millions of "ordinary" citizens get their primary cultural experience from book-of-the-month clubs, Harlequin romances, mass-produced magazines, and popular television shows. In common discourse, the word *cultured* is associated with elite cultural taste.

Sociologists think of cultural taste and participation in the arts as socially structured. First, different groups in society partake of culture in different ways. For example, social class affects the ability of groups to participate in certain forms of culture. Who can afford to buy a $100 symphony ticket? Second, familiarity with different cultural forms stems from patterns of exclusion throughout history, as well as integration into networks that provide information about the arts. As a result, African Americans are much more likely than White Americans to attend jazz concerts and listen to soul, blues, rhythm and blues, and other historically African American musical forms (DiMaggio and Ostrower 1990). Third, popular culture is increasingly disseminated by the mass media, either through television, film, radio, or the Internet; thus, it is buttressed by the interests of big entertainment and information industries who profit from the cultural forms they produce. Each of these factors reveals the social structure of popular and elite culture.

The Influence of the Mass Media

Popular culture is characterized by mass distribution. The term **mass media** refers to those channels of communication that are available to wide segments of the population—the print, film, and electronic media (radio and television) and, increasingly, the Internet. The mass media have extraordinary power to shape public perceptions. In an era when complex issues are reduced to "sound bites" and "photo opportunities," the mass media increasingly have the ability to shape public information.

For most Americans, leisure time is dominated by media. The average person consumes some form of media 71 hours per week—more time than they likely spend in school or at work; thirty-two of these hours are spent watching television (U.S. Census Bureau 2004). Ninety-five percent of all homes in the United States have at least one television—more than have telephone service. Watching television is the most popular leisure activity of Americans; 26 percent say it is their favorite way to spend an evening, compared to nine percent who would rather read and one percent whose favorite evening activity is listening to music (Saad 2002).

Television is the common basis for social interaction in a widely dispersed and diverse national community. Television is a powerful transmitter of culture, but it also portrays a very homogeneous view of culture because, in seeking the widest possible audience, networks and sponsors find the most common ground and take few risks.

THINKING SOCIOLOGICALLY

Watch a particular kind of television show (situation comedy, sports broadcast, children's cartoon, or news program) and, using *content analysis,* make careful written notes about how different groups are depicted in this show. How often are women and men, boys and girls, shown? How are they depicted? You could also observe the portrayal of Asian Americans, Native Americans, African Americans, or Latinos. What do your observations tell you about the cultural ideals that are communicated through *popular culture*?

Television and other forms of mass media have enormous power to shape public opinion and behavior. If you doubt this, observe how familiar certain characters from television sit-coms and dramas are in everyday life. People at work may talk about last night's episode of a particular show or laugh about the antics of their favorite sit-com character. The media is also ubiquitous—present in airports, elevators, classrooms, bars and restaurants, hospital waiting rooms. You may even be born to the sounds and images of television, since they are turned on in many hospital delivery rooms. Television is so ever-present in our lives that there are now homes called "constant television households"—that is, those where television is on most of the time: 42 percent of all U.S. households. Black children are more likely than White or Hispanic children to live in such households (Gitlin 2001). For many families, TV is the "babysitter."

The mass media (and television, in particular) play a huge role in shaping people's perception and aware-ness of social issues. For example, even though crime has actually decreased, the amount of time spent reporting crime in the media has actually increased. Sociologists have found that people's fear of crime is directly related to the time they spend watching television or listening to the radio (Angotti 1997; Chiricos et al. 1997). Similarly, given the media attention to school violence, you would think that such violence is increasing. In fact, throughout the 1990s, there was an actual decline in school violence with fewer students carrying guns and other weapons. Although the number of events with multiple-victim homicides in schools has increased, school-associated violent deaths have decreased (U.S. Department of Education and U.S. Department of Justice 2000; Brener et al. 1999).

Although people tend to think of the news as authentic and true, news is actually manufactured in a complex social process. From a sociological perspective, it is not objective reality that determines what news is presented and how it is portrayed, but commercial interests, the values of news producers, and perceptions of what matters to the public. The media shape our definition of social problems by determining the range of opinion or information that is defined as legitimate and by deciding which experts will be called on to elaborate an issue (Gans 1979; Gitlin 2001).

Race, Gender, Sex, and Class in the Media

Many sociologists have argued that the mass media promote narrow definitions of who people are and what they can be. What is considered beauty, for example, is not universal. Ideals of beauty change as cultures change and depend upon what certain cultural institutions promote as beautiful. Aging is not beauti-

---

A SOCIOLOGICAL EYE ON THE MEDIA
Media Blackout

Suppose that you lived for a few days without use of the mass media that permeates our lives? How would this affect you? In an intriguing experiment, Charles Gallagher (a sociologist at Georgia State University) has developed a research project for students in which he asks them to stage a media blackout in their lives for just forty-eight hours. First, students write a log of the amount of time they spend engaged with the media in the week prior to the blackout. Included is time spent watching television, on the Internet, reading

books and magazines, listening to music, viewing films, even using cell phones—any activity that can be construed as part of the media monopoly on people's time.

Next, students take a forty-eight-hour period during which they eliminate all discretionary time with the media. (Note that you are allowed to do required work and schoolwork during the imposed blackout.) When one of the authors of this book (Andersen) had her students do this experiment, they complained—even before starting—

that they wouldn't be able to do it! But they had to try. What did they find?

First, Andersen's students got a big assist—the week of the assignment came during Hurricane Isabel on the East Coast when many were without power for several days. This did not deter the students from thinking they just *had to have* their DVD players, music, TV, and cell phones! Many of the students said they could only stand being without access to the media for a few hours and couldn't go the full two days without using the

ful, youth is. Light skin is promoted as more beautiful than dark skin, regardless of race, although being tan is seen as more beautiful than being pale. In African American women's magazines, the models typified as most beautiful are generally those with the clearly Anglo features of light skin, blue eyes, and straight or wavy hair. These depictions have fluctuated over time. In the early 1970s, for example, there was a more Afrocentric ideal of beauty with darker skin, Afro hairdos, and African clothing. European facial features are also pervasive in the images of Asian and Latino women appearing in U.S. magazines. The media communicate that only certain forms of beauty are culturally valued. These ideals are not somehow "natural." They are constructed by those who control cultural and economic institutions (Kassell 1995; Wolf 1991; Leslie 1995).

Images of women and racial and ethnic minorities in the media are similarly limiting. Content analyses of television reveal that during prime time, men are a large majority of the characters shown. There has been an increase in women characters depicted in professional jobs, but such images usually show young professional women, suggesting that career success comes early, especially to thin and beautiful women (Signorielli and Bacue 1999). On soap operas, women are cast either as evil, or good but naive. In music videos, women characters appear less frequently, have more beautiful bodies, are more physically attractive, wear more sexy and skimpy clothing, and are more often the object of another's gaze than their male counterparts (Signorielli et al. 1994).

Media images also distort the realities of gender roles for men. In television commercials, men appearing with children without a spouse present are more likely to be shown outside, less likely to be doing household chores, more likely to be shown with boys, and

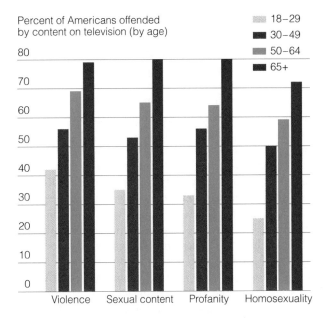

FIGURE 3.2 Percentage of Americans who say they are offended by the content on television (by age)

Data: Jones, Jeffrey M. 2004. "Most Americans Offended by Sex and Violence on Television." *The Gallup Poll*, Princeton, NJ. **www.gallup.com**

seldom shown with infants. Although such portrayals show men as involved in family life, they still project stereotypical norms of fatherhood (Kaufman 1999). Also, whereas nurturing images of men are now more frequent in the media, they most often appear in magazines read by women, not by men (Vigorito and Curry 1998).

Even though African Americans and Hispanics watch more television than Whites do, they are a small proportion of television characters and generally confined to a narrow variety of stereotypical character types. Latinos are often stereotyped as criminals or passionate

media. What did they report happened to them when they were not engaged with the media to the extent that is normal in their lives?

Most reported that they felt isolated during the media blackout—not just from information, but mostly from other people. They were excluded from conversations with friends—about what happened on a given television episode, about film characters or movie stars profiled in magazines and they could not play computer games. One even wrote that without the media she felt

that she had no personality! Overall, without their connection to the media, students were alienated, isolated, and detached—although most also reported that they studied more without the distraction of the media. And, a most interesting finding was that several reported that they were much more reflective during this time and had more meaningful conversations with friends.

Should you try this experiment, you can think about the enormous influence that the mass media have in shaping everyday life—including our

self-concepts and our relationships with other people. A warning: If you try the media blackout, be sure to have some plan in place for having your family and/or friends contact you in case of an emergency!

Sources: Gitlin, Todd. 2002. *Media Unlimited: How the Torrent of Images and Sounds Overwhelms Our Lives.* New York: Metropolitan Books; Personal correspondence, Charles Gallagher, Georgia State University. ●●●

lovers. African American men are most typically seen as athletes, sports commentators, criminals, or entertainers. African American women are shown in domestic or sexual roles or as sex objects. It is difficult to find a single show where Asians are the principal characters. Instead, they are usually depicted in silent roles as domestics or other behind-the-scenes or sidekick characters. Typically, if presented at all, Native Americans are marginal characters stereotyped as warriors or as silent, exotic, and mysterious. Jewish women are generally invisible on popular television programming, except when they are ridiculed in stereotypical roles (Kray 1993).

Class stereotypes also abound with working-class men typically portrayed as ineffectual, even buffoonish (Butsch 1992; Dines and Humez 1995). Working-class men are most likely seen in beer commercials or police shows (Bettie 1995). A recent study of TV talk shows also shows how such shows exploit working-class people. The researcher, Laura Grindstaff, spent half a year working on two popular talk shows, carefully doing participant observation and interviewing staff and guests. She found that guests had to enact stereotypes of their groups to get air time. She argues that, although these shows give ordinary people a place to air their problems and be heard, they exploit the working class—making it seem that others care about them when they do not and making a spectacle of their troubles (Grindstaff 2002; Press 2002).

Recently, there has been increased representation of gays and lesbians in the media—after years of being virtually invisible or only the subject of ridicule. In 2002, the staid *New York Times* began showing lesbian and gay couples in the wedding announcements. Now, as advertisers have sought to expand their commercial markets, there are more gay and lesbian characters being shown on television. This makes gays and lesbians more visible, although critics point out that they are still cast in narrow and stereotypical terms, showing little about real life for gays and lesbians. Nonetheless, cultural visibility for any group is important because it validates people and can influence the public's acceptance and generate support for equal rights protection (Gamson 1998).

Music, film, books, and other industries all play a significant role in molding public consciousness. What images are produced by these cultural forms? You can look for yourself. Try to buy a birthday card that contains neither an age or gender stereotype, or watch rock videos and see how different gender and race groups are portrayed. You will likely find that women are depicted as trying to get the attention of men. Greeting cards ridicule the process of aging, especially for women. Men are portrayed as only interested in golf and fishing. Women's magazines send endless messages about the ideal women's body image and are filled with exhortations to diet, despite also showing highly caloric foods!

Even the new world of cyberspace depicts images and activities based on cultural stereotypes (Ebo 1998).

Do these images matter? Studies find that exposure to traditional sexual imagery in music videos has an effect on college students' attitudes about adversarial sexual relationships (Kalof 1999). Other studies find that even when viewers see media images as unrealistic, they think that others find the images important and will evaluate them accordingly; this has been found especially to be true for young White girls who think boys will judge them by how well they match the media ideal (Milkie 1999). Although people do not just passively internalize media images and do distinguish between fantasy and reality (Currie 1997), such images form cultural ideals that have a huge impact on people's behavior, values, and self-image.

Theoretical Perspectives on Culture

Sociologists study culture in a variety of ways, asking a variety of questions about the relationship of culture to other social institutions and the role of culture in modern life. One important question for sociologists studying the mass media is whether these images have any effect on those who see them. Do the media create popular values or reflect them? The **reflection hypothesis** contends that the mass media reflect the values of the general population (Tuchman 1979). The media try to appeal to the most broad-based audience, so they aim for the most common values to depict images and ideas. Maximizing popular appeal is central to television program development. Media organizations invest huge amounts on market research to uncover what people think and believe and what they will like. Characters are then created with whom people will identify, but they are distorted versions of reality.

The reflection hypothesis assumes that images and values portrayed in the media reflect the values existing in the public, but the reverse can also be true. The ideals portrayed in the media also influence the values of those who see them. Although there is no direct relationship between the content of mass media images and what people think of themselves, clearly these mass-produced images have a significant impact on who we are and what we think.

Culture and Group Solidarity

Many sociologists have examined how culture integrates members into society and social groups. Functionalist theorists think that norms and values create social bonds that attach people to society (see Chapter 1). Culture therefore provides coherence and stability in society.

Robert Putnam examines this idea in his book *Bowling Alone* (2000), in which he argues that there has been a recent decline in civic engagement, defined as participation in voluntary organizations, religious activities, and other forms of public life. As people become less engaged in such activities, there is a decline in the shared values and norms of the society so that social disorder results. Sociologists are debating the extent to which there has been such a decline in public life, but from a functionalist perspective, the point is that participation in a common culture is an important social bond and one that unites society (Etzioni et al. 2001).

Classical theoretical analyses of culture have placed special emphasis on nonmaterial culture—the values, norms, and belief systems of society. Sociologists who use this perspective emphasize the integrative function of culture, that is, its ability to give people a sense of belonging in an otherwise complex social system (Smelser 1992a). In the broadest sense, they see culture as a major integrative force in society, providing a sense of collective identity and commonly shared world views.

Culture, Power, and Social Conflict

Whereas some see culture as producing shared values and group solidarity, others understand culture as driven by economic interests and powerful monopolies. Conflict theorists (see Chapter 1) analyze culture in terms of power in society, viewing culture as dominated by economic interests, with a few powerful groups as the major producers and distributors. With more corporate mergers, a single corporation can control a majority share of television, radio, newspapers, music, publishing, film, and the Internet, as shown in Figure 3.3. As the production of popular culture becomes concentrated in the hands of just a few, there may be less diversity in the content.

Conflict theorists think contemporary culture is produced within institutions that are based on inequality and capitalist principles. As a result, the cultural values and products produced and sold promote the economic and political interests of the few who own or benefit from these cultural industries. This is especially evident in the study of popular culture, which is mass-marketed by entities with a vast economic stake in the dissemination of their products. Conflict theorists conclude that the cultural products most likely to be produced are consistent with the values, needs, and interests of the most powerful groups in society. The evening news, for example, is typically sponsored by major financial institutions, oil companies, and automobile makers. Conflict theorists then ask how this commercial sponsorship influences the content of the news. If the news were sponsored by labor unions, would conflicts between management and workers always be defined as "labor troubles," or might newscasters refer instead to "capitalist troubles" (Lee and Solomon 1990)?

Whether it is books, music, films, news, or other cultural forms, the monopolies in the communications industry have a strong interest in protecting the status quo. As media conglomerates swallow up smaller companies and drive out less efficient competitors, their control over the production and distribution of culture can influence everything from the movies and television, to your daily newspaper, to the books you read in school.

Sociologists refer to the concentration of cultural power as **cultural hegemony** (pronounced "heh-JEM-o-nee"), which is defined as the pervasive and excessive influence of one culture throughout society. Cultural hegemony creates a homogeneous mass culture. It is used by powerful groups to gain the assent of those they rule. The concept of cultural hegemony implies that culture is highly politicized, even if it does not appear so. Through cultural hegemony, those who control cultural institutions can also control people's political awareness because the rulers create cultural beliefs that make the rule of those in power seem inevitable

Table 3.1
Theoretical Perspectives on Culture

Culture according to:			
Functionalism	**Conflict Theory**	**Symbolic Interaction**	**New Cultural Studies**
Integrates people into groups	Serves the interests of powerful groups	Creates group identity from diverse cultural meaning systems	Is ephemeral, unpredictable, and constantly changing
Provides coherence and stability in society	Can be a source of political resistance	Changes as people produce new cultural meaning stystems	Is a material manifestation of a consumer-oriented society
Creates norms and values that integrate people in society	Is increasingly controlled by economic monopolies	Is socially constructed through the activities of social groups	Is best understood by analyzing its artifacts—books, films, television images.

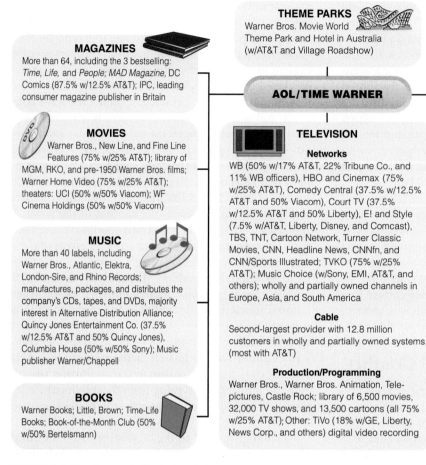

THEME PARKS
Warner Bros. Movie World
Theme Park and Hotel in Australia
(w/AT&T and Village Roadshow)

AOL/TIME WARNER

MAGAZINES
More than 64, including the 3 bestselling:
Time, Life, and *People; MAD Magazine,* DC
Comics (87.5% w/12.5% AT&T); IPC, leading
consumer magazine publisher in Britain

MOVIES
Warner Bros., New Line, and Fine Line
Features (75% w/25% AT&T); library of
MGM, RKO, and pre-1950 Warner Bros. films;
Warner Home Video (75% w/25% AT&T);
theaters: UCI (50% w/50% Viacom); WF
Cinema Holdings (50% w/50% Viacom)

MUSIC
More than 40 labels, including
Warner Bros., Atlantic, Elektra,
London-Sire, and Rhino Records;
manufactures, packages, and distributes the
company's CDs, tapes, and DVDs, majority
interest in Alternative Distribution Alliance;
Quincy Jones Entertainment Co. (37.5%
w/12.5% AT&T and 50% Quincy Jones),
Columbia House (50% w/50% Sony); Music
publisher Warner/Chappell

BOOKS
Warner Books; Little, Brown; Time-Life
Books; Book-of-the-Month Club (50%
w/50% Bertelsmann)

TELEVISION

Networks
WB (50% w/17% AT&T, 22% Tribune Co., and
11% WB officers), HBO and Cinemax (75%
w/25% AT&T), Comedy Central (37.5% w/12.5%
AT&T and 50% Viacom), Court TV (37.5%
w/12.5% AT&T and 50% Liberty), E! and Style
(7.5% w/AT&T, Liberty, Disney, and Comcast),
TBS, TNT, Cartoon Network, Turner Classic
Movies, CNN, Headline News, CNNfn, and
CNN/Sports Illustrated; TVKO (75% w/25%
AT&T); Music Choice (w/Sony, EMI, AT&T, and
others); wholly and partially owned channels in
Europe, Asia, and South America

Cable
Second-largest provider with 12.8 million
customers in wholly and partially owned systems
(most with AT&T)

Production/Programming
Warner Bros., Warner Bros. Animation, Tele-
pictures, Castle Rock; library of 6,500 movies,
32,000 TV shows, and 13,500 cartoons (all 75%
w/25% AT&T); Other: TiVo (18% w/GE, Liberty,
News Corp., and others) digital video recording

INTERNET
America Online, CompuServe,
Netscape, ICQ, and AOL Instant
Messenger; websites include MusicNet
(20% w/20% Bertelsmann, 20% EMI and
RealNetworks), digitalcity, moviefone,
mapquest, and music sites Spinner.com,
Winamp, and SHOUTcast; stakes in
Amazon.com (2%), Dr. Koop (10%),
RoadRunner cable modems (majority
stake w/AT&T and Advance-Newhouse)

SPORTS
Atlanta Braves, Atlanta Hawks,
Atlanta Thrashers, Goodwill
Games, Phillips Arena

OTHER
Time Warner Telecom (37%), Warner
Bros. Studio Stores (75% w/25% AT&T),
licenses rights to DC Comics, Hanna-
Barbera characters, other WB properties
(75% w/25% AT&T); stake in Sportsline
Radio

FIGURE 3.3 The Corporate Control of Culture

Data: The Nation, July 7–14, 2002, pp. 27–30.

and right. As a result, political resistance to the dominant culture is blunted (Gramsci 1971).

Not only does the mass-marketing of culture have
a political effect, but it has an interesting economic
side as well. Those who produce cultural products in
a capitalist society must constantly produce new needs
to expand economic markets. When the mass media
report on changing fashions, it encourages people to
buy more clothes. When radio stations give the latest
recordings extensive air time, it sends listeners out to
buy more culturally current music. Cornel West, a contemporary African American philosopher, argues that
the development of a pervasive market culture like
that of contemporary American society produces an
"addiction to stimulation" (hooks and West 1991: 95).
The purveyors of culture thrive by constantly producing and feeding new needs in the public.

Conflict theorists point out that culture can also be
a source of political resistance. The development of
new cultural beliefs and practices can challenge dominant group control over others. Reclamation of indigenous culture is one way that groups have overthrown
colonial rulers. Within the United States, the American
Indian Movement (AIM) used the resuscitation of traditional Indian culture as a central organizing tool in
its campaign to reassert the independence of Native
Americans.

A final point of focus for sociologists studying culture from a conflict perspective lies in the concept of
cultural capital, which refers to the cultural resources
that are socially designated as worthy (such as knowledge of elite culture) and that give advantages to groups
possessing such capital (Bourdieu 1977; DiMaggio
1982). This idea has been most developed by the French
sociologist Pierre Bourdieu (1984), who sees the appropriation of culture as one way groups maintain their social status. Bourdieu argues that members of the dominant class have distinctive lifestyles that mark their
status in society. Their ability to display this cultural
lifestyle signals their importance to others. That is,
they possess cultural capital. From this point of view,
culture has a role in reproducing inequality between
groups. Those with cultural capital use it to improve
their social and economic position in society. Sociologists have found a significant relationship between cultural capital and grades in schools. Those from the more

well-to-do classes (those with more cultural capital) are able to parlay their knowledge into higher grades, thereby reproducing their social position by being more competitive in school admissions and, eventually, in the labor market (DiMaggio 1982). Likewise, Michèle Lamont's study (1992) of American and French upper middle-class men shows how cultural status is used to mark their moral worth. Lamont argues that the upper middle class uses its cultural capital to proclaim its social value to others.

Culture as a Social Construction

Especially productive when applied to the study of culture has been symbolic interaction theory, a perspective that analyzes behavior in terms of the meaning people give it (see Chapter 1). Symbolic interaction emphasizes the interpretive basis of social behavior, with culture providing the interpretive framework through which behavior is understood. Culture, like all other forms of social behavior, is socially constructed. That is, culture is produced through social relationships and in social groups, such as media organizations.

Symbolic interaction theorists remind us that cultural forms (such as popular culture) are produced by human actors. As such, the images we see and the messages we receive are shaped by the actual activities of people in newsrooms, film studios, chat rooms, and the like. Symbolic interaction theory emphasizes the centrality of human action and human consciousness in shaping culture, as well as emphasizing the interpretive processes through which people both create culture and respond to perceived cultural forms (Denzin 1992). Thus, the mass media, as one form of culture, not only reflects the practices of those who control what is produced, but the media also produces frames through which people perceive their social world. Combining symbolic interaction theory with conflict theory, you can see that those who control the mass media have the power to define what is perceived as real.

New Cultural Studies

In recent years, a new interdisciplinary field known as *cultural studies* has emerged that builds on the insights of the symbolic interaction perspective. Sociologists who work in cultural studies are often critical of classical sociological approaches to culture, arguing that the classical approach has overemphasized the nonmaterial culture of ideas, beliefs, values, and norms. The new scholars of cultural studies find that material culture has increasing importance in modern society (Crane 1994). This includes cultural forms that are recorded through print, film, artifacts, or the electronic media. *Postmodernist theory* has greatly influenced new cultural studies. Recall from Chapter 1 that postmodern-ism is based on the idea that society is not an objective thing. Society is found in the words and images that people use to represent behavior and ideas. Given this orientation, postmodernism calls significant attention to the artifacts produced in modern culture (Walters 1995).

Classical theorists tend to study the unifying features of culture. Cultural studies researchers tend to see culture as more fragmented and unpredictable. To them, culture is a series of images that can be interpreted many ways depending on the viewpoint of the observer. From the perspective of new cultural studies theorists, the ephemeral and rapidly changing quality of contemporary cultural forms reflects the highly technological and consumer-based culture on which the modern economy rests. Their fascination is in part with the illusions that such a dynamic and rapidly changing culture produces.

Cultural Change

In one sense, culture is a conservative force in society. It tends to be based on tradition and is passed on through generations, conserving and regenerating the values and beliefs of society. Culture is also based on institutions that develop an economic interest in maintaining the status quo. Often, people are also resistant to cultural change because familiar ways and established patterns of doing things are hard to give up. Nevertheless, cultures do change. Culture is dynamic, not static, and it develops as people respond to changes in their physical and social environments.

Culture Lag

Culture lag refers to the delay in cultural adjustments to changing social conditions (Ogburn 1922). Some parts of culture may change more rapidly than others; thus, one aspect of culture may "lag" behind another. Rapid technological change is often attended by culture lag, given that some elements of the culture do not keep pace with technological innovation. For example, we have the scientific capability of cloning human beings, but there is strong resistance to doing so because of cultural values about human life.

When culture changes rapidly, or someone is suddenly thrust into a new cultural situation, the result can be **culture shock,** the feeling of disorientation that can occur when one encounters a new or rapidly changed cultural situation. Even returning home after visiting a foreign country can make one's "native" cultural habits seem suddenly strange. Even moving from one cultural environment to another within one's own society can make a person feel out of place and alienated. The greater the difference between cultural settings, the greater the culture shock.

Sources of Cultural Change

There are several causes of cultural change, including changes in the societal conditions, cultural diffusion, innovation, and the imposition of cultural change by an outside agency.

Cultures change in response to changed conditions in the society. Economic changes, population changes, and other social transformations all influence the development of culture. A change in the makeup of a society's population may be enough by itself to cause a cultural transformation. Think about the cultural changes that result from the high rate of immigration in the United States. In most major cities, you can now find Spanish radio stations, and diverse holiday traditions like Cinco de Mayo are becoming more familiar to many in the U.S. public.

Cultures change through cultural diffusion. **Cultural diffusion** is the transmission of cultural elements from one society or cultural group to another. In a world of instantaneous communication, cultural diffusion is swift and widespread. This is evident in the degree to which worldwide cultures have been westernized. Cultural diffusion also occurs when subcultural influences enter the dominant group. Dominant cultures are regularly enriched by minority cultures. One example is the influ-

ence of Black and Latino music on other musical forms; another is the expression *dissing*, a term from Black urban street culture that is now part of the common parlance. Cultural diffusion is one thing that drives cultural evolution, especially in a society such as the United States that is lush with diversity.

Cultures change as the result of innovation. The discovery of new knowledge, including technological advances, leads to cultural developments that would be unlikely to evolve in the absence of research and education. The modern kitchen reflects many such changes, including automatic dishwashers, microwaves, food processors, and other appliances. These innovations have fostered dramatic changes in household organization and cultural practices such as cooking. The popularity of microwaves shows how technological advances match changes in modern life. Harried schedules lead to quick meals. The employment of women outside the home increases the likelihood that a household will include at least two workers. Therefore, the cultural habits of the past, such as the family dinner prepared in the afternoon and shared around a table in the evening, may be disappearing.

Innovations in computing have also had a giant effect on contemporary culture. The smallest laptop or handheld computer today weighs hardly more than a

FORCES OF SOCIAL CHANGE

Fast Food and the Transformation of Culture

Can you imagine the United States without fast food? Probably not. Fast food is so ubiquitous in contemporary culture that it is hard to imagine life without it. Consider this (Schlosser 2001):

- The average person in the United States consumes three hamburgers and four orders of french fries per week.
- Americans spend more money on fast food than on movies, books, magazines, newspapers, videos, music, computers, and higher education combined.
- One in eight workers has at some point been employed by McDonald's.
- McDonald's is the largest private operator of playgrounds in the United States and the single largest purchaser of beef, pork, and potatoes.
- 96 percent of American schoolchildren can identify Ronald McDonald, which is only exceeded by the number who can identify Santa Claus.

Eric Schlosser, who has written about the permeation of society by fast food culture, claims that "a nation's diet can be more revealing than its art or literature" (2001: 3). He relates the growth of the fast food industry to fundamental changes in American society, including the vast entry of women into the paid labor market, the development of an automobile culture, the increased reliance on low-wage service jobs, the decline of family farming, and the growth of agribusiness. One result is a cultural emphasis on uniformity, not to mention increased fat and calories in people's diets.

Think about the influence of fast food on contemporary culture by

John Phillips/Time Pix

making a list of the values associated with fast food. How has the fast food industry changed cultural practices in the United States? What impact does this have in your community?

Source: Schlosser, Eric. 2001. *Fast Food Nation: The Dark Side of the All-American Meal.* New York: Houghton Mifflin. ●●●

© Kathy McLaughlin/The Image Works

few ounces, and its capabilities rival that of computers that filled entire buildings only twenty years ago. Computing infiltrates every dimension of life. It is hard to overestimate the effect of innovation on contemporary cultural change. Technological innovation is so rapid and dynamic that one generation can barely maintain competence with the hardware of the next. Answering machines, cellular phones, fax machines, and handheld computers were glamorous devices just a few years ago. Today, they are as routine as electric can openers. In such a rapidly changing technological world, it is hard to imagine what will be common in just a few years.

Cultural change can be imposed. Change can occur when a powerful group takes over a society and imposes a new culture. The dominating group may arise internally, as in a political revolution, or it may appear from outside the group, perhaps as an invasion. When an external group takes over the society of an indigenous group, as White settlers did with Native American societies, it typically imposes its own culture while prohibiting the indigenous group from expressing its original cultural ways. Manipulating the culture of a group is a way to exert social control. Many have argued that public education in the United States, which developed during a period of mass immigration, was designed to force White, northern European, middle-class values onto a diverse immigrant population that was perceived to be potentially unruly and politically disruptive. Likewise, the schools run by the Bureau of Indian Affairs have been used to impose these dominant group values on Native American children (Snipp 1996).

Resistance to political oppression often takes the form of a cultural movement that asserts or revives the culture of an oppressed group. Thus, cultural expression can be a form of political protest. Nationalist movements that identify a common culture as the basis for group solidarity are an example of political protest. Black nationalism in the United States, called Afrocentrism, is the idea that African Americans should be united by their African heritage and that this should

Technology now infiltrates virtually every aspect of modern culture.

be the basis for cultural identity and political mobilization. Black nationalism was a strong theme in Black protest movements in the 1920s and surfaced again in the 1970s as the Black Power movement emphasized that "Black is beautiful," encouraging African Americans to celebrate their African heritage with Afro hairstyles, African dress, and African awareness. Cultural nationalism has also inspired a similar movement among Latinos called *La Raza Unida,* which means "the race, or the people, united" and promotes solidarity among Chicanos. As people acquire new means of cultural expression, they can begin to question the existing value systems and challenge dominant cultural forms. Cultural change can promote social change, just as social change can transform culture.

Sociology⊛Now™

Reviewing is as easy as ❶❷❸.

1. *Before you do your final review, take the SociologyNow diagnostic quiz to help you identify the areas on which you should concentrate. You will find information on SociologyNow and instructions on how to access all of its great resources on the foldout at the beginning of the text.*

2. *As you review, take advantage of SociologyNow's study videos and interactive Map the Stats exercises to help you master the chapter topics.*

3. *When you are finished with your review, take SociologyNow's posttest to confirm you are ready to move on to the next chapter.*

Chapter Summary

What is culture?

Culture is the complex and elaborate system of meaning and behavior that defines the way of life for a group or society. It is shared, learned, taken for granted, symbolic, and emergent, and it varies from one society to another.

What are the major elements of culture?

The major elements of culture are *language, norms, beliefs,* and *values.* Becoming part of a social group requires learning the special language associated with that group. Language also shapes perceptions of reality.

What is the difference in norms, beliefs, and values?

Norms are rules of social behavior that guide every situation and may be formal or informal. When norms are violated, social sanctions are applied. *Beliefs* are strongly shared ideas about the nature of social reality. *Values* are the abstract concepts in a society that define the worth of different things and ideas.

What is the significance of cultural diversity?

As societies develop and become more complex, *cultural diversity* can appear. The United States has many diverse cultures, with many of its traditions influenced by immigrant cultures and the cultures of African Americans, Latinos, and Native Americans. The *dominant culture* is the culture of the most powerful group in society. *Subcultures* are groups whose values and cultural patterns depart significantly from the dominant culture.

How do sociologists define popular culture?

Popular culture is the beliefs, practices, and objects that are part of everyday traditions, including the mass media. The *mass media* have an increasing influence on cultural values and images, both reflecting and creating cultural values.

What different theories do sociologists use to interpret culture?

Sociological theory provides different perspectives on the significance of culture. *Functionalist theory* emphasizes the influence of values, norms, and beliefs on the whole society. *Conflict theorists* think culture is influenced by economic interests and power relations in society. *Symbolic interactionists* emphasize that culture is socially constructed. This has influenced new *cultural studies,* which interpret culture as a series of images that can be analyzed from the viewpoint of different observers.

How do cultures change?

There are several sources of cultural change, including change in societal conditions, cultural diffusion, innovation, and the imposition of change by dominant groups. As cultures change, *culture lag* can result, meaning cultural adjustments are sometimes not synchronous with each other. Persons who experience new cultural situations may experience *culture shock.*

Key Terms

beliefs 63
counterculture 67
cultural capital 74
cultural diffusion 76
cultural hegemony 73
cultural relativism 57
culture 54
culture lag 75
culture shock 75
dominant culture 64
ethnocentrism 67
ethnomethodology 62
folkways 62
global culture 68
language 59
law 62
mass media 69
material culture 54
mores 62
multiculturalism 68
nonmaterial culture 54
norms 62
popular culture 69
reflection hypothesis 72
Sapir–Whorf hypothesis 59
social sanctions 62
subculture 66
symbols 56
taboos 62
values 63

Researching Society with MicroCase Online

You can see the results of actual research by using the Wadsworth MicroCase® Online feature available to you. This feature allows you to look at some of the results from national surveys, census data, and some other data sources. You can explore this easy-to-use feature on your own, but try this example. Suppose you want to know: *Who is most likely to have seen an X-rated movie in the last year?*

To answer this question, go to http://sociology.wadsworth.com/andersen_taylor4e/, select MicroCase Online from the left navigation bar, and follow the directions there to analyze the following data.

Data file: GSS

Task: Auto-Analyzer

Primary Variable: X-MOVIE? Have you seen an x-rated movie in the last year?

Questions

Once you have your results, answer the following questions:

1. Fill in the table shown, noting which category (or group) was most likely and least likely to have watched an X-rated movie in the last year.

Socio-Demographic Variable	Category Most Likely	Category Least Likely
Religion		
Party		
Education		

2. In the results you found, are people thirty and older less likely to have watched an X-rated movie than those under thirty?

3. Are conservative Protestants the least likely to have watched an X-rated movie?

4. Is there a statistically significant difference in the likelihood of having watched an X-rated movie across different educational levels?

5. Do either race or gender influence the likelihood of one's seeing an X-rated movie?

The Companion Website for Sociology: Understanding a Diverse Society, Fourth Edition

http://sociology.wadsworth.com/andersen_taylor4e/

Supplement your review of this chapter by going to the companion website to take one of the Tutorial Quizzes, use the flash cards to master key terms, and check out the many other study aids you'll find there. You'll also find special features such as GSS Data and Census 2000 information, data and resources at your fingertips to help you with that special project or do some research on your own.

Suggested Readings and Web Resources

Adams, Rebecca G., and Robert Sardiello (eds.). 2000. *Deadhead Social Science: "You Ain't Gonna Learn What You Don't Want to Know."* Walnut Creek, CA: AltaMira Press.
This book is based on these sociologists' participant observations of the Grateful Dead subculture—the Deadheads. Utilizing a symbolic interaction framework, it shows the parallels in sociological thinking and the music of the Grateful Dead. It is a great research study of a contemporary subculture.

Barber, Benjamin R. 1995. *Jihad vs. McWorld: How Globalism and Tribalism Are Reshaping the World.* New York: Random House.
Barber analyzes the cultural conflict between the commercial culture of Western capitalism with the cultural values of traditional, locally based cultures. His discussion is especially intriguing in light of the tensions between fundamentalist Islamic nations and the culture of global capitalism.

Gans, Herbert J. 1999. *Popular Culture and High Culture: An Analysis and Evaluation of Taste.* New York: Basic Books.
This classic work examines class differences in cultural taste and the role of popular and high culture in American society. Gans also provides a sociological analysis of mass culture and its influence in society.

Gitlin, Todd. 2002. *Media Unlimited: How the Torrent of Images and Sounds Overwhelms Our Lives.* New York: Metropolitan Books.
A sociologist who studies the mass media examines how U.S. culture is bombarded with messages from the mass media. This constant stream of information is transforming culture and society, although some resist its increasing influence.

Kilbourne, Jean. 1999. *Can't Buy My Love: How Advertising Changes the Way We think and Feel.* New York: Simon and Schuster.
This book, by the author of the well-known film "Killing Us Softly" examines how advertising demeans women. Kilbourne's analysis of the gendered character of advertisement raises new critical awareness of the power of advertising to influence of self-concepts.

Schlosser, Eric. 2001. *Fast Food Nation: The Dark Side of the All-American Meal.* New York: Houghton Mifflin.
Schlosser analyzes the cultural changes that have both inspired and been brought on by the vast growth of the fast food industry. His fascinating account of the impact of fast food on everyday life is a unique way of understanding contemporary culture.

National Coalition of Television Violence
www.nctvv.org
This Web site provides news and information about violence on television, including recent media stories about television violence.

Media Education Foundation
www.mediaed.org
This is an organization devoted to developing documentary videos for the purpose of encouraging critical discussion of the commercial ownership of the media and the images produced therein. The topics are varied but include such things as images of women, alcohol advertising, and other contemporary topics.

● ● ●

Socialization

During the summer of 2000, scientists working on the human genome project announced that they had deciphered the human genetic code. By mapping the complex structure of DNA (deoxyribonucleic acid) on high-speed computers, scientists identified the 3.12 billion chemical base pairs in human DNA and put them in proper sequence, unlocking the genetic code of human life. Scientists likened this to assembling "the book of life," that is, having the knowledge to make and maintain human beings. The stated purpose of the human genome project is to see how genetics influences the development of disease, but it raises numerous ethical questions about human cloning and the possibility of creating human life in the laboratory. Is our genetic constitution what makes us human? Suppose you created a human being in the laboratory, but left that creature without social contact. Would the "person" be human?

Rare cases of *feral children,* who have been raised in the absence of human contact, provide some clues as to what happens during human development when a person has little or no social contact. One such case, discovered in 1970, involved a young girl given the pseudonym of Genie. When her blind mother appeared in the Los Angeles County welfare office seeking assistance for herself, case workers first thought the girl was six years old. In fact, she was thirteen, although she weighed only fifty-nine pounds and was four feet, six inches tall. She was small and withered, unable to stand up straight, incontinent, and severely malnourished. Her eyes did not focus, she had two nearly complete sets of teeth, and a strange ring of calluses circled her buttocks. She could not talk.

As the case unfolded, it was discovered that the girl had been kept in nearly total isolation for most of her life. The first scientific report about Genie states:

In the house Genie was confined to a small bedroom, harnessed to an infant's potty seat. Genie's father sewed the harness himself; unclad except for the harness, Genie was left to sit on that chair. Unable to move anything except her fingers and hands, feet and toes, Genie was left to sit, tied-up, hour after hour, often into the night, day after day, month after month, year after year (Curtiss 1977: 5).

At night, she was restrained in a handmade sleeping bag that held her arms stationary and placed in a crib. If she made a sound, her father beat her. She was given no toys and was allowed to play only with two old raincoats and her father's censored version of *TV Guide*. (He had deleted everything "suggestive," such as pictures of women in bathing suits.) Genie's mother, timid and blind, was also victimized by her husband. Shortly after the mother sought help, after years of abuse, the father committed suicide.

Genie was studied intensively by scientists interested in language acquisition and the psychological effects of extreme confinement. They hoped that her development would throw some light on the question of nature versus nurture; that is, are people the product of their genes or their social training? After intense language instruction and psychological treatment, Genie developed some verbal ability and, after a year, showed progress in her mental and physical development. Yet, the years of isolation and severe abuse took their toll. Genie now lives in a home for retarded adults (Rymer 1993).

This rare case of a feral child sheds some light on the consequences of life without social contact. Knowing the sequence of the human genome may raise the specter of making human beings in the laboratory, but without society, what would humans be like? Genes may confer skin and bone and brain, but only by learning the values, norms, and roles that culture bestows on people, do they become the social beings that define them in the eyes of others and to themselves. Sociologists refer to this process as socialization—the subject of this chapter. •••

UNDERSTANDING DIVERSITY

My Childhood:
An Interview with Bong Hwan Kim

Childhood is a time when children learn their gender, as well as their racial and ethnic identity. This excerpt from an interview with Bong Hwan Kim, a Korean American man, is a reflection on growing up and learning both Korean and American culture.

I came to the United States in 1962, when I was three or four years old. My father had come before us to get a Ph.D. in chemistry. He had planned to return to Korea afterwards, but it was hard for him to support three children in Korea while he was studying in the United States, and he wasn't happy alone, so he brought the family over. He got a job at a photographic chemicals manufacturing company in New Jersey, where he still works after almost 30 years. He never made it past the "glass ceiling." I view him as a simple person who must have been overwhelmed by the flip side of the American dream they never tell you about. There was never a place for him in America except at home with the family or maybe at the Korean church. Both of my parents were perpetual outsiders, never quite comfortable with American life.

The Bergenfield, New Jersey, community where I grew up was a blue-collar town of about 40,000 people, mostly Irish and Italian Americans. I lived a schizophrenic existence. I had one life in the family, where I felt warmth, closeness, love, and protection, and another life outside—school, friends, television, the feeling that I was on my own. I accepted that my parents would not be able to help me much.

I can remember clearly my first childhood memory about difference. I had been in this country for maybe a year. It was the first day of kindergarten, and I was very excited about having lunch at school. All morning, I could think only of the lunch that was waiting for me in my desk. My mother had made kimpahp [rice balls rolled up in dried seaweed] and wrapped it all up in aluminum foil. I was eagerly looking forward to that special treat. I could hardly wait. When the lunch bell rang, I happily took out my foil-wrapped kimpahp. But all the other kids pointed and gawked. "What is that? How could you eat that?" they shrieked. I don't remember whether I ate my lunch or not, but I told my mother I would only bring tuna or peanut butter sandwiches for lunch after that.

I have always liked Korean food, but I had to like it secretly, at home. There are things you don't show to your non-Korean friends. At various times when I was growing up, I felt ashamed of the food in the refrigerator, but only when friends would come over and wonder what it was. They'd see a jar of garlic and say, "You don't eat that stuff, do you?" I would say, "I don't eat it, but my parents do; they do a lot of weird stuff like that."

My father spelled his name "Kim Hong Zoon." The kids at school made fun of his name. "Zoon?" they would laugh. They called me "Bong" because "Bong Hwan" was too hard to pronounce.

As a child you are sensitive; you don't want to be different. You want to be like the other kids. They made fun of my face. They called me "flat face." When I got older, they called me "Chink" or "Jap" or said "remember Pearl Harbor." In all cases, it made me feel terrible. I would get angry and

The Socialization Process

Socialization is the process by which people learn the expectations of society. **Roles** are learned through the socialization process and are the expected behavior associated with a given status in society (see Chapter 5). Through socialization, people absorb their culture—customs, habits, laws, practices, and means of expression. Socialization is the basis for **identity,** which is how one defines oneself. Identity is both personal and social. To a great extent, it is bestowed by others because people come to see themselves as others see them. Socialization also establishes **personality,** defined as the relatively consistent pattern of behavior, feelings, and beliefs of an individual.

The socialization experience differs for individuals, depending on factors such as race, gender, and class, as well as more subtle factors such as attractiveness and personality. Women and men encounter different socialization patterns as they grow up because each

gender brings with it different social expectations (see Chapter 12). Likewise, growing up Jewish, Asian, Latino, or African American involves different socialization experiences, as the box "Understanding Diversity" shows. In this example, a Korean American man reflects on the cultural habits he learned growing up in two cultures, Korean and American. His comments reveal the strain that can occur when socialization involves competing expectations, a strain that can be particularly acute when a person grows up within different, even if overlapping, cultures.

Through socialization, people internalize cultural expectations, then pass these expectations on to others. *Internalization* occurs when behaviors and assumptions are learned so thoroughly that people no longer question them, but simply accept them as correct. Through socialization, one internalizes the expectations of society. The lessons that are internalized can have a powerful influence on attitudes and behavior. For example, someone socialized to believe that

get into fights. Even in high school, even the guys I hung around with on a regular basis would say, "You're just a Chink" when they got angry. Later, they would say they didn't mean it, but that was not much consolation. When you are angry, your true perceptions and emotions come out. The rest is a façade.

They used to say, "We consider you to be just like us. You don't seem Korean." That would give rise to such mixed feelings in me. I wanted to believe that I was no different from my white (because that is how it is in the original and this is a direct quote) classmates. It was painful to be reminded that I was different, which people did when they wanted to put me in my place, as if I should be grateful to them for allowing me to be their friend.

I wanted to be as American as possible—playing football, dating cheerleaders. I drank a lot and tried to be cool. I had convinced myself that I was "American," whatever that meant, all the while knowing underneath that I'd have to reconcile myself, to try to figure out where I would

fit in a society that never sanctioned that identity as a public possibility. Part of growing up in America meant denying my cultural and ethnic identity, and part of that meant negating my parents. I still loved them, but I knew they could not help me outside the home. Once when I was small and had fought with a kid who called me a "Chink," I ran to my mother. She said, "Just tell them to shut up." My parents would say that the people who did things like that were just "uneducated." "You have to study hard to become an educated person so that you will rise above all that," they would advise. I didn't really study hard. Maybe I knew somehow that studying hard alone does not take anyone "above all that." I have lost contact with everyone from my East Coast life except for a few Korean American friends from the Korean church our family attended. . . .

When I got to college, I experienced an identity crisis. Because I was no longer involved in sports and because I was away from home, I no longer felt special or worthwhile, and I became very depressed. I hated

even getting up in the mornings. Finally, I dropped out of school.

I decided to go to Korea, hoping to find something to make me feel more whole. Being in Korea somehow gave me a sense of freedom I had never really felt in America. It also made me love my parents even more. I could imagine where they came from and what they experienced. I began to understand and appreciate their sacrifice and love and what parental support means. Visiting Korea didn't provide answers about the meaning of life, but it gave me a sense of comfort and belonging, the feeling that there was somewhere in this world that validated that part of me that I knew was real but few others outside my immediate family ever recognized.

Source: Edited by Karin Aguilar-San Juan. Copyright © 1994. *The State of Asian America.* Boston, MA: South End Press. Reprinted with permission of South End Press. ● ● ●

© Keith Meyers/NYT Pictures

Formal and informal learning, through schools and other socialization agents, are important elements of the socialization process. Here deaf children are learning sign language.

homosexuality is morally repugnant is unlikely to be tolerant of gays and lesbians. If such a person, say a man, experiences erotic feelings about another man, he is likely to have deep inner conflicts about his identity. Similarly, someone socialized to believe that racism is morally repugnant is likely to be more accepting of different races. However, people can change the cultural expectations they learn. New experiences can undermine narrow cultural expectations. Attending college often has a liberalizing effect, supplanting old expectations with new ones generated by exposure to the diversity of college life.

Examining the socialization process helps reveal the degree to which our lives are *socially constructed,* meaning that the organization of society and the life outcomes of people within it are the result of social definitions and processes. For example, the expression *tabula rasa* means humans are born as a "blank slate" and values and social attitudes are not inborn, but emerge through the interactions we have with others. From a sociological perspective, what a person becomes results more from social experiences than innate (inborn, or natural) traits, although innate traits have some influence. For example, a person may be born with a great capacity for knowledge, but without a good education, that person is unlikely to achieve his or her full potential and may not be recognized as intellectually gifted.

Socialization as Social Control

The sociologist Peter Berger (1963) pointed out that not only do people live in society, but society also lives in people. Socialization is, therefore, a mode of social control. Because socialized people conform to cultural

expectations, socialization gives society a certain degree of predictability, establishing patterns that become the basis for social order. Although few people match cultural ideals exactly, most of us fit comfortably within society's expectations.

To understand how socialization is a form of social control, imagine that the individual in society is surrounded by a series of concentric circles (see Figure 4.1). Each circle is a layer of social controls, ranging from the most subtle, such as the expectations of others, to the most overt, such as physical coercion and violence. Usually, coercion and violence are not necessary to extract conformity because learned beliefs and the expectations of others are enough to keep people in line. These forces can be subtle, such as when a person feels pressure to conform to others. Even if you disagree, you may experience stress and discomfort if you choose not to conform. People learn through a lifetime of experience that deviating from others' expectations—especially those who matter the most to you—invites peer pressure, ridicule, and other social judgments, reminding you of what is expected.

Conformity and Individuality

Saying that people conform to social expectations does not mean there is no individuality. We are all unique to some degree. Our uniqueness arises from different experiences, different patterns of socialization, the choices we make, and the imperfect ways that we learn our roles. People also resist some of society's expectations. Human beings are not totally passive creatures; they interact with their environment in creative ways. Yet, most people do conform, although to differing degrees. Socialization is profoundly significant, but this does not mean that people are robots. Instead, socialization emphasizes the adaptations people make as they learn to live in society.

Some people conform too much, for which they pay a price. Socialization into men's roles can encourage aggression, independence, and competitiveness. Men's lower life expectancy and higher rate of accidental death

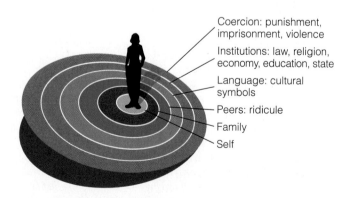

Coercion: punishment, imprisonment, violence

Institutions: law, religion, economy, education, state

Language: cultural symbols

Peers: ridicule

Family

Self

Figure 4.1 Socialization as Social Control

(compared to women) results from some of the risky behaviors associated with men's roles (National Center for Health Statistics 2003; Kimmel and Messner 2003). Women's gender roles carry their own risks. Striving to constantly meet the beauty ideals of the dominant culture can result in feelings of low self-worth and may encourage potentially harmful behaviors, such as smoking or eating less to keep one's weight down. It is not that being a man or woman is inherently bad for your health, but that excessive emphasis on the extremes of gender roles can compromise your physical and mental health.

Socialization and Self-Esteem

Self-esteem is the value a person places on his or her identity. Self-esteem is critical to one's well-being, since poor self-esteem can lead to a number of psychological and social problems. If you devalue yourself—even unconsciously—you are also likely to be unhappy. How much value one sees in oneself is greatly affected by the socialization process and, in a deep way, by how you are seen by society.

This is well illustrated by studies of eating disorders among young men and women. A national study of ninth and twelfth graders examined the eating behaviors of young men and women. Almost two-thirds of the young women (57 percent) and nearly one-third of the young men (31 percent) reported eating disorders—that is, problematic eating behaviors, including using fasting, diet pills, laxatives, vomiting, binge eating, skipping meals, and smoking to control weight. Fear about one's appearance to others was one of the factors associated with this risky behavior, indicative of the influence of the socialization process. At the same time, the study found that some influences stemming from socialization, including having a more positive self-esteem, tended to protect youth from developing eating disorders (Croll et al. 2002). You might ask

yourself what happens in the socialization process that results in these patterns of eating. Note that eating disorders are common among both young men and women, but you might also ask what factors in the socialization process result in young women being more likely to engage in problematic eating.

The Consequences of Socialization

Socialization is a life-long process with consequences that affect how we behave toward others and what we think of ourselves. *Socialization establishes self-concepts.* How we think of ourselves is the result of the socialization experiences we have over a lifetime. Our self-concept is established through the socialization process; see Figure 4.2 for evidence of this.

Socialization creates the capacity for role-taking or, put another way, for seeing ourselves as others see us. Socialization is fundamentally reflective; that is, it involves self-conscious human beings seeing and reacting

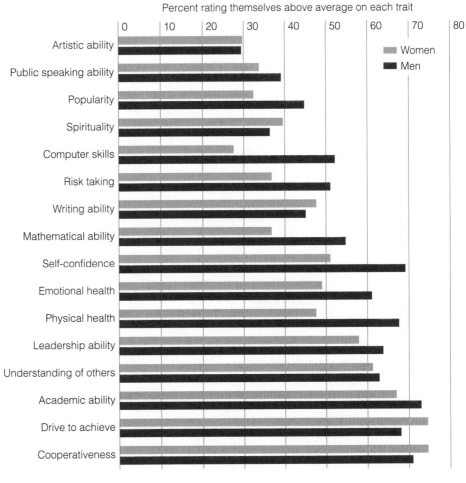

Figure 4.2 Student Self-Concepts*

Source: L. J. Sax, et al. (2003). *The American Freshman. National Norms for Fall 2003.* Los Angeles: Higher Education Research Institute, UCLA. Used by permission.

*Based on national sample of first year college students, Fall 2003

to the expectations of others. The capacity for reflection and the development of identity are ongoing. As we encounter new situations in life, we are able to see what is expected and adapt to the situation accordingly. Of course, not all people do so successfully. This can become the basis for social deviance (explored in Chapter 7) or for many common problems in social and psychological adjustment.

Socialization creates the tendency for people to act in socially acceptable ways. Through socialization, people learn the normative expectations attached to social situations and the expectations of society in general. As a result, socialization creates some predictability in human behavior and brings some order to what might otherwise be social chaos.

Socialization makes people bearers of culture. Socialization is the process by which people learn and internalize the attitudes, beliefs, and behaviors of their culture. At the same time, socialization is a two-way process. A person is not only the recipient of culture, but also the creator of culture who passes cultural expectations on to others. The main product of socialization, then, is society itself. By molding individuals, socializing forces perpetuate the society into which individuals are born. Beginning with the newborn infant, socialization contributes to the formation of a self. As people learn roles and rules, they form a self and become bearers of culture, passing on all they have acquired to others.

Agents of Socialization

Socialization agents are those who pass on social expectations. Everyone is a socializing agent because social expectations are communicated in countless ways and in every interaction people have, intentional or not. When people are simply doing what they consider "normal," they are communicating social expectations to others. When you dress a particular way, you may not feel you are telling others they must dress that way, yet when everyone in the same environment dresses the same way, they clearly convey some expectation about appropriate dress. People feel pressure to become what society expects of them even though the pressure may be subtle and unrecognized.

THINKING SOCIOLOGICALLY

Think about the first week you attended college. What expectations were communicated to you and by whom? Who were the most significant *socialization agents* during this period? Which expectations were communicated formally and which informally? If you were analyzing this experience sociologically, what would some of the most important concepts be to help you understand how one "becomes a college student"?

Socialization does not occur only between individuals, it also occurs in the context of *social institutions.* Recall from Chapter 1 that institutions are established patterns of social behavior that persist over time. Institutions are a level of society above individuals. Many social institutions shape the process of socialization, including, as we will see, the family, the media, peers, religion, sports, and schools.

The Family

For most people, the family is the first source of socialization. Through families, children are introduced to the expectations of society. Children learn to see themselves through their parents' eyes. Thus, how parents define and treat a child is crucial to the development of the child's sense of self.

What children learn in families is certainly not uniform. Even though families pass on the expectations of a given culture, within that culture, families may be highly diverse, as we will see in Chapter 15. Some families may emphasize educational achievement; some may be more permissive, whereas others emphasize strict obedience and discipline. Even within families, children may experience different expectations, based on gender or birth order (being first-, second-, or third-born). Researchers have found, for example, that fathers' and mothers' support for gender equity increases when they have only daughters (Warner and Steel 1999). Living in a family experiencing the strain of social problems such as alcoholism, unemployment, domestic violence, or teen pregnancy also affects how children are socialized. The specific effects of different family structures and processes are the basis for ongoing and extensive sociological research.

As important as the family is in socializing the young, it is not the only socialization agent. As children grow up, they encounter other socializing influences, sometimes in ways that might contradict family expectations. Parents who want to socialize their children in less gender-stereotyped ways might be frustrated by the influence of the media, which promotes highly gender-typed toys and activities to boys and girls. These multiple influences on the socialization process create a reflection of society in us.

The Media

Increasingly, the media are important agents of socialization. The average young person (age 8–19) spends six and three-quarter hours per day immersed in media in various forms, often using multiple media forms simultaneously. Television is the dominant medium, although half of all youth use a computer daily. This is more time devoted to media than any other waking activity (Roberts 2000). The images found in the media are powerful throughout our lifetimes (as we have seen in Chapter 3 on culture), but many worry that their effect during childhood may be particularly deleterious.

Take the issue of violence. When two young boys shot and killed twelve students at Columbine High School in Littleton, Colorado in 1999, the public wondered whether their behavior originated in violent programming on television. The high degree of violence in the media has led to the development of a rating system for televised programming. Analysts estimate that by age eighteen, the average child will have witnessed at least 18,000 simulated murders on television. Programs targeted to children actually contain more violence than other programming (Wilson et al. 2002). Violence in children's programming is frequently shown as humorous and rarely shown to have long-term consequences (National Television Violence Study 1997).

Concerns about violence and the media have been extended to studies of young people and electronic games with marked gender differences in the playing of such games. Boys are more likely to believe that violent video games are more appropriate for boys than girls to play with. Both boys and girls, however, believe that girls who spend more time playing video games are less popular than other girls (Funk and Buchman 1996).

Violence is pervasive in the media, but does it affect behavior? Researchers debate this, but there is strong evidence of the unhealthy effects of violence, especially on children. Violence in the media encourages both antisocial behavior and fear among children and tends to desensitize people to the effects of violence, including having less sympathy for victims of violence (Cantor 2000). Children also tend to imitate the aggressive behavior they see in the media. Violence in the media is not solely to blame for violent behavior in society, however. Children, for example, do not watch television in a vacuum; they live in families where they learn different values and attitudes about violent behavior and they observe the society around them, not just the fictional images they see. Most likely, children are not influenced by the images of televised and filmed violence alone, but also by the broader social context in which they live.

Every 9 seconds	a high school student drops out.
Every 20 seconds	a child is arrested.
Every 37 seconds	a child is born to a mother who is not a high school graduate.
Every 43 seconds	a child is born into poverty.
Every minute	a child is born to a teen mother.
Every 2 minutes	a child is born at low birthweight.
Every 4 minutes	a child is born to a mother who received late or no prenatal care.
Every 4 minutes	a child is arrested for drug abuse
Every 8 minutes	a child is arrested for a violent crime.
Every 19 minutes	a baby dies.
Every 3 hours	a child or youth under 20 is killed by a firearm.
Every 3 hours	a child or youth under 20 is a homicide victim.
Every 5 hours	a child or youth under 20 commits suicide.
Every day	a young person under 25 dies from HIV infection.

Figure 4.3 Moments in America for Children

Source: Children's Defense Fund. 2004. Washington, DC: Children's Defense Fund, Website: **www.childrensdefense.org**

Violence in the media only reflects the violence in society, as would be expected by the reflection hypothesis discussed in Chapter 3.

The media expose us to numerous images that shape our definitions of ourselves and the world around us. What we think of as beautiful, sexy, politically acceptable, and materially necessary is strongly influenced by the media. If every week, as you read a weekly news magazine, someone shows you the new car that will give you status and distinction, or if every weekend, as we watch televised sports, someone tells you that to really

TAKING ON SOCIAL ISSUES:
Media Violence

Concerns about the influence of *media violence* on young people have led many to advocate more restrictions on the contents of the media; including films, television programs, and music. Some want more censorship while others want a stronger system of warning labels, such as the television chip (V chip) that allows parents to make certain programs and channels unavailable to their children. Others argue that violence in the media does not cause violence in society because the media only reflect violence that already exists. Based on the understanding of socialization you have acquired from this chapter, what do you think is the effect of violent images in the media on the socialization experiences of young children? What social policies would you suggest regarding the content of the media?

Taking Action

Go to the Taking Action Exercise on the companion website—at http://sociology .wadsworth.com/andersen_taylor4e/— to learn more about an organization that addresses this topic. •••

have fun you should drink the right beer, it is little wonder that we begin to think that our self-worth can be measured by the car we drive and that parties are seen as better when everyone is drunk. The values represented in the media, whether they are about violence, racist and sexist stereotypes, or any number of other social images, have a great effect on what we think and who we are.

Peers

Peers are those with whom you interact on equal terms. Friends, fellow students, and coworkers are examples of peer groups. Among peers, there are no formally defined superior and subordinate roles, although status distinctions commonly arise in peer group interactions. Peers are enormously important in the socialization process. For children, peer culture is an important source of identity. Through interaction with peers, children learn concepts of self, gain social skills, and form values and attitudes. Without peer approval, most people find it hard to feel socially accepted.

Peer cultures for young people often take the form of cliques, which are friendship circles where members identify with each other and hold a sense of common identity. You have probably witnessed the formation of cliques in school and may even be able to name them. Did your school have "jocks," "preps," "druggies," and so forth? Sociologists studying cliques have found that they are formed based on a sense of exclusive membership, like the in-groups and out-groups we will examine in Chapter 6. Cliques are cohesive, but there is an internal hierarchy with certain group leaders who have more power and status than other members. Interaction

Support from peers and family is an important source of strong self-esteem. Organized peer groups like the Special Olympics can also foster a desire for achievement and enhance one's sense of self-worth.

techniques, like making fun of people, maintain group boundaries and define who's in and who's not (Adler and Adler 1998).

Peer relationships vary among different groups. Girls' peer groups tend to be closely knit and egalitarian; boys' peer groups tend to be more hierarchical, with evident status distinctions between members. Girls are more likely than boys to share problems, feelings, fears, and doubts. Beware of generalizations, however. In some all-girl groups, status distinctions form, and one study found that 40 percent of boys, a substantial fraction, have friendships with a high degree of intimacy. In general, however, boys are more prone to share activities than feelings (Corsaro and Eder 1990; Youniss and Smollar 1985).

As agents of socialization, peers are important sources of social approval, disapproval, and support. This is one reason groups without peers of similar status are often at a disadvantage in various settings, such as women in male-dominated professions or minority students on predominantly White campuses. Being a "token" or an "only," as it has come to be called, places unique stresses on people in settings with relatively few peers from whom to draw support. This is one reason minorities in a dominant group context often form same-sex or same-race groups for support, social activities, and the sharing of information about how to succeed in this environment.

Religion

Religion is another powerful agent of socialization, and religious instruction contributes greatly to the identities children construct for themselves. Children tend to develop the same religious beliefs as their parents. Switching to a faith different from the one in which you were raised is rare (Hadaway and Marler 1993). Even those who renounce their family's religion are deeply affected by the attitudes, self-images, and

Peers are an important agent of socialization. Young girls and boys learn society's images of what they are supposed to be through the socialization process.

beliefs instilled by early religious training. Very often those who disavow religion return to their original faith at some point in their life, especially if they have strong ties to their family of origin and after they form families of their own (Wilson 1994).

Religious socialization shapes the beliefs that people develop. An example comes from studies of people who believe in creationism. Typically, creationists have been taught to believe in creationism over a long period, and are specifically socialized to believe in the creationist view of the world's origin and to reject scientific explanations. Sociological research further finds that socialization into creationist beliefs is more likely to be effective among people from small-town environments where they are less exposed to other influences. Those who believe in creationism are also likely to have mothers who have filled the traditional homemaker's role (Eckberg 1992).

Religious socialization influences a large number of beliefs that guide adults in how they organize their lives, including beliefs about moral development and behavior, the roles of men and women, and sexuality. One's religious beliefs strongly influence belief about gender roles within the family, including men's engagement in housework and the odds that wives will be employed outside the home (Ellison and Bartkowski 2002; Becker and Hofmeister 2001; Scott 2000). Religious socialization also influences beliefs about sexuality, including the likelihood of tolerance for gay and lesbian sexuality (Reynolds 2003; Sherkat 2002). Religion can even influence child-rearing practices. Thus, sociologists have found that conservative Protestants are more likely to use strict discipline in raising children, but they are also more likely to hug and praise their children than are parents with less conservative theological views (Wilcox 1998).

Sports

Most people perhaps think of sports as something that is just for fun or perhaps to provide opportunities for college scholarships and athletic careers, but sports are also an agent of socialization. Through sports, men and women learn concepts of self that stay with them in their later lives. Sports are also where many ideas about gender differences are formed and reinforced (Messner 2002; Dworkin and Messner 1999). For men, success or failure as an athlete can be a major part of a man's identity. Even for men who have not been athletes, knowing about and participating in sports is an important source of men's gender socialization. Men learn that being competitive in sports is considered a part of "manhood." Indeed, the attitude that "sports builds character" runs deep in the culture by supposedly passing on values such as competitiveness, the work ethic, fair play, and a winning attitude. Along with the mili-

tary and fraternities, sports are considered to be where one learns to be a man.

DEBUNKING *SOCIETY'S MYTHS*

Myth: Sports are basically played just for the fun of it.

Sociological perspective: Although sports are a form of entertainment, playing sports is also a source for socialization into gender roles (Messner 2002).

Michael Messner's research on men and sports reveals the extent to which sports shape masculine identity. Messner interviewed thirty former athletes: Latino, Black, and White men from poor, working-class, and middle-class backgrounds. All of them spoke of the extraordinary influence of sports on them as they grew up. Participating in sports is a strong source of gender socialization. Men's sports relationships with peers help form their gender identity (Messner 1992, 2002). Not only are sports a major source of gender socialization, but working-class, African American, and Latino men often see sports as their only possibility for a good career, even though the number of men who succeed in athletic careers is a minuscule percentage of those who hold such hopes.

Sports historically have been less significant in the formation of women's identity, although this is changing, largely as the result of Title IX, which opened more opportunities in athletics to women and girls (see the box, "Forces of Social Change"). Athletic prowess, highly esteemed in men, is not tied to cultural images of "womanliness." Quite the contrary, women who excel at sports are often stereotyped as lesbians and may be ridiculed for not being "womanly" enough. These stereotypes are a form of social control because they reinforce traditional gender roles for women (Blinde and Taub 1992a, 1992b). With athleticism and fitness on the rise as components of female beauty, however, the sociological impact of women's participation in sports will likely change.

Current research finds that women in sports develop a strong sense of bodily competence, which is typically denied to them by the prevailing, unattainable cultural images of women's bodies. Sports also give women a strong sense of self-confidence and encourage them to seek challenges, take risks, and set goals. In addition, women report that sports are frequently the background for forming bonds with other women. Women competing together develop a sense of group power. Women athletes also say that playing sports heightens their awareness of gender inequalities off the playing field, although sports participation has not been shown to correlate with activism on women's issues (Blinde et al. 1993, 1994).

Research in the sociology of sports shows how activities as ordinary as shooting baskets on a city lot, playing on the high school soccer team, or playing touch football on a Saturday afternoon can convey powerful cultural messages about our identity and our place in the world. Sports are a good example of the power of socialization in our everyday lives.

Schools

Once children enter kindergarten or day care, another process of socialization begins. At home, parents are the overwhelmingly dominant source of socialization cues. In school, teachers and other students are the source of expectations that encourage children to think and behave in particular ways. The expectations encountered in schools vary for different groups of students. These differences are shaped by a number of factors, including teachers' expectations for different groups and the resources that different parents can bring to bear on the educational process. The parents of children attending elite, private schools often have more influence on school policies and classroom activities than do parents in low-income communities. In any context, studying socialization in the schools is an excellent way to understand the influence of gender, class, and race in shaping the socialization process.

> ### DEBUNKING *SOCIETY'S MYTHS*
>
> Myth: Schools are primarily places where young people learn skills and other knowledge.
>
> Sociological perspective: A hidden curriculum in schools teaches the students many expectations associated with race, class, and gender relations in society (Martin 1998).

For example, research finds that teachers respond differently to boys than to girls, with boys receiving more of their attention. Even when teachers respond negatively to boys who are misbehaving, they are calling more attention to the boys (American Association of University Women 1992, 1998; Sadker and Sadker 1994). Social class stereotypes also affect teachers' interactions with students. Teachers are likely to perceive working-class children and poor children as less bright and less motivated than middle-class children. Teachers are also more likely to define working-class students as troublemakers (Bowditch 1993). These negative ap-

FORCES OF SOCIAL CHANGE

Women's Athletics and Title IX

Title IX of the Education Amendments of 1972 to the 1964 Civil Rights Act states, "No person in the United States shall on the basis of sex, be excluded from participation in, be denied the benefits of, or be subject to discrimination under any educational program or activity receiving Federal financial assistance." As you can see from Figure 4.4, the impact of Title IX on women's athletic participation has been enormous. In 1972, when Title IX was passed, there were under 30,000 women total in college athletic programs, compared to 170,000 men! Now, in some schools, women's basketball is just as popular (and profitable) as men's, although in many the enthusiasm and financial support for men's sports still far surpasses that for women's athletics.

How has Title IX changed the face of college sports? Some argue that the legal requirements of Title IX hurt men's sports, since some schools have can-

Matt Suess/Getty Images

celed certain men's teams to comply with Title IX. Some supporters of Title IX contend that the huge budgets devoted to men's football and basketball should be reduced to create more gender equity in sports. Data from the National Collegiate Athletic Association show that the share of budgets for women's athletics has increased in recent years, as required by Title IX, and the number of women's sports teams at colleges and universities now exceeds the number of teams for men. But, men's sports still command 65 percent of the $2.7 billion dollars expended on college athletics (NCAA 2002; Suggs 2002).

At the heart of these debates are questions about the role of sports in education.

Title IX is based on the premise that schools must not discriminate in offering educational opportunities to women. You might ask yourself, Is athletic participation about educational opportunity or is it a business? What changes do you think have taken place as the result of Title IX and how do these affect the socialization of women and men in college environments? ···

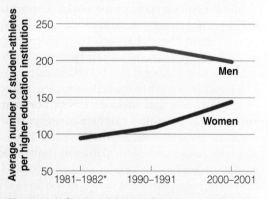

Figure 4.4 Student-Athletes: The Impact of Title IX

Source: National Collegiate Athletic Association. 2002. NCAA Year-by-Year Sports Participation 1982–2001. Indianapolis, IN: NCAA. Website: **www.ncaa.org**

*Data not available prior to 1981.

SOCIOLOGY IN PRACTICE

Building Cooperative Social Relations among Children

Barrie Thorne's research on children's interaction in school found that children establish boundaries between themselves that reflect the race and gender divisions in society. Based on her research findings and her observations of teachers' practices, she offers the following suggestions for improving cross-gender and cross-race relationships among children.

1. When grouping students, use criteria other than gender or race. Thorne notes that, when left to themselves, children will separate by gender and, often, race. Although teachers should not devalue same-gender and same-race relationships, she suggests that teachers should promote more intergroup interaction.

2. Affirm and reinforce the values of cooperation among all children, regardless of social categories. Promoting a sense that the class works together as a group creates more inclusive relationships between different groups.

3. Whenever possible, organize students into small, heterogeneous, and cooperative work groups. Small group instruction tends to promote more cooperative interaction among students. When groups focus on shared goals, they are also more likely to value each other.

4. Facilitate children's access to all activities. Thorne suggests that teachers need to make a point of teaching the same skills to everyone instead of letting children segregate into activities that promote one skill for one group and another skill for another group. Without this intervention, children tend to engage in gender-typed activities.

5. Actively intervene to challenge the dynamics of stereotyping and power. Even when groups are integrated, they may interact based on tensions and inequality. If teachers pretend that race makes no difference, these tensions will simply be reproduced.

Instead, teachers should help students learn to interact with each other across these differences and to explore the nature and meaning of cultural and racial differences and the *dynamics of racism.*

Critical Thinking Exercise

1. Using the guidelines Thorne suggests, identify a group or organization on your campus of which you are a part and make a list of suggestions for facilitating positive cross-race and cross-gender relationships within the group.

2. Observe a classroom where young children are interacting. If you were the teacher, how might you implement the suggestions that Thorne gives for children's interracial interactions?

Source: Thorne, Barrie. 1993. *Gender Play: Girls and Boys in School.* New Brunswick, NJ: Rutgers University Press. •••

---

praisals are *self-fulfilling prophecies,* meaning that the expectations they create often become the basis for actual behavior. Thus, they affect the odds of success for children (see Chapter 16).

Boys also receive more attention in the curriculum than girls. The characters in texts are more frequently boys, the accomplishments of boys are more likely to be portrayed in classroom materials, and boys and men are more typically depicted as active players in history, society, and culture (American Association of University Women 1992; Sadker and Sadker 1994). This is called the *hidden curriculum* in the schools and consists of the informal and often subtle messages about social roles that are conveyed through classroom interaction and classroom materials, roles that are clearly linked to gender, race, and class.

THINKING *SOCIOLOGICALLY*

Visit a local day-care center, preschool, or elementary school and observe children at play. Record the activities they are involved in, and note what both girls and boys are doing. Do you observe any differences between boys' and girls' play? What do your observations tell you about *socialization* patterns for boys and girls?

In schools, boys and girls are often segregated, which has significant sociological consequences. Differences between boys and girls become exaggerated when they are separated based on gender and when they are defined as distinct groups (Thorne 1993). Seating children in separate gender groups or sorting them into play groups based on gender heightens gender differences and greatly increases the significance of gender in the children's interactions with each other. Equally important is that gender becomes less relevant in the interactions between boys and girls when they are grouped together in common working groups, although gender does not disappear altogether as an influence. Barrie Thorne, who has observed gender interaction in schools, concludes from her observations that gender has a "fluid" character and that gender relations between boys and girls can be improved through conscious changes that discourage gender separation. (See the box "Sociology in Practice.")

While in school, young people acquire identities and learn patterns of behavior that are congruent with the needs of other social institutions. School is typically the place where children are first exposed to a hierarchical, bureaucratic environment. School not only teaches children the skills of reading, writing, and other subject areas, but it also trains them to respect authority, be punctual, and follow rules. These skills prepare them for their future lives as workers in organizations that value these traits. Schools emphasize conformity to societal needs, although not everyone internalizes these lessons to the same degree (Bowles and Gintis 1979). Research has found that working-class school children form subcultures in school that resist the dominant culture. Rejecting the mental labor associated with school values, working-class children inadvertently learn the expectations associated with working-class jobs. As a result, the subculture of the school fits the needs of the class system (Willis 1977; MacLeod 1995).

Theories of Socialization

Knowing that people become socialized does not explain how it happens. Different theoretical perspectives explain socialization, including psychoanalytic theory, object relations theory, social learning theory, and symbolic interaction theory. Functionalism and conflict theory—two of the major frameworks of sociological theory—also have particular ways of interpreting the socialization process. Each perspective carries a unique set of assumptions about socialization and its effect on the development of the self. A complete view of socialization benefits from all of these theoretical frameworks.

Psychoanalytic Theory

Psychoanalytic theory originates in the work of **Sigmund Freud** (1856–1939). Among Freud's greatest contributions was the idea that the unconscious mind shapes human behavior. In other words, our behavior may be shaped by forces of which we are not always aware. Freud developed the technique of *psychoanalysis* to help discover the causes of psychological problems in the recesses of troubled patients' minds.

According to Freud, the human psyche has three parts: the id, the superego, and the ego. The **id** consists of deep drives and impulses. Freud was particularly intrigued by the sexual component of the id, which he considered a force in the unconscious mind. The **superego** is the dimension of the self that represents the standards of society. The superego incorporates or internalizes acquired values and norms—in short, society's collective expectations. To maintain an ordered society, people must repress the wild impulses generated by the id. Thus, the id is in permanent conflict with the superego. The superego represents what Freud saw as the inherent repressiveness of society. People cope with the tension between social expectations (the superego)

and their impulses (the id) by developing defense mechanisms, typically repression, avoidance, or denial (Freud 1960/1923, 1961/1930, 1965/1901).

One way people cope with the tension between the superego and the id is through development of the **ego.** The ego plays a balancing act between the id and the superego, adapting the desires of the id to the social expectations of the superego. In one Freudian analysis, "A person with a strong ego is one who is able to accomplish the best, most realistic compromise between the conflicting demands of the id and the superego. To give in to the id is to indulge in infantile and immature behavior. To be totally at the mercy of the superego is to be rigid and repressive" (Cuzzort and King 1980: 27).

The conflict between the id and the superego occurs in the subconscious mind; yet, it shapes human behavior. We get a glimpse of the unconscious mind in dreams and in occasional slips of the tongue—the famous "Freudian slip" that is believed to reveal an underlying state of mind. Someone might intend to say, "There were six people at the party," but instead says, "There were sex people at the party," revealing the unconscious mind. Psychoanalytic theory locates the forces that shape the self in the unconscious mind (Freud 1965/1901). The id demands gratification for biological desires, the superego forces constant awareness of how society perceives one's actions, and the ego negotiates an uneasy peace between the two parties. As the stabilized three-part self interacts with others, the rules of socialization emerge, with the ego bridging the gap between the primal id and the socialized superego.

Freud's work has never been free from controversy, and interpretations suggest different approaches for those in psychotherapy. Some sociologists have criticized Freud's work for not being generalizable because he worked with only a small and nonrepresentative group of clients. Freud has also been criticized by feminist scholars who find his work to be sexist. Feminists have been especially critical of Freud's concept of "penis envy" in women. Freud argued that adult women become jealous and resentful of men because of their unconscious wish to have a penis. Although some have argued that Freud was stating this metaphorically (that is, the penis symbolizes the male power that women resent), many feminists argued that Freud considered women to be sexually and psychologically immature, as well as psychologically maladjusted. Other feminists have adopted a psychoanalytic perspective in their work. Although they reject the sexist dimensions of Freud's arguments, they use the concept of the unconscious to explain numerous dimensions of women's and men's experiences (Chodorow 1994, 1999).

Psychoanalytic theory is an influential and popular way to think about human personality. We often speak of what motivates people, as if motives were internal, unconscious states of mind that direct human behavior. We think we can understand a person if we know what really makes that person tick, as if the behavior we observe is not what actually constitutes the person's being. People also typically think that understanding someone's true self requires knowing about that person's childhood experience. This reflects Freud's contention that personality is fixed at a relatively early age. Psychoanalysis has an enormous influence on how people think about human behavior.

To sum up, the psychoanalytic perspective sees human identity as relatively fixed at an early age in a process greatly influenced by one's family. It sees the development of social identity as an unconscious process, stemming from dynamic tensions between strong instinctual impulses and the social standards of society. Most important, psychoanalysis sees human behavior as directed and motivated by underlying psychic forces that are largely hidden from ordinary view.

Object Relations Theory

Psychoanalytic theory has been modified by a school of thought known as **object relations theory.** Placing less emphasis on biological drives, object relations theorists contend that the social relationships experienced by children determine the development of the adult personality. As in classical Freudian theory, the processes invoked by object relations theory are largely unconscious. Key concepts in object relations theory are *attachment* and *individuation,* the making and breaking of the bond with parents. Infants may be strongly attached to the parent who is their primary caregiver. As infants grow older, they learn to separate themselves from their parents, or individuate, both physically and emotionally. They become free-standing individuals, but early attachments to the primary caregivers persist.

Within sociology, one of the most widely known applications of object relations theory is provided by Nancy Chodorow, a feminist sociologist and practicing psychoanalyst who has used object relations theory to explain how gender shapes personalities. Specifically, Chodorow asks, "Why do women learn to mother and men, generally speaking, do not?" Chodorow's answer connects the process of personality formation directly to the division of labor in the family. The modern family, Chodorow argues, has an *asymmetrical division of labor* where women "mother," and men do not. In addition, she argues, women's work as mothers is devalued by society. It is not defined as real work and does not receive the same social and economic support of work outside the family (Chodorow 1978, 1994).

Chodorow agrees with the basic premise of psychoanalytic theory that personality is shaped by early, unconscious forces. She also agrees with object relations theory in that she sees attachment and individuation as important formative processes. However, she adds

that children identify with their same-sex parent, which means that boys and girls individuate differently. Boys, on the one hand, identify with the father. The asymmetrical division of labor separates the father from the home and full-time nurturing. Therefore, boys form personalities that are emotionally detached, independent, less oriented toward other people. Girls, on the other hand, identify with the mother. The mother's role is one of close attachment to the members of the household. Therefore, girls develop personalities based on attachment and an orientation toward others.

Men and women carry their patterns of separation and attachment from childhood into their adult lives. The unconscious processes associated with individuation and attachment create the gendered personalities that we think of as "masculine" and "feminine." Much research has shown that women are more likely to be oriented toward others and to seek and sustain affiliations, while men are more likely to repress their attachments to others and to be more individualistic in their orientation toward the world (Chodorow 1978).

Although Chodorow based her work on the study of traditional, nuclear families, her work has also been applied in a study of Chicano families. Many Chicano families are characterized by *familism,* defined as large, multigenerational households where there is a high value placed on family unity and much interaction between family and kin (Baca Zinn and Eitzen 2005). Despite the differences between this family form and the nuclear family analyzed by Chodorow, her framework is useful in analyzing the experiences of Chicana mothers and their daughters. Chicana mothers' identities often revolve around family and home, and they are more likely to identify with their daughters than their sons. Yet, Chicanas do not practice exclusive mothering. Instead, mothering figures may include grandmothers, aunts, or godmothers. Young Chicanas identify with their mothers, but they also see themselves as an extension of other women in the family system. In addition, the cultural representation of women within Chicano culture as sacred and self-sacrificing makes Chodorow's point about gender identification and attachment particularly salient for Chicanas (Segura and Pierce 1993).

You might ask how Chodorow's theory would explain personality formation in father-absent families. Chodorow would say that the absence of the father only exacerbates the tendency for boys to develop detached concepts of themselves. Of course, motherhood is not devalued, as Chodorow suggests, in all families, nor among all groups. African American families place a particularly high value on motherhood (Collins 1990). This does not mean Chodorow's theory is wrong. It only suggests that the value we place on mothering and the involvement of men in early child care makes a big difference in children's personality formation. In other words, Chodorow's theory has interesting practical implications. If men were to acquire more mothering skills and participate more in the daily tending of families, men and women in society would probably have less gender-stereotyped personalities.

Social Learning Theory

Whereas psychoanalytic theory places great importance on the unconscious processes of the human mind, **social learning theory** considers the formation of identity to be a learned response to social stimuli (Bandura and Walters 1963). Social learning theory emphasizes the societal context of socialization. Identity is regarded as the result of modeling oneself in response to the expectations of others, not as the product of the unconscious. According to social learning theory, behaviors and attitudes develop in response to reinforcement and encouragement from those around us. Social learning theorists acknowledge the importance of early childhood experience, but they think that the identity people acquire is based more on the behaviors and attitudes of people around them than the interior landscape of the individual.

Early models of social learning theory regarded learning rather simplistically in terms of stimulus and response. People were seen as passive creatures who merely responded to stimuli in their environment. This mechanistic view of social learning was transformed by the work of the Swiss psychologist **Jean Piaget** (1896–1980), who believed that learning was crucial to socialization, but imagination also had a critical role. He argued that the human mind organizes experience into mental categories he called *schema*, which are modified and developed as social experiences accumulate. Schema might be compared with a person's understanding of the rules of a game. Humans do not simply respond to stimulus, but also actively absorb experience and determine what they are seeing to construct a picture of the world.

Piaget proposed that children go through distinct stages of cognitive development as they learn the basic rules of reasoning. They must master the skills at each level before they go on to the next (Piaget 1926). In the initial *sensorimotor stage*, children experience the world only directly through their five senses of touch, taste, sight, smell, and sound. Next comes the *preoperational stage*, in which children begin to use language and other symbols. Children in the preoperational stage cannot think in abstract terms, but they do gain an appreciation of meanings that goes beyond their immediate senses. They also begin to see things as others might see them. The *concrete operational stage* occurs when children learn logical principles regarding the concrete world. This stage prepares them for more abstract forms of reasoning. In the *formal operational stage*, children are able to think abstractly and imagine alternatives to the reality in which they live.

Piaget's work stresses the significance of conscious mental processes in social learning. Socialization in this framework is highly creative and adaptive as young children develop new ways of thinking about the world. The emphasis in social learning theory is on the influence of the environment in socializing people, as well as on human creativity and imagination, because the mind mediates the influence of the environment. Social learning theory holds that behavior can be changed by altering the social environment (Bandura and Walters 1963).

Building on Piaget's model of stages of development, psychologist Lawrence Kohlberg (1969) developed a theory of moral development. Kohlberg interpreted the process of developing moral reasoning as occurring in several stages that were grouped into three levels: the preconventional stage, the conventional stage, and the postconventional stage. In the *preconventional stage*, young children judge right and wrong in simple terms of obedience and punishment, based on their own needs and feelings. In the *conventional stage*, adolescents develop moral judgment in terms of cultural norms, particularly social acceptance and following authority. In the *postconventional stage*, people are able to consider abstract ethical questions, thereby showing maturity in their moral reasoning. In his original research, Kohlberg argued that men reach a higher level of moral development than women because men are concerned with authority, while women remain more concerned with a lower phase of feelings and social opinions.

Kohlberg's work was later criticized by psychologist Carol Gilligan. Gilligan (1982) found that women con-

Social learning theory emphasizes how people model their behaviors and attitudes on that of others.

© Paul Chesley/Stone/Getty Images

ceptualize morality in different terms than men. Instead of judging women by a standard set of men's experiences, Gilligan showed that women's moral judgments are more contextual than those of men. In other words, when faced with a moral dilemma, women are more likely to consider the different relationships affected by any decision, instead of making moral judgments according to abstract principles. Gilligan's research makes an important point, not just about the importance of including women in studies of human development, but also about being careful not to assume that social learning follows a universal course for all groups.

Symbolic Interaction Theory

Social learning theory is primarily represented in sociology by the theoretical perspective known as *symbolic interaction theory,* introduced in Chapter 1. According to symbolic interaction theory, human actions are based on the meanings we attribute to things and these meanings emerge through social interaction (Blumer 1969). To explain further, people learn identities and values through the socialization process as they learn the social meanings that different behaviors imply. Learning to become a good student, then, means taking on the characteristics associated with that role. Because roles are socially defined, they are not real like objects or things, but they are real because of the meanings people give them. As did Piaget, symbolic interactionists understand the human capacity for reflection and interpretation as having an important role in the socialization process.

For symbolic interactionists, meaning is constantly reconstructed as people act within their social environments. The **self** is our concept of who we are and is formed in relationship to others. It is not an interior bundle of drives, instincts, and motives. Because of the importance attributed to reflection in symbolic interaction theory, symbolic interactionists use the term *self,* instead of the term *personality,* to refer to a person's identity. Symbolic interaction theory emphasizes that human beings make conscious and meaningful adaptations to their social environment. Although symbolic interaction interprets childhood as a very influential period in the human life cycle, they see the self as evolving over the life span. Therefore, identity is not something that is unconscious and

hidden from view, it is socially bestowed and socially sustained (Berger 1963).

Two theorists have greatly influenced the development of symbolic interactionist theory in sociology. **Charles Horton Cooley** (1864–1929) and **George Herbert Mead** (1863–1931) were both sociologists at the University of Chicago in the early 1900s (see Chapter 1). Cooley and Mead saw the self developing in response to the expectations and judgments of others in their social environment. Cooley postulated the **looking-glass self** to explain how a person's conception of self arises through reflection about relationships to others (Cooley 1902, 1909). The development of the looking-glass self emerges from (1) how we think we appear to others; (2) how we think others judge us; and (3) the feelings that result from these thoughts. The looking-glass self involves *perception* and *effect*; the perception of how others see us and the effect of others' judgment on us.

How others see us is fundamental to the idea of the looking-glass self. In seeing ourselves as others do, we respond to the expectations others have of us. This means that the formation of the self is fundamentally

The Looking-Glass Self
The looking-glass self refers to the process by which people see themselves as others see them.

Drawing conceptualized by Norman Andersen

a social process that is based in the interaction people have with each other, as well as the human capacity for self-reflection. One unique feature of human life is the ability to see ourselves through others' eyes. People can imagine themselves in relationship to others and develop a definition of themselves, accordingly. From a symbolic interactionist perspective, the reflective process is key to the development of the self. If you grow up with others thinking you are smart and sharp-witted, chances are you will develop this definition of yourself. If others see you as dull-witted and withdrawn, chances are good that you will see yourself this way.

George Herbert Mead agreed with Cooley that children are socialized by responding to other's attitudes toward them. *Roles* are the expectations associated with a given status in society and the basis of all social interaction (see also Chapter 6). When people occupy roles, they are expected to fulfill the expectations associated with those roles. Roles can be thought of as the intersection between the individual and society. By occupying different roles, a person acquires the expectations associated with each role.

Taking the role of the other is the process of imagining oneself from the point of view of another. To Mead, role-taking is a source of self-awareness. As people take on new roles, their awareness of self changes and identity emerges from the roles one plays. He explained this process in detail by examining childhood socialization, which he saw as occurring in three stages: the imitation stage, the play stage, and the game stage (Mead 1934). In each phase of development, the child becomes more proficient at taking the role of the other. In the first stage, the **imitation stage,** children merely copy the behavior of those around them. Role-taking in this phase is nonexistent because the child simply mimics the behavior of those in the surrounding environment without much understanding of the social meaning of the behavior. Although the child in the imitation stage has little understanding of the behavior being copied, the child is learning to become a social being. In Mead's analysis, mimicking behavior is one way that children begin to learn the expectations of others.

In the second stage, the **play stage,** children begin to take on the roles of significant people in their environment, not just imitating them but incorporating their relationship to them. Instead of just mimicking others' behavior, the child now understands more about context and the nuances of meaning that different behaviors represent. Especially meaningful is when children take on the role of **significant others,** those with whom they have a close affiliation. A child pretending to be his or her mother may talk as the mother would. With this behavior the child begins to develop self-awareness, seeing himself or herself as others do.

In the third stage of socialization, the **game stage,** the child becomes capable of taking on a multitude of roles at the same time. These roles are organized in a complex system that gives the child a more general or comprehensive view of the self. In this stage, the child begins to comprehend the system of social relationships in which the child is located. The child not only sees himself or herself from the perspective of a significant other, but also understands how people are related to him or her and to each other. This is the phase where children internalize (incorporate into the self) an abstract understanding of how society sees them.

Mead compared the lessons of the game stage with a baseball game. In baseball, all roles *together* make the game. The pitcher does not just throw the ball past the batter as if they were the only two people on the field. Instead, each player has a specific role, and each role intersects with other roles. The network of social roles and the division of labor in the baseball game is a social system, like the social systems children must learn as they develop a concept of themselves in society.

In the game stage, children learn more than just the roles of significant others in their environment. They also acquire a concept of the **generalized other,** which is the abstract composite of social roles and social expectations. In the generalized other, children have an example of community values and general social expectations that add to their understanding of self, but they do not all learn the same generalized other.

Depending on one's social position (that is, race, class, gender, region, or religion), one learns a particular set of social and cultural expectations. If the self is socially constructed through the expectations of others, how do people become individuals? Mead answered this by saying that the self has two dimensions: the "I" and the "me." The I is the unique part of individual personality; the active, creative, self-defining part. The me is the passive, conforming self; the part that reacts to others. In each person, there is a balance between the I and the me, similar to the conflict Freud proposed between the id and the superego. Mead differed from Freud, however, in his judgment about when identity is formed. Freud felt that identity was fixed in childhood and henceforth driven by internal, not external, forces. In Mead's judgment, social identity is always in flux, constantly emerging (or "becoming") and dependent on social situations. Over time, identity is stabilized as one learns to respond consistently to common situations.

Social expectations associated with given roles learned through the socialization process change as people redefine situations and as social and historical conditions change. For example, as more women enter the paid labor force and as men take on additional responsibilities in the home, the expectations associated with motherhood and fatherhood are changing. Men now experience some of the role conflicts that women have faced in balancing work and family. As the roles of mother and father are redefined, children are learning new socialization patterns. However, traditional gender expectations maintain a remarkable grip. Despite many changes in family life and organization, young girls are still socialized for motherhood, and young boys are still socialized for greater independence and autonomy.

Functionalism and Conflict Theory

Functionalism and conflict theory do not have the specific theories of socialization such as those examined thus far, but each provides some perspective on this process. The concept of social roles stems largely from functionalism. The stability that functionalists emphasize in society stems from the socialization process—that is, people learning the values, norms, and social identities that help to hold society together. As the social institution where people first learn social roles, the family helps stabilize society. To the extent that people learn social identities, they contribute to the overall cohesion in society.

Conflict theorists, given their focus on the role of power in society, do not see socialization in such harmonious terms. Rather, conflict theorists see socialization in the context of inequality and the systems of power that value some social locations more than others. Thus, conflict theorists are interested in the social-ization process as it involves forming identities that are situated within systems of inequality—such as those created by race, gender, age, class, and other social factors. For a conflict theorist, what would be interesting about the socialization process is how people formulate social identities in ways that both reproduce, but also resist, the power relations found in society. Both functionalist theory and conflict theory thus interpret socialization within the larger context of society, seeing how the roles and identities people learn reflect larger social relationships and patterns (see Table 4.1).

Growing Up in a Diverse Society

Socialization makes us members of our society. It instills in us the values of the culture and brings society into our self-definition, our perceptions of others, and our understanding of the world around us. Socialization is not, however, a uniform process, as described by the different examples in this chapter. In a society as complex and diverse as the United States, no two people will have exactly the same experiences. We can find similarities between us, often across vast social and cultural differences, but variation in social contexts creates vastly different socialization experiences. Furthermore,

Table 4.1
Theories of Socialization

How each theory views:	Individual Learning Process	Formation of Self	Influence of Society
Psychoanalytic Theory	The unconscious mind shapes behavior	Self (ego) emerges from tension between the id and the superego	Societal expectations are represented by the superego
Object Relations Theory	Infants identify with the same-sex parent	The self emerges through separating oneself from the primary caretaker	A division of labor in the family shapes identity formation
Social Learning Theory	People respond to social stimuli in their environment	Identity is created through reinforcement and encouragement	Young children learn the logical principles that shape the external world
Symbolic Interaction Theory	Children learn through taking the role of significant others	Identity emerges as the creative self interacts with the social expectations of others	Expectations of others form the social context for learning social roles
Functionalism	Social roles are learned in the family	People internalize social expectations, thus the self contributes to the stability of society	Socialization occurs in social institutions that function to maintain social order
Conflict Theory	Individuals learn social identities in the context of power relationships	People's identities and selves reflect the race, class, and gender relations in society, along with other social influences	The self reflects the needs and interests of the powerful groups in society, although people can also resist these influences

current transformations in the U.S. population are creating new multiracial and multicultural environments where young people grow up. In many communities, schools are being transformed by the large number of immigrant groups entering the school system. In school, where children come into contact with other children from diverse backgrounds, they experience a new context in which to form their social values and learn their social identities.

One task of the sociological imagination is to examine the influence of environmental contexts on socialization. Where you grow up; how your family is structured; what resources you have at your disposal; your racial–ethnic identity, gender, and nationality—all these shape the socialization experience. Growing up White, female, and middle class, is different from growing up Latino, male, and working-class. Even within a given group, there are great differences in socialization patterns. Growing up as an African American, middle-class woman will produce results different from growing up African American and poor. Socialization experiences are shaped by a number of factors that intermingle to form the context for socialization.

To illustrate this, research comparing African American and White childrearing practices finds patterns in parenting practices that are shaped by the gender of the child and the race and social class of the family. Thus, while both Black and White parents give top priority to the long-term goals for their children to get a good education, find a good job, and form a strong and loving family, education and jobs are somewhat more of a priority for Black parents than White parents. Some of the strongest contrasts by race are among upper middle-class parents, where family is a top priority for White parents and education and jobs are the top long-term goals for Black parents. Gender differences are also strongest among upper middle-class parents (both White and Black), where parents of daughters are more likely than parents of sons to emphasize the importance of family as a long-term goal (Hill and Sprague 1999). Many other studies have concluded that gender has a strong influence on parenting. As one illustration, both fathers' and mothers' support for gender equity increases when they have only daughters; this finding is particularly strong for men. Fathers who have only sons are least likely to support policies for gender equity (Warner and Steel 1999).

In a society as diverse as the United States, socialization experiences are multifaceted. From an early age, children learn complex messages about race, including their own racial identity and that of others (see the box "Doing Sociological Research: Children's Understandings of Race.") Consider the increasingly common experience of growing up as a mixed-race person in a society where racial identity is an important marker of how others define you and how you define yourself (Root 1992, 1996; Tizard 1993). In this society, one is presumed to be Black, White, Latino, Asian, *or* Native American. These one-dimensional labels are inadequate for multiracial people. Yet every day they confront

---

DOING SOCIOLOGICAL RESEARCH

Children's Understanding of Race

In a racially stratified society, people learn concepts about race that shape their interaction with others. Sociologists Debra Van Ausdale and Joe Feagin wanted to know how children understand racial and ethnic concepts and how this influences their interactions with other children. These researchers note that much of the social science literature studied older children or looked only at how race shaped children's self-concepts. Prior to this study, most knowledge about children's understandings of race came from experimental studies where children were observed in a laboratory or from psychological tests and interviews with children. Van Ausdale and Feagin wanted to study children in a natural setting, that is, one not contrived for purposes of research. So, they observed children in school,

during which time they systematically observed children's interactions with one another.

They observed three-, four-, and five-year-olds in an urban preschool. Twenty-four of the children were White; nineteen, Asian; four, Black; three, biracial; three, Middle Eastern; two, Latino; and three classified as "other." The children's racial–ethnic designations were provided by parents. The senior researcher (Van Ausdale) was in the classroom all day for five days a week over a period of eleven months. Van Ausdale says she

© Caroline Penn/Corbis

Through the socialization process, young children learn the values of their culture. These values shape their relationships with other people.

observed one to three episodes involving significant racial or ethnic matters each day.

From these observations, the researchers conclude that young children use racial and ethnic concepts to exclude other children from play. Sometimes language is the ethnic marker children use; other times, it is skin color. The children also showed an awareness of negative racial epithets, even though they were attending a school that prided itself on limiting children's exposure to prejudice and discrimination and used a multicultural these labels when they fill out a job application, respond to what people call them, or decide what student organizations to join. A person who is Latino and Black, or Asian and White, or Native American and Black, or any biracial heritage may experience contradictory expectations from the dominant society and the home community. This is compounded by the contradictory expectations that normally occur within either community. In the words of the sociologist Michael Thornton (himself an individual of mixed racial identity), "Individuals are expected to locate themselves 'accurately' within established racial structures . . . , finding where society places them and reconciling this placement with what they want to be. . . . Society defines race as distinctive and homogenous; multiracial people experience it as multidimensional" (1995: 97–98).

The socialization process is clearly patterned by factors such as class, race, gender, religion, regional background, sexual preference, age, and ethnicity, but no single characteristic can define the socialization experience. Certainly, some may be more salient at given times than others. A Latina may be particularly cognizant of her racial identity as she is growing up, but she does not experience it as separate from her gender identity. Both are central to her identity. Although she may be more conscious of one or the other at different times in her life, she is never without the influence of both.

The complexity of belonging to different groups has been described as **social identity complexity**—a term referring to how a person subjectively interprets the interrelationships among multiple group identities. When people perceive a strong overlap between multiple identities, it is easier for them to integrate these multiple identities into a singular definition of themselves (such as in, "I am a woman of color"), but when the multiple sources of their group membership are not perceived as overlapping much, the person may be more likely to identify with just one group. This would be especially the case if one of the sources of their identity is perceived as an "outgroup" (such as an immigrant who might emphasize being American, but reject their prior ethnic identity). Research on social identity complexity finds that the extent to which people can integrate multiple sources of identity into one definition of the self depends in part on the level of stress produced by the different sources of identity, by one's value priorities, and how well outgroups are tolerated in society (Roccas and Brewer 2002).

Socialization Across the Life Course

Socialization begins the moment a person is born. As soon as the sex of a child is known, parents, grandparents, brothers, and sisters greet the infant with expectations that are not the same for a boy as for a girl. Socialization does not come to an end as we reach adulthood; it continues through our lifetime. As we enter new

---

curriculum to teach students to value racial and ethnic diversity. At times, the children also used racial–ethnic understandings to include others by teaching other students about racial–ethnic identities. Race and ethnicity were also the basis for children's concepts of themselves and others. As an example, one four-year-old White child insisted that her classmate was Indian because she wore her long, dark hair in a braid. When the classmate explained that she was not Indian, the young girl remarked that maybe her mother was Indian.

In this and numerous other incidents reported in the research, the children showed how significant racial–ethnic concepts were in their interaction with others. Race and ethnic differences, as the researchers claim, are "powerful identifiers of self and other" (Van Aus-

dale and Feagin 1996: 790). Despite the importance of race in the children's interactions with others, Van Ausdale and Feagin also noted a strong tendency for the adults they observed to deny that race and ethnicity were significant to the children.

Questions to Consider

1. When did you first learn about race? What did you learn? After thinking about these questions, ask yourself what *socialization* experiences and *agents of socialization* affected what you learned. How might what you learned have been different had you been a member of a different racial–ethnic group? (You can get some help on this by talking to people of different racial backgrounds and asking them what and when they first

learned about race.) *Keywords: racial socialization, learning racism, children* and *prejudice*

2. What practices might teachers use to teach students about race in a way that values all groups? Relate your answer to the sociological theories of socialization that are presented in this chapter. *Keywords: teachers* and *race prejudice, reducing racism*

We have included InfoTrac College Edition keywords at the end of each question to make it easier for you to find more to read on these topics. Go to www.infotrac-college.com, an online library, to begin your search.

Source: Van Ausdale, Debra, and Joe R. Feagin. 2000. *The First R: How Children Learn Race and Racism.* Lanham, MD: Rowman and Littlefield. ● ● ●

situations, and even as we interact in familiar ones, we learn new rules and undergo changes in identity.

As we will see in Chapter 14 on age and aging, sociologists examine the experiences that mark the passage from childhood to old age using a **life course perspective** that connects people's personal attributes, the roles they occupy, and the life events they experience to the social and historical context (Stoller and Gibson 2000). This perspective underscores C. Wright Mills' (1959) point that personal biographies are linked to specific social-historical periods. Thus, different generations are strongly influenced by large-scale events such as war, immigration, economic prosperity, or depression, at the same time that the immediate expectations of those around them are part of their socialization. Socialization is an ongoing process, beginning in infancy and continuing through old age.

Childhood

During childhood, socialization establishes one's initial identity and values. In this period, the family is an extremely influential source of socialization, but experiences in school, peer relationships, sports, religion, and the media also have a profound effect. Children acquire knowledge of their culture through countless subtle cues that provide them with an understanding of what it means to live in society.

Socializing cues begin as early as infancy, when parents and others begin to describe their children based on their perceptions. Frequently, these perceptions are derived from the cultural expectations parents have for children. Parents of girls may describe their babies as "sweet" and "cuddly," while boys are described as strong and alert. Even though it is difficult to physi-

MAPPING AMERICA'S DIVERSITY

MAP 4.1 Children in the United States

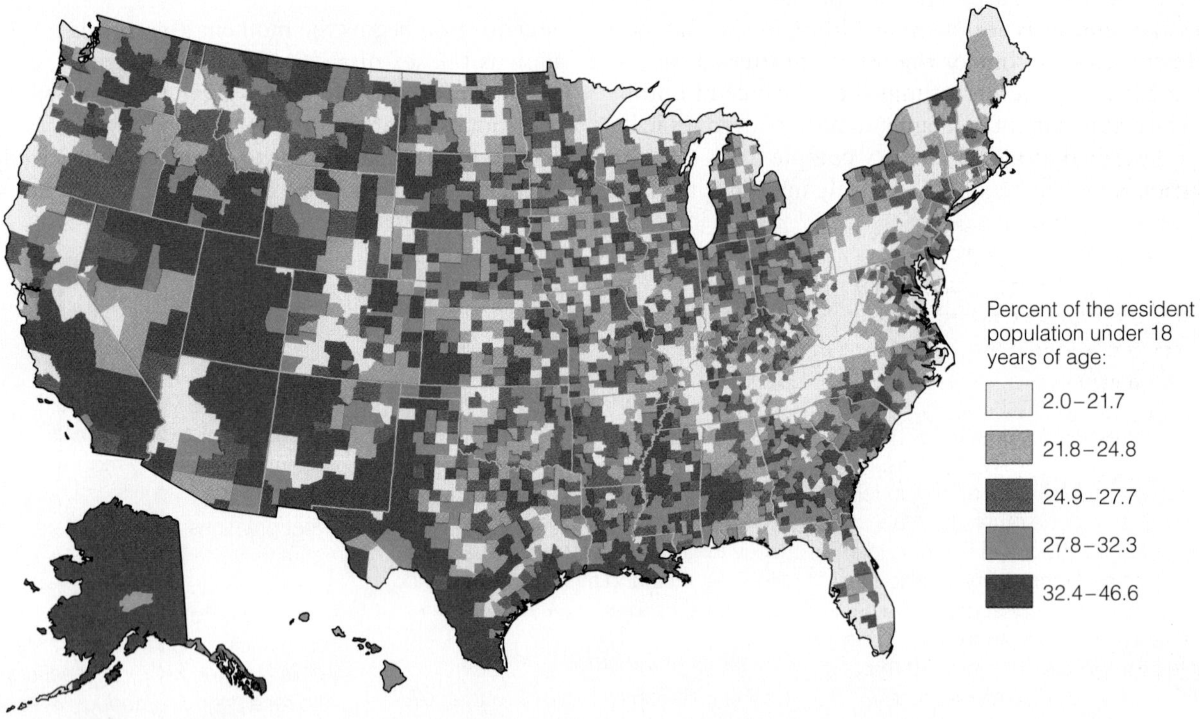

Percent of the resident population under 18 years of age:

- 2.0–21.7
- 21.8–24.8
- 24.9–27.7
- 27.8–32.3
- 32.4–46.6

The socialization of children occurs in the context of specific social influences, including such things as region of residence, national origin, race, social class, and immigrant status. As this map shows, children vary as a proportion of the population in different regions of the United States. Lying behind this distribution of the population are different patterns of immigration and migration, different economic and educational opportunities, and varying levels of social support services for children and their parents. Is your region high or low in the proportion of children and what factors do you think influence this?

Data: From the U.S. Census Bureau. 2004. *American FactFinder.* Website: **www.census.gov**

cally identify baby boys and girls, in this culture parents dress even their infants in colors and styles that typically distinguish one gender from the other.

The lessons of childhood socialization come in a myriad of ways, some more subtle than others. In an example of how gender influences childhood socialization, researchers observed mothers and fathers who were walking young children through public places. Although the parents may not have been aware of it, both mothers and fathers were more protective toward girl toddlers than boy toddlers. Parents were more likely to let boy toddlers walk alone but held girls' hands, carried them, or kept them in strollers. The children were not the only ones learning gender roles. The researchers also observed that when the child was out of the stroller, the mother was far more likely than the father to be pushing the empty stroller (Mitchell et al. 1992). In countless ways, sometimes subtle, sometimes overt, we learn society's expectations.

Much socialization in early childhood takes place through play and games. Games that encourage competition help instill the value of competitiveness throughout someone's life. Likewise, play with other children and games that are challenging give children important intellectual, social, and interpersonal skills. Extensive research has been done on how children's play and games influence their identities as boys and girls (Campenni 1999; Raaj and Rackliff 1998). Generally, the research finds that boys' play tends to be rougher, more aggressive, and involve more specific rules. Boys are also more likely to be involved in group play and girls in more conversational play (Moller et al. 1992). Sociologists have concluded that the games children play significantly influence their development into adults.

Another enormous influence on childhood socialization is what children observe of the adult world. Children are keen observers, and what they perceive will influence their self-concept and how they relate to

VIEWING SOCIETY IN GLOBAL PERSPECTIVE:

MAP 4.2 The World's Children

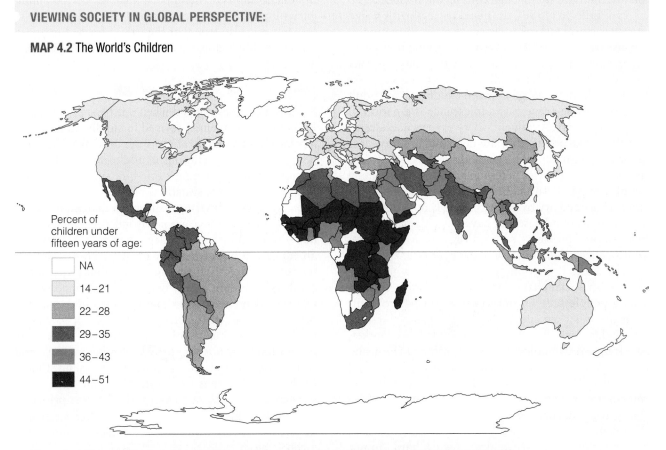

Percent of children under fifteen years of age:

- NA
- 14–21
- 22–28
- 29–35
- 36–43
- 44–51

Throughout the world, the proportion of children as a percentage of the population of a given country tends to be higher in those countries that are most economically disadvantaged and most overpopulated. In such countries, children are also more likely to die young, but are needed to contribute to the labor that families do. What consequences do you think the proportion of children in a given society has for the society as a whole?

Data: From the U.S. Census Bureau. 2004. *Statistical Abstract of the United States 2003.* Washington, DC: U.S. Government Printing Office, p. 845.

others. This is vividly illustrated by research revealing how many adult child abusers were themselves victims of child abuse (Fattah 1994). Children become socialized by observing the roles of those around them and internalizing the values, beliefs, and expectations of their culture.

Adolescence

Only recently has adolescence been thought of as a separate phase in the life cycle. Until the early twentieth century, children moved directly from childhood roles to adult roles. It was only when formal education was extended to all classes that adolescence emerged as a particular phase in life when young people are regarded as no longer children, but not yet adults. There are no clear boundaries to adolescence, although it generally lasts from junior high school until the time one takes on adult roles by getting a job, marrying, and so forth. Adolescence can include the period through high school and extend right up through college graduation.

Erik Erikson (1980), the noted psychologist, stated that the central task of adolescence is the formation of a consistent identity. Adolescents are trying to become independent of their families, but they have not yet moved into adult roles. Conflict and confusion can arise as the adolescent swings between childhood and adult maturity. Some argue that adolescence is a period of delayed maturity. Although society expects adolescents to behave like adults, they are denied many privileges associated with adult life. Until age eighteen they cannot vote; state policies determine the age when young people can purchase alcohol or marry without permission; sexual activity by young people is not condoned. The tensions of adolescence have been blamed for numerous social problems, such as drug and alcohol abuse, youth violence, and the school dropout rate.

The issues that young people face are a good barometer of social change across generations. Today's young people face an uncertain world where adult roles are less predictable than in the past. Marriage later in life, high divorce rates, a volatile labor market, and frequent technological change create a confusing environment for young people (Csikzentmihalyi and Schneider 2000). Studies of adolescents find that, in this context, young people understand the need for flexibility, specialization, and, likely, frequent job change. Although the popular media stereotype adolescents as slackers, most teens are willing to work hard, do not engage in criminal or violent activity, and have high expectations for an education that will lead to a good job. Many, however, find that their expectations are out of alignment with the opportunities that are actually available, creating a paradox of a generation that is "motivated but directionless" (Schneider and Stevenson 1999).

Patterns of adolescent socialization vary significantly by race and social class. National surveys find some intriguing class and race differences in how young people think about work and play in their lives (see Table 4.2). In general, the most economically privileged young people see their activities as more like play than work, whereas those less well off are more likely to define their activities as work. Likewise, White youth (boys especially) are more likely than other groups to see their lives as playful. The researchers interpret these findings to mean that being economically privileged allows you to think of your work as if it were play. Being in a less advantaged position, on the other hand, makes you see the world as more "work-like." This is supported by further findings that young people from less advantaged backgrounds spend more time in activities they define as purposeless (Schneider and Stevenson 1999).

Thus, differences in adolescent socialization are associated with significantly different life paths. The values one forms as a young person have a decided influence on where young people find themselves later in life, but it is also true that the values formed in youth reflect the life chances the young believe are possible.

Adulthood and Old Age

Socialization does not end when one becomes an adult. Building on the identity formed in childhood and adolescence, **adult socialization** is the process of learning new roles and expectations in adult life. Adult socialization involves learning behaviors and attitudes appropriate to specific situations and roles.

Adolescents entering college, to take an example from young adulthood, are newly independent and have new responsibilities. In college, one acquires not just an education but also a new identity. Those who enter college directly from high school may encounter conflicts with their family over their newfound status. Older students who work and attend college may experience difficulties (defined as *role conflict;* see Chapter 5) trying to meet both responsibilities, especially if their family is not supportive. Meeting multiple and conflicting demands may require the returning student to develop different expectations about how much she can accomplish or to establish different priorities about what she will attempt. These changes reflect a new stage in her socialization (Settersten and Lovegreen 1998).

Adult life is peppered with events that may require the adult to adapt to new roles. Marriage, a new career, starting a family, entering the military, getting a divorce, becoming a grandparent, retiring, or coping with death in the family all transform an individual's previous social identity. In today's world, these transitions through the life course are also not as orderly as they were in

Table 4.2
Work and Play among Youth: Percentage of Time Reported Spent in Work, Play, Both, and Neither

	N	% Work	% Play	% Both	% Neither
TOTAL	863	28.5	28.0	9.0	34.5
Gender					
Male	356	29.4	29.6	8.2	32.7
Female	507	27.9	26.9	9.6	35.7
Race/Ethnicity:					
Asian	55	31.4	26.8	11.2	31.5
Hispanic	134	29.1	23.9	7.5	40.6
African American	162	29.3	22.8	7.5	38.4
White	503	27.9	30.9	10.0	31.9
Native American	9	23.9	29.9	10.8	35.4
Parental Education:					
High school	68	27.4	30.4	8.7	33.5
High School graduate	196	28.2	26.7	8.7	36.4
College graduate	128	29.7	28.6	9.7	32.0
Master's	93	26.3	34.6	9.5	29.5
Ph.D.	77	27.6	35.9	9.8	26.7

Ask yourself: What differences do you see in time spent in work by gender? by race/ethnicity? by class (using parental education as an indicator of social class)? Do you think these patterns hold among youth you know? Given what you have observed about young people, how might you explain any differences noted in this table?

Source: Csikzentmihalyi, Mihaly, and Barbara Schneider. 2000. *Becoming Adult: How Teenagers Prepare for the World of Work.* New York: Basic Books, p. 70.

the past. Where there was once a sequential and predictable trajectory of schooling, work, and family roles through one's twenties and thirties, that is no longer the case. Younger generations now experience diverse patterns in the sequencing of work, schooling, and family formation—even returning home—than was true in the past. Moreover, national surveys of young people find that now, by age 25, most young people have not achieved their occupational aspirations. However, studies also find that by the mid-twenties, men are more likely to have moved into higher status occupations, while women tended to move to lower status occupations or leave the labor force altogether. These changes complicate the life course, and people have to make different adaptations to these changing roles (Cooksey and Rindfuss 2001; Rindfuss et al. 1999).

Learning new roles requires **anticipatory socialization,** the learning of expectations associated with a role one expects to enter in the future. One might rehearse the expectations associated with being a professor by working as a teaching assistant, or take a class in preparation for becoming a father, or attend a summer program to prepare for entering college. Anticipatory socialization allows a person to foresee the expectations associated with a new role and learn in advance what is expected in that role.

In the transition from an old role to a new one, individuals often vacillate between their old and new identities. The coming-out process for gays and lesbians supplies an interesting example. *Coming out* is the process of identifying oneself as gay or lesbian. This can be either a public coming out or a private acknowledgment of the homoerotic feelings that one has. Coming out can take years and can be selective. It generally means coming out to a few people at first, people from one's family or friends who are seen as likely to have the most positive reaction. The process is rarely a single event but occurs in stages; the new identity that emerges is not only a new sexual identity, but also a new sense of self (Due 1995).

Do people entering new roles become exactly what is expected of them? Sometimes. Women and minorities who acquire power and prestige are often accused of becoming like White men. White men have long had a near monopoly on positions of power, and their attitudes and behaviors usually define the culture of large organizations. Newcomers may be advised to conform to the same attitudes and behaviors to succeed. How thoroughly they adopt the new role depends on many factors, such as with whom they identify, how they define themselves, and what they receive as rewards and punishments within the organization. Always, in adulthood and earlier, variation exists in the extent to which people internalize new social expectations.

Passage through adulthood involves many transitions. A change in careers requires learning new sets of skills and establishing new relationships. Changes in family relations, such as becoming a grandparent for the first time, produce new identities. Aging, itself, requires a new way of thinking about oneself. In U.S. society, one of the most difficult transitions is the passage to old age. People are taught to fear aging in this society, and many people spend considerable time and money trying to keep looking young. Unlike many societies, ours does not revere the elderly but instead devalues them, making the aging process even more difficult.

Growing old in a society with such a strong emphasis on youth means encountering social stereotypes about the old, adjusting to diminished social and financial resources, and sometimes living in the absence of social support, even when facing some of life's most difficult transitions, such as declining health and the loss of loved ones. Still, many people experience old age as a time of great satisfaction and enjoy a sense of accomplishment connected to work, family, and friends. The degree of satisfaction during old age often depends on the social support networks established earlier in life. Strong social networks have a positive effect on the ability of people to cope with stressful life events. Sociologists have found that older women report higher levels of emotional support than men from such networks. Men tend to experience greater social isolation when they are old—an indication of how the gender expectations one learns earlier in life continue to have an influence over the life course (Calasanti and Slevin 2001).

Rites of Passage

A **rite of passage** is a ceremony or ritual that marks the passage of an individual from one role to another. Rites of passage define and legitimize abrupt role changes that begin or end each stage of life. The ceremonies sur-

Every culture has important rites of passage that mark the transition from one phase in the life course to another. Here different cultural traditions distinguish the rites of passage associated with marriage— a traditional Nigerian wedding (upper left); a young American couple (upper right); a Shinto (Japanese) bride taking a marital pledge by drinking sake (lower left); and a newlywed Orthodox Christian couple in Eritrea (lower right).

rounding rites of passage are often dramatic and infused with awe and solemnity. Examples include graduation ceremonies, weddings, and religious affirmations, such as the Jewish ceremony of the bar mitzvah for boys or the bas mitzvah for girls, confirmation for Catholics, and adult baptism for many Christian denominations.

Formal promotions or entry into certain new careers may also include rites of passage, such as completing police academy training or being handed one's diploma. Such rites are usually attended by family and friends, who watch the ceremony with pride. People frequently keep mementos of these rites as markers of the transition through life's major stages. Bridal showers and baby showers are rites of passage. At a shower, the person who is being honored is about to assume a new role and identity—from young woman to wife or mother (Montemurro 2002). Rites of passage entail public announcement of the new status, for the benefit of both the individual and those with whom the newly anointed person will interact. In the absence of such rituals, the transformation of identity would not be formally recognized, perhaps leaving uncertainty in the rising youngster or the community about the individual's worthiness, preparedness, or community acceptance.

Sociologists have noted that contemporary society has no universally recognized and formalized rite of passage marking the transition from childhood to adulthood. As a consequence, the period of adolescence is attended by ambivalence and uncertainty. As adolescents hover between adult and child status, they may not have the clear sense of identity that a rite of passage can provide.

Although there is no universal ceremony in U.S. culture by which one is promoted from child to adult, some subcultures do mark the occasion. Among the wealthy, the debutante's ball (also called "coming out")

is a traditional introduction of a young woman to adult society. Latinos may celebrate the *quinceañera* (fifteenth birthday) of young girls. As a tradition of the Catholic Church, this rite recognizes the girl's coming of age, while also keeping faith with an ethnic heritage. Dressed in white, she is introduced by her parents to the larger community. Though it was formerly associated with working-class families in the barrios, the *quinceañera* has become popular among affluent Mexican Americans, who may match New York debutante society by spending as much as $30,000 to $50,000 on the event (McLane 1995; Williams 1990).

Resocialization

Most of the transitions people experience in their lives involve continuity with the former self as it undergoes gradual redefinition. Sometimes, however, adults are forced to undergo a radical shift of identity. **Resocialization** is the process by which existing social roles are radically altered or replaced (Fein 1988). This process is likely to occur when people enter institutional settings where the institution claims enormous control over the individual, such as the military, prisons, monastic orders, and some cults (see also Chapter 6 for a discussion of *total institutions*). When military recruits enter boot camp, they are stripped of personal belongings, their heads are shaved, and they are issued identical uniforms. Although military recruits do not discard their former identities, the changes brought about by becoming a soldier can be dramatic and are meant to make the military primary, not one's family, friends, or personal history. The military represents an extreme form of resocialization, where individuals are expected to subordinate their identity to the identity of the group. In such an organization, individuals are interchangeable and group consensus (meaning, in the military, unanimous, unquestioned subordination to higher ranks) is an essential component of group cohesion and effectiveness. Military personnel are expected to act as soldiers, not as individuals. Understanding the importance of resocialization when entering the military helps explain practices such as "the rat line" at the Virginia Military Institute (VMI), where members of the senior class taunt and harass new recruits.

Resocialization often occurs when people enter hierarchical organizations that require them to respond to

Rites of passage, such as this baby shower, mark the passage from one phase of life to another.

© Photodisc Blue/Getty Images

©Getty Images

Hazings are a good example of the rites of passage that often accompany integration or a sense of belonging into a group. This hazing incident became national news when an annual hazing ritual at a suburban Chicago high school in 2003 injured several young women and was broadcast on most major media networks nationwide.

authority on principle, not out of individual loyalty. The resocialization process promotes group solidarity and generates a feeling of belonging. Participants in these settings are expected to honor the symbols and objectives of the organization. Disloyalty is seen as a threat to the entire group. In a convent, for example, nuns are expected to subordinate their own identity to the calling they have taken on, a calling that requires obedience both to God and to an abbess.

Resocialization may involve physically and psychologically degrading new members with the aim of breaking down or redefining their old identity. They may be given menial and humiliating tasks and be expected to act in a subservient manner. Social control may be exerted by peer ridicule or punishment. Fraternities and sororities offer an interesting everyday example of this pattern of resocialization. When fraternities and sororities induct new members, the pledges are often given onerous, meaningless chores and may be forced to behave in comically degrading, sometimes life-threatening, ways. Depending on the traditions of the group, resocialization practices range from mildly sadistic mischief to hazardous behavior, such as hazing. Intense resocialization rituals, whether in jailhouses, barracks, convents, or sorority and fraternity houses serve the same purpose: to impose some sort of ordeal to cement the seriousness and permanence of new roles and expectations.

The Process of Conversion

Resocialization also occurs during what people popularly think of as *conversion*. A conversion is a far-reaching transformation of identity, often related to religious or political beliefs. People usually think of conversion in the context of extreme conversions. Think of John Walker Lindh, a U.S. citizen who joined the Taliban in Afghanistan and was later charged with conspiring to kill Americans abroad and supporting terrorist organizations. He was raised Catholic in an affluent family, but converted to Islam as a teenager, after he traveled to Yemen and Pakistan to study language and the Koran, and was introduced there to the Taliban. He changed not just his ideas, but also his dress. Neighbors described him as being transformed from "a boy who wore blue jeans and T-shirts to an imposing figure in flowing Muslim garb" (Robertson and Burke 2001: A1).

Just as when people join religious cults, Lindh's case is one of extreme conversion, but conversion can also happen in less extreme situations. People may convert to a different religion, thereby undergoing resocialization by changing beliefs and religious practices. Or someone may become strongly influenced by the beliefs of a social movement and can abruptly or gradually change their beliefs and their identity as a result.

The Brainwashing Debate

Extreme examples of resocialization are seen in the phenomenon popularly called *brainwashing*. In the popular view of brainwashing, converts are completely stripped of their previous identities. The transformation is seen as so complete that only deprogramming can restore the former self. Potential candidates for brainwashing include people who enter religious cults, prisoners of war, and hostages. Sociologists have examined brainwashing to illustrate the process of resocialization. As the result of their research, sociologists caution against using the word *brainwashing* when referring to this form of conversion. The term implies that humans are mere puppets or passive victims whose free will can be taken away during these conversions (Robbins 2001, 1988). In religious cults, however, converts do not necessarily drop their former identity.

Sociological research has found that the people most susceptible to cult influence are those who are the most suggestible—primarily young adults who are socially isolated, drifting, and having difficulty performing in their jobs or in school. Such people may choose to affiliate with cults voluntarily. Despite the widespread belief that people have to be deprogrammed to be freed from the influence of cults, many people are able to leave on their own (Robbins 1988). The so-called brain-washing is simply a manifestation of the social influence people experience through interaction with others (see Chapters 6 and 17). Even in cult settings, socialization is an interactive process, not just a transfer of group expectations to passive victims.

Forcible confinement and physical torture can be instruments of extreme resocialization. Under conditions of severe captivity and deprivation, a captured person may come to identify with the captor. This behavior is known as the *Stockholm Syndrome*. In such instances, the captured person has become dependent on the captor. Upon release, the captive frequently needs debriefing. Prisoners of war and hostages may not lose free will altogether, but they do lose freedom of movement and association, which makes prisoners intensely dependent on their captors and therefore vulnerable to the cap-tor's influence. The Stockholm Syndrome can also help explain why some battered women do not leave their abusers. While dependent on their abuser both financially and emotionally, battered women often develop identities that keep them attached to men who abuse them. In these cases, outsiders often think the women should leave instantly, whereas even in the most abusive situations the women may find leaving difficult.

Resocialization involves establishing a radically new definition of oneself. The new identity may seem dramatically different from the former one, but the process by which it is established is much the same as the ordinary socialization process that is a critical part of society. In sum, socialization is an ongoing process over the life course. As people encounter new situations and new relationships, and interact in familiar ones, they continuously observe social expectations and respond to them. Socialization is an abstract process, but one that is realized in concrete, everyday practices.

Sociology ⊛ Now™

Reviewing is as easy as ❶❷❸.

1. *Before you do your final review, take the SociologyNow diagnostic quiz to help you identify the areas on which you should concentrate. You will find information on SociologyNow and instructions on how to access all of its great resources on the foldout at the beginning of the text.*

2. *As you review, take advantage of SociologyNow's study videos and interactive Map the Stats exercises to help you master the chapter topics.*

3. *When you are finished with your review, take SociologyNow's posttest to confirm you are ready to move on to the next chapter.*

Chapter Summary

What is socialization and why is it important?

Socialization is the process by which human beings learn the social expectations of society. Socialization is powerful because it creates the expectations that are the basis for people's attitudes and behaviors. It also shows that society and the people within society are socially constructed. Through socialization, people conform to social expectations while still expressing themselves as individuals.

What are socialization agents?

Socialization agents are those who pass on social expectations. They include the family, the media, peers, sports, religious institutions, and schools. The *family* is usually the first source of socialization. The *media* influence people's values and behaviors. *Peer groups* are an important source of individual identity. Without peer approval, most people find it hard to be socially accepted. *Religion* is also involved in socialization, shaping many of the beliefs and self-concepts people hold. *Sports* pass on many expectations associated with different social roles. *Schools* pass on expectations that are influenced by gender, race, and other social characteristics of people and groups.

What different theories explain the process of socialization?

Several theoretical perspectives are used to understand socialization: psychoanalytic theory, object relations theory, social learning theory, and symbolic interaction theory. *Psychoanalytic theory* sees the self as driven by unconscious drives and forces that interact with the expectations of society. *Object relations theory* explains the development of the self as the result of individuation and attachment in relationship to parenting figures. *Social learning theory* sees identity as a learned response to social stimuli. *Symbolic interaction theory* sees people as constructing the self as they interact with the environment and give meaning to their experience. Charles Horton Cooley described this process as the *looking-glass self*. Another sociologist, George Herbert Mead, described childhood socialization as occurring in stages: *imitation, play,* and *game.* Functionalism and conflict theory emphasize the broader social context in which socialization occurs, with functionalists noting how this process contributes to social order and conflict theorists seeing it in the context of inequality and power relations.

How does socialization differ for different groups in society?

Socialization influences how different groups are valued and value themselves. Studies of race and self-esteem show that more positive images in the dominant culture do enhance people's definitions of themselves.

Does socialization end after childhood?

Socialization continues throughout life, although childhood is an especially significant time for the formation of identity. *Adolescence* is a period when peer cultures have an enormous influence on the formation of people's self-concepts. *Rites of passage* are ceremonies or rituals that symbolize the passage from one role to another. *Adult socialization* involves the learning of specific expectations associated with new roles.

What is resocialization?

Resocialization is the process by which existing social roles are radically altered or replaced. It can take place in an organization that maintains strict social control and demands that the individual conform to the needs of the group or organization.

Key Terms

adult socialization 102	play stage 96
anticipatory socialization 103	psychoanalytic theory 92
ego 92	resocialization 105
game stage 96	rite of passage 104
generalized other 96	roles 83
id 92	self 95
identity 83	self-esteem 85
imitation stage 96	significant others 96
life course perspective 100	social identity complexity 99
looking-glass self 95	social learning theory 94
object relations theory 93	socialization 83
peers 88	socialization agents 86
personality 83	taking the role of the other 96

Researching Society with MicroCase Online

You can see the results of actual research by using the Wadsworth MicroCase® Online feature available to you. This feature allows you to look at some of the results from national surveys, census data, and some other data sources. You can explore this easy-to-use feature on your own, but try this example. Suppose you want to know:

Do age groups differ in whether they think that women should take care of running their homes and leave running the country up to men?

To answer this question, go to http://sociology.wadsworth.com/andersen_taylor4e/, select MicroCase Online from the left navigation bar, and follow the directions there to analyze the following data.

Data file: GSS

Task: Cross-Tabulation

Row Variable: WOMEN HOME

Column Variable: I-AGE

Questions

Once you have your results, answer the following questions:

1. People in which age group are most likely to agree with this statement?
   a. <30
   b. 30–49
   c. 50 and up

2. What percentage of those under the age of 30 agree with this statement?

3. What factors do you think influence this pattern? Could it be that people change over time? Or are the differences due to generational differences? Be sure to use data from the table to support your argument.

The Companion Website for Sociology: Understanding a Diverse Society, Fourth Edition

http://sociology.wadsworth.com/andersen_taylor4e/

Supplement your review of this chapter by going to the companion website to take one of the Tutorial Quizzes, use the flash cards to master key terms, and check out the many other study aids you'll find there. You'll also find special features such as GSS Data and Census 2000 information, data and resources at your fingertips to help you with that special project or do some research on your own.

Suggested Readings and Web Resources

Arana, Marie. 2001. *American Chica: Two Worlds, One Childhood.* New York: Random House.
This memoir is a moving portrayal of a young woman growing up in two very different cultures: Peru and the prairies of Wyoming. It is a fascinating account of the bi-cultural socialization process, a common experience for immigrants.

Due, Linnea. 1995. *Joining the Tribe: Growing Up Gay & Lesbian in the '90s.* New York: Doubleday.
Written for a popular audience and based on many personal narratives, this book examines the sociological dimensions of growing up lesbian or gay.

Messner, Michael A. 2002. *Taking the Field: Women, Men, and Sports.* Minneapolis, MN: University of Minnesota Press.
Messner's research is an excellent overview of the socializing role of sports, particularly as influenced by gender roles.

Riley, Patricia. 1993. *Growing Up Native American: An Anthology*. New York: Morrow.
This is a collection of personal narratives about the experience of growing up Native American. The collection reveals both the influence of growing up in Indian culture and the diversity of cultures among American Indians.

Schneider, Barbara, and David Stevenson. 1999. *The Ambitious Generation: America's Teenagers, Motivated but Directionless*. New Haven: Yale University Press.
This national study of the values of U.S. youth shows how young people are influenced by the context of work and family. It also reveals the social forces in contemporary society that shape the opportunities and aspirations of young people.

Zhou, Min, and Carl L. Bankston, III. 1998. *Growing Up American: How Vietnamese Children Adapt to Life in the United States*. New York: Russell Sage Foundation.
Based on a study of a Vietnamese community in New Orleans, Zhou and Bankston study the children of refugees from Vietnam. Their analysis shows the influence of immigration and resettlement on the socialization experiences of young people.

Annie B. Casey Foundation
www.aecf.org
This is a foundation dedicated to furthering knowledge about and assisting disadvantaged children in the United States. The Web site includes reports, facts, and state-by-state comparisons of various measures of children's well-being.

Children's Defense Fund
www.childrensdefense.org
An organization dedicated to improving children's lives, especially poor and minority children.

• • •

Social Interaction and Social Structure

A parent and teenager argue about what the teen should wear to a party. A police officer has a man spread-eagled against an automobile while searching him for drugs. A group of businesswomen confer over lunch about a recent sale. A congregation listens to a sermon and then prays together. All these highly diverse actions have something in common: They are all regulated, to a greater or lesser degree, by the groups, statuses, roles, and social institutions of society. These elements guide the formation of human society.

Society is more than the sum of the individuals in it. Society takes on a life of its own. This is one of the most fundamental ideas that guides sociological thinking. In this chapter, we examine the different pieces of society, beginning with the study of social interaction—the groups, statuses, roles, and social bonds that people form through social interaction—and proceeding from this close-up level of society (called "microanalysis") to studying the larger forces that hold society together—social institutions and social structures (called "macroanalysis"). As you proceed through the chapter, you will begin to see the complexity of how diverse societies are arranged and held together. • • •

What Is Society?

In Chapter 3, we studied culture as one force that holds society together. Culture refers to the general way of life, to norms, customs, beliefs, and language. Human **society** is a system of social interaction that includes both culture and social organization. Within a society, members have a common culture even though it may include great diversity. In society, people think of themselves as distinct from other societies, maintain ties of interaction, and have a high degree of interdependence. The interaction that they have, whether based on harmony or conflict, is an ingredient of society. That is, social interaction is how human beings communicate with each other, and in so doing, they form a social bond.

Sociologists use the term **social interaction** to mean behavior between two or more people that is given meaning by them. Social interaction involves more than simply acting; it involves communication, which is the conveyance of information to a person by any means. Social interaction may be simple (a word, wave, or threatening gesture) or complex (speaking, organizing a social movement, or forming a family).

Social interaction is the foundation of society, but society becomes more than a collection of individual social actions. Emile Durkheim, the classical sociological theorist, described society as *sui generis,* which is a Latin phrase meaning "a thing in itself, of its own particular kind." To sociologists, seeing society *sui generis* means that society is more than just the sum of its parts. Durkheim saw society as an organism, something comprised of different parts that work together to create a unique whole. Just as a human body is not only a collection of organs but is alive as a whole organism, society is not only a simple collection of individuals, groups, or institutions, but is a whole entity that consists of all these elements plus their interrelationships.

Imagine how a photographer views a landscape. The landscape is not just the sum of its individual parts—mountains, pastures, trees, or clouds—although each part contributes to the whole. The power and beauty of the landscape is that all its parts *relate* to each other, some in harmony, some in contrast, to create a panoramic view. The photographer who tries to capture this view will likely use a wide-angle lens. This method of photography captures the breadth and comprehensive scope of what the photographer sees. Similarly, sociologists try to picture sociology as a whole by seeing its individual parts, but also recognizing the relatedness of these parts and their vast complexity.

From Groups to Institutions: Microanalysis and Macroanalysis

Like photographers, sociologists use different lenses to see the different parts of society. Some lenses are more microscopic; that is, they focus on the smallest, most immediately visible parts of social life, such as specific people interacting together. This is called **microanalysis.** Other views are more "macroscopic"; that is, like a wide-angle lens, they try to comprehend the whole of society, how it is organized and how it changes. This is called **macroanalysis.** Each view provides a distinct vision of society, and both reveal different dimensions of society.

Some sociologists study the patterns of social interactions that are relatively small and less differentiated—the microlevel of society. Studying a small group, such as your friendship group or your family, is an example. A clique (or subgroup) that forms within your own friendship group is another example. Perhaps you are a member of one or more cliques at this time, or perhaps you are a leader within your group. These types of group behavior would be of interest to someone studying the microlevel of society. Another example of microanalysis is the study of interpersonal attraction. What makes people attracted to someone and not attracted to someone else is a topic discussed later in this chapter.

While some sociologists are interested in the microlevels of society, others are interested in the broadest views of society—how it is organized and how it changes. Sociologists who study this macrolevel of society study the large patterns of social interactions that are vast, complex, and highly differentiated by looking at a whole society, or even by comparing different total societies to each other. The broader framework of social problems in U.S. society, such as poverty, homelessness, and urban crime, are all macrolevel problems.

In this chapter, we continue our study of sociology by starting with the microlevel of social life (by studying groups and face-to-face interaction), then continuing through the macrolevel (by studying total social structures). The idea is to help you see the most important ways that social forces influence people's behaviors, proceeding systematically from small processes to larger processes.

Social organization is the term sociologists use to describe the order established in social groups at any level. Specifically, social organization is the order that brings regularity and predictability to human behavior. Social organization is present at every level of interaction, from the smallest groups to the whole society.

Groups

At any given moment, each of us is a member of many groups simultaneously, and we are subject to their influence: family, friendship groups, athletic teams, work groups, office staffs, racial and ethnic groups, and so on. Groups impinge on every aspect of our lives and are a major determinant of our attitudes and opinions on everything from child-care, politics, and the economy to our view on the death penalty.

To sociologists, a **group** is a collection of individuals who:

1. interact and communicate with each other;
2. share goals and norms; and,
3. possess a subjective awareness of themselves as "we"—that is, as a distinct social unit.

To be a group, the social unit in question must possess all three of these characteristics. We will examine the nature and behavior of groups in greater detail in Chapter 6.

In sociological terms, not all social units are groups. *Social categories* are people who are lumped together based on one or more shared characteristics. Examples of social categories are teenagers (an age category), truck drivers (an occupational category), and millionaires (an economic category). Some social categories can also form a social stratum or even a class, as we shall see in Chapter 9. Ethnic and racial groups may be either social categories or groups, depending upon the amount of "we" feeling. When the "we" feeling is high, racial and ethnic categories are groups.

All people nationwide who are watching a particular television program at 8 o'clock Wednesday evening form another distinct social unit, an *audience*. They are not a group, because they do not interact with one another, nor do they necessarily possess an awareness of themselves as "we." If many of the same viewers come together in a television studio, where they interact and develop a "we" feeling, then they would constitute a group.

Finally, *formal organizations* are highly structured social groupings that form to pursue a set of goals. Bureaucracies, such as business corporations or municipal governments, as well as formal associations, such as the Parent-Teacher Association (PTA), are examples of formal organizations. A deeper analysis of bureaucracies and formal organizations appears in Chapter 6.

Statuses

Within groups, people occupy different statuses. **Status** is an established position in a social structure that carries with it a degree of prestige (that is, social value). A status is a rank in society. For example, the position "vice president of the United States" is a status, one that carries very high prestige. "High school teacher" is another status; it carries less prestige than "vice president of the United States" but more prestige than, say, "cab driver." Statuses occur within institutions. "High school teacher" is a status within the education institution. Other statuses in the same institution are "student," "principal," and "school superintendent."

Typically, an individual occupies many statuses simultaneously. The combination of statuses composes the individual's **status set,** which is the complete set of statuses occupied by a person at a given time (Merton 1968). An individual may occupy different statuses in different institutions. Simultaneously, a person may be a daughter (in the family institution), bank president (in the economic institution), voter (in the political institution), and church member (in the religious institution). Each status may be associated with a different level of prestige.

Sometimes the multiple statuses of an individual are in conflict with one another. **Status inconsistency** exists where the statuses occupied by a person bring with them significantly different amounts of prestige, thus differing expectations. For example, someone trained as a lawyer, but working as a cab driver, experiences status inconsistency. Some immigrants from Vietnam and Korea have experienced status inconsistency. Many refugees who had been in high-status occupations in their home country, such as teachers, doctors, and lawyers, could find work in the United States only as grocers, manicurists, and sales clerks—good jobs, to be sure, but jobs of relatively lower status than the jobs they left behind. This status inconsistency thus results from a *downward social mobility*—a concept to which we return in Chapter 9. A relatively large body of research in sociology has demonstrated that status inconsistency (in *addition* to low status) can lead to stress and depression (Taylor et al. 2003; Blalock 1991; Min 1990; Taylor and Hornung 1979; Hornung 1977; Taylor 1973a; Jackson and Curtis 1968; Jackson and Burke 1965).

Achieved statuses are those attained by virtue of independent effort. Most occupational statuses, such as police officer, pharmacist, or boat builder, are achieved statuses. In contrast, **ascribed statuses** are those occupied from the moment a person is born. Your biological sex is an ascribed status. Yet, even ascribed statuses are not exempt from the process of social construction. For most individuals, race is an ascribed status fixed at birth, although an individual with one light-skinned African American parent and one White parent may appear to everyone to be White and may go through life as a White person. This is called "passing," although this term is used less often now than several years ago. Ascribed status is therefore not always perfectly unambiguous, as in the case of individuals who are *biracial* or *multiracial*. Finally, ascribed statuses can arise through other means, which may be beyond the control of the individual, such as severe disability or chronic illness.

Gender, too, although typically thought of as fixed at birth, is a social construct. You can be born female or male (an ascribed status), but becoming a woman or a man is the result of social behaviors associated with your ascribed status. In this sense, gender is a social construction. People who cross-dress, have a sex change, or develop some of the characteristics associated with the other sex are good examples of how gender is achieved. As we will see later, however, you do not have to see these exceptional behaviors to note how gender is achieved. People "do" gender in everyday life. They put on appearances and behaviors that are associated with their presumed gender (West and Zimmerman 1987; West and Fenstermaker 1995; Andersen 2003). In this respect, gender is not a rigidly fixed individual trait but is at least partly created by the process of social interaction.

The line between achieved and ascribed status may be hard to draw. Social class, for example, is determined by occupation, education, and annual income, all of which are achieved statuses. Yet, job, education, and income are known to correlate strongly with the social class of one's parents. Hence, one's social class status is at least partly, though not perfectly, determined at birth. Social class status is an achieved status that includes an inseparable component of ascribed status as well.

Although people occupy many statuses at one time, the person's **master status** usually is dominant, overriding all other features of the person's identity. The master status may be imposed by others, or a person may define his or her own master status. A woman judge, for example, may carry the master status "woman" in the eyes of many. Thus, she is seen not just as a judge, but as a woman judge, thus making gender a master status (Webster and Hysom 1998). Being in a wheelchair is another example of a master status. People may see this, at least at first, as the most salient part of one's identity, ignoring other statuses that define one as a person.

A master status can completely supplant all other statuses in the person's status set. For example, when a person has acquired immune deficiency syndrome (AIDS), that person's health condition becomes the master status. AIDS becomes the defining criterion for the person's status in society. The person then becomes stigmatized by the master status. Not all master statuses are negative; positive-defining master statuses also exist, such as "hero" or "saint."

Roles

A **role** is the expected behavior associated with a particular status. Thus, a role is a collection of expectations that others have for a person occupying a particular status. Statuses are occupied; roles are acted or "played." The status of police officer carries with it numerous expectations for that officer to uphold the law, pursue suspected criminals, assist victims of crimes, and so on. This is the "role" of police officer. Usually, people behave in their role as others expect them to, but not always. When a police officer commits a crime, such as physically brutalizing someone just arrested, he or she has violated the role expectations. Role

expectations may vary according to the role of the observer—whether the person observing the police officer is a member of a minority group, for example.

As we saw in Chapter 4, social learning theory predicts that one learns attitudes and behaviors in response to the positive reinforcements and encouragement that one receives from those around them. This is important in the formation of one's own identity in society. "I am Linda, the waitress"; or, "I am Barry, the guitarist." These identities are often obtained through **role modeling**, a process by which we imitate the behavior of another person we admire and who is in a particular role. A ten-year-old girl or boy who greatly admires the teenage expert skateboarder next door will attempt, through role modeling, to closely imitate the flourishes and "hot dogging" that neighbor performs on the skateboard. As a result, the formation of the ten-year-old's self-identity is significantly influenced.

Just as an individual may occupy several statuses at any one time, an individual will also typically occupy many roles. A person's **role set** includes all the roles occupied by the person at a given time. Thus, a person may be not only Linda the waitress, but also Linda the pianist, painter, and comedian as well. These roles may clash with each other, a situation called **role conflict**, wherein two or more roles are associated with contradictory expectations. Notice that in Figure 5.1, some of the roles diagrammed for this college student may conflict with each other. Can you speculate about which might and which might not?

In U.S. society, one of the most common forms of role conflict arises from the dual responsibilities of job

In role modeling, a person imitates the behavior of an admired other.

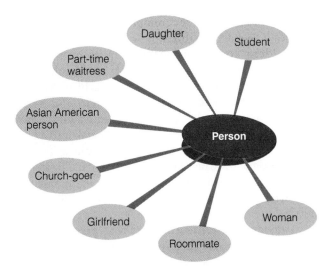

Figure 5.1 Examples of Roles in a College Student's Role Set

and family. The parent role demands extensive time and commitment; so does the job role. Time given to one role is time taken away from the other. Although the norms pertaining to working women and men are rapidly changing, women are more often expected to uphold traditional role expectations and are more likely to be held responsible for minding the family when job and family conflict. The sociologist Arlie Hochschild captured the predicament of today's women when she described the "second shift": A working mother spends time and energy all day on the job, only to come home to the "second shift" of family and home responsibilities. These responsibilities are sometimes delegated to the man of the house, who encounters less well-formed role expectations to take on those responsibilities and who is therefore more likely to leave the jobs undone (Hochschild 1997; Shapiro 1997). Hochschild has further found that the demands of family work, coupled with the demands of jobs, have resulted in a serious time bind for both men and women. She has found that some companies have instituted "family-friendly" policies, designed to reduce the conflicts generated by the "second shift." Ironically, however, in her study, she found that few workers take advantage of such programs as more flexible hours, paid maternity leave, and job sharing—except for the on-site child care that allowed them to work more!

Hochschild's studies point to the conflict between two social roles: family roles and work roles. Her research also illustrates a different sociological concept: **role strain,** a condition that results from a single role that brings conflicting expectations. Different from role conflict, which involves tensions *between* two roles, role strain involves conflicts within a single role. In Hochschild's study, the work role not only has the expectations traditionally associated with work, but also the expectation that one "love" one's work and be as

devoted to it as to one's family. Whether it is role conflict or role strain, the difficulties of managing work and a family have caused some women to "opt out," to use a recently-coined phrase (Barber 2004), meaning leaving one's job and devoting full time to one's family even though one may have obtained advanced education degrees as preparation for the job. Of course, opting out may mean that the woman in question already has sufficient resources, such as wealth, to permit maintaining a certain standard of living after the opting out has taken place. Clearly, many if not most women might find that opting out is not financially feasible and may indeed produce more role strain than not opting out.

The role of student often involves role strain. Students are expected to be independent thinkers, yet often they feel that they are required to simply repeat what a professor tells them. The tension between the two competing expectations is an example of role strain. This is contrasted with the role conflict some students

This excellent photograph shows four aspects of society and social interaction: First, men in a foraging society, hunting tools at the ready; second, socialization (the young boy is being trained in hunting techniques); third, role modeling (the young boy is copying the behavior of the older men); and fourth, the role of gender in social organization (in this society, men are most likely to do the hunting).

encounter when they prepare to pursue their educations away from home. A common experience of first-generation college students, whether Asian, African American, White, Native American, or Latino, is that their families, though proud of their educational achievement and wishing for their success, have the expectation that the student will remain close to home, fulfilling the role of the dutiful child as it has been filled in the family in the past. The student is often even expected to solve complicated family problems that arise when he or she is at college. If the student chooses a school far from home, he or she is liable to experience resistance from the parents. The result is role conflict, which in a sense is a conflict between success and tradition.

Theories About Analyzing Social Interaction

Groups, statuses, and roles form a web of social interaction. The interaction people have with one another is a basic element of society. Sociologists have developed different theories about understanding social interaction. Functional theory, discussed in Chapter 1, is one such theory. Here we detail four others: the social construction of reality, ethnomethodology, impression management, and social exchange. The first three theories come directly from the symbolic interaction perspective.

The Social Construction of Reality

What holds society together? This is a basic question for sociologists, one that, as we will see at the end of this chapter, has long guided sociological thinking. Sociologists note that society cannot hold together without something that is shared—a shared social reality.

Some sociological theorists have argued convincingly that what is shared is, for the most part, socially constructed; namely, there is little reality beyond that which is produced by the process of social interaction itself. This is the principle of *the social construction of reality*: the idea that our perception of what is real is determined by the subjective meaning that we attribute to an experience, a principle that is central to symbolic interaction theory (Berger and Luckmann 1967; Blumer 1969; Jones and Davis 1965; Lamont 2000). In this way, the process of social interaction and the subjective meanings we give to things arising from it determine what we "see" as fact. Hence, there is no objective "reality" in itself. Things do not have their own intrinsic meaning. We subjectively impose meaning on things. (The postmodernist view would argue that objective "reality" exists primarily in our subjective perceptions that exists in the form of communication, culture, text, and other such externalities.)

Children do this routinely—impose meaning on things. Upon seeing a marble roll off a table, the child attributes causation (meaning) to the marble: The marble rolled off the table "because it wanted to." Such perceptions carry into adulthood: The man walking down the street who then accidentally walks smack into a telephone pole at first thought glares at the pole, as though the pole somehow caused the accident! He inadvertently attributes causation and meaning to an inanimate object—the telephone pole (Heider 1958; Taylor et al. 2003).

Considerable evidence exists that people do just that; they will force meaning on something when it serves them to see or perceive what they want to perceive—even if that perception seems to someone else to be contrary to fact. They then come to believe that what they perceived is "fact." A classic and convincing study of this, done by Hastorf and Cantril (1954), was of Princeton and Dartmouth students who watched a film of a basketball game between the two schools. Both sets of students watched the same film. The students were instructed to watch carefully for rule infractions by each team. The results were that the Princeton students reported twice as many rule infractions involving the Dartmouth team than the Dartmouth students saw. The Dartmouth students saw about twice as many rule infractions by Princeton than the Princeton students saw! Remember that they all saw the same film—the same "facts." Each group of students reported what they saw, which was a violation of the rules. We see what "facts" we want to see, as a result of the social construction of reality.

As we saw in Chapter 1, our perceptions of reality are determined by what sociologists call our *definition of the situation*: We observe the context in which we find ourselves and then adjust our attitudes and perceptions accordingly. The sociological theorist W. I. Thomas embodies this idea in his well-known dictum, that *situations defined as real are real in their consequences* (Thomas 1966/1931). The Princeton and Dartmouth students saw different "realities" (different "facts") depending on what college they were attending, and the consequences (the perceived rule infractions) were very real to them.

The definition of the situation is a principle that can affect even so "factual" an event as whether an emergency room patient is perceived to be dead by the doctors there. In his research in the emergency room of a hospital, Sudnow (1967) found that patients who arrived at the emergency room with no discernible heartbeat and no signs of breathing were treated differently by the attending physician depending on the patient's age. A person in his or her early twenties or younger was not immediately pronounced "dead on arrival"

(DOA). Instead, the physicians spent considerable time listening for and testing for a heartbeat, stimulating the heart, examining the patient's eyes, giving oxygen, and administering other stimulations in attempts to revive the patient. If the doctor obtained no lifelike responses, the patient was pronounced dead. Older patients, however, were on the average less likely to receive such extensive procedures. The older person was examined less thoroughly and often pronounced dead on the spot with only a stethoscopic examination of the heart. In such instances, how the physicians defined the situation—how they socially constructed reality—was real in consequence for the patient!

Understanding the social construction of reality helps one see many aspects of society in a new light. Race and gender are significant influences on social experience because people believe them to be so. Society is constructed based on certain assumptions about the significance of race and gender. These assumptions have guided the formation of social institutions, including what work people do, how families are organized, and how power is exercised.

Ethnomethodology

As already discussed in Chapter 3 on culture, our interactions are guided by rules that we follow. These rules are the norms of social interaction. Society cannot hold together without norms, but what rules do we follow? How do we know what these rules or norms are? As we saw earlier, an approach in sociology called ethnomethodology is a clever technique for finding out.

Ethnomethodology (Garfinkel 1967), from "ethno" for "people" and "methodology" for mode of study, is a technique for studying human interaction by deliberately disrupting social norms and observing how individuals attempt to restore normalcy. The idea is that, to study norms, one must first break those norms because the subsequent behavior of the people involved will reveal just what the norms were in the first place.

Ethnomethodology is based on the premise that human interaction takes place within a consensus and, further, that interaction is not possible without this consensus. According to Garfinkel, this consensus will be revealed by people's *background expectancies;* namely, the norms for behavior that they carry with them into situations of interaction. It is presumed that these expectancies are to a great degree shared, and thus studying norms by deliberately violating them will reveal the norms that most people bring with them into interaction. The ethnomethodologist argues that you cannot simply walk up to someone and ask what norms he or she has and uses because most people will not be able to articulate what they are. We are not wholly conscious of what norms we use even though they are shared. Ethnomethodology is designed to "uncover" those norms.

Ethnomethodologists often use ingenious procedures for assessing norms by thinking up clever ways to interrupt "normal" interaction. William Gamson, a sociology professor, had one of his students go into a grocery store where jelly beans, normally priced at that time at 49 cents per pound, were on sale for 35 cents. The student engaged the saleswoman in conversation about the various candies and then asked for a pound of jelly beans. The saleswoman then wrapped them and asked for 35 cents. The rest of the conversation went like this:

Student: Oh, only 35 cents for all those nice jellybeans? There are so many of them. I think I will pay $1.00 for them.

Saleswoman: Yes, there are a lot, and today they are on sale for only 35 cents.

Student: I know they are on sale, but I want to pay $1.00 for them. I just love jellybeans, and they are worth a lot to me.

Saleswoman: Well, uh, no, you see, they are selling for 35 cents today, and you wanted a pound, and they are 35 cents a pound.

Student (voice rising): I am perfectly capable of seeing that they are on sale at 35 cents a pound. That has nothing to do with it. It is just that I personally feel that they are worth more, and I want to pay more for them.

Saleswoman (becoming angry): What is the matter with you? Are you crazy or something? Everything in this store is priced more than what it is worth. Those jelly beans probably cost the store only a nickel. Now do you want them or should I put them back?

At this point, the student became embarrassed, paid the 35 cents, and hurriedly left (Gamson and Modigliani 1974).

The point here is that the saleswoman approached the situation with a presumed consensus, a consensus that becomes revealed by its deliberate violation by the student. The puzzled saleswoman took measures to attempt to normalize the interaction, to even force it to be normal. By so doing, she revealed her expectations, that is, her norms. The ethnomethodological technique used by the student reveals what norms people unconsciously use in everyday interaction and conversation. That is the purpose of ethnomethodology.

Impression Management and Dramaturgy

Another way of analyzing social interaction is to study impression management, a term coined by symbolic interaction theorist Erving Goffman (1959), introduced

Impression management is a technique we all use to manipulate others' perceptions of us. What impressions is this person "giving off"? As a consequence, what perceptions would you have about this person if the two of you were to meet? What impressions might this person be giving in terms of gender? age? social status? politics?

in Chapter 1. **Impression management** is a process by which people control how others will perceive them. A student handing in a term paper late may wish to give the instructor the impression that the student was

not at fault, but that uncontrollable circumstances arose ("my computer hard drive crashed," "my dog ate the last hard copy," and so on). The impression that one wishes to "give off" (to use Goffman's phrase) is that "I am usually a very diligent person, but today—just today—I have been betrayed by circumstances."

Impression management can be seen as a type of *con game*. A person willfully attempts to manipulate the other's impression of himself or herself. Goffman regarded everyday interaction as a series of attempts to "con" the other. Trying in various ways to con the other is, according to Goffman, at the center of much social interaction and social organization in society: Social interaction is just a big con game.

Perhaps this cynical view is not true of all social interaction, but we do in fact present different "selves" to others in different settings. The settings are, in effect, different stages upon which we act as we relate to others. For this reason, Goffman's theory is sometimes called the *dramaturgy model* of social interaction. This is a way of analyzing interaction that assumes the participants are actors on a stage in the drama of everyday social life. People present different faces (give off different impressions) on different stages (in different situations or different roles) with different others. To your mother, you present yourself as the dutiful, obedient daughter, which may not be how you present yourself to a friend. Perhaps you think acting like a diligent student makes you seem like a jerk, so you hide from your friends that you are interested in a class or enjoy your homework. Analyzing impression management

DOING SOCIOLOGICAL RESEARCH

"Doing Hair, Doing Class"

When you begin to study social interaction, you will see that you can study it in many places, including places you would not ordinarily think of as locations for sociological research. Debra Gimlin did this when she began to observe the interaction that takes place in hair salons—interaction that, at least in some salons, she noticed, is often marked by differences in the social class status between clients and stylists.

Her research question was, How do women attempt to cultivate cultural ideals and beauty, and in particular, how is this achieved through the interaction between hair stylists and their clients? She did her research by spending more than 200 hours observing social interaction in a hair salon. She

watched the interaction between clients and stylists, and she conducted interviews with the owner, the staff, and twenty women customers. During the course of her fieldwork, she recorded her observations of the conversations and interaction in the salon, frequently asking questions of patrons and staff. In the salon she studied, the patrons were mostly middle and upper middle class, the stylists, working class; all the stylists were White, as were most of the clients.

"Beauty work," as Gimlin calls it, involves the stylist bridging the gap between those who seek beauty and those who define it; her (or his) role is to be the expert in beauty culture, bringing the latest fashion and technique to

clients. Beauticians are also expected to engage in some "emotion work"; that is, they are supposed to nurture clients and be interested in their lives. They are often put in the position of having to sacrifice their professional expertise to meet clients' wishes.

According to Gimlin, because stylists are typically of lower class status than clients, this introduces an element into the relationship between stylists and clients that stylists negotiate carefully in their routine social interaction. Hairdressers emphasize their special knowledge of beauty and taste as a way of reducing the status differences between them and their clients. They also try to nullify the existing class hierarchy by conceiving an alternative hier-

reveals that we try to con the other into perceiving us as we want to be perceived. The box "Doing Sociological Research" shows how impression management can be involved in many settings, including the everyday world of the hair salon.

A clever study by Albas and Albas (1988) demonstrates just how pervasive impression management is in social interaction. The Albases studied how students interacted with one another when the instructor returned graded papers during class. Some students got good grades ("aces"), others got poor grades ("bombers"), but both employed a variety of devices (cons) to maintain or give off a favorable impression. The aces wanted to show off their grades, but they did not want to appear to be braggarts, so they casually or "accidentally" let others see their papers. In contrast, the bombers hid or covered their papers to hide their poor grades, said they "didn't care" what they got, or simply lied about their grades. Analysis of impression management shows how subject to the influence of others we all are. Although we may protest, claiming we are all "individuals," our individuality is shaped by the numerous social forces we encounter in society.

One thing that Goffman's theory makes clear is that social interaction is a very perilous undertaking. Have you ever been embarrassed? Of course, you have; we all have. Think of a really big embarrassment that you experienced. Goffman defines embarrassment as a spontaneous reaction to a sudden or transitory challenge to our identity: We attempt to restore a prior perception by others of our self. Perhaps you were giving a talk before a class and then suddenly forgot the rest of the talk. Or, perhaps you recently bent over and split your pants. Or perhaps you are a man and barged accidentally into a women's bathroom. All these actions will result in embarrassment, causing you to "lose face."

You will then attempt to *restore face*, that is, eliminate the conditions causing the embarrassment. You thus will attempt to con others into perceiving you as they might have prior to the embarrassing incident. One way to do this is to shift blame from the self to some other, for example, claiming in the first example that the teacher did not give you time to adequately memorize the talk; or in the second example, you claim that you will never buy that particular, obviously inferior brand of pants again; or in the third example, that the sign saying "Women's Room" was not clearly visible. All these represent deliberate manipulations (cons) to save face on your part—to restore the other's prior perception of you.

Social Exchange and Game Theory

Another way of analyzing social interaction is through the social exchange model. The *social exchange model* of social interaction holds that our interactions are determined by the rewards and punishments that we receive from others (Thibaut and Kelly 1959; Homans 1974; Blau 1986; Cook et al. 1988; Levine and Moreland 1998; Taylor et al. 2003). A fundamental principle of exchange theory is that an interaction that elicits

---

archy, not one based on education, income, or occupation (as the usual class hierarchy is), but on the ability to style hair competently. Thus, stylists describe clients as perhaps "having a ton of money" but unable to do their hair or know what looks best on them. Stylists also try to nullify status differences by appearing to create personal relationships with their clients, even though they never see them outside the salon. Stylists become confidantes with clients, often telling them highly personal information about their lives. Gimlin concludes that beauty ideals are shaped in this society by an awareness of social location and cultural distinctions. As she says, "Beauty is . . . one tool women use as they make

claims to particular social statuses" (1996: 525).

The next time you get your hair cut, you might observe the social interaction around you and ask how class, gender, and race shape interaction in the salon or barbershop that you use. Try to get someone in class to corroborate with you so that you can compare observations in different salon settings. In doing so, you will be studying how gender, race, and class shape social interaction in everyday life.

Questions to Consider

1. Would you expect the same dynamic in a salon where men are the clients and the stylists? *Keywords: social interaction, gender interaction*

2. Do Gimlin's findings hold in settings where the customers and stylists are not White or where they are all working class? *Keywords: social class, social status*

3. In your opinion, would Gimlin's findings hold in an African American *men's* barbershop? *Keywords: racial interaction, barbershop*

We have included InfoTrac College Edition keywords at the end of each question to make it easier for you to find more to read on these topics. Go to **www.infotrac-college.com,** an online library, to begin your search.

Source: Gimlin, Debra. 1996. "Pamela's Place: Power and Negotiation in the Hair Salon." *Gender & Society* 10 (October): 505–526.

•••

approval from another (a type of reward) is more likely to be repeated than an interaction that incites disapproval (a type of punishment). According to the exchange principle, one can predict whether a given interaction is likely to be repeated or continued by calculating the degree of reward or punishment inspired by the interaction.

It is often informative to assess social interaction in terms of a difference between reward and punishment. If the reward for an interaction exceeds the punishment for it, then a potential for *social profit* exists and the interaction is likely to occur or continue. If the reward is less than the punishment, then the action will produce a social loss (negative profit) and will be less likely to occur or continue. Exchange theorists analyze human interaction in terms of concepts such as reward, punishment, profit, and loss, in addition to the ratio of inputs to outputs. People want to get from an interaction at least as much as they put into it, resulting in an *equitable* interaction. Calculations of inputs versus outputs constitute a measure of social costs versus rewards. A reward is in effect an approval for conformity; a punishment is a sanction against deviance. Social exchange thus tends to encourage conformity and discourage deviance. In this way, it acts as a force toward cohesion in everyday interaction in society—as a way of "holding society together."

Rewards can take many forms. They can include tangible gains, such as gifts, recognition, and money; or subtle everyday rewards, such as smiles, nods, and pats on the back. Similarly, punishments come in many varieties, from extremes such as public humiliation, beating, banishment, or execution to gestures as subtle as a raised eyebrow or a frown. For example, if you ask someone out for a date and the person says, "Yes," you have gained a reward, and you are likely to repeat the interaction. You are likely to ask the person out again. If you ask someone out, and he or she glares at you and says, "No way!" then you have elicited a punishment that will probably cause you to shy away from repeating this type of interaction with that person.

Social exchange theory has grown partly out of *game theory,* a mathematical and economic theory that predicts human interaction has the characteristics of a "game," namely, strategies, winners and losers, rewards and punishments, and profits and costs (VonNeumann and Morgenstern 1944; Nash 1951; Dixit and Sneath 1997; Kuhn and Nasar 2002). Simply asking someone out for a date indeed has a game-like aspect to it, and you will probably use some kind of strategy to "win" (have the other agree to go out with you) and get "rewarded" (have a pleasant or fun time) at minimal "cost" to you (you don't want to spend a large amount of money on the date; or, you don't want to get into an unpleasant argument on the date). The interesting thing about game theory is that it sees human interaction as

just that: a game. Goffman's impression-management theory also contains a game-like element in its hypothesis that human interaction is a big con game. The mathematician John Nash is one of the inventors of game theory and was featured in the movie, *A Beautiful Mind.*

Exchange theory has been used to probe the perpetuation of racist and sexist attitudes. Research has shown that antiwoman attitudes and antiwoman stereotypes among men persist much longer if they are constantly rewarded by one's social group or are reinforced by other aspects of one's social milieu (Taylor et al. 2003; Levine and Moreland 1998; Cook et al. 1988). In contrast, when such attitudes and stereotypes are punished in social exchange, they are unlikely to persist. If a man exits high school with a few antiwoman attitudes and enters a college that has a strong culture of feminist sophistication, evidence of casual sexism in his interactions is likely to earn quick scorn among disapproving classmates.

Interaction in Cyberspace

When people interact and communicate with one another by personal computers through email, chat rooms, computer bulletin boards, virtual communities, and other computer-to-computer interactions, they are engaging in **cyberspace interaction** (or virtual interaction). In language that has rapidly become popular only within the last few years, one creates a virtual reality as a result of such activity. Virtual reality is the computer control of human perception. What the individual perceives to be real is what is created by and on the computer (Waldhams 2003; Strate et al. 2003; Gross et al. 2002; Kraut et al. 2002; Holmes 1997; Jones 1997; Turkle 1995).

Cyberspace interaction has characteristics that distinguish it from ordinary face-to-face interaction. First, certain kinds of nonverbal communication are eliminated. Thus, when interacting with someone via a chat room, email, or another computer channel, facial expressions such as smiles, laughter, and the raised eyebrow are eliminated. Also eliminated are hand motions, voice pitch, and other nonverbal modes of communication. Second, one is free to become a different self, by creating a new identity or several new identities. One can become a cowboy, a cowgirl, a dragon, or even a sex partner in the realm called *cybersex.* A person can even instantly "become" another gender. Men can interact as (virtual) women, and women as men. While during the 1990s women were significantly less likely to use the Internet for any reason, this gender gap had disappeared by 2000 (Ono and Zavodny 2003). A person with a disability may not feel disabled when interacting

"On the Internet, nobody knows you're a dog."

on the Internet (Smolan and Erwitt 1996). In creating this new virtual self, interaction becomes anonymous, and one's secrets are hidden from the other, including one's appearance, dress, and personal data such as social security number, city and state address, and other characteristics that would otherwise identify one's actual (as opposed to virtual) self, as long as one takes security precautions not to disclose such information.

In this respect, cyberspace interaction is the application of Goffman's principle of *impression management.* The person can put forward a totally different and wholly created self, or identity. One can "give off," in Goffman's terms, any impression that one wishes to and, at the same time, know that one's true self is protected by anonymity. This gives the individual a large and free range of roles and identities from which to choose and that one can become. As predicted by symbolic interaction theory, of which Goffman's is one variety, *the reality of the situation grows out of the interaction process itself.* What one becomes during chat room interaction is a direct outgrowth of the chat room interaction; it is the interaction process that produces the reality. This is a central point of symbolic interaction theory: Interaction creates reality.

Whether cyberspace interaction is ultimately beneficial or harmful to the person has become the subject of much research. Some studies have noted that people can develop extremely close and in depth relationships with each other as a result of their interaction in cyberspace (McKenna 2002; Jones 1997). Another has noted that individuals who have engaged in virtual interaction with each other sometimes establish some other form of contact, such as communication by telephone or by post office mail, and sometimes, they meet face to

face. Yet such a subsequent face-to-face meeting often prompts an end to the relationship (Strate et al. 2003)! Also, many marriages have broken up as a result of cyberspace interaction or even cybersex on the part of a spouse (Fernandes-y-Freitas 1996). Finally, while Internet pornography comes in both soft and hard forms, recent research has shown the persistence of violent forms of pornography, such as violent rape of women by male perpetrators. One such investigation showed at least 31 free Internet sites containing instances of violent rape (Gossett and Byrne 2002).

One researcher who specializes in cyberspace interaction (John Suler, cited in Waldhams 2003) notes that a cyberspace interaction game, where thousands of people interact electronically, can encourage deviant and anti-social behavior. The anonymity offered by cyberspace allows a person to engage in behavior he or she is too afraid to do in real life. Called "griefers," they can pretend to be Mafia figures and harass others by actually hacking into their account, stealing information, and spreading unpleasant rumors about real people through instant messaging. In one case, such rumors were actually circulated about a participant, and the participant suffered harm at work as a result. In such instances, the line between cyberspace and the real world becomes blurred, and the consequences can be unpleasant.

Cyberspace interaction has resulted in a new subculture in our society (often called *cyberculture,* or *virtual culture*)—a totally new social order, including a new social structure as well as a culture (Strate et al. 2003; Turkle 1995). Since the mid-1980s, there has been a transition from the use of large mainframe computers as machines only to perform mathematical and statistical calculations (what Turkle calls the *culture of calculation*) to the use of more user-friendly personal computers (PCs) to perform calculations and provide experiences such as interaction in cyberspace (the *culture of simulation*). This new cyberculture is a true subculture, as defined sociologically (in Chapter 3). It has certain rules or norms, its own language, a set of beliefs, and practices or rituals—in short, all the elements of a culture.

Forms of Nonverbal Communication

Everyday social interaction involves both verbal and nonverbal communication. *Verbal communication* consists of spoken and written language, and it includes a conversation with the person next to you, an exchange of letters, or a telephone conversation between you and a friend overseas. When communication takes place, interaction usually does too, although not always. A television commercial is an example of verbal communication without interaction. The communication is one-way, from the advertiser to you, with no communication from you to the advertiser, unless you write or call them, in which case interaction has occurred.

Nonverbal communication is conveyed by nonverbal means such as touch, tone of voice, and gestures. A punch in the nose is nonverbal communication; so is a knowing glance. A surprisingly large portion of our everyday communication with others is nonverbal, although we are generally only conscious of a small fraction of the nonverbal "conversations" in which we take part. Consider all the nonverbal signals that are exchanged in a casual chat: body position, head nods, eye contact, facial expressions, touching, and so on.

Studies of nonverbal communication, like verbal communication, show it to be much influenced by social forces, including the relationships between diverse groups of people. The meanings of nonverbal communications depend heavily upon race, ethnicity, social class, and particularly, gender, as we shall see. In a society as diverse as the United States, understanding how diversity shapes communication is an essential part of understanding human behavior. Sociologists, psychologists, social psychologists, anthropologists, and linguists have classified nonverbal communication into the following four categories: touch, paralinguistic, body language, and use of personal space, or "proxemics" (Gilbert et al. 1998; Taylor et al. 2003; Ekman 1982; Argyle 1975).

Touch

Touching, also called *tactile communication,* involves any conveyance of meaning through touch. It may involve negative communication (hitting, pushing) as well as positive (shaking hands, embracing, kissing). These actions are defined as positive or negative by the ethnic cultural context. An action that is positive in one culture can be negative in another. Shaking the right hand in greeting is a positive tactile act in the United States, but the same action in East India or certain Arab countries would be an insult. A kiss on the lips is a positive act in most cultures, yet if you were kissed on the lips by a stranger, you would probably consider it a negative act, perhaps even repulsive. The vocabulary of tac-

UNDERSTANDING DIVERSITY

Interaction on the Street

Sociologist Eli Anderson's book *Streetwise* (1990) notes how people use space to interact with strangers in the street. Even the deceptively simple decision to pass a stranger on the street involves a set of mental calculations. Is it day or night? Are there other people around? Is the stranger a child, a woman, a White man, a teenager, or a Black man? Each participant's actions must be matched to the actions and cues of the other. The following field note of Anderson's illustrates how well tuned strangers can be to each other and how capable of subtle gestural communication:

It is about 11:00 on a cold December morning after a snowfall. Outside, the only sound is the scrape of an elderly white woman's snow shovel on the oil-soaked ice of her front walk. Her house is on a corner in the residential heart of the Village, at an intersection that stands deserted between morning and afternoon rush hours. A truck pulls up directly across from the old lady's house. Before long the silence is split by the buzz of two tree sur- geons' gasoline-powered saws. She leans on her shovel, watches for a while, then turns and goes inside. A middle-aged white man in a beige overcoat approaches the site. His collar is turned up against the cold, his chin buried within, and he wears a Russian-style fur-trimmed hat. His hands are sunk in his coat pockets. In his hard-soled shoes he hurries along this east–west street approaching the intersection, slipping a bit, having to watch each step on the icy sidewalk. He crosses the north–south street and continues westward.

A young black male, dressed in a way many Villagers call "street-ish" (white high-top sneakers with loose laces, tongues flopping out from under creased gaberdine slacks, which drag and soak up oily water; navy blue "air force" parka trimmed with matted fake fur, hood up, arms dangling at the sides) is walking up ahead on the same side of the street. He turns around briefly to check who is coming up behind him. The white man keeps his eye on the treacherous sidewalk, brow furrowed, displaying a look of concern and determination. The young black man moves with a certain aplomb, walking rather slowly.

From the two men's different paces it is obvious to both that either the young black man must speed up, the older white man must slow down, or they must pass on the otherwise deserted sidewalk.

The young black man slows up ever so slightly and shifts to the outside edge of the sidewalk. The white man takes the cue and drifts to the right while continuing his forward motion. Thus in five or six steps (and with no obvious lateral motion that might be construed as avoidance), he maximizes the lateral distance between himself and the man he must pass. What a minute ago appeared to be a single-file formation, with the white man ten steps behind, has suddenly become side-by-side, and yet neither participant ever appeared to step sideways at all.

Source: Anderson, Eli. 1990. *Streetwise: Race, Class and Change in an Urban Community.* Chicago, IL: University of Chicago Press, pp. 217–218. •••

tile communication changes with social and cultural context.

Patterns of tactile communication are strongly influenced not only by culture, but also by gender. Parents vary their touching behavior depending upon whether the child is a boy or a girl. Boys tend to be touched more roughly; girls, more tenderly and protectively. This pattern continues into adulthood, where women touch each other more often than do men in everyday conversation.

Touching appears to have different meanings to women and men. Women are on the average more likely to touch and hug as an expression of emotional support, whereas men touch and hug more often to assert power or to express sexual interest (Worchel et al. 2000; Wood 1994). Clearly, there are also instances where women touch to express sexual interest and also dominance, but research shows that in general with some exceptions, for women, touching is a supporting activity; for men, it is a dominance-asserting activity, although the touching and hugging that men do in athletic competition is a supporting activity (Worchel et al. 2000; Wood 1994; Tannen 1990).

Professors, male or female, may pat a man or woman student on the back as a gesture of approval; students will rarely do this to a professor. This shows the effect of status differences. Male professors touch students more often than do female professors. This shows the additional effect of gender. Men often approach women from behind and let their hands rest upon the woman's shoulders; women are less likely to do the same to men. Male doctors touch female nurses more often than nurses touch doctors. Because doctors are disproportionately male and nurses disproportionately female in our society (and most others), this illustrates the combined influences of gender and status (Gilbert et al. 1998; Tannen 1990).

As more women are promoted to high-status positions, we may see changes in how often men and women of higher and lower rank touch each other in the workplace. In the context of a power relationship (that is, one where one person by virtue of his or her status has power over another), these learned patterns can constitute sexual harassment, which is discussed further in Chapter 18.

Paralinguistic Communication

Paralinguistic communication is the component of communication that is conveyed by the pitch and loudness of the speaker's voice, its rhythm, emphasis, and frequency, and the frequency and length of hesitations. In other words, it is not what you say, but how you say it. A baby's cry communicates effectively to its parent. Its pitch, loudness, and frequency convey specific meanings such as hunger, anger, or discomfort to experienced parents. The baby is communicating paralinguistically.

The exact meaning of paralanguage, like that of tactile communication, varies with the ethnic and cultural context. For some people under some circumstances, a pause may communicate emphasis; for others, it may indicate uncertainty. A high-pitched voice may mean a person is excited or that the person is lying. During interactions between Japanese businessmen, long periods of silence often occur. Unlike U.S. citizens, who are experts in "small talk," and who try at all costs to avoid periods of silence in conversation, Japanese people do not need to talk all the time and regard periods of silence as desirable opportunities for collecting their thoughts (Worchel 2000; Fukuda 1994). Unprepared U.S. businesspeople in their first meetings with Japanese executives often think, erroneously, that these silent interludes mean the Japanese are responding negatively to a presentation. More traveled Western executives learn to master the art of paralinguistic communication as practiced in Japan. Even though some find the Japanese mode of conversation highly uncomfortable, getting used to it is a key tool in negotiating successfully with the Japanese. The fate of a deal may depend on a glance, an exhalation, or a smile. Americans often consider paralinguistics to be a minor aspect of the conversation, with much greater attention paid to the verbal transaction; the Japanese, conversely, often consider the paralinguistics to be more important than the verbal communication as a source of information (Mizutami 1990).

People often reveal their true feelings and emotions by paralinguistic slips. Emotions tend to "leak out" even if a person tries to conceal them, hence the term *nonverbal leakage* (Ekman and Friesen 1974; Ekman 1982; Ekman et al. 1988; Gilbert et al. 1998; Taylor et al. 2003; Conniff 2004). People who are lying often betray themselves through paralinguistic expressions of anxiety, tension, and nervousness. Research has shown that when a person is lying, the pitch of his or her voice is higher than when the same person is telling the truth. The difference is small and hard to detect simply by listening, but electronic analysis of vocal pitch can reveal lying with considerable, though not perfect, accuracy.

Body Language

Body language, or more technically *kinesic communication*, involves gestures, facial expressions, body position, and the like. Waving hands, crossed arms, and raised eyebrows all transmit meanings. American Sign Language (ASL), a mode of communication for the hearing-impaired by means of hand signals, uses the principle of body language. Add hundreds of facial expressions and the infinite nuances of eye contact, and body language forms a crucial part of nonverbal communication. Facial expressions are gender-related and reflect dominance patterns in society. White middle-

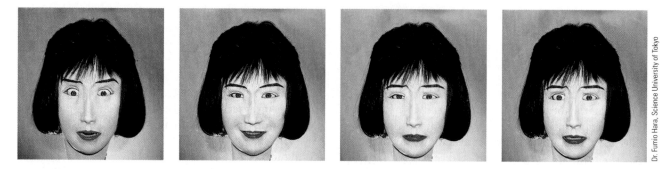

Dr. Fumio Hara, Science University of Tokyo

Facial expressions, a form of nonverbal communication, can be created mechanically by computer, as shown in these photographs. Can you guess what emotions are being conveyed by each of the four different faces?

class women have been socialized to smile often and to do so even if they are not happy. Men are taught not to do this and to generally avoid facial expressions that convey emotion.

Meanings conveyed by body language are usually different in different cultures and ethnic subcultures. Mexicans and Mexican Americans may display the right hand held up, palm inward, all fingers extended, as an obscene gesture meaning "screw you many times over." This provocative gesture has no meaning at all in Anglo (White) society, except insofar as Mexican Amer-

Researcher Paul Ekman (1982; Conniff 2004) has demonstrated that these facial expressions of New Guinea tribesmen are almost identical to those given by college students in the United States to the same four situations depicting anger, happiness, sadness, or disgust. This suggests that some emotions are conveyed by facial expressions in the same way across different cultures, not in different ways. Can you identify which expression goes with which emotion in these four photographs?

© Paul Ekman, from the Nebraska Symposium on Motivation, 1972. Courtesy of the Human Interaction Laboratory, UCSF

ican traditions have been adopted by Anglos—a type of cultural diffusion.

Ignorance of the meanings that gestures have in a society can get you in trouble. Eye contact, a category of body language, is especially prone to misinterpretation. People who grow up in urban environments learn to avoid eye contact on the streets. Staring at someone for only two or three seconds can be interpreted as a hostile act, if done man to man (Anderson 1999, 1990). If a woman maintains mutual eye contact with a strange man for more than merely two or three seconds, she may be assumed by the man to be sexually interested in him. However, in sustained conversation with nonstrangers, women maintain mutual eye contact for longer periods than do men (Romain 1999; Gilbert et al. 1998; Wood 1994).

Conversely, some gestures retain the same meaning across different cultures, ethnic groups, and societies. The hand gestures for "stop," "come here," "go away," and "good-bye" are the same in the United States across racial, socioeconomic, and other group subcultures, and they are the same in other societies as well. The obscene gesture of an extended middle finger, palm inward ("giving the finger," or "flipping the bird") is now virtually universal across all ethnic groups, races, social classes, and both genders both inside and outside the United States. The U.S. gestures for "OK," "shame on you," and "crazy" are also widespread both inside and outside U.S. society (Gilbert et al. 1998). Certain facial expressions are culturally universal, too. The facial expressions for anger, happiness, sadness, and even disgust appear to be recognized in all cultures, from the tribal peoples of New Guinea to the White middle class of the United States. These expressions appear to be used equally by both men and women in these cultures (Conniff 2004; Gilbert et al. 1998; Ekman 1982).

Use of Personal Space

Meaning is conveyed by the amount of space between interacting individuals; this is referred to as **proxemic**

communication. As with much nonverbal communication, the persons involved are generally not conscious of the proxemic messages they are sending. Generally, the more friendly a person feels toward another, the closer he or she will stand. In casual conversation, friends stand closer to each other than strangers. People who are sexually attracted to each other stand especially close, whether the sexual attraction is gay, lesbian, or heterosexual (Taylor et al. 2003).

According to anthropologist E. T. Hall (1966; Hall and Hall 1987), we all carry around us a *proxemic bubble* that represents our personal three-dimensional space. When people we do not know enter our proxemic bubble, we feel threatened and may take evasive action. Friends stand close; enemies tend to avoid interaction and keep far apart. According to Hall's theory, we attempt to exclude from our private space those we do not know or do not like even though we may not be fully aware that we are doing so.

Proxemic interaction varies strongly by cultural differences and by gender. This can be clearly seen when people in different racial and ethnic groups interact. The proxemic bubbles of different groups have different sizes. Hispanic people tend to stand much closer to each other than White middle-class Americans; their proxemic bubble is on the average smaller. Similarly, African Americans also tend to stand close to each other while conversing. Interaction distance is quite large between White middle-class British males—their average interaction distances can be as much as several feet.

This presumably homeless man carries his belongings with him, conspicuously displayed to establish his own particular proxemic bubble—*that area around him representing his own private space, which goes where he goes.*

Black women is very close. Hispanic women also stand close. White men stand fairly far apart, especially older White men, showing the relevance of race, gender, and age. When a Middle Eastern man engages in conversation with a White middle-class U.S. man, the Middle Eastern man tends to move toward the White American, who tends to back away. You can observe the negotiations of proxemic space at cocktail parties or any other setting that involves casual interaction.

Interpersonal Attraction and the Formation of Pairs

Perhaps the most interesting interactions between humans are pairings, including friendships, romances, and sexual pairings. Pairings constitute one of the most fundamental processes in society and go a long way in answering the question, What holds society together? How do pairs form? Are pairings influenced by social structure, or are they as random as chance encounters? You

---

THINKING SOCIOLOGICALLY

Try the following experiment yourself. Enter an empty elevator and go to a far (back) corner. Wait for another person to enter the elevator. If proxemic communication theory is correct, this person will tend to go to a diagonally opposite front corner of the elevator. (This is an illustration of the *proxemic bubble* principle.) Next, instead of staying put in your corner, take a few small steps toward the other back corner. (You will have to get your nerve up a bit to do this!) What does the other person now do? If the theory of proxemics is correct, he or she will take some sort of evasive action, such as move toward the opposite front corner or toward the door. If the person does not move at all, then you should move even closer and observe what the other person does. Repeat the experiment several times, with different people, with people whose gender is different from yours and whose race is different from yours. See what happens.

Proxemic interactions also differ between men and women (Taylor et al. 2003; Romain 1999; Tannen 1990). Women of the same race and culture tend to stand closer to each other in casual conversation than do men of the same race and culture. The space between

will not be surprised to learn that formation of pairs has a strong social structural component; that is, it is patterned and predicted by social forces. It is not a random process, as many might think.

Humans have a powerful desire to be with other human beings; in other words, they have a strong need for *affiliation*. We tend to spend about 75 percent of our time with other people when doing all sorts of activities such as eating, watching television, studying, doing hobbies, working, and so on (Cassidy and Shaver 1999). Overall, women reveal this affiliative tendency somewhat more than men (Basow 1992). People who lack all human contact are rare in the general population, and their isolation is usually rooted in psychotic or schizophrenic disorders. Extreme social isolation at an early age, particularly forced long-term isolation, causes severe disruption of mental and emotional development, as we saw earlier in Chapter 4.

The affiliation tendency has been likened to *imprinting*, a phenomenon seen in newborn or newly hatched animals, who attach themselves to the first living creature they encounter, even if it is of another species (Lorenz 1966). Studies of ducks and squirrels show that once the young animal attaches itself to a human experimenter, the process is irreversible. The young animal seeks the company of the human rather than the company of its own species. A degree of imprinting may be discernible in human infant attachment, but researchers note that in humans the process is more complex, more changeable, and more influenced by social factors (Brown 1986).

Konrad Lorenz, the animal behaviorist, shows that adult ducks that have imprinted on him the moment they were hatched will follow him anywhere, as though he were their mother duck.

Somewhat similar to affiliation is interpersonal attraction, a nonspecific positive response toward another person. Attraction is a factor of ordinary day-to-day interaction and varies from mild attraction (such as thinking your grocer is a "nice person") all the way to deep feelings of love. According to one view, attractions fall on a single continuum ranging from hate to strong dislike to mild dislike to mild liking to strong liking to love. Another view is that attraction and love are two different continua, able to exist separately. By this view, you can like someone a whole lot but not be in love. By the same token, you can feel passionate love for someone (with its associated strong sexual feelings and intense emotion) yet not "like" the person. Have you ever been in love with someone you did not particularly like?

DEBUNKING *SOCIETY'S MYTHS*

Myth: Love is purely an emotional experience that you cannot predict or control.

Sociological perspective: Whom you fall in love with can be predicted beyond chance by such factors as proximity, how often you see the person, how attractive you perceive him or her to be, and whether you are similar (not different) to him or her in social class, race–ethnicity, religion, age, educational aspirations, and general attitudes, including political attitudes and beliefs.

Can attraction be scientifically predicted? Can persons with whom you are most likely to fall in love be identified? The surprising answer to these questions is a loud, although somewhat qualified, yes. Most of us have been raised to believe that love is impossible to measure and certainly impossible to predict scientifically. We think of love, especially romantic love, as ephemeral, mysterious—a lightning bolt. Countless novels and stories support this view, but extensive research in sociology and social psychology suggests otherwise. In a probabilistic sense, love can be predicted beyond the level of pure chance. Let us take a look at some of these intriguing findings.

Proximity

A strong determinant of your attraction toward others is simply whether you live near them, work next to them, or have frequent contact with them. You are more likely to form friendships with people from your own city than with people from a thousand miles away. As was originally shown in a classic study by Festinger et al. (1950), you are more likely to be attracted to someone on your floor, your residence hall, or your apartment building than to someone even two floors down or two streets over.

© Nina Leen/Time Life Pictures/Getty Images

Such is the effect of proximity in the formation of human friendships.

Segal (1974) demonstrated this effect in a study of recruits at a police academy. Seating for classes and seminars at the academy was alphabetical, with recruits sitting beside the person next in the alphabet. Segal found that when the police officers were listed alphabetically, at the end of their training, the list also effectively captured friendship pairings. The major determinant for choosing friends was relative position in the alphabet, because that was who the recruit sat next to in class. Segal found that proximity had a stronger effect than all other factors, including race, socioeconomic background, age, religion, and nationality, although these other factors also had some effect.

Mere Exposure Effect

Our attraction to another is greatly affected by how frequently we see him or her or even that person's picture. Have you ever noticed when watching a movie that the central character seems more attractive at the end of the movie than at the beginning? This is particularly true if you already find the person very attractive when the movie begins. Have you ever noticed that the fabulous-looking person sitting next to you in class looks better every day?

You may be experiencing *mere exposure effect*. The more you see someone in person or in a photograph, the more you like that person. For example, in studies where people are repeatedly shown photographs of the same faces, the more often a person sees a particular face, the more he or she likes that person (Moreland and Beach 1992, Zajonc 1968). There are two qualifications to the effect. First, "overexposure" can result when a photograph is seen too often. The viewer becomes "saturated" and ceases to like the pictured person more with each exposure. (Some celebrities, certainly not all, are careful not to allow themselves to become "overexposed" on talk shows, lest the public tire of them.) Second, the initial response of the viewer can determine how much liking will increase. If one starts out liking someone, seeing that person more will increase the liking for that person. However, if one starts out disliking the pictured person, the amount of dislike tends to remain about the same, regardless of how often one sees the person (Taylor et al. 2003).

Perceived Physical Attractiveness

We hear that "beauty is only skin deep." Apparently that is deep enough. To a surprisingly large degree, the attractions we feel toward people of either gender are based significantly on our perception of their physical attractiveness. A vast amount of research over the years has consistently shown the importance of attractiveness in human interactions. Adults react more leniently to the bad behavior of an attractive child than to the *same* behavior of an unattractive child (Taylor et al. 2003; Berscheid and Reis 1998; Dion 1972). Teachers evaluate cute children of either gender as "smarter" than unattractive children with identical academic records (Worchel et al. 2000; Clifford and Walster 1973). In studies of mock jury trials, attractive defendants, male or female, receive lighter sentences on average than unattractive defendants convicted of the same crime (Gilbert et al. 1998; Sigall and Ostrove 1975).

Of course, standards of attractiveness vary between cultures and between subcultures within the same society. As we saw in Chapter 3, what is highly attractive in one culture may be repulsive in another. In the United States, there is a maxim that you can never be too thin. This cultural belief has been cited as a major cause of eating disorders such as anorexia nervosa and bulimia, especially among White women (Taylor et al. 2003, Wolf 1991), although less so among African American, Hispanic, and Native American women (Thompson 1994). The maxim itself is a source of oppression for women in U.S. society, yet it is clearly culturally relative. In other cultures, plumpness is sexy, as was the case in England and France in the Middle Ages. Among certain African Americans, chubbiness in women is considered attractive. Such women are called "healthy" and "phatt" (not "fat"), which means the same as "stacked," or curvaceous. Similar cultural norms often apply in U.S. Hispanic populations. The skinny woman is considered ugly, not sexy.

Although standards of attractiveness vary from culture to culture, considerable agreement exists *within* a culture about who is attractive. For example, people ranking photographs of men and women of their own race tend to rank the attractiveness of the pictured individuals very similarly (Crandall et al. 2001; Hatfield and Sprecher 1986). This is equally true when men rank photos of women and when women rank photos of men.

Studies of dating patterns among college students show that the more attractive one is, the more likely one will be asked on a date. This finding is consistent across several studies, done over many years (Berscheid and Reis 1998; Speed and Gangestad 1997; Walster et al. 1966). It applies to gay and lesbian dating as well as to heterosexual dating (Cohen and Tannenbaum 2001). However, physical attractiveness predicts only the early stages of a relationship. When one measures relationships that last a while, other factors come into play, primarily religion, political attitudes, social class background, educational aspirations, and race. Perceived physical attractiveness may predict who is attracted to whom initially, but other variables are better predictors of how long a relationship will last (Berscheid and Reis 1998; Hill et al. 1976).

Similarity

"Opposites attract," you say? Not according to the research. We have all heard that people are attracted to their "opposite" in personality, social status, background, and other characteristics. Many of us grow up believing this to be true. However, if the research says one thing about interpersonal attraction, it is that, with few exceptions, we are attracted to those who are similar or even identical to us in socioeconomic status, race, ethnicity, religion, perceived personality traits, and general attitudes and opinions (Taylor et al. 2003; Brehm et al. 2002). "Dominant" people tend to be attracted to other dominant people, not to "submissive" people, as one might otherwise expect. Couples tend to have similar opinions about political issues of great importance to them, such as attitudes about abortion, crime, and urban violence. Overall, couples tend to exhibit strong cultural or subcultural similarity.

There are exceptions, of course. We sometimes fall in love with the exotic—the culturally or socially different. Novels and movies return endlessly to the story of the young, White, preppy woman who falls in love with a Hell's Angel biker, but such a pairing is by far the exception and not the rule. When it comes to long-term relationships, including both friends and lovers (whether heterosexual, gay, or lesbian), humans vastly prefer a great degree of similarity even though, if asked, they might deny it. In fact, the less similar a heterosexual relationship is with respect to race, social class, age, and educational aspirations (how far in school the person wants to go), the quicker the relationship is likely to break up (Silverthorne and Quinsey 2000; Berscheid and Reis 1998; Stover and Hope 1993; Hill et al. 1976).

An especially interesting qualification to all the research on similarity and attraction presents itself, however, in the matter of interracial dating. Although people tend to date within their own race, nationality, or ethnicity, a large number of interracial couples today enjoy long-lasting relationships. Similarity research sheds light on these relationships as well. The research tends to show that for interracial couples, similarity in characteristics other than race—social class background, religion, age, political attitudes, and educational aspirations—tends to predict how long the relationship will last. In general, the more similar the couple is in characteristics other than race, the longer, on average, the interracial relationship will last. The less similar the couple, the shorter the relationship (Berscheid and Reis 1998; Stover and Hope 1993; Eagly et al. 1989).

Most romantic relationships, regrettably, come to an end. On campus, relationships tend to break up most often during gaps in the school calendar, such as Christmas recess and spring vacation. Summers are especially brutal on relationships formed during the academic year. Breakups are seldom mutual. Almost always, only one member of the pair wants to break off the relationship. This sad truth means that the next time someone tells you that their breakup last week was "mutual," you know they are probably lying or deceiving themselves (Taylor et al. 20003; Hill et al. 1976).

Social Institutions and Social Structure

So far, we have been studying the more *microlevels* of society. At the *macrolevel* of society, to which we now turn, sociologists are interested in the role of social institutions in constituting society. Just as sociologists see social forces shaping behavior at the microlevel, so do they see social forces molding society at the macrolevel.

Social Institutions

Societies are identifiable by their cultural characteristics and the social institutions of which they are composed. A **social institution** (or simply, an institution) is an established and organized system of social behavior with a recognized purpose. The term refers to the broad systems that organize specific functions in society. Social institutions are a weave of behaviors, norms, and values. As a whole, institutions are organized to meet various needs in society. The family is an institution that provides for the care of the young and the transmission of culture. Religion is an institution that organizes the sacred beliefs of a society. Education is the institution through which people learn the skills needed to live in the society.

Unlike group behavior, institutions cannot be directly observed, but their impact and structure can be seen nonetheless. An institution is a patterning of social relationships that exist as distinct from individuals or specific groups. Social institutions have an existence all their own. They impinge on the behavior of people and groups, and they persist long after these people and groups are gone. Take the example of education as an institution. You and your classmates form a social group whose behavior can be directly observed. As a group, you are part of a broader educational institution, which is both a specific school and, at the societal level, the institution called education.

The major institutions in society include the family, education, work and the economy, the political institution (or state), religion, and health care, as well as institutions such as the mass media, organized sports, and the military. These all represent institutions; they are all complex structures that exist for explicit reasons. Together, institutions meet certain needs that are necessary to be met for society to exist and proceed. Functionalist theorists have for some time identified these needs (functions) as follows (Aberle et al. 1950; Parsons 1951a; Levy 1949):

1. *The socialization of new members of the society.* This is accomplished by the family and other institutions as well, such as the education institution.

2. *The production and distribution of goods and services.* The economy is generally given as the institution that performs this set of tasks.

3. *Replacement of the membership.* All societies must have a means of replacing its members who die, move or migrate away, or otherwise leave the society. Child-rearing is one of these means.

4. *The maintenance of stability and existence.* As already noted in Chapter 1, one major assertion of functional theory is that certain institutions within a society (such as government, a police force, a military) contribute toward the stability and continuance of the society.

5. *Providing the members with an ultimate sense of purpose.* Societies accomplish this task by having national anthems, patriotism, and the like, in addition to providing basic values and moral codes through institutions such as the family, religion, and education.

Functionalists see these societal needs as universal, although societies do not perform them in the same way or by means of the same institutions. This is what makes societies distinct.

In contrast to functional theory, conflict theory further notes that because conflict is inherent in social existence, the institutions of society do not provide for all its members equally. Some members are provided for better than others, thus demonstrating that institutions affect people with differential power, granting more power to some social groups than to others. Racial and ethnic minorities in a society possess considerably less power and less of society's benefits than does the dominant group. In most societies known to anthropologists and sociologists, women occupy lower social status on the average than do men. There are few exceptions to this. Thus, on average, women have less political power, wealth, and prestige than do men in society. Similarly, power is not equally distributed across social class strata. Generally, the lower one's social class, the less one's political power, influence, and prestige. How institutions affect the person, as well as the benefits that accrue to the individual from the institution, depends upon the person's race–ethnicity, gender, and social class status, among other factors.

Social Structure

At both the micro- and macrolevel, we have seen how social behavior is patterned. Sociologists use the term **social structure** to refer to the organized pattern of social relationships and social institutions that together compose society. The social structure of society is observable in the established patterns of social interaction and social institutions. Social structural analysis is a way of looking at society in which the sociologist analyzes the patterns in social life that reflect and produce social behavior.

Social class distinctions are an example of a social structure. Class shapes the access that different groups have to the resources of society, and it shapes many of the interactions people have with each other. People may form cliques with those who share similar class standing, or they may identify with certain values that are associated with a given class, for example, middle-class values. Class then forms a social structure—one that shapes and guides human behavior at all levels, no matter how overtly visible or invisible this structure is to someone at a given time.

The philosopher Marilyn Frye aptly describes the concept of social structure in her writing. Using the metaphor of a birdcage, she writes that if you look closely at only one wire in a cage, you cannot see the other wires. You might then wonder why the bird within does not fly away. Only when you step back and see the whole cage instead of a single wire do you understand why the bird does not escape. Frye writes:

> It is perfectly obvious that the bird is surrounded by a network of systematically related barriers, no one of which would be the least hindrance to its flight, but all of which, by their relations to each other, are as confining as the solid walls of a dungeon. It is now possible to grasp one reason why oppression can be hard to see and recognize. One can study the elements of an oppressive structure with great care and some good will without seeing or being able to understand that one is looking at a cage and that there are people there who are caged, whose motion and mobility are restricted, whose lives are shaped and reduced (Frye 1983: 4–5).

Frye's analysis focuses on oppressive social structures that confine and exploit people. Oppression and social structure are not the same thing, although many people find existing social structures oppressive. Just as a birdcage is a network of wires, society is a network of social structures, both micro and macro.

THINKING SOCIOLOGICALLY

Using Marilyn Frye's analogy of the birdcage, think of a time when you believed your choices were constrained by *social structure.* When you applied to college could you go anywhere you wanted? What social structural conditions guided your ultimate selection of schools to attend?

What Holds Society Together?

What holds society together? We have been asking this question throughout this chapter. This central question in sociology was first addressed by Emile Durkheim, the French sociologist writing in the late 1800s and early 1900s. He argued that people in society had a **collective consciousness,** defined as the body of beliefs that are common to a community or society and that give people a sense of belonging and a feeling of moral obligation to its demands and values. According to Durkheim, it is collective consciousness that gives groups social solidarity. It gives members of a group the feeling that they are part of one society.

Where does the collective consciousness come from? Durkheim argued that it stems from people's participation in common activities, such as work, family, education, and religion—in short, society's institutions. Goals, values, and beliefs emanate from the institutions of a society, and individuals aligned with common institutions develop a sense of common purpose. This provides solidarity in society.

Mechanical and Organic Solidarity

According to Durkheim, there are two different kinds of social solidarity: mechanical and organic, each the basis of a different form of societal solidarity. **Mechanical solidarity** arises when individuals play similar roles within the society. People feel bonded to the group in less complex societies because everyone in the group is so similar. Individuals in societies marked by mechanical solidarity share the same values and hold the same things sacred. This potent source of cohesiveness is weakened when a society becomes differentiated into more complex systems of work behavior. Contemporary examples of mechanical solidarity are rare because most of the societies of the world have been absorbed in the global trend to greater complexity and interrelatedness. Native American groups prior to European conquest were bound together by mechanical solidarity. Many Native American groups are now trying to regain the mechanical solidarity on which their cultural heritage rests, but they are finding that the superimposition of White institutions on Native American life interferes with the adoption of traditional ways of thinking and being and, hence, prevents mechanical solidarity from gaining its original strength.

In contrast to societies where individuals play the same roles, we find societies marked by **organic solidarity** (also called *contractual solidarity*), in which individuals play a great variety of different roles, and unity is based on role differentiation, not similarity. The United States and other industrial societies are examples. A society built on organic solidarity is cohesive *because of*

its differentiation. Roles are no longer necessarily similar, but they are necessarily interlinked. The performance of multiple roles is necessary for the execution of society's complex and integrated functions.

> ### DEBUNKING *SOCIETY'S MYTHS*
>
> **Myth:** Society is held together because people share common values.
>
> **Sociological perspective:** Some societies are held together by commonly shared values, such as in mechanical solidarity; other societies, marked by organic solidarity, are held together by the interrelated but different roles that are part of a division of labor.

Durkheim described this state as the **division of labor,** defined as the systematic interrelatedness of different tasks that develops in complex societies. The labor force within the contemporary U.S. economy, for example, is divided according to the kinds of work people do. Within any division of labor, tasks become distinct from one another, but they are still woven together into a whole.

The division of labor is a central concept in sociology because it represents how the different pieces of society fall together. The division of labor in most contemporary societies is often marked by gender, race, and class divisions. In other words, if you look at who does what in society, you will see that women and men tend to do different things in society; this is the gender division of labor. This is cross-cut by the racial division of labor, the pattern whereby those in different racial–ethnic groups tend to do different work in society. At the same time, the division of labor is also marked by class distinctions, with some groups providing work that is highly valued and rewarded, and others doing work that is devalued and poorly rewarded. As you will see throughout this book, gender, race, and class intersect and overlap in the division of labor.

Durkheim's thinking about the origins of social cohesion can bring light to contemporary discussions over "family values." Some want to promote traditional family values as the moral standards of society. Is such a thing necessarily good, or even possible? The United States is an increasingly diverse society, and family life differs among different groups. It is unlikely that a single set of family values can be the basis for social solidarity. Groups in this society are bound together through the division of labor, even when the bond of shared values is not strong. Although contemporary society is enormously diverse, it can still have the cohesion of a unified society because its people are interrelated through social institutions.

Gemeinschaft and Gesellschaft

Different societies are held together by different forms of solidarity. Some societies are characterized by what sociologists call **gemeinschaft,** a German word that means "community"; others are characterized as **gesell-schaft,** which means "society" (Tönnies 1963/ 1887). Each involves a type of solidarity or cohesiveness. Those societies that are gemeinschafts ("communities") are characterized by a sense of "we" feeling, a moderate division of labor, strong personal ties, strong family relationships, and a sense of personal loyalty. In gemeinschaft, the sense of solidarity between members of the society arises from personal ties; small, relatively simple social institutions; and a collective sense of loyalty to the whole society. People tend in a gemeinschaft society to be well integrated into the whole, and social cohesion comes from deeply shared values and beliefs (often, sacred values). In such a society, social control need not be imposed externally because control comes from the internal sense of belonging that members share. In general, gemeinschafts tend to be characterized by mechanical solidarity.

In contrast, in societies marked by gesellschaft, an increasing importance is placed on the secondary relationships people have, that is, those that are less intimate and more instrumental, such as work roles instead of family or community roles. Gesellschaft is characterized by less prominence of personal ties, a somewhat diminished role of the nuclear family, and a lessened sense of personal loyalty to the total society. This does not mean that the gesellschaft lacks solidarity and cohesion, for it can be very cohesive, but the cohesion comes from an elaborated division of labor (thus, organic solidarity), greater flexibility in social roles, and the instrumental ties that people have to one another.

Social solidarity under gesellschaft is weaker than in the gemeinschaft society, however. Gesellschaft societies are marked by an elaborate division of labor. Gesellschaft is more likely than gemeinschaft to be torn by class conflict because class distinctions are less prominent, though still present, in the gemeinschaft. Racial–ethnic conflict is also more likely within gesellschaft societies given that the gemeinschaft tends to be ethnically and racially very homogeneous. Often it is characterized by only one racial or ethnic group. This means that conflict between gemeinschaft societies, such as ethnically based wars, can be very high because both groups have a strong internal sense of group identity that may be intolerant of others—for example, Palestinians and Israelis, or Hutus and Tutsis in Rwanda, or Shiite and Sunni Muslims in Iraq.

In summary, complexity and differentiation are what make the gesellschaft cohesive, whereas similarity and unity cohere the gemeinschaft society. In a single society, such as the United States, you can conceptualize the whole society as gesellschaft, with some internal groups marked by gemeinschaft. Our national motto seems to embody this idea: *E pluribus unum*— unity within diversity, although clearly this idealistic motto has been only partly realized.

Types of Societies: A Global View

In addition to comparing how different societies are bound together, sociologists are interested in how social organization evolves in different societies. Over time and across different cultures and continents, societies are distinguished by different forms of social organization. These forms evolve from the relationship of a given society to its environment and from the processes that the society develops to meet basic human needs. Simple things such as the size of a society can also shape its social organization, as do the different roles that men and women engage in as they produce goods, care for the old and young, and pass on societal traditions. Societies also differ according to their resource base, whether they are predominantly agricultural or industrial and whether they are sparsely or densely populated.

Thousands of years ago, societies were small, sparsely populated, and technologically limited. In the competition for scarce resources, larger and more technologically advanced societies dominated smaller ones. Today, we have arrived at a society that is global, with highly evolved degrees of social differentiation and inequality, notably along class, gender, racial, and ethnic lines (Lenski et al. 2001). Sociologists distinguish six types of societies based on the complexity of their social structure, the amount of overall cultural accumulation, and the level of their technology. The various types are *foraging, pastoral, horticultural,* and *agricultural* (these

These Bedouins (a pastoral society) are engaged in an economic transaction in this desert market.

© Penny Tweedie/Corbis

four are called *preindustrial societies*), *industrial,* and *postindustrial societies* (see Table 5.1). Examples of all these types of society can still be found, although all but the most isolated societies are moving toward the industrial and postindustrial forms of societal development.

These different societies vary in the basis for their organization and the complexity of their division of labor. Some, like the foraging societies, are subsistence economies, where men and women hunt and gather food, but accumulate very little. Others, such as the pastoral societies and horticultural societies, develop a more elaborate division of labor as the social roles that are needed for raising livestock and farming become more numerous. With the development of agricultural societies, production becomes more large-scale, and strong patterns of social differentiation sometimes develop in the form of a caste system or slavery.

The key driving forces behind the development of these different societies is the development of technology. All societies utilize technology to assist with human needs. Technology may be as simple as a rough-hewn shovel or as elaborate as computing technology. Either

way, the technology both develops from the society's form and shapes the possibilities for human life.

Preindustrial Societies

A **preindustrial society** is one that directly uses, modifies, and/or tills the land as a major means of survival as a society. There are four kinds of preindustrial societies, listed here by degree of development of their technology: foraging (or hunting-gathering) societies, pastoral societies, horticultural societies; and agricultural societies (see Table 5.1).

In *foraging (hunting-gathering) societies,* the technology enables the hunting of animals and gathering of vegetation. The technology does not permit refrigeration or processing of food, and hence these individuals must search continuously for plants and game. Since hunting and gathering are activities that require large amounts of land, most foraging societies are nomadic; that is, they constantly travel as they deplete the plant supply or follow the migrations of animals. The central institution is the family, which serves as the means

Table 5.1
Types of Societies

		Economic Base	Social Organization	Examples
Preindustrial Societies	*Foraging Societies*	Economic sustenance dependent on hunting and foraging	Gender is important basis for social organization, although division of labor is not rigid; little accumulation of wealth	Pygmies of Central Africa
	Pastoral Societies	Nomadic societies, with substantial dependence on domesticated animals for economic production	Complex social system with an elite upper class and greater gender role differentiation than in foraging societies	Bedouins of Africa and Middle East
	Horticultural Societies	Society marked by relatively permanent settlement and production of domesticated crops	Accumulation of wealth and elaboration of the division of labor, with different occupational roles (farmers, traders, craftspeople, etc.)	Aztecs of Mexico; Incan empire of Peru
	Agricultural Societies	Livelihood dependent on elaborate and large-scale patterns of agriculture and increased use of technology in agricultural production	Caste system develops that differentiates the elite and agriculture laborers; may include system of slavery	American South, pre-Civil War
Industrial Societies		Economic system based on the development of elaborate machinery and a factory system; economy based on cash and wages	Highly differentiated labor force with a complex division of labor and large formal organizations	Nineteenth and most of twentieth century United States and western Europe
Postindustrial Societies		Information-based societies in which technology plays a vital role in social organization	Education increasingly important to the division of labor	Contemporary United States, Japan, and others

of distributing food, training children, and protecting its members. There is usually role differentiation on the basis of gender, although the specific form of the gender division of labor varies in different societies. They occasionally wage war with other clans or similar societies, and spears and bows and arrows are the weapons used. Examples of foraging societies are certain Aborigines of Australia and the Pygmies of Central Africa.

In *pastoral societies,* technology is based on the domestication of animals. Such societies tend to develop in desert areas that are too arid to provide rich vegetation. The pastoral society is nomadic, necessitated by the endless search for fresh grazing grounds for the herds of their domesticated animals. The animals are used as sources of hard work that enable the creation of material surplus. Unlike a foraging society, this surplus frees some individuals from the tasks of hunting and gathering and allows them to create crafts, make pottery, cut hair, build tents, and apply tattoos. The surplus generates a more complex and differentiated social system with an elite or upper class and more role differentiation on the basis of gender. The nomadic Bedouins of Africa and the Middle East are pastoral societies.

In *horticultural societies,* elaborate hand tools are used to cultivate the land, such as the hoe and the digging stick. The individuals in horticultural societies practice ancestor worship and conceive of a deity or deities (God or gods) as a creator. This distinguishes them from foraging societies that generally employ the notion of numerous spirits to explain the unknowable. Horticultural societies recultivate the land each year and tend to establish relatively permanent settlements and villages. Role differentiation is extensive, resulting in different and interdependent occupational roles such as farmer, trader, and craftsperson. The Aztecs of Mexico and the Incas of Peru represent examples of horticultural societies.

The *agricultural society* is exemplified by the pre-Civil War American South. Such societies have a large and complex economic system that is technologically based on large-scale farming using plows harnessed to animals or other sources of energy. Other technological elements such as irrigation, use of the wheel, use of metals, and the ability to write, make such societies considerably advanced technologically. Farms tend to be considerably larger than the cultivated land in horticultural societies. Large and permanent settlements characterize agricultural societies, which also exhibit dramatic social inequalities. A rigid caste system develops, separating the peasants, or slaves, from the controlling elite caste, which is then freed from manual work allowing time for art, literature, and philosophy, activities of which they can then claim the lower castes are incapable.

Industrial Societies

An *industrial society* is one that uses machines and other advanced technologies to produce and distribute goods and services. The Industrial Revolution began only 200 years ago when the steam engine was invented in England, delivering previously unattainable amounts of mechanical power for the performance of work. Steam engines powered locomotives, factories, and dynamos, transforming societies as the Industrial Revolution spread. The growth of science led to advances in farming techniques such as crop rotation, harvesting, and ginning cotton, as well as huge industrial-scale projects such as dams for hydroelectric power. Joining these advances were developments in medicine, new techniques to prolong and improve life, and the emergence of birth control to limit population growth.

Unlike agricultural societies, industrial societies rely upon a highly differentiated labor force and the intensive use of capital and technology. Large formal organizations are common. The task of holding society together, falling more on institutions such as religion in preindustrial societies, now falls more on the high division-of-labor institutions, such as the economy and work, government and politics, and large bureaucracies.

Within industrial societies, the forms of gender inequality that we see in contemporary U.S. society tend to develop. With the advent of industrialization, societies move to a cash-based economy, with labor performed in factories and mills paid on a wage basis and household labor remaining unpaid. This introduced what is known as the *family-wage economy,* one in which families become dependent on wages to support themselves, but work within the family (housework, child care, and other forms of household work) is unpaid and, therefore, increasingly devalued (Tilly and Scott 1978). The family-wage economy is based on the idea that men are the primary breadwinners. Therefore, a system of inequality in men's and women's wages was introduced. This is an economic system that continues today to produce a wage gap between men and women.

Industrial societies tend to be highly productive economically with a large working class of industrial laborers. People become increasingly urbanized as they move from farm lands to urban centers or other areas where factories are located, as we see in Map 5.1 "Viewing Society in Global Perspective: Global Urbanization." Immigration is common in industrial societies, particularly as industries are forming where demand is high for more, cheaply paid labor.

Industrialization has brought many benefits to U.S. society, including a highly productive and efficient economic system, expansion of international markets, extraordinary availability of consumer products, and, for many, a good working wage. Industrialization has, at the same time, also produced some of the most serious

MAP 5.1 Global Urbanization

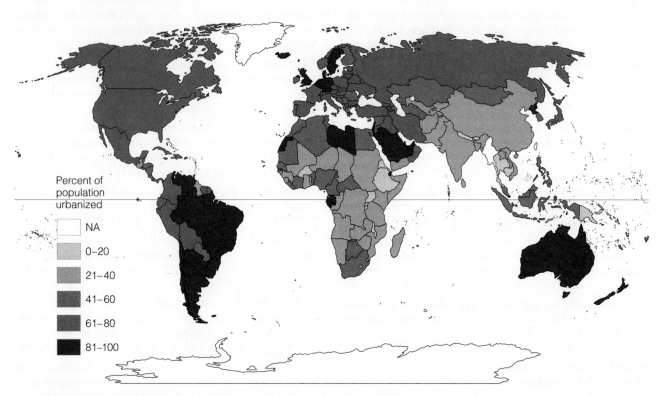

The percentage of the population of a country that is urban is one way of measuring the complexity of a society, but it certainly is not the only one. Can you think of other measures that indicate the complexity of a society?

Data: Population Division of the United Nations Secretariate, 2002. **www.un.org/Depts/unsd/social/hum-set.htm**

social problems that our nation faces: industrial pollution, overdependence on consumer goods, wage inequality and job dislocation for millions, and problems of crowding in urban areas. The portrait of population density depicted in Map 5.2 "Mapping America's Diversity: Population Density in the United States" illustrates where crowding is most extreme. Understanding the process of industrialization and accompanying urbanization is a major avenue for sociological research (see also Chapter 21).

Postindustrial Societies

In the contemporary era, a new type of society is emerging. Whereas most twentieth-century societies can be characterized by their generation of material goods, **postindustrial society** is economically dependent on the production and distribution of services, information, and knowledge (Bell 1973). Postindustrial societies are information-based societies in which technology plays a vital role in the social organization. The United States is fast becoming a postindustrial society, and Japan may

be even further along. In the postindustrial society, many workers provide services such as administration, education, legal services, scientific research, and banking, or they engage in the development, management, and distribution of information, such as computer use and design. Central to the economy of the postindustrial society are highly advanced technologies such as computers, robotics, genetic engineering, and laser technology.

The transition to a postindustrial society has a strong influence on the character of social institutions. Educational institutions acquire paramount importance in the postindustrial society, and science takes an especially prominent place. For some, the transition to a postindustrial society means more discretionary income for leisure activities such as tourism, entertainment, and relaxation (spas, massage centers, and exercise)—at least for those in certain classes. For others, the transition to postindustrialism can mean permanent joblessness or holding more than one job to simply make ends meet. Workers without highly technical skills may be dispossessed in such a society, with millions stuck in low-paid, unskilled work.

MAPPING AMERICA'S DIVERSITY

MAP 5.2 Population Density in the United States

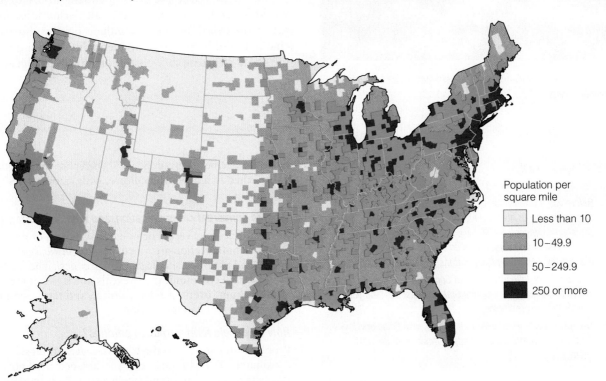

Population per
square mile

☐ Less than 10

▨ 10–49.9

▨ 50–249.9

■ 250 or more

Population density of a state in the United States is measured by the number of people per square mile in that state. Notice which regions are densely populated (for example, the northeastern coast line) and which are sparsely populated (for example, rural areas of the Midwest). How does your own home state compare to other states in population density?

Data: U.S. Census Bureau. 2004. *Statistical Abstract of the United States, 2003.* Washington, DC: U.S. Government Printing Office.
http://www.census.gov/

The United States is suspended between the industrial and postindustrial phases. Manufacturing jobs are still a major segment of the labor force, although they are in decline because most workers are employed in the service sector of the economy (the sector involving the delivery of services and information, not the

TAKING ON SOCIAL ISSUES
A Sense of Community

In their well-received book, *Habits of the Heart* (1996), sociologists Robert Bellah and his colleagues argued that the individualistic orientation of people in the United States has created a society in which people find it difficult to sustain their commitments to others. The authors suggest that the nation needs to develop new traditions—traditions that would unite people through shared community service

activities. Identify one volunteer project in your community, and learn what you can about its purpose, its volunteers, and its effects. At the same time, search the literature on the focus of this project (such as *Habitat for Humanity* and low-income housing or AIDS and the AIDS Caregivers Support Network) and identify the sociological dimensions of this policy issue. Based on what you have learned, would you agree or disagree

with Bellah and his colleagues that service-oriented groups promote a sense of community?

Taking Action

Go to the Taking Action Exercise on the companion website—at http://sociology .wadsworth.com/andersen_taylor4e/— to learn more about an organization that addresses this topic. ● ● ●

An information-based technology is the characteristic of the post-industrial society.

production of material goods). Postindustrial societies are also increasingly dependent on a global economy because goods are still produced, but they tend to be produced in economically dependent areas of the world for consumption in the wealthier nations. As a result, the world, not just individual societies, becomes characterized by poverty and inequality resulting from the social structure that postindustrialism produces.

Chapter Summary

What are the forms of social interaction in a society?

Society is a system of *social interaction* that includes both culture and *social organization*. It involves both *micro*-processes as well as *macro*processes. *Groups* are of varying types, as are social groupings (social categories, audiences, and formal organizations). *Status* is a hierarchical position in a structure; a *role* is the expected behavior associated with a particular status. Roles can conflict, thus producing *status inconsistency, role conflict,* or *role strain,* producing stress.

What theories are there about social interaction?

Social interaction takes place in a society within the context of social structure and social institutions. Social interaction is analyzed in several ways, including the *social construction of reality* (we impose meaning and reality on our interactions with others); *ethnomethodology* (deliberate interruption of interaction to observe how a return to "normal" interaction is accomplished); *impression management* (wherein a person "gives off" a particular impression to "con" the other and to achieve one's goals, as in *cyberspace interaction*); and *social exchange* (wherein one engages in *game-like* reward and punishment interactions to achieve one's goals).

What happens in cyberspace interaction?

In *cyberspace interaction*, a person can manipulate the impression that one gives off, thus creating a new "virtual self" or new identity. This new identity can then

be used to achieve certain goals. Cyberspace interaction has resulted in a new cyberculture, with its own rules and norms.

What is nonverbal communication, and what are its forms?

Interaction can be verbal or nonverbal. Forms of non-verbal communication are *touch, paralinguistic, body language,* and *use of personal space (proxemics)*. The study of nonverbal communication reveals much rich information about diverse gender, cultural, and racial–ethnic variations in human interaction.

How do people form attraction pairings?

People form pairings on the basis of several things: proximity; repeated exposure to another; perceived physical attractiveness; and similarity in age, socioeconomic status, race, ethnicity, religion, perceived personality traits, general attitudes and opinions, and educational aspirations. Similarities over such things attract; opposites do not. Such variables also tend to predict how long an interracial relationship will last.

How is society held together?

According to theorist Emile Durkheim, society, with all its complex social organization and culture, is held together by one of two kinds of cohesion: *mechanical solidarity* (based on individual similarity) or *organic (contractual) solidarity* (based on a *division of labor*). Two other forms of social organization contribute to the cohesion of a society: *gemeinschaft* ("community," based on friendships and loyalties) and *gesellschaft* ("society," based on complexity and differentiation).

What are the types of societies?

Societies vary depending on the complexity of their social structures, their division of labor, and their technologies. From least to most complex, they are *foraging, pastoral, horticultural, agricultural* (these four constitute *preindustrial* societies), *industrial*, and *postindustrial* societies.

Key Terms

achieved status 113
ascribed status 113
collective consciousness 130

cyberspace interaction 120
division of labor 130
ethnomethodology 117

gemeinschaft 131

gesellschaft 131

group 113

impression management 118

master status 114

macroanalysis 112

mechanical solidarity 130

microanalysis 112

organic solidarity 130

paralinguistic communication 123

preindustrial society 132

postindustrial society 134

proxemic communication 124

role 114

role conflict 114

role model 114

role set 114

role strain 115

social institution 128

social interaction 112

social organization 112

social structure 129

society 112

status 113

status inconsistency 113

status set 113

Researching Society with Microcase Online

You can see the results of actual research by using the Wadsworth MicroCase® Online feature available to you. This feature allows you to look at some of the results from national surveys, census data, and some other data sources. You can explore this easy-to-use feature on your own, but try this example. Suppose you want to know:

Having a job is a type of role (called an occupational role). How much in general do people who work like their jobs? Why do some people like their jobs and some not? You can do the analysis yourself!

To answer this question, go to http://sociology.wadsworth.com/andersen_taylor4e/, select MicroCase Online from the left navigation bar, and follow the directions there to analyze the following data.

Data File: GSS

Task: Cross-Tabulation

Column Variable: 1-SEX

Row Variable: LIKE JOB?

Questions

You will get a table that gives the individual's gender (sex) (the columns of the table), and whether a person likes their job (the row variable, which is LIKE-JOB?). When you get these results (these are actual people), answer the following questions:

1. What are the overall percentages (the total percentages—in the right-hand column) of people who are: Very satisfied with their job? Moderately satisfied? Unsatisfied?

2. What is the total number of women in this study? What is the total number of men?

3. In general (look at the table) which gender (women or men) tends to be "very satisfied" with their job? Which gender tends to be "unsatisfied"?

The Companion Website for Sociology: Understanding a Diverse Society, Fourth Edition

http://sociology.wadsworth.com/andersen_taylor4e/

Supplement your review of this chapter by going to the companion website to take one of the Tutorial Quizzes, use the flash cards to master key terms, and check out the many other study aids you'll find there. You'll also find special features such as GSS Data and Census 2000 information, data and resources at your fingertips to help you with that special project or do some research on your own.

Suggested Readings and Web Resources

Goffman, Erving. 1959. *The Presentation of Self in Everyday Life*. Garden City, NY: Doubleday.
Goffman's analysis is the classic analysis of impression management and how people negotiate their individual identity through social interaction.

Putnam, Robert D. 2000. *Bowling Alone: The Collapse and Revival of American Community*. New York: Simon and Schuster.
Putnam argues that our society has experienced a decline in participation in voluntary organizations, religious activities, and other forms of public life, and as a result is undergoing a decline in shared values as a part of culture. Hence, our society is in danger of increased social disorder.

Strate, Lance, Ron L. Jacobson, and Stephen Gibson (eds.). 2003. *Communication and Cyberspace*. Cresskill, NJ: Hampton Press.
A collection of readings covering a wide range of cyberspace topics, for example, who shall control cyberspace, how influence works in cyberspace, and other topics.

Turnbull, Colin. 1972. *The Mountain People*. New York: Simon and Schuster.
The Ik were an isolated people living as a separate society in Uganda. This is a powerful account of how the Ugandan government completely destroyed Ik society by moving its people onto a government-controlled reservation. Comparisons to the historical treatment of Native Americans are made.

The Psychology of Cyberspace
www.rider.edu/users/suler/psycyber/psycyber.html
This site includes an online book about social interaction and cyberspace, with links to other sites exploring the implications of cyberspace for interpersonal interaction.

Center for Social Informatics
www-slis.lib.Indiana.edu/ssi/
The Center for Social Informatics at Indiana University does research on the influence of technology and computerization on society; the website includes information on publications, conferences, and other resources.

Groups and Organizations

Twelve citizens sit together in an elevated enclosure, like a choir loft, and silently watch a drama unfold before them, day after day. They are respectfully addressed by highly paid professionals: lawyers, judges, expert witnesses. Their job, ruling on the innocence or guilt of a defendant or the settlement of a legal claim, was once the prerogative of kings. Their decision may mean freedom or incarceration, fortune or penury, even life or death.

Juries have been the focus of much research, in part because they fill such a vital role in our society and in part because within this curiously artificial, yet intimate, group of random strangers can be found a wealth of interesting sociological phenomena. Jury verdicts and jury deliberations show the same inescapable influences of status, race, and gender that affect the rest of society. Juries are, in some ways, society in miniature.

A vast folklore exists among trial lawyers about jurors and jury performance. During jury selection, in a process called the *voir dire,* lawyers on both sides are entitled to eliminate any potential jurors with no explanation required. Many lawyers who have great

faith in their ability to judge jurors consider jury selection to be the most important part of a trial. By choosing the jurors, they are choosing the verdict. Consider some of the old-fashioned folk wisdom clung to by trial lawyers—with widely varying degrees of accuracy: Farmers believe in strict responsibility, whereas waiters and bartenders are forgiving; avoid wage earners; avoid the clergy; select married women (Belli 1954). A guideline for Dallas, Texas, prosecutors advised against selecting "Yankees . . . unless they appear to have common sense" (Guinther 1988: 54).

High-powered legal teams now make room for a new breed of legal specialist—the *trial consultant* trained in sociological techniques who contributes nothing but juror analysis as part of the jury selection process. Such was the case in the well-publicized murder trial of football player O. J. Simpson for the alleged murder of his ex-wife, Nicole Brown Simpson. According to some researchers, the jury found Simpson not guilty despite strong evidence of his guilt because the prosecutor ignored social science consultants, particularly information regarding the relevance of the racial and gender composition of the

Sociology ⊛ Now™
Reviewing is as easy as ❶ ❷ ❸.

Use SociologyNow to help you make the grade on your next exam. When you are finished reading this chapter, go to the chapter review for instructions on how to make SociologyNow work for you.

jury (Taylor et al. 2003; Toobin 1996). In other trials, simply the size of the jury (six versus twelve members) can make a difference (Saks and Marti 1997).

The level of analysis goes beyond simply identifying the likely bias of a given juror. Juries are *groups,* and groups behave differently from individuals. Understanding group behavior is critical to predicting the performance of a jury. For instance, it is possible to make an educated prediction about who in a jury will become the most influential (Hans and Martinez 1994). Researchers have found that people with high status in society do the most talking in jury deliberations and are thought by other jurors to be the most helpful in reaching a verdict (Cohen and Zhou 1991; Berger and Zelditch 1985).

Factions form during jury deliberations, and if jury analysts expect a difficult decision, they can attempt to influence how fractionalized juries will resolve their disputes based on sociological and psychological data about small-group decision making (Hastie and Pennington 2000; Hastie et al. 1983). For instance, jurors are much less likely to defect from large factions than from small ones. The larger the faction, the less willing a juror will be to defy the weight of group opinion.

What does this say about the state of justice in our legal system, when guilt or innocence depends not only on the legal facts but also on social aspects of the jury? Like society as a whole, and like organizations and bureaucracies, groups are subject to social influences, and understanding these influences can be essential to understanding a process such as the operation of social justice.

This chapter introduces the study of groups and the social processes and dynamics that are characteristic of group and organizational behavior. Whether a relatively small group, such as a jury, or a large organization, such as a major corporation, groups are influenced by sociological forces. •••

Types of Groups

Each of us is a member of many groups simultaneously. We have relationships with family, friends, team members, and professional colleagues in groups. Within these groups are gradations: We are generally closer to our siblings (brothers and sisters) than our cousins; we are intimate with some friends, merely sociable with others. If we count up all of our group associations, ranging from the powerful associations that define our daily lives to the thinnest connections about which we have some feeling, we will uncover connections to literally hundreds of groups.

What is a group? Recall from Chapter 5 that a **group** is a collection of individuals who interact with one another, share goals and norms, and have a subjective awareness as "we." To be considered a group, a social unit must have all three characteristics. The boundaries of the definition are necessarily hazy. Consider three superficially similar examples: The individuals in a line waiting to board a bus are unlikely to have a sense of themselves as a group. A line of prisoners chained together and waiting to board a bus to the penitentiary is more likely to have a thicker sense of common feeling. Finally, no doubt the passengers who overpowered the hijacking pilots on American Airlines Flight 92 on September 11, 2001, sadly subsequently crashing the plane into a Pennsylvania farmland, became a group for a few moments.

As you remember from Chapter 5, certain gatherings are not groups in the strict sense but may be *social categories* (teenagers or truck drivers) or *audiences* (everyone watching the same movie). The importance of defining a group is not to perfectly diagnose whether a social unit is a group—an unnecessary endeavor— but to help us understand the behavior of people in society. As we inspect groups, we are able to identify characteristics that reliably predict trends in the behavior of the group, and even the behavior of individuals in the group.

The study of groups has application at all levels of society, from the attraction between people who fall in love to the characteristics that make some corporations drastically outperform their competitors. The aggregation of individuals into groups has a transforming power, and sociologists understand the social forces that make these transformations possible. In this chapter, we move from the *micro* level of analysis (the analysis of groups and face-to-face social influence) to the *macro* level of analysis (the analysis of formal organizations and bureaucracies).

Dyads and Triads: Group Size Effects

Even the smallest groups are of acute sociological interest. A **dyad** is a group consisting of two people. A **triad** consists of three people. This seemingly minor distinction, first scrutinized by the German sociologist **Georg Simmel** (1858–1918), can have critical consequences for group behavior (Simmel 1902). Simmel was interested in discovering the effects of size on groups, and he found that the mere difference between two and three spawned entirely different group dynamics (the behavior of a group over time).

Imagine two people standing in line for lunch. First one talks, then the other, then the first again. The interaction proceeds in this way for several minutes. Now imagine a third person enters the interaction. The character of the interaction suddenly changes. At any given moment, two people are interacting more with each other than either is with the third. When the third person wins the attention of one of the other two, a new dyad is formed, supplanting the previous pairing. The group, a triad, then consists of a dyad (the pair that is interacting) plus an *isolate.*

Triadic segregation is what Simmel called the tendency for triads to segregate into a pair and an isolate. A triad tends to segregate into a *coalition* of the dyad against the isolate—a two-against-one situation. The isolate then has the option of initiating a coalition with either member of the dyad. This choice is a type of social advantage, leading Simmel to coin the principle of *tertius gaudens,* a Latin term meaning "the third one gains."

Interactions in a triad very often end up as two against one. You may have noticed this principle of coalition formation in your own conversations. Perhaps two friends want to go to a movie you do not want to see. You appeal to one of them to go instead to a minor league baseball game. She wavers and comes over to your point of view—you have formed a coalition of two against one. The friend who wants to go to the movie is now the isolate. He may recover lost social ground by trying to form a new coalition, such as by suggesting a new alternative (going bowling or going to a different movie). This flip-flop interaction may continue for some time, demonstrating another observation by Simmel: A triad is an *unstable* social grouping, whereas dyads are relatively stable. The minor distinction between dyads and triads—one person—has important consequences because it changes the character of the interaction within the group. Simmel is known as the discoverer of **group size effects**—the effects of group number on group behavior.

Primary and Secondary Groups

Charles Horton Cooley (1864–1929), a famous sociologist of the Chicago School of sociology, introduced the concept of the **primary group**—defined as a group consisting of intimate, face-to-face interaction and relatively long-lasting relationships. Cooley had in mind the family and early peer group. In his original formulation, "primary" was used in the sense of "first," the intimate group of the formative years (Cooley 1909/1967). The insight that an important distinction should be made between intimate groups and other groups proved extremely fruitful. Cooley's somewhat narrow concept of family and childhood peers has been elaborated upon over the years to include a variety of intimate relations as examples of primary groups.

Primary groups have a powerful influence on an individual's personality or self-identity. The effect of family on an individual can hardly be overstated. The weight of peer pressure on school children is particularly no-

One of the best examples of the primary group is that consisting of parent and child.

torious. Street gangs are a primary group. The camaraderie formed among Marine Corps units in boot camp is another classic example of a primary group.

In contrast to primary groups are **secondary groups** that are larger in membership, less intimate, and less long-lasting. Secondary groups tend to be less significant in the emotional lives of people. Secondary groups include all the students at a college or university, all the people in your neighborhood, and all the people in a bureaucracy or corporation.

Secondary groups can occasionally take on the characteristics of primary groups. The process can be accelerated in situations of high stress or crisis. When a neighborhood meets with a major catastrophe, people who may know each other only as acquaintances often come to depend on each other and in the process become more intimate. The secondary group of neighbors becomes, for a time, a primary group. This is precisely what happened in otherwise impersonal neighborhoods in New York City near "ground zero" of the September 11, 2001 terrorist attack on the World Trade Center: Thousands of people pitched in to help, and as a result, many primary groups formed. Similar formations of primary groups no doubt took place in many locations in the eastern United States during the power blackout of August 2003.

Primary and secondary groups serve different needs. Primary groups give people intimacy, companionship, and emotional support. These are termed **expressive needs** (also called socioemotional needs). Family and friends share and amplify your good fortune, rescue you when you misbehave, and cheer you up when life looks grim. Primary groups are a major influence on

These are members of a large class of graduating seniors—a secondary group.

social life and an important source of social control. They are also a dominant influence on your likes and dislikes, preferences in clothing, political views, religious attitudes, and other characteristics. Many studies have shown the overwhelming influence of family and friendship groups on religious and political affiliation, as illustrated in the box, "Doing Sociological Research: Sharing the Journey" (Wuthnow 1994).

Secondary groups serve **instrumental needs** (also called task-oriented needs). Athletic teams form to have fun and win games. Political groups form to raise funds and bend the will of the legislature. Corporations form to make profits, and employees join corporations to make a living. Needless to say, intimacies can develop in the act of fulfilling instrumental needs, and primary groups may also devote themselves to meeting instrumental needs. The true distinction between primary and secondary groups is in how strongly the participants feel about one another and how dependent they are on the group for sustenance and identity. Both primary and secondary groups are indispensable elements of society.

> **THINKING *SOCIOLOGICALLY***
>
> Name a *primary group* and a *secondary group* to which you belong. Compare the two groups in terms of their size, degree of intimacy, and the nature of the interaction you have within them. Are these groups primarily *expressive, instrumental,* or both?

Reference Groups

Primary and secondary groups are groups that have members. **Reference groups** are those to which you may or may not belong, but that you use as a standard for evaluating your values, attitudes, and behaviors (Merton and Rossi 1950). Reference groups are generalized versions of role models. They are not "groups" in the sense that the individual interacts within (or in) them.

DOING SOCIOLOGICAL RESEARCH

Sharing the Journey

Modern society is often characterized as remote, alienating, and without much feeling of community or group belongingness. This image of society has been carefully studied by sociologist Robert Wuthnow who noticed that, in the United States, people are looking to small groups as a place where they can find emotional and spiritual support and where they find meaning and commitment, despite the image of society as an increasingly impersonal force.

Wuthnow began his research by noting that, even with the individualistic culture of U.S. society, small groups play a major role in this society. He saw the increasing tendency of people to join recovery groups, reading groups,

spiritual groups, and a myriad of other support groups. Wuthnow began his research by asking some specific questions, including, "What motivates people to join support groups?" "How do these groups function?" and "What do members like most and least about such groups?" His broadest question, however, was to wonder how the wider society is being influenced by the proliferation of small, support groups. To answer these questions, a large research team of fifteen scholars designed a study that included both a quantitative and a qualitative dimension. They distributed a survey to a representative sample of more than 1000 people in the United States. Supplementing the survey were interviews with more than 100 support group

members, group leaders, and clergy. The researchers chose twelve groups for extensive study. Researchers spent six months to three years tracing the history of these groups, meeting with members and attending group sessions.

Based on this research, Wuthnow concludes that the small group movement is fundamentally altering U.S. society. Forty percent of all Americans belong to one or more small groups. As the result of participation in these groups, social values of community and spirituality are undergoing major transformation. People say they are seeking community when they join small groups—whether the group be a recovery group, a religious group, a civic association, or some other small group.

© Joseph Sohm, Chromo Sohm Inc./Corbis

Do you pattern your behavior based on sports stars, musicians, military officers, or business executives? If so, those role models are your reference groups.

Imitation of reference groups can have both positive and negative effects. Members of a Little League baseball team may revere major league baseball players and thus attempt to imitate laudable behaviors such as tenacity and sportsmanship, but young baseball fans are also liable to be exposed to tantrums and fights, tobacco spitting, and scandals such as drug problems and domestic violence that befall many professional athletes. This illustrates that the influence of a single reference group can be both positive and negative.

Research has shown that identification with reference groups can have a strong effect on self-evaluation and self-esteem. Before school desegregation began, it was thought that all-Black schools contributed to a negative self-evaluation among Black students. Desegregation was expected to raise the self-esteem of Black children (Clark and Clark 1947). In some cases, it did, but later research has also found that identification with a positive reference group was more important than desegregation. When racial or ethnic groups were consistently and methodically presented in a positive way, as in later pluralistic and multicultural educational programs designed to increase pride in Black culture, the self-esteem of the children was greater than that of Black children in integrated programs with no pluralistic component. The same has been found for Latino children enrolled in Latino cultural awareness programs. Plainly, the representation of racial and ethnic groups in a society can have a striking positive or negative effect on

children who are acquiring their lifetime set of group affiliations (Zhou and Bankston 2000; Baumeister 1998; Steele 1996, 1992; Steele and Aronson 1995; Banks 1976).

In-Groups and Out-Groups

When groups have a sense of themselves as "us," there will be a complementary sense of other groups as "them." The distinction is commonly characterized as *in-groups* versus *out-groups*. The concept was originally elaborated by the early sociological theorist W. I. Thomas (1931). College fraternities and sororities certainly exemplify "in" versus "out." So do families. The same can be true of the members of your high school class, your sports team, your racial group, your gender, and your social class. Members of the wealthy classes in the United States sometimes refer to one another as PLUs—"people like us" (Graham 1999; Frazier 1957).

Attribution theory is the principle that we all make inferences about the personalities of others (called *dispositional attributions*), such as concluding what another person is "really like." These attributions depend on whether you are in the in-group or the out-group. Thomas F. Pettigrew notes that individuals commonly generate a significantly distorted perception of the motives and capabilities of other people's acts based on whether those people are in-group or out-group members (Pettigrew 1992; Gilbert and Malone 1995). Pettigrew describes this misperception as **attribution error,** meaning errors made in attributing causes for people's behavior to their membership in a particular group,

---

People turn to these small groups for emotional support more than for physical or monetary support.

Wuthnow argues that large-scale participation in small groups has arisen in a social context in which the traditional support structures in U.S. society, such as the family, no longer provide the sense of belonging and social integration that they provided in the past. Geographic mobility, mass society, and the erosion of local ties all contribute to this trend. People still seek a sense of community, but they create it in groups that also allow them to maintain their individuality. In voluntary small groups that are different from the family, you are free to leave if the group no longer meets your needs.

Wuthnow also concludes that these groups represent a quest for spirituality in a society when, for many, traditional religious values have declined. As a consequence, support groups are redefining what is sacred. They also replace explicit religious tenets imposed from the outside with internal norms that are implicit and devised by individual groups. At the same time, these groups reflect the pluralism and diversity that characterizes society. In the end, they buffer the trend toward disintegration and isolation that people often feel in mass societies.

Questions to Consider

1. Are you a member of any support groups? Which ones? *Keywords: small group, support group*

2. Do you think that decline in traditional church membership is at least partly responsible for an increase in membership in support groups? *Keywords: civic participation, voluntary organization*

We have included InfoTrac College Edition keywords at the end of each question to make it easier for you to find more to read on these topics. Go to www.infotrac-college.com, an online library, to begin your search.

Source: Wuthnow, Robert. 1994. *Sharing the Journey: Support Groups and America's New Quest for Community.* New York: Free Press. ●●●

such as a racial group. Attribution error has several dimensions, all tending to favor the in-group over the out-group. In a word, we perceive people in our in-group positively and those in out-groups negatively *regardless* of their actual personal characteristics.

1. When onlookers observe improper behavior by an out-group member, they are likely to attribute the deviance to the *disposition* of the wrongdoer. "Disposition" refers to the perceived "true nature," or "inherent nature," of the person, often considered to be genetically determined. (*Example:* A White person sees a Hispanic person carrying a knife and, without any additional information, attributes this behavior to the "inherent tendency" for Hispanics to be violent. The same would be true if a Hispanic person, without any additional information, assumed that all Whites have an "inherent tendency" to be racists.)

2. When the *same* behavior is exhibited by an in-group member, the common perception is that the act stems from the *situation* of the wrongdoer, not to the in-group member's disposition. (*Example:* A White person sees another White person carrying a knife and concludes, without any additional information, that the weapon must be carried for protection in a dangerous area.)

3. If an out-group member is seen to perform in some laudable way, the behavior is often attributed to a variety of special circumstances, and the out-group member is seen as "the exception."

4. An in-group member who performs in the *same* laudable way is given credit for a worthy personality.

Typical attribution errors include misperceptions between racial groups and also between men and women. If a White policeman shoots a Black or Latino, a White individual, given no additional information, is likely to simply assume that the victim instigated the shooting, whereas a Black individual is more likely to assume that the policeman fired unnecessarily, perhaps because he is dispositionally predisposed to be a racist (Kluegel and Bobo 1993; Bobo and Kluegel 1991).

A related phenomenon has been seen in men's perceptions of women coworkers. Meticulous behavior in a man is perceived positively and is seen by the man as "thorough"; in a woman, the exact *same* behavior is perceived negatively and is considered "picky." Behavior applauded in a man as "aggressive" is condemned in a woman exhibiting the same behavior as "pushy" or "bitchy" (Uleman et al. 1996; Wood 1994).

Social Networks

As already noted, no individual is a member of only one group. Social life is far richer than that. Member-

ship in several groups provides links between groups and many groups often overlap. A **social network** is a set of links between individuals or other social units, such as bureaucratic organizations or even entire nations (Centeno and Hargittai 2003; Mizruchi 1992). Your group of friends, for instance, or all the people on an electronic mail list to which you subscribe are social networks (Wasserman and Faust 1994).

The network of people you are closest to, not those merely linked to you in some impersonal way, is probably most important to you. Numerous research studies indicate that people get jobs via their personal networks more often than through formal job listings, want ads, or placement agencies (Petersen et al. 2000; Granovetter 1995, 1974). This is especially true for high-paying, prestigious jobs. Getting a job is more often a matter of *who* you know than *what* you know, although both are of course important. Who you know, and who they know in turn, is a social network that may have a marked effect on your life and career.

Networks form with all the spontaneity of other forms of human interaction (Granovetter 1973; Mintz and Schwartz 1985; Wasserman and Faust 1994; Fischer 1981; Knoke 1992). One's family usually forms the first social network in a person's life. Later other social networks evolve, within one's neighborhood, professional contacts, and associations formed in fraternal, religious, and volunteer groups. Networks to which you are only *weakly* tied still provide you with access to that network; hence, the sociological principle that there is "strength in weak ties" (Petersen et al. 2000; Montgomery 1992; Granovetter 1973).

Networks based on race, class, and gender form with particular readiness. This has been particularly true of job networks. The person who leads you to a job is likely to have a social background similar to yours. Recent research indicates that the "old boy network"—any network of White male corporate executives—is less important than it used to be. The increasing prominence of women and minorities in business organizations is diminishing the importance of the old boy network. Still, as we will see later in this chapter, women and minorities are considerably underrepresented in corporate life, especially in high-status jobs (Green et al. 1999; Collins-Lowry 1997; Reskin and Padavic 1994; Gerson 1993).

Networks can reach around the world, but how big is the world? How many of us have remarked, "My, it's a small world," upon discovering that someone we just met is a friend of a friend? Research into what has come to be known as the *small world problem* has shown that under certain conditions, networks make the world a lot smaller than you might think.

Original "small world" researchers Travers and Milgram wanted to test whether a document could be routed to a complete stranger more than 1000 miles away using only a chain of acquaintances (Travers and

Milgram 1969; Lin 1989; Kochen 1989; Watts and Strogatz 1998; Watts 1999). The researchers organized an experiment in which approximately 300 "senders" were all charged with getting a document to one "receiver," a complete stranger. The receiver was a male Boston stockbroker. The senders were one group of Nebraskans and one group of Bostonians chosen completely at random. Every sender in the study was given the receiver's name, address, occupation, alma mater, year of graduation, his wife's maiden name, and hometown. They were asked to send the document directly to the stockbroker if they knew him on a first-name basis. Otherwise, they were asked to send the folder to a friend, relative, or acquaintance known on a first-name basis who might be more likely than the sender to know the stockbroker.

How many intermediaries do you think it took, on average, for the document to get through? (Most people estimate from twenty to hundreds.) About one-third of the documents arrived at the target. This was good, considering that the senders did not know the target person. The surprising thing was that the average number of contacts between sender and target was only 6.2! Hence the conclusion that any given person in the country is on average only about "six degrees of separation" from any other person, and in this sense the world "is small."

This original small-world research has recently been criticized on two grounds: First, only one-third of the documents actually arrived at the target person. The 6.2 average intermediaries applied only to these completed chains. Thus two-thirds of the initiated documents never reached the target person at all. For these persons, the world was certainly not "small." Second, the sending chains tended to closely follow occupational, social class, and ethnic lines—just as general network theory would predict (Wasserman and Faust 1994). Thus, the world may indeed be "small," but only for those in your own social network in the first place (Kleinfeld 2002; Watts 1999).

A recent and ongoing study of Black national leaders by Taylor and associates shows that Black leaders form a very closely knit network, one considerably more dense than longer-established White leadership networks (Jackson et al. 1995, 1994; Taylor 1992b; White leadership networks have been examined by Mills 1956; Domhoff 1998; Kadushin 1974; Moore 1979; Alba and Moore 1982). The world is indeed quite "small" for America's Black leadership. Included in the study were members of Congress, mayors, business executives, military officers (generals and full colonels), religious leaders, civil rights leaders, media personalities, entertainment and sports figures, and others. The study found that when considering only personal acquaintances—not indirect links involving intermediaries—one-fifth of the entire national Black leadership network is included: One-fifth of all Black leaders studied

know each other directly, as a friend or close acquaintance. The Black leadership network is considerably more closely connected than White leadership networks. The Black network had greater *density*. Add only one intermediary, the friend of a friend, and almost *three-quarters* of the entire Black leadership network is included: Any given Black leader can generally get in touch with three-quarters of all other Black leaders in the entire country either by knowing them personally (as a "friend") or via only one common acquaintance (as a "friend of a friend"). That is pretty amazing when one realizes that the study is considering the population of Black leaders in the entire country.

Social Influence in Groups

The groups in which we participate exert tremendous influence on us. We often fail to appreciate how powerful these influences are. For example, who decides what you should wear? Do you decide for yourself each morning, or is the decision already made for you by fashion designers, role models, and your peers? Consider how closely your hair length, hair styling, and choice of jewelry has been influenced by your peers. Did you invent your baggy pants, your dreadlocks, or your blue blazer? People who label themselves "nonconformist" often conform rigidly to the dress code of their in-group. This was true of the Beatniks in the 1950s, the hippies of the early 1960s and 1970s, the punk rockers of the 1970s and 1980s, and the grunge culture of the 1990s.

After the rebelliousness of youth has faded, the influences of our parents extend to adulthood. The choices of political party among adults (Republican, Democratic, or Independent) correlate strongly with the political party of their parents—again demonstrating the power of the primary group. Seven out of ten people vote with the party of their parents even though these same people insist that they think only for themselves when voting (Worchel et al. 2000; Taylor et al. 1997; Jennings and Niemi 1974). Furthermore, most people share the religious affiliation of their parents, although they will insist that they choose their own religion, free of any influence by either parent.

We like to think we stand on our own two feet, immune to a phenomenon as superficial as group pressure. After all, it is part of our national motto of individualism. The conviction that one is impervious to social influence results in what social psychologist Philip Zimbardo calls the *Not-Me Syndrome*: When confronted with a description of group behavior that is disappointingly conforming and not individualistic, most individuals counter that some people may be like that, "but not me" (Zimbardo et al. 1977). We all think: "*Other* people yield to group pressure, *but not me*." But sociological experiments often reveal a dramatic gulf between what people think they will do and what they actually do. The conformity study by Solomon Asch discussed in the following section is a case in point.

The Asch Conformity Experiment

Social influences are strong enough to make us behave in ways that would cause us discomfort on later examination. Are they strong enough to make us disbelieve our own senses? In a classic piece of work known as the Asch Conformity Experiment, Solomon Asch showed that even simple objective facts cannot withstand the distorting pressure of group influence (Asch 1951, 1955).

Examine the two figures in Figure 6.1. Which line on the right is equal in length to the line on the left? Line B, obviously. Could anyone fail to answer correctly?

Solomon Asch discovered that social pressure of a gentle sort was sufficient to cause an astonishing rise in the number of wrong answers. Asch lined up five students at a table and administered the test shown in the figure. Unknown to the fifth student, the first four were confederates—collaborators with the experimenter who only pretended to be participants. For several rounds, the confederates gave correct answers to Asch's tests. The fifth student also answered correctly, suspecting nothing. Then the first student gave a wrong answer. The second student gave the same wrong answer. Third, wrong. Fourth, wrong. Then came the fifth student's turn.

In Asch's experiment, fully *one-third* of all students in the fifth position gave the same wrong answer as the confederates at least half the time. Forty percent gave

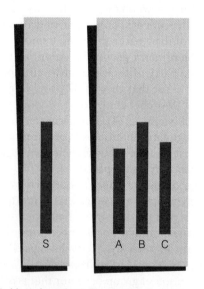

Figure 6.1 Lines from Asch Experiment

Source: Asch, Solomon. 1955. "Opinion and Social Pressure." *Scientific American* 19 (July): 31–35.

"some" wrong answers. Only one in four consistently gave correct answers in defiance of the invisible pressure to conform.

Line length is not a vague or ambiguous stimulus. It is clear and objective; look at Figure 6.1! Wrong answers from one-third of all subjects is a very high percentage. The subjects fidgeted and stammered while doing it, but they did it nonetheless. Those who did not yield to group pressure showed even more stress than those who yielded to the (apparent) opinion of the group.

Would you have gone along with the group? Perhaps, perhaps not. Sociological insight grows when we acknowledge the fact that one-third of all participants will yield to the group. The Asch experiment has been repeated many times over the years, with students and nonstudents, old and young, in groups of different sizes, and in different settings (Worchel et al. 2000; Cialdini 1993; Taylor et al. 1997). The results remain essentially the same. One-third to one-half of the subjects make a judgment contrary to objective fact, yet in conformity with the group.

The Milgram Obedience Studies

What are the limits of social pressure? In terms of moral and psychological issues, judging the length of a line is a small matter. What happens if an authority figure demands obedience—a type of conformity—even if the task is something the test subject finds reprehensible? A chilling answer emerged from the now-famous Milgram Obedience Studies, done from 1960 through 1974 by Stanley Milgram (1974).

In this study, a naive research subject entered a laboratory-like room and was told that an experiment on learning was to be conducted. The subject was to act as a "teacher," presenting a series of test questions to another person, the "learner." Whenever the learner gave a wrong answer, the teacher would administer an electric shock.

The test was relatively easy. The teacher read pairs of words to the learner, such as

blue box
nice house
wild duck

The teacher then tested the learner by reading a multiple-choice answer, such as

blue: sky   ink   box   lamp

The learner had to recall which term completed the pair of terms given originally, in this case, blue box.

If the learner answered incorrectly, the teacher was to press a switch on the shock machine, a formidable-looking device that emitted an ominous hum when activated (see Figure 6.2). For each successive wrong an-

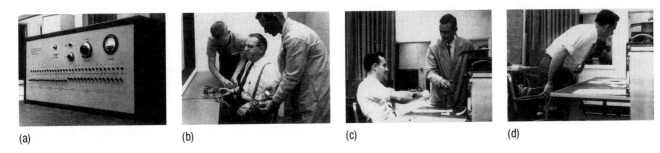

(a) (b) (c) (d)

Figure 6.2 Milgram's Setup

These photographs show how intimidating—and authoritative—the Milgram experiment must have been. Picture (a) shows the formidable-looking shock generator. Picture (b) shows the role player, who pretends to be getting the electric shock, being hooked up. Picture (c) shows an experimental subject (seated) and the experimenter (standing, in lab coat). Picture (d) shows a subject terminating the experiment prematurely, that is, before giving the highest shock level (voltage). A large majority of subjects did not do this and went all the way to the maximum shock level (65 percent of them did).

Source: Milgram, Stanley. 1974. Obedience to Authority: An Experimental View. New York: Harper and Row, p. 25. © Copyright 1965 by Stanley Milgram.

swer, the teacher was to increase the intensity of the shock by 15 volts.

The machine bore labels clearly visible to the teacher: *Slight Shock, Moderate Shock, Strong Shock, Very Strong Shock, Intense Shock, Extreme Intensity Shock, Danger: Severe Shock,* and lastly, *XXX* at 450 volts. As the voltage rose, the learner responded with squirming, groans, then loud screams.

The experiment was rigged. The learner was a confederate, an actor. No shocks were actually delivered. The true purpose of the experiment was to see if any "teacher" would go all the way to 450 volts.

If the subject tried to quit, the experimenter responded with a sequence of prods:

"Please continue."
"The experiment requires that you continue."
"It is absolutely essential that you continue."
"You have no other choice, you *must* go on."

In the first experiment, fully 65 percent of the volunteer subjects went *all the way* to 450 volts on the shock machine.

Milgram himself was astonished. Before carrying out the experiment, he had asked a variety of psychologists, sociologists, psychiatrists, and philosophers to guess how many subjects would go all the way to 450 volts. The opinion of these consultants was that *one-tenth of 1 percent* (or one in a thousand) would do it.

What would you have done? Remember the Not-Me Syndrome. Think about the experimenter as an impressive authority figure in a white lab coat saying, "You have no other choice, you *must* go on." Most people claim they would refuse to continue as the voltage escalates. The significance of this experiment derives in part from how starkly it highlights the difference between what people *think* they will do and what they *actually* do.

Milgram devised a series of additional experiments in which he varied the conditions to find out what might cause subjects *not* to go all the way to 450 volts. He had the learner complain of a heart condition. Still, well over half of the subjects delivered the maximum shock level. Milgram speculated that women might be more humane than men (all prior experiments used only male subjects), so he did the experiment again using only women subjects. The results? Exactly the same. Class background made no difference. Racial and ethnic differences had no detectable effect on compliance rate.

At the time that the Milgram experiments were conducted, the world was watching the trial in Jerusalem of World War II Nazi Adolf Eichmann. Millions of Jews, Gypsies, homosexuals, and communists were murdered between 1939 and 1945 by the Nazi Party, led by Adolf Hitler. As head of the Gestapo's Jewish section, Eichmann oversaw the deportation of Jews to concentration camps and the mass executions that followed. Eichmann disappeared after the war, was abducted in Argentina by Israeli agents in 1961, then transported to Israel, where he was tried and ultimately hanged for crimes against humanity.

The world wanted to see what sort of monster could have committed the crimes of the Holocaust, but a jarring picture of Eichmann emerged. He was slight and mild mannered, not the raging ghoul that everyone expected. The psychiatrists who examined him found him to be sane. He insisted that although he had been a chief administrator in an organization whose product was mass murder, he was guilty only of doing what he was told to do by his superiors. He did not hate Jews, he said. In fact, he had a Jewish half-cousin whom he hid and protected. He claimed, "I was just following orders."

How different was Adolf Eichmann from the rest of us? The political theorist Hannah Arendt dared to suggest in her book *Eichmann in Jerusalem* (1963) that evil on a giant scale is banal. It is not the work of monsters, but an accident of civilization. Arendt argued that to find the villain, we need to look into ourselves.

Other evil figures in history have, of course, existed. For example, Osama bin Laden, the presumed mastermind behind the terrorist attacks on the United States, will no doubt be seen by history as an inherently evil despot. Perhaps his followers complied because they were "following orders."

The Iraqi Prisoners at Abu Ghraib: Research Predicts Reality?

We have just learned that ordinary people will do horrible things to other humans simply because of the influence of the group, or because of an authority figure, or because of a combination of both. This has been the lesson of the Asch studies and the Milgram studies. Recent events in the world have once again shown vividly and clearly how accurate such sociological and psychological experiments are in the prediction of actual human behavior.

In the spring of 2004, it was revealed that American soldiers who were military police guards at a prison in Iraq (the prison was named *Abu Ghraib*) had engaged in severe torture of Iraqi prisoners of war. The torture included sexual abuse of the prisoners—having male prisoners simulate sex with other male prisoners, positioning their mouths next to the genitals of another male prisoner, being forced to masturbate in view of others, and other such acts. Still other acts of torture involved physical abuse such as beatings, stomping on the fingers of prisoners (thus fracturing them), and a large number of other physical acts of torture, including bludgeonings, some allegedly resulting in deaths of prisoners. Such tortures are clearly outlawed by the Geneva Conventions and by clearly stated U.S. principles of war. Both male and female guards participated in these acts of torture, and while most of the Iraqi victims were male, some were female.

The guards later claimed that they were simply following orders, either orders directly given or indirectly assumed. President George W. Bush and Secretary of Defense Donald H. Rumsfeld both claimed that the acts of torture were merely the acts of a "corrupt few" and that the vast majority of American soldiers would never engage in such horrible acts.

Now consider what we know from research. The Milgram studies strongly suggest that many ordinary soldiers—who were not at all "corrupt," at least not more than average—would indeed engage in these acts of torture, particularly if they believed that they were under orders to do so, *or* if they believed that they would

not be punished in any way if they did. The American soldiers must bear a significant portion of the responsibility for their own behavior. Nonetheless, *the causes of the soldiers' behaviors lie not in their personalities* (*their "natures"*), *but in the social structure and group pressures of the situation.*

Evidently, the soldiers (guards) in the Abu Ghraib prison would not even have to receive orders in order to engage in the torture of prisoners. A now classic study of a simulated prison by Haney, Banks, and Zimbardo (1973) shows this quite clearly. In this study, Stanford University students were told by an experimenter to enter a dungeonlike basement. Half were told to pretend to be guards (to role play being a guard) and half were told that they were prisoners (to role play being a prisoner). Which did which was *randomly determined*.

After two or three days, the guards began completely on their own to act very sadistically and brutally toward the prisoners—having them strip naked, simulate sex, act subservient, and so on. Interestingly, the prisoners for the most part did just what the guards wanted them to do, no matter how unpleasant the requested act! The experiment was so scary that the researchers terminated the experiment after six days—more than one week early.

Remember that this study was conducted back in 1973—thirty-one years *before* Abu Ghraib. Yet this simulated prison study predicted quite precisely how both "guards" and "prisoners" act in a prison situation. Group influence effects uncovered by the Asch as well as the Milgram studies took over in both the simulated prison of 1973 as well as the only too real Iraqi prison of 2004.

DEBUNKING *SOCIETY'S MYTHS*

Myth: People are just individuals who make up their own minds about how to behave.

Sociological perspective: The Asch, Milgram, and Simulated Prison experiments show conclusively that people get profoundly influenced by group pressure, often causing them to make up their minds contrary to objective fact and even to deliberately cause harm to another person.

Groupthink

Wealth, power, and experience are apparently not enough to save us from social and group influences. **Groupthink**, as described by I. L. Janis (1982), is the tendency for group members to reach a consensus opinion, even if that decision is downright stupid. Janis reasoned that because major government policies are often the result of group decisions, it would be fruitful to analyze the group dynamics that operate at the highest level of government—for instance, in the Office of the President of the United States. The president makes decisions based on group discussions with his advisers. The president is human and thus susceptible to group influence. To what extent have past presidents and their advisers been influenced by group decision making instead of just the facts?

Janis investigated five ill-fated presidential decisions, all the products of group discussion:

- The decision of the Naval High Command in 1941 *not* to prepare for attack on Pearl Harbor by Japan;
- President Harry Truman's decision to send troops to North Korea in 1951;
- President John F. Kennedy's attempt to overthrow Cuba by launching an invasion at the Bay of Pigs in 1962;
- President Lyndon B. Johnson's decision in 1967 to increase the number of U.S. troops in Vietnam;
- The fateful decision by President Richard M. Nixon's advisers in 1972 to break into Democratic Party headquarters at the Watergate apartment complex, launching the famed "Watergate affair"

All of the preceding were the result of group decisions, and all were absolute fiascoes. The Bay of Pigs invasion was a major humiliation for the United States, a covert outing so ill-conceived it is hard to imagine how it survived discussion by a group of foreign policy experts. Fifteen hundred Cuban exiles trained by the CIA to parachute into heavily armed Cuba landed in a dense, impassable swamp eighty miles from their planned drop zone with inadequate weapons and incorrect maps. A sea landing was demolished by well-prepared, pre-warned Cuban defenders. The fiasco caused the resignation of the then head of the CIA. (In 2004, CIA head George Tenet resigned amid speculation that the Abu Ghraib prison situation was responsible.)

The men who advised President Kennedy to undertake the invasion were not stupid. Many considered them the brightest policy team ever assembled—"the best and the brightest" as they sometimes had been called. Secretary of State Dean Rusk was a past president of the Rockefeller Foundation. Secretary of Defense Robert McNamara was a gifted statistician and past president of the Ford Motor Company. McGeorge Bundy, special assistant for national security, had been dean of Arts and Sciences at Harvard University. How could such a smart team perpetrate such a blunder?

Janis discovered a common pattern of misguided thinking in his investigations of presidential decisions. He surmised that outbreaks of groupthink had several things in common:

1. **An illusion of invulnerability.** "With such a brilliant team and such a nation, how could *any* plan fail," thought those in the group.

2. **A falsely negative impression of those who are antagonists to the group's plans.** Fidel Castro was perceived to be clownish, and Cuban troops were supposed to be patsies. In truth, the defenders at the Bay of Pigs were highly trained commandos.

3. **Discouragement of dissenting opinion.** As groupthink takes hold, dissent is equated with disloyalty. This can discourage dissenters from voicing their objections.

4. **An illusion of unanimity.** In the aftermath, many victims of groupthink recall their reservations, but at the moment of decision, the prevailing sense is that the entire group is in complete agreement.

We now might ask if groupthink influenced the torture of Iraqi prisoners at the Abu Ghraib prison. The actions of the military guards were, it seems at least in part, indirectly or directly the result of high-level group decisions among presidential advisors. Groupthink is not inevitable when a team gathers to make a decision, but it is common and appears in all sorts of groups, from student discussion groups to the highest councils of power (Flowers 1977; McCauley 1989; Aldag and Fuller 1993; Kelley et al. 1999).

Risky Shift

The term *groupthink* is commonly associated with group decision making whose consequences are not merely unexpected but disastrous. Another group phenomenon, **risky shift** (also called *polarization shift*), may help explain why the products of groupthink are frequently calamities. Have you ever found yourself in a group engaged in a high-risk activity that you would not do alone? When you created mischief as a child, were you not usually part of a group? If so, you were probably in the thrall of risky shift—the tendency for groups to weigh risk differently than individuals do.

THINKING *SOCIOLOGICALLY*

Think of a time when you engaged in some risky behavior. What group were you part of, and how did the group influence your behavior? Were you aware of being influenced? How does this illustrate the concept of *risky shift*? Is there more risky shift with more people in the group, thus illustrating *group size effect*?

Risky shift was first observed by James Stoner (1961). Stoner gave study participants descriptions of a situation involving risk, such as one in which an engineer must choose between job security and potentially lucrative but risky advancement. The participant is then asked to decide how much risk the engineer should take. Before performing his study, Stoner believed that individuals in a group would take *less* risk in a group than

© Pohle/Sutcliffe/Sipa Press

Streaking, or running nude in a public place, is more common as a group activity than as a strictly individual one. This illustrates how the group can provide the persons in it with deindividuation, *or merging of self with group. This allows the individual to feel less responsibility or blame for his or her actions and thus convince herself or himself that the group must share in the blame.*

individuals alone, but he found that after his groups had engaged in open discussions, they favored greater risk than before discussion.

His research stimulated hundreds of studies using males and females, different nationalities, different tasks, and other variables (Pruitt 1971; Blaskovitch 1973; Johnson et al. 1977; Hong 1978; Worchel et al. 2000; Taylor et al. 2003). The results are complex. Most but not all group discussion leads to greater risk-taking. In subcultures that value caution above daring, as in some work groups of Japanese and Chinese firms, group decisions are *less* risky after discussion than before. The shift can occur in either direction, driven by the influence of group discussion, but there is generally some kind of shift, in one direction or the other, as opposed to no shift at all (Kerr 1992).

What causes risky shifts? The most convincing explanation is that *deindividuation* occurs. Deindividuation is the sense that one's self has merged with a group. In terms of risk-taking, one feels that responsibility (and possibly blame) is borne not only by oneself but also by the group. This seems to have happened among the American prison guards who tortured prisoners at Iraq's Abu Ghraib prison: Each guard could convince himself or herself that responsibility, hence blame, was to be borne by the group as a whole.

The greater the number of people in a group, the greater the tendency toward deindividuation. In other words, deindividuation is a *group-size effect*. As groups get larger, trends in risk-taking are amplified.

Formal Organizations and Bureaucracies

Groups, as we have seen, are capable of influencing individuals to a very great extent. The study of groups and their effects on the individual represent an example of *microanalysis*, to use a concept introduced in Chapter 5. In contrast, the study of formal organizations and bureaucracies, a subject to which we now turn, represents an example of *macroanalysis*. The focus on groups draws our attention to the relatively small and less complex, whereas the focus now on organizations draws our attention to the relatively large and more structurally complex.

A **formal organization** is a large secondary group, highly organized to accomplish a complex task or tasks and to achieve goals efficiently. Many of us belong to various formal organizations: work organizations, schools, political parties, to name a few. Organizations are characterized by their relatively large size, compared with a small group such as a family or a friendship circle. Often organizations consist of an array of other organizations. The federal government is a huge organization, comprising numerous other organizations, most of which are also vast. Each organization within the federal government is also designed to accomplish specific tasks, be it collecting your taxes, educating the nation's children, or regulating the nation's transportation system and national parks.

Organizations develop their own cultures and routinized practices. The culture of an organization may be reflected in certain symbols, values, and rituals. Some organizations develop their own language and styles of dress. The norms can be subtle, such as men expected to wear long sleeve shirts or women expected to wear stockings, even on hot summer days. It does not take explicit rules to regulate this behavior; comments from coworkers or bosses may be enough to enforce such organizational norms. Some work organizations have instituted a practice called "casual day"—one day per week, usually Friday, when workers can dress less formally.

Organizations tend to be persistent although they are also responsive to the broader social environment where they are located (DiMaggio and Powell 1991). Organizations are frequently under pressure to respond to changes in the society, incorporating new practices and beliefs into their structure as society itself changes. Business corporations have had to respond to increasing global competition. They do so by expanding into new international markets, developing a globally focused workforce, and trimming costs by "downsizing," that is, eliminating workers and various layers of management. Another recent response to increased global competition is outsourcing—having manufacturing tasks ordinarily performed by the home company (such as the manufacture of athletic shoes or soccer balls) performed instead by foreign workers.

Organizations can be tools for innovation, depending on the organization's values and purpose. Rape crisis centers are examples of organizations that originally emerged from the women's movement because of the perceived need for services for rape victims. Rape crisis centers have, in many cases, changed how police departments and hospital emergency personnel respond to rape victims. By advocating changes in rape law and services for rape victims, rape crisis centers have generated change in other organizations as well (Schmitt and Martin 1999; Fried 1994).

Types of Organizations

Sociologists Blau and Scott (1974) and Etzioni (1975) classify formal organizations into three categories, distinguished by their types of membership affiliation: normative, coercive, and utilitarian.

A **normative organization** is one that people join in order to pursue goals that they consider personally worthwhile. They obtain personal satisfaction but no monetary reward for being in such an organization. In many instances, the person joins the normative organization for the social prestige that it offers. Many are service and charitable organizations. Such organizations are often called **voluntary organizations,** and they include organizations such as the Parent-Teacher Association (PTA), Kiwanis clubs, political parties, religious organizations, the National Association for the Advancement of Colored People (NAACP), B'Nai Brith, LaRaza, and other similar voluntary organizations that are concerned with specific issues. Civic and charitable organizations (such as the League of Women Voters) and political organizations (such as the National Women's political caucus) were formed because for decades women were excluded from all-male organizations and political networks. Like other service and charitable organizations, these have been created to meet particular needs, ones that members see as not being served by other organizations.

Gender, class, race, and ethnicity all play a role in who joins what voluntary organizations. Social class is reflected in the fact that many people do not join certain organizations, simply because they cannot afford to join. Joining a professional organization, as one example, can cost hundreds of dollars each year. Those who feel disenfranchised, however, may join grassroots organizations—voluntary organizations that spring from specific local needs that people think are unmet. A tenants' organization may form to protest rent increases or lack of services, or a new political party may emerge from people's sense of alienation from existing party organizations. African Americans, Latinos, and Native Americans have formed many of their own grassroots organizations, in part because of their historical exclusion from traditional White voluntary organizations. African American fraternities and sororities are a classic example (see the box "Understanding Diversity" on the Delta Sigma Theta sorority).

The NAACP, founded in 1909 by W. E. B. Du Bois (recall him from Chapter 1), and the National Urban League are two other large national organizations that have historically fought racial oppression on the legal and urban fronts, respectively. La Raza Unida, a Latino organization devoted to civic activities as well as combating racial–ethnic oppression, has a large membership, with Latinas holding major offices. Such voluntary organizations dedicated to the causes of people of color have in recent years had more women in leadership positions in the organizations than have many standard White organizations. Similarly, Native American voluntary organizations have had increasing numbers of women in leadership positions (Feagin and Feagin 1993; Snipp 1996, 1989).

Coercive organizations are characterized by membership that is largely involuntary. Prisons are an example of organizations that people are coerced to "join" by virtue of their being punished for a crime. Similarly, mental hospitals are coercive organizations. People are placed in them, often involuntarily, for some form of psychiatric treatment. Prisons and mental hospitals are similar in many respects in their treatment of inmates or patients. They both have strong security measures such as guards, locked and barred windows, and high walls (Goffman 1961; Rosenhan 1973). Sexual harassment and sexual victimization are quite common in both prisons and mental hospitals (Andersen 2003; Chesney-Lind 1992).

The sociologist Erving Goffman has described coercive organizations as total institutions. A **total institution** is an organization cut off from the rest of society where individuals who reside there are subject to strict social control (Goffman 1961). Total institutions include two populations: the "inmates" and the staff. Within total institutions, the staff exercises complete power over inmates. Nurses have power over mental patients in the same way that guards have power over prisoners. The staff administers all the affairs of everyday life, including basic human functions such as eating and sleeping. Rigid routines are characteristic of total institutions, thus explaining the common complaint by those in hospitals that they cannot sleep because nurses routinely enter their rooms at night, whether or not the patient needs treatment.

The third type of organization is the **utilitarian organization.** These are large organizations, either for profit or nonprofit, that are joined by individuals for specific purposes, such as monetary reward. Large business organizations that generate profits (in the case of for-profit organizations) and salaries and wages for their employees (as with either for-profit or nonprofit organizations) are these kinds of organizations. Examples of large for-profit organizations include General Motors, Microsoft, Amazon.com, and Procter & Gamble. Examples of large nonprofit organizations that pay salaries to employees are colleges, universities, and the organization that manufactures the Scholastic Assessment Test (SAT), Educational Testing Service (ETS).

Some utilitarian organizations may also be coercive organizations. This is especially the case as various organizations have become increasingly privatized. Mental hospitals may be owned by a large for-profit chain. Although now run by states or the federal government, some prisons have become privatized. Some cities have turned their public schools over to private corporations to see if they can be better managed.

Bureaucracy

As formal organizations develop, many become a **bureaucracy,** a type of formal organization characterized by an authority hierarchy, a clear division of labor, and explicit rules. Bureaucracies are notorious for their un-wieldy size and complexity, as well as for their reputation of being highly impersonal and machinelike in their operation. The federal government is an example of a cumbersome bureaucracy that many believe is ineffective due to its sheer size. Numerous other formal organizations have developed into large bureaucracies: IBM, Disney, and many universities, hospitals, and law firms. Other formal organizations, such as Enron, WorldCom, and Tyco, quickly developed into a large bureaucracy, then subsequently collapsed under fraudulent accounting procedures.

The early sociological theorist **Max Weber** (1947/ 1925) analyzed the classic characteristics of the bureaucracy. These characteristics represent what he called the *ideal type* bureaucracy. This model is rarely seen in reality but defines the typical characteristics of a social form. The characteristics of bureaucracies are

1. *High degree of division of labor and specialization.* The notion of the specialist embodies this criterion. Bureaucracies ideally employ specialists in the various positions and occupations, and these specialists are responsible for a specific set of duties. Job titles and job descriptions define the nature of each such position. Sociologist Charles Perrow (1994, 1986) notes that many modern bureaucracies have hierarchical authority structures and an elaborate division of labor.

UNDERSTANDING DIVERSITY

The Deltas: Black Sororities as Organizations

DELTA Sigma Theta sorority is one of the largest Black women's organizations in the world, with more than 125,000 members. Its membership has included such significant Black women leaders as Mary Church Terrell, Sadie T. M. Alexander, Patricia Roberts Harris, Barbara Jordan, Leontyne Price, and Dorothy Height. Like other Black women's sororities, the unique history of race and gender discrimination gives this organization an identity very different from that of White sororities. In her history of Delta Sigma Theta, whose members are referred to as "Deltas," Paula Giddings, a contemporary African American scholar, writes that organizations:

. . . must be able to adapt to changing environments: in this case, the ever changing exigencies of race relations and the attitudes toward women.

Consequently, Delta has had to alter its purposes and goals throughout the years—and must continue to do so—as well as its internal structure to accommodate them. This makes the Black sorority a particularly dynamic organization.

At the same time, however, changes cannot occur too abruptly or without the consensus of an increasingly diverse constituency. For like other social movement organizations, its viability is dependent on the growth of its membership, which in turn, is largely determined by the number of its members who feel that the sorority's goals are in harmony with their own. This makes it complex as well as concentrated, it invites apathy—the kiss of death for a social movement organization.

The challenge of the sorority, one made all the more difficult by the pathos of the Black women's experience in North America, is to maintain that sense of sisterhood while striving, organizationally, for a more general purpose: aiding the Black community as a whole through social, political, and economic means. As one can see from the rules that govern the social movement organization, this idea can be a difficult one to realize. But the effort to resolve the tension between the goals of organization, and those of the sisterhood, through strengthening social bonds within the context of social action has been an interesting and engaging experiment. It is one that can be seen as a model for Black organizational life, and which adds contour and dimension to the history of Black women in this country.

Source: Giddings, Paula. 1994. *In Search of Sisterhood: Delta Sigma Theta and the Challenge of the Black Sorority Movement.* New York: William Morrow. •••

2. *Hierarchy of authority.* In a bureaucracy, positions are arranged in a hierarchy so that each position is under the supervision of a higher position. The hierarchy is often represented in an *organizational chart*, a diagram in the shape of a pyramid that shows relative rank of each position plus the lines of authority between each. These lines of authority are often called the "chain of command," and they show not only who has authority, but also who is responsible to whom and how many positions are responsible to a given position.

3. *Rules and regulations.* All the activities in a bureaucracy are governed by a set of detailed rules and procedures. These rules are designed, ideally, to cover almost every possible situation and problem that might arise, including hiring, firing, salary scales, rules for sick pay and absences, and the everyday operation of the organization.

4. *Impersonal relationships.* Social interaction in the bureaucracy is supposed to be guided by *instrumental* criteria, such as the organization's rules, rather than by *social–emotional* criteria, such as personal attractions or likes and dislikes. The ideal is based on applying the rules objectively to minimize personal favoritism—such as giving someone a promotion simply because that person is well-liked or firing someone because that person is not well-liked. Of course, as we will see, sociologists have pointed out that bureaucracy has another face—the social interaction that keeps the bureaucracy working and often involves interpersonal friendships and social ties, typically among people taken for granted in these organizations, such as support staff.

5. *Career ladders.* Candidates for the various positions in the bureaucracy should be selected on the basis of specific criteria, such as education, experience, and standardized examinations. The idea is that advancement through the organization becomes a career for the individual.

6. *Efficiency.* Bureaucracies are designed to coordinate the activities of a large number of people in pursuit of organizational goals. Ideally, all activities have been designed to maximize this efficiency. The whole system is intended to keep social–emotional relations and interactions at a minimum and instrumental interactions at a maximum. These characteristics give bureaucracies the reputation of being huge and remote organizations, more intent on generating profit than serving people's needs.

Bureaucracy's Other Face

All the preceding characteristics of Weber's "ideal type" are general defining characteristics. Rarely do bureaucracies meet this exact description. As we will see, sociologists have pointed out that bureaucracy has another face—the social interaction that keeps the bureaucracy working and that often involves interpersonal friendships and social ties, including network ties, typically among those who are taken for granted in these organizations, such as secretaries.

This *informal structure* of social interactions in bureaucratic settings ignore, change, or otherwise bypass the formal structure and rules of the organization. Sociologist Charles Page (1946) coined the term *bureaucracy's other face* to describe this condition.

This other face is the informal culture that evolved over time as a reaction to the formality and impersonality of the bureaucracy. Thus, secretaries "bend the rules a bit" when asked to do something more quickly than usual for a boss they like and slow down or otherwise sabotage work for a boss they do not like. Researchers have noted, for example, that secretaries have more authority than their job titles—and their salaries—suggest. As a way around the cumbersome formal communication channels within the organization, the informal social network or "grapevine" often works better, faster, and sometimes even more accurately. As with any culture, the informal culture in the bureaucracy has its own norms or rules. One is not supposed to "stab friends in the back," as by "ratting on" them to a boss or spreading a rumor about them that is intended to hurt them or get them fired.

Bureaucracy's other face can also be seen in the workplace subcultures that develop, even in the largest bureaucracies. Some sociologists interpret the subcultures that develop within bureaucracies as people's attempts to humanize an otherwise impersonal organization. Keeping photographs of one's family and loved ones in the office, placing personal decorations on one's desk (if permitted), and organizing office parties are ways that people resist the impersonal culture. As with any group, this informal culture can exclude some employees, increasing the isolation they may already feel at work. Gay and lesbian workers may feel left out when other workers gossip about people's dates or discuss family weddings. In male-dominated organizations, women may be left out of the informal banter that may intentionally include inappropriate sexual remarks.

The informal norms that develop within a modern bureaucracy often cause worker productivity to go up or down. The classic Hawthorne studies, so named because they were carried out at the Western Electric telephone plant in Hawthorne, Illinois, in the 1930s (Roethlisberger and Dickson 1939), discovered that small groups of workers developed their own ideas—their own norms—about how much work they should produce each day. If someone produced too many completed tasks in a day, he would make the rest of the workers "look bad" and run the risk of the management raising expectations for the workers' daily production. Because of this, anyone producing too much was informally labeled a "rate buster," and that person was punished by some act, such as punches on the shoulder (called "binging") or by group ridicule ("razzing"). By the same token, someone accused of producing too little was labeled a "chiseler" and punished in the same way. This informal culture of bureaucracy's other face continues today in a manner similar to the culture initially discovered in the early Hawthorne Studies (Perrow 1994, 1986).

Problems of Bureaucracies

Problems have developed in contemporary society that grow out of the nature of the complex bureaucracy. The two problems of risky shift in work groups and the development of groupthink have already been discussed. Two additional problems include a tendency to bureaucratic *ritualism* and the potential for *alienation* on the part of those within the organization.

Ritualism Rigid adherence to rules can produce a slavish following of them, which may not accomplish the purpose for which the rules were originally de-

signed. The rules become an end in themselves rather than a means to an end.

A now classic example of the consequences of *organizational ritualism* has come to haunt us: the explosion of the space shuttle *Challenger* on January 28, 1986, and to our horror, the more recent breakup of yet another space shuttle, the *Columbia*, on February 1, 2003. People in the United States became bound together at the moment of the *Challenger* accident. Many remember where they were and exactly what they were doing when they heard about the tragedy. The failure of the essential O-ring gaskets on the solid fuel booster rockets of the space shuttle caused the catastrophic explosion. It was revealed later that the O-rings became brittle at below-freezing temperatures, which occurred at the launch pad the evening before the *Challenger* took off.

Why did the managers and engineers at NASA (National Aeronautics and Space Administration) allow the shuttle to take off given these prior conditions? The managers had all the information about the O-rings prior to the launch. Furthermore, engineers had warned them against the danger. Sociologist Diane Vaughan

The horror of the explosion of the space shuttle Challenger *in 1986 is seen in the faces of the observers here. All seven astronauts died in the explosion (top right photo). Sociologist Diane Vaughan (1996) attributes the disaster to an ill-formed launch decision based on group interaction phenomena such as risky shift, ritualism, groupthink, and the normalization of deviance. Tragedy struck again in February 2003, when the space shuttle* Columbia *broke up upon reentry into the atmosphere, killing all seven of the astronauts aboard (bottom right photo).*

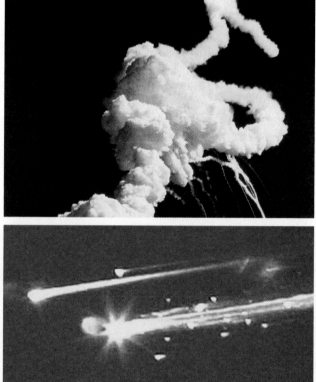

(1996), in a detailed analysis of the decision to launch involving extensive interviews with the managers and engineers who were directly involved, uncovered both risky shift and organizational ritualism within the organization. Vaughan documents the incremental descent into poor judgment supported by a culture of high-risk technology. The NASA insiders, confronted with danger signals, proceeded as if nothing was wrong when they were repeatedly faced with the evidence that something was *very* wrong. They in effect *normalized* their own behavior, so that their actions became acceptable to them, representing nothing out of the ordinary. This is an example of organizational ritualism and what sociologist Vaughan calls the "normalization of deviance."

Unfortunately, history repeated itself on February 1, 2003, when the space shuttle *Columbia*, upon its return from space, broke up in a fiery descent into the atmosphere above Texas, killing all who were aboard. Evidence showed that a piece of hard insulating foam separated from an external fuel tank during launch and struck the shuttle's left wing, damaging it and dislodging its heat-resistant tiles necessary for reentry. The absence of these tiles caused the burn-up upon reentry into the atmosphere. With eerie similarity to the earlier 1986 *Challenger* accident, a recent research report concludes that a "flawed institutional culture" and—citing sociologist Diane Vaughan—a normalization of deviance accompanying a gradual erosion of safety margins were among the causes of the *Columbia* accident (Schwartz and Wald 2003).

No safety rules were broken in either accident. No single individual was at fault. The story is not one of evil intent but of the organizational ritualism in one of the most powerful bureaucracies in the United States. It is a story of rigid group conformity within an organization and of how deviant behavior is redefined, that is, socially constructed, to be perceived as normal. Organizational rituals are so dominant that the means toward goals become the goals themselves. Vaughan's analysis is a powerful warning about the hidden hazards of group conformity in a high-tech age.

Alienation The stresses on rules and procedures within a bureaucracy can result in a decrease in the overall cohesion of the organization. The individual may become psychologically separated from the organization and its goals. This is a state of what is called *alienation*. Alienation results in increased turnover, tardiness, absenteeism, and overall dissatisfaction with the organization.

Alienation can be widespread in organizations where workers have little control over what they do, or are employed on an assembly line, doing the same repetitive action and are treated like machines. Alienation is not restricted to manual labor, however. In organizations where workers are isolated from others, where

they are expected only to implement rules, or where they think they have little chance of advancement, alienation can be common (see Chapter 18). As we will see, some organizations have developed new patterns of work to try to minimize worker alienation and, therefore, enhance their productivity.

The McDonaldization of Society

Sometimes the problems and peculiarities of bureaucracy can affect the total society. Such has been the case with what George Ritzer (2002) has called the *McDonaldization of society*, a term coined from the well-known fast-food chain. In fact, one study (Schlosser 2001) concludes that each month, 90 percent of U.S. children between ages three and nine visit McDonald's! Ritzer noticed that the principles that characterize fast-food organizations are increasingly dominating more and more aspects of U.S. society and societies around the world. "McDonaldization" refers to the increasing and ubiquitous presence of the fast-food model in most organizations that shape daily life. Work, travel, leisure, shopping, health care, education, and politics have all become subject to McDonaldization. Each industry is based on a principle of high and efficient productivity, which translates into a highly rational social organization, with workers employed at low pay, and customers experiencing ease, convenience, and familiarity.

Ritzer argues that McDonald's has been such a successful business model that other industries have adopted the same organizational characteristics. Some have nicknames that associate them with the McDonald's chain: McPaper for *USA Today*, McChild for child-care chains such as Kinder-Care, and McDoctor for the drive-in clinics that deal quickly and efficiently with minor health and dental problems.

Evidence of the "McDonaldization of society" can be seen everywhere, perhaps including on your own campus. Shopping malls, food courts, sports stadiums, even cruise ships reflect this trend toward standardization.

© Billy Barnes/Photo Edit

Ritzer identifies four dimensions of the McDonaldization process: efficiency, calculability, predictability, and control. These characteristics, Ritzer notes, were anticipated long ago by theorist Max Weber:

1. *Efficiency* means that things move from start to completion in a streamlined path. Steps in the production of a hamburger are regulated so that each hamburger is made exactly the same way—hardly characteristic of a home-cooked meal. Business can be even more efficient if the customer does the work once done by an employee, such as using automated teller machines without the personal contact.

2. *Calculability* means there is an emphasis on the quantitative aspects of products sold—size, cost, and the time it takes to get the product. At McDonald's, branch managers must account for the number of cubic inches of ketchup used per day. Sensors on drink machines can cut off the liquid flow to ensure that each drink is exactly the same size. Workers are monitored to determine how long it takes them to complete a trans-

action. Every bit of food and drink is closely monitored by computer, and everything has to be accounted for.

3. *Predictability* is the assurance that products will be exactly the same, no matter when or where they are purchased. Eat an Egg McMuffin in New York, and it will taste just the same as an Egg McMuffin in Los Angeles or Paris!

4. *Control* is the primary organizational principle that lies behind McDonaldization. Behavior of the customers and workers is reduced to a series of machine-like actions. Ultimately, efficient technologies replace much of the work that humans once did. People are also carefully monitored and watched in these organizations, given that uncertainty in human behavior will produce inefficiency and unpredictability.

McDonaldization clearly brings many benefits. There is a greater availability of goods and services to a wide proportion of the population, instantaneous service and convenience to a public with less free time, predictability and familiarity in the goods bought and

MAPPING AMERICA'S DIVERSITY

MAP 6.1 The McDonaldization of Society

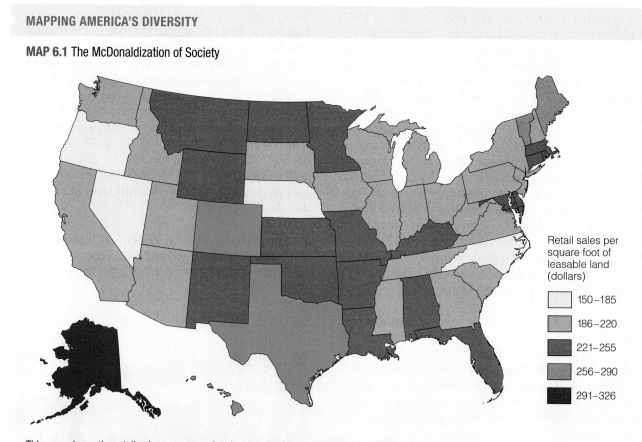

Retail sales per square foot of leasable land (dollars)

150–185
186–220
221–255
256–290
291–326

This map shows the retail sales per square foot by state, within the United States. Notice that some states have a high number of retail sales as a ratio of leasable land and thus a possibly high degree of "McDonaldization"; some states have considerably fewer and thus a relatively low degree of "McDonaldization." Relative to other states, where does your state stand in retail sales per square foot of leasable land?

Data: From the U.S. Census Bureau, 2002. Statistical Abstract of the United States 2002. Washington, DC: U.S. Government Printing Office, p. 650.

sold, and standardization of pricing and uniform quality of goods sold, to name a few. However, this increasingly rational system of goods and services also spawns irrationalities. For example, the majority of workers at McDonald's lack full-time employment, have no worker benefits, have no control over their workplace, and quit on average after four to five months (Schlosser 2001). Ritzer argues that, as we become more dependent on what is familiar and expected, the danger of dehumanization arises. People lose their creativity, and the quality of goods and services is of little concern. This disrupts the fundamental human capacity for error, surprise, and imagination. Even with increasing globalization and the opportunities it provides to expose ourselves to diverse ways of life, McDonaldization has come to characterize other societies, too. A tourist can travel to the other side of the world and taste the familiar Chicken McNuggets or a Dunkin' Donut!

Notice, as you go through your daily life, the extraordinary presence of the process Weber and Ritzer have observed. In what areas of life do you see the process of McDonaldization? How has it influenced the national culture? These questions will help you see how McDonaldization has permeated U.S. society and will help you think about formal organizations as a sociologist would. You might ask where you see evidence of McDonaldization on your campus and in your community.

New Global Organizational Forms and the Japanese Model

The problem of bureaucratization and increasing rationality in social systems has encouraged some people to look beyond Western society bureaucracy for new organizational forms that will potentially avoid some of the problems of dehumanization in bureaucracies that sociologists have noted (Perrow 1994, 1986). The traditional approach to bureaucracy is that, by nature, people hate their jobs and want to avoid work and responsibility. This theory suggested that people do not care about the needs of the organization, only about their personal needs. As a result, according to this perspective, people will work only when they are rigidly controlled and monitored, as with time clocks, and with a system of promotions and salaries. This organizational model lies behind the classical Weberian bureaucracy and, though not so crudely put, still dominates in many work organizations and management perspectives in the United States and parts of western Europe.

More recent bureaucratic perspectives suggest that people are passive and irresponsible only because of their experience with the organization. Many people do have a desire to work as well as to be creative and to take on responsibility. This perspective predicts that an organization can operate efficiently and produce if the self-direction of the individual, called *self-actualization,*

FORCES OF SOCIAL CHANGE

Consumerism

According to sociologist George Ritzer, we all live in a society clogged with McDonald's, Toys "R" Us, Disney World, cybermalls, chain stores, discount houses, cruise ships, gated communities, and the Home Shopping Network. All these structures have pushed us into "hyperconsumption," or *extreme consumerism.* Drawing partly on early sociological theorists Max Weber and Emile Durkheim, Ritzer argues that these "cathedrals of consumption" have created a spectacular and simulated world that has become capable of luring us again and again into new purchases. They have become cathedrals to which we make "pilgrimages" to practice our consumer religion. These new cathedrals of consumption have enticed us to consume far more than we ever did in the past, consume far beyond our needs, giving a picture of the United States (and other countries such as Japan) envisioned decades ago by

the sociologist–economist Thorstein Veblen and his concept of "conspicuous consumption." Even universities entice both parents and students to come and stay with "theme" dorms, fast-food restaurants, souvenir shops, and video arcades. Students become "enchanted" with such cathedrals of consumption, and they are thus more compelled to stay enrolled in the university. The result is that consumption pervades our lives, and we have become consumed by consumption.

Source: Ritzer, George. 1999. *Enchanting a Disenchanted World: Revolutionizing the Means of Consumption.* Thousand Oaks, CA: Pine Forge Press. •••

"Conspicuous consumption," a term used by sociologist-economist Thorstein Veblen, has become increasingly apparent in our materially based culture, even though its "style" has changed, as shown in comparing these two photographs.

© PEMCO/CORBIS

Rick Maiman/CORBIS

A Japanese work group, illustrating close and expressive interaction.

support for the idea that workers are less alienated and experience greater solidarity among themselves in firms that practice more participative management, compared with workers where all power rests with management (Yeats 1991). Participative management and group orientation does not, however, bring workers the same levels of satisfaction as when workers make their own decisions based on their skills and long-term employment (Hodson 1996).

Given the recent collapses of several major bureaucracies in the United States—Enron, WorldCom, and Tyco, for example—no doubt at least some organizational boards of directors will be giving serious consideration to re-tooling and restructuring their organizations along the lines of participative management.

is permitted expression (Argyris 1990). Management's task, under this theory, is to organize things so that people can accomplish their own personal goals while furthering the organization's goals as well.

In recent years, U.S. competition with Japan has spawned interest in Japanese styles of management. Recently, an organizational form used by many Japanese corporations has been adopted by some U.S. companies (Ouchi 1981). This perspective values long-term employment, interpersonal trust, and above all, close personal relationships. In this respect, it advocates the opposite of what Western bureaucracy advocates. The Japanese theory places considerable emphasis upon primary group relationships. It has created a form of work in which small groups of workers meet periodically with managers to discuss ways the organization can work better.

By this style of management, managers and their companies are encouraged to take a long-term orientation toward their problems and products. Personal friendships among coworkers are encouraged. Managers are encouraged to move around to different organizational positions rather than become narrow specialists. They are encouraged to manage their department by "walking around" to develop closer relationships with other employees.

This management style emerges from one particular characteristic of Japanese culture, one that emphasizes lifelong employment in the same company, company loyalty, and complete attachment to one's work. Whether it can be successfully transported to the United States and other Western countries, with their unique cultural traditions, is questionable. United States corporate culture still favors individualism—the exact opposite of the Japanese group-oriented, participative-management approach. Sociological research finds some

Diversity: Race, Gender, and Class in Organizations

The hierarchical structuring of organizations results in the concentration of power and influence with a few individuals at the top. Because organizations tend to reflect patterns within the broader society, this hierarchy, like that of society, is marked by inequality in race, gender, and class relations. Although the concentration of power in organizations is incompatible with the principles of a democratic society (Perrow 1994, 1986), discrimination against women and minorities definitely still occurs. There have been widespread disparities in the promotion rates for White and Black workers, which is a pattern repeated in most work organizations (Eichenwald 1996; McGuire and Reskin 1993; Collins 1996).

Traditionally, within organizations, the most powerful positions are held by White men of upper-class status. Women and minorities, on average, occupy lower positions in the organization. A very few minorities and women do get promoted, but there remains a "glass ceiling" effect, meaning that women and minorities may be promoted but only to a certain level. The glass ceiling acts as a barrier to the promotion of women and minorities into higher ranks of management What are the barriers that prevent more inclusiveness in the higher ranks of organizations?

Sociological research finds that organizations are sensitive to the climate in which they operate. The more egalitarian the environment in which a firm operates, the more equitable its treatment of women and minorities will be. In particular, sociologists have found that those industries with the strongest ties to the federal government tend to be more favorable in their treatment of women and minorities. Put a little differently, the greater the involvement of the federal government in a given industry, the more favorable jobs and earnings are for minority men and women and White women (Beggs 1995).

Other studies find that patterns of race and gender discrimination persist throughout organizations, even when formal barriers to advancement have been removed. Many minorities are now equal to Whites in education, particularly those with organizational jobs that require advanced graduate degrees such as the master of business administration (MBA). Still, White men in organizations are more likely to receive promotions than African American, Hispanic, and Native American workers with the *same* education (DeWitt 1995; Zwerling and Silver 1992). In such cases, lack of promotion of the minority person cannot then be attributed to a lack of education. The same thing often happens to both White and minority women in organizations: Women are less likely to receive promotions than a White male who has the same education. Studies also consistently find that women are held to higher promotion standards than men. For men, the longer they are in a position in an organization, the more likely they will be promoted, but the same is not true for women (Smith 1999). Studies find that women change jobs more frequently within organizations than do men, but these tend to be lateral moves. For men, job changes are more likely to mark a jump from a lower to a higher level in the organization, thus constituting a promotion.

Things work the same way with respect to people being discharged or fired. Studies clearly show that Black federal employees (men and women) are more than *twice as likely* to be dismissed than are their White counterparts (DeWitt 1995; Zwerling and Silver 1992). This disparity occurs regardless of education, occupational category, pay level, type of federal agency, age, performance rating, seniority, or attendance record.

The main reasons cited in the studies for such disparities are lack of network contacts with the "old boy" network, and racial bias within the organization. The studies show conclusively that, with such factors as education, occupational category, pay level, type of federal agency, taken into account, race is still an important reason for the way people are treated in organizations today. The studies strongly suggest that racism thrives in the bureaucracy.

A classic study by Rosabeth Moss Kanter (1977) demonstrated how the hierarchical structure of the bureaucracy negatively affects both minorities and women who are underrepresented in the organization. They feel put "out front" and under the all-too-watchful eyes of their superiors. As a result, they often suffer severe stress (Jackson 2000; Jackson et al. 1994, 1995; Yoder 1991; Spangler et al. 1978; Kanter 1977). This is called "tokenism." Such token representatives find it difficult to gain credibility with not only their superiors, who are for the most part White males, but with their coworkers as well, who often accuse them of getting the position simply because they are women, a minority, or both. It is a widespread phenomenon in universities and colleges that minorities and women are accused of being admitted simply because of their race or their gender, even in instances when this has clearly not been the case. This has stressful effects on the person and shows that tokenism can have very negative consequences.

Social class, in addition to race and gender, plays a part in determining people's place within formal organizations. Middle- and upper-class employees in organizations make higher salaries and wages and are more likely to get promoted than people of less social class status, even for individuals who are of the same race or ethnicity. This even holds for persons coming from families of less social class status but who are themselves as well educated as their middle- and upper-class

Table 6.1 Theoretical Perspectives on Organizations			
	Functionalist Theory	**Conflict Theory**	**Symbolic Interaction Theory**
Central focus	Positive functions (such as efficiency) contribute to unity and stability of the organization	Hierarchical nature of bureaucracy encourages conflict between superior and subordinate, men and women, and people of different racial or class backgrounds	Stresses the role of self in the bureaucracy and how the self develops and changes
Relationship of individual to the organization	Individuals like parts of a machine, irrelevant for their personal characteristics	Individuals are subordinated to systems of power and experience stress and alienation as a result	Interaction between superiors and subordinates forms the structure of the organization
Criticism	Hierarchy can result in dysfunctions such as *ritualism* and *alienation*	De-emphasizes the positive ways that organizations work	Tends to downplay overall social organization

coworkers. This means that their lower salaries and lack of promotion cannot necessarily be attributed to their lack of education. In this respect, their treatment in the bureaucracy only perpetuates the negative effects of the social class system in the United States.

The social class stratification system produces major differences in the opportunities and life chances of individuals and the bureaucracy simply carries these differences forward. Class stereotypes also influence hiring practices in organizations. Personnel officers look for people with "certain demeanors," a code for those who convey middle-class or upper middle-class standards of dress, language, manners, and so forth, which some people may simply be unable to afford.

Patterns of race, class, and gender inequality in organizations persist at the same time that many organizations have become increasingly aware of the need for a more diverse workforce. Responding to the simple fact of more diversity within the working-age population, organizations have developed human relations experts who work within organizations to enhance sensitivity to diversity. Such "diversity training" is now common in most large organizations, a reflection of the significance of diversity in today's society.

DEBUNKING *SOCIETY'S MYTHS*

Myth: Programs designed to enhance the number of women and minorities in organizational leadership are no longer needed because discriminatory barriers have been removed.

Sociological perspective: Research continues to find significant differences in the promotion rates for women and minorities in most organizational settings. Even with the removal of formal discriminatory barriers, organizational practices persist that block the mobility of these workers.

Functional, Conflict, and Symbolic Interaction: Theoretical Perspectives

All three major sociological perspectives—functional, conflict, and symbolic interaction—are exhibited in the analysis of formal organizations and bureaucracies. The functional perspective, based in this case on the early writing of Max Weber, argues that certain functions, called *eufunctions* (meaning positive functions), characterize bureaucracies and contribute to its overall unity. The bureaucracy exists to accomplish these eufunctions, such as efficiency, control, impersonal relations, and a chance for the individual to develop a career within the bureaucracy. As we have seen, however, bureaucracies develop the "other face" (informal interaction and culture, as opposed to formal or bureaucratic interaction and culture), as well as the problems of ritualism and alienation of the person from the organization. These latter problems are called *dysfunctions* (negative outcomes or functions) and have the consequence of contributing to the disunity and lack of harmony in the bureaucracy.

The conflict perspective argues that the hierarchical or stratified nature of the bureaucracy in effect encourages instead of inhibits conflict among the individuals within it. These conflicts are between superior and subordinate, as well as between racial and ethnic groups, men and women, and people of different social class backgrounds. This hampers the smooth and efficient running of the bureaucracy.

The symbolic interaction perspective underlies two management theories, those of Argyris (1990) and Ouchi (1981) discussed earlier. Symbolic interaction

---

TAKING ON SOCIAL ISSUES

Has Racial and Gender Equality Been Achieved in Bureaucracies?

Many organizations have tried to foster a more positive organizational climate for all members by sponsoring diversity workshops and other training efforts designed to make people more aware of racial, ethnic, and gender differences that affect people's interaction in the workplace. Advocates of diversity training say that enhanced awareness and sensitivity toward others makes a more positive environment for all members of the organization. Critics say that, well-meaning as such

programs are, they do not deal with fundamental problems of equity in the workplace and have done little to advance women and racial–ethnic minorities into senior positions in such firms. Finally, research evidence shows that both minorities and women are still quite far away from complete equality in rank, salary, and promotional opportunities in today's bureaucracies.

- How would you go about measuring whether equality for minorities and

women has been attained in an organization?

- In your opinion, what factors (variables) should one consider?

Taking Action

Go to the Taking Action Exercise on the companion website—at http://sociology .wadsworth.com/andersen_taylor4e/— to learn more about an organization that addresses this topic. •••

stresses the role of the self in any group, and especially how the self develops as a product of social interaction. Argyris's theory advocates increased involvement of the self within the organization, as a way of "actualizing" the self and, as a result, reducing the disconnection between individual and organization as well as other organizational problems and dysfunctions. Ouchi's theory argues that increased interaction between superior and subordinate, based on the Japanese organizational model of executives "walking around" and interacting more on a primary group basis, will reduce organizational dysfunctions.

Sociology ⊛ Now™

Reviewing is as easy as ❶❷❸.

1. *Before you do your final review, take the SociologyNow diagnostic quiz to help you identify the areas on which you should concentrate. You will find information on SociologyNow and instructions on how to access all of its great resources on the foldout at the beginning of the text.*

2. *As you review, take advantage of SociologyNow's study videos and interactive Map the Stats exercises to help you master the chapter topics.*

3. *When you are finished with your review, take SociologyNow's posttest to confirm you are ready to move on to the next chapter.*

Chapter Summary

What are the types of groups?

Groups are a fact of human existence and permeate virtually every facet of our lives. Group size is important, as is the otherwise simple distinction between dyads and triads. Groups are of several types. *Primary groups* form the basic building blocks of social interaction in society. *Reference groups* play a major role in forming our attitudes and life goals, as do our relationships with *in-groups* and *out-groups*. *Social networks* partly determine things such as who we know and the kinds of jobs we get. Networks based on race–ethnicity, social class, and other social variables are extremely closely connected, or very dense.

How strong is social influence?

The social influence that groups exert upon us is tremendous, as seen by the Asch conformity experiments. The Milgram experiments demonstrated that the interpersonal influence of an authority figure can cause an individual to act against his or her deep convictions. The recent torture and abuse of Iraqi prisoners of war by American soldiers as prison guards serves as testimony to the powerful effects of both social influence and authority structures.

What is the importance of groupthink and risky shift?

Groupthink can be so pervasive that it adversely affects group decision making. *Risky shift* similarly often compels individuals to reach decisions that are at odds with their better judgment.

What are the types of formal organizations and bureaucracies, and what are some of their problems?

Formal organizations are of several types, such as normative, coercive, or utilitarian. Groups occur within organizations, and thus individuals within organizations are subject to group effects, such as groupthink and risky shift. Weber typified *bureaucracies* as organizations with an efficient division of labor, an authority hierarchy, rules, impersonal relationships, and career ladders. Bureaucratic rigidities often result in organization problems such as ritualism, which may have been at least partly responsible for the space shuttle *Challenger* explosion in 1986, and also the space shuttle *Columbia* explosion in 2003. The *McDonaldization of society* has resulted in greater efficiency, calculability, and control in many industries, probably at the expense of some individual creativity.

What are some new global organizational forms?

Some new global organizational forms have attempted to reduce worker alienation by adapting Japanese styles of management, such as the use of small groups, and participative management, instead of traditional top-down management styles. Whether such practices can thrive in a different cultural context is to date uncertain.

What are some of the problems of diversity in organizations?

Formal organizations tend to perpetuate inequality of race–ethnicity, gender, and social class. Blacks, Hispanics, and Native Americans are less likely to get promoted and more likely to get fired than Whites of comparable education and other qualifications. Women experience similar effects of inequality, especially negative effects of tokenism such as stress and lowered self-esteem. Finally, persons of less than middle-class origins make less money and are less likely to get promoted than a middle-class person of comparable education. Diversity training has been introduced to many organizations as an attempt to combat such problems.

What do the three theoretical perspectives say about organizations?

Functional, conflict, and *symbolic interaction* theories highlight and clarify the analysis of organizations by specifying both organizational functions and dysfunctions (functional theory); by analyzing the consequences of hierarchical, gender, race, and social class conflict in organizations (conflict theory); and finally, by studying the importance of social interaction and integration of the self into the organization (symbolic interaction theory).

Key Terms

attribution error 143
attribution theory 143
bureaucracy 152
coercive organization 151
dyad 140
expressive needs 141
formal organization 150
group 140
group size effect 141
groupthink 148
instrumental needs 142
normative organization 151

primary group 141
reference group 142
risky shift 149
secondary group 141
social network 144
total institution 151
triad 140
triadic segregation 141
utilitarian organization 151
voluntary organization 151

Researching Society with MicroCase Online

You can see the results of actual research by using the Wadsworth MicroCase® Online feature available to you. This feature allows you to look at some of the results from national surveys, census data, and some other data sources. You can explore this easy-to-use feature on your own, but try this example. Suppose you want to know:

Is high communication technology in a country, such as the presence of cell phones, related to some other aspect of communication technology, such as the number of televisions per 1000 people in the country? You can do the analysis yourself!

To answer this question, go to http://sociology.wadsworth .com/andersen_taylor4e/, select MicroCase Online from the left navigation bar, and follow the directions there to analyze the following data:

Data File: NATIONS

Task: SCATTERPLOT

Dependent Variable: CELL PHONE

Independent Variable: TV1000

Questions

You will get a diagram called a scatterplot, where each dot represents a country. One axis of the scatterplott (the X-axis) represents the number of televisions per 1000 persons, and the other axis (the Y-axis) represents the number of cell phone subscribers per 100,000 people. Do you see a pattern among the dots? Answer the following questions:

1. Is the number of cell phones *greater* or *lesser* the greater the number of televisions in a country?

2. How closely is the number of cell phones tied to the number of televisions?

3. What does all this tell you about different types of communication technology in a country (recall that "ism" about how different technologies may be related)? What for example might this tell you about the number of computers with Internet capability in a country?

The Companion Website for Sociology: Understanding a Diverse Society, Fourth Edition

http://sociology.wadsworth.com/andersen_taylor4e/

Supplement your review of this chapter by going to the companion Web site to take one of the Tutorial Quizzes, use the flash cards to master key terms, and check out the many other study aids you'll find there. You'll also find special features such as GSS Data and Census 2000 information, data and resources at your fingertips to help you with that special project or do some research on your own.

Suggested Readings and Web Resources

Collins-Lowry, Sharon M. 1996. *Black Corporate Executives*. Philadelphia, PA: Temple University Press.
Based on extensive interviewing of successful Black executives, this research examines the experiences of some of the first Black managers to succeed in major corporations. Collins-Lowry's work shows the continuing significance of race in structuring experiences within corporate organizations.

Homans, George C. 1992 [1950]. *The Human Group*. New Brunswick, NJ: Thomson Learning.
A classic that combines knowledge about groups with knowledge about formal organizations, this book has been one of the most useful and has served to stimulate research in group interaction over numerous decades.

Kanter, Rosabeth Moss. 1977. *Men and Women of the Corporation*. New York: Basic Books.
This classic study shows the significance of gender, and tokenism, in structuring opportunities for women in corporations. Although more than twenty-five years old, the book's insights continue to be supported by subsequent research on gender dynamics in work organizations.

Ritzer, George. 2002. *The McDonaldization of Society: An Investigation Into the Changing Character of Contemporary Society and Life*, 2nd ed. Newbury Park, CA: Pine Forge Press.
This is an entertaining account of the fast-food approach to bureaucratic organization. The book highlights the influence of this model of bureaucracy upon

many kinds of bureaucracies in American society, even the structure of the community health clinic.

Vaughan, Diane. 1996. *The Challenger Launch Decision: Risky Technology, Culture, and Deviance in NASA.* Chicago: University of Chicago Press.
Vaughan's research explores how ordinary behavior and ritualism within organizations can result in disastrous consequences, such as the Challenger *space*

shuttle explosion in 1986. In many ways, this study foretold the February 1, 2003 breakup upon decent of the space shuttle Columbia.

McDonaldization of Society
www.sociology.net/mcdonald/
This is the home page for George Ritzer's book, and it provides additional information about and links to discussion and illustration of this concept.

● ● ●

Deviance

In the early 1970s, an airplane carrying forty members of an amateur rugby team crashed in the Andes Mountains in South America. The twenty-seven survivors, teenagers and some of their relatives and friends from Uruguay's elite class, were marooned at 12,000 feet in freezing weather and deep snow. There was no food except for a small amount of chocolate and some wine. A few days after the crash, the group heard on a small transistor radio that the search for them had been called off.

Scattered in the snow were the frozen bodies of dead passengers. Preserved by the freezing weather, these bodies became, after a time, sources of food. At first, the survivors were repulsed by the idea of cannibalism—the eating of human flesh—but as the days wore on, they agonized over the decision about whether to eat the dead crash victims. The survivors eventually concluded that they had to eat if they were to live.

In the beginning, only a few ate the human meat, but soon the others began to eat too. One married couple held out the longest, but when they began to think about wanting another child if they found their way home, they too ate. The survivors developed elaborate rules about how, what, and whom they would eat. Some could not bring themselves to cut the meat from the human body but would slice it once someone else had cut off large chunks. They all refused to eat certain parts—the lungs, skin, head, and genitals; no one was expected to eat an immediate friend or relative. The survivors also developed other uses for the bodies, including making warm, insulated "socks" from human skin (Read 1974; Henslin 1993).

After two months, in mid-December, the group sent out an expedition of three survivors to find help. The three men walked for ten days, over the mountains into Chile. The group was eventually rescued, and the world learned of their ordeal. Cannibalism was generally accepted as something they had to do to survive. To many, it even seemed like a triumph over extraordinary hardship. The survivors maintained a sense of themselves as good people, even though their behavior profoundly violated ordinary standards of socially acceptable behavior in most cultures of the world.

Was the behavior of the Andes crash survivors socially deviant? Or was this a normal response to extreme circumstances? Compare the Andes crash with another case of human cannibalism. In 1991 in Milwaukee, Wisconsin, Jeffrey Dahmer plead guilty to charges of murdering at least fifteen men in his home. Dahmer lured the men—eight of them African American, two White, and one a fourteen-year-old Laotian boy—to his apartment, where he murdered

Sociology ⊛ Now™
Reviewing is as easy as ❶❷❸.

Use SociologyNow to help you make the grade on your next exam. When you are finished reading this chapter, go to the chapter review for instructions on how to make SociologyNow work for you.

and dismembered them, then cooked and ate some of their body parts. He boiled the flesh from the heads of those he considered most handsome so that he could save and admire their skulls. Dahmer was seen as a total social deviant, someone who violated every principle of human decency. Even hardened criminals were disgusted. He was killed in the prison bathroom by another inmate in 1994.

Why was Dahmer's behavior considered so deviant when that of the Andes survivors was not? The behavior was the same in each case: eating human flesh. The answer can be found by looking at the situation in which these behaviors occurred. For the Andes survivors, eating human flesh was essential for survival; for Dahmer, however, it was murder. Many people believe Dahmer was simply a sociopath. This psychological explanation has merit, but so does a sociological perspective. From this perspective, the deviance of cannibalism resides not just in the act itself, but also in the social context in which it occurs. That is the essence of sociological explanation: The nature of deviance is not only in the personality of the deviant person, nor is it inherently in the deviant act itself, but instead it is a significant part and product of the social structure. •••

Defining Deviance

Sociologists define **deviance** as *behavior that is recognized as violating expected rules and norms.* Deviance is more than simple nonconformity; it is behavior that departs significantly from social expectations. In the sociological perspective on deviance, there is subtlety that distinguishes it from commonsense understandings of the same behavior.

- *The sociological definition of deviance stresses social context, not individual behavior.* Sociologists see deviance in terms of group processes, definitions, and judgments, not just as unusual individual acts.
- *The sociological definition of deviance recognizes that not all behaviors are judged similarly by all groups.* What is deviant to one group may be normative (nondeviant) to another. Understanding what society sees as deviant also requires understanding the context that determines who has the power to judge some behaviors as deviant and others not.
- *The sociological definition of deviance recognizes that established rules and norms are socially created, not just morally decreed or individually imposed.* Sociologists emphasize that deviance lies not just in behavior itself, but also in the social responses of groups to the behavior.

Strange, unconventional, or nonconformist behavior is often understandable in its sociological context. Consider suicide. Are people who commit suicide crazy, or might their behavior be explained? Are there conditions under which suicide is acceptable? Should someone who commits suicide in the face of a terminal illness be judged differently from a despondent person who jumps from a window? These are the kinds of questions probed by sociologists who study deviance.

Once you have a sociological perspective on deviance, you are likely to see things a little differently when you observe someone behaving in an unusual way. Sociologists sometimes use their understanding of deviance to explain otherwise ordinary events—such as tattooing and body piercing (Irwin 2001; Vail 1999), eating disorders (Sharp et al. 2001; Lovejoy 2001), or drug and alcohol use (Inciardi 2001; Humphries 1999; Logio 1998). Sociologists who examine behavior such as mental illness, suicide, delinquent behavior, and substance abuse as examples of deviance also examine how people respond to social stigmas, such as having a disability, or how people become deviant, such as in prostitution, drug dealing, or crime. Deviant behavior varies in its severity, as well as in how ordinary or unusual the behavior might be.

Sociologists distinguish between two types of deviance: formal and informal. *Formal deviance* is behavior that breaks laws or official rules. Crime is an example. There are formal sanctions against formal deviance, such as imprisonment and fines. *Informal deviance* is behavior that violates customary norms (Schur 1984). Although such deviance may not be specified in law, it is judged to be deviant by those who uphold the society's norms. A good example is the body piercing that is popular among young people. No laws prohibit this practice, yet it violates common norms about dress and appearance and is judged by many to be socially deviant even though it is fashionable for others.

The study of deviance can be divided into the study of why people violate laws or norms and the study of how society reacts. This reaction includes the labeling process by which deviance comes to be recognized. *Labeling theory* is discussed in detail later in the chapter, but it is important to point out here that the meaning of deviance is not just in the breaking of norms or rules; it is also in how people react to those behaviors. This dimension of deviance—the societal reaction to deviant behavior—suggests that social groups in many ways *create* deviance "by making the rules whose infraction constitutes deviance, and by applying those rules to particular people and labeling them as outsiders" (Becker 1963: 9).

The Context of Deviance

Some situations are more conducive than others to creating deviant behavior. Sociologists have underscored that even the most unconventional behavior can be understood if we know the context in which it occurs. Behavior that is deviant in one circumstance may be normal in another, or certain behaviors may be ruled deviant only when performed by certain people. For example, people who break gender stereotypes may be judged as deviant even though their behavior is considered normal for the other sex. Heterosexual men and women who kiss in public are the image of romance. Lesbians and gay men who dare even to hold hands in public are seen as flaunting their sexual orientation.

The definition of deviance can also vary over time. Only recently has date rape been defined as social deviance. Because it was previously not considered deviant, it was not even named. Women were presumed to mean "yes" when they said "no," and men were expected to "seduce" women through aggressive sexual behavior. Even now, women who are raped by someone they know may not think of it as rape. If they do, they may find that prosecuting the offender is difficult because others do not think of it as rape. Studies have found that students think that date rape is justified if the victim was wearing provocative clothing (Wookman and Freeburg 1999; Johnson 1995; Cassidy and Hurrell 1995). People with more traditional attitudes about gender roles are also more likely to excuse men's aggression in date rape and to define the situation as something other than rape. These examples show that the definition of deviance derives not only from what one does, but also from who does it, when, and where.

The sociologist Emile Durkheim argued that one reason acts of deviance are publicly punished is that the social order is threatened by deviance. Judging those behaviors as deviant and punishing them confirms general social standards. Therein lies the value of widely publicized trials, public executions, or the historical practice of displaying a wrongdoer in the pillory, which held the hands and head, or stocks that held the feet. The punishment affirms the collective beliefs of the society, reinforces social order, and inhibits future deviant behavior, especially as defined by those with the power to judge others.

Durkheim argued that societies actually *need* deviance to know what presumably normal behavior is. In this sense, Durkheim considered deviance "functional" for society (Durkheim 1951/1897; Erikson 1966). You could observe Durkheim's point in the aftermath of the terrorist attacks of September 11, 2001. Horrified by the sight of highjacked planes flying into the World Trade Center and the Pentagon, and crashing into the Pennsylvania field, U.S. citizens responded by demonstrating strong patriotism. Durkheim would interpret this as deviance (i.e., the terrorist acts) producing strong solidarity. This was one of Durkheim's most important insights: Instead of breaking up society, deviance produces social solidarity.

DEBUNKING *SOCIETY'S MYTHS*

Myth: Deviance is bad for society because it disrupts normal life.

Sociological perspective: Deviance tends to stabilize society; by defining some forms of behavior as deviant, people are affirming the social norms of groups. In this sense, society actually to some extent *creates* deviance.

Another example of social solidarity comes from a widely publicized stoning that took place in Afghanistan in 1996. A man and a woman who had been caught as adulterers were stoned to death, based on an interpretation of Islamic law by contemporary extremists. Thousands of spectators enthusiastically observed the woman being put in a sandpit with only her chest and head above ground. The man was blindfolded as he faced the Muslim cleric and others who joined in the stoning. After the couple's death, those who witnessed the ritual stoning commented, "It was a good thing—the only way to end this kind of sinning," and "No, I didn't feel sorry for them at all; I was happy to see Sharia [the Muslim code] being implemented" (Burns 1996: 18).

How can such cruelty be seen so enthusiastically? Durkheim would answer that the stoning, by publicly condemning deviant behavior, is how the community reaffirms its values and promotes social solidarity. This contemporary stoning is a classic example of deviance considered to be normal behavior. Likewise, *honor killings* are another example of how social norms can be upheld through publicly witnessed punishment. Honor killings are the murders of women accused of infidelity or even minor violations of social norms, such as flirting. The practice also includes disfiguring women's faces by throwing acid on them. Many places where strict and extreme interpretations of Islamic law are practiced have documented these murders (Mayell 2002).

Some may also remember an image from Afghanistan of a Taliban man firing a rifle into the head of a woman in a blue burqa—an image widely distributed to provide evidence of the Taliban's cruelty toward women. During the rule of the Taliban, Afghan women accused of violating Islamic law were herded into trucks and publicly executed in a large stadium. The image of a rifle being put to a woman's head repulsed and infuriated those opposed to the Taliban regime, including the Afghan women's resistance organization, RAWA (the Revolutionary Association of Women of Afghanistan; see www.rawa.org). Durkheim would say that this, like other public executions, by condemning alleged deviant behavior, promoted social solidarity among the Taliban. At the same time, the fear these executions engendered also enforced Taliban law. Such practices are not restricted to radical, extremist groups. Stoning was also practiced in colonial America.

Public punishments of deviance need not be brutal to generate the social solidarity about which Durkheim wrote. Public ridicule of presumed deviant behavior, such as calling someone a "fag" or teasing a young girl for being a tomboy, are ways of upholding group norms enforcing heterosexual behavior. Likewise, the public display of American flags in the aftermath of terrorist attacks on the United States promoted social solidarity by endorsing patriotic norms and identifying terrorism as deviant behavior.

The Influence of Social Movements

The perception of deviance may also be influenced by social movements, which are networks of groups that organize to support or resist changes in society (see Chapter 22). Smoking, for instance, was once considered glamorous, sexy, and cool. Now, smokers are scorned as polluters and misfits and, despite strong lobbying by the tobacco industry, regulations against smoking have proliferated.

Whereas in 1987 only 17 percent of the public thought that smoking should be banned in restaurants, by 2004 over half (58 percent) thought so (Gallup Organization 2004). The increase in public disapproval

This widely distributed photo of a woman being executed by the Taliban in Afghanistan illustrates the extreme sanctions that can be brought against those defined as deviant by a powerful group. In this case, the photo also mobilized world condemnation of the Taliban regime for its treatment of women.

of smoking results as much from social and political movements as it does from the known health risks. The success of the antismoking movement has come from the mobilization of constituencies able to articulate to the public that smoking is dangerous (Nathanson 1999). Note that the key element here is the ability of the people to mobilize—not just the evidence of risk. In other words, there has to be a social response for deviance to be defined as such; having only scientific evidence of harm is not enough.

Social movements can also be organized to remove the deviant label from certain behaviors. Whereas gay and lesbian behavior traditionally has been defined as deviant, the gay and lesbian movement has encouraged people to see gay and lesbian relationships as legitimate. Mobilization by gays and lesbians has thus challenged the labeling of gays and lesbians as deviant.

Moral entrepreneurs are people who organize a social movement to reform how a behavior is morally perceived and handled. Moral entrepreneurs can create new categories of deviance by imbuing some behaviors with moral value and defining certain groups as deviant (Becker 1963). Public concern about crack mothers provides an example. Sociologist Drew Humphries (1999) argues that the image of crack-addicted mothers as harming innocent babies has created a *moral panic,* in which low-income Black mothers are blamed for an "epidemic" of drug abuse. Although there is real damage to children born of addicted mothers, Humphries argues that moral entrepreneurs use the media to exaggerate the extent of this problem. In the ensuing panic, the underlying causes of the women's drug addiction—namely poverty and inadequate social services—are ignored. Since the mothers, not the social system, are blamed for the drug problem, they—not society—are then expected to change. The point is that deviant categories (i.e., crack mothers and crack babies) are produced by groups who mobilize social movements around specific issues and change how deviant behavior is defined, who is defined as deviant, and how society deals with deviance.

The Social Construction of Deviance

Perhaps because it violates social conventions or because it sometimes involves unusual behavior, deviance captures the public imagination. Commonly, however, people see deviants as crazy, threatening, or sick, and believe deviance results from personality factors. Sociologists do not see deviance in these individualistic terms, rather it is considered to result from social factors.

RAWA/WorldPicture News

Social movements can often call public attention to social issues, such as the protests of minority communities against police brutality.

Deviance, for example, is not necessarily irrational or "sick" and may be a positive and rational adaptation to a situation. Think of the Andes survivors. Was their action irrational, or was it an inventive and rational response to a dreadful situation? Sociological studies of gangs in the United States shed light on this question. The family situations of boys and girls in gangs are often problematic, although in gangs, girls more often than boys tend to be more isolated from their families (Fleisher 2000; Esbensen-Finn et al. 1999). Given the class, race, and gender inequality faced by minority youth, many turn to gangs for the social support they lack elsewhere (Walker-Barnes and Mason 2001; Moore and Hagedorn 1996). For example, many young Puerto Rican girls live in relatively confined social environments where they have little opportunity for educational or occupational advancement. Their community expects them to be "good girls" who are virgins and remain close to their families. Joining a gang is one way to reject these restrictive roles (Messerschmidt 1997; Campbell 1987). Are these young women irrational or just doing the best they can to adapt to their situation? Sociologists interpret their behavior as an understandable adaptation to conditions of poverty, racism, and sexism.

In some subcultures or situations, deviant behavior is encouraged and praised. Deviance may violate social norms, but people do not always disapprove of the behavior. Have you ever been egged on by friends to do something that you thought was deviant, or done something you knew was wrong? Most students know that cheating is wrong, yet many cheat, even openly—perhaps justifying their behavior by claiming that "everybody does it." Many also argue that the reason so many college students drink excessively is that the student subculture encourages them to do so—even though students know it is harmful.

Some behavior patterns defined as deviant are also surprisingly similar to presumably normal behavior. Many people routinely engage in deviant acts, never thinking of themselves as deviant. The practice of employing domestic workers without reporting their wages is deviant—indeed, illegal—but it is commonly done. Have you ever accepted money for work and not reported it to the IRS (Internal Revenue Service)? If so, did you think of yourself as a deviant? Most likely not. Similarly, you might ask if a heroin addict who buys drugs with the only money he has is so different from a business executive who spends a large proportion of his discretionary income on alcohol. Each may establish a daily pattern that facilitates drug use; each may select friends based on shared interests in drinking or taking drugs; and each may become so physically, emotionally, and socially dependent on their "fix" that life seems unimaginable without it. Which of the two is more likely to be considered deviant?

The point is that deviance is both created and defined within the social context. It is not only weird, pathological, irrational, or unconventional behavior. Sociologists who study deviance understand it in the context of social relationships and society, defining deviance in terms of social norms and the social judgments people make about one another. Remember, behavior that is deviant in one context may be perfectly normal in another (for example, men wearing earrings or women wearing boxer shorts). Sometimes deviant behavior can indicate changes taking place in the cultural folkways. Whereas only a few years ago, body piercing and tattooing were associated with gangs and "disrespectable" people, now it is considered fashionable among young, middle-class people—even though to some, it is still a mark of deviance (Irwin 2001).

Psychological Explanations of Deviance

You can see that sociology goes beyond explanations of deviance that root it in the individual personality. Psychological explanations of deviance emphasize individual personality factors as the underlying cause of deviant behavior. For example, from a psychological perspective, violence may be interpreted as the acting out of hostilities toward a parent. Or, a sociopath may be understood as acting out urges rooted in early childhood experiences that make the person dysfunctional later in life.

Sociologists are critical of psychological interpretations not because they are wrong, but because they are incomplete. By locating causes of deviance within individuals, psychological explanations tend to overlook the social context that produced the deviance. Individual motivation simply does not explain the social patterns that sociologists observe in studying deviance. Why is deviance more common in some groups than others? Why are some more likely to be labeled deviant than others, even if they engage in the same behavior? How is deviance related to patterns of inequality in society? The answers to these questions require a sociological explanation. Sociologists do not ignore individual psychology, but integrate it into an explanation of deviance that focuses on the social conditions surrounding deviant behavior.

Sociologists also tend to be critical of biological explanations of deviant behavior. Some of the historically early attempts to explain deviance, particularly criminal behavior, centered on biological explanations. During the early part of the twentieth century—a time when many new immigrant groups were coming into the United States—biological explanations of crime and deviance thrived. Often (though not always) linked to racist and sexist explanations of group differences, these explanations tended to be popular during periods of widespread change in race and gender relations. Even now, when much positive attention is focused on racial and cultural diversity in the population and when gender relations have undergone widespread transformation, there is a resurgence in biological explanations of crime and deviant behavior.

Biological explanations attribute deviance to presumed genetic or biological differences between groups. These explanations reflect a strongly held popular assumption that there is something fundamentally different in biological nature between people who are deviant and people who are not. For example, some have asserted that there may be a genetic predisposition to

DOING SOCIOLOGICAL RESEARCH

Bad Boys

How do young African American boys get associated with images of failure and deviance? This is what Ann Ferguson wanted to know when she began her research on young Black boys in an urban school. She became a participant observer in an urban school where for three and one-half years, she observed how the rules and practices of school discipline "branded" young Black boys as criminally prone.

Ferguson observed children in the fifth and sixth grades. She identified two types of school children, who she calls the *Schoolboys* and the *Troublemakers*. School personnel (teachers, counselors, and staff members) think of the Schoolboys as doing well in school but see the Troublemakers as always in trouble and "at risk." But Ferguson argues that the two groups are fundamentally not different: All come from the same neighborhood and from similar family backgrounds. Yet, Ferguson writes, "As African American males, *schoolboys* were always on the brink of being redefined into the *troublemaker* category" (2001: 10).

When children get into trouble, they are sent to a disciplinary room that Ferguson calls "The Punishing Room." For children, the Punishing Room became a place to escape the drudgery of schoolwork, have fun, and actively challenge school rules. Ironically, it is a place where students proudly acquire a deviant identity—one rooted in their resistance to school authorities. They can make a name for themselves by transgressing school rules. A second room, which students called "The Jailhouse," holds students who are detained after school. The vast majority of students in both "The Punishing Room" and "The Jailhouse" are Black boys. In both places, once labeled as deviant, the boys assert their identities by contesting adult power. As Ferguson writes, "In The Punishing Room, school identities and reputations are constituted, negotiated, challenged and confirmed for African American youth in a process of categorization, reward and punishment, humiliation, and banishment. Children passing through the system are marked and categorized as they encounter state laws, school rules, tests and exams, psychological remedies, screening committees, penalties and punishments, reward and praise. Identities such as worthy, hardworking, devious or dangerous are proffered, assumed, or rejected" (2001: 40–41).

On Ferguson's first day in the school, one school staff member pointed to a young Black boy and said, "That one has a jail-cell with his name on it" (2001: 1). Ferguson's research reveals the extraordinary power of institutions and the labeling process to create, shape, and regulate deviant identities, particularly for young Black boys.

Questions to Consider

1. Make a list of all the different cliques that were part of your high school. What are their names? What characteristics and behaviors were associated with each? What does this tell you about *labeling theory*? **Keywords:** *labeling theory*

2. When you think about the groups who were considered "bad" in your high school, what was the racial, social class, and gender makeup of the group? How did social stereotypes about either race, class, or gender (or all three) influence what people thought about this group? How might students work to overcome such stereotypes? **Keywords:** *race and punishment*

We have included InfoTrac College Edition keywords at the end of each question to make it easier for you to find more to read on these topics. Go to www.infotrac-college.com, an online library, to begin your search.

Source: Ferguson, Ann Arnett. 2001. *Bad Boys: Public Schools in the Making of Black Masculinity.* Ann Arbor, MI: University of Michigan Press.

•••

crime among some racial and social class groups (Gordon 2003; Herrnstein and Murray 1994; Wilson and Herrnstein 1985). But critics of this argument note that there is little scientific proof for such claims and add that biological arguments are typically used only to explain the problems of poor people and minority groups. Seldom are such arguments used to explain the crimes of elites or the middle class, such as tax evasion, embezzlement, or insider stock tading. No one has claimed in the aftermath of white-collar crimes that CEOs and stockbrokers are somehow genetically inferior. Biological explanations of deviance offer easy explanations for complex social problems. Although there are certainly some biological differences between groups in society, attributing complex social phenomena primarily to biological causes oversimplifies and distorts the sociological processes at work.

The Medicalization of Deviance

People commonly interpret acts of deviance, particularly those that are especially harmful, as the behavior of people who are sick or sociopathic. This reaction is what sociologists call the **medicalization of deviance,** referring to explanations of deviant behavior that interpret deviance as the result of individual pathology or sickness. The medical slant may be expressed as a metaphor, such as when deviant behavior is attributed to a "sick" state of mind and where the solution is to "cure" deviance through individual treatments such as intensive therapy for sex offenders (Conrad and Schneider 1992).

THINKING SOCIOLOGICALLY

Ask some of your friends to explain why rape occurs. What evidence of the *medicalization of deviance* is there in your friends' answers?

Like biological explanations, medicalizing deviance emphasizes the physical or genetic roots of deviant behavior. Alcoholism is an example. Certainly alcoholism has serious medical consequences and can be partially understood in medical terms. And there is some evidence of a genetic basis to alcoholism. But viewing alcoholism solely from a medical perspective ignores the social causes that influence the development and persistence of this behavior. Practitioners know that medical treatment alone does not solve the problem. The social relationships, social conditions, and social habits of alcoholics must be altered, or the behavior is likely to recur.

Sociologists criticize the medicalization of deviance for ignoring the effects of social structures on the development of deviant behavior. From a sociological perspective, deviance originates primarily in society, not in individuals. Changing the incidence of deviant behavior requires changes in society in addition to changes in individuals. Most deviance, to most sociologists, is not a pathological state, but an *adaptation to the social structures* in which people live. Family background, social class, racial inequality, and the social structure of gender relations in society produce deviance, and these factors must be considered to explain deviance.

Sociological Theories of Deviance

Sociologists have drawn on several major theoretical traditions to explain deviant behavior, including functionalism, conflict theory, and symbolic interaction theory.

Functionalist Theories of Deviance

Recall that *functionalism* is a theoretical perspective that interprets all parts of society, including those that may seem dysfunctional, as contributing to the stability and continuance of the whole. At first glance, deviance seems dysfunctional for society. Functionalist theorists argue otherwise (see Table 7.1). They contend that deviance is functional *because it creates social cohesion.* Branding certain behaviors as deviant provides a contrast to behaviors that are considered normal, giving people a heightened sense of social order. Norms are meaningless unless there is deviance from the norms,

Deviance can be encouraged in certain peer subcultures, such as the widespread phenomenon of binge drinking among young adults.

Table 7.1
Sociological Theories of Deviance

Functionalist Theory	Symbolic Interaction Theory	Conflict Theory
Deviance creates social cohesion	Deviance is a learned behavior, reinforced through group membership	Dominant classes control the definition of and sanctions attached to deviance
Deviance results from structural strains in society	Deviance results from the process of social labeling, regardless of the actual commission of deviance	Deviance results from inequality in society, including that of class, race, and gender
Deviance occurs when people's attachment to social bonds is diminished	Those with the power to assign deviant labels themselves produce deviance	Elite deviance goes largely unrecognized and unpunished

and deviance is necessary to clarify what society's norms are. Group coherence then comes from sharing a common definition of legitimate behavior and of deviant behavior. The collective identity of the group is affirmed when people defined as deviant are ridiculed or condemned by group members (Erikson 1966).

To give an example, think about how gay men and lesbian women are defined by many people as deviant. Although lesbians and gay men have rejected this label, labeling homosexuality deviant is one way of affirming the presumed normality of heterosexual behavior. Labeling someone else an outsider is, in other words, a way of affirming one's "insider" identity.

Durkheim: The Study of Suicide The functionalist perspective on deviance stems originally from the work of Emile Durkheim. Recall that one of Durkheim's central concerns was how society maintains its coherence, or social order. He saw deviance as functional for society because it produces solidarity among society's members and made a number of important sociological points. First, he criticized the usual psychological interpretations of why people engage in deviance, turning instead to sociological explanations with data to back them up. Second, he emphasized the role of social structure in producing deviance. Third, he pointed to the importance of people's social attachments to society in understanding deviance. Finally, he elaborated the functionalist view that deviance provides the basis for social cohesion. His studies of suicide illustrates these points.

Durkheim was the first to argue that the causes of suicide were to be found in social factors, not individual personalities. Observing that the rate of suicide in a society varied with time and place, Durkheim looked for causes linked to time and place rather than only to emotional stress. Durkheim argued that suicide rates are af-

fected by the different social contexts in which they emerge. He looked at the degree to which people feel integrated into the structure of society and their social surroundings as social factors producing suicide. Building from this, Durkheim analyzed three types of suicide: anomic suicide, altruistic suicide, and egoistic suicide.

Important to Durkheim's studies of deviance is the concept of **anomie,** defined as the condition that exists when social regulations in a society break down. This term refers not to an individual state of mind, but to social conditions. The controlling influences of society are no longer effective, and people exist in a state of relative normlessness. Anomie is reflected in how individuals feel, but its origins are in society. When behavior is no longer regulated by common norms and values, individuals are left without moral guidance (Durkheim 1951/1897; Coser 1977).

Strong ties among the Navajo produce social integration, contributing to the fact that the Navajo have one of the lowest suicide rates of any group in the United States, and also lowest among other Native American Indian tribal groups.

© Terry Eiler/Stock Boston

- **Anomic suicide** occurs when the disintegrating forces in the society make individuals feel lost or alone. Studies of college campuses trace the cause of campus suicides to feelings of loneliness and a sense of hopelessness (Langhinrichsen-Rohling et al. 1998). In addition, studies find that a history of sexual and physical abuse predicts a higher likelihood of suicide among college women (Thakkar et al. 2000; Bryant and Range 1997). You can also use Durkheim's analysis of anomic suicide to understand patterns of race–ethnicity and suicide in the United States (see Figure 7.1). You might expect that suicide rates would be high among minority groups, but as you can see, except for American Indians, they are not. However, suicide rates among young Black and Hispanic men are less than the rate among young White men. (National Center for Health Statistics 2003; Willis et al. 1999). Also, studies find that Asian/Pacific Islander women in the United States have higher suicide rates than Black and Hispanic women, in part, because they are caught between cultural norms emphasizing self-sacrifice and discouraging help-seeking. Coupled with the conditions of poverty, unemployment (or marginal employment), and poor communication with health care providers, Asian/Pacific Islander women face classic conditions of anomie—resulting in an increased risk of suicide (True and Guillermo 1996).

- **Altruistic suicide** occurs when there is excessive regulation of individuals by social forces. An example is someone who commits suicide for the sake of a religious or political cause, such as suicide bombers. As repugnant as you may find the act of someone intentionally killing and injuring others by blowing themselves up—or killing thousands by crashing planes into the World Trade Center and the Pentagon—you can explain these acts in terms of altruistic suicide. Although sociology does not excuse such behavior, it can help explain it. Suicide bombers are so regulated by their extreme beliefs that they are willing to die to achieve their goals. As Durkheim argued, altruistic suicide results when individuals are excessively dominated by the expectations of their social group. People who commit altruistic suicide subordinate themselves to collective expectations, even when death is the result.

- **Egoistic suicide** occurs when people feel completely detached from society. This helps explain the high rate of suicide among the elderly in the United States. People between seventy-five and eighty-four years of age have one of the highest rates of suicide (National Center for Health Statistics 2003; Coren and Hewitt 1999). Ordinarily, people are integrated into society by work roles, ties to family and community, and other social bonds. When these bonds are weakened, the likelihood of egoistic suicide increases. Many elderly people have lost family and social ties, making them most susceptible to egoistic suicide. Suicide is also more likely to occur among people who are not well integrated into social networks (Berkman et al. 2000; Nisbet 1997). Thus, it should not be surprising that women have lower suicide rates than men (see Figure 7.1). Sociologists explain this fact as the result of men's being less embedded in social relationships of care and responsibility than women (Watt and Sharp 2001).

Durkheim's major point is that suicide is a significantly social, not just an individual, phenomenon (see Map 7.1). Recall from Chapter 1 that Durkheim sees sociology as discovering the social forces that influence human behavior. As individualistic as suicide might seem, Durkheim discovered the influence of social structure even here.

Merton: Structural Strain Theory The functionalist perspective on deviance has been further elaborated by the sociologist **Robert Merton** (1910–2003).

Deaths per 100,000 population

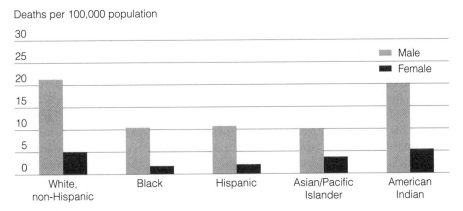

Figure 7.1 Suicide Rates

Source: National Center for Health Statistics. 2003. *Health United States 2003.* Hyattsville, MD: U.S. Department of Health and Human Services, pp. 187–189.

MAP 7.1 Suicide Rates by State

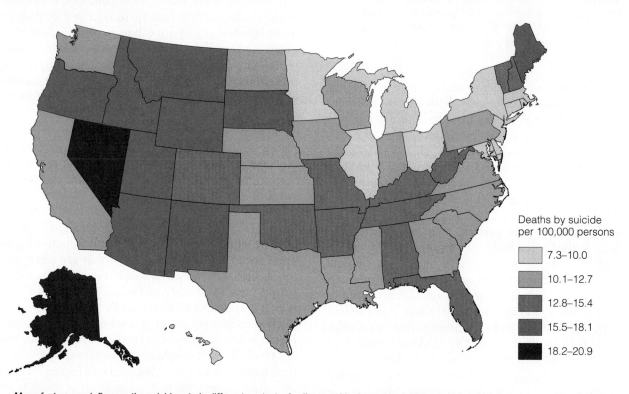

Deaths by suicide per 100,000 persons

	7.3–10.0
	10.1–12.7
	12.8–15.4
	15.5–18.1
	18.2–20.9

Many factors can influence the suicide rate in different contexts. As discussed in the text, suicides can be caused by a multiplicity of structural and cultural factors and sometimes these factors may be differently distributed by state or region. What are some of the social facts about the different states and regions that might affect the different rates of suicide you see in this map?

Data: U.S. Census Bureau. 2004. *Statistical Abstract of the United States, 2003*. Washington, DC: U.S. Government Printing Office.

Merton's **structural strain theory** traces the origins of deviance to the tensions caused by the gap between cultural goals and the means people have to achieve these goals. In society, culture establishes goals for people; social structures provide, or fail to provide, the means for people to achieve those goals. In a well-integrated society, according to Merton, people use accepted means to achieve the goals society establishes. In other words, the goals and means of the society are in balance. When the means are out of balance with the goals, this produces structural strain and deviance is likely to occur. According to Merton, this imbalance, or disjunction, between cultural goals and structurally available means can actually *compel* the individual into deviant behavior (Merton 1968).

To explain further, a collective goal in U.S. society is to achieve economic success. The legitimate means to do so are education and jobs, but not all groups have equal access to those means. The result is structural strain that produces deviance. Poor people are most likely to experience these strains because they internalize the same goals and values of the rest of society but have blocked opportunities for success. Thus, structural strain theory helps explain the moderately high correlation that exists between unemployment and crime.

Figure 7.2 illustrates how strain between cultural goals and structurally available means can produce deviance. *Conformity* is likely to occur when the goals are accepted by the individual and the means toward attaining the goals are made available to the individual via the social structure. If this does not occur, then cultural–structural strain exists and at least one of four possible forms of deviance is likely to result: innovative deviance, retreatism deviance, ritualistic deviance, or rebellion.

Consider the case of female prostitution: The prostitute has accepted the cultural values of dominant society—obtaining economic success and material wealth. Yet if she is poor, then the structural means to attain these goals are less available to her, and turning to prostitution—a type of *innovative deviance* (see Fig-

	Cultural goals accepted?	Institutionalized means toward goal available?
Conformity	Yes	Yes
Innovative deviance	Yes	No
Ritualistic deviance	No	Yes
Retreatism deviance	No	No
Rebellion	No (old goals) Yes (new goals)	No (old means) Yes (new means)

Figure 7.2 Merton's Structural Strain Theory

ure 7.2)—is a likely result. The stockbroker who engages in illegal insider trading constitutes another example of innovative deviance: The cultural goal (wealth) is accepted, but nontraditional means (insider trading) are available and used.

Other forms of deviance also represent this disjunction, or strain, between goals and means. *Retreatism deviance* becomes likely when neither the goals nor the means are available. Examples of retreatism are the severe alcoholic, or the homeless person, or the hermit. *Ritualistic deviance* is illustrated in the case of some eating disorders among college women, such as *bulimia* (purging one's self after eating). The cultural goal of extreme thinness is perceived as unattainable even though the means toward attaining it are plentiful, for example, food, monetary funds, and proper diet methods (Sharp et al. 2000). Finally, *rebellion* as a form of deviance is likely to occur when new goals are substituted for more traditional ones and also new means are undertaken to replace older ones, as by force or armed combat. Many right-wing extremist groups, such as the American Nazi Party, "skinheads," and the Ku Klux Klan (the KKK), are examples of this type of deviance.

Social control theory explains delinquency as the result of weak attachment to social bonds that would support more normative behavior. Instead, social bonds may develop in deviant subcultures, such as a gang.

© Viviane Moos/CORBIS

Social Control Theory

Taking functionalist theory in another direction, Travis Hirschi has developed social control theory to explain the occurrence of deviance. **Social control theory** posits that deviance occurs when a person's (or group's) attachment to social bonds is weakened (Hirschi 1969; Gottfredson and Hirschi 1995, 1990). Most of the time people internalize social norms because of their attachments to others. People care what others think of them and, therefore, conform to social expectations because they accept what people expect. Social control theory, like the functionalist framework from which it stems, assumes the importance of the socialization process in producing conformity to social rules. When that bond is broken, deviance occurs.

Social control theory suggests that most people probably feel some impulse toward deviance at some times, but that the attachment to social norms prevents them from participating in deviant behavior. When conditions arise that break those attachments, deviance occurs. This explains why sociologists find that juveniles whose parents exercise little control over violent behavior and who learn violence from aggressive peers are most likely to engage in violent crimes (Heimer 1997). This helps you understand how two upper middle-class teenaged boys, alienated from the dominant peer culture in their school, could murder twelve high school students and a teacher, plant bombs throughout their school, and then kill themselves at Columbine High School in Littleton, Colorado.

Social control theory has been tested in many different cities and has received considerable research support. For example, cities with high rates of population turnover—where many people are always moving in, out, or around in the city—tend to disrupt personal attachments. This severing of important social bonds tends to correlate with several types of deviance and crime. Burglary, larceny, and rape are thus higher in cities with high population turnover than in cities with less turnover (Crutchfield 1992; Crutchfield et al. 1983).

Functionalism: Strengths And Weaknesses

Functionalism emphasizes that social structure, not just individual motivation, produces deviance. Functionalists argue that social conditions exert actual pressure on individuals to behave in nonconforming ways. Types of deviance are linked to one's place in the social structure, as with Merton's structural strain theory. Functionalists acknowledge that people choose whether to behave in a deviant manner, but they believe that people make their choices from among socially structured options. The emphasis in functionalist theory is on social structure and culture, not individual action. In this sense, functionalist theory is highly sociological.

Functionalists also point out that what appears to be dysfunctional behavior may be functional for the

School violence, although difficult for people to comprehend, can be understood as a form of deviance. One factor leading to the shootings at Columbine High School in 1999 was the formation of a deviant peer group, the Trench Coat Mafia, two of whose members shot and killed twelve students and a teacher before also killing themselves.

society. An example is that most people consider prostitution to be dysfunctional behavior. From the point of view of an individual, that is true—it demeans the women who engage in it, puts them at physical risk, and subjects them to sexual exploitation. From the point of view of functionalist theory, prostitution is functional for the society because prostitution supports and maintains a social system that links women's gender roles with sexuality, associates sex with commercial activity, and defines women as passive sexual objects and men as sexual aggressors. In other words, what appears to be deviant may serve other purposes for society.

Critics of the functionalist perspective argue that it does not explain how norms of deviance are established. Despite its analysis of the ramifications of deviant behavior for society as a whole, functionalism does little to explain why some behaviors are defined as normative and others as illegitimate. Questions such as who determines social norms and upon whom such judgments are most likely to be imposed are seldom asked by those using a functionalist perspective. Functionalists see deviance as having stabilizing consequences in society, but they tend to overlook the injustices that labeling someone deviant can produce. Others would say that the functionalist perspective too easily assumes that deviance has a positive role in society. Functionalists will rarely consider the differential effects that the administration of justice has for various social groups. The tendency in functionalist theory to assume that the system works for the good of the whole too easily ignores the inequities in society and how these inequities are reflected in patterns of deviance. These issues are left for sociologists who work from the perspectives of conflict theory and symbolic interaction.

Conflict Theories of Deviance

Recall that conflict theory emphasizes the unequal distribution of power and resources in society and links

FORCES OF SOCIAL CHANGE

Guns and Violence in America

School violence, shootings in the workplace, drive-by shootings: these are the images of violence in America. Has violence increased over time?

Despite public concerns about violence, the rate of violent crime has decreased. Still, the United States has one of the highest rates of violence among industrialized nations. Violence is not unusual in the United States. The murder rates in the 1930s and 1980s were comparable to those of the early 1990s and, although no official data are available, violence was common in earlier periods as well.

Sociologists emphasize that violence has a social context. It is higher in some regions than others, namely, in the South and in urban areas. Violence is also more likely against certain groups, particularly young African American men and young Hispanic men, for whom homicide is the leading cause of death. Violence is growing more rapidly among youth than any other groups—both as victims and as perpetrators.

What can be done about violence? There is no single answer to such a question. Some suggest that gun control is key to reducing violence; others attribute the cause of violence to family problems. Poverty and unemployment are also strongly related to violence. Some sociologists suggest that the media sensationalizes violence, exaggerating the true extent of violence and creating a "culture of fear" (Best 1999; Glassner 1999). According to sociologist Barry Glassner, politicians, corporations, and advocacy groups profit from creating a culture of fear and use the media to convey a sense that the nation is wracked by crime, drug abuse, and disease. These fears, according to Glassner, divert attention and financial resources from other problems such as poverty, education, and housing—problems that could be addressed with increased resources. •••

the study of deviance to social inequality. Based on the work of Karl Marx (see Chapter 1), conflict theory sees a dominant class as controlling the resources of society and using its power to create the institutional rules and belief systems that support its power. Like functionalist theory, conflict theory is a *macrosocial* approach; that is, both theories look at the structure of society as a whole in developing explanations of deviant behavior.

The economic organization of capitalist societies produces deviance and crime according to conflict theory. Certain groups of people have access to fewer resources in capitalist society and are forced into deviance and crime to sustain themselves. Conflict theorists explain the high rate of economic crimes such as theft, robbery, prostitution, and drug-selling among the poorest groups, as a result of the economic status of these groups. Rather than emphasizing values and conformity as a source of deviance, as do functional analyses, conflict theorists see deviance in terms of power relationships and economic inequality (Grant and Martínez 1997).

The upper class, conflict theorists point out, can better hide their deviance because affluent groups have the resources to mask their deviance and crime. As a result, a working-class man who beats his wife is more likely to be arrested and prosecuted than an upper-class man who engages in the same deviant behavior. In addition, those with greater resources can afford to buy their way out of trouble by posting bail, hiring talented and expensive attorneys, or even resorting to bribes.

Corporate crime is crime committed by the elite within the legitimate context of doing business. Conflict theorists expand our view of crime and deviance by revealing the significance of these crimes. They argue that appropriating profit based on the exploitation of the poor and working class is inherent in the structure of capitalist society. **Elite deviance** refers to the wrongdoing of wealthy, powerful individuals and organizations (Simon 2003). Elite deviance includes tax evasion, illegal campaign contributions, corporate scandals that endanger or deceive the public but profit the corporation or individuals within it, and government actions that abuse public trust. The deceptive accounting practices by Enron Corporation that robbed many workers of their retirement pensions are a good example.

Elite deviance includes what early conflict theorists such as Edwin Sutherland called *white-collar crime* (Sutherland 1940; Sutherland and Cressey 1978). Examples of white-collar crimes are bribery, embezzlement, and antitrust violations—such as the U.S. government's case against the Microsoft Corporation and its founder and CEO (chief executive officer), Bill Gates, for unfairly sabotaging the competition.

According to conflict theory, the ruling groups in society develop numerous mechanisms to protect their

Corporate crime can often have a greater consequence for people in society than do street crimes. An example is the Enron scandal, in which corporate leaders used fraudulent accounting practices that robbed employees of their retirement funds.

interests. Conflict theorists argue that law is created by elites to protect the interests of the dominant class. Therefore, the law that is supposedly neutral and fair in its form and implementation works in the interest of the most well-to-do (Spitzer 1975; Weisburd et al. 2001, 1991). Another way conflict theorists see dominant groups use their power is through the excessive regulation of populations that are a potential threat to affluent interests. The current political support among government leaders for new prison construction is an example. Periodically moving the homeless off city streets, especially when a major political or other elite event takes place, is another example. Conflict theory has also produced analyses of institutions that purportedly "treat" deviants (prisons, mental hospitals, detention homes, for example) but routinely fail those they are intended to help.

Conflict theory emphasizes the significance of social control in managing deviance and crime. **Social control** is the process by which groups are brought into conformity with dominant social expectations. Controlling social deviance is one way that dominant groups control the behavior of others. The least powerful in society are frequently assigned deviant labels and are thus most subject to social control. Social control allows powerful groups to maintain their position while regulating others and managing potential or real dissent.

A dramatic historical example of social control is the treatment of supposed witches during the Middle Ages in Europe and during the early Colonial period in America (Ben-Yehuda 1986; Erikson 1966). In the Middle Ages, the Catholic Church was the preeminent social institution in Europe. Witches were often healers and midwives whose views contradicted the authority of the exclusively patriarchal hierarchy of the church. Witches were seen as agents of Satan and were believed to castrate men and use their organs in satanic rituals. *Witch-hunt* is a term still used today to refer to the aggressive pursuits of those who dissent from prevailing political and social norms.

We do not have to look to past centuries to witness how social control works. Those with the power to define deviance exert the most social control. **Social control agents** are people such as police and mental health workers who regulate and administer the response to deviance. Members of powerless groups may be defined as deviant for even the slightest infraction against social norms, whereas members of other groups may be free to behave in deviant ways without consequence. Oppressed groups have a greater likelihood of being labeled deviant and incarcerated or institutionalized, whether or not they have committed a deviant offense. This labeling of deviance is evidence of the power wielded by social control agents. Poor people and members of racial or ethnic minority groups are more likely to be considered criminals and are therefore more likely to be arrested, convicted, and imprisoned than middle- and upper-class people, even for the same crime. People with physical disabilities are more often labeled stupid—regardless of their actual intellectual acuity. Women are more likely judged as whores, even though prostitution is typically an act involving both women and men. Such is the power of social control.

Conflict Theory: Strengths and Weaknesses

The strength of conflict theory is its insight into the significance of power relationships in the definition, identification, and handling of deviance. Conflict theory links the commission, perception, and treatment of crime to inequality in society. This offers a powerful analysis of how the injustices of society produce crime and result in differing systems of justice for disadvantaged and privileged groups. It is not without its weaknesses, however. Critics of conflict theory point out that laws protect most people, not just the affluent.

In addition, although conflict theory offers a powerful analysis of the origins of crime, it is less effective in

explaining other forms of deviance. For example, how would conflict theorists explain the routine deviance of middle-class adolescents? They might point out that much middle-class deviance is driven by consumer marketing. Profits are made from the accoutrements of deviance—rings in pierced eyebrows, alternative music, "gangsta" rap music, or punk dress, but these economic interests alone cannot explain all the deviance observed in society. As Durkheim argued, deviance is functional for the whole of society, not just those with a major stake in the economic system.

Symbolic Interaction Theories of Deviance

Whereas functionalist and conflict theories are *macrosociological* theories, certain *microsociological* theories of deviance look directly at the interactions people have with one another as the origin of social deviance. *Symbolic interaction theory* holds that people behave as they do because of the meanings people attribute to situations. This perspective emphasizes the meanings surrounding deviance, as well as how people respond to those meanings. Symbolic interaction emphasizes that deviance originates in the interaction between different groups and is defined by society's reaction to certain behaviors. Symbolic interaction theories of deviance originated in the perspective of the Chicago School of sociology, examined earlier in Chapter 1.

W. I. Thomas and the Chicago School W. I. Thomas (1863–1947), one of the early sociologists from the University of Chicago, was among the first to develop a sociological perspective on social deviance. Thomas explained deviance as *a normal response to the social conditions in which people find themselves.* He called this *situational analysis,* meaning that people's actions and the subjective meanings attributed to these actions, including deviant behavior, must be understood in social, not individualized, frameworks. Although some of his early work attributed deviance to biological causes, Thomas was greatly influenced by his women students in the Chicago School (Deegan 1990) and then argued that delinquency was caused by the social disorganization brought on by slum life and urban industrialism. He saw it as a problem of social conditions, not individual character.

Differential Association Theory Thomas's work laid the foundation for a classic theory of deviance: differential association theory. **Differential association theory** interprets deviance, including criminal behavior and white-collar crime, as behavior one learns through interaction with others (Sutherland 1940; Sutherland

and Cressey 1978). Edwin Sutherland argued that becoming a criminal or a juvenile delinquent is a matter of learning criminal ways within the primary groups to which one belongs. To Sutherland, people become criminals when they are more strongly socialized to break the law than to obey it. Those who "differentially associate" with delinquents, deviants, or criminals learn to value deviance. The greater the frequency, duration, and intensity of their immersion in deviant environments, the more likely that they will become deviant.

Consider the career path of con artists and hustlers. Hustlers seldom work alone. Like any skilled workers, they have to learn the "tricks of the trade." A new recruit becomes part of a network of other hustlers who teach the recruit the norms of the deviant culture (Prus and Sharper 1991). Crime also tends to run in families. Rather than seizing on a genetic explanation for crime, sociologists explain that youths raised in deviant families are more likely socialized to become deviant themselves (Miller 1986).

Differential association theory offers a compelling explanation for how deviance is culturally transmitted; that is, people pass on deviant expectations through the social groups and networks in which they interact. This explains how deviance may be passed on through generations or may be learned in particular families or peer groups.

Critics of differential association theory argue that this perspective tends to blame deviance on the values of particular groups. Differential association has been used, for instance, to explain the higher rate of crime among the poor and working class, arguing the cause is that they do not share the values of the middle class. Such an explanation, critics say, is class-biased, both because it overlooks the deviance that occurs among the middle class and elites and because it understates the degree to which disadvantaged groups share the values of the middle class. Disadvantaged groups may share the values of the middle class, but cannot necessarily achieve them through legitimate means (a point, you will remember, made by Merton's structural strain theory.) Still, differential association theory offers a good explanation of why deviant activity may be more common in some groups than others, and it emphasizes the significant role that peers play in encouraging deviance.

Labeling Theory **Labeling theory** interprets the responses of others as the most significant factor in understanding how deviant behavior is both created and sustained (Becker 1963). This theory stems from the work of W. I. Thomas, who wrote, "If men define situations as real, they are real in their consequences" (Thomas 1928: 572). *Labeling* is the assignment or attachment of a deviant identity to a person by others, including by agents of social institutions. Therefore,

the people's reaction, not the action itself, produces deviance as a result of the labeling process. Once applied, the deviant label is difficult to shed.

THINKING SOCIOLOGICALLY

Perform an experiment by doing something deviant for a period. Make a record of how others respond to you, and then ask yourself how *labeling theory* is important to the study of *deviance*. Then take your experiment a step further and ask yourself how people's reactions to you might have differed had you been of another race or gender. *A note of caution:* Do not do anything illegal or dangerous; even the most seemingly harmless acts of deviance can generate strong (and sometimes hostile) reactions or even get you arrested, so be careful in planning your experiment!

Linked with conflict theory, labeling theory shows how those with the power to label someone deviant and to impose sanctions wield great power in determining deviance. When police, court officials, school authorities, experts of various sorts, teachers, and official agents of social institutions apply a label, it sticks. Furthermore, because deviants are handled through complex organizations, bureaucratic workers "process" people according to rules and procedures, seldom questioning the basis for those rules or willing or able to challenge them. Bureaucrats are unlikely to linger over whether someone labeled deviant deserves that label, even though they use their judgments and discretion in deciding whether to apply the label. This leaves tremendous room for all kinds of social influence and prejudice to enter the decision of whether someone is considered deviant (Cicourel 1968; Kitsuse and Cicourel 1963; Margolin 1992; Montada and Lerner 1998).

Once the label of deviant is applied, it is difficult for the deviant to recover a nondeviant identity. Once a social worker or psychiatrist labels a client mentally ill, that person will be treated as mentally ill, regardless of his or her mental state. Pleas by the accused that he or she is mentally sound are typically taken as more evidence of the illness! A person's anger and frustration about the label are taken as further support for the diagnosis. A person need not have engaged in deviant behavior to be labeled deviant.

Convicted criminals are formally and publicly labeled wrongdoers, treated with suspicion ever afterward. Labeling theory helps explain why convicts released from prison have such high rates of *recidivism* (return to criminal activities). The label *criminal* or *ex-con* creates great difficulty in finding legitimate employment.

Labeling theory points to a distinction often made by sociologists between primary, secondary, and tertiary deviance. **Primary deviance** is the actual violation of a norm or law, as when someone breaks a law or violates a norm. **Secondary deviance** is the behavior that results from being labeled deviant, regardless of whether the person has previously engaged in deviance. A student labeled a troublemaker, for example, might accept this identity and move from being merely mischievous to engaging in escalating delinquent acts. In this case, the person at least partly accepts the deviant label and acts in accordance with that role. **Tertiary deviance** occurs when the deviant fully accepts the deviant role but rejects the stigma associated with it, as when lesbians and gays proudly display their identity (Lemert 1972; Kitsuse 1980).

Both social class and the role of prisons in society play an important role in the creation of secondary deviance. Jeffrey Reiman (2002) notes that the prison system in the United States is in effect designed to train and socialize prisoners into a career of secondary deviance and to tell the public that crime is a threat primarily from the poor. He sees the goal of the prison system not as reducing crime, but as impressing upon the public that crime is inevitable and that it originates only from the lower classes. Prisons accomplish this, even if unintentionally, by demeaning prisoners, not training prisoners in marketable skills, and stigmatizing prisoners as different from "decent citizens." As a consequence, the person will never be able to fully pay his or her debt to society. In this respect, the prison system creates the very behavior that it is intended to eliminate.

Deviant Identity Another contribution of labeling theory is the understanding that deviance refers not just to something one does, but also to something one becomes. **Deviant identity** is the definition a person has of himself or herself as a deviant. The formation of a deviant identity, like other identities, involves a process of social transformation in which a new self-image and new public definition of a person emerges. Most often, deviant identities emerge over time (Lemert 1972). A drug addict may not think of herself as a junkie until she realizes she no longer has any nonusing friends. In this example, the development of a deviant identity is gradual, but deviant identities can also develop suddenly. A person who becomes disabled as the result of an accident may be given a deviant label. No longer can that person conform to society's definition of normal behavior. Although the person has done no wrong and has had no choice about his or her condition, society applies a stigma to disability. People respond differently to the disabled person. Someone who enters this status, especially if it happens suddenly, has to adjust to a new social identity. In short, the application by society of a label to a person frequently causes that person to take on a personal identity consistent with the label (Irwin 2001; Vail 1999; Montada and Lerner 1998; Scheff 1984).

> ### THINKING *SOCIOLOGICALLY*
>
> Try an experiment in which you pose as a disabled person for a period of time. (This type of research is "ethnomethodology," discussed in Chapters 3 and 5.) Record how people respond to you. Alternatively, if you are disabled or have a friend whom you can accompany for a period of time, record the responses of others that you observe. How do your observations illustrate the *labeling* process associated with *social stigmas*? How do these reactions affect one's identity?

Deviant Careers In the ordinary context of work, a *career* refers to the sequence of movements a person makes through different positions in an occupational system (Becker 1963). A **deviant career** refers to the sequence of movements people make through a particular subculture of deviance. Deviant careers can be studied sociologically like any other career. People are socialized into new "occupational" roles and encouraged, both materially and psychologically, to engage in deviant behavior. The concept of a deviant career emphasizes that there is a progression through deviance: Deviants are recruited, given or denied rewards, and promoted or demoted.

The concept of a deviant career helps explain why being caught and labeled deviant may actually reinforce, rather than deter, one's commitment to a deviant career. For example, hospitalized mental patients are often rewarded with comfort and attention for "acting sick" but are punished when they act normally—for instance, if they rebel against the boredom and constraints of institutionalization. Acting the role will foster their career as a "mentally ill" person (Scheff 1984, 1966). As with legitimate careers, deviant careers involve an evolution in the person's identity, values, and commitment over time. Deviants, like other careerists, may have to demonstrate their career commitment to their superiors, perhaps by passing certain tests of their mettle, as when a gang expects new members to commit a crime, perhaps even shoot someone.

People's reactions to particular behaviors also sustain deviant careers. This explains why being caught and labeled deviant may reinforce, rather than deter, one's commitment to a deviant career. A first arrest on weapons charges may be seen as a rite of passage that brings increased social status among peers in the gang. Whereas those outside the deviant community may think that arrest is a deterrent to crime, it may encourage a person to continue along a deviant path. Punishments administered by the authorities may even become badges of honor within a deviant community. Thus, labeling a teenager "bad" may encourage the behavior because the juvenile may see the label as a positive affirmation of identity.

Like anyone else, deviants may experience career mobility; that is, they may move up or down in rank among deviants. Male prostitution, for example, is a career organized around a hierarchy of illicit sexual services. Men or boys are recruited into prostitution at different ranks as street hustlers, bar hustlers, or escorts. Some may become specialists in sadomasochism, cross-dressing, or catering to pederasts—those who molest children sexually. Newcomers often acquire mentors who train them in the deviant lifestyle. Someone who "learns the ropes" is more likely to continue in a deviant career (Luckenbill 1986).

© Mark Peterson/Corbis Saba

Some deviance develops in deviant communities, such as the "skinheads" shown here marching in a Ku Klux Klan rally protesting the Martin Luther King, Jr., holiday. Such right-wing extremist groups have become more common in recent years.

Deviant Communities The preceding discussion indicates an important sociological point: Deviant behavior is not just the behavior of maladjusted individuals. It often takes place within a group context and involves group response. Some groups are organized around particular forms of social deviance; these are called **deviant communities.** Like subcultures and countercultures, deviant communities maintain their own values, norms, and rewards for deviant behavior. Joining a deviant community separates that person from conventional society and tends to solidify deviant careers, given that the deviant individual receives rewards and status from the in-group. Disapproval from the out-group may only enhance one's status within. Deviant communities also create a worldview that solidifies the deviant identity of their members. They may develop symbolic systems, such as emblems, forms of dress, publications, and other symbols that promote their identity as a deviant group. Gangs wear their colors; prostitutes have their own vocabulary of tricks and johns; skinheads have their insignia and music. All are examples of deviant communities. Ironically, subcultural norms and values reinforce the deviant label both inside and outside the deviant group, thereby reinforcing the deviant behavior.

Some deviant communities are organized specifically to provide support to those in presumed deviant categories. Groups such as Alcoholics Anonymous, Weight Watchers, and various 12-step programs help those identified as deviant overcome their deviant behavior. These groups, which can be effective, accomplish their mission by encouraging members to accept their deviant identity as the first step to recovery.

The Problem with Official Statistics Because labeling theorists see deviance as produced in significant part by people with the power to assign labels to people, they question the value of official statistics as indicators of the true extent of deviance. Reported rates of deviant behavior are themselves the product of socially determined behavior, specifically the behavior of identifying what or who is deviant. Official rates of deviance are produced by people in the social system who define, classify, and record only certain behaviors as deviant. Labeling theorists are more likely to ask how behavior becomes labeled deviant than they are to ask what motivates people to become deviant (Best 2001; Kitsuse and Cicourel 1963).

In the aftermath of terrorist attacks at the World Trade Center, officials debated whether to count the deaths of thousands as murder or in a separate category of terrorism. The decision would change the official rate of deviance by inflating or deflating the reported crime rate of murder in New York City in that year. In the end, these deaths were not counted in the murder rate.

Official rates of deviance do not necessarily reflect the actual incidence of deviance; they reflect social judgments. Consider suicide. Official reports of suicide rates are based on records typically produced in a coroner's office where someone determines and records the cause of death. Suicide carries a stigma, and staff members in a coroner's office may possess stereotypes about who is likely to commit suicide that influence how a death is recorded. As a result, a designation of "suicide" for upper-class people is less likely. A mentally ill person who kills himself is more likely to have his death recorded as suicide than one not labeled mentally ill. Or a terminally ill middle-class person who takes his or her own life may not be recorded as a suicide in deference to the family. Other factors, such as religious affiliation and nationality, as well as unofficial interference by interested parties, may also influence whether a death is recorded as a suicide. As one sociologist has concluded, "The more socially integrated an individual is, the more he and his significant others will try to avoid having his death categorized as a suicide" (Douglass 1967: 209).

In another example, when AIDS (acquired immune deficiency syndrome) first emerged, it was highly stigmatized because of its perceived association with gay men. Obituaries of AIDS victims seldom noted that the death was due to AIDS. More typically, obituaries reported only that the person died following a "long illness." Labeling theorists also note that official rape rates are underestimates of the actual extent of rape, not only because of victims' reluctance to report, but also because rapes that end in death are classified as murder. Instances of rape are also less likely to be counted as rape by police if the victim is a prostitute, was drunk or on drugs at the time of the assault, or had a prior relationship with the assailant. A number of studies have also shown that whether a police officer lets a drug offender off with only a warning, and whether the incident appears as an arrest in the official records, depends upon the offender's race, demeanor, dress, politeness, and attitude, among other things. This suggests that official records may say more about police behavior than about actual deviances (Babbie 2001; DeFleur 1975). Given such problems, any official statistics must be interpreted with caution.

Labeling Theory: Strengths and Weaknesses
The strength of labeling theory is its recognition that the judgments people make about presumably deviant behavior have powerful social effects. Labeling theory does not, however, explain why deviance occurs in the first place. Labeling theory may illuminate the consequences of a young man's violent behavior, but it does not explain the origins of the behavior. The weakness of labeling theory is that it does not explain why some people become deviant and others do not. Although it focuses on the behaviors and beliefs of officials in the enforcement system, labeling theory does not explain why those officials define some behaviors as deviant or criminal, but not others. This shortcoming in the analysis of deviance has been carefully scrutinized by conflict theorists who place their analysis of deviance within the power relationships of race, class, and gender.

Forms of Deviance

Although deviance takes many forms, the sociology of deviant behavior has focused heavily on subjects such as mental illness, social stigmas, and substance abuse.

The study of crime will be examined in the next chapter. In reviewing different forms of deviance, you will also see how the different sociological theories about deviance contribute to understanding each subject. In addition, you will see how the social structural context of race, class, and gender relationships shape these different forms of deviance. Remember that deviance is not just an individual attribute; it is patterned and supported by social institutions and beliefs. Race, class, and gender are not just individual attributes; they are patterns of relationships supported by social institutions and social ideologies. Consequently, they are an important part of the social context in which different forms of deviance emerge and from which people make judgments about who is deviant and who is not.

Mental Illness

Sociological explanations of mental illness look to the social systems that define, identify, and treat mental illness, even though many typically think of it only in psychological terms. This has several implications for understanding mental illness. Functionalist theory suggests that by recognizing mental illness, society also upholds normative values about more conforming behavior. Symbolic interactionists tell us that mentally ill people are not necessarily "sick" but are the victims of societal reactions to their behavior. Some go so far as to say there is no such thing as mental illness, only people's adverse reactions to unusual behavior. From this point of view, people learn faulty self-images and then are cast into the role of patient when they are treated by therapists. Once someone is labeled a "patient," he or she is forced into the "sick" role, as expected by those who reinforce it, and it becomes extremely difficult to get out of the role (Szasz 1974).

DEBUNKING *SOCIETY'S MYTHS*

Myth: Mental illness is an abnormality best studied by psychologists.

Sociological perspective: Mental illness follows patterns associated with race, class, and gender relations in society and is subject to a significant labeling effect; those who study and treat mental illness benefit from a sociological perspective.

Labeling theory, combined with conflict theory, suggests that people with the fewest resources are most likely to be labeled mentally ill. This is substantiated by data compiled on mental illness. Women, racial minorities, and the poor all suffer higher rates of reported mental illness and more serious disorders than groups of higher social and economic status. Furthermore, research over the years has consistently shown that middle- and upper-class persons are more likely to receive some type of psychotherapy for their illness than poorer individuals and minorities, who are more likely to receive only physical rehabilitation and medication, with no accompanying psychotherapy (Hollingshead and Redlich 1958).

Sociologists give two explanations for the correlation between social status and mental illness. The stresses of being in a low-income group, being a racial minority, or being a woman in a sexist society contribute to higher rates of mental illness. The harsher social environment is also a threat to mental health. However, the same behavior that is labeled mentally ill for some groups may be tolerated in others. For example, behavior considered crazy in a homeless woman who is likely to be seen as deranged may be seen as merely eccentric or charming when exhibited by a rich person. To illustrate this, ask yourself what would have happened to a low-income African American man who exhibited the same violent and threatening behavior that many people reported of John du Pont—the wealthy heir of the du Pont fortune who shot and killed Olympic champion wrestler Dave Schultz on his estate in 1996. For many prior years, people observed du Pont's eccentric and crazy behavior but did nothing about it (Longman 1996).

Patterns of mental illness also reflect gender relations in society. Women have higher rates of mental illness than men, although men and women differ in the kinds of mental illnesses they experience (Horton 1995). Part of this pattern is psychological, and much research has shown a connection between women's learned gender roles and the likelihood of depression, anxiety, and other forms of mental illness. But women also live in stressful conditions, so mental illness is not solely the result of personality factors. Poverty, working environments, unhappy marriages, physical and sexual abuse, and the stress of rearing children all contribute to higher rates of mental illness for women (Elliott 2001). Women's learned gender roles make them more likely than men to seek help when they are distressed, thus producing higher measured rates of mental illness. As labeling theory would predict, gender stereotypes also mean that women are more likely to be labeled mentally ill (Schur 1984). The frequency with which physicians label women's complaints to be "psychologically grounded" is evidence of this fact.

A disproportionate amount of mental illness is also found among racial minority groups in society, pointing again to a correlation between mental health and group status in society (Aponte 1994; Chin 1993). Patterns of mental illness are not the same for all minority groups, however. Mexican Americans have relatively low rates of mental illness and Puerto Ricans have higher rates of mental illness than non-Hispanic Whites. The higher rates of mental illness among African Americans and other racial groups are the result of the stresses of living in a racially conflicted society. White psychiatrists also tend to overdiagnose mental illness among racial minorities. This problem is exacerbated by the availability of only a few minority psychiatrists and psychologists (Williams and Williams 2000). Lower-income people are less able to afford expensive psychiatric care. Consequently, they may delay treatment, and their illness may persist and become aggravated over time. Well-to-do people who can afford private care may be far more likely to recover from mental illness than someone who can only afford to be admitted through the emergency room of a county hospital. Like other forms of deviance, mental illness—its incidence, expression, and treatment—reflects conditions in society.

Social Stigmas

A **stigma** is an attribute that is socially devalued and discredited. Some stigmas result in labeling other people as deviant. The experiences of people who are disabled, disfigured, or in some other way stigmatized are studied in much the same way as other forms of social deviance. Like other deviants, people with stigmas are stereotyped and defined only in terms of their presumed deviance.

Think of how disabled people are treated in society. Their disability can become a **master status** (Chapter 5), a characteristic of a person that overrides all other features of the person's identity (Goffman 1963b). Physical disability can become a master status when other people see the disability as the defining feature of the person. A person with a disability becomes "that blind woman" or "that paralyzed guy." Persons with a particular stigma are often seen to be all alike. This may explain why stigmatized individuals of high visibility are often expected to represent the whole group.

People who are suddenly disabled often have the alarming experience of their new master status rapidly erasing their former identity. They may be treated and seen differently by people they knew before their disability. A master status may also prevent people from seeing other parts of a person. A person with a disability may be assumed to have no meaningful sex life, even if the disability is unrelated to sexual ability or desire. Sociologists have argued that the negative judgments made about people with stigmas tend to confirm the "usualness" of others (Goffman 1963b: 3). For example, when people stigmatize welfare recipients as lazy and undeserving of social support, others are indirectly promoted as industrious and deserving.

Figure 7.3 Cigarette Smoking by Adults

Source: National Center for Health Statistics. 2003. *Health United States 2003.* Hyattsville, MD: U.S. Department of Health and Human Services, p. 214.

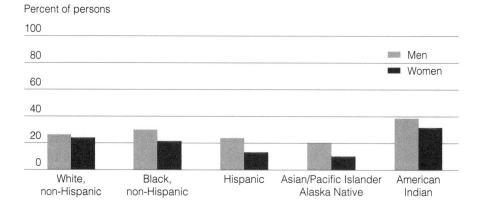

When stigmatized individuals are measured against a presumed norm, they may be labeled, stereotyped, and discriminated against. In Erving Goffman's words, people with stigmas are perceived to have a "spoiled identity." When others see them as deficient or inferior, they are caught in a role imposed by the stigma. The stigmatized individuals may respond by trying to hide their stigma or blame others. What happens, for example, to people who have a sexually transmitted disease? Because this is associated with sexual immorality, people typically react with shame and embarrassment. They may try to conceal that they have the disease (Nack 2000).

Sometimes people with stigmas will bond with others, perhaps even strangers who they believe share their trait. This acknowledgment of "kinship" or affiliation can be as subtle as an understanding look, a greeting that makes a connection between two people, or a favor extended to a stranger presumed to have the same stigma. Such public exchanges are common between various groups that share certain forms of disadvantage, such as people with disabilities, lesbians and gays, or members of other minority groups.

Substance Abuse: Drugs and Alcohol

As with mental illness and stigmas, sociologists study the social factors that influence drug and alcohol use. Who uses what and why? How are users defined by others? These questions guide sociological research on substance abuse.

One of the first things to notice in thinking about drugs and alcohol is to ask why using some substances is considered deviant and stigmatizing while using others is not. How do such definitions of deviance change over time? Until recently, cigarette smoking was considered normative—indeed glamorous and sexy. Now, although smoking is still common (see Figure 7.3), it has become more of a stigma. Some might say this change resulted from the known risks of nicotine addiction. But just knowing the risks of smoking is not enough to define it as deviant behavior. Sociologists study how social groups have mobilized to define smoking as deviant and analyze how the tobacco industry has navigated through the climate of public opinion to maintain the industry's profits (Kall 2002; Brown 2000).

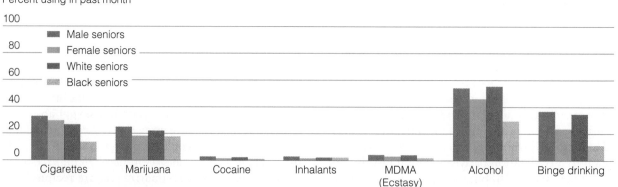

Figure 7.4 Use of Selected Substances by High School Seniors

Source: National Center for Health Statistics. 2003. *Health United States 2003.* Hyattsville, MD: U.S. Department of Health and Human Services, pp. 219–221.

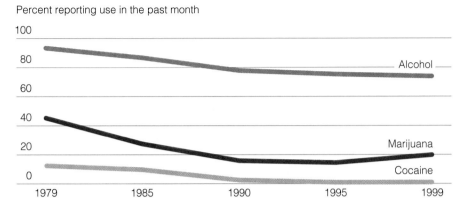

Percent reporting use in the past month

Figure 7.5 Drug and Alcohol Use (persons age 18–25 Years)

Source: National Center for Health Statistics. 2003. *Health United States 2003.* Hyattsville, MD: U.S. Department of Health and Human Services, pp. 216–222.

Even now, however, with increased awareness about the harms of nicotine addiction and more regulation of smoking in public, cigarettes are still largely considered a legitimate drug. Unlike other drugs, cigarettes are publicly marketed, and unless you are under age, you can buy them without fear of arrest. Whatever stigma is attached to cigarette use comes mostly from informal norms that vary in different social settings and among different groups. Young people may be in peer groups where smoking is considered fun-loving and cool. This image is actively promoted by the tobacco industry, which specifically targets young people in the marketing of cigarettes. Recent legislation, however, has limited this practice (Schlosser 2001; Lloyd 1997).

Like cigarettes, alcohol is also a legal drug. Whether one is labeled an alcoholic depends in large part on the social context in which one drinks—not solely the amount of alcohol consumed. Drinking from a bottle in a brown bag on the street corner is considered highly deviant; having martinis at a posh bar is seen as hip.

Sociological understandings challenge views of drug and alcohol use as stemming from individual behaviors that lead to substance abuse. Patterns of use vary by factors such as age, gender, and race, among others. Age is one significant predictor of illegal drug use. Young people are more likely to use marijuana and cocaine and binge drink than are people over age 25; alcohol is most likely to be used by those aged 18–34, although the difference here is less than for other drugs. Although there is much public concern over drug abuse by young people, the extent of use among all age groups has declined quite dramatically since the late 1970s (see Figure 7.5).

TAKING ON SOCIAL ISSUES
Should Drugs Be Decriminalized?

Drug policy in the United States has largely focused on prohibition and enforcement; thus, laws are in place requiring mandatory sentencing for drug possession. An alternative approach would be legalizing drugs like marijuana, cocaine, and heroin, focusing on treatment, not criminalization. The debate over punishment versus decriminalization engages competing values and strong political differences. What does sociological research contribute to this debate—a debate largely muted because of the political climate of zero tolerance?

Researchers Robert MacCoun and Peter Reuter point out that prohibiting drugs has its own harmful results, including increased crime, corruption, and disease (such as from dirty needles). These harms are disproportionately borne by the poor and by racial minorities.

In Holland, where there are no penalties for possessing small amounts of marijuana, use has declined since 1976 when marijuana was decriminalized. There has, however, been increased use by youth in very recent years. This is probably not caused by decriminalization per se; other factors influence increasing usage by young people.

Allowing legal access to drugs eliminates some forms of harm (mostly those associated with enforcement) and reduces some other harms (such as needle sharing and overdoses), but does not eliminate addictive behavior and its impact.

MacCoun and Reuter support legalizing marijuana when used in small amounts. Given their findings—and the public issues surrounding drug use—what sociological considerations do you think should influence the formation of drug policy?

Taking Action

Go to the Taking Action Exercise on the companion website—at http://sociology.wadsworth.com/andersen_taylor4e/—to learn more about an organization that addresses this topic.

Source: MacCoun, Robert J., and Peter Reuter. 2001. *Drug War Heresies: Learning from Other Vices, Times, and Places.* New York: Cambridge University Press. ●●●

MAP 7.2 The Global Fix

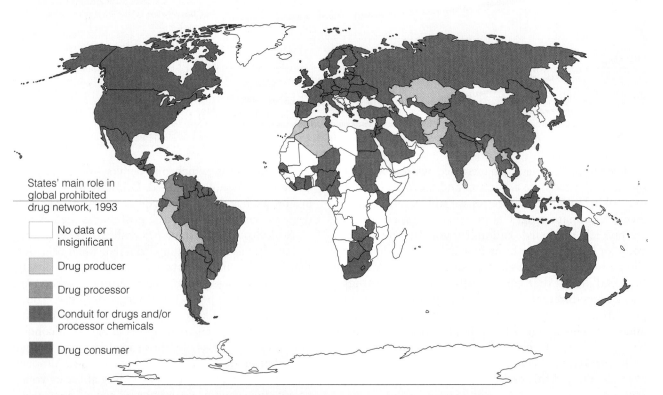

States' main role in global prohibited drug network, 1993

- No data or insignificant
- Drug producer
- Drug processor
- Conduit for drugs and/or processor chemicals
- Drug consumer

This map shows how drug trafficking is organized in a complex global system of production, distribution, and consumption—just like other global commodities. Drug trafficking is thus a wide-ranging global structural phenomenon rather than an individual-based form of deviance. How does seeing this global image of drugs alter the view of drugs from an individual form of deviance to a globally structured phenomenon?

Source: "The Golden Fix (map)," from *State of the World Atlas, New Edition* by Michael Kidron and Ronald Segal. Copyright © 1995 by Michael Kidron and Ronald Segal, text. Copyright © 1995 by Myriad Editions Limited, maps and graphics. Used by permission of Viking Penguin, a division of Penguin Putnam Inc.

Drug and alcohol use also varies substantially by gender and race. Men are more likely than women to be problem drinkers and drug abusers—a pattern that many explain as the result of gender roles that encourage men to be risk-takers. African Americans and Hispanics are less likely to drink than Whites and are far less

SOCIOLOGY IN PRACTICE
Dealing with Binge Drinking

On many college campuses, many students are binge drinkers—drinking an excessive amount on a regular basis. To sociologists, binge drinking is deviant behavior even though it may be common on a given campus. In your sociological judgment, what causes student binge drinking? What are the policies on drinking on your campus, and how effective are they in reducing problems associated with binge drinking?

Critical Thinking Exercise

1. Should binge drinking be defined as deviant behavior? Why or why not? Might it be defined as deviant behavior on some campuses and not others?

2. Whose responsibility should it be to monitor and otherwise deal with binge drinking on your campus? The president? The administration? The students? Some combination of these?

•••

likely to be binge drinkers. Hispanics are less likely than Whites and Blacks to use marijuana but are the group most likely to be using cocaine (National Center for Health Statistics 2003).

Deviance in Global Perspective

As deviance increasingly crosses national borders, understanding crime and deviance now requires a global perspective. Terrorism is a case in point. Worldwide, terrorism has been with us for many years, but was propelled to the public's attention on September 11, 2001. Motivated by political conflicts, often involving ethnic and religious conflict, terrorism has caused some of the world's most violent incidents. Bombings of buildings, airplanes, urban trains and buses, and other targets have become almost commonplace. These expressions of extremist political beliefs stem from the many international conflicts of our current world events. Without understanding the political and economic relations that are the origins for such violence, such deviant acts seem like the crazed behavior of violent individuals. Sociologists in no way excuse such acts, yet they look to the social structural conflicts from which terrorism emerges as the cause of such criminal and deviant behavior. We examine terrorism as a form of crime more completely in the following chapter.

Globalization also means that networks of deviant behavior can more easily flourish across national borders. The same technological developments that ease communication for legitimate business activities also enable illegitimate activities to thrive. Money acquired through illegal activity in one country can easily be transferred to another country. Likewise, transportation systems critical to the international exchange of illegal goods—whether drugs, weapons, or sexual services—link what were once distant and inaccessible places (Binns 2003).

The drug problem well illustrates the globalization of deviance. The map "The Golden Fix" shows how many nations are involved, one way or another, in the international traffic in drugs. Some nations, including the United States, Australia, and parts of western Europe, are vast markets for the consumption of illegal drugs. Other nations, such as Colombia, are known as major drug producers. Still others, such as China, Brazil, and Mexico, play a role in this international drug economy as conduits for drug traffic and production. The person on the streets of New York or Amsterdam who is labeled deviant by virtue of being a "crackhead" is part of an international network of drug production and sales. Going even further, the profits on such drugs may end up in other places like the Cayman Islands, Nigeria, or Switzerland, where the profits from the drug trade are laundered in foreign bank accounts.

Just as globalization is shaping other dimensions of social life, so is it shaping deviant activities. Drug trafficking, the international sex trade, laundering money from corporate deviance into offshore banking accounts all show the global dimensions of deviant behavior. In the next chapter, we examine in detail a particular form of deviance—crime.

Chapter Summary

What is deviance?

Deviance is the behavior recognized as violating expected rules and norms and must be understood in the social context in which it occurs.

What are the major theories of deviance?

Both biological and psychological explanations of deviance, though valuable, place the causes of deviance within the individual person. Early sociological explanations of deviance from the Chicago School, and later sociological explanations, place the causes of deviance within the culture and/or structure of society. *Functionalist theory* sees deviance as functional, thus beneficial, for society because it affirms what is acceptable by defining what is not acceptable. It also attributes deviance to an imbalance between cultural goals and structurally available means to attain the goals. *Symbolic interaction theory* explains deviance as the result of people's perceptions and the meanings people give to various behaviors. *Differential association theory* interprets deviance as learned through social interaction with other deviants. *Labeling theory* argues that society actually creates deviance by noting that some groups have more power than others to assign deviant labels to people. *Conflict theory* explains deviance in the context of unequal power relationships and inequalities in society. Conflict theorists also see powerful groups in society as creating laws and other regulatory mechanisms for protecting dominant group interests.

What are the forms of deviance?

Studies of mental illness and social stigmas reveal some of the sociological factors that produce deviance, relating these phenomena to societal conditions. A stigma is an attribute that is socially devalued and discredited. Those with stigmas such as physical disabilities are often treated like social deviants, even if they are not deviant in any other way. Drug and alcohol users may also experience social stigmas. Patterns of drug and alcohol use vary among different social groups and are defined as deviant only within certain social contexts.

How is deviance global?

Networks of deviant behavior flourish across national borders. International network systems allow for the exchange of illegal goods between countries—for example drugs, money, weapons, or sexual services.

Key Terms

altruistic suicide 173

anomic suicide 173

anomie 172

deviance 166

deviant career 180

deviant community 181

deviant identity 180

differential association theory 178

egoistic suicide 173

elite deviance 177

labeling theory 179

medicalization of deviance 171

moral entrepreneurs 168

primary deviance 179

secondary deviance 180

social control 177

social control agents 178

social control theory 175

stigma 183

structural strain theory 174

tertiary deviance 180

Researching Society with Microcase Online

You can see the results of actual research by using the Wadsworth MicroCase® Online feature available to you. This feature allows you to look at some of the results from national surveys, census data, and some other data sources. You can explore this easy-to-use feature on your own, but try this example. Suppose you want to know:

Have you ever wondered how many married persons have "strayed" (a type of deviant behavior)—that is, had sexual relations with someone other than his or her spouse? And have you ever wondered what factors (variables) predict this behavior?

To answer this question, go to http://sociology.wadsworth .com/andersen_taylor4e/, select MicroCase Online from the left navigation bar, and follow the directions there to analyze the following data.

Data file: GSS

Task: Auto-Analyzer

Primary Variable: EVER STRAY

Question

Once you have your results, answer the following question:

For each of the demographic variables listed, indicate whether there is a significant effect. If so, indicate which category is most likely and least likely to have "strayed."

Socio–Demographic Variable	Is the Overall Effect Significant?	Category Most Likely	Category Least Likely
Religion	Yes No		
Political Party	Yes No		
Education	Yes No		
Sex	Yes No		
Income	Yes No		

The Companion Website for Sociology: Understanding a Diverse Society, Fourth Edition

http://sociology.wadsworth.com/andersen_taylor4e/

Supplement your review of this chapter by going to the companion website to take one of the Tutorial Quizzes, use the flash cards to master key terms, and check out the many other study aids you'll find there. You'll also find special features such as GSS Data and Census 2000 information, data and resources at your fingertips to help you with that special project or do some research on your own.

Suggested Readings and Web Resources

Adler, Patricia A., and Peter Adler. 2000. *Constructions of Deviance: Social Power, Context, and Interaction*, 3rd ed. Belmont, CA: Wadsworth.
   This anthology explores various dimensions of deviant behavior from a labeling theory perspective. Ranging from studies of card sharks to missing children to punks, the book covers a wide range of intriguing sociological analyses of deviant behavior.

Becker, Howard S. 1963. *Outsiders: Studies in the Sociology of Deviance*. New York: Free Press.
   Becker gives a straightforward and important analysis of labeling theory to explain deviance. Using the examples of marijuana users and jazz musicians, among others, to explain labeling theory, he also develops the concepts of deviant careers and moral entrepreneurs. Highly readable and engaging, this book is a classic in the sociology of deviance.

Best, Joel. 1999. *Random Violence: How We Talk About New Crimes and New Victims*. Berkeley, CA: University of California Press.

Best argues that the media present a distorted picture of the extent of violence in the United States by suggesting that violence is random and rampant when it is, in fact, more patterned and more limited than popular images suggest.

Goffman, Erving. 1963. *Stigma: Notes on the Management of Spoiled Identity*. Englewood Cliffs, NJ: Prentice Hall.

Goffman's book is a classic study. Using a symbolic interactionist perspective, he studies social responses to stigmas of various sorts and discusses social responses to stigmas and their effect on the individuals who have stigmas of one sort or another, including physical deformities and devalued social statuses.

Simon, David R. 2003. *Elite Deviance*, 8th ed. Boston, MA: Allyn and Bacon.

By examining the connections between corporate, government, and military institutions, the author explores the causes and consequences of elite deviance and crime, defined as the wrongdoing of wealthy and powerful individuals and organizations.

Bureau of Justice Statistics
www.ojp.usdoj.gov/bjs/
A division of the federal government that compiles information on subjects pertinent to the study of crime and deviance.

National Coalition for the Homeless
www.nationalhomeless.org
This site provides information on the extent of homelessness, as well as bibliographies and other information pertinent to the study of homelessness.

National Institute on Drug Abuse (NIDA)
www.nida.nih.gov
The federal agency that monitors and sponsors research on drug abuse. The site includes extensive data on patterns of drug use, including detailed information from the Monitoring the Future Study—a periodic report on use by high school (and younger) students.

• • •

Crime and Criminal Justice

Crime is all around us, or so most of us think. Crimes vary in their type, seriousness, and victims. Youth gangs are one of the most persistent realities of life in America today. Back in the 1950s and early 1960s, youth gangs busied themselves with the likes of stealing hubcaps from cars and hanging out on street corners. At that time, youth gangs even took on a certain romanticism. Today, youth gangs provide a fertile training ground for more serious crimes, and many adult criminals and mob figures start out in youth gangs.

The impression that youth as well as adult crime is all around us is complicated somewhat by how crime records are collected and kept in our society. Some crimes get reported and become a part of official statistics, and some do not. The keeping of official crime records is subject to sociological influences, as we shall see in this chapter. Whether a person's crime goes on record is to a large extent a matter of sociological forces such as institutional practices, race, ethnicity, social class, and age.

Certain crimes, such as assault and robbery, are more likely to go on record if the person committing them is a person of color, poor or working-class, male, or young. These serious crimes are less likely to get prosecuted and get on record if the person committing them is White, middle-class, and dressed in a suit and tie. If the crime in question is embezzlement, income tax fraud, or insider trading—so-called elite crimes— then the person committing it is more likely to be White, middle-class, and male, and is somewhat less likely to get reported in any official records.

If you are Black or Hispanic and driving at night on an interstate highway in a fancy, late-model car with two or three other Black or Hispanic persons, either male or female, then you stand a roughly eight-in-ten chance of being stopped, questioned, and detained by the police. It does not matter if you are employed in a good job, or are a diligent student, or have committed no crime, and you were simply minding your own business and trying to get home to bed. In a practice called *racial profiling,* the state police may stop and detain you solely on the basis of the color of your skin. They guess that because you are a person of color, the chances are moderately high that you may be up to no good—although, as we shall see, this guess is erroneous.

The police officer who practices racial profiling is probably *not* acting out of personal racial prejudice. He or she is acting on an institutionalized procedure, a procedure that has become part of the social structure. Racial profiling is less an act of a racially prejudiced person and more a result of practices and structures within the criminal justice system. Thus, even racial profiling demonstrates a sociological lesson iterated throughout this book: Acts by humans are often the product of social structure, not just the product of individual personalities.

We will see in this chapter how crime and the criminal justice system in the United States is structured. We will see that crime is not simply a series of random acts, but is patterned and highly predictable. Finally, we will see the role of race, social class, and gender, as well as age, in the occurrence of crime and in the recording and dispensation of crime statistics in this society. •••

Crime and Deviance

We studied deviant behavior in Chapter 7. Crime is a type of deviant behavior, but not all deviant behavior would be called crime. Specifically, **crime** is a type of deviant behavior that violates specific criminal laws. Deviance becomes crime when it is designated by the institutions of society as violating such a law or laws. *Deviance* is behavior that is recognized as violating rules and norms of society. If those rules are formal laws, the deviant behavior would be called crime.

Criminology is the study of crime from a scientific perspective. Criminologists include social scientists such as sociologists who stress the societal causes of crime, psychologists who stress the personality-based causes of crime, and political scientists who view crime as being both caused and regulated by the powerful institutions in society.

Theoretical perspectives on deviance reviewed earlier (in Chapter 7) contribute to our understanding of crime (see Table 8.1). According to the *functionalist perspective,* crime, like other forms of deviance, may be *necessary* to hold society together—a profound hypothesis. By singling out criminals as socially deviant, others are defined as good. The nightly reporting of crime on television is a demonstration of this sociological function of crime. *Conflict theory* suggests that disadvantaged groups are more likely to become criminals than those who are privileged. It also sees the well-to-do as better able to hide their crimes and to be less likely to be punished. The *symbolic interactionist perspective* helps us understand how people learn to become criminals and how being labeled a criminal or ex-criminal can increase the probability that the person will engage in criminal behavior. Each perspective traces criminal behavior to social conditions rather than only to the intrinsic tendencies or personalities of individuals.

Types of Crime

A variety of types of crime exists in our society. Most of us are accustomed to regarding violent crimes as the only real category of crime, but other examples are elite and white-collar crimes such as embezzlement and "victimless" crimes such as gambling. First we address the matter of measuring crime.

Table 8.1		
Sociological Theories of Crime		
Functionalist Theory	**Conflict Theory**	**Symbolic Interaction Theory**
Societies require a certain level of crime in order to clarify norms.	The lower the social class, the more the individual is forced into criminality.	Crime is behavior that is learned through social interaction.
Crime results from social structural strains within society.	Inequalities in society by race, class, gender and other forces tend to produce criminal activity.	Labeling criminals tends to reinforce rather than deter crime.
Crime may be functional to society, thus difficult to eradicate.	Reducing social inequalities in society will reduce crime.	Institutions with power to label, such as prisons, actually produce rather than lessen crime.

Measuring Crime: How Much Is There?

Is crime increasing in America? One would certainly think so from watching the media. Images of violent crime abound and give the impression that crime is a constant threat and that it is on the rise. Data on crime actually show that violent crime peaked in 1990, but actually *decreased* through the 1990s and began to level off before 2000 (see Figure 8.1). Assault and robbery, in particular, decreased quite significantly through the 1990s. Murder and rape have remained more constant, although these too show some decline since the 1990s.

Data about crime come from the Federal Bureau of Investigation (FBI), based on reports from police departments across the nation. The data are distributed annually in the *Uniform Crime Reports*. Data in the Uniform Crime Reports are the basis for official reports about the extent of crime and its rise and fall over time. These data show that, while media reporting of crime has remained high, the officially reported rate of crime has decreased.

A second major source of crime data are the *National Crime Victimization Surveys* published by the Bureau of Justice Statistics in the Department of Justice. These data are based on surveys in which national samples of people are periodically asked if they have been the victim of one or more criminal acts. These surveys also show that violent crime, including rape, assault, robbery and murder, have declined by 15 percent since the 1990s.

Both of these sources of data—the Uniform Crime Reports and the National Crime Victimization Surveys—are subject to the problem of underreporting. About half to two-thirds of all crimes may not be reported to the police, meaning that much crime never shows up in the official statistics. Certain serious crimes, such as rape, are significantly underreported. Victims may be too upset to report a rape to the police or they may believe that the police will not believe a rape has occurred. Equally significant, the victim may not want to undergo the continued emotional stress of an investigation and trial.

Recall from Chapter 7 that certain kinds of noncriminal deviance, such as suicide, are also underreported (especially by upper-income families) because of embarrassment to the deceased person's family. This is one reason crime statistics, and the statistics on deviant behavior such as suicide, may underestimate the true extent of the crime or deviance in question.

Another problem arises in the attempt to measure crime by means of official statistics. The FBI's *Uniform Crime Reports* stresses what it calls **index crimes**, which includes the violent crimes of murder, manslaughter, rape, robbery, and aggravated assault, plus property

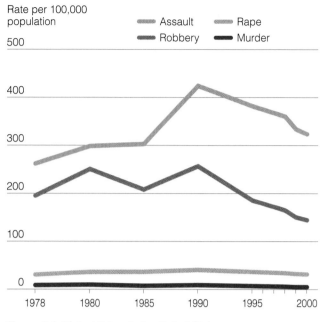

Figure 8.1 Violent Crime in the United States

Source: Federal Bureau of Investigation. 2002. *Uniform Crime Reports.* Washington, DC: U.S. Department of Justice.

crimes of burglary, larceny-theft, and motor vehicle theft. These crimes are committed mostly by individuals who are disproportionately minority and poor people. Statistics based on these offenses do not reflect the crimes that tend to be committed by middle-class and upper-class persons, such as tax violations, insider trading, embezzlement, and other so-called elite crimes. Therefore, the index crimes give a biased picture of crime. The official statistics provide a relatively inflated picture for index crimes but an underreported picture of elite crimes. A final result is, unfortunately, that the public is led to see the stereotyped "criminal" as a lower-class person, most likely African American or Latino, instead of as a middle- or upper-class White person who has committed tax fraud. The official statistics give biased support to the stereotype. This in turn perpetuates the public belief that the "typical" criminal is lower class and minority instead of upper class and nonminority. Criminals, however, can be either.

THINKING *SOCIOLOGICALLY*

Most people have engaged at least once in their lifetime in some relatively petty crime, but crime nonetheless. Have you ever engaged in a minor bit of shoplifting? Have you ever secretly made an obscene phone call? Have you ever stolen anything from school, even a book or supplies? Has anyone you know committed any such crime, even once? Think about how such acts are not reflected in official crime statistics, and consider how these statistics might look if crimes like these were always accurately reported to the police.

Youth Gangs and Crime

American cities have always been a breeding ground for youth gangs. Youth gangs often engage in criminal activities, which to a great extent serve as training for the youth to engage in more crime as adults, thus developing a career of crime. Youth criminals of past years often grow up to participate in organized crime of the present. An especially notorious youth gang of the 1960s and 1970s in Chicago was called the *Blackstone Rangers*. As teenagers they engaged in relatively petty street crimes and street fights, but more recently have evolved into the large and well-organized *El Rukans* gang of organized crime, whose leaders rule the streets from behind the walls of prison. The youth gangs of the 1950s, often romanticized in movies, have graduated to the automatic weapons-carrying *posses* of today's urban environments. Today's youth gangs do not steal hubcaps and soap their neighbors' windows but, in contrast, are more likely to be profit-oriented and to engage in serious crimes such as burglaries, robberies, drug dealing, and other offenses usually associated with hardened adult criminals. Such is the case with the notorious *Bloods* and *Crips* gangs of Los Angeles, gangs that are characterized by an elaborate hierarchical organizational structure, now spread to other cities far beyond just Los Angeles (Jackson 2002; Harris 1998).

The study of urban youth gangs has been of consistent interest to sociologists since the 1920s, when Frederick M. Thrasher published his book called simply *The Gang* (1927). Martin Sanchez Jankowski (1991), consistent with Thrasher and others (Harris 1998; Sanders 1994), notes that urban youth gangs are a relatively small group of young adults who create family-like bonds and hang around together to engage in criminal acts and to protect their "turf," or territory. It is common for the gang member to think of her or his immediate subdivision of the gang as a family. Often new members of such gangs must undergo an initiation ritual, which may involve engaging in a serious fight with one or more gang members or committing some criminal act to demonstrate loyalty to the gang. This act can involve anything from petty theft to assault with a gun and may even involve murder.

Crime committed by youths is usually termed delinquency. One problem in studying juvenile delinquency is that different rules are used for offenses committed by juveniles and different rules are used by different courts. Furthermore, juvenile offenders are placed in vague crime categories such as vagrancy, sexual promiscuity, truancy, and incorrigibility. One reason these vague categories are used is that juvenile court judges are expected to be flexible in judging youths and thus divert them from a life of crime, and the vagueness of the cat-

egories is intended to assist this process. Unfortunately, this vagueness has produced unequal treatment of juveniles depending on their environmental circumstances, social class, race–ethnicity, and gender. For example, being Black or Hispanic and being of lower class status increase a young person's odds of being arrested, whether or not the person has engaged in delinquency (Bond-Maupin et al. 2002; Brownfield et al. 2001). Likewise, probation and parole officers are likely to describe juvenile delinquent girls in sexualized ways, particularly when the girl is a Latina (Leisenring 2002).

Personal and Property Crime

The *Uniform Crime Reports* are subject to the same biases in official statistics mentioned earlier, but they are the major source of information on patterns of crime and arrest, with crimes classified into four categories. **Personal crimes** are violent or nonviolent crimes directed against people. Included in this category are murder, aggravated assault, forcible rape, and robbery. As we saw in Figure 8.1, assault is the most frequently reported personal crime. As already noted, the rate of violent crime has been declining somewhat in recent years.

Property crimes involve theft or change of property, without threat of bodily harm. These include burglary (breaking and entering), larceny (the unlawful taking of property, but without unlawful entry), auto theft, and arson. Property crimes are the most frequent criminal infractions.

Victimless crimes violate laws but are not listed in the FBI's serious crime index. These include the illicit activities of gambling, illegal drug use, and prostitution,

Girls' gangs, like boys', use symbols like common dress and other behavioral mannerisms to signal their affiliated identity.

© Robert Yager/Stone/Getty Images

in which there is no complainant. Enforcement of these crimes is typically not as rigorous as the enforcement of crimes against persons or property, although periodic crackdowns occur, such as the current trend toward mandatory sentencing for drug violations.

Hate crimes refer to assaults and other malicious acts (including crimes against property) motivated by various forms of bias, including but not limited to those based on race, religion, sexual orientation, ethnic or national origin, or disability. The brutal killing of Matthew Shepard in 1998 is an example of a hate crime. Shepard was gay and a student at the University of Wyoming when he was tied to a fence post, tortured, and murdered by two men who espoused hatred toward homosexuals. The number of hate crimes that are reported has increased in recent years, particularly verbal and physical assaults against gays and lesbians. Gays and lesbians of color are particularly susceptible to such assaults. Presently, most states have enacted laws specifically directed against hate crimes as a particular category of crime (Jenness and Broad 2002).

Elite and White-Collar Crime

Sociologists use the term *white-collar* or **elite crime** to refer to criminal activities by persons of high social status who commit their crimes in the context of their occupation (Sutherland and Cressey 1978). White-collar crime includes embezzlement (stealing funds from one's employer), involvement in illegal stock manipulations (insider trading), and a variety of violations of income tax law, including tax evasion. Also included are manipulations of accounting practices to make one's company appear more wealthy, thus artificially increasing the value of the company's stock. These crimes are discussed later in the section on corporate crime.

White-collar crime has traditionally generated less concern in the public mind than street crime. In terms of total dollars, however, white-collar crime is far more consequential for society than street crimes. Scandals involving prominent white-collar criminals come to the public eye occasionally, such as the insider stock sale and subsequent cover-up by media personality Martha Stewart in 2002, for which she was sentenced to five months in prison in 2004. Nonetheless, white-collar crime is generally the least investigated and least prosecuted form of criminal activity.

Some would argue, for example, that tobacco executives are guilty of crimes, given the known causal relationship between smoking and lung disease. From a sociological point of view, one interesting question that stems from studies of crime and deviant behavior is how those who engage in it "normalize" their behavior. (We discussed the "normalization" of deviant behavior in Chapter 6 in connection with the two NASA space shuttle disasters.) Most tobacco executives likely believe they are doing nothing wrong. Rather, they are likely to believe they are pursuing good business practices, even though that particular business may have serious consequences for public health and safety. From the perspective of conflict theory, sociologists argue that class bias lies at the heart of what is perceived as criminal

---

A SOCIOLOGICAL EYE ON THE MEDIA

Images of Violent Crime

The media routinely drive home two points to the consumer: First, that violent crime is always high and may be increasing over time; and second, that there is much *random* violence constantly around us. The media bombard us with stories of "wilding," in which bands of youths kill random victims. Many of us think "road rage" is extensive and completely random. Most of us are now aware of violence in some high schools, where students armed with automatic weapons kill their fellow students. The media vividly and routinely report such occurrences as pointless, random, and probably increasing.

The evidence shows that violent crime in the United States, while it increased during the 1970s and 1980s, nonetheless began to decrease in 1990

and continues to decrease nationally through the present. For example, both robbery and physical assaults have declined dramatically since 1990 (see Figure 8.1). Yet according to one source (Best 1999), the media have consistently given a picture that violent crime has increased during this same period, and furthermore, that the violence is completely unpatterned and random.

No doubt there are occasions when victims are indeed picked at random. But the statistical rule of randomness could not possibly explain what has come to be called *random violence*, a vision of patternless chaos that is advanced by the media. If randomness truly ruled, *then each of us would have an equal chance of being a victim—*

and of being a criminal. This is assuredly not the case. As Best notes, the notion of random violence, and the notion that it is increasing, ignores virtually everything that criminologists, psychologists, sociologists, and extensive research studies know about crime: It is highly patterned and significantly predictable, beyond sheer chance, by taking into account the social structure, social class, location, race–ethnicity, gender, labeling, age, and other such variables and forces in society that affect both criminal and victim. The broad picture then is clearly not conveyed constantly in the media: Criminal violence is not increasing, but decreasing; and it is not random, but highly patterned and even predictable. •••

behavior. Thus, such elite crime is associated with upper-class activity and is less likely to be perceived by the criminal justice system as well as the public as "crime."

Organized Crime and Corporate Crime

The structure of crime and criminal activity in the United States often takes on an organized, almost institutional character. This is crime in the form of mob activity and racketeering and is called *organized crime*. Also, there are crimes committed by bureaucracies, known as *corporate crime*. Both types of crimes are so highly organized, complex, and sophisticated that they take on the nature of social institutions.

Organized Crime

Organized crime is crime committed by organized groups, typically involving the provision of illegal goods and services to others. Organized crime syndicates are often stereotyped as the Mafia, but it can refer to any group that exercises control over large illegal enterprises, such as the drug trade, illegal gambling, prostitution, weapons smuggling, or money laundering. Racial or ethnic ties, as well as family and kinship ties, are the basis for membership, with different groups dominating and replacing each other in different criminal industries. A key concept in sociological studies of organized crime is that these industries are organized in the same kind of hierarchy as legitimate businesses. There are likely to be senior partners who control the profits of the business, workers who manage and provide the labor for the business, and clients who buy the services that organized crime offers. In-depth studies of this underworld are difficult, owing to its secretive nature, although some sociologists have penetrated underworld networks and provided fascinating accounts of how these crime worlds are organized (Block and Scar-

pitti 1993; Carter 1999). Movies such as *The Godfather* and *Goodfellas*, as well as the popular TV series *The Sopranos*, have glamorized organized crime.

The dons, or godfathers, that head crime organizations often lead relatively quiet lives in the suburbs and are good family men who attend religious services and spend time with their children. Women traditionally have been excluded from meaningful leadership roles in organized crime. Organized crime is distinguished from other kinds of crime by its rigid hierarchy of godfathers, bosses, captains, underlings, hit men, and the like. The financial success of organized crime depends upon monopolistic control of prostitution and drug dealing; infiltration of legitimate business monopolies, such as waste and garbage removal; and dependence upon torture and murder for enforcement. *Racketeering*, the extortion of money from legitimate small and large businesses on a regular basis, is another widespread and well-organized undertaking. *Extortion* is accomplished by forcing businesspeople to buy "protection" for their businesses or insisting that they purchase products that they do not want or need (Scarpitti et al. 1997; Carter 1999).

Corporate Crime and Deviance: Doing Well, Doing Time

Corporations—even entire governments—may engage in crime and this behavior can be very costly to society. Sociologists estimate that the costs of corporate crime may be as high as $200 billion every year, dwarfing the take from street crime at roughly $15 billion, which most people imagine to be the bulk of criminal activity. Tax cheaters in business alone probably skim $50 billion a year from the Internal Revenue Service (IRS), three times the value of street crime. Taken as a whole, the cost of corporate crime is almost 6000 times the amount taken in bank robberies in a given year and eleven times the total amount for all theft in a year (Reiman 2002)!

Corporate crime and deviance is wrongdoing that occurs within the context of a formal organization or bureaucracy and is actually sanctioned by the norms and operating principles of the organization. This can occur within any kind of organization—corporate, educational, governmental, or religious. The recent scandals involving sexual assault of youths by Catholic priests, and the attempted cover-ups by assigning offending priests to parishes in different towns or states, constitute examples of organizational crimes. Such practices often become institutionalized in the routine procedures of the organization. Individuals within the organization may actually participate in the criminal behavior with little awareness that their behavior is illegitimate. In fact, their actions are likely defined as in the best

A case in point was the stock trading and accounting practices of Enron Corporation of Houston, Texas. These deviant practices led to the downfall of that organization early in 2002. In the summer of 2001, company executives who found company profits and thus their own personal company stock holdings declining in value, quickly sold off their own stock and pocketed the resulting money before the stocks declined further in value. At the same time, they forbid their own rank-and-file employees from selling their own company stock. The stock held by these unfortunate employees declined to almost nothing over the next several months, thus wiping out the retirement accounts, or "nest eggs," of hundreds of Enron employees. Furthermore, the Enron executives enlisted their own accounting firm, the nationally known firm of Arthur Andersen, to cover up these practices and "cook the books" to conceal the illegal stock transactions. As of this date, the person who initially uncovered these criminal practices, former Enron Vice President Sherron Watkins, has been labeled a "whistle-blower" and has found it difficult to find another job. She has argued that such corporate crimes are so institutionalized within today's organizations that there are sure to be other "Enrons" in the near future (Solomon 2004).

Above, Andrew S. ("Fast Andy") Fastow, former chief financial officer of the Enron Corporation, is being taken to court by FBI agents. David F. Myers, top right, former WorldCom controller, and Scott D. Sullivan, lower right, its former chief financial officer, being arrested by FBI agents for their role in falsely inflating corporate earnings by means of fraudulent corporate accounting.

interests of the organization—business as usual. New members who enter the organization learn to comply with the organizational expectations or leave.

In the 1980s, Beech Nut baby foods proudly claimed that their "nutritionists prepare fresh-tasting vegetables, meats, dinner, fruits, cereals, and juices without artificial flavoring." Not only was there no artificial flavoring in the company's apple juice, but there also were no apples either. Beech Nut was selling sugar water colored brown to resemble apple juice (Ermann and Lundman 1992). No one ordered plant operators to make fake juice. Instead, the owners insisted that the plant make a stronger return on its investment by cutting corners and the pursuit of higher profits resulted in corporate or bureaucratic crime. Most of the people in the production line probably never knew what was happening.

Sociological studies of corporate crime and deviance show that this form of activity is embedded in the ongoing and routine activities of organizations (Punch 1996; Lee and Ermann 1999). They represent cases of what in Chapter 6 was called *normalization of deviance*. Instead of conceptualizing organizational deviance as the behavior of bad individuals, sociologists see it as the result of people in organizations following rules and making decisions in more ordinary ways.

Another example of corporate and accounting malfeasance involved the WorldCom Corporation, a telecommunications company of worldwide repute. The company engaged in a multimillion dollar accounting fraud that disguised mounting losses from early in the year 2000 through the summer of 2002. By incorrectly reporting its operating expenses as though they were capital gains—a type of "cooking the books," or illegal accounting—the company was able to inflate the value of its own stock even though its own finances were rapidly deteriorating. Two former executives of WorldCom, Chief Financial Officer Scott D. Sullivan and Controller David F. Myers, were arrested by the FBI as federal prosecutors presented a point-by-point enumeration of the inner workings of the company's accounting manipulations. Improper inflation put WorldCom's earnings at more than $6.8 billion. Upon the arrest of Sullivan and Myers, the United States Attorney General proclaimed: "Corporate executives who cheat investors, steal earnings and squander pensions will meet the judgment they fear and the punishment they deserve" (Eichenwald 2002; Costello 2004).

Race, Class, Gender, and Crime

Arrest data show a clear pattern of differential arrest concerning race, gender, and class. To sociologists, the central question posed by such data is whether this reflects real differences in the extent of crime among different groups or differential treatment by the criminal justice system. The answer is both (D'Alissio and Stolzenberg 2003).

Certain groups are more likely than others to commit crime given that crime is distinctively linked to patterns of inequality in society. Unemployment, for example, is one correlate of crime, as is poverty. In a direct test of the association between inequality and crime, sociologist Ramiro Martinez Jr. (2002, 1996) explored the connection between rates of violence in Latino communities and the degree of inequality in 111 U.S. cities. His research shows a clear link between the likelihood of lethal violence and the socioeconomic conditions for Latinos in these different cities.

Sociologists use the analysis of socioeconomic conditions to explain the commission of crime, but they also make the important point that prosecution by the criminal justice system is significantly related to patterns of race, gender, and class inequality. We see this in the bias of official arrest statistics, treatment by the police, patterns of sentencing, and studies of imprisonment.

Race, Class, and Crime

One of the most important areas of sociological research on crime is the relationship between crimes committed and social class and race. Arrest statistics show a strong correlation between social class and crime, with the poor more likely than others to be arrested. Does this mean that the poor commit more crimes? To some extent, yes. Sociologists have demonstrated that a strong relationship exists between crime and the social factors of unemployment and poverty. (Scarpitti et al. 1997; Hagan 1993; Britt 1994; Smith et al. 1992). The reason is simple: Those who are economically deprived will often—though not always—see no alternative to crime.

Moreover, the police force tends to be concentrated in lower-income and minority areas. People who live in affluent areas are further removed from police scrutiny and better able to hide their crimes. Middle- and upper-income people may be perceived as less in need of imprisonment because they likely have a job and can find high-status people in the community to testify for their good character. White-collar crime is simply perceived as less threatening than crimes by the poor. Class also predicts that those at the lowest ends of the socioeconomic scale are far more likely to be victims of violent crime, as clearly shown in Figure 8.2.

A parallel to the correlation between class and crime arrest data shows that a strong relationship exists between race and crime. Minorities constitute 25 percent of the population of the United States but are more than 33 percent of the people arrested for property crimes and almost 50 percent of the people arrested for violent crimes. African Americans and Hispanics are more than twice as likely to be arrested for crime than are Whites. Native Americans and Asian Americans are exceptions, with both groups having relatively low rates of arrest for crime (Federal Bureau of Investigation 2003; see Table 8.2).

These data may seem to reinforce racial stereotypes, but sociologists have learned not to take these statistics at face value. Instead, they consider differences in how poor and rich communities are policed and the social origins of crime to explain the differences in criminal behavior among groups. What do they find? Police have wide latitude in deciding when to enforce laws and make arrests. Their discretion is greatest when dealing with minor offenses, such as disorderly conduct. Sociological research has shown that police discretion is strongly influenced by class and race judgments, just as labeling theory would predict, as we saw in Chapter 7 (Ava-

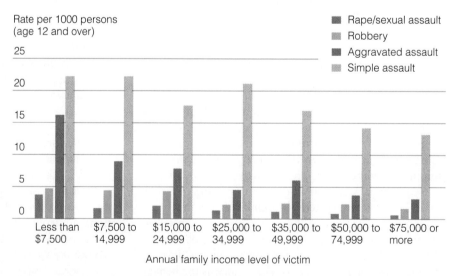

Figure 8.2 Victimization by Crime: A Class Phenomenon

Source: U.S. bureau of Justice Statistics. 2003. *Personal Crimes, 2001.* Washington, DC: U.S. Department of Justice.

Table 8.2 Arrests by Race[1,2]				
Crime	White	Black	American Indian	Asian/Pacific Islander
Total arrests	69.7	27.9	1.2	1.2
Murder	48.7	48.8	1.0	1.5
Forcible rape	63.7	34.1	1.1	1.1
Robbery	44.2	53.9	0.6	1.2
Aggravated assault	63.5	34.0	1.1	1.3
Buglary	69.4	28.4	0.9	1.2
Larcenty-theft	66.7	30.4	1.3	1.6
Motor vehicle theft	55.4	41.6	1.1	1.9
Arson	76.4	21.7	0.9	1.0
Forgery	68.0	30.0	0.6	1.4
Embezzlement	63.6	34.1	0.4	1.9
Vandalism	75.9	21.6	1.4	1.1
Prostitution	58.0	39.5	0.8	1.7
Sex offenses (not rape or prostitution)	74.4	23.2	1.1	1.3
Driving under influence	88.2	9.6	1.3	0.9
Suspicion	69.0	29.6	0.3	1.2

Source: Federal Bureau of Investigation. 2002. *Uniform Crime Reports.* Washington, DC; U.S. Department of Justice, p. 234.
[1]As percentage of all arrests, 2001.
[2]Hispanics appear in any of these categories.

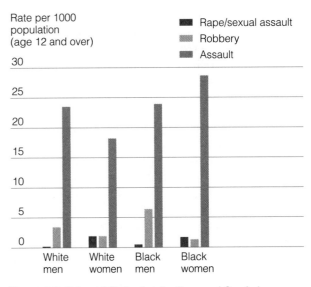

Figure 8.3 Crime Victimization (by Race and Gender)

Source: U.S. Bureau of Justice Statistics. 2003. *Sourcebook of Criminal Justice Statistics.* Washington, DC: U.S. Department of Justice **www.bjs.gov**

kame et al. 1999). The police are more likely to arrest persons they perceive as troublemakers. They are also more likely to make arrests when the complainant is White. Finally, minority communities are policed more intensively, which leads to more frequent arrests of the residents.

A research report on juvenile offenders demonstrates that Black and Latino youths with no prior criminal record are treated far more severely in the juvenile justice system than Whites of comparable social class, also with no prior criminal record. Minority youths are more likely to be arrested, held in jail, sent to juvenile or adult court for trial, convicted, and given longer prison terms. The racial disparities in the juvenile court system are magnified with each additional step into the justice system. In some cases, the racial disparities are stunning. For example, the report notes that 25 percent of arrested White youths are sent to prison but *nearly 60 percent* of arrested Black youths are imprisoned. That is a truly wide racial disparity. The report concludes that these racial disparities do not arise from overt discrimination on the part of prosecutors, judges, and other court personnel, but instead from the stereotypes that these decision makers rely on at each point of the juvenile justice system. Being Black, wearing low-hung baggy pants, and sporting dreadlocks is likely to get a person quickly through the various stages of the juvenile justice system and into prison (Butterfield 2000).

Bearing in mind the factors that affect the official rates of arrest and conviction—the bias of official statistics, the influence of powerful individuals, discrimination in patterns of arrest—there remains evidence that the actual commission of crime varies by race. Why? Again, sociologists find a compelling explanation in social structural conditions and the rise of stereotypes. Racial minority groups are far more likely than Whites to be poor, unemployed, and living in single-parent families. These social facts are all predictors of a higher rate of crime. Note, too, as Figure 8.3 shows, that African Americans are generally more likely to be victimized by crime.

Gender and Crime

Until recently, most sociological research on crime and deviance focused on men. Women's crime was seen as uninteresting or unimportant or studied only with the stereotyped vision of women as accomplices or prostitutes. The development of feminist scholarship within sociology has brought new analyses of women, deviance, and crime (Belknap 2001).

Women commit proportionately fewer crimes than men. Why? Although the number of women arrested for crime has increased slightly in recent years, compared to men the numbers are still small, except for fraud, embezzlement, and prostitution. Some argue that women's lower crime participation reflects their socialization into less risk-taking roles. Others say that women commit crimes that are extensions of their gender roles. This would explain why the largest number of arrests of women are for crimes such as shoplifting, credit card fraud, and passing bad checks.

Nonetheless, women's participation in crime has increased in recent years. Sociologists relate this to several factors. First, changes brought about by the women's movement have made women more likely to be employed in jobs that present opportunities for crimes such as property theft, embezzlement, and fraud. Violent crime by women has also increased notably since the early 1980s, possibly because the images that women have of themselves are changing, making new behaviors possible. The most significant factor in crime by women is related to their continuing disadvantaged status in so-

ciety. Just as crime is linked to socioeconomic status for men, so it is for women (Belknap 2001; Miller 1986).

Despite recent achievements, women on average remain in disadvantaged low-wage positions in the labor market. At the same time, changes in the social structure of families mean that more single women are economically responsible for their children, without the economic support of men. Some women thus may have to turn to illegitimate means for support.

Victimization by crime among women varies significantly by race and age, and women are less likely than men to be victimized by crime. Black women are much more likely than White women to be victims of violent crime. Young Black women are especially vulnerable. Divorced, separated, and single women are more likely than married women to be crime victims. Regardless of their actual rates of victimization, women are more fearful of crime than men. Minority women and widowed, separated, and divorced women are the most fearful. Women's fear of crime increases with age even though

SOCIOLOGY IN PRACTICE
Stopping Date Rape

Date rape, also known as acquaintance rape, has been widely acknowledged as a social problem. Sociological research has exposed a number of myths about rape, including the myth that most rapes are committed by strangers. The research done by sociologists and other social scientists on the subject of rape has been used by organizations that provide information to college students about how they can avoid date rape. The Association of American Colleges (AAC), for example, has produced a pamphlet for students intended to educate them about the occurrence of date rape and to give them strategies to avoid it. Their advice has also been informed by sociological research about gender and power. Understand that victims do not "cause" date rape, but the AAC suggests that women can do things to keep themselves out of precarious situations. Many of their recommendations stem from understanding the dynamics of gender relations. Here are some of their suggestions for women:

• Do not do anything you do not want to do just to be popular or pleasant or to avoid a scene.

• Be aware of situations in which you do not feel comfortable or where you may be at risk—for example, large parties where men greatly outnumber women.

• Avoid putting yourself in a vulnerable position. Have your own transportation or money to get home, if needed.

• Examine your attitudes about money and power. If a date pays for you, does that influence your ability to say "no"?

• Avoid falling for lines like, "You would if you loved me." Someone who loves you will respect your feelings and choices.

• If things start to get out of hand, protest loudly, leave, and go for help.

• Be aware that alcohol and drugs are often involved in acquaintance rape. They compromise your ability to make responsible decisions. If you choose to drink, do not rely on others to take care of you.

• Understand that it is never OK for someone to force himself on you, even if you are drunk or have had sex with the person before. If you do not consent, it is rape.

Critical Thinking Exercise

1. Surveys find that men and women differ significantly in their acceptance of rape myths. Men are more likely than women to think that women's behavior increases their risk of rape. How would you explain this gender difference in acceptance of rape myths and how does it influence the prosecution of date rape?

2. What do you think tend to be social characteristics (in terms of social class, race–ethnicity, education, religion, age, others, if any) of men who are most likely to engage in date rape? What do some of your friends and associates think? To what extent do women and men differ in their answer to this question?

Source: Adapted from Hughes, Jean O'Gorman, and Bernice R. Sandler. 1987. "'Friends Raping Friends.'" Washington, DC: Project on the Education and Status of Women, American Association of Colleges. Also adapted from Greenfield, Laurence A. 1996. *Child Victimizers: Violent Offenders and Their Victims.* Washington, DC: U.S. Bureau of Justice Statistics. •••

the likelihood of victimization decreases with age, a fact that researchers attribute to the elderly's increased sense of vulnerability (Joseph 1997; Wernwrath and Gartrell 1996; Gordon and Riger 1989). Women's fear of crime, sociologist Esther Madriz argues, results from an ideology that depicts women as needing protection from men (Madriz 1997).

For all women, victimization by rape is probably their greatest fear. Although rape is the most under-reported crime, until recently it has been one of the fastest growing. Criminologists explain that this is the result of a greater willingness now to report than in the past *and* an actual increase in the extent of rape (Federal Bureau of Investigation 2003). More than 200,000 rapes (including attempted rapes) are reported to the police annually. Police officials estimate that this is probably only about one-in-four of all rapes committed. Many women are reluctant to report rape because they fear the consequences of questioning by the criminal justice system. Rape victims are least likely to report the assault when they know the assailant, even though a large number of rapes are committed by someone the victim knows.

A disturbingly frequent form of rape is *date rape*—committed by an acquaintance or someone the victim has just met (see the Box "Sociology in Practice: Stopping Date Rape"). The extent of date rape is difficult to measure. The Bureau of Justice Statistics finds that 3 percent of college women experience rape or attempted rape in a given college year and 13 percent report being stalked (Fisher et al. 2001). A substantial amount of research finds that date rape is linked to men's acceptance of various rape myths (such as believing that a woman's "no" means "yes"), the use of alcohol, and the peer support that men receive in some all-male groups and organizations (Ullman et al. 1999; Boeringer 1999; Belknap et al. 1999).

Sociological research shows that rape is clearly linked to gender relations in society. Rape is an act of aggression against women. As feminists have argued, it stems from learned gender roles that teach men to be sexually aggressive. This is reflected in sociological research on convicted rapists who think they have done nothing wrong and who believe, despite having overpowered their victims, that the women asked for it (Scully 1990). Feminists have argued that the causes of rape lie in women's status in society—that women are treated as sexual objects for men's pleasure. The relationship between women's status and rape is also reflected in data revealing who is most likely to become a rape victim. African American women, Asian American women, Latinas, and poor women have the highest likelihood of being raped, as do women who are single, divorced, or separated. Young women are also more likely than older women to be rape victims (U.S. Bureau

of Justice Statistics 2003). Sociologists interpret these patterns to mean that women who are most powerless in society are also most subject to this form of violence.

The Criminal Justice System: Police, Courts, and the Law

Sociological studies consistently find patterns of differential treatment by the institutions that respond to deviance and crime in society. Whether it is in the police station, the courts, or the prisons, the social factors of race, class, and gender are highly influential in the administration of justice in this society. People in the most disadvantaged groups are more likely to be defined and identified as criminal and, having encountered these systems of authority, are more likely to be detained and arrested, found guilty, and punished.

THINKING *SOCIOLOGICALLY*

Although it is unpleasant to contemplate, researchers note that many people, perhaps about 50 percent, have either been involved in or witnessed some form of *family violence* or physical spouse abuse, whether relatively mild or severe. Do you know anyone who has? Have you yourself? Was it reported to the police? To what extent do you think the social class and the race of the participants had something to do with whether it was reported to the police?

The Policing of Minorities

There is little question that minority communities are policed more heavily than White neighborhoods. Moreover, policing in minority communities has a different effect than in White, middle-class communities. To middle-class Whites, the presence of the police is generally reassuring, but for African Americans and Latinos, an encounter with a police officer can be a terrifying experience.

DEBUNKING *SOCIETY'S MYTHS*

Myth: The criminal justice system treats all people according to the neutral principles of law.

Sociological perspective: Race, class, and gender continue to have an influential role in the administration of justice. For example, even when convicted of the same crime as Whites, African American and Latino defendants with the same prior arrest record as Whites are more likely to be arrested, sentenced, and sentenced for longer terms than White defendants.

Numerous studies have also documented the severe treatment that Native Americans, Mexican Americans, and African Americans receive from the police. Whites also get severe treatment, but racial minorities are more *likely* than the rest of the population to be victims of excessive use of force by the police, also called *police brutality*. African Americans are more likely than Whites to be killed by police officers. Studies show that most cases of police brutality involve minority citizens and that usually no penalty is imposed on the officers involved. Moreover, simply showing a disrespectful attitude is just as likely to generate police brutality as posing a serious bodily threat to the police (Lersch and Feagin 1996). Increasing the number of minority police officers has some effect on how the police treat minorities. Cities where African Americans head the police department show a concurrent decline in police brutality complaints, an increase in minority police and

minority recruitment, and in some cases, a decrease in crime. Simply increasing the number of African Americans in police departments does not, however, reduce crime dramatically because it does not change the material conditions that create crime to begin with (Cashmore 1991).

Racial profiling has come to the public's attention relatively recently, although it is a practice that has a long history. Often referred to half in jest by African Americans as the offense of DWB, or "driving while Black," **racial profiling** on the part of a police officer is the use of race alone as the criterion for deciding whether to stop and detain someone, such as the driver of an automobile, on suspicion of committing a crime. Police officers often argue that racial profiling is justified because a high proportion of Blacks and Hispanics commit crimes; an automobile containing Blacks or Hispanics is therefore likely to contain persons who have

---

MAPPING AMERICA'S DIVERSITY

MAP 8.1 The Growth of Imprisonment

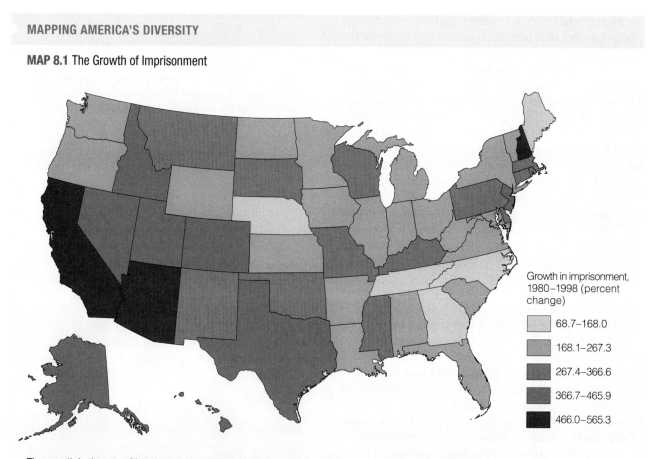

Growth in imprisonment, 1980–1998 (percent change)

	68.7–168.0
	168.1–267.3
	267.4–366.6
	366.7–465.9
	466.0–565.3

The growth in the rate of imprisonment is largely explained by an increase in the number of drug-related arrests, stemming from mandatory sentencing guidelines and "zero tolerance" policies. Sociologists also point out that in the national census count, prisoners are counted in the states where they are incarcerated, not in their home communities, thus potentially drawing resources away from urban and minority communities toward more rural areas where prisons tend to be located (Mauer 1999). What social policies do you think are needed to address this problem?

Source: U.S. Census Bureau. 2004. *Statistical Abstract of the United States,* 2003. Washington, DC: U.S. Government Printing Office.

committed some sort of crime. But in fact, while the overall crime rate for Blacks and Hispanics is indeed higher than that of Whites, race is a particularly bad basis for suspicion, because the vast majority of Blacks or Hispanics, like the vast majority of Whites, do not commit any crimes at all. Annually at least 90 percent of all African Americans are *not* arrested. That means on any given day, there is roughly a 90 percent probability that an African American in a car has *not* committed a crime. That leaves a 10 percent chance for an African American in a car to have committed a crime. Nonetheless, fully *eight out of every ten* automobile searches carried out by state troopers on the New Jersey Turnpike from 1988 through 1998 were conducted on vehicles driven by Blacks and Hispanics. The vast majority of these searches turned up no contraband or crimes of any sort (Kocieniewski and Hanley 2000; Cole 1999).

THINKING SOCIOLOGICALLY

Have you ever been the victim of what you suspected was *racial profiling?* When it happened, were you in a car or on foot? Was it night or day? Think about some of the other circumstances existing at the time, such as the neighborhood you were in or near. What role do you think your own race had, or did not have, in the incident?

Race and Sentencing

What happens once minority citizens are arrested for a crime? On arraignment, bail is set on average higher for Blacks and Latinos than for Whites, and minorities have less success in plea bargaining. Extensive research finds that once on trial, minority defendants are found guilty more often than White defendants. At sentencing, Blacks and Latinos are likely to get longer sentences than Whites for the same crimes, even when they have the same number of prior arrests and the same socioeconomic background. They are also less likely to be released on probation (Steffensmeier and Demuth 2000; Mauer 1999; Chambliss and Taylor 1989; Bridges and Crutchfield 1988).

African American defendants receive longer sentences than White defendants for property and violent crimes, and the disparity between sentences is even greater for serious crimes and crimes in which the victim is White, especially when the crime is rape or murder (Steffensmeier and Demuth 2000; Steffensmeier et al. 1998). Sentencing also differs depending on the racial identity of the judge. A study of Hispanic and White judges found that White judges sentence White defendants less severely than Hispanic defendants; Hispanic judges do not seem to distinguish between defendants based on race (Holmes et al. 1993).

Racial discrimination is particularly evident with regard to the death penalty. Of the over 3500 prisoners currently on Death Row, 44 percent are Black (U.S. Bureau of Justice Statistics 2003). Research shows that when Whites and minorities commit the same crime against a White victim, minorities are more likely to receive a more severe sentence. African American men convicted of raping White women are more likely to receive the death penalty than White men convicted of raping a woman of any race. Someone who kills a White person is also three times more likely to get the death penalty than someone who kills an African American, regardless of the race of the perpetrator (Paternoster and Brame 2003; Keil and Vito 1995).

Prisons: Deterrence or Rehabilitation?

More than half of the federal and state male prisoners in the United States are racial minorities. Blacks have the highest rates of imprisonment, followed by Hispanics, then Native Americans and Asians. Hispanics are the fastest growing minority group in prison (U.S. Bureau of Justice Statistics 2003). Native Americans, though a small proportion of the prison population, are still overrepresented in prisons. In theory, the criminal justice system is supposed to be unbiased, able to objectively weigh guilt and innocence. In reality, the criminal justice system reflects the racial and class stratification and bias in society.

The United States and Russia have the highest rates of incarceration in the world (The Sentencing Project 2003; Mauer 1999; see Figure 8.4). In the United States, the rate of imprisonment has been rapidly growing (see Figure 8.5). By all signs, the population of state and federal prisons continues to grow, with the population in prisons exceeding the capacity of the facilities. The total cost to the nation of keeping people behind bars is approximately $150 billion (U.S. Bureau of Justice Statistics 2003).

The number of people in prison has increased dramatically in recent years, resulting in overcrowding and a huge increase in public spending for the construction of new prisons.

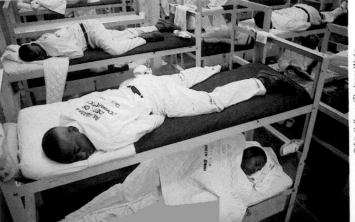

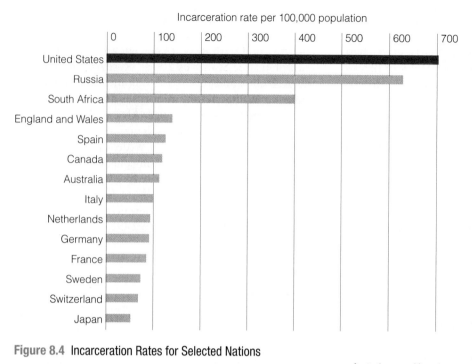

Figure 8.4 Incarceration Rates for Selected Nations

Source: The Sentencing Project. 2003. Washington, DC: **www.sentencingproject.org**
Walmsley, Roy, *World Prison Population List* (2nd ed.). United Kingdom Home Office
Research, Development and Statistics Directorate, July 2000.

to house prisoners, there is financial incentive to keep people in prison longer—at the taxpayers' expense (The Sentencing Project 2004; Bates 1998).

Why is there such growth in the prison population when the crime rate has recently been declining? A major reason for the increasing number of individuals behind bars is the increased enforcement of drug offenses and the mandatory sentencing that has been introduced. Nearly one-quarter of state prisoners are serving a drug sentence. Sixty percent of federal prisoners are serving drug sentences, more than double since 1980. Although the number of drug offenders has grown dramatically, so has the number of violent offenders (U.S. Bureau of Justice Statistics 2003).

The number of women behind bars has also increased at a faster rate than men, although the numbers of women in prison are small by comparison. Women comprise only 8 percent of all state and federal prisoners (U.S. Bureau of Justice Statistics 2003). Like men, three-fourths of the women in federal prisons are there because of drug-related offenses; often they have participated in these crimes by going along with the behavior of their boyfriends (Miller 1986). The typical woman in prison is a poor, young minority who dropped out of high school, is unmarried, and is the mother of two or more children. Fifty-seven percent of

The rate of prison growth in the United States has been so high in recent years that a new trend has emerged: the operation of prisons by private companies. Running corrections systems is now big business, and the "prison market" is expected to more than double in years ahead. For those looking for a business investment, running a prison is a good proposition because, as one business publication has argued, it is like running a hotel with 100 percent occupancy, booked for years in the future!

The privatization of prisons raises new questions for social policy. In the interest of running a profit center, prison managers may overcrowd prisons, reduce staffing, and cut back on food, medical care, or staff training. What may be sound business practice can result in less humane treatment of prisoners—locked away from the eyes of the public. Investigators are beginning to see that there are other costs to privatization. The rate of violence in private prisons is higher than in state facilities, and because the private prisons receive money from the state

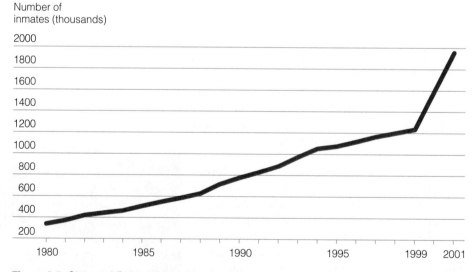

Figure 8.5 State and Federal Prison Population

Source: U.S. Bureau of Justice Statistics. 2002. "Prison Statistics." **www.ojp.usdoj.gov/bjs/prisons.htm**

women in prison are African American. Of all women prisoners, about two-thirds have been victims of sexual abuse. Women prisoners are also more likely than men to be positive for HIV-infection (Greenfield and Snell 1999).

Women in prison face unique problems, in part because they are in a system designed for men and run mostly by men, which tends to ignore the special needs of women. For example, 25 percent of women entering prison are pregnant or have just given birth, but they often get no prenatal or obstetric care. Male prisoners are trained for such jobs as auto mechanics, whereas women are more likely to be trained in relatively lower status jobs such as beauticians or launderers. The result is that few women offenders are rehabilitated by their experience in prison.

The United States is putting offenders in prison at a record pace. Is crime being deterred? Are prisoners being rehabilitated? If the deterrence argument were correct, we would expect that increasing the risk of imprisonment would lower the rate of crime. For example, we would expect drug use to decline as enforcement of drug laws increased. In the past few years, there has been a marked increase in drug law enforcement, but not the expected decrease in drug use. Although the use of drugs did decline slightly, overall there has been an increase in use among Black Americans and inner-city youth—those most likely to feel the crackdown of increased enforcement (Mauer 1999). If drug use is an example, it appears that the threat of imprisonment does not deter crime.

There is also little evidence that the criminal justice system rehabilitates offenders. To use the example of drug use again, only 20 percent who are imprisoned for drug offenses ever receive drug treatment. Although law enforcement is getting "tough on crime," it is doing little to see that offenders do not continue to commit drug crimes once they are released from prison.

In general, prisons rarely seem to deter or rehabilitate offenders. Prisons certainly do nothing to address the societal problems known to promote criminal activity. They concentrate on individual wrongdoers, not on the social structural causes of crime. Although there has been an enormous increase in the number of prisons built and the number of people in prison, it appears that imprisonment is doing little to solve the problem of crime.

If the criminal justice system fails to reduce crime, what does it do? Some sociologists contend that the criminal justice system is not meant to reduce crime but has other functions, namely, to reinforce the stereotype of crime as a threat from the poor and from people of color. The prison experience is demeaning and poorly suited to training prisoners in marketable skills or allowing them to repay their debt to society. In the end, prisons seem, at least in some cases, to refine criminals, not rehabilitate them (see Box, "The Rich Get Richer and the Poor Get Prison").

Courts, the Law, and Minorities

The court system is a pivotal part of social control. Each state has its own guidelines and procedures for its court system. The federal court system has a specific hierarchical structure that determines which cases shall be heard at which level. District courts are the lowest level of the federal court system. There are about ninety district courts, with at least one in each state. The next tier up in the federal legal organization is the U.S. Court of Appeals, also known as circuit courts. There are thirteen courts of appeals in the United States. These courts do not conduct trials but instead review the record of trials in lower courts, hear lawyers' arguments, and decide if errors were made. The highest court is the U.S. Supreme Court, which seats nine justices (one chief justice and eight associate justices). Several specialty courts

---

UNDERSTANDING DIVERSITY

The Rich Get Richer and the Poor Get Prison

Jeffrey H. Reiman (2002) notes that the prison system in the United States, instead of serving as a way to rehabilitate criminals, is in effect designed to train and socialize inmates into a career of crime. It is also designed in such a way as to assure the public that crime is a threat primarily from the poor and that it originates at the lower rungs of society. Reiman notes that prisons contain elements that seem designed to accomplish this.

One can "construct" a prison that ends up looking like a U.S. prison. First, continue to label as criminals those who engage in crimes that have no unwilling victim, such as prostitution or gambling. Second, give prosecutors and judges broad discretion to arrest, convict, and sentence based on appearance, dress, race, and apparent social class. Third, treat prisoners in a painful and demeaning manner, as one might treat children. Fourth, assure that prisoners are not

trained in a marketable skill that would be useful upon their release. And, finally, assure that prisoners will forever be labeled and stigmatized as different from "decent citizens," even after they have paid their debt to society. Once an ex-con, always an ex-con. One has thus socially constructed a U.S. prison, an institution that will continue to generate the very thing that it claims to eliminate.

Source: Reiman, Jeffrey H. 2002. *The Rich Get Richer and the Poor Get Prison.* Boston, MA: Allyn and Bacon. ● ● ●

are also dispersed throughout the court system, for example, bankruptcy court and customs court.

The role of all courts is to interpret and enforce the **law**—the written set of guidelines that determine what is defined as right and wrong in society. The United States uses law to provide guidelines and standards for behavior. The court system is responsible for attempting to ensure that the law is applied to all cases in a fair and just manner. The courts are, however, not the completely neutral parties they are sometimes alleged to be. Social factors, such as the social status of defendants or the wishes of powerful parties, do influence the outcome of legal decisions. At the same time, appealing to the courts is one primary avenue of recourse available to oppressed peoples seeking justice.

All courts balance competing influences in the process of legal decisions. The legal process in the United States is a complicated matter of applying ambiguous legal codes to specific cases. Interpretation plays an important role in case decisions, from interpreting the worth of competing arguments or the relevance of previous rulings to interpreting the wording of the Constitution. The legal process is an eminently subjective system that aspires to objectivity in the name of justice.

Lawyers serve as the spokespersons for each side, and judges serve as moderators and decision makers. Among lawyers, 28 percent are women, 4 percent are Black, 2 percent are Asians, and 4 percent are Hispanic. Among judges, the percentages of women, Blacks, Asians, and Hispanics are even smaller. Decisions that critically affect the lives of the American people, including minorities, lie in the hands of a fairly homogeneous group of individuals (U.S. Department of Labor 2004).

The jury system, intended to ensure the administration of justice by one's peers, has various forms of bias and lack of representation that interfere with fairness. Courts have to take careful measures to ensure representation on juries by different minority groups. Whites have historically been disproportionately represented on juries. Increased attention to fair representation has lessened this problem, although how well diverse groups are included on juries varies enormously in different locales. People who move frequently, immigrants, and some racial–ethnic groups are routinely underrepresented on juries.

The Law and Social Change

Despite problems in the administration of justice, the law remains one of the most effective avenues for addressing the injustices against different groups in U.S. society. This is one irony of the sociology of law: In a society based on legal authority, the legal system may embody racism, sexism, and class injustice within its institutional structure, yet it remains one major method for combating racism, sexism, and class injustice.

One landmark court decision in U.S. history was the 1954 case *Brown v. Board of Education of Topeka, Kansas,* in which the Supreme Court declared segregation was unconstitutional in public facilities such as schools and buses. This decision transformed race relations in the United States because it made formal policies specifying separation by race illegal. Outlawing *de jure* **segregation** (segregation by law) has not, however, eliminated *de facto* **segregation** (segregation in practice). The persistence of racial segregation in the labor market, in housing and education, and in social interaction indicates that the law goes only so far in producing social change. Nevertheless, it is a key element in a long-term historical process that has benefited multiple groups. Despite persistent racial segregation, since *Brown v. Board of Education* there have been many indications of increased racial integration in various dimensions of life (Andersen 2004).

The courts also make decisions that affect seemingly private aspects of life. The Supreme Court decision in *Griswold v. Connecticut* (1965) gave married women the right to birth control. Not until another Supreme Court decision was handed down in 1972 in *Eisenstadt v. Baird* was this right extended to unmarried people. These constitutional cases were followed by *Roe v. Wade* in 1973, in which the Supreme Court gave women the right to choose an abortion under specific federal guidelines. Each law represented a shift in federal policy "from reinforcing women's traditional roles to protecting departures from those roles . . . [and] the U.S. Supreme Court played an especially active role in instituting these changes" (Aliotta 1991: 144). These examples illustrate that the courts and the government can be progressive instruments of social change.

The Supreme Court has also strengthened the protection that lesbians and gays have under the law. In *Romer v. Evans* (1996), the Court ruled that states cannot pass laws that deprive gays and lesbians the equal protection of the law promised under the Fourteenth Amendment to the U.S. Constitution. This decision was the result of a challenge to a law passed in Colorado that banned measures to protect homosexual rights within that state. Presently, gay couples are challenging the movement to legally ban same-sex marriage.

Social change and the law could also involve decriminalizing certain categories of behavior that are now considered crimes. Many criminologists now advocate decriminalizing such "victimless" crimes as gambling and unconventional sexual behavior between consenting adults. One might argue whether such crimes are truly "victimless" as the compulsive gambler is his or her own "victim." Both criminologists and medical researchers now argue that certain narcotics addictions should be treated as medical problems rather than criminal problems, even at the risk of "medicalizing" such forms of deviance (recall this principle from Chapter 7).

The decriminalization of these categories of behavior would enable police forces to devote more time and resources to the investigation of personal and property crimes.

Terrorism As International Crime: A Global Perspective

Crime now crosses international borders and has become global. Terrorism is an example of the globalization of crime. The morning of Tuesday, September 11, 2001, was not expected to be any different from other early fall mornings. However, at approximately 8:45 A.M., the world stood still with images of a civilian airplane crashing into the north tower of the World Trade Center in New York City. Many at that moment assumed that the incident was nothing more than an accident by a pilot who might have lost consciousness or had a mechanical failure. This was a short-lived assumption. At 9:03 A.M., the second plane crashed into the south tower of the World Trade Center, leaving no doubt that this constituted a terrorist attack on the United States. This was confirmed a short time later by reports that another airplane had crashed into the Pentagon, while still another hijacked plane had crashed in Pennsylvania. Virtually everyone in the United States now remembers where they were and what they were doing during those horrible moments.

The FBI defines **terrorism** as the unlawful use of force or violence against persons or property to intimidate or coerce a government or population in furtherance of political or social objectives (Jucha 2002). Thus, terrorism is a crime that violates both international and domestic laws. It is a crime that crosses national borders and its understanding requires a global perspective. The origins of modern terrorism are historically traceable from Western European ideology to the Russian revolution, and then to the incorporation of Russian revolutionary thought in the nationalist struggle in Ireland. Contrary to what many think, the origins of international terrorism go back farther in history than either the Israel-Palestine conflict, or the emergence of Osama bin Laden's terrorist organization, al Qaeda.

Terrorism is linked to other forms of international crime. A case in point was Osama bin Laden's role in the international opium trade. Afghanistan, where bin Laden's terrorist al Qaeda organization was headquartered, was (and still is) the world's largest grower of opium-producing poppies. It is suspected that profit from the international drug trade helped finance the September 11 terrorist attacks. Therefore, a global perspective on crime involves recognizing the global basis of some international crime networks that cross national borders (Binns 2003).

Many nations have long experienced terrorism in the form of bombings, hijackings, suicide attacks, and other terrorist crimes. But the attacks of September 11 focused the world's attention on the problem of terrorism in new ways, including increased fears of **bioterrorism**—the form of terrorism involving the dispersion of chemical or biological substances intended to cause widespread disease and death. Fears of bioterrorism were exacerbated in the United States with a threat in 2002 of the spread of anthrax. This exceptionally deadly poison capable of causing virtually instant death upon inhaling the spores was found in the mail of offices of Congress, the Postal Service, the Supreme Court, and other locations, resulting in the deaths of several persons.

Another form of terrorism and thus cause for international concern is **cyberterrorism,** the use of the computer to commit one or more terrorist acts (**www.cybercrime.gov**). Terrorists may use computers in a number of ways. Data-destroying computer viruses may be implanted in an enemy's computer. Another use would be to employ "logic bombs" that lie dormant for years until they are electronically instructed to overwhelm a computer system. The use of the Internet to serve the needs of international terrorists has already become a reality (Jucha 2002).

Without understanding the political, economic, and social relations from which terrorist groups originate, terrorist acts seem like the crazed behavior of violent individuals. Although sociologists in no way excuse such acts, they look to the social structure of conflicts from which terrorism emerges as the cause of such

Threats of terrorism, such as bioterrorism, have resulted in increased security and countermeasures, particularly in urban areas.

deviant and criminal behavior. Even then, terrorism is not only the work of international extreme groups, as witnessed by the bombing of the Oklahoma City federal buildings in 1995. For this terrorist act, two White,

male U.S. citizens were tried and convicted. Terrorism, whether domestic or international, is best understood not only as individual insanity, but also as a politically, economically, and socially oriented form of violence.

Sociology⊛Now™

Reviewing is as easy as **❶❷❸**.

1. *Before you do your final review, take the SociologyNow diagnostic quiz to help you identify the areas on which you should concentrate. You will find information on SociologyNow and instructions on how to access all of its great resources on the foldout at the beginning of the text.*

2. *As you review, take advantage of SociologyNow's study videos and interactive Map the Stats exercises to help you master the chapter topics.*

3. *When you are finished with your review, take SociologyNow's posttest to confirm you are ready to move on to the next chapter.*

Chapter Summary

What is the difference between deviance and crime?

While deviance is behavior that violates norms and rules of society, *crime* is a type of deviant behavior that violates criminal law. *Criminology* is the study of crime from a scientific perspective.

Is crime accurately measured?

A variety of biases underlie crime statistics, such as over-reporting certain types of crimes among lower social class groups, yet underreporting these crimes at upper-class levels.

What are the types of crime?

Crimes are classified into *personal* versus *property crimes*, street crimes versus *elite crimes*—which are considerably underreported—and *organized* ("mob") *crime*, and *corporate crime*.

How is crime related to race, class, and gender?

In general, crime rates for a variety of crimes are higher among minorities than among Whites, among poorer persons than among middle- or upper middle-class persons, and among men than among women. Women, especially minority women, are more likely to be victimized by serious crimes, such as rape or violence from a spouse or boyfriend. The exceptions are elite crimes, which are higher (and more underreported) among Whites and professional males. These differences reflect both real differences among these groups as well as differential treatment within the justice system. Persons who are minority or poor are far more likely to be arrested and enter the criminal justice system, and they are more likely to be put through trial, sentencing, and longer incarceration than are Whites, even with the same prior criminal record and for the same crime. These

differences are especially prominent within the juvenile justice system. The practice of *racial profiling* of African Americans, and of some Middle Easterners, is a result of unfair differential treatment within the criminal justice system on the basis of skin color.

How is law related to social change?

Although law has often been the basis for excluding people from basic rights, legal change is also a source of social change, such as the *Brown v. Board of Education* (1954) case that ended legally sanctioned segregation, and Supreme Court decisions permitting abortion under certain conditions.

How is crime a global phenomenon?

Crimes of *terrorism*, including *bioterrorism* and *cyberterrorism*, are to be sociologically understood as a product of political, economic, and social structural conditions, as well as internationally based and also domestically based crime networks. In these respects, such crimes are clearly more than the acts of only a crazed small group of individuals, but the result of structural and cultlrural conditions.

Key Terms

bioterrorism 207	index crime 193
corporate crime 196	law 206
crime 192	organized crime 196
criminology 192	personal crime 194
cyberterrorism 207	property crime 194
de facto segregation 206	racial profiling 202
de jure segregation 206	terrorism 207
elite crime 195	victimless crime 194
hate crime 195	

Researching Society with Microcase Online

You can see the results of actual research by using the Wadsworth MicroCase® Online feature available to you. This feature allows you to look at some of the results from national surveys, census data, and other data sources. You can explore this easy-to-use feature on your own, but try this example. Suppose you want to know:

What is the relationship between your family income and whether or not you report that you have ever been arrested (called LAWS in this analysis)? Is there any relationship between these two variables?

To answer this question, go to http://sociology.wadsworth
.com/andersen_taylor4e/, select MicroCase Online from
the left navigation bar, and follow the directions there to
analyze the following data.

Data file: GSS

Analysis Task: Cross Tabulation

Row Variable: LAW5

Column Variable: Income

Questions

The results are displayed in a table showing survey results classified by both family income (columns) and
whether or not the respondent reported they have ever
been arrested (rows). Review the results and answer
the following questions:

1. People in which income group are most likely to say
   yes, they have been arrested?
   a. Low
   b. Middle
   c. High

2. People in which income group are least likely to say
   yes, they have been arrested?
   a. Low
   b. Middle
   c. High

3. Describe the relationship shown in this table between
   family income and the percentage who report they
   have or have not been arrested.

The Companion Website for Sociology: Understanding a Diverse Society, Fourth Edition

http://sociology.wadsworth.com/andersen_taylor4e/

Supplement your review of this chapter by going to the
companion website to take one of the Tutorial Quizzes,
use the flash cards to master key terms, and check out
the many other study aids you'll find there. You'll also
find special features such as GSS Data and Census 2000
information, data and resources at your fingertips to
help you with that special project or do some research
on your own.

Suggested Readings and Web Resources

Anderson, Elijah. 1999. *Code of the Street: Decency,
Violence, and the Moral Life of the Inner City.* New
York: W. W. Norton.
*Based on his ethnographic studies, Anderson pro-
vides a compelling sociological analysis of street life,
its norms or codes, and its relationship to violence
and the social construction of morality.*

Humphries, Drew. 1999. *Crack Mothers: Pregnancy,
Drugs, and the Media.* Columbus, OH: Ohio State
University Press.
*Humphries shows how the media created a drug
scare and moral panic through its exaggerated presentation of cocaine use among low-income women.
She argues that poverty and inadequate social services, not moral failings by women, are the basis
of these social harms.*

Martinez, Ramiro. 2002. *Latino Homicide: Immigration,
Violence and Community.* New York: Routledge.
*This research examines the relationship between class
inequality among Latinos and the levels of violence
within Latino communities.*

Mauer, Marc. 1999. *Race to Incarcerate.* New York:
The New Press.
*Mauer explores the explosion of growth in the prison
industry, including its effects on African Americans
and the poor. The book provides a strong analysis of
why imprisonment has increased and suggests alternative policies for criminal justice.*

Reiman, Jeffrey. 2002. *The Rich Get Richer and the
Poor Get Prison?* 7th ed. Needham Heights, MA:
Allyn and Bacon.
*This critical view of the criminal justice system ex-
plores the racial and class inequalities that the author
argues are at the heart of the nation's crime problem.
He maintains that the criminal justice and prison systems need to be changed dramatically to better deal
with crime, thus moving away from the traditional
emphasis on changing the criminals themselves.*

Bureau of Justice Statistics
www.ojp.usdoj.gov/bjs/
*A division of the federal government that compiles
information on subjects pertinent to the study of
crime and deviance.*

Federal Bureau of Prisons
www.bop.gov
*This site includes data on federal prisons and demo-
graphic data about prison inmates.*

Federal Bureau of Investigation
www.fbi.gov
*The FBI's website includes press releases and infor-
mation about federal crime publications.*

The Sentencing Project
www.sentencingproject.org
*The sentencing project is a private organization dedi-
cated to promoting alternatives to sentencing and
to providing analayses of criminal justice issues for
the public and policy makers. Their website contains
data on various aspects of crime and imprisonment,
including information on privatization of prisons.*

chapter 9

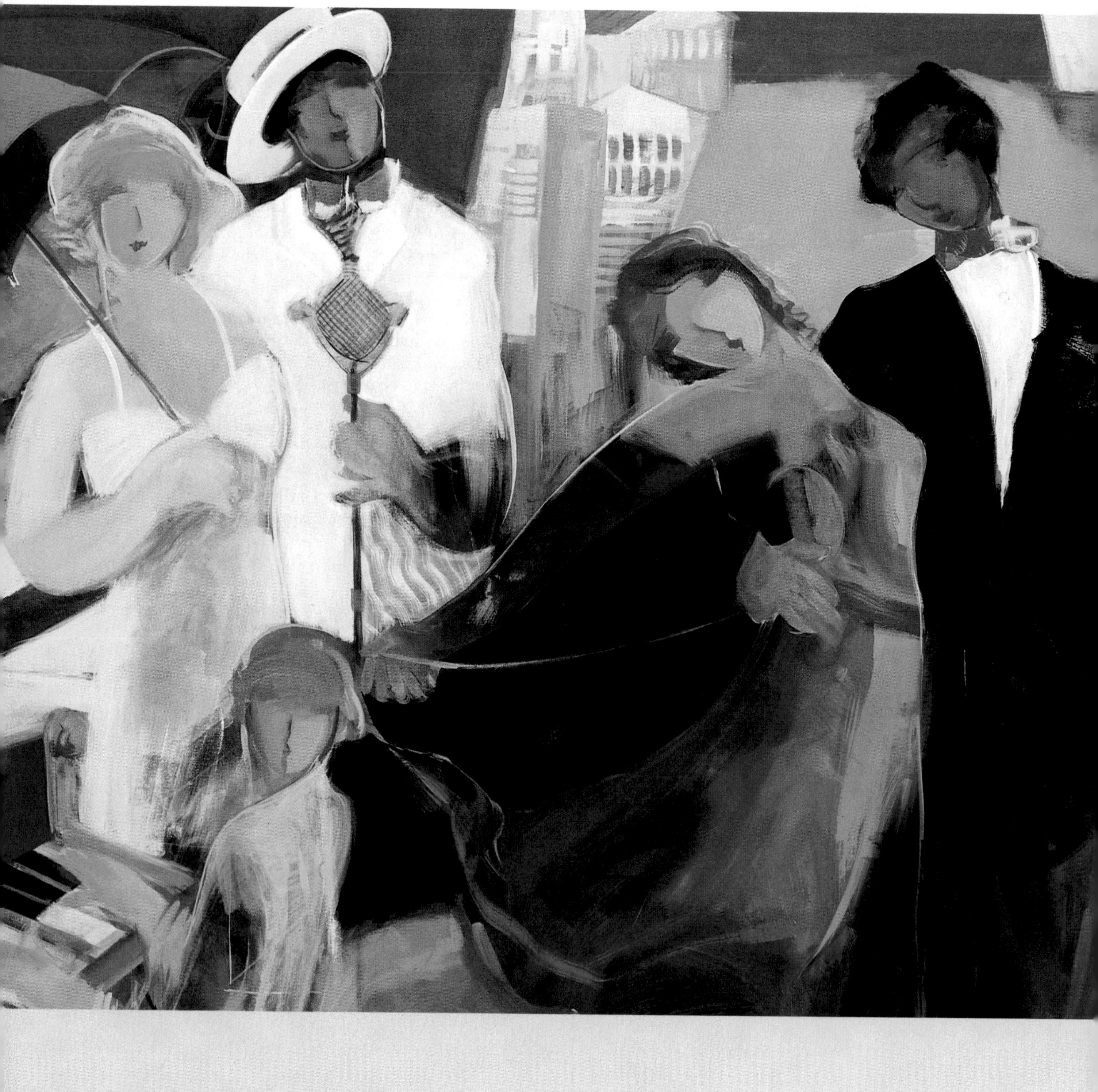

• • •

Social Class and Social Stratification

One afternoon in a major U.S. city, two women go shopping. They are friends—wealthy, suburban women who shop for leisure. They meet in a gourmet restaurant and eat imported foods while discussing their children's private schools. After lunch, they spend the afternoon in exquisite stores—some of them large, elegant department stores; others intimate boutiques where the staff know them by name. When one of the women stops to use the bathroom in one store, she enters a beautifully furnished room with an upholstered chair, a marble sink with brass faucets, fresh flowers on a wooden pedestal, shining mirrors, an ample supply of hand towels, and jars of lotion and soaps. The toilet is in a private stall with solid doors. In the stall there is soft toilet paper and another small vase of flowers.

The same day, in a different part of town, another woman goes shopping. She lives on a marginal income earned as a stitcher in a textiles factory. Her daughter badly needs a new pair of shoes because she has grown out of

last year's pair. The woman goes to a nearby discount store where she hopes to find a pair of shoes for under $15, but she dreads the experience. She knows her daughter would like other new things—a bathing suit for the summer, a pair of jeans, and a blouse. But this summer the daughter will have to wear hand-me-downs because medical bills over the winter have depleted the little money left after food and rent. For the mother, shopping is not recreation but a bitter chore reminding her of the things she is unable to get for her daughter.

While this woman is shopping, she too stops to use the bathroom. She enters a vast space with sinks and mirrors lined up on one side of the room and several stalls on the other. The tile floor is gritty and gray. The locks on the stall doors are missing or broken. Some of the overhead lights are burned out, so the room has dark shadows. In the stall, the toilet paper is coarse. When the woman washes her hands, she discovers there is no soap in the metal dispensers. The mirror before her is cracked. She exits quickly, feeling as though she is being watched.

Two scenarios, one society. The difference is the mark of a society built upon class inequality. The signs are all around you. Think about the clothing you wear. Are some labels worth more than others? Do others in your group see the same marks of distinction and status in clothing labels? Do some people you know never seem to wear the "right" labels? Whether it is clothing, bathrooms, schools, homes, or access to health care, the effect of class inequality is enormous, giving privileges and resources to some and leaving others struggling to get by.

Great inequality divides society. Nevertheless, most people think that in the United States equal opportunity exists for all. The tendency is to blame individuals for their own failures, or attribute success to individual achievement. Many

people think the poor are lazy and do not value work. At the same time, the rich are often admired for their supposed initiative, drive, and motivation. Neither is an accurate portrayal. There are many hard-working individuals who are poor, and rich people have often inherited their wealth rather than earning it themselves.

Observing and analyzing class inequality is fundamental to sociological study. What features of society cause different groups to have different opportunities? Why is there such an unequal allocation of society's resources? Sociologists respect individual achievements but have found the greatest cause for the disparities in material success to be the organization of society. Instead of understanding inequality as the result of individual effort, sociologists thus study the social structural origins of inequality. •••

Social Differentiation and Social Stratification

All social groups and societies exhibit social differentiation. **Status,** as we have seen earlier, refers to a socially defined position in a group or society. **Social differentiation** is the process by which different statuses develop in any group, organization, or society. Think of a sports organization. The players, the owners, the managers, the fans, the cheerleaders, and the sponsors all have a different status within the organization. Together, they constitute a whole social system, one marked by social differentiation.

Status differences can become organized into a hierarchical social system. **Social stratification** is a relatively fixed, hierarchical arrangement in society by which groups have different access to resources, power, and perceived social worth. Social stratification is a system of structured social inequality. To use a sports example again, owners control the resources of the teams. Players earn high salaries, yet do not control the team resources and have less power than the owners and managers. Sponsors (including individuals and corporations) provide the resources on which this system of stratification rests. Fans provide revenue by purchasing tickets to watch the teams play. Sports are a system of stratification where various groups within a sports organization are hierarchically arranged, with some having more resources and power than others.

All societies seem to have a system of social stratification, although they vary in the degree and complexity of stratification. Some societies stratify only along a single dimension, such as age, keeping the stratification system relatively simple. Most contemporary soci-

Table 9.1
Inequality in the United States

- Nearly one in six children in the United States live in poverty: 30 percent of African American, 29 percent of Hispanic, and 12 percent of Asian American children; 9.4 percent of White non-Hispanic children (Proctor and Dalaker 2003).

- Fifteen percent of the U.S. population have no health insurance; the average cost of a day's stay in a hospital is $1,217—two weeks' pay for the average U.S. worker (U.S. Census Bureau 2004).

- When Doris Duke, the tobacco heiress died, she left a $100,000 inheritance to her dog, with the provision that the dog should be fed only imported baby food (Meng 2001).

- One percent of the U.S. population controls 38 percent of the total wealth in the nation; the bottom 20 percent owe more than they own (Rose 2000; Mishel et al. 2003).

- CEOs of major companies earn an average of $13.1 million dollars per year; workers earning the minimum wage make $10,712 per year, if they work 40 hours a week for 52 weeks per year and hold only one job (Lavelle 2001).

- Black family income is 68 percent of White family income—only 9 percent closer than in 1970 (DeNavas-Walt et al. 2003).

- Among women heading their own households, 27 percent live below the poverty line (Proctor and Dalaker 2003).

The class status of different groups can be readily seen by looking at such things as where people shop—
the thrift store or this Ralph Lauren store in Palo Alto, California.

© Tony Freeman/PhotoEdit

© Anne Dowie

THINKING SOCIOLOGICALLY

Take a shopping trip to different stores and observe the appearance of stores serving different economic groups. What kinds of bathrooms are there in stores catering to middle-class clients? the rich? the working class? the poor? Which ones allow the most privacy or provide the nicest amenities? What fixtures are in the display areas? Are they simply utilitarian with minimal ornamentation, or are they opulent displays of consumption? Take detailed notes of your observations and write an analysis of what this tells you about *social class* in the United States.

eties are more complex, with many factors interacting to create social strata. In the United States, social stratification is strongly influenced by class, which is in turn influenced by matters such as one's occupation, income, and education, along with race, gender, and other influences such as age, region of residence, ethnicity, and national origin.

Forms of Stratification: Estate, Caste, and Class

Stratification systems can be broadly categorized into three types: estate systems, caste systems, and class systems. In an **estate system** of stratification, the ownership of property and the exercise of power is monopolized by an elite who have total control over societal resources. Historically, such societies were feudal systems where classes were differentiated into three basic

groups—the nobles, the priesthood, and the commoners. Commoners included peasants (usually the largest class group), small merchants, artisans, domestic workers, and traders. The nobles controlled the land and the resources used to cultivate the land, as well as all the resources resulting from peasant labor.

Estate systems of stratification are most common in agricultural societies. Although such societies have been largely supplanted by industrialization, a few societies still exist that have a small but powerful landholding class ruling over a population that works mainly in agricultural production. Unlike the feudal societies of the European Middle Ages, however, contemporary estate systems of stratification display the influence of international capitalism. It is not knights who conquered lands in war that comprise the "noble class"; it is international capitalists or local elites who control the labor of a vast and impoverished group of people, such as in some Latin American societies where landholding elites maintain a dictatorship over peasants who labor in agricultural fields.

In a **caste system,** one's place in the stratification system is an *ascribed status* (see Chapter 5), meaning it is a quality given to an individual by circumstances of birth. The hierarchy of classes is rigid in caste systems and is often preserved through formal law and cultural practices that prevent movement between classes.

The system of *apartheid* in South Africa was a stark example of a caste system. Under apartheid, the travel, employment, associations, and place of residence of Black Africans were severely restricted. Segregation was enforced using a pass system in which Blacks in White areas were obliged to account for themselves to White authorities. Interracial marriage was illegal, and Black Africans were prohibited from voting. In 1992, following world pressure to abolish apartheid, a Whites-only referendum overwhelmingly endorsed reform of the apartheid system.

Jim Crow segregation in the American South is another example of a caste system. "Jim Crow" refers to the laws and social practices that governed the segregated relations between Blacks and Whites in the American South. From the end of slavery until the enforcement of contemporary civil rights protections, Black Americans in the South were strictly segregated from Whites, attended separate schools and churches, used separate swimming pools and drinking fountains, and were denied the right to vote, among other restrictions. Black people were expected to ride in the back of buses and while Whites referred to Blacks only by first name, Blacks had to address Whites as "Mr.," "Miss," or "Mrs." These demeaning social practices defined the group status of Black Americans as inferior and secondary to White Americans.

In the traditional caste system of India, membership in an Indian caste group was determined by birth. Each group was restricted to certain occupations, and the lowest group, "the untouchables," was prohibited from associating with the others. The Brahmans were the highest caste: nobles who were considered spiritually and socially superior to all other groups. (The term has been transmitted to American culture, where it is used to refer to the old money elite families of New England—the "Boston Brahmans.") The Indian caste system is no longer mandated, but its effects remain. People retain attitudes about the alleged superiority and inferiority of different groups long after formalized distinctions are eradicated, which is one reason social inequalities and prejudices are so tenacious.

In **class systems,** stratification exists, but one's location and rank can change according to individual achievements, even though class is still strongly determined by one's social background. In class systems, class is to some degree *achieved,* that is, earned by the acquisition of resources and power, regardless of one's origins. Class systems are more open than caste systems, because position does not depend strictly on birth. Classes are less rigidly defined than castes, because the divisions are blurred by individuals moving between one class and another.

Despite the potential for movement from one class to another, placement in a class system is still highly dependent on one's social background. *Ascribed status* (i.e., according to birth) is not the basis for social stratification in the United States, yet the class a person is born into has major consequences for that person's life. One's likelihood of achievement is shaped by patterns

of inheritance; access to educational resources; and the financial, political, and social influence of one's family. There is no formal obstacle to movement through the class system, yet achievement is very much influenced by an individual's class of origin.

Defining Class

In common parlance, *class* refers to style or sophistication. In sociological use, **social class** (or **class**) is the

Class differentiation in the United States results in people having different access to society's resources.

social structural position groups hold relative to the economic, social, political, and cultural resources of society. Class determines the access different people have to these resources and puts groups in different positions of privilege and disadvantage. Each class has members with similar opportunities and who tend to share a common way of life. Class also includes a cultural component in that class shapes language, dress, mannerisms, taste, and other preferences. *Class is not just an attribute of individuals; it is a feature of society.*

The social theorist Max Weber described the consequences of stratification in terms of **life chances,** meaning the opportunities that people have in common by virtue of belonging to a particular class. Life chances include the opportunity for possessing goods, having an income, and having access to particular jobs. Life chances are also reflected in the quality of everyday life. Whether you dress in the latest style or wear another person's tossed out clothes, have a vacation in an exclusive resort, take your family to the beach for a week, or have no vacation at all—these life chances are the result of being in a particular class.

To sociologists, life chances stem from social structural arrangements. Put simply, class matters. Class standing determines how well one is served by social institutions. Health care is particularly inadequate for poor and working-class people, who are less likely than others to enjoy private care or have good health insurance (if they have any at all). Class influences access to a high-quality education, critically important because education provides credentials and social networks that pay off over a lifetime. Class is also strongly related to political and social attitudes. Those from higher income brackets are more likely to be Republican than Democrat. Even friendship is influenced by class because friendships arise more frequently within class groups than across them (Allan and Adams 1998; Allan 1998). Whether you can have a certain kind of pet is also a matter of class resources. For some, having a pet means picking up a stray dog who eats cheap dog food or hunts for food scraps. Other people may board their dogs and cats in "pet spas" where there are heated beds and massage therapists for the pets—even after the owner dies (Meng 2001).

Class is a structural phenomenon. It cannot be directly observed. Sociologists cannot isolate and measure social class directly; therefore, they use other indicators to serve as measures of class. Recall from Chapter 2 that an *indicator* is something that represents a concept. By assessing indicators, you can examine concepts that are too abstract to be measured directly. Prominent indicators of class are income, wealth, education,

A SOCIOLOGICAL EYE ON THE MEDIA
Reproducing Class Stereotypes

The media have a substantial impact on how people view the social class system and different groups within it. Especially because people tend to live with and associate with people in their own class, how they see others can be largely framed by the portrayal of different class groups in the media. Research has found this to be true, and in addition, has found that mass media have the power to shape public support for policies for public assistance.

To begin with, the media overrepresent the lifestyle of the most comfortable classes. It is the rare family that can afford the home décor and fashion depicted even in soap operas, ironically most likely watched by those in the working class. Media portrayals, such as those found on television talk shows—and in sports programming as well—tend to emphasize stories of upward mobility. When the working class is depicted, it tends to be shown as deviant, thus reinforcing class antagonism and giving viewers a sense of moral and "class superiority" (Gersch 1999).

Content analyses of the media also find that the poor are largely invisible in the media (Mantsios 2004). Those poor people who are depicted in television and magazines are portrayed as more often Black than is actually the case, leading people to overestimate the actual number of the Black poor. The elderly and working poor are rarely seen (Clawson and Trice 2000; Gilens 1996). Representations of welfare overemphasize themes of dependency, especially when the portrayal is of African Americans. Women are also more likely than men to be represented as dependent (Misra et al. 2003). And, rarely are welfare activists shown as experts; rather, public officials are typically given the voice of authority (Ryan 1996). One result is that the media end up framing the "field of thinkable solutions to public problems" (Sotirovic 2000, 2001), but do so within a context that ignores the social structural context of social issues.

Sources: Gersch, Beate. 1999. "Class in Daytime Talk Television." *Peace Review* 11 (June): 275–281; Sotirovic, Mira. 2001. "Media Use and Perceptions of Welfare." *Journal of Communication* 51 (December): 750–774; Sotirovic, Mira. 2000. "Effects of Media Use On Audience Framing and Support for Welfare." *Mass Communication & Society* 2–3 (Spring–Summer): 269–296; Bullock, Heather E., Karen Fraser, and Wendy R. Williams. 2001. "Media Images of the Poor." *The Journal of Social Issues* 57 (Summer): 229–246; Clawson, Rosalee A., and Rakuya Trice. 2000. "Poverty As We Know It: Media Portrayals of the Poor." *The Public Opinion Quarterly* 64 (Spring): 53–64; Gilens, Martin. 1996. "Race and Poverty in America: Public Misperceptions and the American News Media." *The Public Opinion Quarterly* 60 (Winter): 515–541; Misra, Joy, Stephanie Moller, and Marina Karides. 2003. "Envisioning Dependency: Changing Media Depictions of Welfare in the 20th Century." *Social Problems* 50 (November): 482–504; Ryan, Charlotte. 1996. "Battered in the Media" Mainstream News Coverage of Welfare Reform." *Radical America* 26 (August): 29–41.

occupation, and place of residence. These do not define class by themselves, but they are good measures of class standing. We will see that these indicators also tend to be linked. A good income or family wealth can make it possible to afford a house in a prestigious neighborhood and an exclusive education for one's children. In the sociological study of class, income, wealth, and education are indicators that have enormous value in revealing the contours of the class system. How sociologists conceptualize class depends, however, on the theoretical framework they use for understanding the class system, as we will see in the next section.

Why Is There Inequality?

Stratification occurs in all societies. Why? This question originates in classical sociology in the works of Karl Marx and Max Weber, theorists whose work continues to inform the analysis of class inequality today.

Karl Marx: Class and Capitalism

Karl Marx (1818–1883; see also Chapter 1) provided a complex and profound analysis of the class system under capitalism. Marx's class analysis, developed more than one hundred and fifty years ago, continues to inform sociological analyses and has been the basis for major world change. Marx was specifically interested in how classes formed within the economic system of capitalism and he defined classes in terms of their relationship to the **means of production**—the system by which goods are produced and distributed. In Marx's analysis, two primary classes exist under capitalism: the *capitalist class* (those who own the means of production) and the *working class* (those who sell their labor for wages). There are further divisions within the class system: the *petty bourgeoisie* and the *lumpenproletariat*. The petty bourgeoisie includes small business owners and managers whom you might think of as middle class. They identify with the interests of the capitalist class, but do not own the means of production. The lumpenproletariat are those who have become unnecessary as workers and have become discarded by the economic system. Today, this would include the homeless and other permanently poor people.

Marx thought that with the development of capitalism, the capitalists and the working class would become increasingly antagonistic, referred to as *class struggle*. As class conflicts became more intense, the two classes would become increasingly polarized, with the petty bourgeoisie deprived of their property and dropping into the working class. His analysis is still reflected in contemporary questions about a growing gap between the "haves" and the "have nots," with the rich getting

richer and everyone else getting worse off—an issue further examined later in this chapter.

Much of Marx's analysis boils down to the consequences of a system based on the pursuit of profit. If goods were exchanged at the cost to produce them, no profit would be generated. Capitalist owners want to sell commodities for more than their actual value—more than the cost of production, including materials and labor. Capitalists extract profit by keeping the cost of labor down, so Marx saw capitalists profiting via the exploitation of workers. Marx thought that as profits became increasingly concentrated in the hands of a few capitalists, the working class would become increasingly dissatisfied, leading to the working class revolting and overthrowing the rule of the capitalist class. Class conflict between workers and capitalists, he argued, would inevitably lead to revolution.

In the 1800s when Marx was writing, the middle class (that is, the petty bourgeoisie) was small and consisted mostly of small business owners and managers. Marx saw the middle class as dependent on the capitalist class but also exploited by it, because the middle class did not own the means of production. He saw the middle class as identifying with the interests of the capitalist class, but failing to work in their own best interests because the bourgeoisie falsely believe that they benefit from capitalism. He thought that in the long run the middle class would pay for their misplaced faith when profits became increasingly concentrated in the hands of a few and more of the middle class dropped into the working class.

Not every part of Marx's theory has proved true. He did not foresee the emergence of the large and highly differentiated middle class we have today. Still, his analysis provides a powerful portrayal of the forces of capitalism and the tendency for wealth to belong to a few, while the majority work just to make ends meet. He has also influenced the lives of billions of people under self-proclaimed Marxist systems created in an attempt to overcome the pitfalls of capitalist society.

Max Weber: Class, Status, and Party

Max Weber (1864–1920) agreed with Marx that classes were formed around economic interests, and he agreed that material forces (that is, economic forces) have a powerful effect on people's lives. But, he disagreed with Marx that economic forces are the primary dimension of stratification. Weber saw three dimensions to stratification:

- **class**—the economic dimension;
- **status** (or **prestige**)—the cultural and social dimension;
- **power**—the political dimension.

Weber defined *class* as the economic dimension of stratification. This would include such things as income, property, and other financial assets. Weber understood that a class has common economic interests and that economic well-being was the basis for one's life chances. Obviously, a family with an income of $100,000 per year has more access to the resources of a society than a family living on an income of $40,000 per year. But he thought that social stratification was more than a matter of economics because people are also stratified based on their status and how much power they hold.

Prestige is the judgment or recognition given to a person or group. Weber uses the term *status* (or *prestige*) to refer to the social dimension of stratification. Weber understood that class distinctions are linked to status distinctions: Those with the most economic resources tend to have the highest status in society. In a community, for example, those with the most status may be those who resided there the longest, even if newcomers arrive with more money. Thus, prestige is related to economic standing but may be independent of income. Ministers and priests are accorded high prestige, but they do not typically earn high incomes.

Power is the political dimension of stratification. It is the capacity to influence groups and individuals even in the face of opposition. Power is also reflected in the ability of a person or group to negotiate their way through social institutions. Again, those with great economic resources have more power to influence others, including the power that wealthy individuals and corporations wield in the political process. Those with great power also are better able to negotiate their way through social institutions. A business executive accused of corporate crime can afford expensive lawyers and may go unpunished or, if found guilty, will likely serve a relatively light sentence in pleasant, minimum security facilities. Compare this to the experience of a poor, African American or Latino man wrongly accused of a crime who will not have much power to negotiate his way through the criminal justice system.

Weber had a *multidimensional view of social stratification* because he analyzed the connections between economic, social, and political dimensions of stratification. He pointed out that these different dimensions of stratification are usually related but not necessarily. A person could be high on one or two dimensions, but low on another. A major drug dealer is an example: high wealth (economic dimension) and power (political dimension) but low prestige (social dimension)—at least in the eyes of the mainstream society, even if not in other circles. The point is that stratification does not rest solely on economics. Political power and social judgments are important components of social stratification.

Marx and Weber explain different features of stratification. Both understood the importance of the economic basis of stratification, and they knew the significance of class for determining the course of one's life. Marx saw people as acting primarily out of economic interests, but Weber saw people's position in the stratification system as the result of economic, social, *and* political forces. Together, Marx and Weber provide compelling analyses for understanding the contemporary class structure.

Functionalism and Conflict Theory: The Continuing Debate

Like more contemporary sociologists, Marx and Weber were trying to understand why differences existed in the resources and power held by different groups in society. Why does inequality persist? This has been further studied by sociologists working from these classical traditions.

The Functionalist Perspective on Inequality

Functionalist theory views society as a system of institutions organized to meet society's needs (see Chapter 1). The functionalist perspective emphasizes that the parts of society are in basic harmony with each other. Society is characterized by cohesion, consensus, cooperation, and stability (Parsons 1951a; Merton 1957). Different parts of the social system complement one another and are held together through social consensus and cooperation. To explain stratification, functionalists propose that the roles filled by the upper classes—governance, economic innovation, investment, and management— are essential for a cohesive and smoothly running society; hence, the upper class is rewarded in proportion to their value to the social order (Davis and Moore 1945).

According to the functionalist perspective, social inequality serves an important purpose in society: It motivates people to fill the different positions in society that are needed for the survival of the whole. Functionalists think that some positions in society are more important than others and require the most talent and training. Rewards attached to those positions (such as higher income and prestige) ensure that people will make the sacrifices needed to acquire the training for functionally important positions (Davis and Moore 1945). Social mobility thus comes to those who acquire what is needed for success (such as education and job training). In other words, functionalist theorists see inequality as based on a reward system that motivates people to succeed.

The functionalist perspective is well illustrated by Herbert Gans's (1991/1971) analysis of the functions of poverty. No one typically thinks of poverty as a "good" thing, least of all Gans, but he delineates how poverty sustains an overall social system. Poverty, no matter how unwanted, has both economic and social functions

Table 9.2
Functional and Conflict Theories of Stratification

Interprets	Functionalism	Conflict Theory
Inequality	Inequality serves an important purpose in society by motivating people to fill the different positions in society that are needed for the survival of the whole.	Inequality results from a system of domination and subordination where those with the most resources exploit and control others.
Class structure	Differentiation is essential for a cohesive society.	Different groups struggle over societal resources and compete for social advantage.
Reward system	Rewards are attached to certain positions (such as higher income and prestige) to ensure that people will make the sacrifices needed to acquire the training for functionally important positions in society.	The more stratified a society, the less likely that society will benefit from the talents of all its citizens, because inequality prevents the talents of those at the bottom from being discovered and used.
Classes	Some positions in society are more functionally important than others and are rewarded because they require the greatest degree of talent and training.	Classes exist in conflict with each other as they vie for power and economic, social, and political resources.
Life chances	Those who work hardest and succeed have greater life chances.	The most vital jobs in society—those that sustain life and the quality of life—are usually the least rewarded.
Elites	The most talented are rewarded in proportion to their contribution to the social order.	The most powerful reproduce their advantage by distributing resources and controlling the dominant value system.
Class consciousness/ideology	Beliefs about success and failure confirm status on those who succeed.	Elites shape societal beliefs to make their unequal privilege appear to be legitimate and fair.
Social mobility	Upward mobility is possible for those who acquire the necessary talents and tools for success (such as education and job training).	There is blocked mobility in the system because the working class and poor are denied the same opportunities as others.
Poverty	Poverty serves economic and social functions in society.	Poverty is inevitable because of the exploitation built into the system.
Social policy	Given that the system is basically fair, social policies should only reward merit.	Social policies should support disadvantaged groups by redirecting society's resources for a more equitable distribution of income and wealth.

in society. Poverty ensures that society's "dirty work" will be done, by relegating menial jobs to the poor. In addition, poverty benefits the affluent, by providing a source of cheap domestic labor and the clientele for various illegal activities (drugs, cheap alcohol, prostitution) that benefit wealthy illegal entrepreneurs. Social mobility for some may also be financed at the expense of the poor, such as in the case of slumlords or business owners in poor communities. Another economic function of poverty is that the poor buy or use goods that others do not want, such as secondhand clothes, old automobiles, and surplus food. Without the poor, society would need a different way of disposing of such surplus goods. Having a large poor population also creates jobs for those in social services and charitable foundations and creates a reserve labor force, such as the immigrants who often take society's worst jobs.

Gans thus identifies the social functions that poverty provides. Poverty also produces social deviants—people whose existence can be used to uphold the legitimacy of mainstream social norms. By labeling the poor indolent or unworthy, others uphold a definition of themselves as socially valuable. Thus, the poor provide the basis for status comparisons. The wealthy may congratulate themselves for their charity, even though their largesse may benefit themselves (in the form of tax deductions and social recognition) more than those to whom their philanthropy is directed. Gans's presentation of the functions of poverty makes a strong sociological point: Poverty creates problems for society but is also functional for society because it contributes to society's overall stability. Functionalist theorists might object to the presence of poverty in society, nonetheless, they see its unintended effects.

The Conflict Perspective on Inequality Conflict theory also sees society as a social system, but unlike functionalism, conflict theory interprets society as held together through conflict and coercion. From a conflict-based perspective, society includes competing interest groups, some with more power than others. Different groups struggle over societal resources and compete for social advantage. Conflict theorists argue that those who control society's resources also hold power over others. The powerful are also likely to act to reproduce their advantage, and try to shape societal beliefs to make their privileges appear to be legitimate and fair. In sum, conflict theory emphasizes the friction in society rather than the coherence and sees society as dominated by elites.

Derived largely from the work of Karl Marx, conflict theorists see social stratification as based on class conflict and blocked opportunity. Stratification is a system of domination and subordination in which elites rule while they exploit and control others. The unequal distribution of rewards reflects the class interests of the powerful, not the survival needs of the whole society (Eitzen and Baca Zinn 2004). According to the conflict perspective, inequality provides elites with the power to distribute resources, make and enforce laws, and control value systems. Elites use these powers in ways that reproduce inequality. Others in the class structure, especially the working class and the poor, experience blocked mobility.

From a conflict point of view, the more stratified a society, the less likely that society will benefit from the talents of all of its citizens. Inequality limits the life chances of those at the bottom and prevents their talents from being discovered and used, wasting creativity and productivity.

Implicit in the argument of each perspective is criticism of the other perspective. Functionalism assumes that the most highly rewarded jobs are the most important for society. However, conflict theorists argue that some of the most vital jobs in society—those that sustain life and the quality of life, such as farmers, domestic workers, trash collectors, and a wide range of other laborers—are usually the least rewarded. Conflict theorists also criticize functionalist theorists for assuming that the most talented get the greatest rewards. They point out that systems of stratification tend to devalue the contributions of those left at the bottom and underutilize the diverse talents of all people (Tumin 1953). Functionalist theorists contend that the conflict view of how economic interests shape social organization is too simplistic. Conflict theorists respond by arguing that functionalists overstate the degree of consensus and stability that exists.

The debate between functionalist and conflict theorists raises fundamental questions about how people view inequality. Is it inevitable? Will there always be

poor people? How is inequality maintained? This debate is not just academic. The assumptions made from each perspective frame public policy debates. Whether the topic is taxation, poverty, or homelessness, if people believe that anyone can get ahead by ability alone, they will tend to see the system of inequality as fair and accept the idea that there should be a differential reward system. Those who prefer the conflict view of the stratification system are more likely to advocate programs that emphasize public responsibility for the well-being of all groups and to support programs and policies that result in more of the income and wealth of society going toward the needy.

The Class Structure of the United States

The class structure of the United States is elaborate, arising from the interactions of old wealth, new wealth, intensive immigration, globalization, and the development of new technologies. One can conceptualize the class system as a series of layers, with different class groups arrayed up and down the rungs of the class ladder, each rung corresponding to a different level in the class system. Conceptualized this way, social class is the common position groups hold in a status hierarchy (Wright 1979; Lucal 1994). Class is indicated by factors such as levels of income, occupational standing, and educational attainment. People are relatively high or low on the ladder depending on the resources they have, whether those are education, income, occupation, or any other factors known to influence people's placement (or ranking) in the stratification system (see Figure 9.1). An abundance of sociological research has stemmed from the concept of **status attainment,** the process by which people end up in a given position in the stratification system. Status attainment research describes how factors such as class origins, educational level, and occupation produce class location. It describes the extent to which people are able to move throughout the class system, as we will see in the section on social mobility later in this chapter.

This laddered model of class suggests that stratification in the United States is hierarchical, but somewhat fluid. Different gradients exist in the stratification system, but they are not fixed as they might be in a society where one's class is solely determined by birth. In a relatively open class system such as the United States, people's achievements do matter, although the extent to which people rise rapidly and dramatically through the stratification system is less common than popularly believed. Some people begin from modest origins and amass great wealth and influence (Bill Gates, Oprah Winfrey, and millionaire sports heroes come to

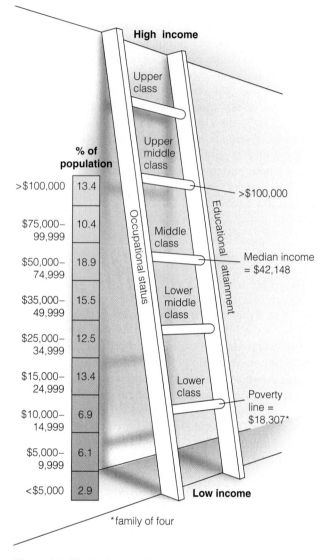

Figure 9.1 The Laddered Model of Stratification

more than the median income; half earn less. In the laddered model of class, those bunched around the median income level are considered middle class. In 2002, median household income in the United States was $42,409 (see Figure 9.2; DeNavas-Walt et al. 2003).

Occupational prestige is a second important indicator of socioeconomic status. **Occupational prestige** refers to the subjective evaluation people give to jobs. To determine occupational prestige, sociological researchers typically ask nationwide samples of adults to rank the general standing of a series of jobs. These subjective ratings provide information about how people perceive the worth of different occupations. People tend to rank Supreme Court justice as one of the most prestigious occupations, followed by occupations such as physician, professor, judge, lawyer, and scientist. In the middle ranges are occupations such as electrician, newspaper columnist, insurance agent, and police officer. The occupations typically considered to have the lowest prestige are farm laborer, maid or servant, garbage collector, janitor, and shoe-shiner (Davis and Smith 1984). These rankings do not reflect the actual worth of people who perform these jobs, but reflect the judgments made about these jobs and their value to society.

The final major indicator of socioeconomic status is **educational attainment,** typically measured as the total years of formal education. The more years of education attained, the more likely a person will have higher class status. The prestige attached to occupations is strongly tied to the amount of education the job requires—the more education people think is needed for a given occupation, the more occupational prestige people attribute to that job (Blau and Duncan 1967; Ollivier 2000; MacKinnon and Langford 1994).

mind), but these are the exceptions, not the rule. Some people also move down in the class system, but as we will see, most people remain relatively close to their class of origin. When people rise or fall in the class system, the distance they travel is usually relatively short, as we will see further in the section on social mobility.

The image of stratification as a laddered system, with different gradients of social standing, emphasizes that one's **socioeconomic status (SES)** is derived from certain factors: income, occupational prestige, and education are the three main measures of socioeconomic status. **Income** is the amount of money a person receives in a given period. Socioeconomic status can be measured in part by a person's income bracket. As we will see, income is distinct from **wealth,** which is the total value of what one owns, minus one's debts. The **median income** for a society is the midpoint of all household incomes. In other words, half of all households earn

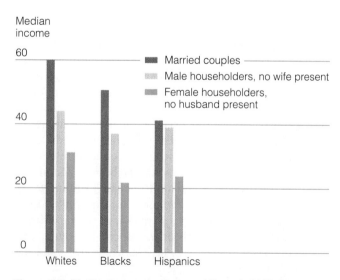

Figure 9.2 Median Income by Race and Household Status

Source: DeNavas–Walt, Carmen, Robert W. Cleveland, and Bruce H. Webster, Jr. 2003. *Income in the United States: 2002.* Washington, DC: U.S. Census Bureau.

MAP 9.1 Median Household Income in the United States

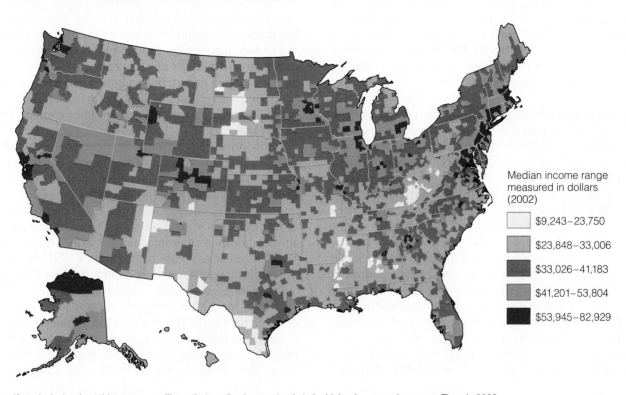

Median income range
measured in dollars
(2002)

$9,243–23,750

$23,848–33,006

$33,026–41,183

$41,201–53,804

$53,945–82,929

If you look closely at this map, you will see that median income tends to be higher in more urban areas. Thus, in 2002, median income inside metropolitan areas was $45,257 and outside such areas, $34,654. What the map does not show, however, are differences within cities. Median income inside central cities is substantially lower ($36,863) than the median income within metropolitan areas, but out of the center city—that is, in suburban areas—it is ($50,717). Given this, what do you conclude about the significance of residence in the structure of the class system?

Data: U.S. Census Bureau. 2004. *American FactFinder.* Website: **www.census.gov**; DeNavas–Walt et al. 2003. *Income in the United States: 2002.* Washington, DC: U.S. Census Bureau, p. 9.

Layers of Social Class

How do sociologists understand the array of classes in the United States? Using the laddered model of stratification, many sociologists describe the class system in the United States as divided into several classes: upper, upper middle, middle, lower middle, and lower class. Each class is defined by characteristics such as income, occupational prestige, and educational attainment. The different class groups are arrayed along a continuum with those with the most money, education, and prestige at the top and those with the least at the bottom.

In the United States, the *upper class* owns the major share of corporate and personal wealth. It includes those who have held wealth for generations as well as those who have recently become rich. Only a very small proportion of people constitute the upper class, but they control vast amounts of wealth and power in the United

States. Those in this class are *elites* who, in Marxist terms, own the means of production. They exercise enormous control throughout society. Most of their wealth is inherited.

Despite social myths to the contrary, the best predictor of future wealth is the family into which you are born. Each year, the business magazine *Forbes* publishes a list of the Forbes 400—the four hundred wealthiest families and individuals in the country. To be on the list published in 2003, your net worth had to be at least $600 million. Bill Gates, the richest person on the list, has an estimated worth of $46 billion. Of all the wealth represented on the Forbes 400 list, most is inherited, although during the late 1990s there was an increase in the number of people on the list with self-created wealth. The four hundred richest Americans have a total net worth that exceeds the gross domestic product of the entire nation of China (*Forbes*, October 8, 2001; *The*

Baltimore Sun, September 24, 1999), and some wealthy individuals can wield as much power as entire nations (Friedman 1999).

Members of the upper class with newly acquired wealth are known as the *nouveau riche.* The "dot-com millionaires" from Internet start-up companies are a good example of this class. Often young, they ushered in a new age of wealth and glamour in the 1990s, although many also lost their fortunes when technology stocks crashed in 2000. Like the old rich, the new rich live lavish lifestyles, often being even more ostentatious than those with old money. At one market in Silicon Valley—home to many of the dot-com millionaires—if you are dissatisfied with an ordinary bottle of apple cider vinegar for 89 cents a pint, in a locked case you can buy a very small bottle of balsamic vinegar for $1500 (Kaplan 1999). Although the nouveau riche may have vast amounts of money, they are often not accepted into "old rich" circles where wealth is not the sole defining characteristic of the upper class. Social connections and family prestige can be as important as money at the pinnacle of the class structure.

The *upper-middle class* includes people with high incomes and high social prestige. They tend to be well-educated professionals or business executives. Their earnings can be high—successful business executives can earn millions of dollars a year. It is difficult to esti-mate exactly how many people fall into this group because of the difficulty of drawing lines between the upper, upper middle, and middle class. The upper middle class is often thought of as "middle class" because their lifestyle sets the standard to which many aspire, but this lifestyle is unattainable by most. A large home full of top-quality furniture and modern appliances, two or three luxurious cars, vacations every year (perhaps a vacation home), high-quality college education for one's children, and a fashionable wardrobe are simply beyond the means of a majority of people in the United States.

Probably the largest group in the class system is the *middle class,* which includes people who fall within a given range above and below the median income figure. It is difficult to pinpoint an exact income bracket that defines the middle class since many people think of themselves as middle class even when their income and lifestyle may differ quite dramatically. Thus, many people earning six-figure incomes think of themselves as middle class, while others earning far less living in much smaller homes also think of themselves as middle class. One reason the middle class is so difficult to define in the United States is that being "middle class" is more than just economic position. Because the United States is an open-class system, many do not want to recognize class distinctions, even though they are real and

DOING SOCIOLOGICAL RESEARCH
The Fragile Middle Class

The hallmark of the middle class in the United States is its presumed stability. Home ownership, a college education for children, and other accoutrements of middle class status (nice cars, annual vacations, an array of consumer goods)—these are the symbols of middle class prosperity. But, the rising rate of bankruptcy among the middle class shows that the middle class is not as secure as it is presumed to be. Personal bankruptcy has risen dramatically in recent years with more than one million filings for bankruptcy per year. How can this be happening in such a prosperous society?

This is the question examined by Teresa Sullivan, Elizabeth Warren, and Jay Lawrence Westbrook in their study of bankruptcy and debt among the middle class. They based their study on an analysis of official records of bank-ruptcy in five states, as well as on detailed questionnaires given to individuals who filed for bankruptcy. Their findings debunk the idea that bankruptcy is most common among poor people. They find that bankruptcy is a middle-class phenomenon representing a cross section of people in this class (meaning that those who are bankrupt are matched on the demographic char-acteristics of race, age, and gender with others in the middle class).

They also debunk the notion that bankruptcy is rising because it is so easy to file. Instead, they find that people in the middle class are simply overwhelmed with debt that they cannot possibly pay. The biggest reason people give for filing for bankruptcy is the loss of income from the instability of

Bankruptcy Cases Per Million Adults

Source: Sullivan, Teresa A., Elizabeth Warren, and Jay Lawrence Westbrook. 2000. *The Fragile Middle Class: Americans in Debt.* New Haven: Yale University Press.

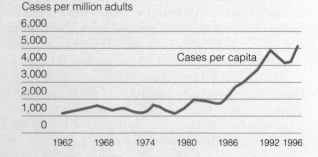

Cases per million adults

pervasive in society. Many who think of themselves as middle class also have a tenuous hold on this class position (see the box "The Fragile Middle Class").

In the hierarchy of social classes, the *lower middle class,* also known as the working class, includes blue-collar workers (those in skilled trades who do manual labor), low-income bureaucratic workers, and many service workers, such as secretaries, hairdressers, waitresses, police, and firefighters. Medium to low income, lower educational attainment, and lower occupational prestige define this class relative to the elite and upper middle class. The term *lower* in this class designation refers to the group's position in the stratification system, but it has a pejorative sound to many people, especially to members of this class. They are unlikely to refer to themselves as lower middle class, and often prefer working class or middle class.

The *lower class* is comprised primarily of the displaced and poor. People in this class have little formal education and are often unemployed or working in minimum wage jobs. People of color and women make up a disproportionate part of this class. The poor include the working poor; 38 percent of the poor hold jobs and 11 percent of them work full-time (Proctor and Dalaker 2003).

Recently, the concept of the underclass has been added to the lower class. The **underclass** includes those with little or no opportunity for movement out of the worst poverty. These people have been left behind by contemporary economic developments and are likely to be unemployed. Often they must turn to public assistance or crime for economic support. According to sociologist William Julius Wilson, structural transformations in the economy have left large groups of people, especially urban minorities, in highly vulnerable positions. Without work and unable to sustain themselves in an economy that has discarded them, these groups form a growing under class—a development that has exacerbated the problems of urban poverty and created new challenges for social policy makers if we are to reverse this trend (Wilson 1996, 1987).

Class Conflict

The model of class just described defines classes in terms of a status hierarchy. Sociologists have also analyzed classes in terms of their structural relationship to other classes and their relationship to the economic system (Vanneman and Cannon 1987; Wright 1985). Derived from conflict theory, this analysis emphasizes power relations in society, interpreting inequality as a result of the unequal distribution of power and resources in society (see Chapter 1). Instead of seeing class simply as a continuum, this perspective sees classes as facing

---

jobs. But, in addition, divorce, medical problems, housing expenses, and credit card debt drive many to bankruptcy court.

Sullivan and her colleagues thus explain the rise of bankruptcy as stemming from structural factors in society that fracture the stability of the middle class. The volatility of jobs under modern capitalism is one of the biggest factors, but add to this the "thin safety net" of no health insurance for many, but rising medical costs. Also, the American dream of owning a home means many people are "mortgage poor"—extended beyond their earning capability.

The United States is also a credit-driven society. Credit cards are routinely mailed to people encouraging them to buy beyond their means. You can now buy virtually anything on credit: cars, clothes, doctor's appointments, enter- tainment, groceries. You can even use one credit card to pay off other credit cards. Indeed, it is difficult to live in this society without credit cards. Increased debt is the result. Many are simply unable to keep up with compounding interest and penalty payments, and debt takes on a life of its own as consumers cannot keep up.

Sullivan, Warren, and Westbrook conclude that increases in debt and uncertainty of income combine to produce the fragility of the middle class. Their research shows that "even the most secure family may be only a job loss, a medical problem, or an out-of-control credit card away from financial catastrophe" (2000: 6).

Questions to Consider

1. How easy is it for students to get a credit card? What forces in society encourage you to use one? Is it more difficult to get by without one? Having answered these questions, why do you think the society has been called a "credit card nation?" *Keywords: student debt, student loan*

2. What evidence do you see in your community of the fragility of the middle class? *Keywords: downward mobility, fragile middle class*

We have included InfoTrac College Edition keywords at the end of each question to make it easier for you to find more to read on these topics. Go to **www.infotrac-college.com**, an online library, to begin your search.

Source: Sullivan, Teresa A., Elizabeth Warren, and Jay Lawrence Westbrook. 2000. *The Fragile Middle Class: Americans in Debt.* New Haven: Yale University Press. ● ● ●

off against each other, with elites exploiting and dominating others. The key idea in this model is that class is not simply a matter of individual levels of income and prestige. Instead, class is defined by the relationship of the classes to the larger system of economic production. In this model, classes are "collectivities of individuals and families with comparable total resources over time" (Perrucci and Wysong 2003: 9). This provides a dynamic, though complex, model of the class structure in which those at the top have stable and relatively secure resources, whereas those at the bottom are less secure but also dominated by those at the top. This is depicted in the double diamond model of class provided by sociologists Robert Perrucci and Earl Wysong (see Figure 9.3). This model emphasizes that no single factor determines one's class standing; rather, it is the relative security of resources that one has over time that shapes class location.

From a conflict perspective, the position of the middle class in society is unique. The middle class, or the *professional-managerial class,* includes managers, supervisors, and professional workers, such as doctors, lawyers, professors, and so forth. Members of this group have substantial control over other people, primarily through their authority to direct the work of others, impose and enforce regulations in the workplace, and determine dominant social values. As Marx argued, the middle class is itself controlled by the ruling class, yet members of this class tend to identify with the interests of the elite. The professional-managerial class, though, is caught in a contradictory position between elites and the working class. Like elites, people in this class have some control over others, but like the working class, they have minimal control over the economic system (Wright 1979). As capitalism progresses, according to conflict theory, more and more people in the middle class drop into the working class, as they are pushed out of managerial jobs into working-class jobs or as professional jobs become organized more along the lines of traditional working-class employment.

Has this happened? Not to the extent Marx predicted whereby he thought that ultimately there would be only two classes—the capitalist and the proletariat—but, to some extent, this is occurring. Classes have become more polarized, with the well-off accumulating even more resources and the middle class seeing their median income falling, measured in constant dollars (Rose 2000). Levels of debt in the middle class also mean that many have a fragile hold on this class position. The loss of a job, a family emergency (such as the death of a working parent), divorce, disability, or a prolonged illness can quickly leave a family with few resources. At the same time, corporate mergers, tax policies that favor the rich, a decline in corporate taxes, and sheer greed continue to concentrate more wealth in the hands of a few. Perhaps the class revolution that Marx predicted has not occurred, but the dynamics of capitalism that he analyzed are unfolding before us.

Recall that Marx defined the *working class* as people who sell their labor for wages. This definition includes blue-collar workers and many white-collar workers. Members of the working class have little control over their own work lives. Instead, they have to take orders from others. This concept of the working class departs from traditional blue-collar definitions of working-class jobs because it includes secretaries, salespeople, and nurses—any group that works under the rules imposed by managers or elites. The middle class may exercise some autonomy at work, but the working class often has little power to challenge decisions of their supervisors, except insofar as they can organize collectively in unions, strikes, or other collective work actions.

Conflict theorists see the poor as under assault by society, in a system of inequality where they are especially vulnerable. Poor through no fault of their own, the poor are still blamed for their own poverty, especially because of belief systems propagated by elites and the middle class.

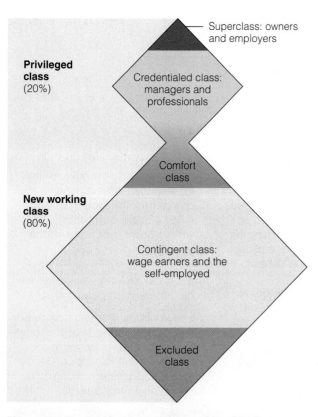

Figure 9.3 The Double Diamond Model of Stratification

Source: Perrucci, Robert, and Earl Wysong. 2003. *The New Class Society: Goodbye American Dream?* 2nd ed. Lanham, MD: Rowman and Littlefield, p. 29. Reprinted by permission of the publisher.

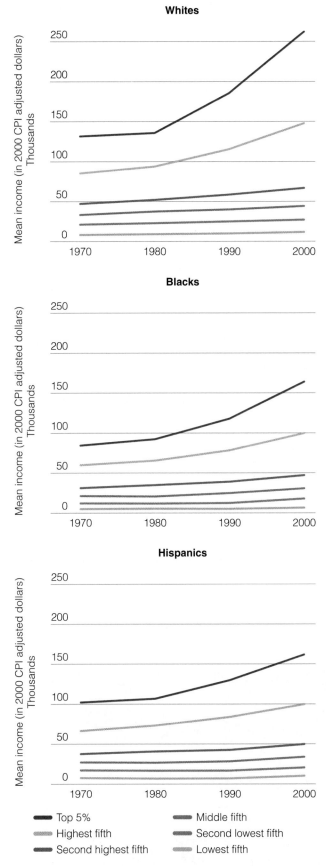

Figure 9.4 Income Growth by Income Group, 1970–2000

Source: U.S. Census Bureau. 2002. *Historical Income Tables—Households.* **www.census .gov/hhes/income/histinc/**

Whether one uses a laddered perspective of class or a class conflict perspective, you can see that the class structure in the United States is a hierarchy. Class position gives people different access to jobs, income, education, power, and social status—all of which bestow further opportunities on some and deprive others. People sometimes move from one class to another, but the class structure is a system with built-in boundaries and judgments of class conflict, such as is reflected in the tax system. The middle and working classes shoulder much of the tax burden for social programs, which produces resentment toward the poor. At the same time, corporate taxes have declined while tax loopholes for the rich have increased—an indication of the privilege that is perpetuated by the class system. In any feature of the class system different sociologists study, they see class stratification as a dynamic process—one involving the interplay of access to resources, group judgments about other groups, and the exercise of power.

The Distribution of Wealth and Income

One thing that is clear about the U.S. class structure is that there is enormous class inequality in this society and it is growing. Elites control an enormous share of the wealth and exercise tremendous control over others. The gap between the rich and the poor is also increasing, while much of the middle class finds its class standing slipping.

Figure 9.4 shows the increasing gap that has developed between the upper classes and everyone else in recent years. These data show that income growth has been greatest for those at the top end of the population—the top 20 percent and the top 5 percent of all income groups, regardless of race. For everyone else, income growth has remained flat. In each racial group, the top 5 and 20 percent have seen the most growth in income, yet even at the top, Black and Hispanic people do not fare as well as the top 5 or 20 percent of Whites.

The top fifth of the population also receives a larger share of total income than the bottom fifth, and that share is growing. High compensation for CEOs of major companies contributes to this gap. For example, the average CEO makes 475 times the salary of the average blue-collar worker. Since 1980, the average pay of working people has increased by 74 percent; for CEOs, 1884 percent (Reingold 2000; AFL-CIO 2002). Income gains have also been augmented by tax reforms that have provided breaks for the highest earners. Add to this the enormous differences in wealth for the elite compared with everyone else and you begin to get a full picture of inequality in the United States. These trends verify the popular adage that the rich are getting richer

Working class heroes: The miners who were dramatically rescued from the Quecreek mine in the summer of 2001 signed a deal to tell their story in a made-for-TV movie. That deal will give them the equivalent of three years' pay each. (They earn about $40,000 per year, counting overtime.) That is a small amount compared to the head of the group that owns the mine: He earns about $18 million a year in salary and bonuses—making in five minutes what it would take the miners to earn in one year!

and the poor, poorer. As the classes become more polarized, the myth that the United States is primarily a middle-class society could be weakened.

When discussing the distribution of resources in society, sociologists make a distinction between wealth and income. *Wealth* is the monetary value of everything one owns, minus debt. It is calculated by adding all financial assets (stocks, bonds, property, insurance, value of investments, and so on) and subtracting all debts. *Income* is the amount of money brought into a household from various sources (wages, investment income, dividends, and so on) during a given period. Unlike income, wealth is cumulative; that is, its value tends to increase through investment, and it can be passed on to the next generation, giving

those who inherit wealth a considerable advantage in accumulating more resources.

THINKING *SOCIOLOGICALLY*

Suppose that you wanted to reduce inequality in the United States. Because you know that the transmission of *wealth* is one basis of *stratification,* would you be willing to eliminate the right to inherit property to achieve greater equality?

To understand the significance of wealth compared to income in determining class position, imagine two college graduates, graduating in the same year, from the same college, with the same major and same grade point average. Imagine further that upon graduation, both get jobs with the same salary in the same organization. Yet, in one case, parents paid all the student's college expenses and gave her a car upon graduation. The other student worked while in school and graduated with substantial debt from student loans. This student's family has no money with which to help support the new worker. Who is better off? Same salary, same credentials, but wealth (even if modest) matters. It gives one person an advantage—one that will be played out many times over as the young worker buys a home, finances her own children's education, and possibly inherits additional assets.

Measures of income are based on annually reported U.S. census data drawn from a sample of the population. But reliable data on the distribution of wealth are difficult to acquire. The figures in the Forbes 400 list of the wealthiest Americans are only estimates, based on publicly disclosed corporate data, shrewd guesswork by market watchers and finance writers, and previous estimates updated by factoring in the overall performance of the economy.

Where is all the wealth? The wealthiest 1 percent own 38 percent of all net worth; the bottom 80 percent

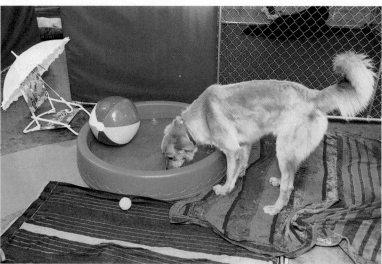

Sociologists have shown that the nouveau riche can be ostentatious in their display of wealth. In this pet care center in San Francisco, even the family pet can engage in conspicuous consumption.

control only 17 percent. The top 1 percent also owns almost half of all stock; the bottom 80 percent own only 4 percent of total stock holdings. Moreover, there has been a sharp increase in the concentration of wealth since the 1980s, and the concentration of wealth is higher in the United States than in any of the other industrialized nations (Mishel et al. 2001). As just one example, John D. Rockefeller is typically heralded as one of the wealthiest men in U.S. history. But if you compare Rockefeller with Bill Gates, in the value of today's dollars, Gates has already surpassed Rockefeller's riches (Myerson 1998).

In contrast to the vast amount of wealth and income controlled by elites, a very large proportion of Americans have hardly any financial assets once debt is subtracted. Nearly one-third have net worth of less than $10,000—hardly enough to handle a major emergency. Another 18 percent have zero or negative net worth, usually because their debt exceeds their assets (Sullivan et al. 2000; Mishel et al. 2001). The American dream of owning a home, a new car, taking annual vacations, and sending one's children to good schools—not to mention saving for a comfortable retirement—cannot be attained by many people. When you see the vast amount of income and wealth controlled by a small segment of the population, a sobering picture of class inequality emerges. Wealth allows you to accumulate assets over generations, giving advantages to subsequent generations that they might not have had on their own.

Popular legends, however, extol the possibility of anyone making it rich in the United States. The well-to-do are admired not just for their style of life, but also for their supposed drive and diligence. Despite the prominence of rags-to-riches stories in American legend, wealth in this society is usually inherited. Some individuals—whether shrewd or lucky—make their way into the elite class by virtue of their own success, but this is rare. The upper class remains overwhelmingly White and Protestant. These elites travel in exclusive social networks that tend to be open only to those in the upper class. They tend to marry other elites, their children are likely to go to expensive schools, and they spend their leisure time in exclusive resorts.

Race has a significant influence on the pattern of wealth distribution in the United States. For every dollar of wealth held by White Americans, Black Americans have only 26 cents (Oliver et al. 1995; Mishel et al. 2001). At all levels of income, occupation, and education, Black families have lower levels of wealth than similarly situated White families. The ability to draw on assets during times of economic stress means that families with some resources are better able

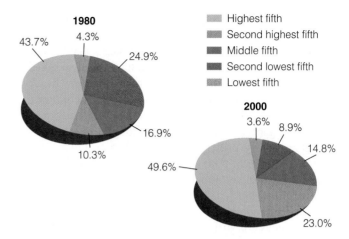

Figure 9.5 Who's Got a Piece of the Pie?

Source: DeNavas–Walt, Carmen, Robert W. Cleveland, and Bruce H. Webster, Jr. 2003. *Income in the United States: 2002.* Washington, DC: U.S. Census Bureau.

to withstand difficult times than are those without assets. Even small assets, such as home ownership or a savings account, provide protection from the crises of increased rent, a health emergency, or unemployment. Because the effects of wealth are intergenerational—that is, they accumulate over time—just providing equality of opportunity in the present does not address the differences in class status that Black and White Americans experience (Oliver and Shapiro 1995).

The ostentatious lifestyle of some can give the appearance that the United States is an open class system.

What explains the disparities in wealth by race? Wealth accumulates over time. Thus, government policies such as discriminatory housing policies, bank lending policies, tax codes, and so forth have disadvantaged Black Americans. These policies have impaired Black Americans in the past from being able to accumulate wealth, resulting in the differing assets Whites and Blacks in general hold now. Even though some of these discriminatory policies have ended, many also continue. Either way, their effects persist, resulting in what sociologists Oliver and Shapiro call the *sedimentation of racial inequality.*

Understanding the significance of how wealth shapes life chances for different groups also decomposes a monolithic view of Hispanics. Cuban Americans and Spaniards are similar to Whites in their wealth holdings; whereas Mexicans, Puerto Ricans, Dominicans, and other Hispanic groups more closely resemble African Americans on the various indicators of wealth and social class. Likewise, one can better understand differences in class status among Asian American groups by carefully considering the importance of not just income, education, and occupation, but also patterns in the net assets of different groups (Conley 1999).

Money and status alone do not tell the whole story of the significance of elites in the United States. Elites wield enormous influence over the political process by funding lobbyists, exerting their social and personal influence on other elites, and contributing heavily to political campaigns. The grip of the upper class on political power is also witnessed by the large number of multimillionaires now in Congress. Those without great wealth are also at a huge disadvantage in financing political campaigns. Studies of elites also find that they are often politically quite conservative (Burris 2000). Even as the elite class becomes more diverse, with more women and minorities in positions of power, sociologists find that women and minorities who make it to these top positions have perspectives and values that do not differ significantly from the White men who predominate (Zweigenhaft and Domhoff 1998; see also Chapter 19).

Many factors have contributed to the declining fortunes of the lower and middle classes in the United States, including the profound effects of national and global economic change. Economic restructuring concentrates the wealth in the hands of only a few. Reductions in state and federal spending have eliminated many government jobs—jobs that have traditionally been the route to middle-class sta-tus for many workers. Job layoffs have left many people out of work or sent former employees (both middle-class and blue-collar) into jobs with lower pay, less prestige, and perhaps no employee benefits. The workers come to expect these circumstances, given their levels of education and experience.

The new economy has had mixed results for different groups (Andersen 2000). Income levels for women have increased, but at the same time have decreased for men, except for the top 20 percent of earners. Since 1980, women in the bottom 20 percent of wage earners have also seen their wages decline. These are aggregate numbers—they reflect trends in the overall population, but these data show the varying impact of economic restructuring on different groups of people (Kilborn and Clemetson 2002; Mishel et al. 2001). Whether these trends reverse or continue remains to be seen.

The tax structure has also distributed benefits unevenly, leading to discontent among the middle class and resistance to social programs that would otherwise be subsidized through federal taxes. Corporations benefit the most from the tax structure, as corporate taxes have decreased dramatically in recent years (see Figure 9.6). While most Americans are paying more in federal tax than ever before (an increase from 13 to 15 cents per dollar of income earned since 1990), corporate taxes since 1990 have fallen from 26 cents on the dollar to 20 cents, yet corporate profits increased 252 percent in that period (Johnston 2000). People at the upper end of the class system have also been able to take advantage of numerous tax benefits and loopholes, reducing their tax burden, while the burden on the middle classes has increased. Understanding the differential impact of changes in the economy is an important part of analyzing the dynamics of social stratification.

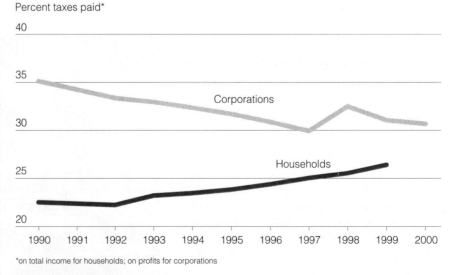

Percent taxes paid*

*on total income for households; on profits for corporations

Figure 9.6 The Tax Burden: For Whom?

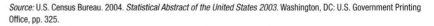

Source: U.S. Census Bureau. 2004. *Statistical Abstract of the United States 2003.* Washington, DC: U.S. Government Printing Office, pp. 325.

Diverse Sources of Stratification

Class is only one basis for stratification in the United States. Age, ethnicity, race, and gender all have a tremendous influence on stratification, as we will also see in subsequent chapters on these topics. For example, analyzing class without also analyzing race and gender can be misleading. Race, class, and gender are overlapping systems of stratification that people experience simultaneously. A working-class Latina, for example, does not experience herself as working class at one moment, Hispanic at another moment, and a woman the next. At any given time, her position in society is the result of her race, class, *and* gender status.

To explain in another way: Class position is manifested differently, depending on one's race and gender, just as gender and race are experienced differently, depending on one's class. At any given moment, race, class, or gender may seem particularly salient in a person's life. For example, a Black middle-class man who is stopped and interrogated by police when driving through a predominantly White middle-class neighborhood may at that moment feel his racial status as his single most outstanding characteristic, but at all times, his race, class, and gender influence his life chances. The social categories of race, class, and gender shape all people's experience in this society—not just those who are disadvantaged (Andersen and Hill Collins 2004).

Class also significantly differentiates group experience within given racial and gender groups. Latinos, for example, are broadly defined as those who trace their origins to regions originally colonized by Spain. The ancestors of this group include both White Spanish colonists and the natives enslaved on Spanish plantations. Today, some Latinos identify as White, others as Black, and others by their specific national and cultural origins. The very different histories of people categorized as Latino are matched by significant differences in class. Some may have been schooled in the most affluent settings; others may be virtually unschooled. Those of upper-class standing may have had little experience with prejudice or discrimination; others may have been highly segregated into barrios and treated with extraordinary prejudice. Latinos who live near each other in the United States and who are the same age and share similar ancestry may have substantially different experiences based on their class standing (Massey 1993). Neither class, race, nor gender, taken alone, can be considered an adequate indicator of different group experiences.

Race and Class

The interaction between race and class can be seen in noting the growth of both the Latino and African American middle class. Although the presence of a significant Black middle class is often seen as a relatively recent phenomenon, the African American sociologist E. Franklin Frazier argued early that the Black middle class dates back to the small numbers of free Blacks in the eighteenth and nineteenth centuries (Frazier 1957). In the twentieth century, the Black middle class expanded to include people who obtained an education and became established enough in industry, business, or a profession to live comfortably or, in a few cases, affluently. These included teachers, doctors, nurses, and businesspeople. The Black middle class also historically included postal workers and railroad porters, occupations generally thought of as working-class, yet Black workers in these trades had relatively high prestige within the Black community. The Black middle class was a class of its own—not comparable to the White middle class, but distinct. Still, wages for Black middle-class and professional workers never matched those of Whites in the same jobs. Furthermore, despite their status within Black communities, members of the Black middle class have been excluded from White schools, clubs, and social settings. The class structure within African American society has existed alongside the White class structure, separate and different.

In recent years, both the African American and Latino middle class have expanded, primarily as the result of increased access to education and middle-class occupations for people of color (Collins 1983; Durant and Louden 1986; Landry 1987; Pattillo-McCoy 1999). This is the result of civil rights legislation, as well as affirmative action policies. The persistence of racial discrimination and the recent arrival of racial groups in the middle class means, however, that their hold on middle-class status is more tenuous than that of many middle-class Whites. For example, many in the Black and Latino middle class work in public sector jobs—positions that depend on continuing government support. During periods of economic recession or political conservatism, when there is considerable pressure to reduce federal spending, the eliminated jobs are likely to cause a significant thinning in the ranks of the Black and Latino middle class (Collins-Lowry 1997; Collins 1983; Silver 1995).

Although middle-class Blacks and Latinos may have economic privileges that other Blacks and Latinos do not have, their class standing does not make them immune to the negative effects of race (Higginbotham 2001). Asian Americans also have a significant middle class, but they have been stereotyped as the most successful minority group because of their presumed educational achievement, hard work, and thrift. This stereotype is referred to as the *myth of the model minority*—a myth that understates the significant obstacles Asian Americans encounter and obscures the high rates of current poverty among many Asian American groups (Lee 1996; Woo 1998).

Mary Pattillo-McCoy's (1999) research on the Black middle class shows the perils associated with this status. Despite recent successes, many in the Black middle class have a tenuous hold on this class status. She questions whether economic gains can be sustained in the context of the new economy where young people have a more difficult time getting a foothold on good jobs. She points out that the Black middle class remains as segregated from Whites as the Black poor. Because of continuing racial segregation in neighborhoods, Black middle-class neighborhoods are typically located next to Black poor neighborhoods, exposing those in the middle class to many of the same risks as those in poverty. Middle-class Black and Latino parents also have to teach their children to avoid victimization by racism. This is not to say that the Black middle class has the same experience as the poor, but McCoy's research challenges the misleading view that the Black middle class "has it all."

Gender and Class

The effects of gender stratification further complicate the analysis of class. In the past, women were thought to derive their class position from their husband or father, but sociologists now challenge this assumption.

Measured by their own income and occupation, the vast majority of women would likely be considered working class. The median income for women, even among those employed full-time, is far below the national median income level. In 2002, when median income for men working year-round and full-time was $39,429, the median income of women working year-round, full-time was $30,203 (DeNavas-Walt et al. 2003). The vast majority of women work in low-prestige and low-wage occupations, even though women and men have comparable levels of educational attainment.

Measuring women's class status is complicated. Many women who have little or no income of their own consider themselves middle class by virtue of their husband's class status. How does one determine the class status of a White woman earning $25,000 per year as a secretary but married to a business executive earning $85,000? The class status of a household may actually differ from an individual woman's class status—something that becomes abundantly clear following a divorce, when women's income typically drops significantly, while men's increases (Peterson 1996a, 1996b; Weitzman 1996; Weitzman 1985).

However hard to measure, class differences between women are highly significant. The problems faced by professional and managerial women are simply not the

UNDERSTANDING DIVERSITY

Latino Class Experience

Latinos in the United States are a diverse population of many different groups, each with many different national origins, histories, and cultural backgrounds. This diversity is represented by the fact that Latinos do not agree among themselves on what they should be called. Some prefer Latino, others Hispanic, and some prefer to be identified by their cultural origins, as in Chicanos, Cuban Americans, or Puerto Riqueños. Some think of themselves as White. Generational differences further add to the diversity among Latinos.

Sociologist Douglas S. Massey reminds us of this diversity in thinking about the diverse class experiences of Latinos. Massey writes that as a result of different histories, Latinos live in different socioeconomic circumstances. He states:

They may be fifth-generation Americans or new immigrants just stepping off the jetway. Depending on when and how they got to the United States, *they may also know a long history of discrimination and repression or they may see the United States as a land of opportunity where origins do not matter. They may be affluent and well educated or poor and unschooled; they may have no personal experience of prejudice or discrimination, or they may harbor stinging resentment at being called a "spic" or being passed over for promotion because of their accent. (1993: 7–8)*

Massey also notes that levels of residential segregation and poverty vary across different Latino groups. As socioeconomic levels rise and as immigrants reach second, third, and later generations within the United States, the degree of segregation for Latinos progressively falls. Puerto Ricans have higher levels of segregation than other Latino groups, as well as some of the highest rates of poverty. The migration of Latinos into this country also contributes to their class position. Some

Latinos are indigenous to the United States; that is, their families owned land in the American Southwest that was colonized by White settlers. Others are recent immigrants who enter the economy as low-wage workers with little opportunity for upward mobility.

Diversity among Latinos has many implications for understanding the experience of this population. Some are caught in the economic underclass, others are middle-class. A few are among the nation's elites. Massey reminds us that factors such as class, historical origins, residential segregation, race, and migration patterns must be carefully analyzed to understand Latino experiences.

Sources: Massey, Douglas S. 1993. "Latino Poverty Research: An Agenda for the 1990s." *Social Science Research Council Newsletter* 47 (March): 7–11; Rodriguez, Clara E. 2000. *Changing Race: Latinos, the Census, and the History of Ethnicity in the United States.* New York: New York University Press. ●●●

same as those experienced by women in low-wage, low-prestige work, even though both groups may encounter sexism in their lives. But each group's experience depends on class, as well as gender and race (Higginbotham and Romero 1997).

Age and Class

Unlike race and gender, age changes over the course of one's life. Still, age is a significant source of stratification in the United States with different age groups experiencing different locations in the stratification system. Just being born in a particular generation can have a significant influence on one's life chances. The current fears of young, middle-class people that they will be unable to achieve the lifestyles of their parents show the effect that being in a particular generation can have on one's life chances. The effect of one's age on life chances is also dependent on the effect of race or gender. Being a young, African American teen, for example, puts you far more at risk of poverty than being an older, White American woman.

Children are the age group most likely to be poor—16.7 percent of those under age 18 are poor in the United States. In the past, the aged were the most likely to be poor. Now, many elderly people are poor (10.4 percent of those sixty-five and over), but far fewer in this age category are poor than was the case not many years ago (see Figure 9.7; Proctor and Dalaker 2003). This shift reflects the greater affluence of the older segments of the population—a trend that is likely to continue as the current large cohort of middle-aged, middle-class Baby Boomers grow older.

Age also interacts with other sources of stratification, particularly race, gender, and marital status. This is shown by differences in the status of people in retirement. Annual retirement income through Social Security for African Americans is 54 percent of that received by White Americans. Women earn between 70 and 81 percent of men's retirement income (the difference varies in different studies). The biggest gap in retirement income is between married and unmarried people. The unmarried claim only 53 percent the Social Security income that the married claim (Hogan and Perrucci 1998). Because Social Security is based on earnings over one's lifetime, these data show how differences in wages prior to retire-ment accumulate over time, affecting the well-being of people long after they have left the labor force.

Class and Cultural Diversity

Class has a cultural, as well as an economic, dimension. Classes are distinguishable not just by access to money, prestige, and power, but sometimes by cultural behaviors and values. These are not uniform within a class, but how you eat, speak, and dress, among other things, can vary by class. Of course, some of these things result from the resources available to class groups. Drinking good wine, for example, is not only an acquired taste; it requires money.

The different cultures and ways of life found among different classes represent a more subjective aspect of class, stretching from friendships and recreation to how different classes communicate. Dress is a marker that people use to make class distinctions. Classes also differ in their language. For example, street language has its own vocabulary, idioms, syntax, intonation, and grammar—much of which would be unintelligible to class elites. Seen on its own terms, street language is efficient and meaningful, its vocabulary rich, but class-based judgments about language and other forms of culture are often based on false assumptions that stem from stereotypes held by the dominant group about subordinate groups. There is a stereotype, for example, that anyone who speaks with a southern accent is ignorant or racist. This belief stems in part from the subordinated social and economic position of the South throughout much of the nation's history.

The cultural dimensions of social class are especially obvious if you think about the experience of entering a class setting different from your own background:

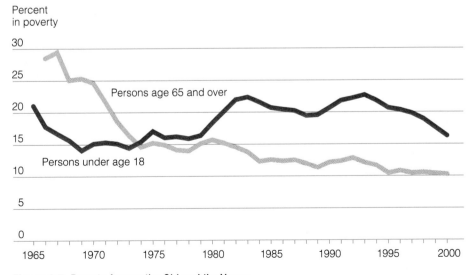

Figure 9.7 Poverty Among the Old and the Young

Source: Proctor, Bernadette D., and Joseph Dalaker. 2003. *Poverty in the United States: 2002.* Washington, DC: U.S. Census Bureau.

You are likely to feel out of place. A middle-class person walking through a working-class community may find the environment uncomfortable and unpleasant. Unfortunately, the politics of class often lead those in the upper and middle classes to make disdainful judgments about the lifestyle of the working class and the poor. At the same time, working-class people who enter middle-class settings often find them inhospitable and alienating and may feel they have to cloak or eliminate their working-class habits to be accepted.

The dominant culture supports White middle-class lifestyles and values more than other class values. As a result, succeeding in the middle-class world usually means abandoning (at least publicly) working-class mannerisms and habits. The United States calls itself the "melting pot," but its dominant culture is distinctively White and middle class. The dominant culture projects the notion that people in the upper classes have more ability, making the entire class system seem legitimate even as it robs people in the working and lower classes of dignity. This is what the sociologists Richard Sennett and Jonathan Cobbs mean when they write of "the hidden injuries of class" (Sennett and Cobbs 1993).

Social Mobility

There is a general belief in the United States that anyone can, by his or her own labor, move relatively freely through the class system. Is this fact or myth? This is an important subject of sociological research. There has been a long-standing argument that Americans are not very conscious of the class system because of the strong cultural belief that upward mobility is possible. Images of opulence also saturate popular culture, making it seem that such material comforts are available to anyone. Many working-class and middle-class Americans focus on getting ahead individually and have little concern for organizing around class interests. The faith that upward mobility is possible ironically perpetuates inequality. If people believe that everyone has the same chances of success, then they are likely to think that whatever inequality exists must be fair.

Class Consciousness

Class consciousness is the perception that a class structure exists, along with the feeling of shared identification with others in one's class—others with whom one perceives common life chances (Centers 1949). Notice that there are two dimensions to the definition of class consciousness: the idea that a class structure exists and one's class identification.

Class consciousness in the United States has been higher at certain times than others. A significant labor movement in the 1920s and 1930s rested on a high degree of class consciousness. But now the formation of a relatively large middle class and a relatively high standard of living mitigate against class discontent. Racial and ethnic divisions also make strong alliances within various classes less stable. The recent trends of growing class inequality could result in greater class consciousness. Researchers find that people in the United States do recognize class divisions and believe that classes are organized around opposing interests. Forty-five percent of the public identifies as working-class; 46 percent as middle-class, 5.4 percent as lower-class and 3.6 percent as upper-class (National Opinion Research Center 2002; Gorman 2000; Vanneman and Cannon 1987), but this has not developed into a significant class-based movement for change.

Class inequality in any society is usually buttressed by ideas that support (or actively promote) inequality. Beliefs that people are biologically, culturally, or socially different can be used to justify the higher position of some groups. If people believe these ideas, the ideas provide legitimacy for the system. Karl Marx used the term **false consciousness** to describe the class consciousness of subordinate classes who internalized the view of the dominant class. Marx argued that the ruling class controls the subordinate classes by infiltrating their consciousness with belief systems consistent with the interests of the ruling class. If people accept these ideas, which justify inequality, they need not be overtly coerced into accepting the roles designated for them by the ruling class.

How much do people identify with their class location? This varies among different classes. There are no direct studies of class consciousness among elites since sociologists rarely include the top stratum of the social hierarchy in their studies. This is not so much the fault of sociologists as it reflects the ability of the elite to isolate themselves from public scrutiny. The upper class is class conscious, however, in the sense that its members are a cohesive group (more so than other class groups) who are well aware of one another and are protective of their common interests (Domhoff 1998, 1970). These are the people who hold institutional power. Outside the elite, the working class is more class conscious than the middle class; that is, working-class people are more likely to perceive that their lives are controlled by others in the higher status classes. Sociologists have found that the single most important determinant of where one sees oneself in the class system is whether one does mental or manual labor (Vanneman and Cannon 1987).

Defining Social Mobility

Social mobility is a person's movement over time from one class to another. Social mobility can be up or down, although the American dream emphasizes upward movement. Mobility can also be either *intergenerational,* occurring between generations, as when a daughter rises above the class of her mother or father; or *intragenerational,* occurring within a generation, as when a person's class status rises as the result of business success or falls with a disaster.

---

*DEBUNKING **SOCIETY'S MYTHS***

Myth: The United States is a land of opportunity where anyone who works hard enough can get ahead.

Sociological perspective: Social mobility occurs in the United States, but less often than the myth asserts and over shorter distances from one class to another; most people remain in their class of origin.

As discussed earlier in this chapter, societies differ in the extent to which social mobility is permitted. Some societies are based on *closed class systems,* in which movement from one class to another is virtually impossible. In a caste system, for example, mobility is strictly limited by the circumstances of one's birth. At the other extreme are *open class systems,* in which placement in the class system is based on individual achievement, not ascription. In open class systems there are relatively loose class boundaries, high rates of class mobility, and weak perceptions of class difference.

The class system in the United States is popularly characterized as an open class system where individual achievement, not birth, is the basis for class placement. Many in the United States revere so-called self-made people and the vast majority, when asked, will say that people have a good chance of improving their standard of living. Most parents also think that their children will be better off than they are—a good indication of the belief in upward mobility (National Opinion Research Center 2002). Many also immigrate to this nation with the knowledge that their life chances are better here than in their country of origins.

The Extent of Social Mobility

How much social mobility exists in the United States? Social mobility is much more limited than people believe. Success stories of social mobility do occur, but research finds that experiences of mobility over great distances are rare, certainly far less than believed. Most people remain in the same class as their parents, and many drop to a lower class. The social mobility that does exist is greatly influenced by education. African Americans, as well as immigrant groups, are often strongly committed to social mobility through education; increases in educational attainment for African Americans account for a considerable portion of the

gains they have made (Smith 1989). But, most of the time, among all groups, people remain in the class where they started. What mobility exists is typically short in distance (Blau and Duncan 1967; Ganzeboom et al. 1991), and recent evidence suggests that mobility between generations may be becoming even more rigid than in the past (Rytina 2000; Gittleman and Joyce 1999).

Social mobility is much more likely to be influenced by factors that affect the whole society than by individual characteristics. Thus, most mobility can be attributed to changes in the occupational system, economic cycles, and demographic factors, such as the number of college graduates in the labor force (Hout 1988; Erikson 1985; Vanneman and Cannon 1987). In sum, social mobility is much more limited than the American dream of mobility suggests.

Upward Mobility People who are *upwardly mobile* are often expected to distance themselves from their origins. This may mean creating some distance from their community of origin and can result in many conflicts with family, with friends, and even within themselves. First-generation college students, for example, often find themselves torn between leaving home to go to school and remaining close to their family and community. Likewise, women who are raised to have greater attachments to family and community may feel pressure from these groups not to move away. White working-class women, for example, are likely to have been socialized to marry, not pursue careers. When working-class women become successful, their families may be ambivalent about their success if success is seen as taking the women away from their family and community origins.

Studies of upward mobility also find that mobility is not just about individual effort, although that clearly plays a role. Most of the time people who are upwardly mobile got help along the way—a willing family, a teacher or mentor, and sometimes just plain luck. This underscores an important sociological point about mobility: It is not just an individual process, but also a collective effort that involves kin and sometimes community (Higginbotham and Weber 1992).

Downward Mobility The attention people in the United States give to upward mobility has obscured the experience of *downward mobility*—movement down in the class system. Downward mobility is becoming more common (Newman 1988; Newman 1993). As income distribution is becoming more skewed toward the top, many in the middle class are experiencing mobility downward. For the first time in American history, many in the middle class are experiencing a decline: Levels of real income (that is, income measured controlling for the value of the dollar) are falling and the cost of living is increasing, especially the cost of housing. Adding to this is the fact that fewer workers are covered by job benefits (health insurance, pension, and the like) so that total compensation levels have also fallen (Mishel et al. 2001).

Doing better than one's parents has long been a hallmark of the American dream—a goal that many people achieved in the twentieth century as the economy of the United States grew. Today, many young people worry that they will be unable to match the lifestyle of their parents. Moreover, many families are finding that just staying in place requires extra effort. Thus, families that are keeping pace with median income are generally those in which wives are working longer hours. Without the greater workload that is falling on women, downward mobility would likely be greater than it is. Consider this: Married-couple families have increased their total working hours an additional six weeks per year since 1989. Husbands worked (on average) 38 more hours per year; wives worked 116; for women, that is fourteen more eight-hour days per year (Mishel et al. 2001). No wonder families feel like things are speeding up!

Poverty

Despite the relatively high average standard of living in the United States, poverty afflicts millions of people. Aside from imposing a grim quality of life on the poor, poverty is also the basis for many of society's problems. Poor health care, failures in the education system, and crime are all related to poverty.

The federal government has established an official definition of poverty used to determine eligibility for government assistance and to measure the extent of poverty in the United States. The **poverty line** is the amount of money needed to support the basic needs of a household, as determined by government. Below this line, one is considered officially poor. To determine the poverty line, the Social Security Administration takes a low-cost food budget (based on dietary information provided by the Department of Agriculture) and multiplies by three, assuming that a family spends approximately one-third of its budget on food. The resulting figure is the official poverty line, adjusted slightly each year for increases in the cost of living. The poverty line also varies by family size and household composition. In 2002, the official poverty line for a family of four (one adult and three children) was $18,307. Although a cutoff point is necessary to administer antipoverty programs, the poverty line can be misleading. A person or family earning $1 above the cut-off point is not officially categorized as poor.

> ### THINKING *SOCIOLOGICALLY*
> Using the current federal *poverty line* ($18,307, pre-tax) as a guide, develop a monthly budget that does not exceed this income level and that accounts for all of your family's needs. For purposes of this exercise, assume that you are a family of four. Base your budget on the actual costs of things in your locale (rent, food, transportation, utilities, child care, clothing, and so forth). Do not forget to account for taxes (state, federal, and local), health care expenses, your children's education, car repairs, and so on. What does this exercise teach you about those who live below the poverty line?

Who Are the Poor?

In 2002, there were about 34.6 million poor people in the United States, representing 12.1 percent of the population. Since the 1950s, poverty has declined in the United States, although it fluctuates depending on the state of the economy. Although the majority of the poor are White, disproportionately high rates of poverty are also found among Asian Americans, Native Americans, Black Americans, and Hispanics. Twenty-four percent of African Americans, 22 percent of Hispanics, 10 percent of Asians and Pacific Islanders, and 10 percent of Whites are poor (Proctor and Dalaker 2003). Among Hispanics, Puerto Ricans have been most likely to suffer increased poverty, probably because of their concentration in the poorest segments of

Families, including women and their children, are the largest portion of the nation's homeless; they are often left homeless because of eviction and/or unemployment.

© Tony Freeman/PhotoEdit

MAPPING AMERICA'S DIVERSITY

MAP 9.2 Poverty in the United States

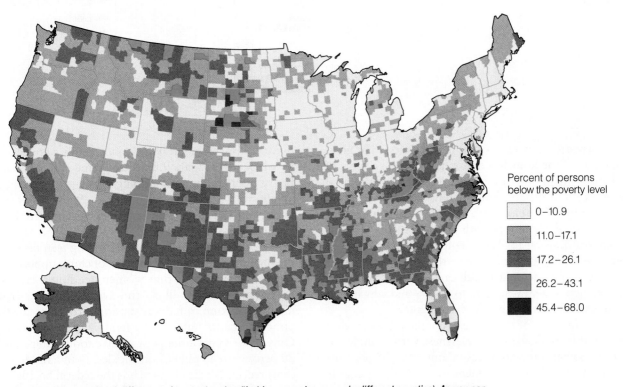

Percent of persons
below the poverty level

	0–10.9
	11.0–17.1
	17.2–26.1
	26.2–43.1
	45.4–68.0

This map shows regional differences in poverty rates (that is, percentage poor in different counties). As you can see, poverty is much higher in the South (13.8 percent) than in the West (12.4 percent), Northeast (10.9 percent), and Midwest (10.3 percent). Various social factors explain different rates of poverty, including regional labor markets, the degree of urbanization, immigration patterns, and population composition, among other factors. What do you think the major causes of poverty are in your region?

Data: U.S. Census Bureau. 2004. *American FactFinder.* Website: **www.census.gov**; Proctor, Bernadette D., and Joseph Dalaker. 2003. *Poverty in the United States: 2002.* Washington, DC: U.S. Census Bureau, p. 6.

the labor market and their high unemployment rates (Tienda and Stier 1996; Hauan et al. 2000). Asian American poverty has also increased substantially since the 1980s, particularly among the most recent immigrant groups, including Laotians, Cambodians, Vietnamese, Chinese, and Korean immigrants; Filipino and Asian Indian families have lower rates of poverty (Lee 1996). The federal government does not include Native Americans in its annual population surveys, but periodic reports indicate a very high rate of poverty among Native Americans—31 percent, higher than any other group (U.S. Census Bureau 1993).

Homelessness For many people, the image of poverty may be the homeless they see on urban streets. Among the poor are thousands of homeless people. It is difficult to estimate the number of homeless people. Depending on how you define and measure homeless-

ness, estimates vary widely. If you count on any given night, there may be 444,000 to 842,000 homeless people (depending on the month measured), but measuring those experiencing homelessness during one year, the estimates jump to 2.3 to 3.5 million people (Urban Institute 2000; National Coalition for the Homeless 2002).

Whatever the actual number, homelessness has substantially increased over the past two decades (National Coalition for the Homeless 2002). Families are the fastest growing segment of the homeless population (about 40 percent). The homeless also includes battered women; elderly, poor men; the disabled; veterans; and AIDS victims. A survey of twenty-seven cities has found that the homeless population is half African American, 35 percent White, 12 percent Hispanic, 2 percent Native American, and 1 percent Asian (U.S. Conference of Mayors 2001).

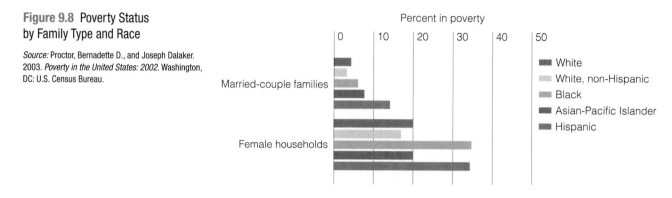

Figure 9.8 Poverty Status by Family Type and Race

Source: Proctor, Bernadette D., and Joseph Dalaker. 2003. *Poverty in the United States: 2002.* Washington, DC: U.S. Census Bureau.

There are many reasons for homelessness. The great majority of the homeless are on the streets because of unemployment and/or eviction. Reductions in federal support for affordable housing have left many with no place to live. Coupled with eroding work opportunities (particularly in jobs with decent benefits) and inadequate housing for low-income people, many people have no choice but to live on the street. Add to that problems of inadequate health care, domestic violence, and addiction, and you begin to understand why homelessness exists (Snow and Anderson 2003).

The diversity of the homeless population makes it impossible to alleviate homelessness with a single solution. Some of the homeless (about 20 to 25 percent) are mentally ill; the movement to get mental patients out of institutional settings has left many without the mental health resources that might help them (National Coalition for the Homeless 2002). Public policy responses to homelessness need to consider the different needs of the various homeless populations. Because homelessness, like other forms of poverty, has social structural causes, simple solutions that target homeless people as deviants are unlikely to have good results.

The Feminization of Poverty The **feminization of poverty** refers to the fact that such a large proportion of the poor are women and children. This results from several factors, mostly the growth of female-headed households and wage inequality between women and men. One-quarter of single-parent, female-headed households were poor; half of poor families are headed by women (see Figure 9.8). At the same time, the number of poor children is alarmingly high—13 percent of White children, 32 percent of Black, 29 percent of Hispanic, and 12 percent of Asian American children (Proctor and Dalaker 2003).

The decline in marriage rates among teen mothers, the high divorce rate, and the lack of child support provided by men means that women heading their own households are less likely to have any income from men. Reduction in federal support programs for the poor also contributes to the feminization of poverty. Only 29 percent of the poor receive food stamps; only 20 percent are in public housing; less than half (46 percent) receive Medicaid, which is the federal health care system for the poor (U.S. Census Bureau 2004). As we will see next, welfare reform has also reduced support for poor women and their children, even though welfare is viewed by some as overly generous and producing dependence.

One marked change in poverty is the growth of poverty in suburban areas (now 8 percent of all poverty). Forty percent of the poor live inside central cities. Within cities, poverty rates are highest in the most racially segregated neighborhoods; the income gap between Latinos and Anglos is higher in the metropolitan areas that have the highest levels of residential segregation (Harris and Curtis 1998; Massey 1999). But the focus on urban poverty should not cause us to forget that 20 percent of the poor live in rural areas.

TAKING ON SOCIAL ISSUES

Is Marriage the Solution to Poverty?

Some, including President George Bush, have suggested that the best solution to poverty and welfare spending is for more poor women to be married. Federal legislation has even included funds for programs promoting marriage in the expenditures for social welfare. Certainly having two incomes in a household raises economic well-being, but what other issues should be addressed in considering marriage as a solution to women's poverty?

Taking Action

Go to the Taking Action Exercise on the Companion Website—at http://sociology.wadsworth.com/andersen_taylor/4e—to learn more about an organization that addresses this topic. •••

Explanations of Poverty

Most agree that poverty is a serious social problem. Far less agreement exists on what to do about it. Two points of view prevail: Some blame the poor for their own condition and some look to social structural causes to explain poverty. The first view, popular with the public and many policy makers, is that poverty is caused by the cultural habits of the poor. According to this point of view, behaviors such as crime, family breakdown, lack of ambition, and educational failure generate and sustain poverty—a syndrome to be treated by forcing the poor to fend for themselves. Sociologists take a more structural view, seeing poverty as rooted in the structure of society, not in the morals and behaviors of individuals. The perspective you take matters because it informs the formation of social policy.

Blaming the Victim: The Culture of Poverty

Blaming the poor for being poor appeals to the myth that success requires only individual motivation and ability. Many adhere to this view and hence have a harsh opinion of the poor. This attitude is reflected in U.S. public policy concerning poverty, which is stingy compared with other industrialized nations. Cheating on welfare is probably far less common than cheating on income taxes, yet welfare recipients are commonly portrayed as lazy and cheating the system. Those who blame the poor for their own plight typically assume that poverty is the result of early childbearing, drug and alcohol abuse, refusal to enter the labor market, and participation in crime. This puts the blame for poverty on individual choices, not on societal problems—blaming the victim, not the society, for social problems (Ryan 1971).

The **culture of poverty** argument contends that poverty is a way of life that is transferred, like other cultures, from generation to generation, where the major causes of poverty are welfare dependency, the absence of work values, and the irresponsibility of the poor. This argument treats poverty as its own cause. The idea that there is a culture of poverty originally came from the work of anthropologist Oscar Lewis, who saw the behaviors of the poor as an adaptation to their marginal position in capitalist, class-stratified societies. Lewis argued that a culture of poverty evolved among the poor as they adapted to their despair and hopelessness (Lewis 1966). The culture of poverty argument has been adapted by policy makers to argue that the causes of poverty are found in the breakdown of major institutions, including the family (Moynihan 1965).

DEBUNKING **SOCIETY'S MYTHS**

Myth: If poor people would only get jobs, they could get out of poverty.

Sociological perspective: Forty-one percent of the poor work; the number of working poor has increased in recent years; 12 percent of the poor work year-round and full-time (U.S. Census Bureau 2002a).

Is there a culture of poverty? To answer this question, we might ask, Is poverty transmitted across generations? Researchers have found only mixed support for this assumption. Children of poor parents have a 16 to 28 percent probability of being poor adults (Rodgers 1995). Many of those who are poor remain poor for only one or two years; only a small percentage of the poor are chronically poor. More often, poverty results from a household crisis, such as divorce, illness, unemployment, or parental death. People tend to cycle in and out of poverty. The public stereotype that poverty is passed through generations is thus not well supported by the facts.

A second question to ask is, Do the poor want to work? The idea that they do not is essential to the culture of poverty thesis. The assumption is that poverty is the fault of the poor and that if they would only change their values and adopt the American work ethic, poverty would go away. What is the evidence for these claims?

A large number of the able-bodied poor *do* work, even if only part-time. Of all poor persons, 38 percent work. And eleven percent of the poor who are of

Despite public stereotypes associating poverty with urban areas, the greatest growth in poverty has been in suburban areas.

© Allen Russell/Index Stock Imagery

working age hold full-time jobs (Dalaker and Proctor 2003). You can see why this is true when you calculate the income of someone working full-time for minimum wage. Someone working forty hours per week, fifty-two weeks per year, at minimum wage will have an income far below the poverty line.

Current policies that force those on welfare to work also tend to overlook how difficult it is for poor people to retain the jobs they get. Prior to welfare reform in the mid-1990s, poor women who went off welfare to take jobs often found they soon had to return to welfare because the wages they earned were not enough to support their families. Leaving welfare often means losing health benefits, but increased expenses. And, the jobs that poor people find often do not lift them out of poverty. Attributing poverty to the values of the poor is both unproved and a poor basis for public policy.

Structural Causes of Poverty From a sociological point of view, the causes of poverty lie in the economic and social transformations taking place in the United States. Careful scholars do not attribute poverty to a single cause. There are many causes. Two of the most important are *the restructuring of the economy,* which has resulted in diminished earning power and increased unemployment, and *the status of women in the family and the labor market,* which has contributed to the overrepresentation of women among the poor. Add to these underlying conditions the diminished social support for the poor in terms of welfare, public housing, and job training and it is little wonder that poverty is so widespread.

The restructuring of the economy has caused the disappearance of manufacturing jobs, traditionally an avenue of job security and social mobility for many workers, especially African American and Latino workers. The working class has been especially vulnerable to these changes. Economic decline in blue-collar sectors of the economy where men have historically received good pay and good benefits has meant fewer men are able to be the sole provider for their families. Most families now need two incomes to achieve a middle-class way of life. The new jobs that are being created fall primarily in occupations that offer low wages and few benefits. They also tend to be filled by women, especially women of color, leaving women poor and men out of work (McCall 2001; Browne 1999; Andreasse 1997), with little chance to get out of poverty. New jobs are also typically located in neighborhoods far away from the poor, creating a mismatch between the employment opportunities and the residential location of the poor. These changes fall particularly hard on young

SOCIOLOGY IN PRACTICE
Welfare Reform: Myths and Realities

Current welfare policy mandates that welfare recipients must work if they are to receive benefits. This policy followed years of debate about welfare, much of which was based on fundamental misperceptions about the welfare system and those who receive welfare. Now there are frequent proclamations that welfare reform has reduced the welfare rolls; thus, many politicians laud welfare reform as a success. Is this so? Much of what is commonly understood about welfare has been proved wrong, based on sociological research. Sociology contributes to the welfare debate by dispelling many of the myths that cloud a deeply opinionated and political discussion.

What do we know about welfare reform from current research?

Myth: The new welfare law passed in 1996 reduced the welfare rolls, indicating the success of welfare reform.

Reality: Fewer people are on welfare now than prior to welfare reform, but the implementation of TANF in 1996 occurred when the economy was strong and former welfare recipients could find jobs. But poverty has risen since 2000 when the economy weakened. Removal of the welfare safety net makes poor people especially vulnerable to economic cycles.

Myth: Women should be encouraged to marry as a way of reducing welfare dependence.
Reality:
- Forcing women to marry encourages women's dependence on men and punishes women for being independent.
- Large numbers of women receiving welfare have been victims of domestic violence; fear of domestic violence is one reason former welfare recipients give for not wanting to marry.

Myth: Mothers on welfare have more children to increase the size of their welfare checks.
Reality:
- No causal relationship exists between the size of welfare benefits and the number of births by welfare recipients.
- "Family cap" policies now in place prohibit increasing welfare benefits with the birth of an additional child.

Myth: Once someone goes on welfare, it creates a cycle of dependency and becomes a way of life.
Reality:
- Most people on welfare remain on welfare for a relatively short period, usually following a household crisis (such as health problems, unemployment, medical disability, domestic violence, or death of a spouse).
- Welfare cycling is the most common pattern; that is, people go on welfare

people, women, and African Americans and Latinos, who are the groups most likely to be among the working poor.

The high rate of poverty among women is strongly related to women's status in the family and the labor market. For White women, divorce can result in poverty. This is less true for minority women because they are more likely than White women to be poor even within marriage (Catanzarite and Ortiz 1996). Women's responsibility for child care also makes working outside the home on a marginal income very difficult. Affordable child care is hard to come by for single mothers who want to work, especially since these women earn wages too low to pay for good quality care. Even though child-care workers receive low wages, the cost of child care consumes a significant proportion of the income most women earn. Many women with children cannot manage to work outside the home because it leaves them with no one to watch their children.

Wages in the United States are also shaped by the **family wage system,** a wage structure historically based on the assumption that men are the breadwinners for families. The family wage system ignores a fundamental change taking place in the economy—men are no longer the mainstay of family income. More women are now dependent on their own earnings, as are their chil-

dren and other dependents. Thus, whereas unemployment is considered a major cause of poverty among men, for women, wage discrimination is a major cause. The median income for all women ($16,614 in 2001) is well below the poverty line. Given this, it is little wonder that so many women are poor.

Poverty has numerous consequences in society, not just for the poor, but also for others. It increases tensions between classes and racial groups. William Julius Wilson, one of the most noted analysts of poverty and racial inequality, has written, "The ultimate basis for current racial tension is the deleterious effect of basic structural changes in the modern American economy on Black and White lower-income groups, changes that include uneven economic growth, increasing technology and automation, industry relocation, and labor market segmentation" (1978: 154). This demonstrates the power of sociological thinking by placing the causes of both poverty and racism in their societal context.

Welfare

Current welfare policy is covered by the 1996 Personal Responsibility and Work Reconciliation Act (PRWRA). This federal policy eliminated the longstanding welfare program titled Aid to Families with Dependent

---

for a relatively short time, exit, but return when jobs cannot sustain them.

Myth: People use their welfare checks to buy things they do not need.
Reality:

- When former welfare recipients find work, their expenses tend to go up. Though they may have increased income, their disposable income remains constant or declines.

- Low-income mothers who buy "treats" for their children (name-brand shoes, a movie, candy, etc.) do so because they want to be good mothers.

Myth: The existence of federal welfare programs encourages people to stay on welfare.
Reality:

- Persistent poverty, especially among children, increases when federal support programs are cut back.

- Poverty dropped during the period when strong federal support programs were available to assist poor people.

Critical Thinking Exercise

1. A number of organizations provide research and encourage public policies and programs designed to assist those in need. What current policies on poverty, welfare, and work are being debated? To get more information, use this sampling of organizations and websites that will provide both policy information and research on welfare and work:

Coalition for Human Needs:
www.chn.org

Institute for Women's Policy Research:
www.iwpr.org

Institute for Poverty Research:
www.northwestern.edu/ipr

Institute for Research on Poverty:
www.ssc.wisc.edu/irp

Joint Center for Poverty Research:
www.jcpr.org

Welfare Information Network:
www.welfareinfo.orgd

Sources: Butler, Amy C. 1996. "The Effect of Welfare Benefit Levels on Poverty Among Single-Parent Families." *Social Problems* 43 (February): 94–115; Edin, Kathyrn, and Laura Lein. 1997. *Making Ends Meet: How Single Mothers Survive Welfare and Low-Wage Work.* New York: Russell Sage Foundation; Harris, Kathleen Mullan. 1996. "Life after Welfare: Women, Work, and Repeat Dependency." *American Sociological Review* 61 (June): 407–426; Larrison, Christopher R., Larry Nackerud, and Ed Risler. 2001. *Journal of Sociology and Social Welfare* 28 (September): 49–69; Mink, Gwendolyn. 2001. "Violating Women: Rights Abuses in the Welfare Police State." *Annals of the American Academy of Political and Social Science* (September): 79–93; O'Campo, Patricia, and Lucia Rojas-Smith. 1998. "Welfare Reform and Women's Health: Review of the Literature and Implications for State Policy." *Journal of Public Health Policy* 19: 420–446. Winship, Scott, and Christopher Jencks. 2002. "The Well-Being of Single Mothers after Welfare Reform, as Measured by Changes in Food Security." *Policy Brief,* Vol 4, No. 7. Joint Center for Poverty Research, Northwestern University. Website: **www.jcpr.org** •••

Children (AFDC), which was created in 1935 as part of the Social Security Act. Implemented during the Great Depression, AFDC was meant to assist poor mothers and their children. It acknowledged that some people are victimized by economic circumstances beyond their control and deserve assistance. For much of its lifetime, this law supported mostly White mothers and their children; not until the 1960s did welfare come to be identified with Black families.

The new welfare policy gives block grants to states to administer their own welfare programs through the program called **Temporary Assistance for Needy Families (TANF)**. TANF stipulates a lifetime limit of five years for people to receive aid and requires all welfare recipients to find work within two years—a policy known as *workfare*. Those who have not found work within two months of receiving welfare can be required to perform community service jobs for free. In addition, welfare policy denies payments to unmarried teen parents under eighteen years of age unless they stay in school and live with an adult. It also requires unmarried mothers to identify the fathers of their children or risk losing their benefits (Mink 1999). These broad guidelines are established at the federal level, but individual states can be more restrictive, as many have been. The very title of the new law, emphasizing personal responsibility and work, suggests that poverty is the fault of the poor. At the heart of welfare reform is the idea that public assistance creates dependence by discouraging people from seeking jobs.

Is welfare reform working? Many studies are finding that low-wage work does not lift former welfare recipients out of poverty (McCrate and Smith 1998). Critics of the current policy also argue that forcing welfare recipients to work provides a cheap labor force for employers and potentially takes jobs from those already employed. In the first few years of welfare reform, the nation was also in the midst of an economic boom; jobs were thus more plentiful. But in an economic downturn, those who are on aid or in marginal jobs can become even more vulnerable to economic distress, particularly given the time limits now placed on receiving public assistance (Albelda and Withorn 2002).

Research done to assess the impact of a changed welfare policy is relatively recent. Politicians brag that welfare rolls have shrunk; of course this would be true when people are denied benefits. Reducing the welfare rolls is a poor measure of the true impact of welfare reform. Because welfare has been decentralized to the state level, studies of the impact of current law have had to be done on a state-by-state basis. These studies show that those who have gone into workfare programs most often earn wages that keep them below the poverty line.

Although some states report that family income has increased, it is slight—and meager. (In Illinois, for example, where one of the most comprehensive studies was done, family income of former welfare recipients increased from only $7,475 to $11,812 following the implementation of TANF.)

There has also been an increase in the number of people evicted from housing because of falling behind on rent. Families also report an increase in other material hardships, such as phones and utilities being cut off. Marriage rates among former recipients have not changed, although more now live with nonmarital partners, most likely as a way of sharing expenses. The number of children living in families without either parent increased, probably because parents had to relocate to find work. In some states the numbers of people neither working nor receiving aid also increased (Lewis et al. 2002; Acker et al. 2002; Bernstein 2002).

The public debate about welfare rages on, often in the absence of informed knowledge from sociological research and almost always without input from the subjects of the debate, the welfare recipients themselves. Although stigmatized as lazy and not wanting to work, those who have received welfare actually believe that it has negative consequences for them, but they say they have no other viable means of support. They typically have needed welfare when they could not find work, or had small children and were without child care. Most were forced to leave their last job because of layoffs or firings, or because the work was only temporary. Few left their jobs voluntarily. Welfare recipients also say that the welfare system makes it hard to become self-supporting because the wages one earns while on welfare are deducted from an already minimal subsistence level. Furthermore, there is not enough affordable day care for mothers to leave home and get jobs. The biggest problem they face in their minds is lack of money. Contrary to the popular image of the conniving "welfare queen," welfare recipients want to be self-sufficient and provide for their families, but they face circumstances that make this very difficult to do (Hays 2003; Edin 1991; Popkin 1990).

Other beneficiaries of government subsidies have not experienced the same kind of stigma. Social Security supports virtually all retired people, yet they are not stereotyped as dependent on federal aid, unable to maintain stable family relationships, or insufficiently self-motivated. Spending on welfare programs is also a pittance compared with the spending on other federal programs. Sociologists conclude that the so-called welfare trap is not a matter of learned dependency, but a pattern of behavior forced on the poor by the requirements of sheer economic survival (Hays 2003; Edin 1991; Edin and Lein 1997).

Chapter Summary

What is social stratification and what forms does it take in society?

Status refers to a socially defined position in a group or society. *Social stratification* is a relatively fixed hierarchical arrangement in society by which groups have different access to resources, power, and perceived social worth. All societies have systems of stratification, although they vary in their composition and complexity. *Estate systems* are those where power and property are held by a single elite class. In *caste systems,* placement in the stratification is by birth, whereas in *class systems,* placement is determined by achievement.

Why is there inequality and what theories have sociologists developed to explain it?

Karl Marx saw class as primarily stemming from economic forces. *Max Weber* had a multidimensional view of stratification, involving economic, social, and political dimensions. Two theoretical perspectives are used in sociology to explain inequality—*functionalism* and *conflict theory.* Functionalists argue that social inequality motivates people to fill the different positions in society necessary for the survival of the whole, claiming that the positions most important for society require the greatest degree of talent or training and are, thus, most rewarded. Conflict theorists see social stratification as based on class conflict and blocked opportunity, criticizing functionalist theory for assuming that the most talented are those who get the greatest rewards and pointing out that those at the bottom of the stratification system are least rewarded because they are subordinated by dominant groups. These perspectives are critical to understanding contemporary debates about social policy.

What is social class?

Social class is the social structural position groups hold relative to the economic, social, political, and cultural resources of society. Social class can be imagined as a hierarchy, where income, occupation, and education are indicators of class, but classes are also organized around common interests and exist in conflict with one another.

Sociological research on the class system shows that the United States is a highly stratified society, with an unequal distribution of wealth and income.

Is social mobility a reality in the United States?

Class consciousness is the awareness that a class structure exists and the identification with others in one's class position. Perception of the class system depends on one's class position. The American dream that people can move up in the class system based on their ability assumes that movement from one class to another is possible. *Social mobility* is the movement between class positions. *Upward social mobility* is less common than is believed, and some people experience *downward social mobility.*

How extensive is poverty in the United States and who is most likely to experience it?

Many factors result in an underestimation of the actual extent of poverty, but poverty is extensive. The majority of the poor are women and their children, but there are diverse groups among the poor. A myth about the poor is that they do not want to work. Such ideas blame the poor for their own situation, reflecting a belief in the *culture of poverty* thesis.

Some people blame the poor for their own failures. Is this true?

The *culture of poverty* thesis is that poverty is the result of the cultural habits of the poor that are transmitted from generation to generation. Sociologists see poverty as caused by social structural conditions, including factors such as unemployment, gender inequality in the workplace, and the absence of support for child care for working parents. In recent years, homelessness has increased. Public debate about poverty focuses on the welfare system. Welfare recipients are stigmatized in ways that other beneficiaries of government support are not.

Key Terms

caste system 213	occupational prestige 220
class 214	
class consciousness 232	poverty line 234
class system 214	prestige 217
culture of poverty 237	social class 214
educational attainment 220	social differentiation 212
	social mobility 232
estate system 213	social stratification 212
false consciousness 232	socioeconomic status (SES) 220
family wage system 239	
feminization of poverty 236	status 212
income 220	status attainment 219
life chances 215	Temporary Assistance for Needy Families (TANF) 240
means of production 216	
median income 220	underclass 223
	wealth 220

Researching Society with MicroCase Online

You can see the results of actual research by using the Wadsworth MicroCase® Online feature available to you. This feature allows you to look at some of the results from national surveys, census data, and some other data sources. You can explore this easy-to-use feature on your own, but try this example. Suppose you want to know:

Does support for spending on social welfare vary by social class?

To answer this question, go to http://sociology.wadsworth .com/andersen_taylor4e/, select MicroCase Online from the left navigation bar, and follow the directions there to analyze the following data.

Data file: GSS

Task: Cross-Tabulation

Row Variable: WELFARE $

Column Variable: INCOME

Questions

Once you have your results, answer the following questions:

1. People in which income group are *most* likely to support more spending on social welfare?

2. What reasons can you think of to explain these differences?

Now analyze the following data to answer the rest of the questions.

Data file: GSS

Task: Auto-Analyzer

Primary Variable: Welfare $

3. Of the demographic groups (variables) included in this analysis, in which groups are there significant differences? List the variable names.

4. Does what you found in question 3 change your answer for question 2? Explain how these data do or do not support your previous explanation.

The Companion Website for Sociology: Understanding a Diverse Society, Fourth Edition

http://sociology.wadsworth.com/andersen_taylor4e/

Supplement your review of this chapter by going to the companion website to take one of the Tutorial Quizzes, use the flash cards to master key terms, and check out the many other study aids you'll find there. You'll also find special features such as GSS Data and Census 2000

information, data and resources at your fingertips to help you with that special project or do some research on your own.

Suggested Readings and Web Resources

Andersen, Margaret L., and Patricia Hill Collins. 2004. *Race, Class, and Gender: An Anthology,* 5th ed. Belmont, CA: Wadsworth.
This anthology explores the intersections of race, class, and gender as systems of stratification. A widely used anthology, the book includes personal narratives, as well as analytical accounts of race, class, and gender in the experiences of different groups.

Ehrenreich, Barbara. 2001. *Nickel and Dimed: On (Not) Getting By in America.* New York: Metropolitan Books.
Ehrenreich, a journalist, spent several months working in low-wage jobs. Her engaging, first-hand account of the indignities of this work shows how millions of workers struggle to make ends meet while performing some of the most socially devalued labor.

Hays, Sharon. 2003. *Flat Broke with Children: Women in the Age of Welfare Reform.* New York: Oxford University Press.
Based on an ethnography in two cities, Hays's analysis shows the impact of current social welfare policies on poor women and their children. Her analysis points to tension in the underlying values of American society as creating a punishing attitude toward women in need of assistance. It is a poignant and sociologically rich analysis of the impact of welfare reform.

Newman, Katherine. 1999. *No Shame in My Game: The Working Poor in the Inner City.* New York: Russell Sage Foundation/Vintage.
Newman's study of Harlem challenges the idea that inner-city communities are only comprised of poor, unemployed people. Using life histories that provide a close-up view of the people's lives in Harlem, she documents the experience of the working poor. It is a compelling and transforming portrait of Harlem and other working-class communities.

Oliver, Melvin L., and Thomas M. Shapiro. 1995. *Black Wealth/White Wealth: A New Perspective on Racial Inequality.* New York: Routledge.
By focusing their analysis on wealth, not just income, Oliver and Shapiro provide a compelling analysis of the source of continuing inequity in the class standing of African Americans and White Americans.

Pattillo-McCoy, Mary. 1999. *Black Picket Fences: Privilege and Peril Among the Black Middle Class.* Chicago: University of Chicago Press.
McCoy's study debunks the idea that the Black middle class now has it made. Instead, she shows how

tenuous the hold on middle-class status is for many and the continuing segregation and racism that marks this experience, even in a context where the middle class is more successful than in the past.

U.S. Census Bureau

www.census.gov

The U.S. Census Bureau publishes numerous reports that provide the best national data on income, poverty, and other measures that depict the class structure of the United States. In addition, the much-used reference book, Statistical Abstract of the United States, *is available online at this site.*

National Coalition for the Homeless

http://nch.ari.net

This site provides information on the extent of homelessness, as well as bibliographies and other information pertinent to the study of homelessness.

The Urban Institute

www.urban.org

The Urban Institute is a nonprofit policy research organization that analyzes society's problems and the efforts to solve them. This website provides analysis and commentary on topics such as welfare reform, urban poverty, and other important issues.

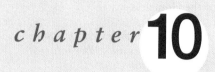

● ● ●

Global Stratification

"It takes a village to raise a child," the saying goes. But it also seems to take a world to make a shirt—or so it seems from looking at the global dimensions of the production and distribution of goods. Try this simple experiment: Look at the labels on your clothing. (If you do this in class, try to do so without embarrassing yourself and others!) What do you see? "Made in Indonesia," "Made in Bangladesh," "Made in Malawi," all indicating the linkage of the United States to systems of production around the world. The popular brand Nike, as just one example, has not a single factory in the United States, although its founder and chief executive officer is one of the wealthiest people in America. Nike products are manufactured mostly in China and Indonesia (Sanders and Kaptur 1997).

Taking your experiment further, ask yourself: Who made your clothing? A young person trying to lift his or her family out of poverty? Might it have been a child? In many areas of the world, one in five children under age fifteen work (International Labour Organization 2002). What countries benefit most from this system of production? Answering these questions reveals much about the interconnection among countries in the *global stratification* system, a system in which the status of the people in one country is intricately linked to the status of the people in others.

Recall from Chapter 1 that C. Wright Mills identified the task of sociology as seeing the social forces that exist beyond the individual. This is particularly important when studying global stratification. The person in the United States (or western Europe or Japan) who thinks he or she is expressing individualism by wearing the latest style is actually located in a global system of inequality whereby the adornments available to that person result from a whole network of forces that produce affluence in some nations and poverty in others.

Dominant in the system of global stratification are the United States and other wealthy nations. Those at the top of the global stratification system have enormous power over the fate of other nations. Although world conflict stems from many sources, including religious differences, cultural conflicts, and struggles over political philosophy, the inequality between rich and poor nations causes much hatred and resentment. One cannot help but wonder what would happen if the differences between the wealth of some nations and the poverty of others were smaller. In this chapter, we examine the dynamics and effects of global stratification. ● ● ●

Sociology ⊛ Now™
Reviewing is as easy as ❶❷❸.

Use SociologyNow to help you make the grade on your next exam. When you are finished reading this chapter, go to the chapter review for instructions on how to make SociologyNow work for you.

Global Stratification

In the world today, there are not only rich and poor people, but also rich and poor countries. Some countries are well-off, some countries are doing so-so, and a growing number of countries are poor and getting poorer. There is, in other words, a system of **global stratification** in which the units are countries, much like a system of stratification within countries in which the units are individuals or families. Just as we can talk about the upper-class or lower-class individuals within a country, we can also talk of the equivalent upper-class or lower-class countries in this world system. One manifestation of global stratification is the great inequality in life chances that differentiates nations around the world. Recall that in Chapter 2 we examined the United Nations human development index. This index and other compilations of international well-being show the great inequities that global stratification brings (see Map 10.1). Simple measures of well-being, including

life expectancy, infant mortality, and access to health services, reveal the consequences of a global system of inequality. And the gap between the rich and poor is sometimes greater in nations where the average person is least well-off. No longer can these nations be understood independently of the global system of stratification of which they are a part.

The effects of the global economy on inequality have become increasingly evident, as witnessed by public concerns about jobs being sent overseas. A coalition of unions, environmentalists, and other groups has also emerged to protest global trade policies that they believe threaten jobs and workers' rights in the United States, as well as contributing to environmental degradation. Such policies also encourage further McDonaldization (see Chapter 6). Thus, popular stores such as The Gap and Niketown have often been targets of political protests because they symbolize the expansion of global capitalism. Protestors see the growth of such stores as eroding local cultural values and spreading the values of

VIEWING SOCIETY IN GLOBAL PERSPECTIVE

MAP 10.1 Rich and Poor

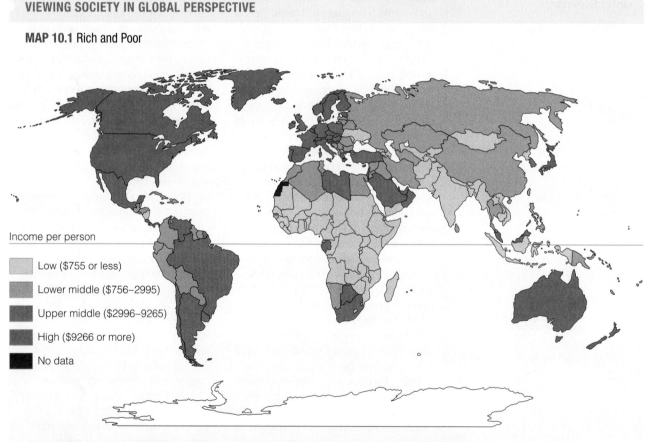

Income per person

- Low ($755 or less)
- Lower middle ($756–2995)
- Upper middle ($2996–9265)
- High ($9266 or more)
- No data

Most nations are linked in a world system that produces wealth for some, poverty for others. The GNI (gross national income), depicted here on a per capita basis for most nations in the world, is an indicator of the wealth and poverty of nations. Identify one of the nations represented here. Would you say this nation is a core, semi-peripheral, or peripheral country (see page 250) in the global economy? Why?

Source: World Bank. 2002. *World Bank Atlas 2002*. Washington, DC: World Bank, p. 42. Used by permission.

unfettered consumerism around the globe. Protests over world trade policies have also emerged in a student-based movement against companies that manufacture apparel with college logos. Calling for fair employment practices, students across the nation have protested the treatment of workers in sweatshops at home and abroad (see the box "Sociology in Practice: Symbols, Sportswear, and Sweatshops").

The relative affluence of the United States means that U.S. consumers have access to goods produced around the world. We can use a simple thing, such as a child's toy, to represent this global system. For many young girls in the United States, Barbie is the ideal of fashion and romance. Young girls may have not just one Barbie, but several, each with a specific role and costume. Cheaply bought in the United States but produced overseas, Barbie is manufactured by those probably not much older than the young girls who play with her and who would need all of their monthly pay to buy just one of the dolls that many U.S. girls collect by the dozens. In China, for instance, where more toys are produced than in any other part of the world, work-

© Marie Dorigny/REA./Corbis Saba

Many common products marketed in the United States are produced in a global economy, sometimes by child laborers.

ers molding Barbie dolls earn 25 cents per hour, and human rights organizations say violations of basic rights are flagrant (Press 1996: 12).

SOCIOLOGY IN PRACTICE
Symbols, Sportswear, and Sweatshops

In the late 1990s, as college students became aware that products bearing college logos were produced in sweatshops, a strong anti-sweatshop movement emerged on many U.S. campuses. Major companies pay universities royalties in exchange for the right to use the campus logo on shirts, hats, jackets, and other items. Students organized teach-ins, sit-ins, and demonstrations, calling for "sweat-free" campuses. Students at the University of Colorado organized a "Why Shop?" week in November 1999, posting a Web page with information about college logos, sweatshops, and garment workers in the United States and abroad. This large, student-led movement was mainly organized through electronic mailing lists and websites where students posted information about sweatshops and the garment industry worldwide. Noting that business was reaping over $2.5 billion in college merchandise sales, students demanded that workers be paid a living wage and that discrimination against women in this industry end. As a result, some universities adopted codes of labor rights ensuring fair practices for those companies

they license to sell products with college logos.

The U.S. General Accounting Office defines a **sweatshop** as a workplace where an employer violates more than one law regarding federal or state labor, industrial homework, occupational safety and health, workers' compensation, or industry regulation. The largest number of sweatshop workers in the United States are immigrant women in the garment industry who work sixty to eighty hours a week, often without minimum wage or overtime pay. The garment industry, though high-tech in the design phase, remains very low-tech in production, where usually women are required to sit at a sewing machine stitching. Work conditions are often dangerous, with blocked exits, unsanitary bathrooms, and poor ventilation. The government estimates 75 percent of U.S. garment shops violate safety and health laws. Workers are often intimidated from organizing or speaking out for fear of reprisal, such as job loss or deportation. Abroad, conditions are just as bad, if not worse. Compensation is extremely low (see Figure 10.1). Sociologists calculate that workers in countries such as

Mexico, Indonesia, or Vietnam can earn as little as 12 cents to 50 cents per hour. Labor costs are a tiny fraction of the retail price finally paid by the consumer. In addition, factories are typically unregulated and pose numerous health hazards for workers. Women working in these settings are sometimes also victims of sexual assault.

Critical Thinking Exercise

1. Where are the products bearing your college labels manufactured and sold? Who profits from the distribution of these goods? Has there been a movement in your community against these practices and, if so, what have been the results?

2. Go to two of the stores that are most popular with college students in your region. Make a list of the countries where their clothing products are made. Given the price of the clothing, what can you infer about global stratification?

For Further Information: Applebaum, Richard, and Peter Dreier. 1999. "The Campus Anti-Sweatshop Movement." *The American Prospect* 46 (September–October): 71.

●●●

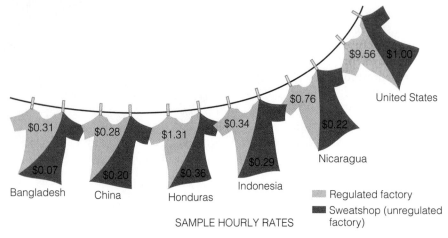

Figure 10.1 Factory Wages: Airing the Dirty Laundry

Data: From UNITE (Union of Needletrades, Industrial and Textile Employees) and Women International.

THINKING *SOCIOLOGICALLY*

Observe the evening news for one full week, noting the countries that are featured in different news stories. What images are shown for each country and what events are covered? What does this tell you about *global stratification*?

The manufacturing of toys and clothing are examples of the global stratification that links the United States and other parts of the world. Workers in the United States may lose jobs when companies export jobs. To see this, note that in 1973, more than 56,000 U.S. workers were employed in toy factories. Now, even

Global stratification often means that consumption in the more affluent nations is dependent on cheap labor in other less affluent nations.

though the market has become more glutted with the latest popular items, only 27,000 workers work in toy factories. The companies that make the toys amass profits, but U.S. workers lose jobs and then blame foreign workers for taking them. But who has gained? Indonesian workers making Barbies earn the minimum wage of $2.25 per day. It would take such a worker a full month to earn the money to buy the Calvin Klein Barbie (Press 1996). But the CEO of Mattel (the company that produces Barbie) earns millions, both in salary and stock options—exceeding the combined salaries of the Mattel workers who produce Barbie dolls in China.

Rich and Poor

One dimension of global stratification between countries is wealth. Enormous differences exist between the wealth of the countries at the top of the global stratification system and the wealth of the countries at the bottom. Although there are different ways to measure the wealth of nations, one of the most commonly used is the **per capita gross national income** (per capita GNI). This measures the total output of goods and services produced by residents of a country each year plus the income from nonresident sources, divided by the size of the population. This does not truly reflect what individuals or families receive in wages or pay; it is simply each person's annual share of their country's income were the proceeds shared equally. Per capita GNI is reliable only in countries that are based on a cash economy.

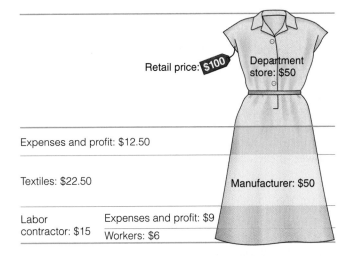

Figure 10.2 The Costs to Produce a Dress

Source: Applebaum, Richard, and Peter Dreier. 1999. "The Campus Anti-Sweatshop Movement." *The American Prospect* (September–October): 71.

It does not measure informal exchanges or bartering in which resources are exchanged without exchanging money. Measures of wealth based on the GN are less reliable among the poorer countries because non-cash transactions do not factor into the GNI calculation and are more common in poor, less industrialized nations.

The per capita GNI of the United States, one of the wealthier nations in the world, was $35,400 in 2002. The per capita GNI in Ethiopia, one of the poorest nations, was $100. Using per capita GNI as a measure of wealth, the average citizen of the United States is 354 times wealthier than the average citizen of Ethiopia. In

Afghanistan, the GNI was estimated to be $735 or less (World Bank 2004). The GNI provides a measure of the relative affluence of those living in the United States.

Which nations are the wealthiest? Figure 10.3 shows data from the World Bank listing the ten richest countries in the world measured by the annual per capita GNI in 2002. Bermuda is the richest country in the world on a per capita basis. The United States ranks sixth. Of course, Bermuda has a tiny population, whereas the United States has a much larger population. Besides the United States, most wealthy countries are in Western Europe. They are all mostly industrialized countries (or those that support them through banking) and, with the exception of Bermuda, Iceland, and the Channel Islands, they are mostly urban. These countries represent the equivalent of the upper class—even though many people within them are poor.

Now consider the ten poorest countries in the world, using per capita GNI as the measure of wealth. Several countries are even poorer than these countries, but they are so poor as to have no reliable statistics. Most of the world's poorest countries are in eastern or central Africa (see Figure 10.3). These countries have not become industrialized, are largely rural, and still depend heavily on subsistence agriculture. They have high fertility rates and rank at the bottom of the global stratification system.

On average, people in the poor countries are much worse off than people in the rich countries. In many poor countries with the largest populations, the life of an average citizen is desperate, and starvation is a growing problem. In a world with a population of nearly six billion, more than three billion—more than half the world's population—live in the poorest forty-five coun-

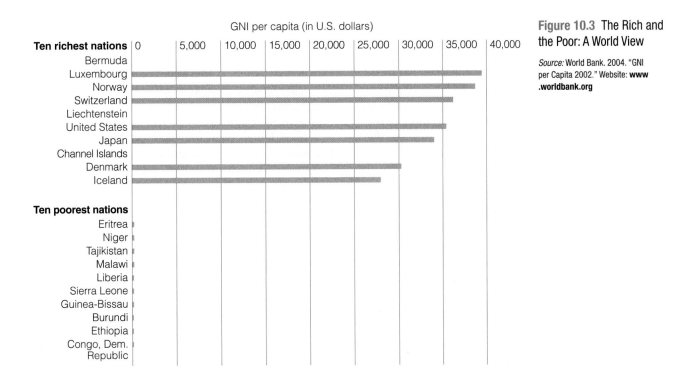

Figure 10.3 The Rich and the Poor: A World View

Source: World Bank. 2004. "GNI per Capita 2002." Website: **www.worldbank.org**

Global stratification means not only that enormous differences exist in the relative well-being of different countries throughout the world, but also that within many nations large numbers of people live in poverty like that of Mazar-i-Sharif refugee camp in Afghanistan, one of the poorer nations of the world.

tries. Often poor nations are rich with natural resources but are exploited for such resources by more powerful nations. We will look more closely at the nature and causes of world poverty later in this chapter.

The Core and Periphery

Global stratification involves nations in a large and integrated network of both economic and political relationships. **Power** is the ability of persons or groups—in this case, countries—to exercise influence and control over others; it is a significant dimension of global stratification. Countries can exercise several kinds of power over other countries, including military and political power. The **core countries** have the most power in the world economic system. These countries that include the powerful nations of Europe, the United States, Australia, and Japan control and profit the most from the world system and thus are the "core" of the world system.

Surrounding the core countries, both structurally and geographically, are the **semiperipheral countries.** The semiperipheral countries (such as Spain, Turkey, and Mexico) are semi-industrialized and, to some degree, represent a kind of middle class. Their middle position is marked by their middleman role in the world economic system—they extract profits from the poor countries and pass those profits on to the core countries.

At the bottom of the world stratification system, in this model, are the **peripheral countries**—the poor, largely agricultural, countries of the world. Even though

the countries are poor, they often have important natural resources that are exploited by the core countries. This exploitation, in turn, keeps them from developing and perpetuates their poverty. Often these nations are politically unstable; though they exercise little world power, political instability can cause a crisis for core nations dependent on their resources. Military intervention by the United States or European nations is often the result. This categorizing system emphasizes the power of each country in the world economic system.

THINKING *SOCIOLOGICALLY*

Look at the labels in your clothes and note where your clothing was made. What does this tell you about the relationship of *core, semiperipheral,* and *peripheral* countries within world systems theory? What further information would reveal the connections among the country where you live and the countries where your clothing is made and distributed?

Another way that countries are sometimes labeled is as first, second, and third world nations. This language grows out of the politics of the Cold War and reflects the political and economic dimensions of global stratification. **First-world countries** consisted of the industrialized capitalist countries of the world, including the United States, New Zealand, Australia, Japan, and the countries of western Europe. They are industrialized with a market-based economy and a democratically elected government. The **second-world countries** are socialist countries without democratically elected governments, including the former Soviet Union, China, Cuba, North Korea, and, prior to the fall of the Berlin Wall, the eastern European nations. During the Cold War, these countries had a communist-based government and a state-managed economy, as some still do. Although less developed than the first-world countries, the second-world countries tried to provide citizens with services such as free education, health care, and low-cost housing, consistent with the principles of socialism, but poverty often prevents them from doing so. **Third-world countries** in this scheme are poor, underdeveloped, largely rural, and have high levels of poverty. The governments of the third-world countries are typically autocratic dictatorships, and wealth is concentrated in the hands of a small elite class.

Because this system of categorization was based on the logic of the Cold War, it has changed. For instance, the oil-rich countries of the Middle East are not part of the first or second world according to this scheme, but since they have considerably more wealth, they do not belong in the same category as the poor countries of Africa and Asia. The collapse of the Soviet Union and the change in the governments of eastern Europe has led to the transformation of almost all the second-world countries. Although some countries still have a communist-based government, many of them, including China, are moving toward a market economic system. Still, like the core and periphery, the terms first-, second-, and third-world are useful in denoting the relationships of different countries to the world economy and global stratification.

Race and Global Inequality

In addition to class inequality, there is a racial component to world inequality, just as there is within many nations. The rich core countries, those that dominate the world system, are largely European, with the addition of the United States and Japan. The populations of these countries are mostly White. The poorest countries of the world are mostly in Africa, Asia, or South America, where the populations are mostly people of color. On average, there are vast differences in life chances and lifestyle between the countries of this world with White populations and those with Black populations.

Many have argued that this pattern reveals the influence of racism in world development. Walter Rodney argues, for example, in a book aptly titled *How Europe Underdeveloped Africa* (1974), that Europeans and North Americans have underdeveloped African nations by exploiting their resources and retarding their economic growth. African nations have historically been overtaken by external nations that profit from the appropriation of natural resources and leave the indigenous population in poverty. Likewise, Latin America has been subjected to exploitation by European and U.S. interests. A rich source of minerals and food, Latin America produces numerous products for wealthier nations. However, the wealthier nations profit more from the consumption and distribution of these goods than the Latin American countries that produce them (Weaver and Chilcote 2000). Western capitalism has exploited the human and natural resources of these regions, with people of color as the primary victims of western colonialism. More than one billion people in the world suffer from malnutrition and hunger. The vast majority of these people are people of color, that is, not of European descent (Uvin 1998).

How did racial inequality evolve on a global basis? Under capitalism, an international division of labor has emerged that is not tied to place, because cheap labor can be found in many countries. The exploitation of cheap labor in the poorest countries has created a global workforce that is mostly people of color; yet, the profits accrue to the wealthy owners, who are mostly White. This has resulted in a racially divided world. Even within the industrialized nations, the availability of low pay in other nations drives labor prices down—or produces unemployment. As a result, racist and xenophobic attitudes (i.e., fear and hatred of foreigners) are created that demonize those who produce the goods on which the capitalist world thrives. Some have argued further that the exploitation of the poor nations has forced an exodus of unskilled workers from the impoverished nations to the rich nations. This has created a flood of third-world refugees into the industrialized nations, increasing racial tensions, fostering violence, and destroying worker solidarity (Sivanandan 1995).

Comparing the United States, South Africa, and Brazil, Anthony Marx (1997) shows how racial groups fare in a country depending on how race is defined, particularly in the early periods of national development. Racial categories are not just based on biological differences (see Chapter 11), but are also constructed by powerful elites according to each country's history. As a result, countries define racial categories differently, and these definitions can have lasting effects on the ability of various racial groups to organize and fight for equality (Fredrickson 2003).

South Africa, the United States, and Brazil each developed different sets of racial categories. Although all three countries have many people of mixed descent, race is defined differently in each place. In South Africa, the particular history of Dutch and English colonialism led to strongly drawn racial categories that defined people as "White," "colored," or "African." In the United States, given its history of slavery, the "one drop" rule was used, which defined anyone with any African heritage as Black, ruling out any category of mixed race.

Brazil is yet a different case. The Brazilian elite declared Brazil a racial democracy at the early stages of national development. Racial differences were thought not to matter. Yet, instead of creating an egalitarian society free of racism, this resulted in Afro-Brazilians being of lower social status, while Euro-Brazilians remain at the highest social status. Thus, color continues to matter because it stratifies people. But, according to Marx, without strict racial categories, Afro-Brazilians are less likely to develop a strong group identity and thus are less likely to experience group solidarity.

However negative racial labels can be, Marx argues that the creation of racial labels may, in the long run, lead to more racial equality because they provide identities around which groups can politically mobilize. An implication of this for the United States is found in the debate over the racial categories. Some have argued not to create new multicultural or mixed-race categories because they might weaken Black American solidarity.

Theories of Global Stratification

Why have so many of the countries in the world fallen into poverty while other countries have developed and become wealthy? How do we understand the geographic distribution of wealth and poverty? Different theoretical perspectives provide answers to these questions. The theories used to explain global stratification are modernization theory, dependency theory, and world systems theory. Each contributes a unique explanation of global inequality (see Table 10.1).

Modernization Theory

Modernization theory views the economic development of countries as stemming from technological change. According to this theory, a country becomes more "modernized" by increased technological development, and this technological development is also dependent on other countries. Modernization theory was initially developed in the 1960s to explain why some countries had achieved economic development and why some had not (Rostow 1978).

Modernization theory sees economic development as a process by which traditional societies become more complex and differentiated. For economic development to occur, modernization theory predicts, countries must change their traditional attitudes, values, and institutions. Economic achievement is thought to derive from attitudes and values that emphasize hard work, saving, efficiency, and enterprise. According to this theory, these values are found in modern (developed) countries but are lacking in traditional societies. Modernization theory suggests that nations remain underdeveloped when traditional customs and culture discourage individual achievement and kin relations dominate.

As an outgrowth of functionalist theory, modernization theory derives some of its thinking from the work of Max Weber. In *The Protestant Ethic and the Spirit of Capitalism* (1958/1904), Weber saw the economic development that occurred in Europe during the Industrial Revolution as a result of the values and attitudes of Protestantism. The Industrial Revolution took place in England and northern Europe, Weber argued, because the people of this area were hard-working Protestants who valued achievement and believed that God helped those who helped themselves.

Modernization theory is similar to the argument of the culture of poverty, which sees people as poor because they have poor work habits, engage in poor time management, are not willing to defer gratification, and do not save or take advantage of educational opportunities (see Chapter 9). Countries are poor, in other words, because they have poor attitudes and poor institutions.

Modernization theory can partially explain why some countries have become successful. Japan is an example of a country that has made huge strides in economic development in part because of a national work ethic (McCord and McCord 1986). But the work ethic alone does not explain Japan's success. In sum, modernization theory may partially explain the value context in which some countries become successful and others do not, but it does not substitute for explanations that also look at the economic and political context of national development. It also rests on an arrogant perspective that the United States and other more economically developed nations have superior values compared to other nations. Critics point out that this perspective blames countries for being poor when other causes of their status in the world may be outside their control. Whether a country develops or remains poor may be the result of other countries exploiting the less powerful. Modernization theory does not sufficiently take into account the interplay and relationships between countries that can affect a country's economic or social condition.

Table 10.1
Theories of Global Stratification

	Modernization Theory	Dependency Theory	World Systems Theory
Economic Development	Arises from relinquishing traditional cultural values and embracing new technologies and market-driven attitudes and values	Exploits the least powerful nations to the benefit of wealthier nations that then control the political and economic systems of the exploited countries	Has resulted in a single economic system stemming from the development of a world market that links core, semiperipheral, and peripheral nations
Poverty	Results from adherence to traditional values and customs that prevent societies from competing in a modern global economy	Results from the dependence of low-income countries on wealthy nations	Is the result of core nations extracting labor and natural resources from peripheral nations
Social Change	Involves increasing complexity, differentiation, and efficiency	Is the result of neocolonialism and the expansion of international capitalism	Leads to an international division of labor that increasingly puts profit in the hands of a few while exploiting those in the poorest and least powerful nations

Developing countries, modernization theory says, are better off if they let the natural forces of competition guide world development. Free markets, according to this perspective, will result in the best economic order. But, as critics argue, markets do not develop independently of government's influence. Governments can spur or hinder economic development, especially as they work with private companies, to enact export strategies, restrict imports, or place embargoes on the products of nonfavored nations.

Dependency Theory

Market-based theories may explain why some countries are successful, but they do not fully explain why some countries remain in poverty or why some countries have not developed. **Dependency theory,** derived from the work of Karl Marx, focuses on explaining the persistence of poverty in the low-income countries as a direct result of their political and economic dependence on the wealthy countries. Specifically, dependency theory argues that the poverty of many countries in the world is a result of their exploitation at the hands of the powerful countries. Marx saw that a capitalist world economy would create an exploited class of dependent countries, just as capitalism within countries had created an exploited class of workers. The development of capitalism led to the European colonization of the world. Many of the poorest nations are former colonies of European powers.

The development schemes of the richest countries (such as the United States and western European nations) have resulted in the underdevelopment and poverty of the poor nations. The economies of these poor countries are often controlled and manipulated by powerful countries, which may also intervene in the political systems of the underdeveloped countries. The struggle for control of oil resources in the Middle East is an example of the process. Although nations like Iraq and Iran are not among the world's poorest countries, the wealth of their natural resource—oil—makes them likely targets for U.S. and western European intervention.

Dependency theory begins by examining the historical development of this system of inequality. As the European countries began to industrialize in the 1600s, they needed raw materials for their factories and they needed places to sell their products. To accomplish this, the European nations colonized much of the world, including most of Africa, Asia, and the Americas. **Colonialism** is a system by which western nations became wealthy by taking raw materials from colonized societies and reaping profits from products finished in the homeland. Colonialism worked best for the industrial countries when the colonies were kept undeveloped to avoid competition with the home country. For example, India was a British colony from 1757 to 1947. During that time, Britain bought cheap cotton from India, made it into cloth in British mills, and then sold the cloth back to India, making large profits. Although India was able to make cotton into cloth at a much cheaper cost than the British, and very fine cloth at that, the British nonetheless did not allow India to develop its cotton industry. As long as India was dependent on Britain, Britain became wealthy and India remained poor.

Under colonialism, dependency was created by the direct political and military control of the poor countries by powerful developed countries. Most colonial powers were European countries, but other countries, particularly Japan and China, had colonies as well. Colonization came to an end soon after the Second World War, largely because it became too expensive to maintain large armies and administrative staffs in distant countries. As a result, according to dependency theory, the powerful countries turned to other ways to control the poor countries and keep them dependent. The powerful countries still intervene directly in the affairs of the dependent nations by sending troops or, more often, by imposing economic or political restrictions and sanctions. But other methods, largely economic, have been developed to control the dependent poor countries, such as price controls, tariffs, and, especially, the control of credit.

The rich industrialized nations are able to set prices very low for raw material produced by the poor countries to prevent the poor countries from accumulating enough profit to industrialize. This form of international control has been called **neocolonialism,** a form of control of the poor countries by the rich countries, but without direct political or military involvement. As a result, the poor, dependent countries must borrow from the rich countries and debt creates only more dependence. Many poor countries are so deeply in debt to the major industrial countries that they must follow the economic edicts of the rich countries that loaned them the money, thus increasing their dependency.

Multinational corporations also play a role in keeping the dependent nations poor, dependency theory suggests. Although their executives and stockholders are from the industrialized countries, multinational corporations recognize no national boundaries and pursue business where they can best make a profit. Multinationals buy resources where they can get the cheapest price, manufacture their products where production and labor costs are the lowest, and sell their products where they can make the largest profits. From one point of view, multinational corporations are only following good business practices. Unfortunately, this strategy usually works to the detriment of the poor countries and to the advantage of the rich countries. Cheap resources and low-wage labor are found in poor countries, yet the profits end up in the core countries. Many people fault companies such as Nike for perpetuating global inequality by taking advantage of cheap overseas labor to make large profits for U.S. stockholders. Another interpretation is that Nike is doing what it should be doing in a market system: trying to make a profit. Nonetheless, dependency theory views the multinationals as maintaining poverty in many parts of the world.

Dependency theory focuses primarily on the connections between specific countries or groups of countries. It faults the powerful countries for causing poverty in the poor countries through colonization, neocolonialism, or the work of multinationals. The theory works well in explaining much of the poverty created in low-income countries in Africa and Latin America. Yet the theory does not explain development in other parts of the world.

One criticism of dependency theory is that many poor countries were never colonies, for example, Ethiopia. The involvement of multinational corporations does not always impoverish nations or increase their dependency. Some nations in which multinationals have built factories to exploit cheap labor have eventually moved up the development ladder by educating their workforce and thus generating more profit. Some have argued that multinationals do as much if not more economic damage to the industrialized nations by pulling out jobs and sending profits overseas. Two countries with the greatest postwar success stories of economic development are Singapore and Hong Kong. Both were British colonies—Hong Kong until 1997—and clearly dependent on Britain, yet they have had successful economic development precisely because of their dependence on Britain. Other former colonies are also improving economically, including India. It is not always the case that dependence leads to exploitation and economic backwardness.

World Systems Theory

Modernization theory examines the factors internal to an individual country, and dependency theory looks to the relationship between countries or groups of countries. Another approach to global stratification is called **world systems theory**. Like dependency theory, this theory begins with the premise that no nation in the world can be seen in isolation. Each country, no matter how remote, is tied in many ways to the other countries in the world. Both dependency theory and world systems theory are derived from *conflict theory* given that they interpret global stratification as a matter of different economic and political power balances between nations. However, unlike dependency theory, world systems theory argues that there is a world economic system that must be understood as a single unit, not as individual countries or groups of countries. This theoretical approach derives to some degree from the work of the dependency theorists and is most closely associated with the work of Immanual Wallerstein in *The Modern World System* (1974) and *The Modern World System II*

(1990). According to this theory, the level of economic development is explained by understanding each country's place and role in the world economic system.

This world system has been developing since the sixteenth century. The countries of the world are tied together in many ways, but of primary importance are the economic connections in the world markets of goods, capital, and labor. All countries sell their products and services on the world market, and they buy products and services from other countries. However, this is not a market of equal partners. Because of historical and strategic imbalances in this economic system, some countries are able to use their advantage to create and maintain wealth, whereas disadvantaged countries remain poor. This has led to a global system of stratification, in which the units are not people, but countries.

World systems theory sees the world divided into three groups of interrelated nations. The three categories were introduced earlier in the chapter. The *core countries* at the center of the world economic system are the rich, powerful industrialized capitalist countries that control the system. Around these nations are the *semiperipheral countries*. These countries occupy an intermediate position in the world system and include such countries as South Korea, Iraq, Iran, Mexico, and Turkey. The semiperipheral countries, which are at a middle level of income and partly industrialized, extract profits from the peripheral countries and pass the profits on to the core countries. The *peripheral countries* are poor, not industrialized, largely agricultural, and manipulated by the core countries that extract resources and profits from them. They are mostly in Africa, Asia, and Latin America.

This world economic system has resulted in a modern world in which some countries have obtained great wealth and other countries have remained poor. The core countries—western European countries, the United States, and Japan—make themselves wealthy by exploiting the resources of the peripheral countries at low prices, according to world systems theory. The core countries control and limit the economic development in the peripheral countries to keep the peripheral countries from developing and thus competing with the core on the world market. This allows the core countries to continue to purchase raw materials at a low price. As a result, peripheral countries suffer low wages, inefficiency, a large dependence on agriculture, poverty, and high levels of inequality, and they depend on the export of raw materials to the core countries.

World systems theory was originally developed to explain the historical evolution of global capitalism, but now also explains how differential profits are attached to the production of goods and services in the world market. Through an **international division of labor** products are produced globally, while profits accrue only for a few. A tennis shoe made by Nike is designed in the United States; uses synthetic rubber made from petroleum from Saudi Arabia; is sewn in Indonesia; is transported on a ship registered in Singapore, which is run by a Korean management firm using Filipino sailors; and is finally marketed in Japan and the United States. At each stage, profits are taken, but at a very different rate. Because of the intense competition in the world economic system, this approach suggests there is a drive to find the countries where wages are lowest and therefore to perpetuate and even deepen the cycle of poverty in the peripheral countries.

World systems theorists call this process a **commodity chain,** which is the network of production and labor processes by which a product becomes a finished commodity (Bonacich et al. 1994). By following a commodity through its production cycle and seeing where the profits go at each link of the chain, one can identify which country is getting rich and which country is being exploited.

The growing phenomenon of international migration is also explained by world systems theory. The international division of labor means that the need for cheap labor in some industrial and developing nations draws workers from poorer parts of the globe. International migration is also the result of refugees seeking asylum from war-torn parts of the world, such as Bosnia, or from countries where political oppression, often against particular ethnic groups, forces people to leave. The development of a world economy thus can

The development of a global economy results in much international migration as diverse groups seek opportunities in new parts of the world. One result is the presence of immigrant enclaves in world cities, such as Pakistani immigrants in London.

cause large changes in the composition of populations around the globe.

World cities, those that are closely linked through the system of international commerce, have emerged (such as London, New York, and Tokyo). Within these cities, families and their surrounding communities often form *transnational communities*—that is, communities

---

DOING SOCIOLOGICAL RESEARCH

Servants of Globalization: Who Does the Domestic Work?

International migration is becoming an increasingly common phenomenon. Women are one of the largest groups to experience migration, often leaving poor nations to become domestic workers in wealthier nations. Rhacel Salazar Parreñas studied two communities of Filipina women—one in Los Angeles, California and one in Rome, Italy—to learn how their experiences were part of the system of global stratification.

She conducted her research through extensive interviewing with Filipina domestics in Rome and Los Angeles, supplementing the interviews with participant observation in church settings, after-work social gatherings, and in employers' homes. The interviews were conducted in English and Tagalog—sometimes a mixture of both.

Parreñas found that Filipina domestics experienced, among other things, many status inconsistencies. They were

upwardly mobile in terms of their home country but were excluded from the middle-class Filipino communities in the communities where they lived. They experienced these feelings of social exclusion, even while they were separated from their own families. Their own families are transnational—that is, they often left family members, including their children, at home as they sought better employment. As part of the process of global restructuring, Filipinas provide some of the labor in more affluent households, even while their own lives are disrupted by new global forces.

Questions to Consider

1. Are there domestic workers in your community who provide child care and other household work for middle- and upper-class households? What is the race, ethnicity, nationality, and gender of these workers? What does

this tell you about the division of labor in domestic labor and its relationship to *immigration* and *race/class/gender stratification*? *Keywords: care work, domestic work and immigration, gender division of labor*

2. Why do you think domestic labor is so underpaid and undervalued? Are there social changes that might result in a reevaluation of the worth of such work? *Keywords: domestic workers, child-care workers*

We have included InfoTrac College Edition keywords at the end of each question to make it easier for you to find more to read on these topics. Go to www.infotrac-college.com, an online library, to begin your search.

Source: Parreñas, Rhacel Salazar. 2001. *Servants of Globalization: Women, Migration and Domestic Work.* Stanford, CA: Stanford University Press. ●●●

that may be geographically distant but socially and politically close. Linked through various communication and transportation networks, transnational communities share information, resources, and strategies for coping with the problems of international migration. Such international migration—sometimes legal, sometimes not—has radically changed the racial and ethnic composition of populations not only in the United States, but also in many European and Asian nations (Rodriquez 1999; Light et al. 1998). Some who migrate internationally are professional workers, but many others remain in the lowest segments of the labor force where, although their work is critical to the world economy, they are treated with hostility and suspicion, discriminated against, and stereotyped as undeserving and threatening. In many nations, including the United States, this has led to numerous political tensions over immigration, even while the emergence of migrant groups in world cities is now a major feature of the urban landscape (White 1998).

There are criticisms of world systems theory. It is very revealing to see the world as an interconnected set of economic ties between countries and to understand that these ties often result in the exploitation of poor countries. However, it is not clear that the system always works to the advantage of the core countries and to the detriment of the peripheral countries. Countries that were once at the center of this world system no longer occupy such a lofty position. England, for example, was once the most powerful nation in the world system; now the United States has that position. Countries like Holland and Italy no longer hold the world power they once had. These countries, though still part of the core, are no longer at the top of the global economy. World systems theory does not in

itself account for changes in the position of countries in the world system.

Critics also point out that the world economic system does not always work to the detriment of the peripheral countries and to the benefit of the core countries. Commodity chain theory finds that peripheral countries often benefit by housing low-wage factories and that the core countries are sometimes hurt when jobs move overseas. And low-wage sweatshops are found in all nations, not just the peripheral countries. In addition, some core countries, for instance, Germany, do not move factories to low-wage countries but improve profitability by increasing efficiency. Nonetheless, world systems theory has provided a powerful tool for understanding global inequality.

Consequences of Global Stratification

It is clear that some nations are wealthy and powerful and some are poor and powerless. What are the consequences of this global stratification system? Table 10.2 shows some basic indicators of national well-being for six nations. You can see there are considerable differences in the quality of life in these different places in the world.

Population

One of the biggest differences in rich and poor nations is population. The poorest countries comprise about three billion people—close to half the world's population. To put this into perspective, three of seven people in the world live on less than one dollar per day; another three billion live on less than $2 per day (World Bank 2004). The poorest countries have the highest birthrates and the highest death rates. The total *fertility rate,* how many live births a woman will have over her lifetime at current fertility rates, shows that in the poorest countries a woman on the average has almost five children over her lifetime. Because of this high fertility rate, the populations of poor countries are growing faster than the populations of wealthy countries and therefore have a high proportion of young children.

In contrast, the richest countries have a total population of approximately 1 billion people—only 15 percent of the world's population. The populations of the richest countries are not growing nearly as fast as the populations of the poorest countries. In the richest countries, women have about two children over their lifetime, and the populations of these countries are growing by only 1.2 percent. Many of the richest countries, including most countries of Europe, are actually experiencing population declines. With a low fertility rate, the rich countries have proportionately fewer children,

Increased awareness of the impact of globalization has generated a protest movement with an unusual alliance between those concerned about the loss of jobs in the United States and those concerned with the impact of globalization on the environment.

© Rachel Epstein/PhotoEdit

Table 10.2
Quality of Life: A Comparative Perspective

	Life Expectancy (in years)	Infant Mortality (per 1000 births)	Percentage of Women Enrolled in Elementary School	Child Malnutrition (percentage under weight)	Access to Safe Water (%)
Afghanistan	43 years	163 births	15%	49%	13%
Iran	69	33	74	11	95
Iraq	61	93	74	n/a	85
Mexico	73	29	100	8	86
United Kingdom	77	7	100	n/a	100
United States	77	7	95	1	100

Source: World Bank. 2002. *World Bank Atlas 2002.* Washington, DC: The World Bank, pp. 28–29.

but they also have proportionately more elderly, which can be a burden on societal resources. The populations in the poorest countries live in mostly rural areas, yet the richest countries are largely urbanized.

Rapid population growth as a result of high fertility rates can make a large difference in the quality of life of the country. Countries with high birthrates are faced with the challenge of having too many children and not enough adults to provide for the younger generation. Public services, such as schools and hospitals, are strained in high-birthrate countries, especially because these countries are poor from the start. However, very low birthrates, as many rich countries are now experiencing, can also lead to problems. In countries with low birthrates, there are often not enough young people to meet labor force needs, and workers must be imported from other countries.

Although the data clearly show that poor countries have large populations and high birthrates and rich countries have smaller populations and low birthrates, does this mean that the large population results in the low level of wealth of the country, or do high fertility rates keep countries poor?

Scholars are divided on the relationship between the rate of population growth and economic development (Cassen 1994; Demeny 1991). Some researchers theorize that rapid population growth and high birthrates lead to economic stagnation keeping a country from developing, thus miring the country in poverty (Ehrlich 1990). However, other researchers point out that some countries with very large populations have become developed (Coale 1986). After all, the United States has the third largest population in the world at 281 million people, and yet is one of the richest and most developed nations in the world. China and India, the two nations in the world with the largest populations, are also showing significant economic development. Scholars now believe large populations and high birthrates can impede economic development in some countries but, in general, fertility levels are affected by

levels of industrialization, not the other way around. As countries develop, their fertility levels decrease and their population growth levels off (Hirschman 1994; Watkins 1987).

Health and Environment

Significant differences are also evident in the basic health standards of countries, determined by their position in the global stratification system. The high-income countries have lower childhood death rates, higher life expectancies, and fewer children born underweight. In addition, most people in the high-income countries, but not all, have clean water and access to adequate sanitation. People born today in wealthy countries can expect to live about seventy-seven years, and women outlive men by several years. Except for some isolated or poor areas of the rich countries, almost all people have access to clean water and acceptable sewer systems.

In the poorest countries, the situation is completely different. Many children die within the first five years of life, people live considerably shorter lives, and fewer people have access to clean water and adequate sanitation. In the low-income countries, the problems of sanitation, clean water, childhood death rates, and life expectancies are all closely related. In many poor countries, drinking water is contaminated from inadequate or nonexistent sewage treatment. Contaminated water is then used to drink, to clean eating utensils, and to make baby formula. For adults, water-born illnesses such as cholera and dysentery sometimes cause severe sickness but seldom result in death. However, children under the age of five, and especially those under the age of one, are highly susceptible to the illnesses carried in contaminated water. Dehydration is brought on by diarrhea and is a common cause of childhood death in poor countries.

Degradation of the environment is a problem that affects all nations—linked in one vast environmental system. But global stratification also means that some nations suffer at the hands of others. Overdevelopment is resulting in deforestation. The depletion of this natural resource is most severe in South America, Africa, Mexico, and southeast Asia. On the other hand, the overproduction of "greenhouse gas"—the emission of carbon dioxide from the burning of fossil fuels—is most severe in the United States, Canada, Australia, parts of western Europe, and Russia—countries that use the most energy.

Although high-income countries have only 15 percent of the world population, together they use more than half of the world's energy. The United States alone uses one-quarter of the world's energy, although it holds only four percent of the world's population (see Figure 10.4). Safe water is also crucial; more than a billion people do not have access to safe water. Moreover, water supplies are declining—a problem that will only be exacerbated by population growth and economic development (see Figure 10.5). The World Bank has in fact warned that the share of the world's population facing water shortages could increase five-fold by the year 2050 (World Bank 2004). Clearly, global stratification has some irreversible environmental effects that are felt around the globe.

Education and Illiteracy

In the high-income nations of the world, education is almost universal, and the vast majority of people have attended school at least at some level. Literacy and school enrollment are now largely taken for granted in the high-income nations. People in these high-income countries without a good education stand little chance of success. In the middle- and lower-income nations, the picture is quite different. Elementary school enrollment, virtually universal in wealthy nations, is not as common in the middle-income nations and even less common in the poorest nations.

How do people survive who are not literate or educated? In much of the world, education takes place outside formal schooling. Just because many people in the poorer countries never go to school does not mean they are ignorant or uneducated. Most education in the

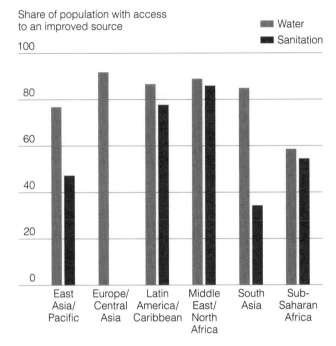

Figure 10.5 **Access to Safe Water and Sanitation**

Source: World Bank. 2004. *World Development Indicators.*
Website: **www.worldbank.org**, p.10.

world takes place in family and religious settings or in other places where elders teach the next generation needed skills and knowledge. Informal education often includes basic literacy and math skills that people in poorer countries need in their daily lives.

Although informal education prepares people for traditional lives, it does not usually give them the skills and knowledge needed to operate in the modern world. Therefore, as poor countries are confronted by the developing world, people may not have the skills and knowledge to adjust to changing world situations, particularly in an increasingly technological world. This can perpetuate their underdeveloped status.

Gender Inequality

The position of a country in the global stratification system also affects gender relations within different countries. Poverty is usually felt more by women than by men. Although gender inequality has not been achieved in the industrialized countries, compared with women in other parts of the world, women in the wealthier countries are much better off.

In the poorer countries, women suffer poorer health and less education. You can see this in something the United Nations calls the gender development index. The **gender development index** is calculated based on gender inequalities in life expectancy, educational attainment, and income for different countries. It provides an indication of the relative well-being of women

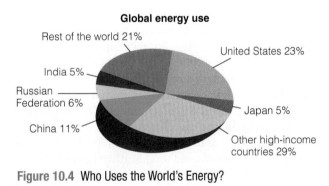

Figure 10.4 **Who Uses the World's Energy?**

Source: World Bank. 2004. *World Development Indicators.*
Website: **www.worldbank.org**, p. 114.

in different nations. In every nation, the gender development index is less than the general human development index. Were there gender equality within a nation, the human development index and gender development index would be the same. Many countries have shown improvement in the gender development index in recent years, including industrialized countries such as Denmark, France, and New Zealand, as well as Estonia, Hungary, and Poland in eastern Europe and the developing countries of Jamaica, Sri Lanka, and Thailand. The diversity of these countries shows that gender equality can be achieved at different income levels and in different stages of national development (United Nations 2000a).

In general, the developed nations have better sanitation, better health, more education, a lower birthrate, and smaller populations. The least developed countries of the world have poor health and sanitation, a much higher death rate, fewer people in school, and a large and growing population. The differences are profound and indicate a markedly unequal world.

War and Terrorism

The consequences of global stratification are also found in the international conflicts that bring war and an increased risk of terrorism. Although global inequality is certainly not the only cause of such problems, it contributes to the instability of world peace and the threat of terrorism. Global stratification generates inequities

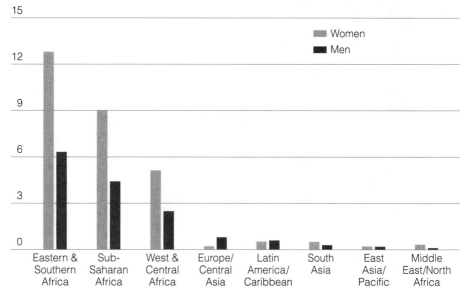

Youth ages 15–24 living with HIV/AIDS

Figure 10.6 AIDS: A Problem for Women and Children

Source: World Bank. 2004. *World Development Indicators.*
Website: **www.worldbank.org**, p.9.

in the distribution of power between nations. Moreover, globalization has created a world-based capitalist class with unprecedented wealth and power. This is a class that now crosses national borders, thus some have defined it as a "transnational capitalist class" (Langman and Morris 2002). Coupled with the enormous poverty that exists, the visibility of this class and its association with western values leads to resentment and conflict. Furthermore, attempts by wealthier nations to control access to the world's natural resources, such as oil, generate much political conflict. Thus, the same power and affluence that makes the United States a leader throughout the world makes it a target by those who resent its dominance.

TAKING ON SOCIAL ISSUES

Global Epidemics

Although disease is usually understood as primarily a medical problem, global epidemics involve much more than the medical treatment of disease. Epidemics like AIDS (acquired immune deficiency syndrome) or SARS (severe acute respiratory syndrome) develop in the context of world poverty, with the poorest nations often being the hardest hit by epidemics (see Figure 10.6). In addition, cultural norms,

technological and scientific resources, the status of women, political institutions, and other sociological factors all influence the ability of the medical community to respond to health crises. Thus, controlling or eliminating disease requires more than medical knowledge. It requires understanding the social and cultural context in which diseases like AIDS or SARS develop. What sociological factors have influenced the spread of

AIDS and SARS in different parts of the world and what can health organizations—or other groups—do to respond?

Taking Action

Go to the Taking Action Exercise on the Companion Website—at http://sociology.wadsworth.com/andersen_taylor4e/—to learn more about an organization that addresses this topic. •••

Global conflicts, including war, often result from the inequality that global stratification produces. Also, competition over the control of natural resources, such as oil, can lead to military intervention by more dominant world powers.

In the Middle East, for example, oil production has created prosperity for some and exposed people in these nations to the values of western culture. When people from different nations, such as those in the Middle East, study at U.S. universities and travel on business or vacations, they are exposed to western values and western patterns of consumption. As one commentator has noted, "Even those who have remained at home have not escaped exposure to western culture. In most of the countries of the modern Middle-east western cultural influences are pervasive. They see western television programs, they watch western movies, they listen to western music, frequently wear western clothes and visit western web sites on their pc's. Even western foods are locally available. McDonald's are now found in many of the major cities" (Bailey 2003: 341). Moreover, the sexual liberalism of western nations and the relative equality of women also add to the volatile mix of nations clashing (Norris and Inglehart 2002).

As a result, some traditional leaders, including religious clerics, define western culture as a source of degeneracy. Countries like the United States where consumerism is rampant then become the target of those who see this as a threat to their traditional way of life (Ehrlich and Liu 2002). In this sense, global stratification and the dominance of western culture are inseparable (Bailey 2002). Understood in this way, terrorism is not just a question of clashing religious values (although that is a contributing factor), but also stems from the global dominance of some nations over others. This is why those who commit atrocious acts, like the flying of jets into the World Trade Center towers, can define themselves as fighting for a righteous cause.

Terrorism can be defined as premeditated, politically motivated violence perpetrated against noncom-

batant targets by persons or groups who use their action to try to achieve their political ends (White 2002). Terrorism can be executed through violence or threats of violence and can be executed through various means—suicide bombs, biochemical terror, cyberterror, or other methods. Because terrorists operate outside the bounds of normative behavior, terrorism is very difficult to prevent. Although rigid safeguards can be put in place, such safeguards also threaten the freedoms that are characteristics of open, democratic societies. The fact that terrorism is so difficult to stop contributes to the fear that it is intended to generate.

Inequality is also connected to the context in which terrorism emerges. A study of al Qaeda terrorists finds that the leaders tend to come from middle-class backgrounds, although they often use those who are young, poorly educated, and economically disadvantaged to carry out suicide missions. Families of suicide bombers often receive large cash payments, at the same time feeling they have served a sacred cause (Stern 2003). This suggests that improving the lives of those who now feel collectively humiliated could provide some protection against terrorism.

World Poverty

One fact of global inequality is the growing presence and persistence of poverty in many parts of the world. There is poverty in the United States, but very few people in the United States live in the extreme levels of deprivation found in some poor countries of the world. In the United States, the poverty level is determined by the yearly income for a family of four that is considered necessary to maintain a suitable standard of living. As mentioned in Chapter 9, the official poverty line in 2002 (for a family of four) was $18,307. By this definition, 34.6 million Americans, or about 12 percent, were living in poverty in 2002 (Proctor and Dalaker 2003). This definition of poverty in the United States identifies **relative poverty.** The households in poverty in the United States are poor compared with other Americans, but when one looks at other parts of the world, an income of $18,307 would make a family very well-off.

The United Nations (UN) measures world poverty in two ways. **Absolute poverty** is the situation in which individuals live on less than $365 a year, meaning that people at this level of poverty live on approximately $1 a day. **Extreme poverty** is defined as the situation in which people live on less than $275 a year; that is, on less than 75 cents a day. There are 600 million people who live at or below this extreme poverty level. Many of these people are in very dire straits, and many are starving and dying.

Money, however, does not tell the whole story because many people in the poor countries do not always deal in cash. In many countries, people survive by rais-

Global stratification results in inequality between nations; the gap between the rich and the poor can be staggering.

ing crops for personal consumption and by bartering or trading services for food or shelter. These activities do not show up in the calculation of poverty levels that use amounts of money as the measure. As a result, the United Nations Development Program also defines what it calls the **human poverty index.** This index is a multidimensional measure of poverty, meant to indicate the degree of deprivation in four basic dimensions of human life: a long and healthy life, knowledge, economic well-being, and social inclusion. Different specific indicators of these four dimensions are used to measure poverty in the industrialized and developing countries and they can vary substantially in such different environments. In the developing countries, the following indicators are used:

- the percentage of people born not expected to live to age forty;
- the adult literacy rate;
- the proportion of people lacking access to health services and safe water;
- the percentage of children under five who are moderately or severely underweight.

In the industrialized countries, the human poverty index is measured by the proportion of people not expected to live to age sixty; the adult illiteracy rate; the incidence of income poverty (because income is the largest source of economic provisioning in industrialized countries); and long-term unemployment rates. Figure 10.7 compares the human poverty index in select developing and industrialized nations (United Nations 2000a).

Who Are the World's Poor?

Using the World Bank's definition of poverty (those whose level of consumption is less than $1 per day),

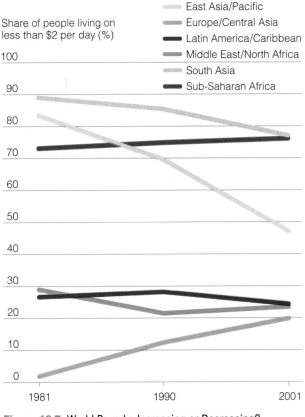

Figure 10.7 World Poverty: Increasing or Decreasing?

Source: World Bank. 2004. *World Development Indicators.*
Website: **www.worldbank.org,** p. 1.

about 28 percent of the world's population live in extreme poverty, but the proportion of people living in poverty is declining, particularly as the result of progress in East Asia, particularly in the People's Republic of China. At the same time, poverty in other areas has increased, including in eastern Europe and Central Asia.

Sub-Saharan Africa has the highest incidence of poverty of anywhere in the world, despite the rich natural resources of this region. Almost half of the population in this region—42 percent—live in poverty. As a result, infant mortality here is high; 151 children per 1000 die before the age of five. Life expectancy is also low and is getting lower because of high death rates from AIDS (World Bank 2004).

The character of poverty differs around the globe. In Asia, the pressures of large population growth leave many without sustainable employment. And, as manufacturing has become less labor-intensive with more mechanized production, the need for labor in certain industries has declined. While new technologies provide new job opportunities, they also create new forms of illiteracy as many people have neither access nor skills to use information technology (see Map 10.2 "The Digital Divide"). In sub-Saharan Africa, the poor live in marginal areas where poor soil, erosion, and continuous warfare create extremely harsh conditions. Political instability and low levels of economic productivity contribute to the high rates of poverty. Solutions to world poverty in these different regions require sustainable economic development, as well as an understanding of the diverse regional factors that also contribute to high levels of poverty.

Women and Children in Poverty

There is no country in the world in which women are treated as well as men. As with poverty in the United States, women bear a larger share of the burden of world poverty. In the poorest areas of the world, the poverty falls particularly hard on the women. For instance, in situations of extreme poverty, women have the burden of much of the manual labor because in many cases the men have left to find work or food. The United Nations Commission on the Status of Women (1996) estimates that women constitute almost 60 percent of the world's population, perform two-thirds of all working hours,

VIEWING SOCIETY IN GLOBAL PERSPECTIVE

MAP 10.2 The Digital Divide

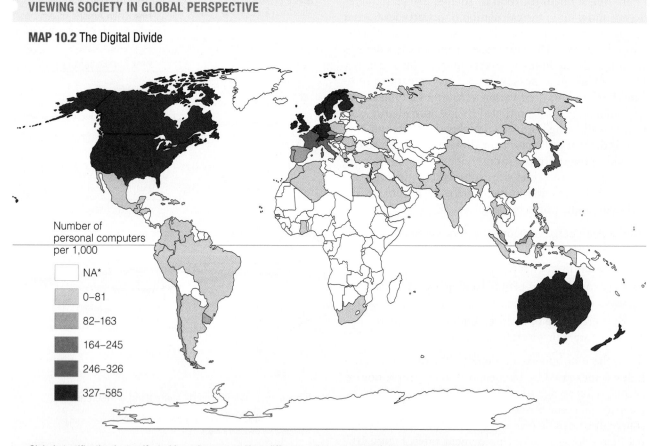

Number of personal computers per 1,000

- NA*
- 0–81
- 82–163
- 164–245
- 246–326
- 327–585

Global stratification is manifested in various ways. Here differences in access to information are depicted by the number of personal computers per 1000 persons in different nations. Access to computers and the Internet exposes you to a wider variety of world opinion. What advantages and disadvantages does access to digital technology provide for a country?

Source: U.S. Census Bureau. 2004. *Statistical Abstract of the United States 2003.* Washington, DC: U.S. Government Printing Office, p. 870.

receive only one-tenth of the world's income, and own less than 1 percent of the world's wealth.

Although women outlive men in most countries, the difference in life expectancy is less in poor countries. In some of the poorest countries, such as Bangladesh, men outlive women. Women suffer greater health risks because of several factors. With fertility rates higher in poor countries, women in poverty experience more pregnancies and childbirth. Poor women, therefore, spend a greater part of their lives pregnant, nursing, and raising small children than do women in the wealthier countries. These factors take a toll on women's health and increase the risks of disease and death. Giving birth is a time of high risk to women, and women in poor countries with poor nutrition, poor maternal care, and the lack of trained birth attendants are at higher risk of dying during and after the birthing event.

High fertility rates are also related to the degree of women's empowerment in society, an often neglected aspect of the discussion between fertility and poverty. Empowering women through providing them employment, education, property, and voting rights can have a strong impact on reducing the fertility rate (Sen 1999, 2000). Societies where women's voices do not count for much are those that have high fertility rates along with social and economic hardships for women, including lack of education, job opportunities, and information about birth control.

Women also suffer in some countries in poverty because of traditions and cultural norms. Most (though

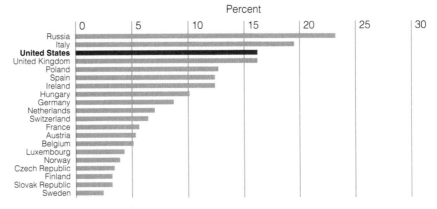

Figure 10.8 Child Poverty in the Wealthier Nations

Source: Vleminckx, Koen, and Timothy Smeeding (eds). 2001. *Child Well-Being, Child Poverty and Child Policy in Modern Nations.* Tonawanda, NY: University of Toronto Press.

not all) poor countries are patriarchal, meaning that men control the society. As a result, in households, women may have to eat after the men. Consequently, women have been found to have higher rates of malnutrition than men, which leads to anemia and other diseases (Doyal 1990). In conditions of extreme poverty, baby boys may be fed before baby girls because boys have higher status than girls. As a result, female infants have a lower rate of survival than male infants.

Children are also severely affected in the countries suffering from poverty (see Figure 10.8). Scholars disagree on the causal relationship between fertility rates and poverty, yet, the highest fertility rates are found in the areas with the greatest poverty. It may seem strange that couples would have more children in situations where many children are dying of starvation. Nonetheless, families in poverty, even extreme poverty, fre-

---

UNDERSTANDING DIVERSITY

War, Childhood, and Poverty

Surgeons were forced to amputate both of Ali Ismaeel Abbas's arms after an errant U.S. bomb slammed into his Baghdad home during the opening phase of the Iraq war. Pictures of the 12-year-old, who lost his parents in the attack, soon appeared on TV screens and in newspapers around the world. Since then, Abbas, who was treated in Kuwait, has come to represent a grim reality: all too often the victims of war are innocent children (McClelland 2003: 20).

In the past ten years alone, UNICEF estimates that over 2 million children have died in war, with even more injured, disabled, orphaned, or forced into refugee camps (Machel 1996). One estimate is that, of all the victims of war, 90 percent are civilian—half of those, children (McClelland 2003). Young children are often exploited as combatants or may be used as human shields.

In the aftermath of war, children are also highly vulnerable to outbreaks of disease. In Iraq, following the war in 2003, many children died of diseases like anemia and diarrhea—diseases

that can be prevented. Children in Iraq were already living under extreme hardship under the regime of Saddam Hussein. Economic sanctions against Iraq during his regime also produced high infant mortality because of food shortages. The United Nations has passed resolutions prohibiting the use of children under 18 in combat and has linked the threat to children from violence with high rates of poverty around the world. Although reducing poverty would not eliminate the threat of war, it would go a long way toward improving children's lives in war-torn regions. ●●●

quently do have many children and often need children as a source of family labor and income.

Children in poverty do not have the luxury of a safe and joyful childhood or an education. Schools often do not exist in poor areas of the world, and families are so poor that children must also work and cannot attend schools. From a very early age, children are required to help the family survive by working or performing domestic tasks such as fetching water. By the time they are five or six, or in some cases even younger, they must be working to support the family. By their early teens, they are usually on their own and must go out in the world to work. In extreme situations, children at a young age work as beggars, young boys and girls are sold to work in sweatshops, and young girls are sold into prostitution by their families. This may seem unusually cruel and harsh by Western standards, but we cannot imagine the horror of starvation and the desperation that many families in the world must feel that would force them to take such measures to survive.

Poverty has become a vicious cycle. In poor countries, families feel they must have more children for their survival, yet having more children perpetuates their poverty. The United Nations estimates that there are 211 million children between the ages of five and fourteen in the paid labor force throughout the world. Most of the children, some 127 million, are in Asia, and 48 million are in sub-Saharan Africa (International Labour Organization 2002).

Some families are so poor that they can no longer care for their children, and the children must go out on their own, even at young ages. Many of these homeless children end up in the streets of the major cities of Asia and Latin America. Latin America has an estimated 13 million street children, some as young as six years old. Alone, they survive through a combination of begging, selling, prostitution, drugs, and stealing. They sleep in alleys or in makeshift shelters. Their lives are harsh, brutal, often short, and they receive no formal education, leaving few prospects for a decent life (Mickelson 2000).

Poverty and Hunger

Malnutrition and hunger are growing problems, given that many people in poverty cannot find or afford food. An estimated 1.1 billion people in the world are so poor that they are unable to obtain enough food to meet their nutritional needs (Gardner and Halweil 1999).

Hunger results when there is not enough food to feed a designated area such as a region, a country, or area of a country. The food supply may not be adequate or households may not be able to afford enough food to feed themselves. Hunger stifles the mental and physical development of children and leads to disease and death. Malnutrition results from a deficiency in the necessary nutrients for healthy living. An estimated 6.6 million children under the age of five die each year

in the developing countries as the result of malnutrition. Although the food supply is plentiful in the world and is increasing faster than the population, malnutrition is still dangerously high (World Health Organization 2002).

DEBUNKING *SOCIETY'S MYTHS*

Myth: There are too many people in the world and simply not enough food to go around.

Sociological perspective: Growing more food will not in itself end hunger. If there were more systems to distribute the world's food, hunger could be reduced.

Why are people starving in the world? Is there not enough food to feed all the people in the world? In fact, there is plenty of food grown in the world. The world's production of wheat, rice, corn, and other grains is sufficient to adequately feed all the people in the world. The problem is that the surplus food in the world does not get to the neediest people. The people who are starving lack the exigencies for obtaining adequate food, such as arable land or jobs that pay a living wage. In the past, many people grew food crops and were able to feed themselves, but today so much of the best land has been taken over by agribusinesses that grow cash crops, such as tobacco or cotton. This has forced subsistence farmers onto marginal lands on the flanks of the desert, where conditions are difficult and crops often do not grow—another example of how the wealthiest nations and businesses profit at the expense of the least well-off.

Some areas of the world have seen a marked decrease in hunger. In China, for instance, 46 percent of the population were chronically underfed in 1970. Just twenty years later, only 16 percent of China's population were underfed. The only region of the globe where hunger has increased is sub-Saharan Africa.

Causes of World Poverty

What causes world poverty? Some think poverty results from overpopulation. Certainly, high fertility rates and poverty are related, but many of the world's most populous countries, India and China, for instance, have large segments of their population that are poor, but these countries have reduced poverty levels, even with very big populations. Poverty is also not caused by people being lazy or uninterested in working. People in extreme poverty work tremendously hard just to survive, and they would work hard at a job if they had one. It is not that they are lazy; it is that there are no jobs for them.

Poverty is caused by a number of factors. The areas where poverty is increasing have a history of unstable governments or, in some cases, virtually no effective government to coordinate national development or plans that might alleviate extreme poverty and starvation. World relief agencies are reluctant to work in or send food to countries where the national governments cannot guarantee the safety of relief workers or the delivery of food and aid to where it should go. Food convoys may be hijacked or roads blocked by bandits or warlords. The world witnessed this problem in the attacks on humanitarian aid agencies, such as the Red Cross, during the war in Iraq.

In many countries with high proportions of poverty, the economies have collapsed and the governments have borrowed heavily to remain in operation. As a condition of these international loans, lenders, including the World Bank and the International Monetary Fund, have demanded harsh economic restructuring to increase capital markets and industrial efficiency. These economic reforms may make good sense for some and may lead these countries out of economic ruin over time, but in the short run, these reforms have placed the poor in a precarious position because the reforms also called for drastically reduced government spending on human services.

Poverty is also caused by changes in the world economic system. Increases in poverty and starvation in Africa and Latin America can be attributed in part to the changes in world markets that favored Asia economically but put sub-Saharan Africa and Latin America at a disadvantage. As the price of products declined with more industrialization in places like India, China, Indonesia, South Korea, Malaysia, and Thailand, commodity-producing nations in Africa and Latin America suffered. In Latin America, the poor have flooded to the cities, hoping to find work, whereas in Africa, they did the opposite, fleeing to the countryside hoping to be able to grow subsistence crops. Governments often had to borrow to provide help to their citizens. Many governments collapsed or found themselves in such great debt that they were unable to help their own people, creating massive amounts of poverty and starvation.

In sum, poverty has many causes. It is now a major global problem that not only affects the billions of people who are living in poverty, but all people on Earth in one way or another, including the increased likelihood of war and violence that poverty brings. In some areas, poverty rates are declining as some countries begin to improve their economic situation. However, in other areas of the world, poverty is increasing, and countries are sinking into financial, political, and social chaos.

The Future of Global Stratification

As we have seen, some countries in the world are well-off, but many other countries in the world are poor, some very poor. This global stratification system has been created over several centuries, but the conditions of extreme poverty and even starvation that we see in the world system at this time are relatively new. Is the world getting better or worse, and what will happen in the future?

There is some good news. In some areas of the world, particularly East Asia but also in Latin America, many countries have shown rapid growth and have emerged as developed countries. These countries are sometimes called the **newly industrializing countries** (NICs), and they include Korea, Malaysia, Thailand, Taiwan, and Singapore. In these countries, individuals have saved and invested, and the governments have invested in social and economic development. Because some of the NICs have large populations, their success demonstrates that economic development can occur in heavily populated countries. China has embarked on an aggressive policy of industrial growth, and India is also improving economically.

Yet for all the success stories, some nations are not making it. In many cases, governments have collapsed or are functioning only at minimal levels, the economy is bankrupt, the standard of living is minimal, and people are starving. In many areas of the world, an increase in ethnic hatred has led to mass genocide and forced millions of refugees from their homes. These situations have increased poverty and hunger.

There has been continued growth of capitalism and of capital markets around the world. With the collapse of the Soviet Union and with China moving rapidly in the direction of capitalism, there has developed a world capitalistic system. The opening of new markets, the increasing global trade, the growth of multinational corporations, and the development of world financial markets will bring prosperity and wealth to many nations and to many individuals. The growing world market economy may allow some of the emerging countries that were once poor to move into the ranks of the rich nations and share in the newly created wealth, but whether this wealth will filter down to the people at the lower levels of society is another question.

The new push to further develop the world as one large capital market will also leave some countries behind, and therefore poverty and hunger will continue in many parts of the world. Market economies create opportunities to become wealthy, both for individuals and for nations. For those who can take advantage of these opportunities, the future looks bright, but many nations and individuals do not have this opportunity. Their conditions are so desperate that they do not have a chance to participate in the world market, except at a great disadvantage.

Globalization is a strong force that will continue to shape the future of most nations. Some see globalization simply as the expansion of western markets and culture into all parts of the world. Western civilization brings new values (including democracy and more equality for women) and certainly new products (movies, clothing styles, and other commercial goods) to other nations, but it also can be a form of *imperialism*—that is, the domination of powerful (in this case, Western) nations over others. Resistance to western globalization and imperialism produces some of the international problems now facing the United States, as evidenced in the hostility of militant, fundamentalist Islamic groups against western power and values.

Globalization has created great progress in the world—including trade, migration, the spread of diverse cultures, the dissemination and sharing of new knowledge, greater freedom for women, travel, and so forth. Moreover, globalization has not simply extended the values and knowledge of western culture. Many of the things we now take for granted in western culture originated in nonwestern cultures. For example, the decimal system—fundamental to modern math and science—originated in India between the second and sixth centuries, and was soon further developed by Arab mathematicians. Western societies certainly get credit for the development of science and technology, but the credit is not theirs alone (Sen 2002).

Sociology ⊛ Now™

Reviewing is as easy as ❶❷❸*.*

1. *Before you do your final review, take the SociologyNow diagnostic quiz to help you identify the areas on which you should concentrate. You will find information on SociologyNow and instructions on how to access all of its great resources on the foldout at the beginning of the text.*

2. *As you review, take advantage of SociologyNow's study videos and interactive Map the Stats exercises to help you master the chapter topics.*

3. *When you are finished with your review, take SociologyNow's posttest to confirm you are ready to move on to the next chapter.*

Chapter Summary

What is global stratification?

Global stratification is a system of inequality of the distribution of resources and opportunities between countries. A particular country's position is determined by its relationship to other countries in the world. The countries in the global stratification system can be categorized according to their *per capita gross national income*, or wealth. The world's countries can also be categorized as *first-, second-, or third-world countries*.

What explanations do sociologists offer for global stratification?

Modernization theory interprets the economic development of a country in terms of its internal attitudes and values. *Dependency theory* draws on the fact that many of the poorest nations are former colonies of European powers, which kept colonies poor and did not allow their industries to develop, thus creating dependency on the colonial powers. *World systems theory* argues that no nation can be seen in isolation and that there is a world economic system that must be understood as a single unit. The economic core—the industrialized countries of Europe, North America, and Japan—exploit the periphery, which is made up of the poor countries of the world.

What are some consequences of global stratification?

High birthrates, high mortality rates, poor health and sanitation, low rates of literacy and school attendance are common in the poorest nations. Although the richest countries constitute only 14 percent of the world's population, they have low birthrates, low mortality rates, better health and sanitation, high literacy rates, and high attendance in schools.

What do sociologists know about world poverty and how is it defined and measured?

Relative poverty means being poor in comparison to others. *Absolute poverty* describes the situation in which people live on less than $365 a year; extreme poverty, on less than $275 per year. Poverty particularly affects women and children.

What can we say about the likely future of global stratification?

The future of global stratification is varied and depends on the country's position within the world economic system. Some countries, particularly those in East Asia—commonly referred to as *newly industrializing countries*—have shown rapid growth and emerged as developed countries. Many nations, though, are not making it. Governments collapse, countries suffer economic bankruptcy, the standard of living plummets, and people starve.

Key Terms

absolute poverty 260
colonialism 253
commodity chain 255
core countries 250
dependency theory 253
extreme poverty 260
first-world countries 250
gender development index 258

global stratification 246
human poverty index 261
international division of labor 254
modernization theory 252
multinational corporation 253
neocolonialism 253

Researching Society with MicroCase Online

You can see the results of actual research by using the Wadsworth MicroCase® Online feature available to you. This feature allows you to look at some of the results from national surveys, census data, and other data sources. You can explore this easy-to-use feature on your own, but try this example. Suppose you want to know:

In what nations is there the greatest gap between rich and poor?

To answer this question, go to http://sociology.wadsworth .com/andersen_taylor4e/, select MicroCase Online from the left navigation bar, and follow the directions there to analyze the following data.

Data file: Nations

Analysis: Mapping

Variable: $ RICH 10%

View: Map

Questions

After viewing the map, return to "Rank Table." This table shows what percent of income the richest ten percent of the population in each nation receives.

1. In what five countries do the richest ten percent receive more than half of all national income?

2. What percent of income is received by the richest ten percent of these countries? (You may want to return to the alphabetical list to answer this question.)

United States _____
Iran _____
Japan _____
Mexico _____
South Africa _____

The Companion Website for Sociology: Understanding a Diverse Society, Fourth Edition

http://sociology.wadsworth.com/andersen_taylor4e/

Supplement your review of this chapter by going to the companion website to take one of the Tutorial Quizzes, use the flash cards to master key terms, and check out the many other study aids you'll find there. You'll also find special features such as GSS Data and Census 2000 information, data and resources at your fingertips to help you with that special project or do some research on your own.

Suggested Readings and Web Resources

Bonacich, Edna, and Richard Applebaum. 2000. *Behind the Label: Exploitation in the Los Angeles Apparel Industry.* Berkeley, CA: University of California Press.
This book examines the growth of sweatshops in the apparel industry, explaining their presence in terms of the rise in global production. The book also shows efforts to eradicate sweatshops.

Chang, Grace. 2000. *Disposable Domestics: Immigrant Women Workers in the Global Economy.* Boston, MA: South End Press.
Chang's book analyzes the patterns of international migration that affect women workers who leave their homelands, providing the care work for families in better-off nations. She links the provision of domestic work to the development of a global economy.

Friedman, Thomas L. 1999. *The Lexus and the Olive Tree.* New York: Farrar Straus & Giroux.
Friedman, an international correspondent for The New York Times, *examines the impact of the global economy on world politics and the character of local cultures.*

Louie, Miriam Ching Yoon. 2001. *Sweatshop Warriors: Immigrant Women Workers Take on the Global Factory.* Cambridge, MA: South End Press.
This book focuses on the organized resistance of women sweatshop workers. Told in their own words, it provides a good analysis of the conditions of this work and what women are doing to improve their situation.

Parreñas, Rhacel Salazar. 2001. *Servants of Globalization: Women Migration and Domestic Work.* Stanford, CA: Stanford University Press.
Parreñas analyzes the migration of Filipina workers into domestic labor in Los Angeles and Rome. Her study shows how this work is linked to a global system of labor and how it transforms family relations for Filipinas.

The United Nations
www.un.org
The United Nations home page includes a wealth of information about social, economic, and political conditions around the world. Their publications on economic and social development are a good source of data about global stratification.

The World Bank
www.worldbank.org
The World Bank site includes publications and other information about the global economy and how different nations are faring.

● ● ●

Race and Ethnicity

You might expect a society based on the values of freedom and equality not to be deeply afflicted by racial conflict, but think of the following situations:

• Eight days after the horrific, catastrophic September 11, 2001 terrorist attacks on New York City's now destroyed World Trade Center and the Pentagon in Washington, DC, a gunman drove into a Chevron gas station in Mesa, Arizona and shot to death the owner, a member of the Sikh religious order who wore a turban. The man who was killed had no known connection with suspected Middle Eastern terrorists, but he had dark skin and wore a turban. Within a few days, more than 200 Sikhs had reported instances of harassment.

• Some years ago in Detroit, Vincent Chin, a Chinese American, was clubbed to death with a baseball bat by a White autoworker and his son for presumably taking away jobs from White workers. Chin died within minutes from multiple skull fractures.

• In 1999, Amadu Diallo, a Black Haitian, standing in front of his own door, reached for his back pocket. Erroneously thinking that he was reaching for

a gun, four White undercover New York police officers fired forty-three separate shots at him in such rapid succession that a witness said it sounded like automatic weapons fire. Nineteen shots hit and killed Diallo instantly.

These ugly incidents, all true, have one thing in common—race. Along with gender and social class, race has fundamental importance in human social interaction, and it is an integral part of social institutions. Of course, the races do not always interact as enemies. Nor is interracial tension always obvious. It can be as subtle as the White person who simply does not initiate interactions with African Americans and Latinos, or the elderly White man who almost imperceptibly leans backward at a cocktail party as a Japanese American man approaches him.

In everyday human interaction, as African American philosopher Cornel West has eloquently noted, race still matters, and matters a lot (West 1993). What is race, and what is ethnicity? Why does society treat racial and ethnic groups differently, and why is there social inequality between these groups? How are these divisions and inequalities able to persist so stubbornly, and how

Sociology ⊛ Now™
Reviewing is as easy as ❶❷❸.

Use SociologyNow to help you make the grade on your next exam. When you are finished reading this chapter, go to the chapter review for instructions on how to make SociologyNow work for you.

extensive are they? These questions fascinate sociologists who research the causes and consequences of racial and ethnic stratification in our society. Just as class stratification differentiates people in society according to the class privileges and disadvantages that they experience, so do racial and ethnic stratification. •••

Race and Ethnicity

Within sociology, the terms *ethnic, race, minority,* and *dominant group* have very specific meanings, different from the meanings these terms have in common usage. These concepts are important to developing a sociological perspective on race and ethnicity.

Ethnicity

An **ethnic group** is a social category of people who share a common culture, for example, a common language or dialect; a common religion; and common norms, practices, customs, and history. Ethnic groups have a consciousness of their common cultural bond. Italian Americans, Japanese Americans, Arab Americans, Polish Americans, Greek Americans, Mexican Americans, and Irish Americans are examples of ethnic groups in the United States, but ethnic groups are also found in other societies, such as the Pashtuns in Afghanistan, or the Shi-

Activities such as this Puerto Rican Day Parade in New York City reflect pride in one's group culture and result in greater cohesiveness of the group.

This band, playing for St. Patrick's Day, an Irish holiday, contains people of varied racial–ethnic backgrounds.

ites and Sunnie in Iraq whose ethnicity is based on religious differences.

An ethnic group does not exist simply because of the common national or cultural origins of the group. Ethnic groups develop because of their unique historical and social experiences. These experiences become the basis for the group's *ethnic identity,* meaning the definition the group has of itself as sharing a common cultural bond. Italian immigrants, for example, did not necessarily think of themselves as a group with common interests and experiences prior to immigration to the United States. Originating from different villages, cities, and regions of Italy, they identified themselves by their family background and community of origin, but the process of immigration and the experiences Italian Americans faced in the United States created a new identity for members of this ethnic group (Waters 2002, 1990; Alba 1990).

The social and cultural basis of ethnicity is also proved by ethnic groups that can develop more or less intense ethnic identification at different points in time. Ethnic identification may grow stronger when groups face prejudice or hostility from other groups. Perceived or real threats from other groups may unite an ethnic

MAPPING AMERICA'S DIVERSITY

MAP 11.1 The New Immigration

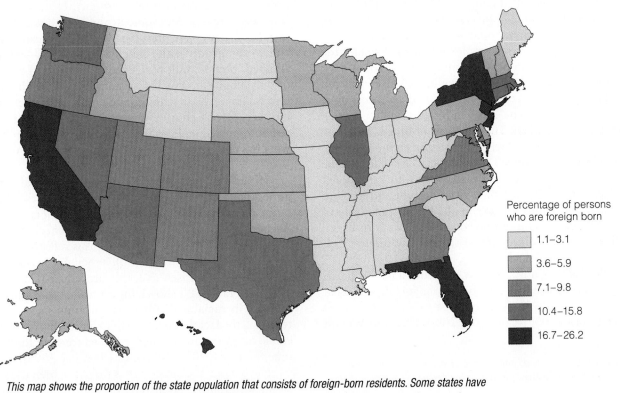

Percentage of persons
who are foreign born

	1.1–3.1
	3.6–5.9
	7.1–9.8
	10.4–15.8
	16.7–26.2

This map shows the proportion of the state population that consists of foreign-born residents. Some states have a high proportion (for example, California, Florida, New York), whereas other states have fewer (for example, Wyoming, Indiana, South Carolina). Where does your state fall in the proportion of foreign-born residents to the total population? Do you see an overall pattern regarding states which have a high proportion of foreign-born immigrants, relative to those that have less? Where do those states tend to be?

Source: U.S. Census Bureau. 2002. American Fact Finder. Website: **www.census.gov**

group around common political and economic interests. Ethnic unity can develop voluntarily or may be involuntarily imposed when ethnic groups are excluded by more powerful groups from certain residential areas, occupations, or social clubs. These exclusionary practices strengthen ethnic identity.

Race

Like ethnicity, race is primarily, though not exclusively, a socially constructed category. A **race** is a group treated as distinct in society on the basis of certain characteristics, some of which may be biological, that have been assigned social importance. Because of presumed biologically or culturally inferior characteristics (as regarded by powerful groups in the society), a race is typically singled out for differential and unfair treatment. It is not biological characteristics per se that de-

fine racial groups, *but how groups have been treated historically and socially.*

Society assigns people racial categories such as Black, White, and so on not by science, logic, or fact, but by opinion and social experience. In other words, how racial groups are defined is a *social* process. This is what is meant when we say that race is "socially constructed." Although the meaning of race begins from perceived biological differences between groups (physical characteristics like skin color, lip form, and hair texture), on closer examination, the assumption that racial differences are purely biological breaks down.

The social categories used presumably to divide racial groups are not fixed and vary from society to society (Washington 2004). Within the United States, laws defining who is Black have historically varied from state to state. North Carolina and Tennessee law historically defined anyone as Black who had even one great-grandparent who was Black (that is, being one-eighth Black). In other Southern states, having any Black ancestry at all (even just one great-great-great-grandparent) defined one as a Black person—the so-called one drop rule (that is, of "Black blood") (Taylor 2006; Malcomson 2000). This "one drop" rule still applies to a great extent today in the United States.

This is even more complex when we consider the meaning of race in other countries. In Brazil, a dark brown–skinned Black person could well be considered White, especially if the person is of high socioeconomic status, showing that "race" in Brazil is in fact *defined by* one's social class. Thus, in parts of Brazil, it is often said that "money lightens" (*o dinheiro embranquence*). In this sense, a category such as social class can become "racialized." In fact, in Brazil, people are considered Black only if they are of African descent and have no discernible White ancestry at all. The vast majority of U.S. Blacks would not be considered Black in Brazil (Surratt and Inciardi 1998; Omi and Winant 1994; Sowell 1983; Blalock 1982b).

Racialization is a process whereby some social category such as a social class or nationality takes on what are *perceived in the society* to be race characteristics (Harrison 2000; Malcomson 2000; Omi and Winant

Tiger Woods, considered among the greatest golfers ever, lines up a putt. He is of Asian American and African American parentage. What race is he?

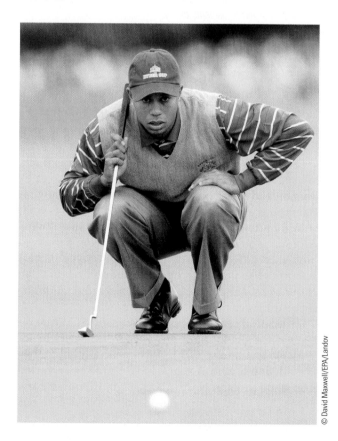

© David Maxwell/EPA/Landov

1994). The experience of Jewish people provides a good example of what it means to say that race is a socially constructed category. Jews are more accurately called an *ethnic group* because of common religious and cultural heritage, but in Nazi Germany, Hitler defined Jews as a "race." An ethnic group had thus become *racialized*. Jews were presumed to be biologically inferior to the group Hitler labeled the Aryans—white-skinned, blonde, tall, blue-eyed people. On the basis of this definition, which was supported through Nazi law, taught in Nazi schools, and enforced by the Nazi military, Jewish people were grossly mistreated: segregated, persecuted, and systematically murdered in what has come to be called the Holocaust during World War II.

Mixed-race people also defy the biological categories that are typically used to define race. Is the child of an Asian mother and an African American father Asian or Black? The current practice of checking several racial categories in the U.S. census reflects this issue (Wright 1994; Waters 1990), although considerable controversy has arisen over this procedure (Harrison 2000). As Table 11.1 shows, the census has dramatically changed its racial and ethnic classification since 1890, reflecting the fact that society's thinking about racial and ethnic categorization has not remained constant through time (Harrison 2000; Mathews 1996; Lee 1993).

The biological characteristics that have been used to define different racial groups vary considerably both within and between groups. Many Asians are lighter skinned than are many Europeans and White Americans, yet regardless of their light skin color, they have been defined in racial terms as yellow-skinned. Some African Americans are also lighter in skin color than some White Americans. Developing racial categories overlooks the fact that human groups defined as races are—biologically speaking—much more alike than they are different.

The differences in biological characteristics presumed to define racial groups seem somewhat arbitrary. Why, for example, do we differentiate people based on skin color and not other characteristics, such as height or hair color or eye color? You might ask yourself how a society based, for example, on the presumed racial inferiority of red-haired people would compare to racial inequalities in the United States. The likelihood is that if one powerful group defined another group as inferior because of certain biological characteristics and then used its power to create social institutions that treated this group unfairly, then a system of racial inequality would result. In fact, *very few specific biological differences exist* between racial groups. Most of the variability in almost all biological characteristics, such as blood type and various body chemicals, is *within* the group, not between various racial groups (Malcomson 2000; Lewontin 1996).

Table 11.1
Comparison of U.S. Census Classifications, 1890–2000

Census Date	White	African American	Native American	Asian American	Other Categories
1890	White	Black Mulatto Quadroon Octoroon	Indian	Chinese Japanese	
1900	White	Black	Indian	Chinese Japanese	
1910	White	Black Mulatto	Indian	Chinese Japanese	Other
1920	White	Black Mulatto	Indian	Chinese Japanese	Other
1930	White	Negro	Indian	Chinese Japanese Filipino Hindu Korean	Mexican Other
1940	White	Negro	Indian	Chinese Japanese Filipino Hindu Korean	Other
1950	White	Negro	American Indian	Chinese Japanese Filipino	Hawaiian Other
1960	White	Negro	American Indian Aleut Eskimo	Chinese Japanese Filipino	Hawaiian Other
1970	White	Negro or Black	Indian (American)	Chinese Japanese Filipino Korean	Hawaiian Other
1980	White	Black or Negro	Indian (American) Eskimo Aleut	Chinese Japanese Filipino Korean Vietnamese	Hawaiian Guamanian Samoan Other
1990	White	Black or Negro	Indian (American) Eskimo Aleut	Chinese Japanese Filipino Korean Asian Indian Vietnamese	Hawaiian Guamanian Samoan Asian or Pacific Islander Other
2000[a,b]	White	Black, African American, or Negro	American Indian or Alaskan Native	Chinese Japanese Filipino Korean Asian Indian Vietnamese Native Hawaiian Samoan Guamanian or Chamarro Other Pacific Islander	Other

[a]In 2000, for the first time ever, individuals could select more than one racial category. Only 2% actually did so.
[b]Hispanics are included under "Other."
Source: Lee, Sharon. 1993. "Racial Classification in the U.S. Census: 1890–1990." *Ethnic and Racial Studies* 16(1): 75–94. U.S. Census Bureau. 2003. "Racial and Ethnic Classification Used in Census 2000 and Beyond." Reprinted by permission of Taylor & Francis Ltd. **www.tandf.co.uk/journals/**.

Different groups use different criteria to define racial groups. To American Indians, being an American Indian depends upon proving Indian ancestry, but this varies considerably from tribe to tribe. Among some American Indians, one must be able to demonstrate 75 percent Indian ancestry to be recognized as American Indian; for other American Indians, demonstrating 50 percent Indian ancestry is sufficient. It also matters whether the government or the people define racial group membership. The government makes Indian tribes prove themselves as tribes through a complex set of federal regulations called the "federal acknowledgment process;" very few are actually given this official status. The criteria for tribal membership as well as definition as Indian or Native American have varied considerably across American history. Thus, as with African Americans, it has been a state or federal government, and not so much the racial–ethnic group *itself*, that has defined who is and is not a member of the group.

Official recognition by the government matters, though, because only those groups officially defined as Indian tribes qualify for health, housing, and educational assistance from the Bureau of Indian Affairs (the BIA) or are allowed to manage the natural resources on Indian lands and maintain their own system of governance (Locklear 1999; Brown 1993; Snipp 1989).

The definition of race emphasizes that what is important about race is not physical or cultural differences, but how definitions of race are created and maintained by the most powerful group (or groups) in society and what these presumed group differences mean in the context of social and historical experience. As a result, to define someone as a particular race is often as much a political question as it is a biological one. Many White Americans probably think of Blacks as a race but do not consciously think of themselves as having a racial identity or experience. Similarly, although Irish Americans probably did not think of themselves as a race, in the early twentieth century, they were defined by more powerful White groups as an inferior "race." At that time, Irish people were not considered by many even to be White (Ignatiev 1995)! In fact, a century ago the Irish were called "Negroes turned inside out," while Negroes (Black people) were called "smoked Irish" (Malcomson 2000).

The social construction of race has been elaborated in a new perspective in sociology known as racial formation theory (Omi and Winant 1994). **Racial formation** is the process by which a group comes to be defined as a race. This definition is supported through official social institutions such as the law and the schools. This concept emphasizes the importance of social institutions in producing and maintaining the meaning of race. It also connects the process of racial formation to the exploitation of so-called racial groups. A good example comes from African American history. During slavery, African Americans were defined as three-fifths of a person for purposes of deciding how slaves would be counted for state representation in the new federal government. The process of defining slaves in this way served the purposes of White Americans, not slaves themselves, and it linked the definition of slaves as a race to the political and economic needs of the most powerful group in society (Higginbotham 1978).

---

*DEBUNKING **SOCIETY'S MYTHS***

Myth: Racial differences are fixed, biological categories.

Sociological perspective: Race is a social concept, one in which certain physical or cultural characteristics take on social meanings that become the basis for racism and discrimination. The definition of "race" varies across cultures within a society and across different societies.

The process of racial formation also explains how groups such as Asian Americans, Native Americans, and Latinos have been defined as races, despite the different experiences and nationalities of the groups composing these three categories. Race, like ethnicity, lumps together groups that may have very different historical and cultural backgrounds, but once so labeled, they are treated as a single entity. This reflects a more general principle in the social sciences called **out-group homogeneity effect,** whereby all members of any out-group are perceived by an individual to be similar or even identical to each other, and differences among them are perceived to be minor or nonexistent (Taylor et al. 2003). This has recently been the case in the United States with Middle Easterners: Lebanese, Iranians, Iraqis, Jordanians, Egyptians, Afghans, and many others are classified as one group and called Middle Easterners, or simply "Arabs."

Minority and Dominant Groups

Minorities are racial or ethnic groups, but not all racial or ethnic groups are minority groups. Irish Americans, for instance, are not now a racial minority as they once were in the early part of the twentieth century. A **minority group** is any distinct group in society that shares common group characteristics and is forced to occupy low status in society because of prejudice and discrimination. A group may be defined as a minority on the basis of ethnicity, race, sexual preference, age, or class status. A minority group is not necessarily a numerical minority, but regardless of the size holds low status relative to other groups in society. In South Africa, Blacks outnumber Whites ten to one, but until Nelson Mandela's election as president and the dramatic change of government in 1994, Blacks were an officially oppressed and politically excluded social minority under the infamous *apartheid* (pronounced "apart-hate" or "apart-hite") system of government. The group that assigns a racial or ethnic group to subordinate status in society is called the **dominant group.**

In general, a racial or ethnic minority group has the following characteristics (Simpson and Yinger 1985):

1. The minority group possesses characteristics of race, ethnicity, social class, sexual preference, age, or religion that are popularly regarded as different from those of the dominant group.

2. The minority group suffers prejudice and discrimination by the dominant group.

3. Membership in the group is frequently ascribed rather than achieved, although either form of status can be the basis for identification as a minority.

4. Members of a minority group feel a strong sense of group solidarity. There is a "consciousness of kind," or "we feeling." This bond grows from common cultural heritage and the shared experience of recipients of prejudice and discrimination.

DEBUNKING *SOCIETY'S MYTHS*

Myth: Minority groups are those with the least numerical representation in society.

Sociological perspective: A minority group is any group, regardless of size, that is singled out in society for unfair treatment and generally occupies lower status in the society.

Racial Stereotypes

Racial and ethnic inequality is peculiarly resistant to change. In society, the inequality produces racial stereotypes, and these stereotypes become the lens through which members of the dominant group perceive members of the minority group.

Stereotypes and Salience

In everyday social interaction, people categorize other people. We all do it. The most common basis for such categorizations are race, gender, and age (Taylor et al. 2003; Worchel et al. 2000). We *immediately* identify a stranger as Black, Asian, Hispanic, White, and so on; as a man or woman; and as a child, adult, or elderly person. Quick and ready categorizations, even from momentary encounters, help us process the huge amounts of information we receive about the people we encounter. We quickly assign people to a few categories, saving ourselves the task of evaluating and remembering every discernible detail about a person. We are

taught to treat each person as a unique individual, but research over the years clearly shows that we do not and, instead, routinely categorize people.

A **stereotype** is an oversimplified set of beliefs about members of a social group or social stratum. Humans tend to categorize individuals of a group based on a narrow range of perceived characteristics. Stereotypes are presumed, usually incorrectly, to describe the "typical" member of some social group.

Stereotypes based on race or ethnicity are called *racial–ethnic stereotypes.* Examples are common: Asian Americans have been stereotyped as overly ambitious, sneaky, and clannish; at the turn of the twentieth century, Asians, particularly the Chinese, were stereotyped as dirty, ignorant, and untrustworthy. African Americans often bear the stereotype of being typically loud, lazy, naturally musical, and so on. Hispanics are stereotyped as lazy, oversexed, and, for men, macho. Jews have been portrayed as materialistic and unethical; Italians as overly emotional, argumentative, and prone to crime; the Irish as political, heavy drinkers, and quarrelsome. Such stereotypes do not apply to the vast majority of the members of the group. Germans were once stereotyped (even by Benjamin Franklin!) as clannish and unable to learn English; Swedes as loud and dirty; Polish as stupid and slow-witted. *No group in U.S. history has escaped the process of categorization and stereotyping.*

The categorization of people into groups and the subsequent application of stereotypes is based on the **salience principle,** which states that we categorize people on the basis of what appears initially prominent and obvious—that is, salient—about them. Skin color is a salient characteristic; it is one of the first things that we notice about someone. Because skin color is so obvious, it becomes a basis for stereotyping. Gender and age are also salient characteristics of an individual and thus serve as notable bases for group stereotyping.

THINKING *SOCIOLOGICALLY*

Observe several people on the street. What are the first things you notice about them? (That is, what is *salient*?) Make a short list of these things. Do these lead you to *stereotype* these people? On what do you base your stereotypes?

The choice of salient characteristics is culturally determined. In the United States, skin color, hair texture, nose form and size, and lip form and size have become salient characteristics. We use these features to categorize people in our minds on the basis of race. In other cultures, religion may be far more salient and take considerable priority over race. In the Middle East, whether one is Muslim or Christian is far more important than skin color. Religion in the Middle East is a salient characteristic and takes considerable priority over race.

Interplay Between Race, Gender, and Class Stereotypes

Alongside racial and ethnic stereotypes, gender and social class are among the most prominent features by which people are categorized. In our society, a complex interplay exists among racial–ethnic, gender, and class stereotypes.

Among *gender stereotypes,* those based on a person's gender, the stereotypes about women are more likely to be negative than those about men. The "typical" woman has been traditionally stereotyped as subservient, flighty, overly emotional, overly talkative, prone to hysteria, and inept at math and science. Many of these are *cultural stereotypes.* They are conveyed and supported by the cultural media—music, TV, magazines, art, and literature. Men, too, are painted in crude unflattering strokes. In the media they are stereotyped as macho, insensitive, and pigheaded and, in situation comedies, as inept. Generally, men are depicted as interested only in having sex with as many women as possible in the shortest time available.

Social class stereotypes are based on assumptions about social class status. Middle- and working-class people stereotype upper-class people as snooty, aloof, condescending, and phony. Some stereotypes held about middle-class people (by both the upper class and the working class) are that they are overly ambitious, striving, and obsessed with "keeping up with the Joneses." Stereotypes about working-class people abound: They are perceived by the upper and middle classes as lazy and unmotivated. Finally, the upper, middle, and working classes perceive lower or *underclass* individuals as inherently violent, dirty, and incapable of improving themselves.

The principle of **stereotype interchangeability** holds that stereotypes, especially negative ones, are often interchangeable from one gender to the other, from one social class to another, and especially from one racial or ethnic group to another, and also from a racial or ethnic group to a social class, and from a social class to a gender. Stereotype interchangeability is sometimes revealed in humor. Ethnic jokes often interchange different groups as the butt of the humor, stereotyping them as dumb and inept. Take the traditional negative stereotype of African Americans as inherently lazy. This stereotype has also been applied in recent history to Hispanics, Polish, Irish, Italians, and other groups. It has even been applied generally to all people perceived as "lower-class." In fact, "laziness" is often used to explain *why* someone is lower-class or poor. Middle-class people are more likely to attribute the low status of a lower-class person to something *internal,* such as lack of willpower or "inherent" laziness (Worchel et al. 2000, Krasnodemski 1996). Lower-class people are more likely to attribute their status to discrimination or poor opportunities, that is, to an *external* societal factor (Taylor et al. 2003; Krasnodemski 1996; Kluegel and Bobo 1993; Bobo and Kluegel 1991).

Similarly, research shows that White Americans by a two-to-one margin blame the "laziness" of urban Black poor for their own predicament, whereas the urban Black poor by a similar margin blame discrimination and racism. The more racially prejudiced the White person, the more likely he or she will blame the lower status of Blacks on internal factors, such as laziness and lack of willpower. Less prejudiced Whites are more likely to attribute status to external causes such as the social structure, racism, and discrimination.

The same kinds of stereotypes have historically been applied to women. Many stereotypes applied to women in literature and the media—they are childlike, overly emotional, and unreasoning—have also been applied to African Americans, people in lower social classes, the poor, and early in the twentieth century, Chinese Americans. A common theme is apparent: Whatever group occupies the lowest social status in society at a given time (whether racial–ethnic minorities, women, or lower-class people) is negatively stereotyped, and often the same negative stereotypes are used among these groups. The stereotype is then used to "explain" the observed behavior of a member of the stereotyped group, serving, incorrectly, as a *justification* for their lower status in society. This in turn subjects the stereotyped group to prejudice, discrimination, and racism, topics we now discuss.

Prejudice, Discrimination, and Racism

Many people use the terms *prejudice, discrimination,* and *racism* as if they all referred to the same thing. Typically, in common parlance, people also think of these terms in reference to individuals, as if the major problems of race were the result of individual people's bad will or biased ideas. Sociologists use more refined concepts to understand race and ethnic relations, distinguishing carefully among prejudice, discrimination, and racism.

Prejudice

Prejudice is the evaluation of a social group and the group members based on conceptions involving both prejudgment and misjudgment that are held despite facts that disprove them (Allport 1954; Pettigrew 1971;

Jones 1997). Thinking ill of people only because they are members of group X is prejudice. The negative evaluation arises solely because the person is seen as a member of group X, without regard to countervailing traits or characteristics the person may have.

Prejudices are usually defined by negative predispositions or evaluations but are occasionally positive. A negative prejudice against someone not in one's own social group is often accompanied by a positive prejudice in favor of someone who *is* in one's own group. Thus, the prejudiced person will have negative attitudes about a member of an *out-group* (any group other than one's own) and positive attitudes about someone simply because he or she is in one's *in-group* (any group one considers one's own).

Most people disavow racial or ethnic prejudice, yet the vast majority of us carry around some prejudices, whether about racial–ethnic groups, men and women, old and young, upper class and lower class, or straight and gay. Virtually no one is free of prejudice. Five decades of research have shown definitively that people who are more prejudiced are also more likely to stereotype others by race or ethnicity, and often by gender, than those who are less prejudiced (Adorno et al. 1950; Jones 1997; Taylor et al. 2003).

Prejudice based on race or ethnicity is called *racial–ethnic prejudice.* In-groups and out-groups in this case are defined along racial or ethnic lines. If you are a Latino and dislike an Anglo only because he or she is White, then this constitutes prejudice. It is a negative judgment (prejudgment) based on race and ethnicity and very little else. In this example, Latino is the in-group and White is the out-group. If the Latino individual attempts to justify these feelings by arguing that "all Whites have the same bad character," then the Latino is using a stereotype as justification for the prejudice. Note that prejudice can be held by any group against another.

Prejudice is also revealed in the phenomenon of **ethnocentrism,** which we examined in Chapter 3 on culture. Ethnocentrism is the belief that one's group is superior to all other groups. The ethnocentric person feels that his or her own group is moral, just, and right and that an out-group—and thus any member of the out-group—is immoral, unjust, wrong, distrustful, or criminal. The ethnocentric individual uses his or her own in-group as the standard against which all other groups are compared.

Prejudice and Socialization Where does racial–ethnic prejudice come from? How do moderately or highly prejudiced people end up that way? People are not born with stereotypes and prejudices. Research shows that they are learned and internalized through the socialization process, including both *primary socialization* (family, peers, and teachers) and *secondary socialization* (such as the media). Children imitate the attitudes of their parents, peers, and teachers. If the parent complains about "Japs taking away jobs" from Americans, then the child grows up thinking negatively about the Japanese, including Japanese Americans. Attitudes about race are formed early in childhood, at about age three or four (Allport 1954; Van Ausdale and Feagin 1996; Feagin 2000). There is a very close correlation between the racial and ethnic attitudes of parents and those of their children. The more ethnically or racially prejudiced the parent, the more ethnically or racially prejudiced, on average, will be the child.

UNDERSTANDING DIVERSITY

Arab Americans: Confronting Prejudice

Whenever violent confrontations in the Middle East reach one of their periodic surges, Marvin Wingfield braces for the worst: ethnic taunts, stereotypes in the mass media, and violence against Arab Americans. Such occurrences have only increased as a result of the terrorist attacks on the World Trade Center and the Pentagon on September 11, 2001.

"The pattern seems to be that whenever there is a crisis in the Mideast, the incidence of hate crimes against Arab Americans increases," says Wingfield, coordinator of conflict resolution for the American-Arab Anti-Discrimination Committee (ADC) in Washington, D.C.

Wingfield notes that these are a manifestation of centuries-old misperceptions in the West about the nature of Arab culture. Colonial arrogance, he says, fostered stereotypes of Arabs as camel-riding hedonists and devious traders.

Another offshoot is the tendency of many White Americans to view broad segments of the population as *homogenous.* References to "the Black community," for example, ignore the diversity among African Americans and their interests. Similarly, "the Arab world" is vast and varied. . . .

The ADC has prepared guidelines for community activists to follow in trying to increase the presence and awareness of Arab culture in school districts. Called "Working With School Systems," it describes various scenarios for involvement, ranging from volunteering at the classroom level to lobbying for districtwide curriculum development.

"Our program is multifaceted," Wingfield says. "It's not a curriculum that can be taken into the schools. Instead, we have many different people doing many different things. We encourage people to take action on their own horizons."

Source: Teaching Tolerance (Spring 1997): 49. This article is reprinted by permission of *Teaching Tolerance*, a publication of the Southern Poverty Law Center. •••

A major vehicle for the communication of racial–ethnic attitudes to both young and old are the media, especially television, magazines, newspapers, and books. For many decades, African Americans, Hispanics, Native Americans, and Asians were rarely presented in the media and then only in negatively stereotyped roles. In the 1950s, the Chinese were shown in movies, magazines, and early television as bucktoothed buffoons who ran shirt laundries. Japanese were depicted as sneaky and untrustworthy. Hispanics were shown as either ruthless banditos or playful, happy-go-lucky characters who took long siestas. American Indians were presented as either villains or subservient characters such as the Lone Ranger's famed sidekick, Tonto. Finally, there is the drearily familiar portrayal of the Black person as subservient, lazy, clowning, and bug-eyed, a stereotypical image that persisted from the late nineteenth century all the way through the 1950s and early 1960s (Thibodeau 1989).

Discrimination

Discrimination is overt negative and unequal treatment of the members of a social group or stratum solely because of their membership in that group or stratum. Prejudice is an attitude, discrimination is behavior, and the two do not always go together. *Racial–ethnic discrimination* is unequal treatment of a person on the basis of race or ethnicity. The discrimination affecting the nation's minorities takes a number of forms; for example, income discrimination and discrimination in housing. Discrimination in employment and promotion (discussed in Chapter 18) and discrimination in education (see Chapter 16) are two other forms of discrimination. Income discrimination is seen in the persistent differences in median income that appear when comparing different racial groups.

Housing discrimination has placed a particular burden on minorities. Many studies have been able to reproduce the situation where two persons identical in nearly all respects (age, education, gender, social class, and other characteristics) both present themselves as potential tenants for the same housing, and if one is White and the other is a minority, the minority person will often be refused housing by a White landlord when the otherwise identical White applicant will not. A minority landlord refusing housing to a White person while granting it to a minority person of similar social characteristics is also discriminating, but reverse discrimination of this sort is far less frequent and far less of a problem in society (Feagin 2000; Feagin and Feagin 1993; Feagin and Vera 1995).

Housing discrimination is illegal under U.S. law. But, banks and mortgage companies often withhold mortgages from minorities based on "red lining," an illegal practice in which an entire minority neighborhood is designated "no-loan." Racial segregation may also be fostered by *gerrymandering,* the calculated redrawing of election districts, school districts, and similar political boundaries to maintain racial segregation. As a result, **residential segregation,** the spatial separation of racial and ethnic groups into different residential areas, continues to be a reality in this country—called "American apartheid" (Massey and Denton 1993).

Income discrimination is evident in the median income of Black and Hispanic families. Although it has increased somewhat since 1950, the income gap between these groups and Whites has remained virtually unchanged since 1967, as can be seen from Figure 11.1. Furthermore, per capita income has grown at a faster rate for Whites. Yet even these median income figures tell only part of the story. Poverty among Blacks decreased from 1960 to 1970, remained fairly steady until the mid-1990s when it decreased again, but then began increasing in 2000. The current poverty rate (the percent of the population below the poverty level; see Figure 11.2) is highest for African Americans (24 percent) and Hispanics (22 percent), and least for White Americans (10 percent) and Asian-Americans (10 percent).

Racism

Prejudice is an attitude (what you *think*); discrimination is behavior (what you *do*). Racism includes both attitudes and behaviors. **Racism** is the perception and treatment of a racial or ethnic group, or member of that group, as intellectually, socially, and culturally inferior to one's own group.

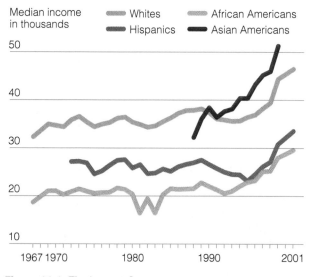

Figure 11.1 The Income Gap

Source: DeNavas-Walt, Carmen et al. 2003 *Income in the United States: 2002.* Washington, DC: U.S. Census Bureau.

There are different **forms of racism.**
Obvious, overt racism, such as physical assaults, from beatings to lynchings, has often
been called *old-fashioned racism,* or *traditional racism* (also *Jim Crow racism*) by researchers (for example, Bobo 1999; Sears
et al. 1997). This form of racism has declined somewhat in our society since the
1950s, though it certainly has not disappeared (Schumann et al. 1997). Racism
can also be subtle, covert, and nonobvious;
this is known as **aversive racism,** another
form of racism (Jones 1997). Consistently
avoiding interaction with someone of another race or ethnicity is an example of
aversive racism. Mischievously imitating
the speech of Blacks, Hispanics, or Asians
even in private conversation is another example of aversive racism. This form of racism is quite common and has remained at roughly the
same level for more than thirty years, with perhaps
a slight increase (Schumann et al. 1997; Kovel 1970;
Dovidio and Gaertner 1986; Katz et al. 1986).

After the Second World War and during the 1950s,
a shift to **laissez-faire racism** occurred. This type of
racism—also called *symbolic racism* by some—involves
several elements:

1. the subtle but persistent negative stereotyping of
   minorities, particularly in the media, which still
   persists to a great degree today;

2. a tendency to blame Blacks, Hispanics, and Native
   American Indians *themselves* for the gap between
   minorities and Whites in socioeconomic stand-

Minorities are more likely to be arrested than Whites for the same
offense, as we saw in Chapter 8. Does this tend to reflect institutional
racism rather than any individual prejudice of the arresting police
officer?

ing, occupational advancement, and educational
achievement;

3. clear resistance to meaningful policy efforts (such
   as affirmative action, discussed later) designed to
   ameliorate American's racially oppressive social
   conditions and practices.

The last element is rooted in perceptions of threat by
the minority to maintaining the status quo (Bobo 1999;
Bobo and Smith 1998).

A close relative of laissez-faire racism is **color-blind
racism**—so named because the individual affected by
this type of racism prefers to ignore legitimate racial–
ethnic, cultural, and other differences and insists that
the race problems in American will go away if only race
is ignored altogether. Accompanying this belief is the
opinion that race is not real. Simply refusing to perceive
any differences at all between racial groups (thus, being "color-blind") is in itself a form of racism. Many
of these types of racism do not necessarily involve explicit or purposeful intent on the part of the nonminority individual to harm the minority person. At least
one researcher (Bonilla-Silva 2001) equates traditional
liberalism with color-blindness—the liberal belief that
racial-cultural differences are irrelevant to people's lives
and their overall well-being.

Institutional racism as a form of racism is negative
treatment and oppression of one racial or ethnic group
by society's existing institutions based on the presumed
inferiority of the oppressed group. It is a form of racism that exists at the level of social structure and is in

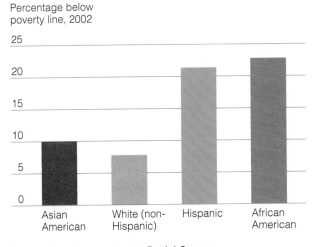

Percentage below
poverty line, 2002

Figure 11.2 Poverty Among Racial Groups

Source: Proctor, Bernadette, and Joseph Dalaker. *Poverty in the United States 2002.*
Washington, DC: U.S. Census Bureau.

Durkheim's sense *external* to the individual—thus, institutional. Key to understanding institutional racism is seeing that dominant groups have the economic and political *power* to subjugate the minority group, *even if they do not have the explicit intent* to show prejudice or discrimination. Power, or lack thereof, accrues to groups because of their position in social institutions, not just because of individual attitudes or behavior. The power that resides in society's institutions can be seen in such patterns as persistent economic inequality between racial groups (which is reflected in high unemployment among minorities, lower wages, and different patterns of job placement).

Racial profiling, already discussed in Chapter 8, is an example of institutional racism. African American and Hispanic people are arrested by police considerably more often than Whites and Asians, although it is not necessarily the fault of the arresting officer as an individual. In fact, an African American or Hispanic wrongdoer is more likely to be arrested than a White man who commits the *same crime,* even when the White man shares the same age, socioeconomic environment, and prior arrest record as the Black or Hispanic.

Institutional racism is also seen in educational institutions, such as in schools assigning Blacks and Latinos to lower tracks than Whites with the *same* ability test scores, as will be seen in Chapter 16. In these instances, racism is a characteristic of the institutions and not necessarily of the individuals within these institutions. *This is why institutional racism can exist even without prejudice being the cause.*

DEBUNKING *SOCIETY'S MYTHS*

Myth: The primary cause of racial inequality in the United States is the persistence of prejudice.

Sociological perspective: Prejudice is one dimension of racial problems in the United States, but institutional racism can flourish even while prejudice is on the decline.

Consider this: Even if every White person in the country lost all of his or her prejudices and even if he or she stopped engaging in individual acts of discrimination, institutional racism would still persist for some time. Over the years, it has become so much a part of U.S. institutions (hence, the term *institutional* racism) that discrimination can occur even when no single person is causing it. Existing at the level of social structure rather than the level of individual attitude *or* behavior, it is external to the individual personality and is thus a

---

DOING SOCIOLOGICAL RESEARCH

American Apartheid

The term *apartheid* was used to describe the society of South Africa prior to the election of Nelson Mandela in 1994. It refers to the rigid separation of the Black and White races. Sociological researchers Massey and Denton argue that the United States is now under a different kind of apartheid, and that it is based on a very rigid residential segregation in the country.

Massey and Denton note that the terms *segregation* and *residential segregation* were rarely used in the American vocabulary in the late 1970s and early 1980s. These terms were spoken little by public officials, journalists, and even civil rights officials. This was because the ills of race relations in America were at the time attributed to causes such as a "culture of poverty" among minorities, or inadequate family structure among Blacks, or too much welfare for minority groups. The Fair Housing Act of 1968, intended to decrease housing discrimination, was thought to have solved the problem.

Nothing could be farther from the truth, according to researchers Massey and Denton, who amassed a large amount of data demonstrating that residential segregation not only has persisted in American society, but that it has actually increased since the 1960s. Most Americans vaguely realize that urban America is still residentially segregated, but few appreciate the depth of Black and Hispanic segregation or the degree to which it is maintained by ongoing institutional arrangements and contemporary individual actions. They find that most people think of racial segregation as a faded notion from the past, one that is decreasing over time. Today theoretical concepts such as the culture of poverty, institutional racism, and welfare are widely debated, yet rarely is residential segregation considered a major contributing cause of urban poverty and the underclass. Massey and Denton argue that their purpose is to redirect the focus of public debate back to race and racial segregation.

Questions to Consider

1. How racially segregated is the neighborhood you were raised in? *Keywords: residential segregation*

2. Was the neighborhood you were raised in racially and/or ethnically homogeneous? *Keywords: institutional racism, racial segregation*

3. For how long has your own neighborhood been racially segregated—or not? *Keywords: hypersegregation*

We have included InfoTrac College Edition keywords at the end of each question to make it easier for you to find more to read on these topics. Go to www.infotrac-college.com, an online library, to begin your search.

Source: Massey, Douglas S., and Nancy A. Denton. 1993. *American Apartheid: Segregation and the Making of the Underclass.* Cambridge, MA: Harvard University Press.

•••

social fact of the sort about which the sociological theorist Emile Durkheim wrote (Chapter 1).

Theories of Prejudice and Racism

Why do prejudice, discrimination, and the various forms of racism exist? Two categories of theories have been advanced to seek to explain why. The first category consists of social psychological theories about prejudice. The second category consists of sociological theories of the types of racism, including institutional racism and discrimination.

Psychological Theories of Prejudice

Two social psychological theories of prejudice are the scapegoat theory and the theory of the authoritarian personality. The **scapegoat theory** argues that, historically, members of the dominant group in the United States have harbored various frustrations in their desire to achieve social and economic success (Feagin and Feagin 1993). As a result of this frustration, they vent their anger in the form of aggression, and this aggression is directed toward some substitute for the original perception of the frustration. Members of minority groups become these substitutes—that is, scapegoats. The psychological principle that aggression often follows frustration (Dollard et al. 1939) is central to the scapegoat principle. For example, a White person who perceived that he was denied a job because too many Mexican American immigrants were being permitted to enter the country would be using Mexican Americans as a scapegoat if he became hostile (thus, prejudiced) toward a specific Mexican American person, even if that person did not have the job in question and had nothing at all to do with the White person not getting the job.

The second theory argues that individuals who possess an authoritarian personality are more likely to be prejudiced against minorities than are nonauthoritarian individuals. The **authoritarian personality** (Adorno et al. 1950) is characterized by a tendency to rigidly categorize other people, as well as tendencies to submit to authority, rigidly conform, be very intolerant of ambiguity, and be inclined to superstition. The authoritarian person is more likely to stereotype or rigidly categorize another person and thus readily place members of minority groups into convenient and oversimplified stereotypes. As a result, the White male authoritarian is likely to stereotype all women as emotional or illogical or all Mexicans as inherently violent. Minorities and women themselves may also reveal such tendencies toward stereotyping if they possess authoritarian personalities. There is some research that links high authoritarianism with high religious orthodoxy and extreme varieties of political conservatism (Bobo and Kluegel 1991; Altemeyer 1988).

Sociological Theories of Prejudice and Racism

Current sociological theory focuses more on explaining the existence of racism and its various forms, particularly institutional racism. Speculation about the existence of prejudice is also a component of these sociological formulations. The three sociological theoretical perspectives considered throughout this text have bearing on the study of racism, discrimination, and prejudice: functionalist theory, symbolic interaction theory, and conflict theory.

Functionalist Theory Functionalist theory argues that, for race and ethnic relations to be functional to society and thus contribute to the harmonious conduct and stability of that society, then racial and ethnic minorities, and women as well, must assimilate into that society. **Assimilation** is a process by which a minority becomes socially, economically, and culturally absorbed within the dominant society. A first step in the functionalist assimilation process is to adopt as much as possible of the language, mannerisms, and goals for success of the dominant society and thus lose or give up much of its own culture. Assimilationism stands in contrast to racial–cultural *pluralism*—the maintenance and persistence of one's culture, language, mannerisms, practices, art, and so on.

Symbolic Interaction Theory Symbolic interaction theory addresses two issues: the role of social interaction in reducing racial and ethnic hostility and how race and ethnicity are socially constructed.

Symbolic interactionism asks: What happens when two people of different racial or ethnic origins come into contact with each other and interact with each other, and how can such interracial or interethnic contact reduce hostility and conflict? **Contact theory** (Allport 1954; Cook 1988) argues that interaction between Whites and minorities will reduce prejudice on the part of both groups—but only if three conditions are met:

1. The contact must be between individuals of equal status; the parties must interact on equal ground. A Hispanic cleaning woman and the wealthier White woman who employs her may interact, but their interaction will not reduce prejudice. Instead, it is likely to perpetuate stereotypes and prejudices on the part of both.

2. The contact between equals must be sustained; short-term contact will not decrease prejudice.

Bringing Whites together with Hispanics or Blacks or Native Americans for short meetings will not erase prejudice. Daily contact on the job, between individuals of *equal job status*, will tend to erase prejudice.

3. Social norms favoring equality must be agreed upon by the participants. Having African Americans and White skinheads interacting on a talk show will probably not decrease prejudice; it might well increase it instead.

Conflict Theory The basic premise of conflict theory is that class-based conflict is an inherent and fundamental part of social interaction. To the extent that racial and ethnic conflict is tied to class conflict, conflict theorists argue that class inequality must be reduced to lessen racial and ethnic conflict in society.

The current "class versus race" controversy in sociology (reviewed in more detail later in this chapter under "Patterns of Racial and Ethnic Relations") concerns the question of whether class (economic differences between races) or race ("caste" differences between races) is more important in explaining inequality and its consequences or whether they are of equal importance. Those focusing primarily on class conflict,

such as sociologist William Julius Wilson (1978, 1987, 1996) argue that class and changes in the economic structure are at times more important than race in shaping the life chances of different groups, although they see both as important. Wilson argues that being disadvantaged in the United States is more a matter of class, although he sees that this is clearly linked to race. Sociologists focusing primarily on the role of race (Willie 1979; Bonilla-Silva 1997; Feagin 2000; Feagin and Feagin 1993) argue the opposite: that race has been and is relatively more important than class—though class is still important—in explaining and accounting for inequality and conflict in society and that directly addressing the question of race forthrightly is the only way to solve the country's racial problems (see Table 11.2).

A recent variety of the conflict perspective is the *intersection perspective*. This perspective refers to the interactive or combined effects of racism, classism (elitism), and gender in the oppression of people. An implication of this theory is that the position of women in society should be studied separately within each racial or ethnic group and within each social class group and then compared, because the observed differences between women and men are not the same within the different racial or class groups. This perspective notes that

SOCIOLOGY IN PRACTICE

Race, the Underclass, and Public Policy

Sociologists use their knowledge in a variety of ways to influence national social policy. One of the most influential to do so is William Julius Wilson, an African American sociologist, currently at the Harvard University School of Government. Wilson's work has been so influential that in 1996 he was listed among *Time* magazine's twenty-five most influential Americans.

Wilson's work about the sociology of the underclass—which he defines to include African Americans, Asian Americans, Latinos, Whites, and others who are concentrated in urban poverty—earned him a place as an adviser to former President Bill Clinton and other government figures. His work came to national prominence with the publication of *The Truly Disadvantaged: The Inner City, the Underclass, and Public Policy* (1987). He argues that deep social and economic transformations have resulted in the formation of an urban underclass, permanently locked in poverty and joblessness. Differing

from both liberals and conservatives, he argues that the primary cause of this persistent poverty is neither overt racial discrimination, as liberals sometimes claim, nor the cultural deficiencies of racial groups, as conservatives sometimes claim.

Instead, Wilson compellingly demonstrates that changes in the global economy have resulted in the concentration of a disenfranchised underclass, permanently jobless and located primarily in inner cities. He argues that the flight of the more prosperous Black middle class from the inner city as traditional heavy industry declines contributes to the concentration of poverty in the urban core.

In his policy-based work, Wilson argues that addressing the needs of the underclass cannot be based on race-specific policies. Arguing with contemporary conservatives, Wilson says he refuses to be intimidated by the rhetoric of conservatives. Instead, he says, "It's quite clear to me that we're going to

have to revise discussion of the need for WPA [Works Progress Administration]-style jobs. Only these more structurally based programs, open to all in need, are likely to garner political support among the majority and to address the deep-seated problems that changes in the global economy have wrought" (*Time*, June 17, 1996: 57).

Critical Thinking Exercise

1. To what extent do you think sociology actually has an effect on national policy? Should it have more or less effect?

2. In your opinion, will there always be an urban underclass? Why or why not?

Source: Wilson, William Julius. 1996. *When Work Disappears: The World of the New Urban Poor.* Chicago, IL: University of Chicago Press; Wilson, William Julius. 1987. *The Truly Disadvantaged: The Inner City, the Underclass, and Public Policy.* Chicago, IL: University of Chicago Press; and "America's Most Influential People." *Time* (June 17, 1996): 56–57.   ●●●

Table 11.2
Comparing Sociological Theories of Race and Ethnicity

	Functionalism	Conflict Theory	Symbolic Interaction
The Racial Order	Has social stability when diverse racial and ethnic groups are assimilated into society	Is intricately intertwined with class stratification	Is based on social construction that assigns groups of people to diverse racial and ethnic categories
Minority Groups	Are assimilated into dominant culture as they adopt cultural practices and beliefs of the dominant group	Have life chances that result from the opportunities formed by the intersection of class, race, and gender	Form identity as the result of sociohistorical change
Social Change	Is a slow and gradual process as groups adapt to the social system	Is the result of organized social movements and other forms of resistance to oppression	Is dependent on the different forms of interaction that characterize intergroup relations

not only are the effects of gender and race intertwined, but they also are intertwined with the effects of class. Class conflict is seen as an integral component of gender and race differences in this society, according to the intersection perspective (Andersen and Collins 2004; Collins 1998, 1990).

Diverse Groups, Diverse Histories

The various racial and ethnic groups in the United States have arrived at their current social condition through histories that are similar in some ways, yet very different in others. Their histories are related because of a common experience of White supremacy, economic exploitation, and political disenfranchisement under the U.S. government. Native Americans first inhabited this continent, but they have been the victims of territorial expansion, armed conquest, genocide, and the destruction of their cultures. They were deprived of their land, forced to live on segregated reservations, or systematically exterminated. African Americans were forcibly abducted from West Africa—transported to this hemisphere under filthy, inhuman conditions, locked in the holds of slave ships, and sold into slavery. Hispanics from both Mexico and Puerto Rico were imported to provide cheap labor. Indigenous Hispanics were also seen as a source of cheap labor. The Chinese were imported to build railroads across the Western frontier and were forced to endure a variety of hardships ranging from hunger and slavelike conditions to murder. Arriving Japanese also met with high prejudice and discrimination. After the United States entered into World War II in the early 1940s against Germany and Japan, Japanese Americans who lived in the United States for several generations suffered the violation of having their property confiscated and their families forcibly moved to concentration camps.

Even White ethnic groups, many of whom arrived as indentured servants during the Colonial era or later expecting religious, political, and social freedom, found instead prejudice and discrimination. The Irish fled domination by the British in Ireland and, in the 1840s, fled from the devastating effects of a famine in Ireland brought on partly by the potato blight. So severe were conditions in Ireland that half of Ireland's population (more than 2 million) disappeared—one million of them dying from starvation and 11.4 million emigrating (Diner 1996). In the past, Italians, Poles, and Jews have been excluded to varying degrees from full participation in U.S. society. They have been denied rights and privileges and have failed to receive full protection under the law. Assimilation of White ethnic groups from southern, central, and eastern Europe was slower than for groups from western and northern Europe (Nagel 1996; Healey 1995; Lieberson 1980).

An historical perspective on each group follows, which will aid in understanding how prejudice, discrimination, and racism have operated throughout the history of U.S. society.

Native American Indians

Native Americans have been largely ignored in the telling of U.S. history. At the time of the Europeans' arrival in 1492, the indigenous population in North America has been estimated at anywhere from one million to ten million people. Native Americans lived here tens of thousands of years before they were "discovered" by Europeans. Discovery quickly turned to conquest, and in the course of the next three centuries, the Europeans systematically drove the Native Americans from their lands, attempting to destroy their ways of life and their various tribal cultures. Native American Indians were subjected to the onslaught of European diseases. Because they lacked immunity to these diseases, the Native Americans suffered a population decline considered

The opening of the National Museum of the American Indian in 2004 (part of the Smithsonian Institution in Washington, DC) was cause for celebration among diverse groups of Native Americans, as well as others.

by some to have been the steepest and most drastic of any people in world history. Native American cultural traditions have nonetheless miraculously survived in many isolated places, despite depletion of the original 500 Indian nations of North America (Nagel 1996; Thornton 1987; Snipp 1989).

At the time of the first European contacts in the 1640s, there was great linguistic, religious, governmental, and economic heterogeneity among Native American tribes. However, between the arrival of Columbus in 1492 and the establishment of the first thirteen colonies in the early 1600s, the ravages of disease and the encroachment of Europeans caused a considerable degree of social disorganization. Sketchy accounts of Indian cultures by early Colonists, fur traders, missionaries, and explorers underestimated the great social heterogeneity among the various Indian tribes and underestimated the devastating effect of the European arrival on Indian society.

By the year 1800, the population of Native Americans had been reduced to a mere 600,000, while wars of extermination against the Indians were being conducted in earnest. Fifty years later, the population had fallen to 300,000. Indians were killed defending their land, or they died of hunger and disease taking refuge in inhospitable country. Four thousand Cherokee died in 1834 on a forced march from their homeland in Georgia to reservations in Arkansas and Oklahoma, a trip memorialized as the Trail of Tears. The Sioux were forced off their lands by the discovery of gold and the new push of European immigration. Their reservation

was established in 1889, and they were designated as wards, subjecting them to capricious and humiliating governmental policies. The following year, the U.S. Army, mistaking Sioux ceremonial dances for war dances, moved in to arrest the leaders. A standoff exploded into violence, during which federal troops killed two hundred Sioux men, women, and children at the famous Wounded Knee massacre.

Today, about 55 percent of all Native Americans live on or near a reservation, which is land set aside for their exclusive use. The other 45 percent live in or near urban areas (U.S. Census Bureau 2004). The reservation system has not served the Indians well. Although a great many Native Americans now live in conditions of poverty, deprivation, and alcoholism, suffering unemployment at more than 50 percent among males, many have fought these ills via self-initiated programs as well as employment in urban areas.

African Americans

The development of slavery in the Americas is related to the development of world markets for sugar and tobacco. Slaves were imported from Africa to provide the labor for sugar and tobacco production and to enhance the profits of the slave owners. It is estimated that somewhere between 20 million and 100 million Africans were transported under appalling conditions to the Americas; 38 percent went to Brazil; 50 percent to the

Novelist and essayist Dorothy West was the oldest living member of the Harlem Renaissance until her death at age ninety-one in 1998.

Langston Hughes (second row from top, third from right), Dorothy West (bottom row, far right), and associates from the Harlem Renaissance aboard the Europa, bound for Russia, June 1932.

nalistic. Whites saw slaves as childlike and incapable of caring for themselves. More recent stereotypes of African Americans as childlike, having been presented endlessly for decades in the media, are traceable directly to the system of slavery.

There is a widespread belief that slaves passively acquiesced to slavery. Scholarship shows this to be false. Instead, the slaves struggled to preserve both their culture and their sense of humanity and to resist, often by open conflict, the dehumanizing and murderous effects of a system that defined human beings as mere property. Slaves revolted against the conditions of enslavement in a variety of ways, from passive means such as work slowdowns and feigned illness to more aggressive means such as destruction of property, escapes, and outright armed rebellion. All the way to the early 1970s, it had been generally believed that active rebellions in the slave community were rare. In fact, they were frequent, numbering in the several hundreds (Myers 1998, Blassingame 1973; Rose 1970).

After slavery was presumably ended by the Civil War (1861–1865) and the Emancipation Proclamation (1863), Black Americans continued to be exploited for their labor. In the South, the system of sharecropping emerged, an exploitative system in which the Black family tilled the fields for White landowners in exchange for a share of the crop. With the onset of World War I and the intensified industrialization of society came the Great Migration of Blacks from the South to the urban North. This massive movement, from 1900 through the 1920s and beyond, had a significant effect on the status of Blacks in society because there was now a greater potential for collective action (Marks 1989).

In the early part of the century, the formation of Black ghettos had a dual effect, victimizing Black Americans with grim urban conditions while also encouraging the development of Black resources, including volunteer organizations, social movements, political action groups, and artistic and cultural achievements. During the period of the 1920s, Harlem in New York City became an important intellectual and artistic oasis for Black America. The Harlem Renaissance gave the nation great literary figures, such as Langston Hughes, Jesse Fausett, Alain Locke, Arna Bontemps, Zora Neale Hurston, Wallace Thurman, and Nella Larson (Bontemps 1972; Rampersad 1986, 1988; Marks and Edkins 1999). At the same time, many of America's greatest musicians, entertainers, and artists came to the fore, such as musicians Duke Ellington, Billie Holiday, Cab Calloway, and Louis Armstrong, and painter Hale Woodruff. The end of the 1920s and the stock market

Caribbean; 6 percent to Dutch, Danish, and Swedish colonies; and only 6 percent to the United States (Genovese 1972; Jordan 1969).

Slavery evolved as a *caste system* in which one caste, the slave owners, profited from the labor of another caste, the slaves. Central to the operation of slavery was the principle that human beings could be *chattel* (property) that is bought and sold. As an economic institution, slavery was based on the belief that Whites were superior to other races, coupled with the belief in a patriarchal social order. The social distinctions maintained between Whites and Blacks were castelike, with rigid categorization and prohibitions, instead of merely classlike, which suggests more pliant social demarcations. Strong remnants of this caste system remain in many parts of the United States to this day.

The slave system also involved the domination of men over women. In this combination of patriarchy and White supremacy, White males presided over their property of White women as well as their property of Black men and women. This in turn led to gender stratification among the slaves themselves, which reflected the White slave owner's assumptions about the relative roles of men and women. Black women performed domestic labor for their masters and their own families. White men further exerted their authority in demanding sexual relations with Black women (Blassingame 1973; Raboteau 1978; Davis 1981; White 1985). The dominant attitude of Whites toward Blacks was pater-

Courtesy of the Schlesinger Library, Radcliffe College

crash of 1929 brought everyone down a peg or two, Whites as well as Blacks, although in the words of Harlem Renaissance writer Langston Hughes, Black Americans at the time "had but a few pegs to fall" (Hughes 1967).

Latinos

Latino Americans include Chicanos and Chicanas (Mexican Americans), Puerto Ricans, Cubans, and other recent Latin American immigrants to the United States. The group also includes Latin Americans who have lived for generations in the United States and are thus not immigrants but whose ancestors were very early settlers from Spain and Portugal in the 1400s. The population of Latin Americans has grown considerably over the past few decades, with the largest increase among Mexican Americans. The terms *Hispanic*, all Spanish-speaking people, and *Latino* or *Latina*, persons of Latin American descent, mask great structural and cultural diversity among the various Hispanic groups. The use of such inclusive terms also tends to ignore important differences in their respective entries into U.S. society: Mexican Americans through military conquest of the Mexican War (1846–1848); Puerto Ricans through war with Spain in the Spanish-American War (1898); and Cubans as political refugees fleeing since 1959 from the communist dictatorship of Fidel Castro, opposed by the U.S. government.

Mexican Americans Before the Anglo (White) conquest began, Mexican colonists had formed settlements and missions throughout the West and Southwest. In 1834, the U.S. government ordered the dismantling of these missions, bringing them under tight governmental control and creating a period known as the Golden Age of the Ranchos. Land then became concentrated into the hands of a few wealthy Mexican ranchers who had been given large land grants by the U.S. government. This economy created a class system *within* the Chicano community, consisting of the wealthy ranchers, mission farmers, and government administrators of the elite class; *mestizos,* who were small farmers and ranchers, as the middle class; a third class of skilled workers; and a bottom class of manual laborers, who were mostly Indians (Mirandé 1985; Maldonado 1997).

With the Mexican-American War of 1846–1848, Chicanos lost claims to huge land areas that ultimately became Texas, New Mexico, and parts of Colorado, Arizona, Nevada, Utah, and California. White cattle ranchers and sheep ranchers enclosed giant tracts of land, thus cutting off many small ranchers, both Mexican and Anglo. It was at this time that Mexicans, as well as early settlers of Mexican descent, became defined as an inferior race that did not deserve social, educational, or political equality. This is an example of the *racial formation* process (Omi and Winant 1994).

Anglos believed that Mexicans were lazy, corrupt, and cowardly, launching stereotypes that would further oppress Mexicans. This belief system was used to justify the lower status of Mexicans and to justify Anglo control of the land, which Mexicans were presumed to be incapable of managing (Moore 1976). As we have noted several times in this chapter, stereotyping has been used in this society as a way of falsely *explaining and justifying* the lower social status of society's minorities.

During the twentieth century, advances in agricultural technology changed the organization of labor in the Southwest and West. Irrigation created year-round production of crops. Consequently, there was a new need for cheap labor to work in the fields, which created more exploitation of migrant workers from Mexico. Migrant work was characterized by low earnings, poor housing conditions, poor health, and extensive use of child labor. Numerous Mexican migrant workers used as field workers, domestic servants, and for other kinds of poorly paid work continues, particularly in the Southwest (Amott and Matthaei 1996).

Puerto Ricans The island of Puerto Rico was ceded to the United States by Spain in 1899. In 1917, the Jones Act extended U.S. citizenship to Puerto Ricans, which eliminated immigration requirements. However, not until 1948 were Puerto Ricans allowed to elect their own governor. In 1952, the United States established the Commonwealth of Puerto Rico, with its own constitution. Following World War II, the first elected governor launched a program known as Operation Bootstrap that was designed to attract large U.S. corporations to the island of Puerto Rico, using tax breaks and other concessions. This program contributed to rapid overall growth in the Puerto Rican economy, although unemployment remained high and wages remained low. Seeking opportunity, unemployed farm workers began migrating to the United States. These migrants were interested in seasonal work, and thus a pattern of temporary migration characterized the early Puerto Rican entrance into the United States (Amott and Matthaei 1996; Rodriquez 1989).

Unemployment in Puerto Rico became so severe that the U.S. government even went so far as to attempt to reduce the population by some form of population control. Pharmaceutical companies experimented with Puerto Rican women in developing contraceptive pills, and the U.S. government encouraged the sterilization of Puerto Rican women. One source notes that by 1974, more than 37 percent of the women of reproductive age in Puerto Rico had been sterilized (Roberts 1997). More than one-third of these women have indicated that they regret being sterilized because they were not informed at the time that the procedure was irreversible.

Cubans Cuban migration to the United States is recent compared to many other Hispanic groups. The largest migration has occurred since the revolution led by Castro in 1959. By 1980, more than 800,000 Cubans—one-tenth of the island population—migrated to the United States. The U.S. government defined this as a political exodus, facilitating the early entrance and acceptance of these migrants. Many of the early migrants had been middle- and upper-class professionals and landowners under the prior dictatorship of Fulencio Batista but had lost their land during the Castro revolution. While in exile in the United States, some worked to overthrow Castro, often with the support of the federal government. Yet, many other Cuban immigrants were of modest means, and like other immigrant groups came seeking freedom from political and social persecution and escape from poverty.

The most recent wave of Cuban immigration came in 1980, when the Cuban government, still under Castro, opened the port of Mariel to anyone who wanted to leave Cuba. In the five months following this pronouncement, 125,000 Cubans came to the United States—more than the combined total for the preceding eight years. The arrival of people from the Mariel boat lift has produced debate and tension, particularly in Florida, a major center of Cuban immigration. The Castro Cuban government had labeled the Cubans fleeing from Mariel as "undesirable" because some had been incarcerated in Cuba. Yet they were actually not much different from prior refugees, such as the "golden exiles" who were professionals with high status (Portes and Rumbaut 1996). Because they were so labeled and because they were forced to live in primitive camps for long periods after their arrival, they have been unable to achieve much social and economic mobility. In contrast, there is a fair degree of success enjoyed by the earlier immigrants, who were on average more educated and much more settled (Portes and Rumbaut 2001, 1996; Amott and Matthei 1996; Pedraza 1996a).

Asian Americans

Like Hispanic Americans, Asian Americans are from many countries and diverse cultural backgrounds and should not be grouped under the single cultural rubric "Asians." They are immigrants from China, Japan, the Philippines, Korea, and Vietnam, and more recently from Cambodia and Laos.

Chinese Attracted by the U.S. demand for labor, Chinese Americans began migrating to the United States during the mid-nineteenth century. In the early stages of migration, the Chinese were tolerated because they provided cheap labor. They were initially seen as good, quiet citizens, but racial stereotypes turned hostile when the Chinese were seen as competing with White California gold miners for jobs. When thousands of Chinese laborers worked for the Central Pacific Railroad from 1865 to 1868, they were relegated to the most diffi-

cult and dangerous jobs, worked longer hours than the White laborers, and for a long time were paid considerably less than the White workers. It was Chinese laborers who performed the dangerous job of planting the stick of dynamite into rock, lighting the short fuse, and racing away, sometimes not in time.

By the turn of the twentieth century, the Chinese were virtually expelled from railroad work and settled in rural areas throughout the western states. As a consequence, anti-Chinese sentiment and prejudice ran high in the West. This ethnic antagonism was largely the result of competition between the White and Chinese laborers for scarce jobs. In 1882, the federal government passed the Chinese Exclusion Act, which banned further immigration of unskilled Chinese laborers. Like African Americans, the Chinese—and Chinese Americans—were legally excluded from intermarriage with Whites (Takaki 1989). The passage of this openly racist act, which was preceded by extensive violence toward the Chinese, drove the Chinese populations from the rural areas into the urban areas of the West. It was during this period that several Chinatowns were established by those who had been forcibly uprooted and who found strength and comfort within enclaves of Chinese people and culture (Nee 1973).

Japanese Japanese immigration to the United States took place mainly between 1890 and 1924. Passage of the Immigration Act of 1924 forbade further immigration of most Asians. Many first-generation immigrants, called *Issei,* who were employed in agriculture or in small Japanese businesses, were from farming families and wished to acquire their own land. However, in 1913, the Alien Land Law of California stipulated that Japanese aliens could lease land for only three years and that lands already owned or leased could not be bequeathed to heirs. The second generation of Japanese Americans, called *Nisei,* were born in the United States of Japanese-born parents and became better educated than their parents, lost their Japanese accents, and in general became more "Americanized"—that is, culturally assimilated. Members of the third generation, called *Sansei,* became even better educated and assimilated, yet they still met with prejudice and discrimination, particularly in areas with the largest populations, as on the West Coast from Washington to southern California (Glenn 1986).

After the Japanese attack on Pearl Harbor in December 1941, Japanese people suffered the maximal indignity of having their loyalty questioned when the federal government, thinking they would side with Japan, herded them into concentration camps. By executive order of President Franklin D. Roosevelt, much of the West Coast Japanese American population, many of them loyal second- and third-generation Americans, had their assets frozen and their real estate confiscated by the government. A media campaign immediately followed labeling Japanese Americans as "traitors" and "enemy aliens." Virtually all Japanese Americans in the United States had been removed from their homes by August 1942, and some remained in relocation camps as late as 1946. Relocation destroyed numerous Japanese families and ruined them financially (Glenn 1986; Kitano 1976; Takaki 1989). In 1986, the U.S. Supreme Court allowed Japanese Americans the right to file suit for monetary reparations. In 1987, legislation was passed awarding $20,000 to each person who had been relocated and also offering an official apology from the U.S. government. One is motivated to contemplate how far this paltry sum and late apology could go in righting what many have argued was the "greatest mistake" the United States has ever made as a government.

Filipinos The Philippine Islands in the Pacific Ocean fell under U.S. rule in 1899 as a result of the Spanish-American War, and Filipinos were allowed to enter the United States freely. By 1934, the islands became a commonwealth of the United States, and immigration quotas were imposed upon Filipinos. More than 200,000 Filipinos immigrated to the United States between 1966 and 1980, settling in major urban centers on the West and East Coasts. More than two-thirds of those arriving were professional workers. Their high average levels of education and skill have eased their assimilation. By 1985, there were more than one million Filipinos in the United States. Within the next thirty years, demographers project that they will become the largest group of Asian Americans in the United States, including Chinese Americans and Japanese Americans (Winnick 1990).

An ESL (English as a second language) class teaches young Vietnamese students. How would this affect their assimilation in an English-speaking country?

Koreans Many Koreans entered the United States in the late 1960s after amendments to the immigration laws in 1965 raised the limit on immigration from the Eastern Hemisphere. The largest concentration of Koreans is in Los Angeles. As many as half of adult Korean Americans are college educated, an exceptionally high proportion. Many of the emigrants were successful professionals in Korea. Upon arrival in the United States, they have been forced to take on menial jobs, thus experiencing the rigors of downward social mobility. This is especially true of those who migrated to the East Coast. However, nearly one in eight Koreans in the United States today owns a business; many are small greengrocers. Many of the stores are in predominantly African American communities and are a source of ongoing conflict between the two cultures. This has fanned negative feeling and prejudice on both sides—among Koreans against African Americans and among African Americans against Koreans (Chen 1991).

Vietnamese Among the more recent groups of Asians to enter the United States have been the South Vietnamese, who began arriving following the fall of South Vietnam to the communist North Vietnamese in 1975. These immigrants, many of them refugees, numbered about 650,000 in the United States in 1975, and approximately one-third settled in California. Many faced prejudice and hostility, resulting in part from the same perception that has dogged many immigrant groups before them: that they were competitors for scarce jobs. A second wave of Vietnamese immigrants arrived after China attacked Vietnam in 1978. As many as 725,000 arrived in the United States, only to face discrimination in a variety of locations. Tensions became especially heated when the Vietnamese became a substantial competitive presence in the fishing and shrimping industries in the Gulf of Mexico on the Texas shore, but many communities have welcomed them, and at last count, 95 percent of all Vietnamese heads of households were employed full time (Kim 1993; Winnick 1990).

Middle Easterners

Since the mid-1970s, immigrants from the Middle East have arrived in the United States from countries such as Syria, Lebanon, Egypt, and Iran. Contrary to popular belief, they speak no single language, follow no singular religion, and are thus ethnically diverse. Some are Catholic, some are Coptic Christian, and many are Muslim. Many are from working-class backgrounds, but many were teachers, engineers, scientists, and other professionals in their homelands. Some have secured employment in their original occupations, but

© Bob Daemmrich/The Image Works

An active Arab-American community exists in Detroit, illustrating both community and residential segregation.

many have undergone the downward mobility and status inconsistency necessitated by taking lower-status jobs—often experienced by present-day immigrant professionals. Like immigrant populations before them, they are forming their own ethnic enclaves in cities and suburbs of this country as they pursue the often elusive American dream (Ansari 1991).

After the terrorist attacks on September 11, 2001, many male Middle Easterners of several nationalities became unjustly suspect in this country and were subjected to severe harassment and racially motivated physical attacks. As noted in Chapter 8, out and out *racial profiling* occurred merely because they had dark skin and wore a turban on their heads. Most of these individuals, of course, probably had no discernible connection with any of the terrorists.

White Ethnic Groups

The story of White ethnic groups in the United States begins in the Colonial era. White Anglo-Saxon Protestants (WASPs), who were originally immigrants from England, Scotland, and Wales, settled in the New World (what is now North America) and were the first ethnic groups to come into contact on a large scale with Native American Indians. WASPs came to dominate the newly emerging society earlier than any other White ethnic group.

The WASPs in the late 1700s regarded the later immigrants from Germany and France as foreigners with odd languages, accents, and customs. For example, Benjamin Franklin thought that the Germans were too clannish and too insistent on keeping their language. He had strong opinions about other groups as well, such as the Dutch, and his opinions no doubt reflected (and influenced) the mood of the times. For example, he sought to bar the Germans, the Dutch, and other nationalities from interaction with the "old stock" Americans, who were largely English. He instituted programs to get the "foreign" groups to assimilate more quickly and become committed to American values, culture, and the political system. The tension between the old stock and the foreigners continued through the Civil War era until 1865, when the national origins of U.S. immigrants began to change (Handlin 1951).

Of all racial and ethnic groups in the United States during that time and since, only WASPs do not think of themselves as a nationality. The WASPs came to think of themselves as the "original" Americans despite the prior presence of Native American Indians, who were in turn described by WASPs as "savages."

The dominance of the WASPs in U.S. society has declined somewhat since 1960, when John F. Kennedy, a Catholic, became the first non-WASP ever to be elected president. Since the late 1960s, there has been increased racial and ethnic awareness for other groups, especially African, Hispanic, Asian, and Native Americans. As a group, women, long discriminated against by the male WASP establishment, also began to assert social and political power, eroding the power of WASPs in the United States. Yet much of that WASP dominance remains, as is evident in the popular use of the terms *race* and *ethnicity* to describe virtually everyone but themselves (Andersen 2003).

Jewish immigrants are questioned at Ellis Island.

There were two waves of migration of White ethnic groups in the mid- and late nineteenth century. The first stretched from about 1850 through 1880 and included northern and western Europeans: English, Irish, Germans, French, and Scandinavians. The second wave of immigration was from 1890 to 1914 and included eastern and southern Europeans: Italians, Greeks, Polish, Russians, and other eastern Europeans. The immigration of Jews to the United States extended for well over a century, but the majority of Jewish immigrants came to the United States during the period from 1880 to 1920.

The Irish arrived in large numbers in the mid-nineteenth century as a consequence of food shortages and massive starvation in Ireland. During the latter half of the nineteenth century and the early twentieth century, the Irish in the United States were abused, attacked, and viciously stereotyped. It is instructive to remember that the Irish, particularly on the East Coast and especially in Boston, underwent a period of severe ethnic oppression. A frequently seen sign posted in Boston saloons of the day proclaimed, "No dogs or Irish allowed." German immigrants were similarly often stereotyped, as were the French and the Scandinavians. It is easy to forget that virtually all immigrant groups came through times of oppression and prejudice, although these periods were considerably longer for some groups than for others. As a rule, where the population density of an ethnic group was greatest, so too was the amount of prejudice, negative stereotyping, and discrimination to which that group was subjected.

More than 40 percent of the world's Jewish population live in the United States, making it the largest community of Jews in the world. Most of the Jews in the United States arrived between 1880 and World War I, originating from the eastern European countries of Russia, Poland, Lithuania, Hungary, and Romania. Jews from Germany arrived in two phases, the first just prior to the arrival of those from eastern Europe, and the second as a result of Hitler's ascension to power in the late 1930s in Germany. Because many German Jews were professionals who also spoke English, they assimilated more rapidly than immigrants from the eastern European countries. Jews from both parts of Europe underwent lengthy periods of extreme anti-Jewish prejudice, **anti-Semitism,** and discrimination, particularly on Manhattan's Lower East Side. Significant anti-Semitism still exists in the United States (Ferber 1999; Simpson and Yinger 1985; Essed 1991).

In 1924, the National Origins Quota Act was passed, one of the most discriminatory legal actions ever taken by the United States in the field of immigration. This act established the first real *ethnic quotas* in the United States, and immigrants were permitted to enter the United States only in proportion to their numbers already in the United States. Thus, ethnics who were already here in relatively high proportions such as English, Germans, French, Scandinavians, and others, mostly western and northern Europeans, were allowed to immigrate in greater numbers than were those from southern and eastern Europe, such as Italians, Poles, Greeks, and other eastern Europeans. Hence, the act discriminated against southern and eastern Europeans in favor of western and northern Europeans. It has been noted that the European groups who were discriminated against by the National Origins Quota Act tended to be those with darker skins, even though they were White and European.

Immigrants during this period were subject to literacy tests and even IQ (intelligence quotient) tests given in English (Kamin 1974). On New York City's Ellis Island, non-English-speaking immigrants, many of them Jews, were given the 1916 version of the Stanford-Binet IQ test *in English,* presumably in order to assess their readiness to be allowed entry into the country. Obviously, non-English-speaking persons taking this test were unlikely to score high. On the basis of this grossly biased test, governmental psychologist H. H. Goddard classified fully 83 percent of the Jews, 80 percent of the Hungarians, and 79 percent of the Italians as "feeble-minded," and hence they should not be permitted to enter the country. It did not dawn on Goddard or the U.S. government that the IQ test, in English, probably did not measure something called "intelligence," as intended, but instead simply measured the immigrant's mastery of the English language (Kamin 1974; Gould 1981; Taylor 2002, 1992a, 1980).

Patterns of Racial and Ethnic Relations

Intergroup contact between racial and ethnic groups in the United States has evolved along a variety of identifiable lines involving specific actions by the dominant White group and the ethnic or racial minority. The character of the contacts has been both negative and positive, obvious and subtle, tragic and helpful. The features and forms of racial and ethnic relations that have received the most attention are assimilation, pluralism, segregation, and the interaction of class and race.

THINKING SOCIOLOGICALLY

Write down your racial–ethnic background and list one thing that people from the same background have positively contributed to U.S. society or culture. Also list one experience (current or historical) in which people from your group have been victimized by society. Discuss how these two things illustrate the fact that racial–ethnic groups both have been *victimized* and have made *positive contributions* to this society. Share your comments with others. What does this reveal to you about the connections between different groups of people and their experience as racial–ethnic groups in the United States?

Assimilation versus Pluralism

A common question raised about African Americans, Hispanics, and Native Americans is, White immigrants made it, why can't they? The question reflects the belief that with enough hard work and loyalty to the dominant White culture of the country, any minority can make it. This is the often-heard assimilation argument that African Americans, Hispanics, and Native Americans need only pull themselves up "by their own bootstraps" to become a success.

This *assimilation perspective* dominated sociological thinking a generation ago and is still prominent in U.S. thought (Portes and Rumbaut 2001; Rumbaut 1996a; Glazer 1970). The assimilationist believes that to overcome adversity and oppression, the minority person need only imitate the dominant White culture as much as possible. In this sense, minorities must assimilate "into" White culture and White society. Many Asian American individuals have followed this pattern and have thus been called the "model minority," but this label ignores the fact that Asians are still subject to prejudice, discrimination, racism, and poverty (Takaki 1989; Lee 1996; Woo 1998).

There are problems with the assimilation model. First, it fails to consider the time that it takes certain groups to assimilate. People from rural backgrounds (Native Americans, Hispanics, African Americans, White Appalachians, some White ethnic immigrants) typically take much longer to assimilate than those from urban backgrounds. Second, the history of the arrival of Black and White ethnic groups was very different, with lasting consequences. Whites came voluntarily; Blacks arrived in chains. Whites sought relatives in the New World; Blacks were sold and separated from relatives. For these and other reasons, the histories of African American and White experience as newcomers can hardly be compared, and their assimilation is hardly likely to follow the same course. Research over the last twenty-five years has thus corrected many other myths about African American history, such as the myth that all African culture was destroyed by slavery, that there is no such thing as Black culture, and that Blacks migrated voluntarily to the United States.

Third, although White ethnic groups did face prejudice and discrimination when they arrived in America, most entered at a time when the economy was growing rapidly and their labor was in high demand. They were thus able to attain education and job skills. In contrast, by the time Blacks migrated to northern industrial areas from the rural South, Whites had already established firm control over labor and used this control to exclude Blacks from better-paying jobs and higher education. Fourth, assimilation is more difficult for people of color because skin color is an especially *salient* characteristic, ascribed and relatively unchange-

able. White ethnics can change their names, but people of color cannot easily change their skin color.

The assimilation model raises the question of whether it is possible for a society to maintain **cultural pluralism,** defined as different groups in society keeping their distinctive cultures while coexisting peacefully with the dominant group. Some groups have explicitly practiced cultural pluralism. The Amish people of Lancaster County in Pennsylvania and also in north central Ohio—who travel by horse and buggy, use no electricity, and run their own schools, banks, and stores—constitute a good example of a relatively complete degree of cultural pluralism.

Cultural pluralism is the opposite of assimilation. Groups that have managed to have cultural pluralism are White ethnic groups, many of whom have also assimilated into different aspects of society. Italians in Little Italy neighborhoods in U.S. cities, for example, maintain some residential segregation, hold on to cultural traditions, and in some cases, speak their own language, but in other ways they are completely assimilated in American life.

Segregation and the Urban Underclass

Segregation refers to the spatial and social separation of racial and ethnic groups. Minorities, who are often believed by the dominant group to be inferior, are compelled to live separately under inferior conditions and are given inferior educations, jobs, and protections under the law. Desegregation has been mandated by law, thereby eliminating *de jure segregation* (legal segregation), yet segregation in fact—*de facto segregation*—still exists, particularly in housing and education.

> ### THINKING SOCIOLOGICALLY
> Using your community or school as an example, what evidence do you see of *racial segregation*? How might a sociologist explain what you have seen?

Segregation has contributed to the creation of an **urban underclass,** a grouping of people, largely minority and poor, who live at the absolute bottom of the socioeconomic ladder in urban areas (Massey and Denton 1993). The rate of segregation of Blacks and Hispanics in U.S. cities is increasing, not decreasing as many assume, thus allowing for less and less interaction among White and Black children and White and Hispanic children (Schmitt 2001; Massey and Denton 1993). Although the proportion of middle-class Blacks has increased in the last two decades, so has the proportion of Blacks, Hispanics, and Native Americans who live in poverty. In 2003, 10 percent of Whites and Asian Americans, 22 percent of Hispanics, and 23 percent of Blacks lived below the official poverty level (see Figure 11.2).

Neighborhoods such as this one in Brooklyn are indicative of residential segregation.

In a seminal study, Wilson (1987) attributes the causes of the urban underclass to economic and social structural deficits in society, while rejecting the "culture of poverty" explanation, which attributes the condition of minorities to their own presumed cultural deficiencies, a view advanced by Lewis (1960, 1966) and Moynihan (1965). The problems of the inner city, such as joblessness, crime, teen pregnancy, welfare dependency, and acquired immune deficiency syndrome (AIDS), are seen to arise from inequalities in the structure of society, and these inequalities have dire behavioral consequences for individuals—in the form of drug abuse, violence, and lack of education (Wilson 1987; Sampson 1987). Wilson argues that the civil rights agendas need to be enlarged and joblessness, the major problem of the underclass, needs to be addressed by fundamental changes in the economic institution.

Wilson (1996) has noted a most dismaying finding: Most adult men in many inner-city, poor neighborhoods are not working in any typical week. In such environments, Wilson argues, people have little chance to gain the educational and social skills that would make them attractive to employers. He advocates government-financed jobs and universal health care as important solutions to such conditions.

The Relative Importance of Class and Race

Which is more important in determining one's chances for overall survival and success in society: social class or race–ethnicity? This is the "class versus caste" controversy. The question has been hotly debated in sociology. One view, put forward by Wilson in his controversial book *The Declining Significance of Race* (1978), contends that class has become more important than race in determining access to privilege and power for

Blacks. In support of his argument, he cites the simultaneous expansion of the Black middle class and the Black urban underclass. Wilson does not argue that race is unimportant, but that the importance of social class is increasing as the importance of race is declining, even though race still remains important. In a more recent work, Wilson (1996) argues that both class and race combine to oppress not only many urban Blacks, but Whites and Hispanics as well.

Among the critics of Wilson's argument, Willie (1979), Farley and Allen (1987), Feagin (1991), Steffensmeier and Demuth (2000), and Bonilla-Silva (2001, 1997) all argue that race is still critically important, which can be seen when one compares Blacks of a given social class with Whites of the same class. Blacks at every level of education earn less than Whites at the same education level. Blacks in blue-collar jobs earn only 80 percent as much in the same jobs as Whites with the same educations and class backgrounds. Blacks are less likely to graduate from college than Whites of comparable family socioeconomic background (Royster 2003).

Finally, as Blalock (1989, 1982b) has noted, in criticizing both the race versus class approaches, there are three contributing influences to be measured: the effects of class, the effects of race, and the effects of the statistical interaction of race and class; that is, the *intersecting* effects of race and class acting together (Collins 1998). A statistical interaction effect is a *double jeopardy* effect of the type discussed in previous chapters. For example, a minority man may earn a lower salary than a White man because of his race; his salary may be lower still because he is of lower-class status; and, finally, the combination of race and class may produce even lower income than one would expect on the basis of race or class alone. When the effects of race and class interact with the effect of gender, this yields a *triple jeopardy* effect. Being Black or Latino or Native American, a woman, and poor subjects one to the maximal interaction effects of this triple jeopardy.

Attaining Racial and Ethnic Equality: The Challenge

The variety of racial and ethnic groups in the United States brings much to the society in cultural richness and diverse patterns of group and interpersonal interaction. But the inequities among these diverse groups also pose challenges for the nation in addressing questions of social justice. Those from other nations often see the United States as a racially divided nation, even though the problems of racial and ethnic conflict are not unique to Americans. Throughout the world, conflicts stemming from racial and ethnic differences are frequently the basis for economic inequities, cultural conflicts, and even war (see Map 11.2).

Civil Rights

The history of racial and ethnic relations in the United States shows several strategies to achieve greater equality. Political mobilization, legal reform, and social policy have been the basis for much social change in race relations, but continuing questions arise about how best to achieve a greater degree of racial justice in this society. Marked by the strong moral and political commitment and courage of participants, the civil rights move-

ment is probably the single most important source for change in race relations in the twentieth century. The civil rights movement was based on the passive resistance philosophy of the Rev. Dr. Martin Luther King Jr., learned from the philosophy of *satyagraha* ("soul firmness and force") of Indian leader Mahatma (meaning "leader") Mohandas Ghandi. This philosophy encouraged resistance to segregation through nonviolent techniques such as sit-ins, marches, and appealing to human conscience in calls for brotherhood, justice, and equality.

The major civil rights movement in the United States intensified shortly after the 1954 *Brown v. Board of Education* decision. In Montgomery, Alabama in 1955, seamstress and National Association for the Advancement of Colored People (NAACP) secretary Rosa Parks, an African American, became famous. By prior arrange-

VIEWING SOCIETY IN GLOBAL PERSPECTIVE

MAP 11.2 Ethnic Conflict Around the World

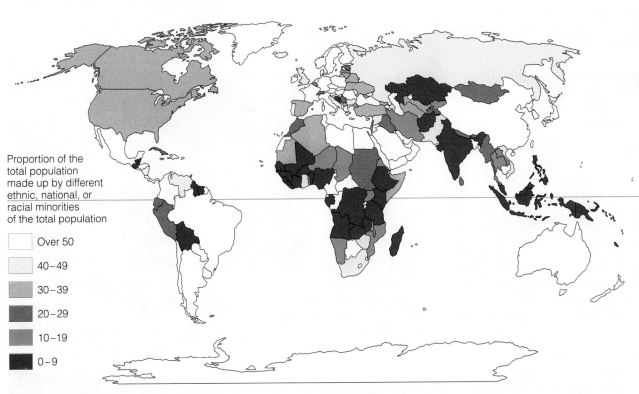

Proportion of the total population made up by different ethnic, national, or racial minorities of the total population

	Over 50
	40–49
	30–39
	20–29
	10–19
	0–9

Countries in the world vary greatly in the proportion of their populations that are made up of different ethnic groups. Some countries (for example, Mexico and Brazil) are ethnically very heterogeneous, whereas other countries (for example, India) are more ethnically homogeneous. In general, the higher the concentration per square mile of a particular ethnic group in an area (state, province, or country), then the higher the hostility, prejudice, and discrimination of the dominant ethnic group against that particular ethnic group. In your opinion, to what extent does the ethnic heterogeneity of a country contribute to ethnic conflict in that country?

Source: *The State of War and Peace Atlas* by Dan Smith. Copyright © 1997 by Dan Smith, text. Maps copyright © 1997 by Myriad Editions Ltd. Used by permission of Penguin, a division of Penguin Putnam, Inc.

ment with the NAACP, she bravely refused to relinquish her seat in the "White only" section on a segregated bus when asked to do so by the White bus driver. At the time, the majority of Montgomery's bus riders were African American, and the action of Rosa Parks initiated the now famous Montgomery bus boycott, led by the young Rev. Martin Luther King, Jr. The boycott, which took place in many cities besides Montgomery, was successful in desegregating the buses, got more African American bus drivers hired, and catapulted Rev. King to the forefront of the civil rights movement. Impetus was given to the civil rights movement and the boycott by the brutal murder in 1954 of Emmett Till, a Black teenager from Chicago, who was killed in Mississippi for whistling at a White woman (see Chapter 3).

The civil rights movement produced many episodes of both tragedy and heroism. In a landmark 1957 decision, President Dwight D. Eisenhower called out the National Guard—after an initial delay—to assist the entrance of nine Black students into Little Rock Central High in Little Rock, Arkansas. Sit-ins followed in which White and Black students perched at lunch counters until the Black students were served. The "freedom rides," organized bus trips from North to South to promote civil rights such as voter registration, forged on despite the murders of Viola Liuzzo, a White Detroit housewife; Andrew Goodman and Michael Schwerner, two White students; and James Chaney, a Black student. The murder of civil rights workers—especially when they were White—galvanized public support for change.

Radical Social Change

As the civil rights movement developed throughout the late 1950s and 1960s, a more radical philosophy of change also developed. More militant leaders grew disenchanted with the limits of the civil rights agenda, which they perceived as slow-moving. The militant Black power movement took its name from the book *Black Power*, published in 1967 by political activist

---

FORCES OF SOCIAL CHANGE
Race Naming

Many have argued for years that people responding to the U.S. census questionnaire should not be limited to checking only one racial category. Instead, they should be allowed to check two or more categories simultaneously—the "multiracial" response. Given this option, a person can self-identify with several racial groups. For the first time ever, the 2000 U.S. Census permitted the individual to select more than one racial category. (Only 2 percent actually did so.) This is useful, for example, if a person has one White parent and one African American parent. That person would then have the option of checking both, thus self-identifying as more than one race instead of, say, Black, as would have been required on prior censuses. An interesting research problem would then be which kinds of persons—in terms of their gender, social class, religion, and so on—tend to choose one option or the other.

Such policy has the potential for major social change in American society. Should self-naming change over time (for example, more people checking more than one racial category who would have checked only Black/African American previously), then the racial composition of the U.S. population itself would change, and could change dramatically, depending upon how many persons elect the multiracial option. Although, as we have already seen, who is of what race is a social construction in society, certain realities do not change, such as who gets state or federal funds that are intended to be allocated on the basis of racial composition of a neighborhood. Arguing that race is primarily a social construction and has no scientific merit only avoids the issue. At least one researcher (Harrison 2000) notes that the new census policy is akin to having Humpty Dumpty fall off the wall: It has created more problems than it was intended to solve. It is now up to the U.S. Census Bureau to find some way of putting Humpty back together again.

The man below (Alison Davis, a Chicago attorney) is African American and has so designated himself for his entire life. What would you designate the person on the left? (She is of Vietnamese and White American parentage.)

© Owen Franken/Corbis

Alison Davis

Source: Harrison, Roderick. 2000. "Inadequacies of Multiple Response Race Data in the Federal Statistical System." Manuscript. Joint Center for Political and Economic Studies and Howard University, Department of Sociology, Washington, DC; and U.S. Census Bureau 2002.

•••

Stokely Carmichael, later Kwame Touré, and Columbia University political science professor Charles V. Hamilton. This movement had a more radical critique of race relations in the United States and saw inequality as stemming not just from moral failures, but from the institutional power that Whites had over Black Americans (Carmichael and Hamilton 1967).

The Black power movement of the late 1960s rejected assimilationism and demanded instead self-determination, cultural pluralism, and self-regulation of Black communities. Militant groups such as the Black Panther Party advocated fighting oppression with armed revolution. The U.S. government acted quickly, imprisoning members of the Black Panther Party and members of similar militant revolutionary groups and, in some cases, killing them outright (Brown 1992).

The Black power movement also influenced the development of other groups who were influenced by the analysis of institutional racism as well as by the assertion of strong group identity. Groups such as *La Raza Unida*, a Chicano organization, encouraged "brown power," promoting solidarity and the use of Chicano power to achieve racial justice. Likewise, the American Indian Movement (AIM) used some of the same strategies and tactics that the Black power movement had encouraged, as have Puerto Rican, Asian American, and other racial protest groups. Elements of Black power strategy were also borrowed by the developing women's movement, and Black feminism developed upon the realization that women, including women of color, shared in the oppressed status fostered by institutions that promoted racism (Collins 1998, 1990). Overall, the Black power movement dramatically altered the nature of political struggle and race and ethnic relations in the United States.

Malcolm X, before breaking with the Black Muslims (the Black Nation of Islam in America) and his religious mentor, Elijah Muhammad, advocated a form of pluralism demanding separate business establishments, banks, churches, and schools for Black Americans. He echoed an earlier movement of the 1920s led by Marcus Garvey's back-to-Africa movement, the Universal Negro Improvement Association (UNIA).

Many leaders of the Black power movement were eliminated, either through assassination or imprisonment or other means, reducing its influence. Its significance, however, remains. Minister Louis Farrakhan, a protege of Malcolm X in the early 1960s, broke from the Black Muslim's Nation of Islam organization to found his own Black Muslim organization. Before his death, even Martin Luther King, the major spokesperson for civil rights, in a more radical analysis of race relations articulated the need for radical economic change to address continuing racial inequality in the United States (Branch 1988). The Black power movement and the movements it spawned changed the nation's consciousness about race and forced academic scholars to develop a deeper understanding of how racism works in society and how fundamental it is to U.S. institutions.

The civil rights movement, which began in the mid-1950s and continued through the 1980s, resulted in significant social, economic, and political gains for Blacks, although gains have been offset by losses (Morris 1999). A major study by the Civil Rights Project at Harvard University shows that schools are now *more segregated* than they were thirty years ago. School desegregation increased from the 1950s through the 1980s, but schools have "resegregated" since. Except in the South and Southwest, White students have little contact with Black and Latino students—hardly racial progress (Frankenberg et al. 2003).

Affirmative Action

A continuing question from the dialogue between a civil rights strategy and more radical strategies for change is the debate between race-specific versus color-blind programs for change. *Color-blind policies* advocate that all groups be treated alike, with no barriers to opportunity posed by race, gender, or other group differences. Equal opportunity is the key concept in color-blind policies. The maxim "equal pay for equal work" is a color- and gender-blind policy, meant to minimize the pay gap in the workplace that is now heavily influenced by race and gender.

Race-specific policies recognize the unique status of racial groups because of the long history of discrimination and the continuing influence of institutional racism. Those advocating such policies argue that color-blind strategies will not work because Whites and other racial–ethnic groups do not start from the same position. Even given equal opportunities, continuing disadvantage produces unequal results. The tension between these two strategies for change is a major source for many of the political debates surrounding race relations now.

Affirmative action, a contested program for change, is a race-specific policy for reducing job and educational inequality that has also had some success. Affirmative action means two things. First, affirmative action means recruiting minorities from a wide base to ensure consideration of groups that have been traditionally overlooked, but at the same time *not* using rigid quotas based on race or ethnicity. Second, affirmative action means using admissions slots (in education) or set-aside contracts or jobs (in job hiring) to assure minority representation. The principal objection, heard from both sides of the racial line, is that either interpretation of affirmative action programs is, in effect, use of a quota.

The Legal Defense Fund (LDF), established by the NAACP (National Association for the Advancement of Colored People) has argued forcefully on legal grounds that the push for affirmative action on the basis of race must continue, though not in the form of rigid quotas. It notes that in the tradition of its founding lawyer and the future member of the U.S. Supreme Court, Thurgood Marshall, the affirmative action policies of the LDF have helped to fundamentally change the compositions of the nation's formerly segregated universities and colleges, as well as greatly expand the educational opportunities for African Americans and other minorities. It notes that the last decade has brought an "assault" on affirmative action, and thus the LDF has had to expend considerable resources in meeting the legal challenge. It argues that standardized tests such as the Scholastic Assessment Test (SAT) are of limited validity, do not adequately predict performance in college, are not as valid for Blacks as for Whites, and are thus not a legitimate basis on which to judge a Black and a White candidate against each other. (We return to this issue of standardized testing in Chapter 16.) Hence, so it argues, there has been no "discrimination" against the dominant majority, as is often maintained by the (White) plaintiffs in the cases. These legal battles are certain to continue for some time.

Recent data have shown that Blacks admitted to selective colleges and universities under affirmative action programs reveal high rates of social and economic success after graduation. For example, the percentage of Blacks graduating from such schools who went on to

Used by permission of Mark Giaimo and the *Naples Daily News.*

graduate school and law school was *higher* than the percentage of Whites from the same schools who did so (Bowen and Bok 1998). This is clearly a benefit of affirmative action in education.

The U.S. Supreme Court decided in 1978 that race could be used as a criterion for admission to undergraduate, professional, and graduate schools or for job recruitment, as long as race is combined with other criteria and as long as rigid racial quotas are not used. Then twenty-five years later, in June of 2003, the U.S. Supreme Court decided two cases that modified its 1978 decision. In *Grutter v. Bollinger,* in a five-to-four decision, the high court decided that, as in the 1978 decision, race could indeed be used as a factor in admissions decisions for the University of Michigan Law School, as long as race was considered along with other

TAKING ON SOCIAL ISSUES
Affirmative Action

Various methods have been suggested for reducing racial inequality in the United States. Some remedies are based on color-blind approaches, such as civil rights laws that prohibit employment practices that exclude people because of race. Other suggested remedies are race-specific, such as *affirmative action* programs that target racial groups for inclusion in employment or college admissions. Proponents of color-blind strategies base their policies on the argument that all groups should be treated alike, regardless of their historical, social, and cultural differences. Critics of this approach say that it cannot address the problems created by past discrimination and the unique experiences of different racial–ethnic groups. What are the arguments for and against each of these positions with regard to solving the problem of inequality among racial groups?

Do you think that race should be used in any way as a criterion for admission to colleges and universities? Do you think that race should be used along with other factors, as long as some form of racial quota is *not* used (as decided by the Supreme Court

in June of 2003 in *Grutter v. Bollinger*)? Do you think that a minority candidate for admission to a college or university should get extra "points" for being a minority (the Supreme Court decided against this in *Gratz v. Bollinger* in 2003)?

Taking Action

Go to the Taking Action Exercise on the Companion Website—at http://sociology .wadsworth.com/andersen_taylor4e/— to learn more about an organization that addresses this topic. •••

factors and the decision to admit or not admit was made on a case-by-case basis—that is, considering many factors, race among them, characterizing any candidate for admission. In a second case (*Gratz v. Bollinger*), a six-to-three decision, the court threw out as unconsti-

tutional any system of assigning favorable "points" to minority candidates seeking admission which would increase their chances for admission. This decision thus ruled out the use of any form of minority quotas, interpreting the point system as a type of quota.

Sociology ⊛ Now™
Reviewing is as easy as ❶❷❸.

1. Before you do your final review, take the SociologyNow diagnostic quiz to help you identify the areas on which you should concentrate. You will find information on SociologyNow and instructions on how to access all of its great resources on the foldout at the beginning of the text.

2. As you review, take advantage of SociologyNow's study videos and interactive Map the Stats exercises to help you master the chapter topics.

3. When you are finished with your review, take SociologyNow's posttest to confirm you are ready to move on to the next chapter.

Chapter Summary

How are race and ethnicity defined?

In virtually every walk of life, race matters. A *race* is a social construction based loosely on physical criteria, whereas an *ethnic group* is a culturally distinct group. A group is *minority* or *dominant* not on the basis of their numbers in a society, but on the basis of which group occupies lower average social status.

What are stereotypes, and how are they important?

Stereotyping and *stereotype interchangeability* reinforce racial and ethnic prejudices and thus cause them to persist in the maintenance of inequality in society. Racial and gender stereotypes are similar in dynamics in society, and both racial and gender stereotypes receive ongoing support in the media. Stereotypes serve to justify and make legitimate the oppression of groups based on race, ethnicity, and gender. Stereotypes such as "lazy" support attributions made to the minority that cast blame on the minority in question, thus, in effect, removing some of the blame from the social structure.

What are the differences among prejudice, discrimination, and racism?

Prejudice is an attitude involving usually negative prejudgment on the basis of race or ethnicity. *Discrimination* is actual behavior involving unequal treatment. *Racism* involves both attitude and behavior. Racism can take several *forms*, such as traditional ("old fashioned," or Jim Crow) racism, *aversive* (subtle) racism, *laissez-faire* (symbolic) racism, *color-blind* racism, and *institutional racism*: unequal treatment, carrying with it notions of cultural inferiority of a minority, which has become

ingrained into the economic, political, and educational institutions of society.

What theories are there about prejudice and racism?

Different theoretical positions have been developed in sociology to explain prejudice and racism and to explain different aspects of race and ethnic relations, including functionalist theories (such as *assimilationism*), symbolic interaction theories (such as *contact theory*), and conflict theories (including an "intersection" perspective).

Do minority groups have similar or different histories?

Historical experiences show that different groups have unique histories, although they are bound by similarities in the prejudice and discrimination they have experienced. Although each group's experience is unique, they are commonly related by a history of *prejudice* and *discrimination*.

What are the general patterns of race–ethnic relations?

Sociologists analyze different forms of racial and ethnic relations, including *assimilation*, *pluralism*, *segregation* and the *urban underclass*, the relative importance of social class versus race (caste), and the question of class–race interaction.

What are the approaches to race–ethnic equality?

The civil rights movement has been the main basis for social change based on an equal rights philosophy. More radical activists have developed an analysis of *institutional racism* and the power relationships on which racism rests. Some programs for change rely on the more traditional *color-blind strategies*, whereas others (such as *affirmative action* programs) propose *race-specific remedies* to address racial inequality. Some argue that only deep-rooted economic change will alleviate persistent racial stratification.

Key Terms

affirmative action 295
anti-Semitism 290
assimilation 281
authoritarian personality 281
aversive racism 279
color-blind racism 279
contact theory 281
cultural pluralism 291

discrimination 278
dominant group 275
ethnic group 270
ethnocentrism 277
forms of racism 279
institutional racism 279
laissez-faire racism 279
minority group 274

Researching Society with MicroCase Online

You can see the results of actual research by using the Wadsworth MicroCase® Online feature available to you. This feature allows you to look at some of the results from national surveys, census data, and other data sources. You can explore this easy-to-use feature on your own, but try this example. Suppose you want to know:

Do you ever wonder nowadays whether Blacks and Whites differ in whether they think there should be laws against intermarriage?

To answer this question, go to http://sociology.wadsworth .com/andersen_taylor4e/, select MicroCase Online from the left navigation bar, and follow the directions there to analyze the data described below.

Data file: GSS

Task: Cross-Tabulation

Row Variable: INTERMAR

Column Variable: HHRACE

Questions

Once you have your results, answer the following questions:

1. Overall, what percent of the people studied felt that there should be laws against racial intermarriage?

2. How many Whites were in this study? Blacks? Others?

3. Look at the numbers and percentages within the table: Is Race of Household related to how they answered the question about intermarriage? You can tell from these numbers!

The Companion Website for Sociology: Understanding a Diverse Society, Fourth Edition

http://sociology.wadsworth.com/andersen_taylor4e/

Supplement your review of this chapter by going to the companion website to take one of the Tutorial Quizzes, use the flash cards to master key terms, and check out the many other study aids you'll find there. You'll also find special features such as GSS Data and Census 2000 information, data and resources at your fingertips to help you with that special project or do some research on your own.

Suggested Readings and Web Resources

Malcomson, Scott L. 2000. *The American Misadventure of Race*. New York: Farrar, Straus, and Giroux. *This book is a readable and comprehensive history of the racial formation process and shows how the names and labels for various racial and ethnic groups have changed over time in the United States. An account of the bizarre history of the U.S. census racial classification is given.*

Omi, Michael, and Howard Winant. 1994. *Racial Formation in the United States*, 2nd ed. New York: Routledge. *This important work advances racial formation theory, which emphasizes the social constructionist and historical processes by which a group comes to be defined as a race and deemphasizes the role of physical or biological traits. The book covers Latinos, American Indians, Asians, African Americans, and other such " racially formed" groups.*

Portes, Alejandro, and Rubén G. Rumbaut. 2001. *Legacies: The Story of the Immigrant Second Generation*. Berkeley, CA: University of California Press. *Focusing on the U.S.-born of foreign-born immigrants, this is a detailed account of these legacy persons and contains rich data on both the similarities and the differences among immigrant groups.*

Snipp, Matthew. 1989. *American Indians: The First of This Land*. New York: Russell Sage Foundation.
This book offers a definitive discussion of how American Indian ancestry is defined and how definitions vary by tribal unit. The role of the U.S. government's Bureau of Indian Affairs (BIA) in this process is discussed.

Yoon, In-Jin. 1997. *On My Own: Korean Businesses and Race Relations in America*. Chicago. IL: University of Chicago Press.
Based on extensive fieldwork among Korean immigrants in Chicago and Los Angeles, this book explores the social problems facing Korean immigrants and examines the tensions between them and African Americans in urban areas.

Racial Legacies and Learning
www.pbs.org/adultlearning/als/race/4.0/index.htm
This site, sponsored by the Public Broadcasting Service (PBS) and the American Association of Colleges and Universities (AAC&U), is directed at fostering dialogue between college campuses and communities in an effort to improve race relations. It includes important resources on cultivating diversity and numerous educational projects designed to improve race relations.

American Indian History and Related Issues
www.csulb.edu/projects/ais
This site features many interesting links to information on Native American history, culture, and sociology.

● ● ●

Gender

Imagine suddenly becoming a member of the other sex. What would you have to change? First, you would probably change your appearance—clothing, hairstyle, and any adornments you wear identify a person's gender. You would also have to change some of your interpersonal behavior. Contrary to popular belief, men talk more than women, are louder, are more likely to interrupt, and are less likely to recognize others in conversation. Women are more likely to laugh, express hesitance, and be polite (Robinson and Smith-Lovin 2001; Anderson and Leaper 1998; Crawford 1995; Cameron 1998). Gender differences also appear in nonverbal communication. Women use less personal space, touch less in impersonal settings (but are touched more), cry more, and smile more—even when they are not necessarily happy (LaFrance 2002; Basow 1992; Lombardo et al. 2001). Researchers find that women and men even write email in a different style, women writing less opinionated mail than men, but also using it to maintain rapport and intimacy (Sussman and Tyson 2000; Colley and Todd 2002). Finally, you might have to change many of your attitudes because men and women differ significantly on many, if not most, social and political issues (see Figure 12.1).

If you are a woman and became a man, perhaps the change would be worth it. You would probably see your income go up (especially if you became a White man). You would have more power in virtually every social setting. You would be far more likely to head a major corporation, run your own business, or be elected to a political office—again, assuming that you were White. Would it be worth it? As a man, you would be far more likely to die a violent death and would probably not live as long (National Center for Health Statistics 2003; U.S. Bureau of Justice Statistics 2004).

If you are a man who becomes a woman, your income would likely drop significantly. Long after passage of the Equal Pay Act in 1963, women still earn 23 percent less than men, counting only those working year-round and full-time (DeNavas-Walt et al. 2003). You would likely become resentful of a number of things because poll data indicate that women are more resentful than men about the amount of money available to live on, the amount of help they get from their mates around the house, how child care is shared, and their appearance. Women also report being more fearful on the streets than men. However, women are more satisfied than men with their roles as parents and with their friendships outside of marriage.

For both women and men, there are benefits, costs, and consequences stemming from the social definitions

Sociology ⊛ Now™
Reviewing is as easy as ➊➋➌.

Use SociologyNow to help you make the grade on your next exam. When you are finished reading this chapter, go to the chapter review for instructions on how to make SociologyNow work for you.

associated with gender. As you imagined this experiment, you may have had difficulty trying to picture the essential change in your *biological* identity—but is this the most significant part of being a man or woman? As we will see in this chapter, nature determines whether you are male or female, but it is society that gives significance to this distinction. Sociologists see *gender* as a social concept. Who we become as men and women is largely shaped by cultural and social expectations. •••

Defining Sex and Gender

Sociologists use the terms *sex* and *gender* to distinguish biological sex identity from learned gender roles. **Sex** refers to biological identity, male or female. For sociologists, the more significant concept is **gender**—the socially learned expectations and behaviors associated with members of each sex. This distinction emphasizes that behavior associated with gender is culturally learned. A person is born male or female, but becoming a man or a woman is the result of social and cultural expectations that pattern men's and women's behavior. Even labeling someone as a man or woman is a cultural decision. Why is this so important to know about someone (Fausto-Sterling 2000)?

From the moment of birth, gender expectations influence how boys and girls are treated. Now that we can identify the sex of a fetus early in a pregnancy, gender expectations may begin before birth. Parents and grandparents may select pink clothes and dolls for baby girls; sports clothing and brighter colors for boys. They have little choice to do otherwise because the clothing styles available on the markets are highly gender-stereotyped. Parents and other adults continue to treat children with gender expectations throughout their childhood. Girls may be expected to cuddle and be sweet, whereas boys are handled more roughly and given greater independence. Fathers and mothers also interact with infants differently, depending on the baby's gender (LaFlamme et al. 2002).

Parents, however, are not the only influence on children's development. Children have gender-stereotyped perceptions of infants, and gender stereotyping increases as children grow into teenagers. Thus, even as they develop their own gender identity, children influence the identities of their peers and the younger children around them (Hibbard and Buhrmester 1998; Garner et al. 1997; Fagot 1995).

The cultural basis of gender is especially apparent when we look at other cultures. Across different cultures, the gender roles associated with masculinity and femininity vary considerably. In Western industrialized societies, people tend to think of masculinity and femininity in dichotomous terms, with men and women perceived as different, even defined as opposites. The view from different cultures challenges this assumption. The Navajo Indians, for example, offer interesting examples of alternative gender roles. Historically, the *berdaches* in Navajo society were anatomically normal men who were defined as a third gender considered to fall between male and female. Berdaches married other men, who were not considered berdaches. Those they married were defined as ordinary men. Moreover, neither the berdaches nor the men

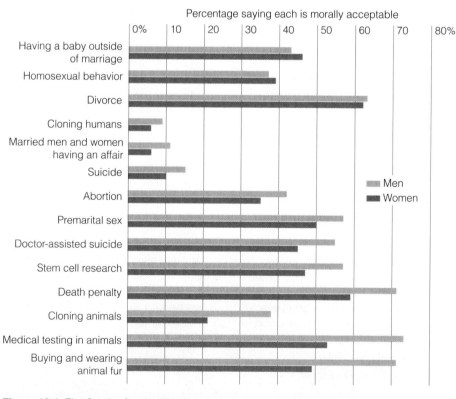

Figure 12.1 The Gender Gap in Attitudes

Data: Saad, Lydia. 2003. "Pondering Women's Issues, Part II." *The Gallup Poll.* Princeton, NJ: The Gallup Organization.
Website: **www.gallup.com**

they married were considered homosexuals, as they would be considered in many contemporary Western cultures (Nanda 1998; Lorber 1994).

There are also substantial differences in the construction of gender across social classes or subcultures within a single culture. Within the United States, as we will see, the experiences of gender vary considerably among different racial and ethnic groups (Baca Zinn et al. 2000). Looking at gender across cultures quickly reveals the social and cultural dimensions of something often popularly defined as biologically fixed.

Sex Differences: Nature or Nurture?

Despite the known power of social expectations, the belief persists that differences between men and women are biologically determined. Biology is, however, only one component in the difference between men and women. The important question is not whether biology or culture is more important in forming men and women, but how biology and culture interact to produce a person's **gender identity** (that is, how one thinks of oneself as a woman or a man).

Biological determinism refers to explanations that attribute complex social phenomena to physical characteristics. The argument that men are more aggressive because of the hormonal difference of the presence of testosterone is a biologically determinist argument. Although people popularly believe that testosterone causes aggressive behavior in men, studies find only a modest correlation between aggressive behavior and testosterone levels. Furthermore, changes in testosterone level (such as by "chemical castration," the administration of drugs that eliminate the production or circulation of testosterone) do not predict changes in men's aggression. In addition, minimal differences exist in the levels of sex hormones between girls and boys in early childhood, yet researchers find considerable differences in aggression exhibited by boys and girls from an early age (Fausto-Sterling 1992).

Biological Sex Identity

A person's sex identity is established at the moment of conception when the father's sperm provides either an X or a Y chromosome to the egg at fertilization. The mother contributes an X chromosome to the embryo. Two X chromosomes make a female; an X and a Y, a male. Normally, genes on the sex-linked chromosomes lead to the formation of male or female genitalia, but sometimes this process is compromised and biological sex identity is unclear, providing fascinating cases for social scientific study.

© Peter Menzel/Stock Boston

How do you react when you see someone and cannot identify them as a man or woman? What does this tell you about the significance of gender in social interaction?

Hermaphroditism is a condition caused by irregularities in the process of chromosome formation or fetal differentiation that produces persons with mixed biological sex characteristics, also known as *intersex persons*. In the most common form of hermaphroditism, the child is born with ovaries or testes, but the genitals are ambiguous or mixed. An example would be a child born with female chromosomes but an enlarged clitoris, making the child appear to be male. Sometimes, a child may be a chromosomal male, but with an incomplete penis and no urinary canal. In other cases of hermaphroditism, physicians typically advise sex reassignment, including reconstruction of the genitals and hormonal treatment.

Case studies of intersexed persons reveal the extraordinary influence of social factors in shaping the person's identity (Preves 2003). Parents of such children are usually advised to have genital reconstruction but also to give the child a new name, a different hairstyle, and new clothes—all intended to provide the child with the social signals judged appropriate to a single gender identity. One physician who has worked on such cases gives the directive to parents that they "need to go home and do their job as child rearers with it very clear whether it's a boy or a girl" (Kessler 1990: 9).

Despite the strong influence of socialization in creating gender, many continue to argue that there are innate (that is, "natural") differences between women and men. This is illustrated by a well-known case involving a pair of male twins. One of the twins had his penis burned off during a routine circumcision by a mismanaged electric current. The boy was then recreated as a girl, including surgical and hormonal treatment. As the child grew up, "she" sometimes imitated male activities—trying to urinate standing up and mimicking the father's shaving—but sometimes the child was feminine, too. The child, however, had difficulty adjusting to her identity and at age fourteen chose to become a man, undergoing extensive surgery to partially restore his penis and have a mastectomy. At age twenty-five, he married a woman and adopted her children.

Some conclude from this case that one's sense of gender identity is fixed at birth, but there is another side to this argument—that gender identity is socially constructed. While the child was growing up as a girl, her peers teased her mercilessly and refused to play with her. She also spent much time having her genitals scrutinized by doctors. By the time the child was a teenager, she was miserable, contemplated suicide, and was then told of what had happened earlier (Diamond and Sigmundson 1997). Although some conclude that this proves the biological basis of gender identity, it also reveals the strong influence of social factors (such as peer ridicule) in managing one's gender identity. Thus, this case shows the strong interplay of culture *and* biology.

Consider the example of *transgendered* people. Transgendered people are those who deviate from the binary (male or female—one or the other) system of gender, including transsexuals, cross-dressers, and others. They do not fit within the normative expectations of gender. Research on transgendered individuals shows that they experience enormous pressure to fit within the normative expectations. When they are young, for example, they may hide their cross-dressing. Those who change their sex as adults report enormous pressure particularly during their transition period, because others expect them to be one sex or the other. Whatever their biological sex, many transgendered people are forced—from fear of rejection and the desire for self-preservation—to manage an identity that falls into just one gender category (Gagne and Tewksbury 1998).

From a sociological perspective, biology alone does not determine gender identity. People must adjust to the expectations of others and the social understanding of what it means to be a man or a woman. A person may remain genetically one sex, socially the other—or perhaps something in between. In other words, there is not a fixed relationship between biological and social outcomes (Fausto-Sterling 2000). If you see men and women only as biologically "natural" states, you miss some fascinating ways that gender is formed in society.

Physical Sex Differences

Physical differences between the sexes do, of course, exist. In addition to differences in anatomy, at birth, boys tend to be slightly longer and weigh more than girls. As adults, men tend to have a lower resting heart rate, higher blood pressure, higher muscle mass and muscle density, and more efficient recovery from muscular activity. These physical differences contribute to the tendency for men to be physically stronger than women. However, the public now routinely sees displays of women's athleticism and expects great performances from both men and women in world-class events, such as the Olympics. Women can achieve high degrees of muscle mass and muscle density through bodybuilding and can win over men in activities that require endurance, such as the four women who have won the Iditarod—the Alaskan dog sled race considered to be one of the most grueling competitions in the world.

Arguments based on biological determinism assume that differences between women and men are "natural" and, presumably, resistant to change. Like biological explanations of race differences, biological explanations of inequality between women and men tend to flourish during periods of rapid social change. They protect the *status quo* (existing social arrangements) by making it appear that the status of women or people of other races is "natural" and therefore should remain as it is. If social differences between women and men were biologically determined, we would also find no variation in gender relations across cultures, but extensive differences are well documented. Moreover, even within the same culture, there can be vast *within gender* differences. That is, the variation on a given trait may be as great, or greater, among women as between women and men (see Figure 12.2.) In sum, we would

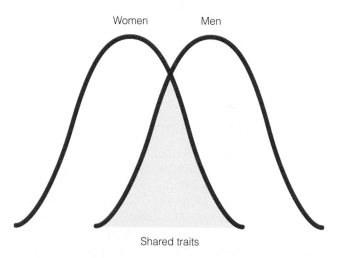

Figure 12.2 Gender: Within and Between Gender Differences
Notice that within gender differences on a given trait can be greater than differences between genders.

not exist without our biological makeup, but we would not be who we are without society and culture.

The Social Construction of Gender

As we saw in Chapter 4, socialization is the process by which social expectations are taught and learned. Through **gender socialization,** men and women learn the expectations associated with their sex. The rules of gender extend to all aspects of society and daily life. Gender socialization affects the self-concepts of women and men, their social and political attitudes, their perceptions about other people, and their feelings about relationships with others. Although not everyone is perfectly socialized to conform to gender expectations, socialization is a powerful force directing the behavior of men and women in gender-typical ways.

Even people who set out to challenge traditional expectations often find themselves yielding to the powerful influence of socialization. Women who consciously reject traditional women's roles may still find themselves inclined to act as hostess or secretary in a group setting. Similarly, men may decide to accept equal responsibility for housework, yet they fail to notice when the refrigerator is empty or the child needs a bath—household needs they have been trained to let someone else notice (DeVault 1991). These expectations are so pervasive that it is also difficult to change them on an individual basis. If you doubt this, try buying clothing or toys for a young child without purchasing something that is gender-typed, or talk to parents who have tried to raise their children without conforming to gender stereotypes and see what they report about the influence of such things as children's peers and the media.

The Formation of Gender Identity

One result of gender socialization is the formation of **gender identity,** one's definition of oneself as a woman or man. Gender identity is basic to our self-concept and shapes our expectations for ourselves, our abilities and interests, and how we interact with others.

Gender identity is formed through social interaction. In all-male groups, for example, boys use more commands and threats than in mixed-sex groups. Boys are less likely than girls to comply with others, and they are more physically aggressive. Researchers interpret these behaviors as forms of domination. Thus, in experimental studies, researchers find that women in subordinate positions tend to smile more than those assigned to dominant roles in the experiment; interestingly, smiling is not associated with subordinate and dominant roles for men, probably evidence of the extent to which

women learn to smile to please others (Mast and Hall 2004). Studies also consistently show that men interrupt in conversation—especially when talking to women (Anderson and Leaper 1998).

> ### THINKING SOCIOLOGICALLY
>
> Try an experiment based on the example of changing genders that opens this chapter. For a period of twenty-four hours note everything you would have to do to change your appearance and behaviors to become the other *gender.* Whenever possible, act in accordance with your new identity and record how others respond to you. What does your experiment tell you about *"doing gender"* and how social interaction supports gender?

Some cautions should be taken when interpreting such studies. First, conventions for reporting research tend to amplify the appearance of gender stereotypes because researchers tend to publish their results only when gender differences are found. As a result, research reports may overemphasize gender differences while denying attention to the many similarities between women and men. Second, the findings of researchers depend highly on how they define the behavior being observed. Men are generally understood to be more physically aggressive than women, but studies have also found that people's gender stereotypes may influence how they perceive aggression in women and men, leading people to overperceive aggression by men and underperceive aggression by women (Stewart-Willliams 2002). Third, no one completely conforms to the expectations passed on through socialization. Our widely different experiences and the creative ways we respond to social expectations are part of our uniqueness as individuals.

Sources of Gender Socialization

As with other forms of socialization, there are different *agents of gender socialization:* family, play, schooling, religious training, and the mass media, to name a few. Gender socialization is reinforced whenever gender-linked behaviors receive approval or disapproval from these multiple influences. Gender socialization is so effective that, as early as eighteen months of age, toddlers have learned to play with presumed gender-appropriate toys (Caldera and Sciaraffa 1998).

Parents and Gender Socialization Parents are one fundamental source of gender socialization. Parents may discourage children from playing with toys that are identified with the other sex, especially when boys play with toys meant for girls. Interviews with preschoolers show that young boys in particular report that their fathers think cross-gender play is bad. As a result, boys are even more likely than girls to choose

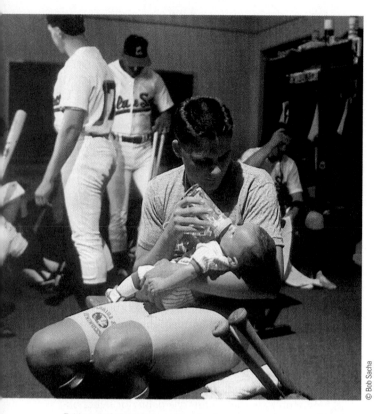

© Bob Sacha

Evidence of changing gender roles for women and men can be seen in many places. Many men have become more involved in child care, thereby changing their gender roles as well.

gender-stereotyped toys (Raaj and Rackliff 1998). Gender norms also seem to be more strictly applied to boys than girls. Boys who engage in behavior associated with girls are more negatively regarded than girls who play or act like boys (Sandnabba and Ahlberg 1999). Girls may be called "tomboys," yet boys who are called "sissies" are more harshly judged. However, although being a tomboy may be acceptable for a girl, beyond a certain age, the same behavior may result in her being considered as unfeminine and labeled a "dyke." Sociological research on tomboys shows that girls who become tomboys do so because they recognize the disadvantages of femininity and the privileges of masculinity. Thus, they are resisting expected gender roles, even while trying to conform within existing gender role choices (Carr 1998).

Expectations about gender are changing, although researchers suggest that the cultural expectations about gender may have changed more than people's actual behavior. Thus, mothers and fathers now report that fathers should be equally involved in child-rearing, but the reality is different. Mothers still spend more time in child-related activities and have more responsibility for children. Furthermore, mothers perceive less father involvement than fathers do. Furthermore, the gap that mothers perceive between fathers' ideal and actual involvement in childrearing is a significant source of mothers' stress (Milkie et al. 2002; Renk et al. 2003).

Gender socialization patterns in families vary within different racial–ethnic groups and in different generations. Latinos have generally been thought to be more traditional in their gender roles, although this varies by generation and by the experiences of family members in the labor force. Thus, although Latinas' parents generally have traditional expectations about how girls behave, the girls themselves have a more complex critique of their gender roles. Gender expectations for Mexican and Puerto Rican American women are more traditional in the older generations. Mexican American men are also more likely than Mexican American women to have traditional views toward women's responsibility for child care (Denner and Dunbar 2004; Gowan and Trevino 1998; Raffaelli and Ontal 2004).

Family experiences also influence patterns of gender socialization. For example, Mexican married couples who migrate to the United States tend to adopt more egalitarian family roles. This is not simply the result of living in a different culture, as many would believe, but is an adaptation to migration itself. Families are often separated during the early phases of their migration. Men may live in bachelor communities for a time, where they often learn to cook and clean for themselves. Once the family is reunited, the men do not necessarily discard these newly learned behaviors (Hondagneu-Sotelo 1992, 1994). Other immigrant groups have similar experiences. In China, women have historically had low rates of labor force participation, but when they migrated to the United States, like other immigrant groups, women's work was necessary to support families. Cultural norms about the desirability of women working then change as the result of actual experience (Geschwender 1992).

Childhood Play and Games From the time they become aware of their surroundings, children are socialized to adopt behaviors and attitudes judged appropriate for members of their sex. Socialization comes not only from parents and other family members, but also from peers. Through play, children learn patterns of social interaction, cognitive and physical development, analytical skills, and the values and attitudes of their culture.

THINKING SOCIOLOGICALLY

Visit a local toy store and try to purchase a toy for a young child that is not gender-typed. What could you buy? What could you not buy? What does this teach you about *gender role socialization*? (If you take a child with you, note what toys he or she wants and does not want. What does this tell you about how effective *gender socialization* is?)

Studies of children's play show numerous gender patterns that have consequences for skills carried into adulthood. Boys are encouraged to play outside; girls, inside. Boys' toys frequently promote the development

Gender socialization is influenced by peers, as well as by parents, the media, schools, and religious institutions.

of militaristic values and tend to encourage aggression, violence, and the stereotyping of enemies—values rarely embodied in girls' toys. Girls' play tends to involve a more communal style of getting along in groups, whereas boys' play tends to emphasize more individualistic style (Hibbard and Buhrmester 1998). Girls also play more cooperatively when they are in same-sex groups (Neppl and Murray 1997). At the same time, boys exert power over girls when they play together, and boys typically establish the conditions of the play activities (Voss 1997).

These gender-typed forms of play are precursors to the behaviors that can appear in adult life. The competitive play of boys socializes them for the hierarchical world they encounter as adults (Lever 1978). The competencies they learn in childhood play are those they need for success in large, competitive, rule-oriented organizations, such as corporations. The greater likelihood that boys will play on team sports and in large groups also teaches boys how to compete and cooperate when they later work in group settings. Although the competencies that women learn as young girls, such as cooperation and flexibility, are important to working in groups, they tend not to be as valued in male-dominated groups and organizations, even though they are no less important.

Gender socialization does not just occur in early childhood. Patterns of play and social expectations continue throughout adolescence and into adulthood. In middle childhood, boys organize their play in definite hierarchical structures, although not all boys enjoy high status in these hierarchies. Girls' groups tend not to have the same single hierarchy that boys' groups have, and they tend not to rest on a singular notion of femininity in the same way that boys' groups rest on a tightly constructed concept of masculinity. Still, both boys and girls have to negotiate these "gender zones," constructing definitions of themselves vis-à-vis the other gender and defining masculinity in terms of power, specifically

in opposition to the powerlessness of girls (McGuffey and Rich 1999). The long-term consequences are the creation of identities that associate masculinity with power over women.

Schools and Gender Socialization Schools have particularly strong influence on gender socialization because of the time children spend in them. As we saw in Chapter 4, teachers often have different expectations for boys and girls. Researchers have found that boys call out answers eight times more often than girls. In general, boys in school get more attention, even if it is negative attention. When teachers of either sex respond more to boys, either positively or negatively, they heighten boys' sense of importance (American Association of University Women 1992, 1998; Sadker and Sadker 1994).

Children's books in schools also communicate gender expectations. Even with publishers' guidelines that discourage stereotyping, textbooks still depict men as aggressive, argumentative, and competitive. Men and boys are also more likely to be in the titles, pictures, and central roles (Tepper and Cassidy 1999; Evans and Davies 2000). Even classic fairy tales emphasize particular beauty ideals for girls and generally suggest that women only gain happiness through finding the right man (Grauerholz and Baker-Sherry 2003).

Classroom experiences influence gender identity by teaching boys and girls different skills. There are small differences in boys' and girls' abilities in the early years. At the preschool level, girls tend to exceed boys in verbal skills, and boys slightly exceed girls in mechanical ability. By the time boys and girls leave school, however, there are great differences in their skills, interests, and abilities. Extensive research on this subject has generally concluded that teacher expectations, classroom interaction, the content of the curriculum, and the representation of men and women as teachers and school leaders all communicate to students that there are different expectations for women and men (American Association of University Women 1998; Sadker and Sadker 1994).

Religion and Gender Socialization Religion is an often overlooked but significant source of gender socialization. The major Judeo-Christian religions in the United States place strong emphasis on gender differences, with explicit affirmation of the authority of men over women. In Orthodox Judaism, men offer a prayer blessing God for not having created them a woman or a slave. The patriarchal language of most Western religions and the exclusion of women from positions of religious leadership in some faiths also signify the lesser status of women in religious institutions.

Any religion, interpreted in a fundamentalist way, can be very oppressive to women, as recent events in many Islamic societies show. One of the important things to know about any religion is that all religious beliefs and texts are subject to interpretation. The most strict believers in any faith tend to hold the most traditional views of women's and men's roles. The influence of religion on gender and attitudes cannot, however, be separated from other factors. For many, religious faith inspires a belief in egalitarian (i.e., equal) roles for women and men. In Christian faith, as well as in Islamic faith, women frequently cite their religious teachings as reason to question and resist sexist practices (Gerami and Lehnerer 2001). Interpretations of religious doctrine can also change over time; as an example, evangelical Protestants' support for shared household roles, women's employment, and women's church leadership has actually declined in recent years (Petersen and Donnenwerth 1997).

The Media and Gender Socialization The media in their various forms (television, film, magazines, music, and so on) communicate strong, some would even say cartoonish, gender stereotypes. Despite some changes in recent years, television, the most pervasive communication medium, continues to depict highly stereotyped roles for women and men. Men outnumber women on television and women are underrepresented in leading roles in film (Eschholz 2002). Women play strong and independent roles on some shows, yet are

FORCES OF SOCIAL CHANGE

The Arrival of Women's Sports

During the Women's World Soccer Cup in 1999, 90,000 people filled the Rose Bowl in Pasadena—the largest crowd ever to attend a women's sporting event. Those who watched the event on television at home saw major corporations (such as Nike) present newly designed television commercials encouraging young girls to be strong and self-confident, to compete, and to develop team spirit. Who could have imagined such change not that many years ago when few women played competitive sports and no women sportscasters were heard in the major media?

Not many years ago, few women were allowed to play competitive sports in school and when they did, it was typically in activities judged appropriate for their gender roles. Now women's basketball on many campuses attracts big audiences (though rarely as big as for men's games). Women play in national professional leagues and are increasingly visible in sports broadcasting.

Sociological research shows that girls who play sports are likely to have higher self-esteem, greater self-confidence, lower rates of school dropout, and more positive body images than girls who do not play sports (Dworkin and Messner 1999). What long-term effects do you think events such as the U.S. women's victory in the World Soccer Match might have for young women and men? More generally, how will women's increased sports participation affect the gender identities of women in your generation?

Further Resources: See Conniff, Ruth. "The Joy of Women's Sports." *The Nation,* August 10–17, 1998, pp. 26–30. ● ● ●

Social change is sometimes apparent by contrasting the social norms of different historical periods.

© AFP/Getty Images

© AP/Wide World Photos

MAPPING AMERICA'S DIVERSITY

MAP 12.1 Women's Participation in High School Sports (as percentage of total number of students participating in sports)

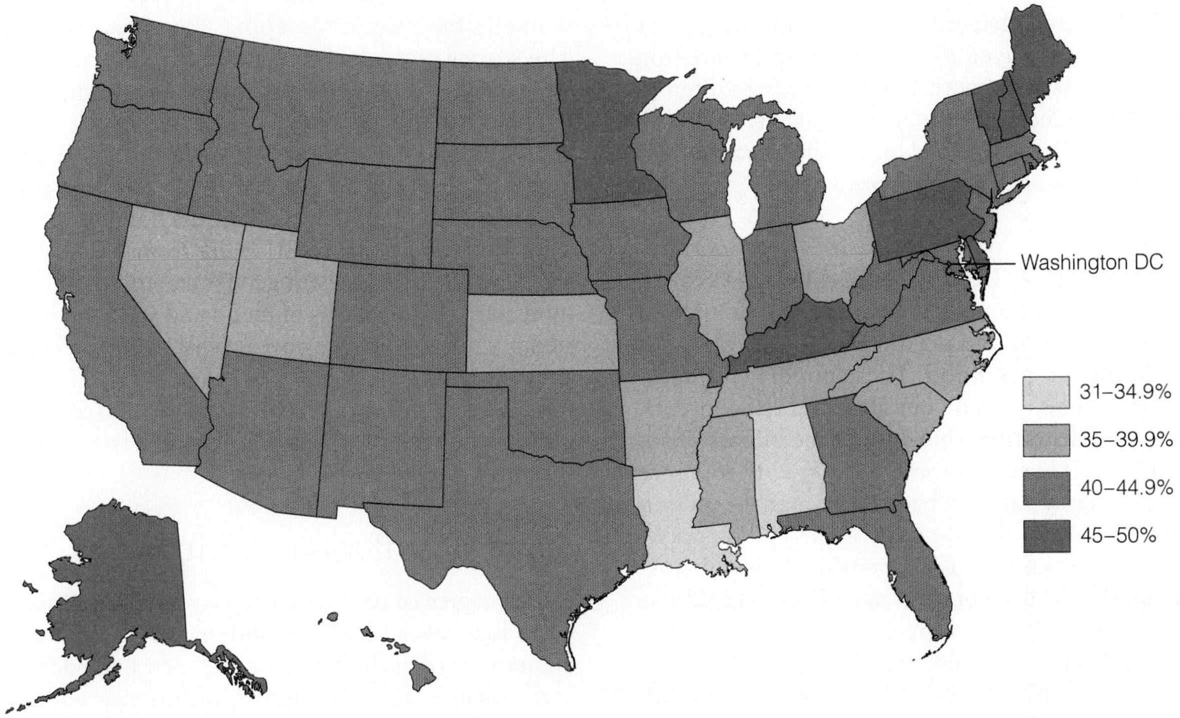

Washington DC

31–34.9%

35–39.9%

40–44.9%

45–50%

In your state, what is the percentage of young women who play sports in high school, relative to young men? What factors might increase or decrease the participation of young women and young men in sports? How would you design a research study to investigate the reasons for gender differences in sports participation? What else would you need to know, beyond what these data show, to identify the factors producing gender disparities in athletic participation?

Source: National Federation of State High School Associations (Indianapolis, IN). 2002. Website: **www.nfhs.org**

still more likely than men to be depicted as sex objects and to appear in situation comedies (Signorielli and Bacue 1999; Lin 1998).

Television also delivers unrealistic portrayals of women and men in terms of age and appearance. The majority of women characters are between the ages of eighteen and thirty-four, although in the actual population only 28 percent of women are in this age group. Women on television are also more likely than men to be shown provocatively dressed—in nightwear, underwear, swimsuits, and tight clothing. Female characters in prime-time shows, regardless of their race, are also depicted as much thinner than typical women, reinforcing cultural ideals of thinness (Fouts and Burggraf 1999). Gender stereotypes on television also cross with racial stereotypes, with White men shown as exercising more authority than either White women or African American men and women. African American men, however, are shown as aggressive and African Ameri-

can women as inconsequential (Coltrane and Messineo 2000).

Social scientists debate the extent to which people believe what they see on television, but research with children shows that they identify with television characters. Children report that they want to be like television characters when they grow up. Boys tend to identify with characters based on their physical strength and activity level; girls relate to perceptions of physical attractiveness (Signorielli 1989; Signorielli et al. 1994). Even with adults, researchers find that there is a link between seeing sexist images and attitudes such as lower acceptance of feminism, more traditional views of women, and attitudes supporting sexual aggression (MacKay and Covell 1997; Garst and Bodenhausen 1997).

Advertisements are another important vehicle for the communication of gender images to the public—one that is especially noted for the communication of idealized, sexist, and racist images of women and men (Cortese 1999). Women in advertisements are routinely shown in poses that would be shocking were the character male. Consider how often women are displayed in ads dropping their pants, skirts, or bathrobe, or squirming on beds. How often are men shown in such poses? Men are now displayed as sex objects in advertising more often than in the past, but not nearly as often as women. The demeanor of women in advertising—on the ground, in the background, or looking dreamily into space—makes them appear subordinate and available to men.

Other parts of popular culture are also a source of gender stereotypes. Greeting cards, CD/DVD covers, books, songs, films, and comic strips all communicate images representing the presumed cultural ideals of womanhood and manhood. These popular products have an enormous effect on our ideas and self-concepts. To take one illustration, think about the impact of romance novels. Harlequin, only one of many romance publishers, is estimated to have more than 14 million loyal readers, with sales in excess of 188 million books per year. In one major bookstore chain, Harlequin novels account for 30 percent of all paperback sales. These romances are marketed to appeal to women, who read them for relaxation and escape. The covers routinely show a feminine, usually White, blonde woman, swoon-

ing in the arms of a dark-skinned man—a man often depicted as savage and ravishing, often suggestive of racial stereotypes of Native American men (Nagel 2003). The plots reflect the powerlessness and depersonalization many women find at work, even while fueling women's fantasies of escaping from these restraints. The stories portray heroines as active and intelligent, struggling to win the recognition and love of their bosses (Rabine 1985).

Mass-marketed products such as this shape our understanding of the possibilities open to ourselves and to others and reflect the values of the dominant culture. Thus popular books like *Men are from Mars, Women are from Venus* reconfirm gender stereotypes and promote narrow definitions of men's and women's life possibilities. Or in romance novels, when women's independence is treated as fantasy or when a love affair with one's boss is shown as the best avenue to recognition, popular culture provides ideological support for the subordination of women in society.

The Price of Conformity

A high degree of conformity to stereotypical gender expectations takes its toll on both men and women. The higher rate of early death among men from accidents and violence can be attributed to the stress and injury

---

A SOCIOLOGICAL EYE ON THE MEDIA:
Cultural Gatekeepers and the Construction of Femininity

Many have noted the distorted images of women that appear in the media. The common argument is that media images present an unrealistic image of women which shapes women's self-concepts and limits their sense of possibilities for their appearance, their relationships, their careers, and so forth. Femininity is defined in the media by *cultural gatekeepers,* those who make decisions about what images to project. Cultural gatekeepers also have to respond to audience criticism. How they do so is an important part of the institutional process by which media images are sustained.

One sociologist, Melissa Milkie, wanted to explore how images of femininity are constructed in the media, particularly when producers encounter criticism from their audience. As readers, girls have protested many of the narrow and limiting images they see

in the media, particularly those portraying girl's bodies. Milkie interviewed ten top editors of leading girls' magazines to find out how they, as cultural gatekeepers, responded to the criticism from girls that the images of girls in teen magazines do not reflect what "real girls" are like.

Milkie found that even the top editors think there are institutional limitations on what they can do to respond to girls' criticism. The editors—who were very sensitive to the criticisms they received—either said there was not much they could do about it or they dismissed the girls' complaints as misguided. They would claim the image was beyond their control—either because of the artistic process, advertisers' needs, or the culture itself. Thus, despite their positions of power, editors believed they could not fully control the images that appear. They pointed to

institutional constraints that, in effect, thwarted efforts for change. Some editors simply dismissed the criticisms as girls' misreading the intent or meaning of an image.

Either way, Milkie's research shows how the organizational complexity of media institutions limits how much change is possible in how images of femininity are constructed. Market forces, advertisers, the values of producers, and the values of the public all intertwine in shaping the decisions of cultural gatekeepers. Milkie also shows, however, that people are not passive about what they see in the media, suggesting that how people respond to images in the media is an important part of the effect of such images in society.

Source: Milkie, Melissa A. 2002. "Contested Images of Femininity: An Analysis of Cultural Gatekeepers' Struggles with the 'Real Girl' Critique." *Gender & Society* 16 (December): 839–859. •••

associated with the cultural definition of masculinity, which includes physical daring and risk-taking. The strong undercurrent of violence in today's culture of masculinity encourages men to engage in behaviors that put them at risk in a variety of ways.

Adhering to gender expectations of thinness for women and strength for men is related to a host of negative health behaviors, including eating disorders, smoking, and steroid abuse. The dominant culture promotes a narrow image of beauty for women—one that leads many women, especially young women, to be disturbed about their body image. Striving to be thin, millions of women engage in constant dieting, fearing being "fat" even when they are well within or below healthy weight standards. Many develop eating disorders by purging themselves of food or cycling through various fad diets—behaviors that can have serious health consequences. Many young women develop a distorted image of themselves, thinking they are "overweight" when they may actually be dangerously thin. And, despite the known risks of smoking, increasing numbers of young women smoke and do so not only because they think it "looks cool," but because they think it will keep them thin. Eating disorders can be related to having a history of emotional trauma, such as sexual abuse, but they also come from the promotion of thinness as an ideal beauty standard for women—a standard that actually puts girls' and women's health in jeopardy (Logio-Rau 1998; Thompson 1994).

The strong undercurrent of violence in today's culture of masculinity encourages men to engage in behaviors that put women and men at risk in a variety of ways. Violence associated with gender roles puts women at risk. Rape, sexual assault, domestic violence, and sexual harassment are all linked to the association of gender with men's power—power that is too frequently manifested in physical and emotional violence against women. Violence against women is endemic in the United States but is also a worldwide problem, as a recent United Nations (UN) report concluded. According to the UN, violence is a matter of women's human rights—a "manifestation of historically unequal power relations between men and women, which have led to domination over and discrimination against women by men and to the prevention of the full advancement of women" (UNICEF 2000: 1). Violence takes many forms, including sexual abuse, rape, wife beating, genital mutilation, honor killings, and rape. Around the world, the UN is working in various ways to reduce violence against women, including some initiatives to help men examine the cultural assumptions about masculinity that promote violence.

© Jacksonville Journal Courier/The Image Works

Despite cultural changes in gender roles, young girls learn stereotypical ideals for feminine beauty early in life.

Men, too, pay the price of overconformity if they too thoroughly internalize gender expectations that they must be independent, self-reliant, and not emotionally expressive. Men's gender socialization discourages intimacy among them, thereby affecting the quality of men's friendships (Basow and Rubenfeld 2003). However, in recent years, men are more likely to say they admire men who show a more sensitive side, and they say they have become better able to express their feelings (Roper Organization 1995). Conformity to traditional gender roles denies women access to power, influence, achievement, and independence in the public world, and denies men the more nurturing, emotional, and other-oriented worlds that women have traditionally inhabited.

Race, Gender, and Identity

Gender identity emerges from the different experiences that we have—experiences that differ not only because of gender, but also other factors such as racial identity. Moreover, people have to form these identities in a context where stereotypes narrowly cast people depending on both their race and gender. Stereotypes are powerful images that can be used to try to justify social injustice, thus Patricia Hill Collins (1990) calls them "controlling images," to emphasize how stereotypes can be part of a system of social control.

As you think about this, you will see that there is interplay between race and gender stereotypes, as we saw in Chapter 11. Gender and race together create stereotypes of different groups of men and women. African American men are stereotyped as hypermasculine and oversexed; African American women as producing too many babies and being "welfare queens." Asian American women are stereotyped as submissive but dainty sex objects (think of the geisha image); Asian American men, as sexless nerds. Jewish American women are stereotyped as rich, spoiled, and materialistic; Jewish men as intellectual, but asexual. Latinos are stereotyped as macho and, like African American men, sexually passionate. Latinas are stereotyped as "hot." White women are stereotyped as madonnas or sluts (also introducing class into the interplay of race and gender, because working-class women are more likely to be seen as "slutty" and upper-class women as frigid and cold).

Because the experiences of race and gender socialization affect each other, men and women from various racial groups have different expectations regarding gender roles. However, the differences are often not the ones people expect based on the gender stereotypes they have absorbed about other groups. For example, when asked to rate desirable characteristics in men and women, White men and women are more likely than Hispanic men and women to select different gender traits for men and women. This runs counter to the idea that Hispanics hold highly polarized views of manhood and womanhood. Comparing Hispanics, Whites, and African Americans, it is African Americans who are most likely to find value in both sexes displaying various traits such as being assertive, athletic, self-reliant, gentle, and eager to soothe hurt feelings (Harris 1994).

African American men and woman also show significant support for feminism and egalitarian views of men's and women's roles, although African American women are somewhat more liberal than African American men (Hunter and Sellers 1998). Asian American women are more likely than Asian American men to value egalitarian roles for men and women (Chia et al. 1994). Sociological research also reveals that Chinese American men actively use many strategies to develop identities that challenge gender- and race-stereotypic views of them (Chen 1999).

Society encourages African American women, as it does White women, to become nurturing and other-oriented, but they are also socialized to become self-sufficient, to aspire to an education, to want an occupation, and to regard work as an expected part of a woman's role. They are also expected to be more independent than White women. This is specific to their experience as African American women (Ladner 1995; Slevin and Wingrove 1998).

African American women are also more likely than White women to reject the gender stereotyping of the dominant culture. These attitudes among African American women are probably related to the examples of their mothers, who were more likely than White women to have been employed and supporting themselves. The image they presented to their daughters and the lessons they passed on would have encouraged self-sufficiency (Wharton and Thorne 1997). The same trends are becoming true for White women as more have entered the labor force.

Among men, too, gender identity is affected by race. Latino men, for example, bear the stereotype of *machismo*—exaggerated masculinity. Machismo is associated with sexist behavior by men, within Latino

UNDERSTANDING DIVERSITY
Racializing Gender/Gendering Race

Gender stereotypes go hand in hand with racial stereotypes. As discussed in the prior chapter, there is interplay between race and gender stereotypes. You see this interplay by analyzing how gender and race together construct stereotypes of different groups of men and women. African American men are stereotyped as hypermasculine and oversexed; African American women as producing too many babies and being "welfare queens." Asian American women are stereotyped as submissive but dainty sex objects (think of the geisha image); Asian American men as sexless nerds. Jewish American women are stereotyped as Jewish American princesses, "JAPS"— rich, spoiled, and materialistic; Jewish men as intellectual, but asexual. Latinos are stereotyped as "macho" and, like African American men, sexually passionate. Latinas are stereotyped as "hot." White women are stereotyped as "madonnas" or "sluts" (also introducing class into the interplay of race and gender, since working-class women are more likely to be seen as "slutty" and upper-class women as frigid and cold.

These stereotypes also show you how racism and sexism shape our views of sexuality and, likewise, how sexuality is used to construct race and gender stereotypes. And as Ruth Atkin and Adrienne Rich (2001) have pointed out, it is probably not accidental that the stereotype of the Jewish American princess—JAP—uses a term that is also associated with anti-Asian racism. Patricia Hill Collins calls these "controlling images"—images that are created by powerful groups and are used to justify race, class, and gender oppression. Whose interests do such stereotypes serve and how do they affect the interactions between different groups?

•••

DOING SOCIOLOGICAL RESEARCH
Succeeding Against the Odds

Given the vast amount of research detailing the obstacles to women's success, how do young women ever establish positive identities for themselves and become socialized to succeed, not fail? Particularly for women of color who must negotiate the stereotypes posed by both gender and race, succeeding can be a challenge. How do they do it? This is what sociologists Heidi Barajas and Jennifer Pierce wanted to know from their study of young Latinas and Latinos.

Barajas and Pierce selected seventy-two Latina and Latino college and high school students who were part of a mentoring program wherein college students mentored high school students who were college bound. The forty-two women and thirty men who participated in the study were from various Hispanic backgrounds, mostly Mexican, Puerto Rican, and Honduran in origin, all of them poor or working-class and second- or third-generation immigrants. Most prior researchers have focused on the high dropout rate from high school among such students. What makes them successful?

Both the women and men in this study reported being made to feel different and inferior at school, but they all expressed a desire to help those in their community. Participating in the mentoring program gave them all an enhanced sense of community awareness. They did not feel they had to give up their racial–gender identity to be successful, as many assimilation theorists would claim (see Chapter 11). The Latinas (women) used their relationships with other Latinas as paths to success, despite the chilly climate they encountered in school. The men, on the other hand, most often found sports to be a source for mentoring—usually by White men. Barajas and Pierce conclude that finding such support, though men and women do so in different ways, provides the kind of encouragement and protective relationships that help Latinas and Latinos navigate their way to success.

Questions to Consider

1. When you were seeking admission to college, what support systems (people or groups) gave you the information and resources you needed? How were these systems related to your social class position, your race, and your gender? *Keywords: class and college admissions, gender and educational equity*

2. Have there been any significant *mentors* in your life? What are characteristics of a mentoring relationship and is it critical to people's success? What sociological functions do mentors provide? *Keywords: mentoring, gender and mentor*

We have included InfoTrac College Edition keywords at the end of each question to make it easier for you to find more to read on these topics. Go to **www.infotrac-college.com**, an online library, to begin your search.

Source: Barajas, Heidi Lasley, and Jennifer L. Pierce. 2001. "The Significance of Race and Gender in School Success among Latinas and Latinos in College." *Gender & Society* 15 (December): 859–878. ••••

culture, and is associated with honor, dignity, and respect (Mirandé 1979). Maxine Baca Zinn argues that machismo is typically misinterpreted within the dominant culture. Although macho behaviors do exist among Latinos, they are not the only way that Latinos interact with women. Researchers find that Latino families are rather egalitarian, with decision making frequently shared by men and women. To the extent that machismo exists, it is not just a cultural holdover from Latin societies, but also can be how men defy their racial oppression (Baca Zinn 1995). The definition of manhood among Latinos is more multidimensional than cultural stereotypes suggest.

African American men also define manhood in ways far more complex than simple stereotypes suggest. Self-determination, responsibility, and accountability to family or community or both are the attributes African American men most commonly associate with manhood. Power over others rates as one of the least important traits (Hunter and Davis 1992). African American men are also more likely than White men to emphasize their importance as the breadwinner, whereas White men are more likely to value the role of being nurturers

as adults (Harris et al. 1994). Men's roles in society, like women's, are conditioned by the social context of their experience. Gender identity is merged with racial identity for all people. For those in racial minority groups, this means that concepts of womanhood and manhood are shaped by the patterns of domination and exclusion that race and gender produce.

Gender Socialization and Homophobia

Homophobia is the fear and hatred of homosexuals (see also Chapter 13). Homophobia plays an important role in gender socialization because it encourages stricter conformity to traditional expectations, especially for men and young boys. Slurs directed against gays encourage boys to act more masculine as a way of affirming for their peers that they are not gay. As a cosequence, homophobia also discourages so-called feminine traits in men, such as caring, nurturing, empathy, emotion, and gentleness. Men who endorse the most traditional male roles also tend to be the most homophobic

(Burgess 2001; Alden 2001; Basow and Johnson 2000; Whitley 2001). In this way, homophobia is one of the means by which socialization into expected gender roles takes place. The consequence is not only conformity to gender roles, but a learned hostility toward gays and lesbians.

Homophobia is a learned attitude, as are other negative social judgments about particular groups. Homophobia becomes embedded in our social definitions of what it means to be a man or a woman. The relationship between homophobia and gender socialization illustrates how socialization contributes to social control. Boys are raised to be "manly" by repressing so-called feminine characteristics in themselves. Being called a "fag or a "sissy" is one of the sanctions that forces conformity into expected gender roles. Similarly, pressures on girls to abandon tomboy behavior are a mechanism by which girls are taught to adopt the behaviors associated with womanhood. Being labeled a lesbian may cause those with a strong love of women to repress this emotion and direct love only toward men. We can see, therefore, how homophobic ridicule, though it may be in the context of play and joking, has serious consequences for both heterosexual and homosexual men and women. Homophobia socializes people into expected gender roles and it produces numerous myths about gays and lesbians—examined in more detail in the next chapter.

Once people internalize societal expectations, they do not challenge or question the status quo. This defines the *social construction of gender:* What appears to be normal or customary is only that which people have been taught is normal. Gender has great significance in society, but the specific forms it takes are learned. Gender is therefore fluid, and because gender expectations are learned, it is possible to redefine and learn them in new ways. Little is inherent in the social definition of women and men as gendered persons that could not be reconsidered and changed.

The Institutional Basis of Gender

The process of gender socialization reveals much about how gender identities are formed, but gender is not just a matter of identity: Gender is embedded in social institutions. This means that institutions are patterned by gender, resulting in different experiences and opportunities for men and women. Sociologists analyze gender as interpersonal expectations as well as a characteristic of institutions. The concept **gendered institution** means that entire institutions are patterned by gender. Gendered institutions are the total pattern of gender relations, which includes the following (Acker 1992):

- stereotypical expectations;
- interpersonal relationships;

- the division of labor along lines of gender;
- the images and symbols that support these divisions; and
- the different placement of men and women in social, economic, and political hierarchies of institutions.

At school, for example, children learn gender roles, but schools also are gendered institutions because they embed specific gender patterns. Seeing institutions as gendered reveals that gender is an attribute of individuals but is a part of the structure of social institutions (Acker 1992: 567).

As an example of the concept of gendered institution, think of what it is like to work as a woman in an organization dominated by men. Women in this situation report that the importance of men in the organization is communicated in subtle ways, while women are treated like outsiders. Important career connections may be made in the context of men's informal interactions with each other—both inside and outside the workplace. Women may be treated as tokens or may think that company policies are ineffective in helping them cope with the particular demands in their lives. These institutional patterns of gender affect men, too, particularly if they try to establish more balance between their personal and work lives. To say then that work institutions are gendered means that, taken together, there is a cumulative and systematic effect of gender throughout the institution.

Good examples of gendered institutions are all-male military academies. Military academies have now been forced to admit women as the result of a Supreme Court decision involving the Virginia Military Institute (VMI) and the Citadel. These institutions had been created with masculinity inherent to the schools' character both in their image of themselves, the behaviors expected of cadets, and the opportunites given (or denied) to women (Kimmel 2000). Other gendered institutions may not be as extreme as this but nonetheless are structured by gendered hierarchies, differences in what men and women do in the institution, and ideas about gender-appropriate behaviors and beliefs. You can see this if you identify an institution with which you are familiar and begin to describe what men and women do in this institution, where they are placed relative to each other, what beliefs support this system, and what happens when someone tries to cross presumed gender lines in the institution.

The concept of gendered institutions also shows the limitation of thinking about gender only in terms of social roles. Gender roles, as we have seen, are the learned patterns of behavior associated with being a man or a woman. Significant as these roles are, the concept of gendered institutions goes further in explaining how gendered patterns persist—even when people themselves try to change their roles. Gender is not only a learned role; it is also part of social structure, just as class and race are social structural dimensions of society. People do not think about the class system or racial inequality in terms of "class roles" or "race roles." Race relations and class relations are far more than matters of interpersonal interaction. Race, class, and gender inequalities are experienced within interpersonal relationships, but they are also more. Just as it would seem strange to think that race relations in the United States are controlled by race-role socialization, it is also wrong to think that gender relations are the result of gender socialization alone. Most people understand that race relations are a matter of systems of privilege and inequality. Likewise, gender is a system of privilege and inequality in which women are systematically disadvantaged relative to men. Gender, like race and class, involves institutionalized power relations between women and men, and it involves unequal access to social and economic resources (Lopata and Thorne 1978; Andersen 2003).

To put the concept simply, socialization and roles cannot explain everything. Socialization affects how women and men choose among options, but the institutional basis of gender determines which options they will choose among. Gender shapes access to economic and political resources, as well as more personal dimensions such as self-definition, relationships with others, and perceived worth. Studying gender as an institutional, or social structural, phenomenon also makes it more clear how gender intersects with systems of race and class relations. Latinas and Native American and African American women, for example, are oppressed by both race and gender and perhaps by class. White men on the whole are accorded more power, prestige, and economic resources than women, but not all men share these advantages equally. As a group, Latino men are disadvantaged relative to White men and relative to some White women (see Figure 12.3). Thinking about gender and its relationship to race and class reveals that gender pervades society beyond the effects of gender socialization. Gender permeates all institutions because it is a strong dimension of how society is organized. Gender stereotypes pervade society, but gendered institutions also create different opportunity structures for women and for men. For this reason, sociologists also study gender as a system of institutional inequality—called gender stratification.

Gender Stratification

Gender stratification refers to the hierarchical distribution of social and economic resources according to gender. Most societies have some form of gender stratification, although the specific form varies from country to country. Comparative research finds that women are more nearly equal in societies characterized by the following (Chafetz 1984):

- Women's work is central to the economy;
- Women have access to education;
- Ideological or religious support for gender inequality is not strong;
- Men make direct contributions to household responsibilities, such as housework and child care;
- Work is not highly segregated by gender; and
- Women have access to formal power and authority in public decision making.

In Sweden, which has a relatively high degree of gender equality, the participation of both men and women in the workforce and the household (including child care and housework) is promoted by government policies. Women also have a strong role in the political system, although women in Sweden still earn less than men and tend to work in occupations different from men. In many countries, women and girls have less access to education than men and boys, although the gap is closing. Two-thirds of the illiterate people in the world are women (United Nations 2000b).

As the preceding list suggests, gender stratification is multidimensional. In some societies, women may be free in some areas of life but not in others. In Japan, for example, women tend to be well-educated and have high labor force participation. Within the family, however, Japanese women have fairly rigid gender roles, but in Japan the rate of violence against women (rape, prostitution, and pornography) is low relative to other

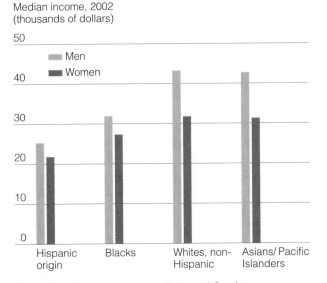

Figure 12.3 Median Income by Race and Gender

Source: U.S. Census Bureau. 2003. *Historical Income Tables–People,* Table P-36A-E. Website: **www.census.gov**

nations, even though women are widely employed as "sex workers" in hostess clubs, bars, and sex joints (Allison 1994). Patterns of gender inequality are most reflected in the wage differentials between women and men around the world, as Figure 12.4 shows.

Gender stratification can be extreme, such as was the case in Afghanistan under the Taliban regime. The Taliban, an extremist militia group, seized power in Afghanistan in 1996, stripping women and girls of basic human rights. Women were banished from the work force, schools were closed to girls, and women who were enrolled in the universities were expelled. Women in Afghanistan were prohibited from leaving their homes unless accompanied by a close male relative. The windows of women's houses were painted black so that women were literally invisible to the public. This extreme segregation and exclusion of women from public life has been labeled **gender apartheid**. Gender apartheid is also evident in other nations, although not so extreme as it was under Taliban rule. But, in Saudi Arabia women are not allowed to drive; in Kuwait, they cannot vote.

Under the rule of the Taliban in Afghanistan, women had to be completely veiled. When the Taliban lost control during the war in 2001, many women shed their burqas, showing their faces in public for the first time in years.

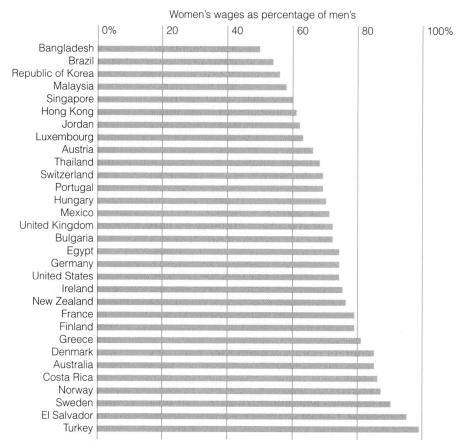

Figure 12.4 The Wage Gap: An International Perspective

Data: From United Nations. 2000. *The World's Women: Trends and Statistics.* Washington, DC: Population Reference Bureau, p. 132.

Sexism and Patriarchy

Gender stratification tends to be supported by beliefs that accept gender inequality. An **ideology** is a belief system that tries to explain and justify the status quo. **Sexism** is an ideology, but it is also a set of institutionalized practices and beliefs through which women are controlled because of the significance given to differences between the sexes. Like racism, sexism distorts reality, making behaviors seem natural when they are rooted in entrenched systems of power and privilege. The idea that men should be paid more than women because they are the primary breadwinners reflects sexist ideology, but when this concept becomes embedded in the wage structure, people no longer must believe explicitly in the original idea for the consequences of sexism to be propagated.

Sexism and racism tend to go hand in hand. Both generate social myths that have no basis in fact, but they justify the continuing advantage of dominant groups over subordinates. A case in point is the belief that women of color are being hired more often and promoted more rapidly than others. This misrepresents the facts. Women rarely take jobs away from men because most women of color work in gender- and race-segregated jobs. The truth is that women, especially women of color, are burdened by obstacles to job mobility that are not present for men, especially White men (Browne 1999). The myth that women of color get all the jobs makes White men seem to be the victims of race and gender privilege. Although there may be occasional individual cases where a woman of color (or a man, for that matter) gets a job that a White man also applied for, the general pattern favors White men.

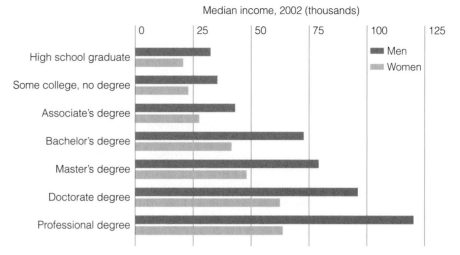

Median income, 2002 (thousands)

Figure 12.5 Education, Gender, and Income

Data: From U.S. Census Bureau. 2003. *Historical Income Tables–People,* Table P28. Website: **www.census.gov**

As an ideology, sexism is part of the structure of society. **Patriarchy** refers to a society or group in which men have power over women. Patriarchy is common throughout the world. In patriarchal societies, husbands have authority over wives in the private sphere of the family, and public institutions are also structured around male power. Men hold all or most positions of public power in patriarchal societies, whether as chief, president, chief executive officer (CEO), or other leadership positions. Forms of patriarchy vary from society to society. In some, it is rigidly upheld in both the public and private spheres. In these societies, women may be formally excluded from voting, holding public office, or working outside the home. In societies such as the contemporary United States, patriarchy may be somewhat diminished in the private sphere (at least in some households), but the public sphere continues to be based on patriarchal relations.

Matriarchy has traditionally been defined as a society or group in which women have power over men. Anthropologists have debated the extent to which such societies exist, but new research finds that matriarchies do exist, though not in the form the traditional definition implies. Based on her study of the Minangkabau—a matriarchal society in West Sumatra (in Indonesia)—anthropologist Peggy Sanday argues that scholars have used a western definition of power that does not apply in nonwestern societies. The Minangkabau define themselves as a matriarchy, meaning that women hold economic and social power, but this is not rule by women since they believe that rule should be by consensus, including men. Thus, matriarchy exists, but not as a mirror image of patriarchy (Sanday 2002).

Women's Worth: Still Unequal

Gender stratification is especially obvious in the persistent earnings gap between women and men. The gap has closed somewhat since the 1960s, when women earned 59 percent of what men earned. Women today who work year-round and full-time still earn, on average, only 73 percent of what men earn. Women with college degrees earn the equivalent of men who have only some college or an Associate's degree (see Figure 12.5). The median income for women working full-time and year-round in 2002 was $30,203; for men, $39,429 (De-Navas et al. 2003).

The income gap between women and men persists despite the increased participation of women in the labor force. The **labor force participation rate** is the percentage of those in a given category who are employed either part-time or full-time. By 2003, 60 percent of all women were in the paid labor force, compared with 74 percent of men. Since 1960, married

Jobs that have historically been defined as "women's work" are some of the most devalued in terms of income and prestige, despite their importance for such things as nurturing children.

women with children have nearly tripled their participation in the labor force. Two-thirds of mothers are now in the labor force, including more than half of mothers with infants. Current projections indicate that women's participation will continue to rise. Men's labor force participation is expected to decline slightly (U.S. Department of Labor 2004).

More women are also now in the labor force—a pattern that has long been true for women of color, but now also characterizes the experience of White women as the labor force participation rates of the two groups has converged. More women are now also the sole supporters of their dependents, given the changes in family patterns in contemporary society. Why then does the pay gap persist?

Explaining the Pay Gap

Laws prohibiting gender discrimination have been in place for more than forty years since the passage of the Equal Pay Act of 1963—the first federal law to require that men and women receive equal pay for equal work. Most employers do not explicitly set out to pay women less than men, but despite good intentions and legislation on the books, differences in men's and women's earnings persist. Why? Research reveals three strong explanations for this continuing difference: overt discrimination, human capital theory, and dual labor market theory.

Overt Discrimination **Discrimination** refers to practices that single out some groups for different and unequal treatment. Despite the progress of recent years, overt and covert discrimination continue to afflict women in the workplace. Much discrimination is covert—that is, only revealed in patterns of differential treatment but not directly observable. Overt discrimination, though, also continues. Men, especially White men, by virtue of being the dominant group in society, have an incentive to preserve their advantages in the labor market. They do so by establishing rules that unequally distribute rewards. Women pose a threat to traditional White male privileges, and men may organize to preserve their power and advantage (Reskin 1988).

The discrimination explanation of the gender wage gap argues that dominant groups will use their position of power to perpetuate their advantage (Lieberson 1980). There is some evidence that this occurs. Historically, White men used labor unions to exclude women and racial minorities from well-paying, unionized jobs, usually in the blue-collar trades. A more contemporary example is seen in the efforts to dilute or repeal legislation that has been developed to assist women and racial–ethnic minorities. These efforts can be seen as an attempt to preserve group power.

Another example of overt discrimination is the harassment that women experience at work, including sexual harassment and other means of intimidation (see Chapter 18). Women who enter traditionally male-dominated professions suffer the most sexual harassment. The reverse seldom occurs for men employed in jobs historically filled by women. Men can be victims of sexual harassment, but the occurrence is rare. Sexual harassment is a mechanism for preserving men's advantage in the labor force—a mechanism that also buttresses the belief that women are sexual objects for the pleasure of men.

Human Capital Theory **Human capital theory** assumes that the economic system is fair and that competitive and wage differences reflect differences in the individual characteristics that workers bring to jobs. Factors such as age, prior experience, number of hours worked, marital status, and education are *human capital variables*. Human capital theory says that the extent to which human beings differ in these variables will influence their worth in the labor market. For example, frequent job turnover or work records interrupted by child rearing and family responsibilities could negatively influence the earning power of women.

Much evidence supports the human capital explanation for the difference in men's and women's earnings because education, age, and experience do influence earnings. However, when we compare men and women at the same level of education, prior experience, and number of hours worked per week, women still earn less than men. Intermittent employment is not as significant in explaining wage differences as human capital theory would lead one to expect (Padavic and Reskin 2002). Although human capital theory explains some of the differences between men's and women's earnings, it does not explain all the differences. Sociologists have looked to other factors to complete the explanation of wage inequality (Browne 1999).

The Dual Labor Market A third explanation of gender differences in earnings is **dual labor market theory,** which contends that women and men earn different incomes because they tend to work in different segments of the labor market. Women tend to work in jobs that employ mostly women and these jobs tend to have low wages and few job benefits. Once an earnings structure is established, it is hard to untangle cause and effect in the relationship between the devaluation of women's work and low wages in certain jobs. As a result, equal pay for equal work may hold in principle, yet it applies to relatively few people because most men and women are not engaged in "equal work."

According to dual labor market theory, the labor market is organized in two sectors: the *primary market*

Although only a small percentage of women work in jobs traditionally defined as for men, a growing number of women are entering such nontraditional fields. Women in such jobs often quickly discover that these are within gendered institutions.

and the *secondary market*. In the primary labor market, jobs are relatively stable, wages are good, opportunities for advancement exist, fringe benefits are likely, and workers are afforded due process. Working for a major corporation in a management job is a good example. Jobs in the primary labor market are often in large organizations where there is general stability, steady profits, and a rational system of management. The secondary labor market is characterized by frequent job turnover, low wages, short or nonexistent promotion ladders, few benefits, poor working conditions, arbitrary work rules, and capricious supervision. Many jobs students take, such as waiting tables, selling fast food, or cooking and serving fast food, fall into this category. Fortunately for students, unlike those permanently stuck in this labor market, these jobs are usually short-term.

Women and racial–ethnic minorities are far more likely to be employed in the secondary labor market than in the primary labor market. Even within the primary labor market, there are two tiers. The first tier consists of high-status professional and managerial jobs with good potential for upward mobility, room for creativity and initiative, and more autonomy. The second tier comprises working-class jobs, including clerical work and skilled and semiskilled blue-collar work. Women and minorities in the primary labor market tend to be in the second tier. These jobs are secure compared with jobs in the secondary labor market, but are more

tenuous and do not have as much mobility, pay, prestige, or autonomy as jobs in the first tier of the primary labor market.

In addition, the *informal sector* of the labor market has even greater wage inequality, no benefits, and little, if any, oversight of employment practices. Individuals may hire such workers as private service workers or under-the-table workers who perform a service for a fee (painting, babysitting, car repairs, and any number of services). Businesses and corporations also employ such workers and can reap huge profits by not paying benefits or providing compensation when workers are sick, injured, or disabled. No formal data have been gathered on the informal sector, because much of it tends to be in an underground economy, but women and minorities likely form a large segment of this market activity.

According to dual labor market theory, wage inequality is a function of the structure of the labor market, not the individual characteristics of workers, as suggested by human capital theory. Because of the dual labor market, men and women tend to work in different occupations, and when working in the same occupation, in different jobs. This is referred to as **occupational segregation,** a pattern in which different groups of workers are separated into different occupations. Occupational segregation can be by gender, class, race, and other factors, as we discuss further in Chapter 18.

Wages tend to be linked to occupational segregation. There is a direct association between the sex and race composition of a given job and the wages paid (Catanzarite 2003; Kmec 2003). In other words, the greater the proportion of women in a given occupation, the lower the pay. Figure 12.6, a scattergram, illustrates this trend. At one extreme, occupations close to 100 percent female (private household child-care workers and dental hygienists) pay only half of what is paid in jobs that are at least 50 percent male. Studies find that workers in jobs requiring nurturing social skills have the lowest pay, even when their education and experience are comparable to workers in other jobs (Kilbourne et al. 1994).

Occupational segregation is exacerbated by race in that the higher the proportion of women of color in a given occupation, the lower the wages. Not surprisingly, the jobs where White men are most prevalent are the best paid (U.S. Department of Labor 2004). The social structural analysis provided by dual labor market theory suggests that women's exclusion from job networks, the size of given industries, and other factors in the environment of work organizations are significant in explaining the wage gap between women and men (Browne 1999).

Each explanation—overt discrimination, human capital theory, and dual labor market theory—contributes to an understanding of the continuing differences

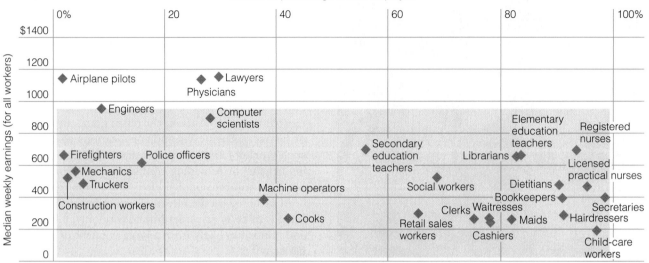

Figure 12.6 Earnings in Selected Occupations

This scattergram relates the percentage of women in selected occupations to the average weekly earnings in these occupations. The scattergram plots the percentage of women in an occupation (on the X axis, or bottom line of the figure) and the average weekly earnings in that occupation for all workers (on the Y axis, or the left side of the figure). Each dot in the scattergram thus represents a particular occupation. Those occupations to the right side are those with the highest percentage of women employed. For example, 97.1 percent of child-care workers are women; average weekly earnings in this occupation in 2003 were $198 per week. On the left side of the scattergram, note that women are 1.4 percent of airplane pilots, who have an average weekly earning of $1,138. After studying the figure carefully, what conclusions would you draw about gender segregation and the wage gap?

Data: From the U.S. Department of Labor. 2004. *Employment and Earnings.* Washington, DC: U.S. Department of Labor.

in pay between women and men. Wage inequality by gender is the result of multiple factors that together operate to systematically disadvantage women in the workplace.

Gender Segregation

Gender segregation refers to the distribution of men and women in different jobs in the labor force. It is a specific form of occupational segregation. Despite several decades of legislation prohibiting discrimination against women in the workplace, most women and men still work in gender-segregated occupations (Wootton 1997; Browne 1999). That is, the majority of women work in occupations where most of the other workers are women, and the majority of men work mostly with men. To this day, more than half of all employed women work as clerical workers, sales clerks, or in service occupations such as food service workers, maids, health-service workers, hairdressers, and child-care workers. Women tend to be concentrated in a small range of occupations, while men are dispersed over a much broader array of occupations (U.S. Department of Labor 2004).

Sociologists use the **index of dissimilarity** to measure the extent of occupational segregation. This measure

indicates the number of workers who would have to change jobs to have the same occupational distribution as the comparison group. By current estimates, at least 53 percent of men (or women) would have to change occupations to achieve occupational balance by gender. This decline of only 5 points from the mid-1980s is progress, but not much (Wootton 1997). Most women, in fact, continue to work in occupations where two-thirds or more of the other workers are women (U.S. Department of Labor 2004). This is especially true for women of color because gender segregation at work is aggravated by race. Women of color tend to be employed in occupations where they are segregated not only from men but also from White women, such as in domestic work in hotels. The occupations with the highest concentrations of women of color are also among the worst paid of all jobs (U.S. Department of Labor 2004).

The Devaluation of Women's Work

Across the labor force, women tend to be located in jobs that are the most devalued, causing some to wonder if the very fact that the jobs are held by women is what devalues the jobs. Why, for example, are elementary

Gender segregation in the labor market intersects with racial segregation, with jobs employing primarily women of color being among the least paid in society.

© John Eastcott/Yva Momatiuk/Woodfin Camp & Associates

school teachers (83 percent women) paid less than airplane mechanics (99 percent men)? The association of elementary school teaching with children and its identification as "women's work" lowers its prestige and economic value. If measured by the wages attached to an occupation, child care is one of the least prestigious jobs in the nation—paying on average only $330 per week in 2004, which averages out to $17,160 per year if you worked every week of the year (U.S. Department of Labor 2004)—an annual income that is below the federal poverty line!

Only a small proportion of women work in occupations traditionally thought to be men's jobs, such as the skilled trades. The representation of women in skilled blue-collar jobs has increased twofold from 2 percent in 1940, to about 4 percent in 2003, still less than one in ten (U.S. Department of Labor 2004). Likewise, very few men work in occupations historically considered to be women's work, such as nursing, elementary school teaching, and clerical work. Men who work in these occupations tend to be more upwardly mobile in their jobs than women who enter fields traditionally reserved for men (Williams 1995, 1992).

DEBUNKING *SOCIETY'S MYTHS*

Myth: Because of affirmative action, Black women are taking jobs away from White men.

Sociological perspective: Sociological research finds no evidence of this claim. On the contrary, women of color work in gender- and race-segregated jobs and only rarely in occupations where they compete with White men in the labor market (Browne 1999; Padavic and Reskin 2002).

Gender segregation in the labor market is so prevalent that most jobs can easily be categorized as those considered to be "men's work" or "women's work." Occupational segregation reinforces the belief that there are significant differences between the sexes. Think of the characteristics of a soldier. Do you imagine someone who is compassionate, gentle, and demure? Similarly, imagine a secretary. Is this someone who is aggressive, independent, and stalwart? The association of each with a particular gender makes the occupation itself "gendered."

Gender Segregation and Gender Identity Perceptions of gender-appropriate behavior influence the likelihood of success at work. Even something as simple as wearing makeup has been linked to women's success in professional jobs (Dellinger and Williams 1997). When men or women cross the boundaries established by occupational segregation, they may be considered gender deviants and possibly stereotyped as homosexual, which questions their "true gender identity." Men who are ballerinas may be stereotyped as effeminate or gay; women in the Marines may be stereotyped as butch. Social practices like these reassert traditional gender identities (Williams 1995, 1989).

As a result, many men and women in nontraditional occupations feel pressure to assert gender-appropriate behavior. Men in jobs historically defined as women's work may feel impelled to emphasize their masculinity, or if they are gay, they may feel even more pressure to not reveal their sexual identity. Such social disguises can make them seem unfriendly and distant, characteristics that can have a negative effect on their professional evaluations. Heterosexual women in male-dominated jobs may feel obliged to squash suspicions that they are lesbians or excessively mannish, whereas lesbian women may be especially wary about having their sexual identity revealed. Studies have found that lesbian women are more likely to be open about their sexual identity at work when they work predominantly with women and have women as bosses (Schneider 1984).

Explanations of Gender Segregation Why is gender segregation still so prevalent, and why do obstacles to mobility at work persist? One explanation is socialization that influences how men and women choose to go into different fields. Women exposed to "masculine" tasks in childhood may be more likely as adults to enter jobs that utilize these skills (Padavic 1991). Conversely, many women shy away from traditionally "male" jobs because they believe that others will disapprove. Gender socialization certainly contributes to why men and women choose the occupations they do, but preference alone does not explain the gender segregation of women and men at work. When given the opportunity, women will move into jobs traditionally defined as men's work (Padavic and Reskin 2002).

A second explanation is that structural obstacles discourage women from entering and advancing in male-dominated jobs. The **glass ceiling** refers to the subtle yet decisive barrier to advancement that women encounter in the workplace. Despite four decades of policies meant to address inequality in the labor market, women and minorities are still substantially blocked from senior management positions (Glass Ceiling Commission 1995). Women who have risen to the top attribute their success to social networks, in addition to their abilities, whereas men attribute their success only to their own individual effort (Davies-Netzley 1998).

THINKING SOCIOLOGICALLY

Identify three women working in a field where men are the majority. Ask them about what influences the possibilities for being promoted within the organization. Do they think a *glass ceiling* exists? Why or why not?

The glass ceiling exists at every level where men and women work together, but research suggests that women have even less chance for advancement in top levels of the organization. Furthermore, African American women have even less chance of advancement than White women—a pattern also experienced by African American men, although not to the same extent as women. Promotion is also less likely for White women and African American men and women in employment positions with a high proportion of women and people of color (Porter 2003; Durbin 2002; Cotter et al. 2001; Baxter and Wright 2000; Maume 1999).

The glass ceiling is explained as a result of gender bias of white male managers. Gender bias shapes the evaluation of women's performance on the job and their perceived abilities. Women are also punished by gender bias when they are assertive—a characteristic normally expected of organizational leaders. But assertiveness in women violates the presumed gender order and reduces the likelihood that others will comply with their directions (Ridgeway 2001).

Before women can encounter the glass ceiling, in the words of one researcher, they must pull away from the sticky floor (Berheide 1992). Others have added that "glass walls" also prevent women from advancing, meaning that they tend to work in parts of work organizations where advancement is less likely. Secretaries, for example, may not be able to move to other professional jobs in an organization.

Despite the difficulties imposed, women continue to move into new areas of work and are gradually advancing to some degree. Change is slow, however, and as we discuss next, women and men struggle with the competing demands of work and family.

Balancing Work and Family

As the participation of women in the labor force has increased, so have the demands of keeping up with work and home life. Although some changes are evident, women continue to hold primary responsibility for meeting the needs of families, as we will see in more detail in Chapter 13. Many men are now much more involved in housework and childcare than has been true in the past, although most of this work still falls on women—a phenomenon that has been labeled "the second shift."

The social speedup that comes from increased hours of employment for both men and women (but especially women), coupled with the demands of maintaining a household, are a source of considerable stress (Hochschild 1989). Women continue to provide most of the labor that keeps households running—cleaning, cooking, running errands, driving children around, and managing household affairs. Although more men are engaged in housework and child care, a huge gender gap in the amount of such work done by women and men remains. Women are also much more likely to be providing care not just for children, but also for their older parents. The strains produced by these demands have for many made the home seem more and more like work, with many women and men reporting that their days at both work and home are harried and that they find work to be the place where they find emotional gratification and social support. In this contest, simply finding time can be an enormous challenge (Hochschild 1997). Little wonder then that women report stress as one of their greatest concerns (Newport 2000).

Gender and Diversity

Gender inequality does not exist in a vacuum. Gender inequality overlaps with race and class inequality. At the same time, the experiences of women in the United States are increasingly affected by global transformations. Understanding the diversity of such experiences is critical to understanding gender.

The Intersections of Race, Class, and Gender

The tendency to think of gender as referring only to White women has been one criticism of the women's movement consistently articulated by women of color. Until recently, White women's lives have tended to be the norm for many research studies, but working-class women of all races and women of color have experiences unique to their class, race, *and* gender position, just as middle-class White women's experience is predicated on their class, race, and gender status. Women have many things in common because of the influence of gender in their lives, but their experiences also vary depending on other factors, including age, sexual orientation, and religion.

Understanding diversity among women and men means thinking about how gender shapes social experiences, but also how it intersects with other social systems. Race, class, and gender together influence all aspects of people's lives. At a given moment in a particular man's or woman's life, either gender, race, or class may feel more salient than the other factors, but together each configures the experiences people have. This is what it means to say that race, class, and gender are different but interrelated dimensions of social structure (Andersen and Hill Collins 2004). Each is manifested differently, depending on a group's location in the nexus of gender, race, and class relations. As an example, one can note that the income of employed women is less than that of employed men. This is true in general, but among Hispanics and African Americans, women's income more nearly approximates that of men in their same racial group. Thus, women's income overall is 74 percent of men's and Black women's income is 83 percent of African American men's income, but only 64 percent of White men's income. Likewise, Hispanic women earn 84 percent of what Hispanic men earn, but only 52 percent of what White men earn (U.S. Census Bureau 2004).

None of this is to say that gender is less significant, only that one has to be careful to understand the unique manifestation of experiences that come from different social locations. Developing an inclusive perspective means trying to understand the multiplicity of experiences, while also comprehending the significance of diverse social factors.

Gender in Global Perspective

Increasingly, the economic condition of women and men in the United States is also linked to the fortunes of people in other parts of the world. The growth of a global economy and the availability of a cheaper industrial workforce outside the United States have meant that U.S. workers have become part of an international division of labor. Companies looking around the world for less expensive labor frequently turn to the cheapest laborers—often women or children. The global division of labor thus has a gendered component, with women workers, usually from the poorest countries, providing a cheap supply of labor for the manufacture of products that are distributed in the richer industrial nations (see Chapters 10 and 18).

Worldwide, women work as much or more than men and do most of the work associated with home, child, and elder care. While women's paid labor has been increasing, their unpaid labor in virtually every part of the world exceeds that of men (see Table 12.1). Often, unpaid labor is considered to be economically unimportant, even though the value of women's unpaid housework in the developed regions has been estimated to be as much as 30 percent of the gross national product (United Nations 1995).

Despite these worldwide trends, women's work situations differ significantly from nation to nation. China is unusual in that there is far greater sharing of household responsibilities than is true in most other nations. In China, both women and men work long hours in paid employment, 82 percent of women and 83 percent of men are in the paid labor force, and women are encouraged to stay in the labor force when they have children. There are extensive child-care facilities in China and a fifty-six-day paid maternity leave. Many work organizations have extended this paid leave to six months, although women can lose seniority rights when they are on maternity leave (something that is illegal in the United States).

In contrast, Japan has marked inequality in the domestic sphere. Women are far more likely to leave the labor force upon marriage or following childbirth. The identities of Japanese women are more defined by their roles at home, although this is changing. Still, compared

Table 12.1

Women's Work Around the World (in weeks)

Country	Total Hours		Paid Work		Unpaid Work	
	Women	Men	Women	Men	Women	Men
Australia	50	48	15	30	35	18
France	46	42	15	26	31	17
Japan	46	42	20	39	26	3
Netherlands	36	36	10	25	26	11
New Zealand	49	48	16	29	33	19
Republic of Korea	40	38	23	36	17	2

Women spend more time on unpaid work than men, but less on paid work.

Source: United Nations, 2000. *World's Women: Trends and Statistics.* New York: United Nations, p. 125. Used by permission.

with women in China, Japanese women more closely resemble the pattern in Britain and, to some extent, the United States.

Work is not the only measure of women's inequality. Throughout the world, women are vastly underrepresented in national parliaments and other forms of government. In only sixteen countries of the world do women represent more than 25 percent of national parliaments. Only seventeen countries have elected a woman president; only twenty-two have had a woman as prime minister (United Nations 2000b) (see Map 12.2).

The United Nations has concluded also that violence against women and girls is a "global epidemic" and one of the most pervasive violations of human rights. Violence against women takes many forms, including rape, domestic violence, infanticide, incest, genital mutilation, and murder (including so-called honor killings, when a woman may be killed to uphold the honor of the family if she has been raped or accused of

adultery). While violence is pervasive, specific groups of women are more vulnerable than others, namely, minority groups, refugees, women with disabilities, elderly women, poor and migrant women, and women living in countries with armed conflict. Statistics on the extent of violence against women are hard to report with accuracy, both because of the secrecy that surrounds many forms of violence and the differences in how nations might report violence. Nonetheless, the United Nations estimates between 20 and 50 percent of women worldwide have experienced violence by an intimate partner or family member (UNICEF 2000b).

Many factors are related to the high rates of violence against women, including cultural norms, women's economic and social dependence on men, and political practices that either provide inadequate legal protection or provide explicit support for women's subordination (as in the actions of the Taliban). In recent years, many groups have organized campaigns to educate the public about violence against women and to initiate a global

VIEWING SOCIETY IN GLOBAL PERSPECTIVE

MAP 12.2 Women in Government

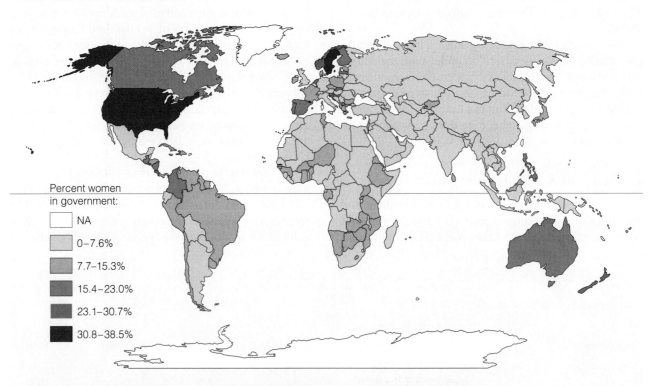

Percent women in government:

- NA
- 0–7.6%
- 7.7–15.3%
- 15.4–23.0%
- 23.1–30.7%
- 30.8–38.5%

Being represented in government is an important mark of citizenship. In most parts of the world, women are politically active, yet they are highly underrepresented in positions of government leadership. Why has there never been a woman president of the United States and what do you think will have to happen before there can be a woman president of the United States? How does this differ from other nations where women have been the major national leader?

Source: From United Nations. 2002. *Fact Sheet on Women in Government.* WomenWatch. Website: **www.un.org/womenwatch**

effort to reduce the harms done to women (UNICEF 2000b).

Theories of Gender

Why is there gender inequality? The answer to this question is important, not only because it makes us think about the experiences of women and men, but also because it guides attempts to address the persistence of gender injustice. The major theoretical frameworks in sociology provide some answers, but feminist scholars have also found that traditional perspectives in the discipline are inadequate to address the new issues that have emerged from feminist research.

The Frameworks of Sociology

The major frameworks of sociological theory—functionalism, conflict theory, and symbolic interaction—provide some answers to the question of why gender inequality exists, although feminist scholars have developed new, additional theories to analyze women's experiences (see Table 12.2). Functionalists, for example, have been criticized for interpreting gender as a fixed role in society. Functionalist theory purported that men fill instrumental roles in society, whereas women fill expressive roles, and the theory presumed that this arrangement worked to the benefit of society (see Chapter 1). Feminists objected to this characterization, arguing that it presumed that sexist arrangements were functional for society. Feminists view limiting women's role to expressive functions and men's to instrumental functions as dysfunctional, both for men and women. Although few contemporary functionalist theorists would make such traditionalist arguments, functionalism emphasizes people's socialization into prescribed roles as the major impetus behind

gender inequality. A functionalist might argue conditions such as wage inequality are the result of choices women make that may result in their inequality but, nonetheless, involve functional adaptation to the competing demands of family and work roles.

Conflict theorists see women as disadvantaged by power inequities between women and men that are built into the social structure, including economic inequity and a disadvantage in political and social systems. According to conflict theorists, wage inequality is produced from the power that men have historically had to devalue women's work and to benefit as a group from the services that women's labor provides. At the same time, conflict theorists have been much more attuned to the interactions of race, class, and gender inequality because they fundamentally see all forms of inequality as stemming from the differential access to resources that dominant groups in society have.

Functionalism and conflict theory tend to be macrosociological theories; that is, they focus on the broad institutional structure of society to explain gender relations. Symbolic interaction is more microsociological in that it tends to focus on direct social interaction as the context for understanding gender. From this perspective, feminist scholars have developed what is known as **doing gender**—a theoretical perspective that interprets gender as something accomplished through the ongoing social interactions people have with one another (West and Zimmerman 1987; West and Fenstermaker 1995). Seen from this framework, people "produce" gender through the interactions they have with one another and through their interpretations of certain actions and appearances. In other words, gender is not just an attribute of different people, as functionalists suggest. Instead, it is constantly recreated through social interaction. When you "act like a man" or "act like a woman," you are constructing gender and reproducing the existing social order.

Table 12.2
Feminist Theory: Comparing Perspectives

	Liberal Feminism	Socialist Feminism	Radical Feminism	Multiracial Feminism	Gendered Institutions	"Doing Gender"
Gender Identity	Learned through traditional patterns of gender role socialization.	Gender division of labor reflects the needs of a capitalist workforce.	Women's identification with men gives men power over women.	Women and men of color form an oppositional consciousness as a reaction against oppression.	Gender is learned in institutional settings that are structured along gender lines.	Gender is an accomplished activity created through social interaction.
Gender Inequality	Inequality is the result of formal barriers to equal opportunity.	Gender inequality stems from class relations.	Patriarchy is the basis for women's powerlessness.	Race, class, and gender intersect to form a matrix of domination.	Organizations reproduce inequity by making gendered activity "business as usual."	People reproduce race and class inequality through assumptions they make about different. groups.
Social Change	Change accomplished through legal reform and attitudinal change.	Transformation of the gender division of labor accompanies change in the class division of labor.	Liberation comes as women organize on their own behalf.	Women of color become agents of feminist change through alliances with other groups.	Change comes through policies that redesign the basic structure of institutions.	People can alter social relations through changed forms of social interaction.

By implication, from this point of view, gender should be relatively easy to change because people need only behave differently. This theory has been criticized by sociologists with a more macrosociological point of view. They say that it underplays the significance of social structure and the economic and political basis for women's inequality. Critics of the "doing gender" perspective also say that it ignores the power differences and economic differences that exist based on gender, along with race and class. They conclude, that although this perspective tells us much about how people reproduce gendered behaviors, it does not explain the structural basis of women's oppression (Collins et al. 1995).

A recent theoretical perspective developed on gender is gendered institutions theory, previously discussed. This perspective sees organizations as gendered because gendered expectations are built into social institutions, without people recognizing the specifically gendered outcomes that result. In this framework, sociologists do not study individual attitudes or roles, but dissect the structural patterns that construct gender in society.

These sociological frameworks have each provided direction in developing an understanding of the significance of gender in society. Feminist scholars, however, do not see functionalism, conflict theory, or symbolic interaction as adequate to address the complexities of women's lives. Although feminist sociologists have been especially influenced by the perspective of conflict theory and, in some cases, by symbolic interaction, they have developed other theoretical frameworks to suggest ways to comprehensively understand women's and, thereby, men's experiences.

Feminist Theory

Feminism has many meanings, but essentially it refers to beliefs and actions that support justice, fairness, and equity for all women, regardless of their race, age, or class. A large proportion of women (30 to 40 percent in various surveys) call themselves feminists; an even larger proportion identify with the major principles of feminism, including gender equality in employment, gender equality in family roles, support of reproductive freedom, support for affirmative action, and support for women in public office (Schnittker et al. 2003; Hall and Rodriguez 2003).

Why the gap? Most say it is because the label "feminist" has been so stigmatized that people are reluctant to claim it as an identity. The media has stereotyped feminists as ugly, man-haters, lesbians, radicals, and other derogatory labels. Still, despite frequent claims in the media that feminism is dead, when you ask people if they support specific goals of feminism, the majority say yes. This is especially true among younger women and among women now middle-aged who were coming of age during the second wave of the feminist movement in the 1960s and 1970s (Peltola et al. 2004; Aronson 2003).

Feminist theory refers to analyses that seek to understand the position of women in society for the purposes of bringing about liberating social changes. Feminist theory assumes that theory is important, not just because it analyzes gender and society, but because it also assumes that change is essential to make women fully equal citizens. Underlying feminist theory is also the idea that gender relations are fundamental to how society has been organized and that understanding how this is so is critical to the success of the feminist movement. Four major frameworks have developed in feminist theory: *liberal feminism, socialist feminism, radical feminism,* and *multiracial feminism* (Andersen 2003).

Liberal feminism argues that inequality for women originates in traditions of the past that pose barriers to women's advancement. It emphasizes individual rights and equal opportunity as the basis for social justice and social reform. The framework of liberal feminism has been used to support many legal changes required to bring about greater equality for women in the United States. Liberal feminists contend that gender socialization contributes to women's inequality because it is learned customs that perpetuate inequality. Liberal feminism advocates the removal of barriers to women's advancement and the development of policies that promote equal rights for women.

Socialist feminism is a more radical perspective that finds the origins of women's oppression in the system of capitalism. Because women constitute a cheap supply of labor, they are exploited by capitalism in much the way the working class is exploited. Some socialist feminists believe capitalism interacts with patriarchy to make women less powerful both as women and as laborers. Socialist feminists are critical of liberal feminism for not addressing the fundamental inequalities built into capitalist–patriarchal systems. To these feminists, equality for women will come only when the economic and political system is changed.

Radical feminism interprets patriarchy as the primary cause of women's oppression. To radical feminists, the origins of women's oppression lie in men's control over women's bodies. They see violence against women, in the form of rape, sexual harassment, wife-beating, and sexual abuse, as mechanisms that men use to assert their power in society. Radical feminists think that change cannot come about through the existing system because that system is controlled and dominated by men. Parting with liberal feminists who see that state reform holds the promise to free women (through legislative action and political participation), radical feminists see "the state as male" and as unlikely to be the source of change on women's behalf.

Most recently, **multiracial feminism** has developed new avenues of theory for guiding the study of race, class, and gender (Andersen and Collins 2004; Baca Zinn and Dill 1996; Collins 1998). Multiracial feminism examines the interactive influence of gender, race, and class, pointing to race as a major factor influencing the different ways the lives of women and men are constructed. Central to this perspective is that there is not a universal experience associated with being a woman. Instead, different privileges and disadvantages accrue to women (and to men) as the result of their location in a racially stratified and class-based society. Much of the new scholarship emerging from multiracial feminism examines the lives of women of color and shows how race, class, and gender *together* shape the experiences of all women (Baca Zinn and Dill 1996).

Feminist theory has developed in the context of the feminist movement. Feminist theory is not theory for theory's sake but is meant to be the basis for programs for social change. Each perspective provides unique ways to look at the experiences of women and men in society. These theoretical orientations have been the bedrock upon which feminists have built their programs of social and political change.

Young women are the group most likely to support feminist goals, although the feminist movement has support across generations, as was apparent at the March for Women's Lives in March 2004 in Washington, DC.

Gender and Social Change

Few lives have been untouched by the transformations that have occurred in the wake of the feminist movement. The women's movement has changed attitudes among both women and men, created new opportunities for women, generated laws that protect women's rights, and spawned organizations that lobby for public policies on behalf of women. Many young women and men now take for granted freedoms that their generation is the first to enjoy—including access to birth control, equal opportunity legislation, and laws protecting against sexual harassment, increased athletic opportunities for women, more presence in political life, and greater access to child care, to name a few. These impressive changes occurred in a relatively short period. How have attitudes and policies changed?

Contemporary Attitudes

One of the most significant results of the feminist movement is the change in people's thinking about women and men (Loo and Thorpe 1998). Only a small minority of people now disapprove of women being employed while they have young children (16 percent of women and 20 percent of men); neither women nor men think it is fair for men to be the sole decision maker in the household. Half of all women and men say the ideal lifestyle is to be in a marriage in which husband and wife share responsibilities, including work, housekeeping, and child care (Roper Organization 1995). The majority of women now want to combine work and families, yet they believe they will be discriminated against in the labor force if they do. Eighty-seven percent of women say that making laws to establish equal pay should be a legislative priority (Greenhouse 2000b).

People's beliefs about appropriate gender roles have evolved as women's and men's lives have changed. Less than half of men (47 percent) now believe that it is best for men to hold the provider role, compared with 69 percent who thought so in 1970. Men's support for women's roles in the family and at work, however, varies across different groups. Not surprisingly, women employed full-time are most supportive of nontraditional gender roles. Homemakers hold the most traditional views (Cassidy and Warren 1997).

Younger men and single men are more egalitarian than older, married men. Among college students, however, women hold more egalitarian views of women's roles than men; although both become less traditional in their views during college, women change more than men (Bryant 2003). There are also racial and ethnic differences in how different groups view gender roles, with minority men usually being more supportive of egalitarian roles than White men. The mothers of minority men are more likely to have been employed, and these men have different educational and employment backgrounds from White men. Their attitudinal differences reflect the economic necessity that minority men attribute to women working (Wilkie 1993; Blee and Tickamyer 1995).

Old attitudes do not die easily, and changed attitudes do not necessarily mean changed behavior. Gender attitudes change as society changes. But also, young people's expectations for being able to "have it all" can be unrealistic. As an example, sociologist Michele Hoffnung surveyed a random sample of college women in their senior year, surveying them again seven years later (in 2001). She found that as seniors, most of the women wanted careers, marriage, and motherhood, with career development being their top priority in their twenties. But at the seven-year point, those who had become mothers had fewer advanced degrees and lower career status than the nonmothers; marriage was not related to career status (Hoffnung 2004).

It is likely that further adjustments in the attitudes of men and women are on the way because it seems unlikely that the roles of men and women will return to old patterns in the future. Attitudes, however, are only part of the problem of persistent gender inequality. Social change requires more than changing individual attitudes; it also means changing social institutions.

Legislative Change

Much legislation is in place that prohibits overt discrimination against women. In addition to the Equal Pay Act of 1963, the Civil Rights Act of 1964, enacted as the result of political pressure from the civil rights movement, banned discrimination in voting and public accommodations and required fair employment practices. Specifically, *Title VII of the Civil Rights Act of 1964* forbids discrimination in employment on the basis of race, color, national origin, religion, or sex. It is almost accidental that prohibitions against sex discrimination are part of this bill. The word *sex* was added as a last-ditch effort by conservative members of Congress who thought the idea of including women was so ludicrous that it would defeat passage of the bill. They argued that a woman's place was in the home and that adoption of the bill would upset "natural" differences between the sexes. Some supporters of the bill also appealed to White racism to promote adoption of the bill, arguing that it would be wrong to elevate the rights of Black women over those of White women (Deitch 1993).

The passage of the Civil Rights Act and Title VII opened up new opportunities to women in employment and education. This was further supported by Title IX (part of the Educational Amendments of 1972). **Title IX** forbids gender discrimination in any educational institution receiving federal funds. Title IX prohibits colleges and universities from receiving federal funds if they discriminate against women in any program, including athletics. Adoption of this bill radically altered the opportunities available to women students and laid the foundation for many coeducational programs that are now an ordinary part of educational life. This law has been particularly effective in opening up athletics to women. In 1997, the U.S. Supreme Court further strengthened Title IX by refusing to rule on a case that supported the principles embedded in Title IX. Still, Title IX has been challenged recently by some who argue that it has reduced opportunities for men in sports. Proponents of maintaining strong enforcement of Title IX have responded by noting the still greater preponderance of men in school sports.

Passage of antidiscrimination policies does not, however, guarantee their universal implementation. Has equality been achieved? In college sports, men still outnumber women athletes by more than two to one, and there is still more scholarship support for male athletes than women athletes. In fact, Title IX allows institutions to spend more money on male athletes if they outnumber women athletes, but it also stipulates that the number of male and female athletes should be roughly proportional to their representation in the student body. Although there has been dramatic improvement in support for women's athletics since the implementation of Title IX, most schools are nowhere near compliance with the law, especially colleges with a large athletics program and a football team (Sigelman and Wahlbeck 1999).

DEBUNKING SOCIETY'S MYTHS

Myth: Title IX results in fewer athletic opportunities for men.

Sociological perspective: Since the enactment of Title IX in 1978, there has been an increase in sports programs both for men and women, although a much greater increase for women. Reductions in men's athletics occurred only in those Division I-A and I-AA schools—with the largest sports budgets. Elimination of sports teams—both women's and men's—is usually the result of budget reductions, not Title IX per se (Sabo 1998).

A strong, legal framework for gender equity in work has been established, yet equity has not been achieved. Because most women work in different jobs from men, the principle of equal pay for equal work does not address all the inequities women experience in the labor market. **Comparable worth** is the principle of paying women and men equivalent wages for jobs involving similar levels of skill, recognizing that men and women tend to work at different jobs. Comparable worth goes beyond equal pay for equal work by evaluating jobs to assess their degree of similarity. In the few places where comparable worth plans have been implemented, women's wages have improved (Blum 1991; Michel et al. 1989; Steinberg 1992; Jacobs and Steinberg 1990).

Many victories in the fight for gender equity are also now at risk. *Affirmative action* is a method for opening opportunities to women and minorities, specifically redressing past discrimination by taking positive measures to recruit and hire previously disadvantaged groups. This method has been effective in opening new opportunities for women, although the national climate is increasingly critical of it. Opponents have argued that it constitutes "reverse discrimination" because it specifically takes race and gender into account. Proponents of affirmative action argue that, as long as the structural conditions of gender and race inequality exist, there is still a need for race- and gender-conscious actions to address persistent injustices—a position supported by the U.S. Supreme Court (see Chapter 11).

One solution to the problem of gender inequality is to have more women in positions of public power. Is increasing the representation of women in existing institutions enough? Without reforming the sexism in the institutions, groups who are already privileged may be the major beneficiaries of change. Feminists advocate restructuring social institutions to meet the needs of all groups, not just those who already have enough power and privilege to make social institutions work for them. The successes of the women's movement demonstrate that change is possible, but only when people are vigilant about their needs and organize to accomplish new results.

Sociology⊛Now™
Reviewing is as easy as ❶❷❸.

1. *Before you do your final review, take the SociologyNow diagnostic quiz to help you identify the areas on which you should concentrate. You will find information on SociologyNow and instructions on how to access all of its great resources on the foldout at the beginning of the text.*

2. *As you review, take advantage of SociologyNow's study videos and interactive Map the Stats exercises to help you master the chapter topics.*

3. *When you are finished with your review, take SociologyNow's posttest to confirm you are ready to move on to the next chapter.*

Chapter Summary

How do sociologists distinguish between sex and gender?

Sociologists use *sex* to refer to biological identity and *gender* to refer to the socially learned expectations associated with members of each sex. *Biological determinism* refers to explanations that attribute complex social phenomena to physical or natural characteristics. Studies of hermaphrodites (those of biologically mixed sex) show that biology alone does not produce gender differences. Biological and social systems are interrelated.

What is gender socialization and why is it significant in understanding women and men?

Gender socialization is the process by which gender expectations are learned. One result of socialization is the formation of gender identity. Gender identity also develops alongside racial identity. *Homophobia* plays a role in gender socialization because it encourages strict conformity to gender expectations.

What is a gendered institution?

A *gendered institution* is one in which the entire institution is patterned by gender. Sociologists analyze gender both as a learned attribute and as an institutional structure.

What is gender stratification and what is the evidence for it?

Gender stratification refers to the hierarchical distribution of social and economic resources according to gender. Most societies have some form of gender stratification, although they differ in the degree and type. Gender stratification in the United States is reflected in the wage differences between men and women.

How do sociologists explain pay inequality between women and men?

Human capital theory explains wage differences as the result of individual differences between workers. *Dual labor market theory* refers to the tendency for the labor market to be organized in two sectors: the primary and secondary market. Jobs in the primary market are higher paying, more prestigious, and provide greater opportunity for advancement than do those in the secondary market. Women and racial minorities tend to be concentrated in the secondary labor market. Continuing *overt discrimination* against women is another way that men protect their privilege in the labor market.

What is gender segregation?

Gender segregation refers to the unequal distribution of men and women in different job categories. Women are segregated both within and across occupations. Numerous structural barriers exist that discourage women's advancement at work. These barriers are popularly referred to as the *glass ceiling*. Gender inequality is a world-wide phenomenon.

What different theories explain the status of women in society?

Different sociological theories have emerged from functionalism, conflict theory, and symbolic interaction to explain the position of women in society. *Liberal feminism, socialist feminism, radical feminism,* and *multiracial feminism* also contribute to our understanding of the status of women.

Are there significant changes in the status of women in society?

Public attitudes about gender relations have changed dramatically in recent years. Women and men are now more egalitarian in their attitudes, although women still perceive high degrees of discrimination in the labor force. A legal framework is in place to protect against discrimination, but legal reform is not enough to create gender equity.

Key Terms

biological determinism 303	human capital theory 318
comparable worth 329	ideology 316
discrimination 318	index of dissimilarity 320
doing gender 325	
dual labor market theory 318	labor force participation rate 317
feminism 326	liberal feminism 327
gender 302	matriarchy 317
gender apartheid 316	multiracial feminism 327
gender identity 303	
gender segregation 320	occupational segregation 319
gender socialization 305	patriarchy 317
gender stratification 315	radical feminism 327
gendered institution 314	sex 302
glass ceiling 322	sexism 316
hermaphroditism 303	socialist feminism 327
homophobia 313	Title IX 328

Researching Society with MicroCase Online

You can see the results of actual research by using the Wadsworth MicroCase® Online feature available to you. This feature allows you to look at some of the results from national surveys, census data, and other data sources. You can explore this easy-to-use feature on your own, but try this example. Suppose you want to know:

Do some people still think that men are better suited for politics than women?

To answer this question, go to http://sociology.wadsworth .com/andersen_taylor4e/, select MicroCase Online from the left navigation bar, and follow the directions there to analyze the following data.

Data file: GSS

Task: Auto-Analyzer

Primary Variable: MEN BETTER

Questions

Once you have your results, answer the following questions:

1. For each of the demographic variables listed, indicate whether there is a significant effect. If so, indicate which category is most likely and least likely to believe men are better suited emotionally for politics than most women.

Socio-Demographic Variable	Overall Effect Significant?		Category Most Likely	Category Least Likely
Religion	Yes	No		
Political Party	Yes	No		
Age	Yes	No		
Education	Yes	No		

2. Using the information you entered in the table in Question 1, describe the person who is most likely to believe men are better suited emotionally for politics than most women.

3. Describe the person least likely to believe men are better suited for politics than women.

4. Are men more likely to believe this than women? Are you surprised?

5. Do African Americans and Whites have different beliefs regarding women's suitability for politics?

The Companion Website for Sociology: Understanding a Diverse Society, Fourth Edition

http://sociology.wadsworth.com/andersen_taylor4e/

Supplement your review of this chapter by going to the companion website to take one of the Tutorial Quizzes, use the flash cards to master key terms, and check out the many other study aids you'll find there. You'll also find special features such as GSS Data and Census 2000 information, data and resources at your fingertips to help you with that special project or do some research on your own.

Suggested Readings and Web Resources

Andersen, Margaret. 2003. *Thinking About Women: Sociological Perspectives on Sex and Gender*, 6th ed. Boston, MA: Allyn and Bacon.
This is a comprehensive review of feminist studies about women. Widely read by undergraduates, it provides an understanding of the sociological analysis of women's lives and a basic introduction to feminist thinking.

Collins, Patricia Hill. 1998. *Fighting Words*. Minneapolis, MN: University of Minnesota Press.
Collins analyzes the experiences of African American women by developing a theory of Black feminism. She also examines contemporary discussions of Afrocentrism and Black women's lives, the contributions of postmodernist theory, and a global perspective on Black feminism.

Craig, Maxine Leeds. 2002. *Ain't I a Beauty Queen? Black Women, Beauty, and the Politics of Race*. New York: Oxford University Press.
Standards of beauty for African American women are affected by gender, race, the beauty culture, and social movements. Craig examines all of these—and their interrelationship—in this fascinating book of how beauty standards for Black women emerge in society.

Espiritu, Yen Le. 1996. *Asian American Women and Men*. Thousand Oaks, CA: Sage.
This comprehensive book examines the construction of gender for Asian American women and men. It includes a discussion of gender and race stereotyping and how immigration policies have affected change in the lives of Asian American women and men.

Freedman, Estelle B. 2002. *No Turning Back: The History of Feminism and the Future of Women*. New York: Ballantine.
This comprehensive book discusses the achievements of the feminist movement and the historical forces that generate it. It also identifies feminist issues for future generations.

Hurtado, Aida. 2003. *Voicing Chicano Feminisms: Young Women Speak Out on Sexuality and Identity*. New York: New York University Press.
Showing the diversity of experiences within the category "Latina," Hurtado examines the complexity of Chicanas' experiences, including within families, religion, work, and sexual relationships. It provides a full view of feminism among young Chicanas.

Kimmel, Michael S., and Michael A. Messner. 2000. *Men's Lives*, 5th ed. Needham Heights, MA: Allyn and Bacon.
This is an excellent collection of articles on men's experiences. It is notable for its inclusion of diverse groups, including racial minority groups and gay men. The subjects covered include families, work, sexuality, friendships, health, and sports.

Institute for Women's Policy Research
www.iwpr.org
This is a Washington-based research organization, working to establish public policies for the benefit of women. The website includes numerous publications from the Institute, which provide data and research on women's employment, poverty and welfare, work and family, and other matters.

The Feminist Majority
www.feminist.org
The Feminist Majority promotes feminist social policies in the United States and abroad. The Web page includes news and information about women's issues as well as a link specifically for students with information about feminist action on different college campuses.

• • •

Sexuality

A visitor from another planet might conclude that people in the United States are obsessed with sex. A glimpse of MTV shows men and women gyrating in sexual movements. A stroll through a shopping mall reveals expensive shops selling delicate, skimpy women's lingerie. Popular magazines are filled with images of women in seductive poses trying to sell every product imaginable. Even bumper stickers brag about sexual accomplishments. People dream about sex, form relationships based on sex, fight about sex, and spend money to have sex.

Ample evidence exists of sexual permissiveness in the United States. Diversity in sexual partnerships is becoming more evident. Sex, usually thought to be a most private form of behavior, has taken on a public life by being at the center of some of the most heated public controversies. Should sex be confined to marriage between two heterosexual partners? Should young people be educated about birth control or just encouraged to abstain from sex? What forms of birth control should be available—and who should have access to it? Sex is clearly a subject that polarizes the public on various issues, such as abortion, teenage pregnancy, birth control, and gay and lesbian rights.

At the same time that this appears to be a very sexually open society, sexual oppression still exists. Sexual orientation (that is, to whom people are attracted) singles out some groups for differential treatment and extreme prejudice. Thus, lesbian, gay, and bisexual people are **minority groups** in our society. Recall that minority groups are people with similar characteristics (or at least perceived similar characteristics) who are treated with prejudice and discrimination (see Chapter 11). Minority groups encounter significant prejudice directed against them and are denied the same rights as dominant groups. Lesbians, gays, and bisexuals have organized to advocate for their rights and to be recognized as socially legitimate citizens, as is reflected in contemporary debates about recognizing gay marriage. Many organizations and municipalities have enacted civil rights protections on behalf of gays and lesbians, typically prohibiting discrimination in hiring against gays. Gays and lesbians won a major civil rights victory in 1996 when the Supreme Court ruled (*Romer v. Evans*) that gays and lesbians cannot be denied equal protection under the law. This ruling overturned the ordinances that many states and municipalities had passed that specifically denied civil rights protections to lesbians and gays (Greenhouse 1996a).

Still, prejudice and discrimination remain, indicating that sexuality is one of

Sociology ⊛ Now™

Reviewing is as easy as ❶❷❸.

Use SociologyNow to help you make the grade on your next exam. When you are finished reading this chapter, go to the chapter review for instructions on how to make SociologyNow work for you.

the bases for social inequality in society. For sociologists, studying sexuality in the context of inequality reveals how sexuality is deeply entrenched in social norms, values, and social structures. Sexuality is a complex social phenomenon. This chapter examines the social significance of sexuality and how sexual experience is shaped by society and culture. Using the sociological imagination, we can see that human sexuality, like other forms of social behavior, is socially structured. •••

Sex and Culture

Sex would seem to be utterly natural. Pleasure and, sometimes, the desire to reproduce are reasons people have sex, but sexual relationships develop within a social context. The social context establishes what sexual relationships mean, how they are conducted, and what social supports are given (or denied) to particular kinds of sexual relationships. *Sexuality is socially defined and patterned.*

Is Sex Natural?

From a sociological point of view, little in human behavior is purely natural, as we have seen in previous chapters. Behavior that appears to be natural is usually that which is accepted by cultural customs and is sanctioned by social institutions. Sex is a physiological experience, but it is not physiology alone that makes sex pleasurable. People engage in sex not just because it feels good, but also because it is socially meaningful, and sexuality is an important part of our social identity. How we enjoy sex is shaped as much by cultural influences as by physical possibilities for arousal and pleasure. Devoid of a cultural context and the social meanings attributed to sexual behavior, people might not find certain behaviors sexy and others a turnoff nor attribute the emotional commitments, psychological interpretations, spiritual meanings, and social significance to sexuality that it has in different human cultures.

If sex is a social phenomenon, is there any biological basis to sexual identity? This question is usually asked in the context of asking whether there is a biological basis to homosexual behavior. Various groups, sometimes including gays and lesbians, claim that there is a biological, or genetic, basis to sexual orientation. This claim is sometimes supported by referring to a so-called gay gene, which presumably directs the sexual orientation of some people. On closer look, however, evidence for a gay gene, is skimpy.

The evidence usually cited to assert a genetic basis to homosexuality was based on a study of gay brothers (identical twins) who were found to have similar DNA markers on one of their X chromosomes. But on closer examination, this research was refuted when it was found that the shared DNA markers found in pairs of gay brothers were no more likely than would be expected by chance (Wickelgran 1999). Moreover, the original study did not control for the environment in which the brothers were raised. They grew up in the same family, so obviously they were raised in the same environment.

A true scientific test of the hypothesis that homosexuality is biologically caused would require a stricter standard of evidence. One could study identical twins who were the offspring of gay parents to see if, controlling for the environments in which they were raised, both turned out to be gay. To date, there are no such studies of biological relatives raised apart (Hamer et al. 1993).

Despite the serious scientific flaws in studies purporting to have found a gay gene, news reports about an alleged biological basis to homosexuality are common. One-third of the public believes that homosexuality is a biological trait (Gallup Organization 1998). Public belief in the biological basis of homosexuality is much stronger than the scientific evidence for such a claim. Far more evidence exists for a social basis for the development of homosexuality, but these studies are rarely, if ever, reported as headline news (Gagnon 1995, 1973; Caulfield 1985; Money 1995, 1988). If the evidence for the social basis of homosexuality received such loud acclaim, the public might be less inclined to think that sexual identity has a biological origin.

Perhaps there is some biological basis to sexual orientation, but the evidence is not yet there. Sociologists would say that, even if a biological influence exists—which is not proven and which many doubt—social experiences are far more significant in shaping sexual identity (Connell 1992; Lorber 1994; Brookey 2001). On balance, no compelling evidence exists for a biological cause of sexual identity. Even if a biological basis for sexual identity were found, social influences interact with biological foundations in that social and cultural environments play an important part in creating the various dimensions of sexual identity. Although most sociologists would not rule out a biological influence, they are far more interested in how sexual identity is constructed through social experiences.

The Social Basis of Sexuality

From a sociological perspective, sexual identity is not innate (that is, inborn). Instead, it develops through social experiences. Biological processes such as hormonal fluctuations, sexual physiology, and perhaps some genetic factors are elements in the biological foundations of sexual desire. The particulars of the biological basis of sexual behavior are much debated, but sociologists conclude that "the social world tends to mold biology as much as biology shapes human's sexuality" (Schwartz and Rutter 1998: 5).

Think about this: In some cultures women do not believe that orgasm exists, even though biologically it does. In the eighteenth century, European and American writers advised men that masturbation robbed them of their physical powers and that instead they should apply their minds to the study of business. These cultural dictates encouraged men to conserve semen on the presumption that its release would lessen men's intelligence or cause insanity (Freedman and D'Emilio 1988; Schwartz and Rutter 1998). Although biology cannot be ignored, social and cultural experiences shape and direct our sexual identities. Moreover, sexual identity is constructed over the life course, not just determined at birth.

Human sexual behavior is situated in a cultural and social context. You can see the social and cultural basis of sexuality in a number of ways:

1. **Human sexual attitudes and behavior vary in different cultural contexts.** Culture affects whether we define certain sexual behaviors as normal or deviant. For example, if the culture allowed men to wear dresses, would there be a category of people known as crossdressers or transvestites? Likewise, you might ask why those who choose same-sex partners are judged to be sexually deviant. From a sociological point of view, it is not because such behavior is unnatural or inherently wrong, but because of the cultural assumptions made about same-sex relationships.

If sex were purely natural behavior, sexual behavior would also be uniform from one society to another, but it is not. Sexual behaviors considered normal in one society may be seen as peculiar in another. For example, the common practice of "French kissing" in the United States is by no means universal. Some societies would consider putting your tongue in another's mouth repulsive (Tiefer 1978). The expression of sexual feeling is influenced by its cultural setting.

Consider the hijras (pronounced HIJ-ras) of India, a religious community of men who are born male but come to think of themselves as neither men nor women; the hijras' role is a third gender identity. Hijras dress as women and may marry men, although they typically live within a hijra communal subculture. Men are not born hijras, but they become so. As male adolescents, they have their penis and testicles cut off in an elaborate and prolonged cultural ritual. This rite of passage marks the transition to becoming a hijra. Hijras perceive that

The hijras of India are a sexual minority group; this "man" is considered to be a "third gender"—neither man nor woman. Hijras provide a good illustration of the socially constructed basis of sexuality.

© Serena Nanda

sexual desire results in the loss of spiritual energy. Their emasculation is seen as proof that they are beyond such desires.

Whereas Western culture emphasizes dichotomous and separate identities for women and men, Hindu religion values the ambiguity of in-between sexual categories, holding that all persons have both male and female principles. Hijras are believed to represent the power of man and woman combined, although they are impotent themselves. At Indian weddings, they may ritually bless the newly married couple's fertility. They also perform at some celebrations following the birth of a male child, a cherished event in Indian society (Nanda 1998). Hijras are not transvestites, hermaphrodites, or sex impersonators, as Westerners might think. Although at times they are treated as social outcasts, they also represent valued ideals. They can be understood only within the context of Hindu culture.

2. **Sexual attitudes and behavior change over time.** Fluctuations in sexual attitudes are easy to document. For example, public opinion polls show that young people today are more permissive in their sexual values than were young people in the past, although tolerance of casual sex has decreased somewhat in recent years (Lyons 2002). Behavioral changes are harder to document

because people in this culture generally consider sex to be a private matter and thus do not always talk frankly about what they do. But, we know that young people are having sex at an earlier age. Now, more people experience sex before marriage than was true in the past, and people have more sex partners over their lifetimes (Janus and Janus 1993; Laumann et al. 1994).

A long-term historical perspective shows that sexual behavior has changed dramatically. During the Victorian period in nineteenth-century England, the prevailing sexual attitudes were very repressive. It was during this time when Sylvester Graham invented the graham cracker, thinking that its blandness would reduce people's carnal desires and make them think pure thoughts. In the same period, Queen Victoria coined the phrases *light* and *dark* meat for chicken and other fowl because she thought mention of the words *breast* and *leg* would cause men to be overcome with lust and faint!

Changes in how the breast is socially defined also show how the meaning of sexuality varies over time. Not until the sixteenth century were women's bosoms considered erotic. Now, breasts are commercialized as advertisers use erotic images of them to sell all kinds of products. Companies also market a huge array of breast products designed to push up, minimize, maximize, augment, or reduce women's breasts. Governments regulate women's breasts through prohibitions against breast-feeding in public. And breasts have been politicized as women sometimes expose their breasts as a symbolic way to protest the oppression of women (Yalom 1997).

3. **Sexual identity is learned.** Like other forms of social identity, sexual identity is acquired through individual socialization and the ongoing relationships we have with other people. Information about sexuality is transmitted culturally and becomes the basis for our identity. Where did you first learn about sex and what did you learn? For some, parents are the source of information about sexual behavior. For many, peers have the strongest influence. Learning about sex is part of the socialization process because it is an important dimension of learning social norms and mores.

Sexual socialization takes place at a very early age, long before young people become sexually active. Even as young children, people learn sexual scripts—often through their talk and play (Thorne and Luria 1986). A **script** is a learned performance. Like an actor, one learns one's part by learning the script that dictates how one is expected to act. **Sexual scripts** teach us what is appropriate sexual behavior for a person of our gender (Schwartz and Rutter 1998; Laws and Schwartz 1977). From an early age, children learn sexual scripts by playing roles, playing doctor as a way to explore their bodies, or hugging and kissing, pretending to be romantic partners. Role-playing teaches children social norms about sexuality, as well as norms about marriage and gender relationships. For instance, sexual talk among young boys socializes them into masculine roles. As

we saw in the study of gender (see prior chapter), the roles learned in youth have a significant impact on our relationships as men and women. This is true of sexual socialization, as it is with other social attitudes and behavior.

THINKING SOCIOLOGICALLY

Observe a popular children's film and describe the *sexual scripts* that are suggested by the film. What do these sexual scripts tell you about the *social construction of sexual identity*?

4. **Social institutions channel and direct human sexuality.** Social institutions, such as religion, education, or the family, define some forms of sexual expression as more legitimate than others. For example, being heterosexual is considered a more privileged status in society than being gay or lesbian. Institutional privileges for heterosexual couples include legal marriage, mutual employee health care benefits, and the option to file joint tax returns. Although many employers now extend benefits to domestic partners, and some states recognize gay marriage, far more institutional supports exist for heterosexual relationships. Numerous laws, religious doctrines, and family and employment policies explicitly and implicitly promote heterosexual relationships. Thus, what we come to see as natural or chosen is promoted by the social structure of society.

Intimate relationships, though highly personal, are also socially constructed.

Laws direct sexuality into socially legitimate forms by defining some forms of sexual behavior as lawful and others as criminal. Most states have laws prohibiting "crimes against nature," which can include oral and anal sex, even though surveys reveal that these practices are common (Laumann et al. 1994). Luckily for those who engage in such practices, these laws are rarely enforced, especially not against heterosexuals.

Sex is also influenced by the economic institutions of society. Sex sells. In the capitalist economy of the United States, sex is used to hawk everything from cars and personal care products to stocks and bonds. Moreover, sex itself is bought and sold as some people, particularly poor women, are forced to sell sexual services for a living. Sex workers are among some of the most exploited and misunderstood workers. Sometimes the economics of sex and the law intersect as quasi-legal sex trades are regulated in the form of red-light districts, selective enforcement, and even outright state licensing.

Although sexuality is generally considered a private matter and many think state authority should have no jurisdiction over private life, governments routinely intervene in people's sexual lives. State agencies may exert their influence on sexuality and reproduction by pressuring poor women who receive state benefits to use birth control or be sterilized. Numerous studies have found that the sterilization of poor and minority women often occurs with inadequate or no consent. Therefore, sterilization rates are highest among poor African American, Puerto Rican, and Native American women (Horton 1995; Roberts 1997).

Public policies also regulate people's sexual and reproductive behaviors in other ways. Prohibiting federal spending on abortion, for example, eliminates reproductive choices for women needing public assistance. Government endorsements of reproductive technologies influence the availability of birth control for men and women. Government funding, or lack thereof, for sex education can influence how people understand sexual behavior. These examples challenge the assumption that sex is a private matter and they show how social institutions can channel sexual behavior.

The Science of Sex

Sex is increasingly capturing the attention of social scientists who want to understand its relationship to social institutions, as well as how groups are treated in society as the result of their sexual orientation and behavior. Studying sex, though, carries some risks because people

---

UNDERSTANDING DIVERSITY
Gay-Bashing and Gender Betrayal

Hate crimes that involve attacks against gays and lesbians have received increased public attention in recent years. Carmen Vázquez sees such crimes as stemming from gender relations.

She writes:

At the simplest level, looking or behaving like the stereotypical gay man or lesbian is reason enough to provoke a homophobic assault. Beneath the veneer of the effeminate gay male or the butch dyke, however, is a more basic trigger for homophobic violence. I call it gender betrayal.

The clearest expression I have heard of this sense of gender betrayal comes from Doug Barr, who was acquitted of murder in an incident of gay bashing in San Francisco that resulted in the death of John O'Connell, a gay man. Barr is currently serving a prison sentence for related assaults on the same night that O'Connell was killed. . . . When asked what he and his friends thought of gay men, he said, *"We hate homosexuals. They degrade our manhood. We was brought up in a high school where guys are football players, mean and macho. Homosexuals are sissies who wear dresses. I'd rather be seen as a football player."*

Doug Barr's perspective is one shared by many young men. I have made about three hundred presentations to high school students in San Francisco, to boards of directors and staff of nonprofit organizations, and at conferences and workshops on the topic of homophobia or "being lesbian and gay." Over and over again, I have asked, "Why do gay men and lesbians bother you?" The most popular response to the question is, "Because they act like girls," or "Because they think they're men." I have even been told, quite explicitly, "I don't care what they do in bed, but they shouldn't act like that."

They shouldn't act like that. Women who are not identified by the relationship to a man, who value their female friendship, who like and are knowledgeable about sports, or work as blue-collar laborers and wear what they wish are very likely to be "lesbian-baited" at some point in their lives. Men who are not pursuing sexual conquests of women at every available opportunity, who disdain sports, who choose to stay at home and be a house-husband, who are employed as hairdressers, designers, or housecleaners, or who dress in any way remotely resembling traditional female attire (an earring will do) are very likely to experience the taunts and sometimes the brutality of "fag bashing."

The straitjacket of gender roles suffocates many lesbians, gay men, and bisexuals, forcing them into closets without an exit and threatening our very existence when we tear the closet open.

Source: Carmen Vazquez. 1992. "Appearances." In *Homophobia: How We All Pay the Price*, Warren J. Blumenfeld (ed.). Boston, MA: Beacon Press, pp. 157–166. ● ● ●

tend to associate sex with frivolity or, worse, with prurient or deviant interests. Thus, those who study sex are often ridiculed and stigmatized, even though the study of sexuality is a growing field of study.

Sexology—the scientific study of sex and sexuality—first developed in the early part of the twentieth century. Early sexologists changed the cultural understanding of sex from something believed to be ordained by religious belief to something subject to social science research. Their work brought profound changes to the understanding of sexuality.

The Influence of Freud

No one is more significant in the scientific study of sex than **Sigmund Freud** (1856–1939). His influence was felt in academic and scientific quarters, as well as among the general public. He opened the door for scholars and scientists to make sex the subject of serious study. His message that the formation of sexual identity is a basic part of personality development still underlies much of the public understanding of sex (see also Chapter 4).

Freud had a developmental model of sexuality that postulated sexual expression as originating in childhood and developing over the lifecycle. The transition to adulthood involves, in Freud's view, several stages of psychosexual development. Freud thought that sexual energy, or *libido*, was the driving force behind all human endeavors, generating the tension and excited state that leads to creativity in artistic and intellectual expression (Freud 1960/1923). Once Freud's writings were translated from German in the 1920s, they became the common language of sexuality among large segments of the public.

Freud's interpretation of women's sexuality assumed that men are more sexually and psychologically mature, an argument that feminists have subsequently criticized for buttressing men's interests. His emphasis on male penetration as required for women to have a "mature" sexual experience has now been rejected. Despite the criticism, Freud established the groundwork for our modern understanding of sexuality. He and other sexologists asserted science, not religion, as the main authority on human sexual behavior.

Havelock Ellis

Havelock Ellis (1859–1939), one of the most influential of the early sexologists, saw sexual dysfunction as pathologically rooted. He often tinged his work with the sexist assumptions of the times. He thought, for example, that lesbian women had turned their emotions upside down by rejecting the proper passive place of female sexuality. Lesbians, he claimed, loved the public world instead of men. He associated lesbianism with insanity, arguing that the professional women emerging during the 1920s and 1930s were particularly prone to this "disease." Ellis regarded anything but heterosexual, monogamous sexuality as "sexual deviance," and this was the prevailing wisdom for many years to come. Now, Ellis's ideas have been largely discredited, but they show how social stereotypes about homosexuality can pervade even seemingly scientific studies of sexuality.

The Kinsey Reports

The first major national surveys of sexual behavior were the Kinsey (1953, 1948) reports published in the 1940s and 1950s. Alfred Kinsey, a zoologist at Indiana University and head of the Institute for Sex Research, was the first to give a detailed picture of the sexual activities of the U.S. public. Kinsey's research was based on a national sample of 11,000 interviews, but his sample was not representative. All the research subjects were White, relatively well-educated, and middle-class. All the interviewers and staff members were White, heterosexual, Anglo-Saxon, Protestant men because Kinsey believed they represented the "yardstick of morality" (Irvine 1990: 44). Kinsey refused to hire interviewers with ethnic names, arguing that this would make rapport with the research subjects difficult. He claimed that he wanted to neither make any judgments about sexual practices nor condemn any behavior, but he would have been wise to have used a more representative sample. Kinsey also did not question the race and class stereotypes prevalent at the time, trusting that the scientific method alone would control for bias. His research had the built-in limitation that it was based on self-reports about sexual habits. Information based on self-reports is notoriously unreliable, especially when the topic is a delicate one, but it was the best information Kinsey could get.

Even with this bias in his sample, the Kinsey reports were nevertheless the first comprehensive and nationally based studies of sexual practices. In the 1940s it was Kinsey who first reported that 33 percent of women and 71 percent of men engaged in premarital intercourse despite public belief to the contrary. Surveys now show these figures have grown to 70 percent of women and at least 80 percent of men having sex before marriage; some estimate that as many as 97 percent of men have sex before marriage (Hunt 1974; Laumann et al. 1994; Lips 1993). Kinsey also found that sexual activities were more varied than was typically believed. Oral sex, a variety of coital positions, and genital touching were common among the couples he interviewed. Kinsey's reports also documented that many women were more sexually adventurous than was expected. Prior to 1920, sex for women was often considered a duty, not a source of pleasure. Kinsey found that women were enjoying sex and experimenting with different sexual practices, although he also found that, generally speaking, women desired less sex than men. Men were also much more likely than women to desire extramarital affairs. Overall, Kinsey reported a weakening of sexual taboos and a more permissive sexual atmosphere.

Kinsey was the first to deliver evidence that a significant proportion of the population was homosexual. He reported that 37 percent of men had experienced homosexual contact resulting in orgasm at some point in their lives. As we will see, the extent of homosexual behavior is still much debated.

When the Kinsey studies were done, widespread societal changes were taking place in the sexual practices and beliefs of the U.S. public. Women were increasingly independent, moving into cities, living on their own, remaining single longer, and working in the public sector. These changes were more pronounced among White, middle-class women—those most likely to have been included in Kinsey's sample. For the White middle class, the moralistic sexual ideology of the Victorian period was giving way to a more secular view of sex. Marriage was increasingly defined as an equal partnership between a man and a woman, not simply a pairing in which the woman was expected to obey the man. The study of sexuality and marriage were more often seen as the province of experts. Marriage manuals instructed women and men on ways to achieve sexual fulfillment. In retrospect, much of their advice now seems antiquated and silly, but Kinsey's findings helped liberate many people from their shame and guilt about their sexual practices. His reports questioned the absolute distinctions made between heterosexual and homosexual people. As a result, he laid the foundation for some of the sexual liberation movements that followed (Bullough 1998).

The Masters and Johnson Studies Kinsey's work provided the groundwork for a second series of important studies of human sexuality—the Masters and Johnson research (1966). The work of William Masters and Virginia Johnson is still the primary basis for understanding physiological sexual response (heart rate, blood pressure, orgasm, and so on). Their work is still the most extensive and thorough study of human sexual response (in a physical sense) ever done.

Like Kinsey's research, the Masters and Johnson study is flawed by a nonrepresentative sample. Their sample included 510 married couples and 57 single people, but they chose only people who they thought were "respectable"—White, middle-class, well-educated men and women. Like Kinsey, Masters and Johnson carried their own race and class biases into the design of their sample. They thought that well-educated people would be better able to visualize the sexual experiments and communicate fine details of sexual response. As a result, their sample was purposefully weighted toward people from higher socioeconomic backgrounds (Irvine 1990).

Masters and Johnson began their studies using prostitutes as their research subjects, believing they would be the only women who would participate in such a project. This turned out to be a boon. In the early stages of the research, the prostitutes, who had numerous sexual experiences, made many recommendations for refining the research and suggested a woman be added to the research team. They became collaborators as much as research subjects, by demonstrating methods for sexual stimulation and techniques for elevating or controlling sexual tensions. Many techniques they demonstrated have since been used in sex therapy programs, marital counseling, and sex manuals (Irvine 1990).

Masters and Johnson's research subjects engaged in a number of explicit sexual activities in the lab, with sexual response observed and measured (Masters and Johnson 1966). They documented the physiological responses that men and women experience during sexual excitement, including during orgasm. Perhaps the greatest significance of their work is that it dispelled a number of prevalent social myths about sexuality—most notably the myth that men and women experience sex in radically different ways. Masters and Johnson defined sex as a natural bodily function, and they asserted that women as well as men had a right to sexual pleasure. Despite the enormously liberating impact of their research, Masters and Johnson were themselves sexual conservatives. They emphasized the importance of sex within marital units only—one reason their research sample was biased so strongly in favor of married couples (Irvine 1990).

The work of the early sexologists laid the foundation for contemporary studies of sexual attitudes and behaviors, the result of which we will see in the following sections.

Sexual Practices of the U.S. Public

Sexual practices are notoriously difficult to document. What we know about sexual behavior is typically drawn from surveys asking about sexual attitudes, not actual behavior. What people say they do may differ significantly from the truth. The most recent, comprehensive survey of the sexual practices of the U.S. public was conducted by a team of sociologists in the early 1990s (Laumann et al. 1994; Michael et al. 1994). Led by Edward Laumann at the University of Chicago, these researchers surveyed a national random sample of nearly 3500 adults. Although their study has limitations because they do not analyze gay, lesbian, or bisexual behavior separately from heterosexual behavior, it is the latest comprehensive study of sexual behavior utilizing careful social science research methods. This study and others tell us several things about the sexual practices of the U.S. public.

1. **Young people are likely to become sexually active in their teenage years.** Sex is rare among very young teenagers, but it is common in the later teenage years (see Figure 13.1). By age nineteen, 85 percent of men

Percentage who have
had sexual intercourse
at different ages

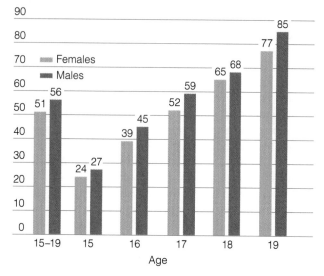

Figure 13.1 Sex Among Teenagers

Source: Alan Guttmacher Institute. 1999. *Teen Sex and Pregnancy.* New York: The Alan Guttmacher Institute. Reproduced with the permission of the Alan Guttmacher Institute.

and 77 percent of women have had sex (Alan Guttmacher Institute 1999b). However, recent reports also indicate a decline in the proportion of young men and women who have had sexual intercourse, particularly among White, Black, and Hispanic young men. At the same time, contraceptive use among young people has increased, resulting in a decline in teen pregnancy rates (Santelli et al. 2000; Alan Guttmacher Institute 2004).

2. Having only one sexual partner in one's lifetime is rare. The common perception that people have more sexual partners than they did in the past seems to be true, according to the Laumann study. Different age groups were asked how many sexual partners they had in their lifetime. Those over fifty years of age typically reported having five or more sexual partners and those thirty to fifty years old reported the same number. Laumann and his associates argued that the similarity in numbers is only because the younger group had less time to "accumulate" partners (Laumann et al. 1994).

Men report more sexual partners than women report, although some of this difference is probably attributable to overreporting by men and underreporting by women. There are no reported differences in the number of sexual partners people have by race, education,

DOING SOCIOLOGICAL RESEARCH

Teens and Sex: Are Young People Becoming More Sexually Conservative?

By the late 1990s several national studies reported a decline in sexual activity among teens. The percentage of sexually active teens has dropped from the early 1990s; rates of teen pregnancy have fallen; teens are having fewer abortions; and the rate of sexually transmitted diseases among teens has declined. Does this herald a growth in sexual conservatism among young people and the success of policies encouraging sexual abstinence?

Sociologists Barbara Risman and Pepper Schwartz reviewed the various studies exploring teen sexuality and concluded that the answer to these questions is more complex than typically assumed. Their review of several national studies found that most of the change in teen sex activity is attributable to changes in boys', not girls', behavior. The number of high school boys who are virgins has increased. Girls' behavior has not changed significantly, except among African American girls whose rates of sexual activity have de-

clined, nearly matching that of White and Hispanic girls. Risman and Schwartz conclude that boys' sexual behavior is then becoming more like girls' behavior, the implication being that boys and girls are now more likely to begin their sexual lives within the context of romantic relationships.

While many declare that the changes in teen sexual behavior mean a decline in the sexual revolution, Risman and Schwartz disagree. Certainly fear of AIDS, education about safe sex, and some growth in conservative values have contributed to changes in teen sexual norms. Risman and Schwartz show that numerous factors influence sexual behavior among teens, just as among adults. They suggest that sexuality is a normal part of adolescent social development and conclude that the sexual revolution—along with the revolution in gender norms—is generating more responsible, not more problematic, sexual behavior among young people.

Questions to Consider

1. Do you think that people in your age group are generally sexually conservative or sexually liberal? What factors influence young people's attitudes about sexuality? *Keywords: adolescent sexuality, gender and sexuality*

2. Following from the above question, what evidence would you need to find out if young people are more sexually liberal or conservative than youth generations in the past? How would you design a study to investigate this question? *Keywords: abstinence, sex education, youth and sexual identity*

We have included InfoTrac College Edition keywords at the end of each question to make it easier for you to find more to read on these topics. Go to www.infotrac-college.com, an online library, to begin your search.

Source: Risman, Barbara, and Pepper Schwartz. 2002. "After the Sexual Revolution: Gender Politics in Teen Dating." *Contexts* 1 (Spring): 16–24. •••

or region of residence, but the average number of partners for people in metropolitan areas is slightly higher than for those living in other areas. About 22 percent of adults report having no sexual partners in the past year. Abstinence from sex is somewhat more prevalent among women than men (Smith 1991).

DEBUNKING *SOCIETY'S MYTHS*

Myth: Over time in a society, sexual attitudes become more permissive.

Sociological perspective: All values and attitudes develop in specific social contexts; change is not always in a more permissive direction (Freedman and D'Emilio 1988).

---

3. **A significant number of people have extramarital affairs.** Most of the public strongly believes in sexual fidelity. About three-quarters say that it is always wrong for a married person to have sex with someone other than the person's spouse. Yet, extramarital affairs are fairly common. Laumann's study found that about 25 percent of married men and 15 percent of married women have extramarital affairs, fairly consistent with other studies (Laumann et al. 1994). There is likely significant misreporting of these data given the secrecy surrounding such behavior. The likelihood of infidelity seems to increase with age—and, presumably, the number of years married (Laumann et al. 1994; Smith 1991).

4. **A significant number of people are lesbian or gay.** Popular estimates suggest about 10 percent of the population is gay. However, precisely measuring the extent of homosexuality is difficult, and reports about it also depend on how the question is asked. Kinsey asked men whether they ever had homosexual contact at some time in their lives. Thirty-seven percent said yes. Current surveys report that 22 percent of men and 17 percent of women would say yes.

Having a homosexual experience at some point in your life, however, is different from having a gay or lesbian identity. Laumann's study distinguished between people who had engaged in any same-sex behavior, people who felt desire for someone of the same sex, and people who identified as homosexual. He found that 2.8 percent of men and 1.4 percent of women identified themselves as homosexual, but 9 percent of men and 4 percent of women reported having had a sexual experience with someone of the same sex. Likewise, more men and women reported having experienced same-sex desire than the number reporting an identity as gay. This helps explain why Laumann reports a smaller percentage of gay people in the population than has been reported in other surveys, such as the Janus survey reporting 9 percent of men and 5 percent of women as gay (Janus and Janus 1993).

Whatever the extent of gay and lesbian experience, it is more common than many think. No doubt, the social stigma associated with being gay or lesbian also causes underreporting. Sociologists have concluded that gays and lesbians are a substantial group in the population that is typically misunderstood, discriminated against, and incorrectly portrayed in much of the popular and traditional social science literature.

5. **For those who are sexually active, sex is relatively frequent.** On average, almost two-thirds of Americans report having sex a few times per month or two to three times per week (Laumann et al. 1994). In some studies, men report having sex more often—a statistical oddity if one presumes that most sex is between men and women. This difference is explained by the tendency for men to overreport both their number of sex partners and how often they have sex (Smith 1991). People in their twenties report having sex most often, confirming popular stereotypes. Contrary, however, to stereotypical images about single life, married and cohabiting people in their twenties have sex more often than singles (Laumann et al. 1994). Comparing groups across racial, religious, or educational lines, there are no significant differences in the frequency of sex. Age and marital status are the most significant predictors of the frequency of sex (Laumann et al. 1994).

Research studies, like those we've cited, describe many facets to sexual behavior, but how do you interpret what this all means? This is the task of sociological theories.

Sexuality and Sociological Theory

Putting sex in a theoretical context may seem abstract and dull (and probably not your idea of a good time!), but thinking about sexuality in an analytical framework reveals how sex is connected to other social processes and sheds light on many of the debates about sex and social policy that frequent public discussion. Sociological theory puts an analytical framework around the study of sexuality, examining its connection to social institutions and current social issues. How do the major sociological theories frame an understanding of sexuality?

Sex: Functional or Conflict-Based?

The major sociological frameworks—functionalist theory, conflict theory, and symbolic interaction—take divergent paths in interpreting the social basis of human sexuality (see Table 13.1). Functionalist theory with its emphasis on the interrelatedness of different parts of society tends to depict sexuality in terms of how it contributes to the stability of social institutions. Functionalist theorists would point out that sexual norms that restrict sex to within marriage encourage the formation of families. Similarly, functionalist theorists might argue that giving legitimacy to heterosexual behavior but not homosexual behavior, maintains a particular form of social organization, one where gender roles are easily differentiated and where nuclear families are the dominant family structure. From this point of view, regulating sexual behavior is functional for society because it prevents the instability and conflict that diverse family structures might generate.

The functionalist perspective is useful in understanding many contemporary public issues. Some have called for a return to family values, meaning a return to more traditional nuclear families and less tolerance for the diverse family forms that have emerged in recent years. In so doing, they are arguing for uniformity in values that they see as necessary for social order.

Conflict theorists see sexuality as part of the power relations and economic inequality in society. Power is the ability of one person or group to influence the behavior of another. Instead of arguing for a return to traditional and commonly shared values, conflict theorists argue that sexual relations are linked to subordination in race, class, and gender inequality that associates sex with power. According to conflict theorists, sexual violence, such as rape or sexual harassment, is the result of power imbalances, especially between women and men. At the same time, because conflict theorists see economic inequality as a major basis for social conflict, they tie the study of sexuality to economic institutions. Conflict theorists, for example, would link such things as sexual trafficking of women and children as prostitutes to poverty, the sexual exploitation of women, and the economics of international development and tourism (Enloe 1989). It is important to note that conflict theorists do not see all sexual relations as oppressive. In connecting sexuality and inequality, conflict theorists are developing a structural analysis of sexuality, not condemning sexual intimacy.

Both functionalist theory and conflict theory are macrosociological theories. That is, they take a broad view of society and see sexuality as the overall social organization of society—they do not tell us much about the social construction of sexual identities. This is where the third major sociological framework, symbolic interaction, is important.

Symbolic Interaction and the Social Construction of Sexual Identity

Symbolic interactionists see sexual identity, whatever its form, as created within specific cultural, historical, and social contexts. Symbolic interaction theory uses a **social construction** perspective to interpret sexual identity as learned, not inborn. To symbolic interactionists, culture and society shape sexual experiences. Patterns

Table 13.1 Theoretical Perspectives on Sexuality				
Interprets:	**Functionalism**	**Conflict Theory**	**Symbolic Interaction**	**Queer Theory**
Sexual norms	Sexual norms are functional for society because they encourage the formation of stable institutions.	Sexual norms are frequently contested by those who are subordinated by dominant groups.	Sexual norms emerge through social interaction and the construction of beliefs.	Sexual norms are easily contested through play and performances that transgress the dominant sex/gender categories.
Sexual identity	Sexual identity is learned in the family; deviant identities contribute to social disorder.	Sexual identity is regulated by individuals and institutions that enforce compulsory heterosexuality.	Sexual identity is socially constructed as people learn the sexual scripts created in society.	Multiple forms of sexual identity are possible and are found by people crossing the ordinarily assumed boundaries.
Changing sexual values	Regulating sexual values and norms is important for maintaining traditional and social stability; too much change results in social disorganization.	Social change comes through the activism of people who challenge dominant belief systems and practices.	Change in sexual value systems evolves as people construct new beliefs and practices over time.	Sexual values can be changed through disrupting categories of the dominant culture long taken for granted.

of social approval and social taboos make some forms of sexuality permissible and others not (Connell 1992; Lorber 1994).

Various forms of sexual identity are possible. *Heterosexuals* are those sexually attracted to members of the other sex; *homosexuals* are attracted to the same sex; and *bisexuals* are attracted to members of both sexes. **Sexual orientation** is how individuals experience sexual arousal and pleasure. Debate about the terms *sexual orientation* and *sexual preference* is further evidence of the significance of social meaning in talking about sexual identities. Although some use the terms interchangeably, gays and lesbians have argued that the term *sexual preference* implies that sexual orientation is a choice, whereas the term *sexual orientation* implies something more deeply rooted in their identity development.

The social construction perspective considers no form of sexual identity more natural than any other. Instead, this perspective explains all forms of sexual identity as created through the self-definitions that people develop through relationships with others. Sexual identity develops from exposure to different cultural expectations and, in the case of heterosexual identity, adherence to dominant cultural expectations. Thus, those who develop heterosexual identities have experiences and receive social supports that reinforce heterosexual identity. Adrienne Rich, a feminist poet and essayist, has called this **compulsory heterosexuality**, the idea that heterosexual identity is not a choice. Instead, she says, institutions define heterosexuality as the only legitimate form of sexual identity and enforce it through social norms and sanctions, including peer pressure, socialization, law, economic policies, and, at the extreme, violence (Rich 1980).

How sexual identity is socially constructed is also revealed by studies of the coming out process. **Coming out**—the process of defining oneself as gay or lesbian—is a series of events and redefinitions in which a person comes to see herself or himself as having a gay identity. In coming out, a person consciously labels that identity either to oneself or others (or both). Coming out is usually not the result of a single homosexual experience. If it were, there would be far more self-identified homosexuals, because researchers find that a substantial portion of both men and women have some form of homosexual experience at some point in their lifetimes. A person may fluctuate among gay, straight, and bisexual identities (Rust 1995, 1993). This indicates, as the social construction perspective suggests, that identity is created, not fixed, over the course of a life.

The development of sexual identity is not necessarily a linear or unidirectional process with persons moving predictably through a defined sequence of steps or phases. Although they may experience certain milestones in their identity development, some people experience periods of ambivalence about their identity and may switch back and forth between lesbian, heterosexual, and bisexual identity over time (Rust 1993). Some people may engage in lesbian or gay behavior but not adopt a formal definition of themselves as lesbian or gay. Certainly many gays and lesbians never adopt a public definition of themselves as gay or lesbian, instead remaining "closeted" for long periods, if not entire lifetimes. Sometimes, one's sexual identity may change. For example, a person who has always thought of himself or herself as a heterosexual may decide at a later time that he or she is gay or lesbian. In more unusual cases, people may undergo a sex change operation, perhaps changing their sexual identity in the process. In the case of bisexuals, a person might adopt a dual sexual identity.

Although most people learn stable sexual identities, over the course of one's life, sexual identity evolves. From a sociological point of view, changing one's sexual identity is not a sign of immaturity, as if someone has missed a so-called normal phase of development. Change is a normal outcome of the process of identity formation (Rust 1993). The plasticity of human sexuality means that it is not fixed but emerges through social experience.

Changing social contexts (including dominant group attitudes, laws, and systems of social control), relationships with others, and even changes in the language used to describe different sexual identities all affect people's self-definition. Political movements can encourage people to adopt a sexual identity that they previously had no context to understand. Further, studies of gay and lesbian academics who are publicly "out" have shown that there are negative consequences in the form of limited job opportunities, fewer promotions, and exclusion from social networks (Taylor and Raeburn 1995). Such a context can discourage some from publicly acknowledging their lesbian or gay identities.

Queer Theory

Related to the social constructionist perspective is a new understanding of sexual identity known as queer theory. **Queer theory** interprets various dimensions of sexuality as thoroughly social and constructed through institutional practices. Queer theory underscores the idea that sexual identity is fluid; that is, it evolves and can change over the life course. Instead of seeing heterosexual or homosexual attraction as fixed in biology, queer theory interprets society as forcing sexual boundaries, or dichotomies, on people.

Queer theory has developed largely through new studies of lesbian and gay experience, as well as studies of transgender, bisexuality, and cross-dressing. Many of these forms of sexual identity have traditionally been thought of in the context of social deviance (see

Chapter 7), but this approach has been challenged by feminist and gay liberation movements, where activists and scholars have argued that there is a broad spectrum of human sexuality and various identities possible. Queer theory challenges the "either/or" thinking that one is either gay or straight and also challenges the assumption that only one form of sexuality is normal and all other forms deviant or wrong. As a result, fascinating new studies of gay, straight, bisexual, and transsexual identities have appeared (Nardi and Schneider 1997; Seidman 1997; Stein and Plummer 1994; Rust 1995).

Queer theory has also linked the study of sexuality to the study of gender, showing how transgressing (or violating) fixed gender categories can reconstruct the possibilities of how all people—men and women, gay, bisexual, or straight—construct their gender and sexual identity. Transgressing gender categories, such as by drag queens, cross-dressers, or other playful demonstrations, can show how sex and gender categories are usually constructed in dichotomous categories (that is, opposite or binary types). By violating these constructions, people are liberated from the social constraints that presumably fixed categories of identity create. Thus, queer theory emphasizes how performance and play with gender categories can be a political tool for deconstructing fixed sex and gender identities (Rupp and Taylor 2003).

Sexual Politics: Diversity and Inequality

Patterns of sexuality reflect the social organization of society. When you understand this, you also see that sex is related to other social factors—such as race, class, and gender—and you see how sexuality is connected to social institutions and social change.

Sexual Politics

Sexual politics refers to the link between sexuality and power. Feminists have argued that one cannot have an equal or satisfying sexual relationship if one is powerless in a relationship or is defined as the property of someone else. But sexual politics does not just refer to power within individual relationships; it also refers to the link between the sexual exploitation of women and the distribution of power in society. This is reflected in the high rates of violence against women, the cultural treatment of women as sex objects, and patterns of sexual harassment at work and in schools (Andersen 2003; see Chapter 18).

The feminist and gay and lesbian liberation movements have put sexual politics at the center of the public's attention by challenging gender role stereotyping

---

A SOCIOLOGICAL EYE ON THE MEDIA

Publicity Traps: Sex on TV Talk Shows

If you even casually peruse television talk shows, you are likely to see a wide array of sexual nonconformists—people who may be bisexual, transgendered, or in some way sexually nonconforming. Joshua Gamson, a sociologist who specializes in the study of media, sexuality, and social change, has studied television talk shows and used them to think about the cultural visibility of nonconforming sexual groups. Lesbians, gays, transgendered people, bisexuals, and others are usually marginalized in society—treated as invisible, seldom recognized, misunderstood, and usually suppressed in public space. But on television talk shows, the public can witness outrageous, boisterous, even wild guests. In fact, in an attempt to build a larger audience (because, after all, this is about profit), television talk shows have become less likely to feature calm, well-educated, distinguished

guests. Instead, they often feature those who least conform to the dominant sexual value system. Gamson has asked, What is the impact of the visibility of sexual nonconformists for gays and lesbians?

Gamson sees no simple answer to this question. On the one hand, showcasing sexual nonconformity in a way that may seem freakish, foul-mouthed, and "abnormal" presents a distorted image of gay life and gives legitimacy to those who think that lesbian, gay, bisexual, transgender, and other diverse sexual lifestyles are deviant. The presentation of people on "the fringe," as Gamson puts it, "makes social acceptability harder to gain by overemphasizing difference, often presented as frightening, pathological, pathetic, or silly" (1998: 33).

But Gamson also sees another dimension to this question. He writes

that, although these images present a distorted image of gay life, the talk shows also make diverse sexual identities public, thereby having the added effect of making sexual nonconformity less shocking and thus, in the long run, more acceptable. In addition, Gamson argues, these portrayals open up public space where challenges to sexual conformity can transform the ordinarily fixed boundaries between gay and straight, normal and "queer." The dilemma comes from the fact that these portrayals are generally not made by or on behalf of gay people. Yet, becoming more visible in a public space such as the media can change people's understanding of who has a right to share public space.

Source: Gamson, Joshua. 1998. "Publicity Traps: Television Talk Shows and Lesbian, Gay, Bisexual, and Transgender Visibility." *Sexualities* 1: 11–41. ●●●

and sexual oppression (D'Emilio 1998). Among other things, this has profoundly changed knowledge of gay and lesbian sexuality. Gay, lesbian, and feminist scholars have argued, and many now concur, that homosexuality is not the result of psychological deviance or personal maladjustment but is one of several alternatives for happy and intimate social relationships. The political mobilization of large numbers of lesbian women and gay men and the willingness of many to make their sexual identity public have also raised public awareness of the civil and personal rights of gays and lesbians. These changes make other changes in intimate relations possible.

The link between sex and power is well illustrated by the scandals involving sexual abuse of young boys by Catholic priests that have recently come to light. Without a sociological perspective, you might think of this behavior solely as the result of certain individuals gone astray by violating strong moral values. But from a sociological perspective—and by using the concept of sexual politics—you can interpret sexual abuse within the church in a different light—one with very different implications for addressing the problem.

Sociologists have shown that certain institutional contexts are more prone to sexual violence than others, namely those that are marked by a highly masculine subculture, exclusive in their membership, and organized in a hierarchical fashion around norms of privacy, secrecy, and silence (Martin and Hummer 1989). The sexual abuse of young boys by priests occurs within such an institutional context. The priesthood within the Catholic Church is highly *patriarchal* (that is, power lies with men). The priesthood is imbued with a sense of *secrecy* (think of the confessional and the private relationship that occurs there between a parishioner, such as a young boy, and a priest). The Catholic Church is an organization marked by a clear hierarchy of *power,* and it is defined as *sacred,* and thus, in some sense, immune from the usual forms of social regulation. These institutional characteristics, one could argue, created a climate wherein abuse could occur and be unreported for many years. Understanding the link between sex and power can also help you understand how sexual abuse occurs within other institutional settings that are marked by similar characteristics (such as the military or even families).

The Influence of Race, Class, and Gender

Sexual behavior follows gendered patterns, as well as patterns established by race and class relations (Schwartz and Rutter 1998). Cultural definitions of what is sexually appropriate differ significantly for women and men.

Gender expectations emphasize passivity for women and assertiveness for men in sexual encounters. The "double standard" is the idea that different standards apply to men and women for sexual behavior. Although this idea is weakening somewhat, men are still stereotyped as sexually overactive; women, less so. Women who openly violate this cultural double standard by being openly sexual are then cast in a negative light as "loose," as if the appropriate role for women is the opposite of loose, say, "secured" or "caged." The double standard forces women into polarized roles as "good" girls or "bad" girls. The belief that women who are raped must have somehow brought it on themselves rests on such images of women as "temptresses." Contrary to popular belief, men do not have a stronger sex drive than women. Men are, however, socialized more often to see sex in terms of performance and achievement, whereas women are more likely socialized to associate sex with intimacy and affection. Men are also more likely to associate sexuality with power, an important point in understanding the high rates of violence against women (see Chapter 12).

Because sex is associated with power, it is tied not only to gender in society, but *sexual politics are also integrally tied to race and class relations in society.* Sexual stereotypes of different racial–ethnic groups illustrate how sexuality is linked to race and class oppression (Collins 2004). Latino women are stereotyped as either "hot" or virgins and Latino men as "hot lovers" and sexually uncontrollable. African American men are stereotyped as threatening and overly virile. Asian American women are stereotyped as compliant and submissive, but passionate, which is a way of keeping them in their place, because to define them as active and independent would be to liberate them from their racial–sexual status. Class relations also produce sexual stereotypes of women and men. Working-class and poor men may be stereotyped as dangerous, whereas working-class women are disproportionately likely to be labeled "sluts."

Class, race, and gender hierarchies have historically been justified based on claims that Black men were highly sexed, lustful beasts whose sexuality had to be controlled by the "superior" Whites and that Black women were sexual animals who were openly available to White men (Nagel 2003). During slavery, for example, the sexual abuse of African American women was one way slave owners expressed their ownership of African American people. Many slave owners saw sexual access to women slaves as their right. Under slavery, the racist and sexist images of Black men and women were also developed to justify the system of slavery. Lynching was another mechanism of social control used to maintain White men's power over this system of extreme racial exploitation. A Black man falsely accused of having had sex with a White woman could be murdered without penalty to his killers. After the end of slavery, and well into the twentieth century, lynching was a method of terrorism used to tell the Black community to keep in its place (Genovese 1972; Jordan 1968). Sexual abuse was also part of the White conquest of American Indians. Historical accounts show that the rape of Indian women by White conquerors was common (Freedman and D'Emilio 1988), as is the rape of women following war and military conquest.

These historical patterns continue to influence contemporary images and national legends. Advertisements commonly depict women of color in bondage, as animals, or as an exotic part of nature (Tuan 1984). These condescending images strongly imply that women of color need to be tamed or can be treated as pets (Tuan 1984; Collins 1990). American Indian women are routinely displayed as using their sexual wiles to negotiate treaties between Whites; few people know that Sacagawea, the Indian woman who helped lead the Lewis and Clark expedition, had been purchased as a slave by a French trader while still a young teen (Nagel 2003). American Indians are part of the cultural mythology, including that about sex (Tuan 1984).

The linkage of sexuality to systems of power and inequality can be seen in the fact that poor women and women of color are the groups most vulnerable to sexual exploitation. These women may have to resort to selling their sexuality if they lack economic resources. Becoming a prostitute, or otherwise working in the sex industry as a topless dancer, striptease artist, pornographic actress, or other sex-based occupation, is often the last resort for women with limited options for income. Women who sell sex are condemned for their behavior, and more so than their male clients. This further illustrates how gender stereotypes mix with race and class exploitation. Why, for example, are women, not usually men, arrested for prostitution? A sociological perspective on sexuality helps one see the social conditions of inequality that shape race, class, and gender relations in our society.

Understanding Gay and Lesbian Experience

The institutional context for sexuality within the United States, as well as other societies, is one where homophobia permeates the culture. **Homophobia** is the fear and hatred of homosexuality. It is manifested in prejudiced attitudes toward gays and lesbians, as well as in overt hostility and violence directed against people suspected of being gay. Homophobia is a learned attitude, as are other forms of negative social judgments about particular groups, and is deeply embedded in people's definitions of themselves. Boys are often raised to be

"manly" by repressing so-called feminine characteristics in themselves. Being called a "fag" or a "sissy" is one peer sanction that socializes a child to conform to particular gender roles. Similarly, verbal attacks on lesbians are a mechanism of social control because ridicule can be interpreted as encouraging social conformity. As we saw in the prior chapter on gender, homophobia is one of the ways that conformity to gender roles is enforced.

THINKING SOCIOLOGICALLY

Keep a diary for one week and write down as many examples of *homophobia* and *heterosexism* as you observe in routine social behavior. What do your observations tell you about how *compulsory heterosexuality* is enforced?

Homophobia produces many myths about gay people. One such myth is that gays have a desire to seduce straight people. This belief falsely leads some to presume that gay men are potential child molesters, even though there is no evidence that homosexual men are more likely to commit sex crimes than heterosexual men. Homosexual men convicted of child molestation are, however, almost seven times more likely to be imprisoned than are heterosexual men who are convicted as child molesters and have the same past criminal record (Walsh 1994).

Fears that gay and lesbian parents will have negative effects on their children are also unsubstantiated by research. The ability of parents to form good relationships with their children is far more significant in children's social development than is their parents' sexual orientation (Stacey and Biblarz 2001).

Other myths about gay people are that they are mostly White men with large discretionary incomes who work primarily in artistic areas and personal service jobs (such as hairdressing). This stereotype prevents people from recognizing that gays and lesbians come from all racial–ethnic groups, some of whom are working-class or poor and who are employed in a wide range of occupations (Gluckman and Reed 1997). Some lesbians and gays are also older people, even though the stereotype defines them as mostly young or middle-aged (Smith 1983). These different misunderstandings reveal just a few of the many unfounded myths about gays and lesbians. Support for these attitudes comes from homophobic attitudes, not from empirical truth.

Heterosexism refers to the institutionalization of heterosexuality as the only socially legitimate sexual orientation. Heterosexism is rooted in the belief that heterosexual behavior is the only natural form of sexual expression and that homosexuality is a perversion of "normal" sexual identity. Heterosexism is reinforced through institutional mechanisms that project the idea that only heterosexuality is normal. Institutions also provide different benefits to people presumed to be heterosexual. Businesses and communities, for example, rarely recognize the legal rights of those in homosexual relationships, although this is changing. Heterosexism is an institutional phenomenon, although people's individual beliefs can reflect heterosexist assumptions. Thus, a person may accept gay and lesbian people (that is, not be homophobic) but still benefit from heterosexual privileges. At the behavioral level, the effect of heterosexist practices can exclude lesbians and gays, such as when coworkers talk about their dating activities assuming that everyone is interested in a heterosexual partner.

In the absence of institutional supports from the dominant culture, lesbians and gays have invented their own institutional support systems. Gay communities and gay rituals, such as gay pride marches, affirm gay and lesbian identity and provide a support system that is counter to the dominant heterosexual culture. Those who remain in the closet deny themselves this support system.

The absence of institutionalized roles for lesbians and gays affects the roles they adopt within relationships. Despite popular stereotypes, gay partners typically do not assume roles as the dominant or subordinate sexual partner. They are more likely to adopt roles as equals. Gay and lesbian couples are also more likely than heterosexual couples to both be employed, another source of greater equality within the relationship. Researchers have also found that the quality of relationships among gay men is positively correlated with the social support the couple receives from other people (Smith and Brown 1997; Metz et al. 1994; Harry 1979).

Few ever question their heterosexual identity as anything other than normal, but when you consider the extent of institutional supports for developing a heterosexual identity, it is not surprising that most take it for granted. In other societies, however, where cultural supports for multiple forms of sexual identity are stronger, distinctions between heterosexual and homosexual behavior may not be so rigid. For example, the *Xaniths* of Oman, a strictly gender-segregated Islamic society, are men who are homosexual prostitutes. Different from male prostitutes in Western culture, they have a specific and valued social function: preserving the sexual purity of women. Women in this society are traditionally prohibited from mingling with men in public or having sex prior to marriage. They are veiled and robed in public. For legal purposes, women are considered lifelong minors. Xaniths are biological males, who mingle freely with women and men but maintain the legal

Many gay and lesbian couples jubilantly took formal vows of marriage following San Francisco's legally recognizing the right to gay unions, although the California Courts later ruled same sex marriage to be invalid.

status of men. They provide sexual outlets for men who are not married, even though the men who have sex with Xaniths are not considered homosexuals. Xaniths themselves are seen as "reverting to manhood" if they later marry a woman (and they often do), but they must provide evidence of having taken her virginity. In the Oman culture, no man is seen as culminating his manhood until he marries and "deflowers" his bride (Lorber 1994). Putting heterosexuality and homosexuality into this cross-cultural perspective helps you see the extent to which sexuality is defined by the cultural context in which it is located.

A Global Perspective on Sexuality

Once you understand the connection between sexuality and systems of inequality within the United States, you can use a similar perspective to develop a global perspective on sexuality. Returning to the point made at the beginning of this chapter, if you thought of sexuality as only a "natural" thing, without a broader sociological perspective, you would probably never stop to think about global dimensions of sexuality. But, sexuality is expressed differently across cultural contexts.

Cross-cultural studies of sexuality show that different sexual norms, like other social norms, develop differently within cultural meaning systems. Take sexual jealousy. Perhaps you think that seeing your sexual partner becoming sexually involved with another would "naturally" evoke jealousy—no matter where it happened. Researchers have found this not to be true. In a study comparing patterns of sexual jealousy in seven different nations (Hungary, Ireland, Mexico, the Netherlands, the United States, Russia, and the former Yugoslavia), researchers found significant cross-national differences in the degree of jealousy when women and men saw their partners kissing, flirting, or being sexually involved with another person (Buunk and Hupka 1987). Such cross-cultural studies can make you more sensitive to the varying cultural norms and expectations that apply to sexuality in different contexts.

Sex is big business and deeply tied to the world economic order (see Map 13.1). As the world has become more globally connected, an international sex trade has flourished. This business is linked to economic development, world poverty, tourism, and the subordinate status of women in many nations. The *international sex trade,* sometimes also referred to as the "traffic in women" (Rubin 1975), is the use of women, worldwide,

as sex workers in an institutional context where sex itself is a commodity. Sex is marketed in an international marketplace, and women as sex workers are used to promote tourism, to cater to business and military men, and to support a huge industry of nightclubs, massage parlors, and teahouses (Enloe 1989). Children, too, may be exploited as child prostitutes, through which boys and girls may be lured into conditions of slavery. Unlike the slavery associated with African Americans in the Americas, children in these circumstances are not seen as a long-term investment, but rather are quickly expendable, disposed of once they are too old, too diseased, or in other ways no longer profitable (Bales 1999; Flowers 2001).

"Sex capitals" where prostitution openly flourishes, such as Thailand, Amsterdam, and other locales, are an integral part of the world tourism industry. In Thailand, for example, men as tourists outnumber women

VIEWING SOCIETY IN GLOBAL PERSPECTIVE

MAP 13.1 The International Sex Trade

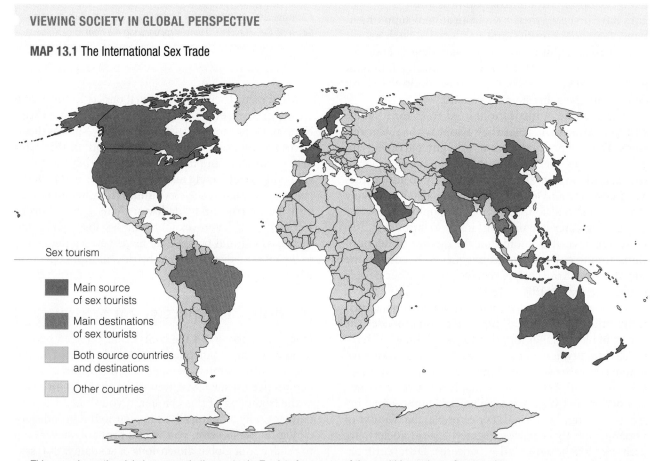

Sex tourism

- Main source of sex tourists
- Main destinations of sex tourists
- Both source countries and destinations
- Other countries

This map shows the global economy in the sex trade. Tourists from some of the wealthier nations often travel to particular parts of the world where sex is promoted as part of the tourism industry. Businessmen flock to Thailand, as an example, to buy sex from prostitutes. Moreover, the trafficking of women and children in the sex industry is controlled by large cartels, just as is the case in legitimate businesses. Still, most people think about prostitution as a matter of morality, not an international business. How do the questions you ask about prostitution change if you think about sex as a business instead a matter of individual morality?

Source: From *The State of Women in the World Atlas* by Joni Seager, copyright © 1997 by Joni Seager, text; © by Myriad Editions, Ltd., maps & graphics. Used by permission of Viking Penguin, a division of Penguin Group (USA), Inc.

The international trafficking of women for sex exploits women—and often children, and puts them at risk for disease and violence.

by three to one. Planeloads of businessmen come to Thailand as tourists, sometimes explicitly to buy sexual companionship. Hostess clubs in Tokyo similarly cater to corporate men. One fascinating study, where the researcher worked as a hostess in a Tokyo club, shows how the men's behavior in these settings is linked to the expression of their gender identity among other men (Allison 1994).

Sociologists see the international sex trade as part of the global economy, contributing to the economic development of many nations and supported by the economic dominance of certain other nations. As with other businesses, the products of the sex industry may be produced in one region and distributed in others. Think, for example, of the pornographic film industry that is centered in southern California but distributed globally. The sex trade is also associated with world poverty. Sociologists have found that the weaker the local economy, the more important sex is to the economy of such places. The international sex trade is also implicated in such problems as the spread of AIDS worldwide, as well as the exploitation of women in parts of the world where women have limited economic opportunities (Altman 2001).

Sex and Social Issues

In studying sexuality, sociologists tap into some of the highly contested social issues of the time. Birth control, reproductive technology, abortion, teen pregnancy, pornography, and sexual violence are all subjects of public concern and public policy. These social issues can generate personal troubles that have their origins in the structure of society—recall the distinction made by C. Wright Mills in Chapter 1 between personal troubles and social issues. Debates about these issues hinge in part on attitudes about sexuality and are shaped by

race, class, and gender relations. Although most people think of sex in terms of interpersonal relationships, sexual norms are deeply intertwined with various social problems in society. These different issues show us how the social structure of society contributes to some of our nation's social problems.

Birth Control

The availability of birth control is now less debated than it was in the not too distant past, but this important reproductive technology has been strongly related to the social position of women (Gordon 1977). Reproduction has been controlled by men who, to this day, have made most of the decisions, legal and scientific, about whether birth control will be available. The social position of women also means that women are more likely seen as responsible for reproduction because that is a presumed part of their traditional role. At the same time, changes in birth control technology have also made it possible for women to change their roles in society. Breaking the link between sex and reproduction has freed women from some traditional constraints.

The right to available birth control is a recently won freedom. It was not until 1965 that the Supreme Court, in *Griswold v. Connecticut*, defined the use of birth control as a right, not a crime. Even then this ruling applied originally only to married people; unmarried people were not extended the same right until 1972, in the Supreme Court decision, *Eisenstadt v. Baird*. Today, birth control is routinely available over the counter, but there is heated debate at the highest levels of public policy. Should access to birth control be curtailed for the young and information about it restricted? Sociological data show that U.S. youths are experimenting with sex at younger ages and therefore increasing the risk of teenage pregnancy and AIDS. Some argue that sexual activity at this age must be controlled and increasing access to birth control and knowledge about how to use it would encourage more sexual activity among the young.

Class and race relations had a strong role in shaping birth control policy. The birth control movement in the United States began in the mid-nineteenth century. During this period, increased urbanization and industrialization removed the necessity for large families, especially in the middle class, because fewer laborers were needed to support the family. The ideology of the birth control movement reflected these conditions. Early feminist activists such as Emma Goldman and Margaret Sanger also saw birth control as a way to free women from unwanted pregnancies and allow them to work outside the home if they chose, instead of being obliged to work in the home in support of a large family. As the birthrate fell among White upper-class and middle-class families in this period, these classes feared that immigrants, the poor, and racial minorities would soon

MAP 13.2 Contraceptive Use Worldwide

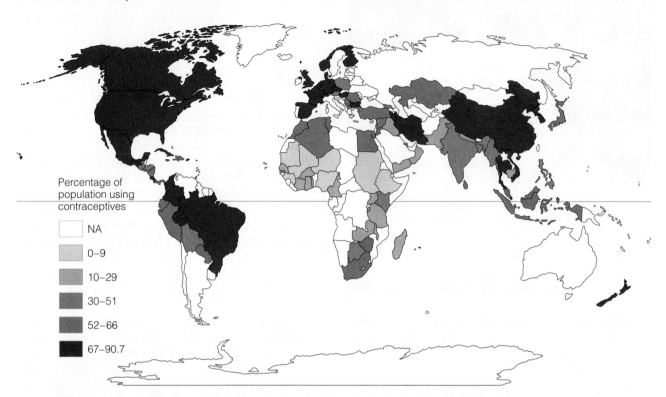

Percentage of
population using
contraceptives

- NA
- 0–9
- 10–29
- 30–51
- 52–66
- 67–90.7

Contraceptive use has been increasing worldwide, increasing as women's levels of education increase. Contraception is also related to the level of social and economic development in different societies, with the most developed regions most likely to have higher rates of contraceptive use. Name other cultural factors that you think influence contraceptive use in the different nations that you see here.

Data: UNICEF. July 2002. *Progress Since the World Summit for Children: A Statistical Review.* Website: **www.childinfo.org/eddb/fertility/dbcontrc.htm**

outnumber them. The dominant class feared "race suicide." They imagined that immigrants and the poor would overpopulate the society, leaving the White middle and upper class as a numerical minority.

The **eugenics movement** of the early twentieth century sought to apply scientific principles of genetic selection through selective breeding to "improve" the offspring of the human race. This movement was explicitly racist and class-based, calling for, among other things, the compulsory sterilization of people the eugenicists thought were unfit. Margaret Sanger made eugenics part of her birth control campaign because she wanted poor women to have fewer children so that their children would not be exploited as workers. Eugenicist arguments, though they make scientific claims, are typically based on popular social stereotypes rather than scientific fact. The arguments appeal to a public that fears the social problems that emerge from racism, class inequality, and other injustices. Instead of attributing problems such as crime and poor educational

achievement to the structure of society, eugenicists put the blame on the genetic composition of certain groups that are least powerful in society.

New Reproductive Technologies

Developments in the technology of reproduction have ushered in new sexual freedoms, but also raise questions for social policy. Reproduction is no longer inextricably linked to biological parents with the advent of practices of surrogate mothering and in vitro fertilization, as well as the new biotechnologies of gene splicing, cloning, and genetic engineering (Rifkin 1998). A child may be conceived through means other than sexual relations between one man and one woman. One woman may carry the child of another. Offspring may be planned through genetic engineering. A sheep can be cloned (genetically duplicated). Are humans next? To whom are such new technologies available? Which groups are most likely to sell reproductive services, which groups to buy? What are the social implications of such changes?

There are no simple answers to such questions, but sociologists would point first to the class, race, and gender dimensions of these issues (Roberts 1997). Poor women are far more likely than middle-class or elite-class women to sell their eggs or offer their bodies as biological incubators. Those who can afford new, costly methods of reproduction may do so at the expense of women whose need for economic support places them in the position of selling their bodies.

Sophisticated prenatal screening now makes it possible to identify fetuses with presumed defects. Socially identifiable groups can now be screened for "genetic disorders." And echoes of eugenics can be heard in the calls for sperm banks reserved for "smart" people, where eager parents could create "designer children" free of defects and stigmatized traits. Might the society then try to weed out those perceived as undesirables—the disabled, certain racial groups, certain sexes? If boys and girls are differently valued, one sex may be more frequently aborted. This is a frequent practice in India and China—two of the most populous nations on earth. State policy in China, for example, because of population pressures, encourages families to have only one child. Because girls are less valued than boys, the aborting and selling of girls is common. This has created a market for the adoption of Chinese baby girls in the United States.

Breakthroughs in reproductive technology raise especially difficult questions for makers of social policy. There are no traditions to guide us on questions that would have been considered outlandish and remote just a few years ago. Our thinking about sexuality and reproduction will have to evolve. Although the concept of reproductive choice is important to most people, choice is conditioned by the constraints of race, class, and gender inequalities in society. Like other social phenomena, sexuality and reproduction are shaped by their social context.

Abortion

Abortion is one of the most seriously contested political issues in recent years. The majority of the U.S. public clearly support abortion rights—41 percent of the public think abortion should be legal in any or most circumstances; another 40 percent think it should be legal in a few circumstances, and 17 percent think it should be illegal in all circumstances (see Figure 13.2). Public support for abortion rights has remained relatively constant over the last twenty years.

The right to abortion was first established in constitutional law by the *Roe v. Wade* decision in 1973. In *Roe v. Wade,* the Supreme Court ruled that at different points during a pregnancy, separate but legitimate rights collide: the right to privacy, the right of the state

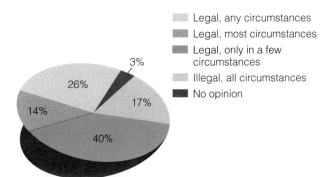

Figure 13.2 Attitudes Toward Abortion

Data: Gallup Organization. 2002a. "Abortion." Princeton, NJ: The Gallup Organization. Website: **www.gallup.com**

to protect maternal health, and the right of the state to protect developing life. To resolve this conflict of rights, the Supreme Court ruled that pregnancy occurred in trimesters. In the first trimester, women's right to privacy without interference from the state prevails; in the second, the state's right to protect maternal health takes precedence; in the third, the state's right to protect developing life prevails. In the second trimester, the government cannot deny the right to abortion, but it can insist on reasonable standards of medical procedure. In the third, abortion may be performed only to save the life or health of the mother. Clearly *Roe v. Wade* has mattered in saving women's lives, as indicated by the dramatic drop in death rates from abortion following this Supreme Court decision (see Figure 13.3). More recently, the Court has allowed states to impose restrictions on abortion, but it has not, to date, overturned the legal framework of *Roe v. Wade.*

Data on abortion show that it occurs across social groups, although certain patterns do emerge. The abortion rate has declined somewhat since 1980, from a rate of 35.9 per 1000 live births to 25.6 in 1999. Among age groups, the most likely group to get abortions are women between age fifteen and nineteen, although the second most likely group is women over forty. Women who have already had one or two live births are also the most likely group to have an abortion (National Center for Health Statistics 2004).

The abortion issue provides a good illustration of how sexuality has entered the political realm. Abortion rights activists and antiabortion activists hold very different views about sexuality and the roles of women. Antiabortion activists tend to believe that giving women control over their fertility breaks up the stable relationships in traditional families. They tend to view sex as something that is sacred, and they are disturbed by changes that make sex less restrictive. Antiabortion activists also tend to believe that making contraception available to teens encourages promiscuity. Abortion rights activists see women's control over reproduction as essential for their independence. They also tend to see

Number of abortion-related deaths

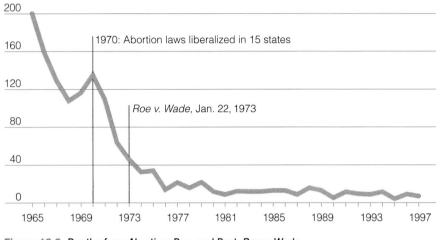

Figure 13.3 Deaths from Abortion: Pre- and Post-*Roe v. Wade*

Source: Alan Guttmacher Institute. 1999. *Trends in Abortion 1973–2000.* New York: The Alan Guttmacher Institute.
Website: **www.agi.usa.org**

sex as an experience that develops intimacy and communication between people who love each other. The abortion debate can thus be interpreted as a struggle over abortion as well as a battle over different sexual values and a referendum on the nature of relationships between men and women (Luker 1984).

Pornography

Little social consensus has emerged about the acceptability and effects of pornography. One purpose of this debate is to define what is obscene. The legal definition of obscenity is one that changes over time and in different political contexts. Public agitation over pornography divides those who think it is solidly protected by the First Amendment, those who want it strictly controlled, those who think it should be totally banned for moral reasons, and those who think it must be banned because it harms women. Who is right? Recall the discussion from Chapter 2 that reported on studies showing that exposure to violent pornography does have an effect on sexual attitudes. After exposure to violent pornography, men are more likely to see victims of rape as responsible for their assault, less likely to regard them as injured, and more likely to accept the idea that women enjoy rape. These effects are found only after exposure to pornography that is both sexually explicit and violent (Donnerstein et al. 1987). Little evidence supports the assertion that exposure to pornography in general increases sexual promiscuity or sexual deviance.

Despite public concerns about pornography, a strong two-thirds majority thinks pornography should be protected by the constitutional guarantees of free speech and a free press. At the same time, people believe that pornography dehumanizes women. Women tend to be

more negative about pornography than men, perhaps because pornography is generally geared toward men. The controversy about pornography is not likely to go away because it taps so many different sexual values among the public.

Teenage Pregnancy

Each year about one million teenage girls (ages 15–19) have babies in the United States, 78 percent of which are unplanned. The United States has the highest rate of teen pregnancy among developed nations (see Figure 13.4), even though levels of teen sexual activity around the world are roughly comparable (see Figure 13.5). Teens account for almost 13 percent of all births. Teen pregnancy has declined since 1990, mostly because of the increased use of birth control and, to some extent, more abstinence by teens. (Analysts find that abstinence accounts for about one-quarter of the decline, birth control, the rest; Alan Guttmacher Institute 2004.)

Although the rate of pregnancy among teens has dropped, so has the rate of marriage for those who be-

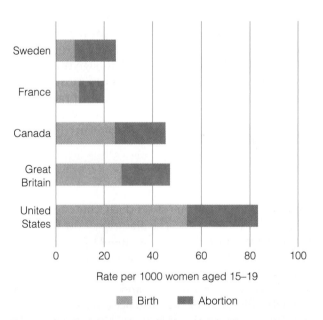

Rate per 1000 women aged 15–19

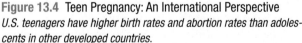

Figure 13.4 Teen Pregnancy: An International Perspective
U.S. teenagers have higher birth rates and abortion rates than adolescents in other developed countries.

Source: Alan Guttmacher Institute. 1999. *Teenagers' Sexual and Reproductive Health.* New York: The Alan Guttmacher Institute. Reproduced with the permission of the Alan Guttmacher Institute.

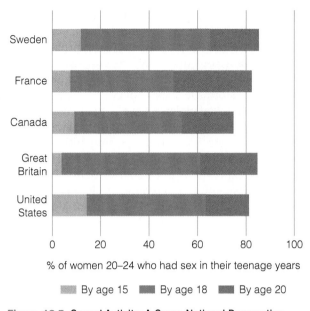

Figure 13.5 Sexual Activity: A Cross-National Perspective
Differences in levels of teenage sexual activity across developed countries are small.

Source: Alan Guttmacher Institute. 1999. *Teenagers' Sexual and Reproductive Health.* New York: The Alan Guttmacher Institute. Reproduced with the permission of the Alan Guttmacher Institute.

come pregnant. Thus, most babies born to teens will be raised by single mothers—a departure from the past when teen mothers more often got married (U.S. Census Bureau 2004). What concerns people about teen parents is that they are more likely to be poor than other mothers, although sociologists have cautioned that this is because teen mothers are more likely poor *before* getting pregnant (Luker 1996). Still, teen parents are among the most vulnerable of all social groups.

Teenage pregnancy correlates strongly with poverty, lower educational attainment, joblessness, and health problems. Teen mothers have a greater incidence of problem pregnancies and are more likely to deliver low-birth-weight babies, a condition associated with a myriad of other health problems. Teen parents face chronic unemployment and are less likely to complete high school than those who delay childbearing. Many teen parents continue to live with their parents, although this is more likely among Black teens than among Whites. Given the higher rates of poverty among African Americans, the costs of adolescent pregnancy then fall most heavily on those families least able to help.

Although teen mothers feel less pressure to marry than they did in the past, if they raise their children alone, they suffer the economic consequences of raising children in female-headed households—the poorest of all income

groups. Teen mothers report that they do not marry because they do not think the fathers are ready for marriage. Sometimes their families also counsel them against marrying precipitously. These young women are often doubtful about men's ability to support them. They want men to be committed to them and their child, but they do not expect their hopes to be fulfilled (Farber 1990). Research shows that low-income single mothers are distrustful of men, especially after an unplanned pregnancy. They think they will have greater control of their household if they remain unmarried. Many also express their fear of domestic violence as a reason for not marrying (Edin 2000). Both Black and White teen parents see marriage as an impediment to their career ambitions, although Black teen mothers are more likely than Whites to continue attending school (Farber 1990; Trent and Harlan 1990).

Why do so many teens become pregnant given the widespread availability of birth control? One-third of teenagers use no contraceptive the first time they have sexual intercourse, typically delaying the use of contraceptives until several months after they become sexually active. A sexually active teenager not using contraceptives has a 90 percent chance of becoming pregnant within the first year of sexual intercourse (Alan Guttmacher Institute 1999a). Teens are more likely to be using contraceptives if they are in a serious, not just a casual, relationship (Manning et al. 2000). In recent years, the percentage of teens using birth control, especially condoms, has increased, although the pill is still the most widely used method (see Figure 13.6; Alan Guttmacher Institute 2004).

Teen pregnancy rates have actually been falling, but the issue remains as one for the challenges of social policies involving sexual behavior and young people.

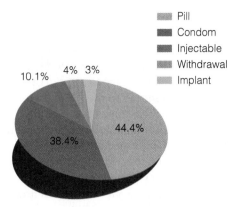

Legend:
- Pill
- Condom
- Injectable
- Withdrawal
- Implant

10.1% 4% 3%
44.4%
38.4%

Figure 13.6 Contraceptive Use Among Teens

Source: Alan Guttmacher Institute. 1999. *Teen Sex and Pregnancy.* New York: The Alan Guttmacher Institute. Reproduced with the permission of the Alan Guttmacher Institute.

Sociologists have argued that the effective use of birth control requires a person to identify himself or herself as sexually active (Luker 1975). Teen sex, however, tends to be episodic. Teens who have sex on a couple of special occasions may not identify themselves as sexually active and may not feel obliged to take responsibility for birth control. Despite the large number of teens initiating sex at an earlier age, social pressure continues to discourage them from defining themselves openly, or even privately, as sexually active.

Teen pregnancy is integrally linked to the gender expectations of men and women in society. Some teen men consciously avoid birth control, thinking it takes away from their manhood. Teen women often romanticize motherhood, thinking that becoming a mother will give them social value they do not otherwise have. For teens in disadvantaged groups, motherhood confers a legitimate social identity on those otherwise devalued by society (Horowitz 1995). Although their hopes about motherhood are not realistic, they indicate how pessimistic the teenagers feel about their lives, when they live in poverty, lack education, and have few good job possibilities (Ladner 1986). That young women romanticize motherhood is not surprising in a culture where motherhood is defined as a cultural ideal for women, but the ideal can seldom be realized when society gives mothers little institutional or economic support.

Sexual Violence

Prior to the development of the feminist movement, sexual violence was largely hidden from public view. One great success of the women's movement has been to identify, study, and advocate better social policies to address the problems of rape, sexual harassment, domestic violence, incest, and other forms of sexual coercion. Feminists have argued that sexual coercion is not just a matter of sexuality, but is a form of power relations shaped by the inequality between women and men in

society. In the thirty years or so that these issues have been identified as serious social problems, volumes of research have been published on these different subjects, and numerous organizations and agencies have been established to serve victims of sexual abuse and to advocate reforms in social policy.

Rape and sexual violence have been discussed in the prior chapters on deviance and the criminal justice system (see Chapters 7 and 8), in keeping with the argument that these are forms of deviant and criminal behavior, not expressions of human sexuality. Likewise, sexual harassment is covered in the chapter on work (Chapter 18), although sexual harassment also occurs in educational institutions. Domestic violence is reviewed in Chapter 15 on families. Here we point out that these various forms of sexual coercion can best be understood (and, therefore, changed) by understanding how social institutions shape human behavior and how social interactions are influenced by social factors such as gender, race, age, and class, among others.

Incest, for example, violates social values so much that it is considered taboo. But we now know that incest is much more common than the taboo suggests. Gender plays a major role in incest, with girls far more likely to be victims of incest than boys and men most likely to be the perpetrators. Most interpret incest as resulting from the intersection of power and gender within family structures (Candib 1999).

Consider too the phenomenon now known as acquaintance (or date) rape. *Acquaintance rape* is forced and unwanted sexual relations by someone who knows the victim (even if only a brief acquaintance) and is common on college campuses, although it is also the most underreported form of rape. Researchers estimate, based on national surveys, about 27 in every 1000 college women will experience either a completed or attempted rape in a given year; only 4 percent of completed rapes and 8 percent of attempted rapes are by perpetrators not known to the victim (Fisher et al. 2000).

Studies of acquaintance rape show that, although rape is an abuse of power, it is related to people's gender attitudes. College men, for example, are more likely than women to see heterosexual relations as adversarial. Men are also more likely to believe in various rape myths (Reilly et al. 1992), such as that women mean yes when they say no, that women's dress is a signal that they want sex, or women precipitate rape by their behavior. Studies also find that acquaintance rape is more likely to occur in organizations set up around a definition of masculinity as competitive, involving alcohol abuse, and where women are defined as sexual prey. This is one explanation given for the high rates of rape in some college fraternities (Martin and Hummer 1989; Stombler and Padavic 1997).

Research on violence against women finds that Black, Hispanic, and poor women are more likely than other women to be victimized by various forms of violence, including rape (U.S. Bureau of Justice Statistics 2002). Hispanic, Black, and White women are, however, equally likely to be victimized by an intimate partner who is an ex-spouse, boyfriend, or husband (Tjaden and Thoennes 2000). Studies also find that Black women are more aware of their vulnerability to rape than are White women and are more likely to organize themselves to resist rape collectively (Stombler and Padavic 1997).

In sum, sociological research on sexual violence shows how strongly sexual coercion is tied to the status of diverse groups of women in society. Rather than explaining sexual coercion as the result of maladjusted men or the behavior of victims, feminists have encouraged a view of sexual coercion that links it to an understanding of dominant beliefs about the sexual dominance of men and the sexual passivity of women. Researchers have shown that those holding the most traditional gender role stereotypes are more tolerant of rapists and less likely to give credibility to victims of rape (Marciniak 1998; Varelas and Foley 1998). Thus, from a sociological point of view, understanding sexual violence requires an understanding of patterns of gender, along with race and class, in society.

Sex and Social Change

Because sexual attitudes and behaviors are social phenomena, they are subject to the forces of social change. Social changes in sexual attitudes and behavior are also connected to other changes in society, particularly changes in family structure, women's and men's roles, even changes in the economic marketplace. Contrary to what many people think, sexual attitudes do not always move in a more liberal direction over time. Attitudes can change and people's behavior adapts to the time. Historians have also documented that social and political movements intended to change sexual ideas and practices thrive during periods when other aspects of society are undergoing rapid change (Freedman and D'Emilio 1988).

DEBUNKING **SOCIETY'S** MYTHS

Myth: Providing sex education to teens only encourages them to become sexually active.

Sociological perspective: Comprehensive sex education actually delays the age of first intercourse; abstinence-only education has not been shown to be similarly effective in delaying intercourse (Kirby 1997; Risman and Schwartz 2002).

In general, public attitudes about sex in the United States today are generally more liberal than they have been in the past, and there is greater tolerance for diverse sexual lifestyles and practices. A huge majority of the public (92 percent) now approve of sex education in the schools—an increase from 65 percent in 1970 (Gallup Organization 2000). The U.S. public has also become much more accepting of gays and lesbians. Fifty-four percent of the public think that homosexuality is an acceptable lifestyle, compared to 38 percent in 1992. A huge majority, 86 percent, think that gays should have equal rights in the workplace (Gallup Organization 2002b). Fifty percent also think that homosexual relations between consenting adults should be legal (Gallup Organization 2003)—a position that was strongly endorsed by the U.S. Supreme Court in 2003 when it ruled that state laws against sodomy were unconstitutional (*Lawrence v. Texas*). This decision essentially says that private sexual relations are a constitutional liberty, a conclusion that was widely interpreted as a major victory for gay rights. This is likely to be tested again if the Supreme Court takes up the case of same-sex marriage—a subject on which the public is greatly divided. Forty-two percent of the public think that gay marriage should be valid; 55 percent do not. But when you follow that question by asking people if they support civil unions between homosexual couples, support increases to 56 percent of the public (Moore and Carroll 2004).

Public opinion is now a mix of greater liberalism along with more conservatism on matters involving sexuality. Although adults are much less likely now than in the past to think that premarital sex is wrong (38 percent now versus 68 percent in 1969), more teens now think premarital sex is morally wrong than teens did in the late 1970s (see the box on page 340, "Doing Sociological Research: Teens and Sex: Are Young People Becoming More Sexually Conservative"; Lyons 2002). College students, generally quite liberal on sexual matters, have also changed their opinions about sexual relationships over time. In 1974, 46 percent agreed strongly or somewhat strongly that it was alright for people who really like each other to have sex even if they have known each other only for a short time. By the 1990s, this had changed to 51 percent, but it has now dropped to 42 percent (Higher Education Research Institute 2002).

Attitudes about sex also differ, depending on various social characteristics. There are significant differences between women and men. Men are more likely than women (62 versus 56 percent) to think that homosexual relations are morally wrong. The gender gap in sexual attitudes may explain some of the personal conflicts that men and women experience about sex (see Figure 13.7). Sexual attitudes are also shaped by age and education. For example, young people under thirty are less likely than older people, especially those over sixty-five, to think that homosexual relations are morally wrong. These differences likely reflect not only the

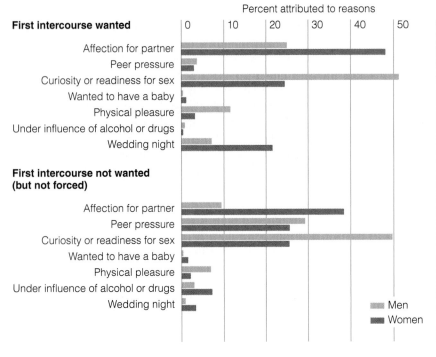

Figure 13.7 Reasons for Sexual Intercourse

Data: From Michael, Robert T., John H. Gagnon, Edward O. Laumann, and Gina Kolata. 1994. *Sex in America: A Definitive Survey.* Boston, MA: Little, Brown, and Company, p. 93.

influence of age, but also historical influences on different generations. People with more education are less likely to think that homosexuality is morally wrong (Newport 2003).

Public opinion on matters such as teen pregnancy, AIDS, child care, women's place in the workplace, and abortion rights tap underlying sexual value systems, often generating public conflicts. In general, sexual liberalism is associated with political liberalism on other social issues. In other words, the meaning that sex has for different groups can shape their support for or resistance to other social issues. The fact that sexual attitudes differ by factors such as gender, age, education, class, religion, and other sociological variables shows that sexual value systems are shaped by society's social and political systems.

The Sexual Revolution: Is It Over?

The **sexual revolution** refers to the widespread changes in roles of men and women and a greater public acceptance of sexuality as a normal part of social development. This movement originated in social and historical changes in the late nineteenth and early twentieth centuries, particularly in how sexuality

came to be understood and how women's and men's roles were emerging. The sexual revolution was also influenced by the development of technology—particularly the availability of the birth control pill and the growth of mass media. The pill opened new possibilities for sexual behavior at the same time that the media widely disseminated new cultural ideals for sexuality.

The sexual revolution has been strongly influenced recently by political movements, especially feminism and gay rights. Many changes associated with the sexual revolution have been changes in women's behaviors. Fewer women today are virgins at marriage, and women are more likely now than in the past to describe their sex lives as active and satisfying. Women are having more sex with more partners. Both men and women initiate sex at a younger age and before marriage. Essentially, the sexual revolution has narrowed the differences in the sexual experiences of men and women.

There is no doubt that there have been major changes in sexual behavior, especially if you take a long view. But, you have to be careful not to assume that there was no sexual freedom nor permissiveness in the past. And current indications are that whatever changes

As gays and lesbians and their allies have mobilized for social justice, they have also fostered pride and celebration, such as demonstrated in this gay rights parade.

AP/Wide World Photos

the sexual revolution ushered in may now be diminishing. The percentage of teens having sex is declining and some are reporting abstinence from sex altogether. But you also have to be careful in interpreting these changes. Careful review of sexual behavior among young people finds that most of the change is in the behavior of boys, not girls, as sociologists Pepper Schwartz and Barbara Risman show in the box, "Doing Sociological Research: Teens and Sex: Are Young People Becoming More Sexually Conservative" (see page 340).

No doubt the sexual revolution has opened up public discourse about sex, brought women new sexual freedoms (as well as dangers), and made possible the public expression of diverse forms of sexual identity. The degree to which the continuing inequities of sexual oppression continue or abate will depend on the continued social activism of those pressing for change and the public and political response to those demands. Social change in sexual behavior and attitudes can come as the result of other changes in society, but in the end is a matter of what people support.

Technology, Sex, and Cybersex

The sexual revolution has been significantly influenced by the development of new technologies. Contraceptives bring new possibilities for sexual freedom. With effective and available contraception like the birth control pill, sex is no longer necessarily linked with reproduction. New sexual norms associate sex (both within and outside marriage) with intimacy, emotional ties, and physical pleasure (Freedman and D'Emilio 1988). These sexual freedoms are not equally distributed among all groups, however. For women, sex is still more closely tied to reproduction than it is for men because women are still more likely to take the responsibility for birth control. One way to see this is to note how many different contraceptive devices are designed for women, whereas only two—the condom and vasectomy—are designed for men.

Contraceptives are not the only technology influencing sexual values and practices. Now the Internet has introduced new forms of sexual relations as many people seek sexual stimulation from pornographic websites or online sexual chat rooms. *Cybersex,* known as sex via the Internet, can transform sex from a personal, face-to-face encounter to a seemingly anonymous relationship with mutual online sex. Researchers have found that often people say they get to know people faster and better from online sex chat rooms (Wysocki 1998). Such Internet encounters can progress to off-line relationships, a phenomenon more common among women than men (Cooper and Scherer 1999). Some argue that this introduces new risks for women because off-line meetings can become dangerous. Two-thirds of those visiting chat rooms are adults masquerading as children (Lamb 1998). This deception introduces new forms of deviance that are difficult to regulate (see the box "Taking on Social Issues: Sex, Censorship, and Cyberspace").

While the sexual revolution has brought new freedoms in the expression of sexuality, it has also ushered in new dangers—not just in the potential for sexual deviance and violence, but also in health risks from sexually transmitted diseases. (For more information on sexually transmitted disease, see Chapter 20.) The AIDS epidemic can be traced in part to these new freedoms because one way it is spread is through sex with multiple partners. Other health risks include venereal diseases, unwanted pregnancies, and, some argue, a new form of addiction to cyberspace sex (Brody 2000).

TAKING ON SOCIAL ISSUES

Sex, Censorship, and Cyberspace

Given widespread access to the Internet, pornography has become more easily available in a seemingly anonymous context. Studies show that one-third of Internet users will visit some sort of sexual site. Some worry that this makes pornography easily accessible particularly to children, which has stimulated debate over how to regulate the Internet. What policies are currently in place regarding pornography and the Internet? What issues arise in balancing concerns about access to pornography with the First Amendment?

Taking Action

Go to the Taking Action Exercise on the Companion Website—at http://sociology.wadsworth.com/andersen_taylor4e/—to learn more about an organization that addresses this topic.

•••

Sexually explicit images are easily available on the Internet, raising new policy questions about balancing the values of First Amendment freedoms of speech and access to information with questions about censorship.

Over time, sex has been used—explicitly or in a suggestive way—to sell all kinds of products—cigarettes, cars, liquor, even computers. Where in the contemporary world do you see evidence of the commercialization of sex? How is this related to the exploitation of women?

Commercializing Sex

While the sexual revolution has brought new freedoms and new dangers in expression of sexuality, it has also resulted in more commercialized sex. Sex has been defined as a commodity. Definitions of sexuality in the culture are heavily influenced by the advertising industry, which narrowly defines what is considered sexy. Thin women, White women, and rich women are all depicted as more sexually appealing in the mainstream media. Images defining "sexy" are also explicitly heterosexual. The commercialization of sex uses women, and increasingly men, in demeaning ways. Poor women are also more likely to have to work in the sex trade for economic survival. Although the sexual revolution has removed sexuality from many of its traditional constraints, the inequalities of race, class, and gender still shape sexual relationships and values.

Chapter Summary

What is the sociological significance of studying sexuality?

Sexuality is deeply entrenched in social norms, values, and social structures. It can be studied within the context of inequality because some groups, particularly gays, lesbians, and bisexual people, are treated as minority groups in society.

What role do culture and biology play in the formation of sexuality?

Sexuality develops within a social and cultural context. Although there is debate about the significance of biology in determining sexual orientation, there is no conclusive evidence for a biological cause to *sexual orientation*. Sociologists are interested in how culture and social institutions direct and channel sexual behavior and attitudes.

How do sociologists use theory to explain human sexuality?

Functionalism interprets cohesion in sexual norms, values, and practices as necessary for the stability of social institutions. *Conflict theorists* interpret sexuality in the context of power relations in society, noting how some forms of sexual relationships are connected to power relations, such as those generated via race, class, and gender inequality. *Symbolic interaction* emphasizes that sexual identity is socially constructed and emerges through learning sexual scripts. *Queer theory* is a new perspective that sees the transgression of assumed boundaries of sexual identity as necessary for challenging the existing social order.

How is sex influenced by diversity and inequality in society?

Patterns of sexuality reflect the social organization of society. In a society organized around race, class, and gender inequality, sexual stereotypes and sexual behaviors will reflect that inequality. This can be seen in how sexual stereotypes are related to the race, gender, and class of different social groups. Contemporary sexual attitudes also vary considerably by diverse social factors such as age, gender, race, and religion. Gays and lesbians experience the brunt of *homophobia* and *heterosexism* in society. A global perspective on sexuality also shows how sex is linked to a worldwide economy in which sex is bought and sold.

How is sexuality related to contemporary social issues?

Sexuality is related to some of the most difficult social problems, including reproductive technologies, pornography, and teen pregnancy. Social issues over birth control, abortion, and new reproductive technologies can be understood by analyzing the sexual, class, and racial politics of society.

How has the sexual revolution changed sexual relations over time?

The *sexual revolution* refers to widespread changes in roles of men and women and a greater acceptance of sexuality as a normal part of social development. Technological changes, such as the development of the pill, have also created new sexual freedoms. Now, sexuality is also being influenced by the growth of cyberspace and its impact on personal and sexual interactions.

Key Terms

coming out 343	script 336
compulsory heterosexuality 343	sexology 338
	sexual orientation 343
eugenics movement 350	sexual politics 344
heterosexism 346	sexual revolution 356
heterosexuality 343	sexual scripts 336
homophobia 346	social construction 342
queer theory 343	

Researching Society with MicroCase Online

You can see the results of actual research by using the Wadsworth MicroCase® Online feature available to you. This feature allows you to look at some of the results from national surveys, census data, and other data sources. You can explore this easy-to-use feature on your own, but try this example. Suppose you want to know:

The last time you had sex, was it with someone you were in an ongoing relationship with?

To answer this question, go to http://sociology.wadsworth .com/andersen_taylor4e/, select MicroCase Online from the left navigation bar, and follow the directions there to analyze the following data.

Data file: GSS

Task: Auto-Analyzer

Primary Variable: RELATESEX

Questions

Once you have your results, answer the following questions:

1. People in which age group are most likely to have had sex outside of an ongoing relationship?

a. <30
b. 30–49
c. 50 and up

2. What percentage of those under the age of 30 had their last sexual experience outside of an ongoing relationship?

3. Are the differences larger or smaller than you would have expected? Do you think the differences are due to changes in sexual mores over time or because people's attitudes change as they grow older?

The Companion Website for Sociology: Understanding a Diverse Society, Fourth Edition

http://sociology.wadsworth.com/andersen_taylor4e/

Supplement your review of this chapter by going to the companion website to take one of the Tutorial Quizzes, use the flash cards to master key terms, and check out the many other study aids you'll find there. You'll also find special features such as GSS Data and Census 2000 information, data and resources at your fingertips to help you with that special project or do some research on your own.

Suggested Readings and Web Resources

Altman, Dennis. 2001. *Global Sex*. Chicago, IL: University of Chicago Press.
  Analyzing sex as big business, Altman discusses the global dimensions of the sex trade. He shows the connection between the growth of the sex trade and national poverty, as well as describing the international networks through which sex is bought and sold.

Collins, Patricia Hill. 2004. *Black Sexual Politics: African Americans, Gender, and the New Racism*. New York: Routledge.
  Collins examines how sexual stereotypes support the maintenance of racism. She demonstrates the strong linkages between race, sexuality, and inequality and illustrates her argument through drawing on Black popular culture.

Freedman, Estelle B. and John D'Emilio. 1988. *Intimate Matters: A History of Sexuality in America*. New York: Harper and Row.
  This is a comprehensive examination of the historical context of sexual behavior and attitudes. The book shows how sexuality is often the basis for political and social conflicts, and it examines the relationship of race, class, and gender inequality to the history of sexual practices and ideologies.

Gold, Jodi, and Susan Villari (eds). 1999. *Just Sex: Students Rewrite the Rules on Sex, Violence, Equality and Activism*. Lanham, MD: Rowman and Littlefield.
  This collection includes the writings of young student activists who have worked on campuses to stop sexual violence. The book includes a review of university policies against gender violence, as well as

personal testimonies and scholarly essays on sex, gender, and violence on university campuses.

Irvine, Janice M. 2002. *Talk about Sex: The Battles over Sex Education in the United States*. Berkeley, CA: University of California Press.
Irvine analyzes the politics surrounding debates about sex education. Her book is a sociological analysis of sexual morality and its relationship to liberal and conservative politics.

Nardi, Peter M., and Beth Schneider. 1997. *Social Perspectives on Lesbian and Gay Studies*. New York: Routledge.
This collection of articles provides a sociological perspective on new research on gay and lesbian issues. It includes classic readings with the work of young scholars who are redefining questions about sexual identity, community, and social change.

Roberts, Dorothy. 1997. *Killing the Black Body: Race, Reproduction and the Meaning of Liberty*. New York: Vintage Books.
Roberts presents compelling evidence about the continuing racism that shapes Black women's reproductive lives. Her analysis of birth control policies and new reproductive technologies shows how historical stereotypes of Black women continue to influence their reproductive freedom.

Schwartz, Pepper, and Virginia Rutter. 1998. *The Gender of Sexuality*. Thousand Oaks, CA: Pine Forge Press.
This review of sociological perspectives on sexuality links the study of gender roles to sexuality.

Stombler, Mindy, Dawn M. Baunach, Elisabeth O. Burgess, Denise Donnelley, and Wendy Simonds (eds). 2004. *Sex Matters: The Sexuality and Society Reader*. Boston: Allyn and Bacon.
This anthology explores numerous sociological dimensions of sex in society. Intended for undergraduate readers, it is an engaging and thoughtful analysis of the social context of sexuality.

Alan Guttmacher Institute
www.agi-usa.org
This is a major national organization that produces numerous useful reports on sexual activity and contraceptive practices.

chapter 14

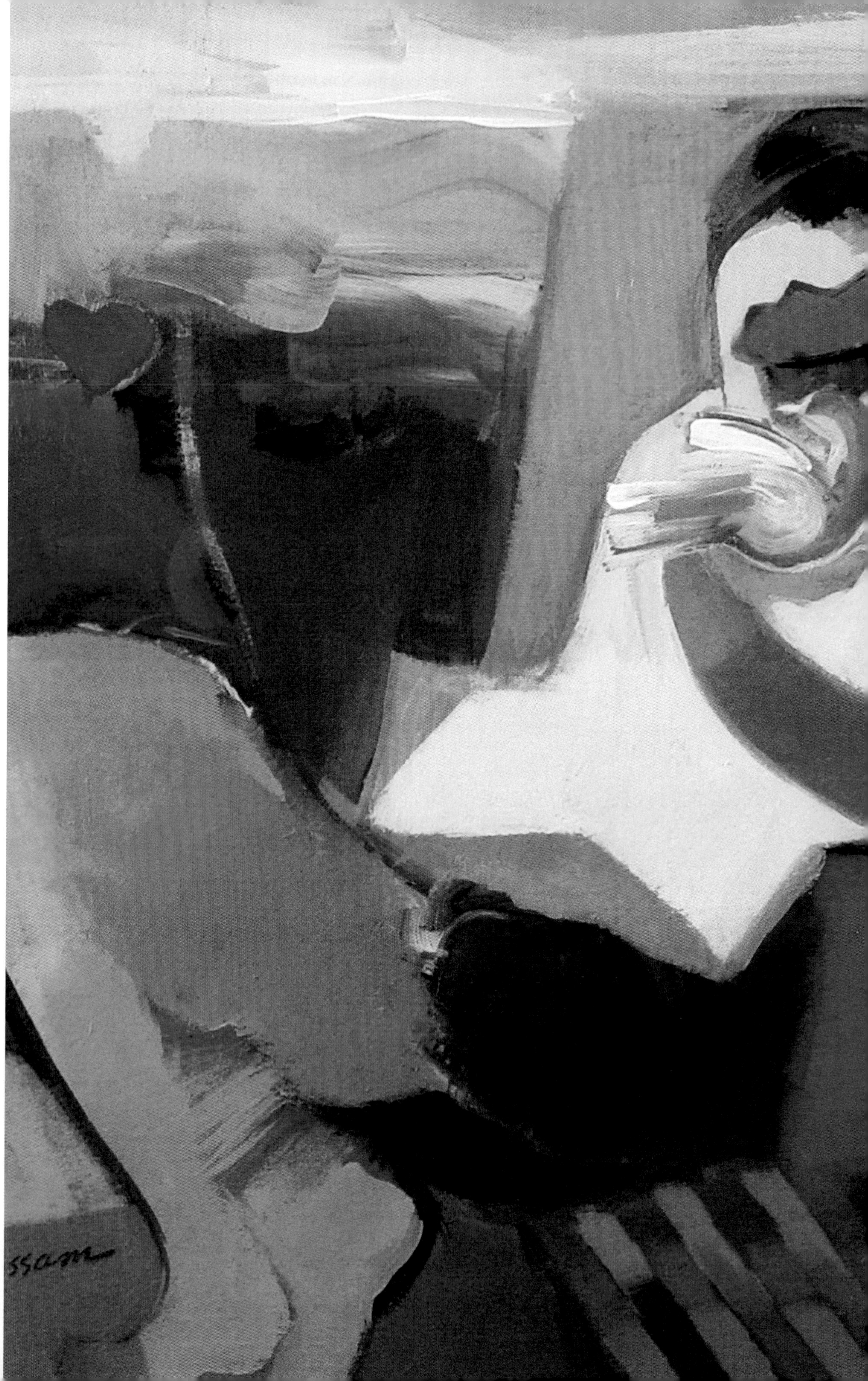

● ● ●

Age and Aging

It is Thanksgiving Day and a family gathers for dinner. Represented at the table are five generations, from great grandmother to the newest great grandchild. Same family, different generations—each facing experiences particular to his or her age group. Great grandma lives alone but worries about how long she will be able to do so and how she will afford extended long-term care, should she need it. Grandmother and grandfather have just returned from an elder hostel where they combined their love of travel and lifelong learning. They are enjoying their retirement but wonder how long their health will allow them to savor their older years. The mother is busily preparing last-minute parts of the meal. She is exhausted from having cooked every evening this week after coming home from her job. The father is watching football with his daughter who is home from college. During commercial breaks, they talk about her concerns about finding a good job when she graduates. Because rents in the city are so high, he wonders if she will decide to live at home again once she graduates, as an increasing number of young adults have done.

The family's teenaged son is in a surly mood. He would rather be hanging out at the local mall. Another son and daughter-in-law are tending their own children, including a newborn. Their children represent the family's hopes for the future, although the parents are concerned about the world these young children will face as they grow older. Reports of violence and overcrowding in the schools and the increasing cost of raising a family make them wonder what the future holds for their children.

Society differentiates people on the basis of age. Thus, different age groups experience different life situations—situations that are further shaped by people's race, class, and gender. Studying the experiences of differently situated generations reveals unique social expectations and different life chances for different age groups in society. Simply being part of a given generation can shape much of your life experience, and being of a certain age also influences the opportunities available to you.

Understanding the sociology of age reveals that age matters, not just to individuals, but also to the structure of society. The age composition of a society makes a difference in the social issues that society faces and in how well social institutions serve different generations of people. Since 1950, for example, the number of those over age fifty in the United States has doubled and is predicted to reach 125 million by the year 2050—one-third of the population (compared to under one-quarter now). The aging of the Baby Boomer

Sociology ⊛ Now™
Reviewing is as easy as ❶❷❸.

Use SociologyNow to help you make the grade on your next exam. When you are finished reading this chapter, go to the chapter review for instructions on how to make SociologyNow work for you.

population—those born in the late 1940s and 1950s—means that this group will eventually become the single largest age group in the country.

The increase in the population of older people is referred to as the *graying of America*. In other words, there are more older people in the population now than at any previous time in history.

Not long ago, the elderly were the group most likely to be poor. Although there still is significant poverty among the elderly, the highest rates of poverty now, unlike in the past, are among children (Proctor and Dalaker 2003). As America is graying, poverty among children in the United States is increasing. These changes in the population structure will pose new dilemmas for providing social services to different population segments. For instance, as the number of older people increases, the demand for Medicare and Social Security benefits increases. (Medicare is most well known for providing health care to those above 65 regard-

less of their income.) Moreover, as poverty among children in the United States increases, the demand for more Medicaid increases. (Medicaid benefits provides health care to low income individuals and families.) How will the younger generation care for themselves let alone their elders?

Under these changing social conditions, sociologists' analyses of age and its relationship to society become even more critical. Some of the questions they address include: How does society's expectations about age influence people's experiences? Does society, for example, shape the social meaning of aging differently for women and for men? What stereotypes have developed about aging, and how are they communicated through popular culture? What opportunities and obstacles do people face in society, depending on their age generation? How will your life be affected by the *graying of America*? To address these questions and more, we begin by first examining the social dimensions of aging. •••

The Social Construction of Aging

It is easy to think that aging is just a natural fact. Despite desperate attempts to hide gray hair, eliminate wrinkles, and reduce middle-aged bulge, aging is inevitable. The skin creases and sags, the hair thins, metabolism slows, and one's bones become less dense and more brittle by losing bone mass. Older persons have slower psychomotor responses and reflexes. The older people get, the less accurate their short-term memory becomes, although they can usually remember events in the distant past with great accuracy.

Social stereotypes of old people as "senile" have assumed that old people inevitably become forgetful and dumb, but this is not necessarily true. Although some measures of intelligence may decline slightly with age, such as speed in solving math problems and speed in overall problem solving, other dimensions of intelligence may actually increase with age. Some artistic abilities, such as painting, have been shown to develop and flourish in later life, while the most prominent gift of age may be the sheer amount of accumulated knowledge and overall "wisdom."

The aging process also has psychological effects, which are linked to physical changes. For example, the hearing loss that accompanies advancing age can have psychological consequences. Many persons over the age of sixty-five suffer enough hearing loss that they think those around them are speaking in muffled tones. People who believe they are being whispered about are likely to find their social situation highly distressing

Famous singer Tina Turner, born Anna Mae Bullock on November 26, 1939, in Nut Bush, Tennessee, clearly demonstrates that chronological age need not limit one's physical appearance. She is sixty years old in this picture.

© Mark Allan/Alpha/Globe Photos

and are liable to succumb to paranoia and depression. Whereas it was once believed that depression was a physical symptom of aging, depression is not an inevitable consequence of aging (Zisook et al. 1994).

Dementia is the term used to describe a variety of diseases that involve some permanent damage to the brain. Dementia usually involves an impaired awareness of one's self and surroundings, memory loss, and tendencies toward delusions and hallucinations. Some forms of dementia may be short-term and are treatable. Others are irreversible and degenerative, the most common of which is Alzheimer's disease.

Alzheimer's disease is a progressive loss of mental ability that involves the degeneration of neurological impulses in the brain. New medical research is revealing more understanding of Alzheimer's disease. Once thought to be rare, Alzheimer's disease is now known to occur among approximately 10 percent of the population over age sixty-five and half of the population over age 85. The number of diagnosed cases is increasing and is expected to continue increasing, both because of the improved ability to diagnose this disease and the larger number of old people in the population (Alzheimer's Association 2004). Alzheimer's disease is not necessarily age-related, although it is most common among older people. The major symptoms of Alzheimer's include gradual changes in a person's mental functioning, specifically a decline in memory (especially short-term memory), learning, attention, and judgment. Someone with Alzheimer's disease may be disoriented in time, have difficulty communicating with others, engage in inappropriate social behavior, and experience changes in personality. These symptoms are typically mild in the beginning but get progressively worse. Because people with Alzheimer's disease typically lose memory of their most recent years, they may not remember their spouses or children—those who are most likely to be caring for them and who therefore may suffer the emotional pain of having been forgotten by the person they love (Reid 1994).

Social Factors in the Aging Process

Important as the physiological changes that accompany aging are, they pale beside the social and cultural aspects of aging. The physiology of aging proceeds according to biological processes. What it *means* to grow older and how people age are social phenomena.

Even the physiological dimensions of aging are influenced by the social context. Take how long people live. **Life expectancy** is the probable number of years a particular group is likely to live on average, given aggregate statistical patterns. It is based on the age at which half the people born in a particular year die (see also Chapter 21). Life expectancy is clearly shaped by several social factors, including gender, race, and social class. On average, women live longer than men, but the life expectancies of minorities of both genders are shorter than those of Whites. Adding in the effects of social class further differentiates life expectancy patterns. The life expectancies are shorter for both men and women who are poor than for men and women who are middle- and upper-class.

The influence of social definitions on aging is illustrated by the distinction in *cognitive age* and *chronological age*—cognitive age being how old one thinks of oneself as being and chronological age one's actual age. Research finds that the older population generally now thinks of itself as younger, reflecting the anti–aging themes in the culture. Those who are most physically active also perceive themselves to be younger than they actually are (Clark et al. 1999; Katz 2001-2002).

Cross-Cultural Dimensions of Aging The social dimensions of aging are also obvious when you look at aging across different cultures. Cultural norms about the meaning of aging, as well as responsibility for caretaking of older persons, shape the treatment of older people in society. In many other societies, the elderly hold very high status and their judgment on important matters is sought and respected. In Samoa, for example, old age is considered to be the best time of life, and the elderly are much revered. In many African nations, social status also increases with age, and the elderly are regularly consulted in acknowledgment of their superior wisdom.

One cannot necessarily conclude that esteem for the elderly is reserved only for less industrialized societies, however, because even in some highly industrialized societies, such as Japan, there is a long-standing tradition of respect for the elderly. This is further reinforced

This grandmother, in her seventies, is teaching origami, the Japanese art of paper folding, to her granddaughter. In Japanese culture, the elderly are generally held in high esteem and are regarded as possessing great knowledge.

© Walter Hodges/Stone/Getty Images

by the Japanese stratification system, which demands that servants, students, and children respect those in roles regarded as superior, such as masters, teachers, or parents. In Asian societies, the Confucian principle of filial duty emphasizes respect for the aged. To call an old man "lao" (as in a man named "Xu" becoming "Xu Lao") is a great compliment. Chinese children are also required to repay their parents with gratitude for bringing them up (Kristof 1996; Palmore and Maeda 1985; Cohen and Eames 1985).

You cannot assume, however, that cultural beliefs alone shape the experience of aging. Important as cultural beliefs are, social changes associated with population change, the degree of urbanization, changing family structures, employment patterns, and even the sheer number of old people in society can change beliefs about aging. As the world undergoes global economic development, differences in how the elderly are perceived and treated can become more similar across cultures (Antonucci et al. 2002; Kim and Maeda 2001). The actual physical health of older persons also affects how much others perceive them to be a burden, even in different cultural settings (Chappell 2003).

Within the United States, you can see the social dimensions of aging by considering how the same phases in the life course are judged differently, according to social factors, such as gender (among others, as we will see). At midlife, for example, both men and women experience changes in their reproductive system. For men, there is a decline in the quantity of sperm produced, although most men remain fertile; some men even produce children when they are quite old. But these biological facts about men's aging have not been imbued with the social meaning that is attached to menopause for women. Social stereotypes about men at midlife portray them as wanting fancy cars, younger women, and generally engaging in more thrill-seeking and youthful behaviors—the so-called midlife crisis.

For women, *menopause* is the biological cessation of ovulation and the menstrual cycle. But menopause is culturally defined as marking old age for women. Menopausal women are stereotyped as unable to control their emotions and highly prone to irritability and depression, despite the evidence that menopause is not related to serious depression among women. To the contrary, some studies find that the majority of women feel happy about the loss of ovulation ability. Studies also find that a majority of menopausal women do not experience the hot flashes and other physical symptoms generally associated with menopause, although some women certainly do (Fausto-Sterling 1992).

Cultural understandings of menopause have also changed over time. For earlier generations, menopause was something one experienced silently. As the Baby Boomer generation has experienced menopause, it has become a widely discussed and recognized experience.

New markets have developed to dispense advice to women about menopause, and women have developed new support networks for sharing their feelings and experiences of menopause. For many women, menopause has been redefined from being a time of sadness and depression to being cause for celebration, humor, and self-affirmation.

Age Stereotypes

The definitions applied to different groups make the experience of growing older highly dependent on one's social circumstances. **Age stereotypes** are preconceived judgments and oversimplified categorizations of beliefs about the characteristics of members of different age groups. Just as the media condition our views on gender and race, they present us with a distorted view of both youth and the elderly. Young people are portrayed as carefree, the elderly as unhappy or evil. Sometimes the elderly are presented as childish, a common and harmful representation called *infantilization of the elderly*. This stereotype also influences how people behave toward the elderly, as shown in how people speak to the elderly in patronizing and infantile tones, as if they were children or did not speak English (Hummert et al. 1998).

Although age stereotypes of the elderly in our society are more pervasive than stereotypes of youth, both groups are burdened by negative preconceptions. Studies routinely find that adults perceive teenagers to be irresponsible, addicted to loud music, lazy, sloppy, and so on. Common stereotyping of the elderly includes that they are forgetful, set in their ways, meddlesome, conservative, inactive, unproductive, lonely, mentally dim, and uninterested in sex .These stereotypes are largely myths, but they are widely believed (see Table 14.1), and they deeply influence how people perceive different age groups.

In one such study, undergraduate students were asked to listen to a lecture, presented via slides of an age- and gender-neutral stick figure with a neutral voice. Students were then provided forms indicating different age and gender conditions (male and female, old and young) and were asked to evaluate the "professor." Students consistently rated the "young male" higher than the "young female"; "young" professors were also rated higher than "old" professors (Arbuckle and Williams 2003). This research shows that both age and gender influence people's perception of others—even, in this case, of the exact same thing!

Other studies of the interaction of age and gender also find that women are viewed as old as much as a decade sooner than men are thought of as old (Stoller and Gibson 2000). Older women are stereotyped as having lost their sexual appeal, while older men are stereotyped as more handsome, "dashing," and desirable.

Table 14.1
The Aged: Myths and Realities

Myth	Reality
Most old people have no interest or capacity for sex.	Although there is some decline in sexual activity as people age, there is less decline in interest than in activity; those who want to be sexually active and are not usually attribute it to the lack of a suitable partner. Still, even after age sixty-five, people report an average of sex 2.5 times a month, compared to 7.1 times for those eighteen to sixty-five (Clements 1996; Calasanti and Slevin 2001).
As people get older, they get depressed.	The majority of older adults experience sound mental health, although this can be greatly affected by the socioeconomic resources available and by one's physical health (Keyes 2002; Martin 2002).
Most old people are dissatisfied with their bodies and think they no longer look good.	Most older people (75–85 percent) are satisfied with their bodies—in part because they use other old people as the standard. Among women, young women are much more dissatisfied with their bodies than are old women (Oberg and Tornstam 1999; Heidrich and Ryff 1993).
Old people are usually senile.	Only a small minority of the elderly can be considered senile; about 10 percent suffer from Alzheimer's disease (Alzheimer's Association 2004).
Most old people live in poverty.	Although poverty among some older persons is a problem, children are the age group most likely to be poor—a pattern that was not true in the recent past (Proctor and Dalaker 2003).
Most old people end up in nursing homes and other institutions.	Less than 5 percent of all elderly are in a nursing home or other institution at any particular time (U.S. Census Bureau 2004).

These social definitions of gender and aging denigrate older women and associate beauty for women only with being young.

Age stereotypes are perpetuated and reinforced through popular culture. Advertisements depict women as needing creams and lotions to hide "the tell-tale signs of aging." Men are admonished to cover the patches of gray that appear in their hair or to use other products to prevent baldness. Entire industries are constructed on the fear of aging that popular culture promotes. Face-lifts, tummy tucks, and vitamin advertisements all claim to reverse the process of aging even though the aging process is a fact of life. On television commercials, older people are not only underrepresented, but women also tend to be shown as younger than men (Ganahl et al. 2003). An analysis of Academy Award–nominated feature films from 1929 to the mid-1990s has also found that older men tend to be depicted as vigorous, employed, and involved in adventure, whereas women of the same age are usually peripheral to the main action or are portrayed as rich dowagers, wives and mothers, or lonely spinsters. Furthermore, the roles for women in film have remained remarkably static over this period of time (Markson and Taylor 2000).

Stereotypes tend to be very persistent, but they can change. Thus, many think that as the Baby Boomer generation ages, images of older people may change too. Yet, although more older people are now seen in the media, they are still underrrepresented and greatly distorted (Bazzini et al. 1997; Carrigan and Szmigin 2000). All this reveals a basic sociological idea: *Perceptions of aging are socially constructed.*

Age Norms

The social significance of aging is also apparent in the expectations people hold for different age groups. **Age norms** are explicit and implicit rules that spell out the expectations society has for the different age strata. Age norms define what you should or should not do according to your age; some behaviors are appropriate for one age category but not for another. Thus, the norms change as you age. Consider this: We have all been admonished at one time or another to "act your age." As one ages, one is expected to act more "mature," more "grown up." Age groupings are treated differently in society, and different expectations, or norms, go along with each. Young people are not supposed to be sexually active and are supposed to stay in school. The elderly are expected to retire from their jobs and be less active and publicly visible than those supposedly in their prime. Many age norms become laws. In most states, you are not allowed to drive until age 16, vote until age 18, and drink until age 21. Children are not permitted to skip school, and they are not supposed to work before their midteens. Until recently, people were required to retire at age sixty-five, although the Age Discrimination in Employment Act has eliminated the mandatory retirement age for many occupations.

Age norms are not fixed. Like other norms, they change as society changes and people adjust to new social conditions. For example, becoming old in this culture has traditionally meant not working and being seen as not contributing to society. These norms are now changing as people recognize that older people can be productive members of society. Of course, the norms have never accurately described the experience of all people.

Many retirees continue to work because they need the economic resources to survive. A study of Mexican Americans has found, for example, that retirement for them often does not come as the culmination of a life of working. Because many Mexican Americans have histories of work marked by unemployment, part-time work, and jobs with few retirement benefits, they are less likely to perceive themselves as retired even when they are no longer employed. This study shows that people's work experiences shape their perceptions of retirement. Women, for example, who no longer work in the paid labor force, often do not perceive themselves as retired because their work as homemakers typically continues (Hatch 1992).

Contemporary social changes have also disrupted the traditional norms that have distinguished different age groups. Older people now return to school and complete college degrees. For many, marriage and childbearing come ten or twenty years later in the life cycle than would have been true years ago. These changes have mixed up the various age norms that apply to a given generation. The increased presence of women in the labor market, especially among the White middle class, has meant a dramatic transformation of the norms associated with aging. Women no longer follow a prescribed life course in which marriage and children come first, and work second. Lifestyles are simply more diverse than ever before, with some women reentering the labor market in middle age, others returning to school after their children are grown, others delaying child rearing until after their careers are well established, and an increasing, though small, number of people forgoing childbearing altogether (Gillespie 2003).

Age norms also differ within groups. Many racial–ethnic communities revere the elderly, valuing their knowledge and perspective and giving them high social status as a result. In Native American cultures, as an example, the elderly have been accorded much respect, as illustrated in the box "Understanding Diversity: Grandparenting Among Native Americans," where the role of Native American grandparents as cultural conservators is described. Similarly, other minority cultures emphasize great respect and care for elder parents although they do not always have the extended family networks that are sometimes assumed (Goodman 1990; Stoller and Gibson 2000).

Age and Social Structure

All societies, including the United States, practice **age differentiation**, the division of labor or roles in a society on the basis of age, although the specific roles given

---

UNDERSTANDING DIVERSITY
Grandparenting Among Native Americans

Social roles develop in the context of cultural traditions and social institutions that vary among different groups. Being a grandparent is an example of a social role where there are different expectations among different racial and ethnic groups. The dominant culture defines grandparents as indulgent, playful, and fun-seeking, but also hands-off. This role differs, however, among various groups in the society, and it is more important for some than for others. African American men, for example, tend to see grandfathering as a more central role in their identity than do most White men (Kivett 1991). Multigenerational households may give grandparents a primary role in child rearing; whereas in other groups, grand-parents are supposed to take an assisting, but hands-off approach to child care. In many cultures, grandparents are also defined as the dispensers of wisdom.

Joan Weibel-Orlando has studied grandparenting among Native American groups and found that, within these cultures, there are diverse roles for grandparents, all of which reflect the high esteem in which old people are held in Native American societies. A few styles she identifies are cultural conservator, ceremonial instructor, and custodian. As cultural conservators, grandparents pass on the traditions of the group, providing cultural continuity and identity for young children as Native Americans. Storytelling can be an important way that this role is enacted, given that stories pass on the cultural beliefs of the group. Grandparents also teach young Native Americans a wide array of ceremonial activities—sun dances, rodeos, powwows, and memorial feasts. Through this instruction, children learn the values of the group. As custodians of young children, Native American grandparents also provide essential household labor. The gender division of labor typically assigns this role to women.

Source: Based on Weibel-Orlando, Joan. 1990. "Grandparenting Styles: Native American Perspectives." In *The Cultural Context of Aging: Worldwide Perspectives,* Jay Sokolovsky (ed.). New York: Bergin and Garvey, pp. 109–124.

•••

to different age groups vary from society to society. It seems that everywhere societies classify people into age categories as infants, children, adults, and elderly—each with specific meanings. But what these categories mean and how people are thus treated within them varies depending on the cultural context.

In some societies, one is not considered a "human" until long after birth. Particularly where infant mortality rates are high, parents may find it too difficult to attach human status to an infant until they have shown signs that they are healthy and likely to live. In the very poor regions of Northeast Brazil, for example, mothers show little attachment to those who are born small and weak; if they die, there is little ceremony and their graves remain unmarked. Because sick infants are believed to be angels who fly to heaven, mother's tears are believed to dampen their wings, risking their flight. Anthropologists interpret this as the mothers' reaction to impoverishment in which they cannot invest attention or emotion in the lives of children who are unlikely to live (Sheper-Hughes 1992; Peoples and Bailey 2003).

On the other end of the life course, there is also great variation in how the old are treated. In many societies, they are given enormous respect. There may be traditions to honor the elders and they may be given authority over decisions in society, as they are perceived as most wise. On the other hand, among some cultures, adults who can no longer contribute to the society because of old age or illness may become perceived as extreme burdens. Among the Comanche Indians, as an example, mourning was reserved only for those who died in their prime because they were seen as a greater loss to the well-being of the community (Peoples and Bailer 2003). You might ask yourself how cultural definitions in this society affect how people grieve for different age groups. Is the death of a young person perceived as somehow more tragic than the death of a very old person? How do people react in each circumstance? Your answer will likely reveal the cultural beliefs surrounding age differentiation in this culture.

In the United States, age differentiation can also be seen in the different rights and privileges people have by virtue of their age. In general, youth in the United States enjoy only a subset of the rights of adults. The rights of youth to drink, drive, own property, work, and get married are all abridged. Some rights and restrictions associated with youth are implausibly contradictory. At age eighteen, for example, you cannot buy a beer in most states, but you can be sent overseas to fight and die for your country. In most cities, you cannot be a police officer or firefighter until you have reached the age of twenty-one. You cannot take a seat in the House of Representatives until you are twenty-five; a senator must be thirty; the president of the United States must be at least thirty-five.

In addition to differentiating roles on the basis of age, societies also produce age hierarchies—systems in which some age groups have more power and better life chances than others. **Age stratification**—the hierarchical ranking of age groups—exists because processes in society ensure that people of different ages differ in their access to society's rewards, power, and privileges. We have seen throughout this book that the United States is stratified into hierarchical groupings on the basis of class, race, gender, earnings, occupational rank, educational attainment, and other variables. Age, too, is the basis for stratification.

Age is an *ascribed status;* it is determined by the day and year you were born. Recall that biological sex and race are also ascribed statuses established at birth. These ascribed statuses remain relatively constant over the duration of a person's life, yet age changes steadily throughout your life. Still, you remain part of a particular generation—something sociologists call an **age cohort**—an aggregate group of people born during the same period. People in the same age cohort share the same historical experiences—wars, technological developments, economic fluctuations—although they might do so in different ways depending on other life factors.

Culture is often transmitted from grandparent to grandchild, as demonstrated by this Iroquois Native American grandfather.

Living through the Second World War, for example, dramatically shaped an entire generation's attitudes and behaviors, as did growing up in the 1960s, and as will being a member of the current youth generation. Yet, within a given cohort, there will be considerable diversity on many dimensions: sexual orientation, gender, race, class, nationality, and ethnicity. How these cohorts are arrayed in a given society at a given time shapes the character of society and the social issues within it.

Sociologists note that there is a continuing interplay between age and social change. In society, successive groups of people grow up, grow old, and die, each group being replaced by the next age group coming up behind them. As people of different ages pass through social institutions, society itself changes. Each age group faces a unique slice of historical time that offers unique challenges and social changes. To understand this, think of some significant facts of life that derive simply from being born at a certain time—from being in an age cohort. In 1900, someone twenty years old would not likely look ahead to retirement because people seldom lived to what we would now consider retirement age. Many twenty-year-old women in 1900 could not expect to live beyond their childbearing years.

Different generations have to grapple with and respond to different social contexts. Someone born just after World War II would, upon graduation from high school or college, enter a labor market that had many jobs available and, for many, expanding opportunities. Now, college graduates face a labor market where entry-level jobs in secure corporate environments are rare and many employees are trapped in low-level jobs with little opportunity for advancement. Many young people worry, as a result, about whether they will be able to achieve even the same degree of economic status as their parents—the first time this has happened in U.S. history. Understanding how society shapes the experiences of different generations is what sociologists mean by saying that age is a structural feature of society.

A Society Grows Old

Never before have so many people in the United States lived so long. This fact, in itself, has a number of implications for how society is organized and the issues to be faced in years to come. Current generations—whether young, middle-aged, or old—will be profoundly influenced by the *graying of America*. Consider the following:

- America's older population will double by 2030, reaching some 70 million (U.S. Census Bureau 2004).
- The older population will become more ethnically and racially diverse. Of those age sixty-five or older now, about 84 percent are non-Hispanic Whites. By 2050, that number will be 64 percent (U.S. Census Bureau 2004).
- Women will continue to outnumber men in old age, especially among the oldest old (U.S. Census Bureau 2004).
- The most rapid growth among the older population is occurring among those who are the oldest old—those 85 years and over (Hetzel and Smith 2001).

To see the magnitude of these changes, recognize that in 1900, only 4 percent of the population was over age sixty-five; by the end of the century, it was 12 percent; and by the year 2025, it will be 18 percent—equal to the current proportion of old people in Florida (see Figure 14.1; U.S. Census Bureau 2004).

Currently structured into society are certain assumptions that guide the expectations and obligations

**Figure 14.1
An Aging Society**

Source: U.S. Census Bureau. 2004. *Statistical Abstract of the United States 2003.* Washington, DC: U.S. Government Printing Office, pp. 13–14.

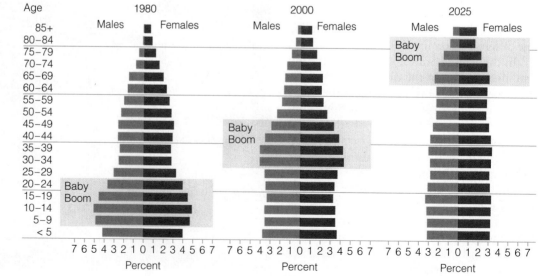

that exist between generations of people. Sociologists speak of these assumptions as the *contract between generations*. Imagine this contract exists between your generation, your parents' generation, and your grandparents' generation. Not a formal contract but a set of social norms and traditions, the contract between generations is the expectation that the first generation (say, your grandparents' generation) will grow up and raise the second generation (your parents' generation), who in turn produce a third generation (your generation). The expectation has been that each generation cares for the next, and the second or third generation will care for the first when they become old.

THINKING SOCIOLOGICALLY

Imagine yourself twenty years from now. How do you think your life will be affected by the *graying of America*?

Currently, however, the shrinking size of families means that the proportion of elderly people is growing faster than the number of younger potential caretakers. Moreover, as life expectancy increases and people live longer, the traditional contract is upset. Women, who shoulder the work of elder care, can expect to spend more years as the care giver of an elderly parent than as the mother of children under eighteen (Watkins 1987). As a result, family members can now expect to spend more time in intergenerational family roles than ever before. This in-between grouping has come to be called the *sandwich generation*, because of the time and resources its members spend with both their parents and their own offspring.

Changes in family structure (discussed further in Chapter 15) further alter the traditional patterns of intergenerational care. Childless couples may not have a younger generation family member to care for them in their older years. Those in single-parent families, particularly women, have the extra burden of caring not only for their own children, but perhaps also for an elderly parent. Men and women in middle age may have to find ways, often with few institutional supports, to care for older, and perhaps ill, parents. Given the geographic mobility that has characterized modern life, they may have to do so over long distances. We may be experiencing is a classic case of *culture lag*, in which the norms of care and support have not changed as rapidly as the composition of the population.

Currently, **Social Security** is one of the older and most successful national social policies. Social Security was first established in 1935 as Old Age, Survivors', and Disability Insurance. It works through a payroll tax placed on the earnings of current workers, employers, and self-employed people. The money is then placed in a federal trust fund drawn on by those currently receiving Social Security. How much you receive depends on your lifetime earnings, even though the fund is being supported through the earnings of current, most likely, younger workers. (There are also death benefits for spouses and children, though most Social Security goes to older, retired people.)

Social Security expenditures are now one-quarter of the entire federal budget for human, physical, and defense resources (U.S. Census Bureau 2004), and there is considerable national debate on the financial sustainability of this system. To explain, in 1945, the Social Security system had thirty-five working people paying into the fund for every recipient drawing upon it; this is a ratio of 35 to 1. By the late 1990s, the ratio of wage earners to recipients was down to 3.2 to 1. With this dramatic drop in the ratio of earners to recipients, most say that without significant changes in Social Security policy, there will not be enough workers to support the number of retirees by about the year 2020 (Kingson and Quadagno 1995). This prediction has led politicians to search for new ideas for funding the Social Security system, such as private investment in the stock market. Add to that the fact that the Social Security trust was drawn from to help pay for some of the war in Iraq and you see the looming problem as the population ages.

For the most part, this social issue has been posed as a matter of fairness between generations. Social Security is an issue used in political campaigns, pitting young against old. Most people assume that more support for other groups can come only at the expense of

The politics of Social Security have raised questions about generational inequity. Are different age groups in competition for federal resources?

Frank Fournier/Contact Press Images

the elderly. This debate is referred to as the question of **generational equity**—whether one age group or generation is unfairly taxed to support the needs and interests of another generation. There are those who argue that as the proportion of individuals over age sixty-five increases, a disproportionate share of the burden of supporting them will fall upon the younger generations. Consequently, so this argument goes, the younger generations are not treated equally and are not treated fairly. This debate has its origins in the aging of the population, budgetary crises in the federal government, growing health care costs, increased poverty among children, and as some sociologists note, a declining faith in social institutions on the part of the public (Kingson and Williamson 1993; Atchley 2000).

This need not, however, be a divisive issue. In other nations there is not necessarily a connection between higher spending for older adults and lower spending for children and the poor; governments can sponsor programs for family allowances, for example, without targeting one group against another (Adams and Dominick 1995; Pampel 1994; Pampel and Adams 1992). Instead,

---

DOING SOCIOLOGICAL RESEARCH

Death's Work: A Sociology of Funeral Homes

"Developing a sociological perspective reveals new ways of looking at seemingly ordinary events. The following analysis of funerals describes how sociologists, working from the perspective of functionalism, would analyze the manifest and latent functions of a funeral.

Most apparent of all the social shock absorbers of death is the funeral. The ritual disposal of the dead and social reintegration of the affected living are two of the few cultural universals known. But the cross-cultural variations in funerary observances are incredible. One can depart New Orleans–style, complete with marching brass bands and humorous graveside eulogies, or one can be the focal point of political protests, as when thousands march in the funeral processions of victims of political oppression.

Perhaps weddings are overrated as social events. It is at funerals that you meet the widest spectrum of people, where you see how many lives can be touched by a single individual. The funeral is the finished picture of a person, providing a ritual occasion when one reflects on the successes and shortcomings of a concluded biography. It also marks the endeavors of a generation: Only the generation of the deceased can provide the frame of reference needed to grasp the principles to which one's biography was dedicated.

Funerals have the clearly apparent, manifest functions of disposing of the body, aiding the bereaved and giving them reorientation to the world of the living, and publicly acknowledging and commemorating the dead while reaffirming the viability of the group. Although

for some cultures and religious groups funerary ritual is explicitly directed for the dead, it is always a rite of passage for the principal survivors, a mechanism for restoring the rent in the social fabric caused by death.

There are less obvious, "latent" functions served by funerals. Symbolically dramatized in funerary ritual are reaffirmations of the extended kinship system. The restrictions and obligations of survivors (such as their dress, demeanor, food taboos, and social intercourse) serve to identify and demonstrate family cohesion. Also dramatized are the economic and reciprocal social obligations that extend from the family to the community and from the community to the broader society. In other words, the social bonds of the living are acted out, remembered, and reinforced in the minds of community members. Not surprisingly, such functions have political significance. When, in 1982, 16,500 aborted fetuses were found in a container at the home of a Los Angeles man who ran a medical laboratory, three years of heated debate over their disposal followed. Antiabortionists sought permission to hold funeral services for the fetuses, claiming they were humans whose social membership had to be ritually reaffirmed, while a prochoice group, represented by the American Civil Liberties Union, argued that the remains were unwanted biological tissue and should be cremated without ceremony.

Also reaffirmed in funerals are the social roles of the living. Claiming that the deceased was a winner in his or her roles reaffirms the system itself by

having produced the opportunities for such a person to even create meaning. Why do we not speak ill of the dead? To speak ill of them is to speak ill of ourselves.

In the homosexual community, as the list of AIDS victims grows, there is the sense that the traditional rituals are insufficient. When another member of New York City's People with AIDS Coalition dies, white helium balloons are released from St. Peter's Episcopal Church in Greenwich Village. Ashes of one partner are sometimes mixed with those of the other who predeceased him and then are dispersed in a place meaningful for the couple. And increasingly, the rainbow flag, the symbol of the annual Gay Pride Parade, is displayed on coffins."

Questions to Consider

1. In your opinion, are funerals necessary, or should they be dispensed with? *Keywords: death and dying, funeral industry*

2. In your opinion, do funerals increase the cohesion among the family of the deceased? *Keywords: burial rites and ceremonies, burial social aspects*

We have included InfoTrac College Edition keywords at the end of each question to make it easier for you to find more to read on these topics. Go to www.infotrac-college.com, an online library, to begin your search.

Source: Kearl, Michael C. 1989. Endings: A Sociology of Death and Dying. New York: Oxford University Press. Copyright © 1989 by Oxford University Press. Used by permission of Oxford University Press.

•••

sociologists suggest that the problems associated with the graying of America could be addressed by having a changed public agenda—one that provides universal access to the basic needs of income, housing, and health care, a change that would require tax increases and some reduction in benefits such as health insurance (Kingson and Quadagno 1995; Kingson and Williamson 1993, 1991).

To date, such a solution has not gained political appeal. The nation's response may be to exacerbate stereotypes of the aged as "greedy geezers" and to blame the victim, rather than try to solve the problem. This debate about generational equity also has the risk of potentially increasing age prejudice in society, while failing to address the problems of the elderly, children, or the poor.

Growing Up/Growing Old: Aging and the Life Course

The phases in the aging process are familiar in name to all of us: childhood, youth and adolescence, adulthood, and old age. Together these strata make up the complete life span. To interpret the life span, sociologists use a **life course perspective** that connects people's personal attributes, the roles they occupy, the life events they experience, and their sociohistorical context to emphasize that personal biography, sociocultural factors, and historical time are interrelated (Stoller and Gibson 2000). Although psychologists and others study the life span, unique to sociology is the connection that sociologists make between individuals and the social contexts in which they live.

As we noted in Chapter 4 on socialization, the transitions to different phases in the life span are often marked by elaborate cultural rituals. These rituals are *rites of passage* that celebrate or memorialize events in the individual's life. Birth, puberty, marriage, and death are each heralded by rites of passage—a christening or baptism, confirmation, the bar mitzvah and bas mitzvah for Jewish boys and girls at puberty. Among Mexican Americans, the *quinceañera* ceremony similarly marks puberty as the transition to adulthood.

The distinction between different phases in the life course is not, however, a rigid one. Without specific social markers to announce the passage from one phase of the life span to another, the distinctions between different phases blur. When, for example, does one become an adult? At age eighteen in most states, you are adult enough to enter the armed services and to vote but not adult enough to drink. The legal drinking age in most states is now twenty-one, although it used to be eighteen in several states. Social change means that young people remain in school longer, marry and bear children

later, and in some cases remain dependent upon a parent or parents for a longer time (see Table 14.2).

Similarly, middle age, according to the traditional definition, was perceived as beginning when one's children grew up and left home. Now some middle-aged people are just beginning college; others may already be grandparents. Given the high rate of teenage pregnancy, some become parents before they would traditionally have been defined as adults. Teenage pregnancy also means that some parents may become grandparents as early as their late twenties or early thirties—a time when other adults may still be in school. Grandparents themselves may still be in school—something that not long ago would have been extremely rare.

The traditional markers of adulthood (education, marriage, work) no longer easily label the change from one phase of life to another. Definitions of age are social products, and there are no rigid demarcations between age categories, merely approximations. Even old age can be seen as divided into two strata—the "young old" (around sixty-five to eighty-five) and the "oldest old" (eighty-five and older). Although the terminology is imprecise, sociologists have extensive knowledge of the general distinctions in the life span: childhood, youth and adolescence, adulthood and middle age, and finally, old age.

Childhood

The United States has typically been defined as a child-centered society. Many news commentators, historians, educators, and sociologists have noted that the United States is "youth-oriented." The high valuation on youth

Table 14.2 Slowing the Transition to Adulthood		
	1980	**2002**
Percentage aged twenty to twenty-one in school	31.0	45.5*
Median age at first marriage		
Women	22.0	25.3
Men	24.7	26.9
Birth rate, women aged fifteen to nineteen (per 1000 women)	53.0	45.3
Percent aged sixteen to nineteen in labor force		
Women	52.9	56.3
Men	60.5	69.7

*For year 2001

Source: U.S. Census Bureau. 2002. "Children's Living Arrangements and Characteristics: March 2002." Washington, DC: U.S. Census Bureau. Website: **www.census.gov**; U.S. Census Bureau. 2004. *Statistical Abstract of the United States 2003.* Washington, DC: U.S. Department of Commerce; U.S. Department of Labor. 2004. *Employment and Earnings.* Washington, DC: U.S. Department of Labor.

in U.S. culture is especially evident in the mass media. The young are depicted in television commercials more often and in a more positive light than any other age group. Today's youth are generally shown as energetic, smart, and attractive. In the United States, more than in most other societies, to be young is revered and to be aged is reviled. Youth is regarded as the prime of life, whereas middle age and old age are regarded as the long decline at the end of life.

Popular images of childhood also depict it as a period of play, fantasy, and freedom from responsibility. It was not always that way. Thinking of childhood as a separate stage of life dates back to the sixteenth and seventeenth centuries in Europe, when changes in the economy began to emphasize the family as a separate unit in society (Aries 1962). At that time, society was well aware of the economic and occupational value of children. In the nineteenth century in the United States, children were used as additional laborers in the family and were given a variety of responsibilities. This is still true today among many immigrant groups, where children work to help support the family. It is also somewhat more pronounced today among the poor, whatever their race or ethnicity. In general, the lower the socioeconomic status of a family, the more likely the family will use its children as a source of labor.

The exploitation of child labor was so pervasive in the nineteenth century that people finally reacted against it. By the middle of the nineteenth century, at least among the middle classes, the idea that children were of limited economic value as laborers began to take shape. Instead of relying on their children to take care of them in later life, parents began to rely more on life insurance, pension plans, and other financial arrangements. All this resulted in the *sentimentalization* of children in the United States, meaning that children were seen as precious but not generally defined as economically useful. Now, instead of seeing children as "functional" to the family, adults report that among the important advantages of having children are the satisfaction of adults' desires for love and affection as well as the pleasures of being part of a family (Hoffman and Norris 1979). This tender attitude toward children no doubt buffers the realization that raising children is an expensive proposition. In return, parents demand from children only that they show respect, return love, and be happy. Parents who use their children to make money, such as parents of successful child actors and entertainers or those who enter young children in baby beauty contests, are viewed with suspicion. As one researcher put it, we live in a culture that defines children as economically useless but emotionally "priceless" (Zelizer 1985).

Still, the image of childhood as a carefree time is not matched by reality for many children. The United States

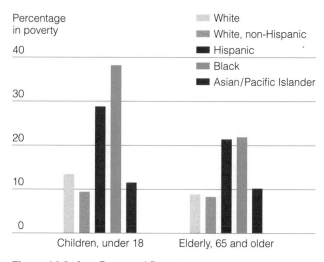

Figure 14.2 Age, Race, and Poverty

Source: Proctor, Bernadette D., and Joseph Dalaker. 2003. *Poverty in the United States 2002.* Washington, DC: U. S. Department of Commerce, pp. 28–32.

is becoming a more dangerous place and violence takes a heavy toll on urban African American and Latino children. More than one-fourth of the homeless are children (National Coalition for the Homeless 2004). A huge number of children live in poverty, with their likelihood of doing so strongly linked to the race or ethnicity of the child. As Figure 14.2 shows, the proportion of Black and Hispanic children under the age of eighteen who are poor greatly exceeds the proportion of White non-Hispanic and Asian children under eighteen who are poor. Furthermore, as shown in Figure 14.3, the proportion of younger people who are racial and ethnic minorities is substantially increasing, raising further concerns about how the nation will be able to support and educate these young people.

Youth and Adolescence

Fast cars, loud music, arguing with parents, facing up to sexuality, dating, what to do when schooling is finished, and questions of "Who am I?"—these are some of the concerns of adolescence. Adolescence is a relatively new category in the life cycle. Until the twentieth century, children moved directly into adult roles. There was no such thing as an in-between adolescent period. Adolescence came to be regarded as a separate stage of life as more people entered and completed high school and the period of formal education became longer. This delayed entry into adult roles, marking the period of adolescence as a particular stage in the life cycle.

The boundaries of adolescence are imprecisely defined, but most regard the lower boundary as the transition from elementary school (sixth grade or its equivalent), or about age twelve, to junior high and high school. What has occurred over time in one institution

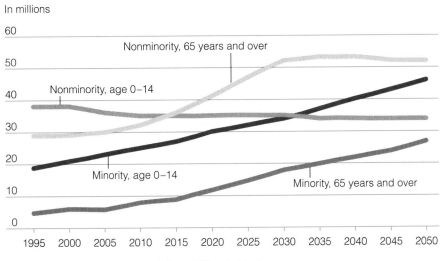

In millions

Figure 14.3 The Growing Population of Minority Youth

Source: Minority Business Development Agency. 1999. *Minority Population Growth 1995 to 2050*. Washington, DC: U.S. Department of Commerce, p. 3.

(education) has structured how an age category is defined in another institution (the family). The upper boundary is usually at entry into some adult role, such as full-time employment, college, marriage, or parenthood, around age twenty. The term *teenager* is often used to describe a person during the period of adolescence. The notion of the preteen, which has come into usage in the last three to four decades, encompasses the ages of about nine through twelve years.

As we saw in Chapter 4, establishing an identity is a central concern in the adolescent period. Youths experiment during this period as they attempt to fit different social roles into a coherent whole. At the same time, adolescents attempt to establish personal autonomy and independence from authority figures such as parents and teachers. Paradoxically, during this period, society attempts to bring the behavior and attitudes of adolescents into line with adult standards while denying adolescents the privileges of adulthood. For example, adolescents may leave school in their midteens and set out on their own, but they will have an extremely difficult time finding steady, gainful employment, mainly on the basis of age. It is an oft-told tale: A young person is denied employment because she does not have "enough experience." She responds, reasonably, "But how can I get any experience if I can't get a job?" One result is very high rate of unemployment for the nation's teens who are looking for work—a particularly acute problem for teens from racial minority groups.

Young people typically try to mark their unique identity through the establishment of *youth subcultures*—relatively distinct habits, customs, norms, and language that define youth in contrast to other generations. Youth subculture is crosscut by African American, Hispanic, and Asian cultures; the combination of youth and racial–ethnic subcultures are encouraged by the consumer markets that promote youth subcultures. Even shopping malls—seemingly places meant primarily for consumption—are places where young people construct their social identities. Hanging out at the mall, a practice common among young people in many cultures, allows the young to assert themselves as public citizens. Even when the older public tries to prevent young people from congregating in malls, the youth assert their right to be there, interpreted by scholars as a way they assert their identity as adults (Matthews et al. 2000).

Some youth subcultures reflect the alienation that young people feel from their families and schools. Many, though certainly not all, young people partake of a variety of "escapist" entertainment, ranging from such mainstream diversions as rock, rap, hip-hop, and videos to the computer underground populated mainly by late adolescents. Such pastimes constitute one element of youth subculture. Another element is style. Analysts of youth subculture define style in terms of its components, which include image, demeanor, and special vocabulary. Image is the impression delivered by hairstyle, jewelry, and dress. Demeanor is communicated by facial expressions of pouting and nonchalance, the shuffling, "cool" walk, traceable to African American and Latino youth in the 1940s and 1950s and now spread, or culturally diffused, to White youth. Also traceable to Black and Hispanic urban youth subculture is the wearing of one's pants so low beneath the waist that they appear to be in defiance of gravity itself. While currently in vogue, this practice was relatively widespread in the mid-1940s, when wearing one's pants "low" was a definite sign of cool. Like many styles, such practices tend to be repeated every generation or two, even though any given generation believes they are the first to discover it.

Within youth subcultures, special vocabularies and manners of speaking, known as *argot*, also define youth autonomy and independence from adults. That which is now described as "hot" (an attractive man or woman, for example) might have been described as "cool" twenty years ago. This tends to confuse adults, as it is supposed to do. Similarly, that which is perceived as good (an especially appealing music video, for example) is now described as "bad." Older people who try to imitate the youth subculture will be seen by young people as silly. Likewise, older people typically view youth subcultures as outlandish or whimsical. With these perceptions both groups implicitly acknowledge the significance that age has in defining our social identity.

Adulthood

The role of adult carries with it more responsibility and more rights and privileges than any other stage in the life cycle. It is also a period of significant change for most people. Moving away from one's parents' home to college, the army, or an apartment or house of one's own are big changes. Currently, the high cost of living has raised the financial threshold on becoming independent, and many young adults are staying at the home of their parents for longer periods, even after entering adult roles.

The four activities for the transition into adulthood of finishing school, getting a job, marrying, and starting a family used to be the norm. Increasingly, however, social conditions make it difficult for many people to progress to adulthood following this traditional path. Many people do not finish high school; some never marry or raise a family; many spend much of their time unemployed. Recent changes in both the sequence of these steps and the length of time between steps have also changed the nature of transition to adulthood.

The percentage of students in school between the ages of twenty and twenty-one has increased since 1980 from 31 percent to about 45 percent today. Further, more persons in the young adult age range (eighteen to twenty-four years old) live with their parents (about 48 percent in 1980 versus 58 percent now). Women are marrying later and having children later. The median age at first marriage for women has increased from 21.8 years to 25.3 years and nearly half of today's married couples have lived together before getting married (U.S. Census Bureau 2004; Sweet and Bumpass 1992). In an earlier era, couples would have married before setting up housekeeping. Today, many marry after doing so, contributing to the increase in the average age of marriage for both men and women. In sum, all such indicators show that the transition to adulthood is occurring at later ages (refer again to Table 14.2).

As adulthood unfolds, the traditional norms of our society suggest that both men and women, but particularly men, should have achieved most of their life goals by the time they reach middle age. As popularly conceived, the *midlife crisis* is a time of trauma during which people, men in particular, become fixated on what they have failed to achieve in their work and family roles or on the things they never attempted. The midlife crisis is reputed to spring from the feeling that something is missing. The media often typify such a man as the conservative accountant or executive who suddenly buys a sportscar and leaves his wife for a younger woman, a futile attempt to stay young. The popular

phrase "empty nest syndrome" also refers to the idea that middle-aged parents will feel lonely—and a little crazy—when their children leave home. Once a presumed syndrome associated only with women, empty nest syndrome is now assumed to affect both mothers and fathers, even though many parents report an enormous sense of freedom when their children leave.

Despite all the talk about the midlife crisis, the transitions experienced at midlife are very often not as painful as the popular image suggests. It is true that during this period, many people rethink their role in society and ask themselves what they have accomplished. They often dislike the physical signs of aging that change their bodies and their appearance, but the bulk of research does not find adults having midlife crises—at least not any more so than in other periods of life. In fact, midlife is experienced as happy and positive by most adults (Keyes 2002).

Retirement

Along with becoming a grandparent, one significant marker of approaching old age is retirement from work. Many look forward to retirement as a time for increased freedom. Some approach retirement with some degree of apprehension and fear because it may symbolize bringing one's life's work to an end. For someone whose career has been the primary basis for identity, retirement can be a difficult period of adjustment. No longer does the person have a predetermined place to be for eight or more hours a day. The social interaction with friends and colleagues at work is absent and the amount of free time increases greatly.

How one experiences retirement and whether one can retire at all are the result of many social factors, although most people view retirement as a satisfying time. This varies significantly, though, depending on such factors as one's physical health, level of education, and economic status. Even whether one's spouse influenced the decision to retire influences people's level of satisfaction, with wives who made independent decisions to retire being among the most satisfied (Smith and Moen 2004). By far the most influential factor on well-being in retirement is education. Studies consistently find that having more education is related to a number of indicators of well-being, including physical health, life satisfaction, well-rounded aging styles, and, of course, one's income during retirement (Murrell and Meeks 2002; Meeks and Murrell 2001; Crosnoe and Elder 2002).

Social factors also influence marital relationships in the retirement years. The transition period to retirement can be especially stressful. Married men and women experience the greatest marital conflict when one spouse remains employed; marital strain is reduced, however, once the couple settles into retirement (Moen et al.

2001). Gender attitudes have a significant effect as well. Husbands and wives with more traditional gender role attitudes tend to report lower marital quality during retirement (Szinovacza 1996). Also, just as gender roles affect patterns of childcare, they also affect caregiving in retirement. Women who have to care for a sick or disabled spouse are five times more likely to retire than are men in the same situation. Researchers have concluded that caregiving responsibilities in the later years actually lead to increased differentiation in gender roles (Dentinger and Clarkberg 2002).

Not surprisingly, the experience of retirement varies, not only because of gender, but due to race and ethnicity as well. Immigrant women are especially disadvantaged in retirement, particularly if they have low levels of education. This puts them at risk of poor health—given the known association between levels of education and patterns of health (Buckley et al. 2000). African American women are the group most likely to have to return to paid employment following retirement (Silverman et al. 1996).

The resources one has available in retirement—both economic and social—are critical to people's well-being, including both their physical and mental health, as well as material well-being. Groups who had reliable patterns of social support in the years prior to retirement are more likely to find such support during retirement (Buckley et al. 2000). But economic resources are critical and these are shaped by gender, race, and social class.

Reflecting the same patterns in employment earnings, women continue to have lower incomes in retirement, even though they now have more working experience than would have been true in the past. Women's lower level of earnings during their employment years mean they have had less chance to build up a good retirement pension (Warren et al. 2001). Women also have higher rates of poverty in old age and less access to pensions—because as we learned, they are more likely to work in jobs without such benefits. Women are thus are more dependent than men on Social Security and government assistance (Morgan 2000 et al. 1999; Gregoire et al. 2002). One difficulty retired women face is that most pension systems are designed with men in mind. Because it was designed in the 1930s, Social Security defined men as wage earners and women as dependents (Hill and Tigges 1995; Myles 1989). Women now tend to receive less in Social Security, which is based on lifetime earnings. In the calculation of Social Security benefits, women who have left the labor force to care for children are disadvantaged.

Retirement is often the beginning of a new life with new activities.

Likewise, research on race and retirement finds that disadvantages in the labor force are reproduced in retirement. Lower levels of job training and benefits mean that minorities have less postretirement income and are more reliant on public retirement funds, such as Social Security, than is true for Whites (Gibson and Burns 1991). Their lower levels of wealth also mean they are less likely to have private retirement investments. Women of color are especially disadvantaged in retirement because they face the influences of both race and gender; they have fewer resources than men (Hogan et al. 2000; Behringer et al. 2000). Race also significantly predicts one's Social Security earnings: African Americans and Latinos are more likely to have been working "off the books" in their employment years, meaning they have fewer assets and benefits compared to Whites (Hogan et al. 1997). In fact, the racial gap in retirement earnings is even greater than the racial gap in employment years (Hogan and Perrucci 1998).

Financial resources are clearly linked to people's sense of life satisfaction, as well as just their ability to get by (Choi 2001). Taking on a job when retired creates larger social networks, which is linked to better health (Slevin and Wingrove 1998), but continuing to work also robs people of the promise of retirement—to live one's later years without the stress and demands of employment. In sum, the experience of retirement, and aging in general, are strongly connected to the social factors that influence other dimensions of life.

Old Age

Aging is not an entirely negative process, but old age is undoubtedly a difficult period, worsened by the inadequacy of social institutions in caring for the aged. Negative images surround those who grow old, especially

Ms. Jeanne Calment died at the age of 122 years, giving her the verified record for being the oldest person to have lived. The odds of living to be over 100 are extremely small, only one out of every 5578 people did so in 2000 (Hetzel and Smith 2001).

older women. Yet, many find being old to be a source of strength, reminding us that just because someone is no longer involved in the labor force does not have to mean that person is unproductive. One sociologist, Jessie Bernard, widely known for her contributions to the sociology of women and the study of families, died in 1996 at the age of ninety-three. Some of her most extraordinary years of research took place after her formal retirement. In one decade, the 1970s, when she was in her seventies, she published nine books and held a sit-in at a hotel bar that refused to serve women.

Although the elderly undeniably face many problems, for many old age is a happy period. Research tends to show that being old looks a lot better to the old than to the young—every age begins to look better as you get closer to it. Studies have found that feelings of optimism actually increase with age. Older people tend to report greater feelings of life satisfaction, although education again is strongly related to emotional well-being among the elderly (Lennings 2000). .

Some say that such findings may be the result of a *cohort effect*—that is, that the particular age group studied, or cohort, has *always* felt positive relative to younger age groups. In other words, a cohort effect occurs when a particular generation of people differs significantly from another because of particular generational experiences. To know whether something is the result of aging per se or a cohort effect, you have to study the same age group over time to see if their feelings change or remain the same.

No doubt one difficult adjustment to make in old age is becoming a widow or widower by losing a life partner—a situation more often faced by women because women, on average, live longer than men. Five times more women than men outlive their spouses. Men who are widowed are more likely to remarry and to do so more quickly. Widowed men tend to know more people and have more acquaintances and work-related contacts than women. Hence, widows tend to suffer more from loneliness. As widowers, men participate in more organizations than women and are more likely to own a car (which translates into a greater ability to get out and around). As the population of widows begins to include more women who held their own jobs and lived more independently, these patterns will likely change (Atchley 2000).

Men have some advantages in widowhood but also experience some disadvantages, mainly that they are much more likely than women to die themselves soon after the death of a spouse, even if they remarry and regardless of their age (Mineau et al. 2002). Widowed men are three times more likely to die in a car accident than comparably aged married men, four times more likely to commit suicide, six times more likely to die of a heart attack, and ten times more likely to die from a stroke (Atchley 2000).

Having social support through extensive friendship and familial networks helps alleviate the stress experienced during widowhood, just as having friends lessens the negative impact of the problems of aging in general. In a study of Puerto Rican families in Boston, Melba Sánchez-Ayéndez (1995) found that strong cultural norms encourage Puerto Rican children to support both parents and grandparents and parents to support older children. The families she studied strongly relied on neighbors, with reciprocity also stressed. As a result, the elderly in Puerto Rican communities, particularly widows, have a wider social support network than is often the case with White widows. This research confirms the importance of social support networks in alleviating the stress of aging.

Elder Care Elder care in the United States is provided primarily in two ways: institutions for the elderly and private care in the home. Most care of older people in the United States is provided informally by families, mostly by women (Aronson 1992). Family members

MAPPING AMERICA'S DIVERSITY

MAP 14.1 Where the Aged Live

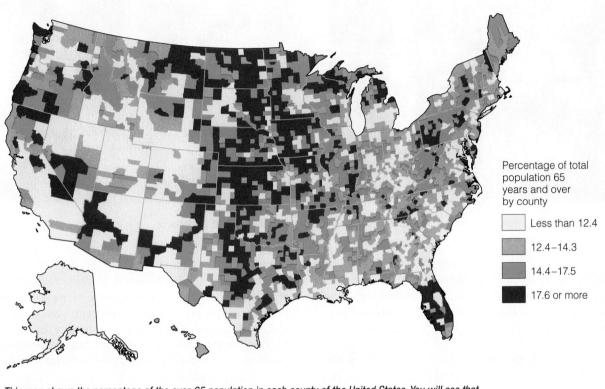

Percentage of total population 65 years and over by county

- Less than 12.4
- 12.4–14.3
- 14.4–17.5
- 17.6 or more

This map shows the percentage of the over-65 population in each county of the United States. You will see that older people are more concentrated in some areas than others. Some older people move to new places, but a large number "age in place," thus it may surprise you to see so many older people in the middle of the country. The Midwest is the region of the country that has the greatest percentage of people 65 and over, relative to the total population in the United States as a whole. What are the implications of having a large aging population for the social services needed in different regions?

Data: Hetzel, Lisa, and Annetta Smith. 2001. "The 65 Years and Over Population: 2000." *Census 2000 Brief.* Washington, DC: U.S. Census Bureau, pp. 5–6.

provide 80 to 90 percent of long-term care for the elderly. Often, this work is taken for granted.

Elder care tends to be confined to the private sphere of households and reflects a gender division of labor that assigns women the responsibility for nurturing others. By contrast, sons and sons-in-law do not face the same expectations to be caregivers as do women. Despite social changes, men are still perceived as having a primary commitment to work and as less able to anticipate and respond to elders' needs. Women also believe they are better at elder care than their husbands and brothers, but with the rapid increase in the older population that lies ahead, these social norms may have to change.

Many elderly people live alone instead of with their families. The highest probability of living alone in old age is found among Black, White, and Hispanic women. At the same time, as noted by Sánchez-Ayéndez (1995), cultural norms among Puerto Rican families containing elderly women are likely to encourage use of informal support networks consisting of extended family and close friends to care for the elderly men as well as women. Race and ethnicity are powerful determinants of household structure, even in addition to the effects of age, sex, marital status, and income. Blacks and Hispanics are more than twice as likely as Whites to live with other persons, not including their spouses (Fields and Casper 2001).

In the past, it was relatively more common among Blacks than Whites for members of the extended family to live together, including grandparents and great grandparents. Now the Black and Hispanic elderly are

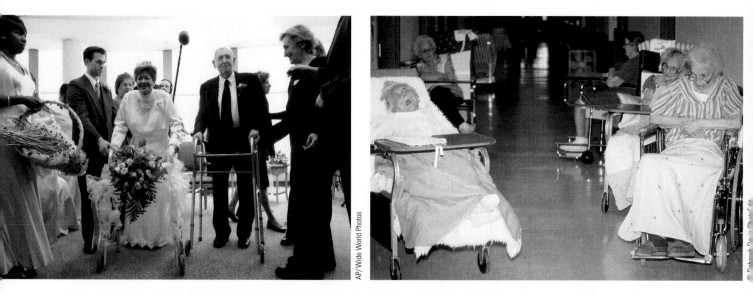

The experience of aging depends greatly on the same factors that influence other phases of life. Social class is one of the strongest such influences.

about as likely to be living alone as the White elderly, suggesting that increased urbanization and some economic advances of Blacks over the last twenty-five years have contributed to these changes (Fields and Casper 2001).

Many elderly continue to live in the homes where they once raised families. Most of these homes are free from mortgage debt, but many were built before 1940 and are now in need of repairs that require more money than the owners can afford. Most elderly are greatly reluctant to be forced from their own homes by economic conditions or other circumstances. Whether voluntary or not, a move from one's home is physically and psychologically disruptive.

A relatively small percentage of the elderly are eventually placed in long-term institutional care. Women are more likely than men to enter nursing homes, largely because of differences in life expectancy and higher rates of chronic illness. African Americans and Hispanics are less likely to enter nursing homes than Whites. Although only a small percentage of all elderly live in nursing homes, many senior citizens worry it will happen to them. The decision to enter a nursing home is almost never made by the elderly person, but by a health-care professional or some other party who assumes control over the care of the older person. Elderly without living children or kin are the most likely to be admitted to a nursing home. Giving up one's residence, separating from one's family, and losing one's independence, coupled with the widespread belief that entering a nursing home is the final step before death, all give the elderly a bleak picture of such institutions (Atchley 2000).

The cost of nursing homes is generally high but nevertheless varies greatly, which perpetuates social class differences. People of higher socioeconomic status are able to take up residence in higher-quality, higher-cost facilities or in modern retirement villages with restaurants, meals-on-wheels, and extensive social programs.

The high cost of nursing homes also forces many, even from the middle class, into poverty. The average annual cost of a nursing home is now about $57,000 per person per year. Given that the median income for all elderly is just over $23,000, it is easy to see that many of those who enter nursing homes also enter poverty (General Electric Financial Survey 2003; DeNavas-Walt et al. 2003).

DEBUNKING *SOCIETY'S MYTHS*

Myth: Nursing homes house most of the nation's elderly people.

Sociological perspective: Family members, most often women, provide almost all the care for elderly people in the United States. Only about 4.5 percent of the elderly live in nursing homes (U.S. Census Bureau 2004).

Medicare and Medicaid are the two federal programs that assist the elderly with health care and the cost of living. **Medicare** is a governmental assistance program established in the 1960s to provide health services for older Americans. **Medicaid** is a governmental assistance program that provides health care assistance for the poor, including the elderly. States can establish the eligibility criteria for Medicare, based on income level. When it comes to long-term care and housing arrangements for the elderly, Medicare and Medicaid stress institutional arrangements, and provide little support for noninstitutional housing. This mitigates against the elderly being able to find more innovative housing arrangements, unless of course they completely pay for such arrangements on their own.

With the recent expansion of Medicaid and Medicare programs, the number of privately owned nursing homes has recently increased compared with those administered by charitable organizations such as churches, the Salvation Army, and others. Federal subsidies contribute to the profitability of nursing homes and make them attractive to investors seeking tax shelters and long-term profits. This creates a problem pervasive throughout the health care industry, as corporate interests and the drive for profitability supplant affordable, humane care as the main interest of the institution. In the medical industry, this situation has focused attention on costs running out of control. The more expensive institutions are generally better run and more closely scrutinized by the kin of the residents. It is the elderly poor in the second-tier institutions who are disproportionately victimized by scams and low-quality care.

Elder Abuse Physical and mental abuse of the elderly has only recently surfaced as a notable social problem. The National Center on Elder Abuse estimates that there are between 820,000 and 1,860,000 abused elders in the United States, but this organization acknowledges the difficulty of gauging the true extent of the problem. Elder abuse is often hidden in the privacy of families or behind institutional doors, and victims are reluctant to talk about their situations. What is known is that reports of elder abuse have increased since the mid-1980s. Whether elder abuse has increased over time or is simply reported more frequently is open to speculation (National Center on Elder Abuse 2003; Teaster 2000).

Why are the elderly abused? One explanation is that having to care for the elderly is very stressful for the caregiver—usually a daughter who may be holding a job as well as caring for the elderly person. Research on elder abuse finds that the abusers are more likely to be women, middle-aged, and the daughter of the victim—in other words, the person most likely to be caring for the older person. Sons, however, are the ones most likely to be engaged in direct physical abuse, accounting for almost half of the known physical abusers. Sometimes the physical abuser is a husband, where the abuse is a continuation of abusive behavior earlier in the marriage. The same factors that affect family life in any generation contribute to the problem of elder abuse (Teaster 2000).

Death and Dying

There is probably nothing sadder than watching a loved one die. At such a time, perhaps the last thing you would think to do is to analyze death sociologically. Still, if you were to engage your sociological imagination, you would find that the behaviors and events surrounding death have a clear sociological character. When it is known someone is going to die, people's behavior changes. As sociologists put it, a person can die a "social death" before biological death. The person dying starts to be treated as if he or she were not there. People may talk about the person in the past tense, and the dying person may be perceived as a nonperson because of his or her physical, emotional, and communicative withdrawal. These tendencies are exacerbated by the dying person's placement in a hospital or nursing home (Glaser and Strauss 1965; Sudnow 1967).

Patterns of social stratification that reveal themselves in life are also apparent in death. Certain groups are more likely to die a violent death than others, namely, African American men, who are seven times more likely to die from homicide as White men. Infant death is also twice as likely to occur among racial minority groups than among White Americans (U.S. Census Bureau 2004). New medical procedures produce new forms of inequality about who will live and die. With a shortage of organs available for heart, kidney, and lung transplants, who receives the transplant reflects social values about whose lives are considered most precious.

When a person dies, he or she dies within social institutions that are organized to handle death. Four-fifths of those who die on a given day die in hospitals. The social organization of death is especially apparent in the funeral home industry—a $16 billion business (Sayre 1994). Within this industry, funeral directors "manage" the death experience for others. No longer referred to as undertakers, these people are seen as professionals who are skilled in the administration of death. As the box "Doing Sociological Research: Death's Work: A Sociology of Funeral Homes" on p. 372 shows, the bereaved are instructed by the funeral director on the ritual about to occur, including what choices to make, how to enter and exit the event, and what accoutrements to choose. Despite the appearance of condolences and sympathy, the funeral director is also a salesperson for an industry with a large market and considerable profit—the average cost of a funeral being around $5000 with no extras (AARP 2004).

Death was not always handled this way. Prior to the twentieth century, death was likely taken care of at home and largely the work of women. Until the emergence in the mid-nineteenth century of undertaking as a profession, women were responsible for the care of

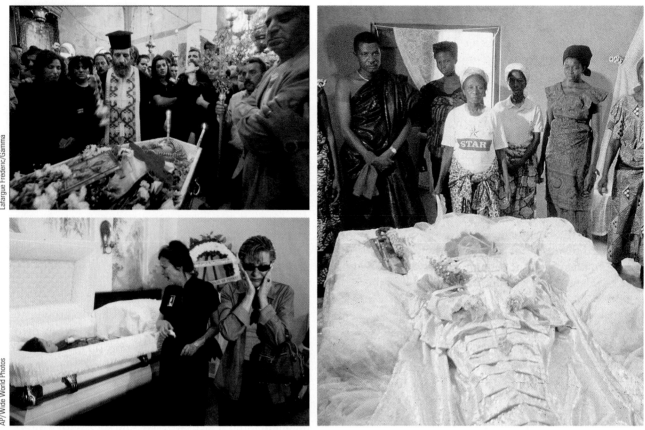

Lafargue Frederic/Gamma

AP/Wide World Photos

© Victor Englebert

*Rituals surrounding death vary in different cultures, such as the funeral of this Palestinian boy (upper left);
the Turkish actress, Derya Arbas in Los Angeles (lower left) and this woman in Ghana (right).*

the dead, particularly preparing the body for burial. Cultural beliefs about women's emotionality and caring supported this role. Women's care for the dead was also justified on religious grounds—society saw women as more devout than men. Because "laying out the body" was considered a sacred act, it logically followed that it should be women's duty. Some women in communities were seen as particularly skilled in this regard. Like midwives, they would be called upon to assist families in laying out the body. Known as "shrouding women," they washed and dressed the body, posed the body in the coffin, and constructed a "cooling board" on which the body laid, giving the impression of a restful sleep (Rundblad 1995).

TAKING ON SOCIAL ISSUES
Privatizing Social Security

When President Bush took office in 2000, he tried to pass legislation that would privatize Social Security—that is, allow people to use funds they would otherwise have in Social Security to invest privately—in stocks, bonds, and other investments. While some say this would allow people potentially to earn more on their investment and re-duce the strain on the federal budget (in support of old people), others say it is not only risky, but would unfairly advantage some groups over others.

Given what you are learning about age stratification, do you think privatizing Social Security is a good solution to the challenge of supporting the growing number of older people?

Taking Action

Go to the Taking Action Exercise on the Companion Website—at http://sociology .wadsworth.com/andersen_taylor4e/—to learn more about an organization that addresses this topic.

A sociological perspective on death includes how social factors like age influence the meaning of and reaction to someone's dying.

Study of the rites of death reveal intriguing patterns in how different groups respond to death. Among some cultural groups, dying is defined as part of living and rituals surrounding death are more likely to celebrate kin. Some groups have created their own death rituals, such as shown by study of gay men in San Francisco coping with the loss of loved ones because of AIDS. Most of these deaths occurred at home, with caregiving partners present at the time of death. The men then developed culturally unique rituals involving multiple memorials, secular (that is, nonreligious) services, and private dispersion of ashes, often with multiple distribution sites. These rituals have been interpreted as reflective of the drive for freedom and self-control that is characteristics of the gay liberation movement (Richards et al. 1999–2000).

The increasing diversity of the U.S. population has implications for dealing with death. People from different cultural backgrounds have different understandings of death—and different traditions of dealing with death. This means that those who work with death—hospital workers, social workers, and others—need to develop more multicultural ways of handling death so that they can be a help to those dealing with this difficult experience (Braun et al. 2000; Seale 2000; Willis 1999).

Euthanasia is the act of killing a severely ill person or allowing the person to die as an act of mercy (see also Chapter 20), such as in physician-assisted suicide. This movement, sometimes referred to as the "requested death movement," is a way of resisting hospital and gov-

ernment control of death (McInerney 2000). The public is divided in its support for euthanasia, with 49 percent saying that physician-assisted suicide is morally wrong and 45 percent saying it is morally acceptable. Interestingly, far fewer support suicide by an individual facing a terminal illness; only 14 percent say this is morally acceptable (Ray 2003). African Americans are less supportive of euthanasia than Whites, a social fact explained by both a higher rate of religious fundamentalism among African Americans as well as their collective fears about giving others the power to end their life. Differences in support for euthanasia are also explained as the result of socioeconomic status and values of political conservatism (MacDonald 1998).

As medical technology has developed, the process of dying can become quite prolonged, with more control over death placed in the hands of medical authorities (Seymour 1999; Braun et al. 2000). As life expectancy has increased, so has the likelihood of living with worsening disabilities. The nation's health and other social services are having to adapt to meet these new challenges. Policymakers and care providers are also increasingly asking what it means to experience a "good death," a good death being defined as one with physical comfort, social support, acceptance of one's fate, and appropriate medical care (Carr 2003). For survivors, a good death can lessen the anger and anxiety experienced after the loss of a loved one—a finding with implications for how social policies can help people cope with death and dying.

The **hospice movement** has emerged in this context. Hospice workers, some of whom are volunteers, provide care for dying people and their families. It is an alternative to hospital-based, technologically controlled death. This movement began as a reaction to the impersonal forms of death that occur in more institutional settings. Instead of letting professional experts control death, hospice workers see control as more appropriately lying with the dying person and his or her family and friends. Ironically, as this movement has developed, many hospice workers have come to be defined as the experts in handling death, and the hospice movement has itself become a large and well-organized industry.

Age, Diversity, and Inequality

Throughout this chapter, we have seen how the experience of aging differs for various groups in society. Aging in itself can result in many problems—physical, psychological, social, and economic. But as we have seen, the effects of aging are compounded by the additional effects of other factors—most notably, class, race, and gender—which shape the experience of aging in both positive and negative ways.

Age Groups as Minorities

Understanding diversity in aging is fundamental to sociological research. Furthermore, concepts from the sociology of race can be used to analyze age stratification. The term *minority group* is one concept that is useful in interpreting age stratification. Recall that sociologists define a minority group as a group with relatively less status and power and fewer social and economic resources than more dominant groups (see Chapter 12). Age minority groups include both the old and the young—those groups with the least power and the poorest life chances because of their age.

Seeing oneself as a minority group can also be the basis for mobilizing for group rights—a movement that is apparent among the nation's older population. Groups such as the American Association of Retired Persons (AARP), the Gray Panthers (a more radical group that has advocated, among other things, intergenerational living arrangements as a means for supporting old people), the Older Women's League (OWL), and other organizations have worked on behalf of the elderly to challenge age stereotypes and define the national agenda for old people. Like civil rights organizations, such groups are important advocates of social programs for the elderly.

Despite certain parallels, the similarities between the aged and racial and ethnic minorities should not be overstated. The aged differ from racial and ethnic minorities in important respects. For one thing, a person is in a racial or ethnic group for life but aged for only part of his or her life. Furthermore, as a group, the elderly have, on average, more political power than other minorities. A high percentage of the elderly vote— more than any other group—and the interests of the elderly are represented by powerful lobbying organizations with large and growing memberships, such as the AARP, which has a constituency of thirty-five million members. Unlike other minorities, the aged also receive far more subsidized support in the form of medical care; insurance programs; senior discounts at supermarkets, restaurants, movies, hotels, and auto rentals; and a variety of other benefits that cumulatively are of great value.

Age Prejudice and Discrimination

Age prejudice refers to a negative attitude about an age group that is generalized to all people in that group. Age prejudice is manifested in the stereotypes that we have seen of different age groups. You have probably heard elderly people taunted as "old geezers." As with any minority group, a range of epithets has evolved to describe the elderly: geezer, curmudgeon, codger, fuddy-duddy, old biddy, old fart. The lobbying efforts of the elderly have also generated new stereotypes of the elderly as politically selfish and grasping, stereotypes that are surprisingly acceptable in social circles that would usually be quick to condemn such prejudice.

Prejudice against the elderly is so prominent in our culture that the myths about them are seen as accurate descriptions. Old people are often stereotyped as being passive victims, even though evidence finds that they often organize on their own behalf (Gubrium and Holstein 2003). Just as for other groups who experience prejudice directed against them, prejudice relegates people to a perceived lower status in society.

In Chapter 11, we defined racial–ethnic discrimination as behavior that singles people out for different treatment. **Age discrimination** is the different and unequal treatment of people based solely on their age. While age prejudice is a covert attitude, age discrimination involves actual behavior. Some forms of age discrimination are illegal. The Age Discrimination Employment Act, first passed in 1967 but amended several times since, protects people from age discrimination in employment. An employer can neither hire nor fire someone based solely on age nor segregate or classify workers based on age. Age discrimination cases have become one of the most frequently filed cases through the Equal Employment Opportunity Commission (EEOC), the federal agency set up to monitor violations of civil rights in employment.

Ageism is a term sociologists use to describe the institutionalized practice of age prejudice and discrimination. More than a single attitude or an explicit act of discrimination, ageism is structured into the institutional fabric of society. Like racism and sexism, ageism encompasses both prejudice and discrimination and is manifested in the structure of institutions. As such, it does not have to be intentional or overt to affect how age groups are treated.

We have seen the consequences of ageism throughout this chapter: stereotypes of old and young that define their social worthiness, marked inequalities in the social and economic resources available to different age groups, and institutional practices that manage how age groups are treated in society. Ageism in society means that, regardless of laws that prohibit explicit discrimination and regardless of individuals' positive efforts on behalf of old and young alike, people's age is a significant predictor of their life chances. Old people are stratified into some segments of society; resources are distributed in society in ways that advantage some age groups and disadvantage others; cultural belief systems devalue the elderly; society's systems of care are inadequate to meet people's needs as they grow old—these are the manifestations of ageism, which is a persistent and institutionalized feature of society.

Quadruple Jeopardy

For some groups, ageism is compounded by other factors. **Quadruple (or multiple) jeopardy** is a phrase referring to the simultaneous effects of being old, minority, female, and poor. The effects of age, race, gender, and social class all intersect for these women. The status of old is lower than that of adult; the status of minority is lower than that of White; the status of women is lower than that of men; and the status of poor is lower than that of middle-class. Anyone who is all four is placed in great jeopardy. This does not mean, however, that those experiencing multiple forms of oppression are passive and weak. Rather, those who are most oppressed often have remarkable resilience and strength (Calasanti and Slevin 2001).

The effects of quadruple jeopardy can be seen in several areas. Income differences between groups reveal the risks of this multiple status. Overall, the economic condition of the elderly has improved somewhat over the last thirty years. In combination with race and gender, however, age brings a particular risk of poverty. Poverty among the aged is more pronounced for African Americans, Latinos, and Native Americans than for Whites, and it is more pronounced for women than men. In 2002, over 22 percent of all elderly Black people and 21 percent of all elderly Hispanics were poor, as contrasted with 9 percent of elderly Whites and 10 percent of Asians/Pacific Islanders. For American Indians, more than *half* the elderly were below the poverty level. Fifteen percent of women over the age of sixty-five are poor, compared with 12 percent of men (Proctor and Dalaker 2003).

Health provides another example of the effects of quadruple jeopardy. Older Black and Hispanic women tend to have higher rates of health problems such as hypertension, diabetes, heart disease, and cancer, compared with older White men and White women. Being poor further increases these rates (National Center for Health Statistics 2003). Elderly Native Americans also have less access to doctors, hospitals, and other health care facilities than any other social or ethnic group. Even travel time to health care providers is substantially longer for Native Americans than for any other group

(Cunningham and Cornelius 1995; Inouye 1993). These differences are perpetuated well into old age and are perpetuated even further for those who are poor.

Age Seen Globally

As discussed earlier (in Chapter 5), societies vary in the extent of their industrialization, from societies that are nomadic or agrarian—the Bedouins of Saharan Africa and many settlements of Siberia—to those that are heavily industrial—the United States, England, France, and Germany—to the postindustrial—Japan. The treatment of the very old in society varies to some

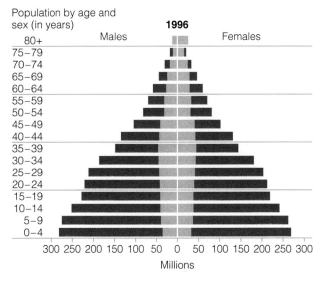

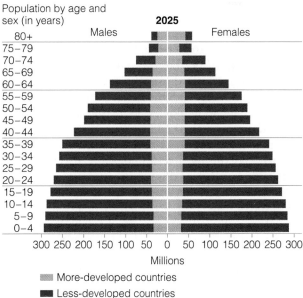

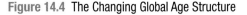

Figure 14.4 The Changing Global Age Structure

Source: U.S. Census Bureau. 2004. *Global Aging in the 21st Century.* Washington, DC: U.S. Census Bureau, p. 3. Website: **www.census.gov**

Figure 14.5
The World's Oldest Old

Source: U.S. Census Bureau. 2004.
Global Aging in the 21st Century.
Washington, DC: U.S. Census Bureau,
p. 6. Website: **www.census.gov**

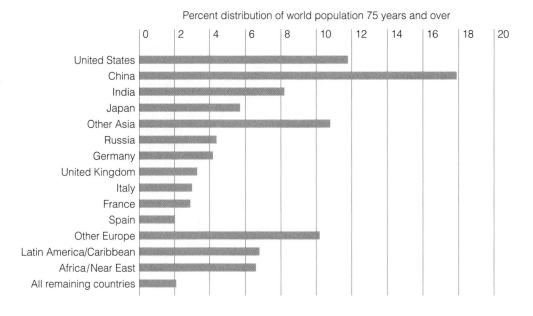

degree depending on the extent of industrialization of the society.

The elderly, on average, have greater social status in nomadic and agrarian societies, although there are exceptions. In such societies, those who have accumulated wealth during their lifetime (such as through raising crops and animals) gain high prestige and also economic and political power that continues into old age, which may be only forty or fifty years of age. When this is a man, he may be regarded as a patriarch and accorded considerable deference in the group, tribe, or town (Cohen and Eames 1985, Stoller and Gibson 2000).

In industrialized societies, advances in medicine and technology tend to result in longer life expectancy. As a consequence, a greater proportion of the population is elderly (Figure 14.5). Increased technology also has the effect of placing those who are relatively young into the workforce, with their education reflecting new and developing technologies. As a result, those who are older and receive their education and training under an earlier technology may be less trained and thus less able to carry out more technologically advanced jobs and tasks. An effect of this process is a surplus of older individuals who are regarded as less economically useful in society (an effect predicted by functionalist theory). This in turn causes the elderly in such a society to be accorded lower social status and prestige. This process results in the elderly having less social status in industrialized societies than in nomadic or agrarian societies.

There is one clear exception to the principle that the elderly have less social status in societies that are more industrialized, and that is the case of Japan. Japan is not only very highly industrialized, but it is also a *postindustrial* society, one that stresses the production and development of services (such as computer engineering and technology) more than goods (such as steel). In contrast to the industrialization principle, the elderly in Japan are accorded very high esteem and social status. Older men, in particular, are more likely than their age counterparts in the United States to remain in their jobs and to stay on as high executives and corporate heads and leaders well into old age. The elderly have been elevated to high status for many decades, even as Japan has moved from being an agrarian society to a highly industrialized and, now, postindustrial society. One explanation may be that the strong cultural tradition in Japan of high prestige for the elderly may have been powerful enough to overcome the effects of industrialization on lowering the status of the elderly. Nonetheless, as some researchers have noted, there is an increasing recent tendency for the elderly in Japan to give up positions of power to younger people. In this respect, Japan may be undergoing social changes that render it somewhat more like the United States and other industrial or postindustrial societies (Treas 1995; Palmore and Maeda 1985).

Explaining Age Stratification

Why does society stratify people on the basis of age? Once again we find that the three main theoretical perspectives of sociological analysis—functionalism, conflict theory, and symbolic interaction—offer different explanations (see Table 14.3). Functionalist sociologists ask whether the grouping of individuals contrib-

Table 14.3
Sociological Theories of Aging

	Functionalism	Conflict Theory	Symbolic Interaction
Age Differentiation	Contributes to the common good of society because each group has varying levels of utility in society	Results from the different economic status and power of age cohorts	Occurs in most societies, but the social value placed in different age groups varies across diverse cultures
Age Groups	Are valued according to their usefulness in society	Compete for resources in society, resulting in generational inequities and thus potential conflict	Are stereotyped according to the perceived value of different groups
Age Stratification	Results from the functional value of different age cohorts	Intertwines with inequalities of class, race, and gender	Promotes ageism, which is institutionalized prejudice and discrimination against old people

utes in some way to the common good of society. From this perspective, adulthood is functional to society because those who are adult and middle aged are seen as the group contributing most fully to society; the elderly are not. Functionalists argue that older people are seen as less useful and therefore granted lower status in society. Youth are in-between. The constraints and expectations placed on youth—prohibited from engaging in a variety of "adult" activities, expected to go to school, not expected to support themselves—are seen to free them from the cares of adulthood and give them time and opportunity to learn an occupation and prepare to contribute to society.

According to the functionalist argument, the elderly voluntarily withdraw from society by retiring and lessening their participation in social activities such as church, civic affairs, and family. **Disengagement theory,** derived from functionalist theory, predicts that as people age, they gradually withdraw from participation in society and are simultaneously relieved of responsibilities. This withdrawal is functional to society because it provides an orderly transition from one generation to the next. The elderly move aside so that the young can step in. The young presumably infuse the roles they take over from the elderly with youthful energy and stamina, while supporting themselves and providing for their families. As the elderly become less useful to society, they are rewarded less. According to the functionalist argument, the diminished usefulness of the elderly justifies their depressed earning power.

Conflict theory assesses the differences between age groups not in terms of what they contribute but in terms of what they want. The focus is on the competition over scarce resources that exists between age groups. Jobs are among the most important scarce resources. Unlike functionalist theory, conflict theory offers an explanation of why both youth and the elderly are assigned lower status in society. Conflict theory argues that barring youth and the elderly from the labor market is a way of eliminating these groups from competition, improving the prospects for workers who are adult and middle-aged. Once removed from competition, both the young and the old have very little power and like other minorities, they are denied access to the resources they need to change their situation. Conflict theory also helps explain the competition between groups that has marked concerns about generational equity, in particular, the questions raised earlier about how the nation will take care of its growing elderly population.

A third approach to age stratification, symbolic interactionism, analyzes the different meanings attributed to social entities; in this case, the different age groups. Interactionists ask what symbolic meanings become attached to the different age groups and to what extent these meanings explain how society ranks them. Symbolic interaction also analyzes how people explain and manage death. Symbolic interaction therefore considers the role of social meanings in understanding the sociology of age. These meanings are cultural products. Age clearly takes on significant social meaning—meanings that vary from society to society for a given age group and that vary within a society for different age groups.

Chapter Summary

What is the social significance of aging?

All societies practice some form of *age differentiation*, and the United States is no exception. *Age stratification* is the hierarchical ranking of different age groups in society. The age structure of the United States is rapidly changing, with a greater proportion of the population consisting of old people. *Age stereotypes* pervade this society and burden the aged, in particular, with negative preconceptions about being old—stereotypes which can actually shorten one's life. Age stereotypes are compounded by the gender stereotypes that define the aging experience for men and women differently. *Age norms* spell out explicit and implicit expectations for different age groups, although these norms change as social conditions change. Age is an ascribed status—established at birth. Individuals in a given *age cohort* share a great deal of historical experiences.

What are some consequences of an aging population?

As the nation's population grows older, the contract between generations that has established expectations for intergenerational care will likely change. Those who are young now can expect to spend a greater portion of their lives as a "sandwich generation," not only raising their own children, but also caring for their elderly parents.

What is the life course perspective?

Sociologists use a *life course perspective* to relate people's life experience to their sociohistorical context. The different phases in the life span include childhood, youth and adolescence, adulthood, and old age. Within each generation, unique life events shape the sociological experience of these age groups. Even death is shaped by sociological factors.

What are both the advantages and disadvantages of aging?

Old age brings with it both the good and the bad. The bad side is that the aged are treated as a social minority and tend to be shunted aside and ignored. The good side is that old age can be, and for many is, a period of great happiness and satisfaction, of renewed interest in education and perhaps a second career, and of increased self-evaluation.

How does age pertain to being a minority in society?

Both the youthful and the aged are treated as minorities in society, especially in industrialized societies. In many nomadic and agrarian societies, the elderly enjoy high social status. *Age discrimination* is the differential treatment of persons on the basis of age alone. *Age prejudice* refers to the tendency to have negative attitudes toward someone on the basis of age. *Ageism* is the institutionalized dimension of age prejudice and discrimination. It need not be conscious or intentional, but ageism shapes the experiences of diverse groups in society. *Quadruple jeopardy* is the combined effect of age, race, gender, and class. Those who are most disadvantaged in society—in income, education, health, and other areas—are those who are old, female, of a racial or ethnic minority, and economically disadvantaged.

What does sociological theory tell us about age discrimination?

In attempting to answer the broad question of why society discriminates against both its youth and its elderly, three sociological theories lend some suggestions: because the elderly are less useful, or "functional," to society (*functionalist theory*); because the elimination of both groups from competition in society frees up those who are adult and middle-aged (*conflict theory*); and because both the youth and the elderly are infantilized via cultural symbols, such as language and popular culture (*symbolic interaction theory*).

Key Terms

age cohort 369	euthanasia 383
age differentiation 368	generational equity 372
age discrimination 384	hospice movement 383
age norms 367	life course perspective 373
age prejudice 384	
age stereotypes 366	life expectancy 365
age stratification 369	Medicaid 380
ageism 384	Medicare 380
Alzheimer's disease 365	quadruple jeopardy 385
dementia 364	
disengagement theory 387	

Researching Society with MicroCase Online

You can see the results of actual research by using the Wadsworth MicroCase® Online feature available to you. This feature allows you to look at some of the results from national surveys, census data, and other data sources. You can explore this easy-to-use feature on your own, but try this example. Suppose you want to know:

Are older people more or less likely to support a person's right to end their own life if the person is tired of living and ready to die?

To answer this question, go to http://sociology.wadsworth
.com/andersen_taylor4e/, select MicroCase Online from
the left navigation bar, and follow the directions there to
analyze the following data.

Data file: GSS

Analysis: Cross-Tabulation

Variable1: SUIC.WISH

Variable 2: AGE

Questions

The table shows you what percentage of three different
age groups support the right to suicide if the person is
tired of living and ready to die.

1. Which age group is most likely to say yes, they support the person's right to end their own life?
   a.   <30
   b.   30–49
   c.   50 and up

2. Which age group is least likely to support the person's
   right to end their own life?
   a.   <30
   b.   30–49
   c.   50 and up

3. Are you surprised by these results? What do you think
   might explain the age differences you have observed?

The Companion Website for Sociology: Understanding a Diverse Society, Fourth Edition

http://sociology.wadsworth.com/andersen_taylor4e/

Supplement your review of this chapter by going to the
companion website to take one of the Tutorial Quizzes,
use the flash cards to master key terms, and check out
the many other study aids you'll find there. You'll also
find special features such as GSS Data and Census 2000
information, data and resources at your fingertips to
help you with that special project or do some research
on your own.

Suggested Readings and Web Resources

Atchley, Robert C. 2000. *Social Forces and Aging,* 9th
ed. Belmont, CA: Wadsworth.
   *This book, a summary of the sociology of aging, is one
   of the best accounts in the field of social gerontology.*

Calasanti, Toni M., and Kathleen F. Slevin. 2001. *Gender, Social Inequalities and Aging.* Walnut Creek, CA:
Altimira Press.
   *Using a feminist perspective, the aurhors examine
   multiple dimensions of the aging experience. They
   also show how understanding old age has been
   distorted through the limiting perspective of men's
   experiences.*

Diamond, Timothy. 1992. *Making Gray Gold: Narratives of Nursing Home Care.* Chicago, IL: University
of Chicago Press.
   *Participant observer Diamond examines profit, care,
   and the exploitation of women of color in a nursing
   home.*

Lyman, Karen A. 1993. *Day In, Day Out with Alzheimer's: Stress in Caregiving Relationships.* Philadelphia, PA: Temple University Press.
   *Based on participant observation at several day-
   care centers for people with Alzheimer's, Lyman
   studies the stress associated with caregiving for Alzheimer's disease. She concludes that the organization
   of Alzheimer's day-care work, that is, how the system is designed, is more stressful for caregivers than
   the care itself.*

Seale, Clive. 1998. *Constructing Death: The Sociology
of Dying and Bereavement.* Cambridge, England:
Cambridge University Press.
   *Placed in the context of social theory, this book
   examines how people construct human social bonds
   even when confronting death. It also demonstrates
   the social construction of bereavement and grief.*

Stoller, Eleanor Palo, and Rose Campbell Gibson.
2000. *Worlds of Difference: Inequality in the Aging
Experience,* 2nd ed. Thousand Oaks, CA: Pine
Forge Press.
   *This anthology explores diverse patterns of aging,
   particularly given the influence of race, class, and
   gender on the aging process. Many of the articles
   are written as personal narratives. Combined with
   a sociological perspective, they provide students
   with a compassionate view of diverse norms and
   social structures that shape the experience of growing old.*

National Institute of Aging
   www.nih.gov/nia
   *The home page of the U.S. government agency, pro-
   vides information on the latest research on aging,
   links to additional sites, and information on funding
   for research on aging.*

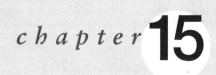

● ● ●

Families

Suppose you were to ask a large group of people in the United States to describe their families. What would they say? Many would describe *divorced* families. Some would describe *single-parent families.* Some would describe *stepfamilies* with new siblings and a new parent from remarriage. Others would describe *gay or lesbian households* in which a single gay parent or a couple of the same sex is raising children. Still others would describe the so-called *traditional family* with two parents living as husband and wife in the same residence as their biological children. Also included would be *adoptive families* and families with foster children. The variety of descriptions would reflect the enormous diversity in families today

Families have become so diverse that it is no longer possible to speak of "the family" as if it were a single thing. The *family ideal*—a father employed as the major breadwinner and a mother at home raising children—has long been the dominant cultural norm, communicated through a variety of sources, including the media, religion, and the law. Few families now conform to this ideal, and the actual number that conformed in the past is probably less than generally imagined (Coontz 1992). Regardless of their form, families now face new challenges—possibly living

on one income or managing family affairs when both parents are employed. Many families also feel that they are under siege by changes in society that are dramatically altering all family experiences.

Families have an enormous influence on our personal lives and the relationships we form. Many live in loving families, where the family provides stability and nurturing care for its members. But the popular image of families as providing refuge from the impersonal, hectic world outside is not shared by everyone. Families can be conflict-ridden and sorrowful. Family affairs are also believed to be private, but as an institution, the family is very much part of the public agenda. Public policies shape family life both directly and indirectly, intentionally and unintentionally, and family life itself shapes the dynamics of behavior in other institutions. Family life is now being openly negotiated in political arenas, corporate boardrooms, and courtrooms, as well as in the bedrooms, kitchens, and living rooms of individual households.

Many view the changes taking place in family life as positive. Women have new options and often greater independence. Fathers are discovering that there can be great pleasure in domestic and child-care responsibilities. Change, however, also brings

Sociology ⊛ Now™
Reviewing is as easy as ❶❷❸.

Use SociologyNow to help you make the grade on your next exam. When you are finished reading this chapter, go to the chapter review for instructions on how to make SociologyNow work for you.

difficulties: balancing the demands of family and employment, coping with the interpersonal conflicts caused by changing expectations, and striving to make ends meet in families without sufficient financial resources. These changes bring new questions to the sociological study of families. •••

Defining the Family

Sociologists address the family as a *social institution*. They are interested in how changes in the society affect families and how families, in turn, affect people's experiences in the society. People commonly believe that families are shaped by the personalities of family members. Individual personalities affect family relationships, and personality is heavily influenced by family experience. Sociologists who study families, however, tend to focus on different issues, such as how families are shaped by the economic, political, social, and cultural institutions of the society.

As a social institution, the family is an established social system that changes and persists over time. Even with social change, the family, like other institutions, is somewhat stable. This does not mean that all families are the same. Studying the family as an institution simply recognizes that families are organized in socially patterned ways. Institutions are "there"; we do not reinvent them each day upon waking, yet they are constantly evolving as people make adaptive changes. Institutions shape and direct our action, and they make it appear that the only options available to us are those deemed acceptable by society. Perhaps you find it hard to imagine living in a family where the husband is expected to have both a wife and a concubine or where children do not live with their parents. The institutional structure of the family in this society does not support such family practices. Institutions shape both the form of families and the expectations we have for family life.

THINKING SOCIOLOGICALLY

Interview three people from different family backgrounds (for example, a White, Black, and Latino student or people from a stepfamily, a female-headed household, and a male-headed household). Ask each about how his or her family is organized. Based on your interviews, how would you describe the *kinship system* of each family? What does each reveal about the *social structure* of families and how they adapt to social change?

Like other institutions, families are shaped by their relationship to systems of inequality in society. Race, class, gender, and age stratification affect how society values certain families. These systems of stratification

Courtesy of Marion and Eilene Hull

Kinship systems vary in different societies, but families usually have the responsibility of caring for different generations.

influence the resources available to different families and influence the power that individual members have within families. Children, for example, have fewer rights than adults; the very old are often relatively powerless within families. Heterosexual privileges also shape the resources available to certain families because our social institutions presume that families will be heterosexual; this sociological insight has also been revealed in the public debates about gay marriage, with advocates of legalizing gay marriage seeking the same rights that other families have.

The family is also intertwined with other social institutions in the society, such as law, religion, and education. Changes in one institution affect other institutions. For example, economic changes can greatly affect families during changes in patterns of employment. The employment status of a family member affects not only the financial condition of the household, but also the social dynamics within the family. Similarly, differences in religious values affect family behaviors, including reproduction, marriage, divorce, and sexual behavior.

The effect of social institutions on the family even extends to things considered private. For example, the family is closely regulated by the government. Why else would one need a license to marry? Legally married, heterosexual couples can file joint tax returns, but in most states gay and lesbian couples who own property together and share expenses and income cannot. In this case, the government is dictating what relationships can be considered a family. The poorest families are those most likely to experience the intrusion of government because for them state intervention in family life can be an everyday fact of life.

Given the diversity among families, how do we define the family? The family has traditionally been defined as a social unit of people related through marriage, birth, or adoption who reside together in officially sanctioned relationships and who engage in economic cooperation, socially approved sexual relations, and reproduction and child rearing (Gough 1984). Not all families fit these conditions, however. Following divorce, for example, the family does not typically share a common residence. Some working parents leave their children in the care of others as they pursue seasonal or regional employment, often far from their families (including across national borders). This pattern is most common among immigrant groups, racial–ethnic families, and now, transnational families, as will be discussed later.

As social scientists have become more aware of the diversity in family life, they have refined their analyses to better show the different realities of family situations. The **family** is now broadly defined to refer to a primary group of people—usually related by ancestry, marriage, or adoption—who form a cooperative economic unit and care for any young (and each other); who consider their identity to be intimately attached to the group; and who are committed to maintaining the group over time (Lamanna and Riedmann 2003: 10). The U.S. Census Bureau, in collecting data about the U.S. population, also refers to **households,** defined as all of the people who occupy a housing unit. This can include family households but is a broader term meant to include the diverse array of living arrangements that characterize modern life.

THINKING SOCIOLOGICALLY

Identify two popular family shows on television. Count how often the *family ideal* is portrayed by these shows. What do your observations reveal about how the family ideal is communicated through the popular media?

Comparing Kinship Systems

Families are part of what are more broadly considered to be kinship systems. A **kinship system** is the pattern of relationships that define people's family relation-

ships to one another. Kinship systems vary enormously across cultures and at different times. Thus, although most societies recognize marriage in some form as a durable relationship between men and women, in some societies, marriage is seen as a union of individuals; in others, marriage is seen as creating alliances between groups. In some societies, maintaining multiple marriage partners may be the norm; in others, marriages are arranged—either by parents or a marriage broker (Croll 1995). At the heart of all these diverse family patterns are the social norms and structures associated with kinship systems.

Kinship systems can generally be categorized by the following features:

- how many marriage partners are permitted at one time;
- who is permitted to marry whom;
- how descent is determined and how property is passed on;
- where the family resides; and,
- how power is distributed (O'Kelly and Carney 1986).

Number of Marriage Partners **Polygamy** is the practice of men or women having multiple marriage partners. Polygamy usually involves one man having more than one wife—technically referred to as **polygyny.** (People commonly use the term polygamy mistakenly when they are actually referring to polygyny.) **Polyandry** is the practice of a woman having more than one husband—an extremely rare custom.

Within the United States, polygynous marriage was once practiced among the Mormons, whose religious doctrine allowed men to have more than one wife. In 1890, polygyny was prohibited by the Mormon Church and had previously been outlawed by the U.S. Congress. The theological basis of polygyny is the Mormon belief that whole families should live as a unit under priestly authority in the afterlife. Mormon theology defines exultation (a state of extreme religious joy) as dependent on the number of children a man fathers and, for women, the number they bear. Under polygyny, men were assured more children than they could have with only one wife. Very few Mormons actually practiced polygyny. Plural marriage had to be approved by the church and was mostly confined to the church leadership. It is practiced now only by a few Mormon fundamentalists (about 2 percent), who do so without official church sanction (Bachman and Esplin 1992; Driggs 1990; Brooke 1998) or legal recognition.

Although polygyny is most commonly associated with Mormons, Old Testament patriarchs also had plural wives. Other groups have practiced this form of marriage in Muslim societies, such as in Morocco,

Tunisia, Egypt, and Kuwait, although mostly among the elite class. Polygyny was also historically linked to high prestige among Muslim men because men with the most wives (and presumably the greatest sexual drive) held the highest social status (Mernissi 1987). Polygyny also has an economic function. It provides wealthy men who can afford multiple wives with a source of cheap labor—their wives. As a result, polygyny has been most common in agrarian societies, where a large and inexpensive labor force is needed—societies where the birthrate is also high.

Monogamy is the practice of a sexually exclusive marriage with one spouse. It is the most common form of marriage in the United States and other Western industrialized nations. In the United States, monogamy is not only a cultural ideal, but is also prescribed through law and promoted through religious teachings. Lifelong monogamy is not always realized, however, as evidenced by the high rate of divorce and extramarital affairs. Many sociologists characterize modern marriage as *serial monogamy,* in which individuals may, over a lifetime, have more than one marriage but maintain only one spouse at a time.

Who Marries Whom? In addition to determining how many marital partners one can have, kinship systems determine whom one can marry. In contemporary U.S. society, people are expected to marry outside their own kin group; in other societies, people may be expected to marry within the kinship network. **Exogamy** is the practice of selecting mates from outside one's group. **Endogamy** is the practice of selecting mates from within one's group. The group may be based on religion, territory, racial identity, and so forth.

In the United States, neither exogamy nor endogamy is mandated by law, although some religious doctrines condemn marriage outside the faith. Even if certain forms of marriage are not explicitly outlawed, society establishes normative expectations for an appropriate marriage partner. The *incest taboo,* generally considered to be universal, for example, is a cultural norm forbidding sexual relations and marriage between certain kin. Research on incest shows, however, that despite this cultural taboo, incest does occur with alarming frequency in this society and others (Russell 1986; Margolis 1996; Adkins and Merchant 1996). Many cultures have a tradition of *arranged marriages,* in which parents (or other elders) make rationally calculated choices about the appropriate marriage partner for their children. Arranged marriages work like business arrangements, perhaps even including a broker, in which a sum of money or property brought to the marriage may be part of the calculation about an appropriate partner.

In general, people in the United States clearly tend to select mates with social characteristics similar to their own—a pattern referred to as **homogamy.** Whether it is class, race, religion, or educational background, people tend to choose partners who have a similar background (Kalmijn 1991). Although no laws in the United States prohibit marriage between people of different social rank, such marriages are relatively infrequent.

Most marriages are also between members of the same racial group. Sociologists have often used public attitudes about intermarriage as being indicative of the public's support for or against integration, because marriage is the ultimate loss of social distance between the so-called races. Attitudes toward intermarriage have changed dramatically during the past few decades. In 1972, 73 percent of White Americans disapproved of intermarriage, compared to 33 percent now; in 1972, 24 percent of Black Americans disapproved of intermarriage, compared to 17 percent now (Schumann et al. 1997). Even with changing attitudes, interracial marriage is infrequent, although increasing. Interracial couples are now about 2 percent of married couples (U.S. Census Bureau 2004).

Some groups are more likely to intermarry than others. African Americans and White Americans are the least likely to marry outside their group. Native Americans are the most likely to intermarry, with approximately one-third of Native Americans marrying outside their group (McLemore 1994). There is also an increase of different Asian Americans (Chinese, Korean, Japanese, and others) marrying each other, in part as recognition of their common identity as Asian American (Shinagawa and Pang 1996).

The growth of interracial marriage is contributing to a more multiracial and multicultural society. The presence of mixed-race people brings new issues into sociological focus. Most obviously, as we saw in Chapter 11, it highlights the inadequacy of seeing race as a fixed category. Mixed-race people must negotiate a complex route to identity formation, but their experience provides us with new perspectives on the meaning of race and culture in society (Root 2001, 1996; Storrs 1999; Waters 1998).

Although interracial marriages are not common, historically a tremendous amount of energy has been put into preventing them. **Antimiscegenation laws** have prohibited marriage between various groups, including between Whites and African Americans and between Whites and Chinese, Japanese, Filipinos, Hawaiians, Hindus, and Native Americans (Takaki 1989). Laws against intermarriage have been less rigidly enforced when women of color married White men (Kennedy 2003).

Antimiscegenation laws are significant not only for how they regulated marriage, but also for the importance they have had in establishing definitions of racial groups. As we saw in Chapter 11, the concept of race is socially constructed. To prohibit marriage across racial groups, courts first had to decide who belonged to what race. Definitions of race varied from state to state. In North Carolina and Tennessee, for example, state laws prohibited the marriage of Whites to persons whose ancestry was one-eighth "Negro." Texas and Georgia prohibited marriage between Whites and a person with any "Negro" ancestry. California enacted a law in 1880 prohibiting marriage between a White person and any "negro, mulatto, or Mongolian" (Takaki 1989: 102). Later, when the goal was prohibiting marriages between Whites and Japanese Americans, anyone considered "Mongolian" was not allowed to marry a White person. In 1934, when a debate arose about whether "Mongolians" included Filipinos, the courts cited an earlier precedent that defined Caucasian as excluding the Chinese, Japanese, Hindus, American Indians, and Filipinos. Marriage between these groups and Whites was outlawed. Soon thereafter, another California law forbidding marriage between Whites and "Malays" (Malaysians) was written to include Filipinos explicitly in this ban (Takaki 1989). These laws were finally declared unconstitutional in 1967—quite recently. The enactment of such restrictions provides examples of how the state regulates marriage according to the social norms of the time.

Property and Descent Kinship systems also shape the distribution of property in society, most notably by prescribing how lines of descent are determined. In **patrilineal kinship systems,** family lineage (or ancestry) is traced through the family of the father. (The prefix *patri* means "of the father.") Offspring in patrilineal systems are typically given the name of the father. **Matrilineal kinship systems** are those in which ancestry is traced through the mother. Among Native American groups, family ancestry is often traced through maternal descent (Allen 1986). Among Jewish people, one is considered Jewish when born to a Jewish mother. Others may convert to Judaism, but orthodox Judaism limits Jewish identity to matrilineal descent. This practice evolved under early Talmudic law as the result of widespread persecution and military conquest of Jews. The prevalence of rape of Jewish women following military conquest made paternity difficult to establish. Thus, a child's proven religious identity was established on the basis of the mother's faith.

Many kinship systems involve a mix of both systems. In **bilateral kinship systems,** descent is traced both through the father and the mother. Bilateral kinship is practiced in the United States, although there is a

UNDERSTANDING DIVERSITY

Interracial Dating and Marriage

Picture this: A young couple, stars in their eyes, holding hands, intimacy in their demeanor. Newly in love, the couple imagines a long and happy life together. When you visualize this couple, who do you see? If your imagination reflects the sociological facts, odds are that you do not imagine this to be an interracial couple. Although interracial couples are increasingly common (and have long existed), people are more likely to form relationships with those of their same race—as well as social class, for that matter. What do sociologists know about interracial dating and marriage?

First, patterns of interracial dating are influenced by both race and gender. Black men (81 percent) are those most likely to say they would date a person of another race, although they and White women are the least likely to say they would marry someone of another race. White women are least likely to say they would date someone of another race. Both Blacks and Whites in interracial relationships report negative responses

from their families. The majority of families are not extremely hostile but put strong pressure on the interracial couple (Majete 1999). And although most Blacks and Whites profess to have a color-blind stance toward interracial marriages, when pressed, they raise numerous qualifications and concerns about such pairings (Bonilla-Silva and Hovespan 2000). Among college students, approval of family and friends is the strongest indicator of one's attitude toward interracial dating. Racial minority students are more accepting of interracial dating than are White students; students in the Greek system of fraternities and sororities are less accepting than those who are not (Khanna et al. 1999).

Regardless of these attitudes, interracial marriage is on the rise, although still a small percentage of marriages formed. Patterns of intermarriage vary by age and other social characteristics. Among African Americans, for example, younger men and women are more likely to intermarry than are older African

Americans. Among Hispanics, Mexican Americans are least likely to marry outside their own group (Sung 1990). The most likely marriages are between Black men and White women and between Hispanics and non-Hispanics. National data on Asian American marriage is limited, but studies find that Asian Americans are increasingly likely to marry other Asian Americans, though often someone of a different Asian heritage (Shinagawa and Pang 1996). The data on interracial dating and marriage show how something seemingly "uncontrollable" like love is indeed shaped by many sociological factors.

Source: Khanna, Nikki D., Cherise Harris, and Rana Cullers. 1999. "Attitudes toward Interracial Dating." Paper presented at the annual meeting of the Society for the Study of Social Problems; Bonilla-Silva, Eduardo, and Mary Hovespan. 2000. "If Two People Are in Love: Deconstructing Whites' Views on Interracial Marriage with Blacks." Paper presented at the annual meeting of the Southern Sociological Society; Majete, Clayton A. 1999. "Family Relationships and the Interracial Marriage." Paper presented at the annual meeting of the American Sociological Association; Shinagawa, Larry, and Gin Yong Pang. 1996. "Asian American Panethnicity and Intermarriage." *Amerasia Journal* 22 (Spring): 127–152.

•••

patrilineal bias in that children commonly take the name of the father. Descent, however, is traced through both the mother and the father. The practice of children taking the father's name is also changing, as more women are keeping their names. Children's names also may be hyphenated with both parent's names, taken from the mother only or, in rare cases, simply made up. Note, however, that even when the mother and father have different names, children most typically are given the name of the father.

Place of Residence Residential patterns are also shaped by kinship systems. In the United States, newly married couples are expected to establish independent households if they can afford to do so. In many other societies, however, and among some groups within the United States, newly married couples often take up residence in the household of one spouse's parents. In **patrilocal kinship systems,** after marriage, a woman is separated from her own kinship group and resides with the husband or his kinship group. In **matrilocal kinship systems,** a woman continues to live with her family of origin. The husband resides with the wife and her family, although he does not give up membership in his own group. **Neolocal residence** is the practice of the new couple establishing their own residence. In most matrilocal societies, the husband retains his importance in the group of his birth and may exercise authority over his sisters and their children.

Who Holds Power? Finally, marriage systems vary according to who holds power in the marriage. A **patriarchy** is a society or group where men have power over women; conversely, in a **matriarchy** women hold power. Patriarchal societies are far more common than matriarchal societies. In **egalitarian societies** men and women share power equally, are equally valued by all societal members, have equal access to resources, and share decision making. Although women may have different roles than men in such societies, both men and women are perceived as contributing to the common good and are judged to have equal social worth (Leacock 1978). Among Eskimos, as an example, women traditionally sewed fishing nets and men fished (as did some women), but the activity of both was culturally defined as critical to economic survival (Hensel 1996).

In many Native American societies, gender roles were far more egalitarian than in the societies of White colonizers. In some Native American groups, women held political power in the sense that they had a significant voice in decisions affecting the group as a whole. In some cases, there were councils of women with an equal role in decision making to councils of men. Each group was seen as having different but complementary responsibilities. The colonists disrupted these forms of gover-

nance, finding that what they called "petticoat government" violated the norms of Christian patriarchy (Allen 1986: 32).

Extended and Nuclear Families

Another important distinction is whether family systems are extended or nuclear. These concepts refer to the whole system of family relationships and whether the family resides in extended or relatively small household units.

Extended Families **Extended families** are the whole network of parents, children, and other relatives who form a family unit. Sometimes extended families, or parts thereof, live together, sharing their labor and economic resources to survive. For example, extended families are common among the urban poor because they develop a cooperative system of social and economic support. Kin, in such a context, may refer to those who are not related by blood or marriage but who are intimately involved in the family support system and are considered part of the family (Stack 1974).

As an example, among African Americans, there are those who sociologists have termed *othermothers.* These are "women who assist bloodmothers by sharing mothering responsibilities" (Collins 1990: 119). Having othermothers is an adaptation to the demands of motherhood and work that characterize African American women's experience, given that they have always been likely to be employed and have families. This dual responsibility in families and in paid labor has meant that African American women have created alternative means of providing family care for children—a situation that the majority of all women now face. An othermother may be a grandmother, sister, aunt, cousin, or a member of the local community, but she is someone who provides extensive child care and receives recognition and support from the community around her.

Extended families are also found at the very top of the socioeconomic scale. For example, family compounds, such as the Kennedy estate in Hyannisport, Massachusetts, serve as community centers for extended kin groups (Baca Zinn and Eitzen 2002). Among the elite, extended family systems preserve inherited wealth, whereas among the poor, extended family systems contribute to economic survival. In sum, extended families provide a means of adaptation to economic conditions that require great cooperation within families.

The system of *compadrazgo* among Chicanos is another example of an extended kinship system. In this system, the family includes godparents, to whom the family feels a connection that equals actual kinship. The result is an extended system of connections between "fictive kin" (not related by birth but considered part of the family) and actual kin that deeply affects family relationships among Chicanos. Cultural traditions among Mexican Americans foster strong ties between families and godparents, but culture alone does not explain the prevalence of extended family networks among Mexican Americans (Baca Zinn and Eitzen 2002). Many of these families have migrated, and they rely on kin support for financial assistance, help finding housing and employment, and for assistance in child care and household work (Angel and Tienda 1982; Baca Zinn and Eitzen 2002). Not all Mexican American families have experienced migration, and middle-class Mexican Americans, like middle-class African Americans, are more likely to live in *nuclear families.*

Nuclear Families The **nuclear family** is comprised of one married couple residing together with their children. Like extended families, nuclear families developed in response to economic and social conditions, particularly industrialization. Before industrialization in Western societies, families were the basic economic unit of society, and large household units produced and distributed goods. Production took place primarily in the home, and all family members were economically vital. No sharp distinction was made between economic and domestic life, because household and production were one, whether in small communities or large plantations and feudal systems where slaves and peasants provided most of the labor. Women's role in the preindustrial family, though still marked by patriarchal relations, was publicly visible and economically valued. Women performed and supervised much of the household work, engaged in agricultural labor, and produced cloth and food. Although the tasks women, men, and children performed might differ, together they were an interdependent unit of economic production.

With industrialization, the transition to wages paid for labor created an economy based on cash instead of domestic production. Families became dependent on the wages that workers brought home rather than goods they were able to produce at home. Single women were among the first to be employed in the factories; the shift to wage labor was accompanied by a patriarchal assumption that men should earn the "family wage," that is, provide for dependents. Thus, men who worked as laborers were paid more than women (offering a great savings for industries employing women), and women became more economically dependent on men. At the same time, men's status was enhanced by having a wife who could afford to stay at home—a privilege seldom accorded to working-class or poor families. The family wage system has persisted and is reflected in the unequal wages of men and women today.

Another result of industrialization was the separation of the family and the workplace. Paid labor was performed in factories and public marketplaces. The shift to factory production moved workers out of the household, which soon created dual roles for women as paid laborers and unpaid housewives. Moreover, the invisibility of women's labor in the home eventually diminished their perceived status.

Racial–ethnic families have developed in the context of disruptions posed by the experiences of slavery, migration, and urban poverty. These experiences have created unique social conditions that affect how families are formed, their ability to stay together, the resources they have, and the problems they face. Chinese American laborers were explicitly forbidden to form families by state laws designed to regulate the flow of labor. Only a small number of merchant families were exempt from the law. Thus, the development of Chinese American families was channeled not only by racial exclusion, but also by class differences.

Under slavery, African American families faced a constant threat of disruption. Marriages among slaves were not officially recognized in law, because slaves were not considered citizens. Slaves formed families nonetheless, although the children of slave parents were the legal property of the slaveholder. Slave owners found it economically advantageous to allow the formation of slave families because the reproduction of new slaves added to their labor force. Slave owners maintained strict control over slave life and, when it served the owners' needs, would sell or separate family members. After slavery was abolished, African Americans faced new challenges in holding families together under new conditions of rural poverty and urban unemployment. African American men could typically find only unskilled or seasonal employment; African American women were most likely employed as domestic workers, a more year-round occupation. As a result, African American women were often the steady providers for their families, resulting in the strong role of women in African American families (Gutman 1976).

Families of other racial–ethnic groups have also been uprooted as workers have had to travel to find jobs. Many Mexicans who had settled in the Southwest, for instance, were displaced when Whites seized Mexican lands during the westward expansion of the United States. The loss of their land disrupted the family and kinship system they had developed, and they became more dependent on the dominant White society for economic survival. In the rapidly industrializing United States of the 1800s, many Mexican Americans were able to find work in the mines opening in the new territories or building the railroads spreading from the East toward the Pacific. Employers apparently thought that they had better control over laborers if their families were not there to distract them, so families were typically prohibited from living with the worker. As a result, prostitution camps were developed, which followed workers from place to place. Some wives also followed their husbands to the railroad and mining camps (Dill 1988).

Poor families often find it necessary for all members to work to meet the economic needs of the household. A family pattern in which three earners span different generations within the family (children, parents, and grandparents, for example), is part of many Cuban American families, one of the most economically successful immigrant groups.

Migration to a new land and exposure to new customs and needs also disrupt traditional family values. Among Korean immigrants, for example, the majority of Korean wives are employed as full-time workers—a change from traditional Korean values, according to which a wife is expected to be totally devoted to her family. Not all changes in customs and values are bad. In this example, Korean women gained new opportunities and the possibility for expanded social roles. Immigrant women who enter the labor force often change their expectations about their gender role, but usually they also find they must continue to perform traditional household tasks. Studies of Korean wives who are employed indicate that, after a while, they begin to feel an acute sense of injustice. Their experience is probably shared by women in other immigrant families (Kim and Hurh 1988: 162).

The experience of Chinese Americans is another example of how social policies shape the experience of family life. The Chinese Exclusion Act of 1882 prohibited the wives and children of resident Chinese laborers from entering the country. The result was an extraordinary sex imbalance in the Chinese American population—26.8 males for every female in 1890. The Chinese

Exclusion Act was buttressed by the Immigration Act of 1924, which prohibited Chinese women to enter the United States, making family formation impossible. As a result, Chinatowns in the early twentieth century were primarily bachelor societies (Glenn 1986; Takaki 1989). These historical examples show how the ability to form and sustain nuclear families is directly linked to the economic, political, and racial organization of society.

Sociological Theory and Families

The complexity of family patterns makes it impossible to understand families from any singular perspective. Is the family a source of stability or change in society? Are families organized around harmonious interests, or are they sources of conflict and differential power? How do new family forms emerge, and how do people negotiate the changes that affect families? These and other questions guide sociological theories of the family.

Sociologists who study the family have used four primary perspectives in their analyses. These are *functionalism, conflict theory, feminist theory*, and *symbolic interaction* (see Table 15.1).

Functionalist Theory and Families

According to functionalist theory, all social institutions are organized to provide for the needs of society. Functionalism also emphasizes that institutions are based on shared values among members of the society. Functionalist theorists interpret the family as filling particular societal needs, including socializing the young, regulating sexual activity and procreation, providing physical care for family members, assigning identity to people, and giving psychological support and emotional security to individuals. According to functionalism, families exist to meet these needs. Marriage is conceptualized as a mutually beneficial exchange wherein women receive protection, economic support, and status in return for emotional and sexual support, household maintenance, and the production of offspring (Glenn 1987). At the same time, in traditional marriages, men get the services that women provide—housework, nurturing, food service, and sexual partnership. Functionalists also see families as providing care for children, who are taught the values that society and the family purport to have.

When societies experience disruption and change, according to functionalist theory, institutions such as the family become disorganized, weakening the social consensus around

The family is the major institution where socialization of children occurs.

© Don Smetzer/Stone/Getty Images

which they have formed. Currently, some functionalists interpret the family as "breaking down" under societal strains, suggesting this breakdown is the result of the disorganizing forces that rapid social change fosters.

Functionalists also note that, over time, other institutions have begun to take on some functions originally performed solely by the family. For example, as children now attend school earlier in life and stay in school for longer periods of the day, schools (and other caregivers) have taken on some functions of physical care and socialization originally reserved for the family. The family's share of these functions has been dwindling, while other institutions have taken on more of the original functions of the family. Functionalists would say that the diminishment of the family's functions produces further social disorganization because the family no longer carefully integrates its members into society. To functionalists, the family is shaped by the template of society, and things such as the high rate of divorce and the rising numbers of female-headed and single-parent households are the result of social disorganization.

Conflict Theory and Families

Conflict theory makes different assumptions about the family as an institution, interpreting the family as a system of power relations that reinforces and reflects the inequalities in society. Conflict theorists are especially interested in how families are affected by class, race, and gender inequality. This perspective sees families as the units through which the privileges and the disadvantages of race, class, and gender are acquired. Families are essential to maintaining inequality in society be-

cause they are the vehicles through which property and social status are acquired (Eitzen and Baca Zinn 2004).

The conflict perspective also emphasizes that families in American society are vital to capitalism because the family produces the workers that capitalism needs. Accordingly, within families, personalities are shaped by adapting to the needs of a capitalist system. Thus, families socialize children to become obedient, subordinate to authority, and good consumers. Those who learn these traits become the workers and consumers that capitalism wants. Families also serve capitalism in other ways—for example, giving a child an allowance teaches the child capitalist habits for earning money.

Whereas functionalist theory conceptualizes the family as an integrative institution—it has the function of maintaining social stability—conflict theorists depict the family as an institution subject to the same conflicts and tensions that characterize the rest of society. Families are not isolated from the problems facing society as a whole. The struggles brought on by racism, class inequality, sexism, homophobia, and other social conflicts are played out within family life.

Feminist Theory and Families

Feminist theory has contributed new ways of conceptualizing the family by focusing sociological analyses on women's experiences in the family and by making gender a central concept in analyzing the family as a social institution. Feminist theories of the family emerged initially as a criticism of functionalist theory. Feminist scholars argued that functionalist theory assumed that the gender division of labor in the household

Table 15.1
Theoretical Perspectives on Families

	Functionalism	Conflict Theory	Feminist Theory	Symbolic Interaction
Families	Meet the needs of society to socialize children and reproduce new members	Reinforce and support power relations in society	Are gendered institutions that reflect the gender hierarchies in society	Emerge as people interact to meet basic needs and develop meaningful relationships
	Teach people the norms and values of society	inculcate values consistent with the needs of dominant institutions	Are a primary agent of gender socialization	Are where people learn social identities through their interactions with others
	Are organized around a harmony of interests	Are sites for conflict and diverse interests of different family members	Involve a power imbalance between men and women	Are places where people negotiate their roles and relationships with each other
	Experience social disorganization ("breakdown") when society undergoes rapid social change	Change as the economic organization of society change	Evolve in new forms as the society becomes more or less egalitarian	Change as people develop new understandings of family life

is functional for society. Feminists have also been critical of functional theory for assuming an inevitable gender division of labor within the family. Feminist critics argue that, although functionalists may see the gender division of labor as functional, it is based on stereotypes about men's and women's roles.

Influenced by conflict theory, feminist scholars see the family as not serving the needs of all members equally. On the contrary, the family is one primary institution producing the gender relations found in society. Feminist theory conceptualizes the family as a system of power relations and social conflict (Glenn 1987; Thorne 1993). In this sense, it emerges from conflict theory but adds the idea that the family is a gendered institution (see Chapter 12).

Symbolic Interaction Theory and Families

Sociologists have also used symbolic interaction theory to understand families. Remember that symbolic interaction emphasizes that the meanings people give to their behavior and the behavior of others is the basis of social interaction. Symbolic interactionists tend to take a more microscopic view of families and might ask how different people define and understand their family experience. They also study how people negotiate family relationships, such as deciding who does what housework, how they will arrange child care, and how they will balance the demands of work and family life.

To illustrate, when two people get married, they form a new relationship that has a specific meaning within society. The newlyweds acquire a new identity to which they must adjust. Some changes may seem very abrupt—a change of name certainly requires adjustment, as does being called a husband or wife. Some changes are more subtle, for example, how one is treated by others and the privileges couples enjoy (such as being a recognized legal unit). Symbolic interactionists see the married relationship as socially constructed; that is, it evolves through the definitions that others in society give it, as well as through the evolving definition of self that married partners make for themselves.

The symbolic interaction perspective emphasizes the construction of meaning within families. Roles within families are not fixed but will evolve as participants define and redefine their behavior toward each other. This perspective is especially helpful in understanding changes in the family because it supplies a basis for analyzing new meaning systems and the evolution of new family forms over time. Each theoretical perspective used to analyze families illuminates different features of family experiences.

A SOCIOLOGICAL EYE ON THE MEDIA

Idealizing Family Life

Cultural norms about motherhood and fatherhood come from many places, but the media is certainly a strong influence on how family ideals—and the ideals for mothers and fathers—are created in society. Media images of the family have certainly changed since the inception of television. In the 1950s, Hollywood Codes—that is, official rules in Hollywood about what could and could not be seen on television—meant that families were always shown as two-parent, marriage intact, father working, and mother staying at home. Parents never talked about sex; indeed, it appeared as though they never had it, because the Hollywood Codes required that scenes of passion could never be explicit and bedrooms, when shown, were supposed to be presented in an "innocent" manner. The codes went so far as to say that the use of bedrooms as a location was wrong in a comedy because their use suggested "sexual laxity and obscenity" (Martin 1970: 283).

Now, family images on television are more diverse: Fathers are shown in parenting roles; some families are divorced or re-blended following divorce; working mothers are typical, not exceptional; and children may be stepchildren. Not all TV families are heterosexual, though most are. Still, the media continues to construct an ideal for family life—one that continues to stereotype men and women in family roles. What does research show about these social constructions?

- Most family characters are middle-class.
- Men appearing with children are most likely to be shown outside, less likely to be portrayed doing family chores; they are also more likely to be seen with boys, not girls.

- Fathers are infrequently seen with infants.
- Fathers are shown playing with, reading to, talking with, and eating with children, but not preparing meals, cleaning house, changing diapers, and so forth.
- Women are disproportionately shown in family settings in the media.

What gender stereotypes do such images project? How do they influence people's views of ideal family roles? How realistically do they portray the actual gender division of labor in families? What race and class images confound these results?

Sources: Kaufmann, Gayle. 1999. "The Portrayal of Men's Family Roles in Television Commercials." *Sex Roles* 41 (September): 439–458; Coltrane, Scott, and Melinda Messineo. 2000. "The Perpetuation of Subtle Prejudice: Race and Gender Imagery in 1990s Television Advertising." *Sex Roles* 42 (March): 363–389.

•••

Diversity Among Contemporary American Families

Today, the family is a rapidly changing institution of society (Lempert and DeVault 2000). Family forms emerge as adaptations to new societal conditions. Central to sociological analyses of the family is that *families are systems of social relationships that emerge in response to social conditions and that, in turn, shape the future direction of society.*

Among other changes in the family, families today are smaller than in the past, with fewer births that are more closely spaced. These characteristics of families vary by social class, region of residence, race, and other factors. Because of longer life expectancy, childbearing and child rearing now occupy a smaller fraction of the adult life of parents. Death, once the major cause of early family disruption, has been replaced by divorce. In earlier periods, death (often from childbirth) was more likely to claim the mother than the father of small children, and men in the past would have been more likely to raise children on their own after the death of a spouse. That trend has reversed, and it is now women who are more likely to be widowed with children (Rossi and Rossi 1990).

Demographic and structural changes have resulted in great diversity in family forms. Compared to thirty years ago, married couples make up a smaller proportion of households; single-parent households have increased dramatically, and divorced and never-married people make up a larger proportion of the population (Fields and Casper 2001; see Figure 15.1). Overall, married-couple families are three-quarters of household types, but this varies significantly by race. Single-parent households (typically headed by women), post-childbearing couples, gay and lesbian couples, and those without children are increasingly common. It is now also becoming common for people to spend more years caring for their elderly parents than they did raising their children. These changes mean that few families actually experience the family ideal often extolled by politicians as the only desirable family form.

Female-Headed Households

Perhaps one of the greatest changes in family life has been the increase in the number of families headed by women. Half of all children can expect to live with only one parent at some point in their lives. The odds of living in a single-parent household are even greater for African American and Latino children (see Figure 15.2), although the number of households headed by women

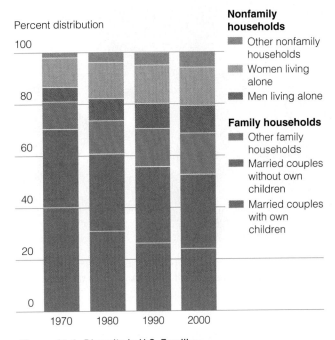

Figure 15.1 Diversity in U.S. Families

Source: Fields, Jason, and Lynne M. Casper. 2001. *America's Families and Living Arrangements: March 2000.* Current Population Reports, P20–537. Washington, DC: U.S. Bureau of the Census, p. 3.

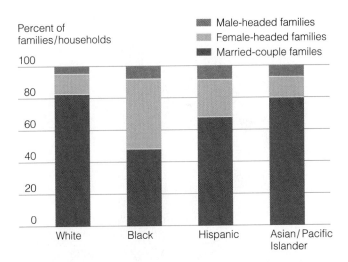

Figure 15.2 Family Structure by Race

Source: Fields, Jason, and Lynne M. Casper. 2001. *America's Families and Living Arrangements: March 2000.* Current Population Reports, P20–537. Washington, DC: U.S. Census Bureau, p. 2.

has risen across all racial groups (Bock 2000; Fields and Casper 2001).

The two primary causes for the growing number of women heading their own households are the *high rate of pregnancy among unmarried teens* and the *high divorce rate,* with death of a spouse also contributing. Although the rate of pregnancy among both White and Black teenagers is lower today than it was in 1960, the

Female-headed families are more common among some groups than others. Twenty-three percent of Hispanic families are headed by women.

© Andrew Rarney/Stock Boston

proportion of teen births that occur outside marriage has increased. Teens who become pregnant are now less likely than in the past to marry so that the number of never-married mothers is now higher (Fields and Casper 2001).

Regardless of race, teen mothers are among the most disadvantaged groups in society. They find it difficult to get jobs, their schooling is often interrupted (if not discontinued), and they are unlikely to receive child support (Kaplan 1996). Teen fathers also are less likely to complete school and therefore they have lower earnings (Nock 1998). The likelihood of poverty is great for teen parents, many of whom come from families that are already poor. (For more discussion on teen pregnancy, see Chapter 13.)

Divorce, the second reason for the increase in the number of female-headed households, also contributes to the growing rate of poverty among women. Most women see a substantial decline in their income in the

MAPPING AMERICA'S DIVERSITY

MAP 15.1 Births to Teenage Mothers

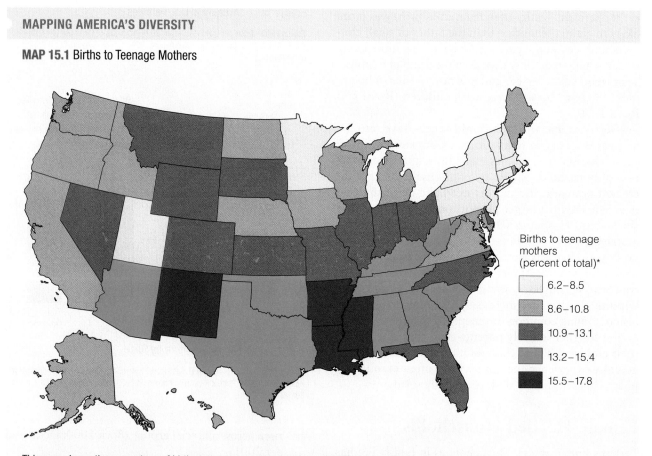

Births to teenage
mothers
(percent of total)*

	6.2–8.5
	8.6–10.8
	10.9–13.1
	13.2–15.4
	15.5–17.8

This map shows the percentage of births to teenage mothers as a percent of all births. Although you cannot see the historic change from this map, it is the case that the rate of teen pregnancy has been declining in the United States, but so has the likelihood of marriage for teen mothers who give birth. These data include births to both married and unmarried teen mothers. If you were trying to explain why the rates of teen pregnancy vary from state-to-state, what factors would you want to examine and why?

Data: From the U.S. Census Bureau. 2004. *Statistical Abstract of the United States, 2003.* Washington, DC: U.S. Government Printing Office, p. 78.

year following divorce (Hoffman and Duncan 1988; Peterson 1996a, 1996b; Weitzman 1985). In addition, women who marry but later divorce are more likely to experience poverty than never-married women (Lichter et al. 2003). Contrary to popular belief, most men also experience a decline in their standard of living after divorce, primarily due to the loss of wives' income. Men's loss, however, is not as great as what women experience (McManus and DiPrete 2001). Following divorce, most men pay very little support, either to their wives or to their children. Following divorce, of those supposed to receive child support payments (almost all of whom are mothers), only two-thirds actually receive any and only about half receive all that is due. The average amount paid amounted to only $3787 in a year (U.S. Census Bureau 2004).

DEBUNKING SOCIETY'S MYTHS

Myth: If a teen woman who gets pregnant would marry the father of the child, a more stable family would be formed.

Sociological perspective: Because the social conditions that produce unstable families (poverty, unemployment, and low wages, to name a few) are more likely among teens, marriage for young people is more unstable than other marriages (Kaplan 1996).

Another cause for the large number of female-headed households among African Americans and Latinos is the economic status of minority men. High unemployment among young African American men makes marriage unlikely because, according to some sociologists, women are not likely to marry men who are economically unstable. What further diminishes the pool of potential African American husbands is their relatively low life expectancy and high rate of incarceration (Thornberry et al. 1997; Wilson 1987).

Many people see the rise of female-headed households as representing a weakening of social values. An alternative view is that the rise of female-headed households reflects the growing independence of women, some making decisions to raise children on their own. Not all female-headed households are women who remain single, however. Many are divorced and widowed women, whose circumstances may be substantially different from those of younger, never-married women.

Some claim that female-headed households—or father-absent families—are linked to problems such as delinquency, the school dropout rate, poor self-image, and other social problems. Sociologists, however, have not found the absence of men to be the only basis for such problems. Instead, economic pressure faced by female-headed households, compared with male-headed families, puts female-headed households under great strain—the threat of poverty being by far the greatest

problem they face (see Figure 15.4 on page 408). As we saw in Chapter 9, one-third of households headed by women live below the poverty line. The problem is not the makeup of households headed by women, it is the fact that they are so likely to be poor (Brewer 1988).

DEBUNKING SOCIETY'S MYTHS

Myth: Father absence is the cause of numerous social problems; if these fathers would just adopt "family values," families would be stronger and children wouldn't get into so much trouble.

Sociological perspective: Research on never-married, poor, noncustodial African American fathers finds that they want to be able to provide for their children; that they spend a lot of time with their children; and they want to be good "daddies." Understanding what fatherhood means to them requires understanding the context in which they live (Hamer 2001).

Although the majority of single-parent families are headed by women, families headed by a single father are also increasing. In most of these families, the father has gained custody of the children following a divorce or after the mother has died. Unlike female-headed households, where there is typically not a man to help with housework and children, single fathers commonly get help from women—either girlfriends, daughters, or mothers (Popenoe 2001; Hilton et al. 2001). Mothering, however, is not exclusively a skill restricted to women. Single fathers report feeling competent as parents. Compared with married fathers, single fathers spend more time with their children, report more sharing of feelings between father and child, are more likely to stay home with children when they are sick, and take a more active interest in their children's out-of-home activities (Risman 1987, 1986).

Despite social stereotypes, many if not most unwed fathers try to take some responsibility for their children, but the likelihood of economic disadvantage, especially among young, unwed fathers, is high. Most young unwed fathers are generally less well educated than other groups, have limited employment prospects, and are more likely to engage in crime (Lerman and Ooms 1993). These conditions make it difficult for these fathers to support their children, particularly if they are young and poor.

Married-Couple Families

The marriage relationship is based on intimacy, commitment, and love—two individuals coming together to form a complex and long-term relationship. How well they can do this, though influenced by their personalities and past histories, is the result of sociological factors, including age at marriage, economic resources, religion, and family histories (Baca Zinn and Eitzen 2002).

Social speedup has fallen particularly on women, although most parents find themselves having to juggle multiple tasks when combining work and family commitments.

For many, marriage is a beneficial and satisfying relationship, evidenced by the fact that married people generally have better mental and physical health than single people (Waite and Gallagher 2000). Married people also tend to have more economic resources than others. On the other hand, marriage can be a source of conflict and disappointment. Moreover, men and women have different experiences within marriage, with the benefits of marriage generally accruing more to men than women. Men get greater health benefits from marriage and as a rule work less, at least when considering both employment outside the home and housework and child care done inside the home.

Among married-couple families, a significant change in recent years has been the increased participation of women in the paid labor force. As we saw in Chapter 9, families sustain a median income level by having both husband and wife in the paid labor force. Although the increase in women's work is most dramatic among White families—minority and poor families have long had women in the labor force—most families are experiencing substantial speedup. This is reflected in longer hours worked—especially among women (Mishel et al. 2003).

Women in particular work a "second shift" of unpaid household work even when they also have paid employment. Arlie Hochschild has studied the effects of the second shift on dual-career families. She finds that women who try to become "supermoms" report feeling numb and out of touch with their feelings. Women develop a variety of strategies for coping with the combined demands of work and family. Some women feign helplessness over certain tasks as a way to get men to do them; others avoid conflict by taking on more work;

others cut back on their expectations or reduce their hours of paid employment (which in turn causes more household stress, given the income loss). Often women turn to women friends and family members for help or, if they have the economic ability, pay other women to work for them. The result is that many find life in the family to be like work, whereas work is where many find personal fulfillment, recognition, and emotional support (Hochschild 1997, 1989; see also Chapter 18).

Women's labor force participation has created other changes in family life. One example is the number of married couples who have *commuter marriages,* an arrangement that typically arises when work requires one partner in a dual-earner couple to reside in a different city much of the time. Commuter marriage tends to be associated with dual-career couples, separated by job transfers or jobs too distant for a daily commute. The common image of a commuter marriage consists of a prosperous professional couple, each holding important jobs, flashing credit cards, and using airplanes like taxis. But commuter marriages also occur among the working class when one or both partners have to seek work in a place distant from the family's residence. Agricultural workers may follow seasonal jobs; although their commute is less glamorous than that of professional spouses, they are commuting nonetheless. When all types of commuter marriages are considered, this form of marriage is more prevalent than typically imagined.

The movement of one or more members of a family in search of employment strongly affects family relations. Migration includes any movement through which people are seeking work in a new geographic area. This can include professional workers, people in the skilled trades, service workers, and domestic and farm laborers. Living in a new culture results, possibly changing the family roles of men and women. Faced with the usual routines of domestic life (cooking, child care, cleaning, and so on) but without each other to depend on, men and women learn new behaviors. Women learn to act more assertively and autonomously; men learn to cook and clean up after themselves. These behavioral changes can also change the expectations that men and women have, making the family more egalitarian when they are reunited (Hondagneu-Sotelo 1994, 1992).

Stepfamilies

Stepfamilies are becoming more common in the United States, matching the rise in divorce and remarriage. Stepfamilies take numerous forms, including married adults with stepchildren, cohabitating stepparents, and stepparents who do not reside together (Stewart 2001).

When two family systems are combined or when new people join an existing family system, stepfamilies can face a difficult period of readjustment. Parents and children discover that they must learn new roles when they become part of a stepfamily. Children accustomed to being the oldest child in the family, or the youngest, may find that their status in the family group is suddenly transformed. New living arrangements may require children to share rooms, toys, and time with people they perceive as strangers. The parenting roles of mothers and fathers suddenly expand to include more children, each with complex needs. Jealousy, competition, and demands for time and attention can create tense relationships within stepfamilies.

The problems are compounded by the absence of norms and institutional support systems for stepfamilies. Even the norms of language are vague in describing stepfamilies. When a woman's former husband remarries, how does the woman refer to the husband's new wife? On visitation weekends, she may take care of the woman's children more than the ex-husband does (Coontz 1992). There are no norms to follow, and people have to adapt by creating new language and new relationships. Many develop strong relationships within this new kinship system. Others find the adjustment extremely difficult, resulting in a high probability of divorce among remarried divorced couples with children (Baca Zinn and Eitzen 2002).

Gay and Lesbian Households

The increased visibility of gay and lesbian households challenges the traditional heterosexual understanding of the family. Although, in most states, lesbians and gays do not have the official supports of social institutions, many gay and lesbian couples form long-term, primary relationships that they define as marriage. Like other families, gay and lesbian couples share living arrangements and household expenses, make decisions as partners, and in many cases, raise children (Dalton and Bielby 2000; Dunne 2000).

Researchers have found that gay and lesbian couples tend to be more flexible and less gender-stereotyped in their household roles than heterosexual couples. Lesbian households, in particular, are more egalitarian than are either heterosexual or gay male couples. Unlike heterosexual couples, for lesbian couples, money has little effect on the balance of power in the relationship or on the couple's feelings about each other. Among gay men, however, the highest earner usually has more power in the relationship. Some researchers have concluded from these results that, regardless of their sexual orientation, men have

been socialized to believe that money equals power. Lesbians, on the other hand, tend to be critical of the power relationships between men and women in marriage and actively construct different household relationships (Jacobs 1997).

At the same time, lesbian families are shaped by some of the same structural features of society that shape heterosexual families. Sociologist Maureen Sullivan has studied lesbian coparents—those who together are raising children—finding that these couples tend to have gender equality within their households, provided both partners are employed. Nonetheless, patterns of employment still shape their relationship. Where one partner is the primary breadwinner and the other the primary caregiver for children, the partner staying at home becomes economically vulnerable, less able to negotiate her needs, and more devalued as a person and daily contributor—just as in heterosexual relationships where women stay home and men are the primary breadwinners. Sullivan's research shows that social structural arrangements, not just the values of domestic partners,

Research finds that children raised in gay or lesbian households develop less gender-stereotyped roles and identities.

© Shelley Gazin/Corbis

shape the ability of individual people to influence family decisions, to be seen as an equal partner, and to have his or her needs satisfied (Sullivan 1996).

Gay and lesbian couples also face unique domestic problems because the dominant culture denies them benefits and privileges such as shared healthcare plans and other benefits that go along with a legally recognized marriage that heterosexual couples receive. An increasing number of businesses now allow same-sex partners to be enrolled in employee benefit plans, and some municipalities give legal recognition to gay marriage, but these remain the exception. Gay and lesbian couples are also frequently the targets of hostility, and children can be affected when custody disputes arise over whether a gay parent has the right to custody. Although the law does not formally deny custody to lesbian mothers or gay fathers, judges tend to take sexual orientation into account in awarding custody, even when there is no evidence it creates adverse effects on the children (Duran-Aydintug and Causey 1996; Crawford and Solliday 1996; Zicklin 1995).

Many people think that being raised in a gay or lesbian family has adverse effects on children, yet research does not support this claim. There is little difference in outcomes for children raised in gay or lesbian households and those raised in heterosexual households. What differences are found result from other factors—not simply the sexual orientation of parents. Thus, the homophobia directed against children in lesbian and gay families can stigmatize them in the eyes of others. But children raised in gay or lesbian families are less likely to develop stereotypical gender roles and are more tolerant and open-minded about sexual matters. They are no more likely than children raised in heterosexual families to become gay themselves (Allen 1997; Stacey and Biblarz 2001). If we lived in a society more tolerant of diversity, the differences that do emerge might be viewed as strengths, not deficits.

Figure 15.3 Acceptance of Gay Marriage

Source: Lyons, Lydia. 2003. "U.S. Next Down the Aisle Toward Gay Marriage?" Princeton, NJ: The Gallup Organization. Website: **www.gallup.com**

Gay and lesbian marriages are producing new social forms and new social debates. Should gay marriage be recognized in law? The Supreme Court of Massachusetts has ruled to recognize gay marriage, setting off a public debate about whether gay marriage should be legal and prompting conservative groups to propose that there should be an amendment to the U.S. Constitution defining marriage as only between a man and a women. Forty-two percent of the American public believes that gay marriages should be recognized as valid and given the same rights as heterosexual marriage; 55 percent disagree. A larger number (48 percent) believe that civil unions between gays should be legally valid, suggesting that the word *marriage* per se is imbued with meaning that many do not want to extend to gay couples (Moore and Carroll 2004). Gay marriage is also more acceptable in the eyes of younger people than to older groups, raising the question of whether over time social support for gay marriages will increase or whether as young people age, they will shift their values (see Figure 15.3). In the meantime, in most states and municipalities, gays and lesbians who form strong and lasting relationships do so without formal institutional support and, as a result, have had to be innovative in producing new support systems.

Singles

Single people (those never married) today are 28 percent of the population. Since 1970, the proportion of younger women (age 20 to 24) who have never been married has doubled; for women age 30 to 34, the number never married has tripled. Men, too, are more likely single—84 percent of those aged 20 to 24 (compared to 55 percent in 1970) and 30 percent of those aged

---

DEBUNKING SOCIETY'S MYTHS

Myth: Children raised in gay and lesbian families are likely to become gay.

Sociological perspective: Children raised in gay or lesbian families are no more likely to become gay than children raised in heterosexual families. The primary problem for children raised in gay and lesbian families is the prejudice directed against them (Stacey and Biblarz 2001).

30 to 34 (compared to 9 percent in 1970; Fields and Casper 2001). The rise in the number of never-married people is partially the result of men and women marrying at a later age. Longer life expectancy, higher educational attainment rates, and cohabitation contribute to this later age for a first marriage.

As sexual attitudes have moved toward greater permissiveness, many people find the same sexual and emotional gratification in single life as they would in marriage. Being single no longer holds the same stigma it once did, especially for women. Single women were once labeled "old maids" and "spinsters." Now they have the image of being carefree, sexually active, unencumbered by family obligations, and free-thinking, even though this stereotype is an inaccurate portrayal of the diversity within single lifestyles. Changes in social attitudes about single life seem to have had some effect on the happiness of singles as well. For many years, sociologists have documented a tendency for married people to report greater happiness than singles. Now, greater diversity in family forms means that personal happiness and satisfaction are found in a variety of lifestyles.

Among singles, Black women are likely to remain single longer than Black men or White women. Although Black women see marriage as a desirable goal, it is not regarded as an end in itself. They expect to work to support themselves and do not assume that someone else will be financially responsible for them. Many White working-class women are also raised to prepare themselves for work; however, somewhat different from Black women, they see their work as contributing to family income, not as the potentially sole source of economic support (Higginbotham and Weber 1992). Single people have been studied less often than married people, yet research is beginning to show that some remain single by choice. Embedded in this decision may be a critique of traditional marriage, a lack of available marriage partners, or pursuit of educational and career plans—a pattern that has been found among Chinese and Japanese American women but may be true of other singles as well (Ferguson 2000).

FORCES OF SOCIAL CHANGE

"Hooking Up": Dating and Mating on College Campuses

There was a time when dating on college campuses was a quite formal activity. There were clear social norms about who dated who, how the date was arranged (by the young man), and what was expected—perhaps a first kiss after several dates, then a more formal stage of courtship prior to marriage, and sex only after marriage. Does this sound like the current dating scene on your campus? Not likely!

A national study of a diverse group of women on eleven different campuses, coupled with a nationally representative sample of 1000 college women interviewed by telephone in 2000–2001, found the following about the current college "dating" scene:

1. "Hooking up"—meaning sexual interaction without commitment—is widespread on college campuses and influences the campus culture, although only a minority of students engage in it (about 40 percent of college women).

2. The meaning of "dating" varies for college women. For some, it means a highly committed relationship, including sex, but rarely going out on "dates." For others, it means just "hanging out." For some, dating and hooking up mean the same thing, but "dating" is not generally used to describe the relationship.

3. "Hooking up" also carries multiple meanings. For some it means kissing, for others it means sexual-genital play but not intercourse; for some, it means sexual intercourse. The vagueness of the term contributes to its becoming a shared cultural phenomenon.

4. Hooking up commonly occurs when both participants are drinking or drunk.

5. Two-thirds of women say hooking up made them feel desirable, but also awkward. Forty percent of women say they have experienced a hookup; 10 percent say more than six times. But women are wary of getting a bad reputation from "hooking up" too much.

6. It is rare for college men to invite college women out on a date; women also report that men do not acknowledge when they have become a couple, although men, not the women, make the decision about when they are committed.

7. Even with the hooking-up culture, the majority of college women still want to meet a spouse while at college, a result that is compromised by the average sex ratio on campus (seventy-nine men for every hundred women).

The sociologists who conducted this study did so for a conservative organization that interpreted the results as interrupting "an important pathway to traditional marriage" (Glenn and Marquist 2001: 7). They recommended more adult involvement in establishing dating norms on college campuses and that men take more initiative in dating.

What is the hooking-up culture on your campus? How do gender relations influence the culture? Given what you see and the preceding results, do you agree with the recommendations of these authors?

Source: Glenn, Norval, and Elizabeth Marquardt. 2001. "Hooking Up, Hanging Out, and Hoping for Mr. Right." New York: Institute for American Values. •••

Finally, a rising number of single people are remaining in their parents' homes for longer periods. Known as the *boomerang generation,* these young people return home in their twenties when they would normally be expected to live independently. The increased cost of living means that many young people find themselves unable to pay their own way, even after marrying or getting an education. They economize by joining the household of their parents. Many pregnant, unmarried women also stay in their parents' homes, forming extended households, often with multiple generations living in a single residence.

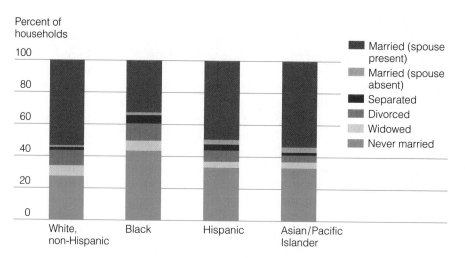

Figure 15.4 Marital Status of the U.S. Population

Data: From Fields, Jason, and Lynne M. Casper. 2001. *America's Families and Living Arrangements: March 2000.* Current Population Reports. Washington, DC: U.S. Bureau of the Census, pp. P20–537.

Cohabitation

Cohabitation (living together) has become common among single people. More than three times as many couples live together without being married now than was true in the 1970s. Some cohabiting couples have never been married; others live together while families are re-forming following divorce. Some couples cohabit because they are critical of the existing norms surrounding marriage (Elizabeth 2000). Estimates are that one-quarter of all children will at some time during their childhood live in a family headed by a cohabiting couple (Graefe and Lichter 1999), indicating how common cohabitation has become.

Those who cohabit tend to have more egalitarian attitudes toward gender roles (Sanchez et al. 1998; Barber and Axinn 1998). However, studies also find that cohabiters tend to develop a division of labor within household roles similar to that of married heterosexual couples. But cohabiters are more likely to remain together when they establish conditions of equality within the relationship (Brines and Joyner 1999).

Marriage and Divorce

Although there is extraordinary diversity of family forms in the United States, the majority of people will marry at some point in their lives (see Figure 15.4). The United States has the highest rate of marriage of any Western industrialized nation, as well as a high divorce rate. Even with a high rate of divorce, a majority of people place a high value on family life. People's opinions about their preferred family lifestyle have changed considerably in recent years. Most people know the importance of both parents working, although few think

it is ideal to have both parents working full time. But the public is not unified in this view. Men (45 percent) are more likely than women (38 percent) to think only one person should work, but most people (69 percent) say it does not matter which parent remains the full-time worker (McComb 2001). Opinion is one thing, reality another, since most families now find it necessary for both parents in two-parent families to work.

Marriage

The picture of marriage as a consensual unit based on intimacy, economic cooperation, and mutual goals is widely shared. But marital relationships also involve a complex set of social dynamics, including cooperation and conflict, different patterns of resource allocation, and a division of labor. For example, the amount of money a person earns can shape that person's relative power within the marriage (such as the ability to influence decisions and the degree of independence held by each partner). Studies have traditionally shown that greater marital power is exercised by the partner with the highest educational and occupational level, but this is not the case when wives earn more than husbands. Women who earn more than their husbands seem to have no greater power in marriage than women in more conventional marriages (Tichenor 1999; Komter 1989). Thus, gender—not just income and occupational level—influences power within marriage.

Such findings reveal that marriage is shaped by systems of social stratification—systems that, although part of the structure of society, have tremendous influence on the social interaction within people's day-to-day lives. Yet, people actively negotiate power relationships

Despite the high divorce rate, marriage is a cultural ideal and most people marry at some point in their lives.

in marriage; how they do so depends on the attitudes and beliefs each partner has about marriage and the resources each brings to the household (Pyke 1994).

Other factors also influence social dynamics within marriage, including the presence of children. Children can enrich marriage relationships, yet the time devoted to caring for children can detract from the quality of a marriage relationship. As a result, as reported in one study of over 6700 married couples, people report higher marital quality in the pre- and post-parental stages of the marriage (Benin and Robinson 1997). This does not mean that marriage suffers because of children, but it points to the time, cost, and stress that raising children involves, which can increase tension within marriage.

The values of partners, as well as the roles they play, influence their experience of marriage. Even with the ideal marriage defined as one based on consensus, harmony, and sharing, men and women have different experiences within marriages (Bernard 1972; Thorne 1993). As marriage roles change, the amount of work people do in families varies significantly for women and men. Women still do far more work in the home and have less leisure time.

Are men more involved in housework? Yes and no. Men report that they do more housework, but they devote only slightly more of their time to housework than in the past. Studies find a large gap between the number of hours women and men give to housework and

child care. Among couples where both partners are employed, only 28 percent share the housework equally. Fathers do more when there is a child in the house under two years of age.

For the most part, the increases in fathers' contributions to household work have been in the amount of child care they provide, not the housework they do. Interestingly, sociologists have found that the allocation of housework is greatly affected by men's and women's experience in their own families of origin; men and women who come from households with a more egalitarian division of labor are likely to carry this into their own relationships (Cunningham 2001). The arrival of a first child significantly increases the gender division of labor in households; women increase the housework they do and lessen their employment. There is far less effect for men (Sanchez and Thomson 1997). The end result is that men have about eleven more hours of leisure per week on average than women do (Press and Townsley 1998).

Despite a widespread belief that young professional couples are the most egalitarian, studies find that there is little difference across social class in the amount of housework that men do (Wright et al. 1992). With regard to race, African American husbands provide a greater share of housework than do White husbands. Latino households have more diversity in gender roles than stereotypes about machismo would lead us to believe (McLoyd 2000). Almost two-thirds (61 percent) of women say the amount of work they have to get done during the day is a cause of stress, and one-half say that they feel resentment, at least sometimes, about how little their mate helps around the house and about their lack of free time (Roper Organization 1995).

Although marriage can be seen as a romantic and intimate relationship between two people, it can also be seen within a sociological context. Marriage relationships are shaped by a vast array of social factors, not just the commitment of two people to each other. You see this especially when examining marital conflicts. Life events, such as the birth of a child, job loss, retirement, and other family commitments, such as elder care or caring for a child with special needs, all influence the degree of marital conflict and stability (Moen et al. 2001; Crowley 1998). As conditions in society change, people make adjustments within their relationships, but how well they can cope within a marriage depends on a large array of sociological—not just individual—factors.

Divorce

The United States leads the world not only in the number of people who marry, but also in the number of people who divorce. More than sixteen million people have divorced but not remarried in the population today; more women are in this group than men, since women are less likely to remarry following a divorce. Since 1960, the rate of divorce has more than doubled, although it has declined recently since its all-time high in 1980 (see Figure 15.5).

You will often hear that one in every two marriages ends in divorce, but this is a misleading statistic. The marriage rate is 8.4 marriages per 1000 people and the divorce rate, 4.0 per 1000 people (U.S. Census Bureau 2004). At first glance, it appears that there are half as many divorces as marriages. But these are divorces out of *all married couples, not just those formed in one year,* so divorce is not as widespread as half of all marriages.

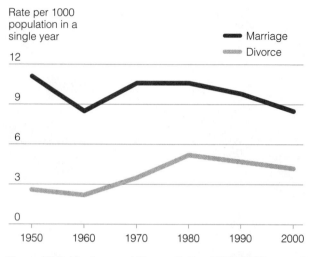

Rate per 1000 population in a single year

— Marriage
— Divorce

Figure 15.5 Marriage and Divorce Rates, 1950–2000

Data: U.S. Bureau of the Census. 2004. *Statistical Abstract of the United States 2003.* Washington, DC: U.S. Department of Commerce, p. 100.

Still, the rate of divorce is high and it has risen since 1950, though fallen again in recent years. The likelihood of divorce is not equally distributed across all groups,

---

DOING SOCIOLOGICAL RESEARCH

Men's Caregiving

Much research has documented the fact that women do the majority of the housework and child care within families. Why? Many have explained it as the result of gender socialization. Women learn early on to be nurturing and responsible for others, while men are less likely to do so. Yet, things are changing and some men are more involved in the "care work" of family life. What explains whether men will be more engaged in family care work?

This is the question that sociologists Naomi Gerstel and Sally Gallagher examined in their study of one hundred eighty-eight married people. They interviewed ninety-four husbands and ninety-four wives, married to each other; the sample was 86 percent White and 14 percent African American but was too small to examine similarities or differences by race. You might expect that men who had attitudes expressing support for men's family responsibilities would be more involved in family care (defined by Gerstel and Gallagher to include elder care, child care, and various household

tasks). But this is not what Gerstel and Gallagher found.

Gender attitudes did not influence men's involvement in caregiving. Rather, the characteristics of the men's families were the most influential determinant of their engagement in housework and child care. Men whose wives spent the most time helping kin and men who had daughters were more likely to help kin; having sons had no influence. But, men with more sisters spend less time helping with elder parents than men with fewer sisters. Furthermore, men's employment (neither hours employed, job flexibility, or job stability) had no effect on their involvement in care work.

Gerstel and Gallagher conclude that it is the social structure of the family, not only gender beliefs, that shape men's involvement in family work. As they put it, "It is primarily the women in men's lives who shape the amount and types of care men provide" (2001: 211). This study shows a most important sociological point: Social structure, not just individual attitudes, is the most significant determinant of social behavior.

Questions to Consider

1. Think about the care work that is done in your family (both your immediate and extended family). Who does it? Is it related to the *social organization* of your family, as Gerstel and Gallagher find in other families? *Keywords: care work, family division of labor, gender and housework*

2. Do you think that men's *gender identity* changes when they become more involved in care work? What hinders and/or facilitates men's engagement in this form of work? *Keywords: men and child care, men and nurturing*

We have included InfoTrac College Edition keywords at the end of each question to make it easier for you to find more to read on these topics. Go to www.infotrac-college.com, an online library, to begin your search.

Source: Gerstel, Naomi, and Sally Gallagher. 2001. "Men's Caregiving: Gender and the Contingent Character of Care." *Gender & Society* 15 (April): 197–217. ●●●

however. Divorce is more likely for couples who marry young (in their teens or early twenties). Second marriages are more likely to end in divorce than first marriages. Divorce also varies by race and class. It is somewhat higher among low-income couples, a fact that reflects the strains that financial problems put on marriages. Divorce is also somewhat higher among African Americans than among Whites, partially because African Americans make up a disproportionate part of lower-income groups. Hispanics have a lower rate of divorce than either Whites or Blacks, probably the result of religious influence. Recently, the divorce rate among Asian Americans has also risen, interpreted as the possible shedding of cultural taboos (Fields 2003; Armas 2003). This explanation seems supported by the fact that Asian Americans born in the United States are more likely to be divorced than Asians who immigrated (McLoyd et al. 2000). Those who come from divorced families are also more likely to be divorced themselves at some point, although children of divorced families are less likely to become married than those from other families (Wolfinger 2003).

THINKING SOCIOLOGICALLY

Identify a small sample of students in your school who have experienced *divorce* in their families. Design a short interview to test the research conclusions reported in this chapter on the effects of divorce on children. Do your results support the conclusions in this earlier research? What additional sociological insights does your research reveal?

A number of factors contribute to the current high rate of divorce in the United States. Demographic changes (shifts in the composition of the population) are part of the explanation. The rise in life expectancy, for example, has an effect on the length of marriages. In earlier eras, people died younger, and thus the average length of marriages was shorter. Some marriages that earlier would have ended with the death of a spouse may be now be dissolved by divorce. Still, cultural factors also contribute to divorce.

In the United States, individualism is highly valued, placing a high value on individual satisfaction within marriage. The cultural orientation toward individualism may predispose people to terminate a marriage in which they are personally unhappy. In other cultural contexts (including years ago in this society), marriage, no matter how difficult, may have been seen as an unbreakable bond, whether one was happy or not. As we have seen in the preceding data, for racial-ethnic groups in the United States, cultural factors can also influence the likelihood of divorce.

Changes in women's roles also are related to the rate of divorce. Women are now less financially dependent on husbands than they once were, even though they still earn less. As a result, the economic interdependence that bound women and men together as a marital unit is no longer as strong. Although most married women would be less well-off without access to their husband's income, they could probably still support themselves. This can make it possible for people to end marriages that they find unsatisfactory.

To people in unhappy marriages, divorce, though painful and financially risky, can be a positive option (Kurz 1995). The belief that couples should stay together for the sake of the children is now giving way to a belief, supported by research, that a marriage with protracted conflict is more detrimental to children than divorce. Although there are periodic public cries about the negative effect of divorce on children, many other factors influence the long-term psychological and social adjustment of children. Few children feel relieved or pleased by divorce; feelings of sadness, fear, loss, and anger are common, along with desires for reconciliation and feelings of conflicting loyalties. But most children adjust reasonably well after a year or so.

Moreover, children's adjustment is influenced most by factors that precede the divorce per se. The single most important factor influencing children's poor adjustment is marital violence and prolonged discord (Stewart et al. 1997; Arendell 1998; Furstenberg 1998; Cherlin et al. 1998; Amato and Booth 1997). The emotional strain on children is significantly reduced if the couple remain amicable. If both parents remain active in the upbringing of the children, the evidence shows that children do not suffer from divorce; especially important is the ability of the mother to be an effective parent after a divorce. This, in turn, is affected by the resources she has and her ongoing relationship with the father (Buchanan et al. 1996; Simons et al. 1996; Furstenberg and Nord 1985).

In the aftermath of divorce, many fathers become distant from their children. Sociologists have argued that the tradition of defining men in terms of their role as breadwinners minimizes the attachment they feel for their children. If the family is then disrupted, they may feel that their primary responsibility, as financial provider, is lessened, leaving them with a diminished sense of obligation to their children. A man may also distance himself as a way of minimizing or avoiding conflicts with his former wife. Salvaging a sense of masculinity may also play a role. Many fathers report feelings of emasculation when they are displaced from their home, especially if being the head of a household was a primary gender identity. In the absence of this, some men may reassert another aspect of their masculine role—independence (Arendell 1992). One consequence of distancing is the limited child support that families receive from fathers after divorce. Only two-thirds of those supposed to receive child support actually receive any; almost half of those receive only a partial amount (U.S. Bureau of the Census 2004).

Family Violence

Generally speaking, the family is depicted as a private sphere where members are nurtured and protected from the influences of the outside world. This situation is true for many, yet families can also be locales for violence, disruption, and conflict. Family violence is a phenomenon that, hidden for many years, has now been the subject of much sociological research. It can affect all age groups—not only partners in relationships, but also children and, as we saw in the prior chapter, older people in the form of elder abuse.

Partner Violence Estimates of the extent of partner violence are difficult to determine and notoriously unreliable because most cases of domestic violence go unreported. The National Violence Against Women Office estimates that 25 percent of women will be raped, physically assaulted, or stalked by an intimate partner in their lifetime. Twenty-two percent experience physical assault; 7–10 percent are raped by intimates; 5 percent will be stalked by an intimate partner. (The num-

Violence against women in the family is the result, sociologists argue, of the differential power of men in the family and women's economic and emotional dependence upon them.

© Ansell Horn

bers do not total to 25 percent since a given person may experience multiple violent events.) Men also experience partner violence, though far less frequently, and women who experience violence are also twice as likely to be injured (Brush 1990; Tjaden and Thoennes 2000; West 1998).

Violence also occurs in gay and lesbian relationships, although silence about the issue may be even more pervasive, given the marginalized status of gays and lesbians. Men living with male intimate partners are just as likely to be raped, assaulted, or stalked as women who live with men, but the incidence of violence against women by women intimates is about half as likely. Researchers conclude that this is because most domestic violence is committed by men, and violence is usually accompanied by emotionally abusive and controlling behavior. Partners who are jealous and dominating are the most likely perpetrators of domestic violence (Renzetti 1992).

One common question about partner (or domestic) violence is why victims stay with their abuser. The answers are complex and stem from sociological, psychological, and economic factors. Victims tend to believe that the batterer will change, but they also find they have few options. They may perceive that leaving will be more dangerous given that violence can escalate when the abuser thinks he (or she) has lost control. Many women are also unable to support their children and their living expenses without a husband's income. Mandatory arrest laws in cases of domestic violence exacerbate this problem because they may, despite their intentions, discourage women from reporting violence for fear their batterer will lose his job (Miller 1997). Despite the belief that battered women do not leave their abuser, however, the majority do leave and seek ways to prevent further victimization (Gelles 1999).

Sociological analyses of violence in the family conclude that women's relative powerlessness in the family is at the root of high rates of violence against women. Two perspectives have been developed to explain violence in the family: The *family violence approach* emphasizes that violence occurs in families because the society condones violence. According to this perspective, violence is endemic in society, and the general acceptance of violence is reflected in family relationships (Straus et al. 1980). The *feminist approach* places inequality between men and women at the center of analyses of violence in the family, arguing that because most violence in the family is directed against women, the imbalance of power between men and women in the family is the source of most domestic violence. The feminist perspective also emphasizes the degree to which many women are trapped in violent relationships because they are relatively powerless within the society and may not have the resources to leave their marriage (Kurz 1989; McCloskey 1996).

Child Abuse Violence within families also victimizes many children who are subjected to child abuse. Not all forms of child abuse are alike. Some consider repeated spanking to be abusive; others think of this as legitimate behavior. Child abuse, however, is behavior that puts children at risk and may include physical violence and neglect. As with battering, the exact incidence of child abuse is difficult to know, but annually 1.5 million children receive preventative services (Sedlak and Broadhurst 1996; National Clearinghouse on Child Abuse and Neglect 2003). Whereas men are the most likely perpetrators of violence against their spouses, with child abuse, women are the perpetrators as frequently as men.

Research on child abuse finds a number of factors associated with the abuse, including chronic alcohol use by a parent, unemployment, and isolation of the family. Sociologists point to the absence of social supports—social services, community assistance, and cultural norms about the primacy of motherhood—as related to child abuse because most abusers have weak community ties and little contact with friends and relatives (Eitzen and Baca Zinn 2004).

Incest Incest is a particular form of child abuse, involving sexual relations between persons who are closely related. A history of incest has been related to a variety of other problems, including drug and alcohol abuse, runaways, delinquency, and various psychological problems, such as the potential for violent partnerships in adult life. Studies find that fathers and uncles are the most frequent incestuous abusers and that incest is most likely in families where mothers are, for one reason or another, debilitated. In such families, daughters often take on the mothering role, being taught to comply with men's demands to hold the family together. Thus, scholars have linked women's powerlessness within families to the dynamics surrounding incest (Herman 1981).

Changing Families/ Changing Society

Like other social institutions, the family is in a constant state of change, particularly as new social conditions arise and as people in families adapt to the individual changes that come from the birth of a new child, the loss of a partner, divorce, migration, and other life events. These changes are what C. Wright Mills referred to as "troubles" (see Chapter 1), although not all of them constitute troubling events. Some may even be happy events. The point is that they are changes that happen at the individual level, such as the birth of a child. These microsociological events are interesting to sociologists, particularly when sociologists see common patterns in people's reactions and adjustments to such situations.

At the macrosociological level, other social changes affect families on a broad scale, and as Mills would have pointed out, many microsociological things that people experience in families have their origins in the broader changes affecting society as a whole. Migration is a good example. Certain economic, social, and political conditions may encourage people to move. How they do so, where they go, what kind of work they seek, and who can go all affect family life. This example shows how the immediate conditions of family life can be understood in the context of larger social patterns.

Global Changes in Family Life

Changes in the institutional structure of families are being affected by the process of globalization. The increasingly global basis of the economy means that people often work long distances from other family members—a phenomenon that happens at all ends of the social class spectrum, yet varies significantly by social class. A corporate executive may accumulate thousands—even millions—of first-class flight miles, criss-crossing the globe to conduct business. A regional sales manager may spend most nights away from a family, likely staying in modestly priced motels, eating in fast-food franchises along the way. Truckers may sleep in the cabs of their tractor-trailers, after logging extraordinary numbers of hours of driving in a given week. Laborers may move from one state to the next, following the pattern of the harvest, living in camps away from families, and being paid by the amount they pick.

These patterns of work and migration have created a new family form—the **transnational family,** defined as families where one parent (or both) live in one country, while other immediate family members (such as children) live in another country. A good example is found in Hong Kong, where most domestic labor is performed by Filipina women, who work on multiple-year contracts, managed by the government, typically on a live-in basis. They leave their children in the Philippines, usually cared for by another relative, and send money home because the meager wages they earn in Hong Kong (the equivalent of about U.S.$325 per month) far exceed the average income of workers in the Philippines (on average, the equivalent of U.S.$1500 per year). This pattern is so common that the average Filipino migrant worker supports five people at home; one in five Filipinos directly depends on migrant workers' earnings (Constable 1997; Parreñas 2001; Hondagneu-Sotelo 2001).

One need not go to other nations to see such transnational patterns in family life. In the United States,

MAP 15.2 Women-Headed Households Around the World

Women-headed
households
(in percent)

☐ NA
☐ 5–12
☐ 13–22
☐ 23–30
☐ 31–38
☐ 39–47

Women-headed households are increasingly common in the industrialized world. What factors do you think re-sult in this family form? As you can see from this map, they are more common in some areas of the world than others, although note that information on this is not available for many regions. What do you think influences the formation of female-headed households?

Source: United Nations. 2000. *The World's Women 2000: Trends and Statistics.* New York: United Nations. Used with permission.

Caribbean women and African American women have had a long history of leaving their children with others as they sought employment in different regions of the country. Many Latinas are experiencing similar pulls on their need to work and their roles as mothers. Central American and Mexican women may come to the United States and work as domestics or in other service jobs, while their children stay behind. Mothers may return to see them whenever they can, or alternatively, children may move for part of the year but spend part of the year with other relatives in a different location. Sociologists studying transnational families find the mothers must develop new concepts of their roles as mothers. Their situation means giving up the idea that biological mothers should raise their own children. Many have expanded their definition of motherhood to include breadwinning, traditionally defined as the role of fathers. Transnational families also create a new sense of home, one not limited to the traditional understanding of "home" as a single place where mothers,

fathers, and their children reside (Hondagneu-Sotelo 2001; Hondagneu-Sotelo and Avila 1997; Das Gupta 1997; Alicea 1997).

Apart from transnational families, the high rate of geographic mobility that emerged in the United States in the late twentieth century means that many U.S. families are geographically separated from their families of origin. Grandparents, sisters and brothers, aunts and uncles rarely live in the same community—a pattern that has profoundly affected systems of care. The extended family is dispersed too widely to provide child care and other forms of family support. Coupled with the presence of women in the labor force, there is a great need for state-sponsored child care. As the Baby Boom cohort grows older, there will also be an increased need in the United States for services for an aging population (see Chapter 14). To sociologists, global developments that produce new work patterns, wars that change the mix of women and men in a given region, and gender relations that change with new patterns of societal development are all fascinating processes that affect many dimensions of life, not the least of which is change in the family as a social institution.

Families and Social Policy

Family social policies are the subject of intense national debate. Should gay marriages be recognized by the state? What responsibility does society have to help parents balance the demands of work and family? What should society do to discourage teen pregnancy? Should marriage be promoted by the government? Many of the issues on the front lines on national social policy engage intense discussions of families. Some claim the family is breaking down. Others celebrate the increased diversity in families. Most people still form intimate partnerships, have children, and form cooperative household arrangements. But new forms of the family are emerging.

Despite the changes that have taken place in the institution of the family, the family ideal persists in public discussions about family values and is a potent element in national policy discussions about the family (Stacey 1996). Many blame the family for the social problems our society faces. Drugs, low educational achievement, crime, and violence are often attributed to a crisis in "family values," as if rectifying these attitudes is all it will take to solve our nation's difficulties. The family is the only social institution that typically takes the blame for all of society's problems. Is it reasonable to expect families to solve social problems? Families are afflicted by most of the structural problems that are generated by racism, poverty, gender inequality, and class inequality. Expecting families to solve the problems that are the basis for their own difficulties is like asking a poor person to save us from the national debt.

Who Cares?: The Spillover of Work and Family The United States is one of the hardest working societies in the world (see also Chapter 18). For those who are employed, hours of work seem to be increasing. At the same time, more employees have responsibility for caring for others—children, parents, or other relatives. This means that the long-assumed 40-hour work week is no longer the experience of most working people. Those employed full time now work an average of more than 43 hours per week, and that is counting only paid employment, not the extra work of unpaid work in the home. With more women in the labor force, this also means that the dual-earner family is now the norm (Moen 2004; Bureau of Labor Statistics 2004).

In this context, who cares for children and others whose needs must be met? The concept of **care work** is relatively new, defined to mean all of the labor that is needed to nurture, reproduce, and sustain people—in other words, work that is critical to the maintenance of social institutions. Care work includes such things as child care, elder care, and even caring for oneself but is a broader concept than just child care and housework. It has been developed to emphasize the actual labor that is required to support daily life (Litt and Zimmerman 2003; Cancian and Oliker 2000).

In this society, care work—in its various forms—falls most heavily on women, and it is work that is typically either unpaid or underpaid and generally devalued, despite its being necessary for the maintenance of human life. Many argue that because care work has historically been the work of women in traditional families, social institutions have not yet emerged to meet the new demands coming from the increased role of women in public life. Work places are not yet "family friendly;" social policies still tend to presume that women will do most of this work. Even when care work is acknowledged, it is not regulated nor counted as are other forms of paid labor. And care work often falls to women of color and immigrant women to perform, such as in the

TAKING ON SOCIAL ISSUES
Gay Marriage

A debate is taking place in the United States about whether gay marriage should be legally recognized. Some courts have supported gay marriage. A substantial portion of the public support gay marriage and even more support civil unions among same-sex couples. Opponents of gay marriage have argued that an amendment to the U.S. Constitution is needed to define marriage as only between a man and a woman.

The argument has been made that denying lesbians and gays the right to marry is similar to earlier bans on interracial marriage. Interracial marriage was banned by laws in many states until the Supreme Court (in *Loving v. Virginia*) ruled in 1967 that such laws were unconstitutional. What similarities do you see between earlier debates about interracial marriage and current debates about gay marriage? Are there differences?

Taking Action

Go to the Taking Action Exercise on the Companion Website—at http://sociology .wadsworth.com/andersen_taylor4e/—to learn more about an organization that addresses this topic. •••

use of domestic workers (often immigrants) who manage households for professional women and men.

For most families, care work—in the form of balancing the multiple demands of work and family—is a daily challenge. Even the traditional family dinner at the end of the day seems to be at risk: Less than half of all families report that they eat together four to six days a week; only one-quarter say they eat together seven days a week and another quarter say zero to three days (Mason 2004). Many people report feeling like they have to do more in less time. Simply coordinating the activities of different family members, getting them where they need to be, and staying in touch requires new solutions and produces new forms of stress. Armed with beepers, cell phones, and elaborate family calendars, parents and children all work to manage the complex affairs of family life. As we saw in Chapter 12 on gender, with more parents employed, having time from one's paid job to care for newborn or newly adopted children, tend to sick children, manage the family's affairs, or care for elderly parents requires great organizational skill, often with little recognition of the importance of this work.

Among industrialized nations, the United States provides the fewest federally supported maternity policies (see Table 15.2). The **Family and Medical Leave Act (FMLA)**, adopted by Congress in 1993, is meant to help address these conflicts. It requires employers to grant employees a total of twelve weeks in unpaid leave to care for newborns, adopted children, or other family members with a serious health condition. The Family and Medical Leave Act is the first law to recognize the need of families to care for children and other dependents. The Family and Medical Leave Act is gender neutral, meaning this right is available to men or women. The policy requires that employers continue their contribution to the employee's benefits during the leave. The employee must also be reinstated to the same or similar position upon return. In addition, the law states that it is not intended to subvert more generous policies, only to establish minimum standards for employers. This law is a major step forward in promoting balance between work and family demands.

A number of conditions, however, limit the effectiveness of the FMLA, not the least of which is that the leave is unpaid, making it impossible for many employed parents. The law covers only employers who have fifty or more employees within a seventy-five-mile radius. Employees must have been employed by the granting employer at least one year or 1250 hours to be eligible. Although it provides twelve weeks of leave, employers may require employees to use other forms of leave first, including sick leave and accrued vacation time. Extensive verification is also required to docu-

---

Table 15.2
Maternity Leave Benefits: A Comparative Perspective

Country	Length of Maternity Leave	Percentage of Wages Paid in Covered Period	Provider of Coverage
Zimbabwe	90 days	6–75%	Employer
Cuba	18 weeks	100%	Social Security
Iran	90 days	66.7% for 16 weeks	Social Security
China	90 days	100%	Employer
Saudi Arabia	10 weeks	50 or 100%	Employer
Canada	17–18 weeks	55% for 15 weeks	Unemployment insurance
Germany	14 weeks	100%	Social Security to a ceiling; employer pays difference
France	16–26 weeks	100%	Social Security
Italy	5 months	80%	Social Security
Japan	14 weeks	60%	Social Security or health insurance
Russian Federation	140 days	100%	Social Security
Sweden	14 weeks	450 days, 100% paid	Social Security
United Kingdom	14–18 weeks	90% for 6 weeks; flat rate thereafter	Social Security
United States	**12 weeks**	**0**	**n/a**

Source: United Nations, 2000. *The World's Women 2000: Trends and Statistics.* New York: United Nations, pp. 140–143. Used with permission.

ment the illness of someone for whom the employee needs to provide care. Finally, it is not yet clear by legal test of this law if it will apply to gay and lesbian workers (Gowan and Zimmermann 1996; Auerbach 1992; Rosenthal 1993). Many workers in firms where there are family-friendly policies worry that taking advantage of these policies will harm their prospects for career advancement (Blair-Loy and Wharton 2002). Currently, only 13 percent of workers have child-care benefits available to them from employers (U.S. Bureau of Labor Statistics 2003). Family leave policies that give parents time off to care for their children or sick relatives are helpful but of little use to people who cannot afford time off work without pay.

Child Care Family leave polices, much as they are needed, also do not address the ongoing issue of child care. For employed parents, child-care expenses require a large portion of their budget. For most families, this is problematic, but it falls especially hard on poor and single-parent families. Low-income families have to pay a higher portion of their earnings for child care (16 percent) than do higher-earning families. Single-parent families need 16 percent of their income (or $258 per month on average) to pay for child care, (6 percent; Giannarelli and Barsimantov 2000). In most urban areas, it now costs as much to send a four-year-old to child care as it does to pay public college tuition (Folbre 2000). At the same time, child-care workers also remain among the poorest-paid workers in the labor force (U.S. Department of Labor 2004). Furthermore, for a job that is so important, specialized training is seldom required and employee benefits for workers are usually nonexistent.

Most families have to find care for more than one child (Harris et al. 2002), and parents struggle to find good and affordable child care for their children. Some rely on relatives for care; others, on paid providers; and, for some, a combination of both. Half of children age three and two-thirds of four-year-olds in the United States now spend much of their time in child-care centers. But the national approach is one of patching together different programs and primarily relying on private initiatives for care. Compare this to France, for example, where participation is voluntary but almost all parents enroll young children in the *école maternelle* system, where a place is guaranteed to every child age three to six. These child-care centers are integrated with the school system and are seen as a form of early education. Moreover, whereas parents in the United States pay child-care costs equivalent to tuition at public universities, in France, child care is seen as a social responsibility and is government-sponsored and free to parents. National norms about whether families are a private or public responsibility clearly shape social policy (Clawson and Gerstel 2002; Folbre 2001).

Child care is one of the lowest paid occupations, despite its importance in the socialization of children.

Elder Care Child care is not the only form of care work. Elder care and caring for other dependents, including those with disabilities, still falls mostly on the shoulders of women. Little wonder that women in their thirties and forties report that time, stress, and money are their most pressing concerns (Newport 2000).

The shrinking size of families means that the proportion of elderly people is growing faster than the number of younger potential caretakers; moreover, as life

Elder care is most often provided by the family, usually by a woman. Given the increase in life expectancy, most adults can now anticipate spending as many years providing elder care as they have providing child care.

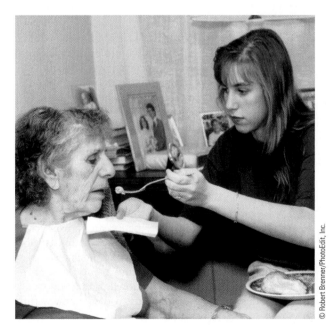

expectancy has increased and people live longer, elder care becomes a greater and greater need. Family members provide 80 to 90 percent of long-term care for the elderly—work that is often taken for granted. Some estimate that for every $120 spent in publicly funded long-term care, families provide $287 of unpaid services (Glazer 1990; Meyer 1994).

Women, who shoulder the work of elder care, can now expect to spend more years as the child of an elderly parent than as the mother of children under eighteen (Watkins 1987). Indeed, young people now can expect to spend more years caring for an elderly parent than raising their own children. The effects of the burden of care are apparent in the stress that women report from this role.

Because families are so diverse, different families need different social supports. Some policies will benefit some groups more than others—one reason that policymakers need to be sensitive to the diversity of family experiences. Policies also need to recognize the gender-specific character of people's experiences within families. Policies can contribute to change in such matters in several ways. For example, legislation can create incentives for employers to allow men to take greater responsibility for family life. The institution of flexible work hours, for instance, can permit more flexible parenting. Social policies cannot solve all the problems that families face, but they can go a long way toward creating the conditions under which diverse family units can thrive.

Chapter Summary

Given the diversity in family forms, how do sociologists define the family?

Because of the diversity of families, it is difficult to define the family as a single thing. *Families* are primary groups of people—usually related by ancestry, marriage, or adoption—who form a cooperative economic unit and care for any young (and each other); who consider their identity to be intimately attached to the group; and who are committed to maintaining the group over time.

What are the different kinships systems that exist in societies?

All societies are organized around a *kinship* system. Different kinship systems are defined by how many marriage partners are allowed, who can marry whom, how descent is determined, family residence, and power relations within the family. *Extended family* systems develop when there is a need for extensive economic and social cooperation. The *nuclear family* arose as the result of Western industrialization that separated production from the home. In the United States, most people marry within

their class and race background, although the number of interracial marriages is increasing.

What sociological theories are used to understand families?

Sociologists use functionalism, conflict theory, feminist theory, and symbolic interaction to explain families. *Functionalism* emphasizes that families have the function of integrating members into society's needs. *Conflict theorists* see the family as a power relationship, related to other systems of power and inequality. *Feminist theory* emphasizes the family as a gendered institution and is critical of perspectives that take women's place in the family for granted. *Symbolic interaction* tends to take a more microscopic look at families, emphasizing such things as how different family members experience and define their family experience.

How are families changing in contemporary society?

One significant change in families has been the increase in *female-headed households,* which are more likely than others to live in poverty. The increase in *women's labor force participation* has also affected experiences within families, resulting in dual roles for women. *Stepfamilies* face unique problems stemming from the blending of different households. *Gay and lesbian households* are also more common and challenge traditional heterosexual definitions of the family. *Single people* make up an increasing portion of the population, in part as a result of the later age when people marry.

How are marriage and divorce shaped by social changes?

The United States has the highest *marriage* and the highest *divorce* rates of any industrialized nation. The majority of people now prefer a lifestyle in which the husband and wife work and share household responsibilities, although many retain more traditional values. The high divorce rate is explained as the result of a cultural orientation toward individualism and personal gratification, as well as structural changes that make women less dependent on men within the family. Following divorce, women experience a decline in economic well-being.

What changes are occurring in family structures as the result of globalization?

Changes at the global level are producing new forms of families—*transnational families*—where at least one parent lives and works in a different nation than the children. Patterns of migration, war, and economic development have a profound effect on the social structure of families. Changing patterns of family relationships create new social policy needs.

How is the family linked to public discussions of social problems and social policy?

The family is often blamed for many social problems the nation experiences. Social policies designed to assist families should recognize the diversity of family forms and needs and the interdependence of the family with other social conditions and social institutions.

Key Terms

anti-miscegenation 394	matriarchy 396
bilateral kinship system 395	matrilineal kinship system 395
care work 415	matrilocal kinship system 396
cohabitation 408	monogamy 394
egalitarian societies 396	neolocal residence 396
endogamy 394	nuclear families 397
exogamy 394	patriarchy 396
extended families 396	patrilineal kinship system 395
family 393	patrilocal kinship system 396
Family and Medical Leave Act (FMLA) 416	polyandry 393
homogamy 394	polygamy 393
household 393	polygyny 393
kinship system 393	transnational family 413

Researching Society with Microcase Online

You can see the results of actual research by using the Wadsworth MicroCase® Online feature available to you. This feature allows you to look at some of the results from national surveys, census data, and other data sources. You can explore this easy-to-use feature on your own, but try this example. Suppose you want to know:

Taking all things together, how would you describe your marriage? Would you say your marriage is very happy, pretty happy, or not too happy?

To answer this question, go to http://sociology.wadsworth .com/andersen_taylor4e/, select MicroCase Online from the left navigation bar, and follow the directions there to analyze the following data.

Data file: GSS

Analysis: Auto-Analyzer

Variable1: HAP.MARRY

Questions:

The tables show you whether various factors (religion, political party, age, education, sex of respondent, and race) are related to whether people describe their marriage as happy.

1. Among all respondents, what percentage report their marriage is very happy?

2. Complete the table using the data in the row "very happy" for the various demographic variables.

Socio-Demographic Variable	Category Most Likely	Category Least Likely	Significant?	
			Yes	No
			Yes	No
			Yes	No
			Yes	No
			Yes	No
			Yes	No
			Yes	No
			Yes	No

3. Describe what characteristics are most likely among people whose marriages are "very happy."

The Companion Website for Sociology: Understanding a Diverse Society, Fourth Edition

http://sociology.wadsworth.com/andersen_taylor4e/

Supplement your review of this chapter by going to the companion website to take one of the Tutorial Quizzes, use the flash cards to master key terms, and check out the many other study aids you'll find there. You'll also find special features such as GSS Data and Census 2000 information, data and resources at your fingertips to help you with that special project or do some research on your own.

Suggested Readings and Web Resources

Baca Zinn, Maxine, and Stanley Eitzen. 2002. *Diversity in American Families*, 6th ed. New York: Allyn and Bacon.
  This is the most comprehensive sociological analysis of families with a focus on the different family experiences across race, class, and gender.

Garey, Anita. 1999. *Weaving Work and Motherhood*. Philadelphia: Temple University Press.
Based on an interview study with women hospital employees in various jobs, Garey examines how working mothers construct their identities. Unlike the prevailing idea that women see work and motherhood as "either/or" choices, Garey demonstrates how women see both as a part of who they are, despite the difficulties of managing these two roles.

Gerstl, Naomi, Dan Clawson, and Robert Zussman (eds). 2002. *Families at Work: Expanding the Boundaries*. Nashville, TN: Vanderbilt University Press.
This anthology explores contemporary issues and social policies in the connection between families and work, including such topics as child care and family labor.

Hochschild, Arlie. 1997. *The Time Bind: When Work Becomes Home and Home Becomes Work*. New York: Metropolitan Books.
Hochschild's study of a company with "family-friendly" policies explores the increasing stresses that time demands at work, for both women and men, put on family life. As a result, family life is more hectic, not the haven that the family ideal purports.

Lareau, Annette. 2003. *Unequal Childhoods: Class, Race, and Family Life*. Berkeley, CA: University of California Press.
Lareau's research compares Black and White, middle-class and working-class families, showing how childhood is treated in each. Her research shows how social class is reproduced in families as children in middle-class homes gain an increasing sense of entitlement, while those in working class families get an increasing sense of constraint.

Weeks, Jeffrey, Brian Heaphy, and Catherine Donovan. 2001. *Same Sex Intimacies: Families of Choice and Other Life Experiments*. New York: Routledge.
Using studies of diverse families from Britain, this book explores the social changes that have influenced family forms among gays, lesbians, and bisexuals. It is a sociological account of how people form intimacy outside of traditional family norms.

Council on Contemporary Families
www.contemporaryfamilies.org
This is a nonprofit organization of family researchers whose purpose is to provide information about the needs of contemporary families. The organization disseminates educational materials and sponsors conferences, as well as responding to family issues addressed in the media.

Family Violence Prevention Fund
www.fvpf.org
This is a nonprofit organization focusing on policy and education about domestic violence. The website includes research information, personal testimonies, and national contacts for related resources.

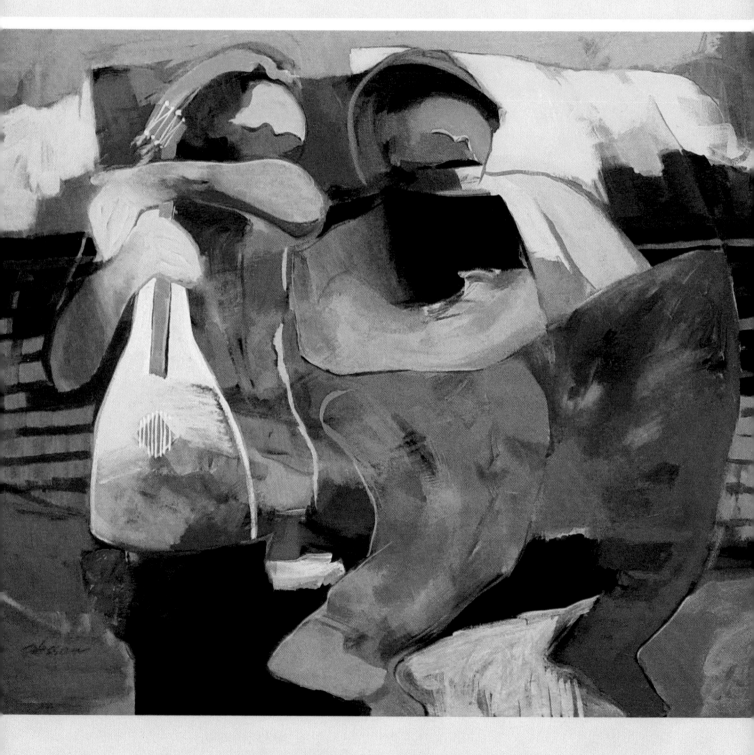

Education

Ollie Taylor (a pseudonym) is a bright eleven-year-old African American. He is growing up in a two-parent family in Boston. Although he receives considerable encouragement, love, and respect from his family, Ollie Taylor is certain that everything he attempts will fail. He is convinced that he is, in his own words, worthless. Ollie's feelings about himself can be traced directly to his school rather than his family or his peers. In Ollie Taylor's case, the feeling of worthlessness was significantly created by the system of tracking: the grouping, or stratifying, of students within the educational system according to their presumed academic ability.

In Ollie's own words:

> The only thing that matters in my life is school, and there they think I am dumb and always will be. I'm starting to think they are right. . . . Upper tracks? Man, when do you think I see those kids? I never see them. . . . If I ever walked into one of their rooms, they'd throw me out before the teacher ever came in. They'd say I'd only be holding them back from their learning (Psychologist Tom Cottle, quoting Ollie Taylor (pseudonym), in Persell 1990: 82).

Stories like Ollie Taylor's are common. His case highlights one of the problems facing the American system of education: Should children be grouped (tracked) according to their ability, allowing them to learn at their most comfortable pace, or should they be grouped simply by age, perhaps impeding the progress of fast learners and setting too quick a pace for the slow learners?

There are both positive and negative consequences to early tracking of students, a system discussed in detail later in this chapter. Ollie Taylor's case exemplifies the negative consequences, but all is not necessarily bleak for Ollie and those like him. Ollie's case also exemplifies the positive as well as the negative possibilities of getting an education in the United States, in general. As an African American, he is six times more likely to graduate from high school now than in 1940. After earning a high school diploma, he will be more likely to get a good job and less likely to join the ranks of the unemployed. If he completes his education, he will be more likely—though certainly not guaranteed—to earn a sufficient wage.

Unfortunately, his chances of dropping out prior to graduation are fairly high. They are considerably higher than for a White man of his age but significantly lower than for a Hispanic young man of his age. If, however, he does graduate, he will stand a better chance of avoiding the grim path of many urban Black and Hispanic male dropouts and

Sociology ⊛ Now™

Reviewing is as easy as ❶❷❸.

Use SociologyNow to help you make the grade on your next exam. When you are finished reading this chapter, go to the chapter review for instructions on how to make SociologyNow work for you.

many White youth as well—a path leading to unemployment and possibly drug abuse, poor health, incarceration, and death at an early age.

The education institution in the United States is capable of sending a person down two entirely different paths. One leads to the perpetuation of racial, ethnic, socioeconomic, and gender inequalities in society; the other lessens these inequalities as the individual goes on to a good job, a decent income, and relative happiness. The system of education in this country spawns both outcomes, making or breaking the life prospects of millions of students like Ollie Taylor. •••

Schooling and Society: Theories of Education

Education in a society is concerned with the systematic transmission of the society's knowledge. This includes teaching formal knowledge such as the "three R's"—reading, writing, and arithmetic—as well as the conveyance of morals, values, and ethics. Education prepares the young for entry into society and is thus a form of socialization. Sociologists refer to the more formal, institutionalized aspects of education as **schooling.**

DEBUNKING **SOCIETY'S MYTHS**

Myth: To get ahead in society, all you need is an education.

Sociological perspective: Education is necessary, but not sufficient, for getting ahead in society. Success depends significantly on one's class origins; the formal education of one's parent or parents; and one's race–ethnicity and gender.

The Rise of Education in the United States

Compulsory education is a relatively new idea. During the nineteenth century, many states had not yet passed laws requiring education for all children. Many, if not most, jobs in the mid-nineteenth century did not require education or literacy. Education was considered a luxury, available only to children of the upper classes (Cookson and Persell 1985), and was prohibited by law for Blacks, even after the Emancipation Proclamation was issued in 1863.

By 1900, compulsory education was established by law in all states, excluding a few Southern states, where Black Americans were still largely denied formal education of any kind (Higginbotham 1978). In the past, state laws in the South and West have also prohibited education for Hispanics, American Indians, and Chinese immigrants. State laws requiring attendance were generally enforced for White Americans at least through eighth grade. Completing education all the way through high school lagged considerably. In 1910, *fewer than 10 percent* of White eighteen-year-olds in the United States graduated from high school. High school attendance rose steadily in the following decades, but by the 1930s, less than half of the eighteen-year-olds in this country had attended high school. It was not until as recently as 1960 that the number of young adults with a diploma approached 50 percent.

Attendance in both high school and college has expanded dramatically since 1960, as shown in Figure 16.1. This figure reveals another dramatic trend—

Ethnic and cultural diversity in the public high school classroom are increasingly common.

© Rick Scott/Index Stock Imagery

high school and college graduation rates have not been equal across racial groups. The high school graduation rate for Whites has increased steadily from 26 percent in 1940 to nearly 90 percent in 2003. For African Americans, the increase has been equally dramatic but only roughly parallel to the increase for Whites, from less than 9 percent in 1940 to about 85 percent in 2003. These trends may certainly be encouraging, but high school graduations for African Americans nevertheless consistently lag behind high school graduations for Whites.

As Figure 16.1 shows, the Hispanic high school graduation rate has fallen behind both the African American and the White and Asian American rates since 1975. College graduations have shown similar trends.

The Functionalist View of Education

All known societies have some type of educational institution of some sort. In the United States, as in other industrialized societies, the education institution is large and highly formalized. In other societies, such as pastoral societies, it may consist simply of parents teaching their children how to till land and gather food. Under these circumstances, the family is both the education institution and the kinship institution.

Why does an education institution exist in the first place? What does it do for society? Functionalist theory in sociology answers these questions by arguing that education accomplishes certain consequences, or functions, for a society. Among these functions are socialization, occupational training, and social control.

Socialization is brought about as the cultural heritage is passed on from one generation to the next. This heritage includes much more than "book knowledge." It also includes moral values, ethics, politics, religious beliefs, habits, and norms—in short, the elements of culture. Schools strive to teach a variety of skills and knowledge, from history, literature, and mathematics to handcrafts and social skills, while also inculcating values such as school loyalty and punctuality. According to functionalist explanations, the importance to society of this kind of socialization explains why an education institution began and grew in society.

Occupational training is another function of education, especially in an industrialized society such as the United States. In the less complex society of the United States prior to the nineteenth century, jobs and training were passed on from father to son or, more rarely, from father or mother to daughter. A significant number of occupations and professions today are still passed on from parent to offspring, particularly among the upper classes (such as a father passing on a law practice to his son) but also among certain highly skilled occupations such as plumbers, ironworkers, and electricians, who pass on both training and union memberships. Modern industrialized societies need a system that trains people for jobs. Most jobs today require at least a high school education, and many professions require a graduate degree.

Social control is also a function of education, although a less obvious one. Such indirect, nonobvious consequences emerging from the activities of institutions are called **latent functions**. Increased urbanization and immigration beginning in the late nineteenth century were accompanied by rises in crime, overcrowding, homelessness, and other urban ills. Consequently, one perceived benefit of compulsory education was that it kept young people off the streets and out of trouble. The more obvious consequence or function of education was job training. The latent function was the social control of deviant behavior. In the 1920s, educa-

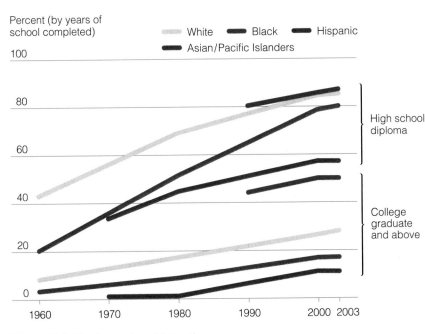

Percent (by years of school completed)

White   Black   Hispanic
Asian/Pacific Islanders

100

80 — High school diploma

60

40 — College graduate and above

20

0

1960   1970   1980   1990   2000 2003

Figure 16.1 The Expansion of Education

Data: U.S. Census Bureau. 2004 *Statistical Abstract of the United States 2003.* Washington, DC: U.S. Government Printing Office. For Asians and Hispanics, data not available prior to 1970.

MAP 16.1 School Dropout Rates

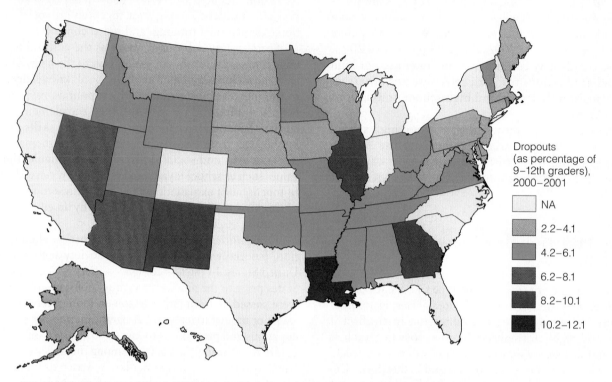

Dropouts
(as percentage of
9–12th graders),
2000–2001

	NA
	2.2–4.1
	4.2–6.1
	6.2–8.1
	8.2–10.1
	10.2–12.1

Notice that states vary quite considerably in their school dropout rates. What is the dropout rate in your home state? Which social factors have had the greatest impact on the dropout rate there? In your opinion, why are the school dropout rates higher in some states than in others? How and in what ways does a high dropout rate hurt a state?

Source: National Center for Educational Statistics. 2004. *Digest of Education Statistics.* Washington, DC: U.S. Department of Education.
Website: **http://nces.ed.gov/quicktables/Detail.asp?Key=1099**

tion also became a way to socialize new immigrants from Italy, Poland, Ireland, and other European countries, "Americanizing" them in the interest of social control (Katz 1987).

The Conflict View of Education

In contrast with functionalist theory that emphasizes how education unifies and stabilizes society, conflict theory emphasizes the disintegrative aspects of education. Conflict theory focuses on the competition between groups for power, income, and social status, giving special attention to the prevailing importance of institutions in the conflict. One intersection of education with group and class competition is shown in the significant correlation between education and class, race, and gender. The unequal distribution of education separates groups. The higher the educational attainment of a person, the more likely that person will be middle- to upper-

class, White, and male. Conflict theorists argue that educational level is a mechanism for producing and reproducing inequality in our society.

According to conflict theorists, educational level can be used as a tool for discrimination via the mechanism of **credentialism,** the insistence upon educational credentials for their own sake, even if the credentials bear little relationship to the intended job (Collins 1979; Marshall 1997). This device can be used by potential employers to discriminate against minorities, working-class people, or women—that is, those who are often less educated, thus less likely to be credentialed because discriminatory practices within the educational system have limited their opportunities for educational achievement.

Although functionalists argue that jobs are becoming more technical and thus require workers with greater education, conflict theorists argue that the reverse is true—most new opportunities appearing today are in categories such as assembly-line work, jobs that are becoming less complex and less technical and therefore require less traditional education or training. Nonetheless, potential employers will insist on a particular degree for the job, even though there should be little expectation that education level will affect job performance. Education is thus used as a discriminatory barrier.

The Symbolic Interactionist View of Education

Symbolic interaction focuses on what arises from the operation of the interaction process during the schooling experience. Through interaction between student and teacher, certain expectations arise on the part of both. As a result, the teacher begins to expect or anticipate certain behaviors, good or bad, from students. Through the operation of the *expectancy effect,* discussed in detail later, the expectations a teacher has for a student can create the very behavior in question. Thus, the behavior is caused by the expectation instead of being simply anticipated by it.

For example, if a White teacher expects Latino boys to perform below average on a math test relative to White students, over time the teacher may act—perhaps unwittingly—in ways that actually encourage the Latino boys to score below average on tests. The teacher might provoke increased stress among Latinos, thus increasing test anxiety, resulting in decreased performance. Therefore, teachers' expectations can affect actual test performance in addition to the effects of students' aptitudes or abilities. Later we shall look at studies that show how the expectancy effect works.

Does Schooling Matter?

How much does schooling really matter? Does more schooling lead to a better job, more annual income, and greater happiness? Is the effect of more education great or small?

Effects of Education on Occupation and Income

One way sociologists measure a person's social class or socioeconomic status (SES) is to determine the person's level of schooling, current income, and type of occupation (see Chapter 9 on class stratification). Sociologists call these the *indicators* of SES. In the general population, there is a fairly strong, though not perfect, relationship between formal education and occupation. Measuring occupations in terms of social status or prestige, we find that the higher a person's occupational status, the more formal education he or she is likely to have received. Overall, we know that, on average, doctors, lawyers, professors, and nuclear physicists spend many more years in school than unskilled laborers such as garbage collectors and shoe shiners. This relationship is strong enough that we can often, though not always, guess a person's approximate educational attainment just by knowing the person's occupation. There are instances of semiskilled laborers (such as taxi drivers) with law degrees or Ph.D.s, but they are relatively rare. Also exceedingly rare is the reverse: the self-educated, self-made individual who completed only the fourth grade and is now the CEO of a major corporation. Even then, at least one researcher (Kasarda 1999) argues that there

Table 16.1
Sociological Theories of Education

	Functionalism	Conflict Theory	Symbolic Interaction
Education in Society	Fulfills certain societal needs for socialization and training; "sorts" people in society according to their abilities	Reflects other inequities in society, including race, class, and gender inequality, and perpetuates such inequalities, by tracking practices for example	Emerges depending on the character of social interaction between groups in schools
Schools	Inculcate values needed by the society	Are hierarchical institutions reflecting conflict and power relations in society	Are sites where social interaction between groups (such as teachers and students) influences chances for individual and group success
Social Change	Means that schools take on functions that other institutions, such as the family, originally fulfilled	Threatens to put some groups at continuing disadvantage in the quality of education	Can be positive as people develop new perceptions of formerly stereotyped groups

is a serious *mismatch* between the skills youths learn today and the skills required to enter the job market.

The connection between income (and jobs) and education is not independent of gender. Gender heavily influences the relationship between income and education. Note from Table 16.2 that although the higher one's education, the higher one's (average) income, it is nonetheless true that the average income for women is less than the average income for men at each educational level. In general, throughout our society, women consistently earn less than men who are of comparable education. The data show clearly that differences persist at *each* level of education. This is because, in general throughout our society, the average woman earns less than a man of the same *or even less* education. Men with professional degrees (law, medicine, and so forth) earn a median annual income of $81,602, whereas a woman with that same education earns only $46,635, about 57 percent of what a man earns. A man with no graduate education but a college-only education earns $54,069, more than a woman with a master's degree. And men with some college but no degree earn more than women with bachelor's degrees (see Table 16.2).

As we already noted, the number of conferred high school diplomas and college degrees have increased rapidly in the last thirty years. This has affected the economic value of a college education. In the past, when few people earned college degrees, college graduates were a scarce and thus valuable commodity. Because far more people earn college degrees today, a college education is no longer the same automatic ticket to success (Pedersen 1997). The French sociologist R. Boudon (1974) noted some time ago that as the level of educa-

tion has risen in industrialized nations, the relative economic advantage of completing college, measured in dollars, has declined. Boudon calls this **educational deflation.** He notes that it applies to all levels of education, not just college. Not only is a college degree worth less now, so is an eighth-grade (junior high) education. Therefore, children of high school dropouts who are themselves dropouts will earn less, on average, and have a more difficult time finding a skilled job than their dropout parents did.

THINKING SOCIOLOGICALLY

What is the amount of (formal) education of your parents? Do they want you to have more education than they had, about the same, or less? What do these questions suggest to you about the relationship between *educational attainment* and *social mobility*?

Effects of Social Class Background on Education and Social Mobility

Education has traditionally been viewed in the United States as the way out of poverty and low social standing—the main route to upward social mobility. The assumption has been that a person can overcome modest beginnings, which starts by staying in school.

Much sociological research has demonstrated that the effect of education upon a person's eventual job and income depends to a great extent—though not entirely—upon the social class that the person was born into. Hence, there is no straightforward relationship between education, occupation, and income. Among White people of the upper class, including those who inherited wealth as well as professionals and high-level managers, social class origin is *more important than education* in determining occupation and income (Blau and Duncan 1967; Jencks et al. 1972; Taylor 1973b; Jencks et al. 1979; Bielby 1981; Cookson and Persell 1985; Persell 1990, 1977; Jencks 1993).

Class and race work together to "protect" the upper class from downward social mobility—and to block the lower classes from too much upward mobility. Education is used by the upper class to avoid downward mobility, by such means as sending their children to elite, private secondary schools. A disproportionately high number of upper-class children attend elite boarding schools and day schools, whereas working-class children are *considerably* underrepresented in such schools (Rendon and Hope 1996; Cookson and Persell 1985).

Among middle-class Whites, education considerably improves the chances of getting middle-class jobs, yet access to upper-class positions is more limited. Among those of lower-class origins, such as unskilled laborers or the chronically unemployed, chances are poor of

Table 16.2 Median Income by Education and Gender (in dollars)		
Level of schooling	Men (percentage in category)	Women (percentage in category)
Less than 9th grade	$14,594 (6.8)	$8,846 (5.9)
9th to 12th grade (no diploma)	19,434 (8.8)	10,330 (8.4)
High school graduate	28,343 (30.7)	15,665 (32.9)
Some college, no degree	33,777 (17.0)	20,101 (17.5)
Associate degree	38,870 (7.5)	22,638 (9.3)
Bachelor's degree	49,985 (18.7)	30,973 (17.8)
Master's degree	61,960 (6.5)	40,744 (6.5)
Professional degree	81,602 (2.1)	46,635 (1.0)
Doctoral degree	72,642 (1.8)	52.181 (0.7)

Source: U.S. Census Bureau. 2004 *Historical Income Tables–People, Table P-16.* Washington, DC: U.S. Department of Commerce, Website: **www.census.gov**

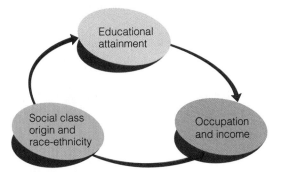

Figure 16.2 Relationship of Social Class, Race–Ethnicity, Education, Occupation, and Income

getting a good education as well as a prestigious job. In sum, education is affected by social class origins, and occupation (and income) is heavily influenced by education but also by social class origins. Individuals with lower-class origins are less likely to get a college education and thus less likely to get a prestigious job. These interrelationships are summarized in Figure 16.2, which shows that social class origin affects occupation and income both directly and indirectly by way of education.

DEBUNKING *SOCIETY'S MYTHS*

Myth: Education is more important than social class in determining one's job and income.

Sociological perspective: Although education has an effect on the job one gets and the income one earns, overall, social class origin is more important than acquired education in determining the prestige of one's job and earned income.

Education, Social Class, and Mobility Seen Globally

It is sometimes argued that because of its educational system, the United States has more occupational and income mobility than other countries, particularly England, Germany, and Japan. In general this is true, but not by much. Until a few years ago, students in England were required at age eleven to take an examination, called the Eleven Plus. A student's score on this examination determined whether he or she was put on a track to prestigious universities such as Oxford or Cambridge or went directly into the labor force from high school. Children of the upper class stood a *far* better chance of scoring high on this examination than did middle- or working-class children, and the average scores of women and minorities, especially Africans and East Indians, were considerably lower than those of upper-class White males.

A similar situation exists in the United States. Students from lower-class families have lower average

scores on exams such as the Scholastic Assessment Test (SAT) and that of the American College Testing (ACT) Program. As shown in Table 16.3, there is a smooth and dramatic increase in average (mean) SAT score with higher family income, for both SAT verbal as well as math scores. In this sense, one's SAT score is a "proxy" measure of one's social class: *Within a certain range, you can guess one's likely SAT score from knowing only the income and social class of one's parents!* As you can see from Table 16.3, each $10,000 increase in family income is worth about 10 to 15 more points on either the SAT verbal or the SAT math tests (thus 20 to 30 points more for combined score)! This is truly ironic, since the multiple-choice SAT was originally designed back in the 1940s as an "objective" test in order to combat the pattern of children from wealthy families having an advantage for admission to college.

With lower test scores come diminished odds of getting into many colleges or universities. African Americans, Latinos, and American Indians score on the average lower than Whites, and women tend to score lower than men on the quantitative (mathematical) sections of the SAT (see Table 16.4). Asian Americans as a group have scored higher than Whites in recent years on the quantitative sections of the SAT but somewhat lower on the verbal sections. Women of any ethnic group score lower on the quantitative sections than the men of the same ethnic group (see Table 16.4). These patterns indicate that the SAT has an effect in the United States similar to that of the Eleven Plus in England, directing the futures of the young according to the results of widely administered exams.

Table 16.3
Average SAT Scores by Family Income

Family Income	Number of Students	SAT Verbal Average Scores	SAT Math Average Scores
<$10,000	34,890	421	443
$10,000–$20,000	70,696	442	456
$20,000–$30,000	86,414	468	474
$30,000–$40,000	101,692	487	489
$40,000–$50,000	86,637	501	503
$50,000–$60,000	89,620	509	512
$60,000–$70,000	77,020	516	519
$70,000–$80,000	72,298	522	527
$80,000–$90,000	95,656	534	540
>$100,000	152,191	557	569
No response	323,451		

Source: College Board, 2001. *College Board Seniors 2001: A Profile of SAT Program Test Takers.* New York: The College Board.

Table 16.4
Average SAT Scores by Ethnicity and Gender

SAT Test Takers Who Described Themselves as:	SAT Verbal Mean Scores			SAT Math Mean Scores		
	2003 Male	2003 Female	2003 Total	2003 Male	2003 Female	2003 Total
American Indian or Alaskan Native	482	481	482	498	467	481
Asian, Asian American, or Pacific Islander	502	498	499	583	548	565
African American or Black	433	436	435	436	419	426
Hispanic or Latino						
Mexican or Mexican American	460	500	455	481	445	461
Puerto Rican	460	452	456	471	437	451
Latin American, South American, Central American, or other Hispanic or Latino	467	457	461	489	451	467
White	530	528	529	551	513	531
Other	509	506	508	538	497	515

Source: College Board, 2003. *College Board Seniors 2003. A Profile of SAT Program Test Takers.* New York: The College Board; Website: **www.collegeboard.com/about/news_info/cbsenior/yr2002/html/links.html**

In Germany, an examination called the *Abitur* is taken during the equivalent of the junior year in high school. A high score on the *Abitur* facilitates admission to a university; a low score inhibits getting into a university. Low-scoring students must take two or three more years of courses and then reapply to a university if they wish to attend.

In Japan, a similar examination, given at age twelve, determines even more rigidly a child's subsequent educational opportunities. Students who wish to continue their education at a college or university must score high enough to gain admission to prep schools. Especially high scores guarantee admission to prestigious prep schools, which are necessary for later admission to the best universities. Low scorers are virtually shut out from prep school admission and thus become ineligible for a university education. In recent years, many parents have begun to send their children to weekend "cram" seminars called *jukos* to prepare for this examination, adding a grueling additional regimen to the already stiff requirements of the Japanese school system. Many Japanese, including educators, are becoming concerned that the extreme competitiveness of this system and the great burden of work being put on students are brutalizing the youngsters.

Overall, the educational system in the United States appears to allow for a bit more social mobility than can be achieved in Germany, possibly England, and certainly Japan. The stratification systems in those countries and others with similar systems are more rigid, or *castelike*, than in the United States. However, there is danger in concluding that the U.S. educational system permits *much* more social mobility. In general, the similarities tend to be more prominent than the differences.

Japanese students and their parents at this juko, or "cram" seminar, vow academic success at a meeting in a hotel in Tokyo over the New Year's holiday.

© Kaku Kurita/Gamma-Liaison

Education and Inequality

In its original nineteenth-century conception, the educational system was to serve as a leveling force in society in the United States—the road to full equality for all citizens regardless of race, social class origin, religion, or gender. Jew and gentile, Black and White, rich and poor, male and female would learn together side by side. Through education, each would learn the ways of others and thus come to understand and respect them. Full equality for humankind was to follow.

Education has indeed reduced many inequalities in society since the turn of the twen-

VIEWING SOCIETY IN GLOBAL PERSPECTIVE

MAP 16.2 Literacy Around the World

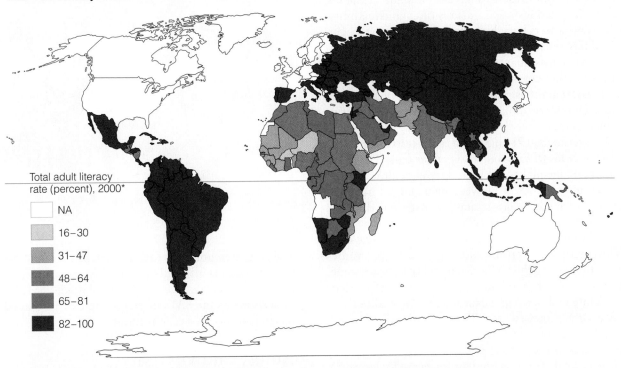

Total adult literacy rate (percent), 2000*

	NA
	16–30
	31–47
	48–64
	65–81
	82–100

The literacy rate of a country is an important index; it indicates the proportion of persons in the country who can read and write the major language of that country. Literacy rate is one type of measure of culture. Notice that on some large continents (such as Africa), literacy rate varies from country to country on the same continent. In your opinion, what social factors would cause this? What problems would this cause when the countries attempt to communicate or have diplomatic relations with one another?

*Percentage of persons age 15 and over who can read and write
Source: From UNICEF, 2003. *State of the World's Children 2003.* Website: **www.unicef.org/sowc03/tables/table4.html**

tieth century. The percentage of high school graduates has risen among Whites and minorities, both male and female, as have certain types of social mobility. Despite continuing inequalities in college enrollments, comparing African Americans, Hispanics, and Whites, the enrollment of all minorities has risen overall. Furthermore, as more minorities and women attend and graduate from two- and four-year colleges, the result has been employment of more minorities in mid-level and high-level jobs. Nonetheless, very many inequalities still exist in U.S. education.

Cognitive Ability and Its Measurement

Since as long ago as classical Greece, humans have sought to measure a "mental faculty" or "intelligence." It is now called **cognitive ability**—the capacity for abstract thinking. Since early in the twentieth century,

educators in our society from preschools to universities have attempted to measure intelligence by means of **standardized ability tests,** such as the SAT, and also by the more traditional IQ ("Intelligence Quotient") tests.

The system of education in the United States has relied heavily upon the idea that intelligence or ability or "potential" is a single, unitary trait. Cognitive ability has been gauged according to the numerical results of the standardized tests. There has been a will to reduce measurements of cognitive ability to a single number, as with IQ tests in the recent past, or perhaps to two numbers, such as the language and math scores of SAT tests.

Standardized cognitive ability tests such as the SAT or IQ tests, intended to measure ability or potential, are not the same as **achievement tests,** which are intended to measure what has actually been learned, in addition to ability or potential. Advanced placement (AP) exams are achievement tests taken before entering college.

Students who score high demonstrate that they have already mastered certain material and can skip those courses in college.

Three major criticisms have been made of the use of standardized tests as measures of cognitive ability; achievement tests have been less criticized. First, cognitive ability tests tend to measure only limited ranges of abilities (such as quantitative or verbal aptitude) while ignoring other cognitive endowments such as creativity, musical abilities, spatial perception, political skill, and even athletic ability ("athleticism" as a cognitive or mental trait; Freedle 2003; Gardner 1999; Lehmann 1999; Sternberg 1988). Second, such tests possess degrees of *cultural bias* and also *gender bias*. They may perpetuate inequality between different cultural or racial groups, as well as between men and women, and also between social class strata. The tests were designed primarily by middle-class White males, and the "standardization" they strive to achieve mirrors middle-class White male populations and cultures. Many studies show that although standardized ability tests are somewhat capable of predicting future school performance for White males, a significant number of studies show less accurate forecasts for the success of minorities, especially Hispanics, African Americans, and American Indians, and they often predict school performance less accurately for women than for men (Taylor 2002, 1992a, 1981; Epps 2002; Fleming and Garcia 1998; Jencks and Phillips 1998; Pennock-Roman 1994; Young 1994; Crouse and Trusheim 1988; Jensen 1980). In other words, the **predictive validity** of the tests—the extent to which the tests accurately predict later college grades—is compromised for minorities, women, and persons of working-class origins.

The third criticism of SAT tests is that their predictive validity even for Whites is not particularly impressive. For example, SAT scores are only modestly accurate predictors of college grades for White persons (Flemming and Garcia 1998; Manning and Jackson

"You're kidding! You count S.A.T.s?"

1984). Grade point average in high school (and class rank as well) is also only a modestly accurate predictor of success in college. High school grades are about as accurate as the SATs in predicting college grades—which is to say, not very accurate.

Ability and Diversity

As already noted, average scores on cognitive ability tests such as the SAT differ by racial–ethnic group, social class, and gender. Overall, Whites score higher on average than minorities; men score higher than women, especially on the math portion; and, in general, the higher a person's social class, the higher the test score. The differences between groups are regarded by experts as primarily environmental in origin, reflecting group differences in years of parental education, social class status, childhood socialization, language, nutrition, and cultural advantages received in the home and during youth. There is no evidence whatsoever that

TAKING ON SOCIAL ISSUES

Athletes and Academics

Does participation in school athletics affect academic performance? The stereotype of "jocks" would say so, although many point out that athletic accomplishment enhances academic performance because it can boost self-esteem and provide encouragement from coaches and teammates. Still, many are concerned about low graduation rates, particularly in the most winning athletic programs. Many worry that Black athletes are exploited by athletics, while others see athletics as providing educational opportunities that might not ever be available to some. Differences also seem to exist in the academic performance of women and men who are student athletes. If you were the athletic director at your campus, what evidence would you cite regarding the connection of academic performance and athletics?

Taking Action

Go to the Taking Action Exercise on the companion website—at http://sociology.wadsworth.com/andersen_taylor4e/—to learn more about an organization that addresses this topic. •••

such *between-group* differences are in any way genetically inherited. Certain *within-group* differences may reflect genetic differences among individuals within the *same* racial or ethnic group, social class, or gender. But even the within-group effect of genes is estimated to be much smaller than the within-group effect of social environment. That is, the effects of social environment are greater than the effect of genes, even though genes do have some effect (Taylor 2002, 1992a, 1980; Jang et al. 1996; Chipuer et al. 1990; Gould 1981; Goldberger 1979; Kamin 1974).

Gender mixes with class in the results of ability tests just as race mixes with class. In the vast majority of societies known to anthropologists and sociologists, including the United States, women have been forced to occupy lower social and economic status than men. Some female-to-male differences in standardized tests are attributable to this status ranking, but the differences are not completely one-sided. Men have tended to score higher in numerical reasoning, spatial perception, and mechanical aptitude, but women have tended to score higher in perception of detail, memory, and certain verbal skills. These trends reflect differences in the childhood socialization of boys and girls as well as differences in societal expectations pertaining to men and women. The tradeoff, however, is not an even one, because our society tends to assign more value to the abilities at which men excel, such as numerical reasoning.

Both women and minorities have been catching up to men on the math part of the SATs in the last few years. The change, coinciding with some social gains on the part of minorities and women over last decade or two, tends to discredit the traditional belief that women, Blacks, and Latinos on average have less mathematical ability than White men. This belief has been used in the past to support the argument that women, as well as minorities, are less fit than men to perform high-level executive jobs that require analytical reasoning and number crunching.

The "Cognitive Elite" and *The Bell Curve* Debate

A few years ago, a book titled *The Bell Curve* was published and caused a major stir among educators, lawmakers, teachers, public officials, policymakers, and the general public. In this book, which contains analyses of great masses of data, authors Richard J. Herrnstein and Charles Murray (1994) argue that not only does the distribution of intelligence in the general population closely approximate a bell-shaped curve (called the *normal distribution*), but also that there is one basic, fundamental kind of intelligence, not several independent kinds of intelligences, that predict how well an individual will do in school and on the job, thus predicting how successful or not a person will be in society.

***DEBUNKING* SOCIETY'S MYTHS**

Myth: Intelligence is mostly determined by genetic inheritance.

Sociological perspective: Intelligence is a complex concept, not easily measured by any one thing and—according to recent research—likely shaped as much by environmental factors as by genetic endowment.

Herrnstein and Murray estimate that intelligence is about 70 percent genetically heritable and only 30 percent the result of social environment. How do they arrive at such a figure? They arrived at the figure by reviewing studies of pairs of identical twins separated early in life for one reason or another and then raised apart. The idea is that because identical twins (as opposed to fraternal twins) are genetic *clones* (exact genetic duplicates) of each other, any similarities that remain between them after their separation must necessarily be the result of their identical genes, not similarities in their social or educational environments. The authors argue that the similarity in intelligence between separated twins is about 70 percent.

Critics, however, point out that some identical twins in the studies cited by Herrnstein and Murray were more separated than others. That is, some were not very separated at all, whereas some were separated for longer periods in their lives and thus had fewer similarities in their social and educational environments. When this is taken into account, it is seen that those twins who were more separated (who attended different schools or were raised in different socioeconomic circumstances) were also less similar in intelligence. In general, *the more separated the twins were, the less similar they were in intelligence.* This shows the effect of their differing social environments over the effect of their identical genes. Some studies show that the similarity in intelligence among truly separated identical twins is only about 50 percent (Chipuer et al. 1990); other studies show it is as low as 30 to 40 percent (Taylor 2003, 2002, 1992a, 1980; Kamin 1974; Jencks et al. 1972).

Another point made by *The Bell Curve* authors is that because intelligence is primarily inherited, and different social classes differ on average in intelligence (with the lower classes having less intelligence), then it follows that the lower classes are on average less endowed with genes for high intelligence, and the upper classes are relatively more endowed with genes for high intelligence. Thus, they reason that the upper- and upper-middle classes constitute a *genetically based* **cognitive elite** in the United States, consisting of people with high IQs, high incomes, and prestigious jobs. The authors strongly imply, but do not state outright, that any two groups presumed to differ in average intelligence, such as Blacks versus Whites or Latinos versus Whites, and any other such minority versus dominant comparison, may differ in genes for intelligence. They imply further that because men and women differ in certain kinds of intelligence (citing the presumed male superiority in math intelligence), they must also differ in genes for this kind of intelligence, and women have less.

The problems with this cognitive-elite argument, largely ignored by the authors, are as follows:

1. Their conclusions tend to ignore the vast number of studies, some of which were discussed earlier, that show that intelligence tests and standardized ability tests are not as accurate a measure of intelligence or cognitive ability of minorities as of Whites, of women as of men, and of individuals of lower socioeconomic status as of individuals of higher status.

2. Their conclusions presume that intelligence is strongly genetically heritable, whereas there is convincing evidence, already noted, that even for Whites the relative contribution of environment may be greater than the relative contribution of genes, even though intelligence is probably the result of some combination of genes and environment.

3. They base a between-group conclusion on a within-group estimate of genetic heritability. Thus, they base their conclusions on women versus men, minority versus White, and lower class versus upper class on heritability results attained on *White men*. The authors ignore a vast scientific literature detailing why one cannot draw conclusions about between-group genetic differences from within-group results (Fischer et al. 1996; Lewontin 1996; 1970; Hauser et al. 1995; Kamin 1995; Taylor 1995, 1992a, 1980; Gould 1994, 1981).

Tracking and Labeling Effects

Over half of America's secondary schools and elementary schools currently use some kind of **tracking** (also called *ability grouping*), separating students according to some measure of cognitive ability (Lucas 1999; Oakes and Lipton 1996; Maldonado and Willie 1996; Oakes 1985). Tracking has been in place for more than seventy years. Starting as early as first grade, children are divided into high-track, middle-track, and lower-track groups. In high school, the high-track students take college preparatory courses in calculus and read Shakespeare. The middle-track students take courses in business administration and typing. The lower-track students take vocational courses in auto mechanics, metal shop, and cooking. While this kind of tracking is now on the decline in the United States, it is still with us in many schools (Hallinan 2003; Lucas 1999; Oakes 1990, 1985).

The basic idea behind tracking is that students will get a better education and be better prepared for life after high school if they are grouped early according to cognitive ability. Tracking is supposed to benefit the gifted, the slow learners, and everyone in the middle. Theoretically, students in all tracks learn faster because the curriculum is tailored to their level of ability and the teacher can concentrate on smaller, more homogenous groups.

The opposite argument is given by advocates of *detracking*. The detracking movement is based on the belief that mixing up students of varying cognitive abilities is more beneficial to students than tracking, especially by the time students get to junior high and high school. Students of high and low ability can thus learn from each other. The high-ability students are not seen as "held back" by students with less ability but as enriched by their presence. Finally, advocates of detracking point out that students in the lower tracks get less teacher attention and simply learn less. They are thus, in effect, penalized for being in the lower tracks. The idea is mix, do not match.

Which approach is better? Most researchers and educators who have studied tracking in detail agree that not *all* students should be mixed together in the same classes. The differences between students can be too great and their needs too dissimilar. Yet some degree of tracking has always had advocates, based on its presumed benefits for all students.

This presumption is under attack. One consistent finding from research on tracking is that students in the higher tracks receive positive effects but that the lower-track students suffer negative effects (Lucas 1999; Owens 1998; Rendon and Hope 1996; Cardenas 1996; Perez 1996; Oakes and Lipton 1996; Oakes 1990, 1985; Gamoran and Mare 1989; Gamoran 1972; Braddock 1988).

First of all, students in the lower tracks learn less because they are, simply, taught less. They are asked to read less and do less homework. High-track students are taught more. Furthermore, they are consistently rewarded for their academic abilities by teachers and administrators. As a result, they find school to be more enjoyable; they have better attendance; and they have higher educational and occupational aspirations. In turn, these advantages increase their academic performance in the classroom and on exams. In contrast, students in the lower tracks are expected to do less well and, as a result, find school relatively less enjoyable, have lower rates of attendance, and lower educational and occupational aspirations. Consequently, their in-class academic performance is less (Hallinan 2003; Owens 1998; Gamoran and Mare 1989; Braddock 1988). One study found that students assigned to low tracks in the eighth grade performed significantly less well in the tenth grade than students who had the *same* eighth grade test scores and social class background but went to untracked schools (Slavin 1993).

THINKING *SOCIOLOGICALLY*

As far as you know, were you in a tracked elementary school? What were the tracks? Describe them. Did you get the impression that teachers devoted different amounts of time to students in different tracks? Did teachers "look down" on those in the lower tracks? What about the students—did they treat some tracks as "better" or "worse" than others (were they perceived as differing in *prestige*)? Based on your recollections, what does this tell you about *tracking* and *social class*?

Both high- and low-track students are subject to **labeling effect:** Once a student is assigned to a particular track and is thereby labeled, the label has a tendency to stick, whether or not it is accurate. Once a student is labeled "gifted" or "high ability," other people—students, teachers, administrators—tend to react in accordance with that label. One is regarded as smart and high achieving. Students labeled "slow" or "low ability" encounter a negative reaction from the same people, including the expectation of low achievement. Even when a student is transferred from one track into another—for example, from a lower track to a higher one as the

result of a recent cognitive ability test—the prior perceptions tend to persist. Teachers and students still think of the youngster as "lower track," and even the recently promoted student may retain the self-perception developed in response to the prior track assignment. It should be noted that getting assigned from a lower to a higher track is more difficult than being downwardly mobile from a higher track to a lower one.

Who gets assigned to which tracks? Research shows that track assignment is not based solely on performance in cognitive ability tests. Social class and race are involved. Students with the *same* test scores often get assigned to different tracks because of differences in their social class and race. Few administrators or teachers consciously and deliberately assign students to tracks based on these criteria, but it occurs nevertheless. Researchers have consistently found that when following two students with *identical scores* on cognitive ability tests, the student of higher social class status is more likely than the student of lower social class status to get assigned to the higher track.

Teacher Expectancy Effect

Similar to the labeling effect of tracking is the **teacher expectancy effect,** which is the effect of teacher expectations on a student's performance, *independent* of the student's ability. What the teacher *expects* students to do affects what they will actually do. The expectations

UNDERSTANDING DIVERSITY

Is Ability Tracking Still Around?

In a detailed quantitative analysis, sociological researcher Samuel R. Lucas argues that ability tracking in high schools in the United States is alive and well. It is a system of stratifying students, presumably according to their academic abilities, which was to have been largely phased out twenty or thirty years ago but has returned in somewhat different forms. It is simply old wine in a new bottle. From the 1940s through the 1970s and early 1980s, most high schools in the United States had roughly three tracks, or ability groupings, based presumably on scores on cognitive ability tests: the college-bound track, a middle general education track, and a vocational track. Some schools allowed students some say about which track they were to be in, and some did not. It became evident that students were being assigned

to tracks more on the basis of their social class and their race–ethnicity than their abilities. Furthermore, women were being unfairly kept from the science tracks or curricula, not on the basis of their math or science ability but because of their gender. They were also being kept from vocational curricula containing the likes of woodworking or auto shop.

Many schools have argued that they have been phasing out tracking for the last twenty years or more, but a system of stratifying students still remains—in the guise of Advance Placement programs (tracks), honors tracks, gifted and talented tracks, and other such strata. Yet the disproportionate absence of Blacks and Hispanics from such strata, as well as the relative absence of women from the math and science strata, suggests that the older

system of tracking, in effect, continues. Lucas further demonstrates how earlier reliance on the concept of "intelligence" and the use of intelligence quotient (IQ) tests has been replaced with other, yet very similar, forms of standardized testing, such as the Preliminary Scholastic Assessment Test (PSAT), Scholastic Assessment Test (SAT), and American College Test (ACT), and other such tests that only serve to rigidify this newer system of academic stratification. He argues that Americans must extricate schools from perpetuating these inequalities and only frank and forward-looking discussion of curriculum reform and other issues will accomplish this.

Source: Lucas, Samuel R. 1999. *Tracking Inequality: Stratification and Mobility in American High Schools.* New York: Teachers College Press.

•••

a teacher has for a student's performance can dramatically influence how much the student learns.

Insights into the teacher expectancy effect come from symbolic interactionist theory. In a classic study that demonstrates this effect, Robert Rosenthal and Lenore Jacobson (1968) told teachers of several grades in an elementary school that certain children in their class were academic "spurters" who would increase their performance that year. The rest of the students were called "nonspurters." The researchers selected the "spurters" list *completely at random*, unbeknownst to the teachers. The distinction had no relation at all to an ability test the children took early in the school year, although the teachers were told (falsely) that it did. At the end of the school year, it was found that all students improved somewhat on the achievement test, yet those labeled spurters made greater gains than those designated nonspurters, especially among first and second graders. This experiment isolates the effect of the label because it is the only difference that distinguishes a randomly selected group. Variations of this clever and revealing study have been conducted many times over, and the results are generally similar despite the fact that the original study by Rosenthal and Jacobson was criticized.

How are expectations converted into performance? By the powerful mechanism of the **self-fulfilling prophecy,** in which merely applying a label has the effect of justifying the label (Cardenas 1996; Darley and Fazio 1980; see Figure 16.3). Recall the quote from early sociologist W. I. Thomas in Chapter 7 : "If men [sic] define situations as real, they are real in their consequences" (Thomas 1928: 572). If a student is defined as a certain type, the student becomes that type. The process unfolds in stages. First, a teacher is told that a student merits a label such as "spurter." Perhaps the designation originates with administrators or comes from the

scoring key of a standardized exam. The teacher's perception of the student is then colored by the label. A student labeled a spurter may be coaxed and praised more often than nonspurters. The student then reacts to the teacher's behavior. Students expected to perform well and encouraged to excel, perform better in class and on exams than other students. Finally, the original prophecy fulfills itself. The teacher observes the behavior of the student, notes the increase in performance, and concludes that the designation "spurter" is affirmed because the so-called spurters perform better by objective measures than the nonspurters. Further praise and encouragement follows. Teachers unaware of the overall effect will not realize that the label itself produced part of the greater performance of the spurters. The entire process tends to work in a similar manner but in opposite direction, if the student is initially labeled "slow" or "non-spurter" (Lucas 1999; Rendon and Hope 1996; Cardenas 1996; Perez 1996; Hallinan 2003, 1994; Gamoran 1972). The self-fulfilling prophecy is diagrammed in Figure 16.3.

Schooling and Gender

Teachers hold different expectations about girls and boys in school, and it therefore comes as no surprise that the gender of the student affects teacher expectations, which affect teachers' actual behaviors, which in turn affect the performance of these children. Tracking, particularly in high school, is significantly dependent upon gender. For example, there is a far greater proportion of young men in the science and math tracks than young women. Throughout schooling, from preschool to graduate or professional school, women are discriminated against in ways that men are not. This consistent and long-term differential treatment has had profound

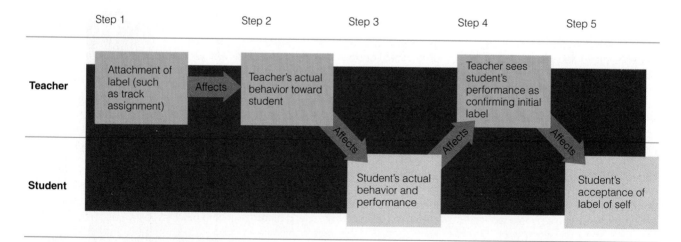

Figure 16.3 The Self-Fulfilling Prophecy

Source: Adapted from Taylor, Shelley E., Letitia Ann Peplau, and David O. Sears. 2003. *Social Psychology,* 11th ed. Englewood Cliffs, NJ: Prentice Hall.

consequences. Girls and boys start out in school roughly equal in skills and confidence, but by the end of high school, gender differences appear in some areas, especially advanced math and science. What happens in between has been documented in a comprehensive report commissioned by the American Association of University Women (AAUW) that summarizes the results of more than 1000 publications and studies. This extensive research shows the following with regard to schooling and gender (American Association of University Women 1998).

1. In general, teachers—women as well as men teachers—pay less attention to girls and women. In elementary school, as well as high school, teachers direct more interaction to boys than to girls. As a result, boys tend to talk more in class, and teachers interact with them more. The difference is particularly notable in math and science classes.

2. On national tests of reading and writing, girls perform equally to boys. But on advanced mathematics and some science tests, differences emerge. The most dramatic gender differences appear on tests with the highest stakes—on the quantitative sections of tests like the Preliminary Scholastic Assessment Test (PSAT), Scholastic Assessment Test (SAT), and Advanced Placement (AP) tests. Researchers explain this as the result of several factors, including course-taking patterns and bias in the tests. With regard to course-taking, young men and women tend to take the same number of math and science courses, but what they take differs substantially. In science, girls are more likely to take biology and chemistry; boys, physics. Girls also tend to end their math studies after a second algebra course; boys are more likely to take trigonometry and calculus, thus giving them an advantage on higher-level math skills (American Association of University Women 1998).

3. Some standardized math and science tests still retain gender bias, despite twenty years or more of effort to weed out gender bias by education specialists and testing organizations such as the Educational Testing Service, manufacturer of the SATs. Bias is especially prevalent in mathematical word problems. Certain problems are gender-typed in the sense that they employ words and concepts more familiar to men than women. For example, an SAT word-math question built on the concept of volume may ask the student to calculate the volume of oil in an automobile crankcase.

Women outnumber men in this high school science class, which is the exception rather than the rule, according to the most recent studies.

The same question could be asked about the volume of a household article, such as a pot on a stove. The question would be the same, but the testing results would likely be different because women (because of gender socialization) are less likely to be able to visualize a car crankcase and will therefore be less comfortable with the question. The phrasing of the question is thus gender-typed. Women tend to score higher on word-math questions of this type when gender typing is neutralized, even though the revised question may be conceptually identical. Subtle bias of this sort is a fixture on the actual SAT exams (Chipman 1991; American Association of University Women 1998).

4. Standardized tests in math tend to underpredict women's actual grades in mathematics. Women tend to do somewhat better in math courses than their test scores would predict.

5. Teachers tend to treat Black women and White women differently. This is particularly true of Black and White girls during the elementary school years. Teachers tend to rebuff Black girls and interact more with White girls. The trend appears, surprisingly, to be independent of the teacher's own race; namely, Black teachers tend to interact more with White girls, just as do White teachers.

6. Many textbooks still tend to either ignore women or stereotype them. In this respect, textbooks are gender-role socializers. In elementary school texts,

male characters greatly outnumber female characters. Boys are portrayed as building things, being clever, and leading others. Girls tend to be shown performing dull tasks and following boys. Even to this day, men are routinely presented as doctors, lawyers, and businesspeople; women as homemakers, librarians, and nurses. There is evidence that such gender typing is decreasing, but the decrease began only very recently, and it has not been eliminated.

7. As girls and boys approach adolescence, their self-esteem tends to drop, with the erosion of self-esteem occurring more quickly among girls than boys. This trend has been noticed in many studies of the social psychology of young men and women. The trend is further exacerbated by the discrimination against women in the classroom, gender bias in standardized tests, and stereotyped presentations of women in presumably authoritative textbooks.

Stereotype Threat Effect

As has already been noted in Chapters 11 and 12, racial and gender stereotypes can affect behavior. To what extent can a negative stereotype one has about one's self affect one's own behavior and academic performance?

As with the self-fulfilling prophecy, to what extent do minorities and women internalize negative stereotypes about themselves and thus show such effects via their behavior and academic performance?

An answer has recently been provided by the research of Claude M. Steele and associates (Steele 1999, 1997, 1992; Steele and Aronson 1995; Aronson et al. 2002; McIntyre 2001; Brown and Josephs 1999). They note that two common stereotypes exist in the United States. First, because on average Blacks perform less well than Whites on tests of math and verbal ability, Blacks must have, or so it is believed, some inherent deficiency in math and verbal abilities relative to Whites. Second, because women perform less well than men on tests of math ability, women must therefore have some inherent deficiency in math ability.

To the extent that Black students in high school or college may believe (internalize) such stereotypes, they may perform less well on a test if they are told that "this is a genuine test of your true ability." This can *activate* the stereotype in the mind of the person so informed and thus increase test anxiety, with the result of lowered test performance. White students who are also told this would be less likely to have the stereotype activated (because the stereotype is not *about* Whites), and thus be less likely to have their test performances lowered— they would be less threatened by the stereotype.

DOING SOCIOLOGICAL RESEARCH

The Agony of Education

What is the experience of African American college students attending school on predominantly White campuses? This question is, according to sociologists Joe Feagin, Hernán Vera, and Imani Nikitah, clouded with misunderstanding.

Because of the many misconceptions that define Black students' experiences, Feagin, Vera, and Nikitah wanted to get firsthand information about the experiences of Black students on White campuses. To do so, they used randomly selected focus groups, a method whereby several people participate in a collective interview conducted by the researchers. Instead of just answering questionnaires or answering interview questions singly, participants in focus groups are able to interact with each other, sometimes discussing particular issues at length and responding to each other's experiences. This method can bring more nuance and subtlety to the data being collected. This team of re-

searchers used focus groups of juniors and seniors at a major state university, supplementing the student data with several focus groups of Black parents whose children were attending college or considering application. The moderator for all the groups was also African American.

The resulting book, *The Agony of Education* (1996), provides a rich analysis of the experience of Black students on White campuses. The students and their parents describe the importance of education within the Black community but also report at length about "the Whiteness of university settings." The students describe being treated as intruders on campus, while White students deny that racism exists. The students describe in poignant detail the stereotyping and discrimination they experience, as well as some of the positive changes that create a more welcoming environment. Throughout the book, Feagin, Vera, and Nikitah use their socio-

logical perspective to understand the students' and parents' experiences and to make recommendations for change.

Questions to Consider

1. Is your campus predominantly White?
2. Are you a minority person? What proportion of students at your campus are minority? *Keywords: Black students on White campuses*
3. Regardless of what ethnicity you are, do you feel you are the recipient of negative stereotyping, by ethnicity and/or gender, on your campus? *Keywords: racism in education*

We have included InfoTrac College Edition keywords at the end of each question to make it easier for you to find more to read on these topics. Go to www.infotrac-college.com, an online library, to begin your search.

Source: Feagin, Joe, Hernán Vera, and Imani Nikitah. 1996. *The Agony of Education.* New York: Routledge. •••

Results show that this is just what happens. Figure 16.4 shows the results for a test of verbal ability based on the GRE (Graduate Record Examination), a test similar to the SAT for college students who contemplate graduate education. Black college students who are simply told that the test is a "genuine" test of their true verbal ability (the *diagnostic condition*) perform less well than Whites who are also told the same thing—even though the Whites and Blacks compared *start out equal* in their average test scores. This is the **stereotype threat effect**. If the groups of both Blacks and Whites are told nothing (the *nondiagnostic condition*), then they perform about the same on the test. Note that nothing is said to the students about Black and White test performance specifically, only that the test was designed to be a "genuine" test of verbal ability.

Stereotype threat appears to operate in the same way with regard to the presumed female–male difference in math ability test performance. As Figure 16.4 shows, when a group of both women and men are told that the math test being given to them is a "genuine" test of their true math ability (the diagnostic condition), the women do much worse than the men. If a group of women and men are told nothing (the nondiagnostic condition), the women and men perform about the same. Other studies also find that merely checking a box on a form indicating "female," such women score lower on math ability tests than women who check no such box (McIntyre et al. 2001; Brown and Josephs 1999).

These results suggest that stereotype threat, as reflected on standardized test performance, might operate for a woman–man comparison in a similar manner as with a Black–White comparison. It suggests that at least part of the long-believed female deficit in math ability may stem simply from what they are told before they take the test and less from inherent differences between women and men in math ability.

Recent evidence shows that an internalized *positive* stereotype can increase test performance. Our culture contains the stereotype that Asians will perform better than anyone else on a cognitive mathematics test, such as the math SAT. Hence, if the stereotype threat principle works, then Asian Americans, when "primed" about their ethnic identity (for example, being required to check off a box identifying themselves as Asian American to the test givers who never actually see them), should get a somewhat *higher* score than Asians not so

"primed." This is precisely what happened in a study by Shih et al. (1999).

We do not know what the result would have been if Asian *women* had to simultaneously indicate both their ethnicity *and* their gender. Checking "Asian" would tend to push the test score up, but checking "female" would tend to push it down. Perhaps their score would end up somewhere in between. Preliminary findings in a study by Lau (2002; Lau and Taylor 2003) do show such results for Asian women checking both "female" and "Asian." Studies such as these not only offer convincing evidence that self-stereotypes can affect test performance, but also that the effect can be negative (score decrease—by activating a negative stereotype) as well as positive (score increase—by activating a positive stereotype).

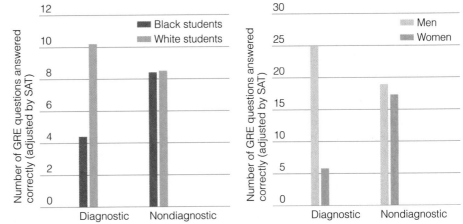

Figure 16.4 Stereotype Threat, Race, and Test Performance

Source: Adapted from Steele, Claude M., and Joshua Aronson. 1995. "Stereotype Vulnerability and African American Intellectual Performance." In *Readings About the Social Animal,* 7th ed., Eliot Aronson (ed.), pp. 409–421; Steele, Claude M. 1999. "Thin Ice: Stereotype Threat and Black College Students." *The Atlantic Monthly* (August): 44–54; and Aronson, Joshua, Carrie B. Fried, and Catherine Good. 2002. "Reducing the Effects of Stereotype Threat on African American College Students by Shaping Theories of Intelligence." *Journal of Experimental Social Psychology* 38: 113–125.

THINKING *SOCIOLOGICALLY*

What are some stereotypes of different racial–ethnic, gender, and class groups in your school or college? How does the concept of *stereotype threat* explain the educational experiences of each group? What evidence of the *expectancy effect* do you see in thinking about the success of different groups?

School Reform

School reform is an ongoing theme in the history of education in the United States. Currently, new challenges face the institution of education. Increasing diversity in the population, economic competition with other nations, inequalities plaguing the schools, and fiscal constraints also pose challenges for those who want a strong system of education in the nation. How should the nation respond to these challenges? Many solutions are debated, and new methods of organizing and delivering education are being developed. Here we examine some of these changes and challenges.

Reducing Unequal Funding

One persistent issue for reform in education is the problem of unequal funding for different school districts within the same city, urban versus suburban schools, public versus private schools, and even greatly differing funding for education across different states within the United States. Some states, such as Connecticut, Minnesota, Alaska, and New York, spend considerably more per pupil on education than states such as Tennessee, Mississippi, or Louisiana.

Major differences in funding often occur between school districts within the same state. School districts are funded primarily through property taxes. Because wealthier school districts contain more expensive real estate, such districts receive more funding. This constitutes an inequity among the school districts. Many aspects of public education depend upon funding, such as textbooks, the availability of computers, smaller class sizes, laboratory equipment, and teacher salaries. Gross inequities in funding translate into inequalities in such necessities. As a consequence, the quality of education, to the extent that such is determined through funding, varies widely from district to district. As racial and ethnic minorities continue to migrate to American cities, schools are now re-segregating with levels of segregation now higher than in the mid-1980s (Frankenberg and Lee 2002).

A proposed alternative to combating the new racial–ethnic resegregation of urban schools is the **voucher system** in education. Vouchers are essentially individual scholarships given to parents that can be used to defray the cost of a child's tuition at any school—public, private, or parochial—with the stipulation that the voucher is not awarded on the basis of religious criteria. The plan draws from public tax funds to pay part or all of the cost of schooling for their children. The voucher system, however, has fallen under legal attack in a number of communities, on the grounds that giving vouchers to attend parochial (religious) schools may threaten the doctrine of separation of church and state as guaranteed by the U.S. Constitution (Meeks et al. 2000).

Back-to-Basics, Charter Schools, and Multiculturalism

During the 1980s and 1990s, two educational reform movements became prominent. One movement stresses a return to a traditional curriculum delivered with traditional methods. The other movement stresses multiculturalism. The two movements are not in opposition but can be complementary (Orfield and Kornhaber 2001; Meeks et al. 2000; Gates 1993).

The **back-to-basics movement** emerged from the dissatisfaction that professional educators felt about declining student discipline, rising functional illiteracy, and teacher incompetence in elementary and high school, all detailed in a widely read 1983 report called *A Nation at Risk*, released by the American Commission for Excellence in Education. In that report, the blame was

FORCES OF SOCIAL CHANGE

Teach for America

Back in 1986, Wendy Kopp was a college senior with an idea, which she wrote up as her senior thesis paper. She wanted to make the world a better place by creating an ethnically diverse corps of dedicated, energetic new teachers who would go out into the poorer communities across the United States and teach children who were at risk of failing.

She was told that the idea would never work, that it was too idealistic and impractical. Even her own senior thesis faculty advisor tried to talk her out of it! But she persisted. Naming her new organization Teach for America (of which she is president), it has now

become the most successful and widespread teacher-recruitment effort for needy communities in the history of the country. The heart of the idea was to train young persons who, like herself, were excited about being trained as teachers in poor, rural as well as urban communities.

She obtained foundation grants to set up the first teacher training institute. The success of this new venture was overwhelming: Over 2500 applications were received from which she and the new organization selected 500 recruits to attend the new training institute. By the year 2000, over 5000 newly trained teachers had completed two years in

classrooms in schools all over the United States. Now, at any given time, there are approximately 1000 corps members on the job in the classroom in the United States. As testimony to the ultimate success and practicality of the program, several thousand graduates of the program have continued in the teaching profession or have become administrators and lawyers in disadvantaged communities. Teach for America has truly been a force of social change in society—and it shows what a dedicated person with just an idea can do.

•••

placed on the education system and the curricula, not upon the students. Many schools have therefore been led to place greater emphasis upon the three R's: reading, writing, and arithmetic. Freedom to choose from a variety of elective courses has been somewhat reduced. Discipline has been increased in some elementary, junior high, and high schools. There is also a push for stiffer standards in grading. This has carried over to colleges and universities, where "grade inflation" has come under recent attack. Grade inflation occurs when an excessive number of high grades are given or when the average grade for a course edges up from the unadmired C level toward the B level once reserved for above-average performance.

Many elementary schools have discontinued the practice of "social promotion" (passing students from one grade to the next regardless of their performance or grades) and have returned to the old practice of requiring students to repeat a year if their performance was unsatisfactory. In addition, many states now require teachers themselves to attain a minimum score on a standardized test called the National Teacher Examination (NTE), produced by the Educational Testing Service (ETS) of Princeton, New Jersey. Finally, the U.S. Department of Education has in the past several years advocated a national program in competency testing for high school seniors.

The **multiculturalism movement** is more recent than the back-to-basics movement and has made considerable progress to date. Its main goal is to introduce more courses and educational materials into school curricula on different cultures, subcultures, and social groups. Thus, programs such as African American Studies, His-

panic or Latino Studies, Jewish Studies, Caribbean Studies, Women's Studies, and Gay and Lesbian Studies have made their way into elementary, high school, or college curricula. The driving principle behind this movement is the belief that traditional curricula tend to stereotype women, minorities, lesbians and gays, and working-class persons, thereby giving an inaccurate picture of these groups and of society.

Multiculturalism in lower grades is supported by African American and Ethnic Studies Programs on college campuses. Established in the late 1960s and early 1970s, some of these programs have been dissolved, but others have thrived, establishing these areas as legitimate fields of academic inquiry. Women's Studies programs have similarly battled and won, and they have now established their institutional legitimacy on many college campuses. Currently, there are more than five-hundred Women's Studies programs across the country. More recently, Latino Studies and Gay and Lesbian Studies have entered the college curricula. The multiculturalism movement has followed on this success with initiatives to make courses in these programs standard requirements for graduation at many schools.

The current interest in the establishment of **charter schools** in some U.S. cities reflects the trend toward. In charter schools, public taxes remain the financier of the school but the responsibility for the running of the school and making policy decisions is delegated to the private sector. In this respect, the movement toward charter schools is a form of *privatization*. They are thus owned and managed by a private entity rather than by a traditional school board. They are chartered to produce noticeable achievement gains, with state and federal funding based on results. They are tuition-free (Meeks et al. 2000).

The multicultural classroom is one place where youth can learn about cultures other than their own.

© Will & Deni McIntyre/Photo Researchers

Home Schooling

Parental dissatisfaction with public schooling has within the last ten to fifteen years led to increasing interest in *home schooling*— parents and tutors educating their own children in the elementary years and junior high school years. Home schooling is seen as an alternative not only to traditional public schools, but as an alternative to what some parents perceive as overburdened schools. Some home school their children for religious reasons, trying to avoid the secularism of the schools and teaching their children academic skills and religious beliefs.

Some estimate that over a million students are home-schooled each year (Cloud and Morse 2004). There are both advantages and disadvantages to home schooling. Among the advantages are that parents, and tutors, can spend considerable time teaching one on one, a situation rarely possible in traditional public schools. As a result, so it is argued, home schooling accelerates the student, who thus ends up ahead of where she or he would have otherwise been in a public school. A recent study claims that home schooled students at the high school level, on the average, outscore students in traditional high schools on the SATs by about 80 points (out of a maximum of 1600). Even William Bennett, former U.S. Secretary of Education, is an advocate of home schooling.

Home schooling has its downside. For example, there is no overarching organization—no leader, no ideology—to guide the home schooling movement. Another disadvantage is that some parents and tutors often leave glaring gaps in the child's knowledge of certain subjects, while still other subjects are over-taught. For example, one home schooled 17-year-old could handle high school level calculus but could not perform simple multiplication or long division. She also lacked other simple arithmetic skills. Going back to pick up such skills is often found to be exceedingly difficult. Other problems often uncovered are low rates of reading speed and emotional problems arising from lack of social interaction with school-age peers. As one home-schooled student sees it, he missed out on this important association with peers and suffered crushing boredom during his years of home schooling from second to eighth grade, noting that his "best friend was the mailman" (Mathews 2004).

No Child Left Behind Act

Fueling the fire of controversy over school reform is the *No Child Left Behind Act,* signed into federal law by President George W. Bush on January 8, 2002. This law was put into place as an attempt to close the "achievement gap" between advantaged and disadvantaged students by means of increased funding to schools in poorer areas, improvement in teacher qualifications, and improved methods for testing students. It was intended to be seen as the first major federal legislation to improve education since the passing of the Elementary and Secondary Education Act of 1965. The intent of the *No Child Left Behind* (NCLB) Act was to assure fulfillment of the most fundamental goals of our educational system—that all children, regardless of race, ethnicity, socioeconomic status/family wealth, gender, religion, or disability, would have equal access to a good education.

Components of the NCLB Act were intended to be the following:

1. increased funding to schools in the poorer areas and states of the country;

2. ensuring that ideally every student would be taught by highly qualified teachers;

3. holding schools that receive federal funds accountable for raising the achievement levels of their students—especially by *disaggregating* their achievement test data, that is, by reporting data school by school rather than by entire school districts. In this way, it would be difficult for a school district to "hide" the data for an under-performing school;

4. providing special test procedures for students with certified disabilities.

While these are laudable goals, it is nonetheless true that after its inception, the NCLB Act has been opposed by many educators, federal legislators, and even many state legislatures. The reasons have been: the administration having provided funds far below what was initially promised to school districts; difficulties in measuring teacher effectiveness fairly across different school districts and states; administering tests that were simply too difficult for disabled students, thus virtually guaranteeing their failure; and a failure in the means by which schools were to have been held accountable. Sadly, at least one analysis (Meier at al. 2004) argues forcibly that the *No Child Left Behind Act* should be called instead the "Many Children Left Behind" Act.

High-Stakes Testing

Prior to the 2002 *No Child Left Behind Act,* in 1997, then-President William Clinton proposed a voluntary nationwide program of reading and mathematics testing. The tests were to be achievement tests, not traditional "ability" tests. Thus, the testing program was to have measured what is actually learned by the student and was not presumed to measure some abstract "potential" or "ability." The tests were "high stakes" in the sense that one's score on the test was to be used for purposes of (a) track assignment, (b) whether or not to promote the student to the next grade, and finally, (c) whether or not to award or withhold the diploma at graduation. Such tests were to be designed for elementary school, middle school, and high school.

The idea was that accountability for educational outcomes should be the shared responsibility of the states, the school districts, educators, parents, and students. Although the ones being tested were the students, nevertheless the responsibility for outcomes was not to be shouldered by the students alone. Considerable data regarding the *validity* of such tests was to be provided, particularly as regarding the matter of whether or not the tests were equally valid for all racial–ethnic groups, social class strata, and for both genders (Huebert and Hauser 1999). To date, such data have yet to be produced. In particular, one recent high-stakes test—the

Texas Assessment of Academic Skills (TAAS), a voluntary test given to fourth and eighth graders in the last several years—is, according to one source, "harmful by its rigid format, its artificial treatment of subject matter, its embodiment of discredited learning theories, [and] its lack of attention to children's cultures and languages" (Orfield and Kornhaber 2001, p. 147).

The Future of American Education: New Technology in the Classroom

Among the tasks of educational institutions in the United States is preparation of the young for a rapidly changing world. Technological changes are taking place, and education is both cause and effect in the process. Education must therefore look toward the future and newly emerging technology and social structure, instead of relying heavily on past curricula and programs. With increased diversity in the schools, the public will need to be vigilant in seeing that all students have the technological skills needed to compete in today's world.

Current projections are that, by the year 2008, approximately 70 million students will be enrolled in public and private schools (U.S. Census Bureau 2002). These demographic trends indicate that the proportions of Blacks and Hispanics in the population will increase substantially and thus place pressure on schools to adapt to the needs of a more racially and ethnically diverse population. The educational system needs to adapt to the changing social, ethnic, and gender diversity of the U.S. population, including developing and using technology to meet diverse needs in the population.

By way of technological change, the increased use of computers in schools will continue to have consequences for both curricular and extracurricular activities. In the same way that the Industrial Revolution of the nineteenth century had a dramatic impact on education, so education is now being changed by the advent of the computer and the "information revolution." Because computers are used widely throughout the occupational spectrum, use in schools has greatly increased and will in all likelihood continue to increase. Current computer use is related to race and class: 65 percent of White students use computers in schools compared with 55 percent of Black and Hispanic students. Looking at usage by social class, students in high-income families are more likely to use computers in school.

Student use of the Internet is also related to race. White students are more likely to have used the Internet in the prior six months, use the Internet on a home PC (personal computer), and use the Internet at other locations (Kerbicov 1998). Differences that once existed by gender have virtually disappeared, although men are more likely than women to use computers in college and beyond.

In current parlance, this "digital divide" is along the lines of race and class and less along the lines of gender. Finally, a digital divide has also developed between professors and students at many two- and four-year colleges, as students become more adept than their professors at searching the net.

Chapter Summary

What is the importance of the education institution?

Education is the social institution that is concerned with the formal transmission of society's knowledge. It is therefore part of the socialization process.

How diverse are college enrollments and graduations?

Although the overall percentage of U.S. high school and college graduates has increased dramatically since the 1950s, the increase has been unequal across racial and ethnic groups. Blacks lag behind Whites, and Hispanics lag behind Blacks in both high school and four-year college graduation rates.

How does education affect, or not affect, occupational and income attainment and mobility?

The number of years of formal education that individuals have has important, but in many ways modest, effects on their ultimate occupation and income. Social class origin affects the extent of educational attainment (the higher the social class origins, the more education ultimately attained) as well as occupation and income (higher social class origin means, probabilistically, both a more prestigious occupation and more income). There is evidence that the social class one is born into has a greater effect on later occupation and income than does educational attainment. This limits social mobility in the United States, but even more so in countries such as Japan.

In what ways has education increased social inequality instead of reduced it?

Although the education system in the United States has traditionally been a major means for reducing racial, gender, and class inequalities among people, the educational institution has also perpetuated these inequalities. Test biases based on culture, language, race, gender, and class have not been substantially reduced in *standardized tests*. Despite scientific evidence to the contrary, books such as *The Bell Curve* argue for the presence of a *cognitive-elite class* based on inherited intelligence.

How do tracking, labeling, and teacher expectancy affect women, minorities, and working-class persons?

Tracking and *labeling* continue to affect minorities, women, and the working classes disproportionately. Women are especially underrepresented in high school science tracks and classes. Students are subject to the *labeling effect,* which affects both teacher expectations and student performance. *Teacher expectancy effects* and the *self-fulfilling prophecy* work to the detriment of minorities, working-class persons, and women. Women are consistently and routinely short-changed by the entire education system in America by a relative lack of interaction from teachers, gender bias on standardized tests, lack of predictive validity in science and math on the standardized tests, differential treatment of minority women by both White as well as minority teachers, negative gender stereotyping in textbooks and readers, and many other aspects.

How does stereotyping affect academic performance?

Stereotype threat affects the actual performance of Blacks as well as women on standardized ability tests. Positive stereotyping (as with Asian Americans) can increase test performance, and negative stereotyping (as with other minorities and women) can decrease it.

What are current issues in school reform?

New social policies regarding schools are fueled by public concerns about educational quality and cost. Charter schools, vouchers, and home schooling are all new ways to school children. Multiculturalism is also needed to reflect the nation's diverse population, as are new technologies needed in an information-based society.

Key Terms

achievement test 431
back-to-basics
    movement 440
charter schools 441
cognitive ability 431
cognitive elite 433
credentialism 426
educational deflation
    428

labeling effect 435
latent functions 425
multiculturalism
    movement 441
predictive validity 432
schooling 424
self-fulfilling prophecy
    436

standardized ability
    test 431
stereotype threat effect
    439

teacher expectancy
    effect 435
tracking 434

Researching Society with MicroCase Online

You can see the results of actual research by using the Wadsworth MicroCase® Online feature available to you. This feature allows you to look at some of the results from national surveys, census data, and other data sources. You can explore this easy-to-use feature on your own, but try this example. Suppose you want to know:

Is the amount of education one has related to job satisfaction?

To answer this question, go to http://sociology.wadsworth .com/andersen_taylor4e/, select MicroCase Online from the left navigation bar, and follow the directions there to analyze the following data.

Data File: GSS

Task: Cross-Tabulation

Row Variable: LIKEJOB?

Column Variable: Education

Questions

Once you have your results, answer the following questions:

1. What percentage of respondents reported they are very satisfied with their job?
    _____%

2. What percentage of respondents reported they are very dissatisfied with their job?
    _____%

3. Which educational group is most satisfied with their jobs?
    _____%

4. Describe the differences you found, if any, among respondents with different levels of education. How do these results meet your expectations?

The Companion Website for Sociology: Understanding a Diverse Society, *Fourth Edition*

http://sociology.wadsworth.com/andersen_taylor4e/

Supplement your review of this chapter by going to the companion website to take one of the Tutorial Quizzes, use the flash cards to master key terms, and check out the many other study aids you'll find there. You'll also find special features such as GSS Data and Census 2000 information, data and resources at your fingertips to help you with that special project or do some research on your own.

Suggested Reading and Web Resources

Crouse, James, and Dale Trusheim. 1988. *The Case Against the SAT.* Chicago, IL: University of Chicago Press.
Based on hundreds of studies, this book concludes that the Scholastic Assessment Test (SAT) is unnecessary and harms students from non-White and low-income backgrounds, as well as some women. The book recommends dropping the SAT and relying on courses taken and course grades until more appropriate tests can be developed.

Feagin, Joe, Hernan Vera, and Imani Nikitah. 1996. *The Agony of Education.* New York: Routledge.
This is an engaging account about the experience of African American college students attending predominantly White institutions. The poignantly told experiences challenge stereotypical notions about Black experience on White campuses and suggest pathways for positive change.

Fischer, Claude S., Michael Hout, Martin Sánchez Jankowski, Samuel R. Lucas, Ann Swidler, and Kim Voss. 1996. *Inequality by Design: Cracking the Bell Curve Myth.* Princeton, NJ: Princeton University Press.
This is a clearly written and comprehensive critique of Richard J. Herrnstein and Charles Murray's (1994) The Bell Curve. *The authors give attention to definitions and measurements of intelligence, problems with the data in the studies used, race and intelligence, types of inequality in society, and other conceptual and methodological issues important to understanding the public debate about* The Bell Curve.

Jencks, Christopher, and Meredith Phillips (eds). 1998. *The Black-White Test Score Gap.* Washington, DC: Brookings Institution Press.
This anthology is a highly readable and well-balanced account of aspects of the Black/White test score gap on the Scholastic Assessment Test (SAT) and other tests. It contains complete discussions of test bias and its different types and forms, the role of socioeconomic background and parenting, teacher expectancy, stereotype threat effect, and other causes of the test score gap. A number of policy issues are also discussed.

Sjulmana, James L., and William G. Bowen. 2001. *The Game of Life: College Sports and Educational Values.* Princeton, NJ: Princeton University Press.
The book addresses a major issue in higher education: the influence of college and university athletics on the fundamental goals of higher education. Based on surveys of athletes, professors, and college and university administrators, the work represents an impressive study of the issues, pro and con.

American Association of University Women
www.aauw.org
This organization, which promotes equity for women and education, regularly produces reports about gender and education. Their reports are often the basis for changes in social policy.

National Center for Education Statistics
www.ed.gov/NCES/index.html
This division of the U.S. government collects and publishes reports on the condition of education, including students, faculty and teaching staff, school expenditures, and other useful compilations of data.

U.S. Census Bureau
www.census.gov
From the U.S. Census Bureau home page, one can find data on educational attainment, including detailed tables showing such things as educational attainment and race, sex, age, income, and family status.

Hassam

• • •

Religion

The profound effect of religion on society and human behavior is easily observed in everyday life. Church steeples dot the landscape everywhere. Invocations to a religious deity occur at the beginning of many public gatherings. Marriage ceremonies are typically held in a religious setting. Millions of people watch religious programming on television and listen to religious broadcasts on the radio. Much of what is driving the world news seems fueled by religious extremism. But even within the United States, intensely felt religious values have held the stage in legislatures, classrooms, and the courts. Some of life's sweetest moments are marked by religious celebration, and some of its most bitter conflicts persist because of unshakable religious conviction. Religion is an integrative force in society, but it is also the basis for many deeply rooted social conflicts.

Religion is a social phenomenon that can be studied just like other social phenomena such as education, work, and family. From a sociological perspective, several questions about religion and its role in society are of paramount interest. How are patterns of religious belief and practice related to social class, race, age, gender, and level of education? How are religious institutions organized? How does religion influence social change? Whether they are personally religious or not, sociologists use their skills of objective analysis to study the social dimensions of religion and answer such questions as these. Studying religion as a sociological phenomenon requires not just faith, but empirical evidence.

Still, one can have a sociological perspective on religion and be a religious person. Many sociologists have strong religious commitments, and some have belonged to the clergy or other religious orders. Many combine their sociological and religious commitments to work for a more just society. In developing a sociological perspective on religion, what is important is one's ability to be nonjudgmental, not religious beliefs per se. The sociological observer must be willing to examine religion in its social and cultural context and to separate the practices of religion from dogma and moral tenets. This may be difficult for people who hold dogmatic religious values or who have strict ideas of right and wrong with no sense of the underlying social dimensions to religious beliefs and practices.

Religion is a social institution, one that is connected to the other institutions of society. People typically learn religious beliefs in the family, and religion and politics become intertwined, as we see in noticing how religion has often driven political debates about such things as abortion, school prayer, and international politics. The economy also

Sociology ⊛ Now™

Reviewing is as easy as ❶❷❸.

Use SociologyNow to help you make the grade on your next exam. When you are finished reading this chapter, go to the chapter review for instructions on how to make SociologyNow work for you.

is intertwined with religion. Many businesses make one-half of their annual profits in the sales season leading up to Christmas. Holidays such as Kwanza and Hanukkah are also commercialized now through the sale of greeting cards, gifts, wrapping paper, and other products. Understanding religion from a sociological perspective helps you see the many social dimensions in this complex phenomenon. •••

Defining Religion

What is religion? Most people think of it as a category of experience separate from the everyday, perhaps involving communication with a deity or communion with the supernatural (Johnstone 1992). One person might define religion as the belief in God, another as the observation of religious rituals—going to a church, temple, or mosque and observing the rituals of religious faith. Another person might define religion as a purely private experience, a matter of what one person believes. But is believing in UFOs (unidentified flying objects) religion? What about believing that the world is round or believing in evolution? Sociologists define **religion** as *an institutionalized system of symbols, beliefs, values, and practices by which a group of people interprets and responds to what they feel is sacred and that provides answers to questions of ultimate meaning* (Glock and Stark 1965: 4; Johnstone 1992: 14). The elements of this definition are

1. **Religion is institutionalized.** Religion is more than just beliefs. It is a pattern of social action organized around the beliefs, practices, and symbols that people develop to answer questions about the meaning of existence. As an institution, religion presents itself as larger than any single individual. It persists over time and has an organizational structure into which members are socialized.

2. **Religion is a feature of groups.** Religion is built around a community of people with similar beliefs. It is a cohesive force among believers because it is a basis for group identity and gives people a sense of belonging to a community or organization. Religious groups can be formally organized, as in the case of large, bureaucratic churches, or they may be more informally organized, ranging from prayer groups to cults. Some religious communities are extremely close-knit, as in convents. Others are more diffuse, such as people who identify themselves as Protestant but attend church only on Easter.

3. **Religions are based on beliefs that are considered sacred.** The **sacred** is that which is set apart from ordinary activity, seen as holy, and protected by special rites and rituals. The sacred is distinguished from what is called the **profane,** that which is of the everyday world and is specifically not religious (Durkheim 1912 [1947]; Chalfant et al. 1987). Religions define what is sacred. Most religions have sacred objects and sacred symbols. These holy symbols are infused with special religious meaning and inspire awe and reverence.

A **totem** is an object or living thing that a religious group regards with special awe and reverence. A statue of Buddha is a totem; so is a crucifix hanging on a wall. Among the Zuni (a Native American religious group), *fetishes* are totems. These are small, beautifully carved animal objects representing different dimensions of Zuni spirituality. A totem is important not for what it is, but for what it represents. The sight of a crucifix may bring tears to the faithful's eyes, and kneeling before a Buddha may inspire deep spiritual meditation. There is nothing inherent in these objects or events that defines them as sacred. Instead, the significance of a totem derives from the sacred meaning that is socially attributed to the object. To a Christian taking communion, a piece of bread is defined as the flesh of Jesus. Eating the bread unites the communicant mystically with Christ. To a nonbeliever, the bread is simply that—a piece of bread (McGuire 2002). Likewise, Native Americans hold certain ground to be sacred and are deeply offended when the holy ground is disturbed by industrial or commercial developers who see only

Jeff Stahler reprinted by permission of Newspaper Enterprise Association, Inc.

potential profit. Because religious beliefs are held so strongly, sacred religious symbols have enormous power and generate strong emotional responses. Certain behaviors can inspire, offend, or anger people, depending on the religious meaning of the behavior.

4. Religion establishes values and moral proscriptions for behavior. A *proscription* is a constraint imposed by external forces. Religion typically establishes proscriptions for the behavior of believers, some quite strict. For example, the Catholic Church defines living together as sexual partners outside marriage as a sin. The lifestyle is condemned as immoral "selfishness" and an "unwillingness to make a lifelong commitment" through marriage (National Council of Catholic Bishops 1992). Often religious believers come to see such moral proscriptions as simply "right" and behave accordingly. Other times, individuals may consciously reject moral proscriptions, although they may still feel guilty when they engage in a forbidden practice.

5. Religion establishes norms for behavior. Religious beliefs establish social norms about how the faithful should behave in certain situations. Worshipers may be expected to cover their heads in a temple, mosque, or cathedral, or to wear certain clothes. Such behavioral expectations may be very strong. The next time you are at a gathering where a prayer is said before a meal, note how many people bow their heads, even though some people present may not believe in the deity being invoked.

6. Religion provides answers to questions of ultimate meaning. The ordinary beliefs of daily life, **secular** beliefs, may be institutionalized, but they are specifically not religious. Science, for example, generates secular beliefs based on particular ways of thinking—logic and empirical observation are at the root of scientific beliefs. Religious beliefs often have a supernatural element. They emerge from spiritual needs and may provide answers to questions that cannot be probed with the profane tools of science and reason. Think of the difference in how religion and science explain the origins of life.

THINKING SOCIOLOGICALLY

For one week, keep a daily log, noting every time you see an explicit or implicit reference to religion. At the end of the week, review your notes and ask yourself how religion is connected to other social institutions. Based on what you have seen, describe what you think is the relationship between the *sacred* and the *secular* in this society.

Religious spirituality takes many forms but produces feelings of awe and reverence among believers.

© Bill Gillette/Stock Boston

The Significance of Religion in U.S. Society

It may surprise you to know that the United States is one of the most religious societies in the world. Two-thirds of Americans think religion can solve all or most of society's problems. Forty-two percent of the population describe themselves as "born again" (Winseman 2004). These indicators reveal some of the significance of religion in contemporary U.S. society and the dominance of Christian beliefs.

For millions of people, religion is the strongest component of their individual and group identity, and often the basis of culture in society. Much of the world's celebrated art, architecture, and music has its origins in religion. In the classical art of western Europe, the Buddhist temples of the East, and the gospel rhythms of contemporary rock are signs of how cultural development is inspired by religion. For many Mexican Americans, the belief that the dead live on is reflected in many cultural artifacts (carvings, paintings, and clay objects), as well as Day of the Dead commemorative celebrations. These religious beliefs ignite feelings of awe and reverence. Even among people who are not particularly religious, the experience of seeing such religious images or entering a beautiful cathedral or temple may be deeply moving.

The Dominance of Christianity

Despite the constitutional principle of the separation of church and state, Christian religious beliefs and practices dominate U.S. culture. Christianity is often treated as if it were the national religion, although there are many religions in the United States. It is commonly said that the United States is based on a *Judeo-Christian* heritage, meaning that our basic cultural beliefs stem from the traditions of the pre-Christian Old Testament of the Bible (the Judaic tradition) and the Gospels of the New Testament (the Christian heritage). The dominance of Christianity is visible everywhere. State-sponsored colleges and universities typically close for Christmas break, not Hanukkah. Christmas is a national holiday, but not Ramadan, the most sacred holiday among Muslims. The dominance of Christianity in national thought is well illustrated by a comment once heard on a major city radio station: "Happy Hanukkah to all our Jewish friends in the audience. Hanukkah is to our Jewish friends what Christmas is to *we Americans*." Clearly, Christian religious traditions are often observed in the culture with little sensitivity to the religious beliefs of other groups.

DEBUNKING SOCIETY'S MYTHS

Myth: The principle of separation of church and state is firmly held in the United States.

Sociological perspective: Although separation of church and state is a constitutional principle, dominant Christian traditions and symbols pervade U.S. culture and are the basis for many of our social institutions.

Measuring Religious Faith

Religiosity is the intensity and consistency of practice of a person's (or group's) faith. Sociologists measure religiosity both by asking people about their religious beliefs and by measuring membership in religious organizations and attendance at religious services. The vast majority in the United States identify themselves as Protestant (48 percent), Catholic (25 percent), Christian (nonspecific, 11 percent), Jewish (3 percent), Mormon (1 percent), or Greek or Russian Orthodox (1 percent), with diverse religions (4 percent) and those with no religious identification (7 percent) constituting the balance (see Figure 17.1). Sixty-four percent say they are members of a church or synagogue, although only

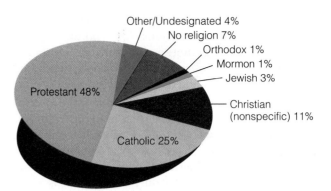

Figure 17.1 Religious Identification in the United States

Source: Carroll, Joseph. 2004. "Religion is 'Very Important' to 6 in 10 Americans." *The Gallup Poll.* Princeton, NJ: Gallup Organization. Used with permission. Website: **www.gallup.com**

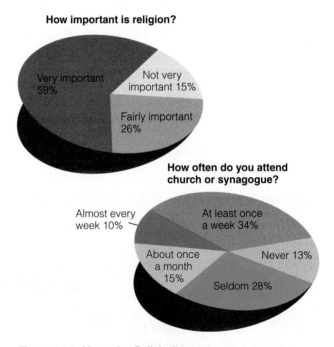

Figure 17.2 Measuring Religiosity

Source: Carroll, Joseph. 2004. "Religion is 'Very Important' to 6 in 10 Americans." *The Gallup Poll.* Princeton, NJ: Gallup Organization. Used with permission. Website: **www.gallup.com**

34 percent attend weekly (Carroll 2004; see Figure 17.2).

THINKING SOCIOLOGICALLY

Go to the library and look up a recent Gallup national opinion poll. Find a question asked on a subject of interest to you (for example, abortion rights, a presidential election, or gay rights) and look to see if opinions of different religious orientations are reported. What does this tell you about the connection between *religiosity* and social and political attitudes? If you were designing the national survey, what questions would you ask?

Sociologists are cautious about the accuracy of such polls, because they are based on self-reports. Measures of religiosity based on observed church attendance can be tricky because people can be religious but observe religion in diverse ways. Nevertheless, the data on religious identification and practice show the great significance of religion in people's lives (Caplow 1998; Hout and Greeley 1998; Woodberry 1998; Hadaway et al. 1998; Presser and Stinson 1998).

Religiosity varies significantly among different groups in society. Church membership and attendance are higher among women than men and more prevalent among older than younger people. African Americans are more likely to belong to and attend church than Whites. On the whole, church membership and attendance fluctuate over time. Membership has decreased slightly since 1940, but attendance has remained largely the same since then. The pattern of religion in the United States is, however, a mosaic one. Large national religious organizations, such as the mainline Protestant denominations, have lost many members, whereas smaller, local congregations have been growing. Changes in immigration patterns have also affected religious patterns in the United States, with Muslims, Buddhists, and Hindus now accounting for several million believers (Niebuhr 1998; Haddad et al. 2003). One of the greatest changes has been a tremendous increase in the number identifying as evangelical Protestants, but Islam has also been one of the fastest growing religions in the

United States in recent years (Dudley and Roozen 2001; see Figure 17.3).

These data indicate the extent of religious identification but tell little about religiosity. To develop a picture of the day-to-day importance of religion, the sociologist must grapple with measures derived from the answers to subjective questions such as, How important is religion in your life? In recent years, the number of people who think that religion is very important in daily life has been relatively steady at about 59 percent; at the same time, about one-quarter of the population believe that religion is largely old-fashioned and out of date. The number who think that religion can answer all or most of today's problems has also been relatively steady at about 61 percent (Carroll 2004).

Forms of Religion

Religions can be categorized in different ways according to the specific characteristics of faiths and how religious groups are organized. In different societies and among different religious groups, the form religion takes reflects differing belief systems and reflects and supports other features of the society. Believing in one god or many, worshiping in small or large groups, and associating religious faith with gender roles all contribute to the social organization of religion and its relationship to the rest of society.

Monotheism and Polytheism

One basic way to categorize religions is by the number of gods or goddesses they worship. **Monotheism** is the worship of a single god. Christianity and Judaism are monotheistic because both Christians and Jews believe in a single god who created the universe. Monotheistic religions typically define god as omnipotent (all-powerful) and omniscient (all-knowing).

Polytheism is the worship of more than one god or deity. Most Native American cultures, for example, do not worship a single god but see harmony in the world derived from the connection among many gods and many peoples, spirits, and the earth (Allen 1986). They believe there are many gods and goddesses, each presiding over different aspects of life. Many Asian religions are polytheistic, defining a number of gods in terms of the phenomena they control. Thus, there may be a goddess of the harvest or god of the sun.

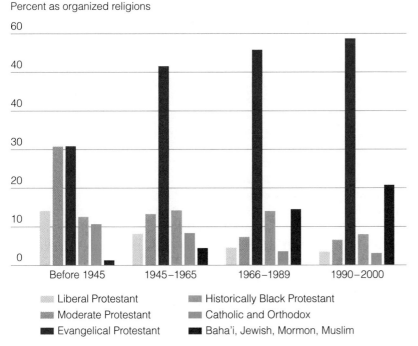

Percent as organized religions

Figure 17.3 The New Face of Religion

Source: Dudley, Carl S., and Roozen, David A. 2001. *Faith Communities Today: A Report on Religion in the United States Today.* Hartford, CT: Hartford Institute for Religion Research, Hartford Seminary.

In some religions—Hinduism, for example—individual worshipers choose deities to whom they feel most near. Within Hinduism, the universe is seen as so vast that it is believed to be beyond the grasp of a single individual, even a powerful god. Hinduism is extraordinarily complex, with millions of gods, demons, sages, and heroes—all overlapping and combined in religious mythology (Grimal 1963).

Patriarchal and Matriarchal Religions

Religions may also be patriarchal or matriarchal. In **patriarchal religions,** the beliefs and practices of the religion are based on male power and authority. Christianity is a patriarchal religion. The dominance of men is emphasized in the role of women in the church, the instruction given on relations between the sexes, and the language of worship itself (as in "Declare *his* glory," Psalms 96:3). The patriarchal character of Christianity has recently been challenged by arguments that God is a genderless spirit and that the patriarchal trend in Christianity propagates injustice.

Matriarchal religions are based on the centrality of female goddesses, who may be seen as the source of food, nurturance, and love, or who may serve as emblems of the power of women (McGuire 2002). In societies based on matriarchal religions, women are more likely to share power with men in the society at large. Because women and men participate more equally in secular life, it is more possible for them to imagine egalitarian divine beings (Ochshorn 1981). Likewise, in highly sexist, patriarchal societies, religious beliefs are likely to be very patriarchal.

Exclusive and Inclusive Religious Groups

Religious groups can also be categorized according to how exclusive or inclusive they are. **Exclusive religious groups** have an easily identifiable religion and culture, including distinctive beliefs and strong moral teachings. Members are expected to conform strictly to religious values and behavioral norms. Black Muslims, for example, expect followers to lead a disciplined life, the strictures of which range from being economically self-reliant (for example, running their own businesses) to following certain dietary restrictions (such as not eating pork).

Exclusive religious groups tend to be small in membership but are coherent, close-knit communities with strongly held beliefs. Groups such as the Jehovah's Witnesses, Assembly of God, Mormons, and Pentecostal and Holiness churches have exclusive orientations. Exclusive groups are the fastest growing category of religious groups in recent years (Roof 1999; Wuthnow 1998).

Inclusive religious groups have a more moderate and liberal religious orientation. They are more *ecumenical,* meaning that they stress interdenominational cooperation and the importance of common religious work. The National Council of Churches, for example, has encouraged participating churches to adopt liberal, activist agendas for social change. This organization provided some support to the civil rights movement in the 1960s. Inclusive groups tend to have a broad-based membership. Their beliefs are often more weakly held, consequently forming a more diffuse religious community. As a result, they are more likely than exclusive groups to make secular compromises. Examples of inclusive religious groups are the more "establishment" churches, such as the United Methodist and United Presbyterian Churches.

Sociological Theories of Religion

The sociological study of religion probes how religion relates to the structure of society. Recall that one basic question sociologists ask is, What holds society together? To people living in a society, the society seems to be both unique and coherent. Coherence comes from both the social institutions that characterize society and the beliefs that hold society together. In both instances, religion plays a key role. From the functionalist perspective of sociological theory, religion is an integrative force in society because it has the power to shape collective beliefs. This perspective was originally developed by Emile Durkheim. In a somewhat different vein, the sociologist Max Weber saw religion in terms of how it supported other social institutions. Weber thought that religious beliefs provided a cultural framework that supported the development of specific social institutions in other realms, such as the economy. From a conflict theory point of view, religion is related to social inequality. Karl Marx was the historical source for the ideas relating religion and social conflict. In the following sections, we examine each of these major perspectives in the sociology of religion.

Emile Durkheim: The Functions of Religion

Emile Durkheim, a classic sociological theorist, argued that religion is functional for society because it reaffirms the social bonds that people have with each other, creating social cohesion and integration. Durkheim believed that the cohesiveness of a society depends on the organization of its belief system. Societies with a unified belief system would be highly cohesive; those with a more diffuse or competing belief system would be less cohesive.

Religious **rituals** are symbolic activities that express a group's spiritual convictions. Making a pilgrimage to Mecca, for example, is an expression of religious faith and a reminder of religious belonging. In Durkheim's view, religious rituals are vehicles for the creation, expression, and reinforcement of social cohesion. Groups performing rituals are expressing their identity as a group. Most religions incorporate rituals as part of the practice of faith, although some are more ritualistic than others. Whether the rituals of a group are highly elaborate or casually informal, they are symbolic behaviors that freshen a group's awareness of its unifying beliefs. Lighting candles, chanting, or receiving a sacrament are behaviors that reunite the faithful and help them identify with the religious group, its goals, and its beliefs (McGuire 2002).

Durkheim assessed the significance of religious ritual within the context of the larger point that a major function of religion is to confer identity on a person. In his classic work *The Elementary Forms of Religious Life*, Durkheim argued that through religion, individuals are able to transcend their individual identities and see themselves as part of a larger group. Durkheim believed that religion binds individuals to the society in which they live by establishing what he called a **collective consciousness**, the body of beliefs that are common to a community or society and that give people a sense of belonging. In many societies, religion establishes the collective consciousness and creates in people the feeling that they are part of a common whole.

Max Weber: The Protestant Ethic and the Spirit of Capitalism

Theorist Max Weber also saw a fit between the religious principles of society and other institutional needs. In his classic work *The Protestant Ethic and the Spirit of Capitalism,* Weber argued that the Protestant faith supported the development of capitalism in the Western world. He began by noting a seeming contradiction: How could a religion that supposedly condemns extensive material consumption coexist in a society such as the United States with an economic system based on the pursuit of profit and material success?

Weber argued that these ideals were not as contradictory as they seemed. As the Protestant faith developed, it included a belief in predestination—one's salvation is predetermined and a gift from God, not something earned. This state of affairs created doubt and anxiety among believers, who searched for clues in the here and now about whether they were among the chosen—called the "elect." According to Weber, material success was taken to be one clue that a person was among the elect and thus favored by God, which drove early Protestants to relentless work as a means of confirming (and demonstrating) their salvation. As it happens, hard work and self-denial—the key features of the **Protestant Ethic**—lead not only to salvation but also to the accumulation of capital. The religious ideas supported by the Protestant Ethic therefore fit nicely with the needs of capitalism. According to Weber, these austere religionists stockpiled wealth, had an irresistible motive to earn more (eternal salvation), and were inclined to spend little on themselves, leaving a larger share for investment and driving the growth of capitalism (Weber 1958/1904).

Karl Marx: Religion, Social Conflict, and Oppression

Durkheim and Weber concentrated on how religion contributes to the cohesion of society. Religion can also be the basis for conflict. This aspect of religion can be seen daily in the headlines of the world's newspapers. Conflict between Protestants and Catholics in Northern Ireland has been the basis for years of social and political warfare. In the Middle East, differences between Muslims and Jews have caused decades of political instability. In Bosnia, the ethnic cleansing perpetrated by the Serbs against the Croats resulted in the annihilation of tens of thousands of people. These conflicts are not solely religious, but religion plays an inextricable part. Certainly religious wars, religious terrorism, and religious genocide have contributed some of the most violent and tragic episodes of world history. Religion is often used as a *justification* for terrorist acts. In the case of the September 11, 2001, terrorist attacks on the World Trade Center Towers and the Pentagon, tape recordings of the pilots just prior to impact revealed the pilots shouting praise to Allah. Thus, the image of religion in history has two incompatible sides: piety and contemplation on the one hand, battle flags and weapons on the other.

The link between religion and social inequality is also key to the theories of Karl Marx. Marx saw religion as a tool for class oppression. According to Marx, oppressed people develop religion to soothe their distress (Marx 1972/1843). The promise of a better life hereafter makes the present life more bearable, and the belief that "God's will" steers the present life makes it easier for people to accept their lot. To Marx, religion is a form of *false consciousness* (see Chapter 9) because it prevents people from rising up against oppression. He called religion the "opium of the people" because it encourages passivity and acceptance of the status quo.

To Marx, religion is an **ideology**—a belief system that legitimates the social order and supports the ideas of the ruling class. When subordinated groups internalize the views of the dominant class, they come to believe the social order that oppresses them is legitimate. Marx thus saw religion as supporting the status quo and being inherently conservative, that is, resisting change and preserving the existing social order. Marx suggested that religion promotes stratification because it generally supports a hierarchy of people on earth and the subordination of humankind to divine authority.

We can use Marx's argument to see how Christianity supported the system of slavery. When European explorers first encountered African people in an attempt to teach them Christianity, they regarded them as godless savages. The brutality of the slave trade was justified by the argument that slaves were being rescued from damnation and exposed to the Christian way of life. Principles of Christianity thus legitimated the system of slavery in the eyes of the slave owners, who could convince themselves they were doing good by enslaving Black people. Slaves were perceived as inferior to their masters because of their allegedly heathen beliefs, but preachers also taught slaves that they would go to heaven if they obeyed their masters. In this way, Christian religion served the economic needs of slavery.

Ironically, this extraordinary system of oppression has produced a strong belief in Christianity on the part of most African American people. While Christianity was used by slave owners to justify slavery, for the slaves it became a wellspring of hope. Slaves had a strong belief in the afterlife and religious meetings offered one of the few legitimate occasions for slaves to congregate in groups. Religious gatherings were frequently the scene for political mobilization and a collective response to oppression (Raboteau 1978; Genovese 1972).

Religion can be the basis for liberating social change. In the civil rights movement in the United States and in Latin American liberation movements, the words and actions of religious organizations have been central to mobilizing people for change. This does not undermine Marx's main point, however, because there remains ample evidence of the role of religion in generating social conflict and resisting social change.

Symbolic Interaction: Becoming Religious

Symbolic interaction theory emphasizes the process by which people become religious. The process can be slow and gradual, as when someone switches to a new religious faith (such as a Christian person converting to Judaism) or it can be more dramatic, as when a person joins a cult or some other extreme religious group. Different religious beliefs and practices emerge in different social and historical contexts and are recognized by symbolic interactionists because these form the systems of understanding that frame religious behavior.

The emphasis on meaning construction that is typical of symbolic interaction also helps explain how the same religion can be interpreted differently by different groups or in different time periods. From this perspective, therefore, religious texts are not simply "truth" but have meaning that arise from the interpretation given to these texts by individuals. Thus, while the Bible may be seen as the literal word of God to some, symbolic interactionists would see this as one of many religious meaning systems. Other religious systems may interpret the same document in a different light.

Each sociological framework identified here reveals a different dimension of religion's role in contemporary society (see Table 17.1) providing a perspective that helps us understand the significance of religion. In the upcoming discussions, you can ask yourself, Is religion giving people a sense of common ground or dividing them into different factions? Is religion a source of oppression or liberation? Although stated in ordinary terms, such questions are the bases for theoretical analyses of religion in society.

World Religions and Religious Diversity

Worldwide, religion is a significant dimension of diverse cultures. In some nations, religion defines the political order and shapes many basic social institutions. Major religious movements, such as the spread of Christianity and the growth of Islam, have defined major events in world history. Across different cultures, many people share in the same faith, although the expression of that faith may vary in different societal contexts.

Table 17.1
Sociological Theories of Religion

	Functionalism	Conflict Theory	Symbolic Interaction
Religion and the Social Order	Is an integrative force in society	Is the basis for intergroup conflict; inequality in society is reflected in religious organizations, which are stratified by factors such as race, class, or gender	Is socially constructed and emerges with social and historical change
Religious Beliefs	Promote order by a sense of *collective consciousness*	Can provide legitimation for oppressive social conditions	Are socially constructed and subject to interpretation; they can also be learned through religious conversion
Religious Practices and Rituals	Reinforce a sense of social belonging	Define in-groups and out-groups, thereby defining group boundaries	Are symbolic activities that provide definitions of group and individual identity

The largest religion in the world, if measured in terms of numbers of followers, is Christianity, with an estimated 1.9 billion adherents, followed by Islam with 1.2 billion adherents. Persons expressing no religion are 759 million; Hindus are 761 million; Chinese folk-religionists are 379 million; Buddhists are 363 million; and other smaller groups include Sikhs, Bahá is, Confucians, and Shintoists. People of Jewish faith are 14 mil-

UNDERSTANDING DIVERSITY

Immigration, Faith, and Community

As Emile Durkheim theorized, religion provides people with a sense of attachment to society. It provides a collective identity and bonds with their community. While this is true for all groups, it has special meaning for immigrant groups, groups who sit between two cultures: their culture of origin and the culture of the host society.

Sociological studies of immigration typically note the significance of religion in providing immigrants a sense of community. Within immigrant communities, the churches, mosques, or temples operate as cultural centers. Not only do the organizations provide religious affiliation, but they may also provide social services and support networks to immigrants that are especially important in adjusting to a new social environment. This is apparent in the account of Charles Ryu, a Korean American, who came to the United States when he was seventeen and here reflects on the importance of the church in Korean communities in the United States:

In America, whatever the reason, the church has become a major and central anchoring institution for Korean immigrant society. Whereas no other institution supported the Korean immigrants, the church played the role of anything and everything—from social service, to education, to learning the Korean language: a place to gather, to meet other people, for social gratification, you name it. The way we think of church is more than in a religious connotation, as a place to go and pray, have worship service, to learn of God and comfort. Your identity is tied so closely to the church you go to. I think almost seventy to eighty percent of Korean Americans belong to church. And it becomes social evangelism. If those who had never considered themselves to be Christian want to be Korean, they go to church. They just go there for social reasons. You are acclimated into the gospel, and you say, I want to be baptized. Your life revolves around the church. It could also become a ghetto in the sense of a Chinatown or a Koreatown. But at the same time, it can be a place, if done properly, where the bruised identity can be healed and affirmed, because living in American society as a minority is a very difficult thing. You are nobody out there, but when you come to church, you are a somebody. The role of the black church in the civil rights era was the same thing.

Probably American churches have become specifically religious. I have my love life, my civic life, my professional life, and I have this church, religious life. But the Korean churches provide a new community of some sort that permeates the social matrix. It is such a solid, close-knit community, because everyone feels in a sense alienated outside, so there is this centripetal force. At the same time, most who come are well established or potentially well established.

Source: Ryu, Charles. 1992. "Koreans and Church." Pp. 162–163 in *Asian Americans,* edited by Joann Lee. New York: The New Press. Reprinted by permission. ●●●

Percent of
population

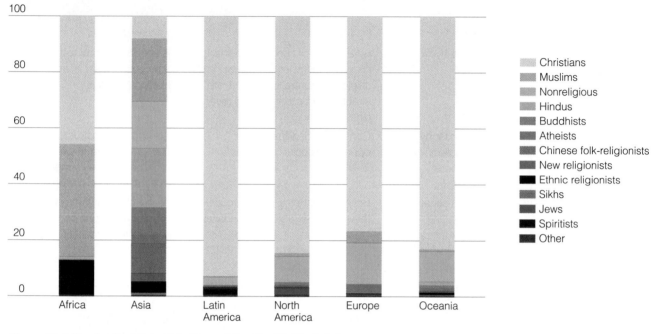

Figure 17.4 Viewing Society in Global Perspective: The World's Religions

Data: From the U.S. Census Bureau. 1999. *Statistical Abstract of the United States, 1999.* Washington, DC: U.S. Government Printing Office, p. 831.

lion, a relatively small number, but nonetheless one of the most significant religious groups (see Figure 17.4).

The United States is more religiously diverse than most societies in the world. Within the United States are well over two hundred various religious groups and denominations. These groups can be divided into a relative handful of broad categories, but the many faiths and denominations that people give as their religious affiliation show great diversity of religious belief within a single country. For many years, the major religious groupings have been Protestants, Catholics, and Jews, but there is enormous diversity in the various faiths observed in the United States, including such groups as Mormons, Greek Orthodox, Muslims, Hindus, Quakers, Mennonites, Swedenborgians, and new spiritual groups such as Eckankar, New Age groups, and others (Moore 2000; Cadge and Bender 2004). Increased immigration in the United States has also influenced the growth of diverse religions in the United States, including traditional Asian religions such as Hinduism and Buddhism. With increased diversity in society, you can expect increasing religious diversity as well (Eck 2002).

Christianity

Christianity developed in the Mediterranean region of Europe. Begun initially as a small cult, it grew rapidly from the years 40 to 350 A.D. to encompass about 56 percent of the population of the Roman empire

(Stark 1996). It has spread throughout the globe, largely as the result of European missionaries who colonized different parts of the world, proselytizing the Christian faith. Christianity now is rapidly spreading in its global influence, especially in Africa, Asia, and Latin America (Jenkins 2002).

Based on the teachings of Jesus, Christianity is a belief in the existence of a Holy Trinity: God, the creator; Jesus, the son of God; and the Holy Spirit, the personal experience of the presence of God. Through the charisma of Jesus, Christians learned that they could be saved by following Christian beliefs and that there would be an afterlife in heaven, where there is an everlasting communion with God. In the United States, Christianity is the dominant religion, although there is great diversity in the different forms of Christianity—both in the United States and worldwide. The two most common forms of Christianity are Protestantism and Catholicism—different in their organizational structure and specific religious practices but sharing the same basic religious beliefs.

Protestants Protestants form the largest religious group in the United States, although there is further diversity within this group. There are two main categories of Protestants: *mainline Protestants* and *conservatives*, also known as fundamentalists. Mainline Protestants include twenty-eight denominations, all belonging to the National Council of Churches. The denominations have

MAP 17.1 The World's Religions

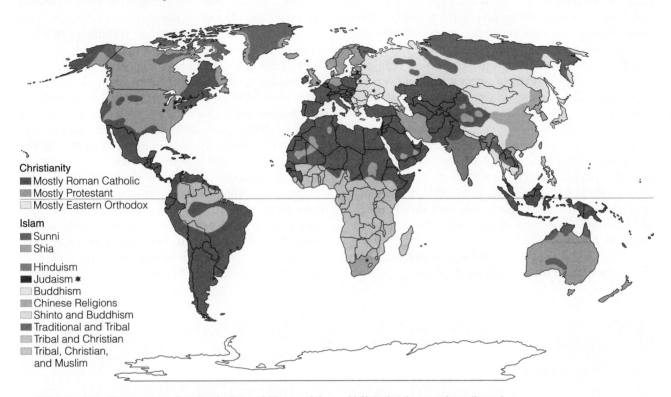

Christianity
- ■ Mostly Roman Catholic
- ■ Mostly Protestant
- □ Mostly Eastern Orthodox

Islam
- ■ Sunni
- ■ Shia

- ■ Hinduism
- ■ Judaism ✷
- □ Buddhism
- ■ Chinese Religions
- □ Shinto and Buddhism
- ■ Traditional and Tribal
- ■ Tribal and Christian
- ■ Tribal, Christian, and Muslim

This map shows the enormous diversity of religious faith around the world. Note that the map shows the major religion in each region, not the only religion. For example, western Europe is denoted as "mostly Roman Catholic" and the United States as "mostly Protestant" when we know that many other religions are prominent in both regions. Nonetheless, given that people tend to "see the world" through the eyes of their own religion, what does this map suggest to you about how religious faith informs people's worldviews? In what ways does this challenge the ethnocentrism that religion can produce?

*The red star indicates cities with large Jewish populations.
Data: Matthews, Warren. 2004. *World Religions.* Belmont, CA: Wadsworth.

a history of intergroup cooperation and social activism. Mainline Protestants tend to be less diligent about religious observance than either Roman Catholics or conservative Protestants. Many do not attend church regularly, even though the majority believe in the existence of God and the divinity of Christ (Chalfant et al. 1987).

Recently, mainline Protestants in the United States have seen declines in church membership and attendance. Two principal reasons for the decline are the general drop in religious affiliation and the growth in the number of conservative Protestants, which include the more dogmatic groups labeled evangelical, fundamentalist, and Pentecostal. These labels apply to a diverse group of religious affiliations centered on the belief that one can be "saved" only through a personal commitment to Christ. The label *fundamentalist* is taken from

the belief that the fundamental tenet of the Christian life claims the Bible is the literal word of God.

In the United States, fundamentalism is the conservative offspring of the evangelical movement, and fundamentalists are sometimes called evangelicals (Stacey and Gerard 1990). Pentecostals—a form of fundamentalism—are part of a worldwide religious movement, begun in the United States around the beginning of the twentieth century. Pentecostals believe in faith healing and a mystical conversion experience in which a follower is able to "speak in tongues," called *glossolalia*. Highly organized for political action, new conservative religious groups have had an enormous influence in society in recent years.

Roman Catholics Another form of Christianity is Catholicism. Worldwide, with its emphasis on Church hierarchy, Roman Catholics identify the Pope as the source of religious authority. Catholicism is a strict hierarchical religious system, with religious values and codes of behavior mandated by the Pope and the Vatican. The Vatican in Rome acts as the system of religious governance, with enormous power to influence the religious norms and beliefs of Catholic followers. The Catholic Church was once the ruling religious and political force in the Western world and is typically perceived as having a strong hold on its members, who are worldwide and cross all social and political boundaries.

In recent decades, the Catholic Church has embraced a number of liberal reforms. No longer, for instance, does the Catholic Church claim to be the sole source of religious truth. Mass is now typically celebrated in secular language, after centuries of being done only in Latin, and lay people have been granted larger roles in church rituals and church affairs.

Many Catholics in the United States hold more liberal attitudes than the official church promulgates. For example, the majority (61 percent) believe that Catholics who remarry are still good Catholics. Sixty-eight percent favor ordaining women, fully 84 percent believe Catholics should be allowed to practice artificial means of birth control, and 58 percent believe the Catholic Church should relax its standards forbidding all abortions under any circumstances, a matter explored more fully in the box "Doing Sociological Research: Diversity and Religious Tradition." Almost one-half (44 percent) also disagree with the official Church position that homosexual behavior is always wrong (Gallup and Lindsay 2000; Lyons 2002). These attitudes produce much social strain between Catholics in the United States and the position of the Vatican.

In the United States, Catholicism has long been associated with immigrant groups who brought Catholicism with them, just as Protestantism in the United States was established by immigrants from other nations. Irish Americans, Italian Americans, and most recently, Latin American immigrants have tended to be of the Catholic faith. As we have seen, recent immi-

DOING SOCIOLOGICAL RESEARCH

Diversity and Religious Tradition

What is it like to belong to a religious faith if you fundamentally disagree with the official teachings of the church? Can you be gay or lesbian and Catholic when the Catholic Church denounces homosexuality? Can you be Catholic and be pro-choice on the question of abortion? How do Catholic women who support the ordination of women keep their faith when the church denies women this opportunity? These are the questions that inspired Michele Dillon's research on Catholic identity.

Dillon studied those who are "institutionally marginalized Catholics"—that is, Catholics who choose to remain Catholic even though their understanding of Catholicism is denounced in church teachings. She also studied members of organizations who are committed to change within the Catholic Church: the Women's Ordination Conference; Catholics for Free Choice; and Dignity, a national association of gay and lesbian Catholics. Using archival data from these organizations, participant observation, questionnaires, interviews, and a content analysis of organizational literature, she learned how people integrate the multiple and seemingly contradictory identities of being

Catholic and being gay, lesbian, pro-choice, or feminist.

Dillon learned that these different groups see church doctrine as a social construct, contingent on specific historical and cultural contexts. This enables them to interpret church doctrine in ways that support their identities and beliefs. For example, gays and lesbians can be critical of official church teachings on sexuality but not reject other aspects of the Catholic faith, and women can interpret religious texts as supporting the ordination of women. By recognizing the socially constructed nature of Catholic doctrine, diverse groups can use church symbols and theology to challenge doctrine and to sustain their own religious faith.

Moreover, such groups see themselves as involved in emancipatory projects wherein they are trying to free themselves from oppressive traditions. Those Dillon studied recognize the differences among people that are part of modern life and see the Catholic Church not as a homogenous and unchanging force, but as an evolving institution that needs to respond to pluralism and diversity.

Michele Dillon's research suggests the need to see people's actions and identities as complex and multifaceted. Even within some of the most traditional institutions, groups can affirm difference while still maintaining a strong sense of group solidarity.

Questions to Consider

1. How does Michele Dillon's research challenge stereotypical views of those who adhere to Catholicism? *Keywords: Catholic identity, women and the priesthood*

2. What various minority groups have to reconcile traditional religious interpretations with more liberal interpretations of religious texts? *Keywords: religion and diversity, gays and the church, feminism and religion*

We have included InfoTrac College Edition keywords at the end of each question to make it easier for you to find more to read on these topics. Go to www.infotrac-college.com, an online library, to begin your search.

Source: Dillon, Michele. 1999. *Catholic Identity: Balancing Reason, Faith, and Power.* New York: Cambridge University Press. •••

gration has increased the number of Catholics in the United States.

Judaism

The Jewish faith is more than 4000 years old. Under Egyptian rule in ancient history, Jewish people endured centuries of slavery. Led from Egypt by Moses in the thirteenth century B.C., Jewish people were liberated and celebrate this freedom in the annual ritual of Passover—one of the holiest holidays. The Jews see themselves as "chosen people," meant to recognize their duty to obey God's laws as revealed in the Ten Commandments.

Although this religion has fewer members worldwide, Judaism has enormous world significance. Its teachings are the source of both Christian beliefs and Islamic beliefs. More than 40 percent of the world's Jewish population now lives in the United States, creating the largest community of Jewish people in the world. Despite this fact, Jews constitute only about 3 percent of the entire population of the United States. The state of Israel, founded in 1948 as a homeland for all Jews following the Holocaust of World War II, has given Jewish people a high profile in international politics.

In the United States, Jewish Americans are a relatively small proportion of the population, and their number is declining as the result of a low birthrate and high rate of interfaith marriage. When a Jewish man marries a non-Jewish woman, their offspring are not considered Jewish; offspring are considered Jewish only if the mother is Jewish. Nevertheless, Jewish people remain a significant religious minority. The periods of greatest Jewish immigration to the United States were in the mid-nineteenth century and between the 1880s and World War I. Many Jews tried to immigrate to the United States in the 1930s when Nazi persecution of Jews in Europe intensified, but the United States turned away many of these refugees, forcing them to return to Europe, where many were killed in the Holocaust (Sklare 1971; Wyman 1984).

There are significant divisions of culture and religious practice within Judaism. Orthodox Jews, a small fraction of the total number of Jews in the United States, adhere strictly to a traditional conception of their religious faith. They observe the biblical dietary laws, honor traditional codes of dress and behavior, and strictly observe the Sabbath, during which they may not travel, carry money, write, work, or do business. Reform Jews have a more secular orientation. Laypersons are typically excused from strict observation of religious proscriptions. In Reform temples, where women and men are treated more equally, prayers are often in English, not Hebrew. In general, there is flexibility of religious practice among Reform Jews. Conservative Judaism falls between Orthodox and Reform in terms of strict-

ness of observation (Sklare 1971). Only 19 percent of all Jewish Americans attend services in a typical week, but most observe the high holidays and hold bar mitzvahs (the ritual ceremony marking the transition from boyhood to manhood) for Jewish boys and sometimes bas mitzvahs (also called bat mitzvahs) for girls.

Islam

Islam is most typically associated with Middle Eastern countries, in part because of the significance of Islamic faith in Iran, Iraq, and Saudi Arabia. But Islamic people are also found in northern Africa, southeastern Asia, and increasingly, in North America and Europe. Estimates are that there are now somewhere between six and seven million Islamic people in the United States, one-third of whom are South Asian in origin, one-third African American, and one-quarter Arab American; the remainder are from diverse origins (Dudley and Roozen 2001).

Followers of Islamic religion are called Muslims. They believe that Islam is the word of God (Allah), revealed in the prophet of Muhammad, born in Mecca in the year 570 A.D. The Koran is the holy book of Islam. The highly traditional or reactionary Islamic fundamentalists adhere strictly to the word of the Koran, believing that they are obligated to defend their faith and that death in doing so is a way of honoring Allah. Not all Muslims take such a fundamentalist view, although they are frequently stereotyped because of the prominence of Islamic fundamentalism in contemporary world conflicts. Many Muslims have a highly patriarchal worldview, with women denied the freedoms that men enjoy. But as with other religions, the degree to which people adhere to this faith varies in different social settings and depends on the institutional supports given to religious strictures by the political state.

As noted, many practicing Muslims in the United States are African Americans who observe the strict dietary habits and prohibitions against many activities, such as consuming pork, alcohol use, drug use, and gambling. Forbidden by religious belief from eating foods identified as traditional food among African Americans (such as collard greens, barbequed pork ribs, and cornbread), Black Muslims have symbolically rejected their slave heritage. Black Muslims (known as the Nation of Islam) were also an important part of the more radical Black protest movement in the 1960s and 1970s, when they promoted the idea of community self-control for African Americans, establishing many of their own institutions, including schools, businesses, and media (Essien-Udom 1962; Lincoln and Mamiya 1990). The emphasis of Black Muslims on self-reliance has earned it a fervent following. Many Black leaders, including Malcolm X and now the Rev. Louis Farrakhan, though controversial, have a huge influence and following.

Hinduism

Unlike Christianity, Judaism, and Islam, Hindu religion is not linked to a singular God, as we have already seen in the discussion of polytheism. Hinduism rejects the idea that there is a single, powerful god. In this religion, god is not a specific entity. Instead, people are called upon to see a moral force in the world and to live in a way that contributes to one's spiritual and moral development. Karma is the principle in Hindu that sees all human action as having spiritual consequences, ultimately leading to a higher state of spiritual consciousness, perhaps found in reincarnation—the rebirth one experiences following death.

Hinduism is deeply linked to the social system of India, because their caste system (see Chapter 9) is seen as stemming from people's commitment to Hindu principles. Those who live the most ideal forms of life are seen as part of the higher caste, with the lower caste ("untouchables") as spiritually bereft, and several castes in between. But the complexity of Hinduism can be seen in the fact that, at the same time it has been used to justify this caste system, Gandhi, a great world leader, used Hindu principles of justice, honesty, and courage to guide one of the most important independence movements in the world. His teachings, as we saw in Chapter 11, were also crucial in the development of the civil rights movement in the United States.

Buddhism

Buddhism is another extremely complex religion—one that, like Hinduism, does not follow a strict or singular theological god. Also arising from the culture of India, Buddhism is most widely practiced in Asian societies, although many countercultural groups throughout the world have adopted some beliefs and practices of Buddhist religion.

The Buddha in Buddhism is Siddartha Gautama, born of the highest caste in India in the year 563 B.C. As a young man, he sought a path of enlightenment, based on travel and meditation. Buddhism thus encourages its followers to pursue spiritual transformation. Many "New Age" spiritual groups in Western society have adopted its focus on meditation. Like Hinduism, Buddhism involves a concept of birth and rebirth through reincarnation. Through seeking spiritual enlightenment, Buddhists see people as relieving themselves of their worldly suffering.

Buddhism now draws many adherents other than its traditional followers. Although the majority of those observing Buddhism are Asian and Asian American, Buddhism appeals to many other people—most typi-

Although Buddhists do not worship a particular deity, Buddhism involves belief in the importance of enlightenment through meditation and, for some, ritual practices.

cally White Americans who are liberal in their cultural orientation and attracted to countercultural trends in religious belief (Kniss 2004; Bromley 2004).

Confucianism

Confucians follow the principles of Confucius, a Chinese philosopher who promoted certain moral practices. Various moral principles are layered on one another, each contributing to the cohesion of a Confucian worldview. Confucianism promotes a disciplined way of life, more of a moral code than a sacred religion as there is not a particular god or set of religious disciples whom Confucians follow. The expression of goodness and social unity is an important principle in Confucian thought.

Eastern religions such as Hinduism, Buddhism, and Confucianism are difficult to describe in concepts drawn from Western thought. As religious belief systems, each rests on philosophical principles different from Western forms of thought. Confucianism is mostly found in China, although migration to other parts of Southeast Asia and North America means that Confucianism is practiced in various societies.

World religions each reveal the power of belief in regulating social behavior—a fact that sociologists find intriguing. Likewise, the association of religion with other social factors is a key part of sociological research on religion. In the next section, we examine some of these associations in the context of the diverse religious beliefs in the United States.

Diversity and Religious Belief

Religious identification varies with a number of social factors, including age, income level, education, and political affiliation. Younger people, for example, are least likely to say religion is important to them. People over age 50 are most likely to say it is important. People in higher income brackets are more likely to identify as Catholic or Jewish than people in lower income brackets, who are more likely to identify as Protestant, although these trends vary among Protestants by denomination. Fundamentalist Protestants, for example, are more likely to come from lower income groups (Gallup and Lindsay 2000).

Race and Religion

Race is a significant marker when examining patterns of religious orientation. African Americans and Hispanics are more likely than Whites to attend church and to say religion is important to them (Gallup and Lindsay 2000). African Americans are more likely to identify as Protestant than are Whites or Hispanics, although the number of African American Catholics is increasing. Many urban African Americans have also become committed Black Muslims.

For many African Americans, religion has been a defense against the damage caused by segregation, bigotry, and discrimination. The church is a central institution in many Black communities, acting simultaneously as a spiritual base and a political and community center. It has served to channel the emotions of an oppressed people and provide emergency support, social networks, and other resources to meet people's needs (Pattillo-McCoy 1998). Combined, these roles have

For African Americans, the Christian church has been a source for faith in social justice.

© C. W. McKeen/The Image Works

made churches among the most important institutions within the African American community (Gilkes 2000).

Religion has also been a strong force in Latino communities. About 70 percent of Latinos identify themselves as Catholic. However, there are a significant and growing number of Latino Protestants in mainline Protestant denominations and in fundamentalist groups. Latinos who are conservative Protestants have greater involvement in religion than those who are Catholic or mainline Protestants. They attend church more often, are more likely to be in a church-related group, and attribute greater importance to religion (Hunt 2000).

Asian Americans have a great variety of religious orientations. One reason for this is that the category "Asian American" is constructed from so many different Asian cultures. Hinduism and Buddhism are common among Asians, but many are also Christian as the result of missionary efforts in Asia. As with all groups whose family histories include immigration, religious belief and practice among Asian Americans frequently changes between generations. The youngest generation may not worship as their parents and grandparents did, although some aspects of the inherited faith may be retained. Within families, the discontinuity with a religious past brought on by cultural assimilation can be a source of tension between grandparents, parents, and children. Within the United States, Asian Americans often mix Christian and traditionally Buddhist, Confucian, or Hindu beliefs, resulting in new religious practices.

Ethnicity and Religion

The great diversity of religious belief in the United States stems from the diverse ethnic and cultural backgrounds of Americans. For many groups, ethnic identity is inseparable from religious identity. The two identities may be so interconnected that someone from an ethnic community who adopts a religion outside the dominant tradition, such as an Italian Jew, may feel ostracized and stigmatized.

For some groups, religion and ethnicity are especially intertwined, and the group may withdraw from society to preserve its distinctiveness. Such groups are known as **ethnoreligious groups**—an extreme form of exclusive religious groups, described earlier in this chapter. Hasidic Jews and the Amish are groups that have limited interaction with the larger society but that have maintained their distinctive faith by withdrawing into their own communities. Hasidic Jews (also called Hasidim) are an ultraconservative group consisting mostly of immigrants who fled persecution in central and eastern Europe. There are approximately 100,000 Hasidic Jews in the United States. Their lives are regimented to maintain strict observation of Jewish tradition. Men and women are segregated outside the home.

Men also arrange their work so that they have no interaction with women. The Hasidim maintain their own schools and shops and their own system of law, under which the rabbi is the ultimate authority on secular and sacred matters (McGuire 2002).

Religious Extremism

Since 9/11 the world has witnessed some of the violent consequences that stem from *religious extremism*—meaning actions and beliefs that are driven by high levels of religious intolerance. Any religion, taken to an extreme, is a dangerous phenomenon because extremists come to believe that it is their sacred duty to impose their beliefs on others and eliminate those having a different worldview. Religious extremism usually becomes the basis for extremely violent behavior.

Religious extremists tend to see the world in simplistic either/or terms—dividing people into either good or evil, us and them, godly or demonic. Such divisive imagery reduces the complexity of human life into simplistic categories—categories that fuel hate and conflict. Thus, religious extremism also produces martyrs and enemies, as if the world's people were divided along some "axis of evil," good people on one side and everyone else on the other (Anthony et al. 2002). When such religious fanaticism is intertwined with the power of a state government, religiously inspired leaders can use the power of the military and government propaganda to wield extraordinary power.

Although people in the United States now associate religious extremism with terrorism, this is not the first time the world has seen the violence and hatred produced by the intolerance of religious fanatics. In Northern Ireland, the decades-long conflict between Catholics and Protestants has been fueled by religious beliefs; in the United States, the religious right (a phrase referring to extreme Christian fundamentalists) has been linked to abortion clinic bombings and the murder of doctors who perform abortions. Throughout the world—both now and in the past—religion has been used to justify horrendous acts, including mass executions and genocide, enslavement, and other heinous crimes against humanity. All religions, taken to an extreme, are dangerous social forces because they can drive adherents to think they are doing sacred work when they are engaging in violent, murderous behavior.

It is easy to see the acts of religious extremists as the work of misled individuals, but those who study religious extremism know that it has social origins. First, religious extremism is learned—usually within a narrowly circumscribed social world, such as the *madrassas*—religious camps in Pakistan (and other areas) where young boys are taught a strict interpretation of Islam combined with instruction that women are a temptation to evil and that they should feel threatened by Westerners, especially Americans. For young boys uprooted from families by war, detached from other social contacts, and with no other education, it is easy to be socialized into a narrow worldview that gives them a cause to fight for (Rashid 2000). Knowing this helps explain how young men in Pakistan and other places could celebrate the death of Americans by cheering and waving guns in the streets. Such behavior can easily emerge when all you are taught is to perceive some other group as your enemy.

Second, religious extremism typically emerges in countries (or subcultures) where people are very poor and without access to education. Young men without an education and with little life opportunity can be attracted to violence as a way of asserting a collective identity (Khashan and Kreidie 2001). The young and uneducated are also more subject to the influence of extremist leaders who may themselves come from a higher social class. Many of the young suicide bombers who have committed horrendous acts of violence come from needy families who gain status and money from having their children do the footwork of terrorists.

Also, where there is a lack of modernism—and especially where people perceive that their traditional way of life is being overtaken by Western influences—some can feel that they must defend a traditional way of life (Pain 2002). And, where most people are poor, frustrations can be channeled into extremist movements that are fueled by highly traditional religious beliefs (Heilman and Kaiser 2002). Add to this economic stagnation and political corruption and you have a powerful social context in which to produce religious extremism (Pant 2001). Thus, it helps to understand religious extremism—and the terrorism it can produce—in the context of global stratification (see Chapter 10).

Finally, religious extremism tends to emerge in societies and subcultures where women are oppressed. Religious extremist movements tend to be highly patriarchal—that is, based on the power of men and the subordination of women. This is true not only in the extremist factions of contemporary Islamic movements, but also in extremist segments of the Christian right in the United States (Antrobus 2002; Ferber 1998). When religious extremism links with militaristic and patriarchal values, it becomes extremely dangerous.

Responding to extremist movements with other forms of religious extremism is likely only to exacerbate such conflicts. A better solution is to educate people, improve the economic conditions under which terrorism thrives, and improve the status of women worldwide.

Religious Organizations

The various religious faiths in the world are organized into **denominations**, meaning religious organizations that unite various congregations into a single adminis-

Gene Robinson, an openly gay Episcopal priest, was ordained as the Bishop by the Episcopalians in New Hampshire in 2003.

trative structure. Thus, within Protestantism, there are various denominations, including Presbyterians, Methodists, Baptists, and others. Denominations can be quite large and fall under the administration of larger national organizations, such as National Baptist Convention—a national organization uniting African American Baptist churches claiming over 25 million members (Taylor 2001). National meetings of such denominational groups can be huge—just like a large political convention—and they are often sites of considerable controversy, such as when the leaders of the Anglican Communion—the international organization of 77 million Episcopalians—decided to allow the election of a gay priest as Bishop of New Hampshire.

Sociologists further study religious organizations in terms of smaller units where members congregate for worship services, including churches, sects, and cults. These are *ideal types* in the sense that Max Weber used the term. That is, the ideal types convey the essential characteristics of some social entity or phenomenon, even though they do not explain every feature of each entity included in the generic category. Sects and cults are not typically part of large, national organizations, nor are all churches.

Churches

Churches are formal organizations that tend to see themselves, and are seen by society, as the primary and legitimate religious institutions. They tend to be integrated into the secular world to a degree that sects and cults are not. They are sometimes closely tied to the state. Churches are organized as complex bureaucracies with division of labor and different roles for groups within. Generally, churches employ a professional, full-time clergy who have been formally ordained following a specialized education. Church membership is renewed as the children of existing members are brought up in the church. Churches may also actively prosely-

tize (that is, actually recruit converts). Churches are less exclusive than cults and sects because they see all in society as potential members (Johnstone 1992).

Recently, especially among African Americans, some churches have become so large that they are now called *megachurches*. These are churches of 2000 or more members. They tend to be located in the suburbs of large cities and are found mostly in the "Sunbelt"—the area including the American Southeast as well as California, Texas, Arizona, and New Mexico. The worship services in such churches are highly planned and carefully polished productions and may even be televised. Many think that the form of such religious worship is congruent with the structure of modern society, allowing great diversity in membership and an organizational style that is formal, complex, and cosmopolitan. Yet, given the emphasis on worship, megachurches can adopt such a modern style, yet not feel alienating. Megachurches are found in many communities, including African American communities where they draw their membership heavily from the African American middle class. African American megachurches tend to take an activist role in their communities, addressing the social problems that afflict many African Americans and providing services that dominant institutions have ignored (Taylor 2001; Thumma 2000; Wuthnow 1988; Vaughan 1993).

Sects

Sects are groups that have broken off from an established church. They emerge when a faction within an established religion questions the legitimacy or purity of the group from which they are separating. Leaders of sects are sometimes laypeople with no formal clerical training, although many sects form as offshoots of existing religious organizations, typically formed from fine points of theological disagreement. Within sects, there tends to be less emphasis on organization (as in churches) and more emphasis on the purity of members' faith. The Shakers, for example, formed by departing from the Society of Friends (the Quakers). They retained some Quaker practices, such as simplicity of dress and a belief in pacifism, but departed from their religious philosophy. The Shakers believed that the second coming of Christ was imminent but that Christ would appear in the form of a woman (Kephart 1993).

Sects typically develop in protest against events or beliefs within the secular world. Different from churches, sects tend to be exclusive, admitting only truly committed members. Sectarians typically refuse to compromise their beliefs in any way, thinking instead that they are practicing the pure doctrine of their faith. Often, they hark back to original principles they believe are being violated by the group from which they parted.

The Amish formed as a sect in 1700 when Swiss Mennonites, descendants of the Anabaptist sect, were led by a Mennonite bishop named Joseph Amman. This group denounced the other Mennonite bishops for not enforcing the Meidung, a practice of shunning, or ignoring, errant members. The Meidung is a strict system of social control, especially so among Mennonites because they typically associate only with other Mennonites. The Amish brought the Meidung with them when they emigrated in the early 1700s to the United States, where they set up exclusive communities that have been remarkably faithful to the present day in living up to a principled mode of life that predates the industrial era (Kephart 1993).

Sects tend to hold emotionally charged worship services. The Shakers, for example, had such emotional services that they shook, shouted, and quivered while "talking with the Lord," earning them their name, from the "Shaking Quakers." The only bodily contact permitted among the Shakers was during the unrestrained religious rituals. Thus, they were celibate (did not have sexual relations) and gained new members only through adoption of children or recruitment of newcomers. Consequently, the Shakers sect has died out.

Cults

Cults, like sects in their intensity, are religious groups devoted to a specific cause or leaders with great **charisma,** a quality attributed to individuals believed by their followers to have special powers (Johnstone 1992). Typically, followers are convinced that the charismatic leader has received a unique revelation or possesses supernatural gifts. Cult leaders are usually men, probably because men are more likely to be seen as having the aggressive and charismatic characteristics associated with heroic leadership. Exceptions include Ann Lee, founder of the Shaker sect in America, and Mary Baker Eddy of the Christian Science movement (Johnstone 1992).

Many cults arise within established religions and sometimes continue to peaceably reside within the parent religion as simply a fellowship of like-minded persons with a particular, often mystical, set of dogma. The term *cult* is also used to refer to a quasi-religious organization, often dominated by a single charismatic individual. A good example was the Heaven's Gate cult, in which ninety members committed group suicide in Rancho Santa Fe, California, in 1997, believing they were going away on spacecrafts that would lead them to a new world (Balch 1995).

Many mainstream religions began as cults. Christian Science is an example of a now well-established religious organization that first evolved as a cult. As cults are developing, it is common for tension to exist between them and the surrounding society. Cults tend

to exist outside the mainstream of society, arising when believers think that society is not satisfying their spiritual needs. People longing for meaningful attachments are who cults tend to attract. Internally, cults seldom develop an elaborate organizational structure but are instead close-knit communities held together by personal attachment and loyalty to the cult leader.

The Branch Davidians were a good example of a cult. Their leader, David Koresh, convinced his followers that he was the Messiah and that they should prepare for a violent day of reckoning that would end in his resurrection and their passage to heaven. He also taught his women followers that it was God's will that he have sex with all of them so that their children could populate the new House of David (Verhovek 1993; Tabor and Gallagher 1995). Seventy-two members of this cult, including Koresh, were killed in the fire that followed an FBI (Federal Bureau of Investigation) assault on the cult's heavily armed compound in Waco, Texas, in 1993.

Becoming Religious

Regardless of affiliation, religion can be a significant part of social identity. Sixty percent of Americans say that religion is very important in their lives; only 15 percent say it is not very important (Carroll 2004). Becoming religious is the result of many experiences, which sociologists group under the concept of **religious socialization,** the process by which one learns a particular religious faith. For most people, religious socialization takes place primarily, or at least initially, within the family or through a process of conversion. Some faiths even insist that a formal conversion ceremony take place to establish commitment to a religious identity. Whatever the process by which one becomes religious, initiation into religion typically involves a period of instruction, learning, and experimentation. This process involves learning group norms and values, just as people learn the norms and values of other groups into which they are socialized. The result of religious socialization is that individuals internalize religious norms, sometimes so thoroughly that they believe the religion they have acquired is the only possible way to think.

Religious socialization can be both formal and informal. *Informal religious socialization* occurs when one observes and absorbs the religious perspective of parents and peers. Interaction with people in religious groups and organizations, such as church-based youth camps, can be an important source of informal religious training. *Formal religious socialization* occurs through explicit religious instruction. An example is the formal religious instruction that Jewish children receive prior to their bar mitzvah or bas mitzvah. The purpose of this religious training is to instill the values, history, and beliefs of the Jewish faith in the young person. Likewise, a child who is becoming socialized into the Christian faith will receive religious training in Sunday school or catechism classes.

Religion and the Family

Most people adopt the faith of the family into which they are born. Typically, children develop some religious identity by the time they are five or six years old (Chalfant et al. 1987). As late as high school age, most

SOCIOLOGY IN PRACTICE
Lessons from Waco

"In 1993, there was a standoff in Waco, Texas between a religious cult, the Branch Davidians, and the FBI. The confrontation ended in a fiery blaze of the Branch Davidian complex, with at least seventy-five people, including the leader, David Karesh, being killed. In the aftermath, sociologist Nancy T. Ammerman, an expert on the sociology of religion, was one of ten experts appointed by the Justice and Treasury Departments to make recommendations to the government on how to avoid such situations in the future. Following the release of the experts' report, Professor Ammerman also reflected on the sociological lessons learned from this experience and offered her insights and sociological imagination to those who wish to understand the nature of groups such as the Branch Davidians. Ammerman also tells us that we learn not just about religious cults from such events, but also about how the U.S. government works. Her conclusions follow:

1. Religious experimentation is pervasive in American history. We simply have been a very religious people. From the days of the first European settlers, there have always been new and dissident religious groups challenging the boundaries of toleration, and the First Amendment to our Constitution guarantees those groups the right to practice their faith. Only when there is clear evidence of criminal wrongdoing can authorities intervene in the free exercise of religion, and then only with appropriately low levels of intrusiveness.

2. New groups almost always provoke their neighbors. By definition, new religious groups think old ways of doing things are at best obsolete, at worst evil. Their very reason for exist-

ing is to call into question the status quo. They defy conventional rules and question conventional authorities.

3. Many new religious movements ask for commitments that seem abnormal to most Americans, commitments that mean the disruption of "normal" family and work lives. While it may seem disturbing to outsiders that converts live all of life under a religious authority, it is certainly not illegal (nor particularly unusual, if we look around the world and back in history). No matter how strange such commitments may seem to many, they are widely sought by millions of others.

4. The vast majority of those who make such commitments do so voluntarily. The notion of "cult brainwashing" has been thoroughly discredited by the academic community, and "experts" who propagate such notions in the courts have been discredited by the American Psychological Association and the American Sociological Association. While there may be real psychological needs that lead persons to seek such groups, and while their judgment may indeed be altered by their participation, neither of those facts constitute coercion.

5. People who deal with new or marginal religious groups must understand the ability of such groups to create an alternative world. The first dictum of sociology is "Situations perceived to be real are real in their consequences." No matter how illogical or unreasonable the beliefs of a group seem to an outsider, they are the real facts that describe the world through the eyes of the insider.

6. People who deal with the leaders of such groups should understand that "charisma" is not just an individual

trait, but a property of the constantly evolving relationship between leader and followers. So long as the leader's interpretations make sense to the group's experience, that leader is likely to be able to maintain authority.

7. Authorities who deal with high-commitment groups of any kind must realize that any group under siege is likely to turn inward, bonding to each other and to their leader even more strongly than before. Outside pressure only consolidates the group's view that outsiders are the enemy. And isolation decreases the availability of information that might counter their internal view of the world. In the Waco case, negotiating strategies were constantly undermined by the actions of the tactical teams. Pressure from encroaching tanks, psychological warfare tactics, and the like only increased the paranoia of the group and further convinced them that the only person they could trust was Koresh."

Critical Thinking Exercise

1. Identify another example of a religious cult, either from the newspaper, observations of your community, or interviewing someone with experience with a cult. How do Ammerman's conclusions help you understand this phenomenon?

2. Have you ever known someone who became fanatical about something (whether religious or otherwise)? Do Ammerman's conclusions help you understand that person's experience and other's reactions to it?

Source: Ammerman, Nancy T. 1994. "Lessons from Waco." *Footnotes* (January): Washington, DC: American Sociological Association.

•••

© Myrleen Cate/Stone/Getty Images

Young people are typically socialized into the religion of their parents. Religious socialization can be a powerful source of social identity and beliefs.

youngsters conform to their parents' religious orientation and levels of church attendance (Willits and Crider 1989).

Religious rituals within the home, the celebration and recognition of religious holidays, attendance at religious ceremonies, and religious instruction inculcate religious values in the young. Sitting at a *seder* (pronounced "say-der"), the Jewish ritual dinner held on the first evening of Passover in commemoration of the exodus of the Israelites from Egypt, young children are socialized into the religious teachings of the Jewish faith. The seder includes a dialogue between the father and the youngest child in the house and other rituals that teach young people the history and beliefs of Jewish people. In Christian households, children learn early about the birth and death of Jesus through enactments of the Nativity story at Christmas and the Passion (the suffering and crucifixion of Jesus) at Easter. In various faiths, religious ceremonies are important family events, such as in christenings and religious confirmations. The *quinceañera* among Mexican Americans recognizes the fifteenth birthday of girls, but it is also tied to religious confirmation, linking the child with the family and the church.

As people grow older, the influence of the family of origin on religious beliefs lessens but does not disappear. It is common for people who dropped out of religious activities during their young adulthood to return to religious practice when they begin raising families (Roof 1993). That influence must compete, however, with new experiences and exposure to new communities. College students, for example, frequently find themselves profoundly questioning the religious identity with which they have grown up. Some completely dissolve their religious faith, and others recommit to religious values later.

Religious Conversion

Becoming religious may be a private or public process, subtle or dramatic. **Conversion** is a transformation of religious identity. It can be slow and gradual, as when someone switches to a new religious faith (such as a Christian person converting to Judaism or an African American Baptist converting to Catholicism). Conversion to a new religious faith typically involves several phases, beginning with *exposure* to a new way of thinking. You may encounter a new experience that causes you to question a prior faith, or someone may recruit you to a new religious organization. This can initiate a period of *questioning* or searching for new understandings, a process symbolic interactionists would see as the social construction of a religious worldview. Fundamentalist Christians describe this as being "born again." A more sociological view would see this as a period of *resocialization* (see Chapter 4) in which original ways of understanding are transformed through the new meanings and new relationships one is forming. Someone who, for example, is converting to Judaism, will learn the tenets of Jewish faith and will likely reorder their understanding of the world accordingly. Conversion can also mean giving up prior religious beliefs, such as people who were once faithful Catholics but later question this faith and become more secular in their orientation.

When a person joins a cult or some extreme religious group, conversion may be more dramatic. Sociologists have been especially interested in these latter forms of conversion, perhaps because they have also captivated the public. In common thought, those who enter extreme religious groups are thought of as brainwashed, manipulated, and coerced. Because such groups proselytize, there is a popular image of religious conversion as involving a radical change in one's identity.

Are people who become converts to religious cults or other extreme groups being brainwashed? The **brainwashing thesis** claims that innocent people are tricked into religious conversion, that religious cults manipulate and coerce people into accepting their beliefs. This popular explanation of extreme religious conversion is seen as analogous to prisoners of war and hostages whose utter dependence on their captors can lead to erosion of the ego and a tendency to take on the captor's perspective. The brainwashing thesis portrays religious converts as "passive, unwilling, or unsuspecting victims of devious but specifiable forces that manufacture conversion" (Wright 1991: 126). This simple explanation that the brainwashing thesis offers for why people join bizarre groups may have led to the popularity of the thesis. Dramatic tales of parents who tried to "rescue" their child from a religious cult feed this popular image of religious conversion. The cults people join are seen as exercising mind control that strips converts of their earlier identities, robs them of free will, and "programs" their minds with cult beliefs.

DEBUNKING *SOCIETY'S MYTHS*

Myth: People who join extreme religious cults are maladjusted and have typically been brainwashed by cult leaders.

Sociological perspective: Conversion to a religious cult is usually a gradual process wherein the convert voluntarily develops new associations with others and through these relationships develops a new worldview. Brainwashing is simply a process of social influence in operation, not evil mind control.

Sociologists suggest an alternative way to understand the process of conversion into religious cults. **Social drift theory** interprets people as moving into such groups gradually, particularly if they have experienced recent personal strains or have become disenchanted with their prior affiliations. Social drift theory emphasizes that conversion is linked to shifting patterns of association, not mind control. People are active participants in the process of their own conversion, not passive creatures "programmed" with new ideas. The social psychological perspective is that conversion comes from simple, though powerful, social influence. People will do practically anything if their friends and an autocratic leader tell them to do it (see Chapter 6). No brainwashing is needed.

The case of the "American Taliban," John Walker Lindh, is a recent example of social drift theory. Lindh, who was raised in the United States, became a member of the Taliban and was later captured by U.S. troops during the invasion of Afghanistan following the terrorist attacks on the United States in 2001. Although many have thought of Lindh as brainwashed by the Taliban forces in Afghanistan, in actuality, Lindh's conversion to Islam and Taliban politics was a long process begun during his preteen and teenage years in the United States.

Most conversions to religious cults do not entail a radical alteration of lifestyle. The process of religious conversion can be broken down into several phases. The first phase typically involves an experience that leads a potential convert to perceive disruption or failure in his or her previous life, allowing the person to be open to a serious change in the social environment. Fervent religious conversion tends to produce withdrawal, thereby reducing other ties, including those to family and friends. An initiate who withdraws to a new and unfamiliar context will be more susceptible to new influences and, in the flush of new experiences, less attentive to inconsistencies and contradictions in the new meaning systems to which they are being exposed. Put

another way, the person may experience some loss of autonomy. How quickly one is converted at this stage depends in large part on how much social pressure is exerted by the new group (Johnstone 1992; Robbins 1988; Wright 1991). In the case of the Heaven's Gate cult, early followers were asked to give up their possessions and to cut off relations with their family and friends. After joining the cult, they were discouraged from contacting outsiders (Balch 1995, 1980). In the case of John Walker Lindh, giving up virtually all of his possessions was voluntary and self-induced, prior to his migration to Afghanistan.

In the second phase of conversion, an emotional bond is created between the initiate and one or more group members. This bond creates in the initiate a greater willingness to adopt the worldview of the new group. As the bond deepens, relationships with people outside the new group are progressively weakened. This is the phase where friends and family members become concerned that they are "losing touch" with the person. Groups often actively manipulate the environment so as to separate new recruits from preexisting ties, intensifying the conditions that promote conversion. A good example of this is the way potential new members are socialized into the Unification Church. Adherents of this religion are popularly known as "Moonies," named after the church founder, the Rev. Sun Myung Moon. Members segregate new recruits from the influence of other social influences by hosting weekend workshops and isolating them in week-long seminars. Recruits are showered with constant attention and kindness by existing members, a heady experience the Unification Church calls "love-bombing" (Long and Hadden 1983; Bromley 1979; Ayella 1998).

The third phase of religious conversion is a period of intense interaction with the new group. Presuming the initiate stays with the group, this can lead to total conversion, as with John Walker Lindh, who interacted on a constant day-to-day basis with members of the Taliban. In the case of the Unification Church, new recruits are given activities that keep them from thinking about other things. They sell flowers on the street, for example, which places them in a new social context and gives them something to do. Initiates are also encouraged to sever ties to their former lifestyles and abandon their material resources—a sacrifice that, not accidentally, benefits the cult, as initiates usually donate their material goods to the organization (Bromley 1979; Long and Hadden 1983; Robbins 1988).

Deconversion also involves a social process of disengagement. Like conversion, it occurs in phases. First, deconversion requires that the emotional attachment to the group be broken. Thenceforth, the process of deconversion is roughly the reverse of conversion. That is, former converts are opened to new ideas and forms of association, eventually breaking from their original group. Sociologists have found that most converts who leave religious cults typically join others, such as the Hare Krishnas and the Unification Church. There is a popular image that religious converts need so-called deprogramming or radical resocialization to rejoin society. However, studies find that most cult defectors (89 percent) are slowly reintegrated into society, achieving complete reintegration in most cases within no more than two years of leaving the cult (Wright 1991).

Social and Political Attitudes: The Impact of Religion

Would it surprise you to learn that religious affiliation affects voting outcomes in Congress? Research has shown that the voting records of Congress's members follow clear religious patterns. Perhaps this is not surprising on religiously charged issues such as abortion and the death penalty, but the pattern also holds up on issues such as defense spending, minimum wage laws, and welfare reform (Fastnow et al. 1999).

Given the strong effect that religion has on people's behavior and beliefs, it is not surprising that a powerful association exists between religious identification and social and political attitudes. Some of these links may be unexpected. For example, contrary to what one might think, Protestants are more likely than Catholics to want to see a reversal of *Roe v. Wade*, the Supreme Court decision upholding women's rights to abortion, probably due in large part to the many fundamentalist Protestants who want this court decision reversed because of their stand against abortion. Jewish Americans, whose politics on many issues tend to be relatively liberal, show the greatest support for abortion rights (Hugick 1992a).

Religious identification is also a good predictor of how traditional a person's gender beliefs will be. Generally speaking, people with deeper religious involvement have more traditional gender attitudes. Religion also affects people's sexual attitudes and behavior. The most religiously devout tend to be the least sexually active and the most conventional in their sexual practices.

The outlooks of different religious groups vary on many social issues. Jewish Americans, Unitarians, and Universalists, and those with no religious preference are generally the most liberal. On the issues of women's roles, race relations, homosexuality, abortion, and premarital sex, the most conservative groups are certain fundamentalist Protestant denominations, Mormons, and Jehovah's Witnesses. Conservative Christians are those most likely to back educational policies supporting the teaching of creationism, prayer in school, and voucher programs (Deckman 2002).

People's explanations of inequality are also strongly related to religion. Among Whites, evangelical Protestants are far more likely to explain racial inequality in individualistic terms than are Whites with less fundamentalist attitudes (Emerson et al. 1999). Religion and race interact, however, with Latino Catholics and White Protestants most likely to believe that this is a just world. African Americans are least likely to believe this, regardless of religious faith (Hunt 2000). The easily discernible trends within groups and the striking differences between groups are part of the reason that religion has played an increasing role in the political life of the country in recent years.

Racial Prejudice

Religion can be the basis for group oppression, and there is ample evidence for this throughout the world and through history. The relationship between religious belief and prejudice has long been of interest to social scientists. Bigotry against religious groups is one way to manifest prejudice, such as in the desecration of Jewish temples and cemeteries that sometimes occurs. Not only are religious groups sometimes targeted by prejudice, but religious belief is also related to the likelihood that someone will or will not be prejudiced. The relationship between religion and prejudice is not a simple one, however.

Researchers have distinguished patterns relating the degree of religious prejudice to the very nature of religious experience. An *extrinsic religious orientation* denotes an exclusionary and highly devout religious attitude, such as that of fundamentalist religious groups. An *intrinsic religious orientation* is more tolerant and open to different forms of religious expression. This is more characteristic of some mainline Protestant churches. A *quest orientation* features a searching attitude toward religion, as is the case with Quakers, Unitarians, and Jews. This is also characteristic of those who join "New Age" spiritual groups. People with a quest orientation tend to be tolerant of ambiguity in general (Adorno et al. 1950; Allport 1966; Griffin et al. 1987; Sapp 1986).

Homophobia

Homophobia (fear and hatred of homosexuals) has also been linked to religious belief. Some people argue that Christianity has encouraged homophobia in society because the Bible is interpreted as prohibiting same-sex sexual relations. Some religious congregations have actively worked to encourage the participation of gays and lesbians. This has sparked controversy in some churches, such as the fight within the United Methodist Church and the United Presbyterian Church about whether to ordain gays and lesbians (Clark et al. 1989).

Protestants are more likely than Catholics to think that a homosexual relationship between consenting adults is an acceptable lifestyle and should be legal. As with racial prejudice, religious orientations that promote intolerance of any kind are likely to promote homophobia.

Anti-Semitism

Anti-Semitism is the belief or behavior that defines Jewish people as inferior and targets them for stereotyping, mistreatment, and acts of hatred. One of the most horrendous acts of anti-Semitism was the Holocaust in Nazi Germany in the 1930s and 1940s, during which Jewish people were held in concentration camps, brutally treated, and millions killed. Today, there are still many Jewish people living who lost all of their family members in the Holocaust; the memories of Holocaust survivors—and the museums that have been created to document this horrendous event in world history—show how destructive the force of any form of group hatred can be (Epstein 1988).

Anti-Semitism is a persistent form of prejudice in the world. Early sociological research found that the people most likely to be anti-Semitic were those with less education and those experiencing downward mobility in the social class system (Adorno et al. 1950; Selznik and Steinberg 1969; Silberstein and Seeman 1959).

Like other forms of intolerance (such as racism, sexism, and homophobia), anti-Semitism is expressed through the beliefs and actions of specific people, but it has its origins in the social structural conditions within society. Some have been surprised by incidents of anti-Semitism occurring, for example, on college campuses, where one expects educated people to be more tolerant and knowledgeable about religious difference. Societal conditions, however, can promote the expression of group hatred. One recurring reason for anti-Semitic acts and beliefs is that the Jewish people are being scapegoated, blamed for problems that other groups might be having. Anti-Semitic prejudices allege that Jewish people control some professions, the media, and the banking in this country even though such elites are overwhelmingly White, Anglo-Saxon Protestants. But if people believe these stereotypes, they may blame Jewish people for the problems of others (Ferber 1998).

Religion and Social Change

Despite the stability of the major faiths, religion, like other social institutions, evolves over time as people adapt to changing social conditions. Religion historically has also had a deep connection to other forms of social change. Although many people think that religion is losing its influence in society, and there has been a decrease in the importance of religion to many people, conservative religious groups have recently displayed an ability to campaign successfully for their own social and political agendas, and religion continues to have an important role in liberation movements around the world.

The New Religious Conservatism

Perhaps one of the most significant changes in religion in recent years is the rise in religious conservatism. There has been a dramatic increase in the number of people who say they are born again or evangelical. In the United States, 42 percent of the population call themselves born again—women being more likely to do so than men and African Americans more likely than Whites. Thinking of oneself as born again is also correlated with education and income. People with no college experience are most likely to think of themselves as born again; those with college degrees, least likely. Evangelicalism seems to appeal most to blue-collar workers and the poor, although its appeal is increasing among the young (Chalfant et al. 1987; Smith 2000; Moore 2000; Winseman 2004).

The evangelical movement consists of diverse groups, including Faith Assemblies of God, Churches of Christ, and the Jehovah's Witnesses, to name only a few. In the past, conservative religious groups had largely distanced themselves from politics. Beginning in the 1970s,

Young people experience religiosity in a variety of ways, such as this group of Billy Graham supporters at the Alamodome in Houston, Texas.

however, conservative activists realized they could have an enormous impact on national politics if they mobilized the growing numbers of conservative Christian groups. Their affiliation with each other and with conservative political causes has resulted in a movement known as the "new Christian right" (Liebman and Wuthnow 1983)—most evident in the presidential election of 2004.

The conservative Christian movement has fueled antiabortion activism, revived the effort to teach creationism in the schools, and supported so-called pro-family legislation that promotes a variety of conservative values. The Christian right sees the changing role of women in society and the influence of the feminist movement as threatening traditional "family values" and undermining what they see as "natural" arrangements between women and men (Gallagher 2003)

The influence of religious conservatism has brought out intense debates about the role of religion in government-supported initiatives. Should faith-based organizations receive government support for work they do in helping people, or does this violate the constitutional separation of church and state? The constitutional issues will ultimately be settled by law. Studies show that faith-based initiatives do engage traditionally disadvantaged groups in civic participation (Wood 2002; Kniss 2003), but many worry that faith-based government initiatives will infuse religion into government, violating the U.S. Constitution. Such debates reveal the considerable impact that religious beliefs have on people's social values and political behavior.

Race, Religion, and Justice

What is the role of religion in social change? Durkheim saw religion as promoting social cohesion; Weber saw it as culturally linked to other social institutions; Marx assessed religion in terms of its contribution to social oppression. Is religion a source of oppression, or is it a source of personal and collective liberation from worldly problems? There is no simple answer to this question. The role of religion in social change is as broad as the spectrum of people who have religious commitments—as broad as society itself. Religious institutions have been a conservative force in society, often supporting homophobia, sexism, and other prejudices and frequently resisting social change, but religion has also been an important part of movements for social justice and human emancipation. For example, religion has played an important role in the development of social movements for the liberation of Latinos. We mentioned earlier the emergence of liberation theology among Catholic clergy in Latin America. Liberation theologians have used the prestige and organizational resources of the Catholic Church to develop a consciousness of oppression in poor peasants and working-class people.

> ### DEBUNKING SOCIETY'S MYTHS
>
> **Myth:** Most deeply religious people are intolerant of social change.
>
> **Sociological perspective:** Some religious affiliations and belief systems are more likely to promote progressive social change than others. Protestant fundamentalist groups are typically conservative in their social and political attitudes, whereas other groups have used their religious commitments as the basis for the promotion of civil rights and other progressive social issues.

One central theme of African American religiosity has been liberation from oppression. African American spirituals are rich with the symbolism of struggle and redemption from bondage. During slavery, churches served as "stations" on the Underground Railroad that smuggled Blacks out of slave states, and they were meeting places where social change was mapped out.

TAKING ON SOCIAL ISSUES

Separation of Church and State: Faith-Based Initiatives

A long-standing issue in the United States is the constitutional principle of the separation of church and state. In recent years, many policy questions have centered on questions of public support for religious observances, such as support for school prayer, civic displays of religious symbols, and the use of public funds for religious schools. Now there is the question of whether public funds should be used to support the activities of faith-based organizations. At the heart of these policy debates are questions about how to observe the constitutional separation of church and state while also allowing diverse groups the freedom of religious expression. What risks are there to supporting faith-based initiatives? What are the arguments, pro and con, regarding public support for faith-based programs?

Taking Action

Go to the Taking Action Exercise on the Companion Website—at http://sociology.wadsworth.com/andersen_taylor4e/—to learn more about an organization that addresses this topic. •••

Religion can be interpreted as producing social conflict, as well as promoting social justice.

African American churches had a prominent role during the mobilization of the civil rights movement (Marx 1967/1867; Morris 1984). Churches served as headquarters for protestors and clearinghouses for information on protest strategies and organizational tactics. They also supplied the infrastructure of the developing Black protest movements of the 1950s and 1960s, and the moral authority of the church was used to reinforce the appeal to Christian values as the basis for racial justice. Militant Black leaders have tended to be less religiously oriented than more moderate activists. Emphasizing community self-determination and critical of the White power structure, Black Muslims have fueled some of the more radical forms of Black protest, but the Muslim faith also has a conservative edge, especially in its patriarchal orientation, which has often limited women's leadership and independence.

Women and Religion

Generally speaking, women are more religiously devout than men. Women tend to say religion is very important to them and they are more likely to be members of churches or synagogues (Winseman 2002). Women's work in religious organizations, whether churches, synagogues, mosques, or religious clubs, has also sustained religious practice. Women have raised much of the money that has supported religious organizations, organized the social services provided by religious organizations, and passed on religious traditions to children. For many religious organizations, the movement of women into the paid labor force has resulted in a shortage of the volunteer labor that women have histori-

cally provided to religious organizations. In the Catholic Church, for example, there is concern that the population of nuns will not be replenished as current nuns grow old. As women now have other opportunities, the limited resources nuns are given, especially relative to priests, makes this a less attractive occupation for young women.

Women have also had to bear being denied the right to full participation in many churches. Some churches refuse to ordain women as clergy, and in the past women were denied admission to divinity schools. The public now generally supports the ordination of women, even among a majority of Roman Catholics. A slight majority opposes this step among evangelical Protestants. Across all denominational groups, men are more supportive of female ordination than women, although there is a small difference among Protestants (Roper Organization 1993; Bedell and Jones 1992; Jelen 1989).

African American women have tended to be more equal to men within African American churches than White women have been in White churches. African American women make up the largest percentage of African American church congregations, typically about 75 percent, and they have been far more likely than White women to assume positions of leadership in the ministry. The *sanctified church* is a term used to refer to Holiness and Pentecostal churches in the African American community. It has elevated women to positions of leadership, rejected the patriarchal organization of mainstream churches, and encouraged more feminist models of religious organization and practice (Gilkes 2000, 1985).

Traditional religious images of women have been the basis of many gender stereotypes. For example, although women are now rabbis in some Reform congregations, Orthodox Jewish men are supposed to thank God every morning that they are not women and are prohibited from associating with women who are menstruating. Among Christians, the New Testament of the Bible encourages women to be subordinate to their husbands, reflecting a presumed natural order in which women are subordinate to men as men are subordinate to God. Many feminists believe that a great deal of *misogyny* (hatred of women) has been inspired by the biblical creation story in which Woman, in the form of Eve, is responsible for the fall of Man. This biblical legend is the archetype, the original mold, of the stereotype of women as seductresses leading men into sin. Two polarized images of women developed within Christianity: (1) women as temptresses, witches, and whores and (2) women as madonnas. The net result is that religion has been a powerful source of the subordination of women in society.

Religion in Decline?

There has been a general trend in life in the United States of a decline in religious participation. Since the 1970s, church membership has declined, and fewer Americans are likely to say that religion is a significant influence in their lives. People are more likely to believe that science will solve the world's problems than they are to see religion as the solution. Does this mean that religion is losing its force in society in the United States?

Secularization is the process by which religious institutions, behavior, and consciousness lose their religious significance, a change in the basic organization of society. Secularization is not "antireligion." It refers to the process by which society becomes increasingly complex, bureaucratized, fragmented, and impersonal. In a secular society, people are no longer bound by sacred principles. A shift occurs from religious to secular control of all institutions, including education and the family. With secularization, people spend less time in religious activities. With secularization, the religious regulation of everyday life is supplanted by other, less spiritual imperatives. Max Weber referred to this trend as the **rationalization of society,** by which he meant

that society is increasingly organized around rational, empirical, and scientific forms of thought.

Although for many, the rationalization of society liberates them from the constraints that religion can impose, it leaves many feeling that they have no spiritual attachments. Without strong religious belief systems, society can seem adrift, as we see in the words of many social commentators who say the United States is in a moral crisis, is overly materialistic, and has no strong spiritual cohesion. As society becomes more secularized, traditional religious values lose their strength, and various forms of religious experimentation are likely to develop. Without spiritual attachment, people may seek out new means of psychological and social expression. Thus, in the contemporary United States, there are many who would describe themselves as "spiritual" but not religious, meaning that they express their faith in more inward ways, often not belonging to a religious organization and also drawing from multiple beliefs and practices of faith (Fuller 2002). One consequence is the development of new religious movements that counter the trend toward secularization. This helps explain the popularity of Rastafarianism, Eckankar, New Age, and assorted other religious groups and movements (Wuthnow 1994). These groups are attractive to those who seek meaningful ties and a sense of community.

FORCES OF SOCIAL CHANGE
Marketing Religion

Where do the forces of capitalism and religion intersect? Most people likely think of religion as a relatively private matter that should be untainted by economic forces such as the pursuit of profit, supply and demand, or consumerism. Yet, many sociologists now describe religion in the United States in terms of a "spiritual marketplace" (Roof 1999; Wuthnow 1998), where religion is bought and sold and people shop for spirituality as they would other commodities.

Evidence for this can be found in the growth of religious bookstores, an explosion in the number of religious self-help books that are on the market, even a growing connection between religion and the entertainment industry. Think, for example, of prime-time shows such as *Touched by an Angel* or the

proliferation of Christian radio stations that promote new Christian music groups. Even diet fads, such as the Hallelujah Diet, can be seen as part of a market-driven culture where profits can be made from faith.

Sociologists interpret these developments as reflecting a "quest culture" wherein more people, particularly younger people, are searching for a spiritual inner life, while moving away from participation in formal religious organizations. In documenting the expansion of the production and consumption of religion, sociologists see

the character of religion in this culture as being redrawn. Reflecting cultural themes of individualism and self-help, this transformation of religion means that people may become more spiritual but less drawn to traditional religious organizations.

Further Resources: Roof, Wade Clark. 1999. *Spiritual Marketplace.* Princeton, NJ: Princeton University Press; Wuthnow, Robert. 1998. *After Heaven: Spirituality in America Since the 1950s.* Berkeley, CA: University of California Press; Cimino, Richard P., and Don Lattin. 1998. *Shopping for Faith: American Religion in the New Millennium.* San Francisco: Jossey-Bass. •••

The place of religion in society is complex and full of contradictions. Yes, society is becoming more secularized, but in ways that perhaps Weber could not imagine, religion is still extremely influential in society. In other words, society is not just secular or religious, but is an intriguing mixture of both.

Chapter Summary

How do sociologists define religion?

Religion is an institutionalized system of symbols, beliefs, values, and practices by which a group of people interprets and responds to what they feel is sacred and which provides answers to questions of ultimate meaning. The *sacred* is that which is set apart from ordinary activity, is seen as holy, and is protected by special rites and rituals. *Totems* are sacred objects that people regard with special awe and reverence.

What form does religion take in different societies?

Religions may be *polytheistic* or *monotheistic, patriarchal* or *matriarchal. Exclusive religious groups* have a highly identifiable set of religious beliefs and a distinctive religious culture. *Inclusive groups* are more moderate and emphasize the importance of common religious work. In the United States, Christianity dominates the national culture, even though the U.S. Constitution specifies a separation between church and state. *Religiosity* is the measure of the intensity and practice of religious commitment.

How does sociological theory interpret the role of religion in society?

Emile Durkheim understood religions and religious rituals as creating social cohesion. *Max Weber* saw a fit between the ideology of the *Protestant Ethic* and the needs of a capitalistic economy. Religion is also related to social conflict. *Karl Marx* saw religion as supporting societal oppression and encouraging people to accept their lot in life. Symbolic interactionists see religious faith as socially constructed and have used this to explain people's religious belief systems.

What is the character of religious affiliation in the United States?

The United States is a diverse religious society. Protestants, Catholics, and Jews make up the major religious faiths in the United States. Patterns of religious faith and participation vary by age, income level, education, ethnicity, and race.

How do people become religious?

People learn religious faith through *religious socialization,* which can be formal or informal. The family is a major source of religious socialization. Religious *conversion* involves a dramatic transformation of religious identity. Individuals proceed through several phases to learn to identify with a new group and lose other existing social ties.

What are the different forms of religious organization?

Churches are formal religious organizations. They are distinct from *sects,* which are religious groups that have withdrawn from an established religion. *Cults* are groups that have also rejected a dominant religious faith, but they tend to exist outside the mainstream of society.

What relationship is there between religious beliefs and other social and political attitudes?

Religiosity is related to a wide array of social and political attitudes. Racial prejudice, homophobia, and *anti-Semitism* are all linked to patterns of religious affiliation. Religious extremism develops in particular social contexts and can fuel dangerous behaviors, such as terrorism.

What changes have characterized religious behavior in recent years in the United States?

There has been an enormous growth in conservative religious groups. Evangelical groups have been highly influential, particularly through their use of the electronic media as a means of communication and their affiliation with conservative political causes. *Secularization* is the process in society by which religious institutions, action, and consciousness lose their social significance. Religion in the United States has become more secular.

Key Terms

anti-Semitism 469
brainwashing thesis 466
charisma 464
churches 463

collective consciousness 453
conversion 466
cult 464

Researching Society with Microcase Online

You can see the results of actual research by using the Wadsworth MicroCase® Online feature available to you. This feature allows you to look at some of the results from national surveys, census data, and other data sources. You can explore this easy-to-use feature on your own, but try this example. Suppose you want to know:

How often do you take part in the activities and organizations of a church or place of worship other than attending services?

To answer this question, go to http://sociology.wadsworth .com/andersen_taylor4e/, select MicroCase Online from the left navigation bar, and follow the directions there to analyze the following data.

Data file: GSS

Analysis: Auto-Analyzer

Variable 1: RELG.ACTIV

Questions

Once you have your results, answer the following questions:

1. People reporting to follow which religion are most likely to have taken part in religious activities?
   a. Liberal Protestants
   b. Conservative Protestants
   c. Catholics
   d. Jews
   e. No religion

2. What percentages of those in each political party shown often take part in religious activities?
   ____ % Democrats
   ____ % Republicans
   ____ % Independents

3. How do African Americans and Whites compare in their participation in religious activities? How might you explain these differences?

4. Is gender a significant factor in predicting people's participation in religious activities? Why do you think this is so?

The Companion Website for Sociology: Understanding a Diverse Society, Fourth Edition

http://sociology.wadsworth.com/andersen_taylor4e/

Supplement your review of this chapter by going to the Companion Website to take one of the Tutorial Quizzes, use the flash cards to master key terms, and check out the many other study aids you'll find there. You'll also find special features such as GSS Data and Census 2000 information, data and resources at your fingertips to help you with that special project or do some research on your own.

Suggested Readings and Web Resources

Gallagher, Sally K. 2003. *Evangelical Identity and Gendered Family Life.* New Brunswick, NJ: Rutgers University Press.
This study of evangelical Christians examines how religion influences values about the family and how values about gender are also part of contemporary discussions of religion and the family.

Gilkes, Cheryl Townsend. 2000. *"If It Wasn't for the Women . . . ," Black Women's Experience and Womanist Culture in Church and Community.* Mary Knoll, NY: Orbis Books.
Gilkes examines the significant role of African American women in churches, including how churches work for social justice.

Haddad, Yvonne Yazbeck, Jane I. Smith, and John L. Esposito (eds.). 2003. *Religion and Immigration: Christian, Jewish, and Muslim Experiences in the United States.* Walnut Creek, CA: AltaMira Press.
These different articles explore the diversity of religious faith in America and how it is changing as the result of immigration.

Lincoln, C. Eric, and Lawrence H. Mamiya. 1990. *The Black Church in the African-American Experience.* Durham, NC: Duke University Press.
This is a comprehensive analysis of the influence of diverse religious groups and experiences in the history of African American people. It also shows the centrality of religious faith in movements for racial justice.

Manning, Christel. 1999. *God Gave Us the Right: Conservative Catholic, Evangelical Protestant, and Orthodox Jewish Women Grapple with Feminism.* New Brunswick, NJ: Rutgers University Press.
Based on ethnographic studies of women in these different conservative faiths, Manning analyzes how

women in conservative religions construct a positive identity for themselves even though their religion defines women's role as submissive. She shows how the women construct a definition of gender that enables them and does not victimize them. She uses a sociological perspective to unravel questions about identity, religious faith, gender, and tradition.

Wuthnow, Robert. 1998. *After Heaven: Spirituality in America Since the 1950s.* Berkeley, CA: University of California Press.
Wuthnow connects patterns of religious practice to the changing character of American society. He argues that people feel "rootless" and therefore use religion to anchor themselves in a complex and mobile society. His analysis is a good sociological example of the connections among personal beliefs, behavior, and institutional change.

Academic Info Religion
www.academicinfo.net/religindex.html
This site is a reference to numerous Internet resources on religion. It includes links to sites on diverse religious faiths, religious studies programs, directories of religion in the United States, new religious movements, and special topics, such as women and religion.

The Center for the Study of American Religion, Princeton University
www.princeton.edu/~nadelman/csar/csar.html
The Center for the Study of American Religion encourages the academic study of religion; the website contains information about the center's activities and links to numerous other sites for the study of American religion.

Economy and Work

Every day millions of people get up, get dressed, and go to work. Another several million wish they could find work. Many people who work wish they earned more or had jobs with better working conditions. Work has an enormous effect on most aspects of your life, including relationships with your family, how much power you have, and the resources available to you. Work establishes the routines of daily life—what you wear, with whom you associate, what hazards you face, and even what time you get out of bed. For some, work is a satisfying experience that provides the opportunity to be creative and pursue talents and interests; for others, work is a source of frustration and exploitation. Any way you look at it, the significance of work in people's lives is hard to overestimate.

When thinking about work, most people think in terms of what work they might do, how much they can earn, whether their work will be satisfying, and whether their job will give them opportunities for advancement. Students, for example, may see their education as bringing them work opportunities, but they may also worry about what work they will find and whether their work will utilize their talents and bring them a lifetime of rewards. These are individual issues; sociologists study the social forces that shape people's experiences. They examine the social processes that are changing the world of work, including globalization of the economy, technological change, and the diversity of the workforce. What work is available to different groups, indeed if one will be working at all, is shaped by the convergence of these forces in people's lives. This situation places the concerns of individual workers into a broader social structural context. ●●●

Sociology ⊛ Now™
Reviewing is as easy as ❶❷❸.

Use SociologyNow to help you make the grade on your next exam. When you are finished reading this chapter, go to the chapter review for instructions on how to make SociologyNow work for you.

Economy and Society

To understand work, you first have to see it in the context of the broad social institution known as the economy. All societies are organized around an economic base. The **economy** of a society is the system by which goods and services are produced, distributed, and consumed. How the economic structure of a society is organized shapes how work is done and who performs it. This has to be understood first in terms of the historic transformation of economic systems from agriculturally based societies to industrial and, now, postindustrial societies.

The Industrial Revolution

In Chapter 5, we discussed the evolution of different societies. A significant change was, first, the development of agricultural societies and, later, the far-ranging impact of the *Industrial Revolution.* Now, the Industrial Revolution is giving way to the growth of *postindustrial societies*—a development in the economic system with far-reaching consequences for how societies are organized.

The development of agricultural societies followed the development of technologies that enabled the large-scale production of food. The invention of the plow, for example, allowed people to form settlements organized around farm production in large fields and replaced the more labor-intensive methods of gathering small amounts of food. This change in production meant that societies could settle in one place, forming

markets for the sale and distribution of goods, thereby radically changing how society was organized and how people worked. Agricultural production remains a vast part of the world economy, although now it has been changed by the processes of industrialization—probably the most significant historical development affecting the social organization of work, prior to the invention of the silicon chip.

The Industrial Revolution is usually pinpointed as beginning in mid-eighteenth century Europe, soon thereafter spreading throughout other parts of the world, which has led to numerous social changes. The creation of factories, as we saw in Chapter 15, separated work and family by relocating the place where labor was performed. The Industrial Revolution also transformed the consumption of energy and natural resources with the large-scale use of coal, steam, and later electrical power, the basis for running the machinery needed for production. Industries became highly specialized; workers would repeat the same action many times over the course of a working day—involved in only one step of the production, not the total process. (This is something that, as we will see in the later discussion of worker alienation, has an enormous impact on workers' attitudes.) Another economic change of the Industrial Revolution was the creation of a cash-based economy: Laborers are paid a cash wage and goods are sold, not for exchange, but for their cash value. All in all, the social relations created by the Industrial Revolution are hard to overestimate.

We still live in a society that is largely industrial but quickly giving way to a new kind of social organization: postindustrial society. Where industrial societies are primarily organized around the production of goods, **postindustrial societies** are organized around the provision of services. Thus, the United States has moved from being a manufacturing-based economy to a service-based economy. The provision of services pertains to a wide range of economic activities now common in the labor market, including, for example, banking and finance, retail sales, hotel and restaurant work, and health care, to name but a few. The service economy also includes parts of the vastly expanded information technology industry, primarily software design and the exchange of information (through publishing, video production, and the like), not the assembly of electronics. Information technology forms the core of a postindustrial society because it is the mechanism through which most services are delivered and organized. Whereas the Industrial Revolution was once seen as the source for broad-scale social change, now the *Information Revolution* is probably one of the

The Industrial Revolution transformed labor, moving it from the household to the factory—or other sites where work was mechanized and oriented to mass production.

© Mystic Seaport, Rosenfeld Collection, Mystic CT, neg. 1634

greatest sources for social and economic change in the future (see also Chapter 23).

Comparing Economic Systems

There are different ways that an economy can be organized, depending on the cultural values and principles of a given society. **Capitalism,** the basis for the U.S. economy, is an economic system based on the principles of market competition, private property, and the pursuit of profit. Under capitalism, the means of production are privately owned. (Recall from Chapter 9 that the *means of production* refers to the system by which goods are produced and distributed.) To say that in capitalist societies, some people own the means of production does not simply mean that people own property; it means some people control the natural resources and own the industries in which goods are produced and sold. Within capitalist societies, stockholders together own corporations—each owning a share of the corporation's wealth. Under capitalism, owners keep the profit from the revenue that is generated. Profit is created by selling a product for more than the cost of creating it; thus, owners pay workers less than the value of what they produce. Under capitalism, workers produce the goods and provide the services, whereas owners disproportionately consume goods and reap the profits. This class relationship is what defines the system of capitalism.

The capitalist basis of society in the United States shapes the character of the nation's other institutions. Health care institutions, for example, are administered on a profit-based system. In other industrialized societies, health care is regarded as a human right that is paid for and administered by state agencies—a more socialist model. Other institutions are also shaped by capitalism. Even public school systems in some cities (such as the Edison schools in Philadelphia) are now being run by private companies, as are many prisons. The United States is not purely a capitalist society, however, in that government subsidies support some industries. Some people say government bailouts for failing industries or government support for agribusinesses provides socialism for corporate interests.

Socialism is an economic institution characterized by state ownership and management of the basic industries; that is, the means of production are the property of the state. In many nations, the global forces of capitalism mix with socialist principles. Many European nations, for example, have strong elements of socialism. Sweden supports numerous state-run social services, such as health care, education, and social welfare programs, but its industry is capitalist. Likewise, the state in Great Britain has historically owned the basic industries of the country, such as the railroads, mines, and the communications industry; most other industrial entities are privately owned. Still, the influence of socialist values in Britain can be witnessed in things such as the national support for the BBC (British Broadcasting System). Compare this with the privately owned, though publicly regulated, communications industries in the United States (such as CBS, NBC, ABC, Fox, and CNN).

Other nations are more strongly socialist, although they are not immune from the penetrating influence of global capitalism. The People's Republic of China, formerly a strongly socialist society, is currently undergoing transformation to a mix of socialist and capitalist principles. This change occurs with state encouragement of a market-based economy, the introduction of privately owned industries, and increased engagement in the international capitalist economy. Many developing nations have pledged socialist principles, but socialism in the developing world has frequently met with considerable hostility from the capitalist world powers. In a world context dominated by capitalism, it is difficult for any nation not to become part of this global economy.

Communism is sometimes described as socialism in its purest form. In pure communism, industry cannot be the private property of owners. Instead, the state is the sole owner of the systems of production. A critical feature of communist economics has been the centralization of the economy in which administrators declare prices, quotas, and production goals for the entire country. This is perhaps the most striking difference between communism and capitalism. Under capitalism, market forces are permitted to dictate these decisions.

Communist philosophy argues that capitalism is fundamentally unjust because powerful owners take more from laborers (and society) than they give and use their power to maintain the inequalities between workers and owners. Recall that Marx thought that capitalism would inevitably be overthrown when workers worldwide united against the system that exploited them. Class divisions were supposed to be erased at that time, along with private property and all forms of inequality. Communism should not be confused with *totalitarianism*—a political system in which powerful elites exercise total control over the population—although the communist nations that have existed have typically also been totalitarian states. People commonly refer to communism to mean any repressive regime, but this is not the true meaning of communism as an economic system.

The Changing Global Economy

As capitalism has spread throughout the world, a **global economy** has been created. Of course, trade between nations has long been a feature of economic relations, but globalization now encompasses more than world trade. In a global economy, economic transactions, including investment, production, management, markets, labor, information, and technology, cross and penetrate national borders, and nations become increasingly interdependent (Altman 2001; Carnoy et al. 1993). Economic events in one nation now can have major reverberations throughout the world, so when the economy in Brazil, Japan, Argentina, or Russia is unstable, the effects are felt worldwide. **Multinational corporations**—those that draw a large share of their revenues from foreign investments and conduct business across national borders—have become increasingly powerful, generating profits for a few, while spreading their influence around the globe.

The global economy links the lives of millions of Americans to the experiences of other people throughout the world. College students who worry about getting a good job after graduation do so in an economy where manufacturing jobs are being exported to other nations and where U.S. industries face stiff competition from other parts of the world. You can see the global economy in everyday life: Status symbols such as high-priced sneakers are manufactured for just a few cents in China. The Barbie dolls that young girls accumulate, though inexpensive by U.S. standards, are made by workers in Indonesia where, as the box "Sociology in Practice: Toys Are Not U.S." explains, it would take a month's wages for the Indonesian worker who makes the doll to buy it herself. Such are the sociological forces now guiding the experience of work both in the United States and abroad.

An obvious consequence of the global economy for U.S. workers is the increased transfer of jobs overseas. When the government produces data that show the number of manufacturing jobs in the United States has declined as a percentage of the total labor force, this does not mean that manufacturing is ceasing—only that most of the work is performed overseas. For business owners, moving jobs overseas reduces the cost of labor because wages for workers are extremely low in countries where labor is situated. In China, as an example, the young women who make Barbie dolls earn less than $1.99 per day—and frequently migrate thousands of miles from their homes to do so (Macmillan 1996; Press 1996). Soccer balls that sell in the United States for

SOCIOLOGY IN PRACTICE

Toys Are Not U.S.

Using the sociological perspective brings everyday objects into new light. For example, every year millions of young girls in the United States ask for a Barbie doll for Christmas. What do Barbie dolls reveal about the sociological structure of work and the global economy? Made by Mattel, Barbie is the ultimate American "dream girl," but how many people who buy Barbies know that the doll is manufactured by workers not much older than the ones who play with her and who would need all of their monthly pay to buy just one of the dolls that U.S. girls collect by the dozens?

The toys that many U.S. kids play with are manufactured through a growing global division of labor, and Barbie is not the only example. Whether it is soccer balls, *Toy Story* action figures, or *Monsters, Inc.* plastic characters, toys are made in China, Indonesia, Mexico, and other parts of the world where work conditions are poor and hazardous and where workers' rights are routinely ignored. In China, where more toys are produced than in any other country, workers molding Barbie dolls may earn as little as 25 cents per hour; human rights organizations say violations of basic rights are flagrant. As one journalist writes, "Behind the glitter of F.A.O. Schwartz and Toys 'R' Us, the toy industry is a showcase for the injustices at the heart of the global economy" (Press 1996: 12).

The manufacturing of toys is a classic example of the global assembly line resulting in job loss in the United States. In 1973, more than 56,000 U.S. workers were employed in toy factories. Now, as the market has become more glutted with the latest popular items, only 27,000 U.S. workers work in toy factories. The companies that make the toys are amassing record profits. At whose expense is this happening? U.S. workers have lost jobs and tend to blame foreign workers for taking them. But who has gained? In 1995, the CEO of Mattel earned $7 million and an additional $23 million in stock options— far more than the combined salaries of the 11,000 Mattel workers in China. Indonesian workers making Barbies earn the minimum wage of $2.25 per day. It would take such a worker a full month to earn the money to buy the Calvin Klein Barbie (Press 1996). The next time you look at the labels in your local toy stores, think about the global restructuring that brings entertainment to U.S. children.

Critical Thinking Exercise

1. Take a walk through one of your local malls. What products do you see that have been developed on the global assembly line? What does this tell you about the evolving character of the global economy?

2. What different styles of Barbie are currently being produced? How do the images being sold produce gender and race stereotypes? Is there a link between the images produced and the division of labor that produces the dolls?

Source: Press, Eyal. 1996. "Barbie's Betrayal." *The Nation* (December 30): 11–16; Macmillan, Jerry. 1996. "Santa's Sweatshop." *U.S. News & World Report* (December 16): 50–60. •••

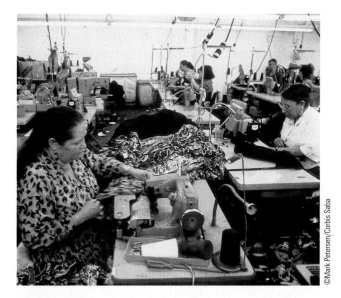

©Mark Petersen/Corbis Saba

Immigrant women and women of color are the typical employees in sweatshops in the United States, such as this sweatshop in New York City's Chinatown.

$150 are manufactured in Sri Lanka and Haiti, where not only are wages low, but labor and human rights are also routinely ignored. Many products sold in the United States are manufactured in sweatshops abroad, where young women and children are the majority of the workers.

In the global economy, research and management are controlled by the most developed countries, and assembly-line work is performed in nations with less privileged positions in the global economy (Ward 1990). A single product, such as an automobile, may be built from parts made all over the world—the engine assembled in Mexico, tires manufactured in Malaysia, electronic parts constructed in China. The relocation of manufacturing to wherever labor is cheap and management is strong has led to the emergence of the *global assembly line,* a new international division of labor in which research and development is conducted in the United States, Japan, Germany, and other major world powers, while the assembly of goods is done primarily in underdeveloped and poor nations—mostly by women and children. The development of the global assembly line has contributed to a significant trade imbalance, with the United States increasingly importing the goods it consumes.

Manufacturing products overseas suits the interest of capitalist economies because labor in the underdeveloped nations is cheaper and typically nonunionized. The relative absence of state regulations governing work conditions and terms of employment makes the transfer of work overseas even more attractive to profit-seeking corporations. The enhancement of corporate profits made possible by purchasing cheap labor in the poorest parts of the world comes partly at the expense of U.S. workers, however, who are faced with a shrinking domestic job market.

Within the United States, the development of a global economy has also created anxieties about foreign workers, particularly among the working class. Because it is easier to blame foreign workers for increases in unemployment in the United States than it is to understand the complex processes that have produced this phenomenon, U.S. workers have been prone to **xenophobia,** the fear and hatred of foreigners. Campaigns to "Buy American" reflect this trend, although the concept of buying American is increasingly antiquated in a multinational economy. When buying a product from a U.S. company, the likelihood is that the parts, if not the product itself, were built overseas. In a global economy, distinctions between U.S. and foreign businesses blur. Moreover, the label "Made in the U.S.A." does not necessarily mean that the product was made by well-paid workers in the United States. In the garment industry, sweatshop workers, many of whom are recent immigrants and most of whom are women, are likely to have stitched the clothing that bears such a label. Moreover, these workers are likely to have done so under exploitative conditions. For example, recent surveys of the garment industry in Los Angeles, where 25 percent of all women's outerwear is made, found that 96 percent of the garment firms were in violation of health and safety regulations for workers; 61 percent were violating wage regulations (Bonacich and Applebaum 2000; Louie 2001).

The development of a global economy is part of the broad process of **economic restructuring,** which refers to the contemporary transformations in the basic structure of work that are permanently altering the workplace. This process includes the changing composition of the workplace, deindustrialization, the use of enhanced technology, and the development of a global economy. Some changes are *demographic*—that is, resulting from changes in the population. The workforce is becoming more diverse, with women and people of color becoming the majority of those employed. Other changes are driven by *technological developments*. For example, the economy is now based less on its earlier manufacturing base and more on service industries, where the primary business is not the production of goods but the delivery of services (banking, health care, provision of food, or the like). All these developments are happening within a global context.

A More Diverse Workplace

Projections about the labor force indicate that women and people of color will be an even larger proportion of the workforce in the years ahead. Asian Americans and Hispanics are the groups with the most rapidly growing labor force participation. Women are also increasing their participation in the labor force more than men. A big change in the labor force over the next several years will be the age of workers. With the aging of Baby Boomers, the workforce will be older than ever before (Bowman 1999; Dohm 2000; Fullerton 1999).

These changes in the social organization of work and the economy are creating a more diverse workforce, but much of the growth in the economy is projected to be in service industries where, for the better jobs, education and training are required. People without these skills will not be well positioned for success. Manufacturing industries, where racial minorities have historically been able to get a foothold on employment, are in decline. New technologies and corporate layoffs have reduced the number of entry-level corporate jobs, which recent college graduates have always used as a starting point for career mobility. Many college graduates are employed in jobs that do not require a college degree. College graduates, however, do still have higher earnings than those with less education, but the declining value of the dollar means that workers at various levels of educational attainment have seen a loss in the buying power of their income.

Deindustrialization

Deindustrialization refers to the transition from a predominantly goods-producing economy to one based on the provision of services (Harrison and Bluestone 1982). The goods are still produced, but fewer workers in the United States are required to produce the goods because machines can do the work people once did, and many goods-producing jobs have moved overseas. Different from traditional manufacturing jobs, such as the manufacture of cloth or automobiles, service-based industries are based on the delivery of a product or provision of a particular service.

Deindustrialization is most easily observed by looking at the decline in the number of jobs in the manufacturing sector of the U.S. economy since World War II. At the end of the war, the majority of workers (51 percent) in the United States were employed in manufacturing-based jobs. Now, the majority (at least 75 percent) are employed in what is called the service sector (U.S. Department of Labor 2004; Wilson 1978). The service sector includes two segments: service delivery (such as food preparation, cleaning, or child care) and information processing (such as banking and finance, computer operation, or clerical work). The service delivery segment consists of many low-wage, semiskilled, and unskilled jobs, and it provides employment to numerous women and people of color.

Deindustrialization has led to **job displacement,** the permanent loss of certain job types that occurs when employment patterns shift. For example, job displace-

ment occurs when the manufacturing base shrinks as the data-handling industry grows. Job displacement creates **structural unemployment**—that which occurs because of changes at the societal level, different from the unemployment that results from individual performance, the seasonality of work, or individual business failure.

Whole communities can be affected by deindustrialization, and the human costs can be severe. For example, increased food production in South America, Mexico, and other international sites can cause farmers to lose their income, even their land. Likewise, residing in an industrial city that depends on a single industry leaves workers vulnerable if that industry shuts down. When industries close, workers may be forced to leave the community altogether, although many find it difficult to do so. Moving means leaving the support systems provided by family and friends. And, workers who own their homes may find it difficult to move after a plant closing because the real estate market is likely to be depressed.

Deindustrialization has transformed whole cities, such as those that were heavily dependent on a single industry. (Think of the steel towns or the automobile-manufacturing cities of Detroit, Flint, Lansing, Cleveland, and Akron.) Some may rebound by investing in new, high-tech industries, but this can still leave workers without jobs if they do not have the skills required by the new work. Many find they have to take jobs at lower pay and without job benefits.

Minority workers in particular have been severely affected by these changes—one result being the high rates of unemployment in inner cities that traditionally depended on manufacturing jobs. African Americans who migrated to northern cities in the mid-twentieth century and found skilled blue-collar work now find themselves deeply affected by the decline in manufacturing work. Latinos, too, have more recently been affected by the same changes. As a result, there are extremely high rates of joblessness and dim prospects for economic recovery in many urban communities; in these locations, teens and new workers are especially disadvantaged and experience a very high rate of unemployment (see Figure 18.1; Wilson 1996).

Deindustrialization has also resulted in major changes in labor unions. Traditionally, unions have been strong among blue-collar workers; now much union activity is from the service industry, where women and people of color predominate. Unionized workers have better pay and better benefits than nonunion workers, even in the same occupations. The decline of manufacturing industries has resulted in a decline in union membership over the years. Still, 13 percent of all workers are union members—Black men and women being the most likely to be unionized. Unions are also increasingly found among workers other than blue-collar

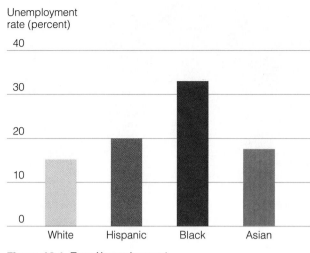

Unemployment
rate (percent)

Figure 18.1 Teen Unemployment

Data: U.S. Department of Labor. 2004. *Employment and Earnings.* Washington, DC: U.S. Government Printing Office, p. 196–200.

Table 18.1		
Top Ten Occupations by Job Growth, 2002–2012		
	Number of jobs*	Percent growth
Registered nurses	623	27
Postsecondary teachers	603	38
Retail salespersons	596	15
Customer service representatives	460	24
Food preparation and serving workers, including fast food	454	23
Cashiers, except gaming	454	13
Janitors and cleaners	414	18
General and operations managers	376	19
Waiters and waitresses	367	18
Nursing aides, orderlies, and attendants	343	25

Source: Bureau of Labor Statistics. 2004. "Occupations with Largest Job Growth, 2002–2012." Washington, DC: U.S. Department of Labor. Website: **www.bls.gov/emp/emptab4.htm**

manufacturing workers. For example, 43 percent of those in education, training, and library occupations are unionized, compared to 22 percent of those in construction and maintenance occupations (U.S. Department of Labor 2004). This reflects a significant trend in union membership—that unions are now much more diverse than in the past and have the potential to represent different groups of workers than was their historic tradition.

Although labor unions are historically identified with White, working-class men, increasingly unions are the province of a diverse population of workers, including women and people of color in the service industry—such as these AFSCME (American Federation of State, County, and Municipal Employees) workers.

With deindustrialization, job growth is greatest in professional and administrative positions with high educational requirements and in jobs that require advanced technological skills (Wilson 1996). At the same time, job growth in a service-based economy is at the lower end of the occupational system—jobs requiring little education, skill, or training (see Table 18.1; Braddock 2000). Highly educated and highly trained people are best positioned to take advantage of the desirable new jobs; most people find themselves in far less competitive positions, dependent on low-wage service occupations.

Technological Change

Rapidly changing technologies are also drastically changing work, including how it is organized, who performs it, and how much it pays. A highly influential technological development of the twentieth century was the invention of the semiconductor. Some have argued that the computer chip has as much significance for social change as the invention of the wheel and the steam engine in earlier times. Computer technology makes possible many workplace transactions that would have seemed like science fiction just a few years

AP/Wide World Photos

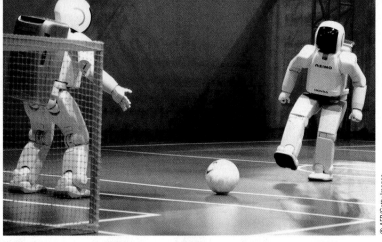

Automation means that machines can now supply the labor originally provided by human workers, such as this robot.

ago. Electronic information can be transferred around the world in less than a second. Employees work for corporations based on another continent. Offshore data entry is a perfect example of the confluence of low wages and high technology. A young woman in Southeast Asia may key in the information for a U.S. citizen's health care claim or two typists in the Caribbean may key in the manuscript of a book with a machine checking for errors.

Increasing reliance on the rapid transmission of electronic data has produced *electronic sweatshops,* a term referring to the back offices found in many industries, such as airlines, insurance firms, mail-order houses, and telephone companies, where workers at computer terminals process hundreds or thousands of transactions in a day. During this time, workers' performance is monitored by the very computer at which they work. Telephone operators may be given a few seconds to root out a number, the speed of the transaction recorded by computer. Computers can measure how fast cashiers ring up groceries and how fast ticket agents book reservations. Supervisors may monitor employee phone calls and e-mail. Records derived from computer monitoring then become the basis for job evaluations (Kilborn 1990). The electronic monitoring of workers conjures images of "Big Brother" invisibly watching workers' actions and is the focus of debates about workers' rights to privacy.

Automation is the replacement of human labor by machines. Robots can perform repetitive tasks once done by human workers. For example, robots now perform 98 percent of the spot-welding on Ford automobiles. Some sophisticated robots are capable of highly complex tasks, enabling them to assemble finished products or flip burgers in fast-food restaurants, replacing the cooks. Robots are expensive to buy, but employers can calculate their advantage given that "the robot hamburger-flipper would need no lunch or bathroom breaks, would not take sick days, and most certainly would neither strike nor quit" (Rosengarten 2000: 4).

Technological innovation in the workplace is a mixed blessing. Automation eliminates many repetitive and tiresome tasks and makes possible rapid communication and access to information. But critics worry that increasing dependence on technology also makes workers subservient to machines. The human tasks on assembly lines, for example, are paced by the speed of conveyor belts, robots farther up the line, and so on. Automation has also completely eliminated many jobs, and not only in manufacturing. Elevator operators, once a common sight in department stores, are now found in only the most elite office or retail settings.

The benefits of technological advancement are many, to be sure—new forms of work become possible, including work that can be done at home or across long distances; new products and markets are created; new outlets for talent and ingenuity emerge. The benefits associated with technological change are, however, distributed unevenly across social classes. Professional workers may find that technology brings them new options and opportunities, but many working-class people with fewer resources may simply be displaced.

Technology also changes the character of work organizations. In automated workplaces, there is often a less hierarchical and formal structure compared with less automated workplaces. But computerized organizations also have a more bifurcated workforce and higher levels of race and sex segregation (Burris 1998).

The rise of technologically based work has other consequences for workers. Any group without the educational or technological skills needed to compete in a technologically advanced labor force will be in a severely disadvantaged position (Wilson 1987). This situation is worsened by the tendency of high-tech industries to move out of old, urban manufacturing districts to suburbs where land is less expensive, unions are less pervasive, and nearby universities supply a pool of young, educated workers and highly skilled academics. This erodes the job base of center cities, leaving many people in the city impoverished. Thus, racial minorities, who once moved to cities to find industrial employment, accrue a disproportionate share of the disadvantages emerging from these trends. Suburban residents, on the other hand, benefit.

A further consequence of the growth of technology is deskilling. *Deskilling* occurs when the level of skill required to perform a certain job declines over time. Deskilling may result when a job is automated or when a more complex job is fragmented into a sequence of easily performed units. As work roles become deskilled, employees are paid less and have less control over their tasks. Less mental labor is required of workers and jobs become routinized and boring. Deskilling contributes to polarization of the labor force. The best jobs require ever-greater levels of skill and technological knowledge, whereas at the bottom of the occupational hierarchy, people stuck in deadend positions with no skills development become alienated from their work (Braverman 1974; Apple 1991).

The Impact of Economic Restructuring

Economic restructuring has numerous effects. Just as the workplace is becoming more diverse, there is growing inequality between the different groups in the labor market. For some people, there is too much work; for others, too little. Those who are employed are now working longer hours. Wages are flat. Economists calculate that since 1979, workers have increased their annual working hours by three weeks per year. The increase in hours worked is highest among working- and middle-class workers. Furthermore, whereas men have increased their working hours by 4 percent since 1979, women have increased theirs by 42 percent—or, measured another way, 44 more eight-hour days per year (Mishel et al. 2003).

THINKING SOCIOLOGICALLY

Identify a job you have once held (or currently hold) and make a list of all the ways that workers in this segment of the labor market are being affected by the various dimensions of *economic restructuring: demographic changes, globalization of the economy,* and *technological change.* What does your list tell you about how people's individual work experiences are shaped by *social structure?*

At the same time, wages (controlling for the value of the dollar) have been stagnant, especially for men, younger workers, blue-collar and service workers, and those without college degrees. Among women, wages have increased, especially among upper earners, but have fallen for those in the lowest-income groups. Given the stagnancy of wages and the longer hours worked, it is not surprising that many people feel they are working harder to simply stay in place (Jacobs and Gerson 1998; Mishel et al. 2003).

Economic restructuring has also meant a change in the location of jobs. As we have learned, many jobs have moved from central cities to the suburbs. Compounded with residential segregation, this has serious consequences for urban minorities. **Mismatch theory** is an argument that specific groups are disadvantaged in the labor market by the combination of residential segregation and the movement of jobs to suburban areas. In particular, the movement of jobs from center city areas has a disproportionately negative effect on minority workers, especially minority women and young Black and Latino men (Thompson 1997; Sarbaugh et al. 1999; Stoll 1998). Most transportation systems are not designed to move inner cities residents into the suburbs

where jobs are likely to be found. Moreover, for workers with children, having jobs located far from home and school can pose serious problems for child care.

Economic restructuring also involves *downsizing*—a euphemism for laying off workers to reduce corporate expense. In addition to job layoffs hurting the working classes, employers have shrunk the layers of management, thereby decreasing the overall number of managers; and downgraded the rank and salary of managers, thereby eliminating many career ladders that allowed job mobility in the past. Companies now use more temporary workers and independent contractors who are not paid benefits.

Contingent workers do not hold regular jobs, but their employment is dependent on demand. These workers include those who contract independently with employers, temporary workers, on-call workers (those called in only when needed), the self-employed, part-time workers, and day laborers. Contingent workers are estimated now to comprise 30 percent of the labor force—a huge increase over the past thirty years and the most rapidly growing sector of the economy. Although these jobs bring flexibility and autonomy, contingent workers earn less, are less likely to receive health insurance or pensions, and have less job security than workers in regular full-time jobs—even when comparing workers with similar educational, personal, and job characteristics (Hudson 1999; Kalleberg et al. 2000).

Women are more likely than men to be employed in contingent jobs; the employment of women is also concentrated in the least desirable of such jobs—those with the lowest pay and least likelihood of providing benefits. Whites are more likely to be independent contractors or self-employed, whereas Blacks and Hispanics are more likely to be found in temporary and part-time work (Hudson 1999; Barker and Christensen 1998; Cook 2000). Contingent workers are heterogeneous in terms of skill and educational background (think of part-time or temporary faculty members, day laborers, and temporary clerical workers), yet they have tenuous status in the labor market. They are rarely organized in unions and highly subject to fluctuations in the economy and, therefore, are among the most vulnerable of all workers. That is why such workers suit employers' interests. Changes associated with economic restructuring may be good business decisions, but this comes at great human cost.

Theoretical Perspectives on Work

The major theoretical perspectives of sociology provide frameworks for understanding the social structural forces that are transforming work. Each viewpoint—

functionalist theory, conflict theory, and symbolic interaction—offers a unique analysis of work and the economic institution of which it is a part (see Table 18.2).

Functionalist Theory

Functionalist theorists interpret the work and the economy as a functional necessity for society. Certain tasks must be done to sustain society, and the organization of work reflects the values and other characteristics of a given social order. Functionalists argue that when society changes too rapidly, as is the case with new technological and global developments, work institutions undergo social disorganization—perhaps alienation, unemployment, or economic anxiety—as social institutions try to readjust and develop new forms that will again bring social stability.

Functionalist theory also calls attention to the cultural values that are widely shared about work. In the United States, the work ethic, for example, is a strong and central cultural value. People place a high value on the work ethic, meaning the belief that hard work is a moral obligation. As Max Weber noted, the work ethic stems from the Protestant belief that hard work is a sign of moral stature, and prosperity is a sign of God's favor (see also Chapter 17). Those perceived not to value work, and therefore not to have a strong work ethic, are judged as moral failures and are blamed for their own lack of success. This cultural value is the crux of stereotypes about the "undeserving poor"—the belief that the poor have become so because of their own failures and refusal to internalize the values of diligence and hard work.

At the same time, people who are most admired for success tend to be thought of as hard workers, even if their success comes largely through inheritance. But once a principle with the primacy of the work ethic becomes embedded in the value system of a culture, such contradictions tend to be ignored.

Conflict Theory

Conflict theorists view the transformations taking place in the workplace as the result of inherent tensions in the social systems, tensions that arise from the power differences between groups vying for social and economic resources. Class conflict is then a major element of the social structure of work, and conflict theorists see class inequality (and its relationships to gender and race inequality) as the source of unequal rewards that workers receive for work.

In addition, conflict theorists analyze the fact that some forms of work are more highly valued than others, both in how the work is perceived by society and how it is rewarded. Mental labor has been more highly valued than manual labor. In addition, work performed outside the home is typically judged to be more valuable than work performed inside the home. Given the stratification based on class, race, gender, and age in this society, generally speaking, the work most highly valued has been that done by White, middle-class, older men. Sociologists debate whether this occurs because the most highly valued jobs are reserved for this group, or whether the fact that this group performs these jobs results in their being more highly valued. It is a question of which comes first, but the point from conflict theory

Table 18.2			
Theoretical Perspectives on Work			
	Functionalism	**Conflict Theory**	**Symbolic Interaction**
Defines Work	As functional for society because work teaches people the values of society and integrates people within the social order	As generating class conflict because of the unequal rewards associated with different jobs	As organizing social bonds between people who interact within work settings
Views Work Organizations	As functionally integrated with other social institutions	As producing alienation, especially among those who perform repetitive tasks	As interactive systems within which people form relationships and create beliefs that define their relationships to others
Interprets Changing Work Systems	As an adaptation to social change	As based in tensions arising from power differences between different class, race, and gender groups	As the result of the changing meaning of work resulting from changed social conditions
Explains Wage Inequality	As motivating people to work harder	As reflecting the devaluation of different classes of workers	As producing different perceptions of the value of different occupations

is that the prestige attributed to different jobs follows along lines of race, class, gender, and age, among other factors.

Judgments made about the value of different forms of work have also been turned into judgments about the value of different groups of workers. Thus, men who perform manual labor have been devalued as workers, including White men. Work done by Latinos, Native Americans, African Americans, and Asian Americans has been some of the most demanding, yet least rewarded, work in society (see the box "Understanding Diversity: Native American Women and Work.") The labor of the White working class has also been undervalued. Historically, White ethnic groups such as the Irish, Polish, and Italians performed some of the most onerous work. Their contributions as workers have also been devalued as the result of prestige being typically reserved for the most dominant group.

Symbolic Interaction Theory

Symbolic interaction brings a different perspective to the sociology of work. Symbolic interaction theorists are interested in the meaning people give to work, as well as the actual interactions that people have in the workplace. Thus, some classic studies have examined how new workers learn their new roles and how a worker's identity is shaped by social interaction in the workplace (Becker et al. 1961). Some studies using this perspective also analyze the creative ways that people deal with routinized jobs. People may create elaborate and exaggerated displays of routine tasks to bring some

human dimension to otherwise dehumanizing labor (Leidner 1993).

Another way to use the insights of symbolic interaction theory is to think about how work is defined in society. Most people think of work as an activity for which a person gets paid, but does this definition devalue work that people do without pay? Unpaid jobs such as housework, child care, and volunteer activities make up much of the work done in the world. If you define work as productive human activity that creates something of value—either goods or services—you see that work takes many forms. It may be paid or unpaid. It may be performed inside or outside the home. It may involve physical or mental labor, or both.

Influenced by feminist studies, sociologists recognize that housework and other forms of unpaid labor are an important part of the productivity of a society. Much productive labor is done within families, whether paid or unpaid. This includes *reproductive labor,* or *care work,* referring to the care provided for children and adults within families (including the provision of food, clothing, and nurturing). As we have seen (in Chapter 15), some nations are far ahead of the United States in recognizing the value of this productive work. Germany, for example, provides fourteen weeks of paid maternity leave for women and a monthly allowance for mothers of small children who have left the paid labor force. In Norway, employed women may take eighteen weeks maternity leave, with either parent allowed to take twenty-six extra paid weeks (United Nations 2000b). By contrast, the Family and Medical Leave Act (FMLA) in the United States provides twelve weeks

UNDERSTANDING DIVERSITY
Native American Women and Work

Whether they live on or off the reservations, American Indian men and women live in difficult times, struggling daily against white control, the threat of cultural extinction, and the challenges posed by poverty. . . . Still, American Indians are growing in strength, cultural pride, unity, and militancy. . . .

Native American women's labor force participation [has risen] sharply . . . to 55 percent (in 1990). Those who held fulltime, year-round jobs earned 83 percent as much as White women—but these jobs were hard to come by. More than 70 percent of American Indian women held part-time or part-year jobs—the highest rate for any racial-ethnic group

of women—and American Indian women faced the highest unemployment rate (13 percent) of the racial-ethnic groups. . . .

Native American women have made substantial progress into the more highly paid, masculine jobs in the lower tier of the primary sector. . . . In the upper tier of the primary labor market, Indian women [have] made real inroads into sales jobs in commodities and finance, and into executive, managerial, and administrative jobs. . . . Many American Indian women have attained primary-sector jobs in the public sector, which [employs] three-fifths of American Indian women. . . .

Despite some gains in the primary sector, over two-thirds of American Indian women were employed in the secondary labor market in 1990, compared to three-fifths of European American women and just under one-third of European American men. . . . Repressive, inaccessible, and inadequate education bears much of the blame for this low occupational status, along with discrimination by employers and fellow employees and the stagnation of the reservation economy.

Source: Amott, Teresa, and Julie Matthaei. 1996. *Race, Gender and Work: A Multicultural Economic History of Women in the United States.* Cambridge, MA: South End Press, pp. 59–61. Used with permission. ● ● ●

of unpaid leave—but only in certain organizations and only after the employee has used up other vacation and sick leave.

Another way to think about work is through the concept of emotional labor (Hochschild 1983). **Emotional labor** is work that is specifically intended to produce a desired state of mind in a client. Many jobs require some handling of other people's feelings. Emotional labor involves putting on a false front before clients, and is performed in jobs where inducing or suppressing a feeling in the client is a primary work task. Airline flight attendants are an example—their job is to please the passenger and to make passengers feel as though they are guests in someone's living room, not objects catapulting through the sky at 500 miles per hour! Emotional labor, like other work, is done for wages. It is supervised and evaluated. Workers are trained to produce the desired effect among clients.

Many jobs require the performance of emotional labor. In other words, producing a particular state of mind in the client is part of the product being sold. In a service-based economy, emotional labor is a growing part of the work that people do, although it is seldom recognized as real work. Emotional labor also makes the production of emotion a commodity—a product created for profit and consumed. This can result in the "commercialization of human feeling," meaning that the production and management of emotional states of mind is increasingly seen as a commodity to be bought and sold in the marketplace (Hochschild 1983).

In sum, each theoretical perspective reveals different dimensions to the sociological study of work—either in the organization of work within systems of stratification, in studying the values and meanings associated with work, and in analyzing the interaction people have with one another at work. You can also see that these different theoretical perspectives can be combined to explain particular subjects. Thus, the work ethic in U.S. culture can be understood as a shared cultural value (functionalism), as reflected in judgments about various groups of people (symbolic interaction), and as differentially applied depending on the group's class, race, age, or gender (conflict theory).

Characteristics of the Labor Force

Data on characteristics of the U.S. labor force are typically drawn from official statistics reported by the U.S. Department of Labor. In 2003, the labor force included approximately 146 million people—66 percent of the working age population, that is those sixteen years of age and older (U.S. Department of Labor 2004). The specific characteristics of those in the labor force have also changed considerably.

Who Works and How Much?

Employment varies significantly for different groups in the population. Hispanic men are the most likely group to be employed; Hispanic women are the least likely (see Figure 18.2). It is difficult to assess changes in the employment of Hispanics, Asian Americans, and Native Americans over time, because the Department of Labor did not collect separate data on these groups prior to 1980. Even now the broad categories "Black" and "White" make it difficult to see the unique experience of diverse groups in the labor force. Hispanics can also appear in both the "Black" and "White" categories, depending on their self-identification. Pacific Islanders, Native Americans, and Alaskan natives may be grouped as "other," a category that is seldom reported in summary tables. Thus, making comparisons among different groups is difficult, without more detailed information.

Despite these limitations, several trends can be discerned. A dramatic change in the labor force since World War II has been the increase in the number of women employed from 35 to 60 percent of all women since 1948. Because women of color have historically had a high rate of employment, the biggest increase is among White women. Although Black and White women are now equally likely to be employed (U.S. Department of Labor 2004).

Increases in employment among women have been greatest for women with young children. Sixty-three percent of married women with children under six years of age and 68 percent of single women with children under six are now employed (U.S. Census Bureau 2004). This has caused significant changes in family life (see Chapter 15) and created new demands for services

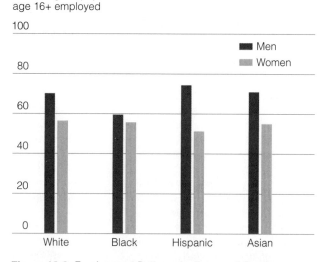

Percentage of population
age 16+ employed

Figure 18.2 Employment Patterns by Race and Gender

Data: U.S. Department of Labor. 2004. *Employment and Earnings.* Washington, DC: U.S. Department of Labor, p. 196–200.

traditionally provided free of charge by women in the home—day care, fast food, laundry services, and house-cleaning—activities that contribute to the growth of a service-based economy. Growth in these jobs has also generated an underground economy of workers, primarily women of color, who provide domestic labor at wages below the federal poverty line.

The increase in employment of women with children has also created new demands on the workplace. Some organizations have responded by creating on-site child-care facilities or facilities to tend children who are sick to allow their parents to come to work. The Family and Medical Leave Act of 1993 means many more employers provide maternity and paternity leave—95 percent of medium- to large-sized companies. This is compared with only 27 percent who did so before the FMLA legislation passed. However, in smaller firms, only 19 percent now provide leave—indicative of the limitations in the law (Waldfogel 1999). However, most employers provide only unpaid leave, thus few workers are able to take advantage of this policy. The work-

ers most likely to take such leaves are married, White women (Gerstel and McGonagle 1999). There are also significant class differences in who takes such leaves (see Table 18.3), with the working class being least likely to be able to do so.

While the employment of women has been increasing, employment for men has been falling from 87 percent of men in the labor force in 1951 to 74 percent now (U.S. Department of Labor 2004). This decline is expected to continue, while the presence of women in the labor force is expected to keep growing. Black, Hispanic, and Asian American employment is also predicted to grow faster than White employment, with Hispanics predicted to have the greatest increase (Kutscher 1995).

These changes in the employment status of diverse groups reflect the economic restructuring of the workplace and, as conflict theorists would point out, produce strife between different groups competing for the jobs that are left. Jobs where White men have predominated are declining, especially in the manufacturing

Table 18.3
Employee Benefit Programs

Benefit (measured as percent who participate)	All Employees	Professional, Technical, and Related Employees	Clerical and Sales Employees	Blue-Collar and Service Employees
Medium and large private establishment (100 or more employees)				
Paid time off				
Holidays	89%	89%	91%	88%
Vacations	95	96	97	94
Personal leave	20	23	33	13
Funeral leave	81	84	85	76
Jury duty leave	87	92	89	83
Military leave	47	60	50	38
Family leave	2	3	3	1
Unpaid family leave	93	95	96	91
Medical Health Care Benefits	76	79	78	74
Small private establishments (fewer than 100 employees)				
Paid time off				
Holidays	80%	86%	91%	71%
Vacations	86	90	95	79
Personal leave	14	21	18	8
Funeral leave	51	60	60	42
Jury duty leave	59	74	68	47
Military leave	18	25	22	12
Family leave	2	3	3	1
Unpaid family leave	48	53	52	43
Medical Health Care Benefits	56	67	60	49

Table covers full-time employment only.
Source: Bureau of Labor Statistics. 1999. *Employee Benefits in Medium and Large Private Establishments.* Washington, DC: U.S. Department of Labor, p. 6;
Bureau of Labor Statistics. 1999. *Employee Benefits in Small Private Establishments.* Washington, DC: U.S. Department of Labor, p. 6.

sector; jobs in niches of the labor market segregated by race and gender are those most likely to increase. The popular belief that women and minorities are taking jobs from White men derives from this fact. Note, however, that women are not taking the jobs where White men have predominated in the past. These jobs simply are not as numerous, and much growth in the labor market is in menial jobs, where women and minorities are likely to be employed. Beliefs that foreign workers are taking U.S. jobs, that women and people of color are taking jobs from White men, and debates about the role of immigrant labor in the U.S. economy are the result of social structural transformations—that is, sociological factors lying behind statistical patterns.

Workers in the United States work more hours per year than workers in other industrialized nations.

Data also indicate that, unlike the image of the "9 to 5" job, a large proportion of workers work at nonstandard times—that is working hours that fall outside of the Monday through Friday daytime hours. Forty percent of workers work at nonstandard times—weekends, nights, and shifting hours. Currently, one-third of all employed persons work on Saturday, Sunday, or both. Men are more likely to work nonstandard hours than women and racial minorities are more likely to do so than White workers.

With more women in the labor force, working at nonstandard hours can accommodate family care, although it takes a heavy toll on family relationships. Couples may report rarely seeing each other as they juggle working hours, child care, and other family chores. Shift work, night work, and changing work hours also lead to various health problems, with sleep deprivation an increasing health risk for many (Presser 2004). Studies also find large numbers of workers reporting the "spillover" of work into their home lives—with work making people too tired to enjoy other parts of life or job worries distracting them when they are at home (Moen 2004).

The Value of Work: Earnings and Prestige

One pattern that sociologists investigate is differences in earnings and the value attached to different jobs. There is extensive documentation that earnings are highly dependent on race, gender, and class (see Figure 18.3). White men earn the most, with a gap between men's and women's earnings among all racial groups. Occupations in which White men are the numerical majority also tend to pay more than occupations in which women and minorities are a majority of workers. Not all men benefit equally, however, from the earnings hi-

erarchy, as White professional men earn more than men in working-class occupations. CEOs have the highest pay of all—a whopping 150 times what workers earn. White professional women are also disadvantaged relative to men at the same occupational level but are far more advantaged than Black and Hispanic men and White, working-class men.

Why are there such disparities in earnings? Again, theory puts the facts into perspective. According to functionalist theories, workers are paid according to their value—value derived from the characteristics they bring

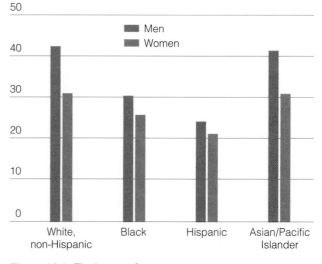

Median income, year-round, full-time workers, 2001

■ Men
■ Women

White, non-Hispanic · Black · Hispanic · Asian/Pacific Islander

Figure 18.3 The Income Gap

Source: U.S. Census Bureau. 2001. *Income: Detailed Historical Tables.* Website: **www.census.gov/hhes/income/histinc/p36a.html**

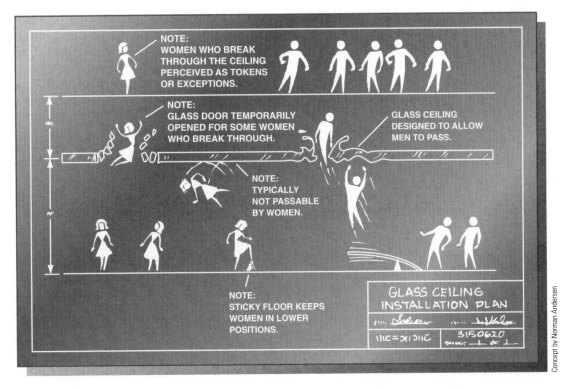

The Glass Ceiling
Whimsically depicted here as an architectural drawing, the glass ceiling refers to the structural obstacles still inhibiting upward mobility for women workers. Most employed women remain clustered in low-status, low-wage jobs that hold little chance for mobility. Those who make it to the top often report being blocked and frustrated by patterns of exclusion and gender stereotyping.

to the job: education, experience, training, and motivation to work. From this point of view, the high wages and other rewards associated with some jobs are the incentive for people to spend long years in training and garnering experience. To functionalists, differential wages are a source of motivation and a means to ensure that the most talented people fill jobs essential to society. Functionalists see wage inequality as reflecting the differently valued characteristics (education, years of experience, training, and so forth) that workers bring to a job.

DEBUNKING *SOCIETY'S MYTHS*

Myth: The racial gap in earnings persists because many Blacks and Hispanics are not motivated to work; the effect of discrimination has been eliminated by equal employment policies.

Sociological perspective: The proportion of the racial gap in earnings resulting from discrimination has increased, in part because of the government's retreat from antidiscrimination policies (Cancio et al. 1996).

Conflict theorists disagree with this point of view, arguing that many talented people are thwarted by the systems of inequality they encounter in society. Thus, far

from ensuring that the most talented will fill the most important jobs, conflict theorists note that some essential jobs are the most devalued and underrewarded. From a conflict perspective, wage inequality maintains race, class, and gender inequality. They say that differences in pay reflect the devaluation of certain classes of workers, and wage inequality primarily reflects power differences between groups in society.

The value of work is not just measured by how much people make. Subjective factors also weigh in as judgments about the worth of various jobs. **Occupational prestige** is the perceived social value of an occupation in the eyes of the general public. Sociologists find a strong correlation between occupational prestige and the race and gender of people employed in given jobs, where the influence of race is stronger than that of gender. Thus, African American and Latino men are found disproportionately in jobs that have the lowest occupational prestige. White men and Asian American men hold the jobs with the highest occupational prestige, followed by White women, Asian American women, African American women, and Latinas (Xu and Leffler 1992; Stearns and Coleman 1990).

With regard to gender, three patterns are apparent. First, women receive less prestige for the same work as men. Second, the gender composition of the job and its occupational prestige are linked: Jobs that employ mostly women are lower in prestige than those that employ more men. Jobs often lose their prestige as many women enter a given profession. Likewise, the prestige of jobs increases as more men enter the field. Finally, men and women assign occupational prestige differently. Women give occupations where most of the incumbents are women a higher prestige ranking than men do (Bose and Rossi 1983; Tyree and Hicks 1988).

Unemployment and Joblessness

The U.S. Department of Labor regularly reports data on the **unemployment rate,** the percentage of those not working but officially defined as looking for work. Currently, almost 8.7 million people are officially un-

employed (U.S. Department of Labor 2004). The official number does not, however, include all people who are jobless. It includes only people who meet the official definition of a job seeker—someone who has actively sought to obtain a job during the prior four weeks and who is registered with the unemployment office. Several categories of job seekers are excluded from the list, including people who earned money at any job during the week prior to the data being collected, even if it was just a single day's work with no prospect of more employment. Also excluded are people who have given up looking for full-time work, either because they are discouraged, have settled for part-time work, are ill or disabled, or cannot afford the child care needed to allow them to go to work, among other reasons. Because the number of job-seeking people excluded from the official defini-

VIEWING SOCIETY IN GLOBAL PERSPECTIVE

MAP 18.1 Women and Global Unemployment

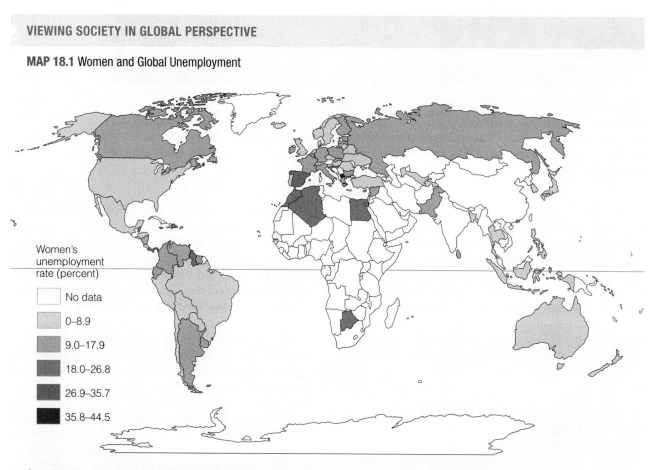

Women's
unemployment
rate (percent)

☐ No data
0–8.9
9.0–17.9
18.0–26.8
26.9–35.7
35.8–44.5

One of the first things you will notice about this map is the large number of countries in Africa, the Middle East, and Asia where data on women's unemployment is not available. This may be because women's work in many of these countries is informal and thus not captured by official statistics. In those areas of the world where unemployment is reported, where do you see the highest rate of unemployment for women? What are some of the consequences of women's unemployment for society? For the women themselves and for their families? What factors would you want to study were you to compare the causes of women's unemployment in different nations?

Data: Population Reference Bureau. 2000. *Women's World.* Washington, DC: Population Reference Bureau, p. 138. Used with permission.

Young minority workers are the most likely group to be unemployed.

tion of unemployment can be large, the official reported rates of unemployment seriously underestimate the extent of joblessness in the United States.

People most likely to be undercounted in unemployment statistics are the groups for whom unemployment runs the highest—the youngest and oldest workers, women, and members of racial minority groups. These groups are least likely to meet the official criteria of unemployment because they are more likely to have left jobs that do not qualify them for unemployment insurance. Migrant workers and other transient or mobile populations are also undercounted in the official statistics. People who work only a few hours a week are excluded, even though their economic position may be little different from someone with no work at all. Workers on strike are also counted as employed, even if they receive no income while they are on strike. Official unemployment rates also take no notice of **underemployment**—the condition of being employed at a skill level below what would be expected given a person's training, experience, or education. It can also include working fewer hours than desired. A laid-off autoworker flipping hamburgers at a fast-food restaurant is underemployed; so is a Ph.D. who drives a taxi for a living.

Unemployment is routinely used as an indicator of the state of the economy. A 4 to 5 percent unemployment rate is considered full employment. Thus, even in a "good" economy, many people are without work. Unemployment among African Americans, Native

Americans, Puerto Ricans, and Mexican Americans is currently at a level considered to be a major economic depression. As we have seen, globalization, reliance on technology, economic restructuring, and the move to a service economy fall especially hard on minority workers (see Figure 18.4).

Unemployment is not an abstraction, however. It is linked in people's lives tied to a host of social problems, including crime, alcoholism, poverty, and poor health to name some. Unemployment affects not only the worker, but also the whole social network of the person. When a parent becomes unemployed, children may have to leave school to work. Unemployment causes strains on marriages and friendships. It disrupts the identity a person gets from work and separates people from the social mechanisms that integrate them into society, as symbolic interaction theory and functionalist perspectives on work would explain.

Popular explanations of unemployment attribute it to individual failings. This reflects the belief that anyone who works hard enough can succeed thus locating the cause of unemployment in personal motivation. Sociologists explain unemployment instead by looking at structural problems in the economy, such as rapidly changing technology that reduces the need for human labor, discriminatory employment practices, deindustrialization, corporate restructuring, and the export of jobs overseas, where cheap labor is abundant. Instead of blaming the victim for joblessness, sociologists examine how changes in the social organization of work contribute to unemployment.

Some also argue that unemployment for White men is the result of hiring more women and minorities. Yet, the employment of women and minorities does not

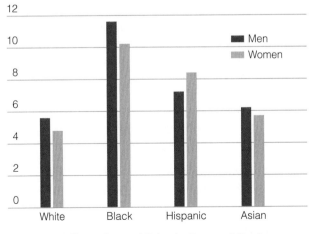

Figure 18.4 Unemployment Rates by Race and Gender

Data: U.S. Department of Labor. 2004. *Employment and Earnings.* Washington, DC: U.S. Department of Labor, pp. 196–200.

cause White male unemployment if for no other rea-son than race and gender segregation. Women are typi-cally not employed in the same jobs as men. Racial seg-regation also concentrates racial minorities in certain segments of the labor force. The fact is that men and women of color have the highest unemployment rates; White women's unemployment matches that of White men (U.S. Department of Labor 2004).

Diversity in the U.S. Occupational System

From a sociological perspective, jobs are organized in a system increasingly being defined by the diversity of groups who make up the system of work in society. As we will see, people are distributed in patterns that re-flect the race, class, and gender organization of society. Fundamental to sociological analyses of the occupa-tional system is the idea that individual and group ex-periences at work are the result of social structures, not just individual attributes. There is, for example, a rough correlation between the desirability of given jobs in the occupational system and the social status of the group most likely to fill those jobs. When sociologists study the occupational system, they acknowledge the importance of individual attributes such as level of edu-cation, training, and prior work experience in predict-ing the place of a worker in the occupational system, but they see a greater importance in the societal pat-terns that structure the diverse experiences of workers.

The Division of Labor

The **division of labor** is the systematic interrelatedness of different tasks that develops in complex societies (see also Chapter 5). When different groups engage in dif-ferent economic activities, a division of labor exists. In a relatively simple division of labor, one group may be responsible for planting and harvesting crops, whereas another group is responsible for hunting game. As the economic system becomes more complex, the division of labor becomes more elaborate.

THINKING *SOCIOLOGICALLY*

Look around you at the different occupations in your school. What evidence do you see of a *race, class, or gender division of labor*? How do you think these dif-ferent factors intersect in the experience of workers in this organization?

In the United States, the division of labor is affected by gender, race, class, and age, to name the major axes of stratification. The *class division of labor* can be ob-

served by looking at the work done by people with dif-ferent educational backgrounds, education being a fairly reliable indicator of class. People with more education tend to work in higher-paid, higher-prestige occupa-tions. Education, income, and social prestige are thus indicators intricately linked to the class division of la-bor. Class also leads to perceived distinctions in the value of manual and mental labor. People doing men-tal labor (management and professional positions) tend to be paid more and have more job prestige than people doing manual labor. Therefore, class produces stereo-types about the working class. Manual labor is also presumed to be the inverse of mental labor, meaning it is presumed to require no thinking. By extension, work-ers who perform manual labor may be assumed not to be very smart, regardless of their intelligence.

The *gender division of labor* refers to the different work that women and men perform in society. In soci-eties with a strong gender division of labor, the belief that some activities are "women's work" (secretarial) and others, "men's work" (construction) contributes to the inequality between women and men; cultural ex-pectations usually place more value, both social and economic, on men's work. This helps explain why li-brarians and social workers are typically paid less than electricians, despite the likely differences in their edu-cational level. The traditional gender division of labor has also been the basis for women not being paid for the labor they contribute within families. As long as some forms of work are presumed to be what women are "supposed" to do, the gender division of labor ap-pears to be natural instead of based on the social defi-nitions attributed to women's and men's work.

There is also a *racial division of labor* in the United States, seen in the pattern of people from different racial and ethnic groups working in different jobs. The labor performed by racial–ethnic minority groups has often been the lowest paid, least prestigious, and most ardu-ous work. Today's racial division of labor is rooted in the racial patterns of the past. In the mid- and late nine-teenth century, Chinese immigrants were used to pro-vide a cheap supply of labor on the sugar plantations of Hawaii and in mines and railroads on the U.S. main-land (see also Chapter 11). The Chinese laborers came to be called *coolies*, a term originally meaning "bitter strength" and "slave." The Chinese were denied legal rights, paid substandard wages, and assigned to live in labor camps. Chinese laborers were pitted against White workers. Chinese workers were also used as strikebreakers—a tactic employers used to defuse the development of a strong labor movement. These early divisions predicated a long history of racial conflict between Asian and White workers and solidified the racial division of labor in the American West (Takaki 1993, 1989).

The labor history of other racial groups reflects similar patterns of exclusion and segregation. Mexicans who had settled and owned property in what is now the U.S. Southwest were robbed of their land during Anglo conquest, culminating in the Mexican-American War of 1846–1848. Mexican Americans displaced from their land were forced to seek work as seasonal agricultural workers, domestic workers, and miners. Like Asian Americans, they faced terrible hardships, doing backbreaking work for very little pay. Frequently, they lived in company towns where housing was barely fit for human habitation. Women and children provided additional unpaid labor. Families often lived in *peonage*—a system of labor in which families lived as tenants on an employer's land and were bound to an employer by debt. Unable to pay rent, families accrued debts to their landlord that they were expected to pay off with their labor. The peonage system is very much like feudalism, a system in which a large peasant class occupies lands held by a small number of elites and works in exchange for its subsistence. Under peonage, workers are seldom able to pay off their "debt" to their landlords because it builds up faster than what they earn, so they work for little more than sustenance over very long periods (Amott and Matthaei 1996).

The racial, class, and gender divisions of labor intersect, creating unique work experiences for different groups. For example, following the abolition of slavery, African American women found work as domestics—work that, though severely underpaid, was readily available year-round. In the South, African American men often found only seasonal work as agricultural laborers or day laborers. Thus, African American women often became the primary breadwinners, even in stable, married-couple families. The history of Chinese Americans shows a different pattern, yet one that again reflects the structural intersections of race, class, and gender. In the nineteenth and early twentieth centuries, restrictive immigration laws that prohibited Chinese women from joining their husbands in the United States created a shortage of women in Chinese American communities. The Chinese maintained what has been called a "split-household family system," wherein wives and children remained in China while the husband worked in the United States. As a result, Chinese American men filled certain roles that otherwise would have been considered women's work, such as domestic worker, launderer, and food service worker (Amott and Matthaei 1996; Glenn 2002, 1986).

In short, the racial and gender division of labor has historically segregated women and people of color into different and inferior jobs. The race and gender division of labor intersects with the class-based division of labor and age-based divisions. Certain jobs are rarely performed by middle- and upper-class people. Likewise, young workers typically find themselves confined to a niche in the labor market—such as in the fast-food industry. All these patterns reflect the cross-cutting arrangements of race, class, gender, and age stratification.

The Dual Labor Market

The division of labor is reflected in the segregation that can be observed in contemporary work patterns. **Occupational distribution** describes the pattern by which workers are located in the labor force. The U.S. Department of Labor classifies workers into five broad categories according to the kind of work involved: (1) managerial and professional; (2) technical, sales, and administrative support; (3) service and office occupations; (4) natural resources, construction, and maintenance occupations; and, (5) production, transportation, and material moving occupations. Each broad grouping contains a number of specific jobs.

Workers are dispersed throughout the occupational system in patterns that vary by race, class, and gender (see Figure 18.5). Women are most likely to work in sales and office occupations, primarily because of their heavy concentration in clerical work. This is now true for both White women and women of color. White men are most likely to be found in managerial and professional jobs, whereas African American and Hispanic men are most likely to be employed in production,

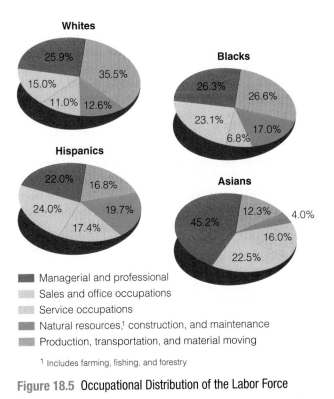

Figure 18.5 Occupational Distribution of the Labor Force

Data: U.S. Department of Labor. 2004. *Employment and Earnings.* Washington, DC: U.S. Government Printing Office, p. 207–208.

transportation, and material-moving occupations—an area of the labor market that includes such jobs as machine operators, laundry workers, and packagers. These jobs are some of the least well paid and least prestigious in the occupational system (U.S. Department of Labor 2004).

Several changes are noticeable in these patterns over time; first is the decline in the number of Black women in private domestic work. In 1960, 38 percent of all Black women in the labor force were employed as private domestic workers; now, fewer than 1 percent. This work is now done most often by recent immigrant women (often undocumented) from Latin and Central America, Southeast Asia, and the Caribbean. They work for professionals and other middle-class working people, earn low wages, have few benefits (if any), and constitute the underground economy of unreported and untaxed income (Hondagneu-Sotelo 2001).

A second major change in occupational distribution is the greater number of racial minorities in professional and managerial work, such as in business, law, and municipal management. Now, 31 percent of Black women and 21 percent of Black men work as professionals or managers, compared with 7.4 percent in 1960. Even when racial segregation was most rigid, there was a professional and managerial segment among African Americans, but before the civil rights movement, Black professionals worked primarily in Black-owned enterprises serving primarily Black communities. Now, Black professionals are most likely to work in predominantly White organizations or in government work (Collins-Lowry 1997). Almost half of Asian American men and women work in the professional and managerial category; among Hispanics, 21 percent of women and 14 percent of men are professionals or managers (U.S. Department of Labor 2004). The entry of significantly higher numbers of minorities into these occupations has expanded the number of minorities in the middle class and increased diversity in the workplace.

Race and gender segregation in the labor market mean that women of color are concentrated in occupations where most other workers are also women of color.

© Fritz Hoffmann/The Image Works

Over time, the number of women employed in skilled trades has also increased, although the number remains low—about 10 percent in skilled trades now, compared with 1 percent in 1970 (U.S. Department of Labor 2004). The women most likely to be employed in such jobs are African American women, low-income mothers, women who previously held lower-rank jobs and disliked their previous job, and women who can work other than during the day. From this information, we can conclude that women are attracted to these jobs not just because they prefer them to other work, but also because they offer better opportunities than the available alternatives (Padavic 1992).

Occupational segregation refers to the separation of workers into different occupations on the basis of social characteristics such as race and gender, as we have seen in Chapter 12 on gender. Most occupations reveal some degree of occupational segregation, meaning there is a discernible tendency for the workers in a particular occupation to be of a particular gender or race or both. Women of color, for example, experience occupational segregation by both race and gender. That is, they are most likely to work in occupations where most other workers are also women of color.

There is also *internal segregation* within the broad occupational categories. Women in the professional category, for example, are primarily concentrated in specialties that are gender-typed, such as nursing, social work, elementary and secondary school teaching, and library work. Within the most prestigious occupations,

---

DEBUNKING *SOCIETY'S MYTHS*

Myth: Because of programs such as affirmative action, women and minorities are taking jobs away from White men.

Sociological perspective: Although women and minorities have made many gains because of programs such as affirmative action, women and minority workers still are clustered in occupations that are segregated by race and gender. There is little evidence that the progress of these groups has been at the expense of White men (Hartmann 1996).

women are still a minority—14 percent of engineers and architects, 30 percent of physicians, 28 percent of lawyers, 24 percent of CEOs (chief executive officers of companies); however, their numbers in these fields are growing (U.S. Department of Labor 2004).

Men in predominantly female professions, such as nursing, school teaching, library work, and social work, do not encounter the same discrimination facing women who enter male-dominated professions. They may encounter prejudice and stereotypes, but they experience real advantages, too. Men are more likely to be hired in occupations that are predominantly female than are women in occupations that are predominantly male. Men in predominantly female jobs are also more likely than women in the same field to be promoted and advantageously treated. Men typically do not experience the same negative working climate that many women do in male-dominated organizations (Williams 1995, 1992). For all groups, however, there are economic penalties for being employed in occupations that have disproportionate numbers of women or minorities. The penalty is greatest in jobs where gender segregation is dramatic and long-established, unionization is uncommon, and performance criteria are ambiguous (Baron and Newman 1990).

Ongoing segregation in the labor market is a major explanation for the inequalities that are found in earnings, as well as other job benefits. **Dual labor market theory** views the labor market as comprising two major segments: the *primary labor market* and the *secondary labor market* (see Table 18.4). The primary labor market includes jobs with relatively high wages, benefits, stability, good working conditions, opportunities for promotion, job protection, and due process for workers (meaning workers are treated according to established rules and procedures that are allegedly fairly administered). Blue-collar and service workers in the primary labor market are often unionized, leading to better wages and job benefits. High-level corporate jobs and unionized occupations fall into this segment of the labor market.

The secondary labor market is characterized by low wages, few benefits, high turnover, poor working conditions, little opportunity for advancement, no job protection, and arbitrary treatment of workers. See Table 18.4 for examples of jobs that fall into each segment. Many service jobs, such as waiting tables, nonunionized assembly work, and domestic work, which primarily employ women and minorities, are in the secondary labor market (Browne 1999; McCall 2001).

Table 18.4
The Segmented Labor Market: Examples of Occupations

Primary Labor Market	Secondary Labor Market	Underground Economy
Upper tier	**Upper tier**	
High status professional specialties (physicians, lawyers, professors, engineers, and so on)	Waitresses, bartenders	Prostitution
	Hairdressers, cosmetologists	Con artists
	Sales workers, cashiers	Unreported domestic labor
Business executives	Clerical workers, clerks	Undocumented workers
Supervisors	Machine operators	Sweatshop labor
Farm managers		Thieves
		Drug dealers
Lower tier	**Lower tier**	
Lower status professional specialties (teachers, librarians, social workers, actors)	Service work (cooks, maids, janitors, child care)	
Middle managers	Private household workers	
Protective service (police and fire)	Farm labor	
Truck driving	Food counter workers	
Mechanics		
Precision craft, repair, production (electricians, plumbers, welders, construction, and so on)		
Technicians		
Health assessment and training		

Adapted From: Amott, Teresa, and Julie Matthaei. 1996. *Race, Gender and Work: A Multicultural History of Women in the United States.* Boston, MA: South End Press, p. 344; U.S. Department of Labor. 2004. *Employment and Earnings.* Washington, DC: U.S. Department of Labor.

In addition to the two major segments of the labor market, the *underground economy* includes unreported domestic labor; illegal work such as drug dealing, prostitution, and confidence games, or "scams"; and so-called under the table work—work that is not reported in official data on the labor market. Workers are often forced into the underground economy because of their inability to secure good, legitimate jobs. Even when full-time, minimum wage jobs are available, some may find more lucrative employment in the illegitimate, underground segment of the economy.

The placement of women and minorities in the most devalued segments of the labor market is a major factor in gender and racial inequality. Workers who benefit from this arrangement may organize to exclude other workers from the better jobs (Reskin 1988). For example, women and minorities historically have been excluded from labor unions, which preserved the better jobs for White men. Today, exclusion in the dual labor market remains a less visible means by which certain groups are excluded from the best jobs.

THINKING SOCIOLOGICALLY

Think about the labor market in the region where you live. What racial and ethnic groups have historically worked in various segments of this labor market? Are they in the *primary* or *secondary labor markets*?

Conflict theorists would argue that the dual labor market favors capitalist owners, who find it advantageous to encourage conflict between different groups of workers. Segmenting women and minorities into niches in the labor market creates antagonisms between groups, for example, White working-class men and minority groups, who see themselves as competing for the limited jobs. When people perceive some other group to be the cause of their problems, they are less likely to blame the structure of the labor market or the structure of the economy for difficulties they experience (Bonacich 1972).

Work and Immigration

The contemporary labor force is also being shaped by the employment of recent immigrants (Rumbaut 1996a, 1996b; Pedraza and Rumbaut 1996). Popular wisdom holds that the bulk of new immigrants are illegal, poor, and desperate, but the data on immigration show otherwise. The proportion of professionals and technicians among legal immigrants exceeds the proportion of professionals in the labor force as a whole. Note, however, that this figure is based on formal immigration data that exclude illegal immigrants, most of whom are working-class. Still, people who migrate are usually not the poorest or most downtrodden in their home country. Even illegal immigrants tend to have higher levels of education and occupational skill than the typical worker in their homeland. Among immigrants can be found both the most educated and the last educated segments of the population (Rumbaut 1996a, 1996b).

What happens when migrants arrive? Immigrants in professions tend to enter at the bottom of their occupational ladder, but compared with less skilled immigrants, they are more likely to succeed economically. Less skilled immigrants do not fare so well. Immigrants have higher unemployment than native-born workers, and they earn significantly less per week than other groups (Meisenheimer 1992). Enormous variation exists in the well-being of diverse immigrant groups, depending in large part on the circumstances under which they enter the United States and the resources they bring with them (Pedraza and Rumbaut 1996).

Perhaps surprising to many is that the majority of immigrants are women. Women are especially numerous among today's largest immigrant streams: those from Central and South America, the Caribbean, Southeast Asia, and Europe (Donato 1992). Women immigrants are some of the poorest immigrants and typically concentrated in just a few occupations—domestic work, the garment industry, family enterprises, and skilled service occupations, such as nursing (Pedraza 1996).

Some immigrant groups have found work through the development of *ethnic enclaves,* areas, typically urban, in which there is a concentration of ethnic entrepreneurs. Ethnic enclaves arise when a significant number of immigrants with business experience and access to labor and capital locate in a particular area. The source of labor is often recent immigrants from the same country of origin, as well as family members and people recruited from social networks. Examples of ethnic enclaves are Little Havana in Miami, where Cuban entrepreneurs own many of the businesses, the Chinatowns of many major American cities, and the Dominican neighborhoods in New York City. In several large cities, Koreans have become the primary merchants in low-income areas. Ethnic enclaves can begin when immigrants disadvantaged in the labor market are pushed into self-employment. A few businesses formed to serve the immigrant community can be the start of an ethnic enclave that eventually includes its own banks, insurance firms, and other large-scale businesses (Portes and Rumbaut 1996).

Disability and Work

Not too many years ago, people did not think of those with disabilities as a social group. Instead, disability was thought of as an individual frailty or perhaps a stigma. Sociologist **Irving Zola** (1935–1994) was one of the first to suggest that people with disabilities face issues similar to minority groups. Instead of using a medical model that treats disability like a disease and sees individuals as impaired, conceptualizing disabled people as a minority group enabled people to think about the social, economic, and political environment that disabled people face. Instead of seeing disabled people as pitiful victims, this approach emphasizes the group rights of the disabled, illuminating such things as access to employment and education (Zola 1993, 1989).

Now people with disabilities have legal protections similar to those afforded to other minority groups. Key to these rights is the Americans with Disabilities Act (ADA), adopted by Congress in 1990. Building on the Civil Rights Act of 1964 and earlier rehabilitation law, the ADA protects disabled persons from discrimination in employment and stipulates that employers and others (such as schools and public transportation systems) must provide disabled persons with "reasonable accom-

modation." Disabled persons must be qualified for the jobs or activities for which they seek access, meaning that they must be able to perform the essential requirements of the job or program without accommodation to the disability. For students, reasonable accommodation includes the provision of adaptive technology, exam assistants, and accessible buildings.

The law, which applies to state and local governments, as well as other employers, prohibits organizations with fifteen or more employees from discriminating against job applicants who are disabled or current employees who become disabled. The ADA also requires public facilities and transportation to be accessible to the disabled (although airlines are excluded from this requirement).

Not every disabled person is covered by ADA, and recent court decisions have restricted the application of the ADA. To be considered disabled, a person must have a condition or the history of a condition that im-

---

DOING SOCIOLOGICAL RESEARCH

All in the Family:
Children of Immigrants and Family Businesses

A common pattern among Asian American immigrants is to establish a small family business, utilizing the labor of family members to establish and run the business. Owning a family business is part of the American dream, thus idealized as promoting entrepreneurial and family values, including the value of hard work, devotion to family, and self-sacrifice. But what kind of life does this produce for the children of immigrant families? This is what sociologist Lisa Park wanted to know in her research on the children of Asian American entrepreneurs.

Park, a child of such entrepreneurs herself, conducted her study by interviewing Korean American and Chinese American high school and college students who grew up in small family businesses. She utilized several qualitative research methods, including in-depth individual interviews, focus group interviews, and participant observation in family stores.

Although many teens hold jobs—often to help support their families, sometimes for their own spending money—Park found that the life of Asian American teenagers who are

sons and daughters of entrepreneurs was difficult. Her research subjects reported wanting to have the freedom to be bored that they associated with White, middle-class teens. They were critical of the popular images of teens in families as only worrying about their hair, their clothes, and their friends.

In contrast to the stereotype of Asian American families as close-knit and conflict-free, Park found that teens in these families report feeling overworked and with little family "quality time." Some of these young people have worked since they were six years old. Often parents set up a small "home away from home" in the family restaurant or business—a small table in the corner where children can read, color, or watch television. But her subjects also reported being burdened and not able to have a normal childhood. They often had to do all the household chores while parents worked, as well as serving customers or helping in other ways.

Park's research shows immigration shapes family structures and affects children's lives. She challenges simplistic views of the family business as an ideal way of life and helps us under-

stand some of the social forces that are part of the immigrant experience and that shape the lives of some Asian American youth.

Questions to Consider

1. Are there businesses in your community run by immigrant families? If so, what evidence of children's involvement in the business do you see, and how do you think this influences the children? *Keywords: immigrant children, ethnic enclave*

2. In addition to immigrant businesses, how are children involved in the work that their parents do? In what ways does this influence the children's identities, values about work, future opportunities? *Keywords: immigrant labor, immigrant worker*

We have included InfoTrac College Edition keywords at the end of each question to make it easier for you to find more to read on these topics. Go to www.infotrac-college.com, an online library, to begin your search.

Source: Park, Lisa Sun-Hee. 2002. "A Life of One's Own." *Contexts* 1 (Summer): 56–57. •••

The Americans with Disabilities Act provides legal protection for disabled workers, establishing rights to reasonable accommodations by employers and access to education and jobs.

pairs a major life activity. The law specifically excludes the use of illegal drugs as a cause of disability and excludes pregnancy as a disability—ironic because pregnancy is treated as a disability by other federal laws for purposes of maternity leave.

Passage of the law followed extensive documentation of pervasive discrimination against the disabled in employment, transportation, and public accommoda-

tions. Despite change in the law, disabled people report that they are often treated with prejudice. Studies have even found that employers discourage employees with disabilities from making requests for accommodations; one-third of requests are denied (Harlan and Robert 1998).

With more women in the labor force and more continuing to work while they are pregnant, new questions are also arising about how to treat pregnant workers. Currently, the law stipulates that pregnancy is to be treated like any other disability. That is, states may require employers to grant up to four months of unpaid leave to women who are pregnant, and women who return to work following pregnancy must be permitted to return to their old job or its equivalent. These laws reversed a long-term historical trend in which employed women who became pregnant were forced to leave the workforce. The Family and Medical Leave Act of 1993 provides legal protection against this form of discrimination.

Common myths that women who become pregnant drop out of the labor force are simply untrue. Because most women work out of economic necessity, their maternity leaves are generally short. Family circumstances are a strong predictor of whether a woman will return to work following a birth, and how soon. Contrary to what one might expect, mothers with a spouse or another adult in the household return to work more quickly than those without other adults in their household. The greater the proportion of the family income a mother provides, the sooner she returns to work. There are no differences by race in these patterns (Wenk and Garrett 1992).

Gays and Lesbians in the Workplace

The increased willingness of lesbians and gay men to be open about their sexual identity has resulted in more attention being paid to their experience in the workplace. Surveys find that a large majority (85 percent)

TAKING ON SOCIAL ISSUES
Americans with Disabilities

The Americans with Disabilities Act extends various rights to those with documented disabilities. Employers, as well as schools and municipalities, must provide "reasonable accommodations" to allow disabled people to work, learn, and have access to public facilities.

This law has made a tremendous difference in opening opportunities for the disabled. What does your school provide for disabled students? What programs and policies might you recommend to further enable the education and employment of disabled people?

Taking Action

Go to the Taking Action Exercise on the Companion Website—at http://sociology .wadsworth.com/andersen_taylor4e/— to learn more about an organization that addresses this topic. •••

MAPPING AMERICA'S DIVERSITY

MAP 18.2 Regional Differences in Disabled Among Working-Age Population

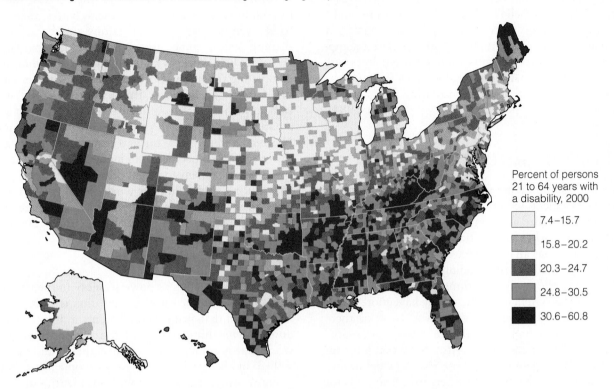

Percent of persons
21 to 64 years with
a disability, 2000

☐	7.4–15.7
▨	15.8–20.2
▨	20.3–24.7
▨	24.8–30.5
■	30.6–60.8

This map shows the percentage (in different counties) of persons of working age (age 21 to 64 years old) who have a disability. You will see that there are some regions of the country that have a particularly high percentage of disabled people. What do you think explains this pattern and how might it affect the needs of the labor force in those parts of the country? For example, do you think it is possible that the rate of disability might be related to the poverty rate in different regions or to the proportion of minority populations? If so, how would you explain this?

Data: U.S. Census Bureau. 2004. *American FactFinder.* Website: **http://factfinder.census.gov**

believe that lesbians and gay men should have equal rights in job opportunities. The public has also become far more accepting of gays and lesbians in various occupations. Thus, 78 percent of Americans now say that gays should be hired as doctors, compared with only 44 percent who thought so in the 1970s. Seventy-two percent also think gays should be employed in the armed forces and be included in the president's cabinet; almost two-thirds (63 percent) think they should be employed as high school teachers (up from 36 percent in 1989). The majority (54 percent) think they should be employed as elementary school teachers and clergy (Newport 2001; Mazzuca 2002).

These changes in opinion show improvement in the public's acceptance of lesbians and gays, although homophobia in the workplace is still widespread. Any negative experience in the workplace can affect self-esteem, productivity, and economic and social well-

being. Studies of gay and lesbian workers find that they fear they will suffer adverse career consequences if coworkers know they are gay. This leads many to "pass" as heterosexual at work or keep their private lives secret, but being closeted at work puts lesbian and gay workers at a disadvantage. Shielding themselves from antagonism or rejection may make them appear distant and isolate them from social networks. These behaviors can negatively affect their performance reviews because fellow workers may find them unfriendly, boring, and withdrawn. Research finds that the relationships of lesbian and gay employees with their coworkers are less stressful when the employees are "out" (Schneider 1984). The work organization is also improved for their employees when employee assistance programs encourage open communication, policies are sensitive to their needs, and discriminatory practices can be identified and stopped (Sussal 1994).

Most workers, including gays and lesbians, think it is desirable to integrate work and social life, but one-third of gays keep their social life separate from work, trying not to discuss personal matters with coworkers—even when they believe that such discussions make things at work go more smoothly. Lesbians and gay men at work encounter numerous heterosexist assumptions as coworkers discuss their personal lives and families; this can make gays and lesbians fearful of revealing their identity, further isolating gays and lesbians from interpersonal networks (Schneider 1984).

Worker Safety

Safe and sanitary conditions, good pay, opportunities for advancement, freedom from harassment, and policies that recognize human dignity are among the foundations of worker satisfaction. The size of the work organization has an effect on worker satisfaction, with workers in smaller firms tending to be more satisfied. Workers in these organizations perceive a less rigid stratification system and believe they have more possibility for job advancement. When workers think there is room to grow on the job, they tend to be more satisfied (Free 1990; Kelley 1990). Also, work that is intrinsically rewarding and challenging, that provides for advancement, and that makes workers feel responsible for their achievements results in greater job satisfaction.

The same elements of work life that promote worker satisfaction are the cause of dissatisfaction when they are absent. As the routinization of work increases, so does worker dissatisfaction. When unemployment goes up, worker discontent rises as well because workers in a weak labor market feel they have no alternatives to the job they are doing, whether they like their job or not. In general, women, minorities, and working-class men are often the least satisfied with their jobs—likely a reflection of their position in the race, class, and gender hierarchies at work. For women, being treated as equals is important. Women also value social exchanges at work, more than men do, and they see socializing as an important part of developing trusting relationships (Congdon 1990).

Management and administrative style is a major factor in creating job satisfaction. Highly authoritarian styles—that is, those where power is highly concentrated and blind authority to power is expected—breed worker discontent. Under such systems, workers may try to sabotage the operation of a workplace, although they have to do so subversively, such as by holding work slowdowns.

Because workers today experience the stress brought on by longer working hours and the demands of family and personal life, some employers have responded by providing more personal services within the workplace. They may be the exception, yet some large corporations, in particular, now provide exercise facilities, child-care centers, dry cleaners, banks, take-out food, even hairdressers, nail salons, on-site massage therapy, and, in some cases, Bible study groups. These amenities help workers maintain their work while also completing the many chores associated with personal life. Critics say, however, that as much as this "new company store" eases some demands on workers, it also creates the potential for public life to become more "barren" and to widen the gap between the haves and the have-nots. If people begin living all their life at the office, what happens to neighborhoods, community organizations, and volunteer work? Such benefits also work to the advantage of employers because they encourage people to work longer hours (Useem 2000).

Worker Alienation

Alienation is a feeling of powerlessness and separation from one's group or society. Alienation is a more specific concept than just general discontent. It is the boredom and meaninglessness of routine and repetitive work that is controlled, overspecialized, and underskilled. In alienated work, people become like robots, going through the motions without identity or commitment to the process or product.

The concept of worker alienation was first developed by Karl Marx, who believed that in a capitalist society, workers would always feel alienated because they do not control their labor. Moreover, the value of the goods workers produce is always greater than the amount workers are paid, so the workers are estranged from the profits of their labor. Alienation, Marx thought, was inherent in the system of capitalism.

The sociologist Robert Blauner produced a classic study of worker alienation. Like Marx, he defined alienation as the feeling of separation from one's labor (Blauner 1964). Worker alienation, Blauner argued, was especially intense in work such as automobile assembly lines, where workers have little control over their labor. Alienation can occur, however, in any job where workers have little control over repetitive tasks. Blauner emphasized the performance of fragmented and repetitive tasks as a major source of alienation; something typical of many jobs in the industrial workplace and increasingly of middle-class jobs involving lots of paper work and repetitive labor. On assembly lines, where workers engage in only one step of the production process, alienation is widespread. One worker may put rivets in the same part of an airplane all day long. Doing only fragments of the production process and never seeing the full product of one's labor leads to feelings of powerlessness and dehumanization. Workers may come to think of themselves as machines—as mechanized and routinized as robots. With the rise of automation, feelings of alienation and dissatisfaction have increased, but any kind of work can be alienating if workers view their tasks as meaningless and out of their control. Alienation is also exacerbated by workers having little control over the work environment, when they are isolated from other workers, experience little group cooperation or interaction at work, are dissatisfied with low pay and low status, and find stagnancy in the job market.

Alienation leads to feelings of powerlessness. Having no sense of the relationship between one's work and the overall productive process can produce a sense that one's work is without value or meaning. One consequence of alienation may be organizational deviance, such as deliberate destruction of company property, theft, or embezzlement (see Chapter 8). Alienated workers are likely to feel no loyalty to the organization that employs them. Their productivity is likely to be low. Their discontent can be measured not just in human unhappiness, but also in lost economic potential.

Sexual Harassment

Sexual harassment is defined as unwanted physical or verbal sexual behavior that occurs in the context of a relationship of unequal power and is experienced as a threat to the victim's job or educational activities (Martin 1989; Saguy 2003). Two primary forms of sexual harassment are recognized in the law. *Quid pro quo sexual harassment* forces sexual compliance in exchange for an employment or educational benefit. A professor who suggests to a student that going out on a date or having sex would improve the student's grade is engaging in quid pro quo sexual harassment. The second form of sexual harassment recognized by law is the creation of a *hostile working environment* in which unwanted sexual behaviors are a continuing condition of work. This kind of sexual harassment may not involve outright sexual demands but includes unwanted behaviors such as touching, teasing, sexual joking, and comments that contribute to what is termed, in law, a hostile working environment.

Sexual harassment was first made illegal by Title VII of the Civil Rights Act of 1964, which identified sexual harassment as a form of sex discrimination; in 1986, the U.S. Supreme Court upheld the principle that sexual harassment violates federal laws against sex discrimination in *Meritor Savings Bank v. Vinson*. The law defines sexual harassment as discriminatory because it makes sex a condition of employment or education.

By creating a hostile working environment, sexual harassment makes productive work difficult and discourages educational and work advancement. Fundamentally, sexual harassment is an abuse of power by which perpetrators use their position to exploit subordinates. Federal law makes employers liable for financial damages if they do not have policies appropriate for handling complaints or have not educated employees about their paths of redress.

The true extent of sexual harassment is difficult to estimate. Most surveys indicate that as many as one-half of all employed women experience some form of sexual harassment at some time. Men have also been victims of sexual harassment, although far less frequently than women (probably less than 3 percent of all cases). Same-sex harassment also occurs; when men are harassed by other men, they react with more severe consequences than do men who have been harassed by women (DuBois et al. 1998). The typical harasser is male, older than his victim, and of the same race and ethnicity. He is also likely to have harassed others in the past. There is some evidence that women of color are more likely to be harassed than White women (Gruber 1982). Many major legislative cases defining sexual harassment as illegal (including the landmark case of *Meritor Savings Bank v. Vinson*) involved African American women plaintiffs and White male harassers (Martin 1989; Rubin and Borgers 1990; McKinney 1994; Welsh 1999).

Sexual harassment occurs in every kind of work setting, from factories to schools and colleges, but some organizational settings are more prone to sexual harassment than others. A strong predictor of sexual harassment is a high male-to-female ratio (Martin 1989). Thus, women in male-dominated workplaces, including professions and the skilled trades, are particularly vulnerable to sexual harassment (Mansfield et al. 1991).

Some sociologists have argued that sexual harassment reflects men's resentment of women entering relatively high-paying fields once reserved for men. It is, thus, a mechanism men use to maintain their dominance in the workplace. Some men say they engage in this activity to show their interest and affection, but others say it is as a response to their perception of unfair treatment or is a demonstration of their anger (McKinney 1992). Men are more likely than women to see sexual harassment as normal and tolerable. Consequently, they are less likely to think of harassment as doing something wrong (Reilly et al. 1992). Women, on the other hand, are more likely to see inappropriate behavior as sexual harassment and to judge it more severely than men (Loredo et al. 1995).

Sexual harassment tends to be underreported. Most studies have found that typically neither women nor men are aware of the proper channels for reporting sexual harassment. Women are also less likely to report sexual harassment than are men who are victims, mostly because women believe that nothing will be done to stop the behavior.

The consequences of sexual harassment are economic, emotional, and sometimes physical. Workers and students who have been sexually harassed report feeling helpless, fearful, and powerless. Some victims lose their ambition and self-confidence, and they take a negative view of their work. Among students, the experience of being harassed creates self-doubt, anxiety, confusion, and distrust of all male faculty in general, feelings that certainly interfere with learning and achievement (Van Roosmalen and McDaniel 1998; Cortina et al. 1998; Benson and Thomson 1982; Bremer et al. 1991). Many women students respond by leaving fields of study in which they have experienced harassment.

Occupational Health and Safety

Sometimes people face outright dangers at work, including disability, injury, and occasionally death. The most dangerous occupations are those where the majority of workers are working-class men. Traffic accidents, equipment accidents, and homicides at work are the most common cause of fatal work injuries—thus, truckers, construction workers, and agricultural workers face some of the greatest risks. Men are more likely than women to be killed at work, probably reflective of the gender segregation in hazardous occupations (Bureau of Labor Statistics 2002).

Occupations that employ numerous immigrant workers are especially hazardous for employees. Workers in iron foundries have the highest injury rate of all U.S. industries. Airplane pilots have the highest rate of fatality at work, followed by farm workers, construction laborers, and truck drivers. Taxicab drivers—another occupation where immigrant labor is common—run the highest risk of homicide at work: forty times the national average. In sweatshops where recent immigrants are likely to be working, exposure to lead, as-

bestos, and other hazardous substances creates serious health risks. Language barriers often contribute to workers' lack of information about hazards they face on these jobs (Hawkins 1996; Cooper 1997).

Although the technological revolution has brought the promise of a "clean" workplace in which people perform mental labor while machines do the dirty and dangerous work once performed by humans, whether technologically–based jobs are truly as clean as believed is increasingly in doubt. New hazards of the technological workplace include toxic chemicals, nuclear hazards, repetitive motion disorders, and problems associated with prolonged use of video display terminals. Although technological developments can enhance productivity and efficiency, they also have a downside by causing physical problems, such as carpal tunnel syndrome, eye strain, and back problems from sitting at a monitor, and ironically the possibility of more paper trash as people write less and print more (Heim 1990). Occupational hazards of the past also continue to exist. In modern-day sweatshops, workers (usually recent immigrant women) labor outside the reach of regulatory protections, operating unsafe equipment in unsafe facilities.

Hazards to reproductive health can also occur in the workplace, and some employees have gone so far as to exclude all women, pregnant or not, from occupations deemed risky. Feminists argue that such legislation is discriminatory because it targets only women when men can be equally vulnerable to reproductive injuries. Feminists also note that it is no accident that the occupations deemed too risky to the reproductive health of women are traditionally male-dominated jobs with relatively high pay. The argument is that both men and women should be protected from reproductive harm. Seldom are women excluded from hazardous occupations that have been identified as women's work—operating-room nurses, for example. Policies intended to regulate occupational risks have been clouded by assumptions about appropriate jobs for women and men. The simple truth is that all workers need to be safe and secure in their work environments.

Chapter Summary

How do sociologists define the economy as an institution?

The *economy* is the system on which the production, distribution, and consumption of goods and services are based.

How has the economy in Western societies developed over time and what other economic systems are found in different societies?

Western economies have evolved from agricultural systems to industrial and postindustrial economies, affecting change in other parts of society. The different economic systems include capitalism, socialism, and communism. *Capitalism* is an economic system based on the pursuit of profit, market competition, and private property. *Socialism* is characterized by state ownership of industry. *Communism* is the purest form of socialism.

What defines contemporary economic restructuring?

The contemporary economy is increasingly *global,* includes a more diverse workforce, and is marked by *deindustrialization* and technological change. Together these changes are profoundly affecting the social organization of work, including who works and how.

What theories do sociologists use to understand work?

Conflict theorists study the tensions generated by power differences as groups compete for social and economic resources. *Functionalist theory* emphasizes that different rewards for important jobs in society motivate workers to pursue these areas. *Symbolic interactionists* study the meaning systems that shape people's behavior and identity at work.

What are some of the characteristics of the contemporary workforce?

In recent years, the employment of women has been increasing, whereas that of men has been decreasing. The value of work is reflected both in earnings and the prestige attached to different occupations—both of which are influenced by the dynamics of inequality. Official *unemployment rates* underestimate the true extent of joblessness. Women and racial minorities are the most likely to be unemployed.

What concepts describe the social organization of the labor market?

The U.S. occupational system is characterized by a *dual labor market* with jobs in the primary sector carrying better wages and working conditions and those in the secondary labor market paying less and providing fewer job benefits. Women and minorities are disproportionately employed in the secondary labor market. Patterns of *occupational distribution* also show tremendous race and gender segregation.

How is the labor market being influenced by the diversity of those within it?

The labor market is being transformed by the high rate of immigration. Recent immigrants are often in the lowest paying and least prestigious jobs. Some groups form ethnic enclaves—areas of ethnically owned businesses that serve predominantly ethnic communities. Disabled workers are now protected in law by the Americans with Disabilities Act. Homophobia in the workplace also negatively affects the working experience of gays and lesbians.

What factors affect work satisfaction and occupational safety?

Work satisfaction is contingent on the characteristics of work organizations. Workers in more routinized jobs tend to be less satisfied than others. Worker *alienation* is the feeling of powerlessness and separation that results when workers have little control over the products of their labor. *Sexual harassment* is the unequal imposition of sexual requirements in the context of a power relationship; it affects primarily women workers. Health and safety hazards also affect worker health. The greater participation of women in the labor force has also meant that pregnancy is now protected by law.

Key Terms

alienation 502	job displacement 482
automation 484	mismatch theory 485
capitalism 479	multinational
communism 479	corporations 480
contingent workers 485	occupational
deindustrialization 482	distribution 495
division of labor 494	occupational prestige 491
dual labor market theory 497	occupational segregation 496
economic restructuring 481	postindustrial society 478
economy 478	sexual harassment 503
emotional labor 488	socialism 479
global economy 479	

Researching Society with MicroCase Online

You can see the results of actual research by using the Wadsworth MicroCase® Online feature available to you. This feature allows you to look at some of the results from national surveys, census data, and other data sources. You can explore this easy-to-use feature on your own, but try this example. Suppose you want to know:

Does the number of workers covered by labor unions vary in different parts of the country?

To answer this question, go to http://sociology.wadsworth .com/andersen_taylor4e/, select MicroCase Online from the left navigation bar, and follow the directions there to analyze the following data.

Data file: States

Analysis: Mapping

Variable 1: %UNIONS-01

Click on: Continue

Questions

Once you have your results, answer the following questions:

1. What regions of the country have the highest percentage of unionized workers? What regions have the lowest?

2. The phrase "The Rust Belt" refers to the region of the country where the economy was once heavily based on manufacturing and where labor unions were common. These are some of the areas now hardest hit by unemployment resulting from the decline of manufacturing jobs. How do you think this affects the social, political, and economic life of these regions?

The Companion Website for Sociology: Understanding a Diverse Society, Fourth Edition

http://sociology.wadsworth.com/andersen_taylor4e/

Supplement your review of this chapter by going to the Companion Website to take one of the Tutorial Quizzes, use the flash cards to master key terms, and check out the many other study aids you'll find there. You'll also find special features such as GSS Data and Census 2000

information, data and resources at your fingertips to help you with that special project or do some research on your own.

Suggested Readings and Web Resources

Amott, Teresa L., and Julie A. Matthaei. 1996. *Race, Gender, and Work: A Multicultural History of Women in the United States,* 2nd ed. Boston, MA: South End Press.
Based on an analysis of the intersections of race, class, and gender in the labor market, this book presents the histories of the work of women of color and White women.

Chang, Grace. 2000. *Disposable Domestics: Immigrant Women Workers in the Global Economy.* Cambridge, MA: South End Press.
Chang argues that immigrant women workers and former welfare recipients are used as reserve supply of labor and are pitted against each other by employers seeking cheap labor. Her book also documents the position of immigrant women as domestic workers in the contemporary economy.

Harley, Sharon, and the Black Women and Work Collective. 2002. *Sister Circle: Black Women and Work.* New Brunswick, NJ: Rutgers University Press.
This collection of writing by Black women workers—activists and scholars—examines various dimensions of the experiences of Black women workers.

Hochschild, Arlie. 1997. *The Time Bind: When Work Becomes Home and Home Becomes Work.* New York: Metropolitan Books.
Hochschild's study examines two related developments: the increased pressure of work in the home and the primary attachments many people find at work. This reversal in primary and secondary roles results, she argues, from the impact of social speedup and the new organization of work and family.

Williams, Christine L. 1995. *Still a Man's World: Men Who Do Women's Work.* Berkeley, CA: University of California Press.
By focusing on men who work in traditionally female jobs, Williams demonstrates how gender structures opportunities in different occupations.

Wilson, William Julius. 1996. *When Work Disappears.* Chicago, IL: University of Chicago Press.
Wilson's book is an important analysis of the effect of economic restructuring on the family and work experience of Black Americans. His recommendations for social policy in this area have also been the basis for federal planning on urban problems.

U.S. Department of Labor

www.dol.gov

As the federal agency responsible for overseeing issues related to labor and employment, the Department of Labor maintains a site with statistical data on employment, as well as such services as a job bank in the public employment sector.

U.S. Bureau of Labor Statistics

www.bls.gov

The Bureau of Labor Statistics home page provides current information on the U.S. economy, press releases on matters pertinent to the study of work, data on work and employment, and links to numerous data sources on employment.

Government and Politics

Picture a couple lounging together on a public beach on a national holiday. They imagine themselves married someday, perhaps having children who they will send to public school. Does the government have anything to say about the couple's children? They may decide to use birth control until they are ready for children. If so, their choice of birth control methods will be limited to those the government allows. What if they cannot have children? They might consider adopting—if state agencies judge them acceptable as parents.

As you imagined this couple, who did you picture? Were they the same race? Today, they would be legally able to marry no matter what their races. Not so long ago, the law would have forbidden the marriage if the marriage were interracial. Is this a man and a woman or is this a same-sex couple? If they are lesbian or gay, despite being in love and wanting to form a lifelong relationship, in most states they would be prevented by law from marrying.

This couple is probably giving no thought to how much their life is influenced by the state at that very moment. Permission to walk on the public beach comes from the state; the day off to celebrate a holiday is sanctioned by the state; the drive to the beach took place in a car registered with the state and inspected by the state, driven by a motorist licensed by the state. The range of things regulated by the state is simply enormous, and yet many are never noticed.

If so much state attention seems oppressive, consider life without state regulations. People assume that when they go to work, they will get paid. Some employers would be sure to discover the profitability of withholding paychecks, were it not for laws requiring that employers meet their obligation to pay employees in exchange for work, and further requiring that wages be fair, that hiring be done without regard to race or gender, and that the workplace be safe. Without the state, a person who is robbed would have little recourse except personal revenge. In the form of law and the judicial system, the state defines good and evil, right and wrong.

Because not all conflicts in society are so well defined, the state also regulates disputes between people who are well-intentioned, yet bitterly opposed to each other's wishes. The state steps in at all levels, from arbitrating disputes

Sociology ⊛ Now™
Reviewing is as easy as ❶❷❸.

Use SociologyNow to help you make the grade on your next exam. When you are finished reading this chapter, go to the chapter review for instructions on how to make SociologyNow work for you.

between two parties (as in a lawsuit or divorce) all the way up to negotiating and defining class, race, and gender relations in society. Imagine the United States with no government, no laws to regulate people's behavior, no organized police or security force to defend the nation, and no prisons. Imagine the infrastructure with no paved roads, no expensive bridges, no satellites, no flood-control programs, no weather forecasting, no printed money. Imagining society without the state is like imagining the individual without socialization, as we did in Chapter 4. Remove the state from society and what is left? •••

Defining the State

In sociological usage, the **state** is the organized system of power and authority in society. The state is an abstract concept that refers to all those institutions that represent official power in society, including the government and its legal system (including the courts and the prison system), the police, and the military. The state regulates many societal relations, ranging from individual behavior to interpersonal conflicts to international affairs.

Theoretically, the state exists to regulate social order, although it does not always do so fairly or equitably. The guarantee of life, liberty, and the pursuit of happiness, as promised by the Declaration of Independence, when examined carefully, is not evenly distributed by the state. Less powerful groups in society may see the state more as an oppressive force than as a protector of individual rights. They may still turn, however, to the state to rectify injustice. For example, when African Americans sought to end segregation, they looked for legal reform enacted through the state.

The state has a central role in shaping class, race, and gender relations in society and in determining the rights and privileges of different groups (see the box Understanding Diversity: "American Indians and State Policy"). The involvement of the state may include the resolution of management and labor conflicts (such as in airline strikes), congressional legislation determining the benefits for different groups (such as the Americans with Disabilities Act of 1990), or Supreme Court decisions interpreting the U.S. Constitution. The state also supports the basic institutional structure of society because, through its laws, the state determines what institutional forms will receive societal legitimacy. Laws regulating family relationships emanate from the state. Thus, families without legitimate sanctions do not receive the same benefits and rights as other families. The state determines how people are selected to govern other people and dictates the system of governance they use. The state configures economic institutions by regulating economic policy. The state also maintains security forces, such as the police and the military, to protect the citizenry from criminals as well as internal and external threats.

Sociological analyses of the state focus on several different issues. One issue is how the state is related to inequality in society. State policies can have greatly different impacts on different groups, as we will see later in this chapter. Another issue explored by sociological theory is the connection between the state and other social institutions—the state and religion, the state and the family, and so on. Finally, a central topic is the state's role in maintaining social order, a basic question in sociological theory, as we have seen repeatedly since Chapter 1. Some theorists see the state as regulating society through coercion and power; others emphasize the role that consensus plays in the maintenance of public order. The issue of how power is exercised within the state is a subject of intense and continuing sociological debate and research.

The Institutions of the State

A number of institutions make up the state, including the government, the legal system, the military, and the police. The *government* creates laws and procedures that regulate the actions and behaviors of society. The *military* is the branch of government responsible for defending the nation against domestic and foreign conflicts. The *court system* is designed to punish wrongdoers and adjudicate disputes. Court decisions also determine the guiding principles or laws of human interaction. *Law* is a fundamental type of formal social control that outlines what is permissible and what is forbidden. The *police* are responsible for enforcing law at the local level and for maintaining public order. The *prison system* is the institution responsible for punishing people who have broken the law. Recall from Chapter 8, however, that courts and the law regulate civil behavior as well as criminal behavior.

The State and Social Order

Throughout this book, we have seen that a variety of social processes contribute to order in society, including the learning of cultural norms (socialization), peer pressure, and the social control of deviance. Each plays a part in producing social order, but none so explicitly and unambiguously as the official system of power and authority in society. In making laws, the state clearly decrees if actions are legitimate or illegitimate. Punishments for illegitimate actions are spelled out, and systems for administering punishment are maintained. The state also influences public opinion through its power to regulate the media and, in some cases, by circulating **propaganda,** information disseminated by a group or organization (such as the state) intended to justify its own power. Censorship is another means by which the state can direct public opinion. The movement to censor sexually explicit materials on the Internet is an example of state-based censorship.

The state's role in maintaining social order is also apparent in how it manages dissent. If those in power perceive protest movements as a challenge to state authority or threaten the disruption of society, the movement may be repressed through state action. Options available to the state range from surveillance through imprisonment all the way to military force. Witness the use of increased surveillance that has occurred in the aftermath of 9/11 via security screenings at airports, increased powers to intercept e-mail and voice mail via the Patriot Act of 2001, even more cameras at traffic intersections. Federal troops may be called upon by the state to quell riots and urban uprisings. As the system

UNDERSTANDING DIVERSITY

American Indians and State Policy

Although most people think of the state as neutral in its administration of policy, the experience of American Indians shows just how deeply the state influences group experience. C. Matthew Snipp, an American Indian sociologist who studies American Indians, has identified five historical periods characterizing the role of the U.S. government vis-à-vis American Indians: removal, assimilation, the Indian New Deal, termination/relocation, and self-determination. During each period, state policy had profound effects on the organization of Indian life—and, now, whether certain groups can even be considered "Indian."

In the first period, the U.S. government used its military forces and various pieces of legislation to forcibly remove dozens of Indian tribes from the eastern half of the United States to so-called Indian Territory. As White citizens moved westward, a very bloody period in American history occurred, resulting in both genocide and one of the largest forced migrations in history. Spanning more than half a century that began in the mid-1800s, removal forced thousands of Cherokees, Creeks, Choctaws, Seminoles, and other Indian groups from their homes to face mass death and drastic relocation.

Near the end of the nineteenth century, after tribes were moved to the Indian Territory, the government adopted the goal of isolating American Indians on reservations and trying to force them out of their cultural ways. American In-

dians were forced into boarding schools designed to indoctrinate Indian children and teach them that their culture was inferior to that of Euro-Americans. They were forbidden to wear native clothes, speak their own language, or practice their traditional religions. The 1887 General Allotment Act mandated that tribal lands would be allotted to individual American Indians, with surplus lands sold on the open market. This caused the loss of about 90 million acres of Indian land (approximately two-thirds of all Indian land held in 1887).

The third period, the Indian New Deal, was implemented in the 1930s along with other New Deal policies of the Roosevelt administration. Providing some relief from the Great Depression, the Indian Reorganization Act of 1934 allowed for tribal self-government. However, these laws have been an ongoing source of conflict because critics say it also forced an alien form of government on tribal groups.

Following World War II, in the termination and relocation period, the U.S. government tried unsuccessfully to abolish all reservations. The Bureau of Indian Affairs (which had been established in 1824) encouraged American Indians to move to cities, but many returned to reservations when outside the reservation, they found only seasonal employment.

In the 1960s, American Indian leaders made self-determination their major priority. The U.S. government

responded by passing the American Indian Self-Determination and Education Assistance Act (1975), allowing American Indians to oversee the affairs of their own communities without federal intervention. This has resulted in some reservations having their own police forces and being able to levy taxes and, in the case of the Onondaga tribe in New York, to issue passports that are internationally recognized. Smaller reservations and those with limited resources are still, however, dependent on Bureau of Indian Affairs' services.

Today, only Indian groups officially recognized by the U.S. government can be considered Indians. Without recognition, there is no access to the limited aid provided by the Bureau of Indian Affairs. The process of formal recognition is a difficult one because petitioners must do years of research to document their heritage and negotiate their way through a maze of federal regulations. The absence of written records among most American Indian groups complicates this search for verification. In 1993, of the 143 groups seeking recognition, only 23 cases were resolved, with only 8 officially granted Indian status (Brown 1993).

Source: Adapted from Snipp, C. Matthew. 1996. "The First Americans: American Indians." Pp. 390–401 in *Origins and Destinies: Immigration, Race, and Ethnicity in America,* edited by Sylvia Pedraza and Rubén G. Rumbaut. Belmont, CA: Wadsworth.

•••

of institutionalized power and authority in society, the state has the dual role of protecting its citizens and ensuring the preservation of society. Different states work in different ways—some explicitly protecting the status quo, others more revolutionary or more totalitarian in operation. Even in a democratic state, such as the United States, the state typically protects the interests of those with the most power, leaving the least powerful groups in society vulnerable to oppressive state action.

Global Interdependence and the State

On an international level, increasingly strong ties exist between the state and the global economy. The interdependence of national economies means that political systems are also elaborately entangled—a phenomenon that can be observed daily in the newspaper. Political tensions in what may once have been a remote part of the world have reverberations around the globe. Furthermore, the phenomenal rise of information technology means that political developments in one part of the world can be watched and heard just about everywhere. What will be the effect of globalization on the state institutions of diverse nations?

Some argue that increased economic interdependence will mean that nations will move to adopt similar institutions, including similar forms of government, law, and state rule. The European Union (EU) is an example of interdependence—an alliance of separate nations established to promote a common economic market and develop a political union within western Europe. The European Community now includes France, Belgium, Italy, Luxembourg, the Netherlands, Germany, the United Kingdom, Denmark, Ireland, Greece, Portugal, Spain, Austria, Finland, and Sweden. These nations have adopted shared laws and shared currency (the euro) that simplify trade and the movement of people across national boundaries. The United States has also entered similar agreements, such as NAFTA (North American Free Trade Agreement), which eliminates many restrictions on trade between the United States, Mexico, and Canada.

Does the permeability of national borders produced through such agreements result in a more democratic or less democratic world? Critics differ. Some say that the greater interdependence exposes authoritarian regimes and creates pressures for an international system of law where democratic rights are respected and human rights protected. Others argue that the strong alliance produced between economic interests, corporate power, and the state shifts global political power, favoring the most dominant nations and impoverishing and marginalizing others, while also destroying their local traditions. Such analysts worry that the increasing similarity produced by global interdependence will produce a *monoculture* where everything will look alike—same hotels, same clothes, same music, same stores (Mander and Goldsmith 1996).

Either way, the process of globalization is clearly having profound effects on the character of states and their relationships to each other. A good example is the World Trade Organization (WTO)—created in 1994 to monitor and resolve trade disputes. Member nations may challenge decisions made by local and national governments, with the WTO Council in Geneva resolving disagreements. The creation of this organization has produced a system that transcends individual nation-states by formalizing and strengthening the rules that govern many aspects of world trade. This represents the trend, in an increasingly international economy, toward the creation of a single state—at least in the sense of a singular governing body that has the potential to have authority over many aspects of people's lives around the world. This system of global governance is not without

TAKING ON SOCIAL ISSUES

Civil Liberties and National Security

In the aftermath of terrorist attacks on the United States, heightened concerns about national security have resulted in increased surveillance of people's actions. A debate has ensued regarding the balance of national security with the protection of civil liberties—the hallmark of a democratic and free society. What factors weigh in on each side of this issue? Can there be restrictions of freedom of movement without compromising civil liberties? Should civil liberties be compromised even if it means more risk? Would you give up your rights to privacy and freedom of movement if it meant less risk? What sociological factors must be considered in this debate? For example, should citizens who have the characteristics associated with terrorists be subjected to greater surveillance?

Taking Action

Go to the Taking Action Exercise on the Companion Website—at http://sociology.wadsworth.com/andersen_taylor4e/—to learn more about an organization that addresses this topic. •••

problems, however. Critics say it results in the pervasive dominance of capitalist interests.

Power and Authority

The concepts of power and authority are central to sociological analyses of the state. **Power** is the ability of one person or group to exercise influence and control over others. The exercise of power can be seen in relationships ranging from the interaction of two people (husband and wife, police officer and suspect) to one nation threatening or dominating other nations. Sociologists are most interested in how power is structured within societies—who has it, how it is used, and how it is built into institutionalized structures, such as the state. Power can be structured into social institutions, for example, when men as a group have power over women as a group. In the United States, a society that is heavily stratified by race, class, and gender, power is structured into basic social institutions in ways that reflect these inequalities. Sociologists also understand that power institutionalized at the societal level influences the social dynamics within individual and group relationships.

The exercise of power may take the form of persuasion or coercion. For example, a group may be encouraged to act a certain way based on a persuasive argument. A strong political leader may persuade the nation to support a military invasion or a social policy. Alternatively, power may be exerted by sheer force. Between persuasion and coercion are many gradations. Generally speaking, groups with the greatest material resources will likely have the advantage in transactions involving power, but this is not always the case. A group may by sheer size be able to exercise power, or groups may use other means to exert power, such as armed uprisings or organized social protests.

Power can be legitimate—that is, accepted by the members of society as right and just—or it can be illegitimate. **Authority** is power that is perceived by others as legitimate, emerging from the exercise of power and the belief of constituents that the power is legitimate. People who accept the status quo as a legitimate system of authority perceive the guardians of law to be exercising *legitimate power*. In the United States, the source of the president's domestic power is his status as commander in chief of the armed forces as well as the belief by most people that his power is legitimate. The law is also a source of authority in the United States. *Coercive power* is achieved through force, often against the will of the people being forced. Those people may be a few dissidents or most of the citizenry of an entire nation. A dictatorship often relies on its ability to exercise coercive power through its control of the military or the state police. The ability to maintain power through force may be enough to keep someone in power even if he or she lacks legitimate authority, whereas someone lacking in legitimate authority *and* coercive power is unlikely to remain in power long.

THINKING *SOCIOLOGICALLY*

Observe the national evening news for one week, noting the people featured who have some kind of *authority*. Listing each of them and noting their area of influence, what form of authority would you say each represents: *traditional, charismatic,* or *rational-legal*? How is the kind of authority a person has related to his or her position in society (that is, race, class, gender, occupation, education, and so on)?

Types of Authority

Max Weber (1864–1920), the German classical sociologist, postulated that there are three types of authority in society: traditional, charismatic, and rational-legal (Weber 1978/1921). **Traditional authority** stems from long-established patterns that give certain people or groups legitimate power in society. A monarchy is an example of a traditional system of authority. Within a monarchy, kings and queens rule, not necessarily because of their appeal or because they have won elections, but because of long-standing traditions within the society.

Charismatic authority is derived from the personal appeal of a leader. Charismatic leaders are often believed to have special gifts, even magical powers, and their presumed personal attributes inspire devotion and obedience. Charismatic leaders often emerge from religious movements (see Chapter 17), but they come from other realms also. John F. Kennedy was for many a charismatic president, admired for the vigorous image he projected and his political skill. Charismatic leaders may mobilize large numbers of people in the name of lofty principles, as in the case of Dr. Martin Luther King, Jr. In the case of cults, charismatic leaders may inspire such loyalty among their followers that the group solidifies around what others would consider preposterous beliefs. Because the foundation of charismatic power rests on the qualities of a single individual, when that person leaves or dies, the movement he or she inspired may quickly dissipate.

Rational-legal authority stems from rules and regulations, typically written down as laws, procedures, or codes of conduct, and is the most common form of authority in the contemporary United States. People obey not because national leaders are charismatic or because social traditions are followed, but because an unquestioned system of authority is established by legal rules and regulations.

Under rational-legal authority, rules are formalized. Authority is based not on the personal appeal of charismatic or traditional leaders but on the written rules. Rulers gain legitimate authority through having been elected or appointed in accordance with society's rules. In systems based on rational-legal authority, the rules are upheld by means of state agents such as the police, judges, social workers, and other state functionaries to whom power is delegated.

Weber wrote that modern societies would be increasingly based on rational-legal authority, as opposed to the traditional institutions of monarchies and local and regional custom. We can see evidence of Weber's far-sightedness in the proliferation of regulations governing daily life and in the fading authority of folklore as a source of authority. Consider the explosive growth in the population of lawyers, who tend to the legalities governing all important transactions in our society and who act as advocates for official resolution of disputes. At one time, their role was filled by appeals to local custom or the adjudication of a wise person whose judgment both parties either trusted or were obliged to accept based on the high regard of others. As societies modernize, they become more rationalized in their systems of authority. Weber acknowledged the societal gain of rational authority over superstition or oppressive custom, but he also warned of the cost in terms of diminished social intimacy and group feeling.

Bureaucracies, such as this Department of Motor Vehicles, are organized according to hierarchical and rule-driven forms of social organization.

The Growth of Bureaucratic Government

According to Weber, rational-legal authority leads inevitably to the formation of bureaucracies. As we saw in Chapter 6, a **bureaucracy** is a type of formal organization characterized by an authority hierarchy, a clear division of labor, explicit rules, and impersonality. Bureaucratic power is established by the accepted legitimacy of the rules, not personal ties to individual people. The rules may change, but they do so via formal—that is, bureaucratic—procedures. People who work within bureaucracies are selected, trained, and promoted based on how well they apply the rules. People who make the rules are unlikely to be the same people who administer them.

Bureaucracies are hierarchical, and the bureaucratic leadership may be remote. Power in bureaucracies is dispersed downward through the system to people who carry out the bureaucratic functions. It is an odd feature of bureaucracy that those with the least power to influence how the rules are formulated—people at the bottom of the bureaucratic hierarchy—are very often the most adamant about strict adherence to the rules.

In principle, bureaucracies are highly efficient modes of organization, although the reality is often very different. There is a tendency within bureaucracies to proliferate rules, often to the point that the organization becomes ensnared in its own bureaucratic requirements. Thus, as an example, procurement of desktop computers by the government has become so overburdened by bureaucratic requirements for review, comparison, bidding, and approval that by the time the federal government approves a purchase, the chosen models are sometimes already obsolete.

An early critic of bureaucracies, the German sociologist Robert Michels argued that there is an **iron law of oligarchy** in bureaucracies (Michels 1962/1911). He noted that the formal organization of bureaucracies tends to evolve into a system where a small elite become increasingly powerful. Those at the top of bureaucratic organizations tend to become enchanted with their elite status and then make decisions that mostly protect and reinforce their own power. This leads to rule by a few in such formal bureaucracies.

Within bureaucracies, administrators also tend to become so focused on rules and regulations that the actual work of the system bogs down. As bureaucratic systems become ever larger, people feel themselves becoming smaller—they feel powerless in the face of a powerful system. Now, voice mail used to answer calls can be seen as the epitome of bureaucratic organization: predetermined and fixed paths for conducting a transaction; timed and supposedly efficient use of the organization's time; no human discretion—indeed, no human interaction. The next time you get stuck in an endless voice mail loop, you might think about how sociological concepts of bureaucracy at least explain your frustrating predicament.

Within bureaucracies, personal temperament and individual discretion are not supposed to influence the application of rules. Of course, we know that the face of the bureaucracy is not always perfectly stony. As we saw in Chapter 6, bureaucracy has another face. Bureaucratic workers frequently exercise discretion in applying rules and procedures. Some who encounter bureaucracies know how to "work the system," perhaps by personalizing the interaction and making a willing accomplice of the bureaucrat in dodging bureaucratic stipulations, or perhaps by using knowledge of some rules to evade other rules. People without privileged relationships or privileged information are continually disadvantaged by their inability to negotiate within the system. Despite the supposed impersonal administration in bureaucracies, people may receive widely different treatment from bureaucrats, who may favor some persons while acting prejudiced against others based on their race, gender, age, or other characteristics.

Our picture of the state so far—bureaucratic, powerful, omnipresent—presents an important question to be addressed by the sociological imagination: Does the state act in the interests of its different constituencies, or does it merely reflect the needs of a select group who sit at the top of the pyramid of state power? This question has spawned much sociological study and debate, and it has resulted in several different theoretical models of state power.

Theories of Power

How is power exercised in society? Four different theoretical models have been developed by sociologists to answer this question: the *pluralist model*, the *power elite model*, the *autonomous state model*, and *feminist theories of the state*. Each model begins with a different set of assumptions and arrives at different conclusions.

The Pluralist Model

The **pluralist model** interprets power in society as coming from the representation of diverse interests of various groups in society. This model assumes that in democratic societies, the system of government works to balance the various interest groups in society. An **interest group** can be any constituency in society organized to promote its own agenda, including large, nationally based groups such as the American Association of Retired People (AARP) or the National Rifle Association (NRA); groups organized around professional and business interests, such as the American Medical Association (AMA) and the Tobacco Institute; or groups that concentrate on a single political or social goal, such as Mothers Against Drunk Driving (MADD). According to the pluralist model, interest groups achieve power and influence through their organized mobilization of concerned people and groups.

> ### THINKING SOCIOLOGICALLY
> Using a major daily newspaper, for a period of two weeks, follow a political story on a subject of interest to you. As you follow this story, make a list of the different *interest groups* with a stake in this issue. Which groups exercise the most *power* in determining the outcomes of this issue and why? What does your example tell you about how the political process operates?

The pluralist model has its origins in functionalist theory. This model sees the state as benign and representative of the whole society. No particular group is seen as politically dominant. Instead, the pluralist model

The pluralist model of the state sees interest groups as attempting to wield power to influence political decision-making.

© Najlah Feanny/Corbis Saba

sees power as broadly diffused across the public. Groups that want to effect a change or express their point of view need only mobilize to do so (Block 1987). The pluralist model also suggests that members of diverse ethnic, racial, and social groups can participate equally in a representative and democratic government. This theoretical model assumes that power does not depend upon social status or wealth. Different interest groups compete for government attention and action, with equality of political opportunity for any group that organizes to pursue its interests (Harrison 1980).

According to pluralism, special-interest groups are the link between the people and the government. Interest groups form when a group of people who share a belief in an issue organize themselves to get the attention of the government. They compete with other interest groups in shaping public policy, each group using any influence it can garner to encourage policies favorable to its interests.

One resource of special-interest groups is sheer numbers. A special-interest group with a large membership can influence a politician by threatening to support another candidate. Another tool interest groups use is money. A small interest group with a great deal of money can wield a disproportionate amount of influence. The principle mechanisms for converting money into political influence are campaign contributions, lobbying, and propaganda. The pluralist model sees special-interest groups as an integral part of the political system, even though they are not an official part of government. In the pluralist view, interest groups make government more responsive to the needs and interests of different people, an especially important function in a highly diverse society (Berberoglu 1990).

The pluralist model helps explain the importance of **political action committees** (PACs), groups of people who organize to support candidates they feel will represent their views. In 1974, Congress passed legislation enabling employees of companies, members of unions, professional groups, and trade associations to support political candidates with money they raise collectively. The number of political action committees has now grown to almost 4000. PACs have enormous influence on the political process. One measure of their growing influence is the tremendous increase in financial contributions that PACs have made to political campaigns over the last two decades—in excess of $300 million in 2004 (Federal Election Commission 2004; see Figure 19.1). Campaign finance reforms now limit the amount of money PACs can give to specific candidates, but special-interest groups, including large corporations, unions, and other organized groups, have circumvented these restrictions through the use of so-called soft money. *Soft money* cannot be given directly to candidates but can be used by political parties for a whole host of "party-building" expenses (advertisements ad-

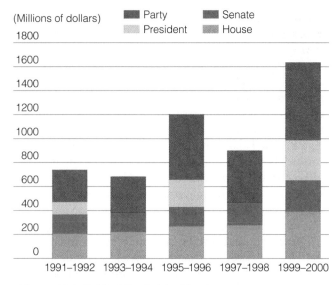

Figure 19.1 Political Fundraising Trends

Source: The Center for Responsive Politics. 2000. "Historical Fundraising Trends." Reprinted with permission. **www.opensecrets.org**

vocating support for particular issues, office overhead, voter registration drives, and other behind-the-scenes activities), thus freeing other funds for direct contributions. Through soft money, interest groups take advantage of loopholes in campaign finance laws and thus continue to wield enormous influence (see Table 19.1).

How realistic is the pluralist model in explaining how political power works? Interest groups certainly are influential in the political process, although there is not as much equality between them as the pluralist model tends to assume. Special-interest groups representing powerful organizations and constituencies are able to mobilize resources that smaller groups cannot. Still, in a diverse society, the formation of interest groups is critical to the political process if any degree of equity is to be achieved.

The Power Elite Model

The **power elite model** originated in the work of **Karl Marx** (1818–1883) and conflict theory. According to Marx, the dominant or "ruling" class controls all the major institutions in society. The state itself is simply an instrument by which the ruling class exercises its power. The Marxist view of the state emphasizes the power of the upper class over the lower classes—the small group of elites over the rest of the population. The state, according to Marx, is not a representative, rational institution but an expression of the will of the ruling class (Marx 1972/1845).

Marx's theory was elaborated much later by **C. Wright Mills** (1956), who analyzed the power elite. Mills attacked the pluralist model, arguing that the true power structure consists of people well positioned

Table 19.1
Top PAC Donors (Soft Money to National Parties), 2003–2004

Organization	Total	Percent to Republicans	Percent to Democrats
National Association of Realtors	$3,763,083	52%	48%
Laborers Union	$2,580,750	14%	86%
National Auto Dealers Association	$2,461,300	73%	27%
National Beer Wholesalers Association	$2,160,000	77%	23%
Association of Trial Lawyers of America	$2,043,999	7%	93%
United Parcel Service	$1,988,959	73%	23%
Int'l Brotherhood of Electrical Workers	$1,959,480	5%	95%
American Medical Association	$1,881,891	80%	20%
United Auto Workers	$1,858,200	1%	98%
National Association of Home Builders	$1,837,700	65%	35%
Credit Union National Association	$1,824,023	58%	42%
Carpenters and Joiners Union	$1,800,560	26%	74%
Machinists/Aerospace Workers Union	$1,794,500	1%	99%
SBC Communications	$1,655,366	68%	32%
American Federation of Teachers	$1,615,800	3%	97%
Wal-Mart	$1,606,000	78%	22%

Source: Center for Responsive Politics. 2005. "Top 20 PAC Contributors to Federal Candidates, 2003–2004." Web site: **www.opensecrets.org**

in three areas: the economy, the government, and the military. Sharing common beliefs and goals, the power elite shape political agendas and outcomes in the society along the narrow lines of their own collective interests. This small and influential group at the top of the pyramid of power holds the key positions in the major state institutions, giving it a great deal of power and control over the rest of society.

The power elite model posits a strong link between government and business, a view that is supported by the strong hand government takes in directing the economy and by the role of military spending as a principal component of U.S. economic affairs. Since World War II, huge conglomerates have emerged whose holdings are spread over many different industries, whose employees number in the hundreds of thousands, and whose political interests support armies of lobbyists. Corporations have an intense interest in public policy given that government regulates how business is conducted.

The power elite model also emphasizes how power overlaps between influential groups. **Interlocking directorates** are organizational linkages created when people in elite circles sit on the boards of directors of a number of different corporations, major companies, universities, and foundations at the same time. People drawn from the same group receive most of the major government appointments. Thus, the same relatively small group of people tends to the interests of all these

organizations and the interests of the government. These interests naturally overlap and reinforce one another.

The power elite model sees the state as part of a structure of domination in society, one in which the state is simply a piece of the whole. Members of the upper class do not need to occupy high office themselves to exert their will, as long as they are in a position to influence those who are in power (Domhoff 1998). The majority in the power elite are White men, which according to the power elite model means that the interests and outlooks of White men dominate the national agenda.

DEBUNKING *SOCIETY'S MYTHS*

Myth: Increasing the diversity of people in positions of power brings new values to political institutions.

Sociological Perspective: Women and minorities who make it into positions of political power tend to come from backgrounds similar to White male elites and feel pressure to gain legitimacy by sharing values similar to those already in power (Zweigenhaft and Domhoff 1998).

Is the power elite model the best description of power in the United States? In some ways the model is hard to argue against because the influence of powerful groups is so obvious. Two strong criticisms have been leveled against the power elite model. One critique is that the model assumes too readily that there is a unity of interests among elites. In fact, according to critics, the most powerful people in society hold widely divergent views on many political issues—for example, there is no single power elite position on abortion or environmental protection. A second criticism is that the power elite model fails to acknowledge how well public interest groups have been able to make themselves heard. One of the largest PACs (measured by expenditures) is Emily's List, which is devoted to supporting candidates strong on women's issues (Federal Election Commission 2004). According to critics, the power elite model does not explain how a group such as Emily's

List can rally behind candidates who earlier would have been excluded from the power elite. Of course, at the same time, the power elite model recognizes that even groups such as Emily's List operate within elite circles (see the box Doing Sociological Research: "Diversity in the Power Elite").

The Autonomous State Model

A third view of power developed by sociologists, the **autonomous state model,** interprets the state as its own major constituent. From this perspective, the state develops interests of its own, which it seeks to promote

---

DOING SOCIOLOGICAL RESEARCH
Diversity in the Power Elite

As society has become more diverse, has it made a difference in the makeup of the power elite? Various groups—women, racial–ethnic groups, lesbians, and gays—have vied for more representation in the halls of power, but have their efforts succeeded? If they make it to power, does this change the corporations, military, or government— the major institutions composing the power elite?

Sociologists Richard L. Zweigenhaft and G. William Domhoff examined these questions by analyzing the composition of boards of directors and chief executive officers (CEOs) of the largest banks and corporations in the United States, as well as analyzing Congress, presidential cabinets, and the generals and admirals who form the military elite. In addition, they examined the political party preferences and the political positions of people found among the power elite. Do women and minorities bring new values into power, thereby changing society as they move into powerful positions, or do their values match those of the traditional power elite or become absorbed by a system more powerful than they are? Zweigenhaft and Domhoff's study also looks at whether those who do make it into the power elite are within the innermost circles or whether they are marginalized.

They find that women, Jews, gays, lesbians, Black Americans, and Hispanics have become more numerous within the power elite, but only to a

small degree. The power elite is still overwhelmingly White, wealthy, Christian, and male. Women and other minorities who make it into the power elite also tend to come from already privileged backgrounds, measured by their social class and education. Among African Americans and Latinos, skin color continues to make a difference, with darker-skinned Blacks and Hispanics less likely to achieve prominence, compared with lighter-skinned people. Furthermore, Zweigenhaft and Domhoff find that the perspectives and values of women and minorities who rise to the top do not differ substantially from their White male counterparts. Some of this is explained by the common class origins of those in the power elite. The researchers also attribute the managing of one's identity to avoid challenging the system as a sorting factor that perpetuates the dominant worldview and practices of the most powerful.

The authors of this study conclude that "the irony of diversity" is that greater diversity may have strengthened the position of the power elite because its members appear to be more legitimate through their inclusion of those previously left out. But by including only those who share the perspectives and values of those already in power, little is actually changed. Clarence Thomas, Sandra Day O'Connor, Ruth Bader Ginsburg, and Colin Powell may make it seem that women and African Americans have it made, but as

long as they support the positions of White male elites, the power elite goes undisturbed.

Questions to Consider

1. Think of a time when you were a *minority* in a group because of some social characteristic (such as age, race, gender, religion, and so forth). Were your views alike or different from the other group members? If they were different, what would it have taken to make your views prevail? What does this reveal to you about Zweigenhaft and Domhoff's research on *diversity in the power elite*? *Keywords: tokenism, diversity and power*

2. Who are some of the prominent members of the power elite who are women and/or people of color? What evidence do you see of their values? Are they similar to or different from the values of others in the power elite? *Keywords: class privilege, corporate elite, power elite*

We have included InfoTrac College Edition keywords at the end of each question to make it easier for you to find more to read on these topics. Go to www.infotrac-college.com, an online library, to begin your search.

Source: Zweigenhaft, Richard L., and G. William Domhoff. 1998. Diversity in the Power Elite: Have Women and Minorities Reached the Top? New Haven, CN: Yale University Press. •••

independent of outside interests and the public that it allegedly serves. The state is not simply reflective of the needs of dominant groups, as Marx and power elite theorists would contend. Autonomous state theory sees the state as a network of administrative and policing organizations, each with its own interests (Domhoff 1990), such as maintenance of its complex bureaucracies and protection of its special privileges (Evans et al. 1985; Skocpol 1992; Rueschmeyer and Skocpol 1996). Autonomous state theory actually builds from the sociological discussion of bureaucracy originally proposed by Max Weber.

The interests of the state may intersect at times with the interests of the dominant class or the members of society as a whole, but the major concern of the state is maintaining the status quo and upholding its own interests in competition with other states. State policies are created by independent state managers who represent their own interests.

Autonomous state theorists note that states tend to grow over time, possibly including expansion beyond their original boundaries. The North American Free Trade Agreement (NAFTA), for example, is a case of the United States expanding its state interests by regulating business not only in the United States, but also in Mexico and Canada.

The huge government apparatus now in place in the United States is a good illustration of autonomous state theory. The government provides a huge array of social support programs, including Social Security, unemployment benefits, agricultural subsidies, public assistance, and other economic interventions intended to protect citizens from the vagaries of a capitalist market system (Collins 1988). The purpose of these programs is to serve people in need. Autonomous state theory argues that the government has grown into a massive, elaborate bureaucracy run by bureaucrats more absorbed in their own interests than in meeting the needs of the people. As a consequence, government can become paralyzed in conflicts between revenue-seeking state bureaucrats and those who must fund them. This can lead to revolt against the state, as in the tax revolts appearing sporadically throughout the country (Lo 1990; Collins 1988).

When people ask why the government, for all its hugeness, seems to be incompetent to deal with the needs of its citizens, the autonomous state model explains that the government is too busy tending to its own problems. Critics of this theory say, however, that by focusing on the state alone, the autonomous state model overlooks the degree to which the state supports the interests of big business. Like the pluralist and power elite models, the autonomous state model contributes to our understanding of the state, but it is not a total explanation of state power.

Feminist Theories of the State

Feminist theory diverges from the preceding theoretical models by seeing men as having the most important power in society. Pluralist theorists see power as widely dispersed through the class system. Power elite theorists see political power directly linked to upper-class interests. Autonomous state theorists see the state as relatively independent of class interests. Feminist theory begins with the premise that an understanding of power cannot be sound without a strong analysis of gender (Haney 1996).

Some feminist theorists argue that all state institutions reflect men's interests. They see the state as fundamentally patriarchal, its organization embodying the principle that men are more powerful than women. Feminist theories of the state conclude that, despite the presence of a few powerful women, the state is devoted primarily to men's interests and, moreover, the actions of the state will tend to support gender inequality (Blankenship 1993). One historical example would be laws denying women the right to own property once they married. Such laws protected men's interests at the expense of women.

The argument that "the state is male" (MacKinnon 1983: 644) is easily observed in powerful political circles. Despite the recent inclusion of more women in powerful circles and the presence of some notable women as major national figures, the most powerful members are men. The U.S. Senate is 86 percent men. Groups that exercise state power, such as the police and military, are predominantly men. Moreover, these institutions are structured by values and systems that can be described as culturally masculine—that is, based on hierarchical relationships, aggression, and force.

Comparing Theories of Power

These four models each see power in a different way (see Table 19.2). *Pluralist theory* sees interest groups compete in a struggle for power. *Power elite theory* sees power as stemming from the top down (the power elite model). *Autonomous state theory* sees the power of the state as feeding on itself. And *feminist theory* sees men as holding power in all social institutions. Each perspective makes its own contribution and complements the others to give a more complete picture of how power works.

In its own way, each theory addresses the same question: Does power move from the top down or from the bottom up? The pluralist theory sees power rising from the bottom up in the form of interest groups that organize to express their needs or use their vote to influence state policy. Power elite theory and feminist theory see power operating from the top down, with the most

Table 19.2
Theories of Power in Society

Interprets . . .	Pluralism	Power Elite	Autonomous State	Feminist Theory
The State	As representing diverse and multiple groups in society	As representing the interests of a small, but economically dominant, class	As taking on a life of its own, perpetuating its own form and interests	As masculine in its organization and values (that is, based on rational principles and a patriarchal structure)
Political Power	As derived from the activities of interest groups and as broadly diffused throughout the public	As held by the ruling class	As residing in the organizational structure of state institutions	As emerging from the dominance of men over women
Social Conflict	As the competition between diverse groups that mobilize to promote their interests	As stemming from the domination of elites over less powerful groups	As developing between states, as each vies to uphold its own interests	As resulting from the power men have over women
Social Order	As the result of the equilibrium created by multiple groups balancing their interests	As coming from the interlocking directories created by the linkages among those few people who control institutions	As the result of administrative systems that work to maintain the status quo	As resulting from the patriarchal control that men have over social institutions

powerful groups in society (either the power elite or elite men) using the state to enforce their will. The model of the autonomous state falls somewhere in between, with the state seen to operate in its own interest, sometimes supporting the interests of the people or the elites, sometimes not.

Government: Power and Politics in a Diverse Society

The terms *government* and *state* are often used interchangeably. More precisely, the government is one of several institutions that make up the state. The **government** includes state institutions that represent the population, making rules that govern the society. The government in the United States is a **democracy,** meaning that it is based on the principle of representing all people through the right to vote. The actual makeup of the government, however, is far from representative of society. Not all people participate equally in the workings of government, either as elected officials or voters, nor do their interests receive equal attention. Women, the poor and working class, and racial–ethnic minori-

ties are less likely to be represented by government than are White middle- and upper-class men. Sociological research on political power has demonstrated large, persistent differences in the political participation and representation of different groups in society.

> ### DEBUNKING *SOCIETY'S MYTHS*
> **Myth:** In democratic societies, all people are equally represented through the vote.
> **Sociological perspective:** Although democracies are based on principles of equal representation, in practice, race, class, and gender influence the power that diverse groups have in the political system.

Democracies, despite their flaws, are the most representative form of government and depend on the full participation of all citizens to meet their promise. Contrast this with other forms of government: monarchies and dictatorships. A **monarchy** is a form of government characterized by having a head of state who rules for life and where authority tends to be inherited. Monarchies usually involve royalty, as in the case of Britain, where the royal family inherits the rule of the land. In actuality, Britain is both a monarchy and a democracy since the traditional authority of the royal family is being replaced and supplemented with rational-legal authority in the form of Parliament and a prime minister.

Also common in today's world are **dictatorships** where power resides in the rule of one person who acquired power through force and maintains it usually through having control of the military. In most known cases of dictatorships, the ruler is also a man—showing a relationship between patriarchy and dictatorships. Saddam Hussein was a dictator in Iraq—ruling through the force of terror and the loyalty of his Baath party.

Diverse Patterns of Political Participation

One would hope that in a democratic society all people would be equally eager to exercise their right to vote; that is far from the case. Among democratic nations, the United States has one of the lowest voter turnouts (see Figure 19.2). In the 2004 presidential election, the percentage of eligible voters who went to the polls was only 61—an abnormally high turnout. A turnout of 50 percent or less is typical of most national elections. Voter turnout in elections is even lower (see Figure 19.3).

The group most likely to vote is older, better educated, and financially better off than the average citizen.

A U.S. soldier watches as a statue of Iraq's President Saddam Hussein falls in central Baghdad, April 9, 2003.

In sociological terms, age, income, and education are the strongest predictors of whether someone will vote. Despite the fact that almost all young people say it is important for people in a democracy to vote (97 percent said so in a recent national poll), only about one-third of young people (those between age 18 and 29) actually vote in presidential elections. Analysts interpret this as the result of cynicism because large numbers of young people think their vote won't matter (Rosenberg 2004). After the close election of 2000, many thought that young people would see that their vote mattered and that they would turn out in higher numbers in the 2004 election. Turnout among young people (aged 18–24) in 2004 did increase to 52 percent, up from 42 percent in 2000.

Social class also influences voting patterns; the higher a person's social class, the higher the likelihood that she or he will be a voter. As a result, those in the upper social classes are far more likely to have their voices heard in the political process. Social class also influences who votes how, as Table 19.3 shows. The upper classes are more likely to vote for the Republican party; the working class, Democrats.

The requirements of voter registration to register each time you move may discourage many from voting (Piven and Cloward 1988). The National Voter Reg-

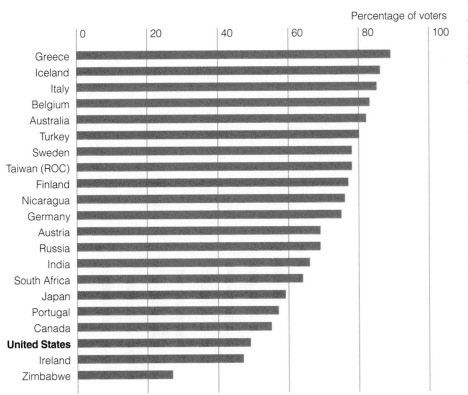

Figure 19.2 Voter Participation in Democratic Nations

Source: International Institute for Democracy and Electoral Assistance. 2001. "Voter Turnout." **www.idea.int/vt/country_view.cfm**

istration Act of 1993 intended to provide greater flexibility in how one registers to vote by including "motor voter" procedures that can register you to vote when you apply for a driver's license. To date, however, voting turnout has not significantly increased since implementation of this policy (refer again to Figure 19.3).

There is also significant variation in voting patterns by race. Overall, African Americans are less likely to vote than Whites, a fact reflecting the disproportionate numbers of African Americans in the poor and working classes—groups that typically are less likely to vote than middle- and upper-class people. During the late 1960s, however, controlling for factors such as age, education, and income, African Americans were more likely to vote than White Americans, probably due to the heightened sense of the importance of voting generated by the civil rights movement and voter registration drives mobilized during this period. Beginning in the 1970s, the likelihood of voting converged among Blacks and Whites of similar social class, education, and age (Ellison and Gay 1989). Problems with the disenfranchisement of African American voters continue, however, as was apparently the case in the close election of 2000, when thousands of African Americans had their names purged from voter lists and many were turned away from polling places after waiting in long lines.

A subjective factor influencing the voting behavior of African Americans is that they are, in general, less

Following the 2000 presidential election, many groups organized demonstrations to protest voting irregularities, especially as they disenfranchised many African American voters.

trusting and more alienated from politics than their White counterparts. The more that Black Americans and Latinos identify themselves as groups with distinct needs and interests, the more likely they will participate in the political process, reflecting the long-standing sociological finding that a strong sense of ethnic identity is

Figure 19.3
Voter Turnout in U.S. Elections

Source: U.S. Census Bureau. 2004. *Statistical Abstract of the United States, 2003.* Washington, DC: U.S. Government Printing Office, p. 269.

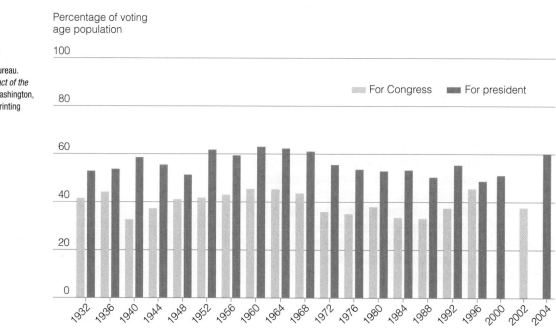

linked to increased political participation. Research on Latinos also finds that Latinos are less alienated from politics when they are well represented among legislators (Welch and Sigelman 1993; Pantoja and Segura 2003).

Social factors also influence whom people vote for, as Table 19.3 shows, and race is one major influence. African Americans and Latinos, with the exception of Cuban Americans, tend to be Democratic; Cuban Americans are disproportionately Republican. One reason that both Latinos and African Americans are more likely to support the Democratic Party and its candidates is that Democrats historically have been more committed to government action that promotes social and economic equality. The allegiance of Latino and Black voters to Democratic candidates is also explained in part by socioeconomic factors such as age, gender, income, education, and religion. All these factors, regardless of racial–ethnic identity, are linked to political behavior. Even without controlling for these factors, Latinos are more liberal than Anglos and generally more likely than other groups to support government spending for various purposes (Welch and Sigelman 1993).

Ethnicity also influences voting patterns. People tend to vote for people of their own ethnic group. Sociologists have also found that ethnicity can outweigh socioeconomic factors such as occupation, income, and education in predicting voting patterns. White Catholics, for example, have traditionally aligned with the Democratic Party, although that link has been weakened somewhat by social mobility among some Catholics that has influenced voting patterns (Legge 1993; Smith 1993).

Gender is also a major social factor influencing political attitudes and behavior. The **gender gap** refers to the differences in political attitudes and behavior between women and men. One aspect of the gender gap is that women are more likely than men to identify and vote as Democrats and to have liberal views on a variety of social and political issues. For many decades, men were more likely to vote than women and, when asked, women reported voting the same as their husbands or fathers. In recent years, women have been as likely to vote as men, but there are significant differences in their political outlooks. The gender gap is widest on issues involving violence and the use of force. Women, for example, are more likely than men to be peace-seeking and to support gun control. The gender gap is also evident on so-called compassion issues—women are more likely than men to support government spending for social service programs that aid the young, the old, and the disadvantaged. They are also more liberal on issues such as militarism, gay and lesbian rights, and feminism (Saad 2002; Wilcox et al. 1996; Eliason et al. 1996; Jackson et al. 1996).

Table 19.3
The 2004 Presidential Election: Who Voted How?

	George W. Bush	John Kerry
By gender		
Women	48%	51%
Men	51	48
By race		
White	59	41
Black	11	88
Hispanic/Latino	44	53
Asian	44	56
By age		
18 to 29 years old	45	54
30 to 44 years old	53	46
45 to 59 years old	51	48
60 or over	54	46
By education		
No high school diploma	49	50
High school graduate	52	47
Some college	54	46
College graduate	52	46
Postgraduate study	44	55
By income		
>$15,000	36	63
$15,000–30,000	42	57
$30,000–50,000	49	50
$50,000–75,000	56	43
$75,000–100,000	55	45
$100,000–150,000	57	42
$150,000–200,000	58	42
$200,000 or more	63	35
By religion		
Protestant	59	40
Catholic	52	47
Jewish	25	74
None	31	67
Gay, lesbian, bisexual	**23**	**77**

Source: Edison/Mitofsky National Survey, as reported in *The New York Times*, November 4, 2004, p. P4; www.cnn.com

Finally, as Map 19.1 shows, region can affect voting patterns. This map shows the outcome of the 2004 presidential election by weighting the proportion of votes cast for each candidate (red for George W. Bush and blue for John Kerry). The more red a county, the higher the proportion of votes cast for Bush and vice

MAP 19.1 The 2004 Election

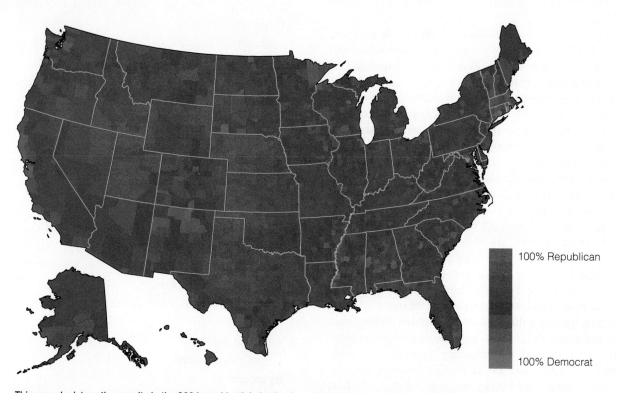

100% Republican

100% Democrat

This map depicts voting results in the 2004 presidential election by weighting each county according to what proportion voted for George W. Bush (red) and the proportion for John Kerry (blue). What does it suggested to you about regional differences in the political choices people make? Furthermore, as you look closely at the shades of red and blue in different areas, what sociological factors can you identify that might have been pertinent to people's voting decisions?

Source: Conceived by Robert Vanderbei, Princeton University, based on national election results and U.S. census data.

versa. The map shows strong regional differences. (See also the map caption.)

Political Power: Who's in Charge?

Although democratic government is supposed to be representative, the class, race, and gender composition of the ruling bodies in this country indicate that this is hardly the case. Most of the members of Congress are White men (see Figure 19.4). They are well educated, from upper middle- or upper-income backgrounds, and have an Anglo-Saxon Protestant heritage. One-third of the people in Congress are lawyers; another third are businesspeople and bankers. The remaining members come from professional occupations. Very few senators or representatives were blue-collar workers before coming to Congress (Ornstein et al. 1996; Saunders 1990; Freedman 1997).

Many members of Congress are millionaires, and many have large financial interests in the industries they regulate. Simply getting into politics requires a substantial investment of money. The average cost to run a Senate campaign is $2.5 million; a successful run for the House of Representatives costs nearly $500,000. In the 2004 presidential election, George W. Bush spent a record $339 million and John Kerry, $299 million, much of it coming from PACs associated with business and other organizations.

Electioneering has become a high-tech, highly expensive business. A record amount was spent on the national elections in 2004—much of it on media advertising, but also on overnight polls, computer modeling of the electorate, and communications systems— all of which add up to huge campaign expenses. Many lawmakers report that the need to constantly raise money for campaigns distracts them from their jobs as

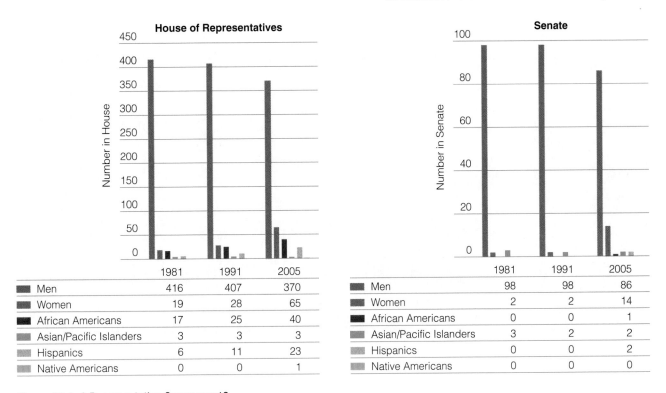

Figure 19.4 A Representative Government?

Source: U.S. Census Bureau. 2004. *Statistical Abstract of the United States, 2003.* Washington, DC: U.S. Government Printing Office, p. 263.

legislators. Meanwhile, spending big money does not guarantee that a candidate will win an election. It is just the entry fee to get into the game. Little wonder that political action committees (PACs) have become such a fixture on the campaign scene. Their influence, based on the sheer growth in their numbers, is on the rise. PACs also shift their contributions, depending on which party they see as likely to win (Salant and Cloud 1995).

All candidates in state and national elections depend on contributions from individuals and groups to finance their election campaigns. Wealthy families and individuals are among the largest campaign contributors to presidential elections (Allen and Broyles 1991). The largest individual contributors typically have interests in the same industries that fund political action committees. Much of the money given by individuals and PACs goes to incumbents, who historically have had an overwhelming edge in elections and who already sit on the committees where public policy is hammered out. This picture of elites and business interests funneling money to candidates, who return to the same donors for more money when the next campaign rolls around, has shaken the faith of many Americans in their political system. National surveys show that 30 percent have a great deal or quite a lot of confidence in Congress; half (46 percent) have confidence in the Supreme Court (Saad 2004).

Women and Minorities in Government

Although there have been some gains in the numbers of women and minorities in government, they remain a substantial minority—both at the federal and state levels (Zweingenhaft and Domhoff 1998). In the Senate of the 109th Congress (convening in 2005), there are fourteen women, one African American, two Latinos, two Asian Americans, and no Native American (refer again to Figure 19.4). The first Native American to be sent to the Senate in more than sixty years was elected only as recently as 1992.

Each minority group is vastly underrepresented in Congress. Note that only 14 percent of the senators are women, whereas women comprise 51 percent of the population. Thirteen percent of the population is Black, with only one black senator (Barack Obama from Illinois). Hispanics comprise 13 percent of the population but there are only two Hispanic senators, and in the House the Hispanic members comprise only 5 percent. There is a long way to go before we have a truly representative government, and there is no guarantee that gains made in one election will be sustained in the next.

Researchers offer several explanations for why women and racial–ethnic groups continue to be underrepresented in government. Certainly, prejudice plays

The world's leadership: Where are the women?

a role. It was not long ago, in the 1960 Kennedy–Nixon election, that the first Catholic president was elected. In the 2000 election, Joseph I. Lieberman was the first Jewish candidate to appear on a major national ticket. Gender and racial prejudice run just as deep in the public mind, although the percentage of Americans who say they would vote for a woman for president has climbed to 92 percent (compared to 53 percent in 1969), a substantial number (42 percent) also say they think a man would make a better president than a woman; 39 percent of women and 22 percent of men think a woman would make a better president (Simmons 2001).

Prejudice alone, however, cannot account for the lack of representation. Social structural causes are a major factor in the successful elections of women and people of color. First, women and people of color are better represented in local political office. Women and minority candidates receive a great deal of political support from local groups, but at the national level, they do not fare as well. The power of incumbents, most of whom are White men, disadvantages any new office seeker (Darcy et al. 1994).

A central question for sociologists is whether the inclusion of previously excluded groups in government will make a difference. In other words, do numbers matter? The participation of minorities does alter the outcomes of elections; and the political outlook of Latinos, African Americans, and women differs from that of White men. Studies of Black elected officials show that they have different political agendas from White officials. Thus, an increase in Black representation in

government does have a potential effect on political outcomes (Scavo 1990). The presence of women and minorities in government brings new perspectives on old issues and new attention on issues otherwise overlooked. Women legislators are more likely than men to support feminist issues, such as the Equal Rights Amendment, government-subsidized child care, and abortion rights (Mandel and Dodson 1992).

The Military

The military arm of the state is among the most powerful and influential social institutions in almost all societies. In the United States, the military is the largest single employer, and it accounts for a large portion of the federal budget (see Figure 19.5). Approximately 2.56 million men and women serve in the U.S. military, 1.3 million on active duty and the rest in the reserves. This does not include the many hundreds of thousands employed in industries that support the military or the civilians who work for the Department of Defense and other militarily affiliated agencies (Department of Defense 2003).

Highly ranked in the value system of the United States is *militarism,* the pervasive influence of military goals and values throughout the culture. The militarism in our culture is evident in many ways. The toys children play with often inculcate militaristic values. Military garb falls in and out of fashion. The plots of each summer's biggest blockbuster movies almost always re-

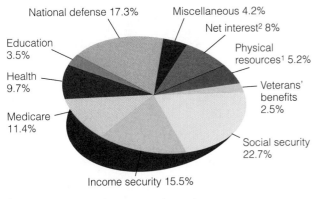

National defense 17.3% Miscellaneous 4.2%

Net interest[2] 8%

Education 3.5% Physical resources[1] 5.2%

Health 9.7%

Veterans' benefits 2.5%

Medicare 11.4%

Social security 22.7%

Income security 15.5%

[1] Includes energy, natural resources, environment, commerce, housing, transportation, and community development

[2] Includes international affairs, agriculture, justice, and general government

Figure 19.5 Military Spending and Federal Budget Outlays

Source: U.S. Census Bureau. 2004. *Statistical Abstract of the United States, 2003.* Washington, DC: U.S. Government Printing Office, p. 322.

Military values permeate American culture, including its fashion.

volve around a burly character hugely gifted in the military arts, often a veteran of the armed forces, who generally stocks up on the latest weaponry. Fighter jets flash over the Super Bowl. Military brass sparkles in parades. Being the strongest nation, militarily, in the world is frequent in the national political rhetoric. Most people in the United States, 75 percent, say they have quite a lot or a great deal of confidence in the military (Saad 2004).

The Military as a Social Institution

Institutions are stable systems of norms and values that fill certain functions in society. The military is a social institution whose function is to defend the nation against external (and sometimes internal) threats. A strong military is often considered an essential tool for maintaining peace, although the values that promote preparedness in the armed forces are perilously close to the warlike values that lead to military aggression against others (see Map 19.2).

The military is a strict hierarchical and formal social institution. People who join the military are explicitly labeled with rank and, if promoted, pass through a series of well-defined levels, each with clearly demarcated sets of rights and responsibilities. There is an explicit line between officers and enlisted personnel, and officers have numerous privileges that others do not. People in the higher ranks are also entitled to absolute obedience from the ranks below them, with elaborate rituals created to remind dominants and subordinates of their status. As in other social institutions, military enlistees are carefully socialized to learn the norms of the

culture they have joined. Military socialization places a high premium on conformity. New recruits have their hair cut short to look alike, they are issued identical uniforms, and they are allowed to retain very few of their personal possessions. They must quickly learn new codes of behavior that are strictly enforced.

Like other hierarchical institutions, the military embeds inequality into its structure. For those who are working-class or poor, the military may be the best hope for social mobility—a funded education, learning technical skills, and travel. Yet, this promise is also met with increased risk since those from the upper classes who serve in the military are most likely to be officers and in noncombat positions—if they serve at all. Thus, during the Iraq War, only one member of Congress—those who make the decision to go to war—had a child serving in the military. Blacks, Hispanics, and poor Whites also perceive that they have borne most of the burden for fighting and dying in the war—a perception partially substantiated in fact. As of May 2004, Blacks made up 14 percent of casualties in Iraq and Hispanics, 12 percent—close to their proportion in the population at large. But between 1980 and now, the proportion of Blacks killed in all military conflicts is 19 percent, exceeding their proportion in the population. At the same

MAP 19.2 Military Expenditures per Capita

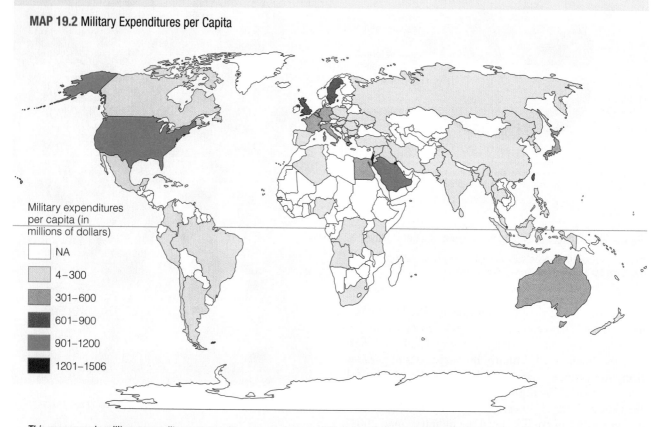

Military expenditures per capita (in millions of dollars)

☐ NA
☐ 4–300
☐ 301–600
☐ 601–900
☐ 901–1200
☐ 1201–1506

This map reveals military expenditures per capita, compared across different countries. Israel, Kuwait, Singapore, and the United States rank highest in military expenditures per capita. Why do you think the United States has such high military expenditures? (Note that this was true even before the current military conflicts in the Middle East). In what ways do government expenditures represent the values of a given culture? How do they reflect the different degrees of global political power?

Data: U.S. Census Bureau. 2004. *Statistical Abstract of the United States, 2003.* Washington, DC: U.S. Government Printing Office, p. 878.

time, social class figures prominently in the casualty count, with Whites from small, mostly poor rural areas a disproportionately large percentage of the U.S. casualties in Iraq (Neal 2004).

These patterns are not without historical precedent. During the Vietnam War, a program known as Project 100,000 was promoted by President Lyndon B. Johnson to recruit and rehabilitate 100,000 young men—mostly southern, Black, and poor men who had previously been rejected from service for failing to meet the mental and physical requirements. By 1969, the soldiers recruited through Project 100,000 had twice the ratio of soldiers killed in action than any other unit (MacPherson 1984).

Many find the military to be a distinctively masculine institution. Not only are men the majority in the military, but the organization itself rests on masculine cultural traits such as aggression, competition, hierar-

chy, and violence. Soldiers are often abused if they fail to live up to the masculine image of the military. Homophobic and misogynistic attitudes are used to enforce highly masculine codes of behavior during military socialization. New recruits may derisively be called "sissies" and told that if they fail to live up to military ideals, it is because they are effeminate or homosexual.

Military life is obviously very different from life off the base, yet the military is intimately linked to other institutions in the society. Most notably, a strong connection exists between the military and corporate America. The **military–industrial complex** is the term used to describe the linkage between business and military interests. The link is so strong that the military supports many of the basic research and development projects in the nation. From university research labs to corporate research institutes and centers, the military funds much of the basic scientific and technological knowledge produced in this country. Many have argued that research and the knowledge produced from corporate and government sponsorship should be adapted to more humanistic social service values.

Race and the Military

A striking feature of the military as a social institution is the increased representation of racial minority groups and women within the armed forces. Picture a U.S. soldier. Who do you see? At one time, you would have almost certainly pictured a young, White male, possibly wearing army green camouflage and carrying a weapon. Today, the image of the military is much different. You are just as likely to picture a young African American man in military blues with a stiffly starched shirt and a neat and trim appearance, or perhaps a woman wearing a flight helmet in the cockpit of a fighter plane.

In 1973, the all-male draft of civilians into the Army was ended and an all-volunteer force was initiated in the U.S. armed services. The idea behind the all-volunteer force was that the Army would be made more professional because the people who joined would be choosing to make the military their career. In fact, the greatest change in having an all-volunteer force has been a tremendous increase in the representation of racial minority groups in the armed services. Decisions to enlist are affected by a number of societal factors, including existing employment opportunities, possibilities for educational attainment, and the military values held at the time by the public. Military life is chosen far more often by African Americans, and to some extent Hispanics, than by Whites. The number of minorities in the military became an issue again with regard to the war in Iraq. Some suggested reinstituting the draft so that military would more fairly represent the whole population and the risks of being in military service would not fall so disproportionately on certain groups—the poor and the nation's minorities.

African Americans have served in the military for almost as long as the U.S. armed forces have been in existence. Except for the Marines, which desegregated in 1942, the armed forces were officially segregated until 1948, when President Harry S. Truman signed an executive order banning discrimination in the armed services. Although much segregation continued following this order, the desegregation of the armed forces is often credited with having promoted more positive interracial relationships and increased awareness among Black Americans of their right to equal opportunities.

Thirty-four percent of military personnel are racial minorities—20 percent African American, 8 percent Hispanic, and 6 percent other racial minorities (U.S. Department of Defense 2002). For groups with limited opportunities in civilian society, joining the military seems to promise an educational and economic boost. Is this realized? Sociologists have found that African Americans in the military have higher levels of interpersonal satisfaction and work satisfaction than do Whites. African American and Latino recruits are also more likely than Whites to complete their first term of service and remain in the military beyond their first term. Some evidence shows that minorities in the military are promoted at least as fast as Whites, although the research on this issue has yielded mixed results.

Within the military, there is equal pay for equal rank. African Americans and Latinos, however, are overrepresented in lower-ranking support positions within the armed forces. Often, they are excluded from the higher-status, technologically based positions—those most likely to bring advancement and higher earnings both in the military and beyond. Despite the visibility of General Colin Powell, at the highest ranks in the military, there are few African Americans, Latinos, Asian Americans, or Native Americans. For example, although Blacks comprise 20 percent of the military, they are only 8 percent of officers; Hispanics comprise only 4 percent of officers. Officers among Native Americans and Asians/Pacific Islanders are even more rare (U.S. Department of Defense 2000; Moskos and Butler 1996).

For both Whites and racial minorities, serving in the military leads to higher earnings relative to one's nonmilitary peers. The economic payoff is greatest for African Americans and Latinos, relative to people with

The men and women who serve in the Armed Forces, such as this young man in Iraq, are often separated from families and loved ones for long periods of time.

comparable background, education, experience, and ability. The earnings difference between military and civilian groups is substantial (Phillips et al. 1992).

The military has long been seen as an institution in which race relations are better than among the general public. There is some truth to this. Whites in the armed forces are more likely to say they have a close personal friend who is Black than are Whites in general (82 percent compared with 59 percent). While the military has created a more racially tolerant environment than many other institutions, problems remain. African American and Latino military personnel are more likely than Whites to say that the military pays too little attention to discrimination within the military (Scarville et al. 1999).

Women in the Military

Military academies in the United States did not open their doors to admit women until 1976. Since then, there have been profound changes in the armed services, although the hostile reaction to women who have entered these academies shows how fierce resistance to the inclusion of women in the military can be. Shannon Faulkner, the young woman who won the right to enter The Citadel, had to embark on a two-and-one-half-year legal struggle to gain admission. Once admitted, she was harassed and ridiculed. When she succumbed to exhaustion during the first week of intense physical training and chose to leave, many cadets whooped and shouted in glee, leaving many to argue that none of the men had been forced to endure such tribulations. Later, two other women left The Citadel after serious hazing in which, among other things, the women were sprinkled with nail polish remover and their clothes set on fire.

The Supreme Court ruling in 1996 (*United States v. Virginia*) that women cannot be excluded from state-supported military academies such as The Citadel and the Virginia Military Institute (VMI) was a landmark decision that opened new opportunities for those women who want the rigorous physical and academic training that such academies provide (Kimmel 2000).

The incorporation of women into the military has changed the look of the military, but it is still a highly masculine institution in terms of personnel and policies. The first female uniformed personnel appeared in 1901 as nurses. By the end of World War I, 34,000 women served in the Army and Navy Nurse Corps. Nursing is indicative of the roles women filled in the military. As military personnel, women were confined to "women's work," namely, clerical work and nursing. (In the 1960s, the phrase *typewriter soldiers* was used to refer to women in the armed services, nearly all of whom held clerical positions.) Serving behind a desk, instead of in the field, effectively removed women from the possibil-

ity of achieving rapid promotion or high rank (Becraft 1992a; Holm 1992).

The marginalization of women in the military has been justified by the popular conviction that women should not serve in combat. The traditional belief that men are protectors and women are dependents in need of protection has led many to believe that women, especially mothers and wives, should stay on home soil where they can safely carry out their "womanly" duties. Despite these beliefs, women have fought to defend this country. Women are about 10 percent of the forces serving in the Afghanistan and Iraq wars (The Women's Research and Education Institute 2004). In 1992, the Defense Authorization Act allowed women to fly aircraft in combat, traditionally a position reserved for men only. The increased visibility of women soldiers during the Persian Gulf War also shifted public opinion in favor of women serving in combat positions. Thirty-eight percent now think women *should* get combat assignments, with another 45 percent thinking they should if they so choose (Carlson 2003).

The involvement of women in the military has reached an all-time high in recent years. Now, almost 200,000 women are on active duty, with an additional 151,000 in the reserves, not including the Coast Guard and its reserves. The Air Force has the highest proportion of women (18 percent), followed by the Army (15 percent), the Navy (13 percent), and the Marines (6 percent). Minority women are overrepresented in the military, relative to their percentage in the general population. Minority women are a greater proportion of women in the military (29 percent) than are minority men as a proportion of men in the military (18 percent; U.S. Department of Defense 2000).

Gender relations in the military extend beyond just women who serve. The experience of military wives, for example, is greatly affected by their husbands' employment as soldiers. Frequent relocation means that military wives are less competitive in the labor market. They have lower labor force participation and earn less than their "nonmilitary" women peers at the same educational level (Enloe 1993; Moskos 1992; Payne et al. 1992). Sociologists have also found that the presence of a military base depresses wages and employment for all women in the surrounding community (Booth et al. 2000).

Gays and Lesbians in the Military

According to the *"don't ask, don't tell"* policy, military recruiting officers cannot ask about sexual preference, and individuals who keep their sexual preferences a private matter shall not be discriminated against. Those who publicly reveal that they are gay or lesbian can be expelled from the military based on their sexual orientation. Although the "don't ask, don't tell" policy does not explicitly permit the military to exclude and discharge service people on the grounds of sexual orientation, nor does it explicitly acknowledge the civil rights of gays and lesbians. Gay soldiers and potential recruits have filed several lawsuits against the U.S. military. Whether gays and lesbians will ultimately be able to live openly as homosexuals while pursuing military careers remains unclear.

THINKING *SOCIOLOGICALLY*

The military is generally defined as the institution that provides defense for the society and maintains the *public order.* Sociologists have argued that the military, like other social institutions, is a *gendered institution.* Similarly, there is a racial and class order in the military. Discuss what this means using examples from current events.

Prejudice, homophobia, and discrimination continue against lesbian women and gay men in the military. Seventy percent of the public believe gays should be allowed to serve openly; only 9 percent of the public do not think gays should be allowed to serve under any circumstances (Newport 2002). Ironically, studies of student attitudes about gays in the military show that the "don't ask, don't tell" policy may have increased support for gays in the military among the general public. A survey of college students taken before and after initiation of this policy, has found that students see gay soldiers' careers as more promising and heterosexuality as less of an advantage after this policy generated so much public attention (Pesina et al. 1994).

DEBUNKING *SOCIETY'S MYTHS*

Myth: Letting lesbians and gays serve in the military erodes morale and weakens the armed services.

Sociological perspective: Similar arguments were historically made to exclude Black Americans from military service. This belief is an ideology that serves to support the status quo.

Supporters of the ban on gays in the military often use arguments similar to those used before 1948 to defend the racial segregation of fighting units. They claim that the morale of soldiers will drop if they are forced to serve alongside gay men and women, that national security will be threatened, and that having known homosexuals serving in the military will upset the status quo and destroy the fighting spirit of military units. Ominous reference to the "tight quarters" where recruits live and work implies that gays and lesbians are seen as unable to control their sexuality. Fear of acquired immune deficiency syndrome (AIDS) and the connection of this disease to the gay population further inflame the feeling against lesbians and gays.

The long-standing policy against gays in the military has kept homosexuality hidden, but not nonexistent. Even the military has had to admit that there have always been some gays and lesbians in all branches of the U.S. armed forces. Surveys have found that antigay bias in the military is common, and disparaging comments and harassment are widespread. Despite some increased tolerance of gays and lesbians in the military, homophobia is a pervasive part of military culture (Myers 2000; Becker 2000).

The government policy on gays in the military is an example of how state institutions can have different effects on individuals within society. Lesbians and gays, like heterosexual women and members of racial minority groups, are discriminated against by the military. The policies and recruitment practices of the armed forces affect members of these groups differently than they affect members of the dominant group in society. The phrase used to recruit military soldiers, "Be all that you can be," seems a hollow promise in an institution that forbids a segment of its people from admitting who they are.

Sociology ⊛ Now™
Reviewing is as easy as ❶❷❸.

1. *Before you do your final review, take the SociologyNow diagnostic quiz to help you identify the areas on which you should concentrate. You will find information on SociologyNow and instructions on how to access all of its great resources on the foldout at the beginning of the text.*

2. *As you review, take advantage of SociologyNow's study videos and interactive Map the Stats exercises to help you master the chapter topics.*

3. *When you are finished with your review, take SociologyNow's posttest to confirm you are ready to move on to the next chapter.*

Chapter Summary

What is the state?

The *state* is the organized system of power and authority in society. It comprises different institutions, including the government, the military, the police, the law and the courts, and the prison system. The purpose of the state is to protect its citizens and preserve society, but it often protects the status quo, sometimes to the disadvantage of less powerful groups in the society. *Revolutions* occur when the state breaks down, either from an overthrow or the total transformation of state institutions.

How do sociologists define power and authority?

Power is the ability of a person or group to influence another. *Authority* is power perceived to be legitimate. There are three kinds of authority—*traditional authority,* based on long-established patterns; *charismatic authority,* based on an individual's personal appeal or charm; and *rational-legal authority,* based on the authority of rules and regulations (such as law). The United States is primarily built on a system of rational-legal authority. *Bureaucracies* typically flourish in a system of rational-legal authority. The growth of bureaucracy is especially notable in contemporary governmental institutions.

What theories explain how power operates in the state?

Sociologists have developed four theories of power. The *pluralist model* sees power as operating through the influence of diverse interest groups in society. The *power elite model* sees power as based on the interconnections between the state, industry, and the military. The *autonomous state* model sees the state as an entity in itself that operates to protect its own interests. *Feminist theorists* argue that the state is patriarchal, representing primarily men's interests. Each theory reveals a different aspect of how power operates.

How well does the government represent the diversity of the U.S. population?

An ideal democratic government would reflect and equally represent all members of society. The makeup of American government does not reflect the diversity of the general population. African Americans, Latinos, Native Americans, Asians, and women are underrepresented within the government. Political participation also varies by a number of social factors, including income, education, race, gender, and age. The large sums of money necessary to run campaigns make politicians *susceptible* to the influence of *political action committees*. One consequence of the current structure of government is that many Americans have little confidence in the government to solve the nation's problems.

What role does the military play in the state?

The military is a social institution that is the nation's system of defense. Militaristic values permeate the culture, emphasizing violence and aggression. Through military socialization, new recruits must learn the norms and values of military culture. The military is strongly linked to other social institutions in the society. African Americans and Latinos are overrepresented in the military, in part because of the opportunity the military purports to offer groups otherwise disadvantaged in education and the labor market. Women also have an increased presence in the military.

Key Terms

authority 513
autonomous state model 518
bureaucracy 514
charismatic authority 513
democracy 520
dictatorship 521
gender gap 523
government 520
interest group 515
interlocking directorate 517
iron law of oligarchy 514
military–industrial complex 528
monarchy 520
pluralist model 515
political action committees 516
power 513
power elite model 516
propaganda 511
rational-legal authority 513
state 510
traditional authority 513

Researching Society with MicroCase Online

You can see the results of actual research by using the Wadsworth MicroCase® Online feature available to you. This feature allows you to look at some of the results from national surveys, census data, and other data sources. You can explore this easy-to-use feature on your own, but try this example. Suppose you want to know:

Does social class affect one's political party identification?

To answer this question, go to http://sociology.wadsworth .com/andersen_taylor4e/, select MicroCase Online from the left navigation bar, and follow the directions there to analyze the following data.

Data file: GSS

Analysis: Cross-Tabulation

Row Variable: I-INCOME

Column Variable: PARTY

Questions:

Once you have your results, answer the following questions:

1. Democrats are most likely to be in which income group?
   a. $0k–24.9k
   b. $45k–49.9k
   c. $50k +

2. Republicans are most likely to be in which income group?
   a. $0k - 24.9k
   b. $45k - 49.9k
   c. $50k +

3. Independents are most likely to be in which income group?
   a. $0k - 24.9k
   b. $45k - 49.9k
   c. $50k +

4. In what ways do you think this affects the political issues the different parties tend to promote?

The Companion Website for Sociology: Understanding a Diverse Society, *Fourth Edition*

http://sociology.wadsworth.com/andersen_taylor4e/

Supplement your review of this chapter by going to the Companion Website to take one of the Tutorial Quizzes, use the flash cards to master key terms, and check out the many other study aids you'll find there. You'll also find special features such as GSS Data and Census 2000 information, data and resources at your fingertips to help you with that special project or do some research on your own.

Suggested Readings and Web Resources

DeLoria, Vine. 1992. *American Indian Policy in the Twentieth Century.* Norman, OK: University of Oklahoma Press.
DeLoria, a major American Indian writer and activist, explores state policies toward American Indians in the twentieth century.

Domhoff, G. William. 1998. *Who Rules America?* Mountain View, CA: Mayfield Publishing.
This is an updated version of a classic book, detailing the power elite in the United States.

Enloe, Cynthia. 2000. *Maneuvers: The International Politics of Militarizing Women's Lives.* Berkeley, CA: University of California Press.
Enloe examines the effects of militarization on women's lives, both women in the military (includ-ing military wives) and women who are not. Her analysis of the military as a gendered institution also shows how pervasive militarization is within society.

Moskos, Charles C., John Allen Williams, and David R. Segal. 1999. *The Postmodern Military: Armed Forces After the Cold War.* New York: Oxford University Press.
This book examines the changes occurring in the U.S. military as a result of the end of the Cold War. It includes a comparative perspective from a team of international sociologists.

Shilts, Randy. 1993. *Conduct Unbecoming: Gays & Lesbians in the U.S. Military.* New York: St. Martin's Press.
This book provides a historical overview of gays and lesbians in the military and analyzes public and government response to this issue.

Federal Election Commission
www.fec.gov
This site includes information on elections, campaign finance, and the various voting systems used in local, state, and federal elections.

Center for Responsible Politics
www.opensecrets.org
A comprehensive site providing information on campaign financing, including which organizations and which people contribute the most to different parties and candidates.

Roll Call
www.rollcall.com
An organization that tracks political news and provides commentary on current political events.

● ● ●

Health and Health Care

Imagine the following scenarios:

• A 55 year old woman has been diagnosed with a terminal illness. She and her husband have accumulated $300,000 of debt trying to pay for medical bills. Having depleted all of her savings, they sell their home and move from Ohio to Connecticut to live with their oldest daughter and family while the wife is in hospice care. While there, the husband develops a condition that requires an expensive drug treatment, which they cannot afford, so he forgoes treatment and dies soon after his wife.

• A young woman, recently graduated from college, works at two jobs—waiting tables at a local restaurant most nights and weekends and working in the local mall during the day. She wants to return to school for a Master's degree but can't save enough for the tuition, and her family is unable to help. Meanwhile, she is earning a minimal income, with occasional good nights from tips, but she drives an older car and can't seem to get ahead. She knows she should carry health insurance, but her employers don't provide it, and she can't afford the monthly premiums. She decides she is young and healthy, so

she takes the risk of not having coverage. One night, driving home from work, a drunk driver hits her car and she is hospitalized and permanently disabled.

The first account is based on an actual case, taken from the notes of a physician at Yale-New Haven Hospital, where over the course of one week, she reported several such scenarios (Pollitt 2004). The second scenario, though not based on an actual case, reflects the reality of health care for many young people—the group least likely to be covered by health insurance. Both scenarios reflect fundamental topics in the sociology of health and health care— that with all of the modern advances in health care and despite the fact that the United States has one of the most scientifically sophisticated health care systems in the world, many people are without adequate care.

Health care as an institution in society is in crisis. The public wonders: Should there be universal health care supported by the government? Why are costs for medical insurance so high, and why don't all people have health insurance? Why are prescription drugs cheaper in Canada than the very same drugs sold in the United States?

Sociology ⊛ Now™
Reviewing is as easy as ❶❷❸.

Use SociologyNow to help you make the grade on your next exam. When you are finished reading this chapter, go to the chapter review for instructions on how to make SociologyNow work for you.

Indeed, when the public is surveyed about the most pressing health care issues, the cost of health care and access to good health care are foremost on the public's mind (Jones 2003).

Sociological studies of health and health care explore these issues and also show how health—seemingly a physical reality—is also a deeply social phenomenon. Social facts as basic as how long one lives are the result of one's position in society. Even the diseases people die from are correlated with factors like gender, race, and social class. And for many people, the inequalities that mark other social institutions are also reflected in who has access to good health care and who does not. This chapter reviews the sociology of health and health care, showing how social processes affect the most basic fact of our lives—our physical and mental health. •••

The Emergence of Modern Medicine

The highly technological, scientific, corporate-based health care that now characterizes modern medicine in the United States was not always how health care was delivered. In colonial times, American physicians received their training in western Europe. Their competitors in the healing arts included alchemists, herbalists, ministers, faith healers, and even barbers. Treatments were a combination of folk wisdom, superstition, tried-and-true regimens, and sometimes dangerous quackery. A simple scratch, once infected, could easily cost a limb or a life. Ambiguous maladies such as "the fever" were a common cause of death. Etiologies (causes) of disease were believed to include everything from "bilious humors" to demonic possession. Even the best-trained practitioners did not know to sterilize instruments, and a common intervention was the "bleeding" of patients (namely, drawing "bad blood" from the patient, sometimes in amounts large enough to seriously weaken the victim and sometimes by drilling holes in the skull). Needless to say, the cure was often worse than the disease.

The mentally ill, in particular, were often thought to be possessed or, worse, witches. Once labeled a witch, one had little hope of shedding the label, and the consequences could be as horrendous as burnings or drownings. The vast majority of people labeled witches were women. Sociological researchers have shown that these women were generally older, unmarried, and childless, and they were thus defined as of little use to the local community (Erikson 1966). Defining them as possessed by demons and thus incurable was one way of achieving their permanent elimination from the community.

By the start of the nineteenth century, advances in biology and chemistry ignited a century of explosive growth in medical knowledge. One insight of the scientific revolution of the mid-1800s was the *germ theory,* the idea that many illnesses were caused by microscopic organisms, or germs. The theory is now considered scientific fact, but at the time the notion that something called germs caused illness was a hotly debated topic.

The development of new technologies has transformed the delivery of health care. The "house call" by a physician is now a rarity, although it is being renewed in a few communities. Many new medical technologies, such as this heart bypass procedure assisted by robots, would have been unimaginable not many years ago.

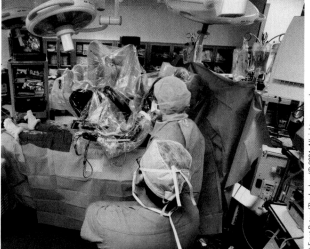

During the late 1800s, germ theory established itself as a foundation of medicine. Doctors were able to show that tactics such as isolating infected people and sealing infected wells could stop the spread of illness by stopping the spread of germs. Relentless study and research transformed medicine into a science. Coincidentally, the social prestige of medicine greatly increased, contributing to the status of physicians, who had formerly enjoyed a more modest social standing.

The year 1847 saw the founding of the American Medical Association (AMA). After half a century of successful campaigns to sweep away rivals in the healing arts—to have alternative therapies delegitimated or outlawed—the AMA emerged as the most powerful organization in U.S. health care. As one example, prior to the development of obstetrics and gynecology as medical specialties, most babies were born at home, delivered by midwives who were considered the experts on birth. When the American Medical Association was founded in the mid-nineteenth century, the AMA actively campaigned to eliminate midwives from practice—a phenomenon that is still reflected today in the social control of childbirth by medical doctors (Barker-Benfield 1976).

It was in the late 1800s that the image of medicine as an upper-class profession took hold. A medical education was expensive, and medical schools drew upon White, male, urban populations for their students. People trained as physicians were entitled to take their place in the upper social strata. Herbalists and faith healers more frequently were raised in the rural lower class and generally remained there. As the ranks of the medical profession swelled with wealthy Whites, African Americans and Hispanics became proportionately more affiliated with older folk practices and midwifery. This overall trend continued through the early part of the twentieth century. Even today, folk practices have adherents among rural, lower-class Whites and among rural and urban, lower-class Blacks, Hispanics, and Native Americans (Starr 1982).

Not only are social class and race–ethnicity interwoven with the development of modern medicine in the United States, but so too is gender. At the outset of the twentieth century, the male-dominated medical profession vigorously opposed gender equality within their profession. Prevailing medical opinion at the time labeled the differences between the genders both natural and unchangeable. Men were seen as inherently rational and scientific; women, as dominated by emotions and incapable of rigorous scientific thought (Smith-Rosenberg and Rosenberg 1984). Vestiges of these beliefs still persist in the medical profession today.

Women were expected to devote their time and energy to childbearing instead of professions such as science and medicine. One physician in 1890 stated that God had "in creating the female sex . . . taken the uterus and built up a woman around it" (Smith-Rosenberg and Rosenberg 1984: 13). Such beliefs kept women out of medical school because physicians warned that too much thinking tended to interfere with a woman's ability to have children.

Physicians in the 1800s had strong opinions about female sexuality. Women were supposed to have no interest in sex beyond the reproductive function. The female orgasm, when it was acknowledged, was likely to be regarded as a type of disorder. The male orgasm was considered natural, and regular sexual relations were understood to be necessary for men. This view was promoted by the American Medical Association well into the 1920s. Contraception and abortion were opposed because it was believed they would encourage increased sexual activity among women, thus endangering their health.

Specialization in Medicine

Since World War II ended in 1945, there has been tremendous growth in the medical establishment. Along with the huge growth in medicine came increased specialization. Today, specialists (about 80 percent of physicians) greatly outnumber general practitioners (about 20 percent). Physicians who specialize, particularly in fields like surgery, radiology, and obstetrics/gynecology, generally enjoy high incomes (Weitz 2001; U.S. Census Bureau 2004).

Specialization has its advantages in that it provides expert care to those who have access to experts. One of the hallmarks of the U.S. health care system is that medical care is based on a vast research enterprise where experts are learning more and more about the causes, consequences, and treatment of various diseases. On the other hand, specialization also adds to the cost of health care and can envelop the patient in a confusing and usually highly bureaucratic system of health care delivery. One result is that people who develop a major illness often report that dealing with the health care system can be as difficult as confronting the disease.

Excessive bureaucratization adds to the alienation of patients, making them feel that their own health is largely out of their control. Long waits for medical attention are normal, even in the emergency room. Or when facing a major disease, much time can pass as patients wait for appointments with one expert or another. Meanwhile, the disease persists. Specialization and bureaucratization also mean that the health care system is burdened by endless forms for both physician and patient, including paperwork to enter individuals into the system, authorize procedures, dispense medicines, monitor progress, and process payments.

The Role of Government in Medicine

At the heart of many current public discussions about health care is the role of government in health care. The U.S. government has for some time sought to have some form of widespread guaranteed health service, at least for certain categories of people, such as veterans, the poor, and the elderly. The **Medicare** program, begun in 1965 under the administration of President Lyndon B. Johnson, provides medical care in the form of medical insurance covering hospital costs for all individuals who are age 65 or older. The Medicare program does not cover physician costs incurred outside the hospital, but there are programs available that do so, although the patient usually must pay a portion of the cost.

Medicaid is a governmental program that provides medical care in the form of health insurance for people of any age who are poor, on welfare, or disabled. The program is funded through tax revenues. The costs covered per individual vary from state to state because the state must provide funds to the individual in addition to the funds that are provided by the federal government. The Medicare and Medicaid programs together are as close as the United States has come to the ideal of universal health insurance. More recent attempts at the establishment of universal coverage have not been successful to date. Health care issues, such as a patient's bill of rights, the cost of prescription drugs, and the availability of health insurance and care continue to be dominant themes in national politics (Weitz 2001; Starr 1995).

At the core of these discussions are different perspectives on the role of government in health care. Opponents of universal health care argue that government regulation will lead to greater bureaucratization in the

MAPPING AMERICA'S DIVERSITY

MAP 20.1 Persons Not Covered by Health Insurance

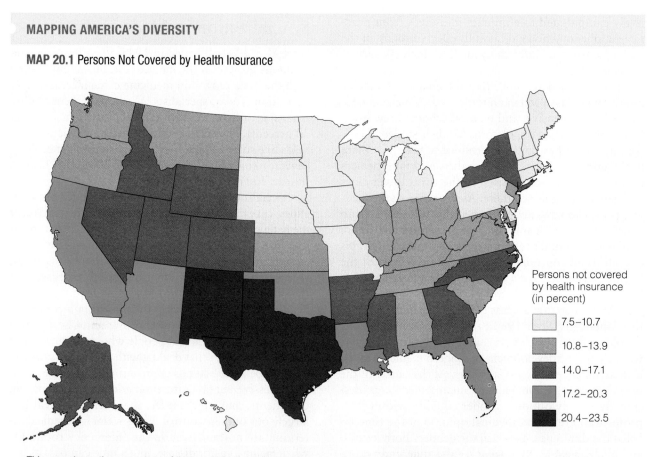

Persons not covered by health insurance (in percent)

- 7.5–10.7
- 10.8–13.9
- 14.0–17.1
- 17.2–20.3
- 20.4–23.5

This map shows the percentage of the population in each state that was not covered by health insurance in 2001. Texas has the highest percentage of people uncovered (23.5 percent), followed by New Mexico (20.7 percent), then California (19.5 percent). Iowa has the lowest percentage of those not covered by health insurance, followed by Rhode Island and Wisconsin (both at 7.7 percent uncovered). Were you to design a study to explain state-by-state variation in coverage, what explanatory factors would you want to examine?

Data: From the U.S. Census Bureau. 2004. *Statistical Abstract of the United States, 2003.* Washington, DC: U.S. Department of Commerce, p. 114.

health care system and lack of individual choice about providers. Proponents argue that the current system of private health care benefits primarily the more affluent and, of course, private companies that control access to health care. What role government will play in the future in insuring and serving the different health needs of the U.S. population remains to be seen.

Corporate Control of Health Care

There is little doubt that corporate interests are increasingly dominating how medicine is practiced. From pharmaceutical companies to hospitals, health care is increasingly being delivered in an organizational form where corporate interests dominate. This has profound implications for both patients and health care providers. Physicians have to be as concerned with maintaining a profitable business as they are with patients and the provision of care. Patients may feel overwhelmed by the complexity of dealing with business interests when they are facing a serious illness. And everyone complains about decisions being run by corporate and insurance regulations, instead of by patients and their doctors.

As medical care has become increasingly based in a corporate system, physicians may become as concerned with legal regulations as they are with preventative care. Indeed, doctors report feeling a loss of medical autonomy as they have had to respond to increased regulation. Some even argue that the elaboration of corporate rules encourages fraud as health care workers attempt to subvert the corporate control (Draper and Karst 1999).

As health care corporations have become larger, the management of health care is also increasingly done by those with a background in business, not medicine. Thus, the corporate structure of health care emphasizes productivity, cost efficiency, and rationalization (McKinley and Stoeckle 1998; McKinlay and Marceau 2002). Such managerial needs may actually conflict with physicians' work as medical practices have to manage staff schedules, coordinate large staffs, contract with third-party purchasers and providers, and deal with insurance regulations. Because corporations are modeled on a profit-based system, there are also consequences for health care workers, many of whom are among the lowest paid workers in the occupational system (U.S. Department of Labor 2004; Scherzer 2002).

Given the corporate structure of most U.S. social institutions, it is unlikely that this trend will reverse. Analysts project that this will mean increasing regulation, continued importance of the "bottom line" over quality health care, and less control by medical professionals over the delivery of health care.

Theoretical Perspectives on Health Care

A deeper understanding of the nation's health care system and its problems can be extracted by applying the three major theoretical paradigms of sociology: functionalism, conflict theory, and symbolic interaction theory.

The Functionalist View of Health Care

Functionalism argues that any institution, group, or organization can be interpreted by looking at its positive and negative functions in society. Positive functions contribute to the harmony and stability of society. The positive functions of the health care system are the prevention and treatment of disease. Ideally, this would mean the delivery of health care to the entire population without regard to race–ethnicity, social class, gender, age, or any other characteristic. At the same time, the health care system is currently notable for a number of negative functions, ones that contribute to the disharmony and instability of society.

Functionalism also emphasizes the systemic way that various social institutions are related to each other—together forming the relatively stable character of society. You can see this with regard to how the health care system is entangled with government, through such things as federal regulation of new drugs and procedures. The government is also deeply involved in health care through scientific institutions such as the National Institute of Health—a huge government agency that funds new research on various matters of health and health care policy. As a social institution, health care is also one of the nation's largest employers and, thus, is integrally tied to systems of work and the economy.

Functionalist theory primarily examines the institutionalized system of health care and studies how different institutional forms of health care benefit society as a whole.

The Conflict Theory View of Health Care

Conflict theory stresses the importance of social structural inequality in society. From the conflict perspective, the inequality of available resources inherent in our society is responsible for the unequal access to medical care. Minorities, the poor and working class, and the elderly, particularly elderly women, have less access to the health care system in the United States than

Whites, the middle and upper classes, and the middle-aged. To the contrary, functionalists argue that relatively greater access of the middle and upper classes to medical care is good for society, because the upper classes are more beneficial ("functional") to society.

Conflict theorists are also interested in how illness and death are distributed across the various groups in society. Although heart disease and cancer are the most likely causes of death for both women and men, the third leading cause of death for men is unintentional injury. Gender patterns in the likelihood of disease are further exacerbated by race. Among Native American, Asian American, African American, and Latino men, homicide is one of the ten leading causes of death. In fact, for Latinos and African American men, it is the fifth leading cause of death. Thus, conflict theorists are interested in how inequality is related both to the likelihood of death and disease, as well as to the likelihood of treatment.

In addition, conflict theorists are especially critical of the corporate control of health care and associate the drive for corporate profits with the rising costs of health care. Conflict theorists would also examine inequality within health care employment patterns. Health care institutions employ some of the highest paid professionals, but they are also workplaces where various service workers (such as orderlies, cooks, etc.) are among the least prestigious and lesser paid occupations.

Symbolic Interaction and Health Care

Symbolic interaction theory holds that illness is, in part, socially constructed. For example, the definitions of illness and wellness are culturally relative. That is, sickness in one culture may be considered wellness in another culture. Sickness can be time-dependent as well.

A physical condition considered optimal in one era may be defined as sickness at another time in the same culture. At the turn of the twentieth century, a healthy woman was supposed to be plump; a thin woman would be suspect of being unhealthy. Similarly, the health care system itself has a socially constructed aspect. The ways we behave toward the ill, toward doctors, and toward innovative ventures such as health maintenance organizations (HMOs) are all social creations.

One way to think about a symbolic interactionist interpretation of illness is the concept of the **sick role**—a pattern of behavioral expectations defined as appropriate for one who is ill. Just as people play other roles (as men, as mothers, as teachers), so might they play the sick role in particular ways. Some people may engage in a sick role even when they are perfectly healthy, even coming to believe in the part they play. This would be one way to explain hypochondria—the belief that one is sick even when not. Or when actually sick, different people play sick roles in different ways—some dramatizing their condition, others withdrawing.

DEBUNKING **SOCIETY'S MYTHS**

Myth: Illness is strictly a physical thing.

Sociological perspective: What is defined as illness, even physical illness, is heavily influenced by culture. Also, the likelihood of experiencing a particular illness is strongly influenced by social factors, such as gender, race, and class (Ruzek, Olesen, and Clarke 1997).

People also treat sick people in specific ways—something that symbolic interactionists find fascinating. Medical practitioners frequently subject patients to *infantilization*, treating them like children—even adult patients—and talking to them with "baby talk." The patient is assigned a role that is heavily dependent on the physician and the health care system, much as an

Table 20.1 Theoretical Perspectives on the Sociology of Health	Functionalism	Conflict Theory	Symbolic Interaction
Central Point	The health care system has certain functions, both positive and negative.	Health care reflects the inequalities in society.	Illness is partly socially constructed.
Fundamental Problem Uncovered	The health care system produces some negative functions.	Excessive bureaucratization of the health care system and privatization lead to excess cost.	Patients are patronized and infantilized.
Policy Implications	Policy should decrease negative functions of health care system for minority groups, the poor, and women.	Policy should improve access to health care for minority racial–ethnic groups, the poor, and women.	Doctors, nurses, and other medical personnel should periodically take the sick role of the patient, as an instructional device.

infant is dependent on its parents. Doctors and nurses may begin patronizing the patient from the initial greeting, a condescending, "How are *we* today?" Physicians will commonly address women patients by their first names more often than men patients, yet all patients virtually always address physicians as "Doctor." The patronizing and infantilizing of patients is common in emergency rooms, where minority patients are infantilized even more often than others (Weitz 2001; González 1996).

Symbolic interaction also studies how the interaction between a physician and patient can be "managed." During a gynecological exam, for example, women may feel uncomfortably vulnerable when they lie partially naked on an examination table with their heels in elevated stirrups and their legs open. Strong social norms say that when a man touches a woman's genitals, it is an act of intimacy. Yet the gynecological examination is supposed to be completely impersonal. How gynecologists negotiate this apparent contradiction and construct the gynecological exam as "normal" would be a subject for a symbolic interaction study (Scully 1994; Emerson 1970; see Table 20.1).

Together, the different theoretical perspectives in sociology each contribute unique ideas to understanding different dimensions of health and health care. No one is complete, but all inform the sociological study of health care behavior and institutions.

Health, Diversity, and Social Inequality

Generally speaking, the citizens of the United States are quite healthy in relation to the rest of the world. However, there are wide discrepancies among people in the United States in terms of longevity, general health, and access to health care, with the least advantaged people including primarily minorities, the lower classes, and for a number of ailments, women.

Prominent among the problem areas in the U.S. health care system are

- **unequal distribution of health care by race–ethnicity, social class, or gender.** Health care is more readily available and more readily delivered to White or middle-class individuals in urban and suburban areas than to minorities and the poor. Particularly serious is the lack of health care delivery to Native American populations. On average, women tend to receive a lesser quality of health care than men, even though they tend to utilize the health care system more.
- **unequal distribution of health care by region.** Each year, many people in the United States die because they live too far away from a doctor, hospital, or emergency room. Medical offices and hospitals are concentrated in cities and suburbs. They are much less likely to be located in isolated rural areas. Rural people in Appalachia and some parts of the South and Midwest may travel a hundred miles or more to get to the doctor or an emergency room.
- **inadequate health education of inner-city and rural parents.** Many inner-city and rural parents do not understand the importance of immunizing their children against smallpox, tuberculosis, and other illnesses, and they are often suspicious of immunization programs. This hesitancy is reinforced by the depersonalized and inadequate care inner-city and rural residents often encounter when care is available at all.

Health can be affected by personal factors, such as dietary and hygienic habits, but what interests sociologists the most is how health is shaped by the social characteristics of human groups. Factors like one's race, class, and gender—as well as age, as we saw in Chapter 14, can have as great an impact on health as the dietary and health habits of given individuals.

DEBUNKING **SOCIETY'S MYTHS**

Myth: The health care system works with the best interests of clients in mind.

Sociological perspective: The health care system is structured along the same lines as other social institutions, thus reflecting similar patterns of inequality in society.

Race and Health Care

Being in a racial or ethnic minority in the United States has a great influence on one's physical and mental health. This is so not only in the diseases people experience, but also in a group's access to the health care system. You can see the impact of race easily by looking simply at the dramatic differences in life expectancies for White Americans compared with other groups (as well as in differences in life expectancies between men and women). White men can now expect to live to 75 years of age (on average), whereas African American men have a life expectancy of only 68.6 years. White women can expect to live longer than White men, 80.2 years. African American women can expect to live more than seven years longer than African American men, 75.5 years, which is the same life expectancy as White men but noticeably less than White women (see Figure 20.1). Even medical researchers note the strong effect that race, especially when in combination with living in a poor community, has on health. In a study of health in poor, Black communities, the *New*

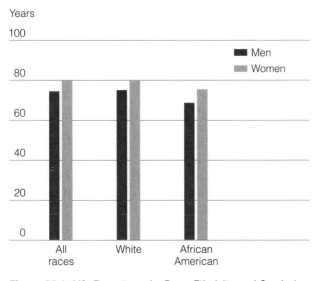

Figure 20.1 Life Expectancy by Race, Ethnicity, and Gender*

*Data not available for other groups.
Sources: National Center for Health Statistics. 2002. *National Vital Statistics Report*,
Vol. 50, No. 6, March 21, 2002. Website: **www.cdc.gov/nchs/fastats/lifeexpec.htm;**
National Center for Health Statistics. 2004. *Health United States 2003.* Washington, DC:
U.S. Department of Health and Human Services.

England Journal of Medicine has concluded that socio-economic characteristics of communities—which would include income, education, occupation, and race—are related to the incidence of disease, death, and, in particular, coronary risk due to stress (Marmot 2001; Roux et al. 2001).

Among women, African Americans are more likely than Whites to fall victim to diseases such as cancer, heart disease, stroke, and diabetes. Death of the mother for complications during pregnancy or childbirth is four times higher among African American women than among White women. And though White and Black women are about equally likely to get breast cancer, the death rate for breast cancer among African American women is considerably greater than for White women. This reflects the fact that White women are more likely to get high-quality care, and get it more rapidly, than African American women. Alaskan Native, American Indian, Asian American, and Hispanic women are less likely to die from breast cancer than either Black or White women (National Center for Health Statistics 2003).

Indian Health Services has the federal responsibility for the health of the Native American population, but because more than half of American Indians do not live on reservations, they do not have access to these services. Overall, the health of Native Americans is poorer than for other groups. Geographic isolation, poverty, and suspicion of the medical profession toward traditional healing practices are all said to contribute toward the poorer health of Native Americans (Office of Minority Health 2004).

Hispanics, like African Americans, Native Americans, and other minorities, are significantly less healthy than Whites. For example, Hispanics contract tuberculosis at a rate seven times that of Whites. The other indicators of health, such as infant mortality, reveal a picture for Hispanics similar to that of African Americans and Native Americans. Hispanics are less likely than Whites to have a regular source of medical care, and when they do, it is likely to be a public health facility or an outpatient clinic. Because of language barriers as well as other cultural differences, Hispanics are less likely than other minority groups to use available health services, such as hospitals, doctors' offices, and clinics (National Center for Health Statistics 2003).

One of the challenges for the health care system in a society with increasing diversity is responding to the different cultural orientations of various groups in society. Immigrants, for example, who may have limited English language skills and may come from cultures with very different health care practices, may be especially confused by the practices within the U.S. health care system (Suro 2000). Developing the ability for greater cross-cultural administration of health care will likely continue to be a challenge in the future. Also, because the majority of physicians are White, patients of a different racial or ethnic background will likely feel some social distance between themselves and the health care provider, contributing to a reluctance to seek care (Malat 2001). At the same time, researchers find that patients are much more satisfied with their care when their physician is of the same race as they are (Laveist and Nuru-Jeter 2002). Such findings indicate the continuing significance that race and color have in shaping people's health.

Social Class and Health Care

In the United States, social class has a pronounced effect on health and the availability of health services. Health depends in no small part on wealth. In general, the higher one's social class, the longer one will live (Jacobs and Morone 2004). Likewise, the lower the social class status of the person or family, the less access they have to adequate health care. People with higher incomes, when asked to rate their own health, also tend to rate themselves as healthier than people with lesser incomes. The effects of social class are nowhere more evident than in the distribution of health and disease, showing up dramatically in the rates of infant mortality, stillbirths, tuberculosis, heart disease, cancer, arthritis, diabetes, and a variety of other illnesses. The reasons lie partly in personal habits that are themselves partly dependent on one's social class. For example, people with lower socioeconomic status smoke more, and smoking is the major cause of lung cancer and an important contributor to cardiovascular disease.

Social circumstances such as poor living conditions, elevated levels of pollution in low-income neighborhoods, and lack of access to health care facilities all contribute to the high rate of disease among lower social classes. Another contributing factor is the stress caused by financial troubles. Research has consistently shown correlations between psychological stress and physical illness (Worchel et al. 2000; Taylor et al. 2000; Jackson 1992; Thoits 1991; House 1980). The poor are more subject to psychological stress than the middle and upper classes, and it shows up in their comparatively high level of illness. Hypertension, depression, and stress all contribute to poorer health.

With the exception of the elderly, now subsidized by Medicare, low-income people also remain largely outside the mainstream of private health care. Nearly 41 million Americans—14.5 percent of the population—have no health insurance at all (see Map 20.1, "Mapping America's Diversity: Persons Not Covered by Health Insurance" and Figure 20.2). The main sources for medical care for people without insurance are hospital emergency rooms and health department clinics, resulting in vastly overcrowded emergency rooms in the inner cities, often called the "doctor's office of the poor." In emergency rooms, treatment is given only for specific critical ailments and rarely is there any follow-up care or comprehensive treatment.

Sociologists have found that during interactions between health care providers and poor patients, Black and Hispanic poor in particular are likely to be *infantilized* and to receive health counseling that is incorrect, incomplete, or delivered in inappropriate language not likely to be understood by the patient (Weitz 2001; González 1996). The symbolic interaction perspective has also noted that this tendency has been attributed to an attitude among health counselors that the poor are charity cases who should be satisfied with whatever they get because they are probably not paying for their own care (Diaz-Duque 1989).

The connection between health and social class does not fall solely on the poor; however, Middle and working class patients also worry about the effect that medical emergency will have on household finances. A major disease, accident, disability, or other health emergency can rapidly deplete a family's finances and leave even relatively well-off families with huge medical bills and depleted financial margins.

Gender and Health Care

Gender also affects people's health, with women and men experiencing different effects of their gender status. On average, women live longer than men, although older women are more likely than older men to suffer from stress, overweight, hypertension, and chronic illness. Hypertension is more common among men than women until age fifty-five, when the pattern reverses. This may reflect differences in the social environment experienced by men and women, with women finding their situation to be more stressful as they advance toward old age. Under the age of thirty-five, men are more likely to be overweight than women; after that age, women are more likely to be overweight. Women also have a higher likelihood of developing chronic disease than men, although men are more likely to be disabled by disease (National Center for Health Statistics 2003).

Researchers cite differences in male and female roles and cultural practices to explain the variation in life expectancy between women and men. Traditionally, men's occupational roles call for more travel and more exposure to other people, both major sources of infection. Men also smoke more, and despite what tobacco executives often dispute, cigarette smoking definitely causes cancer and cardiovascular disease. In recent years, though, the proportion of smokers who are women

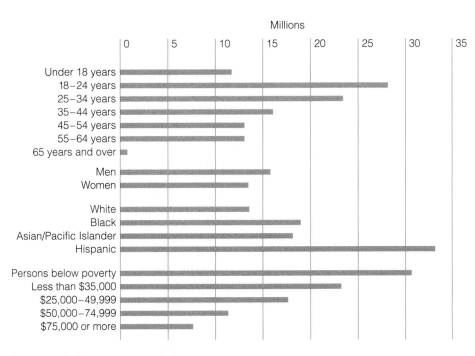

Figure 20.2 Persons Without Health Insurance

Source: U.S. Census Bureau. 2004. *Statistical Abstract of the United States, 2003.* Washington, DC: U.S. Printing Office, p. 114.

has increased, with a convergence in the smoking rates of men and women (National Center for Health Statistics 2003). Men are also more likely than women to quit smoking. These changes in smoking rates are expected to appear in the future as higher death rates for females from diseases linked to smoking.

It is typically assumed that the work-oriented, hard-driving lifestyle associated with men's traditional role as the primary source of financial support tends to produce elevated levels of heart disease and other health problems. In general, this is true. However, the role of women in society is changing, with associated changes in women's health. The health of women and men varies with their social circumstances. People who are "tokens" in the workplace—primarily women and Blacks, and especially Black women—suffer more stress in the form of depression and anxiety than "nontokens," or women and Blacks who work in places where there is nothing exceptional about their presence (Jackson et al. 1995). Research has also found that perceiving that there is discrimination against you is significantly related to both physical and mental health (Pavalko et al. 2003). Those who experience discrimination by both gender and race are therefore the most vulnerable.

THINKING SOCIOLOGICALLY

You have seen here that illness is related to social factors. Try observing this the next time you are in a medical office (outside a college campus, preferably). Who is in the waiting room? How many men and how many women? What is the racial composition of those who are waiting? Are there any indications that the group is relatively homogeneous with respect to social class, as indicated by dress, manner, or speech? What do your observations tell you about *social inequality* and health care?

Global Dimensions of Health

Despite the health problems faced in the United States, the fact remains that we are a healthy nation relative to many other nations around the world. In societies that are largely pastoral or horticultural, the lack of technology severely limits both the development of health care systems and the delivery of health care to the populations of these countries. Such is the case in many third-world countries, where poverty also limits access to good health care. As in the United States, elites in third-world countries receive better medical attention than people who are poor and possibly even near starvation.

There is some evidence that the degree of income inequality within a nation also affects the overall health of the public. In other words, the more unequal a country's wealth distribution, the less healthy its population

will be (Wilkinson 1992). As income inequality within a nation increases, so do rates of infant mortality, lack of health care insurance, more low-birth-weight babies, and so forth (Kaplan and Lynch 1997). Thus, inequality affects people's health both directly (in the form of stress and disease) and indirectly (in the form of income inequality impairing public health).

Abject poverty in much of the world, particularly in third-world regions and areas of extreme poverty in the United States, cuts life expectancy to far below the 75 to 80 years experienced by the wealthy, elite, and middle-class populations in the United States and elsewhere. Life expectancy in the African countries of Niger, Chad, Ethiopia, and Somalia is barely 45 years of age. It is the same in Afghanistan and in Laos and Cambodia. Serious illness goes hand in hand with severe poverty, and lowered life expectancy is the result. Such is also the case in many ghettoes and barrios in the United States, although considerably less is written about such conditions. Drinking water contaminated by poor sewage breeds many infectious diseases that kill. The diseases of tuberculosis and pneumonia, diseases that were the leading causes of death in the United States nearly a century ago, today kill people in poorer areas of the world (Lenski et al. 1998).

Doctors and other medical personnel are rare in impoverished areas; thus, the world's poorest people, in dire need of medical care, may never be seen by a physician. There are fewer than five doctors per 100,000 people in African countries such as Niger, Chad, and Ethiopia, Guyana in South America, and Guatemala in Latin America. As a result, infant mortality is high, and 10 percent of the children in poor societies die within one year of birth. Health care and poverty are intimately related: Poverty breeds disease, which makes it less likely for people to be able to work, which in turn produces more poverty and more disease and premature death. The absence of health care professionals and facilities only exacerbates this vicious cycle of poverty and disease.

Social Patterns of Health and Disease

Epidemiology is the study of all the factors—biological, social, economic, and cultural—associated with disease in society. **Social epidemiology** is the study of the effects of social, cultural, temporal, and regional factors in disease and health. Social factors affect any number of diseases, including who is likely to get them, who is likely to survive them, and how they are treated. In this section, we examine some select phenomena that show the social epidemiology of various aspects of health.

VIEWING SOCIETY IN GLOBAL PERSPECTIVE

MAP 20.2 World Infant Mortality

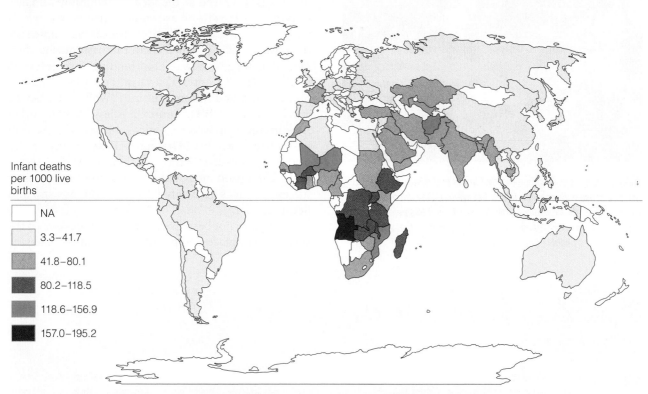

Infant deaths
per 1000 live
births

	NA
	3.3–41.7
	41.8–80.1
	80.2–118.5
	118.6–156.9
	157.0–195.2

The infant mortality rate is computed as the number of deaths of children under one year of age per 1000 live births in a given year. Infant mortality is strongly correlated with poverty rates in a given country, although within any country, the rate can vary among different populations. In the United States, for example, the infant mortality rate among inner-city, poor African Americans exceeds that in many third world countries, although you cannot see that variation on a map such as this that reports only country-wide rates. Which four countries shown have the highest rate of infant mortality? Name four of the lowest. What do you think can be done to improve the survival rate of infants in those countries where mortality is high?

Data: From the U.S. Census Bureau. 2004. *Statistical Abstract of the United States, 2003.* Washington, DC: U.S. Department of Commerce, p. 847.

Eating Disorders and Obesity

From the early 1900s until the mid-1940s, thinness was associated with poverty and hunger. If you were skinny, that meant you were in bad health. From the late 1950s through the present, however, a positive value has been placed on being thin. Role models in our society, such as movie stars and fashion models, have firmly established that thin is "in," especially for young women. Millions of young women have attempted to copy the slender, high-fashion look. An increased incidence of *anorexia nervosa* has been one result (Weitz 2001; Williams and Collins 1995; Taylor et al. 2002).

Anorexia nervosa, or anorexia for short, is an eating disorder characterized by compulsive dieting. Vic-

tims of this illness starve themselves, sometimes to death. While suffering from this disorder, they do not typically define themselves as ill, because they tend to see themselves as overweight, even though they are dangerously thin. A related malady, **bulimia,** is an eating disorder characterized by alternating between binge eating and purging (induced vomiting) to avoid gaining weight (see also the box, Doing Sociological Research: "Beauty and Health").

Like many other diseases, anorexia and bulimia have social as well as biological causes. A majority of people suffering from the disease are young, White women from well-to-do families, most often two-parent families. Many behavioral scientists have noted that anorexics have generally been pressured excessively by

© Nina Berman/Sipa Press

Anorexia nervosa is common mostly, but not exclusively, among young women. Those with anorexia falsely believe that they are overweight, although they are exceedingly thin. Many interpret this potentially life-threatening condition as encouraged by a culture that advertises being thin as a beauty ideal for women.

their parents to be high achievers. Other scientists have detected a link between anorexia and the socially constructed ideals of beauty in our society, in which slenderness is displayed as the ideal of femininity. Images of bodily "perfection" are emblazoned across television, magazines, and billboards. Researchers note that these social values, which encourage compulsive dieting, are comparable to the footbinding once practiced in China (see Chapter 1) and other forms of female mutilation found in some cultures (Wolf 1991; Levine 1987; Chernin 1991). Anorexia is less likely to afflict African American women, Latinas, and lesbians. According to one researcher (Thompson 1994), many people in these groups overeat, rather than self-starve, something Thompson interprets as a reaction to oppressive life experiences associated with racism, sexism, and homophobia.

Men have not been exempt from the pressure of the social values placed on physical appearance. Since the

DOING SOCIOLOGICAL RESEARCH

Beauty and Health

Beauty is only skin deep, so the saying goes. This implies that beauty is solely a matter of appearance, that is, how we appear to others. Certainly many people spend an enormous amount of time and money trying to appear beautiful. Some such efforts are harmless, but can beauty affect your health? This is the question asked by Jane Sprague Zones in her research on beauty myths and women's health.

Zones identifies the beauty myth as the ideals set for women's appearance in Western culture—ideals that increasingly pressure women to conform to narrow standards. The greatest impact of internalizing such narrow expectations is on women's self-esteem. Focusing too much attention on one's weight, fitness, and physical appearance can lead women never to be satisfied with their bodies and to have nagging self-doubts about their appearance. With models appearing as petite figures, size six or smaller, women may be driven to starve themselves or to take drastic measures to reduce their weight. Thus, not only is internalizing the beauty myth harmful to one's self-esteem and mental health, but it also can be harmful to your physical health.

Likewise, trying to appear eternally youthful may lead you to buy endless products promising to eliminate wrinkles. Or you may dye your hair, chisel your nose, minimize or maximize your breast size through surgery, or use numerous products promising eternal youth. The Consumer Product Safety Commission has noted many consumer-related health problems from the use of some cosmetics. And the National Cancer Institute has documented a greater risk of cancer of the lymph system among women who use hair coloring. The damage is not only to individuals. Hydrocarbons in hairspray are a primary contributor to pollution in the air and ground, thereby affecting the health of the earth.

Restricted images of beauty can not only harm one's physical and mental health, but also can lead to labeling people with visible physical disabilities as unattractive. Likewise, not being able to afford the array of products associated with being beautiful leads people to make invidious class distinctions about who is beautiful and who is not.

Zones concludes that appearance and beauty are powerful forces in

women's lives and that both personal and societal transformation are needed to overcome the deleterious impact of narrow beauty ideals on women's health.

Questions to Consider

1. To what extent are men affected by beauty standards and by the advertising of male "beauty" products? *Keywords: beauty myth, men and appearance*

2. To what extent is women's health harmed by society's attention to beauty? *Keywords: anorexia, beauty and health, culture of thinness*

We have included InfoTrac College Edition keywords at the end of each question to make it easier for you to find more to read on these topics. Go to www.infotrac-college.com, an online library, to begin your search.

Source: Zones, Jane Sprague. 1997. "Beauty Myths and Realities and Their Impact on Women's Health." Pp. 249–275 in *Women's Health: Complexities and Differences,* edited by Sheryl B. Ruzek et al. Columbus, OH: Ohio State University Press. •••

1940s and especially from the mid-1970s on, a persistent male physical ideal has been the rippling physique of the body builder or weight lifter. Young men have been urged by the media and their peers to "pump iron" for the perfect body, thus presumably to gain pride, muscle mass, and the adoration of women (Logio 1998).

Many athletes, professional and amateur, have been goaded by the competition for physical size to use **anabolic steroids,** powerful hormones that stimulate the growth of muscle. Used widely (despite dire warnings by physicians), steroids will build muscle as advertised, but they can also shrivel the testicles and cause impotence, hair loss, heart arrhythmia, liver damage, strokes, and very possibly some forms of cancer. Lyle Alzado was a huge, fast, monstrously strong football star in the National Football League and later a Hollywood actor. As he slowly succumbed to brain cancer in his thirties, Alzado freely admitted to prolonged abuse of anabolic steroids, which he believed caused his fatal disease.

More recently, obesity also has become defined as a public health problem. Obesity has traditionally been considered a matter of individual habits, but in 2004 the officials in the federal Medicare program changed their policy to include obesity as a disease (and thus various weight loss programs are now eligible for Medicare funding). This is a good example of how something once defined as a matter of individual habit can become subsumed under the **medical model**—that is, a framework that interprets some condition as purely a physiological or medical issue.

As obesity has become defined as a medical problem, analysts tend to emphasize the physiological causes and consequences of being overweight. But a sociological perspective on obesity would also look to cultural factors as a source of obesity. One might note the increasing reliance on fast food, which is high in fat and calories and generally low in nutrient value. Further, the increased consumption of fast food follows from a lifestyle where people are hurried, overworked, and seldom take the time to prepare meals with fresh ingredients. Business interests also promote the consumption of more and more fast food, and one must also note that having an overweight population is a sign of the relative affluence of the population.

Furthermore, patterns of being overweight also vary by the same social factors that predict other forms of good and bad health. Under the age of thirty-five, men are more likely than women to be overweight. Hispanic men are more likely overweight than other men, and Black women are the race–gender group most likely to be overweight. Weight is also related to educational attainment, with those with less education being far more likely to be overweight than those with higher educational attainment. Education and income are also related to physical inactivity, with less educated and poorer people more likely to be physically inactive than others (U.S. Census Bureau 2004). In other words, although obesity is now being interpreted as a medical problem, a sociological perspective reveals much about the social origins of this phenomenon.

Smoking and Tobacco

The Centers for Disease Control estimate that about 450,000 people die each year as a direct result of smoking—representing 20 percent of all deaths. Nonsmokers exposed to secondhand cigarette smoke have a higher risk of smoking-related disease, including death, than nonsmokers who are not exposed. In fact, a study of more than 32,000 healthy women who never smoked found that regular exposure to other people's smoking doubled the risk of heart disease for the nonsmoker.

Smoking and the tobacco industry have become a major national issue. When tobacco companies began directing smoking campaigns at children, using commercials and billboards (such as the "Joe Camel" campaign of several years ago), the federal government passed legislation forbidding such advertising. Especially targeted were ad campaigns that carried special appeal to women and girls, as well as ad campaigns centered in urban areas with large minority populations. Furthermore, state laws mandating a smoke-free environment

The national campaign to stop smoking has often focussed on young people—a group highly susceptible to media representations.

EPA/MICHAEL REYNOLDS/Landov

in restaurants, waiting rooms, and all professional buildings, have been enacted on a large scale, including the entire state of California. The effect has been to lower the overall rates of smoking in the United States, although cigar smoking has shown a recent increase. To compensate for lost sales resulting from the antismoking trend in the United States, tobacco companies are advertising and selling more in other countries—particularly in third-world countries with low average incomes—where there is little governmental regulation of tobacco advertising and sales.

AIDS and Sexually Transmitted Diseases

Approximately fifty sexually transmitted diseases have been medically diagnosed. The four major STDs are syphilis, gonorrhea, genital herpes, and AIDS. Less frequent are the more esoteric diseases such as lymphogranuloma venerium (LGV), which devastates the body with open sores and has no known cure. The incidence of all STDs increased during the sexual revolution, when sex was thought of simply as a pleasant way to express affection. Little fear was harbored about contracting a venereal disease because most, such as syphilis or gonorrhea, were known to be medically curable, and the remaining diseases (such as LGV) were thought of as too rare to worry about. With the subsequent dramatic rise in STDs, especially in AIDS, the late 1980s and early 1990s witnessed a new revolution in contrast to the sexual revolution. The new counterrevolution caused people to reevaluate the nature of sex, the possibly fatal risks involved in unprotected sexual activity, and sexual behavior in general (Laumann et al. 1994; Alan Guttmacher Institute 1994).

Syphilis and gonorrhea have been around for a long time. Both are caused by microorganisms transmitted through sexual contact involving the mucous membranes of the body. It is virtually impossible to get either syphilis or gonorrhea any other way. If untreated, syphilis causes damage to major body organs, blindness, mental deterioration, and death. Each result was seen all too dramatically in the infamous Tuskegee "experiment" done on a group of Black men in the twentieth century. Untreated gonorrhea can cause sterility in both women and men. Both diseases are quickly curable with penicillin or other appropriate antibiotic medication.

Genital herpes (Herpes Simplex II) is more widespread than either syphilis or gonorrhea and affects roughly 30 million people in the United States alone. That represents one person in seven. Although genital herpes can remain dormant for the life of the infected individual, it is nonetheless to date, incurable. Genital herpes began to get attention in the early 1980s, when concern with the risks of STDs generally was on the rise. A person with genital herpes may have no symptoms or may experience blisters in the genital area and a fever as well. Genital herpes is not fatal to adults, but it can be fatal to infants born through vaginal delivery, but not via cesarean section.

People tend to regard sexually transmitted diseases as not merely diseases but as punishment for being immoral. As a result, people who contract an STD become negatively stigmatized, sometimes resulting in their not seeking treatment. This shows the power that social influences can have on the treatment of disease.

A **stigma** occurs when an individual is socially devalued because of some malady, illness, misfortune, or similar attribute (see Chapter 7). A stigma is viewed as a relatively permanent characteristic of the stigmatized individual; the negative attribute of the stigmatized person is expected to persist, with no cure in sight (Goffman 1963b; Jones et al. 1986; Epstein 1996). When the disease AIDS first appeared in the early 1980s, it was mostly associated with gay men and was heavily stigmatized. The result was that the federal government (during the Reagan administration) largely ignored this spreading problem and devoted little in research funds to identify its causes. The stigma associating AIDS with gay men and the resulting delay by the medical and scientific community in researching treatment likely cost many lives (Shilts 1988).

THINKING SOCIOLOGICALLY

Have you or has anyone you know ever contracted a *sexually transmitted disease*? What was your reaction? What do your observations tell you about *stigma*?

AIDS is the term for a category of disorders that result from a breakdown of the body's immune system. HIV, the virus that causes AIDS, was first identified in 1981. The incubation period between infection with HIV and the development of AIDS can stretch longer than ten years. Thus, one can be infected with HIV and yet not have full-blown AIDS for up to ten years. It is not the actual HIV infection that causes death; rather, it is caused by a complex of severe illnesses that thrive in the absence of a working immune system, such as pneumonia, certain cancers, and a number of other illnesses rare enough that their presence is judged to be diagnostic of AIDS. Since the 1980s, the disease has spread rapidly, with over 830,000 cases reported in the United States since 1981.

Over 34 million adults and children worldwide are infected with HIV, 14.8 million of whom are women. Among new cases of AIDS worldwide, half are women. The global AIDS epidemic among women is overwhelmingly the result of heterosexual contact—almost entirely so in Africa and South and Southeast Asia. Analysts have argued that the high rate of transmission to women worldwide results from women's financial dependence on men. Because of such dependence, women may have little control over when and with whom they have sex. Many women have to exchange sex for financial support and in highly patriarchal cultures, women are not expected, nor allowed, to make decisions about sex. If they refuse sex or request condom use, they risk abuse and violence—or they may be suspected of infidelity, which can also put them at great risk. These facts mean that treating the worldwide AIDS epidemic requires that an analysis of gender relations in different cultures must be a part of the solution to this health epidemic (World Health Organization 2000). World leaders have also severely criticized the United States for limiting its international funding of AIDS-prevention programs to only those that promote abstinence (Associated Press 2004).

The AIDS disease is transmitted through the exchange of bodily fluids, particularly blood and semen. A little more than one-third of all new cases of AIDS are the result of male-to-male sexual contact, followed by intravenous drug use (among both women and men). Although AIDS initially affected primarily White, gay men, now one-third of new AIDS cases are women, the largest share of whom acquire AIDS through heterosexual contact. The second most frequent cause of transmission for women and men is intravenous drug use (U.S. Census Bureau 2004). In the United States, AIDS has hit inner-city minority communities disproportionately hard. The contextual problems of poverty, poor health, inadequate health care, drug and alcohol abuse, and violence have to be taken into account in explaining the high rate of AIDS in these communities (see Figure 20.3). Because things like drug use vary by race,

social class, and gender, this points to different interventions for different communities.

AIDS can be understood in medical terms, and new treatment protocols promise to bring more effective treatment for those who contract the disease. But to reduce the incidence of AIDS requires a sociological perspective that takes into account the social networks and social norms that contribute to the transmission of this disease. Social norms affecting the age of first intercourse, number of sexual partners, drug use, and homosexual and bisexual sexual practices influence the spread of AIDS. There are in any given society groups who are more likely to use drugs, have sex with multiple partners, and engage in risky sexual behavior (such as not using condoms or other "safe sex" practices). A cavalier attitude toward AIDS can result in death; young people, in particular, may be the most vulnerable since they are more likely single, more likely to have multiple sexual partners, and are often relaxed about safe sex methods (see Figure 20.4).

Understanding social networks can also contribute to reducing the spread of AIDS. Sociologists suggest that because HIV/AIDS is primarily transmitted as a result of social activities (such as sexual relations and sharing of needles used for drug injections), the social networks and relationships in which transmission occurs are critical in understanding this disease. Networks may be so dense and interconnected, for example, that members interact only with each other, as in a well-defined homosexual or drug-injecting community. Or the network may be diffuse and loosely tied together, as with prostitutes and their occasional clients, or among young adults that date casually. The identification and characterization of these networks is critical for predicting

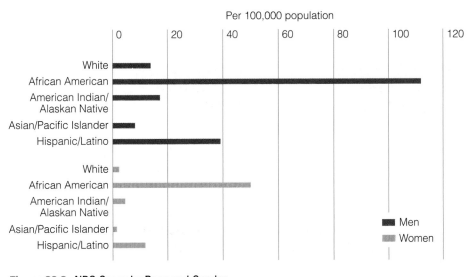

Figure 20.3 AIDS Cases by Race and Gender

Data: National Center for Health Statistics. 2003. *Health United States 2003.* Washington, DC: U.S. Department of Health and Human Services, p. 200.

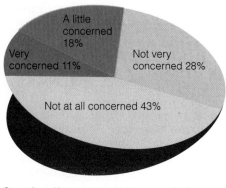

Question: How concerned are you that you, yourself, will get AIDS?

Figure 20.4 Teen Concerns about AIDS (ages 13 to 17)

Source: McMurray, Colleen. 2004. "Teens Still View AIDS as Urgent Health Problem." *The Gallup Poll.* Princeton, NJ: The Gallup Organization. Website: **www.gallup.com**

the spread of infection, focusing and shaping interventions. It was precisely this network analysis by sociologists at the CDC (Centers for Disease Control and Prevention) in the early 1980s that produced epidemiological mapping of gay men and helped target safe sex education to the most at-risk subgroups within that community (American Sociological Association 1993). As with other forms of illness, AIDS can be understood from a physiological perspective, but a socio-logical perspective adds to our understanding of the social origins and transmission of this terrible disease.

Disability

The disability rights movement has generated significant change in how society treats people with disabilities. Whereas the disabled once would have had little access to public facilities and to opportunities for work and education, the disability rights movement has challenged these old practices and defined the disabled as a social minority with rights—such as to public transportation, accommodations for various needs, and equal protection under the law.

The Americans with Disabilities Act, passed by Congress in 1990 (see also Chapter 18 and the box Understanding Diversity: "The Americans with Disabilities Act") protects disabled persons from discrimination and requires employers and other institutional providers (such as schools) to provide "reasonable accommodation" so that disabled people can be productive citizens. This can include various technologies to help people be mobile or able to read class assignments. The disability rights movement challenged the medical model of interpreting disability by redefining disability as another form of difference that society could accommodate. Thus, even the language people use to describe disability has evolved to terms with more positive con-

UNDERSTANDING DIVERSITY
The Americans with Disabilities Act

The Americans with Disabilities Act (ADA), passed by Congress in 1990, prohibits discrimination against disabled persons (see also Chapter 18). In 1999, the U.S. Supreme Court restricted the definition of disability to exclude disabilities that can be corrected with devices such as eyeglasses or with medication. The decision was based on cases in which people had been denied employment because they did not meet a health standard required by the employer, even though with correction, the standard was met. One case involved two nearsighted women denied jobs as airline pilots because, without glasses, their vision did not meet the airline company's standard of 20/40. With glasses, their vision was 20/20. Another case involved a man whose high blood pressure was above the federal standard for driving trucks. He was denied a job as a trucker—even

though with medication, his pressure was regulated to an acceptable level. The Court's decision denied these people the right to claim discrimination.

The decision raises several interesting sociological questions about disabilities and civil rights. Writing for the majority, Sandra Day O'Connor argued that the law requires people to be assessed based on each individual's condition, not as members of groups affected in a particular way. The disability rights movement would argue that disabled people are a minority group with certain civil rights. The issue being mediated by the courts involves competing definitions of what constitutes a disability. This point intrigues symbolic interactionists, who would note the role of socially constructed meanings in legal negotiations. Furthermore, as one lawyer who drafted the original ADA law has pointed out, according to

the original decision, someone may be disabled enough to be excluded from a job, but not disabled enough to claim discrimination.

This raises interesting questions about the rights of employers to establish physical and medical standards for certain jobs, even if those standards result in the exclusion from employment for some groups of people. You might ask yourself how you would balance the diverse and competing interests of disabled people, employers, and the courts if it were your responsibility to decide whether the exclusion of a disabled individual from a particular job constitutes discrimination. What are the implications of your argument for cases involving race or gender discrimination?

Additional Resources: Smith, Bonnie G., and Beth Hutchinson (eds.). 2004. *Gendering Disability.* Piscataway, NJ: Rutgers University Press. •••

The disability rights movement has opened up new opportunities to those who face the challenge of a disability.

notations than *disabled*. Some speak of the *physically challenged* or the *differently abled*—all terms that are seen as more empowering than what *disability* connotes.

Certainly anyone can become disabled. A sudden accident, a crippling disease, a congenital condition, a stroke or the simple fact of aging can leave someone with impaired abilities. But as you would predict from a sociological framework, the likelihood of disability is socially structured (see the box A Sociological Eye on the Media: "Images of Disability in the Media"). Black Americans are 50 percent more likely than White Amer-

icans or Hispanics to have chronic conditions that result in the limitation of activity. Women and men are about equally likely to be disabled. However, Whites are more likely than Black Americans and men more likely than women to be receiving Social Security when disabled—an indication of the inequality that pervades society (U.S. Census Bureau 2004).

Mental Illness

Mental illness is a complex phenomenon that also has social roots. Some (certainly not all) of what we define as mental illness in our society may owe as much to the labeling effect as to psychological or physical conditions. Certainly, how the mentally ill are perceived by the medical profession and the public depends in significant part on the label that is attached to various behaviors (Best 2001; Jones et al. 1986; Thoits 1985; Scheff 1984; Szasz 1974; Scott 1969). If we are told that someone is "weird," then we are likely to perceive the person's behavior as weird no matter what the person does.

A landmark study by David L. Rosenhan (1973) demonstrated how profoundly the definition of illness in our society is subject to the effects of labeling. Rosenhan and his graduate students contrived to get themselves admitted to a mental hospital. Working in alliance with a psychiatrist connected with the hospital, they pretended to hear voices and complain of hallucinations. They were successful in getting themselves

A SOCIOLOGICAL EYE ON THE MEDIA

Images of Disability in the Media

Advertising plays a huge role in the cultural definition of people. The media's emphasis on beauty and bodily perfection has long meant that disabled people have been excluded—or grossly stereotyped—in most advertisements and other media outlets. But with an increased consciousness on the part of marketers about the size of the disabled market, advertising images of disabled people have changed in recent years. No longer are disabled people simply ignored or portrayed as pitiful.

From a conflict theory point of view, you could argue that this is just another form of exploitation as profit-based businesses seek to expand their busi-

ness market. But, at the same time, as symbolic interaction theory would point out, ideas about people are socially constructed. If the media—with their power to produce cultural understandings of different groups—begin to change their portrayal of disabled people, then social perceptions of the disabled in society may change, and the stigma associated with disability can be lessened.

Those who have studied images of disabled people in the media observe that the most typical forms of disability seen in the media are wheelchair users and deaf people. And as with any other form of advertising, only the "pretty peo-

ple" are shown. Thus, advertisements showing people in wheelchairs often use able-bodied people to pose in the picture.

As you observe the media, to what extent do you think disabled people are shown in advertising? Are the images positive or negative? Do more positive images of disabled people enhance integration of disabled people into society, or does the use of pretty and handsome models only narrow the range of acceptance of the disabled?

Source: Haller, Beth A., and Sue Ralph. 2001. "Profitability, Diversity, and Disability Images in Advertising in the United States and Britain." *Disability Studies Quarterly* 21 (Spring). Website: **www.dsq-sds.org**

• • •

admitted to the hospital and thus became *pseudo-patients*. Once admitted, they ceased completely to fake symptoms of mental illness; that is, they simply acted normally. Nevertheless, they found that in their day-to-day interaction with doctors and psychiatric nurses, they were treated just like any of the other patients. The staff failed to recognize that the imposters were not ill.

During their stays, the researchers discovered that patients were routinely dehumanized and infantilized. In fact, the hospital staff routinely used "baby talk" to address the researchers and the other patients. Particularly illuminating was the fact that the *real* mental patients in the hospital began to suspect that Rosenhan and his students, the pseudopatients, were fakes who should not be in the hospital! The real patients were able to pick out the pseudopatients, but the professionals, trained to identify those who are mentally ill, were not able to do so. This and other studies lead sociologists to conclude that at least *part* of what is defined as mental illness in this society is socially constructed.

Mental illness is also not distributed uniformly through society, again revealing the social basis of this health problem. Poverty, because it is stress-producing, is significantly related to the likelihood of mental illness (McLeod and Nonnemaker 2000). Racial and gender discrimination also produce mental illness because those who experience discrimination often experience high rates of hypertension, depression, and other forms of mental disorders as a consequence. Interestingly, researchers have found that strongly identifying with one's racial–ethnic group can reduce the depression that comes from perceived discrimination. A strong ethnic identity, involvement in ethnic communities, and a general sense of ethnic pride are related to significantly lower rates of depression (Mossakowski 2003). In general, having a strong sense of self-esteem and self-worth lessens the chance that one will experience mental illness (Markowitz 2001). Low self-esteem is a problem that particularly affects women, especially teenage girls. Depression, anxiety, and possibly self-destructive behavior can result.

Euthanasia

Euthanasia is the act of killing a severely ill person as an act of mercy. There are two forms of euthanasia. Negative euthanasia, sometimes called passive euthanasia, involves the withholding of treatment ("pulling the plug") with the knowledge that doing so will produce the death of the patient, such as may be stipulated in a living will. The second form, positive or "active" euthanasia, involves killing the severely ill person who would otherwise live, though in constant pain, coma, or other extreme conditions, as an act of mercy. Posi-

tive euthanasia is for this reason often called "mercy killing." In most states, positive euthanasia is now considered a form of murder and thus is not permitted under law.

A pressing problem facing the medical profession and society is the issue of whether people have a right to die. Does the terminally ill individual have a right to take his or her own life? Does the terminally ill person have the right to instruct someone else, perhaps a loved one, to do so? What is the definition of "terminally ill?" The issue is a moral, ethical, and legal one.

Many thousands of men and women in the United States exist in a state that is often described as "brain dead." In such instances, the person has a flat EEG (electroencephalogram), indicating no brain activity, and has no reflexes or spontaneous breathing. Many others are suffering from some terminal illness that causes them great pain, and they wish to end their own lives. There is thus a dilemma of patient's rights, or the question of the person's right to take his or her own life, versus the principles of the medical profession, which dictate that all forms of medical care should be given to the patient to sustain life. Not only has this resulted in the ethical and legal question of just what constitutes death, but it has also caused the courts over the last two decades to wrestle with the question of whether a chronically ill patient who wishes to take his or her own life may legally do so.

One such case is that of Terry Schiavo, age 40, who more than a decade ago collapsed from a chemical imbalance brought on by an eating disorder. She was severely brain-damaged and in a vegetative state. Her husband, Michael Schiavo, wanted to remove her feeding tube thereby ending her life, as he says she would have wanted. But her parents objected on religious grounds and had the Florida legislature and Governor Jeb Bush intervene. The legislature passed a law, "Terri's Law," applying only to her and allowing the governor to have the feeding tube reinserted after the husband had a court order that he could have it removed. The Florida Supreme Court subsequently ruled the law unconstitutional, though her fate is still uncertain in that the decision could be appealed to the U.S. Supreme Court.

As a result of such court cases, the medical profession has established two guidelines. First, the physician must clearly explain to the patient all the medical options available to sustain life. If the patient is in a coma or otherwise incapacitated and is not capable of understanding such explanations, then the physician will explain the options that are available to close members of the patient's family. If this is done, then a terminally ill patient, or a close family member, has the right to refuse what is called "heroic" treatment that might prolong life, or a coma, but offer no hope of recovery.

Second, the physician may honor the *living will* of the patient: a statement made by the patient, while still in possession of mental faculty, of whether heroic treatment should be given in the case of severe incapacity. Terri Schiavo had no such written document. Most physicians now recognize the validity of the living will and will abide by it, with the backing of law in most states. If no living will has been made, then the physicians and hospital are required to give treatment intended to sustain life. Nonetheless, even in such instances where a living will is absent, the patient or close family member may confer with the physician about discontinuing heroic treatment, and the physician may comply.

The Health Care System in the United States

Currently, the cost of medical care in the United States is more than 14 percent of our gross national product, making health care the nation's third leading industry (U.S. Census Bureau 2004). The cost of health care has at times risen at twice the annual rate of inflation. The United States tops the list of all countries in per-person expenditures for health care (see Figure 20.5). Other countries that spend considerably less money deliver health care that is at least comparable to that of the United States. Sweden and France spend roughly half as much per capita as the United States, and Great Britain spends a bit more than a third as much, yet these countries have national health insurance programs that cover virtually their *entire* populations.

The high cost of health care is a pressing social issue. Consumers complain about limited choice through HMOs, doctors worry over the cost of malpractice, businesses complain about the high cost, and insurance companies complain that they cannot recover their costs. There are many dimensions to the health care system. In the remaining section, we examine just a few: the high cost, including the cost of malpractice insurance,

the development of HMOs, the emergence of alternative forms of medicine, and the debate over universal health care.

The Cost of Health Care

The greatest contributors to skyrocketing health care costs are the soaring costs of hospital care and the rise in fees for the services of physicians. The services of specialists in particular are expensive. Health care must be paid for, and in the United States, the structure in place for paying the doctor's bill is the fee-for-service principle. Under this arrangement, the patient is responsible for paying the fees charged by the physician or hospital. Fortunate patients will be adequately insured and will be able to pass on their health expenses to the insurance company. Many people are not insured, however, especially people with lower incomes. This group must reckon with large bills by drawing on their own limited resources. Overall, 41 million people in the United States are without health insurance (see Map 20.1). A third source of payment is the government, which is increasingly the ultimate payer of health care expenses.

Another culprit is the third-party payment system. With medical care increasingly dominated by large corporate interests, hospitals, pharmaceutical companies,

Figure 20.5 Health Expenditures: A Comparative International View

Source: U.S. Census Bureau. 2004. *Statistical Abstract of the United States, 2003.* Washington, DC: U.S. Printing Office, p. 848.

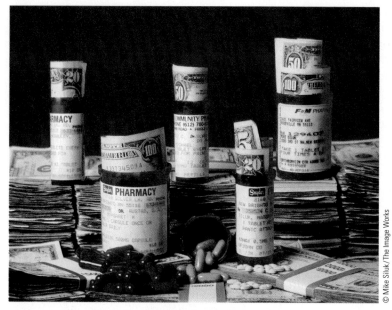

The high cost of prescription drugs is indicative of the problems generated in a profit-based health care system.

and other health care providers are interested in making a profit. Insurance companies, too, are based on a profit system. With so many corporate interests vying for a share of the profits, there is little pressure to keep the cost of care down, except from consumers who probably have the least power in this profit-driven system. Of course, health care is expensive. Most diagnostic procedures require expensive equipment; the research and development that goes into developing new drug treatments is expensive; and someone has to pay the salaries of the millions of health care workers. Still, one has to wonder if a profit-based system is not largely responsible for the high cost of health care.

Medical Malpractice

A rising number of patients are suing their physicians. In response, physicians are increasing the amount of their malpractice insurance to protect themselves. Annual insurance premiums for physicians have risen tenfold in the past decade and can be as high as $150,000 per year for physicians in specialties such as radiology, anesthesiology, and surgery. The extremely high cost of these insurance premiums is simply passed along to consumers (patients), contributing to the rise in the overall cost of health care. Some physicians have even had to shut down their practices because the cost of malpractice insurance was more than they could afford.

The U.S. public has traditionally accorded physicians high social status and high incomes, but the pub-

lic is beginning to question the privileged status of physicians (Pescosolido et al. 2001). Several specific reasons have been put forward for client revolts against the medical profession. Attorneys claim that the primary cause of the malpractice dilemma is a declining standard of medical care and a rising incidence of medical negligence. Another argument is that as a result of specialization, doctors now fail to establish old-fashioned rapport with patients, and thus patients are less trusting of their physicians and more likely to turn hostile. A third notion is that the high cost of medical care and resentment about the incomes of physicians has generated animosity in patients who are quick to show dissatisfaction, especially when medical episodes have an unfortunate outcome. Because so much of the system is impersonal, the physician may simply be the most easily seen target.

Contributing to the public's increasing mistrust of the medical profession is the incidence of medical errors and the medical profession's response to them. A doctor's actions often mean the difference between life and death for patients. On occasion, a doctor's own actions can become the direct cause of the death or further disability of the patient. When a suspected medical error occurs, a staff meeting at the involved hospital is called, but a number of researchers have noted that such meetings seem designed more to downplay or cover up medical errors than to account for them or prevent further errors (Weitz 2002; Bosk 1979).

Adding to the problem is the eagerness with which lawyers take on malpractice suits. This stems in part from the numerous lawyers available and in part from the vast profits that can be reaped from successful cases. Malpractice suits are generally conducted on a contingency fee basis. If the lawyer chooses to accept the case, the patient is required to pay little or nothing initially. Lawyers who work on contingency generally get one-third of any eventual settlement as their payment and nothing if the case is lost.

The problem of malpractice may actually contribute to the rising cost of health care. To protect themselves against potential malpractice suits, doctors may practice **defensive medicine,** which entails ordering excessively thorough tests, X rays, MRIs, and so on at the least indication that something might be amiss. This is to try to ensure that nothing is missed and also to document that the highest possible level of care was given. Although this extra attentiveness may contribute marginally to favorable medical outcomes, it contributes mightily to the overall cost of health care.

A Response to the Problem: HMOs

Health maintenance organizations (HMOs) are private clinical care organizations that provide medical services in exchange for a set membership fee and hence have direct responsibility and control over the costs incurred.

Health maintenance organizations are an innovation in health care, first begun shortly after World War II. In 1986, there were approximately 26 million HMO subscribers in the United States, a number that has risen to more than 55 million today. The staffs in HMOs can include anywhere from several doctors to several hundred doctors, along with physicians' assistants, nurses, medical technologists, and other personnel. People who join HMOs are assigned to a physician who administers care and, when necessary, gives referrals to specialists affiliated with the HMO. The doctors in an HMO earn salaries rather than fees. The membership fee paid by subscribers to the HMO ultimately pays for all services.

The elimination of both the fee-for-service system and the third-party insurer can drive costs down in several ways. Physicians have an incentive to give the most economical levels of care, while presumably retaining their motive to maintain a high standard of care because they will want to secure the reputation of the HMO and avoid malpractice suits. Presumably, the corporate structure of HMOs offers the economy of scale and organization enjoyed by other profit-oriented corporations, but it can increase the bureaucratic delivery of health care and contribute to patients' feeling like the system is far beyond their control.

Critics of HMOs argue that they decrease the rights of physicians to determine treatments and they limit the rights of patients to choose their own doctors or seek out treatments from specialists. The American Medical Association has also argued in a number of forums that HMOs are inclined to pay too much attention to cost containment and not enough to patient welfare (Scheid 2003). The problem with HMOs—and health care—has created a call for major reform in the health care system, but to date there is not a new national plan, even though public concern is high.

Alternative Health Care

As will be seen in Chapter 22, people often form social movements that seek to bring about some kind of change in society. The recent "New Age" movements, which advocate meditation, vegetarianism, yoga, and other holistic practices, are examples of alternative health care movements. This overall movement has sprung up in response to the general health crisis in the United States. **Holistic medicine,** which emphasizes

Although medicine has been dominated by Western values, many have come to appreciate alternative therapies, including such ancient Asian practices as acupuncture and herbal therapies.

the person's entire mental and physical state, the integration of the two, and the person's physical and social environment, is an example of alternative medicine that has developed in recent years (Reischer 1992).

Practitioners of holistic medicine advocate the treatment of the entire mental and physical person, but they do not necessarily reject the remedies of traditional medicine such as surgery and drugs. They do, however, caution against the use of such measures alone on the grounds that they result in overspecialization. The consequence is a narrow focus on isolated symptoms, not the individual as an integrated whole, and on specific diseases, not disease as a part of the entire concept of health.

The way each person's health is affected by physical and social environments concerns practitioners of holistic medicine. For example, the holistic practitioner recognizes that a variety of illnesses can result from the stresses of one's lifestyle, or from work-related stress, or from poverty. Another approach of holistic medicine is to decrease the dependency of the patient upon the physician by shifting some of the responsibility of cure from the physician to the patient, such as by advocating health-promoting behaviors involving diet, exercise, and the use of organic foods. As a consequence, holistic medicine takes an active role in combating environmental pollution.

The alternative health care movement rests on the assumption that individuals can be responsible for their own well-being and that individuals have some measure of control over the prevention of illness or recovery from bad health. As a result, many people in the United States have begun to pay more attention to their health. The recent growth in the number of health clubs and gyms is one indication of this. However, although individual responsibility for health is important, one cannot ignore the deeply structural realities that produce good or bad health within society—phenomena that extend beyond the ability of single individuals to control.

As many people have become aware of the importance of good health, a major industry has developed to provide assorted health services.

The Universal Health Care Debate

Periodically in the United States, plans for the reorganization of the health care system are suggested. The suggested plans are designed to combat the major problems of health care, including high costs, access, and improved health care for less well-off Americans. Universal health insurance for all Americans is suggested as part of such intended programs.

The core of health care reform was to convert the entire system to a model called **managed care**. Under this plan, all individuals in the United States would belong to a complex of managed care organizations, rather like HMOs, that would use their collective bargaining force to drive down the cost of health insurance, while accepting responsibility for operating their own facilities in an economical manner and continuing to provide high-quality care. Everyone would join the managed

care complex. However, individuals would still be free to retain their personal physicians if they so chose, as long as their physicians met government-stipulated criteria. The plan was intended to achieve the advantages of socialized medicine systems as administered in Great Britain, Canada, and other places where everyone is entitled to see a doctor when they need one, while retaining elements of the profit motive in the system.

THINKING *SOCIOLOGICALLY*

The age group least likely to be covered by *health insurance* is comprised of those aged 18 to 24 (see Figure 20.2). What social factors are involved in young people not having health care coverage?

TAKING ON SOCIAL ISSUES

Universal Health Care

National debate on the desirability of universal health care engages some of our basic social values. Should health care insurance be a government responsibility or left in the hands of private companies? What rights do patients have relative to the power of drug companies and physicians to control treatment plans? These and other questions engage many of the value conflicts over the right to privacy, the role of government in people's daily lives, and the power of corporations over individuals. Would you support a federally funded program that would provide health care insurance to all who need it? Why or why not, and how does your answer reflect an underlying value system?

Taking Action

Go to the Taking Action Exercise on the Companion Website—at http://sociology .wadsworth.com/andersen_taylor4e/— to learn more about an organization that addresses this topic. •••

A managed care plan for health care reorganization as well as universal health insurance was defeated in Congress in 1994. To date, no detailed proposals for universal health care have been supported by the federal government, although it is commonly debated by politicians. Public concern over the issue remains high and the promise of a working proposal remains part of the national conscience.

Sociology ⊛ Now™

Reviewing is as easy as ❶❷❸.

1. *Before you do your final review, take the SociologyNow diagnostic quiz to help you identify the areas on which you should concentrate. You will find information on SociologyNow and instructions on how to access all of its great resources on the foldout at the beginning of the text.*

2. *As you review, take advantage of SociologyNow's study videos and interactive Map the Stats exercises to help you master the chapter topics.*

3. *When you are finished with your review, take SociologyNow's posttest to confirm you are ready to move on to the next chapter.*

Chapter Summary

How has the modern system of medicine emerged?

What was once based more on folk wisdom and perhaps superstitious healing practices has developed into a highly specialized and scientifically based enterprise. Moreover, whereas health care was once delivered largely through one-on-one contact between a patient and physician, it is now largely controlled through profit-based corporations.

What theories in sociology inform the analysis of health and health care?

The three major theoretical perspectives in sociology each contribute to our understanding of health and health care. *Functionalism* emphasizes the interconnections between health care and other institutions; this is a major contributor to the stability of society. *Conflict theorists* emphasize the inequalities associated with access to health care and the occurrence of disease. They also see the profit-based system of health care as contributing to the high cost of health care. *Symbolic interactionists* study such things as how sickness is socially defined and how people act out the sick role.

How does the system of inequality influence health in the United States?

Race–ethnicity, social class, and gender are major factors in determining people's health and access to health care. Life expectancy, disease rates, and the utilization of the health care system vary for different racial–ethnic groups. Class status also influences health and health care. The lower one's social class status, the greater are one's chances of various diseases and the more likely one will be uninsured and/or receive poor health care. Gender also influences health. Women on average live longer than men but tend to be treated differently in the health care system because there is still a tendency for this male-dominated profession to regard the problems of women as "special." At the global level, the economic status of different nations is related to the health of their populations.

How do social factors influence various forms of illness?

Social epidemiology is the study of the role of social and cultural factors in disease and health. Diseases such as eating disorders, sexually transmitted diseases, smoking-related diseases, disability, and mental health disorders all have social dimensions. Understanding the social basis of health and health care is critical in the treatment of disease. Social factors also play a role in ethical issues, such as *euthanasia*.

What is the health care crisis in the United States?

Modern medicine in the United States is a highly structured, specialized, high-status profession. High costs have resulted in a policy crisis today in the health care system in the United States, including the failure of some medical practices because of the high cost of malpractice insurance. The growth of *health maintenance organizations* (HMOs) provides a new organizational model of health care, one that can be more efficient but also more bureaucratic. Despite the need for better health care coverage, universal health insurance has not yet been established by the federal government.

Key Terms

anabolic steroids 547	holistic medicine 555
anorexia nervosa 545	managed care 556
bulimia 545	Medicaid 538
defensive medicine 554	medical model 547
epidemiology 544	Medicare 538
euthanasia 552	sick role 540
health maintenance organization (HMO) 555	social epidemiology 544
	stigma 548

Researching Society with MicroCase Online

You can see the results of actual research by using the Wadsworth MicroCase Online feature available to you. This feature allows you to look at some of the results from national surveys, census data, and other data sources. You can explore this easy-to-use feature on your own, but try this example. Suppose you want to know:

What social factors are associated with whether or not people are covered by health insurance?

To answer this question, go to http://sociology.wadsworth .com/andersen_taylor4e/, select MicroCase Online from the left navigation bar, and follow the directions there to analyze the following data.

Data File GSS

Task: Auto-Analyzer

Primary Variable: HLTHINSR

Questions

Once you have your results, answer the following questions:

The tables show you whether various factors (religion, political party, age, education, sex of respondent, and race) are related to whether people are covered by health insurance.

1. Among all respondents, what percentage report they are covered by health insurance?

2. Complete the table using the data in the row "Yes" for the following demographic variables.

Socio-Demographic Variable	Category Most Likely	Category Least Likely	Significant?
Age			Yes No
Education			Yes No
Income			Yes No
Party			Yes No
Race			Yes No
Region			Yes No
Religion			Yes No
Sex			Yes No

3. Describe what characteristics are most likely among people who have health insurance.

The Companion Website for Sociology: Understanding a Diverse Society, Fourth Edition

http://sociology.wadsworth.com/andersen_taylor4e/

Supplement your review of this chapter by going to the Companion Website to take one of the Tutorial Quizzes, use the flash cards to master key terms, and check out the many other study aids you'll find there. You'll also find special features such as GSS Data and Census 2000

information, data and resources at your fingertips to help you with that special project or do some research on your own.

Suggested Readings and Web Resources

Morgen, Sandra. 2002. *Into Our Own Hands: The Women's Health Movement in the United States, 1969–1990.* Piscataway, NJ: Rutgers University Press.
Morgen chronicles the development of the women's health movement as it emerged from feminism and challenged the system of medical control—and male authority—over women's bodies. It is a good illustration of how grassroots movements can challenge the medical establishment.

Scheff, Thomas. 1984. *Being Mentally Ill: A Sociological Theory,* 2nd ed. New York: Aldine.
This classic is one of the best explanations and applications of labeling theory as applied to the study of mental illness as a social construction. The work discusses how patients are in effect rewarded for continuing their sickness and punished for trying to get well.

Shilts, Randy. 1988. *And the Band Played On.* New York: Penguin.
This book offers an extremely moving analysis of the public policy response to the AIDS crisis. The book documents how the association of AIDS with homosexual sex slowed scientific research and inhibited a strong policy response to the growing threat of AIDS.

Starr, Paul. 1982. *The Social Transformation of American Medicine.* New York: Basic Books.
Winner of a Pulitzer Prize, this book chronicles the emergence of modern medicine over two centuries of development, emphasizing the sociological context in which medicine has emerged.

Thompson, Becky. 1994. *A Hunger So Wide and So Deep: American Women Speak Out on Eating Problems.* Minneapolis: University of Minnesota Press.
Thompson's book develops the interesting thesis that the American culture of thinness affects White women differently than it affects African American women, Latinas, and lesbians. White women are somewhat more likely to undereat and become anorexic, but the latter groups engage in what Thompson calls strategies of self-preservation as a result of oppression and are thus more likely to engage in overeating than self-starvation.

Centers for Disease Control and Prevention (CDC)
www.cdc.gov
A national research center aimed at preventing and controlling disease, the CDC also publishes national data on the health status of the population.

National Institutes of Health (NIH)

www.nih.gov

The national agency dedicated to research that will improve the nation's health, the NIH provides access to a large number of governmental documents about health on its website.

World Health Organization

www.who.int/en/

An organization dedicated to improvement of health on a global basis.

• • •

Population, Urbanization, and the Environment

The study of population is a most important topic as we begin the new millennium. A great many issues that are of central concern to the entire country are driven by population. Population growth and density are heavily responsible for the major policy issues of the day such as overcrowding in cities, traffic jams, long lines at markets and other stores, environmental pollution, family planning, diminishing resources, and food shortages. Population issues are always high on the federal government's list of problems to be solved, and the U.S. Census Bureau debates with Congress methods for accurately counting the number of people who constitute the population of the United States. The federal as well as the state governments endlessly debate who should be permitted to immigrate into the country and who should not.

Finally, the terrorist attacks on the World Trade Center towers, and on the Pentagon, on September 11, 2001, presumably by foreign persons residing in the United States, have—rightly or wrongly—resulted in stricter immigration policies.

The population of the United States is presently more than 280 million. At the current rate of growth, the country will reach almost 300 million by the year 2025. The population has more than doubled since 1946, when it stood at about 132 million people. In that year, a spike in the number of births began that lasted until 1964—from the end of World War II to the beginning of the Vietnam War. The Baby Boom is the name given to that crop of 75 million babies currently representing nearly one-third of all the people in the United States. This generation, who in the tumultuous 1960s declared "don't trust

Sociology ⊛ Now™
Reviewing is as easy as ❶❷❸.

Use SociologyNow to help you make the grade on your next exam. When you are finished reading this chapter, go to the chapter review for instructions on how to make SociologyNow work for you.

anyone over thirty," is now beginning to reach their midfifties and starting to have grandchildren.

Babies are born every year to about 70 in 1000 American women aged fifteen to forty-five, across all social classes and racial–ethnic groups. Planning around children greatly affects the goals that adults set for themselves. For example, if you want to go directly into a career after college, you may choose to postpone having a child. Having a career and a child at the same time is a heavy strain—but many people do it. Graduate or professional school may also cause you to postpone the decision to have a child. That first job or overseas assignment may cause you to put off children yet again. The decision of whether to have or adopt children is among the most important that you will ever make, affecting not only your own life, but that of your child, the rest of your family, and, ultimately, society.

The educational and occupational structure of the United States has been affected by the decisions of so many peo-ple to have children. Today, as globalization, economic restructuring, and a myriad of other changes are causing young people to feel insecure about their future prospects, many twenty-something couples are deciding, at least for the present, not to have children. Their decisions will ripple forward in the years to come, as fewer youngsters are enrolled in grade school in the next few years, and in college two decades down the road. Inevitably, there will follow a decline in the number of middle-aged persons to take care of the elderly in their old age.

In this chapter, we examine the nature of human populations. How do births, deaths, and migrations affect society? How are the male and female populations different? Who in our society is most likely to die young or die old? How do demographic hazards such as accidental death or exposure to toxic waste differ across racial and ethnic groups? Finally, we will address a question that many people find most chilling of all: Are there too many people on this planet? •••

Demography and the U.S. Census

The study of population is called demography. **Demography** is the scientific study of the current state and changes over time in the size, composition, and distribution of populations. This field of sociology draws on huge bodies of data generated by a variety of sources, including the U.S. Census Bureau.

A **census** is a head count of the entire population of a country, usually done at regular intervals. The census conducted every ten years by the U.S. Census Bureau, as required by the Constitution, not only attempts to enumerate every individual, but also obtains information on gender, race, ethnicity, age, education, occupation, and other social factors.

The 2000 U.S. census is estimated to have missed or undercounted a small percentage of the country's population. Among those most likely to be missed by the census are the homeless, immigrants, minorities who live in ghetto neighborhoods, and other people of low social status. The constitutional requirement for a census was installed to assure fair apportionment of representatives for each state in the U.S. House of Representatives. However, the undercounting of minorities and the underclass tends to leave these groups underrepresented in government. The estimated undercount for the entire U.S. population is only about 2 percent, yet the overall undercount for African Americans nationally has been estimated to be as high as 20 percent, and for Hispanics, as high as 25 percent (National Urban League 2003; NAACP Legal Defense and Education Fund 2003; Harrison 2000).

The U.S. Congress and the U.S. Census Bureau have hotly debated the following issue: Should people be allowed to select multiracial (or "mixed race") responses on the census questionnaire, as they are now, by checking more than one category for African American, Hispanic, Non-Hispanic White, Native American, Asian, Eskimo, and Aleut? Or should individuals be limited to checking one category only? Use of this multiple response option gives individuals who define themselves as mixed race an opportunity to so designate themselves. One argument against this option is that it subtracts from the number of people indicating some of these categories, such as African Americans, Hispanics, and Native Americans, thus only further undercounting them (Harrison 2000).

Another body of data used in demography is vital statistics. **Vital statistics** include information about births, marriages, deaths, migrations in and out of the country, and other fundamental quantities related to population. From the census and vital statistics gathered from a wide variety of sources, we can create a picture of the U.S. population—who we are, how we are changing, and even who we will be in the future.

Diversity and the Three Basic Demographic Processes

The total number of people in a society at any given moment is determined by three variables: births, deaths, and migration. These three variables show different patterns for different racial and ethnic groups, different social strata, and both genders. Births add to the total population, and deaths subtract from it. Migration into a society from outside the society is **immigration,** which adds to the population; whereas **emigration,** the departure of people from a society (also called out-migration), subtracts from the population.

The population of the entire world is increasing at a rate of about 270,000 people per day, or just fewer than 200 people per minute. The world's population does not increase in a linear fashion—the curve on a graph of population does not rise in a straight line with the same number of people added each year. Instead, the population grows exponentially, with an upward-accelerating curve, as dramatically shown in Figure 21.1. An ever-increasing number of people are added each year. At the present rate of growth, the world's population will double in about forty years.

Birthrate

The **crude birthrate** of a population is the number of babies born each year for every 1000 members of the population or, alternatively, the number of births divided by the total population, times 1000:

$$\text{Crude birthrate (CBR)} = \frac{\text{Number of births}}{\text{Total population}} \times 1000$$

The crude birthrate for the total world population is approximately 27.1 births per 1000 people. Different countries and subgroups within a country can have dramatically different birthrates. For example, the country with the highest birthrate in the world is Niger, with 50.4 births per 1000 people. The lowest birthrate is found in Italy and Belarus, with birthrates of only 9 births per 1000 people (U.S. Census Bureau 2004).

The crude birthrate reflects what is called the *fertility* of a population, which is live births per number of women in the population. Fertility is different from *fecundity,* which is the potential number of children in a population that could be born (per 1000 women) if every woman reproduced at her maximum biological capacity during the childbearing years. Current demographic estimates are that a majority of women could have a maximum average of nearly twenty children. Fertility rates are not as high as fecundity rates because few women in any given population reproduce according to the theoretical biological maximum.

The overall birthrate for the United States is about 16.5 births per 1000 people, compared with the all-time high rate of 27 births per 1000 people in 1947, the start of the Baby Boom following World War II. The rate varies according to racial–ethnic group, region, socioeconomic status, religion, and other factors. Overall, we find that for different racial–ethnic groups, the birth-

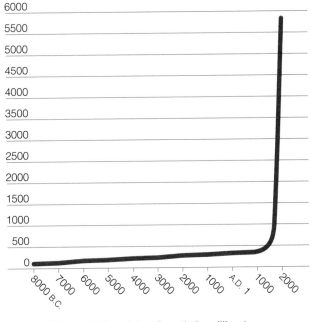

Figure 21.1 World Population Growth (in millions)

Data: From the Population Reference Bureau. 2003. Washington, DC. Reprinted with permission. Website: **www.prb.org**

rates in the United States are as follows (National Center for Health Statistics 2003):

Group	Birthrate per Thousand
Whites	14.1
African Americans	17.6
Latinos (Mexican Americans, Puerto Ricans, and other Hispanic groups combined)	24.0
Native American Indians	17.1
Asian Americans	17.8

In general, minority groups tend to have somewhat higher birthrates than White nonminority groups, and lower socioeconomic groups tend to have higher birthrates than those higher on the socioeconomic scale. Native American Indians have a birthrate higher than the national average, even though their birthrate has *declined* steadily over the past fifty years.

The effects of birthrates are somewhat cumulative. For example, minorities tend to be overrepresented on the lower end of the socioeconomic scale, compounding the likelihood of a high birthrate. Similarly, religious and cultural differences can make themselves felt. For example, Catholics have a higher birthrate than non-Catholics of the same socioeconomic status; Hispanic Americans have a higher likelihood of being Catholic. Thus, the combination of multiple factors can contribute to the higher birthrate among Hispanic Americans. Because minorities tend to have higher birthrates, assuming that present migration rates continue and that death rates do not outstrip the birthrates, the United States will have a significantly greater proportion of minorities, thus a relatively lower proportion of Whites, in the coming years.

Given current birthrates, the population of Latinos in the United States will double by the year 2030, and the population of African Americans will double by the year 2025. This increase has great implications for local and national policy. As a result, local, state, and federal governments must account for these projected increases when adopting future legislation and programs.

Death Rate

The **crude death rate** of a population is the number of deaths each year per 1000 people, or the number of deaths divided by the total population, multiplied by 1000:

$$\text{Crude death rate (CDR)} = \frac{\text{Number of deaths}}{\text{Total population}} \times 1000$$

The crude death rate can be an important measure of the overall standard of living of a population. In general, the higher the standard of living enjoyed by a country, or a group within the country, the lower the death rate. The death rate also reflects the quality of medicine and health care. Poor medical care, which goes along with a low standard of living, will correlate with a high death rate.

Another measure that tends to reflect the standard of living in a population is the **infant mortality rate.** This is the number of deaths per year of infants less than one year old for every 1000 live births. In the United States, the overall infant mortality rate is about 7 infant deaths for every 1000 live births. The highest infant mortality rates among 77 countries throughout the world are in Angola in Africa (nearly 200 infant deaths per 1000 live births) and Afghanistan (150 per 1000; U.S. Census Bureau 2004).

Infant mortality rates are important to compare across racial–ethnic groups and across social class strata because they are a good indicator of the overall quality of life and the chances of survival for members of each group. Inadequate health care and facilities cause higher infant mortality rates; consequently, the greater infant mortality among minorities and those in lower socioeconomic strata in the United States suggests lack of adequate health care and access to health facilities. Many other factors cause higher infant mortality, a measure of the chances of the very survival of members of the population, such as presence of toxic wastes, malnutrition of the mother, inadequate food, and outright starvation.

The **life expectancy** of a population or group is defined as the average number of years that the group can expect to live. In the United States, life expectancy has gone from forty years of age in the year 1900 to seventy-seven years of age today. That means that people born in the twenty-first century can expect to live, on average, until they are about seventy-seven years of age.

Although a life expectancy of seventy-six years might seem high, the truth is that the United States does not compare very well with other developed nations in either life expectancy or infant mortality. Table 21.1 indicates the United States ranks near the *bottom* among industrialized nations in life expectancy, including a lower rate than Japan, Canada, and the Netherlands. The picture is similar regarding infant mortality. Interestingly, Russia also has a low life expectancy and high infant mortality rate.

Life expectancy and infant mortality both vary with gender, race–ethnicity, and social class. Women on average live longer than men, not only in the United States but also throughout most of the world. Women have higher rates of numerous illnesses than men, but they still live longer. Women survive illnesses more frequently than men, and they are more resistant to physical stress.

African Americans, Hispanics, and Native Americans all have shorter life expectancies than Whites. White women live on average five years longer than African American women, and White men on average live seven years longer than African American men. African Americans are twice as likely as Whites of comparable age to die from diseases such as hypertension or the flu, as are Hispanics. Both groups are also considerably more likely than Whites to die of acquired immune deficiency syndrome (AIDS).

Table 21.1
Life Expectancy and Infant Mortality
(by country, industrialized countries only)

Country	Life Expectancy	Infant Mortality Rate
Japan	80.8	3.9
Australia	79.9	5.0
Canada	79.6	5.0
Italy	79.1	5.8
France	78.9	4.5
Spain	78.9	4.9
Netherlands	78.4	4.4
United Kingdom	77.8	5.5
Germany	77.6	4.7
United States	**77.3**	**6.8**
China	71.6	28.1
Russia	67.3	20.1

Source: U.S. Census Bureau. 2004. *Statistical Abstract of the United States, 2003.* Washington, DC: U.S. Bureau of the Census.

Sadly, these racial and ethnic differences are even more striking when it comes to infant survival. African American babies are almost *thirty* times more likely to contract AIDS than White babies, and Hispanic babies, twenty-five times more likely. Native Americans are ten times more likely than Whites to get tuberculosis, and nearly seventy times more likely to get dysentery (National Urban League 2004). Once again, we see the dramatic effects of gender and race–ethnicity, as well as class, this time on the odds of avoiding the devastation of chronic disease and surviving to a relatively old age. Those odds are considerably lower in the United States if you are poor and minority.

DEBUNKING *SOCIETY'S MYTHS*

Myth: High infant mortality and a high death rate for young adults are problems in underdeveloped countries but not in the United States.

Sociological perspective: Both infant mortality and adult death rates are significantly greater among lower-class than among middle-class persons in the United States, greater among people of color (Latinos, Native Americans, and African Americans) than among Whites of the same social class, and greater among men than among women.

Migration

Joining the birthrate and death rate in determining the size of a population is migration in and out. Migration affects society in many other ways as well. For example, Israel, since its establishment in 1948, has experienced considerable growth in its population, primarily the result of a tremendous migration of Jews from Europe and the United States. These migrants tend to be younger on average than the rest of the Israeli population, and their arrival therefore has certain direct consequences, such as increasing the birthrate, which is higher among the young than among older Israelis.

Migration can also occur within the boundaries of a country. In the 1980s, internal migration by African Americans, Hispanics, Asians, and Pacific Islanders within the borders of the United States has occurred at a rate unmatched since World War I and the great Black migration from the South to the North in the 1920s. During that era, Blacks migrated from the South to major urban areas in the North, such as Chicago, New York, Detroit, and Cleveland. The recent pattern of mi-

© 2004 Gwendolyn Knight Lawrence/Artists Rights Society (ARS), New York.

gration for African Americans has been not only from the South, but also from the major Northern urban centers to the West, the Southwest, and back to the South. This migration has carried many African Americans into areas of the country that were previously all White and has frequently been associated with the increased presence of African Americans at institutions such as military bases or universities (Barringer 1993).

Among Hispanics, migration patterns have been traditionally linked to the agriculture industry, but more recently, they are linked to industries such as meatpacking, textiles, and other industries centered in urban areas (Portes and Rumbaut 2001; Edmondson 2000; Barringer 1993). Although Mexican Americans have traditionally settled in the West and Southwest, recent movement has taken them to such places as Michigan, Washington state, and New England. Farm workers from Mexico have settled near the beet fields of northern Minnesota, tripling the Hispanic population there since 1980. Puerto Rican Americans have migrated northward in increasing numbers to rural, urban, and suburban areas. In the Yankee towns of New England that were once heavily White, such as Lynn and Lowell, Massachusetts, the Hispanic population has grown 180 percent since 1980 (Rumbaut and Portes 2001; Portes and Rumbaut 2001).

As shown in this painting by African American artist Jacob Lawrence, in the 1920s from virtually every Southern town, African Americans left by the thousands to go North to enter industry. This has come to be known as the Great Black Migration in America.

Asians in the last decade have migrated to a variety of destinations within the country. Many Vietnamese have joined the fishing industry on Louisiana's Gulf Coast. The Lutheran Church resettled hundreds of Laotians in Wisconsin. In Smyrna, Tennessee, where Nissan has built a truck plant, the Asian population has more than doubled. Finally, many Asians, African Americans, and Hispanics have joined Whites in migrating from urban to suburban areas.

Population Characteristics

The composition of a society's population can reveal a tremendous amount about the society's past, present, and future. The populations of many nations show a striking imbalance in the number of men and women of certain ages, an indication that many men are apparently missing. The explanation can be found by looking back in time to when the missing men were the right age for military service—the demographic vacancy will usually coincide with a major war from which many young men failed to return. The demographic data thus become a record of national history. World War II was responsible for killing many men, and some women, in the early 1940s—people who were in their twenties. As a result, this age category in the United States census revealed a shortage of persons, and considerably fewer men than women, even as the entire population of the country aged.

Sociologists put together data about population characteristics to develop pictures of the population in slices or as a whole. Important characteristics of populations are sex ratio, age composition, the age–sex population pyramid, and age cohorts. Here we examine these main approaches to describing the population.

Sex Ratio and the Population Pyramid

Two factors that affect the composition of a population are its sex ratio and its age–sex pyramid. The **sex ratio** (also called *gender ratio*) is the number of males per 100 females, or the number of males divided by the number of females, times 100.

$$\text{Sex ratio} = \left[\frac{\text{Number of males}}{\text{Number of females}} \right] \times 100$$

A sex ratio above 100 means that there are more males than females in the population; below 100, more females than males. The sex ratio could just as easily have been defined as the number of females per 100 males. A ratio of exactly 100 means the number of males and females are equal. In almost all societies more boys

are born than girls, but males have a higher infant mortality rate and a higher death rate after infancy, which results in more females than males in the overall population. In the United States, approximately 105 males are born for every 100 females, thus giving a sex ratio for live births of 105. After factoring in male mortality, the sex ratio for all ages for the entire country ends up being 94—there are 94 males for every 100 females.

The *age composition* of the U.S. population is presently undergoing major changes. More and more people are entering the sixty-five-and-over age bracket. This trend is known as the *graying of America,* as we saw in Chapter 14. The elderly will soon become the largest population category in our society. As our society gets grayer, its older members will have more influence on national policy and a greater say in matters such as health care, housing, and other areas in which the elderly have traditionally experienced age discrimination.

Sex and age data are often combined in a graph called an **age–sex pyramid** (or age–gender pyramid), which represents the age and gender structure of a society (see Figure 21.2). The left side of Figure 21.2 shows the age–sex pyramid for the United States. Note that there are slightly more males than females in the younger age ranges (due to the higher birthrate for males), a trend that reverses in the middle-age brackets. At the upper range of the age scale, women outnumber men. The pronounced bulge near the middle of the pyramid represents the Baby Boom generation. As these Baby Boomers age, the bulge will rise toward the top of the pyramid, to be replaced underneath by future birth trends. This restructuring of the population pyramid will necessitate the restructuring of society's institutions to serve an aging population. Marketing will have to be directed more toward people over age sixty-five; considerably more funds will be needed for health care; and recreational facilities for those who are older will have to be greatly expanded. (Because this bulge in the population is continually moving forward, much like food in a large snake, demographers have sometimes jokingly referred to the Baby Boomers as "the pig in the python.")

The birthrate of a society or group tends to rise and fall in line with the shape of the age–sex pyramid. The right side of Figure 21.2 shows the age–sex pyramid for Mexico. A pyramid of this shape is more characteristic of developing nations than of the industrial powers. It is wide at the bottom and narrow at the top, suggesting a high birthrate and a death rate that increases rapidly with age. Relatively few males or females survive to fill the elderly age categories. Countries with a high birthrate tend to have a high proportion of women in their childbearing years. The birthrate is a statistic of great consequence for the future. Because the children being born will themselves grow up and have children, a high birthrate in the present can be

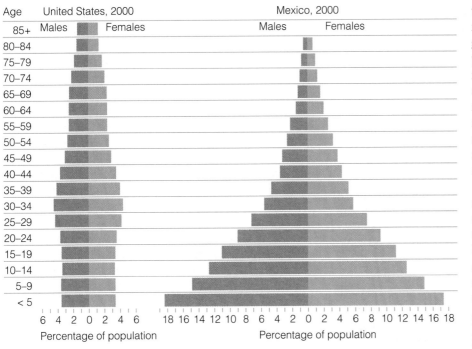

Figure 21.2 A Comparison of Two Age–Sex Pyramids

Source: Weeks, John R. 2004. *Population: An Introduction to Concepts and Issues,* 9th ed. Belmont, CA: Wadsworth.

projected to indicate a high birthrate in future generations. Within the United States, the population pyramid for African Americans tends to be shaped somewhat more like the one on the right in Figure 21.2, also true for Mexican Americans and Puerto Rican Americans.

THINKING SOCIOLOGICALLY

The *age–sex pyramid* is a graph showing the age and gender structure of a society's population. How would the age–sex pyramid for a society look if its men aged eighteen to thirty-five had a greater *crude death rate* than they would otherwise have had because of a major war or natural disaster? What if the women in the society were killed off in the same way? What would such dramatic reductions in the numbers of both women and men mean for the future of the society?

Cohorts

A birth cohort, or more simply, a **cohort,** consists of all the persons born within a given period. A cohort can include all persons born within the same year, decade, or longer. Over time, cohorts either remain the same size or get smaller owing to deaths, but they never grow larger. If we have knowledge of the death rates for this population, we can predict fairly accurately the size of the cohort as it passes through the stages of life from infancy through adulthood to old age. This enables us

to predict, for example, how many people will enter the first grade at age six between the years 2010 and 2015, how many will enroll in college, and how many will arrive at retirement decades down the road. Social entities such as schools and pension funds can make preparations on the basis of such predictions.

To see how dramatically a single cohort can affect a society over time, consider the effect of the cohort born in the United States between 1946 and 1964—the Baby Boom cohort. Many of the parents of students reading this book are Baby Boomers. The Baby Boom has been a significant demographic event in U.S. history, along with the great Black migration from the late 1800s through the 1920s, the simultaneous massive immigration of Europeans and other groups, and current migration patterns of Hispanic groups (Portes and Rumbaut 2001; Rumbaut and Portes 2001; Weeks 2004; Kennedy 1989; Ehrlich 1968).

The Baby Boom cohort, now comprising nearly one-third of the entire population of the United States, has had a major impact on the practices, politics, preferences, and culture of our society. Raised in the relatively permissive late 1950s and 1960s, watching the likes of *Howdy Doody* and *The Mickey Mouse Club* on television, they became a large part of the "Greed Generation" of the 1990s. Dr. Benjamin Spock's book of down-to-earth advice about how to raise healthy children, *Baby and Child Care,* became the greatest bestseller of the twentieth century, with more than 30 billion copies sold. Among its principles was the suggestion that parents should communicate with their children and forego physical punishments when possible. For encouraging this brand of "permissiveness," Dr. Spock has often been blamed at one time or another for nearly every social problem at the time, including the youth rebellions on campuses of the 1960s, protests against the Viet Nam War, and other practices often associated with that generation, such as experimentation with drugs like marijuana, LSD, and cocaine.

As they begin to pass age sixty-five in 2010, the Baby Boom cohort will greatly increase the ranks of the elderly. One effect will be an increase in political clout for those over age sixty-five. A heavy burden will befall those born between 1965 and 1975 because they will be the main contributors to the Social Security fund just as the fund will be required to meet the needs of the giant

Boomer population bulge as it comes to rest in its retirement years (Weitz 2004; Kennedy 1989; Robey 1982).

Theories of Population Growth Locally and Globally

Among the major problems facing modern-day civilization is the specter of uncontrolled population growth. As noted earlier, the population of the world increases by about 270,000 people every day. Some view overpopulation as an epochal catastrophe that endangers the future of the world. Other people dispute whether the problem exists at all, pointing out that there is no scientific consensus on the *carrying capacity* of the planet, meaning the number of people the planet can support on a sustained basis, and that technological advances, which have dependably met our needs in the past, can be counted on to do so in the future as the number of mouths to feed continues to grow.

Is the world overpopulated? Is the United States? What can we expect from the future? If the less optimistic scenarios turn out to be accurate, these could be the most important questions facing humankind.

Malthusian Theory

Humans, like other animals, can survive and reproduce only when they have access to the means of *subsistence,* meaning the necessities of life, such as food and shelter. Human and animal populations have in common that they decline and die off when encountering checks on population growth such as famine, disease, and war. In the face of a daunting environment, humans have managed to thrive, and their population has doubled many times over. The period of doubling gets shorter and shorter.

Thomas R. Malthus, a clergyman born in 1766 in Scotland and educated in England, pondered the realities of life on earth and assembled his observations into a chilling depiction of disastrous population growth called **Malthusian theory,** the idea that a population tends to grow faster than the subsistence needed to sustain it. Malthus declared that the earth must be near the limits of its ability to support so many humans and that the future must inevitably hold catastrophe and famine. In his *Essay on the Principle of Population* published in 1798 (Malthus 1798), he propounded his gloomy views, and he continued to revise and extend his theory until his death in 1834.

Malthus noted that populations tend to grow not by *arithmetic increase,* which adds the same number of new individuals each year, but by *exponential in-*

crease, in which the number of individuals added each year increases, with the larger population generating an even larger number of births with each passing year. Arithmetic increase would cause a population to double in size at a decelerating rate. Exponential increase causes a population to double more quickly. It took 200 years for the population of the world to double from half a billion in 1650 to one billion in 1850. The increase from two billion to four billion took forty-five years in the twentieth century. The mathematical power of doubling is remarkable when applied to population. If we were to start with just one couple and imagine a lineage that doubled itself each generation by having four children, a mere thirty-two generations (roughly 600 years) would result in a population of 8.4 billion—considerably more than today's world population of approximately 6 billion. To put Malthus's fears in perspective, in the United States at the time Malthus was writing, the average number of births per couple was seven!

Malthus hypothesized that unchecked doubling of any population would swiftly spawn enough people to carpet the earth many times over. Clearly, something prevents populations, human and other, from doubling every generation. Malthus reasoned that two forces were at work to keep population growth in check. First, the growth in the amount of food produced tends to be only arithmetic and not exponential. Second, Malthus surmised that there were three major *positive checks* on population growth—famine, disease, and war. In Malthus's time, disease could reach apocalyptic scales. The outbreak of bubonic plague in Europe from 1334 to 1354 eliminated a third of the entire population; a smallpox epidemic in 1707 wiped out *three-fourths* of the entire populations of Mexico and the West Indies. Wars had a giant appetite for European men, with deaths in battle causing semipermanent gaps in the population pyramids of European populations. Along with positive checks on population growth, Malthus acknowledged what he called *preventive checks,* such as sexual abstinence, but he knew that sexual abstinence was unlikely to be the mechanism to halt uncontrolled population growth.

Malthusian theory predicted rather well the population fluctuations of many agrarian societies, such as Egypt from about 500 A.D. through Malthus's own lifetime. However, Malthus failed to foresee three revolutionary developments that have derailed his predicted cycle of growth and catastrophe. In agriculture, technological advances have permitted farmers to work larger plots of land and to grow more food per acre, resulting in subsistence levels higher than Malthus would have predicted. In medicine, science has fought off diseases such as the bubonic plague that Malthus expected to periodically wipe out entire nations. Finally, the development and widespread use of contraceptives in many

Jacob Van Oost the Younger, *Saint Macavius of Ghent Succors the Plague Victims*, the Louvre, Paris. Photo © Scala/Art Resource

Bubonic plague (called the Black Death) severely decreased the entire population in Medieval Europe.

countries have kept the birthrate at a level lower than Malthus would have thought possible.

The technological victories of the twentieth century have not completely erased the specter of Malthus. Even though outbreaks such as AIDS can be medically controlled somewhat, the worldwide epidemic warns us that nature can still hurl catastrophes our way. Heart-rending pictures of bloated, starving babies remind us that famine continues to destroy human populations in some parts of the world just as it has for thousands of years. Overall, Malthus's theory has served as a warning that subsistence and natural resources are limited. The Malthusian doomsday has not occurred, but some believe that Malthus's warning was not in error, just premature.

Demographic Transition Theory

An alternative to Malthusian theory is the demographic transition theory, developed initially in the early 1940s by Kingsley Davis (1945) and extended by Ansley Coale (1974, 1986) and others (Weeks 2004). **Demographic transition theory** proposes that countries pass through a predictable and consistent sequence of population patterns linked to the degree of technological develop-

ment in the society, ending with a situation in which the birthrates and death rates are both relatively low. Overall, the population level is predicted to eventually stabilize, with little subsequent increase or decrease over the long term.

Population change involves three main stages, according to demographic transition theory (see Figure 21.3). Stage 1 is characterized by a high birthrate and high death rate. The United States during its Colonial period was in this stage. Women began bearing children at younger ages, and it was common for a woman to have twelve or thirteen children—a very high birthrate. However, infant mortality was also high, as was the overall death rate owing to primitive medical technique and unhealthy sanitary conditions.

> ### DEBUNKING *SOCIETY'S MYTHS*
>
> **Myth:** The average number of children per family in the United States may have decreased a little since Malthus's time (around 1800) but not by much.
>
> **Sociological perspective:** The average number of children has been about 2.1 per family in the United States for the last twenty years or so [which is close to zero population growth (ZPG) level], but it was seven children per family in Malthus's time! This may have seemed high even for Malthus, which is one reason he became so concerned with the effects of overpopulation.

Stage 2 in the demographic transition is characterized by a high birthrate but a declining death rate, increasing the overall level of the population. The United States entered Stage 2 in the second half of the nineteenth century as industrialization took hold in earnest. The norms of the day continued to encourage large families, and thus high birthrates, while advances in medicine and public sanitation whittled away at the infant mortality rate and the overall death rate. Life expectancy increased, and the population grew in size.

The characteristics of Stage 2 did not apply across all social groups or social classes. Minorities at the time (Blacks, and especially Native Americans and Chinese in the Midwest and West) were less likely to benefit from medical advances, and the infant mortality and overall mortality rates for Blacks, Native Americans, and Chinese remained high, while life expectancy remained considerably shorter than for the White population. Lack of access to quality medical care was particularly devastating to Native Americans, who had very high death rates for all ages during the 1800s. Demographic transition theory is not completely accurate for different racial–ethnic groups within the same society.

Stage 3 of the demographic transition is characterized by a low birthrate and low death rate. The overall level of the population tends to stabilize in Stage 3. Medical advances continue, and the general prosperity

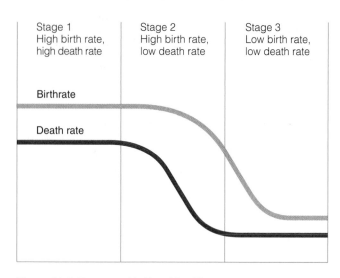

Stage 1	Stage 2	Stage 3
High birth rate, high death rate	High birth rate, low death rate	Low birth rate, low death rate

Birthrate

Death rate

Figure 21.3 Demographic Transition Theory

of the society is reflected in lowered death rates. Cultural changes also take place, such as a lowering of the family size most people consider ideal. The United States entered this stage prior to World War II, and with the notable exception of the Baby Boom, it has exhibited Stage 3 demographics since the 1940s. Again, however, the trend does not apply equally to all groups in U.S. society. Some groups, such as African Americans and Hispanics, continue to have high infant mortality rates and higher death rates owing to the disproportionate impact of disease upon these groups, as well as a disproportionate lack of access to health resources. American Whites are further along into Stage 3 than the rest of U.S. society.

Demographic transition theory has received some criticism primarily because it tends to be based on heavily industrialized countries with a White majority (Coale 1986). In this respect, the theory is *ethnocentric*. It argues that industrialized countries in the past have gone from high birth and death rates to low birth and death rates, with the demographic "transition" phase in between. Therefore, all countries, and large ethnic groups within countries, should go through these three phases and thus experience a decline in birth rate as well. The theory carried with it the presupposition that nations and ethnic groups of color were considerably less industrialized and were thus at Stage 1 with high birth and death rates. Yet it is not true that such groups are necessarily nonindustrialized. Thus, such nations and groups can be industrialized yet still be at Stage 1.

Nearly forty years after the theory was developed, it was discovered that individuals can certainly make their own decisions about how many children to have and when to have them. In other words, people often behave according to their own perception of rational choice, instead of being driven by some theoretical principle over which they are presumed to have no con-

trol. This is why **rational choice theory** was used to counter the predictions of demographic transition theory—namely, that humans are capable of making cost-benefit analyses of what to do and how many children to have (Weeks 2004).

The "Population Bomb" and Zero Population Growth

In 1968, a modern-day Malthus appeared in the form of Paul Ehrlich, a biology professor at Stanford University. His book, *The Population Bomb* (Ehrlich 1968), was the first in a series of writings in which Ehrlich argued that many dire earlier predictions of Malthus were not far from wrong. The growth in world population, according to Ehrlich, was a time bomb ready to go off in the near future, with dismal consequences. Supporting his argument with scientific research (a commodity used rather loosely in the work of Malthus, whose reasoning depended heavily on speculative guesswork about how many people the planet could and should bear), Ehrlich stated that the sheer mathematics of population growth worldwide were sufficient to demonstrate that world population could not possibly continue to expand at its present rates.

Worldwide population growth has outgrown food production and massive starvation must inevitably follow, according to Ehrlich. He went beyond Malthusian theory, however, to state that the problem transcended food production and was in fact a problem of the environment. At the time, Ehrlich was among the first modern thinkers to argue that the quality of the environment, especially the availability of clean air and water, was a critical factor in the growth and health of populations.

Many dire predictions Ehrlich made back in 1968 have come true. In a later book Ehrlich wrote with his wife, Anne, *The Population Explosion* (Ehrlich and Ehrlich 1991), the authors pointed to a variety of disasters that confirmed the predictions of the 1968 work and even of Malthus: mass starvation in parts of Africa; starvation in the United States among Black and Hispanic populations; increased homelessness in cities, especially among minorities; acid rain; rampant extinctions of plant and animal species; the irrecoverable destruction of environments such as the rain forests; and global warming.

Ehrlich has stated that only limits on population growth can avert disaster. He has advocated the implementation of Malthusian preventive checks by the government, for example supporting organizations such as zero population growth (ZPG). ZPG is dedicated to reaching the **population replacement level,** a state in which the combined birthrate and death rate of a population simply sustains the population at a steady level,

Table 21.2
A Comparison of Demographic Theories

	Malthusian Theory	Demographic Transition Theory	Zero Population Growth (ZPG) Theory
Main Point	A population grows faster than the subsistence (food supply) needed to sustain it.	Populations go through predictable stages ("transitions") from high birth and death rates to a stable population with low birth and death rates.	Achievement of zero population growth solves the Malthusian problem of unchecked population growth.
"Positive" Checks on Population Growth	Famine, disease, and war are likely.	Famine, disease, and war are moderately likely.	Famine, disease, and war are unlikely.
"Preventive" Checks on Population Growth	Sexual abstinence.	Sexual abstinence, birth control, and contraceptive methods.	Sexual abstinence, birth control, and contraceptive methods.
Predictions for the Future	Pessimistic, despite positive and preventive checks, a population will ultimately outstrip its food supply.	Optimistic, given technology and medical advances in a population.	Very optimistic; zero population growth has already been achieved in the United States and other countries.

called the *equilibrium level*. To achieve zero population growth, and thus population equilibrium, couples would have to limit themselves to approximately two children per family. By 1980, the United States had reached the replacement level of reproduction with an average number of 2.1 children per family, without considering race, ethnic, and class differences.

THINKING SOCIOLOGICALLY

What is the one major demographic change that would have to take place for a society to achieve *zero population growth* (ZPG)? Can you think of the advantages of ZPG? Any disadvantages? What effect would ZPG likely have on the *demographic transition* process? What is a major sociological barrier that would cause a population to resist adopting a governmental policy of birth control?

Despite these advances, worldwide population growth is still out of control. Population-related maladies such as urban crime, unemployment, and increasing homelessness remain a scourge of worldwide populations as well as in the United States, adding an updated twist to the positive checks (meaning factors that reduce population size) of the past, such as disease and war, which are still with us. Pollution of the environment remains a major threat. The Ehrlichs deliver a final message: The population bomb must inevitably explode. What remains in doubt is whether we will arrive at that historic moment having made humane preparations, perhaps mitigating the problem with preventive checks such as a lowered birthrate; or whether the population crisis will end tragically with a realization of Malthus's vision of people starving worldwide and disease ram-

paging through a weakened population (Ehrlich and Ehrlich 1991; see Table 21.2).

Checking Population Growth

As early as the 1950s, most countries had accepted that population growth was a problem that must be addressed. By the 1980s, countries representing 95 percent of the world's population had formulated some policies aimed at stemming population growth. However, consensus does not exist on how population should be controlled. For example, many religious and political authorities have argued against the use of birth control. Efforts to encourage the use of contraceptives among rapidly growing populations have therefore had mixed support, whereas other attempts to curb population growth, such as encouraging changes in social habits, generally meet with little success.

Family Planning and Diversity

Many governments, including that of the United States, make contraceptives available to individuals and families. Doing so, however, is not always consistent with the beliefs and cultural practices of all groups in the society. Catholics are taught that it is acceptable to use natural means of birth control (such as the "rhythm method") but are forbidden to use contraceptive devices such as an IUD (intrauterine device) or the pill. Many Catholics choose to use contraceptives anyway. Some studies, as already noted in Chapter 17, have shown that the *majority* of Catholics in the United States practice forms of birth control forbidden by their church.

Governmental programs that advocate contraception can be successful only if couples choose to have smaller families. This is most likely to occur only if the wider culture supports that decision. Creating a new image of the ideal family size is a central concern of people involved in the family planning movement.

Birthrate and family size are known to be related to the overall level of economic development of a country—including the economic status of certain ethnic groups within a country. In general, as countries become more economically developed, their birthrates and average family size drop (though not always), as predicted by demographic transition theory. In the more industrialized countries, people generally prefer smaller families. Developing countries tend to have higher birthrates and tend to give greater cultural value to large families. The same is true of ethnic and racial minorities living in developed nations but not entirely assimilated, such as many Hispanics in the United States. Large families are seen as a demonstration of potency, a source of needed labor, and as a preparation for old age, when offspring will be expected to care for their elders. Before contraceptive devices such as condoms, the IUD, and the pill will be widely adopted, cultural values in favor of undisturbed fertility must be countered (Vanlandingham 1993; Westoff et al. 1990).

A large-scale study has called into question the assumed relationship between economic development and family size in demographic transition theory (Stevens 1994). The study shows that countries in Stage 2, such as Bangladesh, which had a declining death rate because of rapid economic and medical developments but retained a high birthrate, have lately become very receptive to birth control programs. As a result, the birthrate as well as the population level have begun to decline, even in the absence of Western-style advanced development. Contraceptive programs have worked in other non-Western countries as well, such as Thailand and Colombia.

Population Policy and Diversity

Findings such as those in Bangladesh have an important bearing on population policy elsewhere, including in the United States. Family planning programs offer great potential for achieving large declines in birthrates. In fact, in underdeveloped, overpopulated countries where such programs can have the most effect, the demand for family planning resources surpasses the supply.

Some cultural resistance is evident from some U.S. racial and ethnic groups to government-sponsored contraceptive programs, including some Hispanic groups and some African American groups. Their argument is that decreasing the birthrate by means of contraceptive methods is genocidal and threatens the survival of members of these groups. If such programs are sponsored by the federal government, they are perceived as direct attempts by the government to reduce the number of minority people in the country. To the extent that the government promotes such programs disproportionately in Hispanic or African American communities and promotes them less aggressively in traditionally White or upper-class areas, then the programs are perceived to be racist. As a consequence, the individual is less likely to be convinced to adopt a contraceptive method or device.

A field worker in Bangladesh explains birth control pills and their use to village women.

At the borderline of policy and social custom are still other cultural barriers against the use of contraceptive methods, such as the belief among some groups (some young urban men, for instance) that condoms are unmasculine. Still, the popularity of condoms and other means of contraception is rising. Two-thirds of sexually active U.S. women in their childbearing years use some form of birth control—an increase from about half in the 1980s. African Americans, Hispanics, and low-income women had the largest increase in birth control usage. The greatest increase has been increased use of condoms and newer methods, including injectables and implants (Alan Guttmacher Institute 2004). Globally, the availability of birth control is hindered by Bush administration policies prohibiting funding to organizations that provide abortion services.

© Dilip Mehta/Contact Press Images

Urbanization

The growth and development of cities, or centers of human activity with high degrees of population density, is a relatively recent occurrence in the course of human history. Scholars locate the development of the first city at around 3500 B.C. (Flanagan 1995). The study of the urban, the rural, and the suburban is the task of *urban sociology,* a subfield of sociology that examines the social structure and cultural aspects of the city in comparison to rural and suburban centers. These comparisons involve what urban sociologist Gideon Sjøberg (1965) calls the *rural–urban continuum,* those structural and cultural differences that exist as a consequence of differing degrees of social structural complexity. **Urbanization** is the process by which a community has the characteristics of city life and the "urban" end of the rural–urban continuum.

Urbanization as a Lifestyle

Early German sociological theorist **Georg Simmel** (1950/1905) argued that urban living had profound social psychological effects, meaning that social structure could affect the individual. Urban life has a quick pace and is stimulating, but as a consequence of this intense style of life, the individual becomes insensitive to surrounding people and events. The urban dweller tends to avoid emotional involvement, which according to Simmel was more likely to be found in rural communities. Interaction tends to be characterized as economic rather than social, and close, personal interaction is frowned upon and discouraged. Yet urban dwelling can increase the likelihood of other ills: Early theorist Emile Durkheim noted that the suicide rate per 10,000 people was greater in more urbanized areas than in rural areas (Durkheim 1951/1897).

The sociologist Louis Wirth (1938), focusing his studies on Chicago in the 1930s, also argued that the city was a center of distant, cold interpersonal interaction and that, as a result, the urban dweller experienced alienation, loneliness, and powerlessness. One positive consequence of all this, according to both Wirth and Simmel, was the liberating effect that arises from the relative absence of close, restrictive ties and interactions. Thus, city life offered the individual a certain feeling of freedom.

A contrasting view of urban life is offered by Herbert Gans (1982/1962), who studied Boston in the late 1950s and concluded that many city residents develop strong loyalties to others and are characterized by a sense of community. Such subgroupings he referred to as the *urban village,* which is characterized by several "modes of adaptation." These adaptations include *cosmopolites,* who are typically students, artists, writers, and musicians, who together form a tightly knit community and choose urban living to be near the city's cultural facilities. A second category are the *ethnic villagers,* who live in ethnically and racially segregated neighborhoods. Today's *urban underclass* would encompass what Gans called the *trapped,* individuals who are unable to escape from the city because of extreme poverty, homelessness, unemployment, and other familiar urban ills.

Race, Class, and the Suburbs

The impact of race and class can clearly be seen in the distinction between city and suburb. Today, only about one-fourth of African Americans live in suburban areas. Echoing the earlier *urban villagers* principle of Gans, closely knit communal subgroups form in the suburbs, but the subgroupings tend to form on the basis of class and race, with class, according to some, being more important than race (DeWitt 1995). In the suburbs, one chooses one's neighbors and friends on the basis of educational and occupational similarity in addition to race.

Pleasant though the suburbs can be, it is also true that people of color, particularly African Americans, often become as segregated there as they do in cities (Feagin and Feagin 1993; Massey and Denton 1993). Racial segregation persists not only in suburban neighborhoods (Chicago suburbs are a case in point) but in schools as well. Although some people may argue that moving to the suburbs and taking up residence is a matter of personal choice, ample evidence shows that residential segregation is maintained by landlords, homeowners, and White realtors who engage in steering people of color to segregated neighborhoods (Feagin and Feagin 1993). The practice of *redlining* by banks, which renders it impossible for a person of color to get a mortgage loan for a specific property, further intensifies residential segregation. Finally, these practices serve only to encourage further segregation in the realm of interpersonal interaction.

The New Suburbanites

The United States ended the twentieth century as it began it—in a great wave of immigration. The 1924 National Origins Quota Law encouraged immigration from northern and western Europe (England, France, Germany, Switzerland, and the Scandinavian countries) but discouraged immigration from eastern and southern Europe (Greece, Italy, Poland, Turkey, and Eastern Europe, the latter notably affecting Ashkanazi Jews, among others; see Chapter 11). Despite this law, millions of Eastern Europeans successfully made the journey to Ellis Island and thence to the U.S. mainland, only to face prejudice, discrimination, and the accusation that they were taking jobs that would otherwise have gone to the already present White majority.

At the end of the twentieth century, once again U.S. shores received millions from abroad, and once again prejudice and discrimination were part of the picture, but often in more subtle forms, though sometimes flaring up. Neighborhoods are now invigorated and culturally enriched by mosques or Buddhist temples; by whole neighborhoods of Vietnamese Catholics, Koreans, or Asian Indians; or by war refugees from Somalia and Bosnia and other nations. The most prominent immigrants in suburban neighborhoods are Hispanic Americans and Asian Americans, two groups that presently comprise most of the country's foreign-born population (Edmondson 2000).

One long-term consequence of the current immigration settlement is what at least one demographer (William Frey of the University of Michigan) calls *the new demographic divide*. Today's immigrants settle in a relatively small number of big cities and their suburbs, mostly located on either the East or West Coast. Yet, many parts of the United States are still feeling the effects of new immigration trends. In cities and suburbs, there are a number of consequences. For example, for immigrant groups who are youthful and have a relatively high birthrate, schooling has become a big political issue. But in other areas with a higher proportion of older immigrants, tax cuts for the elderly and Medicare are the major issues.

The flow of immigrants can change from year to year, but the largest number of immigrants to the United States recently have been from Mexico, the Philippines, China, and countries of the former Soviet Union. Each group brings with it its own culture, politics, and differing ages. A consequence of the presence of these new suburbanites across the country is that Whites now in suburbia often are outnumbered by minorities such as African Americans, Asians, Hispanics, and other new immigrant groups. As a result, many whites refer to themselves as the "minority majority." One consequence is that the "minority majority" may perceive itself to be in competition with the new immigrants for jobs and other resources. In this respect, relations between Whites and the new immigrants may come to resemble the history of race and ethnic relations of the 1950s and early 1960s in the United States (Edmondson 2000).

Ecology and the Environment

It should be apparent by now that population size has an important social dimension. Social forces can cause changes in the size of populations, and population changes can transform society. Values are acquired from one's culture, including values about what family size is considered ideal or what degree of crowding in a city is considered tolerable. The values people hold will affect the number of children they want, the places they want to live, and even the degree to which they consider population control an urgent problem.

Population density is determined by the number of people per unit of area, usually per square mile. As population density rises to high levels, as it has in today's cities, the familiar problems of urban living appear, such as high rates of crime and homelessness. Interacting with these problems are crises of the physical environment, such as air and water pollution, acid rain, and the growing output of hazardous wastes. Humans sometimes forget that, like all other creatures, they are intimately dependent on their physical environment.

Human ecology is the scientific study of the interdependencies that exist between humans and their physical environment. A **human ecosystem** is any system of interdependent parts that involves human beings in interaction with one another and the physical environment. A city is a human ecosystem; so is a rural farmland community. In fact, the entire world is a human ecosystem. Human ecology is the study of human ecosystems. Two fundamental and closely related problems confront our present ecosystems: overpopulation and the destruction or exhaustion of natural resources (Weeks 2001; Hawley 1986).

Vanishing Resources

In all ecosystems, whether human, animal, or plant, organisms depend on one another as well as the physical environment for their survival. Plants use carbon dioxide and give off oxygen, which all humans and animals need to survive. Terrestrial creatures metabolize oxygen and produce carbon dioxide, which is then used by plants, completing a natural cycle. Humans and other animals require nutrients they can get only by eating plants. When living beings die, they decompose and provide nutrients to the soil that are taken up by plants, completing another cycle.

The examination of ecosystems has demonstrated two things. First, the supply of many natural resources is finite. Second, if one element of an ecosystem is disturbed, the entire system is affected. For much of the history of humankind, the natural resources of the earth were so abundant in relation to the amounts used by humans that they may as well have been infinite. No more. Some resources, such as certain fossil fuels, are simply nonrenewable and will be depleted soon. Other resources, such as timber or seafood, are renewable provided we do not plunder the sources of supply so recklessly that they disappear. We have made this ecological blunder many times before. Finally, some of earth's natural resources like air and water are so abundant that they *still* seem infinite, but at this stage of our technological development, we are learning that our powers extend to such heights and depths that we can even destroy the near-infinite resources.

Concerning our planet as a whole, our gaseous wastes are gnawing away at the ozone layer, and our buried chemical wastes are trickling into the water table and creating underground pools of poison. One study (Barlow and Clarke 2002) actually notes that pollution has damaged the Earth's surface water so badly that the United States is mining underground water reserves faster than nature can replenish them. This has led to attempts to privatize water, as in Alltandra Township, South Africa, thus cutting the water supply to those in this poverty-stricken community who cannot afford it.

One of the best demonstrations of how each part of the ecosystem affects all the other parts was seen in the use of DDT in the United States during the 1940s and 1950s. *Dichloro-Diphenyl-Trichloroethane* (DDT) was sprayed on plants by farmers and suburbanites to kill a variety of pests. After just a few years, DDT had seeped through the soil, into the groundwater, on into the seas, then into fish, and finally into birds that ate the fish. Birds that accumulated DDT in their systems appeared healthy enough, but their numbers plummeted. It was soon discovered that DDT caused a disastrous brittleness in the eggs laid by the birds, decimating succeeding generations. The chemical also found its way into the human food supply, with the dangerous consequence of the contamination of human breast milk (Carson 1962; Ehrlich and Ehrlich 1991). By the time DDT was identified as a major environmental hazard, tremendous damage had already been done.

If a growing population is a problem of the developing world, then shrinking resources are a problem of the industrialized world. The United States alone uses more than 40 percent of the world's aluminum and coal as well as about 30 percent of its platinum and copper (Ehrlich et al. 1977; Ehrlich and Ehrlich 1991). Real estate development takes over millions of acres of farmland each year. In the western and southwestern United States, the groundwater supply is being depleted at a rapid pace. We are racing through our non-renewable natural resources and destroying much that should be renewable. However, some activists have begun to claim that the environmental situation, though perilous, is improving (Simon 1995). There is evidence to support that view, but even optimists who wish to show that things are getting better are quick to point out that "better" is not sufficient if the situation is bad enough to begin with (Montagne 1995). According to some ob-

servers, within forty years we will reach the end of the world's supply of lead, silver, tungsten, and mercury, mainstays of heavy and high-tech manufacturing, including the computer hardware industry.

Environmental Pollution

It would be a bitter irony if we managed to avoid exhausting the resources of the planet, only to find that what we had conserved was too degraded by pollution to be useful. The most threatening forms of pollution are the poisoning of the planet's air and water. Air pollution is not only ugly and uncomfortable, it is deadly. The skies of all major cities around the world are stained with pollution hazes, and in cities that rest within geological basins, such as Mexico City and Los Angeles, the concentrations of pollutants can rise so high that pollution-sensitive individuals cannot leave their homes. The numbers of respiratory cases in hospitals rise and fall with the passing of weather systems that cause the pollutants to concentrate or disperse.

Water pollution has an especially insidious side to it—most people never see how much waste is dumped into the world's waterways. They are not present far off shore when the interior walls of vast tankers, acres upon acres of fouled surface area, are rinsed with hot seawater and the waste flushed into the ocean. Small factories dump waste invisibly into canals that feed streams that lead to the sea.

This power plant in Moscow, Russia, spews pollution into the environment, a problem affecting nearly all major industrial cities throughout the world.

The leading air and water polluters are the United States, Japan, Russia, and Poland. When the reign of secrecy in the former Soviet Union (Russia) ended, bloodcurdling stories of environmental vandalism emerged. Nuclear-powered ships at the ends of their useful lives were blithely sunk—contaminated reactors, spent fuel, and all—in Antarctic seas. In the farther reaches of the giant Russian territory (which crosses eleven time zones), areas were designated as open dumping sites for toxic wastes and then sealed off. Similarly, a toxic quarantine area exists downwind of the infamous Chernobyl nuclear power plant that blew up in 1986, spewing radioactive poisons over Europe in amounts never accurately determined, but now appearing to be far in excess of what was once thought to be the worst case imaginable. For comparison, the Three Mile Island nuclear accident that occurred in the United States in 1979 released 15–20 curies of radioactive iodine-131 into the environment; Chernobyl is now estimated to have released as much as *50 million* curies of the same dangerous isotope. Incredibly, the Chernobyl nuclear complex was not shut down until November 2000.

Most pollutants released into the air come from the exhaust systems of motor vehicles, which emit fumes containing carbon monoxide, a highly toxic substance.

Also found in exhaust fumes are nitrogen oxides, the substances that give smog its brownish yellow tinge. The action of sunlight causes nitrogen oxides to combine with hydrocarbons also emitted from exhausts, forming a host of health-threatening substances.

On the industrial side, the Environmental Protection Agency (EPA) has estimated that hazardous and cancer-causing pollutants released into the air by industry are responsible for approximately 12,000 or more deaths a year. Electric utility companies and other industries often burn low-grade fossil fuels that emit harmful sulfur dioxide. When mixed with other chemicals normally present in the air, sulfur dioxide turns into sulfuric acid, which gets carried back to earth through precipitation in droplets of *acid rain*. Acid rain can change the acidity of lakes, soil, and forests so severely that they no longer support life.

A daunting international issue has focused on a group of chemicals called chlorofluorocarbons (CFCs), which are used as coolant in refrigerators, used in the manufacture of plastics, and used as an aerosol propellant. CFCs released into the air find their way to the ozone layer in the upper atmosphere, where they eliminate the highly reactive ozone. The ozone layer is a shield that blocks dangerous ultraviolet light, and as

UNDERSTANDING DIVERSITY
Environmentalism and Social Justice

Robert Bullard notes that the environmental movement in the United States has emerged with agendas that focused on areas such as wilderness and wildlife preservation, resource conservation, pollution abatement, and population control. Over the years, environmentalism has shifted from a "participatory" to a "power" strategy, where the active environmental movement is focused on litigation, political lobbying, and technical evaluation instead of mass mobilization. Concern about the environment cuts across racial and class lines, yet environmental activism and actual participation have been most pronounced among individuals who have above-average education and greater access to economic resources, primarily middle- and upper middle-class Whites. Mainstream environmental organizations were late in broadening their base of support to include Blacks and other minorities, the poor, and working-class persons. Pollution in minority and poor neighborhoods has provided a major impetus for the many environmentalists to embrace equity issues confronting the poor in this country and in the countries of the third world.

An abundance of documentation shows that Blacks, lower-income groups, and working-class persons are subjected to a disproportionately large amount of pollution and other environmental stressors in their neighborhoods as well as in their workplaces. However, these groups have been only recently involved in the nation's environmental movement. Problems facing the Black community have been topics of much discussion in recent years, and race has not been eliminated as a factor in the allocation of community amenities.

Pollution is exacting a heavy toll in health and environmental costs on Black communities across the nation. There are few studies that document, for example, the way Blacks cope with environmental stressors such as municipal solid waste facilities, hazardous waste landfills, toxic waste dumps, chemical emissions from industrial plants, and on-the-job hazards that pose extreme risks to their health. Coping in this case is seen as a response to stress. Coping strategies employed by individuals confronted with a stressor are of two general types: *problem-focused coping,* which is individual or group efforts to directly address the problem, and *emotion-focused coping,* which is the effort to control one's psychological response to the stressor. The decision whether to take direct action or to tolerate a stressor often depends on how individuals perceive their ability to do something about or have an impact on the stressful situation.

Source: Bullard, Robert. 1994. *Dumping in Dixie: Race, Class, and Environmental Quality.* Boulder, CO: Westview Press. •••

MAPPING AMERICA'S DIVERSITY

MAP 21.1 Total Hazardous Waste Sites by State

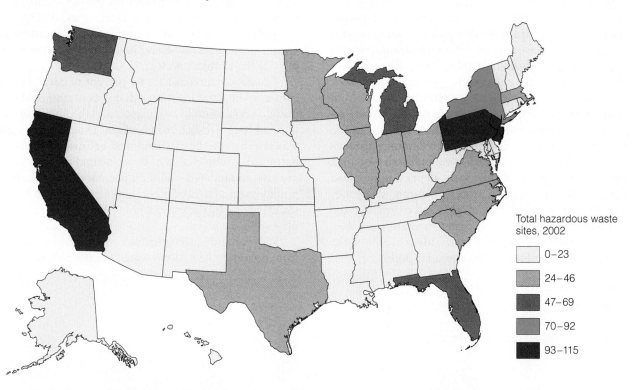

Total hazardous waste sites, 2002

- 0–23
- 24–46
- 47–69
- 70–92
- 93–115

The total number of waste sites in a state varies greatly from state to state. Can you pick out your home state on this map? What is the approximate number of hazardous waste sites in it?

Data: U.S. Census Bureau. 2004. *Statistical Abstract of the United States, 2003.* Washington, DC: U.S. Government Printing. **http://www.census.gov/**

this shield is destroyed, more ultraviolet light gets through, causing an increase in sunburn, skin cancer, and other illnesses. In 1987, a treaty named the Montreal Protocol on Substances that Deplete the Ozone Layer called for drastic reduction in the use of CFCs, but the alternatives are expensive, and developing nations are being called upon to decrease refrigeration, which they are reluctant to do.

Related to the problem of ozone depletion is the **greenhouse effect.** As the sun's energy pours onto the earth, some is reflected from the earth's surface back out again. Of the reflected energy, a portion is captured by carbon dioxide in the earth's atmosphere, while the rest radiates into space. If the amount of solar energy trapped by carbon dioxide rises, the temperature in the atmosphere goes up, a process called *global warming*. Small changes in the average temperature of the earth can have dramatic consequences. A few degrees' difference can cause melting in the arctic regions, which raises the level of the sea, which can affect water, land, and weather systems worldwide. The exact effects are unpredictable.

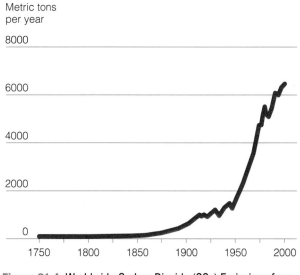

Figure 21.4 Worldwide Carbon Dioxide (CO_2) Emissions from Burning Fossil Fuels, 1751–2000 (metric tons per year)

Source: U.S. Census Bureau. 2004. *Statistical Abstract of the United States, 2003.* Washington, DC: U.S. Department of Commerce.

In the face of warnings about water pollution, many people take comfort in the vastness of the ocean—three-quarters of the earth is covered with water. However, only a tiny sliver of the planet's water is usable by humans. Nearly all the water on earth is seawater—too salty too drink. Of the fresh water on earth, most is locked in the polar ice caps. Of the remainder, most is inaccessible groundwater. All that is left and available to us is the trickle of the planet's river systems and its lakes, and this fragile supply we are polluting. Since about 1955, it has been clear that water is not the infinite resource it was once thought to be (Rice 1986). The nation's rivers and lakes have long been dumping grounds for heavy industry. Yet these same industries—paper, steel, automobile, and chemical—depend upon clean water for their production processes, during which they take water from the rivers and lakes and return it heated up and polluted. The difference in temperature can alter aquatic habitats and kill aquatic life, earning it the name **thermal pollution.** The chemical pollutants that industry discharges into rivers, lakes, and the oceans include solid wastes, sewage, nondegradable byproducts, synthetic materials, toxic chemicals, and even radioactive substances. Add to this the polluting effects of sewage systems of towns and large cities, detergents, oil spills, pesticide runoff, runoff from mines, and even wastes from large cattle and hog farms, then the enormity of the problem is clear.

The EPA estimates that 63 percent of rural Americans may be drinking water that is contaminated as a result of agricultural waste runoff and the improper disposal of toxic substances in landfills. Thousands of rural water wells have been abandoned because of contamination. Households served by municipal water systems are also endangered. Fully 20 percent of the country's public water systems do not meet the minimum health standards set by the government (Weeks 2004; Shaberoff 1988).

Federal and state statutes now prohibit industry from polluting the nation's water, but the pollution con-

VIEWING SOCIETY IN GLOBAL PERSPECTIVE

MAP 21.2 Energy Consumption Per Capita

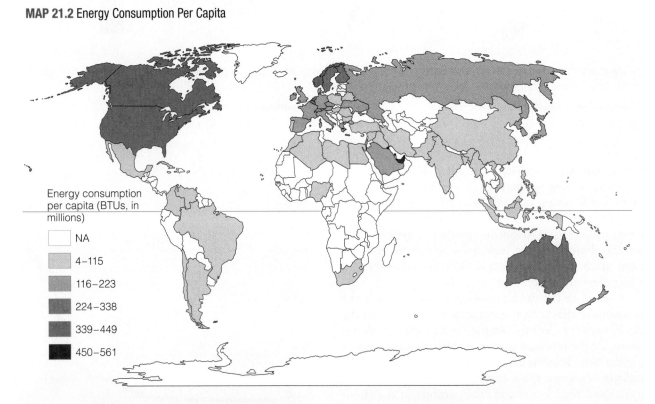

Energy consumption per capita (BTUs, in millions)

- NA
- 4–115
- 116–223
- 224–338
- 339–449
- 450–561

As you can see from this map, the United States and Canada are among the highest consumers of energy in the world. For the United States to reduce its energy consumption and thus become less dependent on foreign energy sources, what lifestyle changes would Americans need to make? Do you think there are conditions in which they would be encouraged to do so and why might they resist?

Data: From the U.S. Census Bureau, 2004. *Statistical Abstract of the United States, 2003.* Washington, DC: U.S. Energy Information Administration, p. 847.

tinues. Why? The answer is economic, political, and sociological. Industries that contribute to a vigorous economy have traditionally met little interference from the government. Only recently, and only because of public awareness and outrage, has the government started cracking down on major polluters. Nevertheless, the government often chooses a look-the-other-way attitude. The problem was addressed in 1993 by the inauguration of project GLOBE (Global Organization for a Better Environment). Administered by the Congressional Institute for the Future, GLOBE is an international organization attempting to stop all forms of pollution, particularly water pollution, through a combined effort by several major polluters, including the United States and Japan. Despite such national and international efforts, federal or local legal actions against major polluters often end with the corporation paying a relatively painless fine and continuing to pollute as appeals and further proceedings for new violations wind their way slowly through the cumbersome regulatory and judicial systems.

Many argue that of all the environmental problems facing the United States today, the most urgent is the dumping of hazardous wastes, if only for the sheer noxiousness of some of the materials being dumped. It is estimated that since 1970, the production of toxic wastes increased fivefold (Weeks 2004). This dramatic increase in the amount and variety of hazardous waste production is traceable to new and profitable industrial technologies. The public creates great demand for products that inevitably produce hazardous wastes, such as insecticides and other useful poisons; products requiring mercury or lead (both acutely toxic when released into the environment); a variety of dyes, pigments, and paints; and an endless list of specialty plastics whose manufacture produces dangerous byproducts.

Environmental Racism and Classism

Adding to the social problem of toxic wastes is the fact that wastes are dumped with disproportionate frequency in areas with high concentrations of minorities, particularly American Indians, Hispanics, and African Americans, as well as persons of lower socioeconomic status (Pellow 2002; Holmes 2000; Boer et al. 1997; Pollock and Vittas 1995; Bullard 1994b; Williams 1987). One study determined that it was "virtually impossible" that dumps were being placed so often in minority and lower SES (socio-economic status) communities by chance (Bullard 1994a, 1994b). The same study found that communities with the greatest number of toxic dumps had the highest concentration of non-White residents. Such communities also tended to fall below the national average economically and educationally. **Environmen-**

© Ted Spiegel/Corbis

These are townspeople protesting a toxic waste site located next to an Hispanic neighborhood.

tal racism consists of the dumping of toxic wastes with disproportionate frequency at or very near areas with high concentrations of minorities.

One study of households throughout Florida found that Native American, Hispanic, and particularly African American populations are disproportionately found to reside closer to toxic sources than are Whites. This pattern is *not* explainable by social class differences alone. That is, when comparing communities of the same socioeconomic characteristics but different racial–ethnic compositions, Native Americans, Hispanics, and African Americans of a given socioeconomic level are still closer to toxic dumps than are Whites of the *same* socioeconomic level (Holmes 2000; Pollock and Vittas 1995).

DEBUNKING SOCIETY'S MYTHS

Myth: Environmental pollution is in fact more common in or near economically poor areas, but race has nothing to do with it.

Sociological perspective: Even when areas of the same low socioeconomic status but different racial compositions are compared, those with a greater percentage of minorities are on average closer to polluted areas than those with a lower percentage of minorities.

Concept by Norman Andersen

Environmental Racism

Environmental racism refers to the pattern whereby people living in predominantly minority communities are more likely exposed to toxic dumping and other forms of pollution. Nuclear waste and testing in the American Southwest, for example, have been located in areas predominantly inhabited by Native Americans. In other areas, African Americans and Latinos are exposed to the effects of industrial waste.

These studies found that a greater proportion of ethnic minority households are closer to toxic sites (TRIs, or toxic release inventories), even when considering that all the households are of comparable low-income status. The researchers note that amount of exposure to the harmful effects of pollution is directly related to physical distance from a toxic facility. An *environmental inequity* exists among ethnic groups, with African Americans, Hispanics, and Native Americans bearing a greater share of the burden of exposure to the harmful effects of environmental pollution.

In a similar study of Los Angeles, Boer and associates (Boer et al. 1997) found that lower social class areas were significantly nearer toxic waste dumps, as were areas that were predominantly Latino or African

American. Generally, they found that communities most affected by toxic waste dumps were working-class communities or people of color located near industrial areas.

Among the largest commercial hazardous waste landfills in the nation is that located in Emelle, Alabama, where Blacks make up nearly 80 percent of the population. In Scotlandville, Louisiana, another large toxic landfill in the country, Blacks make up 93 percent of the population. In Kettleman City, California, the site of another large toxic landfill, more than 78 percent of the residents are Hispanic. In many cases, siting landfills was linked to the economic interests of both residents and corporations. Companies with profit in mind negotiated favorable deals with the residents of these communities to permit the dumping of wastes, often misrepresented as nontoxic, in exchange for jobs and other economic incentives, which were often slow in coming (Bullard 1994b).

From the 1950s through the 1970s, the Navajo population of Shiprock, New Mexico, was exposed to waste from uranium mining and dumping that was 90 to 100 times more radioactive than the level permissible by law. Kerr-McGee, the corporation involved, was forced out of the area in the early 1970s. When it left, it simply abandoned the site, leaving seventy acres of radioactive mine tailings (the residue from the separation of ores). An even worse situation developed during the same period in Laguna, New Mexico, involving the Pueblo Indians. Anaconda Copper, a subsidiary of the Atlantic-Richfield Corporation, virtually wrecked the traditional Pueblo economy by recruiting the community's youth for hazardous jobs even as it contaminated their environment with the wastes from uranium mining. A high rate of cancer deaths in both communities serves as testimony to the horrible consequences of carelessly discarding toxic wastes (Churchill 1992).

THINKING SOCIOLOGICALLY

The *human ecosystem* can be adversely affected by environmental hazards such as toxic wastes. Have you ever witnessed the effects on a population of a major environmental hazard disaster such as Love Canal in upstate New York, or Three Mile Island? Did you grow up in or near such an area, or know someone who did? From your observations, are toxic waste dumps more likely to be in or near areas that are working-class or heavily occupied by people of color? And as shown by research? Quickly survey a few women you know, then a few men, and find out how concerned they are about environmental issues. Do the women on average show more concern, as shown by past research?

Feminism and the Environment

Women and men do not regard environmental issues equally. In general, women tend to be more concerned with issues of environmental risk, and this has important policy implications: Lack of attention on the part

DOING SOCIOLOGICAL RESEARCH
New Findings Challenge Demographic Transition

As demographic transition theory argues, it used to be assumed that economic development was the best way for a poor country to reduce its population growth. New studies suggest that a country such as Bangladesh can cut its birthrate significantly if it aggressively promotes the adoption of modern contraceptive methods—without waiting for the reduction that traditionally comes with higher living standards. Presumably, such a strategy may work with certain U.S. populations, although cultural resistance to contraceptive methods may lessen the likelihood of their adoption.

In what some experts are calling a reproductive revolution, birthrates are falling in countries presumably too poor for economic development that would stabilize rapid population growth (as it did in Europe and North America early in the twentieth century), suggesting that contraceptives can work.

Demographic transition theory, predicated on the experience of the industrialized world before modern con-

traceptives, says that in a preindustrial economy, people tend to have many children but that high death rates hold the population down. As a country industrializes, living conditions improve, the practice of medicine advances, and life expectancy increases, but the birthrate remains high and the population soars. Only as education spreads and people find that too many children are an economic liability do birthrates drop, the theory goes. Thus begins an era of low mortality, low fertility, and stable but not runaway population growth—the condition of the industrialized world today.

Bangladesh is a perfect example of how the concept linking the fertility rate to developing economies has been disproved. This South Asian country, one of the world's poorest and most densely populated, has a traditional agrarian economy in which most families still depend on children for economic security. Yet Bangladeshi fertility rates declined significantly between 1970 and 2001, from 7 to 2.8 children per woman. In

the same period, the use of contraception among married women of reproductive age rose to 40 percent from 3 percent.

Questions to Consider
1. How and in what way is demographic transition an "ethnocentric theory"? *Keywords: demographic transition*
2. How and in what ways does *rational choice theory* contradict demographic transition theory? *Keywords: fertility rate; population growth*

We have included InfoTrac College Edition keywords at the end of each question to make it easier for you to find more to read on these topics. Go to www.infotrac-college.com, an online library, to begin your search.

Source: Office of Population Research, Princeton University, 2004; Stevens, William K. 1994. "Poor Lands Success in Cutting Birthrate Upsets Old Theories." *The New York Times*, January 2, 1994, pp. 1, 8; U.S. Census Bureau. 2004. *Statistical Abstract of the United States, 2003.* Washington, DC: U.S. Government Printing Office. •••

of local and federal governments to environmental issues can be interpreted as lack of attention to policy that differentially affects women. In this respect, it is a feminist issue.

In one study (Bord and O'Connor 1997), women and men were both asked a set of detailed questions pertaining to their perceptions of risk to themselves from environmental hazards. Women consistently showed more concern than men for environmental issues and perceived themselves to be at considerably more risk from environmental hazards than men. For example, women were more likely than men to perceive that abandoned waste sites cause cancer, could produce miscarriages, and cause other health problems. They were also more likely than men to feel that waste sites posed dangers to trees, fish, and other wildlife. Women were also more likely than men to perceive dangers in global warming. They were more likely to predict, as a result, coastal flooding from polar ice meltdown, loss of forests, increased killing off of certain animal species, increases in hurricanes and tornadoes, and increased air and water pollution.

The issue is not, for example, whether global warming will definitely result in such calamities, because there is room for debate on the issue. The issue is that women feel more vulnerable than men to the risks posed by such environmental problems (Bord and O'Connor 1997; Blocker and Eckberg 1997), and as a consequence, women are more concerned that policy makers will act to reduce these risks.

Environmental Policy

Environmental policy of the United States government over the last thirty years or so has been affected by what has come to be called the "environmental movement." This social movement (see Chapter 22) consists of various loosely organized groups such as the Earth Day group, NIMBY ("Not in my back yard"), the Sierra Club, and other similar organizations. Rachel Carson's seminal book titled *The Silent Spring* (Carson 1962) served as a significant impetus to the environmental movement. The book is a scathing indictment concerning the pesticide chemical DDT and its far-reaching

SOCIOLOGY IN PRACTICE
The Office of Population Research

The Office of Population Research at Princeton University is the oldest population research center in the country. Founded in 1936, it has trained many persons, from undergraduate degree recipients to doctoral recipients, who have professional positions in the field of demography in the United States as well as in developing countries. Many of these jobs are concerned with the reduction of population growth, and the Office of Population Research engages in distribution and training in the use of contraceptives.

The Office provides many services other than undergraduate and graduate degree programs. Information about reproductive health is provided, as is information on *emergency* contraception. It provides extensive data archives and produces a research journal, *Population Index,* on the Web. Research programs housed in the Office cover areas such as aging, fertility and fecundity, health and mortality, marriage and the family, demographic methods, migration, and the environment.

A current ongoing project on the environment is housed in Mexico. It is engaged in a cooperative effort with a grassroots community action organization in Michoacan, Mexico, called ORCA. The purpose of this cooperative effort is to introduce alternative and environmentally positive technologies into the community and then engage in research on their effects. These technologies involve the following:

- latrines that do not contaminate the groundwater with human wastes;
- stoves (for cooking) and kilns (for pottery) that reduce levels of indoor air pollution through improved ventilation mechanisms;
- elimination of the need for high-priced chemical fertilizers and pesticides;
- introduction of alternative crop rotation strategies.

These four projects all form an integrative framework within the existing organizations. They function as a unit. Each project learns from the experiences of the other projects and thus provides feedback to the benefit of the ecology and inhabitants of the region.

Critical Thinking Exercise

1. What are some of the ways you can think of to attempt to overcome cultural resistance to contraception and encourage a group or society to adopt some method of contraception? To what extent might it depend upon the type of contraception advocated (for example, the condom versus the pill)?

2. Do you think it is a good idea to try to get any culture to decrease the birthrate by adopting some form of contraception? Do you think such attempts are a threat to the survival of the group or society in question?

Source: Office of Population Research, Princeton University. Website: **http://opr.princeton.edu** •••

polluting effects from toxic runoff, including the severe reduction of many bird populations and even the contamination of human breast milk. The Earth Day mobilization in 1970, repeated again in 1990, received tremendous support from the U.S. public (nearly 70 percent according to one survey; Dunlop and Mertig 1992).

The infamous Love Canal debacle is often cited as a recent spur to the movement. During the 1980s, in an area of upper New York state near Niagara Falls called Love Canal, new homes were discovered to have been built on what had been large toxic waste dumps and landfills. The pollution was so severe that many of the homeowners discovered a dark gray chemical ooze issuing from their backyard lawns (Levine 1982). The homeowners abandoned their properties, which remain unoccupied and unsalable to this day.

In the past three decades, federal and local agencies have made concerted efforts to bring the problems of environmental pollution under control. The main lines of attack have been stiffer antipollution laws and the encouragement of alternative technologies.

Antipollution laws have been resisted by industry because they require expensive adaptations of manufacturing processes, and resisted by unions for fear that the added expense to industry would cost jobs. Despite these points of opposition, many antipollution laws have been passed since the late 1960s and have won great public support. In the early 1980s, then-President Ronald Reagan's administration relaxed antipollution standards on the grounds that they were too costly for industry, and such relaxed policies generally continued under both Bush presidencies—President George H. W. Bush as well as President George W. Bush. Important questions still persist involving the balance of interests of big business and environmental protection.

FORCES OF SOCIAL CHANGE
Suburbanization

The growth of suburbs since World War II has been so pronounced that, in many ways, it defines the American dream—a house of one's own, good schools, family stability, two cars in the driveway, and green lawns. The process of suburbanization has been profoundly influenced by the growth of the automobile industry after World War II and the expansion of state and federal highways. How has suburbanization transformed society, and has the promised dream been realized?

Suburbanization has created massive urban sprawl as most large cities have expanded over formerly rural areas. Once remote, these areas now are densely populated, placing strains on natural resources and increasing pollution. Witness the political struggles over water in cities of the Southwest and southern California or the drying of the Everglades in southern Florida.

The reality of suburbanization may also be very different from the American dream. Ideals of family togetherness are challenged by the fact that longer commuting times and long distances from work may result in the isolation of family members from others—a development exacerbated by an emphasis on home-centered entertainment and home maintenance (Miller 1995).

The suburbs have also contributed to racial segregation, both in housing and schools. As more White, middle-class families have moved to the sub-

urbs, urban schools have suffered from a weakened tax base. "White flight" has then increased the residential segregation of African Americans and, now, Latinos. Moreover, as racial minorities have moved into the suburbs, they have tended to live in considerably poorer neighborhoods than is the case for Whites. Access to public services reflects this racial disparity.

Such conclusions show the connection between the growth of suburbs and the race and class structure of society in the United States. You might ask yourself how suburbanization has changed the region where you live and how the growth of suburbs might affect the future for diverse groups in your region of the country.

Resource: The Urban Institute. 2003. Website: **www.urban.org**

Suburban sprawl in places such as the San Fernando Valley (pictured in 1930 and the 1990s) has transformed not only the natural environment but also social relationships.

Several automakers have recently released "hybrid" (gas and electric) pollution-reducing automobiles.

The development of new technologies in the last thirty years or so has played a major role in the reduction of certain kinds of pollution. Since the early 1970s, emissions controls for automobile exhaust have been widely installed. Electric cars are once again enjoying a vogue, as is the search for alternative fuels to replace gasoline, such as methanol or methane derived from the fermentation of human and animal waste products.

Globalization: Population and Environment in the Twenty-First Century

The U.S. census predicts that the world's population will increase from the 6 billion it is now to 7.9 billion by the year 2020. Even more upsetting is the United Nation's revised predictions concerning when the world's population would stabilize. A few years ago, the United Nations Division on Population estimated that the world population would stabilize as it reached 9 billion. That estimate has been revised to 10 billion, with a high estimate of as much as 14 billion.

Sociologists predict that the United States will continue to experience increasing suburban development, with accompanying increases in heavy industry and thus additional pollution. The rate at which people are leaving the centers of today's cities will slow somewhat. A major concern that today's sociologists have for the future is the effect that a changing planet will have upon our lifestyle (Weitz 2004; Brown et al. 2000; Palen 1992; Logan and Molotch 1987) and the reverse—the effect our lifestyle will have on the planet.

For the first time in our history, our own lifestyles may threaten our very existence. Perhaps, as philosopher Matthew Arnold once said, humankind "carries with it the seeds of [its] own destruction."

One analyst, Jeremy Rifkin, admittedly combining speculation with only a dash of science fiction, foresees a depressing outcome to our abuse of the environment:

> "The year—2035. In an effort to hold back the rising sea water, massive dikes have been built around New Orleans, New York, and Miami. Phoenix is baking in its third week of temperatures of 155 degrees. Decades of drought have laid waste to the once fertile Midwest farm belt. Hurricanes batter the Gulf Coast and forest fires continue to blacken thousands of acres across the country." (Rifkin 1989)

Ecological concern has stimulated such developments as the field of **ecological demography,** which combines the studies of demography and ecology (Namboodiri 1988). This field monitors experimentation

Courtesy of American Honda Motor Co., Inc.

TAKING ON SOCIAL ISSUES
Environmental Racism

Environmental racism has become a cause for concern among people worried about the impact of pollution and waste on the nation's already most disadvantaged groups. What sociological factors influence environmental racism, and what groups have developed to promote social action and social policy on this issue? Are there such organizations in your community or state, and what communities in your region are most vulnerable to environmental hazards?

Taking Action

Go to the Taking Action Exercise on the Companion Website—at http://sociology.wadsworth.com/andersen_taylor4e/—to learn more about an organization that addresses this topic. •••

with alternative fuels, fertilizers, and pesticides; efforts at the recycling of toxic wastes; and protection for the ozone layer. It also observes the use and success of alternative technologies, such as solar, wind, and geothermal power. The development of such alternative technologies offers some hope for deflecting Rifkin's dire predictions.

As the twenty-first century dawns, earth seems to be shrinking, giving ecology a worldwide, or global, dimension. **Ecological globalization** is now upon us—the worldwide dispersion of problems and issues involving the relationships between humans and the physical and social global environment. Goods, money, people, ideas, and pollution are traveling around the world on an unprecedented scale. With the upward soaring of international trade in fish and timber comes the internationalization of environmental issues. Environmental problems are escalating on the international political agenda, at times occupying the attention of diplomats as much as the issues of arms control and war. Fears arising from the introduction of *genetically modified organisms* (GMOs) permeate the governments of many nations. Many governments are ill-suited for managing environmental problems that transcend their borders, whether via air, water, or international commerce. Yet global environmental governance is still in its infancy, and nations and economies are coming to the realization that international coordination in ecological globalization has become necessary for any significant likelihood of global human survival.

Sociology ⊛ Now™
Reviewing is as easy as ❶❷❸.

1. *Before you do your final review, take the SociologyNow diagnostic quiz to help you identify the areas on which you should concentrate. You will find information on SociologyNow and instructions on how to access all of its great resources on the foldout at the beginning of the text.*

2. *As you review, take advantage of SociologyNow's study videos and interactive Map the Stats exercises to help you master the chapter topics.*

3. *When you are finished with your review, take SociologyNow's posttest to confirm you are ready to move on to the next chapter.*

Chapter Summary

What is demography?

Demography is the study of population, a field that focuses upon three fundamental processes, which together determine the level of population at a given moment: births, deaths, and migrations. We noted that the United States ranks very low in *life expectancy* and high in *infant mortality*, relative to other Western countries.

How is diversity relevant?

Diversity is of great significance because both *infant mortality* and *life expectancy* are not equal across all races and social classes, nor for men and women. Women have a greater life expectancy than men in virtually all countries and at all social class levels. However, the lower one's social class, the lower one's life expectancy, regardless of gender, and the greater the infant mortality. Minority group individuals, especially African Americans, Hispanics, and American Indians, all have lower life expectancies and higher infant mortality than Whites.

What about current migration patterns?

Current migration patterns show that in addition to overall movement from city to suburb by Whites and people of color, including the "new suburbanite" populations from other countries and cultures, large portions of populations, such as the underclass, remain stuck in central cities. The study of a *cohort*, such as Baby Boomers, showed how a demographic cohort can both use and produce cultural change.

What is the Malthusian problem, and why is it important?

Malthusian theory, still relevant today, warns us about the dangers of exponential population growth along with only arithmetic increases in food and natural resources. To avoid the Malthusian positive checks of famine and war, population control can be, and is being, instituted by means of programs of family planning and birth control. The level of zero population growth (ZPG), and replacement-level birthrates, can be achieved and already has been in many parts of the United States. Whether methods of contraception advocated by the government are adopted, either in the United States or in other countries, depends on the culture of the group in question. Some Latino groups and some African Americans consider advocacy of contraception for people of color, without equally rigorous contraception advocacy for middle-class Whites, to border on genocide.

What are the current problems pertaining to urbanization, human ecology, and the environment?

Any society is an *ecosystem* with interacting and interdependent forces, consisting of human populations, natural resources, and the state of the environment. Depletion of one natural resource affects many other things in the ecosystem. The dumping of toxic wastes is a very serious problem in our society, especially when toxic dumps are found more frequently—as they are—in or very near African American, Hispanic, and American

Indian communities. Such practices constitute what some researchers call *environmental racism*. Surveys have shown that environmental risks are of more concern to women than to men. As a consequence, environmental policy has more impact on women than on men. *Ecological globalization* shows that environmental issues are now international in scope and are interconnected between countries around the globe.

Key Terms

age–sex (age–gender) pyramid 566	human ecology 574
census 562	human ecosystem 574
cohort 567	immigration 563
crude birthrate 563	infant mortality rate 564
crude death rate 564	life expectancy 564
demographic transition theory 569	Malthusian theory 568
demography 562	population density 574
ecological demography 584	population replacement level 570
ecological globalization 585	rational choice theory 570
emigration 563	sex ratio 566
environmental racism 579	thermal pollution 578
greenhouse effect 577	urbanization 573
	vital statistics 562

Researching Society with MicroCase Online

You can see the results of actual research by using the Wadsworth Microcase® Online feature available to you. This feature allows you to look at some of the results from national surveys, census data, and other data sources. You can explore this easy-to-use feature on your own, but try this example. Suppose you want to know:

Which states in the United States are the most dense in population in terms of persons per square mile? Which states are the least dense? Which are somewhere in the middle?

To answer this question, go to http://sociology.wadsworth .com/andersen_taylor4e/, select MicroCase Online from the left navigation bar, and follow the directions there to analyze the following data.

Data file: States—The 50 United States

Task: Mapping

Variable: DENSITY—POPULATION PER SQUARE MILE

Questions

Once you have your results, answer the following questions:

1. Which states are the most dense? Which states are the least dense?

2. Which states have a moderate amount of density?

3. Locate your home state on the map. What is the approximate density of your home state?

4. To what extent do you think many modern environmental problems, such as pollution, are related to population density? Are the more densely populated states those with more pollution?

The Companion Website for Sociology: Understanding a Diverse Society, Fourth Edition

http://sociology.wadsworth.com/andersen_taylor4e/

Supplement your review of this chapter by going to the Companion Website to take one of the Tutorial Quizzes, use the flash cards to master key terms, and check out the many other study aids you'll find there. You'll also find special features such as GSS Data and Census 2000 information, data and resources at your fingertips to help you with that special project or do some research on your own.

Suggested Readings and Web Resources

Brown, Lester R. 2000. *State of the World: A World-watch Institute Report on Progress Toward a Sustainable Society.* New York: W. W. Norton.
This is a collection of articles discussing environmental issues and risks globally.

Bullard, Robert. 1994. *Dumping in Dixie: Race, Class, and Environmental Quality.* Boulder, CO: Westview Press.
Bullard's book is a readable and glaringly clear account of how race as well as class is connected with toxic waste disposal. It is a major source of hard evidence of environmental racism and classism.

Carson, Rachel. 1962. *The Silent Spring.* New York: Knopf.
This book is the classic on how pollution from DDT affected the environment, birds, other animals, and humans. The book represents a major phase in the environmental movement.

Pellow, David. 2002. *Garbage Wars: The Struggle for Environmental Justice in Chicago.* Cambridge, MA: MIT Press.
This book establishes a framework for understanding environmental racism, using a case study of the City of Chicago.

Rumbaut, Ruben G., and Alejandro Portes (eds.). 2001. *Ethnicities: Children of Immigrants in America.* New York: Russell Sage Foundation.

A readable collection of essays and studies of second-generation ethnics in America, focusing on urban patterns, struggles, and attempts of families to sustain homeland culture in their children.

Environmental Protection Agency

www.epa.gov

The EPA is the government agency responsible for monitoring the environment and developing public policy to protect it.

Greenpeace

www.greenpeace.org

An activist organization that uses nonviolent action to protect the environment.

World Resources Institute

www.wri.org

An international organization dedicated to protecting the earth's environment.

● ● ●

Collective Behavior and Social Movements

When was the last time you were in a large crowd? Perhaps it was a concert where people clapped, danced, and called for an encore at the show's end. Or maybe it was a sports event where fans cheered for their team and booed the other. Did fans begin doing the wave or stand and gesture to "YMCA"? Perhaps you have marched on Washington on behalf of an issue you feel is important. Or perhaps you have joined some local organization promoting a cause that you feel strongly about, such as protecting the environment, demonstrating against war, or protesting something that people perceive to be wrong and want changed. In any of these cases, you would have been participating in what sociologists call *collective behavior*.

Collective behavior is defined as behavior that occurs when the usual social conventions are suspended and people collectively establish new norms of behavior in response to an emerging situation (Turner and Killian 1988: 3). This can include crowd behavior, such as in the preceding examples, but it can also be more sustained and organized than crowds typically are. Thus, collective behavior includes the study of **social movements,** groups that act with continuity and organization to promote or resist change (Turner and Killian 1988).

Collective behavior occurs when something out of the ordinary happens and people respond by establishing new behavioral norms, such as people spontaneously placing numerous shrines on the sidewalks of New York following the terrorist attack on September 11, 2001. Although collective behavior may emerge spontaneously in response to a unique situation, it is not entirely unpredictable. Established patterns of collective behavior exist, even when

Sociology ⊗ Now™
Reviewing is as easy as ❶❷❸.

Use SociologyNow to help you make the grade on your next exam. When you are finished reading this chapter, go to the chapter review for instructions on how to make SociologyNow work for you.

unusual, unpredicted events generate such behavior. Collective behavior thus includes crowds, riots, disasters, and social movements, as well as other forms of mass action, such as fads and fashion.

Sociologists who study collective behavior are interested in how even unique and idiosyncratic events are socially structured and how collective behavior generates social change. As we will see, some of the phenomena defined as collective behavior are whimsical and fun, such as fads and some kinds of crowds. Other collective behaviors can be terrifying, such as a riot that gets out of control. Whether whimsical or awesome, collective behavior is innovative, sometimes revolutionary, and it is this feature that links collective behavior to social change. •••

Characteristics of Collective Behavior

Most days, life is more or less predictable, but that can change in a moment. When something unusual happens or when a gathering becomes focused on a specific event, collective behavior is spawned. Collective behavior can also emerge when people think something unusual will occur, even if it never does. Thus, included in the study of collective behavior are such things as cults that emerge when people think spaceships are about to land on earth or that some other supernatural event will occur at some designated time. Such was the case with the Heaven's Gate cult, whose members collectively committed suicide in 1997, thinking that they would be getting on a spaceship in the tail of Halley's Comet and find a new world, free of the perceived restrictions in this world.

Collective behavior can also occur as the result of natural and manmade disasters. Blizzards, hurricanes, floods, and earthquakes all create situations in which people develop new ways of behaving in the face of extreme and unusual circumstances. Now sociologists are also increasingly interested in the collective behavior that results from terrorism. As with responses to disasters, amid the devastation of a terrorist event, groups of neighbors or other affected groups, as well as organized response teams, come together to help one another meet their immediate needs and try to reestablish their ways of living before the disaster (Kreps 1994).

To understand collective behavior, it helps to understand what the different forms of collective behavior have in common.

1. **Collective behavior always represents the actions of groups of people, not individuals.** The action of a lone gunman who opens fire in a post office is not collective behavior, because it is the action of only one person. However, groups that gather at the scene to observe the emergency response are engaged in collective behavior, as are people who organize to publicly protest a nuclear power plant. Collective behavior has some of the same characteristics as other forms of social behavior. It is rooted in relationships between people (Weller and Quarantelli 1973) and involves group norms, such as the expectation that people in a baseball stadium stand up and sing "Take Me Out to the Ballgame" in the middle of the seventh inning.

2. **Collective behavior involves new or emergent relationships that arise in unusual or unexpected circumstances.** Collective behavior often falls outside everyday institutional social behavior. The behavior of people who commute to work together is not considered collective behavior, because commuting is an ordinary part of their everyday life. However, if traffic comes to a halt because a barge has plowed into a drawbridge up ahead and everyone gets out of their cars to look, their actions constitute collective behavior because they are all responding to the same unusual circumstance.

Collective behavior arises when uncertainty in the environment creates the need for new forms of action. People may undertake tasks that are new to them (Weller and Quarantelli 1973), perhaps doing things they never imagined themselves capable of before. For example, when a community is struck by a flood, earthquake, or hurricane, suddenly nothing can be taken for granted—not food, water, transportation, electricity, or shelter. Collective behavior emerges to meet the new and immediate needs that people face. Following a hurricane, people may form neighborhood work groups to clear debris, or they may organize teams to stack sandbags in the threat of a flood. Following the collapse of the World Trade Center, there were numerous reports of people helping each other to exits, as well as heroic efforts by firefighters and other rescue workers who quickly organized—even at risk of death—to save as many people as they could.

DEBUNKING *SOCIETY'S MYTHS*

Myth: In the face of disasters and other unexpected events, people behave irrationally and outside of the normal social influences.

Sociological perspective: When faced with unexpected events, people develop norms to guide their behavior, often drawing on prior social behavior and knowledge to guide new interactions (Turner and Killian 1988).

---

3. **Because of its emergent character, collective behavior captures the more novel, dynamic, and changing**

elements of society to a greater degree than other forms of social action. Fads, such as Yu-Gi-Oh among young children, rollerblading, or body piercing among young adults, introduce something new into everyday social life. Although fads may stretch across several seasons, the collective behavior it involves is usually short-lived. Social movements are a collective behavior that typically develops over a longer period; crowds are more ephemeral. Both long-term and short-term incidents of collective behavior can, however, transform society. The American Indian takeover of Alcatraz Island in 1968 followed by a seventy-one-day occupation at Wounded Knee, for example, transfixed the nation and called attention to the demands of American Indians. Actions demanding political response can be the basis for long-term and short-term social change.

4. **Collective behavior may mark the beginning of more organized social behavior.** Collective behavior often precedes the establishment of formal social movements. People who spontaneously organize to protest something may develop structured ways of sustaining their protest. One of the best historical examples comes from the community protest at Love Canal. When the New York state health commissioner announced in 1978 that a toxic waste dump site located at Love Canal in Niagara Falls, New York, presented a "great and imminent peril to the health of the general public residing at or near the said site" (Levine 1982: 28), within days, Lois Gibbs and other housewives residing in the area came together to form the Love Canal Home Owner's Association (Gibbs 1982). The action of this small neighborhood group was an important first step in the development of social movements to protest the dumping of toxic waste, a precursor to today's environmental justice movement (Thomas 1995). Since then, numerous comparable social movements can be found—many of which involve minority communities threatened by the high level of environmental hazards in their communities. The *environmental justice movement* includes a wide array of Native American, African American, Latino, and other communities that have organized to protest the dumping and pollution that imperils their neighborhoods (Pellow 2002; Bullard 1994a).

5. **Collective behavior is patterned behavior, not the irrational behavior of crazed individuals.** Patterned behavior is activity that is relatively coordinated among the participants. For example, crowd members may all be focused on the same thing, such as a rock band. They attend concerts with their friends and progress to and from the concert site in a more or less orderly fashion. During episodes of collective behavior, people may follow new guidelines of social behavior, but they do follow guidelines. Even episodes of panic, which may appear to be asocial and disorganized, follow a relatively orderly pattern.

A good example of collective behavior caused by panic is the actions of the passengers on the flight that was overtaken by terrorists on September 11, 2001 and then crashed in a field in Pennsylvania. Although we will never know exactly what transpired on that flight, we do know that passengers, faced with the threat from terrorists, organized to thwart the terrorists' plan. In doing so, they engaged in collective behavior—quickly changing from a collection of mostly unrelated individuals to an organized social group.

6. **Many forms of collective behavior appear to be highly emotional, even volatile.** Episodes of collective behavior often exhibit the more emotional side of life, as when hundreds of vigils were quickly organized to grieve the death of thousands killed on 9/11. Not all emotional behavior is collective behavior, however, nor is all collective behavior deeply emotional. Parents who grieve over a dead child are emotional but not necessarily acting collectively. But if a group of parents gather at the site of a school bus accident, weeping over lost lives, this is collective behavior. Emotionality per se does not define collective behavior; what defines collective behavior is its spontaneous character.

7. **During collective behavior, people communicate extensively through rumors.** Lacking communication channels or distrusting the channels available, people

Collective behavior is spontaneous, as when people created memorials like this one on New York's sidewalks as a tribute to those lost on 9/11/01.

use rumors to define an ambiguous situation (Turner et al. 1986). In the popular view, rumors are rapidly circulated information of questionable accuracy. From a sociological point of view, *rumors* are the information transmitted by participants in collective behavior as they try to make sense out of an ambiguous situation. Rumors are transmitted by people who are piecing together information about a story with facts that are partly obscured. Rumoring is a collective activity and most common when there is inadequate information to interpret a problematic situation or event (Dahlhamer and Nigg 1993: 2). Rumor is present in almost all instances of collective behavior. Ambiguity is a key component of such behavior, and rumors thrive on ambiguity (Turner and Killian 1988). Rumors are important because they convey information, and they meet social-psychological needs in people faced with uncertainty (Shibutani 1966).

8. **Finally, collective behavior is often associated with efforts to achieve social change.** This can mean promoting change or resisting it. Members of protest groups and social movements are always trying to effect change

in the political, cultural, or value structures of society. The civil rights movement and *La Raza* (a Latino rights movement) are examples of social movements promoting change in the racial order of society. Reactionary movements like the militia movement and White supremacist movements are attempts to resist change toward greater racial equality.

Some people may believe the stereotype of collective behavior is irrational and antisocial, but collective behavior involves patterned group activity that is spontaneous and arises when people see the need for new forms of action (Goode 1992). Collective behavior includes different forms of behavior, such as crowds, riots, panics, and other emergent forms of collective action, as we will see next.

Crowds

Crowds are one form of collective behavior. Sociologists differ in their conceptions of crowds, but most agree that crowds share several characteristics. First, *crowds involve groups of people coming together in face-to-face or visual space with one another.* This is true of other forms of social behavior as well, but crowds are *transitory*. They form when groups come together for a specific transient event (a sports event, a spiritual retreat, a rock concert, or a riot). The same group of people will probably never reconvene. Crowds are *volatile*. As crowds develop, their behavior may change suddenly. The behavior of people in crowds is different from their behavior in other social settings. Finally, crowds usually have a sense of *urgency*. They are focused intensely on a single event.

THINKING SOCIOLOGICALLY

The next time you find yourself in a *crowd,* think about the sociological characteristics of the situation. Resist the pull of the crowd and instead ask: What is the character of the crowd? Is there a division between the crowd and onlookers? Are there permeable boundaries? What norms are guiding the behavior of different participants in the crowd? Answering these and other questions will help you see the *social structure* of the crowd.

You can find countless examples of crowds, differing in their purposes and behavior, but all following observable sociological patterns. Within crowds, people think they are acting as individuals, but like other forms of group behavior, they are being shaped by the collective action of others. People in crowds seem to take on a collective identity, and it may even be difficult to distinguish individual and group behavior. Crowds seem to act as one even though there may be great diversity within them.

The Social Structure of Crowds

Crowds have a discernible social structure. Any one crowd is usually a particular size, with participants packed together in a particular density, and people in the crowd more or less connected to one another (Goode 1992). Crowds are usually "circular," surrounding the object of the crowd's attention. The people closest to the crowd's center of interest are the *core* of the crowd and show the greatest focus on the object of interest. At the outer edges of the crowd, attention is less focused. People there are more likely to be talking to their friends, playing Frisbee, or participating in other activities.

DEBUNKING SOCIETY'S MYTHS

Myth: Crowds are disorganized groups whose behavior is less orderly and predictable than other social groups.

Sociological perspective: Crowds have a social structure that is observable in the patterns of crowd behavior (Goode 1992).

The boundaries crowds have are more or less permeable. Some crowds make movement in and out of them very difficult (Milgram and Toch 1968). Other crowds have movable boundaries, such as crowds at outdoor concerts where people may walk in and out of the area nearest the stage and wander the grounds, perhaps tailgating, watching others in the crowd, or even engaging in illegal activity on the edges of the crowd.

Early social theorists such as **Gustave LeBon** (1895) described crowds as having mental unity or a "group mind." LeBon thought that people in crowds were highly suggestible and the usual controls on people's behavior disappeared as people took on a single way of acting and thinking. His term for this was **contagion theory**. This depiction of crowds, however, overlooks the social factors that influence crowd behavior. Crowds are distinctly social, not individualistic, forms of behavior. Although much of the behavior in crowds may be impromptu, crowds are socially organized and have a social structure.

Sociologists Ralph Turner and Lewis Killian, two influential collective behavior theorists, developed emergent norm theory to describe how crowds can be *both* emergent forms of behavior *and* socially organized. **Emergent norm theory** postulates that people faced with an unusual situation can create meanings that define and direct the situation. Remember that norms are the common understandings people develop to guide their behavior. During collective behavior, new norms emerge. The interactions people have with one another and the cues they get from group leaders or members of the crowd encourage the formation of new norms that will guide the entire group. Emergent norm theory emphasizes that group norms govern collective behavior, but the norms that are obeyed are newly created as the group responds to its new situation. Although collective behavior can sometimes appear to be without social organization, emergent norm theory emphasizes that members of the group do follow norms—they just may be created on the spot (Turner and Killian 1988).

THINKING *SOCIOLOGICALLY*

Think of the last time you were in a situation where something unexpected happened (a disaster, emergency, or other sudden and unanticipated event). What norms guided people's behavior in this situation? What forms of *collective behavior* developed? How would *emergent norm theory* explain what happened?

Within crowds, as elsewhere, norms guide behavior, but just because a guiding norm exists, not every person in the crowd is doing the same thing. In fact, crowds may exhibit a division of labor. At a sports event, one fan section may lead a cheer; other sections of the crowd may respond. Crowds may also generate *bystander crowds*, groups that may be physically present, such as the crowd of onlookers at a protest march, or that may be as remote as a mass media audience. In fact, crowd actions or protests are often explicitly set up for a media audience. Protesters planning a demonstration may expend great effort lining up media coverage, and during the protest, the participants may come to life for the cameras, then settle down when the media are gone.

Expressive crowds are groups whose primary function is the release or expression of emotion. More than individual people or ordinary groups typically show, expressive crowds tend to exhibit high levels of feeling. The object on which the crowd is focused is seen by participants with deep, even religious or awesome, feelings (Goode 1992: 142). Although participants are moved by any emotion, the most common are collective grief (as when citizens file past a slain leader lying in state) or joy (as in a victory celebration or religious revival). Expressive crowds can instill a permanent change in the mood and behavior of participants (Turner and Killian 1988: 97). Religious revivals, for example, often provide more than a cathartic moment and members often come away feeling substantially renewed or changed.

People who have seen the AIDS Memorial Quilt have likely felt the power of an expressive crowd. The AIDS quilt is a giant mosaic of panels designed as a memorial to the thousands of children, women, and men who have died from AIDs. Each panel is created by friends and family members of an AIDS victim. The panels are pieced together to form a patchwork quilt representing all the lives lost to this fatal disease. The unfolding of this huge quilt is a moving event. Often, one or two songs sung by a gay or lesbian chorus open the unfolding ceremony. Members of the crowd quietly file past sections of the quilt, many holding hands or grieving the passing of so many people. This is one example of many that show the power that crowd behavior can have on people's deepest feelings.

The Influence of Social Control Agents

Social control agents (police, chaperones, and other authorities) are present in most crowd situations because crowds are generally believed to easily get out of control. Social control agents at a rock concert, for example, may simply establish a perimeter for the crowd, as when police block fans from rushing the stage. Sometimes the behavior of the social control agents may be the cause of a crowd action, particularly if they overreact to crowd behavior. Police action has been a precipitating factor in student riots in some college towns when police have arrested people at large parties where hundreds, sometimes thousands, of students have gathered. Often the presence of the police sparks the riot—particularly if they are perceived as overly aggressive.

Panic

A *panic* is behavior that occurs when people in a group suddenly become concerned for their safety. Panics may be triggered by physical, psychological, social, or even financial danger (Lang and Lang 1961). In the popular conception (often found in disaster movies), people caught in a panic are divested of all socially acquired characteristics and become irresponsible, emotional, and dangerous (Quarantelli 1978; Clarke 2002). In reality, even during panics, more social structure exists than the popular image suggests. The sinking of the *Titanic* in 1912 is a good example. When this supposedly unsinkable ship struck an iceberg and began to sink, social class, gender, and age guided the pattern of escape. Of the 2200 people onboard, 600 escaped in lifeboats, with almost all the women and children traveling in first class surviving (including the few men who pretended to be women). Of the people traveling in third class, 45 percent of the women and

70 percent of the children died (Lord 1956). Getting people into lifeboats was carefully managed by crew who played their roles as expected, carefully executing an escape plan highly structured by class, gender, and age.

Social ties endure during panics. People tend to flee in groups, often stopping to look out for one another. People help their spouses, friends, and coworkers and develop cooperative forms of behavior to find safe exits. Some people even die as the result of strong ties to others, because they refuse to abandon others or they return to threatened areas in search of loved ones (Miller 2000). Headlong flight without regard for others is actually rare. We know, for example, that in the World Trade Center on 9/11, people for the most part tried to leave in an orderly fashion. There were numerous instances of people helping each other, including the firefighters who, even prior to imminent death, methodically tried to assist the injured (Tierney 2002; Clarke 2002).

Three main factors characterize panic-producing situations. First, there is a *perceived threat*. The threat may be physical, psychological, or a combination of both, and it is usually perceived as so imminent that there is no time to do anything but flee. The second characteristic of panic is a sense of *possible entrapment*. Panic occurs when people perceive that if they do not act fast, they may miss their chance to achieve a certain goal. In situations where persons know that they are completely trapped, as in airplane crashes or subway fires, there is often no evidence of panic behavior. It seems that only when people think there are a limited number of routes does panic ensue (Quarantelli 1954). The final

Even in panics, people tend to form social ties, working with others and behaving in a relatively orderly fashion.

AP/Wide World Photos

characteristic of panic is a *failure of front-to-rear communication*. People at the rear of the crowd, especially when they perceive themselves to be unfairly disadvantaged in reaching their goal, exert strong physical or psychological pressure to advance toward the goal, whether it be an emergency exit or an entrance to a stadium. In instances where people are trampled to death, physical pressure from the rear is usually the single most important factor (Turner and Killian 1988: 81).

Riots

Popular conceptions of riots are similar to conceptions of crowds. The assumption is that a mob mentality exists and rioters lose their will and their ability to be rational and are highly suggestible to the actions of leaders. One need look no further than popular descriptions of riots as "mobs" and rioters as "giddy looters" to see that this conception of riots is alive and well. Sociological analyses of riots are less inflamed and more analytical. Sociologists see riots as a multitude of small crowd actions spread over a particular geographic area, where the crowd is directed at a particular target (Stark et al. 1974).

Riots occur when groups of people band together to express a collective grievance or when groups are provoked by anger or excitement. Some riots can be anticipated, as when Los Angeles exploded in 1992 following the acquittal by an all-White jury of four police officers accused of beating Rodney King, who is Black. Similarly, riots occurred in 1980 in Miami as the result of longheld grievances held by the Black community against the police (Porter and Dunn 1984). Riots have also frequently erupted following highly contested sporting events, when large crowds of people are gathered together. Such situations are highly prone to rioting because of the excitability of the crowd, each loyal to one team or the other; any triggering event—a fight or a strong police response—can quickly erupt into a riot.

Riots are made up of many different crowd formations and are likely to consist of different actions dispersed over a potentially wide area. Fires, looting, rock throwing, and simple crowd assemblies can be dispersed over many square miles. In the Watts riots of Los Angeles in 1965 and the 1992 Los Angeles riots, some parts of the city experienced more intense riot activity, and the type of crowd action varied in different places. Crowd actions did not spread in connected paths. This means that crowd actions do not spread like wildfire; instead, mini-riots pop up throughout the city (Stark et al. 1974). Studies of riots debunk the notion that people in riots are consumed by a mob mentality that spreads through

Although the stereotype of looters is that they are lone actors, behavior during riots is socially organized and follows predictable sociological patterns such as targeting the property of those identified as outsiders or oppressors.

a crowd. There are also variations in crowd activities during riots throughout the day and evening. Most riot activity occurs in the evening and late at night, suggesting that riot behavior is linked to other social routines such as work and leisure time. Instead of rioters being possessed by a mob mentality, their activity is influenced by the social conditions in their lives.

Even looting, which seems motivated by individual wants, is organized along social lines. Looters choose selected targets; they often act as groups; and they behave in the context of communities or social groups that give considerable support to their actions (Quarantelli and Dynes 1970). This is quite contrary to images of looters as crazed, out solely to produce mayhem, and acting from individual greed.

Why Do Riots Occur?

Some people think that riots occur because of unruly individual behavior. Rioters have also been characterized as the "criminal element," the dregs of society. Sociologists, however, have shown that this is not true. In a classic study to understand who participates in riots, George Rudé (1964) examined police, prison, hospital, and judicial records; poll books; petitions; and parish registries of births and deaths for several historical periods. He discovered that people who rioted were likely to have no prior arrest records, to live in "fixed abodes," and to have steady jobs. He also found that different kinds of people are at different riots. During the riots leading to the French Revolution, women were more likely to be present at bread riots. Likewise, workers may participate in industrial disputes, but not in other types of riots (Rudé 1964).

Studies such as Rudé's help us understand individual participation in riots. Explanations of riots that focus on individual attitudes and states of mind fall into the category of **convergence theory**, which explains riots by focusing on the participants in riots, assuming that rioters are acting on predispositions and attitudes. One problem with this theory is that people have many attitudes and predispositions, and there is no one-to-one correspondence between membership in a particular group and participation in a riot. If the focus is on the individual or even the group that riots, attention is shifted away from the organizations or groups that contributed to the conditions to begin with. Studies of Latino residents in south-central Los Angeles after the 1992 riots found few significant differences among different Latinos that would explain why some looted and rioted and others did not (Hayes-Bautista et al. 1993). Rather than focusing on the individual participants to explain riots, it is more productive to look at the social conditions that cause riots to erupt.

Sociologists have found, for example, that the characteristics of urban areas can make some cities more prone to riots than others. First, riots are more likely to occur in cities with economic deprivation of racial–ethnic minority groups, including low levels of educational attainment, low median income, high unemployment, and poor housing conditions. Second, riots are most likely to occur in cities where grievances of the rioting group have not been addressed. This may be a reaction to an unsympathetic or unresponsive city government (Lieberson and Silverman 1965). Third, the rapid influx of new populations (through migration or immigration) is a common characteristic of cities where riots take place. Fourth, whether a group has the resources to initiate and sustain rebellious activity influences the development of riots (Carter 1992; Spilerman 1976).

All four conditions have been identified as underlying causes of the 1992 Los Angeles riots. The economic deprivation of African Americans and Latinos in Los Angeles, coupled with competition between Latino, Korean, and African Americans for jobs and housing have been shown to be primary causes (Baldassare 1994; Herman 1995). Moreover, participants in the riots defined themselves as "protestors" and "freedom fighters," concerned with specific grievances, including poverty, unemployment, police brutality, and racial discrimination (Murty et al. 1994). Tension between Korean Americans and African Americans has been popularly identified as one cause of the riots, but researchers have found that the majority of both Korean Americans and African Americans identify institutional racism and the poor living conditions for impoverished groups in Los Angeles as the primary causes of intergroup tensions (Stewart 1993).

Even with the underlying social structural conditions of poverty (such as unemployment and poor housing), riots are typically sparked by precipitating events, such as a confrontation with the police. For the 1992 Los Angeles riot, the precipitating event was the verdict acquitting four police officers of criminal charges in the beating of Rodney King. To rioters, rioting may seem like the only way to express the injustice they feel since they are often powerless to effect change in legitimate ways. As the Rev. Dr. Martin Luther King, Jr., said, "A riot is the language of the unheard" (Hampton 1987).

Competition theory argues that conflicts between different groups can erupt into riots when these groups must compete for limited resources, for example, as when groups compete with others for jobs, housing, and other resources (Olzak and Shanahan 1996; Olzak et al. 1996; Shanahan and Olzak 1999). Competition theory alone does not predict that riots will occur. Background conditions, such as the degree of inequality in a city, patterns of residential segregation by race, and the contraction of job opportunities provide a context where the likelihood of riots increases. Still, for a riot to occur, there must be a precipitating incident and situational determinants, such as the time of day or the number of people present in a public space (Carter 1990; Carter 1986). Riots are most likely to develop when social control mechanisms, such as the police or community leaders, fail to quell escalating crowd actions.

For riots to occur, a number of precipitating factors must be present. Precipitating factors alone, such as an instance of police brutality, do not spark riots. Other conditions must also be present, such as longstanding conditions of deprivation or unaddressed grievances. Situational factors also allow riots to develop, such as the presence of people in the area who may assemble and take collective action (McCarthy and McPhail 1996; McPhail 1994; Tierney, K. 2000).

What Stops Riots?

Eventually, riots stop. What brings riots to an end? First, the original goals of the protest groups may have been satisfied. For example, if a race riot breaks out because political grievances have not been addressed, a satisfactory response to those grievances will bring the riot to a close.

Second, the actions of social control agents may end violence. Law enforcement officials and politicians frequently argue that the proper way to end a riot is to increase repressive measures. Following the 1965 Watts riots, Congress responded by developing anti-riot legislation, stepping up law enforcement training in riot control, and building up stocks of tactical weapons to disperse crowds in riot situations (Kelly and Isaac 1984).

Third, riots and violence may end when the political situation changes. A government may become more responsive to the needs of groups that have been in revolt, or government may become more repressive, making the human cost of a riot (in loss of life or imprisonment) too great for the potential gain (Oberschall 1979).

Finally, some have argued that discontent can be regulated by the expansion of relief services. According to this argument, the expansion of the welfare state quiets discontent by making people most likely to rebel dependent on federal subsidies. In so doing, discontent is managed by the seemingly benevolent actions of the state (Piven and Cloward 1971).

Collective Preoccupations

Collective preoccupations are forms of collective behavior wherein many people, over a relatively broad social spectrum, engage in similar behavior and have a shared definition of their behavior as needed to bring social change or to identify their place in the society (Tierney 1994). Fads, fashion, hysterical contagions, and scapegoating are all collective preoccupations. Many of the more interesting, newsworthy, or novel parts of life comprise collective preoccupations. Beanie Babies and Digimon represent some of the more creative and whimsical parts of human life. Collective preoccupations can be harmless, such as the fad of streaking (running naked) on college campuses, or horrifying, such as the persecution of the Jews in Nazi Germany. Although collective preoccupations are popularly seen as weird or eccentric, like other forms of collective behavior, they are social in nature. It is the social part of these episodes that sociologists are interested in explaining.

The distinctly social phenomena of collective preoccupations take on many different forms but share a number of features. First, they often begin within a small group of people involved in face-to-face interaction. Usually, the people must deal with an ambiguous situation from which they attempt to derive meaning. For example, UFO sightings are often reported in local newspapers. Groups of friends and family members may read these stories and come up with explanations for an unidentified flying object they have seen. Explanations may range from identifying them as experiments, advertising gimmicks for car dealerships or tires, or exploratory vehicles from Mars (Miller 1985). The point is that groups of people together form a collective definition of the situation to explain an ambiguous event. To become a fad, fashion, craze, or hysterical contagion, the collective definition of the situation must spread be-

yond the initial group. This usually occurs through the mass media, preexisting friendship networks, or organizational ties (Aguirre et al. 1988).

Most collective preoccupations also involve some aspect of social change. Social change may bring on the respective preoccupation, as when groups defined as outsiders are targeted for scapegoating. The social change also may be of a very limited sort, as is the case with fads, when a large number of people behave for a brief period in a curious manner.

The various collective preoccupations provide opportunities for participants to belong to a group while differentiating themselves from other groups. The initial collective definition of the situation among group members may bring about a feeling of belonging. However, groups may use their definitions to exclude others or differentiate themselves from other groups. They become members of the in-group and others become part of the out-group who are persecuted by the in-group.

Fads

Fads represent change that has a less consequential impact than other social change. Fads do not usually fundamentally transform society (Goode 1992) but are "an amusing mass involvement defined as of little or no consequence and in which involvement is brief" (Lofland 1985: 68). They may be products (scooters, hula hoops, yo-yos), activities (streaking, raves), words or phrases (yo!, whatever, cool), or popular heroes (Harry Potter, Barbie). Fads provide a sense of unity among their participants and a sense of differentiation between participants and nonparticipants. Fads usually represent a departure from the mainstream. Streaking on college campuses, for instance, was novel in its time. Sometimes there is a negative judgment applied to fads. For example, piercing the face or other parts of the body is seen by some as morally reprehensible. More often, however, fad behavior is seen by all as harmless fun.

Despite their seemingly idiosyncratic nature, even fads follow certain social norms. Streaking, to continue the example, was defined in the student subculture and by most school authorities as harmless fun, but its acceptability was confined to certain attitudes and certain locations on campus; for example, the nudity of streaking was considered nonsexual, and it was taboo to streak in classes when tests were being given (Aguirre et al. 1988).

Fads are initially created within a small group that defines a particular product, phrase, or activity as meaningful and desirable to possess or to say. A *latent period* is followed by the *breakout period,* during which the product, phrase, or activity spreads to other groups via friendship networks and mass media. In this period, small groups take up the product, phrase, or activity and affirm it as something meaningful to take on (Goode 1992: 356). Commonly, commercial interests latch on to fads (or even create them) by massmarketing the item and manipulating public demand. In the *peaking period,* the use of the new item is defined as a fad, and people enthusiastically adopt it. The spread of a fad may depend on the success of commercial marketing (such as marketing popular film characters or baseball heroes through fast-food chains) or the spread may unintentionally result from the attempts of social control agents such as police or officials to regulate the activity. For example, on some campuses, school administrators saw streaking as intolerably deviant. Stiff sanctions sometimes led to confrontations, even riots (Aguirre et al. 1988). Ironically, such penalties only seem to make the activity more appealing. In the *decline period,* the fad quickly fades (Miller 1985), no longer defined as meaningful or desirable. Participants looking back at their participation in a fad may be bewildered that they ever found it meaningful.

Crazes are similar to fads except that they tend to represent more intense involvement for participants. Participants are "more or less encompassingly involved for periods of time" (Lofland 1985: 640). The craze takes up more of the participant's life. Crazes tend to involve fewer people than fads because they require more of a commitment. They can take place in any arena of life, but most often appear in the religious, political, economic, and expressive or aesthetic realms (Smelser 1963).

People involved in crazes tend to be highly focused on the craze behavior. They may seem fanatical, devoted to the craze above all else. Whether it is rollerblading, a new dance craze, or a financial craze, crazes consume participants' attention. Participants may endure considerable inconvenience to pursue the craze, such as waiting in long lines, traveling long distances, or absorbing large expenses. A mountain biker may spend $1500 on a bike and then augment it with a constant stream of expensive add-ons. The most desirable bikes are built of a space-age alloy, just beyond the purchaser's price range, known in biker parlance as "unobtainium." To people not addicted, the craze behavior may seem bizarre, senseless, or immoral. Crazes usually peak and then decline, although some people may become lifelong devotees of the craze behavior (Turner and Killian 1988).

Fashion

More institutionalized than fads, **fashion** has traditionally been considered a form of collective behavior because it constantly introduces something novel into the society. People wear clothing for protection and concealment of their bodies, but they also wear fashionable clothing to feel a part of a group and to differentiate themselves from others (Simmel 1904; Veblen 1953/1899). Particular kinds of clothing and adornment can give people a feeling of acceptance. Nuns may wear habits to indicate their commitment to a vow of celibacy and service, but the wearing of the habit is also a symbol recognized by others that they belong to a group. Likewise, young men and women may wear sports clothing or particular brand items (such as Tommy Hilfiger clothing or Nikes) to signify their belonging to a given group and social class, or they may wear clothing in a certain way, such as loose jeans barely hanging on one's hips or wearing caps backward. Hairstyles, clothing, jewelry, and other adornments all symbolize an identity people wish to convey to others.

Along with providing a sense of group identity, fashion is important in differentiating groups. Around the turn of the twentieth century in the United States, when industrialization and technological progress were creating a new class of wealthy people, people used personal adornment and dress as a way to make other people aware of their wealth.

Crazes, such as the popular desire for Furbies, a child's toy, can produce collective behavior when huge crowds gather to purchase a scarce resource. Such crazes can be produced by commercial manufacturers with an interest in profiting from the mass reactions.

© AP/Wide World Photos

Economist and sociologist **Thorstein Veblen** referred to this purchasing and displaying of goods to symbolize wealth and status as *conspicuous consumption*. A fully equipped, dark-colored sports utility vehicle, such as the Cadillac Escalade, is an example of conspicuous consumption. Fashions often emerge when the fads of marginalized subcultures are adopted by the fashion industry and then recycled to elite groups. Hip-hop fashion first emerged as a style among inner-city, low-income Black youth and was captured by the fashion industry, which then marketed wide, baggy jeans, caps, and oversized shorts to elite and middle-class markets. A cycle of fashion develops when the style of low-status groups trickles up to high-status groups. The style then becomes a status symbol and is sold widely to the middle class. Marginalized subcultures then may develop new styles that, if appropriated by high-status groups, create another cycle of fashion. The recent women's fashion of tight, cropped shirts, animal skin prints, low-cut jeans, and high-heeled boots—a style originally associated with streetwalkers (prostitutes)—is a case in point.

Because fashion differentiates groups, it is also a means of marking inequality between groups. Designer fashion labels, often displayed on the outer part of clothing, communicate material status. Within bureaucratic organizations, the status of different groups is also marked by apparel. Thus, workers may wear uniforms, but management wears suits.

Hysterical Contagions

Hysterical contagions involve the spread of symptoms of an illness among a group, usually one in close contact, when there is no physiological disease present. For example, a mass contagion closed a Tennessee school for two weeks in 1998. More than 170 students sought emergency treatment, but neither a virus, chemical, nor other toxin could ever be identified as causing the mass outbreak. Doctors concluded that this was a case of mass contagion with no physical cause (Jones 2000).

An episode of hysterical contagion usually begins with one person exhibiting physiological symptoms such as nausea, stomach cramps, itching, or uncontrollable trembling. That person receives sympathetic attention as people accept that some genuine physical or biological agent has caused the symptoms. People begin to talk about the case. Soon other people begin to experience similar symptoms, then also receive attention for their illness. When the outbreak becomes more widespread, authorities attempt to locate and eliminate the source of the problem. When they are unable to find a biological cause for the outbreak, speculation begins that the episode is a case of mass hysteria (Gehlen 1977). The word *hysterical* connotes frenzied behavior, yet hysterical contagions follow a consistent pattern that can be accounted for by social factors.

Hysterical contagion is most likely to occur when it provides a way of coping with a situation that cannot be handled in the more usual ways. Workers, for example, who think it is not legitimate to take a day off from work just because they need a rest may become part of a hysterical contagion because being "sick" is then validated by others in the group. These diligent people may believe they are actually sick (Gehlen 1977; Kerckhoff and Back 1968). If others participate in the sickness, there is less risk of being viewed as a slouch or malingerer. This is not to say that people are faking. Like other types of collective preoccupations, hysterical contagion results when people try to cope with an ambiguous situation. It usually means redefining how we feel and what causes those feelings.

Political contexts also shape how hysterical contagions are defined. During a hysterical contagion in a West Bank school in Israel in 1983, the contagion began in a girls' school, soon spread to other schools and, eventually, the whole community. Hospital doctors, working without laboratory facilities, thought the girls had been poisoned by gas. At the height of the epidemic, rumors circulated among the Arab population that poison gas was being used to diminish the fertility of the schoolgirls, a ploy that would depopulate the West Bank area, then under Israeli occupation. Only later were external medical experts able to establish that there was no gas and no poison (Hefez 1985).

Like other types of collective preoccupations, organization and communication networks are crucial to the spread of hysterical contagion. Friendship networks are often the crucial communication link, explaining why hysterical contagion typically occurs in schools and work organizations (Kerckhoff and Back 1968; Gehlen 1977).

Scapegoating

Often, collective preoccupations lead to **scapegoating,** which occurs when a group collectively identifies another group as a threat to the perceived social order and incorrectly blames the other group for problems they have not caused. The group so identified then becomes the target of negative actions that can range from ridicule to imprisonment, extreme violence, and even death. Sometimes scapegoating involves one person blamed by another person or group for a perceived problem. When scapegoating becomes a collective phenomenon—an entire group blamed by another group—it is a collective preoccupation.

Racial minority groups and other groups perceived by the dominant group to be a threat are commonly the victims of scapegoating. This often occurs when major social changes are taking place in society; especially when older values are being replaced by new ones. You probably noticed the scapegoating of Arab Americans following the events of 9/11. Arab Americans were racially profiled at the same time that symbolic displays of patriotism were flourishing.

To become a collective preoccupation, scapegoating must gain the support of the media, government, or other influential organizations. Anti-Arab sentiment in the United States, for example, has been fanned by unflattering government and media depictions of Arab leaders defined as enemies of the American people. This has the effect of generating bad will against all Arab people in the United States, as if all Arabs were terrorists, even when most are not.

When people defined as outsiders are perceived as taking jobs—or other privileges—away from presumed insiders, scapegoating can occur. Thus, immigrants get blamed when jobs are scarce; foreigners are scapegoated when U.S. businesses keep profits up by hiring labor in other countries. Ethnic violence sometimes results, reflecting the mass hysteria that fear of the "outsider" can create (Jenkins 1983).

International conflicts or internal strife may exacerbate the tension between the dominant culture and perceived outgroups. War also produces strong feelings of national identification. Scapegoats perceived as "other" may be singled out and targeted as enemies. Such was the case during World War II when more than 110,000 Japanese Americans unjustly feared as potential saboteurs were interned in concentration camps in the United States.

Social Movements

A social movement is an organized social group that acts with continuity and coordination to promote or resist change in society or other social units. Social movements are the most organized form of collective behavior, and they tend to be the most sustained. They often have a connection to the past, and they tend to become organized in coherent (sometimes even bureaucratic) social organizations.

Social movements usually involve large numbers of people, but unlike a crowd, the individuals who make up a social movement are dispersed over time and space. ACT-UP (AIDS Coalition to Unleash Power), for example, comprises hundreds of local chapters that participate in "zap actions" across the United States, that is, single episodes of dramatic behavior intended to call attention to the AIDS cause. Unlike a crowd, not all the members of ACT-UP interact with each other face-to-face. Although ACT-UP chapters are diffuse

(like collective preoccupations, they are spread across several groups), they are more enduring than a fad or an instance of hysterical contagion. Social movements, then, share some characteristics of crowds and riots and some features of more established organizations.

Like crowds, social movements can give the impression that all of their members are unified around a single goal or ideology. Although movements focus on a shared goal, they can be internally diverse and generate different public reactions (see Figure 22.1). Members of a given group within a movement also may have divergent ideas about the movement's goals and tactics. This can lead to tensions within movements as different groups vie for dominance, but it can also lead to increased vitality because social movements (especially those on a large scale) can often benefit from the diverse backgrounds, ideas, and interests of members.

The women's movement is a good example of a vital, diverse social movement. Within women's movement organizations, there are different styles of activism—some of them more formal and mainstream than others (see the box Forces of Social Change: "Feminist Generations" for a discussion of the changes among radical feminists over the past thirty years). Class and racial differences among women also create distinct organizational needs, depending on the group's particular focus (Poster 1995). The diverse organizational ideologies, activities, and structures of the women's movement have likely contributed to its vitality over the years, as well as its ability to respond to changing social conditions. At the same time, however, diversity within the women's movement also produces conflicts over whose interests the movement is most likely to represent. Women of color, in particular, have been critical of the women's movement for primarily representing White, middle-class women's experiences, even though women of color tend to hold strong feminist values and have organized many groups that speak to their needs.

Social movements, such as the disability rights movement, can raise public awareness and result in new forms of social behavior.

Thomas Frey/DPA/Landov

Question: There are many social movements that try to have an impact on policy-making in our nation. For each of the following social movements, please tell me how much of an impact you think it has had on our nation's policies—a great deal, a moderate amount, a slight amount, or none at all.

Question: Regardless of how much impact, if any, each movement has had, please tell me if you personally agree or disagree with its goals. As I read each one, please tell me if you strongly agree, somewhat agree, somewhat disagree, or strongly disagree with its goals.

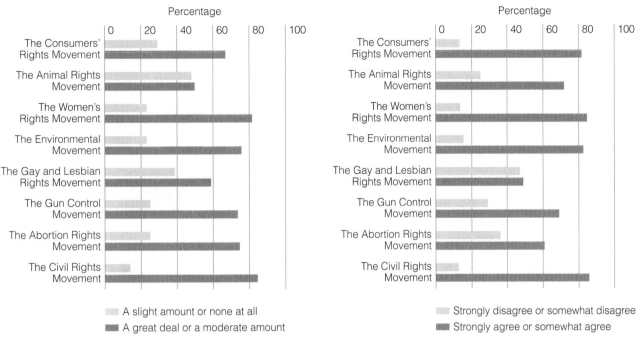

Figure 22.1 The Influence of Social Movements: What the Public Perceives

Source: Dunlap, Riley E. 2000. "Americans Have Positive Image of the Environmental Movement." The Gallup Poll. Princeton, NJ: The Gallup Organization, April 18. Website: **www.gallup.com**

Social movements thrive on spontaneity, unlike everyday organizations. They often must swiftly develop new strategies and tactics in the quest for change. During the civil rights movement, students improvised the technique of sit-ins, which quickly spread to other protests because the technique succeeded in gaining attention for the activist's concerns. Social movements are different from other forms of collective behavior, however, by containing routine elements of organization and lasting for longer periods. Short-lived collective behaviors nevertheless have a role in social movements. Thus, within the environmental movement, as in others, periodic demonstrations involve crowd behavior—or generate fads, such as carrying around refillable coffee mugs or water bottles.

Types of Social Movements

There are three broad types of social movements: personal transformation movements, social change movements, and reactionary movements. **Personal transformation movements** aim to change the individual. Instead of pursuing social change, they focus on the development of new meaning within individual lives (Klapp 1972). The New Age movement is a personal transformation movement that defines mainstream life as stress-

ful and overly rational and promotes relaxation and spiritualism as an emotional release and route to expanded perceptions. New Age music, crystals, massage therapy, and meditation are intended to restore the New Age person to a state of unstressed wholeness. Like many other social movements, New Ageism is supported by an array of dues-charging organizations and commercial products, from tapes and crystals to sessions with the spirits of the dead, retreats in yoga ashrams, and guided tours of Native American holy sites.

Religious and cult movements are also personal transformation movements. The rise in evangelical religious movements can be explained in part by the need people have to give clear meaning to their lives in a complex and sometimes perplexing society. In personal transformation movements, participants adopt a new identity—one they use to redefine their lives, both current and former states.

Social change movements aim to change some aspect of society. Examples are the environmental movement, the gay and lesbian movement, the civil rights movement, the animal rights movement, the religious right movement, and the Green Party. All seek social change, although in distinct and sometimes oppositional ways. Some movements want radical change in existing social institutions; others want a retreat to a former way

FORCES OF SOCIAL CHANGE
Feminist Generations

Different historical periods shape the collective identity of people who participate in social movements, especially social movements such as feminism that have endured over a long time. How has the women's movement changed the identity of its participants from different generations?

Sociologist Nancy Whittier has studied feminist activists of the "second wave" of the women's movement from the 1960s and 1970s, as well as the generation of younger women growing up in the 1990s—a group typically thought of as less influenced by feminism. Using the concept of political generations, Whittier argues that coming of age in different historical periods produces different generational groups with different perspectives. Participation in the women's movement in the 1960s and 1970s was, for women now largely in their middle years, a deeply transformative experience. Having developed their political consciousness in a historical period when they were immersed in the women's movement, these women have had to sustain their feminist beliefs even during times when the dominant culture was extremely hostile to feminist thought and activism. Their experience within the feminist movement, however, forged links between them that still inform their work and progressive attitudes toward change.

Radical feminists of the 1960s and 1970s still use the perspective they gained from participation in the women's movement in numerous phases of their work and lives. In addition, these feminists find that, despite myths about the postfeminist generation, an extensive feminist culture has developed that now draws in younger women. They may differ in outlook and feminist perspective from their foremothers, but still see themselves as forging new identities that are creating new forms of feminist thinking.

Whittier's research provides a strong case study of a social movement in a particular place, as it evolves with changing social conditions. In the end, she

sees feminism as an enduring social movement, yet one that develops in response to particular historical conditions and as a reflection of the changing membership within the movement. Going beyond the glib pronouncements that feminism is dead, Whittier shows how feminism has evolved as a social movement and suggests what it may become in its next wave.

Source: Whittier, Nancy. 1995. *Feminist Generations: The Persistence of the Radical Women's Movement.* Philadelphia, PA: Temple University Press. •••

---

of life or even a move to an imagined past (or future). Social movements use a variety of tactics, strategies, and organizational forms to achieve their goals. The civil rights movement in the early 1950s used collective action in sit-ins, mass demonstrations, and organizational activities to overturn statutes that supported the "separate but equal" principle of segregation. These efforts culminated in the Supreme Court decision *Brown v. Board of Education,* which declared the "separate but equal" doctrine unconstitutional (Morris 1999).

Movements also form alliances with other movements, and many social change movements involve a vast network of diverse groups organized around broadly similar, but also unique, goals. The Asian American movement, for example, includes Asian American women's organizations such as Pan Asia and the National Network of Asian and Pacific Women, Asian workers' groups such as the Six Companies and the Chinese Consolidated Benevolent Association, and other groups of Asian activists (Wei 1993; Espiritu 1992; see the box Understanding Diversity: "Asian American Panethnicity").

Social change movements may be *norm focused* (trying to change the prescribed way of doing things) or they may be *value focused* (trying to change a fundamental idea or something everyone holds dear; Turner and Killian 1988). Often, they are both. The civil rights movement tried to change the law of the land and the attitudes of its people at the same time. The broad charter of the civil rights movement spawned offspring movements devoted to the special interests of such groups as Chicanos, feminists, and lesbians and gays, among many others.

Social change movements may be either reformist or radical. **Reform movements** seek change through legal or other mainstream political means, typically working within existing institutions. **Radical movements** seek fundamental change in the structure of society. Although most movements are primarily one or the other, within a given movement there may be both reformist and radical factions. In the environmental movement, for example, the Sierra Club is a classic reform movement that lobbies within the existing political system to promote legislation protecting the environment. Greenpeace is a more radical group that sometimes uses tactics that disrupt the activities the group finds objectional, such as the killing of whales. The distinction between reform and radical movements is not absolute. Social movements can contain elements of both and may change their orientation in midstride upon meeting success or failure. Moreover, whether a group is defined as radical or reformist is to a great degree a matter of public perception. Just as one person's rebel is another's freedom fighter on the world stage, the definition of a social movement often depends on its social legitimacy and its ability to control how it is defined in the media and other public forums (Killian 1975).

Reactionary movements organize to resist change or to reinstate an earlier social order that participants perceive to be better and are reacting against contemporary changes in society. The New Right provides a case in point. It has emerged in opposition to such changes such as the legalization of abortion, the high divorce rate, women's liberation, and greater freedom for gays and lesbians. To the New Right, these changes symbolize a decline in moral values and traditions (Esterberg 2003). Thus, the New Right is organized to resist the social changes that participants in this conservative movement find deeply objectionable.

UNDERSTANDING DIVERSITY

Asian American Panethnicity: Building Identity from Social Movements

Groups known as Asian American originally emigrated from many different countries and did not necessarily think of themselves as "Asian." Coming from China, Korea, Japan, and, more recently, Southeast Asian countries and India, many immigrants originally based their identity not so much on a country of origin as on a home province or district. Sociologist Yen Le Espiritu (1992) argues that the identity as Asian American—what she calls panethnicity—is a new identity, stemming from the activism of Asian Americans in contemporary social movements.

Asian groups immigrating early to the United States often distanced themselves from other Asians who they perceived to be stigmatized in the United States. For example, upon immigrating to the United States, many Japanese thought of themselves as superior to the Chinese. They saw the Chinese as despised in American society and did not want to share that status. Soon, however, the Japanese experienced the same exclusion and prejudice that Chinese Americans did. Other Asians, notably Filipinos and Koreans, then distanced themselves from the Japanese, not immediately recognizing that they, too, would be subject to racism.

Several changes in the society modified this pattern of identification, resulting in Asian American panethnicity. The first change was demographic. Immigration restrictions in the United States throughout the early half of the twentieth century resulted in American-born Asians outnumbering immigrant Asian groups. As Asian Americans became a native-born community, differences in their language and culture blurred. Second, third, and subsequent generations born in the United States developed a stronger identity as American than as the nationality of their previous generations. The native-born also began to see themselves as having more in common with each other than with either White Americans or Asians living outside the United States.

The second major impact on Asian American identity was historical—namely, World War II. Before then, Asian Americans had already been segregated within the United States in Chinatowns, Koreatowns, Little Tokyos, and other residentially segregated communities. The war exacerbated White fears and hatred of all Asian groups. The internment of Japanese Americans resulted from fears of a "yellow peril" (alleged overpopulation of Asians); in addition, the hatred toward Japan heightened racism against all Asian Americans.

The recent development of social movements among Asian Americans is the third and most important factor in generating an identity as Asian American. Influenced by the Black power movement of the 1960s, Asian Americans rejected the term *Oriental* because of its connotations of passivity and deviousness. Asian pride, like African American pride, is reflected in the names of new organizations such as the Yellow Seed, the Asian American Political Alliance, and Asian Women United of California. Learning from the Black protest movement, Asian American students and faculty developed programs in Asian American Studies to study their history and culture. The feminist movement also enhanced awareness among Asian American women of their common identities.

Espiritu's research makes an important sociological point that ethnic identity, like other forms of group identity, stems from the relationships people have with one another and the meaning given to those relationships. Her research demonstrates the importance that social movements play in forging these identities.

Source: Espiritu, Yen Le. 1992. *Asian American Panethnicity: Bridging Institutions and Identity.* Philadelphia, PA: Temple University Press. ●●●

Reactionary movements, such as the militia movement, are those that try to reinstate a perceived status quo.

THINKING SOCIOLOGICALLY

Identify a *social movement* in your community or school that you find interesting. By talking to the movement leaders and participants and, if possible, examining any written material from the movement, how would you describe this movement in sociological terms? Is it *reform, radical,* or *reactionary*? How have the available resources helped or hindered the movement's development? What tactics does the movement use to achieve its goals? How is it connected to other social movements?

Origins of Social Movements

How do social movements start? At least four elements are necessary: *a preexisting communication network, a preexisting grievance, a precipitating incident,* and *the ability to mobilize.*

Social movements do not typically develop out of thin air. For a movement to begin, *there must be a preexisting communication network* (Freeman 1983a). Communication can be informal, as between neighbors, organizations, or other group networks, but to initiate a movement, activists need some way to communicate with the people who will become part of the new movement.

The importance of previously established networks in the emergence of new social movements is well illustrated by the beginnings of the civil rights movement, considered by most people to be December 1, 1955, the day Rosa Parks was arrested in Montgomery, Ala-

bama, for refusing to give up her seat on a municipal bus to a White man. She is typically understood as simply too tired to give up her seat that day, but Rosa Parks had been an active member of the movement against segregation in Montgomery. The local chapter of the National Association for the Advancement of Colored People (NAACP), of which Parks was secretary, had already been organizing to boycott the Montgomery buses. As a result, Rosa Parks was the one chosen to be the plaintiff in a test case against the bus company. Because she was soft-spoken, middle-aged, and a model citizen, NAACP leaders believed her to make a credible and sympathetic plaintiff in what they had long hoped would be a significant legal challenge to segregation in public transportation. When Parks refused to give up her seat, the movement stood ready to mobilize.

News of Rosa Parks's arrest spread quickly via networks of friends, kin, church, and school organizations. A small group, the Women's Political Council, had previously discussed plans to announce a boycott. The arrest of Rosa Parks presented an opportunity for implementation. One member of the Women's Political Council, Jo Ann Gibson Robinson, called a friend who had access to the mimeograph machine at Alabama State College. In the middle of the night, Robinson and two of her students duplicated 52,500 leaflets calling for a boycott of the segregated bus system, to be distributed throughout Black neighborhoods (Robinson 1987).

In addition to preexisting communications networks among movement participants, movements are also fueled by the news media, which bring public attention to the movement's cause. Especially when movements use dramatic tactics, such as a massive march on Washington, news photos provide a means for information, even if distorted, to travel far and wide. Now, with the extraordinarily widespread availability of television, the Internet, cellular phones, and standard radio transmissions, the news of movements around the world spreads quickly.

The Internet in particular has become increasingly important for the mobilization of social movements. Communication via the Internet was an important organizing tactic for the diverse groups that participated in the March for Women's Lives in Washington, DC, in April 2004, where close to one million people demonstrated on behalf of reproductive choice for women. The Internet has also been a tool of White supremacist groups for recruiting new members and promoting their views. This is especially effective because of the appeal of the Internet to young people and the anonymity it can provide.

© Cynthia Howe/Corbis Sygma

Celebrities bring visibility to social movements such as here when Angelica Houston marched against the Iraq war with millions of others on Feburary 15, 2003.

Celebrities can also advance the cause of social movements, although sociologists have found that celebrities typically involve themselves in less controversial movements that already have widespread popular support. The entry of celebrities into social movements can unintentionally hurt a movement's development because celebrities tend to depoliticize the movement (Meyer and Gamson 1995). Their endorsement and presence, nonetheless, brings visibility to a movement, as for example when Ron Reagan, the son of the former president Ronald Reagan, spoke at the Democratic National Convention in support of the movement endorsing stem cell research to combat disease

For a movement to begin, *there must also be a perceived sense of injustice among the potential participants, or a strong desire for change.* Preestablished communication networks come into play when they are used to express the collective sense of wrong or right from which movements develop. Networks of friends, families, coworkers, and other groups whose members have something in common

provide the channels through which people organize and articulate the developing ideas of an emerging movement (Freeman 1983a). The environmental justice movement, as an example, has emerged largely from grassroots organizations formed in communities where residents have organized to clean up a toxic waste site, close a polluting industry, or protect children from a perceived threat to their health and safety (Krauss 1994). The organizations may develop communications networks through community meetings, newsletters and flyers, or discussions with family members, friends, and coworkers (see Figure 22.2).

With a communication network in place and a preexisting sense of injustice among potential movement participants, the conditions are present for the emergence of a social movement. The next step is *a precipitating factor that translates the sense of perceived grievance into action.* In the environmental justice movement, action may be triggered by the sight of trucks dumping refuse in an empty lot or a plant spewing stinking gases into the air.

The final condition needed to get a movement started is *the ability of groups to mobilize.* **Mobilization** is the process by which social movements and their leaders secure people and resources for the movement. Even with all other conditions present for a movement to develop, movements cannot continue if they cannot mobilize people. Grievances exist all the time in society, but they alone, even in the presence of a communications network and a precipitating incident, do not explain the emergence of social movements. For example, a person might think he or she has been wronged by a particular company and might communicate this grievance to networks of friends and associates. Those people might even share the same grievance, but until they are able to mobilize on behalf of their collective

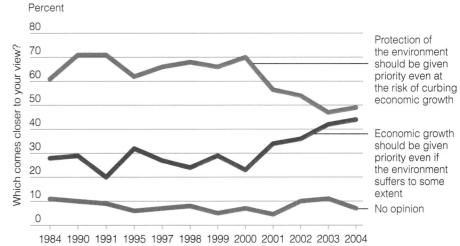

Figure 22.2 Environment and Economy: Competing and Changing Views

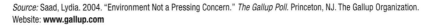

Source: Saad, Lydia. 2004. "Environment Not a Pressing Concern." *The Gallup Poll.* Princeton, NJ. The Gallup Organization. Website: **www.gallup.com**

sense of a need for change, there will be no social movement.

Movement leaders need to coordinate action between the different groups that constitute a movement. Sometimes this means forming new groups and organizations; it can also mean utilizing existing groups on behalf of the new movement. In the Montgomery bus boycott, for example, Black churches were critical to mobilizing the movement. Churches provided volunteers, meeting places, leadership, and a network of information sharing critical to the movement's success. In addition, a transportation committee was formed to ferry people who would normally ride the buses (Robinson 1987). This group may have been the most important factor in mobilizing the boycott. Making an individual decision to stay off the bus was much harder if you did not know how you were going to get to work, especially when you also could not be sure that others would heed the boycott. The transportation system guaranteed both transportation and moral support and made it possible for the yearlong boycott to succeed.

The Organization of Social Movements

As social movements develop, they quickly establish an organizational structure. The shape of the movement's organization may range from formal, bureaucratic structures to decentralized, interpersonal, and egalitarian arrangements. Many social movements combine both. For example, urban tenant movements typically have three organizational levels: groups of tenants in a single building, neighborhood organizations comprising tenant groups, and federations of neighborhood organizations. Tenant groups in a single building are a relatively weak bargaining unit, so to increase their bargaining power, they often unite with the tenant groups in other buildings to form a neighborhood organization. Larger, more bureaucratically organized federations are made up of groups from a larger geographic area that represent the interests

of tenants in relations with the city and state. This broader organization may also educate the general public on the overall goals of the tenant movement (Lawson 1983).

Many social movements are organized as large bureaucracies. The National Organization for Women (NOW), Amnesty International, Greenpeace, and the Jewish Defense League are examples of large-scale bureaucratic organizations. In fact, the success of a social movement can sometimes be seen in the extent to which it becomes institutionalized. As a social movement becomes more established, it is likely to become more bureaucratically organized, often resulting in a large, national organization with formal membership and leadership. This affirms Max Weber's prediction that social movements gradually rationalize their structures to efficiently meet the needs of members and the goals of the organization.

Other social movements are often made up of organizations that are more decentralized, interpersonal, and loosely connected. They may even comprise a vast network of organizations with no central command or decision-making structure. Student organizations protesting anti-affirmative action policies on college campuses are an example. With no national office to coordinate and fuel their movement, leaders may rely on personal networks with students on other campuses or other student newspapers. Increasingly, the Internet provides a rich source for movement communication because it closes the distances between locations where movement activity takes place. Over the course of a movement's life, various organizations and groups may

Social movements often use highly visible tactics to promote their cause, as in the March for Women's Lives in Washington DC in April 2004.

grow, proliferate, fuse, contract, or die (Gerlach and Hine 1970).

Links between social movements arise in a number of ways. Different groups may participate in joint activities, members of a group may switch to other groups or be enrolled in several organizations at once, and individuals in different groups may have friendship ties with each other (Gerlach and Hine 1970). For example, the women's movement comprises numerous groups and organizations, some of which have widely divergent philosophies and organizational structures. To the outsider, a movement may appear to be a single, unified group, but it is often more diverse on the inside than it appears to the causal observer.

The advantages of informal organization over more bureaucratic structures include greater resilience and flexibility (Gerlach and Hine 1970). Among the disadvantages are, first, that the movement cannot act rapidly as a whole. Second, no one person can legitimately speak for the entire movement, and there may be no one who can give clear direction to the movement's principles and plans (Dwyer 1983). Finally, fragmentation within a movement can be the cause of a movement's demise.

Strategies and Tactics

Social movements choose their political and social strategies based on a number of variables: the resources

---

MAPPING AMERICA'S DIVERSITY

MAP 22.1 Hate Groups in the United States

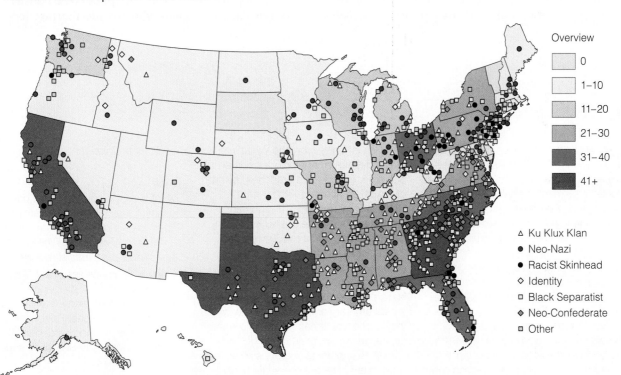

Overview
- 0
- 1–10
- 11–20
- 21–30
- 31–40
- 41+

△ Ku Klux Klan
● Neo-Nazi
● Racist Skinhead
◇ Identity
▢ Black Separatist
◆ Neo-Confederate
▢ Other

The Southern Poverty Law Center, originally founded to protest and monitor the activities of the Ku Klux Klan, now monitors numerous hate groups throughout the country, estimating that there are over 750 such groups currently active. The Center also provides extensive teaching materials to promote tolerance and interracial understanding. You can see that different kinds of hate groups populate various regions of the United States. Many of them involve interracial hatred, such as the Ku Klux Klan, Neo-Nazis, and Skinheads. But they also include other extremist groups. The website of the Southern Poverty Law Center has an interactive version of this map, where you can click on a given location and learn more about the hate groups and statewide legislation to outlaw hate crimes. Based on the map presented here, what part of the country has the highest concentration of hate groups? What factors do you think lead to the presence of hate groups?

Source: Taken from Southern Poverty Law Center. 2004. Used by permission. Website: **www.splcenter.org**

available to them, the constraints on their actions, the organizational structure of the movement, and the expectations the movement has about potential targets of its actions—such as the media, social control agencies, the public at large, and other audiences (Freeman 1983b). The kinds of resources available to a group may heavily influence the tactics they use. Poor people are hardly in a position to mobilize high-powered, high-priced Washington lobbyists. Instead, they are more likely to use disruptive tactics such as occupying buildings and blocking traffic. Their bodies are the resources most readily available to them (Piven and Cloward 1977). Whether the movement is organized more bureaucratically or more interpersonally may influence the kinds of strategies or tactics used.

The Ku Klux Klan uses a variety of strategies in their hate campaigns. Women of the Klan draw on their familial and community ties to form "poison squads," groups that use social networks to spread rumor or slander and organize consumer boycotts. Men use their work connections and participate in the more visible and public strategies of disrupting and corrupting elections, and terrorizing minority citizens (Blee 1991, 2002). The Klan also plays to stereotypes about sexuality, gender, and class in creating a movement ideology that promotes White supremacy while targeting Jews, African Americans, gays and lesbians, and other supposed "out-groups" (Daniels 1997; Ferber 1998; see the box Doing Sociological Research: "White Man Falling").

Sometimes the strategies that movements use can backfire, causing them to lose support. The strategy of antiabortion activists of blocking access to abortion clinics is a case in point. In a study of public attitudes toward abortion that extended over a three-year period in Buffalo, New York, researchers discovered that the militant activities of Operation Rescue during the second year of the study decreased public support for their position (Ansuini et al. 1994). Although there have been no comparable studies done on a national basis, one could surmise that the widely publicized murders of workers and physicians who provide abortion services

DOING SOCIOLOGICAL RESEARCH

White Man Falling

How can some groups hate and fear Jewish and African American people so much that they would organize a social movement to promote this hate? This is what the sociologist Abby Ferber wanted to know when she began the difficult project of studying White supremacist movements. Ferber conducted her research by analyzing the newsletters and other publications of numerous White supremacist organizations. Reflecting on the difficulty of this task as a White, educated, Jewish American woman, she says, "Reading white-supremacist literature is a profoundly disturbing experience, and even more difficult if you are one of those targeted for elimination" (1999:6).

She found that the development of such movements stems from White groups wanting to maintain clear boundaries around what they perceive as "whiteness." Emphasizing "racial purity," these groups perceive interracial relationships, especially interracial sexuality, as the ultimate threat to White power relations.

White supremacist movements have a long history, which includes the eugenics movement and the Nazi movement in Germany. Now the White supremacist movement includes numerous organizations and branches, some of which are affiliated with far right Christian organizations. The primary organizations include the Ku Klux Klan, neo-Nazis, the militia movement, and the Christian Identity Movement—an organization that also links other groups such as Posse Comitatus and Aryan Nations. Some of these organizations are paramilitary organizations; that is, they are heavily armed and believe in war as the way to win their cause.

White supremacist groups have proliferated in recent years, although they are still estimated to include a very small percentage of people. Groups within this movement have a shared ideology. Ferber concludes that we have to understand White supremacy by understanding how those in the mainstream culture define racial and gender dichotomies. By challenging the constructed meaning of race and gender as opposite categories, we can debunk the systems of power that are built around such conceptions and, thereby, challenge the grounds on which White supremacist movements are created.

Questions to Consider

1. What groups in your community are most vulnerable to hate crimes? In what ways do you think hate crimes directed against this group are related to how race and/or gender are understood? *Keywords: White backlash, white supremacist movements*

2. Many hate groups assume that racial, religious, and other differences between people are somehow rooted in biological differences. How does a sociological perspective debunk this idea? *Keywords: anti-Semitism, eugenics, neo-Nazism*

We have included InfoTrac College Edition keywords at the end of each question to make it easier for you to find more to read on these topics. Go to www.infotrac-college.com, an online library, to begin your search.

Source: Ferber, Abby. 1998. *White Man Falling: Race, Gender, and White Supremacy.* Lanham, MD: Rowman and Littlefield; Ferber, Abby. 1999. "What White Supremacists Taught a Jewish Scholar About Identity." *The Chronicle of Higher Education* (May 7): B6–B7. ••••

by those in the antiabortion movement could weaken support for their cause.

Constraints on social movements may include the values, past experiences, reference groups, and expectations of the groups. The value systems of many groups do not allow them to use tactics that harm people, but they may be willing to destroy property. Peace groups, for example, are unlikely to use terrorism or weapons, given their movement's philosophy of nonviolence. Movements sometimes use tactics for dramatic effect, as when women in the environmental movement staged a bare-breasted protest against the logging of a grove of ancient redwood trees in Oregon. Calling their action "striptease for the trees," the activists dramatized the environmental movement by standing with naked breasts in front of logging trucks. Finally, the appearance of opportunities for action, the perceived response of social control agencies, and the expected effect that a group's action will have on the public all contribute to shaping the strategies the group will use.

Dramatic tactics are often used by social movements to bring attention to their causes. Here, La Tigresa protests the logging of redwood forests, calling attention to the environmental movement.

Theories of Social Movements

The question driving sociological theories of social movements is, What makes social movements emerge? Several theories have been developed to answer this question, including resource mobilization theory, political process theory, and new social movement theory (see Table 22.1).

Resource mobilization theory focuses on how movements gain momentum by successfully garnering resources, competing with other movements, and mobilizing their available resources (Marx and McAdam 1994; McCarthy and Zald 1973). Money, communication technology, special technical or legal knowledge, and people with organizational and leadership skills are all examples of resources that can be used to organize a social movement (Zald and McCarthy 1975). Interpersonal contacts are an important resource a group can mobilize because they provide a continuous supply of new recruits, as well as money, knowledge, skills, and other kinds of assistance (Snow et al. 1986). Sometimes,

social movement organizations acquire the resources of other organizations.

Resource mobilization theory is used to explain the development of the civil rights movement, which relied heavily on Black churches and colleges for resources such as money, leadership, meeting space, and administrative support (Morris 1984). In other words, it mobilized existing resources on behalf of its own cause. Resource mobilization theory also notes how movements are connected to each other. The gay and lesbian movement, as an example, has used many of the strategies of the violence against women movement in developing its own campaign to halt hate crimes against gay men and lesbian women (Jenness and Broad 1997).

| Table 22.1 | | | |
| Sociological Theories of Social Movements | | | |
	Resource Mobilization Theory	**Political Process Theory**	**New Social Movement Theory**
How do social movements start?	People garner resources and organize movements by utilizing such things as money, knowledge, skills.	Movements exploit social structural opportunities, such as economic crises and wars.	New forms of identity are created as people participate in social movements.
What does the theory emphasize?	Linkages among groups within a movement.	Vulnerability of political system to social protest.	Interconnection between social structural and cultural perspectives.

Structural strain theory, developed by sociologist Neil Smelser (1982), interprets movements as arising when various tensions make society conducive to people organizing for social change. People may feel deprived of some resource, but conditions have to exist that allow them to mobilize for change. One such condition is that there has to be the spread of some explanation or framework by which people interpret the need for action, as we will see below in the discussion of social movement *frames*. In addition, there is usually some precipitating factor that sparks the growth of a movement. This then leads to mobilization, as we have seen in the discussion of resource mobilization theory.

Political process theory posits that movements achieve success by exploiting a combination of internal factors, such as the ability of organizations to mobilize resources, and external factors, such as changes occurring in the society (McAdam 1982). Some structural conditions provide opportunities for collective action. A war, pressure from international parties, demographic shifts, or an economic crisis may create the possibility for people who challenge the social order to mobilize a movement (McAdam 1999). Political process theory stresses the vulnerability of the political system to social protest. For example, after the 2000 presidential election, African American voters, along with others mobilized to protest the irregularities in voting procedures that had occurred in minority districts in Florida. African Americans have been protesting such impediments to voting for years, but the particularly egregious irregularities in the Bush/Gore election crystallized sentiment that was already present. The subsequent mobilization of disenchanted voters initiated some reforms in the voting process.

Resource mobilization theory and political process theory are social structural explanations. Other explanations of social movements are more cultural in focus. Cultural explanations of social movements focus on the importance of meaning systems in mobilizing people for collective action. People will not organize within social movements unless they develop a shared definition of the situation that gives meaning to their action. Explanations within a cultural perspective on social movements then focus on the collective process by which peo-

ple interpret a perceived grievance or attribute. Through the creation of meaning systems, people create a sense of shared identity—a crucial step in the mobilization of collective action.

Sociologists use the concept of *framing* to explain the process for collective action. **Frames** are schemes of interpretation that allow people in groups to perceive, identify, and label events within their lives that can become the basis for collective action (Snow et al. 1986; Snow and Benford 1988; Goffman 1974). You can think of this as collective sense-making. The framing process emphasizes that social movements do not emerge unless people have a shared understanding of the causes of a perceived injustice or a shared definition of their opposition (see Figure 22.3). As an example, framing can help you understand the emergence of Ralph Nader as a movement leader. He has framed corporate domination as the major source of political and social problems in the nation. This frame then provides the lens through which people have been mobilized in support of Nader.

A social movement frame can also be heightened when opposing movements engage in action that further crystallizes the frame's definition. For example, the actions of the police may sharpen a frame that tells movement participants that their grievance is legitimate. For example, a strong police response to an anti-war demonstration may only accentuate the movement's perception that the government and its authorities are acting unjustly.

Frames can also help us understand how movements change or break down. During the civil rights move-

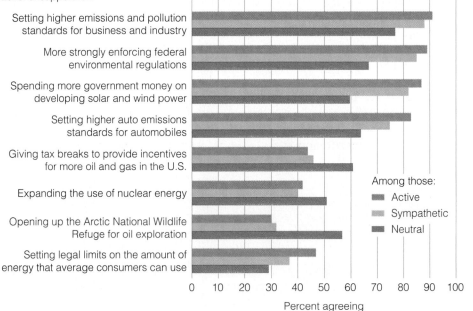

Figure 22.3 The Environmental Movement: Activists and Sympathizers

Source: Crabtree, Steve. 2001. "Environmental Activists: Extreme to Mainstream." *The Gallup Poll.* Princeton, NJ: The Gallup Organization. Website: **www.gallup.com**

ment, participants shared a definition of the situation that Jim Crow laws were the main obstacle to Black social progress. When this singular frame broke down as more radical forms of protest emerged and as laws were passed to eliminate old forms of segregation, the movement subsided.

Cultural explanations also add another dimension to the study of social movements by highlighting the expressive aspects of social movements (Johnston and Klandermans 1995). Many social movements try to use dramatic social actions to call attention to their cause. Often, these actions are creative and exciting to those involved. Movements can stage "performances," as when they march on Washington or otherwise dramatize their cause through some creative demonstration. One example of dramatic action is in the battered women's movement, wherein participants hang an article of clothing on a publicly displayed clothesline and each piece of clothing includes a battered woman's name. The gay and lesbian movement has also used "performances" to contest the sexual categories that allegedly divide "straight" and "gay" people. Playful expressions of sexual identity provide a tactic for protesting sex and gender roles; in this way, cultural expression becomes a tool for social protest (Rupp and Taylor 2003).

Cultural and social structural explanations of social movements come together in what is called **new social movement theory** linking culture, ideology, and identity conceptually to explain how new identities are forged within social movements. Whereas resource mobilization theory emphasizes the rational basis for social movement organization, new social movement theorists are especially interested in how identity is socially constructed through participation in social movements (Gamson 1995b; Larana et al. 1994; Calhoun 1994). This new development in social movement theory links structural explanations with cultural and social-psychological theories, investigating how social movements provide foundations upon which people construct new identities (Gamson 1992; Gamson 1995a; Morris and Mueller 1992).

Each theoretical viewpoint is helpful in explaining different aspects of how movements organize and, in some cases, how they fail. One team of sociologists has examined the demise of the American Indian Movement (AIM)—a specific organization prominent in the Native American liberation movement for several years. These sociologists argue that AIM ultimately failed to reach its goal of building a pan-tribal sense of identity among American Indians because of its inability to mobilize resources in the face of strong government repression. AIM was a threat to both government and corporate interests because of its focus on reclaiming energy resources on lands Native Americans argued were rightfully theirs. The authors conclude that resource mobilization theory, political process theory, and new social movement theory each explain different dimensions of

SOCIOLOGY IN PRACTICE

Understanding Terrorism as a Social Movement

The promise of sociology has been that it can help people understand events in everyday life within a broader social and historical context. The concepts of sociology thus often help us understand contemporary affairs, even when the event may be troubling or even frightening. Such is the case with understanding al Qaeda—the international organization responsible for the terrorist acts against the United States.

Al Qaeda can be understood as a *transnational social movement*—a social movement organized across national borders. It is also a *reactionary social movement*—one that is organized to resist the changes associated with a more modern world order. Thus, it uses its extreme methods of terrorism to try to destroy the symbols of moder-

nity, such as the World Trade Center, a symbol of global capitalism and Western values.

Resource mobilization theory also can be used to explain how al Qaeda developed. The movement utilized modern systems of technology, including the Internet, to communicate with its different organizational cells. Members also developed special technical knowledge to plan their attacks. And they did so under the leadership of an effective leader, Osama bin Laden. The ethnic conflicts in various nations where al Qaeda operated also provided a steady supply of new recruits—people who would become "true believers," that is, fanatics about their cause.

Although al Qaeda is one of the most inhumane and dangerous social

movements to have developed in recent history, we can understand its social organization and mobilization through this sociological perspective. Such an analysis does not excuse terrorism or the work of al Qaeda, but it can help experts and others understand how such movements develop and thus how to take action to combat them.

Critical Thinking Exercise

1. What social conditions promote the development of transnational social movements?

2. Are there other sociological concepts that you would find useful in understanding how terrorism develops and how nations might protect themselves from such developments? ●●●

this movement (Stotik et al. 1994). This is a good illustration of how sociological theory can be used to explain contemporary social movement activity.

Diversity, Globalization, and Social Change

Social movements are a major source of social change. Around the world, as people have organized to protest oppressive forms of government, the absence of civil rights, or economic injustices, social change often results. What would the United States be like had the civil rights movement not been inspired by Mahatma Gandhi's liberation movement in India? How would contemporary politics be different had the African National Congress and other movements for the liberation of Black South Africans not dismantled apartheid? How is the world currently affected by the development of a more fundamentalist Islamic religious movement in the Middle East?

These and countless other examples show the significance of collective behavior and social movements for the many changes affecting our world. Sometimes collective behavior and social movements can be the basis for revolutionary events—those that change the course of world history. Other times, the persistence of social movements more slowly changes the world—or a particular society. Persistence has been the case for the civil rights movement in the United States—a movement that not only has transformed American society, but has also since inspired similar movements throughout the world. Likewise, the women's movement is now a global movement, although the particular issues for women vary from place to place.

Another way to think about globalization and social movements is through the concept of a **transnational social movement,** in which an organization crosses national borders, such as the reactionary, terrorist group al Qaeda. Although al Qaeda is not typically thought of as a social movement, analyzing it as such helps explain how al Qaeda works and the impact it has on world affairs. For example, al Qaeda is organized in small "cells" that are small networks of people affiliated with this reactionary movement. As we see in the box Sociology in Practice: "Understanding Terrorism as a Social Movement," concepts from the sociology of social movements help explain how this terrorist organization works.

In the United States, the most significant social movements are those associated with the nation's diverse population. The women's movement, the civil rights movement, the gay and lesbian movement are all major sources of activism in contemporary society, and these movements have generated some of the most transformative changes in the nation's social institutions. Despite the persistence of class, race, and gender inequality, no longer is segregation legally mandated. Equal opportunity laws, at least in theory, protect diverse minority groups from discriminatory treatment. Large segments of the public have become more conscious of the harmful effects of racism, sexism, and homophobia in society. All these changes can be attributed to the successful mobilization of diverse social movements.

The nation's changing and diverse population will continue to be a powerful influence in shaping social activism in the future. As we will see in Chapter 23, different sources of social change, including the growing racial and ethnic diversity within the population, will likely shape society of the future.

Chapter Summary

What is collective behavior?

Collective behavior occurs when normal conventions cease to guide people's behavior, and people establish new patterns of interaction and social structure. Collective behavior is patterned, not irrational. *Rumors* are a mechanism people use to communicate during collective behavior. Collective behavior is often associated with efforts to promote change.

How do sociologists analyze crowd behavior?

Crowds are groups of people who temporarily come together for a specific reason. They are usually a particular size and density, and they are more or less connected to one another. Social control agents can also influence the behavior of crowds. *Panic* is spontaneous behavior that occurs when people become highly concerned about their safety and security. Despite its image as disorganized, panic is socially structured.

What is emergent norm theory?

Emergent norm theory suggests that people create meanings in unusual situations to guide their behavior.

How do sociologists analyze riots?

The popular conception of riots is that they are the action of unruly mobs, but sociologists explain *riots* by analyzing the social conditions that generate them. In urban riots, these factors can include competition, background conditions, precipitating factors, situational determinants, and the failure of social control mechanisms.

What are collective preoccupations?

Collective preoccupations are forms of collective behavior in which people engage in similar behavior and have a shared understanding of that behavior as necessary to bring about change or identify their place in society. *Fads*, fashion, *crazes, hysterical contagions*, and *scapegoating* are types of collective preoccupations.

What is a social movement?

Social movements are organized social groups that have some continuity and coordination and that exist to promote or resist social change. There are different kinds of social movements, including *personal transformation movements*, *social change movements*, and *reactionary movements*.

How do social movements start?

Social movements start when there is a preexisting communication network, a collective sense of grievance, a precipitating factor initiating the movement, and mobilization of a group of people. *Resource mobilization theory* suggests that social movements develop when people can compete for and gain resources needed for mobilization. *Political process theory* suggests that large-scale social changes, such as industrialization or urbanization, provide the conditions that spawn social movements. Both resource mobilization theory and political process theory are social structural explanations that place the origins of social movements in societal conditions, not individual needs or psychological wishes. Cultural perspectives on social movements emphasize the *framing* process as important to creating a shared definition of reality for movement participants. *New social movement theory* links social structural and cultural explanations, emphasizing that social movements are places where people construct their identities.

How do globalization and diversity affect social movements?

Social movements can be the basis of revolutionary change. Some movements originating in one nation also spill over to affect movements in another. *Transnational social movements* are those whose organizational struc-tures cross national borders. Some of the most profound changes in the United States have come as the result of social movements from the nation's diverse population.

Key Terms

collective behavior 589	new social movement theory 611
collective preoccupations 597	personal transformation movements 601
competition theory 596	political process theory 610
contagion theory 593	radical movements 602
convergence theory 596	reactionary movements 603
craze 598	reform movements 602
emergent norm theory 593	resource mobilization theory 609
expressive crowds 593	scapegoating 599
fad 597	social change movements 601
fashion 598	social movement 589
frame 610	transnational social movement 612
hysterical contagion 599	
mobilization 605	

Researching Society with MicroCase Online

You can see the results of actual research by using the Wadsworth MicroCase® Online feature available to you. This feature allows you to look at some of the results from national surveys, census data, and other data sources. You can explore this easy-to-use feature on your own, but try this example. Suppose you want to know:

Have you done any voluntary activity in the past 12 months in any of the following arenas—helping political parties, political movements, election campaigns, etc.?

To answer this question, go to http://sociology.wadsworth .com/andersen_taylor4e/, select MicroCase Online from the left navigation bar, and follow the directions there to analyze the following data.

Data File: GSS

Task: Auto-Analyzer

Primary Variable: VOLWKPOL

Questions

Once you have your results, answer the following questions:
The tables show you whether various factors (religion, political party, age, education, sex of respondent, and race) are related to whether people have volunteered in political activities.

1. How does the General Social Survey define voluntary activity?

2. Among all respondents, what percentage of the sample had volunteered in political activities?

3. Complete the table using the data in the row "YES" for the following demographic variables.

Socio-Demographic Variable	Category Most Likely	Category Least Likely	Significant?
Age			Yes No
Education			Yes No
Income			Yes No
Party			Yes No
Race			Yes No
Region			Yes No
Religion			Yes No
Sex			Yes No

4. Describe the characteristics most likely among people who volunteered in political activities.

The Companion Website for Sociology: Understanding a Diverse Society, Fourth Edition

http://sociology.wadsworth.com/andersen_taylor4e/

Supplement your review of this chapter by going to the Companion Website to take one of the Tutorial Quizzes, use the flash cards to master key terms, and check out the many other study aids you'll find there. You'll also find special features such as GSS Data and Census 2000 information, data and resources at your fingertips to help you with that special project or do some research on your own.

Suggested Readings and Web Resources

D'Emilio, John. 1998. *Sexual Politics, Sexual Communities: The Making of a Homosexual Minority in the United States, 1940–1970.* Chicago, IL: University of Chicago Press.
D'Emilio analyzes the growth of the gay and lesbian rights movement, showing how the movement broke down the invisibility of gays and lesbians and brought a new analysis of gay identity as socially constructed to the public eye.

Freeman, Jo. 1999. *Waves of Protest: Social Movements Since the Sixties.* Lanham, MD: Rowman and Littlefield.
This anthology analyzes the sociological and political structure of contemporary social movements, looking especially at how movements mobilize, how they are organized, how they use strategies and tactics, how they change consciousness, and how they decline.

Jenness, Valerie, and Kendal Broad. 1997. *Hate Crimes: New Social Movements and the Politics of Violence.* New York: Aldine de Gruyter.
The authors use social movement theory to explain why hate crimes have recently become defined as a social problem, when violence has long been directed against Jews, racial minorities, the disabled, and gays and lesbians.

McAdam, Doug. 1988. *Freedom Summer.* New York: Oxford University Press.
McAdam studies a sample of the volunteers who worked in the voter registration drive of 1964, known as "Freedom Summer." His study of these movement participants—many of them White, northern college students—analyzes why they went, what happened, and what they learned from this historic experience.

Morris, Aldon D. 1984. *The Origins of the Civil Rights Movement.* New York: The Free Press.
This book is a classic study of the history and sociology of the civil rights movement. It utilizes resource mobilization theory to explain the emergence of the movement.

Environmental Justice Resource Center
www.ejrc.cau.edu
A center located at Clark University in Atlanta that has numerous online resources for those interested in how environmental issues particularly affect minority communities.

Southern Poverty Law Center
www.splcenter.org
An organization long devoted to monitoring and abolishing hate groups in the United States.

● ● ●

Social Change in Global Perspective

Technological innovations can transform an entire society. Consider the following: A gigabyte of information can travel from China to the United States in less time than it will take you to read this paragraph. At the same time, you could be sending a fax in the opposite direction while also sipping coffee that was grown in Columbia, roasted in Seattle, distributed across the nation, and delivered to your home via express mail. Was all this imaginable fifty years ago? Not really. What will the next fifty years bring?

We are quick to think of such changes as progress, but sometimes people in a changing society wonder whether the things gained from progress are not overbalanced by the things lost. The Alaskan pipeline and related innovations have brought considerable wealth to the communities of northern Alaska, and many Eskimo entrepreneurs have grown prosperous from the changes. Others have fared less well. Some researchers have attributed increases in suicide and alcohol-related deaths to changes that simply came too rapidly in the social structure and culture of the Eskimos (Klausner and Foulks 1982; Simons 1989).

Technological advances have recently come to the Kaiapo people of Brazil. They are noted for their striking body paint and elaborate ceremonial dress. Recently, many Kaiapo have become wealthy as a result of gold mining and tree harvesting in their region. Yet, a favorite topic of conversation among the elder members of this society is whether their newfound wealth is a blessing or curse. For example, the availability of television has caused the youth among the Kaiapo to retreat from traditional rituals like the evening campfire, where all members of the village, old and young, would sit to tell stories and dispense philosophy. One elder lamented that whereas the night was previously the time when the old taught the young the ways of life, television had "stolen the night," bringing an end to a practice that had many benefits for the young (Simons 1989).

Sociology ⊛ Now™

Reviewing is as easy as ❶❷❸.

Use SociologyNow to help you make the grade on your next exam. When you are finished reading this chapter, go to the chapter review for instructions on how to make SociologyNow work for you.

These examples are studies in the effects of social change. What is social change? What causes it? What has the power to launch deep changes in norms, habits, practices, beliefs, gender roles, racial and ethnic relations, and class distinctions? This chapter examines the causes and consequences of societies in change. •••

What Is Social Change?

Social change is the alteration of social interactions, institutions, stratification systems, and elements of culture over time. Societies are always in a state of flux. Some changes are rapid, such as those brought about by desktop computers in little more than ten to fifteen years. Other changes are more gradual, such as the increase of urbanization that characterizes the contemporary world. Sometimes people adapt quickly to change, as is happening in the development of electronic communication. Other times people resist change or are slow to adapt to new possibilities. Decades of effort to promote contraceptive methods in overpopulated, developing nations have garnered only the most sluggish gains (Goldman et al. 1989). The speed of social change varies from society to society and from one time to another within the same society.

As societies become more complex, the pace of change increases. In U.S. society, this truism can be seen by comparing the rate of change in the 1950s with the rate in the 1990s (Lenski et al. 1998). Most people would agree that the United States in the 1950s was far more predictable and staid than today's United States as the twenty-first century gets underway.

Microchanges are subtle alterations in the day to day interaction between people. A fad "catching on" is an example of a microchange. Fads and other microchanges often spread rapidly across the nation. Take the popularity of bungie jumping. Although not as widespread as some prior fads, this highly dangerous recreation is one in a group of "extreme sports" that have recently become popular across the country. Bungie jumping has caused a number of serious injuries and deaths, but it has also provided thrilling footage for soft-drink commercials, probably accounting for why a large number of youths have suddenly developed a taste for putting themselves in bone-smashing danger. Another such fad is extreme motocross racing, where the participants on motorcycles leap over bumps and hills with such speed that driver and motorcycle become airborne and separate from each other. Although the overall change in the structure of society caused by fads is small, some minor effects may persist. Skateboarding has had several rises and falls as a fad, starting in the early 1960s, and in that time has never faded completely out of the repertoire of youthful recreations—an example of how a microchange can persist.

Macrochanges are gradual transformations that occur on a broad scale and affect many aspects of society. In the process of *modernization*, societies absorb the changes that come with new times and shed old ways. One frequently noted trend accompanying modernization is that societies become more differentiated socially, including greater differentiation in social rank, divisions of labor, and so on. The effects of the fast-food industry and its impact on social structure exemplify a macrochange (see the discussion of McDonaldization in Chapter 6).

In the United States, the quick rise of the computer through all its generations, from vacuum tube to microchip, is another example of a macrochange that has dramatically changed society. Not many years ago, who would have imagined that you could surf the Web to the extent possible today? As recently as 1990, few people had heard of the Internet, much less used it (Friedman 1999).

The Kaiapo people of Brazil wear colorful formal dress. Technology from outside this society (TV, guns) presently threatens the persistence of such cultural practices.

© "Disappearing World," Granada TV/The Hutchison Library

© Mystic Seaport, Rosenfeld Collection, Mystic CT

© Paul L. Sutton/Duomo

Social change is sometimes apparent by contrasting the social norms of different historical periods.

Now it is a daily presence in the lives of millions. The macrochange to a digital culture was swift, but some macrochanges can take generations. Whatever time they require, macrochanges represent deep and pervasive changes in social structure and culture.

THINKING *SOCIOLOGICALLY*

A fad is an example of a *microchange*. What fads have you noticed lately among your friends or acquaintances? Are they completely new, or were they started a year or two ago and are simply continuing? How long do you think these fads will last? Sometimes a microchange such as a fad or craze can endure for many years, in which case it has an increased chance of causing a *macrochange*, a change on a broader scale. Name two fads of the past that resulted in macrochanges.

Large or small, fast or slow, social change generally has the following characteristics in common.

1. **Social change is uneven.** The different parts of a society do not all change at the same rate; some parts lag behind others. This is the principle of *culture lag*, a term coined by early sociological theorist **William F. Ogburn** (1922) and first described in Chapter 3. Recall that culture lag refers to the delay that occurs between

the time that social conditions change and the time that cultural adjustments are made. Often the first change is a development in material culture (for example, hardware technology), which is followed some time later by a change in nonmaterial culture (for example, the habits and norms of the culture). The symptoms of culture lag can be seen in the uneven dissemination of computer capability. Some organizations and bureaucracies adopt state-of-the-art hardware and software more quickly than others, leaping ahead of their colleagues and competitors. Even within single organizations, change occurs unequally, with older employees tending to adapt to new technology more slowly than younger members.

2. **The onset and consequences of social change are often unforeseen.** The inventors of the atomic bomb in the early 1940s could not predict the vast changes in the character of international relations that were to come, including a cold war that lasted until the demise of the Soviet Union in the early 1990s. Television pioneers, who envisioned a mode of mass communication more compelling than radio, could not know that television would become such a dominant force in determining the interests and habits of youth as well as the activities and structure of the family. The notion of culture lag is present in both of these examples: A change in material culture (invention of the atomic bomb, invention

of television) precedes later changes in nonmaterial culture (international relations, youth culture, fads, and family structure).

3. **Social change often creates conflict.** Change often triggers conflicts along racial–ethnic lines, social class lines, and gender lines. Terrorism—both in the United States and the world—focuses attention on the deep conflicts that exist worldwide in political, ethnic, and religious divisions. These conflicts not only produce international tension, but often drive the world events that generate social change.

4. **The direction of social change is not random.** Change has direction relative to a society's history. A populace may want to make a good society better, or it may rebel against a status quo regarded as unendurable. Change may be wanted or resisted, but in either case, when it occurs, it takes place within a specific social and cultural context.

Social change cannot erase the past. As a society moves toward the future, it carries along its past, its traditions, its institutions (Lenski et al. 1998; McCord 1991; McCord and McCord 1986). A generally satisfied populace that strives to make a good society better obviously wishes to preserve its past, but even when a society is in revolt against a status quo that is intolerable, the social change that occurs must be understood in the context of the past as much as the future.

Theories of Social Change

As we have seen, social change may occur for different reasons. It may occur quickly or slowly, may be planned or unplanned, and may represent microchange or macrochange. Different theories of social change emphasize different aspects of the change process. Three main lines of contention in social change theory are functionalist theories, conflict theories, and cyclical theories. Later in this chapter, we consider three additional global theories: modernization theory, world systems theory, and dependency theory.

FORCES OF SOCIAL CHANGE

A Nation Changed?

Following the terrorist attacks on the United States on September 11, 2001, many people wondered if U.S. society was changed forever. Perhaps it was. The spontaneous creation of shrines, the efforts of relief workers, and the organization of candlelight vigils and other memorials demonstrate a major principle of collective behavior: Even under tragic conditions, human beings form meaningful relationships with each other.

However, the bonds of solidarity that emerged among many Americans after 9/11 also reveal a less noble sociological process: The creation of in-group solidarity also produces antagonism toward out-groups, as evidenced by the ethnic profiling of Arab Americans and, in some cases, hate crimes directed against those thought to be Arab Americans. Such acts also show how prejudice can lead people to falsely generalize to all people in a minority group.

Many other sociological processes became visible as a result of these attacks. In our increasingly technologically reliant society, the media play a key role in shaping public opinion and transmitting information. Years from now, few people will forget the televised images of planes crashing into two treasured national symbols. And the use of cell phones, particularly the last minute calls from victims, will haunt many memo-

ries. The surge of patriotic displays also constructed a sense of national identity that may not have been felt so keenly before. For years to come, sociologists will be analyzing many facets of this tragedy, including its long-term impact on civil rights and the response of social institutions (such as the military, transportation systems, the economy) to such a disaster.

Current generations will now likely always remember what they were doing when they first learned of the attacks. The power of this memory reminds us of an important sociological point that C. Wright Mills made years ago: People's individual lives are shaped by the social forces of history. •••

DOING SOCIOLOGICAL RESEARCH

An Insider's Account of a Social Movement and Social Change

Todd Gitlin, a sociologist and journalist, participated in the student rebellions of the 1960s. As a social movement, the rebellions of the 1960s were a force that stimulated social change in this society, and though some of these changes have faded, others remain. Gitlin found that from 1967 to 1969, attitudes on college campuses leaned strongly, even radically, against the Vietnam War. In the fall of 1969, 69 percent of students nationwide called themselves "doves"—twice as many as in the spring of 1967. More and more students seemed to realize that the U.S. government was not about to end the war, and they concluded a few years behind SDS (Students for a Democratic Society) that the war was not merely an isolated "mistake," but instead part of a detailed national plan.

At that time, the percentage of students who agreed with the statement, "The war in Vietnam is pure imperialism" jumped from 16 percent in the spring of 1969 to 41 percent in April 1970. The number of students strongly disagreeing fell from 44 to 21 percent. The percentage of students calling themselves radical or far left was 4 percent in the spring of 1968 and rose to 8 percent in the spring of 1969 and to 11 percent in the spring of 1970. Nonviolent as well as violent protests accelerated throughout the year, 731 of them involving police arrests, 410 involving damage to property (ROTC buildings and the like), and 230 involving violence to persons. The infamous killing of four students by the Ohio National Guard at Kent State University in the spring of 1970 still remains clear in the national memory.

Questions to Consider

1. In retrospect, do you think the United States should have entered the war in Vietnam? Why or why not? *Keywords: anti-war movements*
2. With respect to present U.S. war policy, do you consider yourself to be a "hawk," a "dove," or neither? *Keywords: student protest; student movements*

We have included InfoTrac College Edition keywords at the end of each question to make it easier for you to find more to read on these topics. Go to www.infotrac-college.com, an online library, to begin your search.

Source: Gitlin, Todd. 1987. *The Sixties: Years of Hope, Days of Rage.* New York: Bantam Books. ●●●

Functionalist and Evolutionary Theories

Recall from previous chapters that functionalist theory builds upon the postulate that all societies, past and present, possess basic elements and institutions that perform certain functions permitting a society to survive and persist. A *function* is a consequence of some social element that contributes to the continuance of a society.

The early theorists **Herbert Spencer** (1882) and **Emile Durkheim** (1964/ 1895) both argued that as societies move through history, they become more complex. Spencer argued that societies moved from "homogeneity to heterogeneity." Durkheim similarly argued that societies moved from a state of *mechanical solidarity,* a cohesiveness based on the similarity among its members, to *organic solidarity* (also called *contractual solidarity*), a cohesiveness based on difference—a division of labor that exists among its members joins them together because each is dependent on the others for the performance of specialized tasks (see Chapter 6). Societies thus move from a condition of relative undifferentiation to higher social differentiation through the creation of specialized roles, structures, and institutions.

According to functional theorists, societies move from structurally simple, homogeneous societies, such as foraging or pastoral societies where members engage in largely similar tasks, to structurally more complex, heterogeneous societies, such as agricultural, industrial, and postindustrial societies. Great social differentiation exists in heterogeneous society and there is extensive division of labor among people who perform many specialized tasks. The consequence (or function) of increased differentiation and division of labor is a higher degree of stability and cohesiveness in the society, brought about by the realities of mutual dependence according to functional theorists (Parsons 1951a, 1966).

Evolutionary social theories of social change are a branch of functionalist theory. One variety called **unidimensional evolutionary theory,** now well out of favor, argued that societies follow a single evolutionary path from simple and relatively undifferentiated societies to more complex and highly differentiated societies, with the more differentiated societies perceived as more civilized. Early theorists such as Lewis H. Morgan (1877) labeled this difference a distinction between primitive and civilized, an antiquated notion that has been severely criticized. There is no reason to suppose that an undifferentiated society is necessarily more primitive than a more differentiated one. Furthermore, in these earlier theories, there are no firm definitions for the terms *primitive* or *civilized.* Nevertheless, the notion that some societies are primitive continues to persist today.

Unidimensional theories of social change fell out of favor because social change occurs in several dimensions and affects a variety of institutions and cultural elements. Meeting the need for a theory that better matches what is actually observed, **multidimensional evolutionary theory** (also called **neoevolutionary theory**) argues that the structural, institutional, and cultural development of a society can follow many evolutionary paths simultaneously, with the different paths all emerging from the circumstances of the society in question.

A formulation of multidimensional evolutionary theory is that of Gerhard Lenski and associates (1998). Lenski gives a central role to technology, arguing that technological advances are significantly (though not wholly) responsible for other changes, such as alterations in religious preference, the nature of law, the form of government, and relations between races and genders. The role of technology is presented as central, yet other relationships among institutions continue to be important. For example, changes in technology, such as advances in computer hardware and software, can produce changes in the legal system by creating a need for new laws to deal with computer crimes—child pornography on the Internet, for example.

In support of the overall argument that social change is in fact evolutionary—cumulative and not easily reversible—Lenski and his associates point out that many agricultural societies have transformed into industrial societies throughout history. Few have made the reverse trip from industrial to agricultural, although certain countercultural groups have tried, such as the hippie communes of the 1960s and 1970s. On the other hand, Lenski also argues that social advances can be reversed. For example, a cataclysm such as an earthquake or flood can humble a technologically advanced society. Following a natural disaster, especially in the developing world, residents may find themselves foraging for food if the elaborate infrastructure supporting urban life is destroyed. The devastating tsunamis in 2004 that killed thousands of people and destroyed vast areas in twelve different nations of southeast Asia changed society in seconds. The long-term changes brought on by such a vast tragedy are yet to be known.

DEBUNKING *SOCIETY'S MYTHS*

Myth: Societies change in a linear, directed fashion from primitive to civilized.

Sociological perspective: Social change can occur in several directions at roughly the same time, thus giving more weight to a multidimensional change theory than to a unidimensional one. Furthermore, the terms *primitive* and *civilized* are of limited usefulness and are out of favor as concepts, in that they imply some value judgment about the relative sophistication of diverse cultures.

Unlike the early theories of Spencer, Durkheim, and Parsons, newer functionalist theories emphasize the role of racial–ethnic, social class, and gender differences in the process of social change (Lenski et al. 1998; Alexander and Colomy 1990; McCord 1991; McCord and McCord 1986). The earlier theories made the implicit assumption that European and American societies, predominantly White, were more evolved or advanced. Societies largely comprising people of color were usually assumed to be less evolved and more primitive. This bias was often projected onto analyses *within* a society. For example, in the United States, Native American and Latino cultures were often seen as less advanced. Older functionalist theories also supported the notion that less advanced peoples were less intelligent, a postulate completely rejected by later functionalists, as well as by social and behavioral researchers in general. The new functionalism rejects the "primitive" versus "civilized" dichotomy and its implicit commentary on racial groups and considers relations between racial–ethnic, social class, gender, and other groups as an important part of any society, regardless of its stage of development or evolution.

Conflict Theories

Karl Marx, the founder of conflict theory (Marx 1967/1867), was himself influenced by the early functionalist and evolutionary theories of Herbert Spencer. Marx agreed that societies change and that social change has direction; the central principle in Spencer's social evolutionary theory. However, Marx placed greater emphasis on the role of economics than did functionalist and evolutionary theorists. He argued that societies could "advance" and that advancement was to be measured in terms of movement from a class society to one without class. Marx believed that, along the way, class conflict was inevitable.

As noted earlier in this book, the central notion of conflict theory is that conflict is built into social relations (Dahrendorf 1959). For Marx, social conflict, particularly between the two major social classes in any society—working class versus upper class, proletariat versus bourgeoisie—was not only inherent in social relations but the driving force behind all social change. Marx believed that the most important causes of social change were the tensions between various social groups, especially those defined along social class lines. The main reason for tension was that different classes had different access to power, with the relatively lower class carrying less power. Although the groups to which Marx originally referred were social classes, subsequent interpretations expand on Marx to include conflict between any socially distinct groups that receive unequal privileges and opportunities (Marx, G. T. 1967; Rodney

1974). However, be aware that the distinction between class and other social variables is necessarily murky. For example, conflict between Whites and minorities is at least partly (but not wholly) class conflict because minorities are disproportionately represented among the less well-off classes.

There is far more to racial and ethnic conflict in the United States than class differences alone. Many cultural differences exist between Whites and between and among Native Americans, Latinos, Blacks, and Asians. Furthermore, there are cultural differences *within* broadly defined ethnic groups. We have pointed out earlier in this book that there are broad differences in norms and heritage between Chinese Americans, Japanese Americans, Vietnamese Americans, and so on, all of whom are often grouped rather coarsely as Asian Americans. The central idea of conflict theory is the notion that social groups will have competing interests no matter how they are defined and conflict is an inherent part of the social scene in any society.

A central theme in Karl Marx's writing is that revolution and dramatic social change would come about when class conflict led inevitably to a decisive social rupture. Marx predicted that the capitalist class would progressively eliminate or absorb competitors and relentlessly pursue profits while squeezing the wages of the working class and crushing dissent. Discontent among the working classes was supposed to blossom into a recognition that the common enemy of the worker was the capitalist class. The workers would then join in revolution, overthrow the system of capitalism, eliminate privately owned property, and establish a new economic system that would exist for the good of all.

Although the worldwide revolution predicted by Marx has never come to pass, his highly refined analyses of class-related conflict have advanced our understanding of social change, and his work continues to be of interest. However, Marx seems to have overemphasized the role of economics in the network of social tensions he observed, while ignoring the importance of other relevant factors related to class.

Sociologist **Theda Skocpol** (1979) has noted that in France, Russia, and China—countries where major revolutions have occurred—serious internal conflicts between social classes were combined with major international crises that the elite social classes proved unable to resolve before they were overthrown. The French Revolution, begun in 1789, erupted in a period when the newly arisen capitalist class was asserting itself worldwide against the old monarchies. While France was bankrupt from the many wars of the seventeenth and eighteenth centuries, the country intervened in the American Revolution. The Russian Revolution occurred while Russia was flattened from its disastrous defeat in the First World War, and the communist revolution in

Martin Luther King, Jr., speaks during the 1963 march on Washington, a major event in the civil rights movement, which resulted in major social changes in the United States.

China occurred as the world was still putting out the flames of the Second World War. In each case, internal social change was linked to relations between entire societies in different parts of the globe, as well as to relations among social classes within a society.

The collapse of the Soviet Union (now Russia) offers some support for Skocpol's hypothesis. Years of international trade sanctions and rejection of its currency by foreign traders had eroded the economic foundations of the Soviet Union, and headlong military competition with the United States helped drive the Soviet Union into bankruptcy. Although economic considerations were paramount in the fall of the Soviet Union, the relationships *between* societies, and not just among classes within a society, were crucial in undoing the former superpower. Nevertheless, conflicts among Russians along religious, social class, ethnic, and regional lines also contributed to the social and cultural disintegration of the Soviet state.

Ethnic, racial, and religious and tribal differences join social class differences as major causes of conflict within and between countries. Cases in point include recent horrors of *ethnic cleansing*, vicious battles between ethnic groups in which one group attempts to annihilate the other, as Serbians in the early 1990s attempted to do to Muslims in Bosnia and attempted in the late 1990s to do to Albanians in Kosovo. Prior to the end of their rule in Afghanistan in 2002, the Taliban's systematic slaughter of many of Afghanistan's minorities

Table 23.1
Theories of Social Change

	General Theories			Global Theories		
	Functionalist/ Evolutionary Theory	**Conflict Theory**	**Cyclical Theory**	**Modernization Theory**	**World Systems Theory**	**Dependency Theory**
How do societies change?	Societies change from simple to complex and from an undifferentiated to a highly differentiated division of labor.	Conflict is inherent in social relations, and society changes from a class-based to a classless society.	Societies develop in cycles from idealistic culture to sensate culture.	Societies become more homogenized as the result of technological change.	Unequal political and economic relationships between nations result in some ("core nations") becoming more advanced than others ("peripheral nations").	The most successful nations control the development of less powerful nations, which become dependent on them.
What is the primary cause of social change?	Technology.	Economic conflict between social classes.	Necessity for growth.	Technology and global development.	Growth of international capitalism.	Economic inequality in the global economy.

provides another example of ethnic cleansing among groups within the same society.

Cyclical Theories

Cyclical theories of social change invoke patterns of social structure and culture that are believed to recur at fairly regular intervals. Cyclical theories build on the idea that societies have a life cycle, like seasonal plants, or at least a life span like humans. **Arnold J. Toynbee,** a social historian and a principal theorist of cyclical social change, argues that societies are born, mature, decay, and sometimes die (Toynbee and Caplan 1972). For at least part of his life, Toynbee believed that Western society was fated to self-destruct as energetic social builders were replaced by entrenched minorities who ruled by force and under whose sterile regimes society would wither. Some believe that societies become decrepit, only to be replaced by more youthful societies. This belief is typified in Oswald Spengler's famous work, *The Decline of the West* (1932), which held that western European culture was already deeply in decline, following a path Spengler believed was observable in all cultures.

Sociological theorists Pitrim Sorokin (1941) and, more recently, Theodore Caplow (1991) have argued that societies proceed through three different phases or cycles. In the first phase, dubbed the *idealistic culture,* the society wrestles with the tension between the ideal and the practical. An example would be the situation captured in Gunnar Myrdal's classic work, *An American Dilemma* (1944), in which our nation declared a belief in equality for all, despite intractable racial, class, and gender stratification.

The second phase, *ideational culture,* emphasizes faith and new forms of spirituality. The strong religious institution of the Puritans in Colonial America is one example. Another more current example is the New Age spirituality movement, which stresses nontraditional techniques of meditation and the seeking out of small support groups for close interpersonal interaction in a journey toward self-fulfillment and spiritual peace (Wuthnow 1994).

The third phase is *sensate culture,* which stresses practical approaches to reality and involves the hedonistic and the sensual ("sex, drugs, and rock and roll"). Sorokin may have foreseen the hedonistic elements of popular culture in the 1960s and 1970s, elements indicative of sensate culture. According to the theory, when a society tires of the sensate, the cyclical process begins again with the society seeking refuge in idealistic culture. The emphasis begun in the late 1980s and continuing now for a return to "family values," meaning older and more traditional values, is an example of a return to idealistic culture, presumably as a response to a prior perceived sensate culture.

The Causes of Social Change

Social and cultural change is a broad subject, and the causes of social change are varied and shifting. The major changes are reviewed here: revolution; mobilization as collective behavior and social movements (examined in detail in Chapter 22); cultural diffusion; inequalities in race, ethnicity, social class, and gender; technological innovation; changes in population, including migration; and war and terrorism.

Revolution

A **revolution** is the overthrow of a state or the total transformation of central state institutions. A revolution thus results in far-reaching social change. Numerous sociologists have studied revolutions and identified the conditions under which revolutions are likely to occur. Revolutions can sometimes break down a state as the result of conflict between an oppressive state and various disenfranchised groups. An array of groups in a society may be dissatisfied with the status quo and organize to replace established institutions. Dissatisfaction alone is not enough to produce a resolution, however. The opportunity must exist for group to mobilize en masse. Thus, revolutions can result when structured opportunities are created, such as through war or an economic crisis, or mobilization through a social movement, as we saw in Chapter 22. Groups mobilize to challenge authorities when they are able to garner the resources that enable the challenge to develop (Kurzman 1996; Tilly 1978, 1975).

Social structural conditions that often lead to revolution can include a highly repressive state—so repressed that a strong political culture develops out of resistance to state oppression. A major economic crisis can also produce revolution, as can the development of a new economic system, such as capitalism, that transforms the world economy. Many have noted that displacement of agriculturally based systems that came with the development of capitalism resulted in state revolutions in western European nations (Skocpol 1979).

People typically think of revolutions as producing more democratic and open political systems, such as the French and American Revolutions, but revolutions can also produce a more repressive state, such as the Iranian Revolution that brought conservative Shiites to rule. Whether it is a revolution to gain freedom or absolute power, revolution is the total transformation of state institutions.

Mobilizing People for Change

Social change does not develop in the abstract. Change comes from the actions of human beings. Collective behavior and social movements are ways that people organize to promote, or in some cases, to resist change. As noted in Chapter 22, *collective behavior* occurs when normal conventions cease to guide people's behavior, and people establish new patterns of interaction and so-cial structure. Collective behavior is often associated with efforts to promote change. Forms of collective behavior, such as fads, crazes, and fashions, can initiate social change and may produce more sustained efforts at change, such as the development of social movements.

Social movements are organized and persistent forms of collective behavior. The purpose of a social movement is often to initiate or vigorously resist social change. Examples abound: the civil rights movement, the women's movement, the environmental movement, and the antiwar movement, just to name a few.

DEBUNKING *SOCIETY'S MYTHS*

Myth: Rebellious social movements such as the Black Power movement were simply people blowing off steam. They do not result in long-lasting changes in society.

Sociological perspective: Social movements of this type are one of the several major causes of long-lasting social change, resulting in enduring structural and cultural changes in society.

Cultural Diffusion

Cultural diffusion (as noted in Chapter 3) is the transmission of cultural elements from one society or cultural group to another. Cultural diffusion can occur by means of trade, migration, mass communications media, and social interaction. The anthropologist Ralph Linton (1937) alerted us some time ago to the fact that many things people often regard as "American" originally came from other lands—cloth developed in Asia, clocks were invented in Europe, coins originated in Turkey, and much more.

Cultural diffusion can occur from one culture in a society to another in the same society. Barbecued ribs were originally eaten by Black slaves in the United States after the ribs were discarded by White slaveowners who preferred meatier parts of the pig. They are now a delicacy enjoyed throughout the United States by virtually all ethnic and racial groups. One contemporary theorist, Robert Ferris Thompson (1993), points out that an exceptionally large range of elements in material and nonmaterial culture that originated in Africa have diffused throughout virtually all groups and subcultures in the United States, including aspects of language, dance, art, dress, decorative styles, and even forms of greeting. These examples all illustrate cultural diffusion not only from one place to another (West Africa to the United States), but also diffusion across time, from a community in the past to many diverse ethnic groups in the present.

Similarly, the immigration of Latino groups into the United States over time has dramatically altered U.S. culture by introducing new food, music, language, slang, and many other cultural elements (Muller and

Espenshade 1985). By a similar token, popular culture in the United States has diffused into many other countries and cultures: Witness the adoption of American clothing styles, rock, rap, and Big Macs in countries such as Japan, Germany, Russia, and China. In grocery shops worldwide, from the rain forests of Brazil to the ice floes of Norway, you can find the Coca-Cola logo.

At one time, it was thought that slavery killed off most institutions and cultural elements that the slaves brought to the Americas (Herskovits 1941). Extensive research over the past three decades now demonstrates that elements of culture carried from Africa by Black slaves continue to survive among African Americans and, thanks to cultural diffusion, among many other groups as well. A *step show* is an energetic, highly rhythmic, group choreography performed as a special event by predominantly Black fraternities and sororities in the United States. Researchers have traced these performances to traditional West African and central African group dances (Thompson 1993; Gates 1992, 1988). The step show has recently been noticed by non-Black

Here members of a predominantly African American fraternity, Kappa Alpha Psi (KAΨ) put on a "step show," a highly rhythmic and energetic dance form with roots in slave society as well as West Africa.

© Terry Wild Studio

students at universities and colleges all over the country, with a few White groups, fraternal, sororal, and otherwise, taking up "steppin'."

Many religious practices among African Americans are traceable to Africa. As noted by religious historian Albert J. Raboteau (1978), the religious singing styles of the slaves, influenced by their African heritage, were characterized by polyrhythms, syncopation, slides (glissandos) from one musical note to another, and repetition. That African lineage is clearly detectable in jazz, rock, rhythm and blues, and rap music, having diffused far beyond the African American community today.

Signifyin' is an interpersonal game in which a person scores with insults, often of a sexual nature. Signifyin' began in the African American urban community and is presently diffusing into White youth culture. The roots of signifyin' are planted in Africa, where a godlike symbolic figure called *Eshu,* a "trickster," or "signifyin' monkey," insulted all with his wit and cleverness (Gates 1992, 1988). A closely related game, engaged in by virtually all hip urban youth, including Latinos, Whites, and Blacks, is a highly sexist game called "playing the dozens" (or "running the dozens") and involves signifyin' about (or "on") someone's mother.

Playing the dozens can cross over from fun to serious, in which case it can become dangerous. Playing the dozens and signifyin' are not new or unique to the United States. Similar games can be found in certain Arab and other Middle Eastern cultures. Complete dramas of the Restoration period in seventeenth-century Europe were constructed around duels of sly insult called "raillery." Cultural diffusion and cultural invention can be seen working together when tracing a cultural practice such as playing the dozens, which may draw on several heritages, depending on who taught who how to play.

THINKING SOCIOLOGICALLY

Cultural diffusion can be a source of social change. What words or phrases can you think of that have probably diffused recently into broader culture as the result of recent immigration? To what extent is such diffusion likely to result in social change in the broader society?

Inequality and Change

Inequalities between people on the basis of class, ethnicity, gender, or other social structural characteristics can be a powerful spur toward social change. As noted in Chapter 22, social movements may blossom into full-blown revolution if the underlying tension is great enough. An example of the mechanism of change can be seen when inequalities between the middle class and the urban underclass produce governmental initiatives, such as increased education for the poor, designed to

MAP 23.1 Technological Penetration

Cellular phones
per 1000 people

	NA
	1–193
	194–388
	389–582
	583–776
	777–969

Technological development is a major source of cultural change in any society. Cell phones, for example, are now commonplace in the United States and other nations. What cultural changes inspire the use of cell phones? And what cultural changes does the increased use of cell phones then create?

Data: U.S. Census Bureau. 2004. *Statistical Abstract of the United States 2003.* Washington, DC: U.S. Government Printing Office. Website: **www.census.gov**

reduce this inequality. Social inequalities thus become the causes of social change.

Culture itself can contribute to the persistence of social inequality and thus becomes a source of discontent among the individuals in the society. Inequalities within the educational system often have a cultural basis. For example, a poor child in the United States, having adopted a language useful in the ghetto, is at a disadvantage in the classroom, where standard English is used. Culturally specific linguistic systems, such as urban Black English, or Ebonics (Dillard 1972; Harrison and Trabasso 1976; DiAcosta 1998), are generally not adopted by schools. This may serve to strengthen the inequalities between the poor and the privileged—unless the child is bicultural and can speak both standard English and Ebonics, which many African Americans can do.

Compounding the problem of inequality, a Hispanic or Black child may be criticized by his or her peers for "acting White" if the child studies hard (Fordham 1996). A female student may shrink from studying mathemat-

ics because she has received the culturally transmitted message that adeptness at mathematics is not feminine (Shih et al. 1999; American Association of University Women 1998; Sadker and Sadker 1994). The perpetuation of inequalities of class, race, and gender can stoke the desire for social change on the part of the disadvantaged groups.

Technological Innovation and the Cyberspace Revolution

Technological innovations can be strong catalysts of social change. The historical movement of societies from agrarian to industrialized has been tightly linked to the emergence of technological innovations and inventions. Inventions often come about specifically to answer some need already present in the society that promises to deliver great rewards. The water wheel promised agrarian societies greater power to raise crops despite dry weather, while also saving large amounts of time

TAKING ON SOCIAL ISSUES

Bringing in Social Change

Social change can be purposely created in a society or in a specific group or culture within society. As suggested in several earlier chapters, this change may occur, for example, by introducing a type of contraceptive device or a new technology, such as electronic computers or cell phones. What specific changes do you think cell phones, for instance, have had on things like family relationships, people's use of time, the character of social interaction, and so on?

Taking Action

Go to the Taking Action Exercise on the Companion Website—at http://sociology .wadsworth.com/andersen_taylor4e/— to learn more about an organization that addresses this topic.

and labor. One could trace a direct line from the water wheel to the large hydroelectric dams that power industrialized societies, and along the way one could find evidence of how each major advance changed society.

In today's world, the most obvious technological change is the rise of the digital computer, and the subsequent development since the 1980s of desktop computing. The result is massive social and cultural changes in the United States. The advent of the electronic computer has massively transformed our entire society and all its institutions. What has come to be called the *cyberspace* revolution began with vacuum tube, mainframe computers in the 1950s and early 1960s, followed by the transistorized computer of the mid-1960s and then the integrated circuit computers of the late 1960s and 1970s. The revolution was accelerated by the advent of the PC (personal computer) and desktop computing, which was in turn made possible by the invention and rapid spread of the microchip. The invention of the microchip has had incalculable effects on society. What can now be stored in a microchip memory the size of a wristwatch would have required in the late 1960s a transistorized computer the size of a small auditorium, and in the early 1960s, a vacuum tube computer the size of a building.

The invention and development of the Internet and the resulting communication in "cyberspace" is of special significance. It includes both communication between persons and communication between persons and computers. Unique in its vastness and lack of a central location, the Internet has very rapidly become so much a part of human communication and social reality that it pervades and has transformed every one of society's institutions—educational, economic, political, familial, and religious.

Few institutional structures have not been transformed by the cyberspace revolution. Both the mainframe computer and the PC continued to transform and shorten processes of mathematical and statistical calculation well into the 1980s. In the move from what Sherry Turkle (1995) has called the *culture of calculation* to the *culture of simulation,* the computer user via the Internet develops a new self, a new identity. One person can communicate with another via email and chat rooms with the protection of anonymity (recall the discussion of this in Chapter 5). As one user has noted, "you are what you pretend to be". Thus the revolution in cyberspace has not only transformed society's institutions, but it also has begun to transform the nature of the self and how we as individuals define ourselves.

The path by which technology is introduced into society often reflects the predominant cultural values in that society. Some cultural values may prevent a technological innovation from changing a society. Anthropologists have noted that new technologies introduced into a nonindustrial, agrarian society very often meet with resistance even though the new technology might greatly benefit the society. The Yanomami, a nonindustrial agrarian society existing deep in the rain forests of South America, live without electricity, automobiles, guns, and other items of material culture associated with industrialized societies. The Yanomami place great positive cultural value upon their way of hunting and engaging in war. Although the introduction of guns to their culture might help them in the hunt and in wars against their neighbors, they remain deeply attached to their bows and poison-tipped arrows.

Recent research has shown, however, that steel was introduced to their society, along with other objects in the 1970s, very possibly by an anthropologist who was studying the Yanomami. This changed the society dramatically and enabled steel-tipped spears and other tools of war to be more deadly. Perhaps even more significant, the anthropologist *himself* may have become a source of major social and cultural change in the very society he was studying. There are those who now argue that the changes wrought have made the Yanomami worse off than they were before these changes (Tierney, P. 2000).

Population and Change

Another cause of social and cultural change is change in the population (population and the environment were discussed in Chapter 21). Limitations placed on the population by the natural environment can greatly influence the nature of social relationships. In Japan, a small country with a large population, crowding is a fact of life that affects how people interact with one another. In Japanese cities, bus drivers negotiate streets that U.S. bus drivers would consider far too narrow for even a small bus. Japanese subways are packed so tightly that white-gloved "pushers" must squeeze commuters bodily into subway cars. Riders on the subway are so tightly packed that their entire bodies are in constant contact, a situation that in most parts of the United States would be considered taboo because such close bodily contact has sexual overtones. In Japan, the contact is not considered sexual (although there are of course exceptions), nor is it considered a violation of the other's personal space, as it would be in the United States. This is one example of how population density can affect the nature of interpersonal relations among people.

Immigration is having profound effects on the overall ethnic and racial composition of the United States. Roughly three-quarters of a million immigrants each

Population density can affect social interaction and cultural norms, as illustrated here in Japan, where a subway worker (a "pusher") causes close physical contact among subway riders.

FORCES OF SOCIAL CHANGE

Shock and Social Change: The 2004 Tsunami

How does one comprehend the changes that come as the result of a disaster so colossal as the tsunamis that struck in the Indian Ocean on December 26, 2004? Hundreds of thousands of people killed, twelve nations stretching from the African Coast to Indonesia devastated, human life and property wiped out in an instant . . . what will be the consequences for the many societies affected by this event?

In the largest natural disaster in memory, whole communities were eliminated. Even the earth's geography was changed as the shock of the undersea earthquake that caused the tsunamis actually shifted the location of various islands and other land masses by 98 feet. But the largest toll was that paid by human society. Families lost loved ones, frequently young children who could not withstand the force of the waves. Thousands of families lost multiple members. In some areas, particularly on the islands of Andaman and Nicoba, as many as one-quarter of the entire population were killed.

In the immediate aftermath of this disaster, as sociologists would predict, people organized to help the victims. The massive relief effort required extensive coordination—across nations and across organizations—to get food, water, medical treatment, and shelter to those in need. Because of the influence of mass media, millions of people around the world observed the disaster and its aftermath—on television and on web "blogs." Horrified by the tremendous loss of life, people rallied to donate millions of dollars, much of the fundraising aided by web sites well-poised to generate donations by those in their network. Indeed, the presence of a global media brought images of this disaster into the homes of people across thousands of miles, showing once again the enormous power of the media to influence human behavior—in this case, toward humanitarian aid.

What will be the long-term effect on social change? People generally think of social change as slowly evolving, a process that takes place in society over many years and involving complex so-

cial interactions and institutional adjustments. But a natural disaster like the tsunami also shows how social change can come in an instant, requiring massive reconstruction, not only of the basic needs for human survival, but the long term needs of rebuilding entire societies—their communication and transportation systems, its schools, health care institutions, even their cultural rituals. Following the tsunami, people of various faiths had to quickly adapt to create make-shift funerals and mass disposal of corpses, even when cultural norms would ordinarily require different burial rites. It will be years before we will know how—indeed whether—all of the human societies struck by this disaster recreate themselves, but what is certain is that, even in the face of massive human tragedy, people will create new social bonds and, perhaps, new social forms, showing the tenacity and resiliency of social life. •••

year take up permanent residence in the United States (Edmondson 2000; Espenshade 1995; U.S. Census Bureau 2004). As we noted in Chapter 21, approximately one-half of all immigrants are from Mexico or other Latin American countries, and about one-fourth are from Asian countries. The U.S. population is at present approximately 11 percent Hispanic; by the year 2025, that figure will be about 18 percent. Currently about 4 percent of the U.S. population is of Asian ancestry. By the year 2025, that figure will be about 7 percent. By the year 2050, it is expected that Hispanics will be 25 percent of the U.S. population and Asians 9 percent. These population shifts will cause major changes in society's institutions. The structure of the economy will change, as will the ethnic mix in education, the ethnic complexion of jobs, and the strength of the influence of Hispanic culture.

As demographic variables, age and gender exert influences upon social change in society. The age structure of a society can create social changes, and the changes that occur can affect age groups in different ways.

Current generations—whether young, middle-aged, or old—will be profoundly influenced by the *graying of America* (discussed in Chapter 21), a phrase referring to the fact that the proportion of old people in the population is increasing dramatically. Consider the following:

- By the year 2015, 27 percent of the population will be age 55 and older (U.S. Census Bureau 2004).
- The proportion of the population classified as the "oldest old," those over the age of 85, will continue to increase.
- Women will continue to outnumber men, among the old as well as among the oldest old.

How well we adapt to the graying of America may partially be the result of how carefully we can analyze the potential effects of population changes. The *contract between generations* is the expectation that the first generation (say, your grandparents' generation) will grow up and raise the second generation (your parents' generation), who in turn will produce a third generation (your generation). Then, later in life, the children and the grandchildren take care of their parents and grandparents once they are too old to care for themselves.

War, Terrorism, and Social Change

War and severe political conflict results in large and far-reaching changes for both the conquering society, or the region within a society (as in civil war), and for the conquered. The conquerors can impose their will upon the conquered and restructure many of their institutions, or the conquerors can exercise only minimal changes.

The United States' victory over Japan and Germany in World War II resulted in changes in each of the three societies. The war transformed the United States into a mass-production economy that affected family structure and education. Father-absence increased and women not previously employed joined the workforce while men of college age went off to war in large numbers. Many who returned from the war were educated under a scholarship plan called the GI Bill.

The war also transformed Germany in countless ways, given the vast physical destruction brought on by U.S. bombs and the worldwide attention brought to anti-Semitism and the Nazi Holocaust. The cultural and structural changes in Japan were extensive, as well. The decimation of the Jewish population in Germany, as well as in other nations throughout Europe, resulted in the massive migration of Jews to the United States. More recently, the war in Vietnam has also resulted in numerous social changes, including a high degree of immigration to the United States from Vietnam.

As we noted in Chapter 8, terrorism is a type of crime—the use of force or violence to coerce a government or population in the furtherance of political or social objectives. Terrorism is of global significance, and social changes resulting from terrorism occur not only in that society but in other societies as well. Threats of bioterrorism, for example, have resulted in recent microchanges in the United States (families wrapping their houses in plastic and duct tape) and in other countries as well. As we further noted in Chapter 8, Osama bin Laden had a role in the international opium trade. His al Qaeda organization was headquartered in Afghanistan, the world's largest grower of opium-producing poppies. The profits from this trade may well have financed the 9/11 terrorist attacks on the World Trade Center and the Pentagon in the United States.

Given that our world is ever more globally connected, terrorist events in any one country will more immediately impact other societies as well. The terrorist attack in Beslan, Russia, in the summer of 2004, is a case in point. Nearly three hundred young children lost their lives when their school was bombed, allegedly by Chechyn rebels. Many students were assassinated by gunfire. Some were systematically beheaded prior to the bombing. Others were murdered through the severing of arms and legs. These were elementary-school-aged children. Widely reported by the international media, the horror of this terrorist attack was felt everywhere, further underscoring the global nature of contemporary terrorism, war, and politics. This unspeakable event can, and no doubt will, effect social changes within Russia and in other countries as well. In fact, as a direct result, the U.S. Department of Education and elementary schools throughout the United States are evaluating stepped-up security measures in the schools, measures likely to result in changes in how education—a social institution—is structured in this country.

Modernization

As societies grow and change, they become in a general sense more modern. Sociologists use the term *modernization* in a specific sense. **Modernization** is a process of social and cultural change that is initiated by industrialization and followed by increased social differentiation and division of labor. Societies can, of course, experience social change in the absence of industrialization. Modernization is a specific type of social change that industrialization tends to bring about. The change toward an industrialized society can have both positive and negative consequences—improved transportation and a higher gross national product or pollution, elevated stress, and increases in certain kinds of job discrimination.

There are three general characteristics of modernization (Berger et al. 1974). First, *modernization is typified by the decline of small, traditional communities.* The individuals in foraging or agrarian societies live in small-scale settlements with their extended families and neighbors. The primary group is prominent in social interaction. With industrialization comes an overall decline in the importance of primary group interactions and an increase in the importance of secondary groups, such as colleagues at work. Second, *with increasing modernization, a society becomes more bureaucratized.* Interactions come to be shaped by formal organizations. Traditional ties of kinship and neighborhood feeling decrease, and the members of the society tend to experience feelings of uncertainty and powerlessness. Third, *there is a decline in the importance of religious institutions.* With the mechanization of daily life, people begin to feel that they have lost control of their own lives and may respond by building new religious groups and communities (Wuthnow 1994).

From Community to Society

The German sociologist **Ferdinand Tönnies,** who died in 1936, formulated a theory of modernization that still applies to today's societies (Tönnies 1963/1887). As noted in Chapter 5, Tönnies viewed the process of modernization as a progressive loss of **gemeinschaft** (German for "community"), a state characterized by a sense of fellow feeling, strong personal ties, and sturdy primary group memberships, along with a sense of personal loyalty to one another. Tönnies argued that the Industrial Revolution, with its emphasis on efficiency and task-oriented behavior, destroyed the sense of community and personal ties associated with an earlier rural life, substituting feelings of rootlessness and impersonality. At the

crux of this was a society organized on the basis of self-interest, where division of labor is high and personal feelings of belonging are low. This conflict caused the condition of **gesellschaft** (German for "society"), a social organization characterized by a high division of labor, less prominence of personal ties, the lack of a sense of community among the members of society, and the absence of a feeling of belonging—maladies often associated with modern urban life.

According to Tönnies, the United States was characterized by gemeinschaft through the year 1900. Life was mainly rural, characterized by families that had lived for generations in villages, and one's work was closely tied to the family. In terms of gender roles, patriarchy was prominent because most women's lives were centered on the home and very few women held jobs outside the home. There was no radio, no television, and few telephones. As a result, family members were dependent on each other for entertainment, information, and support. Despite the relative intimacy of the gemeinschaft, social interaction tended to remain within both racial–ethnic and social class boundaries. Mass transportation was not yet developed, and people tended to base their lives in their own town. These characteristics of the United States at the turn of the twentieth century are preserved in some communities today, such as the earlier mentioned Amish living in parts of New York, Pennsylvania, and Ohio—a classic example of a present-day gemeinschaft.

The United States has become a society marked by gesellschaft: Social interaction has become less intimate and less emotional, although certain primary groups such as the family and the friendship group, still permit strong emotional ties. However, Tönnies noted that the role of the family is considerably less prominent in

These Amish women in Lancaster County, Pennsylvania, illustrate gemeinschaft ("community") social organization as they work on a quilt.

These subway riders illustrate the impersonality of gesellschaft *("society") as they ride to work.*

a gesellschaft than in a gemeinschaft. In the large cities that characterize the gesellschaft, people live among strangers, and the people one passes on the street are unfamiliar. In a gemeinschaft, most people one encounters have been seen before. The level of interpersonal trust is considerably less in a gesellschaft. Social interaction tends to be even more confined within ethnic, racial, and social class groups. To find personal contact and to satisfy the need for intimate interaction, individuals often join groups such as small church groups, training groups, or personal awareness groups (Wuthnow 1994).

Mass Society and Bureaucracy

According to contemporary theorists **Ralf Dahrendorf** (1959) and **Peter L. Berger** and colleagues (1974), modernization has produced what they call a *mass society,* one in which industrialization and bureaucracy reach exceedingly high levels. In the mass society, the change from gemeinschaft to gesellschaft is accelerated, and the breakup of primary, family, and kinship ties is particularly pronounced. The government and its functions expand to the extent that much of one's personal life falls under government management, including tasks that were previously performed by family. Care for the elderly, for example, may be placed in the hands of unfamiliar, faceless bureaucrats who run elder-care facilities and administer financial benefits for the aged.

Dahrendorf, Berger, and other mass society theorists argue that not only have we moved from gemeinschaft to gesellschaft, with all the attendant negatives described

by Tönnies, but also that bureaucracies have obtained virtually complete control of the individual's life. As people moved from town to city over the course of the twentieth century, divisions of labor became more pronounced, and social differentiation increased in the workplace, education, government, and other institutions. It became more common to identify people by the personal attributes of their job ("He's John, a lawyer") and their gender ("She's Ms. Blackburn, a judge") instead of their kinship ("She's a Smith") or their home town ("He's from Mantua, Ohio"), which is more commonly done in the gemeinschaft. The importance of mass media increased. Newspapers, television, magazines, radio, and movies took on more prominent roles in society (Starr 2004). People became more mobile geographically, and thus less dependent on neighbors and kin. All these changes worked together to increase the feeling that most people in one's immediate environment are strangers.

The rise of large government is a major part of the overall increased bureaucratization of social life. In the preindustrial societies of both the United States and Europe, government may have been only a clergyman, a nobleman, a justice of the peace, or a sheriff. Industrialization allowed government to expand at the national, state, and local levels, thus becoming more complex and bureaucratized. Government demonstrates an eagerness to involve itself in many aspects of life formerly left to community standards or private resolution—regulating working conditions; setting wages and salaries; establishing standards for products and medicines, health care, and the care of the poor, as well as all sorts of intimate behaviors. Most political and social power today resides in large bureaucracies, leaving the individual a diminishing degree of control over his or her own life.

Social Inequality, Powerlessness, and the Individual

Another product of modernization, along with mass society, is pronounced social stratification, according to theorists such as **Karl Marx** (1967/1867) and **Jürgen Habermas** (1970). In their view, the personal feelings of powerlessness that accompany modernization are the result of social inequalities related to race, ethnicity, class, and gender stratification. Marx argued that inequalities are the inevitable product of the capitalist system. Habermas argued that inequalities are the cause of social conflict.

The social structural conditions that arise from modernization, such as increased social stratification, are felt at the level of the individual. Building a stable personal identity is difficult in a highly modernized society that presents the individual with complex and conflicting choices about how to live. Many individuals flounder between lifestyles in their search for personal stability and a sense of self. According to Habermas, individuals in highly modernized environments are more likely than their less modernized peers to experiment with new religions, social movements, and lifestyles in search of a fit with their conception of their own "true self." These individual responses to social structural conditions reveal how the social structure can affect personality.

Social theorist **David Riesman** (1970/1950) argued that three main orientations of personality can be traced to social structural conditions:

- **other-directedness**—wherein the behavior of the individual is guided by the observed behavior of others and is characterized by rigid conformity and attempts to "keep up with the Joneses;"
- **inner-directedness**—wherein the individual is guided by internal principles and morals and is relatively impervious to the superficialities of those around her or him;
- **tradition-directedness**—strong conformity to long-standing and time-honored norms, practices, and styles of life.

According to Riesman, modernization tends to produce other-directedness. Less modernized gemeinschafts, such as horticultural or agricultural societies, such as the Amish people discussed earlier, tend to produce

In the early twentieth century, without the benefit of modern air conditioning, families on the lower east side of New York City sometimes slept outdoors for relief from the summer's heat.

tradition-directedness. The inner-directed, because they are guided by internal instead of external forces, are less likely to sway with the presence or absence of modernization.

If modernization tends to produce other-directedness, then anyone who happens to be inner-directed or tradition-directed in a highly modernized and rapidly changing society, such as the United States, is likely to be seen as a deviant person. The other-directed person, in contrast to the inner- and tradition-directed persons, is highly flexible, capable of rapid personal change, and more open to the influences of group pressures, changing styles, and shifting interests. These qualities can leave the other-directed individual in the highly modernized society stranded and searching for his or her "true self."

THINKING *SOCIOLOGICALLY*

Did you grow up in what was primarily a *gemeinschaft* (community), such as a rural community, or in what was primarily a *gesellschaft* (urban or suburban environment)? Do you remember how your community or neighborhood changed over time? What were some of those changes— in population, ethnic composition, the presence of offices and office buildings, the nature of interpersonal relationships? Would you say that your community encouraged *inner-, other-,* or *tradition-directedness*?

The poignant question, Who am I? can give rise to feelings of individual powerlessness. The influential social theorist **Herbert Marcuse** (1964) has argued that modernized society fails to meet the basic needs of people, among them the need for a fulfilling identity. In this respect, modern society and its attendant technological advances are not stable and rational, as is often argued, but unstable and irrational. The technological advances of modern society do not increase the feeling that one has control over one's own life, but instead reduce that control and foster feelings of powerlessness. These feelings of powerlessness can, at least according to one recent researcher, give rise to such cultural phenomena as the proliferation in our society of self-help manuals and books—covering dieting, improving your memory, increasing personal power, and a host of other self-help prescriptions (Whelan 2004).

This powerlessness leads to the *alienation* of the individual from society—the individual experiences feelings of separation from his or her group or society. Alienation is more likely to affect those who have traditionally been denied access to power, such as racial minorities, women, and the working class. The alienation of these individuals from the highly modernized, technological society is, in Marcuse's view, a pressing problem of civilization today. Marcuse argues that despite the popular view that technology is supposed to yield efficient solutions to the world's problems, it may be more accurate to say that technology is a primary cause of many problems of modern society.

Global Theories of Social Change

Globalization refers to the increased interconnectedness and interdependence of different societies around the world. No longer can the nations of the world be viewed as separate and independent societies. The irresistible trend in the twentieth century was for societies to develop deep dependencies on each other, with interlocking economies and social customs. In Europe, this trend has proceeded as far as developing a common currency, the *euro*, for all nations participating in the newly constructed common economy.

If the world is becoming increasingly interconnected, does this mean that we are moving toward a single, homogeneous culture—the culture that futurist Marshall McLuhan once called the "global village" (Griswold 1994)? Are some cities, such as the likes of New York, London, and Tokyo, becoming *world cities*—cities that themselves connect entire societies—as some (such as Sassen 1991) have argued? In such formulations, electronic communications, computers, and other developments would erase the geographic distance between cultures, and eventually the cultural differences themselves. In a competing view, greater interconnectedness among societies may *magnify* the cultural differences between interacting groups by making the groups more aware of the incompatibilities between them.

In fact, both processes take place. As societies become ever more interconnected, cultural diffusion between them creates common ground, while cultural differences may become more important as the relationships among nations become more intimate. The different perspectives on globalization are represented by three main theories (discussed in more detail in Chapter 10 on global stratification) that we will briefly review here: modernization theory, world systems theory, and dependency theory (see Table 23.1).

Modernization Theory

Strongly influenced by functionalist theories of social change, **modernization theory** states that global development is a worldwide process affecting nearly all societies touched by technological change. The theory argues that more advanced technology results in greater differentiation, thus more modernization. Several societies having undergone technological change thus become more homogeneous with respect to each other in terms of differentiation and complexity.

Modernization theory traces the beginnings of globalization to western Europe and the United States. Technological advances in these countries propelled them ahead of the less developed nations of the world that were left to adopt the new technologies years after Europe and the United States. Homogenization resulted, and developing nations were shaped in the mold of the Western nations, which had modernized first.

Some proponents of modernization theory, such as William McCord and Arline McCord, reject the assumption that only western European countries and the United States have led technological globalization and its resultant homogenization (McCord 1991; McCord and McCord 1986). The McCords argue that non-Western societies, most notably Japan, have also been leaders in modernization. As a result, the Japanese culture has profoundly influenced other countries and cultures with its emphasis on the importance of small friendship groups in the workplace and a traditional work ethic. According to the McCords, the examples of Japan and other technological leaders, such as Taiwan and South Korea, have added to the impetus of global economic growth.

World Systems Theory

Formulated by theorist Immanuel Wallerstein (1989, 1979, 1974), **world systems theory** argues that all nations are members of a worldwide system of unequal political and economic relationships that benefit the developed and technologically advanced countries at the expense of the less technologically advanced and less developed. Less developed nations are thus shortchanged in the world system. As discussed in Chapter 10, this has resulted in a worldwide (global) system of stratification—stratification of entire countries.

Wallerstein divides the world system into two camps. **Core nations,** such as the United States, England, and Japan, produce goods and services both for their own consumption and for export. The core nations import raw materials and cheap labor from the **noncore nations** (or *peripheral nations*), situated in Africa, Latin America, South America, and parts of Asia. These nations occupy lower positions in the global economy, thus showing a stratification of the global economy. Certain populations in the noncore nations suffer exploitation as a result. Witness the use of children in parts of Malaysia, Singapore, and Latin America as laborers manufacturing shirts, soccer balls, and blankets. With these manufactured goods, the noncore nations end up contributing to the wealth of the core nations.

Dependency Theory

Closely allied with Wallerstein's world systems theory is **dependency theory,** derived from the work of Karl Marx, which maintains that highly industrialized nations tend to imprison developing nations in dependent relationships rather than spurring the upward mobility of developing nations with transfers of technology and business acumen (Rodney 1981, 1974; Reich 1991). Dependency theory sees the highly industrialized core nations as transferring only those narrow capabilities it serves them to deliver. Once these unequal relationships are forged, core nations seek to preserve the status quo because they derive benefits in the form of cheap raw materials and labor from the noncore, or peripheral, nations. In this sense the core nations actively *prevent* upward social and economic mobility both within and among the developing noncore nations.

Borrowing dependency is a form of dependent relationship. Former Secretary of Labor Robert Reich (1991) has noted that core nations have been willing to lend money to noncore nations, but often on terms such as high interest rates that put severe economic strain on the noncore nations. This may sometimes require interventions such as wage and price freezes in the developing societies to maintain solvency. The hardship produced falls disproportionately on the lower social classes in the noncore country. The upper classes are less affected, and occasionally they benefit extravagantly.

As economist Reich (1991) and sociologists McCord and McCord (1986, also McCord 1991) have noted, the network of dependency is complicated by the fact that in today's global economic system, it is often difficult to determine just who owns what. For example, in the early 1990s, General Motors in the United States owned almost half the stock of Isuzu in Japan. Japanese companies own a number of large American enterprises in the automotive and entertainment businesses. This type of worldwide interdependence further connects nations into a homogeneous global economy, but it also facilitates the dependency of the less developed nations, especially those in Latin America and Africa.

International takeovers often spawn humorous incidents. In the early 1990s, a construction company in California turned down a low bid from a Japanese company (Haimatsu) to spite them for a prior deal and accepted a higher bid from John Deere, the tractor maker. As it turned out, Haimatsu was a U.S.-owned company and John Deere was Japanese-owned!

Diversity, Globalization, and Social Change

Issues of race, class, and gender—in other words, diversity issues—have played a major role in social change, both as causes of change and as social institutions

---

A SOCIOLOGICAL EYE ON THE MEDIA

The Media As A Source of Social Change

We have all no doubt wondered whether the major media of our age (TV, magazines, newspapers, and Internet) not only report significant social events but actually become a cause of social change themselves. Does the way in which something gets reported cause significant social, institutional, structural, and psychological changes in a society? Are any resulting changes only *microchanges,* or do *microchanges* caused by the media sometimes occur?

Media treatment of certain subjects no doubt results in certain kinds of social changes. The TV series special *Roots* about a Black American slave family's historical beginnings in Africa, a program aired in the mid-1970s, is still talked about and has resulted in long-term changes in how people of all races now study and trace their own family histories. The assassination of President John F. Kennedy over forty years ago on November 22, 1963, was dramatically presented on television on that day and for years after. Probably as a result of the vividness of the event as displayed in the media, Americans became more aware of their own vulnerabilities and convinced of the frailties in the security systems that surround even the highest office in the land. Finally, the media image of the two aircraft striking the world Trade Center towers on September 11, 2001, has become firmly ingrained in the memories of virtually the entire U.S. population, causing many to think of our society as irrevocably changed in numerous ways yet to be fully known. •••

directly affected by change. Tensions and aspirations arising from diversity can produce major changes within the structure and culture of a society. Social change can also affect the relations *between* societies. In this respect, globalization has had dramatic effects on the entire world and thus on the question of diversity within as well as between societies.

Among the most welcome and long-lasting effects of diversity is the erosion of the tendency to regard the technology of developing nations as more primitive than the technologies of the dominant Western countries. It is certainly true that the technological base in many developing nations is far behind that of Western countries, but it is now fully recognized that some technologies of the future are being developed overseas. In Singapore and Taiwan, state-of-the-art computer products are prototyped by people of color in dust-free clean rooms. Japan was among the pioneers of low-volume, high-efficiency, high-quality steelmaking. The growing army of software writers in India has become a resource drawn upon by the entire world. The distinction between "civilized nation" and "primitive nation" has lost what little meaning it ever had.

Globalization and diversity also mean that as social movements develop in one part of the world, they are rapidly communicated to other parts of the world. What would the United States be like had the civil rights movement, via Martin Luther King, Jr., not been inspired by Ghandi's liberation movement in India? How would contemporary politics be different had the African National Congress and other movements for the liberation of Black South Africans not dismantled apartheid? Would the world be so conscious of human rights issues in China had they not witnessed the killing of students demonstrating in Tiananmen Square in 1989?

Likewise, the civil rights movement in the United States has not only transformed U.S. society, but has since inspired similar movements throughout the world. And the women's movement is now a global movement, although the particular issues for women vary from place to place. Finally, the terrorist attacks on September 11, 2001, have already been felt in virtually every corner of the globe.

Diversity as a cause of change is exemplified by the effects of immigration into a country and the resultant changes in society. The influx of Latinos into the United States has in a relatively short time resulted in profound cultural changes throughout the nation. From east to west, new tastes can be seen in eating habits, music, language, literature, and many other venues. The changes in the United States brought about by African Americans, beginning as long ago as the slavery era, have been documented throughout this book.

The effects of diversity upon cultural lag are of particular interest. New cultural elements (language, music, technology) introduced to a nation by its racial and ethnic minorities are often adopted by the dominant culture of that nation. The adoption may occur quickly, or even instantly, as when White youths adopt African American dance steps. At other times, it can take decades or longer for particular cultural transfers to take hold. The length of the cultural lag may depend on a number of factors, such as the "foreignness" of the minority culture, the degree to which the minority culture has already been assimilated, the ongoing state of relations between the minority and the dominant group, and a variety of other variables that can be as trivial as the vagaries of fashion or as deep-rooted as historical ethnic animosities.

Chapter Summary

What is social change, and what are its types?

Social change is the process by which social interaction, the social stratification system, and entire institutions in a society change over time. Social change can take place quickly, as within one to several years, or it can take longer, sometimes as long as centuries. It can involve *microchanges* (smaller changes, usually rapid, such as a fad "catching on") or *macrochanges* (technological innovation and modernization).

What are the theories of social change?

Functionalist theories (and *evolutionary theories*) predict that societies move or evolve from the structurally simple to the structurally complex. *Unidimensional evolutionary theory* predicts that societies evolve along a path from simpler, socially undifferentiated societies to more complex, highly differentiated ones. *Multidimensional evolutionary theory* predicts that societies follow not one but several different paths in the process of social change. Conflict theories predict that social conflict is an inherent part of any social structure and that conflict between social class strata, or racial–ethnic groups that occupy dramatically differing social strata, can bring about social change. *Cyclical theories* such as those of Arnold J. Toynbee and Pitrim Sorokin predict that certain patterns of social structure and culture recur in a society at different times.

What are some of the causes of social change?

The causes of social change vary widely. Revolution is a major cause. Some others are mobilization of people for change, as with social movements (for example, the civil rights movement); *cultural diffusion,* or the adoption by the dominant culture of some cultural element introduced by another group via trade, migration, mass communication, or social interaction (including many cultural elements introduced to American society by Native American, Hispanic, Black, and Asian groups); social inequality (such as the many changes in gender relations in the United States brought about by women's political activism against long-practiced gender inequalities); technological innovation (for example, the vast social changes brought about by software and hardware innovations in the computing industry); population factors (such as social changes brought about by increases or decreases in population density or by the in- or out-migration of one or more racial or ethnic groups); and finally, war and terrorism, notably including some worldwide changes following the September 11, 2001, terrorist attacks.

What is modernization and what are its aspects?

Modernization is a complex set of processes initiated by industrialization. Modernization results from the decline of small, traditional communities (*gemeinschafts*) and the change to generally larger, more differentiated and impersonal societies (*gesellschafts*), thus causing the individual to seek out more intimate interaction, as through small church groups or personal awareness groups. With modernization comes increased bureaucratization of a society, thus increasing intrusion of governmental bureaucracies into the lives of the individual. Modernization tends to produce *other-directedness*, where one's behavior is strongly guided by the perceived behavior of others.

What theories of globalization pertain to social change?

Globalization refers to the increased economic, political, and social interconnectedness among different societies in the world. The *modernization theory* of globalization states that global development among societies is a worldwide process not confined to any one society, such that technological advances in one society affect other

societies. *World systems theory,* formulated by Immanuel Wallerstein, predicts that societies are members of a worldwide system of inequality that benefits the more technologically developed societies (*core nations*) at the expense of the less technologically developed countries (*noncore,* or peripheral, nations). *Dependency theory* notes that industrialized core nations tend to keep noncore nations economically dependent upon them, and this retards the economic development and upward mobility of the noncore nations.

Key Terms

core nations 634
cultural diffusion 625
cyclical theory 624
dependency theory 635
evolutionary social theory 621
gemeinschaft 631
gesellschaft 631
globalization 634
inner-directedness 633
macrochange 618
microchange 618
modernization 631
modernization theory 634

multidimensional evolutionary theory 622
neoevolutionary theory 622
noncore nations 634
other-directedness 633
revolution 625
social change 618
tradition-directedness 633
unidimensional evolutionary theory 621
world systems theory 634

Researching Society with MicroCase Online

You can see the results of actual research by using the Wadsworth MicroCase® Online feature available to you. This feature allows you to look at some of the results from national surveys, census data, and some other data sources. You can explore this easy-to-use feature on your own, but try this example. Suppose you want to know: *Has the public's view on legalizing marijuana changed over time, say from about 1970 to 2000?*

To answer this question, go to sociology.wadsworth .com/andersen_taylor4e/, select MicroCase Online from the left navigation bar, and follow the directions there to analyze the following data.

Data file: U.S. Trends 1970–2000

Analysis Task: Historical Trends

Select Variable: Variable 1: GRASS?—%SAYING MARIJUANA SHOULD BE LEGAL

Questions

Once you have your results, answer the following questions:

1. In this time period, the percent of the public that favored legalizing marijuana ("grass") increased two times; when was this?

2. In this time period, there was a long decline in the percent of the public that favored legalizing marijuana. What period was this?

3. Do you think these changes reflect major social changes (*macrochanges*) in our society? How so?

The Companion Website for Sociology: Understanding a Diverse Society, Fourth Edition

http://sociology.wadsworth.com/andersen_taylor4e/

Supplement your review of this chapter by going to the Companion Website to take one of the Tutorial Quizzes, use the flash cards to master key terms, and check out the many other study aids you'll find there. You'll also find special features such as GSS Data and Census 2000 information, data and resources at your fingertips to help you with that special project or do some research on your own.

Suggested Readings and Web Resources

Gelb, Joyce, and Marian L. Palley (eds). 1994. *Women of Japan and Korea: Continuity and Change.* Philadelphia, PA: Temple University Press.
Focusing on two Asian nations, this anthology details how the social standing of women is dependent on a society's economic development.

Gitlin, Todd. 1980. *The Whole World Is Watching: Mass Media in the Making and Unmaking of the New Left.* Berkeley, CA: University of California Press.
This is still the best insider account of the student protests of the 1960s and the role of the media in both supporting and then neutralizing them. Gitlin describes how one source of social change (the 1960s student movement) is affected by another source of social change (the media).

Griswold, Wendy. 1994. *Cultures and Societies in a Changing World.* Thousand Oaks, CA: Pine Forge Press.
This is an overview of cultural change, with special reference to race–ethnicity, class, and gender. Culture and meaning in several social change theories are reviewed, including functionalism and conflict theory, with a discussion of the notion of global culture.

Scott, Catherine V. 1995. *Gender and Development: Rethinking Modernization and Dependency Theory.* Boulder, CO: Lynne Rienner Publishers.
Scott demonstrates that prevailing theories about development, dependency, capitalism, and socialism are anchored in social constructions of gender differences. She critically evaluates both modernization theory and dependency theory, plus other theories of social change.

Tierney, Patrick. 2000. *Darkness in El Dorado: How Scientists and Journalists Devastated the Amazon.* New York: W. W. Norton and Co.
Tierney shows how anthropologists and others brought unintended social and cultural change to the Yanomami tribespeople of the Amazon region in South America. These changes included instruments of war and general social unrest. The book is a major treatise on how scientists studying social change can themselves become the agents (causes) of change.

Progress of Nations
www.unicef.org
A report from UNICEF detailing various aspects of worldwide social change.

The World Factbook
www.cia.gov/cia/publications/factbook/index.html
A collection of data prepared by the Central Intelligence Agency (CIA) that provides detail on various countries around the world.

Glossary

absolute poverty the situation in which people do not have enough resources for basic survival

achieved status a status attained by effort

achievement test tests designed to measure what has been learned, not ability or potential

adult socialization the process of learning new roles and expectations in adult life

affirmative action programs in education and job hiring that recruit minorities over a wide range but do not use rigid quotas, or those that use admissions slots (quotas) for minorities in education and set aside contracts in the economy

age cohort an aggregate group of people born during the same time period

age differentiation the division of labor or roles in a society on the basis of age

age discrimination different and unequal treatment of people based solely on their age

age norms rules that spell out, implicitly or explicitly, the expectations that society has for different age strata

age prejudice a negative attitude about an age group that is generalized to all people in that group

age–sex (age–gender) pyramid a graphic representation of the age and gender structure of a society

age stereotypes preconceived judgments about what different age groups are like

age stratification the hierarchical ranking of age groups in society

ageism the institutionalized practice of age prejudice and discrimination

alienation the feeling of powerlessness and separation from one's group or society

altruistic suicide the type of suicide that can occur when there is excessive regulation of individuals by social forces

Alzheimer's disease a degenerative form of dementia that involves neurological changes in the brain

anabolic steroids muscle mass enhancing drugs

anomic suicide the type of suicide occurring when there are disintegrating forces in the society that make individuals feel lost or alone

anomie the condition existing when social regulations (norms) in a society break down

anorexia nervosa; anorexia a condition characterized by compulsive dieting resulting in self-starvation

anticipatory socialization the process of learning the expectations associated with a role one expects to enter in the future

anti-Semitism the belief or behavior that defines Jewish people as inferior and that targets them for stereotyping, mistreatment, and acts of hatred

applied sociology the use of sociological research and theory in solving real human problems

ascribed status a status determined at birth

assimilation process by which a minority becomes socially, economically, and culturally absorbed within the dominant society

attribution error erroneously attributing the causes of someone's behavior to only their membership in some out-group

attribution theory the principle that dispositional attributions are made about others (what the other is "really like") under certain conditions, such as outgroup membership

authoritarian personality a personality characterized by a tendency to rigidly categorize people and to submit to authority, rigidly conform, and be intolerant of ambiguity

authority power that is perceived by others as legitimate

automation the process by which human labor is replaced by machines

autonomous state model a theoretical model of the state that interprets the state as developing interests of its own, independent of other interests

aversive racism a subtle, nonobvious, or covert form of racism

back-to-basics movement push among some professional educators to stress the "basics" of reading, writing, and arithmetic and the "canon" of classic literature

beliefs shared ideas held collectively by people within a given culture

bilateral kinship system a kinship system in which descent is traced through the father and the mother

biological determinism explanations that attribute complex social phenomena to physical characteristics

bioterrorism the dispersion of chemicals or biological substances intended to cause widespread disease and death

brainwashing thesis claims that innocent people are tricked into religious conversion, that religious cults manipulate and coerce people into accepting their beliefs

bulimia an eating disorder characterized by alternate binge-eating followed by purging or induced vomiting

bureaucracy a type of formal organization characterized by an authority hierarchy, a clear division of labor, explicit rules, and impersonality

capitalism an economic system based on the pursuit of profit and the sanctity of private property

care work all of the labor that is needed to nurture, reproduce, and sustain people, which is critical to the maintenance of social institutions

caste system a system of stratification (characterized by low social mobility) in which one's place in the stratification system is determined by birth

census a count of the entire population of a country

charisma a quality attributed to individuals believed by their followers to have special powers

charismatic authority authority derived from the personal appeal of a leader

churches formal organizations that see themselves and are seen by society as primary and legitimate religious institutions

class see social class

class consciousness the awareness that a class structure exists and the feeling of shared identification with others in one's class with whom one perceives common life chances

class system a system of stratification in which location and rank can change according to individual achievements, even though one's class is still strongly determined by social background

coercive organization an organization characterized by membership that is largely involuntary

cognitive ability the capacity for abstract thinking

cognitive elite term used to describe the upper classes in society, based on the premise that they possess a genetically based high intelligence

cohabitation the practice of living together outside of marriage

cohort; birth cohort all persons born within a given time period

collective behavior behavior that occurs when the usual conventions are suspended and people collectively establish new norms of behavior in response to an emerging situation

collective consciousness the body of beliefs that are common to a community or society and that give people a sense of belonging

collective preoccupations forms of collective behavior wherein many people over a relatively broad social spectrum engage in similar behavior and have a shared definition of their behavior as needed to bring social change or to identify their place in the society

colonialism system by which western nations became wealthy by taking raw materials from colonized societies and reaping profits from products finished in the homeland

color-blind racism ignoring legitimate racial–ethnic, cultural, and other differences between groups

coming out the process of openly defining oneself as gay or lesbian

commodity chain the network of production and labor processes by which a product becomes a finished commodity; by following the commodity chain, it is evident which countries gain profits and which ones are being exploited

communism an economic system in which the state is the sole owner of the systems of production

comparable worth the principle of paying women and men equivalent wages for jobs involving similar levels of skill

competition theory an explanation of riots as resulting from conflicts between different groups who compete for limited resources

compulsory heterosexuality the idea that heterosexual identity is not a choice but is created by institutions that treat heterosexuality as the only legitimate form of sexual identity

concept any abstract characteristic or attribute that can be potentially measured

conflict theory a theoretical perspective that emphasizes the role of power and coercion in producing social order

contact theory the theory that prejudice will be reduced through social interaction with those of different race or ethnicity but of equal status

contagion theory the idea (from Gustave LeBon) that people in crowds are highly suggestible and that the crowd takes on a single way of acting and thinking

content analysis the analysis of meanings in cultural artifacts such as books, songs, and other forms of cultural communication

contingent workers those who do not hold regular jobs but whose employment is dependent on demand, including those who contract independently with employers, temporary workers, on-call workers, the self-employed, part-time workers, and day laborers

controlled experiment a method of collecting data that can determine whether a given factor causes something independently of other factors

convergence theory a theory of rioting that focuses on the participants and presupposes that rioters are acting on predispositions and attitudes

conversion a transformation of religious identity

core countries; core nations within world systems theory, those nations that are more technologically advanced

corporate crime wrongdoing that occurs within the context of a formal organization or bureaucracy and is sanctioned by the organization's norms

correlation a statistical technique that analyzes patterns of association between pairs of sociological variables

counterculture subculture created as a reaction against the values of the dominant culture

craze form of collective behavior with very intense involvement for participants

credentialism the insistence upon educational credentials only for their own sake

crime one form of deviance; specifically, behavior that violates criminal laws

criminology the study of crime from a social scientific perspective

cross tabulation a table showing the relationship between two variables

crude birthrate the number of live births each year for every 1000 members of the population

crude death rate the number of deaths each year for every 1000 members of the population

cult a religious group devoted to a specific cause or charismatic leader

cultural capital cultural resources that are socially designated as being worthy (such as knowledge of elite culture) and that give advantages to groups possessing such capital

cultural diffusion the transmission of cultural elements from one society or culture to another

cultural hegemony the pervasive and excessive influence of one culture throughout society

cultural pluralism pattern whereby groups maintain their distinctive culture and history

cultural relativism the idea that something can be understood and judged only in relationship to the cultural context in which it appears

culture the complex system of meaning and behavior that defines the way of life for a given group or society

culture lag the delay in cultural adjustments to changing social conditions

culture of poverty the argument that poverty is a way of life and, like other cultures, is passed on from generation to generation

culture shock the feeling of disorientation that can come when one encounters a new or rapidly changed cultural situation

cyberspace interaction an interaction between people through the medium of computer networks

cyberterrorism the use of the computer to commit one or more terrorist acts

cyclical theory (of social change) from Arnold Toynbee, the theory that certain patterns of social structure and culture recur at different times in the same society

data the systematic information that sociologists use to investigate research questions

data analysis the process by which sociologists organize collected data to discover what patterns and uniformities are revealed

debunking the process of looking behind the facades of everyday life

deductive reasoning a form of reasoning in which specific hypotheses, or predictions, are derived from general principles

de facto segregation segregation "in fact" but not necessarily by law

defensive medicine practiced by physicians who order extra precautionary tests on a patient in an effort to fend off a lawsuit by the patient

deindustrialization the transition from a predominantly goods-producing economy to one based on the provision of services

de jure segregation segregation as defined by law

dementia term used to describe a variety of diseases that involve some permanent damage to the brain

democracy system of government based on the principle of representing all people through the right to vote

demographic transition theory the theory that societies pass through phases based on economic development, which affects the birth and death rates

demography the scientific study of population

denomination religious organizations that unite various congregations into a single administrative structure

dependency theory the global theory that maintains that industrialized nations hold less industrialized nations in a dependency, thus exploitative, relationship that benefits the industrialized nations at the expense of the less industrialized ones, whose upward mobility in the global economy is prevented

dependent variable the variable that is a presumed effect

deviance behavior that is recognized as violating expected rules and norms

deviant career the sequence of movements people make through a particular system of deviance

deviant community groups that are organized around particular forms of social deviance

deviant identity the definition a person has of himself or herself as a deviant

dictatorship form of government where power resides in the rule of one person who acquires power through force and maintains it usually through having control of the military

differential association theory theory that interprets deviance as behavior one learns through interaction with others

discrimination overt negative and unequal treatment of the members of some social group or stratum solely because of their membership in that group or stratum

disengagement theory part of functionalist theory, which predicts that as people age, they gradually withdraw from participation in society

diversity the variety of group experiences resulting from the social structure

division of labor the systematic interrelation of different tasks that develops in complex societies

doing gender a theoretical perspective that interprets gender as something that is accomplished through the ongoing social interactions people have with one another

dominant culture the culture of the most powerful group in society

dominant group the group that assigns a racial or ethnic group to subordinate status in society

dominative racism obvious and overt racism, sometimes called "old-fashioned" racism

double deprivation the situation in poor countries where women suffer because they are in poverty and because they are women

dramaturgical model a perspective that sees society like a stage (that is, a drama) wherein social actors are "on stage," projecting and portraying social roles to others

dual labor market theory theoretical description of the occupational system defining it as divided into two major segments: The primary labor market and the secondary labor market

dyad a group consisting of two people

ecological demography a field combining the study of demography (population) with the study of human ecology

ecological globalization the worldwide dispersion of problems and issues involving the relationships between humans and the physical and social global environments

economic restructuring contemporary transformations in the basic structure of work that are permanently altering the workplace, including the changing composition of the workplace, deindustrialization, the use of enhanced technology, and the development of a global economy

economy the system on which the production, distribution, and consumption of goods and services is based

educational attainment the total years of formal education

educational deflation the decline in the value of a college education arising from increases over time in the number of persons graduating from college; the decline in value of a bachelor's degree

egalitarian societies societies or groups where men and women share power

ego the part of the self representing reason and common sense

egoistic suicide the type of suicide that occurs when people feel totally detached from society

elite crime crimes committed primarily by those in the upper class in the context of their "ordinary activities," such as tax evasion, embezzlement, etc.

elite deviance the wrongdoing of wealthy and powerful individuals and organizations

emergent norm theory theory of collective behavior postulating that, when people are faced with an unusual situation, they create meanings that define and direct the situation

emigration (vs. immigration) migration of people from one society to another (also called out-migration)

emotional labor work that is intended to produce a desired emotional effect on a client

empirical refers to something that is based on careful and systematic observation

endogamy the practice of selecting mates from within one's group

Enlightenment the period in seventeenth- and eighteenth-century Europe characterized by faith in the ability of human reason to solve society's problems

environmental racism the dumping of toxic wastes with disproportionate frequency at or very near areas with high concentrations of minorities

epidemiology the study of all factors—biological, social, economic, and cultural—that are associated with disease and health

estate system a system of stratification in which the ownership of property and the exercise of power is monopolized by an elite or noble class, which has total control over societal resources

ethnic group a social category of people who share a common culture, such as a common language or dialect, a common religion, and common norms, practices, and customs

ethnocentrism the belief that one's in-group is superior to all out-groups

ethnomethodology a technique for studying human interaction by deliberately disrupting social norms and observing how individuals attempt to restore normalcy

ethnoreligious group an extreme form of an exclusive religious group

eugenics movement a social movement in the early twentieth century that sought

to apply scientific principles of genetic selection to "improve" the offspring of the human race

euthanasia the act of killing a severely ill person or allowing the person to die, as an act of mercy

evaluation research research assessing the effect of policies and programs

evolutionary social theory a theory of social change that predicts that societies change in a single direction over time

exclusive religious group religious groups with an easily identifiable religion and culture, including distinctive beliefs and strong moral teachings; they have little tolerance for diversity

exogamy the practice of selecting mates from outside one's group

expressive crowds crowds whose primary function is the release or expression of emotion

expressive needs needs for intimacy, companionship, and emotional support

extended families families in which a large group of related households live together

extreme poverty the situation in which people live on less than per year $275 U.S.

fad form of collective behavior that involves a novel, though usually short-lived, change

false consciousness the thought resulting from subordinate classes internalizing the view of the dominant class

family a primary group of people—usually related by ancestry, marriage, or adoption—who form a cooperative economic unit and care for any young (and each other); who consider their identity to be intimately attached to the group; and who are committed to maintaining the group over time

Family and Medical Leave Act (FMLA) federal law requiring employers to grant employees a total of twelve weeks of unpaid leave to care for newborn or newly adopted children or a family member with serious health needs

family wage system a wage structure based on the assumption that men are the breadwinners for families

fashion form of collective behavior wherein something novel is introduced into society

feminism beliefs and actions that attempt to bring justice, fairness, and equity to all women, regardless of their race, age, class, sexual orientation, or other characteristics

feminization of poverty the process whereby a growing proportion of the poor are women and children

field research research which usually involves the participation of the researcher with the people or group(s) being studied

first-world countries industrialized nations based on a market economy and with democratically elected governments

folkways the general standards of behavior adhered to by a group

formal organization a large secondary group organized to accomplish a complex task or set of tasks

forms of racism types of racism such as old-fashioned racism, aversive racism, laissez-faire racism, color-blind racism, and others

frame schemes of interpretation that allow people in groups to perceive, identify, and label events within their lives that can become the basis for collective action

functionalism a theoretical perspective that interprets each part of society in terms of how it contributes to the stability of the whole

game stage the stage in childhood when children become capable of taking a multitude of roles at the same time

gemeinschaft German for *community,* a state characterized by a sense of fellow feeling among the members of a society, including strong personal ties, sturdy primary group memberships, and a sense of personal loyalty to one another; associated with rural life

gender socially learned expectations and behaviors associated with members of each sex

gender development index a calculation based on gender inequalities in life expectancy, educational attainment, and income for different countries

gender gap term referring to the differences in women's and men's political attitudes and behavior

gender identity one's definition of self as a woman or man

gender segregation the distribution of men and women in different jobs in the labor force

gender socialization the process by which men and women learn the expectations associated with their sex

gender stratification the hierarchical distribution of social and economic resources according to gender

gendered institution the idea that whole institutions are patterned by gender

gendered racism the principle that the effects of racism and sexism are inseparable

generalization a claim that a finding represents something greater than the specific observations on which the finding is based

generalized other the abstract composite of social roles and social expectations

generational equity the question of whether one age group or generation is unfairly taxed to support the needs and interests of another generation

gesellschaft German for *society,* a form of social organization characterized by a high division of labor, less prominence of personal ties, the lack of a sense of community among the members, and the absence of a feeling of belonging; associated with urban life

glass ceiling popular concept referring to the limits that women and minorities experience in job mobility

global culture diffusion of a single culture throughout the world

global economy term used to refer to the fact that all dimensions of the economy now cross national borders

global stratification the systematic inequalities between and among different groups within nations that result from the differences in wealth, power, and prestige of different societies relative to their position in the international economy

globalization increased economic, political, and social interconnectedness and interdependence among societies in the world

government those state institutions that represent the population and make rules that govern the society

greenhouse effect a rise in the earth's surface temperature caused by heat trapped by excess carbon dioxide in the atmosphere

gross national income the total output of goods and services produced by residents of a country each year plus the income from nonresident sources, divided by the size of the population

group a collection of individuals who interact and communicate, share goals and norms, and who have a subjective awareness as "we"

group size effect the effect upon a person of groups of varying sizes

groupthink the tendency for group members to reach a consensus at all costs

hate crime assaults and other malicious acts directed against gays, the disabled, and racial, ethnic, or religious groups

health maintenance organization (HMO) a cooperative of doctors and other medical personnel who provide medical services in exchange for a set membership fee

hermaphroditism a condition produced when irregularities in chromosome formation or fetal differentiation produce persons with mixed sex characteristics

heterosexism the institutionalization of heterosexuality as the only socially legitimate sexual orientation

holistic medicine medical treatment directed toward a person's entire mental and physical state

homogamy the pattern by which people select mates with similar social characteristics to their own

homophobia the fear and hatred of homosexuality

hospice movement movement to provide in-home care for terminally ill patients as an alternative to hospitalization

household term used by the U.S. census to refer to all persons (may or may not be related) occupying a housing unit

human capital theory a theory that explains differences in wages as the result of differences in the individual characteristics of the workers

human ecology the study of the interdependence between humans and their physical environment

human ecosystem a system involving interaction of humans with the physical environment

human poverty index a multidimensional measure of poverty, meant to indicate the degree of deprivation in four basic dimensions of human life: a long and healthy life, knowledge, economic well-being, and social inclusion

humanitarianism the principle that human reason can successfully direct social change for the betterment of society

hypothesis a statement about what one expects to find in research

hysterical contagion collective phenomenon wherein symptoms of an illness spread among a group, even though there is no physiological disease present

id the part of the personality that includes various impulses and drives, including sexual passions and desires, biological urges, and human instincts

identity how one defines oneself

ideology a belief system that tries to explain and justify the status quo

imitation stage the stage in childhood when children copy the behavior of those around them

immigration (vs. emigration) the migration of people into a society from outside (also called in-migration)

impression management a process by which people control how others perceive them

inclusive religious group religious groups with a moderate and liberal religious orientation and tolerance for diversity

income the amount of money brought into a household from various sources during a given year (wages, investment income, dividends, and so on)

independent variable a variable treated as the presumed cause of a particular result

index crime largely "street" crimes of a serious nature such as armed robbery or drug dealing

index of dissimilarity a measure used to indicate the number of workers who would have to change jobs to have the same occupational distribution as the comparison group

indicator something that points to or reflects an abstract concept

inductive reasoning a logical process of building general principles from specific observations

infant mortality rate the number of deaths per year of infants under age 1 for every 1000 live births

informant a group member secretly in alliance with the researcher, as an aid to the researcher in studying the group

inner-directedness a condition wherein the individual's behavior is guided by internal principles and morals

institution see social institution

institutional racism racism involving notions of racial or ethnic inferiority that have become ingrained into society's institutions

instrumental needs emotionally neutral, task-oriented (goal-oriented) needs

interaction see social interaction

interest group a constituency in society organized to promote its own agenda

interlocking directorate organizational linkages created when the same people sit on the boards of directors of a number of different corporations

international division of labor system of labor whereby products are produced globally, while profits accrue only to a few

intervening variable a variable caused by the independent variable and which in turn causes the dependent variable

iron law of oligarchy process wherein a small elite become increasingly powerful as bureaucratic leaders become enchanted with their elite status and make decisions that protect their power

issues problems that affect large numbers of people and have their origins in the institutional arrangements and history of a society

job displacement the permanent loss of certain job types that occurs when employment patterns shift

kinesic communication communication by means of body motion; body language

kinship system the pattern of relationships that define people's family relationships to one another

labeling effect the effect of educational track (role) assignment as distinct from the effect of cognitive ability

labeling theory a theory that interprets the responses of others as most significant in understanding deviant behavior

labor force participation rate the percentage of those in a given category who are employed

laissez-faire racism negative stereotyping of minorities; and blaming minorities themselves for economic, occupational, and educational lack of achievement

language a set of symbols and rules that, put together in a meaningful way, provides a complex communication system

latent functions indirect, nonobvious consequences (functions) emerging from the activities of institutions

law the written set of guidelines that define what is right and wrong in society

liberal feminism a feminist theoretical perspective asserting that the origin of women's inequality is in traditions of the past that pose barriers to women's advancement

life chances the opportunities that people have in common by virtue of belonging to a particular class

life course perspective sociological framework for studying aging that connects people's personal attributes, the roles they occupy, the life events they experience, and their sociohistorical context

life expectancy the probable number of years a particular group is likely to live on average, given aggregate statistical patterns

looking-glass self the idea that people's conception of self arises through reflection about their relationship to others

macroanalysis analysis of large processes and structures, such as institutions

macrochange a social change that is relatively gradual and that broadly affects many aspects of a society

Malthusian theory after T. R. Malthus, the principle that a population tends to grow faster than the subsistence (food) level needed to sustain it

managed care a health care system wherein all individuals are members of an HMO (health maintenance organization), with the intent of reducing fees and insurance costs

manifest functions the stated and open goals of social behavior

market research a type of evaluation research, the purpose of which is to evaluate the sales potential of some product or service

mass media channels of communication that are available to very wide segments of the population

master status some characteristic of a person that overrides all other features of the person's identity

material culture the objects created in a given society—its buildings, art, tools, toys, print and broadcast media, and other tangible objects

matriarchal religion religions based on the centrality of female goddesses, who may be seen as the source of food, nurturance, and love or who may serve as emblems of the power of women

matriarchy a society or group in which women have power over men

matrilineal kinship system kinship systems in which family lineage (or ancestry) is traced through the mother

matrilocal kinship system kinship systems in which women continue to live with their families of origin after marriage

mean the sum of a set of values divided by the number of cases from which the values are obtained; an average

means of production the system by which goods are produced and distributed

median the midpoint in a series of values that are arranged in numerical order

median income the midpoint of all household incomes

Medicaid governmental assistance program for the poor

medical model a framework that interprets a given condition as purely a physiological or medical issue

medicalization of deviance explanations of deviant behavior that interpret deviance as the result of individual pathology or sickness

Medicare governmental assistance program for the elderly

microanalysis analysis of processes and structures of social interaction, such as types of interactions or groups

microchange a subtle alteration in the day-to-day interaction between people

military–industrial complex term used to describe the linkage between business and military interests

minority group any distinct group in society that shares common group characteristics and is forced to occupy low status in society because of prejudice and discrimination

miscegenation the mixing of races through marriage

mismatch theory argument that a group's disadvantage in the labor market results from the combination of residential segregation in center cities and the movement of jobs to suburban areas

mobilization the process by which social movements and their leaders secure people and resources for the movement

mode the value that appears most frequently in a set of data

modernization a process of social change initiated by industrialization and followed by increased social differentiation and division of labor

modernization theory a view of globalization in which global development is a worldwide process affecting nearly all societies that have been touched by technological change

monarchy form of government characterized by having a head of state who rules for life and where authority tends to be inherited

monogamy the marriage practice of a sexually exclusive relationship with one spouse at a time

monotheism the worship of a single god

moral entrepreneurs people who organize a social movement to reform how a particular behavior is morally perceived and handled

mores strict norms that control moral and ethical behavior

multiculturalism modes of thinking that view society through the plural experiences of its diverse membership

multiculturalism movement the push to introduce into the elementary, high school, and college curricula more courses on different and diverse subcultures and groups, ethnic groups, and gender studies

multidimensional evolutionary theory (of social change) a theory predicting that over time societies follow not one but several evolutionary paths

multinational corporation companies that draw a large share of their revenues from foreign investments and that conduct business across national borders

multiracial feminism form of feminist theory noting the exclusion of women of color from other forms of theory and centering its analysis in the experiences of all women

neoevolutionary theory see multidimensional evolutionary theory

neolocal residence the practice whereby newly wedded couples establish their own residence

new social movement theory a theory about social movements linking culture, ideology, and identity conceptually to explain how new identities are forged within social movements

newly industrializing countries (NICs) countries that have shown rapid growth and have emerged as developed countries

noncore (peripheral) nations within world systems theory, those nations that are less technologically advanced

nonmaterial culture the norms, laws, customs, ideas, and beliefs of a group of people

normative organization an organization that people join in order to pursue goals they consider worthwhile

norms the specific cultural expectations for how to act in a given situation

nuclear families families in which married couples reside together with their children

object relations theory a psychoanalytic theory of socialization arguing that social relationships children experience early in life determine the development of their personality

occupational distribution the pattern by which workers are located in the occupational system

occupational prestige the subjective evaluation people give to jobs as better or worse than others

occupational segregation the pattern by which workers are separated into different occupations on the basis of social characteristics such as race and gender

organic metaphor refers to the similarity early sociologists saw between society and other organic systems

organic (contractual) solidarity unity based on role differentiation, not similarity

organizational crime wrongdoing that occurs within an organizational context and that is sanctioned by the norms and operating principles of the organization

other-directedness a condition wherein the individual's behavior is guided by the behavior of others

out-group homogeneity effect the tendency to perceive members of an out-group as identical in various characteristics

paralinguistic communication communication that is conveyed by voice pitch, loudness, rhythm, emphasis, frequency, and other such nonverbal elements

participant observation a method whereby the sociologist becomes both a participant in the group being studied and a scientific observer of the group

patriarchal religion religion in which the beliefs and practices of the religion are based on male power and authority

patriarchy a society or group in which men have power over women

patrilineal kinship system kinship systems in which family lineage (or ancestry) is traced through the father

patrilocal kinship system kinship systems in which, following marriage, women are separated from their families of origin and reside with the husband's kinship group

peers those of similar status

percentage parts per hundred

peripheral countries poor countries, largely agricultural, having little power or influence in the world system

personal crime violent or nonviolent crimes directed against people

personal transformation movements social movements that aim to change the individual

personality the relatively consistent pattern of behavior, feelings, and beliefs in a given person

play stage the stage in childhood when children begin to take on the roles of significant people in their environment

pluralist model a theoretical model of power in society as coming from the representation of diverse interests of different groups in society

policy research research intended to produce results for social policy

political action committees (PACs) groups of people who organize to support candidates they feel will represent their views

political process theory explanation of social movements positing that movements achieve success by exploiting a combination of internal factors

polyandry a marriage practice of a woman having more than one husband

polygamy a marriage practice in which men or women can have multiple marriage partners

polygyny the marriage practice of a man having more than one wife

polytheism the worship of more than one deity

popular culture the beliefs, practices, and objects that are part of everyday traditions

population a relatively large collection of people (or other unit) that a researcher studies and about which generalizations are made

population density the number of persons per square mile

population replacement level a condition wherein the combined birthrate and death rate sustains the population at a steady level

positivism a system of thought in which accurate observation and description is considered the highest form of knowledge

postindustrial society society organized around the provision of services

postmodernism a theoretical perspective based on the idea that society is not an objective thing but is found in the words and images—or discourses—that people use to represent behavior and ideas

poverty line the figure established by the government to indicate the amount of money needed to support the basic needs of a household

power a person or group's ability to exercise influence and control over others

power elite model a theoretical model of power positing a strong link between government and business

predictive validity the extent to which a test accurately predicts later college grades, or some other criterion such as likelihood of graduating

pre-industrial society one that directly uses the land as a major means of survival

prejudice the negative evaluation of a social group, and individuals within that group, based upon conceptions held about the group despite facts that contradict them

prestige the perceived value placed upon different groups or people

primary deviance the violation of a norm or law

primary group a group characterized by intimate, face-to-face interaction and relatively long-lasting relationships

probability the likelihood that a specific behavior or event will occur

profane that which is of the everyday, secular world and is specifically not religious

propaganda information disseminated by a group or organization (such as the state) intended to justify the power of dominant groups

property crime crimes involving theft of property without bodily harm

Protestant Ethic belief that hard work and self-denial lead to salvation

proxemic communication meaning conveyed by the amount of space maintained between interacting individuals

psychoanalytic theory a theory of socialization positing that the unconscious mind shapes human behavior

quadruple jeopardy phrase referring to the simultaneous effects of being old, minority, female, and poor

qualitative research research that is somewhat less structured yet focused on a question being asked; it is more interpretive and tends to have greater depth than quantitative research

quantitative research research that uses statistical methods

queer theory a perspective that interprets various dimensions of sexuality as thoroughly social and constructed through institutional practices

race a social category, or social construction, that we treat as distinct on the basis of certain characteristics, some biological, that have been assigned social importance in the society

racial formation process by which groups come to be defined as a "race" through social institutions such as the law and schools

racial profiling the use of race alone as the criterion to decide whether police stop and detain someone, such as the driver of an automobile, on suspicion of committing a crime

racialization a process whereby some social category such as social class or nationality takes on what are perceived in the society to be race characteristics

racism the perception and treatment of a racial or ethnic group, or member of that group, as intellectually, socially, and culturally inferior to one's own group

radical feminism feminist theoretical perspective that interprets patriarchy as the primary cause of women's oppression

radical movements social movements that seek fundamental change in the structure of society

random sample a sample that gives everyone in the population an equal chance of being selected

rate parts per a given number (for example, per 10,000, per 100,000)

rational choice theory the theory that humans are capable of making cost-benefit analyses of what to do and how many children to have

rational-legal authority authority stemming from rules and regulations, typically written down as laws, procedures, or codes of conduct

rationalization of society term used by Max Weber to describe society being increasingly organized around legal, empirical, and scientific forms of thought

reactionary movements social movements organized to resist change or to reinstate an earlier social order that participants perceive to be better

reference group any group (to which one may or may not belong) used by the individual as a standard for evaluating her or his attitudes, values, and behaviors

reflection hypothesis the idea that the mass media reflect the values of the general population

reform movements social movements that seek change through legal or other mainstream political means, by working within existing institutions

relative poverty a definition of poverty that is set in comparison with an established standard

reliability the likelihood that a particular measure would produce the same results if the measure were repeated

religion an institutionalized system of symbols, beliefs, values, and practices by which a group of people interprets and responds to what they feel is sacred and that provides answers to questions of ultimate meaning

religiosity the intensity and consistency of practice of a person's (or group's) faith

religious socialization the process by which one learns a particular religious faith

replication study research that is repeated exactly but on a different group of people at a different time

research design the overall logic and strategy used in a research project

residential segregation the spatial separation of racial and ethnic groups in different residential areas

resocialization the process by which existing social roles are radically altered or replaced

resource mobilization theory theory of how social movements develop that focuses on how movements gain momentum by successfully garnering organizational resources

revolution the overthrow of a state or the total transformation of central state institutions

risky shift the tendency for group members, after discussion and interaction, to engage in riskier behavior than they would while alone

rite of passage ceremony or ritual that symbolizes the passage of an individual from one role to another

ritual symbolic activities that express a group's spiritual convictions

role the expected behavior associated with a given status in society

role conflict two or more roles associated with contradictory expectations

role model a person we admire and whose behavior we imitate

role set all the roles occupied by a person at a given time

role strain conflicting expectations within the same role

sacred that which is set apart from ordinary activity, seen as holy, and protected by special rites and rituals

salience principle categorizing people on the basis of what initially appears prominent about them

sample any subset from a population that a researcher studies

Sapir–Whorf hypothesis a theory that language determines other aspects of culture because language provides the categories through which social reality is defined and perceived

scapegoat theory argument that dominant group aggression is directed toward a minority as a substitute for frustration with some other problem

scapegoating process whereby a group collectively identifies another group as a threat to the perceived social order and incorrectly blames the other group for problems they have not caused

schooling the formal, institutionalized aspects of education

scientific method the steps in a research process, including observation, hypothesis testing, analysis of data, and generalization

script a learned performance of a social role

second-world countries socialist countries with state-managed economies and typically without a democratically elected government

secondary deviance behavior that results from being labeled deviant, regardless of whether the person has previously engaged in deviance

secondary group a group that is relatively large in number and not as intimate or enduring as a primary group

sect groups that have broken off from an established church

secular the ordinary beliefs of daily life that are specifically not religious

secularization the process by which religious institutions, behavior, and consciousness lose their religious significance

segregation the spatial and social separation of racial and ethnic groups

self our concept of who we are, as formed in relationship to others

self-esteem the value a person places on his or her identity

self-fulfilling prophecy the process by which merely applying a label changes behavior and thus tends to justify the label

semiperipheral countries semi-industrialized countries that represent a kind of middle class within the world system

sex used to refer to biological identity as male or female

sex ratio the number of males per 100 females

sexism a system of practices and beliefs through which women are controlled and exploited because of the significance given to differences between the sexes

sexology the scientific study of sex

sexual harassment unwanted physical or verbal sexual behavior that occurs in the context of a relationship of unequal power and that is experienced as a threat to the victim's job or educational activities

sexual orientation the manner in which individuals experience sexual arousal and pleasure

sexual politics the link feminists argue exists between sexuality and power and between sexuality and race, class, and gender oppression

sexual revolution the widespread changes in men's and women's roles and a greater public acceptance of sexuality as a normal part of social development

sexual scripts the ideas taught to us about what is appropriate sexual behavior for a person of our gender

sick role a pattern of expectations of behaviors defined in society as appropriate for one who is ill

significant others those with whom we have a close affiliation

social action behavior to which people give meaning

social change the alteration of social interaction, social institutions, stratification systems, and elements of culture over time

social change movements movements that aim to change some aspect of society

social class the social structural, hierarchical position groups hold relative to the economic, social, political, and cultural resources of society

social construction a theoretical perspective that explains sexual identity as created and learned within a cultural, social, and historical context

social control a process by which groups and individuals within those groups are brought into conformity with dominant social expectations

social control agents those who institutionally regulate and administer responses to deviance, such as the police and mental health workers

social control theory theory that explains deviance as the result of the weakening of social bonds

Social Darwinism the idea that society evolves to allow the survival of the fittest

social differentiation the process by which different statuses in any group, organization, or society develop

social drift theory interprets people as moving into religious cults gradually, particularly if they have experienced recent personal strains or have become disenchanted with their prior affiliations

social epidemiology the study of the effects of social and cultural factors upon disease and health

social facts social patterns that are external to individuals

social identity complexity a term referring to how a person subjectively interprets the interrelationships among multiple group identities

social institution an established and organized system of social behavior with a recognized purpose

social interaction behavior between two or more people that is given meaning

social learning theory a theory of socialization positing that the formation of identity is a learned response to social stimuli

social mobility a person's movement over time from one class to another

social movement a group that acts with some continuity and organization to promote or resist change in society

social network a set of links between individuals or other social units such as groups or organizations

social organization the order established in social groups

social sanctions mechanisms of social control that enforce norms

social stratification a relatively fixed hierarchical arrangement in society by which groups have different access to resources, power, and perceived social worth; a system of structured social inequality

social structure the patterns of social relationships and social institutions that comprise society

socialism an economic institution characterized by state ownership and management of the basic industries

socialist feminism a feminist theoretical perspective that interprets the origins of women's oppression as lying in the system of capitalism

socialization the process through which people learn the expectations of society

socialization agents those who pass on social expectations

society a system of social interactions that includes both culture and social organization

socioeconomic status (SES) a measure of class standing, typically indicated by income, occupational prestige, and educational attainment

sociological imagination the ability to see the societal patterns that influence individual and group life

sociology the study of human behavior in society

standardized ability test tests given to large populations and scored with respect to population averages

state the organized system of power and authority in society

status an established position in a social structure that carries with it a degree of prestige

status attainment the process by which people end up in a given position in the stratification system

status inconsistency exists when the different statuses occupied by the individual bring with them significantly different amounts of prestige

status set the complete set of statuses occupied by a person at a given time

stereotype an oversimplified set of beliefs about the members of a social group or social stratum that is used to categorize individuals of that group

stereotype interchangeablility the principle that negative stereotypes are interchangeable from one racial group (or gender or social class) to another

stereotype threat effect a decrease in ability test score resulting from the stress and fear of confirming a negative racial, ethnic, or gender stereotype

stigma an attribute that is socially devalued and discredited

structural strain theory a theory that interprets deviance as originating in the tensions that exist in society between cultural goals and the means people have to achieve those goals

structural unemployment loss of work that occurs because of changes at the societal level

subculture the culture of groups whose values and norms of behavior are somewhat different from those of the dominant culture

superego the dimension of the self representing the standards of society

sweatshop a workplace where an employer violates more than one law regarding federal or state labor, industrial homework, occupational safety and health, workers' compensation, or industry regulation

symbolic interaction theory a theoretical perspective claiming that people act toward things because of the meaning things have for them

symbols things or behavior to which people give meaning

taboos behaviors that bring the most serious sanctions

taking the role of the other the process of imagining oneself from the point of view of another

teacher expectancy effect the effect of the teacher's expectations on the student's actual performance, independent of the student's ability

Temporary Assistance for Needy Families (TANF) federal program by which grants are given to states to administer welfare

terrorism premeditated, politically motivated violence perpetrated against noncombatant targets by persons or groups who use their action to try to achieve their political ends

tertiary deviance deviance that occurs when the deviant fully accepts the deviant role, but rejects the stigma associated with it

thermal pollution the heating up of the earth's rivers and lakes as a result of the chemical discharges of heavy industry

third-world countries countries that are poor, underdeveloped, largely rural, and with high levels of poverty; typically, governments in such countries are autocratic dictatorships and wealth is concentrated in the hands of a small elite

Title IX federal law that prohibits gender discrimination in any educational institution that receives federal funds

total institution an organization cut off from the rest of society in which individuals are subject to strict social control

totem an object or living thing that a religious group regards with special awe and reverence

tracking grouping, or stratifying, students in school on the basis of ability test scores

traditional authority authority stemming from long-established patterns that give certain people or groups legitimate power in society

tradition-directedness conformity to long-standing and time-honored norms and practices

transnational family families where one parent (or both) live in one country while other immediate family members live in other countries

transnational social movement a social movement whose organization crosses national boundaries

triad a group consisting of three people

triadic segregation the tendency for a triad to separate into a dyad and an isolate

troubles privately felt problems that come from events or feelings in one individual's life

Tuskegee Syphilis Study an unethical study of about 400 syphilis-infected African American men who went untreated for the disease for forty years, from 1932 to 1972, even though a cure (penicillin) was discovered in the early 1950s

underclass those with little or no opportunity for movement out of the worst poverty

underemployment a term used to describe being employed at a level below what would be expected, given a person's level of training or education

unemployment rate the percentage of those not working but officially defined as looking for work

unidimensional evolutionary theory a theory that predicts that societies over time follow a single path from simple and structurally undifferentiated to more complex and structurally differentiated

urban underclass a grouping of people, largely minority and poor, who live at the absolute bottom of the socioeconomic ladder in urban areas

urbanization is the extent to which a community has the characteristics associated with city life

utilitarian organization an organization, either profit or nonprofit, that is joined by individuals for specific purposes, such as monetary reward

validity the degree to which an indicator accurately measures or reflects a concept

values the abstract standards in a society or group that define ideal principles

variable something that can have more than one value

verstehen the process of understanding social behavior from the point of view of those engaged in it

victimless crimes crimes that violate law but are not listed in the FBI's serious crime index

vital statistics information about births, deaths, marriages, and migration

voluntary organization organizations that people join to pursue goals they consider personally worthwhile

wealth the monetary value of what someone owns, calculated by adding all financial assets (stocks, bonds, property, insurance, value of investments, and so on) and subtracting debts; also called net worth

world cities cities closely linked through the system of international commerce

world systems theory theory that capitalism is a single world economy and a worldwide system of unequal political and economic relationships that benefits the developed and technologically advanced countries at the expense of other countries

xenophobia the fear and hatred of foreigners

References

Aberle, David F., Albert K. Cohen, A. Kingsley Davis, Marion J. Levy Jr., and Francis X. Sutton. 1950. "The Functional Prerequisites of a Society." *Ethics* 60 (January): 100–111.

Acker, Joan. 1992. "Gendered Institutions: From Sex Roles to Gendered Institutions." *Contemporary Sociology* 21 (September): 565–569.

Acker, Joan, Sandra Morgen, and Lisa Gonzales. 2002. "Welfare Restructuring, Work & Poverty: Policy from Oregon." Working Paper, Eugene, OR: Center for the Study of Women in Society.

Adams, Paul, and Gary L. Dominick. 1995. "The Old, the Young, and the Welfare State." *Generations* 19 (Fall): 38–42.

Adkins, Lisa, and Vicki Merchant. 1996. *Sexualizing the Social: Power and the Organization of Sexuality.* New York: St. Martin's Press.

Adler, Patricia, and Peter Adler. 1998. *Peer Power: Preadolescent Culture and Identity.* New Brunswick, NJ: Rutgers University Press.

Adorno, T. W., Else Frenkel-Brunswik, D. J. Levinson, and R. N. Sanford. 1950. *The Authoritarian Personality.* New York: Harper and Row.

AFL-CIO. 2002. "Executive Pay Watch." Website: www.afl-cio.org

Aguilar-San Juan, Karin (ed.). 1994. *The State of Asian America.* Boston, MA: South End Press.

Aguirre, Benigno, Enrico Quarantelli, and Jorge L. Mendoza. 1988. "The Collective Behavior of Fads: The Characteristics, Effects and Career of Streaking." *American Sociological Review* 53 (August): 569–589.

Aitchison, Jean. 1997. "Free or Ensnared? The Hidden Nets of Language." Pp. 75–86 in *On Freedom: A Centenary Anthology,* edited by Eileen Barker. New Brunswick, NJ: Transaction.

Alan Guttmacher Institute. 1999a. "Facts in Brief: Contraceptive Use." New York: Alan Guttmacher Institute. Website: http://www.agi-usa.org

Alan Guttmacher Institute. 1999b. "Facts in Brief: Teen Sex and Pregnancy." New York: Alan Guttmacher Institute. Website: http://www.agi-usa.org

Alan Guttmacher Institute. 2002. *Teenagers' Sexual and Reproductive Health.* New York: Alan Guttmacher Institute. Website: www.agi-usa.org

Alan Guttmacher Institute. 2004. "Fact Sheets." Website: www.agi-usa.org

Alba, Richard. 1990. *Ethnic Identity: The Transformation of Ethnicity in the Lives of Americans of European Ancestry.* New Haven, CT: Yale University Press.

Alba, Richard, and Gwen Moore. 1982. "Ethnicity in the American Elite." *American Sociological Review* 47 (June): 373–383.

Albas, Daniel, and Cheryl Albas. 1988. "Aces and Bombers: The Post-Exam Impression Management Strategies of Students." *Symbolic Interaction* 11: 289–302.

Albelda, Randy, and Ann Withorn (eds.). 2002. *Lost Ground: Welfare Reform, Poverty, and Beyond.* Boston, MA: South End Press.

Aldag, Ramon J., and Sally R. Fuller. 1993. "Beyond Fiasco: A Reappraisal of the Groupthink Phenomenon and a New Model of Group Decision Processes." *Psychological Bulletin* 113: 533–552.

Alden, Helena L. 2001. "Gender Role Ideology and Homophobia." Paper presented at the annual meeting of the Southern Sociological Society, Atlanta, GA.

Alexander, Jeffrey C., and Paul Colomy (eds.). 1990. *Differentiation Theory and Social Change: Comparative and Historical Perspectives.* New York: Columbia University Press.

Alicea, Marixsa. 1997. "'What Is Indian About You?' A Gendered, Transnational Approach to Ethnicity." *Gender & Society* 11 (October): 597–626.

Aliotta, Jilda M. 1991. "The Unfinished Feminist Agenda: The Shifting Forum." *Annals of the American Academy of Political and Social Sciences* 515: 140–150.

Allan, Graham. 1998. "Friendship, Sociology and Social Structure." *Journal of Social and Personal Relationships* 15 (October): 685–702.

Allan, Graham, and Rebecca G. Adams (eds.). 1998. *Placing Friendship in Context.* New York: Cambridge University Press.

Allen, Katherine R. 1997. "Lesbian and Gay Families." Pp. 196–218 in *Contemporary Parenting: Challenges and Issues,* edited by Terry Arendell. Thousand Oaks, CA: Sage.

Allen, Michael Patrick, and Philip Broyles. 1991. "Campaign Finance Reforms and the Presidential Campaign Contributions of Wealthy Capitalist Families." *Social Science Quarterly* 72 (December): 738–750.

Allen, Paula Gunn. 1986. *The Sacred Hoop: Recovering the Feminine in American Indian Traditions.* Boston, MA: Beacon Press.

Allison, Anne. 1994. *Nightwork: Sexuality, Pleasure, and Corporate Masculinity in a Tokyo Hostess Club.* Chicago, IL: University of Chicago Press.

Allport, Gordon W. 1954. *The Nature of Prejudice.* Reading, MA: Addison-Wesley.

Allport, Gordon W. 1966. "The Religious Context of Prejudice." *Journal for the Scientific Study of Religion* 5: 447–457.

Altemeyer, Bob. 1988. *Enemies of Freedom: Understanding Right-Wing Authoritarianism.* San Francisco: Jossey-Bass.

Altman, Dennis. 2001. *Global Sex.* Chicago, IL: University of Chicago Press.

Altman, Laurence K. 1991. "Many Hispanic Americans Reported in Ill Health and Lacking Insurance." *The New York Times* (January 9): A16.

Alzheimer's Association. 2002. "One in Ten Families Affected by Alzheimer's Disease." Chicago, IL: Alzheimer's Association. Website: www.alz.org

Alzheimer's Association. 2004. Website: www.alz.org

Amato, Paul R., and Alan Booth. 1997. *A Generation at Risk: Growing Up in an Era of Family Upheaval.* Cambridge, MA: Harvard University Press.

American Association of Retired Persons. 2004. Website: www.aarp.org

American Association of University Women. 1992. *How Schools Shortchange Girls.* Washington, DC: American Association of University Women.

American Association of University Women. 1998. *Gender Gaps: Where Schools Still Fail Our Children.* Washington, DC: American Association of University Women.

Ammerman, Nancy T. 1994. "Lessons from Waco." *Footnotes* (January): Washington, DC: American Sociological Association.

Amnesty International. 2004. Website: www.amnesty.org

Amott, Teresa, and Julie Matthaei. 1996. *Race, Gender and Work: A Multicultural Economic History of Women in the United States.* Cambridge, MA: South End Press, pp. 59–61.

Amott, Teresa L., and Julie A. Matthaei. 1996. *Race, Gender, and Work: A Multicultural History of Women in the United States,* 2nd ed. Boston, MA: South End Press.

Andersen, Margaret L. 2000. *Thinking About Women: Sociological Perspectives on Sex and Gender,* 5th ed. Boston, MA: Allyn and Bacon.

Andersen, Margaret L. 2001. "Restructuring for Whom: Race, Class, Gender and the Ideology of Invisibility." *Sociological Forum* 16 (June): 181–201.

Andersen, Margaret L. 2003. *Thinking About Women: Sociological Perspectives on Sex and Gender.* 6th ed. Boston, MA: Allyn and Bacon.

Andersen, Margaret L., and Patricia Hill Collins (eds.). 2001. *Race, Class, and Gender: An Anthology,* 4th ed. Belmont, CA: Wadsworth.

Andersen, Margaret L., and Patricia Hill Collins. 2004. *Race, Class, and Gender: An Anthology,* 5th ed. Belmont, CA: Wadsworth.

Anderson, Elijah. 1990. *Streetwise: Race, Class and Change in an Urban Community.* Chicago, IL: University of Chicago Press.

Anderson, Elijah. 1976. *A Place on the Corner.* Chicago, IL: University of Chicago Press.

Anderson, Elijah. 1999. *Code of the Street.* New York: W. W. Norton.

Anderson, Kristin J., and Campbell Leaper. 1998. "Meta-Analyses of Gender Effects of Conversational Interruption: Who, What, When, Where, and How." *Sex Roles* 39 (August): 225–252.

Andreasse, Arthur. 1997. "Evaluating the 1995 Industry Employment Projections." *Monthly Labor Review* 120 (September): 9–31.

Angel, Ronald, and Marta Tienda. 1982. "Determinants of Extended Household Structure: Culture Pattern or Economic Need?" *American Journal of Sociology* 87 (May): 1360–1383.

Angotti, Joseph. 1997. "Content Analysis of Local News Programs in Eight U.S. Television Markets." Miami, FL: University of Miami Center for the Advancement of Modern Media.

Ansari, Maboud. 1991. "Iranians in America: Continuity and Change." Pp. 119–142 in *Rethinking Today's Minorities,* edited by Vincent N. Parillo. New York: Greenwood Press.

Ansuini, Catherine G., Juliana Fiddler-Woite, and Robert S. Woitaszek. 1994. "The Effects of Operation Rescue on Pro-Life Support." *College Student Journal* 28 (December): 441–445.

Anthony, Dick, Thomas Robbins, and Steven Barrie-Anthony. 2002. "Cult and Anticult Totalism: Reciprocal Escalation and Violence." *Terrorism and Political Violence* 14 (Spring): 211–239.

Antoucci, Toni C., Corann Okorodudu, and Hiroko Akiyama. 2002. "Well-being among Older Adults on Different Continents." *Journal of Social Issues* 58 (Winter): 617–626.

Antrobus, Peggy. 2002. "Feminism as Transformational Politics: Towards Possibilities for Another World." *Development* 45 (June): 46–52.

Aponte, Harry J. 1994. *Bread and Spirit: Therapy with the New Poor, Diversity of Race, Culture, and Values.* New York: W. W. Norton.

Apple, Michael W. 1991. "The New Technology: Is It Part of the Solution or Part of the Problem in Education?" *Computers in the Schools* 8 (April/October): 59–81.

Applebaum, Richard, and Peter Dreier. 1999. "The Campus Anti-Sweatshop Movement." *The American Prospect* 46 (September–October): 71.

Arbuckle, Julianne, and Benne D. Williams. 2003. "Students' Perceptions of Expressiveness: Age and Gender Effects on Teacher Evaluations." *Sex Roles* 49 (November): 507–516.

Arendell, Terry. 1992. "After Divorce: Investigations into Father Absence." *Gender & Society* 6 (December): 562–586.

648

Arendell, Terry. 1998. "Divorce American Style." *Contemporary Sociology* 27 (May): 226–228.

Arendt, Hannah. 1963. *Eichmann in Jerusalem: A Report on the Banality of Evil.* New York: Viking Press.

Argyle, Michael. 1975. *Bodily Communication.* London: Methuen.

Argyris, Chris. 1990. *Overcoming Organizational Defenses: Facilitating Organizational Learning.* Boston, MA: Allyn & Bacon.

Aries, Phillippe. 1962. *Centuries of Childhood.* New York: Vintage Books.

Armas, Genaro C. 2003. "Asian American Divorce Rate Up." *San Francisco Chronicle:* A4.

Aronson, Jane. 1992. "Women's Sense of Responsibility for the Care of Old People: 'But Who Else is Going to Do It?'" *Gender & Society* 6 (March): 8–29.

Aronson, Joshua, Carrie B. Fried, and Catherine Good. 2002. "Reducing the Effects of Stereotype Threat on African American College Students by Shaping Theories of Intelligence." *Journal of Experimental Social Psychology* 38: 113–125.

Aronson, Pamela. 2003. "Feminists or 'Postfeminists'? Young Women's Attitudes toward Feminism and Gender Relations." *Gender & Society* 17 (December): 903–922.

Asch, Solomon. 1951. "Effects of Group Pressure upon the Modification and Distortion of Judgments." In *Groups, Leadership, and Men,* edited by H. Guetzkow. Pittsburgh, PA: Carnegie Press.

Asch, Solomon. 1955. "Opinions and Social Pressure." *Scientific American* 19 (July): 31–35.

Associated Press. 2004. "AIDS Meeting Warns of Global Dangers." Associated Press, July 16, 2004.

Atchley, Robert C. 2000. *Social Forces and Aging,* 9th ed. Belmont, CA: Wadsworth.

Atkin, Ruth, and Adrienne Rich. 2001. "J.A.P.—Slapping: The Politics of Scapegoating." Pp. 201–204 in *Race, Class and Gender: An Anthology,* edited by Margaret L. Andersen and Patricia Hill Collins. Belmont, CA: Wadsworth Publishing.

Auerbach, Judy. 1992. "The Family/Medical Leave Bill." Presentation at midyear meeting of Sociologists for Women in Society, Raleigh, NC.

Avakame, Edem F., James J. Fyfe, and Candace McCoy. 1999. "Did You Call The Police? What Did They Do? An Empirical Assessment of Black's Theory of Mobilization of Law." *Justice Quarterly* 16 (December): 765–792.

Ayella, Marybeth. 1998. *Insane Therapy: Portrait of a Psychotherapy Cult.* Philadelphia, PA: Temple University Press.

Babbie, Earl. 2003. *The Practice of Social Research,* 10th ed. Belmont, CA: Wadsworth.

Baca Zinn, Maxine. 1995. "Chicano Men and Masculinity." Pp. 33–41 in *Men's Lives,* 3rd ed., edited by Michael S. Kimmel and Michael A. Messner. Needham Heights, MA: Allyn and Bacon.

Baca Zinn, Maxine, and Bonnie Thornton Dill. 1996. "Theorizing Difference from Multiracial Feminism." *Feminist Studies* 22 (Summer): 321–331.

Baca Zinn, Maxine, and D. Stanley Eitzen. 2000. *Diversity in American Families,* 4th ed. New York: HarperCollins.

Baca Zinn, Maxine, Pierrette Hondagneu-Sotelo, and Michael A. Messner. 2000. *Gender Through the Prism of Difference.* Needham Heights, MA: Allyn and Bacon.

Bachman, Daniel, and Ronald K. Esplin. 1992. "Plural Marriage." *Encyclopedia of Mormonism,* Vol. 3. New York: Macmillan.

Bailey, Garrick. 2002. *Humanity: An Introduction to Cultural Anthropology,* 6th ed. Belmont, CA: Wadsworth.

Balch, Robert W. 1980. "Looking Behind the Scenes in a Religious Cult: Implications for the Study of Conversion." *Sociological Analysis* 41: 137–143.

Balch, Robert W. 1995. "Waiting for the Ships: Disillusionment and Revitalization of Faith in Bo and Peep's UFO Cult." Pp. 137–166 in *The Gods Have Landed: New Religions From Other Worlds,* edited by James R. Lewis. Albany, NY: SUNY Press.

Baldassare, Mark (ed.). 1994. *The Los Angeles Riots: Lessons for the Urban Future.* Boulder, CO: Westview Press.

Bales, Kevin. 1999. *Disposable People: New Slavery in the Global Economy.* Berkeley, CA: University of California Press.

Bandura, A., and R. H. Walters. 1963. *Social Learning and Personality Development.* New York: Holt, Rinehart, and Winston.

Banks, W. Curtis. 1976. "White Preference in Blacks: A Paradigm in Search of a Phenomenon." *Psychological Bulletin* 83: 1179–1186.

Barajas, Heidi Lasley, and Jennifer L. Pierce. 2001. "The Significance of Race and Gender in School Success among Latinas and Latinos in College." *Gender & Society* 15 (December): 859–878.

Barber, Benjamin R. 1995. *Jihad vs. McWorld: How Globalism and Tribalism Are Reshaping the World.* New York: Random House.

Barber, Jennifer S., and William G. Axinn. 1998. "Gender Role Attitudes and Marriage Among Young Women." *Sociological Quarterly* 39 (Winter): 11–31.

Barber, Melvin, Leslie Inniss, and Emmit Hunt. 1998. *African American Contributions to Sociology.* Belmont, CA: Wadsworth.

Barker, Kathleen, and Kathleen Christensen. 1998. *Contingent Work: American Employment Relations in Transition.* Ithaca, NY: ILR Press.

Barker-Benfield, G. J. 1976. *Horrors of the Half Known Life.* New York: Harper and Row.

Barlow, Maude, and Tony Clarke. 2002. "Who Owns Water?" *The Nation* (September 2–9): 11–14.

Baron, James N., and Andrew E. Newman. 1990. "For What It's Worth: Organizations, Occupations, and the Value of Work." *American Sociological Review* 55 (April): 155–175.

Barringer, Felicity. 1993. "Minorities on the Move, Often Unpredictably." *The New York Times* (June 6): sec. 4, p. 4.

Barton, Paul E. 1999. "Too Much Testing of the Wrong Kind, Too Little of the Right Kind in K–12 Education." Princeton, NJ: Educational Testing Service, Policy Information Center.

Basow, Susan A. 1992. *Gender: Stereotypes and Roles,* 3rd ed. Pacific Grove, CA: Brooks/Cole.

Basow, Susan A., and Kelly Johnson. 2000. "Predictors of Homophobia in Female College Students." *Sex Roles* 42 (March): 391–404.

Basow, Susan A., and Kimberly Rubenfeld. 2003. "'Troubles Talk': Effects of Gender and Gender-Typing." *Sex Roles* 48 (February): 183–187.

Bates, Eric. 1998. "Private Prisons." *The Nation* (January 5): 11–17.

Baumeister, Roy F. 1998. "The Self." Pp. 680–740 in D. T. Gilbert, S. F. Fiske, and G. Lindzey, eds., *The Handbook of Social Psychology,* Vol. 1. New York: McGraw Hill.

Baxter, Janeen, and Eric Olin Wright. 2000. "The Glass Ceiling Hypothesis: A Comparative Study of the United States, Sweden, and Australia." *Gender & Society* 14 (April): 275–294.

Bazzini, Doris G., William D. McIntosh, Stephen M. Smith, Sabrina Cook, and Caleigh Harris. 1997. "The Aging Woman in Popular Film: Underrepresented, Unattractive, Unfriendly, and Unintelligent." *Sex Roles* 36 (April): 531–543.

Bean, Frank D., and Marta Tienda. 1987. *The Hispanic Population of the United States.* New York: Russell Sage Foundation.

Becker, Elizabeth. 2000. "Harassment in the Military Is Said to Rise." *The New York Times* (March 10): 14.

Becker, Howard S. 1963. *Outsiders: Studies in the Sociology of Deviance.* New York: Free Press.

Becker, Howard S., Blanche Geer, Everett C. Hughes, and Anselm L. Strauss. 1961. *Boys in White: Student Culture in Medical School.* Chicago, IL: University of Chicago Press.

Becraft, Carolyn H. 1992. Pp. 8–17 in *Women in the Military,* edited by E. A. Blacksmith. New York: H. W. Wilson.

Bedell, Kenneth B., and Alice M. Jones (eds.). 1992. *Yearbook of American and Canadian Churches.* Nashville, TN: Abingdon Press.

Beggs, John J. 1995. "The Institutional Environment: Race and Gender Inequality in the U.S. Labor Market." *American Sociological Review* 60 (August): 612–633.

Behringer, Autumn, Janet M. Wilmoth, Richard Hogan, and Carolyn C. Perrucci. 1999. "Employment Income Inequality among Older U.S. Workers: Gender, Race, and Class Comparisons." Paper presented at the Society for the Study of Social Problems, Chicago, IL.

Belknap, Joanne. 2001. *The Invisible Woman.* Belmont, CA: Wadsworth.

Belknap, Joanne, Bonnie S. Fisher, and Francis T. Cullen. 1999. "The Development of a Comprehensive Measure of the Sexual Victimization of College Women." *Violence Against Women* 5 (February): 185–214.

Bell, Daniel. 1973. *The Coming Crisis of Post-industrial Society.* New York: Basic Books.

Bellah, Robert (ed.). 1973. *Emile Durkheim on Morality and Society: Selected Writings.* Chicago, IL: University of Chicago Press.

Bellah, Robert, Richard Madsen, William M. Sullivan, Ann Swidler, and Steven M. Tipton. 1996. *Habits of the Heart: Individualism and Commitment in American Life.* Berkeley, CA: University of California Press.

Belli, M. M. 1966 [1954]. *Modern Trials.* Indianapolis, IN: Bobbs-Merrill (Suppl.).

Bendick, Marc Jr., Lauren E. Brown, and Kennington Well. 1999. "No Foot in the Door: An Experimental Study of Employment Discrimination against Older Workers." *Journal of Aging and Social Policy* 10: 5–23.

Benedict, Ruth. 1934. *Patterns of Culture.* Boston, MA: Houghton Mifflin.

Benin, Mary Holland, and Linda B. Robinson. 1997. "Marital Happiness across the Family Life Cycle: A Longitudinal Analysis." Paper presented at the American Sociological Association, Toronto.

Benson, Donna J., and Gregory Thomson. 1982. "Sexual Harassment on a University Campus: The Confluence of Authority Relations, Sexual Interest, and Gender Stratification." *Social Problems* 29 (February): 236–251.

Ben-Yehuda, Nachman. 1986. "The European Witchcraze of the Fourteenth–Seventeenth Centuries: A Sociologist's Perspective." *American Journal of Sociology* 86: 1–31.

Berberoglu, Berch. 1990. *Political Sociology: A Comparative/Historical Approach.* New York: General Hall.

Berger, Peter L. 1963. *Invitation to Sociology: A Humanistic Perspective.* Garden City, NY: Doubleday Anchor.

Berger, Peter L., Brigitte Berger, and Hansfried Kellner. 1974. *The Homeless Mind: Modernization and Consciousness.* New York: Vintage Books.

Berger, Peter L., and Thomas Luckmann. 1967. *The Social Construction of Reality: A Treatise in the Sociology of Knowledge.* Garden City, NY: Anchor Books.

Berheide, Catherine W. 1992. "Women Still 'Stuck' in Low-Level Jobs." *Women in Public Services: A Bulletin for the Center for Women in Government* 3 (Fall). Albany, NY: Center for Women in Government, State University of New York.

Berkman, Lisa F., Thomas Glass, Ian Brissette, and Teresa E. Seeman. 2000. "From Social Integration to Health: Durkheim in the New Millennium." *Social Science and Medicine* 51 (September): 843–857.

Berkowitz, L. 1974. "Some Determinants of Impulsive Aggression: The Role of Mediated Associations with Reinforcements for Aggression." *Psychological Review* 81 (March): 165–176.

Bernard, Jessie. 1972. *The Future of Marriage.* New York: Bantam.

Bernstein, Nina. 2002. "Side Effect of Welfare Law: The No-Parent Family." *The New York Times,* July 29: A1.

Berscheid, Ellen, and Harry R. Reis. 1998. "Attraction and Class Relationships." Pp. 193–281 in *The Handbook of Social Psychology,* 4th ed., edited by Daniel T. Gilbert, Susan T. Fiske, and Gardner Lindzey. New York: Oxford University/McGraw-Hill.

Best, Joel. 1999. *Random Violence: How We Talk About New Crimes and New Victims.* Berkeley, CA: University of California Press.

Best, Joel. 2001. *Damned Lies and Statistics: Untangling Numbers From the Media, Politicians, and Activists.* Berkeley, CA: University of California Press.

Best, Joel. 2004. *More Damned Lies and Statistics, How Numbers Confuse Public Issues.* Berkeley, CA: University of California Press.

Bettie, Julie. 1995. "Class Dismissed? Roseanne and the Changing Face of Working-Class Iconography." *Social Text* 45 (Winter): 125–149.

Bian, Yanjie. 1997. "Bringing Strong Ties Back In: Indirect Ties, Bridges, and Job Search in China." *American Sociological Review* 63: 366–385.

Bielby, William T. 1981."Models of Status Attainment." Pp. 3–26 in *Research in Social Stratification and Mobility,* edited by Donald Treiman and Robert Robinson. Greenwich, CT: JAI Press.

Binns, Allison. 2003. *White Gold, Weed, and Blow: The Drug Trades of Afghanistan, Columbia, and Mexico in Comparative Historical Perspective.* Senior Thesis, Princeton University, Princeton, NJ.

Blackwell, James E., and Morris Janowitz. 1974. *Black Sociologists: Historical and Contemporary Perspectives.* Chicago, IL: University of Chicago Press.

Blake, C. Fred. 1994. "Footbinding in Neo-Confucian China and the Appropriation of Female Labor." *Signs* 19 (Spring): 676–712.

Blalock, Hubert M., Jr. 1982. *Race and Ethnic Relations.* Englewood Cliffs, NJ: Prentice Hall.

Blalock, Hubert M., Jr. 1989. "Race Versus Class: Distinguishing Reality from Artifacts." *National Journal of Sociology* 3 (Fall): 127–142.

Blalock, Hubert M., Jr. 1991. *Understanding Social Inequality.* Newbury Park, CA: Sage.

Blankenship, Kim. 1993. "Bringing Gender and Race In: U.S. Employment Discrimination Policy." *Gender & Society* 7 (June): 204–226.

Blaskovitch, J. 1973. "Blackjack and the Risky Shift." *Sociometry* 36 (March): 42ff.

Blassingame, John. 1973. *The Slave Community: Plantation Life in the Antebellum South.* New York: Oxford University Press.

Blau, Peter M. 1986. *Exchange and Power in Social Life,* revised ed. New Brunswick, NJ: Transaction.

Blau, Peter M., and Marshall W. Meyer. 1987. *Bureaucracy in Modern Society,* 3rd ed. New York: Random House.

Blau, Peter M., and Otis Dudley Duncan. 1967. *The American Occupational Structure.* New York: Wiley.

Blau, Peter M., and W. Richard Scott. 1974. *On the Nature of Organizations.* New York: Wiley.

Blauner, Robert. 1964. *Alienation and Freedom: The Factory Worker and His Industry.* Chicago, IL: University of Chicago Press.

Blee, Kathleen. 1991. *Women of the Klan: Racism and Gender in the 1920s.* Los Angeles: University of California Press.

Blee, Kathleen. 2002. *Inside Organized Racism: Women in the Hate Movement.* Berkeley, CA: University of California Press.

Blee, Kathleen M., and Ann R. Tickamyer. 1995. "Racial Differences in Men's Attitudes About Women's Gender Roles." *Journal of Marriage and the Family* 57 (February): 21–30.

Blinde, Elaine M., and Diane E. Taub. 1992a. "Homophobia and Women's Sport: The Disempowerment of Athletes." *Sociological Focus* 25 (May): 151–166.

Blinde, Elaine M., and Diane E. Taub. 1992b. "Women Athletes as Falsely Accused Deviants: Managing the Lesbian Stigma." *Sociological Quarterly* 33 (Winter): 521–533.

Blinde, Elaine M., Diane E. Taub, and Lingling Han. 1993. "Sport Participation and Women's Personal Empowerment: Experiences of the College Athlete." *Journal of Sport and Social Issues* 17 (April): 47–60.

Blinde, Elaine M., Diane E. Taub, and Lingling Han. 1994. "Sport as a Site for Women's Group and Social Empowerment: Perspectives from the College Athlete." *Sociology of Sport Journal* 11 (March): 51–59.

Block, Alan, and Frank R. Scarpitti. 1993. *Poisoning for Profit: The Mafia and Toxic Waste in America.* New York: William Morrow.

Block, Fred. 1987. *Revising State Theory.* Philadelphia, PA: Temple University Press.

Blocker, T. Jean, and Douglas Lee Eckberg. 1997. "Gender and Environmentalism: Results from the 1993 General Social Survey." *Social Science Quarterly* 78 (December): 841–858.

Blum, Nancy S. 1991. "The Management of Stigma by Alzheimer Family Caregivers." *Journal of Contemporary Ethnography* 20 (October): 263–284.

Blumer, Herbert, 1969. *Studies in Symbolic Interaction.* Englewood Cliffs, NJ: Prentice Hall.

Bobo, L., and R. A. Smith. 1998. "From Jim Crow Racism and Laissez-Faire Racism: An Essay on the Transformation or Racial Attitudes in America." Pp. 182–220 in W. Katkin, N. Landsmand, and A. Tyree (eds.), *Beyond Pluralism: Essays on the Conception of Groups and Group Identities in America.* Urbana, IL: University of Illinois Press.

Bobo, Lawrence D. 1999. "Prejudice as Group Position: Microfoundations of a Sociological Approach to Racism and Race Relations." *Journal of Social Issues* 55 (3): 445–492.

Bobo, Lawrence, and J. R. Kluegel. 1991. "Modern American Prejudice: Stereotypes, Social Distance, and Perceptions of Discrimination Toward Blacks, Hispanics, and Asians." Paper presented before the American Sociological Association, Cincinnati, Ohio.

Bock, Jane D. 2000. "Doing the Right Thing?: Single Mothers by Choice and the Struggle for Legitimacy." *Gender & Society* 14 (February): 62–86.

Boer, J. Tom, Manuel Pastor Jr., James L. Sadd, and Lori D. Snyder. 1997. "Is There Environmental Racism? The Demographics of Hazardous Waste in Los Angeles County." *Social Science Quarterly* 78 (December): 793–810.

Boeringer, Scott B. 1999. "Associations of Rape Supportive Attitudes with Fraternal and Athletic Participation." *Violence Against Women* 5 (January): 81–90.

Bonacich, Edna. 1972. "A Theory of Ethnic Antagonism: The Split Labor Market." *American Sociological Review* 37 (October): 547–559.

Bonacich, Edna, and Richard P. Applebaum. 2000. *Behind the Label: Inequality in the Los Angeles Apparel Industry.* Berkeley, CA: University of California Press.

Bonacich, Edna, Lucie Cheng, Norma Chinchilla, and Paul Ong (eds.). 1994. *Global Production: The Apparel Industry in the Pacific Rim.* Philadelphia, PA: Temple University Press.

Bonilla-Silva, Eduardo. 2001. *White Supremacy and Racism in the Post-Civil Rights Era.* Boulder, CO: Lynne Rienner.

Bonilla-Silva, Eduardo, and Mary Hovespan. 2000. "If Two People Are in Love: Deconstructing Whites' Views on Interracial Marriage with Blacks." Paper presented at the Annual meetings of the Southern Sociological Society, New Orleans, LA.

Bontemps, Arna (ed.). 1972. *The Harlem Renaissance Remembered.* New York: Dodd Mead.

Booth, Bradford, William W. Falk, David R. Segal, and Mady Wechsler Segal. 2000. "The Impact of Military Presence in Local Labor Markets and the Employment of Women." *Gender & Society* 14 (April): 318–332.

Bord, Richard J., and Robert E. O'Connor. 1997. "The Gender Gap in Environmental Attitudes: The Case of Perceived Vulnerability to Risk." *Social Science Quarterly* 78 (December): 831–840.

Bose, Christine E., and Peter H. Rossi. 1983. "Gender and Jobs: Prestige Standings of Occupations as Affected by Gender." *American Sociological Review* 48 (June): 316–330.

Bosk, Charles L. 1979. *Forgive and Remember: Managing Medical Failure.* Chicago, IL: University of Chicago Press.

Boudon, Raymond. 1974. *Education, Opportunity, and Social Inequality.* New York: Wiley.

Bourdieu, Pierre. 1977. *Reproduction in Education, Society, Culture.* Beverly Hills, CA: Sage.

Bourdieu, Pierre. 1984. *Distinction: A Social Critique of the Judgement of Taste,* translated by Richard Nice. Cambridge, MA: Harvard University Press.

Bowditch, Christine. 1993. "Getting Rid of Troublemakers: High School Disciplinary Procedures and the Production of Dropouts." *Social Problems* 40 (November): 493–510.

Bowen, William G., and Derek Bok. 1998. *The Shape of the River: Long-Term Consequences of Considering Race in College and University Admissions.* Princeton, NJ: Princeton University Press.

Bowles, Samuel, and Herbert Gintis. 1976. *Schooling in Capitalist America.* New York: Basic Books.

Bowman, Charles. 1999. "BLS Projections to 2008: A Summary." *Monthly Labor Review* 122: 3–4.

Braddock, Jomills H. 1988. *National Education Longitudinal Study of 1988.* Washington, DC: U.S. Department of Education, National Center for Educational Statistics.

Branch, Taylor. 1988. *Parting the Waters: America in the King Years, 1954–1963.* New York: Simon and Schuster.

Braun, Kathyryn L., James H. Pietsch, and Patricia Blanchette (eds.). 2000. *Cultural Issues in End-Of-Life Decision Making.* Thousand Oaks, CA: Sage.

Braverman, Harry. 1974. *Labor and Monopoly Capital.* New York: Monthly Review Press.

Bremer, Barbara A., Cathleen T. Moore, and Ellen F. Bildersee. 1991. "Do You Have to Call It 'Sexual Harassment' to Feel Harassed?" *College Student Journal* 25 (September): 256–268.

Brener, Nancy D., Thomas R. Simon, Etienne G. Krug, and Richard Lowry. 1999. "Recent Trends in Violence-Related Behaviors Among High School Students in the United States." *JAMA* 282 (August 4): 440–446.

Brewer, Rose. 1988. "Black Women in Poverty: Some Comments on Female-Headed Families." *Signs* 13 (Winter): 331–333.

Bridges, George, and Robert Crutchfield. 1988. "Law, Social Standing and Racial Disparities in Imprisonment." *Social Forces* 66 (June): 699–724.

Briggs, Adam, and Paul Cobley. 1999. "'I Like My Shit Sagged': Fashion, 'Black Musics' and Subcultures." *Journal of Youth Studies* 2 (October): 337–352.

Brines, Julie, and Kara Joyner. 1999. "The Ties That Bind: Principles of Cohesion in Cohabitation and Marriage." *American Sociological Review* 64 (June): 333–355.

Brinkerhoff, Merlin B., and Marlene M. MacKie. 1984. "Religious Denominations' Impact Upon Gender Attitudes: Some Methodological Implications." *Review of Religious Research* 25 (June): 365–378.

Britt, Chester L. 1994. "Crime and Unemployment Among Youths in the United States, 1958–1990: A Time Series Analysis." *American Journal of Economics and Sociology* 53 (January): 99–109.

Brody, Jane. 2000. "Cybersex Gives Birth to a Psychological Disorder." *The New York Times* (May 16): D7, D10.

Bromley, David G. 1979. *"Moonies" In America: Cult, Church, and Crusade.* Beverly Hills, CA: Sage.

Bromley, David G. 2004. "Asian Spirituality." *Contexts* 3 (Spring): 4–6 (Letters).

Bronson, Po. 1999. *The Nudist on the Late Shift.* New York: Random House.

Brooke, James. 1998. "Utah Struggles with a Revival of Polygamy." *The New York Times* (August 23): A12.

Brookey, Robert Alan. 2001. "Bio-Rhetoric, Background Beliefs, and the Biology of Homosexuality." *Argumentation and Advocacy* 37: 171–183.

Brooks-Gunn, Jeanne, Wen-Jui Han, and Jane Waldfogel. 2002. "Maternal Employment and Child Cognitive Outcomes in the First Three Years of Life: The NICHD Study of Early Child Care." *Child Development* 73 (July-August): 1052–1072.

Brown, Clare. 2000. "'Judge Me All You Want': Cigarette Smoking and the Stigmatization of Smoking." Paper presented at the annual meeting of the Society for the Study of Social Problems, Washington, DC.

Brown, Cynthia. 1993. "The Vanished Native Americans." *The Nation* 257 (October 11): 384–389.

Brown, Elaine. 1992. *A Taste of Power: A Black Woman's Story.* New York: Pantheon.

Brown, Lester R., Christopher Flavin, and Hilary French. 2000. *State of the World 2000: A Worldwatch Institute Report on Progress Toward a Sustainable Society.* New York: W. W. Norton.

Brown, Ryan P., and Robert A. Josephs. 1999. "A Burden of Proof: Stereotype Relevance and Gender Differences in Math Performance." *Journal of Personality and Social Psychology* 76 (No. 2): 246–257.

Brown, Ursula M. 2001. *The Interracial Experience: Growing Up Black/White Racially Mixed in the United States.* Westport, CT: Praeger.

Browne, Angela, and Kirk R. Williams. 1993. "Gender, Intimacy, and Lethal Violence: Trends from 1976 through 1987." *Gender & Society* 7 (March): 78–98.

Browne, Irene (ed.). 1999. *Latinas and African American Women at Work: Race, Gender, and Economic Inequality.* New York: Russell Sage Foundation.

Brownfield, David, Ann Marie Sorenson, and Kevin M. Thompson. 2001. "Gang Membership, Race, and Social Class: A Test of the Group Hazard and Master Status Hypotheses." *Deviant Behavior* 22 (January–February): 73–89.

Brush, Lisa D. 1990. "Violent Acts and Injurious Outcomes in Married Couples: Methodological Issues in the National Survey of Families and Households." *Gender & Society* 4 (March): 56–67.

Bryant, Alyssa N. 2003. "Changes in Attitudes toward Women's Roles: Predicting Gender-Role Traditionalism among College Students." *Sex Roles* 48 (February): 131–142.

Bryant, Susan L., and Lillian M. Range. 1997. "Type and Severity of Child Abuse and College Students' Lifetime Suicidality." *Child Abuse and Neglect* 21 (December): 1169–1176.

Buchanan, Christy M., Eleanor Maccoby, and Sanford M. Dornbusch. 1996. *Adolescents After Divorce.* Cambridge, MA: Harvard University Press.

Buckley, Cynthia, Jacqueline L. Angel, and Dennis Donahue. 2000. "Nativity and Older Women's Health: Constructed Reliance in the Health and Retirement Study." *Journal of Women and Aging* 12: 21–37.

Bullard, Robert D. 1994a. *Unequal Protection: Environmental Justice and Communities of Color.* San Francisco: Sierra Club Books.

Bullard, Robert D. 1994b. *Dumping in Dixie: Race, Class, and Environmental Quality.* Boulder, CO: Westview.

Bullock, Heather E., Karen Fraser, and Wendy R. Williams. 2001. "Media Images of the Poor." *The Journal of Social Issues* 57 (Summer): 229–246.

Bullough, Vern. 1998. "Alfred Kinsey and the Kinsey Report: Historical Overview and Lasting Contributions." *The Journal of Sex Research* 35: 12–131.

Bullough, Vern L. 1993. *Cross Dressing, Sex, and Gender.* 1993. Philadelphia, PA: Temple University Press.

Bureau of Labor Statistics. 2002. *Fatal Occupational Injuries by Worker Characteristics and Event or Exposure, United States 2001.* Website: **www.bls.gov/iif/oshwc/cfoi/cftb0151.pdf**

Burgess, Samuel H. 2001. "Gender Role Ideology and Anti-Gay Opinion in the United States." Paper presented at the annual meeting of the Southern Sociological Association, Atlanta, GA.

Burris, Beverly H. 1998. "Computerization of the Workplace." *Annual Review of Sociology* 24: 141–157.

Burris, Val. 2000. "The Myth of Old Money Liberalism: The Politics of the *Forbes* 400 Richest Americans." *Social Problems* 47 (Summer): 360–378.

Burt, Cyril. 1966. "The Genetic Determination of Differences in Intelligence: A Study of Monozygotic Twins Reared Together and Apart." *British Journal of Psychology* 57: 137–153.

Bushman, B. P. 1998. "Primary Effects of Media Violence on the Accessibility of Aggressive Constructs in Memory." *Personality and Social Psychology Bulletin* 24: 537–545.

Butler, Amy C. 1996. "The Effect of Welfare Benefit Levels on Poverty Among Single-Parent Families." *Social Problems* 43 (February): 94–115.

Butsch, Richard. 1992. "Class and Gender in Four Decades of Television Situation Comedy: Plus ça Change . . . ." *Critical Studies in Mass Communication* 9: 387–399.

Butterfield, Fox. 2000. "Racial Disparities Seen As Pervasive in Juvenile Justice." *The New York Times* (April 26): 1.

Buunk, Bram, and Ralph B. Hupka. 1987. "Cross-Cultural Differences in the Elicitation of Sexual Jealousy." *Journal of Sex Research* 23 (February): 12–22.

Cadge, Wendy, and Courtney Bender. 2004. "Yoga and Rebirth in America: Asian Religions are Here to Stay." *Contexts* 3 (Winter): 45–51.

Calasanti, Toni M., and Kathleen F. Slevin. 2001. *Gender, Social Inequalities, and Aging.* Walnut Creek, CA: AltaMira Press.

Caldera, Yvonne M., and Mary A. Sciaraffa. 1998. "Parent-Toddler Play with Feminine Toys: Are All Dolls the Same?" *Sex Roles* 39 (November): 657–668.

Calhoun, Craig. 1994. *Social Theory and the Politics of Identity.* Cambridge, MA: Blackwell.

Cameron, Deborah. 1998. "Gender, Language, and Discourse: A Review Essay." *Signs* 23 (Summer): 945–974.

Campbell, Anne. 1987. "Self Definition by Rejection: The Case of Gang Girls." *Social Problems* 34 (December): 451–466.

Campbell, S. 1972. "The Multiple Function of the Criminal Defense *voir dire* in Texas." *American Journal of Criminal Law* 1: 255–282.

Campenni, C. Estelle. 1999. "Gender Stereotyping of Children's Toys: A Comparison of Parents and Nonparents." *Sex Roles* 40 (January): 121–138.

Cancian, Francesca, and Stacey Oliker. 2000. *Caring and Gender.* Thousand Oaks, CA: Pine Forge Press.

Cancio, A. Silvia, T. David Evans, and David J. Maume Jr. 1996. "Reconsidering the Declining Significance of Race: Racial Differences in Early Career Wages." *American Sociological Review* 61 (August): 541–556.

Candib, Lucy M. 1999. "Incest and Other Harms to Daughters across Culture: Maternal Complicity and Patriarchal Power." *Women's Studies International Forum* 22 (March–April): 185–201.

Cantor, Joanne. 2000. "Media Violence." *Journal of Adolescent Health* 27 (August): 30–34.

Caplow, Theodore. 1991. *American Social Trends.* New York: Harcourt Brace Jovanovich.

Caplow, Theodore. 1998. "Comment: The Case of the Phantom Episcopalians." *American Sociological Review* 63 (February): 112–113.

Cardenas, Jose A. 1996. "Ending the Crisis in the K-12 System." Pp. 51–70 in *Educating a New Majority: Transforming America's Educational System for Diversity,* edited by Laura I. Rendon and Richard O. Hope. San Francisco: Jossey-Bass.

Carlson, Darren K. 2000. "Majority of Americans Still Oppose the Idea of Legalized Marijuana." *The Gallup Poll* (December 12). Princeton, NJ: Gallup Organization. Website: **www.gallup.com**

Carlson, Darren K. 2003. "Public OK With Gays, Women in Military." *The Gallup Poll.* Princeton, NJ: Gallup Organization. Website: **www.gallup.com**

Carmichael, Stokely, and Charles V. Hamilton. 1967. *Black Power: The Politics of Liberation in America.* New York: Vintage Books.

Carnoy, Martin, Manuel Castells, Stephen S. Cohen, and Fernando Henrique Cardoso. 1993. *The New Global Economy in the Information Age.* University Park, PA: Pennsylvania State University Press.

Carr, C. Lynn. 1998. "Tomboy Resistance and Conformity: Agency in Social Psychological Gender Theory." *Gender & Society* 12 (October): 528–553.

Carr, Deborah. 2003. "A 'Good Death' for Whom? Quality of Spouses' Death and Psychological Distress among Older Widowed Persons." *Journal of Health and Social Behavior* 44 (June): 215–232.

Carrigan, Marylyn, and Isabelle Szmigin. 2000. "Advertising in an Ageing Society." *Ageing and Society* 20 (March): 217–233.

Carroll, J. B. 1956. *Language, Thought and Reality: Selected Writings of Benjamin Lee Whorf.* Cambridge, MA: MIT Press.

Carroll, Joseph. 2004. "Religion is 'Very Important' to 6 in 10 Americans." *The Gallup Poll.* Princeton, NJ: Gallup Organization. Website: **www.gallup.com**

Carson, Clayborne. 1981. *In Struggle: SNCC and the Black Awakening of the 1950's.* Cambridge, MA: Harvard University Press.

Carson, Clayborne, David J. Garrow, Vincent Harding, and Darlene Clark Hine (eds.). 1987. *Eyes on the Prize: America's Civil Rights Years.* New York: Penguin.

Carson, Rachel. 1962. *The Silent Spring.* New York: Knopf.

Carter, Gregg Lee. 1986. "The 1960s Black Riots Revisited: City Level Explanations of Their Severity." *Sociological Inquiry* 56 (Spring): 210–228.

Carter, Gregg Lee. 1990. "Collective Violence and the Problem of Group Size in Aggregate-Level

Studies." *Sociological Focus* 23 (October): 297–300.

Carter, Gregg Lee. 1992. "Hispanic Rioting During the Civil Rights Era." *Sociological Forum* 7 (2): 301–323.

Carter, Timothy S. 1999. "Ascent of the Corporate Model in Environmental-Organized Crime." *Crime, Law and Social Change* 21: 1–30.

Cashmore, Ellis. 1991."Black Cops Inc." Pp. 87–108 in *Out of Order: Policing Black People,* edited by Ellis Cashmore and Eugene McLaughlin. New York: Routledge.

Casper, Lynne M., Sara McLanahan, and Irwin Garfinkel. 1994. "The Gender-Poverty Gap: What We Can Learn from Other Countries." *American Sociological Review* 59 (August): 594–605.

Cassen, Robert. 1994. "Population and Development: Old Debates, New Conclusions." *U.S.–Third World Policy Perspectives* 19: 282.

Cassidy, Linda, and Rose Marie Hurrell. 1995. "The Influence of Victim's Attire on Adolescents' Judgments of Date Rape." *Adolescence* 30 (Summer): 319–323.

Cassidy, Margaret L., and Bruce O. Warren. 1997. "Family Employment Status and Gender Role Attitudes: A Comparison of Women and Men College Graduates." *Gender & Society* 10 (June): 312–329.

Catanzarite, Lisa. 2003. "Race-Gender Composition and Occupational Pay Degradation." *Social Problems* 50 (February): 14–27.

Catanzarite, Lisa, and Vilma Ortiz. 1996. "Family Matters, Work Matters? Poverty Among Women of Color and White Women." Pp. 121–140 in *For Crying Out Loud: Women's Poverty in the United States,* edited by Diane Dujon and Ann Withorn. Boston, MA: South End Press.

Caulfield, Mina Davis. 1985. "Sexuality in Human Evolution: What Is 'Natural' in Sex?" *Feminist Studies* 11 (Summer): 343–364.

Cavendar, Gray, Lisa Bond-Maupin, and Nancy C. Jurik. 1999. "The Construction of Gender in Reality Crime TV." *Gender & Society* 13 (October): 643–663.

Centeno, Miguel A. and Eszter Hargittai. 2001. "Defining A Global Geography." *The American Behavioral Scientist* 44 (10): 1545–1560.

Centers, Richard. 1949. *The Psychology of Social Classes.* Princeton, NJ: Princeton University Press.

Chafetz, Janet. 1984. *Sex and Advantage.* Totowa, NJ: Rowman and Allanheld.

Chagnon, Napoleon A. 1968. *Yanomami: The Fierce People.* New York: Holt, Rinehart, and Winston.

Chalfant, H. Paul, Robert E. Beckley, and C. Eddie Palmer. 1987. *Religion in Contemporary Society,* 2nd ed. Palo Alto, CA: Mayfield.

Chambliss, William J. 1989. "State Organized Crime." *Criminology* 27 (May): 183–208.

Chambliss, William J. and Howard F. Taylor. 1989. *Bias in the New Jersey Courts.* Trenton, NJ: Administrative Office of the Courts.

Chang, Jung. 1991. *Wild Swans: Three Daughters of China.* New York: Simon and Schuster.

Chappell, Neena L. 2003. "Correcting Crosscultural Stereotypes: Aging in Shanghai and Canada." *Journal of Cross-Cultural Gerontology* 18 (June): 127–147.

Chen, Anthony S. 1999. "Lives at the Center of the Periphery, Lives at the Periphery of the Center: Chinese American Masculinities and Bargaining with Hegemony." *Gender & Society* 13 (October): 584–607.

Chen, Elsa, Y. F. 1991. "Conflict Between Korean Greengrocers and Black Americans." Senior thesis, Princeton University, Princeton, NJ.

Cherlin, Andrew J., Frank F. Furstenberg, P. Lindsay Lansdale-Chase, Kathleen E. Kiernan, Philip K. Robins, Donna-Ruane Morrison, and Julien O. Teitler. 1991. "Longitudinal Studies of Effects of Divorce on Children in Great Britain

and the United States." *Science* 252 (June 7): 1386–1389.

Cherlin, Andrew J., P. Lindsay Chase-Lansdale, and Christine McRae. 1998. "Effects of Parental Divorce on Mental Health Throughout the Life Course." *American Sociological Review* 63 (April): 219–249.

Chernin, Kim. 1991. *The Obsession.* New York: Harper Colophon Books.

Chia, Rosina C., Jamie L. Moore, Ka Nei Lam, C. J. Chuang, and B. S. Cheng. 1994. "Cultural Differences in Gender Role Attitudes Between Chinese and American Students." *Sex Roles* 31 (July): 23–30.

Chin, Jean Lau. 1993. *Diversity in Psychotherapy: The Politics of Race, Ethnicity, and Gender.* Westport, CT: Praeger.

Chipman, Susan. 1991. "Word Problems: Where Bias Creeps In." Unpublished research report, Office of Naval Research, Arlington, VA.

Chipuer, H. M., M. J. Rovine, and R. Plomin. 1990. "LISREL Modeling: Genetic and Environmental Influences on IQ Revisited." *Intelligence* 14: 11–29.

Chiricos, Ted, Sarah Eschholz, and Marc Gertz. 1997. "Crime, News and Fear of Crime: Toward an Identification of Audience Effects." *Social Problems* 44 (August): 342–357.

Chodorow, Nancy. 1978. *The Reproduction of Mothering: Psychoanalysis and the Study of Gender.* Berkeley, CA: University of California Press.

Chodorow, Nancy. 1994. *Femininities, Masculinities, Sexualities: Freud and Beyond.* Lexington, KY: University of Kentucky Press.

Chodorow, Nancy J. 1999. *The Power of Feelings.* New Haven, CT: Yale University Press.

Choi, Namkee G. 2001. "Relationships between Life Satisfaction and Postretirement Employment among Older Women." *International Journal of Aging and Human Development* 52: 45–70.

Churchill, Ward. 1992. *Struggle for the Land.* Toronto: Between the Lines.

Cialdini, R. B. 1993. *Influence: Science and Practice,* 3rd ed. New York: Harper Collins.

Cicourel, Aaron V. 1968. *The Social Organization of Juvenile Justice.* New York: Wiley.

Cimino, Richard P., and Don Lattin. 1998. *Shopping for Faith: American Religion in the New Millennium.* San Francisco: Jossey-Bass.

Clark, Kenneth B., and Mamie P. Clark. 1947. "Racial Identification and Preference in Negro Children." Pp. 602–611 in *Readings in Social Psychology,* edited by T. M. Newcomb and E. L. Hartley. New York: Henry Holt.

Clark, Sylvia D., Mary M. M., and Leon G. Schiffman. 1999. "The Mind-Body Connection: The Relationship among Physical Activity Level, Life Satisfaction, and Cognitive Age among Mature Females." *Journal of Social Behavior and Personality* 14 (June): 221–240.

Clarke, J. Michael, Joanne Carlson Brown, and Lorna M. Hochstein. 1989. "Institutional Religion and Gay/Lesbian Oppression." *Marriage and Family Review* 14: 265–284.

Clarke, Lee. 2002. "Panic: Myth or Reality?" *Contexts* 1 (Fall): 21–26.

Clawson, Rosalee A., and Rakuya Trice. 2000. "Poverty As We Know It: Media Portrayals of the Poor." *The Public Opinion Quarterly* 64 (Spring): 53–64.

Clements, Mark. 1996. "Sex After 65." *Parade Magazine* (March 17): 4–6.

Clifford, N. M., and Elaine Walster. 1973. "Research Note: The Effects of Physical Attractiveness on Teacher Expectations." *Sociology of Education* 46: 248–258.

Coale, Ansley. 1974. "The History of the Human Population." *Scientific American* 231 (3): 40–51.

Coale, Ansley. 1986. "Population Trends and Economic Development." Pp. 96–104 in *World Population and the U.S. Population Policy: The*

Choice Ahead, edited by Jane Menken, New York: W. W. Norton.

Cohen, Adam B. and Ilana J. Tannenbaum. 2001. "Lesbian and Bisexual Women's Judgments of the Attractiveness of Different Body Types." *The Journal of Sex Research* 38 (August): 226–232.

Cohen, Bernard P., and Xueguang Zhou. 1991. "Status Processes in Enduring Work Groups." *American Sociological Review* 56 (April): 179–188.

Cohen, Eugene N., and Edwin Eames. 1985. *Cultural Anthropology.* Boston, MA: Little Brown.

Cohen, Stanley. 1993. "Human Rights and Crimes of the State: Culture of Denial." *The Australian and New Zealand Journal of Criminology* 26 (July): 97–115.

Cole, David. 1999. "The Color of Justice." *The Nation* (October 11): 12–15.

Cole, Johnnetta B. 1988. *Anthropology for the Nineties.* New York: Free Press.

College Board. 2002. *College Bound Seniors 2002: A Profile of SAT Program Test Takers.* New York: College Board.

Colley, Ann, and Zazie Todd. 2002. "Gender-Linked Differences in the Style and Content of E-mails to Friends." *Journal of Language and Social Psychology* 21 (December): 380–393.

Collins, Patricia Hill. 1990. *Black Feminist Thought: Knowledge, Consciousness and the Politics of Empowerment.* Cambridge, MA: Unwin Hyman.

Collins, Patricia Hill. 1998. *Fighting Words: Black Women and the Search for Justice.* Minneapolis, MN: University of Minnesota Press.

Collins, Patricia Hill. 2004. *Black Sexual Politics: African Americans, Gender, and the New Racism.* New York: Routledge.

Collins, Patricia Hill, Lionel A. Maldonado, Dana Y. Takagi, Barrie Thorne, Lynn Weber, and Howard Winant. 1995. "On West and Fenstermaker's 'Doing Difference.'" *Gender & Society* 9 (August): 491–505.

Collins, Randall. 1979. *The Credential Society.* New York: Academic Press.

Collins, Randall. 1988. *Theoretical Sociology.* San Diego, CA: Harcourt Brace Jovanovich.

Collins, Sharon M. 1983. "The Making of the Black Middle Class." *Social Problems* 30 (April): 369–382.

Collins, Sharon M. 1989. "The Marginalization of Black Executives." *Social Problems* 36: 317–331.

Collins-Lowry, Sharon M. 1997. *Black Corporate Executives: The Making and Breaking of a Black Middle Class.* Philadelphia, PA: Temple University Press.

Coltrane, Scott, and Melinda Messineo. 2000. "The Perpetuation of Subtle Prejudice: Race and Gender Imagery in 1990s Television Advertising." *Sex Roles* 42 (March): 363–389.

Congdon, David C., 1990. "Gender, Employment, and Psychosocial Well-Being." *Journal of Sociology and Social Welfare* 17 (September): 101–121.

Coniff, Richard. 2004. "Reading Faces." *Smithsonian* (January): 44–50.

Conley, Dalton. 1999. *Being Black, Living in the Red: Race, Wealth, and Social Policy in America.* Berkeley, CA: University of California Press.

Connell, R. W. 1992. "A Very Straight Gay: Masculinity, Homosexual Experience, and the Dynamics of Gender." *American Sociological Review* 57 (December): 735–751.

Conniff, Ruth. 1998. "The Joy of Women's Sports." *The Nation* (August 10–17): 26–30.

Conrad, Peter, and Joseph W. Schneider. 1992. *Deviance and Medicalization: From Badness to Sickness,* expanded ed. Philadelphia, PA: Temple University Press.

Constable, Nicole. 1997. *Maid to Order in Hong Kong: Stories of Filipina Workers.* Ithaca, NY: Cornell University Press.

Cook, Christopher D. 2000. "Temps Demand a New Deal." *The Nation* (March 27): 13–19.

Cook, Karen S., Karen A. Hegtvedt, and Toshio Yamagishi. 1988. "Structural Inequality, Legitimation, and Reactions to Inequality in Exchange Networks." Pp. 291–308 in *Status Generalization: New Theory and Research,* edited by M. Webster, Jr., and M. Foschi. Stanford, CA: Stanford University Press.

Cook, S. W. 1988. "The 1954 Social Science Statement and School Desegregation: A Reply to Gerard." Pp. 237–256 in *Eliminating Racism: Profiles in Controversy,* edited by D. A. Taylor. New York: Plenum.

Cooksey, Elizabeth C., and Ronald R. Rindfuss. 2001. "Patterns of Work and Schooling in Young Adulthood." *Sociological Forum* 16 (December): 731–755.

Cookson, Peter W., Jr., and Caroline Hodges Persell. 1985. *Preparing for Power: America's Elite Boarding Schools.* New York: Basic Books.

Cooley, Charles Horton. 1902. *Human Nature and Social Order.* New York: Scribners.

Cooley, Charles Horton. 1967 [1909]. *Social Organization.* New York: Schocken Books.

Coontz, Stephanie. 1992. *The Way We Never Were.* New York: Basic Books.

Cooper, Alvin, and Coralie R. Scherer. 1999. "Sexuality on the Internet: From Sexual Exploration to Pathological Expression." *Professional Psychology Research and Practice* 30. Website: **www.sex-centre.com/Internetsex_Folder/ MSNBC_Study_pp.htm**

Cooper, Marc. 1997. "The Heartland's Raw Deal: How Meatpacking Is Creating a New Immigrant Underclass." *The Nation* (February 3): 11–17.

Coren, Stanley, and P. L. Hewitt. 1999. "Sex Differences in Elderly Suicide Rates: Some Predictive Factors." *Aging and Mental Health* 3 (May): 112–118.

Corsaro, William A., and Donna Eder. 1990. "Children's Peer Cultures." *Annual Review of Sociology* 16: 197–220.

Cortese, Anthony J. 1999. *Provocateur: Images of Women and Minorities in Advertising.* Lanham, MD: Rowman and Littlefield.

Cortina, Lilia M., Suzanna Swam, Louise F. Fitzgerald, and Craig Walo. 1998. "Sexual Harassment and Assault: Chilling the Climate for Women in Academia." *Psychology of Women Quarterly* 22 (September): 419–441.

Coser, Lewis. 1977. *Masters of Sociological Thought.* New York: Harcourt Brace Jovanovich.

Costello, Mark. 2004. "Throwing Away The Key." *The New York Times Magazine* (June 6): 41–43.

Cotter, David A., Joan M. Hermsen, Ovadia Smith, and Reeve Vanneman. 2001. "The Glass Ceiling Effect." *Social Forces* 80 (December): 655–682.

Crane, Diana (ed.). 1994. *The Sociology of Culture: Emerging Theoretical Perspectives.* Cambridge, MA: Blackwell.

Crawford, Isiaah, and Elizabeth Solliday. 1996. "The Attitudes of Undergraduate College Students toward Gay Parenting." *Journal of Homosexuality* 30: 63–77.

Crawford, Mary. 1995. *Talking Difference: On Gender and Language.* Thousand Oaks, CA: Sage.

Croll, Elisabeth. 1995. *Changing Identities of Chinese Women.* London: Zed Books.

Croll, Jillian, Dianne Neumark-Sztainer, Mary Story, and Marjorie Ireland. 2002. "Prevalence and Risk and Protective Factors Related to Disordered Eating Behaviors Among Adolescents: Relationship to Gender and Ethnicity." *Journal of Adolescent Health* 31 (August): 166–175.

Crosnoe, Robert, and Glen H. Elder, Jr. 2002. "Successful Adaptation in the Later Years: A Life Course Approach." *Social Psychology Quarterly* 65 (December): 309–328.

Crouse, James, and Dale Trusheim. 1988. *The Case Against the SAT.* Chicago, IL: University of Chicago Press.

Crowley, Sue M. 1998. "Men's Self-Perceived Adequacy as the Family Breadwinner: Implications for Their Psychological, Marital, and Family Well-Being." *Journal of Family and Economic Issues* 19 (Spring): 7–23.

Crutchfield, Robert. 1992. "Anomie and Alienation." Pp. 95–100 in *Encyclopedia of Sociology,* Vol. 1, edited by Edgar F. Borgatta and Marie L. Borgatta. New York: Macmillan.

Csikszentmihalyi, Mihaly, and Barbara Schneider. 2000. *Becoming Adult: How Teenagers Prepare for the World of Work.* New York: Basic Books.

Cunningham, Mick. 2001. "The Influence of Parental Attitudes and Behaviors on Children's Attitudes toward Gender and Household Labor in Early Adulthood." *Journal of Marriage and Family* 63 (February): 111–122.

Cunningham, Peter J., and Llewellyn J. Cornelius. 1995. "Access to Ambulatory Care for American Indians and Alaska Natives: The Relative Importance of Personal and Community Resources." *Social Science and Medicine* 40 (February): 393–407.

Currie, Dawn. 1997. "Decoding Femininity: Advertisements and Their Teenage Readers." *Gender & Society* 11 (August): 453–477.

Curtiss, Susan. 1977. *Genie: A Psycholinguistic Study of a Modern-Day "Wild Child."* New York: Academic Press.

Cuzzort, Ray P., and Edith W. King. 1980. *20th Century Social Thought,* 3rd ed. New York: Holt, Rinehart, and Winston.

D'Alessio, Stewart J., and Lisa Stolzenberg. 2003. "Race and the Probability of Arrest." *Social Forces* 81 (June): 1381–1397.

D'Emilio, John. 1998. *Sexual Politics, Sexual Communities: The Making of a Homosexual Minority in the United States, 1940–1970.* Chicago, IL: University of Chicago Press.

Dahlhamer, James, and Joanne M. Nigg. 1993. "An Empirical Investigation of Rumoring: Anticipating Disaster Under Conditions of Uncertainty." Paper presented at the annual meeting of the Southern Sociological Society, Chattanooga, TN.

Dahrendorf, Ralf. 1959. *Class and Class Conflict in Industrial Society.* Stanford, CA: Stanford University Press.

Dalaker, Joseph. 1999. *Poverty in the United States, 1998.* Washington, DC: U.S. Government Printing Office.

Dalaker, Joseph, and Bernadette D. Proctor. 2000. *Poverty in the United States, 1999.* Washington, DC: U.S. Government Printing Office.

Dalton, Susan E., and Denise D. Bielby. 2000. "'That's Our Kind of Constellation': Lesbian Mothers Negotiate Institutionalized Understandings of Gender Within the Family." *Gender & Society* 14 (February): 36–61.

Daniels, Jessie. 1997. *White Lies: Race, Class, Gender, and Sexuality in White Supremacist Discourse.* New York: Routledge.

Darcy, R., Susan Welch, and Janet Clark. 1994. *Women, Elections, and Representation.* Lincoln, NE: University of Nebraska Press.

Darley, John M., and R. H. Fazio. 1980. "Expectancy Confirmation Processes Arising in the Social Interaction Sequence." *American Psychologist* 35: 867–881.

Das Gupta, Monisha. 1997. "'A Chambered Nautilus': The Contradictory Nature of Puerto Rican Women's Role in the Construction of a Transnational Community." *Gender & Society* 11 (October): 627–655.

Datesman, Susan K., and Frank R. Scarpitti (eds.). 1980. *Women, Crime, and Justice.* New York: Oxford University Press.

Davies-Netzley, Sally Ann. 1998. "Women Above the Glass Ceiling: Perceptions on Corporate Mobility and Strategies for Success." *Gender & Society* 12 (June): 339–355.

Davis, Angela. 1981. *Women, Race, and Class.* New York: Random House.

Davis, James A., and Tom Smith. 1984. *General Social Survey Cumulative File, 1972–1982.* Ann Arbor, MI: Inter-University Consortium for Political and Social Research.

Davis, Kingsley, and Wilbert E. Moore. 1945. "Some Principles of Stratification." *American Sociological Review* 10 (April): 242–247.

Deckman, Melissa. 2002. "Holy ABC's: The Impact of Religion on Attitudes about Education Policies." *Social Science Quarterly* 83 (June): 472–487.

Deegan, Mary Jo. 1988. "W.E.B. Du Bois and the Women of Hull-House, 1895–1899." *The American Sociologist* 19 (Winter): 301–311.

Deegan, Mary Jo. 1990. *Jane Addams and the Men of the Chicago School, 1892–1918.* New Brunswick, NJ: Transaction.

DeFleur, Lois. 1975. "Biasing Influences on Drug Arrest Records: Implications for Deviance Research." *American Sociological Review* 40 (March): 88–103.

Deitch, Cynthia. 1993. "Gender, Race, and Class Politics, and the Inclusion of Women in Title VII of the 1964 Civil Rights Act." *Gender & Society* 7 (June): 183–203.

Dellinger, Kristen, and Christine L. Williams. 1997. "Makeup at Work: Negotiating Appearance Rules in the Workplace." *Gender & Society* 11 (April): 151–177.

Demeny, Paul. 1991. "Tradeoffs Between Human Numbers and Material Standards of Living." Pp. 408–421 in *Resources, Environment, and Population: Present Knowledge, Future Options,* edited by Kingsley Davis and Michail S. Bernstam. New York: Population Council.

DeNavas-Walt, Carmen, Robert W. Cleveland, and Brice H. Webster, Jr. 2003. *Income in the United States: 2002.* Washington, DC: U.S. Census Bureau.

Denner, Jill, and Nora Dunbar. 2004. "Negotiating Femininity: Power and Strategies of Mexican American Girls." *Sex Roles* 50 (March): 301–314.

Dentinger, Emma, and Marin Clarkberg. "Informal Caregiving and Retirement Timing among Men and Women: Gender and Caregiving Relationships in Late Midlife." *Journal of Family Issues* 23 (October): 857–879.

Denzin, Norman. 1992. *Symbolic Interaction and Cultural Studies: The Politics of Interpretation.* New York: Blackwell.

DeVault, Marjorie. 1991. *Feeding the Family: The Social Organization of Caring as Gendered Work.* Chicago, IL: University of Chicago Press.

DeWitt, Karen. 1995. "Blacks Prone to Job Dismissal in Organizations." *The New York Times* (April 20): A19.

DiAcosta, Diego. 1998. "A Sociolinguistic Study of Two Baptist Churches." Senior thesis, Princeton University, Department of Sociology, Princeton, NJ.

Diamond, Milton, and H. K. Sigmundson. 1997. "Sex Reassignment at Birth: Long-Term Review and Clinical Implications." *Archives of Pediatric and Adolescent Medicine* 151 (March): 298–304.

Diaz-Duque, Ozzie F. 1989. "Communication Barriers in Medical Settings: Hispanics in the United States." *International Journal of the Sociology of Language* 79: 93–102.

Dill, Bonnie Thornton. 1988. "Our Mothers' Grief: Racial Ethnic Women and the Maintenance of Families." *Journal of Family History* 13 (October): 415–431.

Dillard, J. L. 1972. *Black English: Its Historical Usage in the United States.* New York: Random House.

Dillon, Michele. 1999. *Catholic Identity: Balancing Reason, Faith, and Power.* New York: Cambridge University Press.

DiMaggio, Paul. 1982. "Cultural Capital and School Success: The Impact of Status Culture Participation on the Grades of U.S. High School Students." *American Sociological Review* 47: 189–201.

DiMaggio, Paul, and Francie Ostrower. 1990. "Participation in the Arts by Black and White Americans." *Social Forces* 68 (March): 753–778.

DiMaggio, Paul J., and Walter W. Powell. 1991. "Introduction." Pp. 1–38 in *The New Institutionalism in Organizational Analysis,* edited by W. W. Powell and P. J. DiMaggio. Chicago, IL: University of Chicago Press.

Diner, Hasia. 1996. "Erin's Children in America: Three Centuries of Irish Immigration to the United States." Pp. 161–171 in *Origins and Destinies: Immigration, Race, and Ethnicity in America,* edited by Silvia Pedraza and Rubén Rumbaut. Belmont, CA: Wadsworth.

Dines, Gail, and Jean M. Humez. 2002. *Gender, Race, and Class in Media,* 2nd ed. Thousand Oaks, CA: Sage.

Dion, K. K. 1972. "Physical Attractiveness and Evaluating Children's Transgressions." *Journal of Personality and Social Psychology* 24: 285–290.

Dixit, Avinash K., and Susan Sneath. 1997. *Games of Strategy.* New York: W. W. Norton.

Dohm, Arlene. 2000. "Gauging the Labor Force Effects of Retiring Baby-Boomers." *Monthly Labor Review* 123 (July): 17–28.

Dollard, John, Neal E. Miller, Leonard W. Doob, O. H. Mowrer, and Robert R. Sears. 1939. *Frustration and Aggression.* New Haven, CT: Yale University Press.

Domhoff, G. William. 1970. *The Higher Circles: The Governing Class in America.* New York: Random House.

Domhoff, G. William. 1990. *The Power Elite and the State: How Policy Is Made in America.* New York: Aldine de Gruyter.

Domhoff, G. William. 1998. *Who Rules America: Power and Politics in the Year 2000,* 3rd ed. Mountain View, CA: Mayfield.

Donato, Katharine. 1992. "Understanding U.S. Immigration: Why Some Countries Send Women and Others Send Men." Pp. 159–184 in *Seeking Common Ground: Multidisciplinary Studies of Immigrant Women in the United States,* edited by Donna Gabaccia. Westport, CT: Praeger.

Donnerstein, E., and L. Berkowitz. 1981. "Victim Reactions in Aggressive Erotic Films as a Factor in Violence Against Women." *Journal of Personality and Social Psychology* 41 (January): 710–724.

Donnerstein, Edward, Daniel Linz, and Steven Penrod. 1987. *The Question of Pornography: Research Findings and Policy Implications.* New York: Free Press.

Douglass, Jack D. 1967. *The Social Meanings of Suicide.* Princeton, NJ: Princeton University Press.

Dovidio, J. F., and S. L. Gaertner (eds.). 1986. *Prejudice, Discrimination, and Racism.* New York: Academic Press.

Doyal, I. 1990. "Hazards of Hearth and Home." *Women's Studies International Forum* 13: 501–517.

Draper, Elaine Alma, and Kenneth Karst. 1999. "Liability and Responsibility in the Work of Company Doctors." Paper presented at the annual meeting of the American Sociological Association, Chicago, IL.

Driggs, Ken. 1990. "After The Manifesto: Modern Polygamy and Fundamentalist Mormons." *Journal of Church and State* 32 (Spring): 367–389.

DuBois, Cathy L., Deborah E. Knapp, Robert H. Faley, and Gary A. Kustis. 1998. "An Empirical Examination of Same and Other Gender Sexual Harassment in the Workplace." *Sex Roles* 9–10 (November): 731–749.

Du Bois, W. E. B. 1901. "The Freedmen's Bureau." *Atlantic Monthly* 86: 354–365.

Dudley, Carl S., and David A. Roozen. 2001. *Faith Communities Today: A Report on Religion in the United States Today.* Hartford, CT: Hartford Institute for Religion Research, Hartford Seminary.

Due, Linnea. 1995. *Joining the Tribe: Growing Up Gay & Lesbian in the '90s.* New York: Doubleday.

Duneier, Mitchell. 1999. *Sidewalk.* New York: Farrar, Strauss, and Giroux.

Dunlop, Riley E., and Angela G. Mertig. 1992. *American Environmentalism: The U.S. Environmental Movement, 1970–1990.* Philadelphia, PA: Taylor and Francis.

Dunne, Gillian A. 2000. "Opting into Motherhood: Lesbians Blurring the Boundaries and Transforming the Meaning of Parenthood and Kinship." *Gender & Society* 14 (February): 11–35.

Duran-Aydintug, Candan, and Kelly A. Causey. 1996. "Child Custody Determination: Implications for Lesbian Mothers." *Journal of Divorce and Remarriage* 25: 55–74.

Durant, Thomas J., Jr., and Joyce S. Louden. 1986. "The Black Middle Class in America: Historical and Contemporary Perspectives." *Phylon* 47 (December): 253–263.

Durbin, Susan. 2002. "Women, Power, and the Class Ceiling: Current Research Perspectives." *Work, Employment, and Society* 16 (December): 755–759.

Durkheim, Emile. 1897 [1951]. *Suicide.* Glencoe, IL: Free Press.

Durkheim, Emile. 1947 [1912]. *Elementary Forms of Religious Life.* Glencoe, IL: Free Press.

Durkheim, Emile. 1950 [1938]. *The Rules of Sociological Method.* Glencoe, IL: Free Press.

Durkheim, Emile. 1964 [1895]. *The Division of Labor in Society.* New York: Free Press.

Dworkin, Shari L., and Michael A. Messner. 1999. "Just Do . . . What? Sport, Bodies, Gender." Pp. 341–361 in *Revisioning Gender,* edited by Myra Marx Ferree, Judith Lorber, and Beth B. Hess. Thousand Oaks, CA: Sage.

Dwyer, Lynn E. 1983. "Structure and Strategy in the Antinuclear Movement." Pp. 148–161 in *Social Movements of the Sixties and Seventies,* edited by Jo Freeman. New York: Longman.

Eagly, A. H., R. D. Ashmore, M. G. Makhijani, and L. C. Longo. 1989. "What Is Beautiful Is Good, But: A Meta-Analysis." *Psychological Bulletin* 110: 109–128.

Ebo, Bosah (ed.). 1998. *Cyberghetto or Cybertopia? Race, Class, and Gender on the Internet.* Westport, CT: Praeger.

Eck, Diane L. 2002. *A New Religious America: How a "Christian Country Has Become the World's Most Religiously Diverse Nation.* San Francisco: Harper.

Eckberg, Douglass Lee. 1992. "Social Influences on Belief in Creationism." *Sociological Spectrum* 12 (April–June): 145–165.

Edin, Kathryn. 1991. "Surviving the Welfare System: How AFDC Recipients Make Ends Meet in Chicago." *Social Problems* 38 (November): 462–472.

Edin, Kathryn. 2000. "What Do Low-Income Single Mothers Say About Marriage?" *Social Problems* 47 (February): 112–133.

Edin, Kathryn, and Laura Lein. 1997. *Making Ends Meet: How Single Mothers Survive Welfare and Low-Wage Work.* New York: Russell Sage Foundation.

Edmondson, Brad. 2000. "Immigration Nation." *Preservation* (January–February): 31–38.

Ehrlich, Paul. 1968. *The Population Bomb.* New York: Ballantine Books.

Ehrlich, Paul. 1990. *The Population Explosion.* New York: Ballantine Books.

Ehrlich, Paul R., and Anne H. Ehrlich. 1991. *The Population Explosion.* New York: Touchstone/Simon and Schuster.

Ehrlich, Paul R., and Jianguo Liu. 2002. "Some Roots of Terrorism." *Population and Environment* 24 (November): 183–192.

Ehrlich, Paul R., Anne H. Ehrlich, and John P. Holdren. 1977. *Ecoscience: Population, Resources, Environment.* San Francisco: Freeman.

Eichenwald, Kurt. 1996. "The Two Faces of Texaco." *The New York Times* (November 10): sec. 3, p. 1ff.

Eichenwald, Kurt. 2002. "Two Ex-Officials at WorldCom Are Charged in Huge Fraud." *The New York Times* (August 2): A1, A11.

Eitzen, D. Stanley, and Maxine Baca Zinn. 2004. *In Conflict and Order,* 10th ed., Boston, MA: Allyn and Bacon.

Ekman, Paul. 1982. *Emotion in the Human Face,* 2nd ed. Cambridge, MA: Cambridge University Press.

Ekman, Paul, and W. V. Friesen. 1974. "Detecting Deception from the Body or Face." *Journal of Personality and Social Psychology* 29: 288–298.

Ekman, Paul, W. V. Friesen, and M. O'Sullivan. 1988. "Smiles When Lying." *Journal of Personality and Social Psychology* 54: 414–420.

Eliason, Michele J., and Salome Raheim. 1996. "Categorical Measurement of Attitudes About Lesbian, Gay, and Bisexual People." *Journal of Gay and Lesbian Social Services* 4: 51–65.

Elizabeth, Vivienne. 2000. "Cohabitation, Marriage, and the Unruly Consequences of Difference." *Gender & Society* 14 (February): 87–110.

Elliott, Marta. 2001. "Gender Differences in the Causes of Depression." *Women and Health* 22: 163–177.

Ellison, Christopher G., and David A. Gay. 1989. "Black Political Participation Revisited: A Test of Compensatory, Ethnic Community, and Public Arena Models." *Social Science Quarterly* 70 (March): 101–119.

Ellison, Christopher G., and John P. Bartkowski. 2002. "Conservative Protestantism and the Division of Household Labor among Married Couples." *Journal of Family Issues* 23 (November): 950–985.

Emerson, Juan P. 1970. "Behaviors in Private Places: Sustaining Definitions of Reality in Gynecological Examinations." Pp. 74–97 in *Recent Sociology,* Vol. 2, edited by H. P. Dreitzel. New York: Collier Books.

Emerson, Michael O., Christian Smith, and David Sikkink. 1999. "Equal in Christ, But Not in the World: White Conservative Protestants and Explanations of Black-White Inequality." *Social Problems* 46 (August): 398–417.

Enloe, Cynthia. 1989. *Bananas, Beaches, and Bases: Making Feminist Sense of International Politics.* Berkeley, CA: University of California Press.

Epps, Edgar T. 2002. "Race, Class and Educational Opportunity: Trends in the Sociology of Education." In *2001 Race Odyssey: African Americans and Sociology.* edited by Bruce R. Hare. Syracuse, NY: Syracuse University Press.

Epstein, Helen. 1988. *Children of the Holocaust: Conversations with Sons and Daughters of Survivors.* New York: Penguin.

Epstein, Steven. 1996. *Impure Science: AIDS, Activism, and the Politics of Knowledge.* Berkeley, CA: University of California Press.

Erikson, Eric. 1980. *Identity and the Life Cycle.* New York: W. W. Norton.

Erikson, Kai. 1966. *Wayward Puritans: A Study in the Sociology of Deviance.* New York: Wiley.

Erikson, Robert. 1985. "Are American Rates of Social Mobility Exceptionally High? New Evidence on an Old Issue." *European Sociological Review* 1 (May): 1–22.

Ermann, M. David, and Richard J. Lundman. 1992. *Corporate and Governmental Deviance.* New York: Oxford University Press.

Esbensen-Finn, Aage, Elizabeth Piper Deschenes, and Thomas L. Winfree, Jr. 1999. "Differences between Gang Girls and Gang Boys: Results from a Multisite Survey." *Youth and Society* 31 (September): 27–53.

Espenshade, Thomas J. 1995. "Unauthorized Immigration to the United States." *Annual Review of Sociology* 21: 195–216.

Espiritu, Yen Le. 1992. *Asian American Panethnicity: Bridging Institutions and Identity.* Philadelphia, PA: Temple University Press.

Essien-Udom, E. U. 1962. *Black Nationalism: A Search for an Identity in America.* Chicago, IL: University of Chicago Press.

Esterberg, Kristin. 2003. "The New Right." Pp. XXX in *Encyclopedia of American Lesbian, Gay, Bisexual, and Transgender History and Culture.* New York: Charles Scribner's Sons.

Esterberg, Kristin G. 2003. "New Right." *Encyclopedia of Lesbian, Gay, Bisexual and Transgendered History in America.* New York: Charles Scribner's Sons.

Etzioni, Amatai. 1975. *A Comparative Analysis of Complex Organization: On Power, Involvement, and Their Correlates,* rev. ed. New York: Free Press.

Etzioni, Amatai, John Wilson, Bob Edwards, and Michael W. Foley. 2001. "A Symposium on Robert D. Putnam's *Bowling Alone: The Collapse and Revival of American Community.*" *Contemporary Sociology* 30 (May): 223–230.

Evans, Lorraine, and Kimberly Davies. 2000. "No Sissy Boys Here: A Content Analysis of the Representation of Masculinity in Elementary School Reading Textbooks." *Sex Roles* 42 (February): 255–270.

Evans, Peter B., Dietrich Ruesschemeyer, and Theda Skocpol. 1985. *Bringing the State Back In.* Cambridge, MA: Cambridge University Press.

Fagot, Beverly I. 1995. "Psychosocial and Cognitive Determinants of Early Gender-Role Development." *Annual Review of Sex Research* 9: 1–31.

Farber, Naomi. 1990. "The Significance of Race and Class in Marital Decisions Among Unmarried Adolescent Mothers." *Social Problems* 37 (February): 51–63.

Farley, Reynolds. 1984. *Blacks and Whites: Narrowing the Gap?* Cambridge, MA: Harvard University Press.

Farley, Reynolds, and Walter R. Allen. 1987. *The Color Line and the Quality of Life in America.* New York: Russell Sage Foundation.

Fastnow, Chris, J. Tobin Grant, and Thomas J. Rudolph. 1999. "Holy Roll Calls: Religious Tradition and Voting Behavior in the U.S. House." *Social Science Quarterly* 80 (December): 687–701.

Fattah, Ezzat A., 1994. "The Interchangeable Roles of Victim and Victimizer." *HEUNI Papers* 3: 1–26.

Fausto-Sterling, Anne. 1992. *Myths of Gender: Biological Theories About Women and Men.* New York: Basic Books.

Fausto-Sterling, Anne. 2000. *Sexing the Body: Gender Politics and the Construction of Sexuality.* New York: Basic Books.

Feagin, Joe R. 1991. "The Continuing Significance of Race: Antiblack Discrimination in Public Places." *American Sociological Review* 56 (February): 101–116.

Feagin, Joe R. 2000. *Racist America: Roots, Future Realities, and Racial Reparations.* New York: Routledge.

Feagin, Joe R., and Clairece B. Feagin. 1993. *Racial and Ethnic Relations,* 4th ed. Englewood Cliffs, NJ: Prentice Hall.

Feagin, Joe R., and Vera Hernán. 1995. *White Racism.* New York: Routledge.

Feagin, Joe R., Hernán Vera, and Nikitah Imani. 1996. *The Agony of Education: Black Students at White Colleges and Universities.* New York: Routledge.

Federal Bureau of Investigation. 2000. *Uniform Crime Reports.* Washington, DC: U.S. Department of Justice.

Federal Bureau of Investigation. 2002. *Uniform Crime Reports.* Washington, DC: U.S. Department of Justice.

Federal Election Commission. 2001. "PAC Financial Activity through December 31, 2000." Website: **www.fec.gov**

Fein, Melvyn L. 1988. "Resocialization: A Neglected Paradigm." *Clinical Sociology* 6: 88–100.

Ferber, Abby L. 1998. *White Man Falling: Race, Gender, and White Supremacy.* Lanham, MD: Rowman and Littlefield.

Ferber, Abby. 1999. "What White Supremacists Taught a Jewish Scholar about Identity." *The Chronicle of Higher Education* (May 7): 86–87.

Ferguson, Ann Arnett. 2001. *Bad Boys: Public Schools in the Making of Black Masculinity.* Ann Arbor, MI: University of Michigan Press.

Ferguson, Susan J. 2000. "Challenging Traditional Marriage: Never Married Chinese American and Japanese American Women." *Gender & Society* 14 (February): 136–159.

Fernald, Anne, and Hiromi Morikawa. 1993. "Common Themes and Cultural Variations in Japanese and American Mothers' Speech to Infants." *Child Development* 64 (June): 637–656.

Fernandes-y-Freitas, Rosa L. 1996. "The Minitel and the Internet: New Houses for Our Old Phantoms." *Societies* 51 (February): 49–57.

Fields, Jason, and Lynn M. Casper. 2001. *America's Families and Living Arrangements: March 2000.* Current Population Reports. Washington, DC: U. S. Census Bureau.

Fischer, Claude. 1981. *To Dwell Among Friends: Personal Networks in Town and City.* Chicago, IL: University of Chicago Press.

Fischer, Claude S., Michael Hout, Mártin Sánchez Jankowski, Samuel R. Lucas, Ann Swidler, and Kim Voss. 1996. *Inequality by Design: Cracking the Bell Curve Myth.* Princeton, NJ: Princeton University Press.

Fisher, Bonnie S., Frances T. Cullen, and Michael G. Turner. 2001. *Sexual Victimization of College Women.* Washington, DC: Bureau of Justice Statistics.

Flanagan, William G. 1995. *Urban Sociology: Images and Structure.* Needham Heights, MA: Allyn and Bacon.

Fleisher, Mark. 2000. *Dead End Kids: Gang Girls and the Boys They Know.* Madison, WI: University of Wisconsin Press.

Fleming, Jacqueline, and Nancy Garcia. 1998. "Are Standardized Tests Fair to African Americans? Predictive Validity of the SAT in Black and White Institutions." *Journal of Higher Education* 69 (September–October): 471–495.

Flowers, M. L. 1977. "A Laboratory Test of Some Implications of Janis' Groupthink Hypothesis." *Journal of Personality and Social Psychology* 35 (December): 888–896.

Flowers, R. Barri. 2001. "The Sex Trade Industry's Worldwide Exploitation of Children." *Annals of the American Academy of Political and Social Science* 575 (May): 147–157.

Folbre, Nancy. 2000. "Universal Childcare: It's Time." *The Nation* (July 3): 21–24.

Fordham, Signithia. 1996. *Blacked Out: Dilemmas of Race, Identity, and Success at A. I. High.* Chicago, IL: University of Chicago Press.

Fouts, Gregory, and Kimberley Burggraf. 1999. "Television Situation Comedies: Female Body Images and Verbal Reinforcements." *Sex Roles* 40 (March): 473–481.

Frank, A. G. 1969. *Latin America: Underdevelopment or Revolution?* New York: Monthly Review Press.

Frankbgerg, Erika, and Chungmei Lee. 2002. *Race in American Public Schools: Rapidly Resegregating School Districts.* Cambridge, MA: The Civil Rights Project, Harvard University. Website: **www.civilrightsproject.harvard.edu**

Frankenberg, Erica, Chungmei Lee, and Gary Orfield. 2003. "A Multiracial Society with Segregated Schools: Are We Losing the Dream?" Civil Rights Project, Harvard University. Website: **www.civilrightsproject.harvard.edu**

Frazier, E. Franklin. 1957. *The Black Bourgeoisie.* New York: Collier Books.

Fredrickson, George M. 2003. *Racism: A Short History.* Princeton, NJ: Princeton University Press.

Free, Marvin D. 1990. "Demographic, Organizational and Economic Determinants of Work Satisfaction: An Assessment of Work Attitudes of Females in Academic Settings." *Sociological Spectrum* 10 (Winter): 79–103.

Freedle, Roy O. 2003. "Correcting the SATs Ethnic and Social Class Bias: A Method of Re-Estimating SAT Scores." *Harvard Educational Review* 73 (Spring): 1–43.

Freedman, Allan. 1997. "Lawyers Take a Back Seat in the 105th Congress." *Congressional Quarterly* 55 (January 4): 27–30.

Freeman, Jo. 1983a. "On the Origins of Social Movements." Pp. 8–33 in *Social Movements of the Sixties and Seventies,* edited by Jo Freeman. New York: Longman.

Freeman, Jo. 1983b. "A Model for Analyzing the Strategic Options of Social Movement Organizations." Pp. 193–210 in *Social Movements of the Sixties and Seventies,* edited by Jo Freeman. New York: Longman.

Freud, Sigmund. 1960 [1923]. *The Ego and the Id,* translated by Joan Riviere. New York: W. W. Norton.

Freud, Sigmund. 1961 [1930]. *Civilization and Its Discontents,* translated by James Strachey. New York: W. W. Norton.

Freud, Sigmund. 1965 [1901]. *The Psychopathology of Everyday Life,* translated by Alan Tyson and edited by James Strachey. New York: W. W. Norton.

Fried, Amy. 1994. "'It's Hard to Change What We Want to Change': Rape Crisis Centers as Organizations." *Gender & Society* 4 (December): 562–583.

Friedman, Thomas L. 1999. *The Lexus and the Olive Tree.* New York: Farrar, Straus, and Giroux.

Frye, Marilyn. 1983. *The Politics of Reality.* Trumansburg, NY: Crossing Press.

Fukuda, Mari. 1994. "Nonverbal Communication Within Japanese and American Corporations." Unpublished manuscript, Princeton University, Princeton, NJ.

Fuller, Robert C. 2002. *Spiritual, But Not Religious: Understanding Unchurched America.* New York: Oxford University Press.

Fullerton, Howard F. 1999. "Labor Force Projections to 2008: Steady Growth and Changing." *Monthly Labor Review* 122: 19–32.

Funk, Jeanne B., and Debra D. Buchman. 1996. "Children's Perceptions of Gender Differences in Social Approval for Playing Electronic Games." *Sex Roles* 35 (August): 219–331.

Furstenberg, Frank. 1998. "Relative Risk: What Is the Family Doing to Our Children?" *Contemporary Sociology* 27 (May): 223–225.

Furstenberg, Frank F., Jr. 1990. "Divorce and the American Family." *Annual Review of Sociology* 16: 379–403.

Furstenberg, Frank F., Jr., and Christine Winquist Nord. 1985. "Parenting Apart: Patterns of Childrearing After Marital Disruption." *Journal of Marriage and the Family* 47 (November): 898–904.

Gagné, Patricia, and Richard Tewksbury. 1998. "Conformity Pressures and Gender Resistance Among Transgendered Individuals." *Social Problems* 45 (February): 81–101.

Gagnon, John. 1973. *Sexual Conduct: The Social Sources of Human Sexuality.* Chicago, IL: Aldine.

Gagnon, John. 1995. *Conceiving Sexuality: Approaches to Sex Research in a Postmodern World.* New York: Routledge.

Gallagher, Charles. 2003. "Color-Blind Privilege: The Social and Political Functions of Erasing the Color Line in Post Race America." *Race, Gender and Class:* 10(4): 22–37.

Gallagher, Sally K. 2003. *Evangelical Identity and Gendered Family Life.* New Brunswick, NJ: Rutgers University Press.

Gallup, George, Jr., and D. Michael Lindsay. 2000. *Surveying the Religious Landscape: Trends in U.S. Beliefs.* Harrisburg, PA: Morehouse.

Gallup Organization. 2000. "Tobacco and Smoking." *The Gallup Poll.* Princeton, NJ: The Gallup Organization. Website: **www.gallup.com**

Gallup Organization. 2002a. "Abortion." Princeton, NJ: Gallup Organization. Website: **www .gallup.com**

Gallup Organization. 2002b. "Homosexual Relations." Princeton, NJ: Gallup Organization. Website: **www.gallup.com**

Gamoran, Adam. 1972. "The Variable Effects of High School Tracking." *American Sociological Review* 57: 812–828.

Gamoran, Adam, and Robert D. Mare, 1989. "Secondary School Tracking and Educational Inequality: Compensation, Reinforcement, or Neutrality?" *American Journal of Sociology* 94: 1146–1183.

Gamson, Joshua. 1995a. "Must Identity Movements Self Destruct? A Queer Dilemma." *Social Problems* 42 (August): 390–407.

Gamson, Joshua. 1995b. "Featured Essay." *Contemporary Sociology* 24 (May): 294–298.

Gamson, Joshua. 1998. "Publicity Traps: Television Talk Shows and Lesbian, Gay, Bisexual, and Transgender Visibility." *Sexualities* 1 (February): 11–42.

Gamson, William A. 1992. "The Social Psychology of Collective Action." Pp. 53–76 in *Frontiers in Social Movement Theory,* edited by Aldon D. Morris and Carol McClurg Mueller. New Haven, CT: Yale University Press.

Gamson, William, and Andre Modigliani. 1974. *Conceptions of Social Life: A Text-Reader for Social Psychology.* Boston, MA: Little Brown.

Ganahl, Dennis J., Thomas J. Prinsen, and Sara Baker Netzley. 2003. "A Content Analysis of Prime Time Commercials: A Contextual Framework of Gender Representation." *Sex Roles* 49 (November): 545–551.

Gans, Herbert. 1979. *Deciding What's News: A Study of the CBS Evening News, NBC Nightly News, Newsweek and Time.* New York: Pantheon.

Gans, Herbert. 1982 [1962]. *The Urban Villagers: Group and Class in the Life of Italian Americans.* New York: Free Press.

Gans, Herbert J. 1991 [1971]. "The Uses of Poverty." Pp. 263–270 in *People, Plans, and Policies: Essays on Poverty, Racism, and Other National Urban Problems,* edited by Herbert J. Gans. New York: Columbia University Press.

Gans, Herbert J. 1999. *Popular Culture and High Culture: An Analysis and Evaluation of Taste.* New York: Basic Books.

Ganzeboom, Harry B. G., Donald J. Treiman, and Wout C. Ultee. 1991. "Comparative Intergenerational Stratification Research: Three Generations and Beyond." *Annual Review of Sociology* 17: 277–302.

Gardiner, R. Allen, and Beatrice T. Gardiner. 1969. "Teaching Sign Language to a Chimpanzee." *Science* 165: 664–672.

Gardner, Gary, and Brian Halweil. 1999. *Underfed and Overfed: the Global Epidemic of Malnutrition.* New York: Worldwatch Institute. Website: **www.worldwatch.org**

Gardner, Howard. 1999 [1993]. *Frames of Mind: The Theory of Multiple Intelligences.* New York: Basic Books.

Garfinkel, Harold. 1967. *Studies in Ethnomethodology.* Englewood Cliffs, NJ: Prentice Hall.

Garner, Pamela W., Shannon Robertson, and Gail Smith. 1997. "Preschool Children's Emotional Expressions with Peers: The Roles of Gender and Emotion Socialization." *Sex Roles* 36 (June): 675–691.

Garrow, David J. 1981. *The FBI and Martin Luther King.* New York: W. W. Norton.

Garst, J., and G. V. Bodenhausen. 1997. "Advertising's Effects on Men's Gender Role Attitudes." *Sex Roles* 36 (May): 551–572.

Gates, Henry Louis, Jr. 1988. *The Signifying Monkey: A Theory of African-American Literary Criticism.* New York: Oxford University Press.

Gates, Henry Louis, Jr. 1992. "Integrating the American Mind." Pp. 105–120 in *Loose Canons: Notes on the Culture Wars,* edited by Henry Louis Gates, Jr. New York: Oxford University Press.

Gates, Henry Louis, Jr. 1993. "The End of Civilization As We Know It." Eberhard Faber Lecture Series, Princeton University, Princeton, NJ.

Gehlen, Frieda. 1977. "Toward a Revised Theory of Hysterical Contagion." *Journal of Health and Social Behavior* 19 (March): 27–35.

Gelles, Richard J. 1999. "Family Violence." Pp. 1–24 in *Family Violence: Prevention and Treatment,* edited by Robert J. Hampton, Thomas P. Gallota, Gerald R. Adams, Earl H. Potter III, and Roger P. Weissberg. Newbury Park, CA: Sage.

General Electric Financial Survey. 2003. Website: **www.ge.com**

Genovese, Eugene. 1972. *Roll, Jordan, Roll: The World the Slaves Made.* New York: Pantheon.

Gerami, Shahin, and Melodye Lehnerer. 2001. "Women's Agency and Houshold Diplomacy: Negotiating Fundamentalism." *Gender & Society* 15 (August): 556–573.

Gerlach, Luther P., and Virginia Hine. 1970. *People, Power, Change: Movements of Social Transformation.* Indianapolis, IN: Bobbs-Merrill.

Gersch, Beate. 1999. "Class in Daytime Talk Television." *Peace Review* 11 (June): 275–281.

Gerson, Kathleen. 1993. *No Man's Land: Men's Changing Commitments to Family and Work.* New York: Basic Books.

Gerstel, Naomi, and Katherine McGonagle. 1999. "Job Leaves and the Limits of the Family and Medical Leave Act: The Effects of Gender, Race, and Family." *Work and Occupations* 26 (November): 510–534.

Gerstel, Naomi, and Sally K. Gallagher. 2001. "Men's Caregiving: Gender and the Contingent Character of Care." *Gender & Society* 15 (April): 197–217.

Gerth, Hans, and C. Wright Mills (eds.). 1946. *From Max Weber: Essays in Sociology.* New York: Oxford University Press.

Geschwender, James. 1992. "Ethnicity and the Social Construction of Gender in the Chinese Diaspora." *Gender & Society* 6 (September): 480–507.

Gibbs, Lois. 1982. *Love Canal: My Story.* Albany, NY: State University of New York Press.

Gibson, Rose C., and Cheryl J. Burns. 1991. "The Health, Labor Force, and Retirement Experiences of Aging Minorities." *Generations* 15 (Fall–Winter): 31–35.

Giddings, Paula. 1994. *In Search of Sisterhood: Delta Sigma Theta and the Challenge of the Black Sorority Movement.* New York: William Morrow.

Gilbert, D. T., and P. S. Malone. 1995. "The Correspondence Bias." *Psychological Bulletin* 117: 21–38.

Gilbert, Daniel R., Susan T. Fiske, and Gardner Lindzey (eds.). 1998. *The Handbook of Social Psychology,* 4th ed. New York: Oxford University/McGraw-Hill.

Gilens, Martin. 1996. "Race and Poverty in America: Public Misperceptions and the American News Media." *The Public Opinion Quarterly* 60 (Winter): 515–541.

Gilkes, Cheryl Townsend. 1985. "'Together and in Harness': Women's Traditions in the Sanctified Church." *Signs* 10 (Summer): 678–699.

Gilkes, Cheryl Townsend. 2000. *"If It Wasn't for the Women . . .": Black Women's Experience and Womanist Culture in Church and Community.* Mayknoll, NY: Orbis Books.

Gillespie, Rosemary. 2003. "Childfree and Feminine: Understanding the Gender Identity of Voluntary Childless Women." *Gender & Society* 17 (February): 122–136.

Gilligan, Carol. 1982. *In a Different Voice: Psychological Theory and Women's Development.* Cambridge, MA: Harvard University Press.

Gimlin, Debra. 1996. "Pamela's Place: Power and Negotiation in the Hair Salon." *Gender & Society* 10 (October): 505–526.

Gitlin, Todd. 1983. *Inside Prime Time.* New York: Basic Books.

Gitlin, Todd. 1987. *The Sixties: Years of Hope, Days of Rage.* New York: Bantam Books.

Gittleman, Maury, and Mary Joyce. 1999. "Have Family Income Mobility Patterns Changed?" *Demography* 36 (August): 299–314.

Glaser, B. G., and Anselm Strauss. 1965. *Awareness of Dying.* Chicago, IL: Aldine de Gruyter.

Glass Ceiling Commission. 1995. *Good for Business: Making Full Use of the Nation's Human Capital.* Washington, DC: U.S. Government Printing Office.

Glassner, Barry. 1999. *Culture of Fear: Why Americans Are Afraid of the Wrong Things.* New York: Basic Books.

Glassner, Barry, and Rosanna Hertz. 2003. *Our Studies, Ourselves: Sociologists' Lives and Work.* New York: Oxford University Press.

Glazer, Nathan. 1970. *Beyond the Melting Pot: The Negroes, Puerto Ricans, Jews, Italians, and Irish of New York City.* Cambridge, MA: MIT Press.

Glazer, Nona. 1990. "The Home as Workshop: Women as Amateur Nurses and Medical Care Providers." *Gender & Society* 4: 479–499.

Glenn, Evelyn Nakano. 1986. *Issei, Nisei, War Bride: Three Generations of Japanese American Women in Domestic Service.* Philadelphia, PA: Temple University Press.

Glenn, Evelyn Nakano. 1987. "Gender and the Family." Pp. 348–380 in *Analyzing Gender: A Handbook of Social Science Research,* edited by Beth Hess and Myra Marx Ferree. Newbury Park, CA: Sage.

Glenn, Evelyn Nakano. 2002. *Unequal Freedom: How Race and Gender Shaped American Citizenship and Labor.* Cambridge, MA: Harvard University Press.

Glenn, Evelyn Nakano, Grace Chang, and Linda Rennie Forcey. 1993. *Mothering: Ideology, Experience, and Agency.* New York: Routledge.

Glenn, Norval, and Elizabeth Marquardt. 2001. "Hooking Up, Hanging Out, and Hoping for Mr. Right." New York: Institute for American Values.

Glock, Charles, and Rodney Stark. 1965. *Religion and Society in Tension.* Chicago, IL: Rand McNally.

Gluckman, Amy, and Betsy Reed (eds.). 1997. *Homo Economics: Capitalism, Community, and Lesbian and Gay Life.* New York: Routledge.

Goffman, Erving. 1959. *The Presentation of Self in Everyday Life.* Garden City, NY: Doubleday.

Goffman, Erving. 1961. *Asylums: Essays on the Social Situation of Mental Patients and Other Inmates.* Garden City, NY: Anchor.

Goffman, Erving. 1963a. *Behavior in Public Places.* New York: Free Press.

Goffman, Erving. 1963b. *Stigma: Notes on the Management of Spoiled Identity.* Englewood Cliffs, NJ: Prentice Hall.

Goffman, Erving. 1967. *Interaction Ritual.* Chicago, IL: Aldine de Guyter.

Goffman, Erving. 1974. *Frame Analysis.* Cambridge, MA: Harvard University Press.

Goldberger, Arthur S. 1979. "Heritability." *Econometrica* 46: 327–347.

Goldman, Noreen, Yorenzo Moreno, and Charles F. Westoff. 1989. *Peru Experimental Study: An Evaluation of Fertility and Child Health Information.* Princeton, NJ: Office of Population Research.

Goldner, Ellen J., and Safiya Henderson-Holmes. 2001. *Race and (E)Racing Language: Living with the Color of Our Words.* Syracuse, NY: Syracuse University Press.

González, Tina Esther. 1996. "Social Control of Medical Professionals, Cultural Notions of Pain, and Doctors as Social Scientists." Junior thesis, Princeton University, Princeton, NJ.

Goodall, Jane. 1990. *Through a Window: My Thirty Years with the Chimpanzees of Gombe.* Boston, MA: Houghton Mifflin.

Goode, Erich. 1992. *Collective Behavior.* Fort Worth, TX: Harcourt Brace Jovanovich.

Goodman, Catherine Chase. 1990. "The Caregiving Roles of Asian American Women." *Journal of Women and Aging* 2: 109–120.

Gordon, Linda. 1977. *Woman's Body/Woman's Right.* New York: Penguin.

Gordon, Margaret T., and Stephanie Riger. 1989. *The Female Fear.* New York: Free Press.

Gorman, Thomas J. 2000. "Cross-Class Perceptions of Social Class." *Sociological Spectrum* 20 (January–March): 93–120.

Gottfredson, Michael R., and Travis Hirschi. 1990. *A General Theory of Crime.* Stanford, CA: Stanford University Press.

Gottfredson, Michael R., and Travis Hirschi. 1995. "National Crime Control Policies." *Society* 32 (January–February): 30–36.

Gough, Kathleen. 1984. "The Origin of the Family." Pp. 83–99 in *Women: A Feminist Perspective,* 3rd ed., edited by Jo Freeman. Palo Alto, CA: Mayfield.

Gould, Stephen J. 1994. "Curveball." *The New Yorker* 70 (November 28): 139–149.

Gould, Stephen Jay. 1981. *The Mismeasure of Man.* New York: W. W. Norton.

Gowan, Mary, and Melanie Trevino. 1998. "An Examination of Gender Differences in Mexican-American Attitudes Toward Family and Career Roles." *Sex Roles* 38 (June): 1079–1093.

Gowan, Mary A., and Raymond A. Zimmermann. 1996. "The Family and Medical Leave Act of 1993: Employee Rights and Responsibilities, Employer Rights and Responsibilities." *Employee Responsibilities and Rights Journal* 9 (March): 57–71.

Graefe, Deborah-Roempke, and Daniel T. Lichter. 1999. "Life Course Transitions of American Children: Parental Cohabitation, Marriage, and Single Motherhood." *Demography* 36 (May): 205–217.

Graham, Lawrence Otis. 1999. *Our Kind of People: Inside America's Black Upper Class.* New York: Harper Collins.

Gramsci, Antonio. 1971. *Selections from the Prison Notebooks of Antonio Gramsci,* edited by Quintin Hoare and Geoffrey Nowell. London: Lawrence and Wishart.

Granovetter, Mark. 1973. "The Strength of Weak Ties." *American Journal of Sociology* 78 (May): 1360–1380.

Granovetter, Mark. 1974. *Getting a Job: A Study of Contacts and Careers.* Cambridge, MA: Harvard University Press.

Granovetter, Mark S. 1995. "Afterward 1994: Reconsiderations and A New Agenda." Pp. 139–182 in *Getting a Job,* 2nd ed. Chicago, IL: University of Chicago Press.

Grant, Don Sherman II, and Ramiro Martinez, Jr. 1997. "Crime and the Restructuring of the U.S. Economy: A Reconsideration of the Class Linkages." *Social Forces* 75 (March): 769–799.

Grauerholz, Liz, and Lori Baker-Sherry. 2003. "The Pervasiveness and Persistence of the Feminine Beauty Ideal in Children's Fairy Tales." *Gender & Society* 17 (October): 711–726.

Greeley, Andrew M., and Michael Hout. 1999. "Americans Increasing Belief in Life After Death." *American Sociological Review* 64 (December): 813–835.

Green, Gary Paul, Leam M. Tigges, and Daniel Diaz. 1999. "Racial and Ethnic Differences in Job Search Strategies In Atlanta, Boston, and Los Angeles." *Social Science Quarterly* 80 (June): 263–290.

Greenfeld, Lawrence A., and Tracy L. Snell. 1999. *Women Offenders.* Washington, DC: U.S. Bureau of Justice Statistics.

Greenfeld, Lawrence A. 1996. *Child Victimizers: Violent Offenders and Their Victims.* Washington, DC: U.S. Bureau of Justice Statistics.

Greenhouse, Linda. 1996. "Gay Rights Laws Can't Be Banned, High Court Rules." *The New York Times* (May 21): A1ff.

Greenhouse, Steven. 2000. "Poll of Working Women Finds Them Stressed." *The New York Times* (March 10): A15.

Gregoire, Thomas K., Keith Kilty, and Virginia Richardson. 2002. "Gender and Racial Inequities in Retirement Resources." *Journal of Women & Aging* 14: 25–39.

Grieco, Elizabeth M., and Rachel C. Cassidy. 2001. *Overview of Race and Hispanic Origin.* Washington, DC: U.S. Census Bureau.

Griffin, Glenn, A. Elmer, Richard L. Gorsuch, and Andrea Lee Davis. 1987. "A Cross-Cultural Investigation of Religious Orientation, Social Norms, and Prejudice." *Journal for the Scientific Study of Religion* 26 (September): 358–365.

Grimal, Pierre (ed.). 1963. *Larousse World Mythology.* New York: Putnam.

Grindstaff, Laura. 2002. *The Money Shot: Trash Class, and the Making of TV Talk Shows.* Chicago, IL: University of Chicago Press.

Griswold, Wendy. 1994. *Cultures and Societies in a Changing World.* Thousand Oaks, CA: Pine Forge Press.

Gruber, James E. 1982. "Blue-Collar Blues: The Sexual Harassment of Women Autoworkers." *Work and Occupations* 3 (August): 271–298.

Gubrium, Jaber F., and James A. Holstein (eds.). 2003. *Ways of Aging.* Malden, MA: Blackwell.

Guinther, J. 1988. *The Jury in America.* New York: Facts on File.

Guterman, Lila. 2003. "As the Rich Get Richer, Do People Get Sicker?" *The Chronicle of Higher Education* (November 28): A22–23.

Gutman, Herbert G. 1976. *The Black Family in Slavery and Freedom, 1750–1925.* New York: Vintage Books.

Habermas, Jürgen. 1970. *Toward a Rational Society: Student Protest, Science, and Politics.* Boston, MA: Beacon Press.

Hadaway, C. Kirk, and Penny Long Marler. 1993. "All in the Family: Religious Mobility in America." *Review of Religious Research* 35 (December): 97–116.

Hadaway, C. Kirk, Penny Long Marler, and Mark Chaves. 1998. "Reply: Overreporting Church Attendance in America: Evidence That Demands the Same Verdict." *American Sociological Review* 63 (February): 122–130.

Haddad, Yvonne Yazbeck, Jane I. Smith, and John L. Esposito (eds.). 2003. *Religion and Immigration: Christian, Jewish, and Muslim Experiences in the United States.* Walnut Creek, CA: AltaMira Press.

Hagan, John. 1993. "The Social Embeddedness of Crime and Unemployment." *Criminology* 31 (November): 465–491.

Haines, Rebecca J. 1999. "Break North: Rap Music and Hip-Hop Culture in Canada." Pp. 54–88 in *Ethnicity, Politics, and Public Policy: Case Studies in Canadian Diversity,* edited by Harold Troper and Morton Weinfeld. Toronto: University of Toronto Press.

Hall, Edward T. 1966. *The Hidden Dimension.* New York: Doubleday.

Hall, Edward T., and Mildred Hall. 1987. *Hidden Differences: Doing Business with the Japanese.* New York: Anchor Press/Doubleday.

Hall, Elaine J., and Marnie Salupo Rodriguez. 2003. "The Myth of Postfeminism." *Gender & Society* 17 (December): 878–902.

Haller, Beth A., and Sue Ralph. 2001. "Profitability, Diversity, and Disability Images in Advertising in the United States and Britain." *Disability*

Studies Quarterly 21 (Spring). Website: **www.dsq-sds.org**

Hallinan, Maureen T. 1994. "Tracking: From Theory to Practice." *Sociology of Education* 67: 79–84.

Hallinan, Maureen T. 2003. *Ability Grouping and Student Learning.* Washington, DC: Brookings Papers on Educational Policy.

Hamer, Dean H., Stella Hu, Victoria L. Magnuson, Nan Hu, and Angela M. L. Pattatucci. 1993. "A Linkage Between DNA Markers on the X Chromosome and Male Sexual Identification." *Science* 261 (July): 321–327.

Hamer, Jennifer. 2001. *What It Means to Be Daddy.* New York: Columbia University Press.

Hampton, Henry. 1987. "Revolt and Repression." *Eyes on the Prize.* Film. Boston, MA: Blackside Productions.

Handlin, Oscar. 1951. *The Uprooted.* Boston, MA: Little Brown.

Haney, C., C. Banks, and P. G. Zimbardo. 1973. "Interpersonal Dynamics in A Simulated Prison." *International Journal of Criminology and Penology* 1: 69–97.

Haney, Lynne. 1996. "Homeboys, Babies, Men in Suits: The State and the Reproduction of Male Dominance." *American Sociological Review* 61 (October): 759–778.

Hans, Valerie, and Ramiro Martinez. 1994. "Intersections of Race, Ethnicity, and the Law." *Law and Human Behavior* 18 (June): 211–221.

Harding, Sandra G. 1998. *Is Science Multicultural? Postcolonialisms, Feminisms, and Epistemologies.* Bloomington, IN: Indiana University Press.

Harlan, Sharon, and Pamela M. Robert. 1998. "The Social Construction of Disability in Organizations: Why Employers Resist Reasonable Accommodation." *Work and Occupations* 25 (November): 397–435.

Harris, Allen C. 1994. "Ethnicity as a Determinant of Sex-Role Identity: A Replication Study of Item Selection for the Bem Sex Role Inventory." *Sex Roles* 31 (August): 241–273.

Harris, Darryl B. 1998. "The Logic of Black Urban Rebellions." *Journal of Black Studies* 28 (3): 373–383.

Harris, Diana K., Gary A. Fine, and Thomas C. Hood. 1992. "The Aging of Desire: Playboy Centerfolds and the Graying of America; A Research Note." *Journal of Aging Studies* 6: 301–306.

Harris, Fred R., and Lynn Curtis (eds.). 1998. *Locked in the Poorhouse: Cities, Race, and Poverty in the United States.* Lanham, MD: Rowman and Littlefield.

Harris, Ian, José B. Torres, and Dale Allender. 1994. "The Responses of African American Men to Dominant Norms of Masculinity Within the United States." *Sex Roles* 31 (December): 703–720.

Harris, Kathleen Mullan. 1996. "Life after Welfare: Women, Work, and Repeat Dependency." *American Sociological Review* 61 (June): 407–426.

Harris, Kathleen Mullan, R. Kelly Raley, and Ronald R. Rindfuss. 2002. "Family Configurations and Child-Care Patterns: Families with Two or More Preschool–Age Children." *Social Science Quarterly* 83 (June): 455–471.

Harris, Marvin. 1974. *Cows, Pigs, Wars, and Witches: The Riddles of Culture.* New York: Vintage Books.

Harrison, Bennett, and Harry Bluestone. 1982. *The Deindustrialization of America: Plant Closings, Community Abandonment, and the Dismantling of Basic Industry.* New York: Basic Books.

Harrison, Deborah, and Tom Trabasso. 1976. *Black English: A Seminar.* Hillsdale, NJ: Lawrence Erlbaum.

Harrison, Reginald. 1980. *Pluralism and Corporatism.* London: George Allen and Unwin.

Harrison, Roderick. 2000. "Inadequacies of Multiple Response Race Data in the Federal Statistical System." Manuscript. Washington, DC: Joint Center for Political and Economic Studies and Howard University, Department of Sociology.

Harry, Joseph. 1979. "The Marital 'Liaisons' of Gay Men." *The Family Coordinator* 28 (October): 622–629.

Hartmann, Heidi. 1996. "Who Has Benefited from Affirmative Action in Employment?" Pp. 77–96 in *The Affirmative Action Debate*, edited by George E. Curry. Reading, MA: Addison-Wesley.

Hastie, Reid, Steven D. Penrod, and Nancy Pennington. 1983. *Inside the Jury.* Cambridge, MA: Harvard University Press.

Hastorf, Albert, and Hadley Cantril. 1954. "They Saw a Game: A Case Study." *Journal of Abnormal and Social Psychology* 40 (2): 129–134.

Hatch, Laurie Russell. 1992. "Gender Differences in Orientation Toward Retirement from Paid Labor." *Gender & Society* 6 (March): 66–87.

Hatfield, Elaine S., and S. Sprecher. 1986. *Mirror, Mirror: The Importance of Looks in Everyday Life.* Albany, NY: State University of New York Press.

Hauan, Susan M., Nancy S. Landale, and Kevin T. Leicht. 2000. "Poverty and Work Effort among Urban Latino Men." *Work and Occupations* 27 (May): 188–222.

Hauser, Robert M., Howard F. Taylor, and Troy Duster. 1995. "The Bell Curve." *Contemporary Sociology* 24 (March): 149–161.

Hawkins, Dana. 1996. "The Most Dangerous Jobs." *U.S. News & World Report* 121 (September 23): 40–42.

Hawley, Amos H. 1986. *Human Ecology: A Theoretical Essay.* Chicago, IL: University of Chicago Press.

Hayes-Bautista, David E., Werner O. Schink, and Maria Hayes-Bautista. 1993. "Latinos and the 1992 Los Angeles Riots: A Behavioral Sciences Perspective." *Hispanic Journal of Behavioral Sciences* 15 (November): 427–448.

Hays, Sharon. 2003. *Flat Broke with Children: Women in the Age of Welfare Reform.* New York: Oxford University Press.

Healey, Joseph H. 1995. *Race, Ethnicity, Gender, and Class: The Sociology of Group Conflict and Change.* Thousand Oaks, CA: Pine Forge Press.

Healy, Sherry. 1986. "Growing to Be an Old Woman." Pp. 38–62 in *Women and Aging*, edited by Jo Alexander, et al. Corvallis, OR: Calyx Books.

Hearnshaw, Leslie. 1979. *Cyril Burt: Psychologist.* Ithaca, NY: Cornell University Press.

Hefez, Albert. 1985. "The Role of the Press and the Medical Community in the Epidemic of 'Mysterious Gas Poisoning' in the Jordan West Bank." *American Journal of Psychiatry* 142 (7): 833–837.

Heider, Fritz. 1958. *The Psychology of Interpersonal Relations.* New York: Wiley.

Heidrich, Susan M., and Carol D. Ryff. 1993. "The Role of Social Comparison Processes in the Psychological Adaptation of Elderly Adults." *Journal of Gerontology: Psychological Sciences* 48: 127–136.

Heilman, Brice E., and Paul J. Kaiser. 2002. "Religion, Identity, and Politics in Tanzania." *Third World Quarterly* 23 (August): 691–709.

Heim, Michael. 1990. "The Dark Side of Infomania." *Electric Word* 18 (March-April): 32–36.

Heimer, Karen. 1997. "Socioeconomic Status, Subcultural Definitions and Violent Delinquency." *Social Forces* 75 (March): 799–833.

Helgesen, Sally. 1990. *The Female Advantage: Women's Ways of Leadership.* New York: Doubleday.

Hensel, Chase. 1996. *Telling Our Selves: Ethnicity and Discourse in Southwestern Alaska.* New York: Oxford University Press.

Henslin, James M. 1993. "Doing the Unthinkable." Pp. 253–262 in *Down to Earth Sociology,* 7th ed., edited by James M. Henslin. New York: Free Press.

Herman, Judith. 1981. *Father-Daughter Incest.* Cambridge, MA: Harvard University Press.

Herman, Max A. 1995. "A Tale of Two Cities: Testing Explanations for Riot Violence in Miami, Florida, and Los Angeles, California." Paper presented at the Annual Meeting of the American Sociological Association, Washington, DC.

Herrnstein, Richard J., and Charles Murray. 1994. *The Bell Curve: Intelligence and Class Structure in American Life.* New York: Free Press.

Hersh, Seymour M. 2004. "Torture At Abu Graib." *The New Yorker* (May 10): 42–47.

Herskovits, Melville J. 1941. *The Myth of the Negro Past.* New York: Harper and Brothers.

Hetzel, Lisa, and Annetta Smith. 2001. "The 65 Years and Over Population: 2000." *Census 2000 Brief.* Washington, DC: U.S. Census Bureau.

Heubert, Jay P., and Robert M. Hauser (eds.). 1999. *High Stakes: Testing for Tracking, Promotion, and Graduation.* Commission on Behavioral and Social Sciences and Education, National Research Council. Washington, DC: National Academy Press.

Hibbard, David R., and Duane Buhrmester. 1998. "The Role of Peers in the Socialization of Gender-Related Social Interaction Styles." *Sex Roles* 39 (August): 185–202.

Higginbotham, A. Leon. 1978. *In the Matter of Color: Race and the American Legal Process.* New York: Oxford University Press.

Higginbotham, A. Leon. 1998. "Breaking Thurgood Marshall's Promise." *The New York Times Magazine* (January 18): 28–29.

Higginbotham, Elizabeth. 2001. *Too Much to Ask: Black Women in the Era of Integration.* Chapel Hill, NC: University of North Carolina Press.

Higginbotham, Elizabeth, and Lynn Weber. 1992. "Moving Up with Kin and Community: Upward Social Mobility for Black and White Women." *Gender & Society* 6 (September): 416–440.

Higginbotham, Elizabeth, and Mary Romero (eds.). 1997. *Women and Work: Exploring Race, Ethnicity, and Class.* Thousand Oaks, CA: Sage.

Higher Education Research Institute. 2002. *American Freshman Survey.* Los Angeles: University of California.

Hill, C. T., Z. Rubin, and L. A. Peplau. 1976. "Breakups Before Marriage: The End of 103 Affairs." *Journal of Social Issues* 32: 147–168.

Hill, Dana, Carol Davis, and Leann M. Tigges. 1995. "Gendering Welfare State Theory: A Cross-National Study of Women's Public Pension Quality." *Gender & Society* 9 (February): 99–119.

Hill, Jane H., and Bruce Mannheim. 1992. "Language and World View." *Annual Review of Anthropology* 21: 381–406.

Hill, Shirley A. 1994. *Managing Sickle Cell Disease in Low-Income Families.* Philadelphia, PA: Temple University Press.

Hill, Shirley A., and Joey Sprague. 1999. "Parenting in Black and White Families: The Interaction of Gender with Race and Class." *Gender & Society* 13 (August): 480–502.

Hilton, Jeanne M., Stephan Desrochers, and Edther L. Devall. 2001. "Comparison of Role Demands, Relationships, and Child Functioning in Single-Mother, Single-Father, and Intact Families." *Journal of Divorce and Remarriage* 35: 29–56.

Hirschi, Travis. 1969. *Causes of Delinquency.* Berkeley, CA: University of California Press.

Hirschman, Charles. 1994. "Why Fertility Changes." *Annual Review of Sociology* 20: 203–223.

Hochschild, Arlie. 1983. *The Managed Heart: Commercialization of Human Feelings.* Berkeley, CA: University of California Press.

Hochschild, Arlie. 1997. *The Time Bind: When Work Becomes Home and Home Becomes Work.* New York: Metropolitan Books.

Hochschild, Arlie Russell, with Anne Machung. 1989. *The Second Shift: Working Parents and the Revolution at Home.* New York: Viking Books.

Hodson, Randy. 1996. "Dignity in the Workplace Under Participative Management." *American Sociological Review* 61 (October): 719–738.

Hoecker-Drysdale, Susan. 1992. *Harriet Martineau, First Woman Sociologist.* New York: St. Martin's Press.

Hoffman, Bruce. 1995. "'Holy Terror': The Implications of Terrorism Motivated by a Religious Perspective." *Studies in Conflict and Terrorism* 18 (October–December): 271–284.

Hoffman, Lois W., and Jean D. Norris. 1979. "The Value of Children in the United States: A New Approach to the Study of Fertility." *Journal of Marriage and the Family* 41 (August): 583–569.

Hoffman, Saul D., and Greg J. Duncan. 1988. "What *Are* the Economic Consequences of Divorce?" *Demography* 25 (November): 641–645.

Hoffnung, Michele. 2004. "Wanting It All: Career, Marriage, and Motherhood During College-Educated Women's 20s." *Sex Roles* 50(May): 711–723.

Hofstadter, Richard. 1944. *Social Darwinism in American Thought.* Philadelphia, PA: University of Pennsylvania Press.

Hogan, Richard, and Carolyn C. Perrucci. 1998. "Producing and Reproducing Class and Status Differences: Racial and Gender Gaps in U.S. Employment and Retirement Income." *Social Problems* 45 (November): 528–549.

Hogan, Richard, Carolyn C. Perrucci, and Janet H. Wilmoth. 2000. "Gender Inequality in Employment and Retirement Income: Effects of Marriage, Industrial Sector, and Self-Employment." *Advances in Gender Research* 4: 27–54.

Hogan, Richard, Meesook Kim, and Carolyn C. Perrucci. "Racial Inequality in Men's Employment and Retirement Savings." *The Sociological Quarterly* 38 (Summer) 431–438.

Hollingshead, August B., and Frederick C. Redlich. 1958. *Social Class and Mental Illness: A Community Study.* New York: Wiley.

Holm, Maj. Gen. Jeanne. 1992. *Women in the Military.* Novato, CA: Presidio Press.

Holmes, David (ed.). 1997. *Virtual Politics: Identity and Community in Cyberspace.* Thousand Oaks, CA: Sage.

Holmes, Malcolm D., H. M. Hosch, H. C. Dauditel, D. A. Perez, and J. B. Graves. 1993. "Judges' Ethnicity and Minority Sentencing: Evidence Concerning Hispanics." *Social Science Quarterly* 74 (September): 496–506.

Holmes, Schuyler D. 2000. "Environmental Racism and Classism in Toxic Waste Dumping in Cleveland, Ohio." Junior thesis, Princeton University, Princeton, NJ.

Homans, George, 1961. *Social Behavior: Its Elementary Forms.* New York: Harcourt Brace Jovanovich.

Homans, George. 1974. *Social Behavior: Its Elementary Forms,* rev. ed. New York: Harcourt Brace Jovanovich.

Homes, G. P. 1988. "Chronic Fatigue Syndrome: A Working Case Definition." *Annals of Internal Medicine* 108 (March): 387–389.

Hondagneu-Sotelo, Pierrette. 1992. "Overcoming Patriarchal Constraints: The Reconstruction of Gender Relations Among Mexican Immigrant Women and Men." *Gender & Society* 6 (September): 393–415.

Hondagneu-Sotelo, Pierrette. 1994. *Gendered Transitions: The Mexican Experience of Immigration.* Berkeley, CA: University of California Press.

Hondagneu-Sotelo, Pierrette. 2001. *Doméstica: Immigrant Workers Cleaning and Caring in the Shadows of Affluence.* Berkeley, CA: University of California Press.

Hondagneu-Sotelo, Pierrette, and Ernestine Avila. 1997. "I'm Here, but I'm There': The Meanings of Latina Transnational Motherhood." *Gender & Society* 11 (October): 548–571.

Hong, Laurence K. 1978. "Risky Shift and Cautious Shift: Some Direct Evidence on the Culture-

Value Theory." *Social Psychology* 41 (December): 342–346.

hooks, bell, and Cornel West. 1991. *Breaking Bread: Insurgent Black Intellectual Life.* Boston, MA: South End Press.

Hornung, Carlton A. 1977. "Social Status, Status Inconsistency and Psychological Stress." *American Sociological Review* 42 (August): 623–638.

Horowitz, Ruth. 1995. *Teen Mothers: Citizens or Dependents?* Chicago, IL: University of Chicago Press.

House, James S. 1980. *Occupational Stress and the Mental and Physical Health of Factory Workers.* Ann Arbor, MI: Survey Research Center.

Hout, Michael. 1988. "More Universalism, Less Structural Mobility: The American Occupational Structure in the 1980s." *American Journal of Sociology* 93 (May): 1358–1400.

Hout, Michael, and Andrew Greeley. 1998. "Comment: What Church Officials' Reports Don't Show: Another Look at Church Attendance Data." *American Sociological Review* 63 (February): 113–118.

Hudson, Ken. 1999. "No Shortage of 'Nonstandard' Jobs." *Briefing Paper.* Washington, DC: Economic Policy Institute, December.

Hughes, Jean O'Gorman, and Bernice R. Sandler. 1987. "Friends Raping Friends." Washington, DC: Project on the Education and Status of Women, American Association of Colleges.

Hughes, Langston. 1967. *The Big Sea.* New York: Knopf.

Hugick, Larry. 1992. ". . . But Public Seeks Middle Ground on Abortion, Supports Pennsylvania Law." *The Gallup Poll Monthly* (January): 6–9.

Hummert, Mary Lee, Jaye L. Shaner, Teri A. Garstka, and Clark Henry. 1998. "Communication with Older Adults: The Influence of Age Stereotypes, Context, and Communicator Age." *Human Communication Research* 28 (September): 124-151.

Humphries, Drew. 1999. *Crack Mothers: Pregnancy, Drugs, and the Media.* Columbus, OH: Ohio State University Press.

Hunt, M. 1974. *Sexual Behavior in the 1970s.* Chicago, IL: Playboy Press.

Hunt, Matthew. 2000. "Status, Religion, and the 'Belief in a Just World': Comparing African Americans, Latinos, and Whites." *Social Science Quarterly* 81 (March): 323–343.

Hunter, Andrea, and James Davis. 1992. "Constructing Gender: An Exploration of Afro-American Men's Conceptualization of Manhood." *Gender & Society* 6 (September): 464–479.

Hunter, Andrea, and Sherrill L. Sellers. 1998. "Feminist Attitudes Among African American Women and Men." *Gender & Society* 12 (February): 81–99.

Hunter, Herbert M., and Sameer Y. Abraham (eds.). 1987. *Race, Class, and the World System: The Sociology of Oliver C. Cox.* New York: Monthly Review Press.

Hurlbert, Jeanne S., Valerie A. Hanes, and John J. Beggs. 2000. "Core Networks and Tie Activation: What Kinds of Routine Networks Allocate Resources in Nonroutine Situations?" *American Sociological Review* 65 (August): 598–618.

Hyde, J. S. 1984. "Children's Understanding of Sexist Language." *Developmental Psychology* 20: 697–706.

Ignatiev, Noel. 1995. *How the Irish Became White.* New York: Routledge.

Inciardi, James A. 2001. *The War on Drugs: The Continuing Saga of the Mysteries and Miseries of Intoxication, Addiction, Crime, and Public Policy,* 3rd ed. Boston, MA: Allyn and Bacon.

Inglehart, Ronald, and Wayne E. Baker. 2000. "Modernization, Cultural Change, and the Persistence of Traditional Values." *American Sociological Review* 65 (February): 19–51.

Inouye, Daniel K. 1993. "Our Future Is in Jeopardy: The Mental Health of Native American

Adolescents." *Journal of Health Care for the Poor* 4 (1): 68.

International Labour Organization. 2002. *Every Child Counts: Estimates on Child Labour.* Website: **www.ilo.org**

Irvine, Janice M. 1990. *Disorders of Desire: Sex and Gender in Modern American Sexology.* Philadelphia, PA: Temple University Press.

Irwin, Katherine. 2001. "Legitimating the First Tattoo: Moral Passage through Informal Interaction." *Symbolic Interaction* 24 (March): 49–73.

Jackson, Elton F., and Peter J. Burke. 1965. "Status and Symptoms of Stress: Additive and Interaction Effects." *American Sociological Review* 30 (August): 556–564.

Jackson, Elton F., and Richard F. Curtis. 1968. "Conceptualization and Measurement in the Study of Social Stratification." Pp. 112–149 in *Methodology in Social Research,* edited by H. M. Blalock, Jr., and A. B. Blalock. New York: McGraw-Hill.

Jackson, Linda A., Ruth E. Fleury, and Donna M. Lewandowski. 1996. "Feminism: Definitions, Support, and Correlates of Support Among Male and Female College Students." *Sex Roles* 34 (May): 687–693.

Jackson, Pamela B. 2000. "Stress and Coping Among Black Elites in Organizational Settings." Unpublished manuscript. Indiana University, Sociology Dept., Bloomington, IN.

Jackson, Pamela B., Peggy A. Thoits, and Howard F. Taylor. 1994. "The Effects of Tokenism on America's Black Elite." Paper read before the American Sociological Association, Los Angeles.

Jackson, Pamela B., Peggy A. Thoits, and Howard F. Taylor. 1995. "Composition of the Workplace and Psychological Well-Being: The Effects of Tokenism on America's Black Elite." *Social Forces* 74 (December): 543–557.

Jackson, Pamela Brayboy. 1992. "Specifying the Buffering Hypothesis: Support, Strain, and Depression." *Social Psychology Quarterly* 55: 363–378.

Jackson, Reginald M. 2002. *Curb Servin': An Analysis of Los Angeles Black Gang Culture.* Senior thesis, Princeton University, Princeton, NJ.

Jacobs, Jerry A., and Kathleen Gerson. 1998. "Who Are the Overworked Americans?" *Review of Social Economy* 56 (Winter): 442–459.

Jacobs, Jerry A., and Ronnie J. Steinberg. 1990. "Compensating Differentials and the Male-Female Wage Gap: Evidence from the New York State Comparable Worth Study." *Social Forces* 69 (December): 430–469.

Jacobs, Lawrence R., and James A. Morone. 2004. "Health and Wealth." *The American Prospect* (June): A20–21.

Jacobs, Michael P. 1997. "Do Gay Men Have a Stake in Male Privilege? The Political Economy of Gay Men's Contradictory Relationship to Feminism." Pp. 165–184 in *Homo Economics: Capitalism, Community, and Lesbian and Gay Life,* edited by Amy Gluckman and Betsy Reed. New York: Routledge.

Jacoby, Susan. 1999. "Great Sex: Our First Ever Sex Survey Is In; What Do The Numbers Say?" *Modern Maturity* 42 (September): 40.

Jang, Kerry L., W. J. Lively, and Philip A. Vernon. 1996. "Heritability of the Big Five Personality Dimensions and Their Facets: A Twin Study." *Journal of Personality* 64: 577–589.

Janis, Irving L. 1982. *Groupthink: Psychological Studies of Policy Decisions and Fiascos,* 2nd ed. Boston, MA: Houghton Mifflin.

Janus, Samuel S., and Cynthia L. Janus. 1993. *The Janus Report on Sexual Behavior.* New York: Wiley.

Jelen, Ted G. 1989. "Weaker Vessels and Helpmeets: Gender Roles, Stereotypes, and Attitudes Toward Female Ordination." *Social Science Quarterly* 70 (September): 579–585.

Jencks, Christopher. 1993. *Rethinking Social Policy: Race, Poverty, and the Underclass.* New York: Harper Collins.

Jencks, Christopher, and Meredith Phillips (eds.). 1998. *The Black-White Test Score Gap.* Washington, DC: Brookings Institution Press.

Jencks, Christopher, Marshall Smith, Henry Ackland, Mary Jo Bane, David Cohen, Herbert Gintis, Barbara Heyns, and Stephan Michelson. 1972. *Inequality: A Reassessment of the Effect of Family and Schooling in America.* New York: Basic Books.

Jencks, Christopher, Susan Bartlett, Mary Corcoran, James Crouse, David Eaglesfield, Gregory Jackson, Kent McClelland, Peter Mueser, Michael Olneck, Joseph Schwartz, Sherry Ward, and Jill Williams. 1979. *Who Gets Ahead? The Determinants of Economic Success in America.* New York: Basic Books.

Jenkins, J. Craig. 1983. "The Transformation of a Constituency into a Movement: Farm Worker Organizing in California." Pp. 52–71 in *Social Movements of the Sixties and Seventies,* edited by Jo Freeman. New York: Longman.

Jenkins, Philip. 2002. *The Next Christendom: The Coming of Global Christianity.* New York: Oxford University Press.

Jenness, Valerie, and Ryken Grattet. 2004. *Making Hate a Crime: from Social Movement to Law Enforcement.* NY: Russell Sage Foundation.

Jenness, Valerie, and Kendal Broad. 1997. *Hate Crimes: New Social Movements and the Politics of Violence.* New York: Aldine de Gruyter.

Jenness, Valerie, and Kendal Broad. 2002. *Hate Crimes: New Social Movements and the Politics of Violence.* New York: Aldine de Gruyter.

Jennings, M. K., and R. G. Niemi. 1974. *The Political Character of Adolescence.* Princeton, NJ: Princeton University Press.

Jensen, Arthur R. 1980. *Bias in Mental Testing.* New York: Free Press.

Johnson, Kim K. P. 1995. "Attributions About Date Rape: Impact of Clothing, Sex, Money Spent, Date Type, and Perceived Similarity." *Family and Consumer Sciences Research Journal* 23 (March): 292–310.

Johnson, Norris R., James G. Stember, and Deborah Hunter. 1977. "Crowd Behavior as Risky Shift: A Laboratory Experiment." *Sociometry* 40 (2): 183–187.

Johnston, David Cay. 2000. "Corporations' Taxes Are Falling Even as Individual Burden Rises." *The New York Times* (February 20): A1ff.

Johnston, Hank, and Bert Klandermans. 1995. "The Cultural Analysis of Social Movements." Pp. 3–24 in *Social Movements and Culture,* Vol. 4 in *Social Movements: Protests and Contention,* edited by Hank Johnston and Bert Klandermans. Minneapolis, MN: University of Minnesota Press.

Johnstone, Ronald L. 1992. *Religion in Society: A Sociology of Religion,* 4th ed. Englewood Cliffs, NJ: Prentice Hall.

Jones, Edward E., and Keith E. Davis. 1965. "From Acts to Dispositions: The Attribution Process in Person Perception." Pp. 219–266 in *Advances in Experimental Social Psychology,* edited by Leonard Berkowitz. New York: Academic Press.

Jones, Edward E., Amerigo Farina, Albert H. Hastorf, Hazel Markus, Dale T. Miller, and Robert A. Scott. 1986. *Social Stigma: The Psychology of Marked Relationships.* New York: W. H. Freeman.

Jones, James. 1993. *Bad Blood: The Tuskegee Syphilis Experiment,* rev. ed. New York: Free Press.

Jones, James M. 1997. *Prejudice and Racism,* 2nd ed. New York: McGraw-Hill.

Jones, Jeffery A. 2003. "Health Care Costs, Access Viewed as Most Urgent U.S. Health Problems." *The Gallup Poll.* Princeton, NJ: Gallup Organization. Website: **www.gallup.com**

Jones, Ray, Audrey J. Murrell, and Jennifer Jackson. 1999. "Pretty Versus Powerful in the Sports

Pages: Print Media Coverage of U.S. Women's Olympic Gold Medal Winning Teams." *Journal of Sport and Social Issues* 23 (May): 183–192.

Jones, Steven G. 1997. *Virtual Culture: Identity and Communication in Cyber-Society.* Thousand Oaks, CA: Sage.

Jones, T. F. 2000. "Mass Psychogenic Illness Attributed to Toxic Exposure at a High School." *The New England Journal of Medicine* 342 (22): 96–101.

Jones, Timothy, and George Gallup, Jr. 2000. *The New American Spirituality: Finding God in the Twenty-First Century.* Colorado Springs, CO: Chariot Victor.

Jordan, June. 1981. *Civil Wars.* Boston, MA: Beacon Press.

Jordan, Winthrop D. 1968. *White Over Black: American Attitudes Toward the Negro 1550–1812.* Chapel Hill, NC: University of North Carolina Press.

Jordan, Winthrop D. 1969. *The White Man's Burden: Historical Origins of Racism in the United States.* New York: Oxford University Press.

Joseph, Janice. 1997. "Fear of Crime Among Black Elderly." *Journal of Black Studies* 27 (May): 698–717.

Jucha, Robert. 2002. *Terrorism.* Belmont, CA: Wadsworth.

Kall, Denise. 2002. "Smoking Gun or Organizational Haze? The Tobacco Companies' Response to Cancer Research." Senior thesis, University of Delaware, Newark, DE.

Kalleberg, Arne L., Barbara F. Reskin, and Ken Hudson. 2000. "Bad Jobs in America: Standard and Nonstandard Employment Relations and Job Quality in the United States." *American Sociological Review* 65 (April): 256–278.

Kalmijn, Matthijs. 1991. "Status Homogamy in the United States." *American Journal of Sociology* 97 (September): 496–523.

Kalof, Linda. 1999. "The Effects of Gender and Music Video Imagery on Sexual Attitudes." *Journal of Social Psychology* 139 (June): 378–385.

Kamin, Leon J. 1974. *The Science and Politics of IQ.* Potomac, MD: Lawrence Erlbaum.

Kamin, Leon J. 1995. "Behind the Curve." *Scientific American* 272 (February): 99–103.

Kanter, Rosabeth Moss. 1977. *Men and Women of the Corporation.* New York: Basic Books.

Kaplan, David A. 1999. *The Silicon Boys and the Valley of Dreams.* New York: William Morrow.

Kaplan, Elaine Bell. 1996. *Not Our Kind of Girl: Unraveling the Myths of Black Teenage Motherhood.* Berkeley, CA: University of California Press.

Kaplan, John W., and George A. Kaplan. 1997. "Understanding How Inequality in the Distribution of Income Affects Health." *Journal of Health Psychology* 2 (July): 297–314.

Kaplan, Richard L. 2000. "Blackface in Italy: Cultural Power among Nations in an Era of Globalization." Paper presented at the annual meeting of the American Sociological Association, Washington, DC.

Kasarda, Jack. 1999. "Industrial Restructuring and the Changing Location of Jobs." In *State of the Union: America in the 1990s: Economic Trends,* edited by Reynolds Farley. New York: Russell Sage Foundation.

Kassell, Scott. 1995. 'Afrocentrism and Eurocentrism in African American Magazine Ads.' Unpublished senior thesis, Princeton University, Princeton, NJ.

Katz, I., J. Wackenhut, and R. G. Hass. 1986. "Racial Ambivalence, Value Duality, and Behavior." Pp. 35–60 in *Prejudice, Discrimination, and Racism,* edited by J. F. Dovidio and S. L. Gaertner. New York: Academic Press.

Katz, Michael. 1987. *Reconstructing American Education.* Cambridge, MA: Harvard University Press.

Katz, Stephen. 2001–2002. "Growing Older without Aging? Positive Aging, Anti-Ageism, and Anti-Aging." *Generations* 25 (Winter): 27–32.

Kaufman, Gayle. 1999. "The Portrayal of Men's Family Roles in Television Commercials." *Sex Roles* 41 (September): 439–458.

Kaufman, Gayle, Hiromi Taniguchi, and Glen Elder, Jr. 2001. "Gender Role Attitudes and Marital Happiness in Later Life." Paper presented at the annual meeting of the Southern Sociological Society, Atlanta, GA.

Kearl, Michael C. 1989. *Endings: A Sociology of Death and Dying.* New York: Oxford University Press.

Keil, Thomas J., and Gennaro F. Vito. 1995. "Factors Influencing the Use of 'Truth in Sentencing' Law in Kentucky Murder Cases: A Research Note." *American Journal of Criminal Justice* 20 (Fall): 105–111.

Kelley, Maryellen R. 1990. "New Process Technology, Job Design, and Work Organization: A Contingency Model." *American Sociological Review* 55 (April): 191–208.

Kelly, J. R., J. W. Jackson, and S. L. Huston-Comeaux. 1999. "The Effects of Time Pressure and Task Differences on Influence Modes and Accuracy in Decision-Making Groups." *Personality and Social Psychology Bulletin* 23: 10–22.

Kelly, Joan B., and Judith S. Wallerstein. 1976. "The Effects of Parental Divorce: Experiences of the Child in Early Latency." *American Journal of Orthopsychiatry* 46 (January): 20–32.

Kelly, William R., and Larry Isaac. 1984. "The Rise and Fall of Urban Racial Violence in the U.S.: 1948–1979." *Research in Social Movements: Conflict and Change* 7: 203–230.

Kendall, Lori. 2002. "'Oh No! I'm a Nerd!': Hegemonic Masculinity on an Online Forum." *Gender & Society* 14 (April): 256–274.

Kennedy, Randall. 2003. *Interracial Intimacies: Sex, Marriage, Identity and Adoption.* New York: Pantheon.

Kennedy, Robert E. 1989. *Life Choices: Applying Sociology,* 2nd ed. New York: Holt, Rinehart, and Winston.

Kephart, W. H. 1993. *Extraordinary Groups: An Examination of Unconventional Life,* rev. ed. New York: St. Martin's Press.

Kephart, William M. 1987. *Extraordinary Groups: An Examination of Unconventional Life Styles.* New York: St. Martin's Press.

Kerbicov, Boris. 1998. "Computer Use by Race and Social Class." Junior research project, Princeton University, Princeton, NJ.

Kerckhoff, Alan C., and Kurt W. Back. 1968. *The June Bug: A Study of Hysterical Contagion.* New York: Appleton-Century-Crofts.

Kerr, N. L. 1992. "Issue Importance and Group Decision Making." Pp. 68–88 in *Group Process and Productivity,* edited by S. Worchel, W. Wood, and J. A. Simpson. Newbury Park, CA: Sage.

Kessler, Suzanne J. 1990. "The Medical Construction of Gender: Case Management of Intersexed Infants." *Signs* 16 (Autumn): 3–26.

Keyes, Corey L. M. 2002. "The Mental Health Continuum: From Languishing to Flourishing in Life." *Journal of Health and Social Behavior* 43 (June): 207–222.

Khanna, Nikki D., Cherise Harris, and Rana Cullers. 1999. "Attitudes toward Interracial Dating." Paper presented at the annual meeting of the Society for the Study of Social Problems, Chicago, IL.

Khashan, Hilal, and Lina Kreidie. 2001. "The Social And Economic Correlates of Islamic Religiosity." *World Affairs* 1654 (Fall): 83–96.

Kilborn, Peter. 1990. "Workers Using Computers Find a Supervisor Inside." *The New York Times* (Dec. 23): A1ff.

Kilbourne, Barbara Stanek, George Farkas, Kurt Beron, Dorothea Weir, and Paula England. 1994. "Characteristics on the Wages of White

Women and Men." *American Journal of Sociology* 100 (November): 698–719.

Killian, Lewis. 1975. *The Impossible Revolution, Phase 2: Black Power and the American Dream.* New York: Random House.

Kim, Elaine H. 1993. "Home Is Where the *Han* Is: A Korean American Perspective on the Los Angeles Upheavals." Pp. 215–235 in *Reading Rodney King/Reading Urban Uprising,* edited by Robert Gooding-Williams. New York: Routledge.

Kim, Ik Ki, and Daisaku Maeda. 2001. "A Comparative Study on Sociodemographic Changes and Long-term Health Care Needs of the Elderly in Japan and South Korea." *Journal of Cross-Cultural Gerontology* 16 (September): 237–255.

Kim, Kwang Chung, and Won Moo Hurh. 1988. "The Burden of Double Roles: Korean Wives in the USA." *Ethnic and Racial Studies* 11 (April): 151–176.

Kimmel, Michael. 2000. "Saving the Males: The Sociological Implications of the Virginia Military Institute and The Citadel." *Gender & Society* 14: 494–516.

Kimmel, Michael S., and Michael A. Messner. 2003. *Men's Lives,* 6th ed. Boston, MA: Allyn and Bacon.

Kingson, Eric R., and Jill Quadagno. 1995. "Social Security: Marketing Radical Reform." *Generations* 19 (Fall): 43–49.

Kingson, Eric R., and John B. Williamson. 1991. "Generational Equity or Privatization of Social Security." *Society* 28 (September–October): 38–41.

Kingson, Eric R., and John B. Williamson. 1993. "The Generational Equity Debate: A Progressive Framing of a Conservative Issue." *Journal of Aging and Social Policy* 5: 31–53.

Kinsey, Alfred C., W. B. Pomeroy, C. E. Martin, and P. H. Gebbard. 1948. *Sexual Behavior in the Human Male.* Philadelphia, PA: W. B. Saunders.

Kinsey, Alfred C., W. B. Pomeroy, C. E. Martin, and P. H. Gebbard. 1953. *Sexual Behavior in the Human Female.* Philadelphia, PA: W. B. Saunders.

Kirby, Douglas. 1997. "No Easy Answers: Research Findings on Programs to Reduce Teen Pregnancy." Washington, DC: National Campaign to Prevent Teen Pregnancy. Website: **www.teenpregnancy.org**

Kitano, Harry. 1976. *Japanese Americans: The Evolution of a Subculture,* 2nd ed. New York: Prentice Hall.

Kitsuse, John I. 1980. "Coming Out All Over: Deviants and the Politics of Social Problems." *Social Problems* 28 (October): 1–13.

Kitsuse, John I., and Aaron V. Cicourel. 1963. "A Note on the Uses of Official Statistics." *Social Problems* 11 (Fall): 131–139.

Kivett, V. R. 1991. "Centrality of the Grandfather Role Among Older Rural Black and White Men." *Journal of Gerontology* 46: 250–258.

Klapp, Orin. 1972. *Currents of Unrest: Introduction to Collective Behavior.* New York: Holt, Rinehart, and Winston.

Klausner, Samuel Z., and Edward F. Foulks. 1982. *Eskimo Capitalists: Oil, Alcohol, and Politics.* Totowa, NJ: Allanheld.

Kleinfeld, Judith S. 2002. "The Small World Problem." *Society* 39:61-66.

Kluegel, J. R., and Lawrence Bobo. 1993. "Dimensions of Whites' Beliefs About the Black-White Socioeconomic Gap." Pp. 127–147 in *Race and Politics in American Society,* edited by P. Sniderman, P. Tetlock, and E. Carmines. Stanford, CA: Stanford University Press.

Kmec, Julie A. 2003. "Minority Job Composition and Wages." *Social Problems* 50 (February): 38–59.

Kniss, Fred. 2003. "Church and State." *Contexts* 2 (Spring): 62–63.

Kniss, Fred. 2004. "Asian Spirituality." *Contexts* 3 (Spring): 5–6.

Knoke, David. 1992. *Political Networks: The Structural Perspective*. New York: Cambridge University Press.

Kochen, M. (ed.). 1989. *The Small World*. Norwood, NJ: Ablex Press.

Kocieniewski, David, and Robert Hanley. 2000. "Racial Profiling Was the Routine, New Jersey Finds." *The New York Times* (November 28): 1.

Kohlberg, Lawrence. 1969. "Stage and Sequence: The Cognitive-Developmental Approach to Socialization." Pp. 347–480 in *Handbook of Socialization and Research*, edited by D. A. Goslin. Chicago, IL: Rand McNally.

Komter, Aafke. 1989. "Hidden Power in Marriage." *Gender & Society* 3 (June): 187–216.

Kovel, Jonathan. 1970. *White Racism: A Psychohistory*. New York: Pantheon.

Krasnodemski, Memory. 1996. "Justified Suffering: Attribution Theory Applied to Perceptions of the Poor in America." Unpublished senior thesis, Princeton University, Princeton, NJ.

Krauss, Celene. 1994. "Women of Color on the Front Line." Pp. 256–271 in *Unequal Protection: Environmental Justice and Communities of Color*, edited by Robert D. Bullard. San Francisco: Sierra Club Books.

Kray, Susan. 1993. "Orientalization of an 'Almost White' Woman: The Interlocking Effects of Race, Class, Gender and Ethnicity in American Mass Media." *Critical Studies in Mass Communication* 10 (December): 349–366.

Kreps, Gary A., 1994. "Disasters as Nonroutine Social Problems." Paper presented at the International Sociological Association, Madrid, Spain.

Kuhn, Harold W., and Sylvia Nasar. 2002. *The Essential John Nash*. Princeton, NJ: Princeton University Press.

Kurz, Demie. 1989. "Social Science Perspectives on Wife Abuse: Current Debates and Future Directions." *Gender & Society* 3 (December): 489–505.

Kurz, Demie. 1995. *For Richer for Poorer: Mothers Confront Divorce*. New York: Routledge.

Kurzman, Charles. 1996. "Structural Opportunity and Perceived Opportunity in Social Movement Theory: The Iranian Revolution of 1979." *American Sociological Review* 61 (February): 153–170.

Kutscher, Ronald E., 1995. "Summary of BLS Projections to 2005." *Monthly Labor Review* 118 (November): 3–9.

Ladner, Joyce A. 1986. "Teenage Pregnancy: Implications for Black Americans." Pp. 65–84 in *The State of Black America 1986*, edited by James D. Williams. New York: National Urban League.

Ladner, Joyce A. 1995 [1971]. *Tomorrow's Tomorrow: The Black Woman*, rev. ed. Garden City, NY: Doubleday.

Laflamme, Darquise, Andrée Pomerleau, and Gérard Macuit. 2002. "A Comparison of Fathers' and Mothers' Involvement in Childcare and Stimulation Behaviors During Free-Play with their Infants at 9 and 15 Months." *Sex Roles* 47 (December): 507–518.

LaFrance, Marianne, 2002. "Smile Boycotts and Other Body Politics." *Feminism & Psychology* 12 (August): 319–323.

Lamanna, Mary Ann, and Agnes Riedman. 2003. *Marriage and Families: Making Choices in a Diverse Society*. Belmont, CA: Wadsworth.

Lamb, Michael. 1998. "Cybersex: Research Notes on the Characteristics of the Visitors to Online Chat Rooms." *Deviant Behavior* 19 (April–June): 121–135.

Lamont, Michele. 2000. "Meaning-Making in Cultural Sociology: Broadening Our Agenda." *Contemporary Sociology* 29 (July): 602–607.

Lampard, Richard, and Kay Peggs. 1999. "Repartnering: The Relevance of Parenthood and Gender to Cohabitation and Remarriage among the Formerly Married." *British Journal of Sociology* 50 (September): 443–465.

Landry, Bart. 1987. *The New Black Middle Class*. Berkeley, CA: University of California Press.

Lang, Kurt, and Gladys Lang. 1961. *Collective Dynamics*. New York: Thomas Y. Crowell.

Langhinrichsen-Rohling, Jennifer, Peter Lewinsohn, Paul Rohde, John Seeley, Candice M. Monson, Kathryn A. Meyer, and Richard Langford. 1998. "Gender Differences in the Suicide-Related Behaviors of Adolescents and Young Adults." *Sex Roles* 39 (December): 839–854.

Langman, Lauren, and Douglas Morris. 2002. "Internetworked Social Movements: The Promises and Prospects for Global Justice." Paper presented at the International Sociological Association, Brisbane, Australia.

Larana, Enrique, Hank Johnston, and Joseph R. Gusfield (eds.). 1994. *New Social Movements: From Ideology to Identity*. Philadelphia, PA: Temple University Press.

Larrison, Christopher R., Larry Nackerud, and Ed Risler. 2001. "A New Perspective on Families that Receive Temporary Assistance for Needy Families (TANF)." *Journal of Sociology and Social Welfare* 28 (September): 49–69.

Lau, Bonnie. 2002. *Stereotype Threat: Minority Education, Intelligence Theory, Affirmative Action, and an Empirical Study of Women in the Quantitative Domain*. Senior thesis, Princeton University, Princeton, NJ.

Lau, Bonnie, and Howard F. Taylor. 2003. "Ethnic and Gender Stereotype Threat Among Asian Women in Quantitative Test Performance." Manuscript, Department of Sociology, Princeton University.

Laumann, Edward O., John H. Gagnon, Robert T. Michael, and Stuart Michaels. 1994. *The Social Organization of Sexuality: Sexual Practices in the United States*. Chicago, IL: University of Chicago Press.

Laveist, Thomas A., and Amani Nuru-Jeter. 2002. "Is Doctor-Patient Race Concordance Associated with Greater Satisfaction with Care?" *Journal of Health and Social Behavior* 43 (September): 296–306.

Lavelle, Louis. 2001. "Executive Pay." *Business Week*, (April 16). Website: **www.businessweek .com**

Laws, Judith Long, and Pepper Schwartz. 1977. *Sexual Scripts: The Social Construction of Female Sexuality*. Hinsdale, IL: Dryden Press.

Lawson, Ronald. 1983. "A Decentralized But Moving Pyramid: The Evolution and Consequences of the Structure of the Tenant Movement." Pp. 119–132 in *Social Movements of the Sixties and Seventies*, edited by Jo Freeman. New York: Longman.

Leacock, Eleanor. 1978. "Women's Status in Egalitarian Society." *Contemporary Anthropology* 19: 247–275.

LeBon, Gustave. 1895 [1960]. *The Crowd*. New York: Viking Press.

Lee, Matthew T., and M. David Ermann. 1999. "Pinto 'Madness' as a Flawed Landmark Narrative: An Organizational and Network Analysis." *Social Problems* 46 (February): 30–47.

Lee, Sharon M. 1993. "Racial Classification in the U.S. Census: 1890–1990." *Ethnic and Racial Studies* 16 (1): 75–94.

Lee, Sharon M. 1994. "Poverty and the U.S. Asian Population." *Social Science Quarterly* 75 (September): 541–559.

Lee, Stacey J. 1996. *Unraveling the "Model Minority" Stereotype: Listening to Asian American Youth*. New York: Teacher's College Press.

Legge, Jerome S., Jr. 1993. "The Persistence of Ethnic Voting: African Americans and Jews in the 1989 New York Mayoral Campaign." *Contemporary Jewry* 14: 133–146.

Lehmann, Nicholas. 1999. *The Big Test: The Secret History of the American Meritocracy*. New York: Farrar, Straus, and Giroux.

Leidner, Robin. 1993. *Fast Food, Fast Talk: Service Work and the Routinization of Everyday Life*. Berkeley, CA: University of California Press.

Lemert, Edwin M. 1972. *Human Deviance, Social Problems, and Social Control*. Englewood Cliffs, NJ: Prentice-Hall.

Lempert, Lora Bex, and Marjorie DeVault. 2000. "Guest Editors' Introduction: Special Issue on Emergent and Reconfigured Forms of Family Life." *Gender & Society* 14 (February): 6–10.

Lenin, Vladimir. 1939. *Imperialism, the Highest Stage of Capitalism*. New York: International Publishers.

Lennings, C. J. 2000. "Optimism, Satisfaction, and Time Perspective in the Elderly." *International Journal of Aging and Human Development* 51: 167–181.

Lenski, Gerhard, Jean Lenski, and Patrick Nolan. 1998. *Human Societies: An Introduction to Macro-Sociology*, 8th ed. New York: McGraw-Hill.

Lenski, Gerhard, Jean Lenski, and Patrick Nolan. 2001. *Human Societies: An Introduction to Macro-Sociology*, 9th ed. New York: McGraw-Hill.

Lerman, Robert I., and Theodora J. Ooms. 1993. *Young Unwed Fathers: Changing Roles and Emerging Policies*. Philadelphia, PA: Temple University Press.

Lersch, Kim-Michelle, and Joe R. Feagin. 1996. "Violent Police-Citizen Encounters: An Analysis of Major Newspaper Accounts." *Critical Sociology* 22: 29–49.

Leslie, Michael. 1995. "Slow Fade to ?: Advertising in Ebony Magazine, 1957–1989." *Journalism and Mass Communication Quarterly* 72 (Summer): 426–435.

Levin, J., and W. Levin. 1980. *Ageism: Prejudice and Discrimination Against the Elderly*. Belmont, CA: Wadsworth.

Levine, Adeline. 1982. *Love Canal: Science, Politics, and People*. Lexington, MA: Lexington Books.

Levine, Felice J., and Katherine J. Rosich. 1996. *Social Causes of Violence: Crafting a Science Agenda*. Washington, DC: American Sociological Association.

Levine, John M., and Richard L. Moreland. 1998. "Small Groups." Pp. 415–469 in *The Handbook of Social Psychology*, 4th ed., edited by Daniel T. Gilbert, Susan T. Fiske, and Gardner Lindzey. New York: Oxford University/McGraw-Hill.

Levine, Lawrence. 1984. "William Shakespeare and the American People: A Study in Cultural Transformation." *American Historical Review* 89 (1): 34–66.

Levy, Marion J. 1949. *The Structure of Society*. Princeton, NJ: Princeton University Press.

Lewin, Tamar. 2002. "Study Links Working Mothers to Slower Learning." *The New York Times* (July 17): A14.

Lewis, Amanda E. 2003. *Race in the Schoolyard: Negotiating the Color Line in Classrooms and Communities*. New Brunswick, NJ: Rutgers University Press.

Lewis, Dan A., Amy Bush Stevens, and Kristen Shook Slack. 2002. *Illinois Families Study, Welfare Reform in Illinois: Is the Moderate Approach Working?* Evanston, IL: Institute for Poverty Research, Northwestern University.

Lewis, David Levering. 1993. *W. E. B. DuBois: Biography of a Race*. New York: Henry Holt.

Lewis, George H. 1989. "Rats and Bunnies: Core Kids in an American Mall." *Adolescence* 24 (Winter): 881–889.

Lewis, Oscar. 1960. *Five Families: Mexican Case Studies in the Culture of Poverty*. New York: Basic Books.

Lewis, Oscar. 1966. "The Culture of Poverty." *Scientific American* 215 (October): 19–25.

Lewontin, Richard. 1996. *Human Diversity.* New York: W. H. Freeman.

Lewontin, Richard C. 1970. "Race and Intelligence." *Bulletin of the Atomic Scientists* 26: 2–8.

Lichter, Daniel T., Deborah Roempke Graefe, and J. Brian Brown. 2003. "Is Marriage a Panacea? Union Formation among Economically Disadvantaged Unwed Mothers." *Social Problems* 50 (February): 60–86.

Lieberson, Stanley. 1980. *A Piece of the Pie: Black and White Immigrants Since 1880.* Berkeley, CA: University of California Press.

Lieberson, Stanley, and Arnold Silverman. 1965. "The Precipitant and Underlying Conditions of Race Riots." *American Sociological Review* 30 (December): 887–898.

Liebman, Robert C., and Robert Wuthnow. 1983. *The New Christian Right.* Hawthorne, NY: Aldine de Gruyter.

Light, Ivan, Rebecca Kim, and Connie Hum. 1998. "Globalization, Vacancy Chains, or Migration Networks?: Immigrant Employment and Income in Greater Los Angeles, 1970–1990." Paper presented at the International Sociological Association, Montreal.

Lin, Carolyn A. 1998. "Uses of Sex Appeals in Prime-Time Television Commercials." *Sex Roles* 38 (March): 461–475.

Lin, Nan. 1989. "The Small World Technique as a Theory Construction Tool." Pp. 231–238 in *The Small World,* edited by M. Kochen. Norwood, NJ: Ablex Press.

Lincoln, C. Eric, and Lawrence H. Mamiya. 1990. *The Black Church in the African-American Experience.* Durham, NC: Duke University Press.

Linton, Ralph. 1937. *The Study of Man.* New York: Appleton-Century.

Litt, Jacquelyn S., and Mary Zimmerman. 2003. "Global Perspectives and Carework: An Introduction." *Gender & Society* 17 (April): 156–165.

Lloyd, Barbara, Kevin Lucas, and Madeline Fernbach. 1997. "Adolescent Girls' Constructions of Smoking Identities: Implications for Health Promotion." *Journal of Adolescence* 20 (February): 43–56.

Lo, Clarence. 1990. *Small Property Versus Big Government: Social Origins of the Property Tax Revolt.* Berkeley, CA: University of California Press.

Locklear, Erin M. 1999. "Where Race and Politics Collide: The Federal Acknowledgement Process and Its Effects on Lumsee and Pequot Indians." Senior thesis, Princeton University, Princeton, NJ.

Lofland, John. 1985. *Protest: Studies of Collective Behavior and Social Movements.* New Brunswick, NJ: Transaction.

Logan, John, and Harvey L. Molotch. 1987. *Urban Fortunes: The Political Economy of Place.* Berkeley, CA: University of California Press.

Logio, Kim. 1998. "Here's Looking At You, Kid: Race, Gender and Health Behaviors Among Adolescents." Ph.D. diss., University of Delaware, Newark, DE.

Lombardo, William K., Gary A. Cretser, and Scott C. Roesch. 2001. "For Crying Out Loud—The Differences Persist into the '90s." *Sex Roles* 45 (December): 529–547.

Long, Theodore E., and Jeffrey K. Hadden. 1983. "Religious Conversion and the Concept of Socialization: Integrating the Brainwashing and Drift Models." *Journal of the Scientific Study of Religion* 22 (March): 1–14.

Longman, Jere. 1996. "Left Alone in His Frigid House, DuPont Heir Is Seized by Police." *The New York Times* (January 29): A1.

Loo, Robert, and Karran Thorpe. 1998. "Attitudes Toward Women's Roles in Society: A Replication After 20 Years." *Sex Roles* 39 (December): 903–912.

Lopata, Helene Z., and Barrie Thorne. 1978. "On the Term 'Sex Roles.'" *Signs* 3 (Spring): 718–721.

Lorber, Judith. 1994. *Paradoxes of Gender.* New Haven, CT: Yale University Press.

Lord, Walter. 1956. *A Night to Remember.* New York: Bantam.

Loredo, Carren, Anne Reid, and Kay Deaux. 1995. "Judgments and Definitions of Sexual Harassment by High School Students." *Sex Roles* 32 (January): 29–45.

Lorenz, Konrad. 1966. *On Aggression.* New York: Harcourt.

Louie, Miriam Ching Yoon. 2001. *Sweatshop Warriors: Immigrant Women Workers Take on the Global Factory.* Cambridge, MA: South End Press.

Lovejoy, M. 2001. "Disturbances in the Social Body: Differences in Body Image and Eating Problems among African American and White Women." *Gender & Society* 15 (Part 2): 239–261.

Lucal, Betsy. 1994. "Class Stratification in Introductory Textbooks: Relational or Distributional Models?" *Teaching Sociology* 22 (April): 139–150.

Lucas, Samuel R. 1999. *Tracking Inequality: Stratification and Mobility in American High Schools.* New York: Teachers College Press.

Luckenbill, David F. 1986. "Deviant Career Mobility: The Case of Male Prostitutes." *Social Problems* 33 (April): 283–296.

Luker, Kristin. 1975. *Taking Chances.* Berkeley,CA: University of California Press.

Luker, Kristin. 1984. *Abortion and the Politics of Motherhood.* Berkeley, CA: University of California Press.

Luker, Kristin. 1996. *Dubious Conceptions: The Politics of Teenage Pregnancy.* Cambridge, MA: Harvard University Press.

Lyons, Linda. 2002. "Church Reform: Women in the Clergy." *The Gallup Poll.* Princeton, NJ: Gallup Organization. Website: **www.gallup.com**

Lyons, Linda. 2002. "Teen Attitudes Contradict Sex-Crazed Stereotype." *The Gallup Poll.* (January 29). Website: **www.gallup.com**

MacCoun, Robert J., and Peter Reuter. 2001. *Drug War Heresies: Learning from Other Vices, Times, and Places.* New York: Cambridge University Press.

MacDonald, William L. 1998. "The Difference between Blacks' and Whites' Attitudes toward Voluntary Euthanasia." *Journal for the Scientific Study of Religion* 37 (September): 411–426.

Machel, Graca. 1996. *Impact of Armed Conflict on Children.* New York: UNICEF/United Nations.

MacKay, N. J., and K. Covell. 1997. "The Impact of Women in Advertisements on Attitudes Toward Women." *Sex Roles* 36 (May): 573–583.

MacKinnon, Catherine. 1983. "Feminism, Marxism, Method, and the State: An Agenda for Theory." *Signs* 7 (Spring): 635–658.

MacKinnon, Neil J., and Tom Langford. 1994. "The Meaning of Occupational Prestige Scores: A Social Psychological Analysis and Interpretation." *Sociological Quarterly* 35 (May): 215–245.

Mackintosh, N. J. 1995. *Cyril Burt: Fraud or Framed?* New York: Oxford University Press.

MacLeod, Jay. 1995. *Ain't No Makin' It: Aspirations and Attainment in a Low-Income Neighborhood.* Boulder, CO: Westview Press.

Macmillan, Jeffrey. 1996. "Santa's Sweatshop." *U.S. News & World Report* (December 16): 50ff.

MacPherson, Myra. 1984. *The Long Time Passing: Vietnam & the Haunted Generation.* New York: Doubleday.

Madriz, Esther. 1997. *Nothing Bad Happens to Good Girls: Fear of Crime in Women's Lives.* Berkeley, CA: University of California Press.

Majete, Clayton A. 1999. "Family Relationships and the Interracial Marriage." Paper presented at the annual meeting of the American Sociological Association, Chicago, IL.

Malat, Jennifer. 2001. "Social Distance and Patients' Rating of Healthcare Providers." *Journal of Health and Social Behavior* 42 (December): 36–372.

Malcomson, Scott L. 2000. *The American Misadventure of Race.* New York: Farrar, Straus, and Giroux.

Maldonado, Lionel A. 1997. "Mexicans in the American System: A Common Destiny." In *Ethnicity in the United States: An Institutional Approach,* edited by William Velez. Bayside, NY: General Hall.

Maldonado, Lionel A., and Charles V. Willie. 1996. "Developing A 'Pipeline' Recruitment Program for Minority Faculty." Pp. 330–371 in *Educating a New Majority: Transforming America's Educational System for Diversity,* edited by Laura I. Rendon and Richard O. Hope. San Francisco: Jossey-Bass.

Malkin, Amy R., Kimberlie Wornian, and Joan C. Chrisler. 1999. "Women and Weight: Gendered Messages on Magazine Covers." *Sex Roles* 40 (April): 647–655.

Malthus, Thomas Robert. 1798. *First Essay on Population, 1798.* London: Macmillan.

Mandel, Ruth, and Debra Dodson. 1992. "Do Women Officeholders Make a Difference?" Pp. 144–177 in *American Women, 1992–93: A Status Report,* edited by Paula Ries and Anne Stone. New York: W. W. Norton.

Mander, Jerry, and Edward Goldsmith (eds.). 1996.*The Case against the Global Economy: And For a Turn toward the Local.* San Francisco: Sierra Club Books.

Manning, Wendy, Monica Longmore, and Peggy Giordano. 2000. "The Relationship Context of Contraceptive Use at First Intercourse." *Family Planning Perspectives* 32 (May–June): 104–110.

Manning, Winton H., and Rex Jackson. 1984. "College Entrance Examinations: Objective Selection or Gatekeeping for the Economically Privileged." Pp. 189–220 in *Perspectives on Bias in Mental Testing,* edited by Cecil R. Reynolds and Robert T. Brown. New York: Plenum.

Mansfield, Phyllis-Kernoff, Patricia Barthalow-Koch, Julie Henderson, Judith R. Vicary, Margaret Cohn, and Elaine W. Young. 1991. "The Job Climate for Women in Traditionally Male Blue-Collar Occupations." *Sex Roles* 25 (July): 63–79.

Mantsios, Gregory. 2001. "Media Magic: Making Class Invisible." Pp. 332–342 in *Race, Class, and Gender: An Anthology,* edited by Margaret L. Andersen and Patricia Hill Collins. Belmont, CA: Wadsworth.

Marciniak, Liz-Marie. 1998. "Adolescent Attitudes Toward Victim Precipitation of Rape." *Violence and Victims* 12 (Fall): 287–300.

Marcuse, Herbert. 1964. *One-Dimensional Man.* Boston, MA: Beacon Press.

Margolin, Leslie. 1992. "Deviance on Record: Techniques for Labeling Child Abusers in Official Documents." *Social Problems* 39 (February): 58–70.

Margolis, Diane Rothbard. 1996. "Victimized Daughters: Incest and the Development of the Female Self." *Gender & Society* 10 (August): 488–489.

Markowitz, Fred E. 2001. "Modeling Processes in Recovery from Mental Illness: Relationships Between Symptoms, Life Satisfaction, and Self-Concept." *Journal of Health and Social Behavior* 42 (March): 64–79.

Marks, Carole. 1989. *Farewell, We're Good and Gone: The Great Black Migration.* Bloomington, IN: Indiana University Press.

Marks, Carole, and Deana Edkins. 1999. *The Power of Pride: Stylemakers and Rulebreakers of the Harlem Renaissance.* New York: Crown.

Markson, Elizabeth W., and Carol A. Taylor. 2000. "The Mirror Has Two Faces." *Ageing and Society* 20 (March): 137–160.

Marlowe and Company. 1995. "How Schools Shortchange Girls." Research report. New York: Marlowe.

Marmot M. 2001. "Inequalities in Health." *New England Journal of Medicine* 345 (July 12): 134–136.

Marshall, Susan. 1997. *Splintered Sisterhood: Gender & Class in the Campaign Against Woman Suffrage.* Madison, WI: University of Wisconsin Press.

Martin, Karin A. 1998. "Becoming a Gendered Boy: Practices of Preschools." *American Sociological Review* 63 (August): 494–511.

Martin, Olga Johanna. 1970 [1930]. *Hollywood's Movie Commandments.* New York: Arno Press.

Martin, Patricia Yancey, and Robert Hummer. 1989. "Fraternities and Rape on Campus." *Gender & Society* 3 (December): 457–473.

Martin, Peter. 2002. "Individual and Social Resources Predicting Well-Being and Functioning in the Later Years: Conceptual Models, Research, and Practices." *Ageing International* 27 (Spring): 3–29.

Martin, Susan Ehrlich. 1989. "Sexual Harassment: The Link Between Gender Stratification, Sexuality, and Women's Economic Status." Pp. 57–75 in *Women: A Feminist Perspective,* 4th ed., edited by Jo Freeman. Palo Alto, CA: Mayfield.

Martinez, Ramiro. 1996. "Latinos and Lethal Violence: The Impact of Poverty and Inequality." *Social Problems* 43 (May): 131–146.

Martinez, Ramiro. 2002. *Latino Homicide: Immigration, Violence and Community.* New York: Routledge.

Marx, Anthony. 1997. *Making Race and Nation: A Comparison of the United States, South Africa, and Brazil.* Cambridge, MA: Cambridge University Press.

Marx, Gary T. 1967. "Religion: Opiate or Inspiration of Civil Rights Militancy Among Negroes." *American Sociological Review* 32 (February): 64–72.

Marx, Gary T., and Douglas McAdam. 1994. *Collective Behavior and Social Movements: Process and Structure.* Englewood Cliffs, NJ: Prentice Hall.

Marx, Karl. 1967 [1867]. *Capital.* Edited by F. Engels. New York: International Publishers.

Marx, Karl. 1972 [1843]. "Contribution to the Critique of Hegel's *Philosophy of Right.*" Pp. 11–23 in *The Marx-Engels Reader,* edited by Robert C. Tucker. New York: W. W. Norton.

Marx, Karl. 1972 [1845]. "The German Ideology." Pp. 110–164 in *The Marx-Engels Reader,* edited by Robert C. Tucker. New York: W. W. Norton.

Mason, Heather. 2004. "Empty Seats: Fewer Families Eat Together." *The Gallup Poll.* Princeton, NJ: Gallup Organization. Website: **www.gallup.org**

Massey, Douglas S. 1993. "Latino Poverty Research: An Agenda for the 1990s." *Social Science Research Council Newsletter* 47 (March): 7–11.

Massey, Douglas S., and Nancy A. Denton. 1993. *American Apartheid: Segregation and the Making of the Underclass.* Cambridge, MA: Harvard University Press.

Massey, Douglas S. 1999. "America's Apartheid and the Urban Underclass." Pp. 125–139 in, *Race and Ethnic Conflict: Contending Views on Prejudice, Discrimination, and Ethnoviolence,* edited by Fred L. Pincus and Howard J. Ehrlich. Boulder, CO: Westview.

Mast, Marianne Schmid, and Judith A. Hall. 2004. "When Is Dominance Related to Smiling? Assigned Dominance, Dominance Preference, Trait Dominance, and Gender as Moderators." *Sex Roles* 50 (March): 387–399.

Masters, William H., and Virginia E. Johnson. 1966. *Human Sexual Response.* Boston, MA: Little Brown.

Mathews, Jay. 2004. "Correcting Misconceptions About Home Schooling." *The Washington Post* (July 27).

Mathews, Linda. 1996. "More Than Identity Rests on a New Racial Category." *The New York Times* (July 6): 1–7.

Matthews, Hugh, Mark Taylor, Barry Percy-Smith, and Melanie Limb. 2000. "The Unacceptable Flaneur: The Shopping Mall as a Teenage Hangout." *Childhood* 7 (August): 279–294.

Mauer, Mark. 1999. *Race To Incarcerate.* New York: New Press.

Maume, David J., Jr. 1999. "Glass Ceilings and Glass Escalators: Occupational Segregation and Race and Sex Differences in Managerial Promotions." *Work and Occupations* 26 (November): 483–509.

Mayell, Hillary. 2002. "Thousands of Women Killed for Family 'Honor.'" *National Geographic* (February 12). Website: **www.nationalgeographic.com**

Mazur, Alan. 1968. "The Littlest Science." *The American Sociologist* 3 (August): 195–200.

Mazzuca, Josephine. 2002. "More Accepting of Homosexuals—Canada or U.S.?" *The Gallup Poll.* Princeton, NJ: Gallup Organization. Website: **www.gallup.com**

McAdam, Doug. 1982. *Political Process and the Development of Black Insurgency.* Chicago, IL: University of Chicago Press.

McAdam, Doug. 1999. *Political Process and the Development of Black Insurgency, 1930–1970,* 2nd ed. Chicago, IL: University of Chicago Press.

McCall, Leslie. 2001. *Complex Inequality: Gender, Class, and Race in the New Economy.* New York: Routledge.

McCarthy, John D., and Clark McPhail. 1996. "Images of Protest: Dimensions of Selection Bias in Media Coverage of Washington Demonstrations, 1982 and 1991." *American Sociological Review* 3 (June): 478–499.

McCarthy, John, and Mayer Zald. 1973. *The Trend of Social Movements in America: Professionalism and Resource Mobilization.* Morristown, NJ: General Learning Press.

McCauley, C. 1989. "The Nature of Social Influence in Groupthink: Compliance and Internalization." *Journal of Personality and Social Psychology* 57 (August): 250–260.

McClelland, Susan. 2003. "A Grim Toll on the Innocent." *Maclean's* (May 12): 20.

McCloskey, Laura Ann. 1996. "Socioeconomic and Coercive Power Within the Family." *Gender & Society* 10 (August): 449–463.

McComb, Chris. 2001. "Few Say It's Ideal for Both Parents to Work Full Time Outside the Home." Princeton, NJ: Gallup Organization.

McCord, William. 1991. "The Asian Renaissance." *Society* 28 (September–October): 50–61.

McCord, William, and Arline McCord. 1986. *Paths to Progress: Bread and Freedom in Developing Societies.* New York: W. W. Norton.

McCrate, Elaine, and Joan Smith. 1998. "When Work Doesn't Work: The Failure of Current Welfare Reform." *Gender & Society* 12 (February): 61–81.

McGrew, W. C. 1992. *Chimpanzee Material Culture.* New York: Cambridge University Press.

McGuffey, C. Shawn, and B. Lindsay Rich. 1999. "Playing in the Gender Transgression Zone: Race, Class, and Hegemonic Masculinity in Middle Childhood." *Gender & Society* 13 (October): 608–627.

McGuire, Gail M., and Barbara F. Reskin. 1993. "Authority Hierarchies at Work: The Impacts of Race and Sex." *Gender & Society* 7 (December): 487–506.

McGuire, Meredith. 1997. *Religion: The Social Context.* Belmont, CA: Wadsworth.

McInerney, Fran. 2000. "'Requested Death': A New Social Movement." *Social Science and Medicine* 50 (January): 137–154.

McIntyre, Rusty B., Rene M. Paulson, and Charles G. Lord. 2003. "Alleviating Women's Mathematics Stereotype Threat Through Salience of Group Achievements." *Journal of Experimental Social Psychology* 39: 83–90.

McKinlay, John B., and John D. Stoeckle. 1998. "Corporatization and the Social Transformation of Doctoring." *International Journal of Health Services* 18: 191–205.

McKinlay, John B., and Lisa D. Marceau. 2002. "The End of the Golden Age of Doctoring." *International Journal of Health Services* 32: 379–416.

McKinney, Kathleen. 1992. "Contrapower Sexual Harassment: The Offender's Viewpoint." *Free Inquiry in Creative Sociology* 20 (May): 3–10.

McKinney, Kathleen. 1994. "Sexual Harassment and College Faculty Members." *Deviant Behavior* 15: 171–191.

McLane, Daisann. 1995. "The Cuban-American Princess." *The New York Times Magazine* (February 26): 42.

McLemore, Dale. 1994. *Racial and Ethnic Relations in America,* 4th ed. Boston, MA: Allyn and Bacon.

McLeod, Jane D., and James M. Nonnemaker. 2000. "Poverty and Child Emotional and Behavioral Problems: Racial/Ethnic Differences in Processes and Effects." *Journal of Health and Social Behavior* 41 (June): 137–161.

McManus, Patricia A., and Thomas A. DiPrete. 2001. "'Losers and Winners' The Financial Consequences of Separation and Divorce for Men." *American Sociological Review* 66 (April): 246–268.

McPhail, Clark. 1994. "The Dark Side of Purpose: Individual and Collective Violence in Riots." *Sociological Quarterly* 35 (February): 1–32.

Mead, George Herbert. 1934. *Mind, Self, and Society.* Chicago, IL: University of Chicago Press.

Meeks, Suzanne, and Stanley A. Murrell. 2001. "Contribution of Education to Health and Life Satisfaction in Older Adults Mediated by Negative Affect." *Journal of Aging and Health* 13 (February): 92–119.

Meier, Deborah, Alfie Kohn, Linda Darling-Hammond, Theodoe R. Sizer, and George Wood. 2004. *Many Children Left Behind: How the No Child Left Behind Act is Damaging Our Children and Our Schools.* Boston, MA: Beacon Press.

Meisenheimer, Joseph R., III. 1992. "How Do Immigrants Fare in the U.S. Labor Market?" *Monthly Labor Review* 115 (December): 3–19.

Meng, Susan. 2001. "Pet Sanctuary." *Forbes* (October 8): 236.

Merlo, Joan M., and Kathleen Maurer Smith. 1994. "The Feminine Voice of Authority in Television Commercials: A Ten-Year Comparison." Paper presented at the annual meeting of the American Sociological Association, Los Angeles.

Mernissi, Fatima. 1987. *Beyond the Veil: Male-Female Dynamics in Modern Muslim Society.* Bloomington, IN: Indiana University Press.

Merton, Robert K. 1938. "Social Structure and Anomie." *American Sociological Review* 3: 672–682.

Merton, Robert K. 1957. *Social Theory and Social Structure.* New York: Free Press.

Merton, Robert K. 1972. "Insiders and Outsiders: A Chapter in the Sociology of Knowledge." *American Journal of Sociology* 78 (July): 9–47.

Merton, Robert, and Alice K. Rossi. 1950. "Contributions to the Theory of Reference Group Behavior." Pp. 279–334 in *Studies, Scope and Method of "The American Soldier."* New York: Free Press.

Messerschmidt, James W. 1997. *Crime as Structured Action: Gender, Race, Class and Crime in the Making.* Thousand Oaks, CA: Sage.

Messner, Michael A. 1992. *Power at Play: Sports and the Problem of Masculinity.* Boston, MA: Beacon Press.

Messner, Michael A. 1996. "Studying Up on Sex." *Sociology of Sport Journal* 13 (September): 221–237.

Messner, Michael A. 2002. *Taking the Field: Women, Men, and Sports.* Minneapolis, MN: University of Minnesota Press.

Metz, Michael A., B. R. Rosser-Simon, and Nancy Strapko. 1994. "Differences in Conflict-Resolution Styles Among Heterosexual, Gay, and Lesbian Couples." *Journal of Sex Research* 31: 293–308.

Meyer, David S., and Joshua Gamson. 1995. "The Challenge of Cultural Elites: Celebrities and Social Movements." *Sociological Inquiry* 65 (Spring): 181–206.

Meyer, Madonna Harrington. 1994. "Gender, Race, and the Distribution of Social Assistance: Medicaid Use Among the Elderly." *Gender & Society* 8 (March): 8–28.

Michael, Robert T., John H. Gagnon, Edward O. Laumann, and Gina Kolata. 1994. *Sex in America: A Definitive Survey.* Boston, MA: Little Brown.

Michel, Robert T., Heidi I. Hartmann, and Brigid O'Farrell (eds.). 1989. *Pay Equity: Empirical Inquiries.* Washington, DC: National Academy Press.

Michels, Robert. 1962[1911]. *Political Parties: A Sociological Study of the Oligarchic Tendencies of Modern Democracy.* New York: Collier Books.

Mickelson, Roslyn Arlin (ed.). 2000. *Children on the Streets of the Americas: Globalization, Homelessness, and Education in the United States, Brazil, and Cuba.* New York: Routledge.

Milgram, Stanley. 1974. *Obedience to Authority: An Experimental View.* New York: Harper and Row.

Milgram, Stanley, and Hans Toch. 1968. "Collective Behavior: Crowds and Social Movements." Pp. 507–610 in *Handbook of Social Psychology,* 2nd ed., edited by Gardner Lindzey and Elliot Aronson. Reading, MA: Addison-Wesley.

Milkie, Melissa A. 1999. "Social Comparisons, Reflected Appraisals, and Mass Media: The Impact of Pervasive Beauty Images on Black and White Girls' Self-Concepts." *Social Psychology Quarterly* 62 (June): 190–210.

Milkie, Melissa A. 2002. "Contested Images of Femininity: An Analysis of Cultural Gatekeepers' Struggles with the 'Real Girl' Critique." *Gender & Society* 16 (December): 839–859.

Milkie, Melissa A. 2002. "Gendered Division of Childrearing: Ideals, Realities, and the Relationship to Parental Well-being." *Sex Roles* 47 (July): 21–38.

Miller, David. 1985. *Collective Behavior.* Belmont, CA: Wadsworth.

Miller, David L. 2000. *Introduction to Collective Behavior and Collective Action,* 2nd ed. Prospect Heights, IL: Waveland Press.

Miller, Eleanor. 1986. *Street Women.* Philadelphia, PA: Temple University Press.

Miller, Laura J. 1995. "Family Togetherness and the Suburban Ideal." *Sociological Forum* 3 (September): 393–418.

Miller, Susan L. 1997. "The Unintended Consequences of Current Criminal Justice Policy." Talk presented at Research on Women Series, University of Delaware, Newark, DE.

Mills, C. Wright. 1956. *The Power Elite.* New York: Oxford University Press.

Mills, C. Wright. 1959. *The Sociological Imagination.* New York: Oxford University Press.

Min, Pyong G. 1990. "Problems of Korean Immigrant Entrepreneurs." *International Migration Review* 24 (Fall): 436–455.

Mineau, Geraldine P., Ken R. Smith, and Lee L. Bean. 2002. "Historical Trends of Survival among Widows and Widowers." *Social Science and Medicine* 54 (January): 245–254.

Miner, Horace. 1956. "Body Ritual Among the Nacirema." *American Anthropologist* 58: 503–507.

Mink, Gwendolyn (ed.). 1999. *Whose Welfare?* Ithaca, NY: Cornell University Press.

Mink, Gwendolyn. 2001. "Violating Women: Rights Abuses in the Welfare Police State." *Annals of the American Academy of Political and Social Science* (September): 79–93.

Mintz, Beth, and Michael Schwartz. 1985. *The Power Structure of American Business.* Chicago, IL: University of Chicago Press.

Mirandé, Alfredo. 1979. "Machismo: A Reinterpretation of Male Dominance in the Chicano Family." *The Family Coordinator* 28: 447–449.

Mirandé, Alfredo. 1985. *The Chicano Experience.* Notre Dame, IN: Notre Dame University Press.

Mishel, Lawrence, Jared Bernstein, and Heather Boushey. 2003. *The State of Working America 2002–03.* Washington, DC: Economic Policy Institute.

Misra, Joy, Stephanie Moller, and Marina Karides. 2003. "Envisioning Dependency: Changing Media Depictions of Welfare in the 20th Century." *Social Problems* 50 (November): 482–504.

Mitchell, G., Stephanie Obradovich, Fred Herring, Chris Tromberg, and Alysson L. Burns. 1992. "Reproducing Gender in Public Places: Adults' Attention to Toddlers in Three Public Locales." *Sex Roles* 26 (September): 323–330.

Mizruchi, Mark S. 1992. *The Structure of Corporate Political Action: Interfirm Relations and Their Consequences.* Cambridge, MA: Harvard University Press.

Mizutami, Osamu. 1990. *Situational Japanese.* Tokyo: The Japan Times.

Moen, Phyllis. 2004. "The New 'Middle' Work Force." Minneapolis, MN: The Life Course Center, University of Minnesota/Ithaca, NY: Bronfenbrenner Life Course Center and Cornell Careers Institute, Cornell University. Website: www.lifecourse.cornell.edu/newmiddleworkforce.pdf

Moen, Phyllis, Jungmeen E. Kim, and Heather Hofmeister. 2001. "Couples' Work/Retirement Transitions, Gender, and Marital Quality." *Social Psychology Quarterly* 64 (March): 55–71.

Moller, Lora C., Shelley Hymel, and Kenneth H. Rubin. 1992. "Sex Typing in Play and Popularity in Middle Childhood." *Sex Roles* 26 (April): 331–353.

Money, John. 1988. *Gay, Straight, and InBetween: The Sexology of Erotic Orientation.* New York: Oxford University Press.

Money, John. 1995. *Gendermaps: Social Constructionism, Feminism and Sexosophical History.* New York: Continuum.

Montada, L., and M. Lerner, Jr. (eds.). 1998. *Responses to Victimization and Beliefs in a Just World.* New York: Plenum.

Montemurro, Beth. 2002. "You Go 'Cause You Have To': The Bridal Shower as a Ritual of Obligation." *Symbolic Interaction* 25: 67–92.

Montgomery, James D. 1992. "Job Search and Network Composition: Implications of the Strength-of-Weak-Ties Hypothesis." *American Sociological Review* 57 (October): 586–596.

Moore, David W. 2000. "Two of Three Americans Feel Religion Can Answer Most of Today's Problems." *The Gallup Poll Monthly* (March): 53–61.

Moore, David W., and Joseph Carroll. 2004. "Support for Gay Marriage/Civil Unions Edges Upward." *The Gallup Poll.* Princeton, NJ: Gallup Organization. Website: www.gallup.com

Moore, Gwen. 1979. "The Structure of a National Elite Network." *American Sociological Review* 44 (October): 673–692.

Moore, Joan. 1976. *Hispanics in the United States.* Englewood Cliffs, NJ: Prentice Hall.

Moore, Joan W., and John M. Hagedorn. 1996. "What Happens to Girls in Gangs?" Pp. 205–218 in *Gangs in America,* edited by C. Ronald Huff. Thousand Oaks, CA: Sage.

Moore, Robert B. 1992. "Racist Stereotyping in the English Language." Pp. 317–328 in *Race, Class, and Gender: An Anthology,* 2nd ed., edited by Margaret L. Andersen and Patricia Hill Collins. Belmont, CA: Wadsworth.

Moreland, Richard L., and Scott R. Beach. 1992. "Exposure Effects in the Classroom: The Development of Affinity Among Students." *Journal of Experimental Social Psychology* 28: 255–276.

Morgan, Leslie A. 2000. "The Continuing Gender Gap in Later Life Economic Security." *Journal of Aging and Social Policy* 11: 157–165.

Morgan, Lewis H. 1877. *Ancient Society, or Researches in the Lines of Human Progress, From Savagery Through Barbarism to Civilization.* Cambridge, MA: Harvard University Press.

Morgan, Mary Y. 1987. "The Impact of Religion on Gender-Role Attitudes." *Psychology of Women Quarterly* 11 (September): 301–310.

Morris, Aldon. 1984. *The Origins of the Civil Rights Movement: Black Communities Organizing for Change.* New York: Free Press.

Morris, Aldon D. 1999. "A Retrospective on the Civil Rights Movement: Political and Intellectual Landmarks." *Annual Review of Sociology* 25: 517–539.

Morris, Aldon, and Carol McClurg Mueller (eds.). 1992. *Frontiers in Social Movement Theory.* New Haven, CT: Yale University Press.

Moskos, Charles. 1992. "Right Behind You, Scarlett!" Pp. 40–54 in *Women in the Military,* edited by E. A. Blacksmith. New York: H. W. Wilson.

Moskos, Charles C., and John Sibley Butler. 1996. *All That We Can Be: Black Leadership and Racial Integration The Army Way.* New York: Basic Books.

Mossakowski, Krysia N. 2003. "Coping with Perceived Discrimination: Does Ethnic Identity Protect Mental Health?" *Journal of Health and Social Behavior* 44 (September): 318–331.

Muller, Thomas, and Thomas J. Espenshade. 1985. *The Fourth Wave: California's Newest Immigrants.* Washington, DC: Urban Institute Press.

Mullin, C. R., and D. Linz. 1995. "Desensitization and Resensitization to Violence Against Women." *Journal of Personality and Social Psychology* 69: 449–459.

Murrell, Stanley A., and Suzanne Meeks. 2002. "Psychological, Economic, and Social Mediators of Education-Health Relationship in Older Adults." *Journal of Aging and Health* 14 (November): 527–550.

Murty, Komanduri S., Julian B. Roebuck, and Gloria R. Armstrong. 1994. "The Black Community's Reactions to the 1992 Los Angeles Riot." *Deviant Behavior* 15 (March): 85–104.

Myers, Steven Lee. 2000. "Survey of Troops Finds Antigay Bias Common in Service." *The New York Times* (March 24): 1.

Myers, Walter D. 1998. *Amistad Affair.* New York: NAL/Dutton.

Myerson, Allen R. 1998. "Rating the Bigshots: Gates vs. Rockefeller." *The New York Times* (May 24): 4.

Myles, John. 1989. *Old Age in the Welfare State: The Political Economy of Public Pensions,* rev. ed. Lawrence, KS: University Press of Kansas.

Myrdal, Gunnar. 1944. *An American Dilemma: The Negro Problem and Modern Democracy,* 2 vols. New York: Harper and Row.

NAACP Legal Defense and Education Fund. 2004. *Annual Report.* Washington, DC: NAACP Legal Defense Fund.

NAACP Legal Defense Fund. 2002. *Annual Report.* Washington, DC: NAACP Legal Defense Fund.

Nack, Adina. 2000. "Damaged Goods: Women Managing the Stigma of AIDS." *Deviant Behavior* 21: 95–121.

Nagel, Joane. 1996. *American Indian Ethnic Renewal: Red Power and the Resurgence of Identity and Culture.* New York: Oxford University Press.

Nagel, Joane. 2003. *Race, Ethnicity, and Sexuality: Intimate Intersections, Forbidden Frontiers.* New York: Oxford University Press.

Namboodiri, Krishnan. 1988. "Ecological Demography: Its Place in Sociology." *American Sociological Review* 53 (August): 619–633.

Nanda, Serena. 1998. *Neither Man Nor Woman: The Hijras of India.* Belmont, CA: Wadsworth.

Nardi, Peter M., and Beth Schneider. 1997. *Social Perspectives on Lesbian and Gay Studies.* New York: Routledge.

Nash, John F. 1951. "Non-Cooperative Games." *Annals of Mathematics* 54: 286–295.

Nathanson, Constance A. 1999. "Social Movements as Catalysts for Policy Change: The Case of Smoking and Guns." *Journal of Health Politics, Policy, and Law* 24 (June): 421–488.

National Center for Health Statistics. 2003. *Health United States.* Rockville, MD: U.S. Department of Health and Human Services.

National Center for Health Statistics. 2003. *National Vital Statistics Report* 52 (February 12).

National Coalition for the Homeless. 2002. *Facts about Homelessness.* Website: **www .nationalhomeless.org**

National Collegiate Athletics Association. 2002. *NCAA Year-by-Year Sports Participation, 1982–2001.* Indianapolis, IN: National Collegiate Athletics Association. Website: **www.ncaa.org**

National Council on Crime and Delinquency. 2000. "Justice System Facing A Crisis of Legitimacy in Communities of Color." Website: **www.nccd-crd.org**

National Opinion Research Center. 2002. *General Social Survey.* Website: **www.norc.uchicago.edu**

National Television Violence Study. 1997. *National Television Violence Study.* Thousand Oaks, CA: Sage.

National Urban League. 2002. *The State of Black America.* New York: National Urban League.

Neal, Terry M. 2004. "Bush, Blacks, and Iraq." *The Washington Post* (May 20).

Nee, Victor. 1973. *Longtime Californ': A Documentary Study of an American Chinatown.* New York: Pantheon Books.

Neppl, Tricia K., and Ann D. Murray. 1997. "Social Dominance and Play Patterns Among Preschoolers: Gender Comparisons." *Sex Roles* 36 (March): 381–394.

Newman, Katherine. 1999. *No Shame in My Game: The Working Poor in the Inner City.* New York: Russell Sage Foundation/Vintage Books.

Newman, Katherine S. 1988. *Falling from Grace: The Experience of Downward Mobility in the American Middle Class.* New York: Free Press.

Newman, Katherine S. 1993. *Declining Fortunes: The Withering of the American Dream.* New York: Basic Books.

Newport, Frank. 2000. "Women's Most Pressing Concerns Today Are Money, Family, Health, and Stress." *The Gallup Poll Monthly* (March): 40–41.

Newport, Frank. 2001. "American Attitudes Toward Homosexuality Continue to Be More Tolerant." *The Gallup Poll* (June 4): Princeton, NJ: Gallup Organization.

Newport, Frank. 2002. "Homosexuality." Princeton, NJ: Gallup Organization. Website: **www .gallup.com**

Newport, Frank. 2003. "Six Out of 10 Americans Say Homosexual Relations Should Be Recognized As Legal." *The Gallup Poll.* Princeton, NJ: Gallup Organization. Website: **www.gallup .com**

Niebuhr, Gustav. 1998. "Makeup of American Religion Is Looking More Like Mosaic, Data Say." *The New York Times* (April 12): 14.

Nieves, Evelyn. 2000. "Many in Silicon Valley Cannot Afford Housing, Even at $50,000 a Year." *The New York Times* (February 20): A16.

Nisbet, Paul A. 1997. "Suicide Opportunity as a Function of Effective Network Size." Paper presented at the annual meeting of the American Sociological Association, Toronto.

Nock, Steven L. 1998. "The Consequences of Premarital Fatherhood." *American Sociological Review* 6 (April): 250–263.

Norris, Pippa, and Ronald Inglehart. 2002. "Islamic Culture and Democracy: Testing the 'Clash of Civilizations' Thesis." *Comparative Sociology* 1: 235–263.

O'Campo, Patricia, and Lucia Rojas-Smith. 1998. "Welfare Reform and Women's Health: Review of the Literature and Implications for State Policy." *Journal of Public Health Policy* 19: 420–446.

O'Kelly, Charlotte G., and Larry S. Carney. 1986. *Women and Men in Society,* 2nd ed. Belmont, CA: Wadsworth.

O'Neil, John. 2002. "Parent Smoking and Teenage Sex." *The New York Times* (September 3): F7.

O'Rand, Angela M., and John C. Henretta. 1999. "Labor Markets and Occupational Welfare in the United States." Pp. 131–157 in *Age and Inequality: Diverse Pathways through Later Life,* edited by Angela O'Rand and John C. Henretta. Boulder, CO: Westview Press.

Oakes, Jeannie. 1985. *Keeping Track: How Schools Structure Inequality.* New Haven, CT: Yale University Press.

Oakes, Jeannie. 1990. "Multiplying Inequalities: The Effects of Race, Social Class and Tracking on Opportunities to Learn Mathematics and Science." Santa Monica, CA: RAND for National Science Foundation.

Oakes, Jeannie, and Martin Lipton. 1996. "Developing Alternatives to Tracking and Grading." Pp. 168–200 in *Educating A New Majority: Transforming America's Educational System for Diversity,* edited by Laura I. Rendon and Richard O. Hope. San Francisco: Jossey-Bass.

Oberg, Peter, and Lars Tornstam. 1999. "Body Images among Women of Different Ages." *Ageing and Society* 5: 629–644.

Oberschall, Anthony. 1979. "Protracted Conflict." Pp. 45–70 in *Dynamics of Social Movements,* edited by Mayer N. Zald and John McCarthy. New Brunswick, NJ: Transaction.

Ochshorn, Judith. 1981. *The Female Experience and the Nature of the Divine.* Bloomington, IN: Indiana University Press.

Office of Population Research. Princeton University. 2004. Website: **http://opr.princeton.edu**

Ogburn, William F. 1922. *Social Change with Respect to Cultural and Original Nature.* New York: B. W. Huebsch.

Ollivier, Michele. 2000. "'Too Much Money Off Other People's Backs': Status in Late Modern Societies." *Canadian Journal of Sociology* 25 (Fall): 441–470.

Olzak, Susan, and Suzanne Shanahan. 1996. "Deprivation and Race Riots: An Extension of Spilerman's Analysis." *Social Forces* 74 (March): 931–961.

Olzak, Susan, Suzanne Shanahan, and Elizabeth H. McEneaney. 1996. "Poverty, Segregation, and Race Riots: 1960–1993." *American Sociological Review* 61 (August): 590–613.

Omi, Michael, and Howard Winant. 1994. *Racial Formation in the United States,* 2nd ed. New York: Routledge.

Orfield, Gary and Mindy L. Kornhaber (eds). 2001. *Raising Standards or Raising Barriers? Inequality and High-Stakes Testing in Public Education.* New York: Century Foundation Press.

Ornstein, Norman J., Thomas E. Mann, and Michael J. Malbin. 1996. *Vital Statistics on Congress, 1995–1996.* Washington, DC: Congressional Quarterly

Ouchi, William. 1981. *Theory Z: How American Business Can Meet the Japanese Challenge.* Reading, MA: Addison-Wesley.

Owens, Sarah E. 1998. "The Effects of Race, Gender, Their Interactions, and Selected School Variables upon Educational Aspirations and Achievements." Senior thesis, Princeton University, Department of Sociology, Princeton, NJ.

Ozawa, Martha N., and Young Choi. 2002. "The Relationship between Pre-Retirement Earnings and Health Status in Old Age: Black-White Differences." *Journal of Gerontological Social Work* 38: 19–37.

Padavic, Irene. 1991. "Attractions in Male Blue-Collar Jobs for Black and White Women: Eco-

nomic Need, Exposure, and Attitudes." *Social Science Quarterly* 72 (March): 33–49.

Padavic, Irene. 1992. "White Collar Values and Women's Interest in Blue-Collar Jobs." *Gender & Society* 6 (June): 215–230.

Padavic, Irene, and Barbara Reskin. 2002. *Women and Men at Work,* 2nd ed. Thousand Oaks, CA: Sage.

Page, Charles H. 1946. "Bureaucracy's Other Face." *Social Forces* 25 (October): 89–94.

Pain, Emil. 2002. "The Social Nature of Extremism and Terrorism." *Social Sciences* 33: 55–68.

Palen, John J. 1992. *The Urban World.* New York: McGraw-Hill.

Pampel, F. C. 1994. "Population Aging, Class Context, and Age Inequality in Public Spending." *American Journal of Sociology* 100: 153–195.

Pampel, F. C., and P. Adams. 1992. "The Effects of Demographic Change and Political Structure on Family Allowance Expenditures." *Social Service Review* 66: 524–546.

Pant, Girijesh. 2001. "Islamic Resurgence and Neoliberal Economic Reforms in West Asia." *International Studies* 38 (October–December): 323–340.

Pantoja, Adrian D., and Gary M. Segura. 2003. "Does Ethnicity Matter? Descriptive Representation in Legislatures and Political Alienation among Latinos." *Social Science Quarterly* 84 (June): 441–460.

Park, Lisa Sun-Hee. 2002. "A Life of One's Own." *Contexts* 1 (Summer): 56–57.

Park, Robert E., and Ernest W. Burgess. 1921. *Introduction to the Science of Society.* Chicago, IL: University of Chicago Press.

Parreñas, Rhacel Salazar. 2001. *Servants of Globalization: Women, Migration and Domestic Work.* Stanford, CA: Stanford University Press.

Parsons, Talcott (ed.). 1947. *Max Weber: The Theory of Social and Economic Organization.* New York: Free Press.

Parsons, Talcott. 1951a. *The Social System.* Glencoe, IL: Free Press.

Parsons, Talcott. 1951b. *Toward a General Theory of Action.* Cambridge, MA: Harvard University Press.

Parsons, Talcott. 1966. *Societies: Evolutionary and Comparative Perspectives.* Englewood Cliffs, NJ: Prentice Hall.

Parsons, Talcott. 1968 [1937]. *The Structure of Social Action.* New York: Free Press.

Paternoster, Raymond, and Robert Brame. 2003. "An Empirical Analysis of Maryland's Death Sentencing System with Respect to the Influence of Race and Legal Jurisdiction." Website: **www .urhome.umd.edu/newsdesk**

Patterson, Francine. 1978. "Conversations with a Gorilla." *National Geographic* 154 (October): 438–465.

Pattillo-McCoy, Mary. 1998. "Church Culture as a Strategy of Action in the Black Community." *American Sociological Review* 63 (December): 767–784.

Pattillo-McCoy, Mary. 1999. *Black Picket Fences: Privilege and Peril Among the Black Middle Class.* Chicago, IL: University of Chicago Press.

Pavalko, Eliza K., Krysia N. Mossakowski, and Vanessa J. Hamilton. 2003. "Does Perceived Discrimination Affect Health? Longitudinal Relationships between Work Discrimination and Women's Physical and Emotional Health." *Journal of Health and Social Behavior* 43 (March): 18–33.

Payne, Deborah M., John T. Warner, and Roger D. Little. 1992. "Tied Migration and Returns to Human Capital: The Case of Military Wives." *Social Science Quarterly* 73 (June): 324–339.

Pedersen, D. 1997. "When An A Is Average." *Newsweek* (March 3): 64.

Pedraza, Sylvia. 1996. "Origins and Destinies: Immigration, Race, and Ethnicitiy in American History." Pp. 1–20 in *Origins and Destinies: Immigration, Race, and Ethnicity in America,* edited by Sylvia Pedraza and Rubén G. Rumbaut. Belmont, CA: Wadsworth.

Pedraza, Sylvia, and Rubén G. Rumbaut (eds.). 1996. *Origins and Destinies: Immigration, Race, and Ethnicity in America*. Belmont, CA: Wadsworth Publishing Company.

Pellow, David N., 2002. *Garbage Wars: The Struggle for Environmental Justice in Chicago*. Cambridge, MA: MIT Press.

Peltola, Pia, Melissa Milkie, and Stanley Presser. 2004. "The 'Feminist' Mystique: Feminist Identity in Three Generations of Women." *Gender & Society* 18 (February): 122–144.

Pennock-Roman, Maria. 1994. "College Major and Gender Differences in the Prediction of College Grades." College Board Research Report No. 94–2, ETS Research Reports No. 94–24.

Peoples, James, and Garrick Bailey. 2003. *Humanity: An Introduction to Cultural Anthropology*, 6th ed. Belmont, CA: Wadsworth.

Perez, Denise N. 1996. *A Case for Derailment: Tracking in the American Public School System as an Obstacle to the Improvement of Minority Education*. Senior thesis, Princeton University, Department of Sociology, Princeton, NJ.

Perrow, Charles. 1986. *Complex Organization: A Critical Essay*, 3rd ed. New York:

Perrow, Charles. 1994. "The Limit of Safety: The Enhancement of a Theory of Accidents." *Journal of Contingencies and Crisis Management* 22: 212–220.

Perrucci, Robert, and Earl Wysong. 2003. *The New Class Society: Goodbye American Dream?* 2nd ed. Lanham, MD: Rowman and Littlefield.

Persell, Caroline Hodges. 1977. *Education and Inequality: The Roots and Results of Stratification in America's Schools*. New York: Free Press.

Persell, Caroline Hodges. 1990. *Understanding Society: An Introduction to Sociology*. New York: Harper and Row.

Pescosolido, Bernice A., Elizabeth Grauerholz, and Melissa A. Milkie. 1997. "Culture and Conflict: The Portrayal of Blacks in U.S. Children's Picture Books Through the Mid- and Late-Twentieth Century." *American Sociological Review* 62 (June): 443–464.

Pescosolido, Bernice A., Steven A. Tuch, and Jack K. Martin. 2001. "The Profession of Medicine and the Public: Examining Americans' Changing Confidence in Physician Authority from the Beginning of the 'Health Care Crisis' to the Era of Health Care Reform." *Journal of Health and Social Behavior* 42 (March): 1–16.

Pesina, Maria D., Daryl L. Hitchcock, and Beth Menees Rienzi. 1994. "The Military Ban Against Gay Males: University Students' Attitudes Before and After the Presidential Decision." *Journal of Social Behavior and Personality* 9 (September): 499–506.

Petersen, Trond, Ishak Saporta, and Mark-David L. Seidel. 2000. "Offering A Job: Meritocracy and Social Networks." *American Journal of Sociology* 106 (November): 763–816.

Peterson, Richard R. 1996a. "A Re-Evaluation of the Economic Consequences of Divorce." *American Sociological Review* 61 (June): 528–536.

Peterson, Richard R. 1996b. "A Re-Evaluation of the Economic Consequences of Divorce: Reply to Weitzman." *American Sociological Review* 61 (June): 539–540.

Pettigrew, Thomas F. 1971. *Racially Separate or Together?* New York: McGraw-Hill.

Pettigrew, Thomas F. 1985. "New Black-White Patterns: How Best to Conceptualize Them?" *Annual Review of Sociology* 11: 329–346.

Pettigrew, Thomas F. 1992. "The Ultimate Attribution Error: Extending Allport's Cognitive Analysis of Prejudice." Pp. 401–419 in *Readings About the Social Animal*, edited by Elliott Aronson. New York: Freeman.

Phillips, Robert L., Paul J. Andrisani, Thomas N. Daymont, and Curtis L. Gilroy. 1992. "The Economic Returns to Military Service: Race-

Ethnic Differences." *Social Science Quarterly* 73 (June): 340–359.

Piaget, Jean. 1926. *The Language and Thought of the Child*. New York: Harcourt.

Piven, Frances Fox, and Richard A. Cloward. 1971. *Regulating the Poor*. New York: Pantheon.

Piven, Frances Fox, and Richard A. Cloward. 1977. *Poor People's Movements: Why They Succeed, How They Fail*. New York: Vintage Books.

Piven, Frances Fox, and Richard A. Cloward. 1988. *Why Americans Don't Vote*. New York: Pantheon.

Pollitt, Katha. 2004. "Desperately Seeking Health Insurance." *The Nation* (June 21): 9.

Pollock, Philip H., and M. Elliot Vittas. 1995. "Who Bears the Burdens of Environmental Pollution: Race, Ethnicity, and Environmental Equity in Florida." *Social Science Quarterly* 76 (June): 294–310.

Popenoe, David. 2001. "Today's Dads: A New Breed?" *The New York Times* (June 19): A22.

Popkin, Susan. 1990. "Welfare: Views from the Bottom." *Social Problems* 37 (February): 64–79.

Port, Bruce, and Marvin Dunn. 1984. *The Miami Riot of 1980: Crossing the Bounds*. New York: Simon and Schuster.

Porter, Jennie Lee. 2003. "An Investigation of the Glass Ceiling in Corporate America: The Perspective of African-American Women." *Dissertation Abstracts International: The Humanities and Social Sciences* 297-A.

Portes, Alejandro. 2002. "English-Only Triumphs, But the Costs Are High." *Contexts* 1 (Spring): 10–15.

Portes, Alejandro, and Rubén G. Rumbaut. 1996. *Immigrant America: A Portrait*, 2nd ed. Berkeley, CA: University of California Press.

Portes, Alejandro, and Ruben G. Rumbaut. 2001. *Legacies: The Story of the Immigrant Second Generation*. Berkeley, CA: University of California Press.

Poster, Winifred. 1995. "The Challenges and Promises of Class and Racial Diversity in the Women's Movement: A Study of Two Women's Organizations." *Gender & Society* (December): 659–679.

Press, Eyal. 1996. "Barbie's Betrayal." *The Nation* (December 30): 11–16.

Press, Julie E., and Eleanor Townsley. 1998. "Wives' and Husbands' Reporting: Gender, Class, and Social Desirability." *Gender & Society* 12 (April): 188–218.

Presser, Harriet. 2004. "The Economy That Never Sleeps." *Contexts* 3 (Spring): 42–49.

Presser, Stanley, and Linda Stinson. 1998. "Data Collection Mode and Social Desirability Bias in Self-Reported Religious Attendance." *American Sociological Review* 63 (February): 137–145.

Preves, Sharon E. 2003. *Intersex and Identity: The Contested Self*. New Brunswick, NJ: Rutgers University Press.

Proctor, Bernadette D., and Joseph Dalaker. 2003. *Poverty in the United States: 2002*. Washington, DC: U.S. Census Bureau.

Pruitt, D. G. 1971. "Choice Shifts in Group Discussion." *Journal of Personality and Social Psychology* 20 (December): 339–360.

Prus, Robert, and C. R. D. Sharper. 1991. *Road Hustler*. New York: Kauffman and Greenery.

Punch, Maurice. 1996. *Dirty Business: Exploring Corporate Misconduct; Analysis and Cases*. Thousand Oaks, CA: Sage.

Putnam, Robert D. 2000. *Bowling Alone: The Collapse and Revival of American Community*. New York: Simon and Schuster.

Pyke, Karen. 1994. "Women's Employment as a Gift or Burden? Marital Power Across Marriage, Divorce, and Remarriage." *Gender & Society* 8 (March): 73–91.

Quarantelli, Enrico. 1954. "The Nature and Conditions of Panic." *American Journal of Sociology* 60 (3): 267–275.

Quarantelli, Enrico. 1978. "The Behavior of Panic Participants." Pp. 141–146 in *Collective Be-

havior and Social Movements*, edited by Louis Genevie. Itasca, IL: F. E. Peacock.

Quarantelli, Enrico L., and Russell R. Dynes. 1970. "Property Norms and Looting: Their Patterns in Community Crisis." *Phylon* 31 (Summer): 181–193.

Raaj, Tarja, and Christine L. Rackliff. 1998. "Preschoolers' Awareness of Social Expectations of Gender: Relationships to Toy Choices." *Sex Roles* 38 (May): 685–700.

Rabine, Leslie W. 1985. "Romance in the Age of Electronics: Harlequin Enterprises." *Feminist Studies* 11 (Spring): 39–60.

Raboteau, Albert J. 1978. *Slave Religion: The "Invisible Institution" in the Antebellum South*. New York: Oxford University Press.

Raffaelli, Marcela, and Lenna L. Ontai. 2004. "Gender Socialization in Latino/a Families: Results from Two Retrospective Studies." *Sex Roles* 50 (March): 287–299.

Rampersad, Arnold. 1986. *The Life of Langston Hughes: Vol. I: 1902–1941. I, Too, Sing America*. New York: Oxford University Press.

Rampersad, Arnold. 1988. *The Life of Langston Hughes: Vol. II: 1941–1967. I Dream a World*. New York: Oxford University Press.

Rashid, Ahmed. 2000. *Taliban: Militant Islam, Oil, and Fundamentalism in Central Asia*. New Haven, CT: Yale University Press.

Ray, Julie, 2003. "The Legality of Life and Death." Princeton, NJ: *The Gallup Poll*. Website: **www .gallup.com**

Read, Piers Paul. 1974. *Alive: The Story of the Andes Survivors*. Philadelphia, PA: J. B. Lippincott.

Reich, Robert. 1991. *The Work of Nations: Preparing Ourselves for 21st Century Capitalism*. New York: Knopf.

Reid, Frances. 1994. "Complaints of a Dutiful Daughter." Film. New York: Women Make Movies.

Reilly, Mary Ellen, Bernice Lott, Donna Caldwell, and Luisa DeLuca. 1992. "Tolerance for Sexual Harassment Related to Self-Reported Sexual Victimization." *Gender & Society* 6 (March): 122–138.

Reiman, Jeffrey. 2002. *The Rich Get Richer and the Poor Get Prison*, 7th ed. Boston, MA: Allyn and Bacon.

Reingold, Jennifer. 2000. "Executive Pay." *Business Week* (April 17). Website: **www.businessweek.com**

Reinisch, June Hanover. 1990. *The Kinsey Institute New Report on Sex: What You Must Know to Be Sexually Literate*. New York: St. Martin's Press.

Reischer, Erica. 1992. *Alternative Medicine in the United States*. Senior thesis, Princeton University, Princeton, NJ.

Rendon, Laura I., and Richard O. Hope (eds.). 1996. *Educating a New Majority: Transforming America's Educational System for Diversity*. San Francisco: Jossey-Bass.

Renk, Kimberly, Rex Roberts, Angela Roddenberry, Mary Luick, Sarah Hillhouse, Cricket Meehan, Arazais Oliveros, and Vicky Phares. 2003. "Mothers, Fathers, Gender Role, and Time Parents Spend with their Children." *Sex Roles* 48 (April): 305–315.

Renzetti, Claire. 1992. *Violent Betrayal: Partner Abuse in Lesbian Relationships*. Newbury Park, CA: Sage.

Reskin, Barbara. 1988. "Bringing the Men Back In: Sex Differentiation and the Devaluation of Women's Work." *Gender & Society* 2 (March): 58–81.

Reskin, Barbara F., and Irene Padavic. 1988. "Supervisors as Gatekeepers: Male Supervisors' Responses to Women's Integration in Plant Jobs." *Social Problems* 35 (December): 536–550.

Reynolds, Ariel Catherine. 2003. "Religiosity as a Predictor of Attitudes toward Homosexuality." Paper presented at the Annual Meetings of the Southern Sociological Society, New Orleans, LA.

Rice, Berkeley. 1986. "Water Shocks of the '80s." *Across the Board* (March): 17.

Rich, Adrienne. 1980. "Compulsory Heterosexuality and Lesbian Existence." *Signs* 5 (Summer): 631–660.

Richards, T. Anne, Judith Wrubel, and Susan Folkman. 1999–2000. "Death Rites in the San Francisco Gay Community: Cultural Developments of the AIDS Epidemic." *Omega* 40: 335–350.

Ridgeway, Cecilia L. 2001. "Gender, Status, and Leadership." *The Journal of Social Issues* 4 (Winter): 637–655.

Riesman, David. 1970 [1950]. *The Lonely Crowd: A Study of the Changing American Character.* New Haven, CT: Yale University Press.

Rifkin, Jeremy. 1989. *Entropy: Into the Greenhouse World,* rev. ed. New York: Bantam Books.

Rifkin, Jeremy. 1998. "The Biotech Century: Human Life as Intellectual Property." *The Nation* 266 (April 13): 11–19.

Rindfuss, Ronald R., Elizabeth C. Cooksey, and Rebecca L. Sutterlin. 1999. "Young Adult Occupational Achievement: Early Expectations versus Behavioral Reality." *Work and Occupations* 26 (May): 220–263.

Risman, Barbara. 1986. "Can Men 'Mother'? Life as a Single Father." *Family Relations* 35 (January): 95–102.

Risman, Barbara. 1987. "Intimate Relationships from a Microstructural Perspective: Men Who Mother." *Gender & Society* 1 (March): 6–32.

Risman, Barbara, and Pepper Schwartz. 2002. "After the Sexual Revolution: Gender Politics in Teen Dating." *Contexts* 1 (Spring): 16–24.

Ritzer, George. 1999. *Enchanting a Disenchanted World: Revolutionizing the Means of Consumption.* Thousand Oaks, CA: Pine Forge Press.

Ritzer, George. 2002. *The McDonaldization of Society: An Investigation Into The Changing Character of Contemporary Social Life,* 3rd ed. Thousand Oaks, CA: Pine Forge Press.

Robbins, Thomas. 1988. *Cults, Converts, and Charisma: The Sociology of New Religious Movements.* Beverly Hills, CA: Sage.

Robbins, Thomas. 2001. "Combating 'Cults' and 'Brainwashing' in the United States and Western Europe: A Comment on Richardson and Introvigne's Report." *Journal for the Scientific Study of Religion* 20 (June): 169–175.

Roberts, Donald F. 2000. "Media and Youth: Access, Exposure, and Privatization." *Journal of Adolescent Health* 27 (August Suppl.): 8–14.

Roberts, Dorothy. 1997. *Killing the Black Body: Race, Reproduction and the Meaning of Liberty.* New York: Vintage Books.

Robertson, Tatsha, and Garrance Burke. 2001. "Fighting Terror: Concerned Family;" "Shock, Worry for Family of U.S. Man Captured with Taliban." *The Boston Globe* (December 4): A1.

Robey, Bryant. 1982. "A Guide to the Baby Boom." *American Demographics* 4 (September): 16–21.

Robinson, Dawn T., and Lynn Smith-Lovin. 2001. "Getting a Laugh: Gender, Status, and Humor in Task Discussions." *Social Forces* 80 (September): 123–158.

Robinson, Jo Ann Gibson. 1987. *The Montgomery Bus Boycott and the Women Who Started It.* Knoxville, TN: The University of Tennessee Press.

Roccas, Sonia, and Marilynn B. Brewer. 2002. "Social Identity Complexity." *Personality and Social Psychology Review* 6: 88–106.

Rodgers, Joan R. 1995. "An Empirical Study of Intergenerational Transmission of Poverty in the United States." *Social Science Quarterly* 76 (March): 178–194.

Rodney, Walter. 1981. *How Europe Underdeveloped Africa,* rev.ed. Washington, DC: Howard University Press.

Rodriguez, Clara E. 1989. *Puerto Ricans: Born in the U.S.A.* Boston, MA: Unwin Hyman.

Rodriguez, Clara E. 2000. *Changing Race: Latinos, the Census, and the History of Ethnicity in the*

United States. New York: New York University Press.

Rodriquez, Nestor P. 1999. "Globalization, Autonomy, and Transnational Migration: Impacts on U.S. Intergroup Relations." *Research in Politics and Society* 6: 65–84.

Roethlisberger, Fritz J., and William J. Dickson. 1939. *Management and the Worker.* Cambridge, MA: Harvard University Press.

Rollins, Judith. 1985. *Between Women: Domestics and Their Employers.* Philadelphia, PA: Temple University Press.

Romain, Suzanne. 1999. *Communicating Gender.* Mahwah, NJ: Lawrence Erlbaum.

Roof, Wade Clark. 1993. *A Generation of Seekers.* San Francisco: Harpers.

Roof, Wade Clark. 1999. *Spiritual Marketplace.* Princeton, NJ: Princeton University Press.

Roof, Wade Clark, and William McKinney. 1987. *American Mainline Religion: Its Changing Shape and Future.* New Brunswick, NJ: Rutgers University Press.

Roosevelt, Margot. 2000. "Yanomami: What Have We Done To Them?" *Time* (October 2): 77–78.

Root, Maria P. P. (ed.). 1996. *The Multiracial Experience: Racial Borders as the New Frontier.* Thousand Oaks, CA: Sage.

Root, Maria P. P. 2001. *Love's Revolution: Interracial Marriage.* Philadelphia, PA: Temple University Press.

Roper Organization. 1993. *The Gallup Poll.* Storrs, CT: Roper Organization.

Roper Organization. 1995. *The 1995 Virginia Slims Opinion Poll: A 25-Year Perspective of Women's Lives.* Storrs, CT: Roper Organization.

Rose, Peter I. (ed.). 1970. *Americans from Africa: Slavery and Its Aftermath.* New York: Atherton.

Rose, J. Stephen. 2000. *Social Stratification in the United States: The New American Profile Poster.* New York: New Press.

Rosenau, Pauline Marie. 1992. *Post-Modernism and the Social Sciences: Insights, Inroads, and Intrusions.* Princeton, NJ: Princeton University Press.

Rosenberg, Yuval. 2004. "Lost Youth." *American Demographics* (March): 17–19.

Rosengarten, Danielle. 2000. "Modern Times." *Dollars & Sense* (September): 4.

Rosenhan, David L. 1973. "On Being Sane in Insane Places." *Science* 179 (January 19): 250–258.

Rosenthal, Elin E. 1993. "New Benefits and New Rules for Family and Medical Leave." *CUPA Benefits Alert* (February 11). Washington, DC: College and University Personnel Association.

Rosenthal, Robert, and Lenore Jacobson. 1968. *Pygmalian in the Classroom: Teacher Expectations and Pupils' Intellectual Development.* New York: Holt, Rinehart, and Winston.

Rossi, Alice (ed.). 1973. *The Feminist Papers: From Adams to deBeauvoir.* New York: Columbia University Press.

Rossi, Alice S., and Peter H. Rossi. 1990. *Of Human Bonding: Parent-Child Relations Across the Life Course.* New York: Aldine de Gruyter.

Rostow, W. W. 1978. *The World Economy: History and Prospect.* Austin, TX: University of Texas Press.

Roux, A. V. D., S. S. Merkin, D. Arnett, L. Chambless, M. Massing, F. J. Nieto, P. Sorlie, M. Szklo, H. A. Tyroler, and R. L. Watson. 2001. "Neighborhood of Residence and Incidence of Coronary Heart Disease." *New England Journal of Medicine* 345 (July 12): 99–106.

Royster, Deidre A. 2003. *Race and the Invisible Hand: How White Networks Exclude Black Men from Blue Collar Jobs.* Berkeley, CA: University of California Press.

Rubin, Gayle. 1975. "The Traffic in Women." Pp. 157–211 in *Toward an Anthropology of Women,* edited by Rayna Reiter. New York: Monthly Review Press.

Rubin, Linda J., and Sherry B. Borgers. 1990. "Sexual Harassment in Universities During the 1980s." *Sex Roles* 23 (October): 397–411.

Rudé, George. 1964. *The Crowd in History, 1730–1848.* New York: Wiley.

Rueschmeyer, Dietrich, and Theda Skocpol. 1996. *States, Social Knowledge, and the Origins of Modern Social Policies.* Princeton, NJ: Princeton University Press.

Rumbaut, Rubén G. 1996a. "Origins and Destinies: Immigration, Race, and Ethnicity in Contemporary America." Pp. 21–42 in *Origins and Destinies: Immigration, Race, and Ethnicity in America,* edited by Sylvia Pedraza and Rubén G. Rumbaut. Belmont, CA: Wadsworth.

Rumbaut, Rubén. 1996b. "Prologue." Pp. xvi–xix in *Origins and Destinies: Immigration, Race, and Ethnicity in America,* edited by Silvia Pedraza and Rubén Rumbaut. Belmont, CA: Wadsworth.

Rumbaut, Rubén G., and Alejandro Portes (eds.). 2001. *Ethnicities: Children of Immigrants in America.* New York: Russell Sage Foundation.

Rumbo, Joseph. 2001. "Wiggers, Oreos, and "Keeping It Real": Negotiating Racial and Cultural Boundaries in Popular Music." Paper presented at the Society for the Study of Social Problems, Anaheim, CA.

Rundblad, Georgeanne. 1995. "Exhuming Women's Premarket Duties in the Care of the Dead." *Gender & Society* 9 (April): 173–193.

Rupp, Leila J., and Verta Taylor. 2003. *Drag Queens at the 801 Cabaret.* Chicago, IL: University of Chicago Press.

Rust, Paula C. 1993. "'Coming Out' in the Age of Social Constructionism: Sexual Identity Formation Among Lesbian and Bisexual Women." *Gender & Society* 7 (March): 50–77.

Rust, Paula. 1995. *Bisexuality and the Challenge to Lesbian Politics.* New York: New York University Press.

Ryan, Charlotte. 1996. "Battered in the Media: Mainstream News Coverage of Welfare Reform." *Radical America* 26 (August): 29–41.

Ryan, William. 1971. *Blaming the Victim.* New York: Pantheon.

Rymer, Russ. 1993. *Genie: A Scientific Study.* New York: Harper Collins.

Rytina, Steven. 2000. "Is Occupational Mobility Declining in the United States?" *Social Forces* 78 (June): 1227–1276.

Ryu, Charles. 1992. "Koreans and Church." Pp. 162–163 in *Asian Americans,* edited by Joann Lee. New York: New Press.

Saad, Lydia. 2002. "There's No Place Like Home to Spend an Evening, Say Most Americans." *The Gallup Poll.* Princeton, NJ: Gallup Organization. Website: **www.gallup.com**

Saad, Lydia. 2004. "Military Still Americans' Top-Rated Institution." *The Gallup Poll.* Princeton, NJ: Gallup Organization. Website: **www.gallup.com**

Sabo, Don. 1998. "Women's Athletics and the Elimination of Men's Sports Programs." *Journal of Sport and Social Issues* 22 (February): 27–31.

Sadker, Myra, and David Sadker. 1994. *Failing at Fairness: How America's Schools Cheat Girls.* New York: Scribners

Saguy, Abigail. 2003. *What Is Sexual Harassment? From Capitol Hill to the Sorbonne.* Berkeley, CA: University of California Press.

Saks, M. J., and M. W. Marti. 1997. "A Meta-Analysis of the Effects of Jury Size." *Law and Human Behavior* 21: 451–467.

Salant, Jonathan D., and David S. Cloud. 1995. "To the '94 Election Victors Go the Fundraising Spoils." *Congressional Quarterly* 53 (April 15): 1055–1059.

Sanchez, Laura, and Elizabeth Thomson. 1997. "Becoming Mothers and Fathers: Parenthood, Gender, and the Division of Labor." *Gender & Society* 11 (December): 747–773.

Sanchez, Laura, Wendy D. Manning, and Pamela J. Smock. 1998. "Sex-Specialized or Collaborative Mate Selection? Union Transitions Among

Cohabitors." *Social Science Research* 27 (September): 280–304.

Sanchez, Lisa Gonzalez. 1999. "Reclaiming Salsa." *Critical Studies* 13 (April): 237–250.

Sánchez-Ayéndez, Melba. 1995. "Puerto Rican Elderly Women: Shared Meanings and Informal Supportive Networks." Pp. 260–274 in *Race, Class, and Gender: An Anthology*, 2nd ed., edited by Margaret L. Andersen and Patricia Hill Collins. Belmont, CA: Wadsworth.

Sanday, Peggy Reeves. 2002. *Women at the Center: Life in a Modern Matriarchy*. Ithaca, NY: Cornell University Press.

Sanders, Bernie, and Marcy Kaptur. 1997. "Just Do It, Nike." *The Nation* (December 8): 6.

Sanders, Clifford. 1999. *Understanding Dogs: Living and Working with Canine Companions*. Philadelphia, PA: Temple University Press.

Sanders, William B. 1994. *Gangbangs and Drive-Bys: Grounded Culture and Juvenile Gang Violence*. Hawthorne, NY: Walter de Gruyter.

Sandnabba, N. Kenneth, and Christian Ahlberg. 1999. "Parents' Attitudes and Expectations About Children's Cross-Gender Behavior." *Sex Roles* 40 (February): 249–263.

Santelli, John S., Laura Duberstein Lindberg, Joyce Abma, Clea Sucoff McNeely, and Michael Resnick. 2002. "Adolescent Sexual Behavior: Estimates and Trends from Four Nationally Representative Surveys." *Family Planning Perspectives* 32 (July–August): 156–165.

Santiago, Anna M. 1995. "Intergenerational and Program-Induced Effects of Welfare Dependency: Evidence from the National Longitudinal Survey of Youth." *Journal of Family and Economic Issues* 16 (Fall): 281–306.

Sapir, Edward. 1921. *Language: An Introduction to the Study of Speech*. New York: Harcourt Brace.

Sapp, Gary L. 1986. "Religious Orientation and Moral Judgment." *Journal for the Scientific Study of Religion* 25 (December): 208–214.

Sarbaugh, Thompson, Marjorie Thompson-Lyke, Harold Wolman, and Marie Olson. 1999. "Organizational Mismatch in Urban Labor Markets: The Detroit Case." *Social Science Quarterly* 80 (March): 19–36.

Saunders, Laura. 1990. "America's Richest Congressmen." *Forbes* 145 (February 19): 44–45.

Saunders, William B. 1994. *Gangbangs and Drive-Bys: Grounded Culture and Juvenile Gang Violence*. New York: Aldine de Gruyter.

Sayre, Alan. 2004. "Death No Longer a Sure Thing for Funeral Industry." *Review Appeal*. Website: **www.reviewappeal.midsouthnews.com**

Scarville, Jacquelyn, Scott B. Button, Jack E. Edwards, Anita R. Lancaster, and Timothy W. Elig. 1999. *Armed Forces Equal Opportunity Survey*. Arlington, VA: Defense Manpower Data Center. Website: **www.dmdc.osd.mil/surveys/**

Scavo, Carmine. 1990. "Racial Integration of Local Government Leadership in Southern Small Cities: Consequences for Equity Relevance and Political Relevance." *Social Science Quarterly* 71 (June): 362–372.

Schacht, Steven P. 1996. "Misogyny on and off the 'Pitch': The Gendered World of Male Rugby Players." *Gender & Society* 10 (October): 550–565.

Scheff, Thomas J. 1966. *Being Mentally Ill: A Sociological Theory*. Chicago, IL: Aldine, de Gruyter.

Scheff, Thomas. 1984. *Being Mentally Ill: A Sociological Theory*, 2nd ed. New York: Aldine, de Gruyter.

Scheid, Teresa L. 2003. "Managed Care and the Rationalization of Mental Health Services." *Journal of Health and Social Behavior* 44 (June): 142–161.

Scheper-Hughes, Nancy. 1992. *Death without Weeping: The Violence of Everyday Life in Brazil*. Berkeley, CA: University of California Press.

Scherzer, Teresa. 2002. "Division of Labor, or Labor Divided? Health Care Workers, Health Care Work, and Labor-Management Relations." *Dissertation Abstracts International, A: The Humanities and Social Sciences* (June): 4352–A.

Schlosser, Eric. 2001. *Fast Food Nation: The Dark Side of the All-American Meal*. New York: Houghton Mifflin.

Schmitt, Eric. 2001. "Segregation Growing Among U.S. Children." *The New York Times* (May 6): 28.

Schmitt, Frederika E., and Patricia Yancey Martin. 1999. "Unobtrusive Mobilization by an Institutionalized Rape Crisis Center: 'It Comes From the Victims.'" *Gender & Society* 13: 364–384.

Schneider, Barbara, and David Stevenson. 1999. *The Ambitious Generation: Motivated but Directionless*. New Haven, CT: Yale University Press.

Schneider, Beth. 1984. "Peril and Promise: Lesbians' Workplace Participation." Pp. 211–230 in *Women Identified Women*, edited by Trudy Darty and Sandee Potter. Palo Alto, CA: Mayfield.

Schnittker, Jason, Jeremy Freese, and Brian Powell. 2003. "Who Are Feminists and What Do They Believe? The Role of Generations." *American Sociological Review* 68 (August): 607–622.

Schumann, Howard, Charlotte Steeh, Lawrence Bobo, and Maria Krysan. 1997. *Racial Attitudes in America: Trends and Interpretations*. Cambridge, MA: Harvard University Press.

Schur, Edwin M. 1984. *Labeling Women Deviant*. New York: Random House.

Schwartz, Pepper, and Virginia Rutter. 1998. *The Gender of Sexuality*. Thousand Oaks, CA: Sage.

Scott, Robert. 1969. *The Making of Blind Men*. New York: Russell Sage Foundation.

Scott, Tracy Lee. 2000. "What's God Got to Do with It? Protestantism, Gender, and the Meaning of Work in the U.S." *Dissertation Abstracts International, A: The Humanities and Social Sciences* (April): 3813A–3814A.

Scully, Diana. 1990. *Understanding Sexual Violence: A Study of Convicted Rapists*. Boston, MA: Unwin Hyman.

Scully, Diana. 1994. *Men Who Control Women's Health: The Miseducation of Obstetrician-Gynecologists*. New York: Teachers College Press.

Seale, Clive. 2000." Changing Patterns of Death and Dying." *Social Science and Medicine* 51 (September): 917–930.

Sears, David O., Letitia Anne Peplau, Jonathan L. Freedman, and Shelley E. Taylor. 1988. *Social Psychology*. Englewood Cliffs, NJ: Prentice Hall.

Sedlak, Andrea J., and Diane D. Broadhurst. 1996. *Executive Summary of the Third National Incidence Study of Child Abuse and Neglect*. Washington, DC: National Clearinghouse on Child Abuse and Neglect. Website: **nccanch .acf.hhs.gov**

Segal, Mady W. 1974. "Alphabet and Attraction: An Unobtrusive Measure of the Effect of Propinquity in a Field Setting." *Journal of Personality and Social Psychology* 30: 654–657.

Segura, Denise A., and Jennifer L. Pierce. 1993. "Chicana/o Family Structure and Gender Personality: Chodorow, Familism, and Psychoanalytic Sociology Revisited." *Signs* 19 (Autumn): 62–91.

Seidman, Steven. 1994. "Symposium: Queer Theory/Sociology: A Dialogue." *Sociological Theory* 12 (July): 178–187.

Selznik, Gertrude J., and Stephen Steinberg. 1969. *The Tenacity of Prejudice: Anti-Semitism in Contemporary America*. New York: Harper and Row.

Sen, Amartya. 1999. *Development as Freedom*. New York: Knopf.

Sen, Amartya. 2000. "Population and Gender Equity." *The Nation* (July 24–31): 16–18.

Sen, Amartya. 2002. "How to Judge Globalism." *The American Prospect*, (Winter/Suppl.): A2–A6.

Sennett, Richard, and Jonathan Cobb. 1993. *The Hidden Injuries of Class*. New York: W. W. Norton.

Settersten, Richard A., Jr., and Loren D. Lovegreen. 1998. "Educational Experiences throughout Adult Life: New Hopes or No Hope for Life-Course Flexibility?" *Research on Aging* 20 (July): 506–538.

Seymour, Jane Elizabeth. 1999. "Revisiting Medicalization and 'Natural' Death." *Social Science and Medicine* 49 (September): 691–704.

Shaberoff, Philip. 1988. "Water Supplies in Ground Held Generally Safe." *The New York Times* (October 8): 1, 6.

Shanahan, Suzanne, and Suzanne Olzak. 1999. "The Effects of Immigrant Diversity and Ethnic Competition on Collective Conflict in Urban America: An Assessment of Two Moments of Mass Migration, 1869–1924 and 1965–1993." *Journal of American Ethnic History* 19 (Spring): 40–64.

Shapiro, Laura. 1997. "Book Review: 'The Time Bind.'" *Newsweek* 129 (April 28): 6.

Sharp, Susan F., Toni L. Terling-Watt, Leslie A. Atkins, Jay Trace Gilliam, and Anna Sanders. 2000. "Purging Behavior in a Sample of College Females: A Research Note on General Strain Theory and Female Deviance." *Deviant Behavior* 22: 171–188.

Sherkat, Darren E. 2002. "Sexuality and Religious Commitment in the United States: An Empirical Examination." *Journal for the Scientific Study of Religion* 41 (June): 313–323.

Shibutani, Tomatsu. 1961. *Society and Personality: An Interactionist Approach to Social Psychology*. Englewood Cliffs, NJ: Prentice Hall.

Shibutani, Tomatsu. 1966. *Improvised News*. Indianapolis, IN: Bobbs-Merrill.

Shih, Margaret, Todd L. Pittinsky, and Nalini Ambady. 1999. "Stereotype Susceptibility: Identity Salience and Shifts in Quantitative Performance." *Psychological Science* 10 (January): 80–83.

Shilts, Randy. 1988. *And the Band Played On*. New York: Penguin.

Shinagawa, Larry, and Gin Yong Pang. 1996. "Asian American Panethnicity and Intermarriage." *Amerasia Journal* 22 (Spring): 127–152.

Sigall, H., and N. Ostrove. 1975. "Beautiful But Dangerous: Effects of Offender Attractiveness and Nature of Crime on Judicial Judgment." *Journal of Personality and Social Psychology* 31: 410–414.

Sigelman, Lee, and Paul J. Wahlbeck. 1999. "Gender Proportionality in Intercollegiate Athletics: The Mathematics of Title IX Compliance." *Social Science Quarterly* 80 (September): 518–538.

Signorielli, Nancy. 1989. "Television and Conceptions About Sex Roles: Maintaining Conventionality and the Status Quo." *Sex Roles* 21: 341–360.

Signorielli, Nancy, and Aaron Bacue. 1999. "Recognition and Respect: A Content Analysis of Prime-Time Television Characters Across Three Decades." *Sex Roles* 40 (April): 527–544.

Silberstein, Fred B., and Melvin Seeman. 1959. "Social Mobility and Prejudice." *American Journal of Sociology* (November): 258–264.

Silver, Hilary. 1995. *Federal Discharge Rates, Final Report: Minority/Non-Minority*. Washington, DC: Office of Personnel Management.

Silverman, Myrna, Esther Skirboll, and Joy Payne. 1996. "An Examination of Women's Retirement: African American Women." *Journal of Cross-Cultural Gerontology* 11 (December): 319–334.

Silverthorne, Zebulon A., and Vernon L Quinsey. 2000. "Sexual Partner Age Preferences of Homosexual and Heterosexual Men and Women." *Archives of Sexual Behavior* 29 (1): 67–76.

Simmel, Georg. 1902. "The Number of Members as Determining the Sociological Form of the Group." *The American Journal of Sociology* 8 (July): 1–46.

Simmel, Georg. 1904. "Fashion." *International Quarterly* 10: 541–558.

Simmel, Georg. 1950 [1905]. *The Sociology of Georg Simmel,* edited by Kurt Wolff. New York: Free Press.

Simmons, Wendy W. 2001. "Majority of Americans Say More Women in Political Office Would Be Positive for the Country." *The Gallup Poll.* Princeton, NJ: Gallup Organization. Website: **www.gallup.com**

Simon, David R. 2001. *Elite Deviance,* 7th ed. Boston, MA: Allyn and Bacon.

Simon, Julian L. (ed.). 1995. *The State of Humanity.* Cambridge, MA: Blackwell.

Simons, Marlise. 1989. "The Amazon's Savvy Indians." *The New York Times Magazine* (February): 36ff.

Simons, Ronald L., and Associates. 1996. *Understanding Differences Between Divorced and Intact Families.* Newbury Park, CA: Sage.

Simpson, George Eaton, and J. Milton Yinger. 1985. *Racial and Cultural Minorities: An Analysis of Prejudice and Discrimination,* 5th ed. New York: Plenum.

Sivananadan, A. 1995. "La trahison des clercs. (Racism)." *New Statesman and Society* 8: 20–22.

Sjøberg, Gideon. 1965. *The Preindustrial City: Past and Present.* New York: Free Press.

Sklare, Marshall. 1971. *America's Jews.* New York: Random House.

Skocpol, Theda. 1979. *States and Social Revolutions: A Comparative Analysis of France, Russia, and China.* New York: Cambridge University Press.

Skocpol, Theda. 1992. *Protecting Soldiers and Mothers: The Origins of Social Policy in the United States.* Cambridge, MA: Belknap Press.

Slavin, Robert E. 1993. "Ability Grouping in the Middle Grades: Achievement Effects and Alternatives." *The Elementary School Journal* 93 (5): 535–552.

Slevin, Kathleen F., and C. Ray Wingrove. 1998. *From Stumbling Blocks to Stepping Stones: The Life Experiences of Fifty Professional African American Women.* New York: New York University Press.

Smelser, Neil J. 1963. *Theory of Collective Behavior.* New York: Free Press.

Smelser, Neil J. 1992a. "Culture: Coherent or Incoherent." Pp. 3–28 in *Theory of Culture,* edited by R. Münch and N. J. Smelser. Berkeley, CA: University of California Press.

Smelser, Neil J. 1992b. "The Rational Choice Perspective: A Theoretical Assessment." *Rationality and Society* 4: 381–410.

Smith, A. Wade. 1989. "Educational Attainment as a Determinant of Social Class Among Black Americans." *Journal of Negro Education* 58 (Summer): 416–429.

Smith, Barbara. 1983. "Homophobia: Why Bring It Up?" *Interracial Books for Children Bulletin* 14: 112–113.

Smith, Bonnie G., and Beth Hutchison (eds.). 2004. *Gendering Disability.* Piscataway, NJ: Rutgers University Press.

Smith, Christian. 1998. *American Evangelicism: Embattled and Thriving.* Chicago, IL: University of Chicago Press.

Smith, Christian. 2000. *Christian America: What Evangelicals Really Want.* Berkeley, CA: University of California Press.

Smith, Deborah B., and Phyllis Moen. 2004. "Retirement Satisfaction for Retirees and Their Spouses: Do Gender and the Retirement Decision-Making Process Matter?" *Journal of Family Issues* 25 (March): 262–285.

Smith, Heather. 1999. "The Promotion of Women and Men." Ph.D. diss. University of Delaware, Newark, DE.

Smith, James. 1998. "Race and Ethnicity in the Labor Markets: Trends over the Short and Long Run." Paper presented at the National Academy of Sciences, Research Conference on Racial Trends in the United States, Washington, DC.

Smith, M. Dwayne, Joel A. Devine, and Joseph F. Sheley. 1992. "Crime and Unemployment: Effects Across Age and Race Categories." *Sociological Perspectives* 35 (Winter): 551–572.

Smith, Richard B., and Robert A. Brown. 1997. "The Impact of Social Support on Gay Male Couples." *Journal of Homosexuality* 33: 39–61.

Smith, Robert B. 1993. "Social Structure and Voting Choice: Hypotheses, Findings, and Interpretations." Paper presented at the annual meeting of the American Sociological Association, Miami, FL.

Smith, Tom W. 1991. "Adult Sexual Behavior in 1989: Number of Partners, Frequency of Intercourse and Risk of AIDS." *Family Planning Perspectives* 23 (May–June): 102–107.

Smith, Tom W. 1992. "Changing Racial Labels: From 'Colored' to 'Negro' to 'Black' to 'African American.'" *Public Opinion Quarterly* 56: 496–512.

Smith-Rosenberg, Carroll, and Charles Rosenberg. 1984. "The Female Animal: Medical and Biological Views of Woman and Her Role in Nineteenth Century America." Pp. 12–27 in *Women and Health in America,* edited by Judith W. Leavitt. Madison, WI: University of Wisconsin Press.

Smolan, Rick, and Jennifer Erwitt. 1996. *24 Hours in Cyberspace.* New York: Macmillan.

Snipp, C. Matthew. 1989. *American Indians: The First of This Land.* New York: Russell Sage Foundation.

Snipp, C. Matthew. 1996. "The First Americans: American Indians." Pp. 390–403 in *Origins and Destinies: Immigration, Race, and Ethnicity in America,* edited by Sylvia Pedraza and Rubén G. Rumbaut. Belmont, CA: Wadsworth.

Snow, David A., and Robert D. Benford. 1988. "Ideology, Frame Resonance, and Participant Mobilization." *International Social Movements Research* 1: 197–217.

Snow, David A., E. Burke Rochford, Jr., Steven K. Worden, and Robert D. Benford. 1986. "Frame Alignment Processes, Micromobilization and Movement Participation." *American Sociological Review* 78 (August): 464–481.

Snow, David, and Leon Anderson. 2003. "Street People." *Contexts* 2 (Winter): 12–17.

Solomon, Deborah. 2004. "Questions for Susan Watkins: Life After Whistle-Blowing." *The New York Times Magazine* (June 6): 25.

Sorokin, Pitrim. 1941. *The Crisis of Our Age.* New York: Dutton.

Sotirovic, Mira. 2000. "Effects of Media Use On Audience Framing and Support for Welfare." *Mass Communication & Society* 2–3 (Spring–Summer): 269–296.

Sotirovic, Mira. 2001. "Media Use and Perceptions of Welfare." *Journal of Communication* 51 (December): 750–774.

Southern Poverty Law Center. 1997. *Teaching Tolerance* (Spring): 49.

Sowell, Thomas. 1983. *The Economics and Politics of Race: An International Perspective.* New York: William Morrow.

Spangler, Eve, Marsha A. Gordon, and Ronald Pipkin. 1978. "Token Women: An Empirical Test of Kanter's Hypothesis." *American Journal of Sociology* 84: 160–170.

Spencer, Herbert. 1882. *The Study of Sociology.* London: Routledge.

Spengler, Oswald. 1932. *The Decline of the West.* New York: Knopf.

Spilerman, Seymour. 1976. "Structural Characteristics of Cities and the Severity of Racial Disorders." *American Sociological Review* 41 (October): 771–793.

Spitzer, Steven. 1975. "Toward a Marxian Theory of Deviance." *Social Problems* 22: 638–651.

Sprock, June, and Carol Y. Yoder. 1997. "Women and Depression: An Update on the Report of the APA Task Force." *Sex Roles* 36 (March): 269–303.

Stacey, Judith. 1996. *In the Name of the Family.* Boston, MA: Beacon Press.

Stacey, Judith, and Susan Elizabeth Gerard. 1990. "'We are Not Doormats: The Influence of Feminism on Contemporary Evangelicals in the United States.'" Pp. 98–117 in *Uncertain Terms: Negotiating Gender in American Culture,* edited by Faye Ginsburg and Anna Lowenhaupt Tsing. Boston, MA: Beacon Press.

Stacey, Judith, and Timothy J. Biblarz. 2001. (How) Does the Sexual Orientation of Parents Matter?" *American Sociological Review* 66 (April): 159–183.

Stack, Carol. 1974. *All Our Kin: Strategies for Survival in a Black Community.* New York: Harper Colophon Books.

Stark, Margaret J., Abudu, Walter J. Raine, Stephen L. Burbeck, and Keith K. Davison. 1974. "Some Empirical Patterns in a Riot Process." *American Sociological Review* 39 (December): 865–876.

Stark, Rodney. 1996. *The Rise of Christianity: A Sociologist Reconsiders History.* Princeton, NJ: Princeton University Press.

Starr, Paul. 1982. *The Social Transformation of American Medicine.* New York: Basic Books.

Starr, Paul. 1995. "What Happened to Health Care Reform"? *The American Prospect* (20/Winter): 20–31.

Starr, Paul. 2004. *The Creation of the Media: Political Origins of Modern Communications.* New York: Basic Books.

Stearns, Linda Brewster, and Charlotte Wilkinson Coleman. 1990. "Industrial and Labor Market Structures and Black Male Employment in the Manufacturing Sector." *Social Science Quarterly* (June): 285–298.

Steele, Claude M. 1992. "Race and the Schooling of Black Americans." *The Atlantic Monthly* 269 (April): 68–78.

Steele, Claude M. 1996. "A Burden of Suspicion: How Stereotypes Shape the Intellectual Identities and Performance of Women and African-Americans." Paper read before the Princeton Conference on Higher Education, Princeton, NJ.

Steele, Claude M. 1997. "A Threat in the Air: How Stereotypes Shape Intellectual Identity and Performance." *American Psychologist* 52 (June): 613–629.

Steele, Claude M. 1999. "Thin Ice: Stereotype Threat and Black College Students." *The Atlantic Monthly* (August): 44–54.

Steele, Claude M., and Joshua Aronson. 1995. "Stereotype Vulnerability and African American Intellectual Performance." Pp. 409–421 in *Readings About the Social Animal,* 7th ed., edited by Elliot Aronson. New York: W. H. Freeman.

Steffensmeier, Darrell, and Stephen Demuth. 2000. "Ethnicity and Sentencing Outcomes in U.S. Federal Courts: Who Is Punished More Harshly?" *American Sociological Review* 65 (October): 705–729.

Steffensmeier, Darrell, Jeffrey Ulmer, and John Kramer. 1998. "The Interaction of Race, Gender and Age in Criminal Sentencing: The Punishment Cost of Being Young, Black, and Male." *Criminology* 36 (November): 763–797.

Stein, Arlene, and Ken Plummer. 1994. "'I Can't Even Think Straight': Queer Theory and the Missing Sexual Revolution in Sociology." *Sociological Theory* 12 (July): 178–187.

Steinberg, Ronnie. 1992. "Gendered Instructions: Cultural Lag and Gender Bias in the Hay System of Job Evaluation." *Work and Occupations* 19 (November): 387–424.

Stern, Jessica. 2003. *Terror in the Name of God: Why Religious Militants Kill.* New York: Ecco.

Stern, Kenneth S. 1996. *A Force upon the Plain.* New York: Simon and Schuster.

Sternberg, Robert J. 1988. *The Triarchic Mind: A New Theory of Human Intelligence.* New York: Penguin.

Stevens, Ann Huff. 1994. "The Dynamics of Poverty Spells: Updating Bane and Ellwood." *The American Economic Review* 84 (May): 34–37.

Stevens, William K. 1994. "Poor Lands Success in Cutting Birthrate Upsets Old Theories." *The New York Times* (January 2): 1, 8.

Stewart, Abigail J., Anne P. Copeland, Nia Lane Chester, Janet E. Malley, and Nicole B. Barenbaum. 1997. *Separating Together: How Divorce Transforms Families.* New York: Guilford Press.

Stewart, Ella. 1993. "Communication Between African Americans and Korean Americans: Before and After the Los Angeles Riots." *Amerasia Journal* 19 (Spring): 23–53.

Stewart, Susan D. 2001. "Contemporary American Stepparenthood: Integrating Cohabiting and Nonresident Stepparents." *Population Research and Policy* 20 (August): 345–364.

Stewart-Williams, Steve. 2002. "Gender, the Perception of Aggression and the Overestimation of Gender Bias." *Sex Roles* 46 (March): 177–189.

Stoll, Michael A. 1998. "When Jobs Move, Do Black and Latino Men Lose? The Effect of Growth in Job Decentralization on Young Men's Jobless Incidence and Duration." *Urban Studies* 12 (December): 2221–2239.

Stoller, Eleanor Palo, and Rose Campbell Gibson. 2000. *Worlds of Difference: Inequality in the Aging Experience.* Thousand Oaks, CA: Pine Forge Press.

Stombler, Mindy, and Irene Padavic. 1997. "Sister Acts: Resisting Men's Domination in Black and White Fraternity Little Sister Programs." *Social Problems* 44 (May): 257–275.

Stoner, J. A. F. 1961. *A Comparison of Individual and Group Decisions Involving Risk.* Master's thesis, MIT, Cambridge, MA.

Storrs, Debbie. 1999. "Whiteness as Stigma: Essentialist Identity Work by Mixed-Race Women." *Symbolic Interaction* 22: 187–212.

Stotik, Jeffrey, Thomas E. Shriver, and Sherry Cable. 1994. "Social Control and Movement Outcome: The Case of AIM." *Sociological Focus* 27 (February): 53–66.

Stover, R. G., and C. A. Hope. 1993. *Marriage, Family, and Intimate Relationships.* New York: Harcourt Brace Jovanovich.

Straus, Murray, Richard Gelles, and Suzanne Steinmetz. 1980. *Behind Closed Doors: Violence in the American Family.* Garden City, NY: Doubleday.

Sudnow, David N. 1967. *Passing On: The Social Organization of Dying.* Englewood Cliffs, NJ: Prentice Hall.

Suggs, Welch. 2002. "Title IX at 30." *The Chronicle of Higher Education* (June 21): A38–42.

Sullivan, Maureen. 1996. "Rozzie and Harriet? Gender and Family Patterns of Lesbian Coparents." *Gender & Society* 12 (December): 747–767.

Sullivan, Teresa A., Elizabeth Warren, and Jay Lawrence Westbrook. 2000. *The Fragile Middle Class: Americans in Debt.* New Haven, CT: Yale University Press.

Sumner, William Graham. 1906. *Folkways.* Boston, MA: Ginn.

Sung, Betty Lee. 1990. "Chinese American Intermarriage." *Journal of Comparative Family Studies* 21 (Autumn): 337–351.

Suro, Roberto. 2000. "Beyond Economics." *American Demographics* (February): 48–55.

Sussal, Carol M. 1994. "Empowering Gays and Lesbians in the Workplace." *Journal of Gay and Lesbian Social Services* 1: 89–103.

Sussman, N. M., and D. H. Tyson. 2000. "Sex and Power: Gender Differences in Computer-Mediated Interactions." *Computers in Human Behavior* 16 (July): 381–394.

Sutherland, Edwin H. 1940. "White Collar Criminality." *American Sociological Review* 5 (February): 1–12.

Sutherland, Edwin H., and Donald R. Cressey. 1978. *Criminology,* 10th ed. New York: Lippincott.

Sweet, James A., and Larry L. Bumpass. 1992. "Young Adults' Views of Marriage, Cohabitation, and Family." Pp. 143–170 in *The Changing American Family: Sociological and Demographic Perspectives,* edited by Scott J. South and Stewart E. Tolnay. Boulder, CO: Westview Press.

Switzer, J. Y. 1990. "The Impact of Generic Word Choices: An Empirical Investigation of Age- and Sex-Related Differences." *Sex Roles* 22: 69–82.

Szasz, Thomas S. 1974. *The Myth of Mental Illness.* New York: Harper and Row.

Szinovacz, Maximiliane. 1996. "Couples' Employment/Retirement Patterns and Perceptions of Marital Quality." *Research on Aging* 18 (June): 243–268.

Szinovacz, Maximiliane E. 2000. "Changes in Housework after Retirement: A Panel Analysis." *Journal of Marriage and the Family* 62 (February): 78–92.

Tabor, James D., and Eugene V. Gallagher. 1995. *Why Waco? Cults and the Battle for Religious Freedom in America.* Berkeley, CA: University of California Press.

Takaki, Ronald. 1989. *Strangers from a Different Shore: A History of Asian Americans.* New York: Penguin.

Takaki, Ronald. 1993. *A Different Mirror: A History of Multicultural America.* Boston, MA: Little Brown.

Takaki, Ronald. 2002. *Debating Diversity: Clashing Perspectives on Race and Ethnicity in America.* New York: Oxford University Press.

Tatum, Beverly. 1997. *Why Are All The Black Kids Sitting Together in the Cafeteria?* New York: Basic Books.

Taylor, Howard F. 1973a. "Linear Models of Consistency: Some Extensions of Blalock's Strategy." *American Journal of Sociology* 78 (March): 1192–1215.

Taylor, Howard F. 1973b. "Playing the Dozens with Path Analysis: Methodological Pitfalls in Jencks et al. *Inequality.*" *Sociology of Education* 46(4): 433–450.

Taylor, Howard F. 1980. *The IQ Game: A Methodological Inquiry into the Heredity Environment Controversy.* New Brunswick, NJ: Rutgers Press.

Taylor, Howard F. 1981. "Biases in *Bias in Mental Testing.*" *Contemporary Sociology* 10: 172–174.

Taylor, Howard F. 1992. "Intelligence." Pp. 941–949 in *Encyclopedia of Sociology,* edited by E. F. Borgatta and M. L. Borgatta. New York: Macmillan.

Taylor, Howard F. 1995. "Symposium on 'The Bell Curve.'" *Contemporary Sociology* 24 (March): 153–158.

Taylor, Howard F. 2002. "Deconstructing the Bell Curve: Racism, Classism, and Intelligence in America." In *2001 Race Odyssey: African Americans and Sociology,* edited by Bruce R. Hare. Syracuse, NY: Syracuse University Press.

Taylor, Howard F. 2003. "Inequality and the Bell Curve: Analyzing the Heritability and Race-Gender Bias of Cognitive Test Scores." Colloquium paper presented at Princeton University, Department of Sociology, Princeton, NJ.

Taylor, Howard F. 2006. "Defining Race." In *Race and Ethnicity in U.S. Society: The Changing Landscape,* edited by Elizabeth Higginbotham and Margaret L. Andersen. Belmont, CA: Wadsworth.

Taylor, Howard F., and Carlton A. Hornung. 1979. "On A General Model for Social and Cognitive Consistency." *Sociological Methods and Research* 7 (February): 259–287.

Taylor, Mark V. C. 2001. "Young Adults and the Appearance of Religion." Pp. 161–174 in *The State of Black America 2001,* edited by Lee A. Daniels. New York: National Urban League.

Taylor, Shelley E., Letitia Ann Peplau, and David O. Sears. 1997. *Social Psychology,* 9th ed. Englewood Cliffs, NJ: Prentice Hall.

Taylor, Shelley E., Letitia Anne Peplau, and David O. Sears. 2003. *Social Psychology,* 11th ed. Upper Saddle River, NJ: Prentice Hall.

Taylor, Verta, and Leila J. Rupp. 1993. "Women's Culture and Lesbian Feminist Activism: A Reconsideration of Cultural Feminism." *Signs* 19 (Autumn): 32–61.

Taylor, Verta, and Nicole C. Raeburn. 1995. "Identity Politics as High-Risk Activism: Career Consequences for Lesbian, Gay and Bisexual Sociologists." *Social Problems* 42 (May): 252–274.

Teaster, Pamela B. 2000. *A Response to the Abuse of Vulnerable Adults: The 2000 Survey of State Adult Protective Services.* Washington, DC: National Center on Elder Abuse. Website: **www.elderabusecenter.org**

Tepper, Clary A., and Kimberly Wright Cassidy. 1999. "Gender Differences in Emotional Language in Children's Picture Books." *Sex Roles* 40 (February): 265–280.

Tewksbury, Richard. 1994. "Gender Construction and the Female Impersonator: The Process of Transforming 'He' to 'She.'" *Deviant Behavior* 15: 27–43.

Thakkar, Reena R., Peter M. Gutierrez, Carly L. Kuczen, and Thomas R. McCanne. 2000. "History of Physical and/or Sexual Abuse and Current Suicidality in College Women." *Child Abuse and Neglect* 24 (October): 1345–1354.

The Sentencing Project. 2000.Washington, DC. Website: **www.sentencingproject.org**

The Sentencing Project. 2001. "Prison Privatization and the Use of Incarceration." Washington, DC: The Sentencing Project. Website: **www.sentencingproject.org**

The Women's Research and Education Institute. 2004. "Women in the Military." The Women's Research and Education Institute. Website: **www.wrei.org**

Thibaut, John W., and Harold T. Kelley. 1959. *The Social Psychology of Groups.* New York: Wiley.

Thibodeau, R. 1989. "From Racism to Tokenism: The Changing Face of Blacks in *New Yorker* Cartoons." *Public Opinion Quarterly* 53: 482–494.

Thoits, Peggy A. 1985. "Self-Labeling Processes in Mental Illness: The Role of Emotional Deviance." *American Journal of Sociology* 91: 221–249.

Thoits, Peggy A. 1991. "On Merging Identity Theory and Stress Research." *Social Psychology Quarterly* 54: 101–112.

Thomas, John K. 1995. "Review: Ecopopulism: Toxic Waste and the Movement for Environmental Justice." *Rural Sociology* 60 (Spring): 151–152.

Thomas, W. I. 1931. *The Unadjusted Girl.* Boston, MA: Little Brown.

Thomas, W. I. 1958 [1918, 1919]. *The Polish Peasant in Europe and America.* New York: Dover.

Thomas, W. I. 1966 [1931]. "The Relation of Research to the Social Process." Pp. 289–305 in *W. I. Thomas on Social Organization and Social Personality,* edited by Morris Janowitz. Chicago, IL: University of Chicago Press.

Thomas, William I., with Dorothy Swaine Thomas. 1928. *The Child in America.* New York: Knopf.

Thompson, Becky. 1994. *A Hunger So Wide and So Deep: American Women Speak Out on Eating Problems.* Minneapolis, MN: University of Minnesota Press.

Thompson, Mark A. 1997. "The Impact of Spatial Mismatch on Female Labor Force Participation." *Economic Development Quarterly* 2 (May): 138–145.

Thompson, Robert Ferris. 1993. *Slash of the Spirit.* New York: Random House.

Thomson, Rob, and Tamar Murachver. 2001. "Predicting Gender from Electronic Discourse." *The British Journal of Social Psychology* 40 (June): 193–208.

Thornberry, Terence P., Carolyn A. Smith, and Gregory J. Howard. 1997. "Risk Factors for Teenage Fatherhood." *Journal of Marriage and the Family* 59 (August): 505–522.

Thorne, Barrie. 1993. *Gender Play: Girls and Boys in School.* New Brunswick, NJ: Rutgers University Press.

Thorne, Barrie, and Zella Luria. 1986. "Sexuality and Gender in Children's Daily Worlds." *Social Problems* 33 (February): 176–190.

Thornton, Michael. 1995. "Is Multiracial Experience Unique? The Personal and Social Experience." Pp. 95–99 in *Race, Class, and Gender: An Anthology,* 2nd ed., edited by Margaret L. Andersen and Patricia Hill Collins. Belmont, CA: Wadsworth.

Thornton, Russell. 1987. *American Indian Holocaust and Survival: A Population History.* Norman, OK: University of Oklahoma Press.

Tichenor, Veronica Jaris. 1999. "Status and Income as Gendered Resources: The Case of Marital Power." *Journal of Marriage and the Family* 61 (August): 638–650.

Tiefer, Leonore. 1978. "The Kiss." *Human Nature* (July): 28–37.

Tienda, Marta, and Haya Stier. 1996. "Generating Labor Market Inequality: Employment Opportunities and the Accumulation of Disadvantage." *Social Problems* 43 (May): 147–165.

Tierney, Kathleen. 1994. "Making Sense of Collective Preoccupations: Lessons from Research on the Iben Browning Earthquake Prediction." Pp. 75–95 in *Collective Behavior and Society,* edited by Gerald Platt and Chad Gordon. Greenwich, CT: JAI Press.

Tierney, Kathleen. 2000. Personal correspondence, University of Delaware, Newark, DE.

Tierney, Kathleen. 2002. "Group Dynamics under Pressure." *Talk of the Nation,* National Public Radio, July 31.

Tierney, Patrick. 2000. *Darkness in El Dorado: How Scientists and Journalists Devastated the Amazon.* New York: W. W. Norton.

Tilly, Charles. 1975. *The Formation of National States in Europe.* Princeton, NJ: Princeton University Press.

Tilly, Charles. 1978. *From Mobilization to Revolution.* Reading, MA: Addison-Wesley.

Tilly, Louise, and Joan Scott. 1978. *Women, Work, and Family.* New York: Holt, Rinehart, and Winston.

Time Magazine (eds.) 1996. "America's Most Influential People." *Time* (June 17): 56–57.

Tizand, Barbara. 1993. "Racism in Children's Lives: A Study of Mainly-White Primary Schools." *Journal of Child Psychology and Psychiatry and Allied Disciplines* 34 (Nov): 1484–1485.

Tjaden, Patricia, and Nancy Thoennes. 2000. *Extent, Nature, and Consequences of Intimate Partner Violence.* Washington, DC: U.S. Bureau of Justice Statistics.

Tönnies, Ferdinand. 1963 [1887]. *Community and Society (Gemeinschaft and Gesellschaft).* New York: Harper and Row.

Toobin, Jeffrey. 1996. "The Marcia Clark Verdict." *The New Yorker* (September 9): 58–71.

Toubia, Nahid, and Susan Izett. 1998. *Female Genital Mutilation: An Overview.* Geneva: World Health Organization.

Toynbee, Arnold J., and Jane Caplan. 1972. *A Study of History.* New York: Oxford University Press.

Travers, Jeffrey, and Stanley Milgram. 1969. "An Experimental Study of the Small World Problem." *Sociometry* 32: 425–443.

Trent, Katherine, and Sharon L. Harlan. 1990. "Household Structure Among Teenage Mothers in the United States." *Social Science Quarterly* 71 (September): 439–457.

True, Reiko Homma, and Tessie, Guillermo. 1996. Pp. 94–120 in *Race, Gender, and Health,* edited by Marcia Bayne-Smith. Thousand Oaks, CA: Sage.

Tschann, Jeanne M., Janet R. Johnson, Marsha Kline, and Judith S. Wallerstein. 1989. "Family Process and Children's Functioning During Divorce." *Journal of Marriage and the Family* 2 (May): 431–444.

Tschann, Jeanne M., Janet R. Johnson, Marsha Kline, and Judith S. Wallerstein. 1990. "Conflict, Loss, Change and Parent-Child Relationships: Predicting Children's Adjustment during Divorce." *Journal of Divorce* 13: 1–22.

Tuan, Yi-Fu. 1984. *Dominance and Affection: The Making of Pets.* New Haven, CT: Yale University Press.

Tuchman, Gaye. 1979. "Women's Depiction by the Mass Media." *Signs* 4 (Spring): 528–542.

Tumin, Melvin M. 1953. "Some Principles of Stratification." *American Sociological Review* 18 (August): 387–393.

Turkle, Sherry. 1995. *Life on the Screen: Identity in the Age of the Internet.* New York: Simon and Schuster.

Turner, Ralph, and Lewis Killian. 1988. *Collective Behavior,* 3rd ed. Englewood Cliffs, NJ: Prentice Hall.

Turner, Ralph, Joanne M. Nigg, and Denise Paz. 1986. *Waiting for Disaster: Earthquake Watch in California.* Berkeley, CA: University of California Press.

Turner, Terence. 1969. "Tchikrin: A Central Brazilian Tribe and Its Symbolic Language of Body Adornment." *Natural History Magazine* 78 (October): 50–59.

Tyree, Andrea, and R. Hicks. 1988. "Sex and the Second Moment of Occupational Basic Distributions." *Social Forces* 66 (June): 1028–1037.

U.S. Bureau of Justice Statistics. 2000a. *Criminal Victimization 2000.* Washington, DC: U.S. Department of Justice.

U.S. Bureau of Justice Statistics. 2000b. *Sourcebook of Criminal Justice Statistics 2000.* Washington, DC: U.S. Department of Justice.

U.S. Bureau of Justice Statistics. 2002a. "Prisons." U.S. Department of Justice. Website: **www.ojp.usdoj.gov/bjs/**

U.S. Bureau of Justice Statistics. 2002b. *Sourcebook of Criminal Justice Statistics 2002.* Washington, DC: U.S. Department of Justice.

U.S. Bureau of Labor Statistics. 2003. *National Compensation Survey: Occupational Wages in the United States, July 2002.* Washington, DC: U.S. Department of Labor.

U.S. Census Bureau. 1993. *We the . . . First Americans.* Washington, DC: U.S. Government Printing Office.

U.S. Census Bureau. 1999. *Statistical Abstract of the United States, 1998.* Washington, DC: U.S. Government Printing Office.

U.S. Census Bureau. 2000. *Statistical Abstract of the United States, 2001.* Washington, DC: U. S. Government Printing Office.

U.S. Census Bureau. 2001. *Living Arrangements of Children Under 18 Years Old: 1960 to Present.* Website: **www.census.gov/population/socdemo/hh-fam/tabCH-1.txt**

U.S. Census Bureau. 2002. *Poverty: Detailed Historical Tables.* Washington, DC: U.S. Census Bureau. Website: **www.census.gov**

U.S. Census Bureau. 2004. *Historic Income Tables-People.* Website: **www.census.gov**

U.S. Census Bureau. 2004. *Statistical Abstract of the United States 2003.* Washington, DC: U.S. Government Printing Office.

U.S. Conference of Mayors. 2001. *A Status Report on Hunger and Homelessness in America's Cities: 2001.* Washington, DC: U.S. Conference of Mayors.

U.S. Department of Defense. 2000. "Minorities in Uniform." *Defense Almanac.* Washington, DC: U.S. Department of Defense. Website: **www.defenselink.mil**

U.S. Department of Defense. 2002. Population Representation in the Military Services, Fiscal Year 2000. Washington, DC: U.S. Department of Defense. Website: **www.defenselink.mil**

U.S. Department of Education and U.S. Department of Justice. 2000. *1999 Annual Report on School Safety Report.* Washington, DC: U.S. Department of Education/U.S. Department of Justice.

U.S. Department of Labor. 2000. *Employment and Earnings.* Washington, DC: U.S. Department of Labor.

U.S. Department of Labor. 2001. *Employment and Earnings.* Washington, DC: U.S. Department of Labor.

U.S. Department of Labor. 2004. *Employment and Earnings.* Washington, DC: U.S. Department of Labor.

Uleman, J. S., L. S. Newman, and G. B. Moskowitz. 1996. "People as Flexible Interpreters: Evidence and Issues from Spontaneous Trait Inference." Pp. 211–279 in *Advances in Experimental Social Psychology,* Vol. 28, edited by Mark P. Zanna. Boston, MA: Academic Press.

Ullman, Sarah E., George Karabatsos, and Mary P. Koss. 1999. "Alcohol and Sexual Assault in A National Sample of College Women." *Journal of Interpersonal Violence* 14 (June): 603–625.

UNICEF. 2000. *Domestic Violence Against Women and Girls.* New York: United Nations.

United Nations Commission on the Status of Women. 1996. *Report of the World Conference of the United Nations Decade for Women.* Copenhagen: United Nations.

United Nations. 2000a. *Human Development Report.* New York: United Nations.

United Nations. 2000b. *The World's Women, 1995: Trends and Statistics.* New York: United Nations.

Urban Institute, The. 2000. *A New Look at Homelessness in America.* Washington, DC: The Urban Institute. Website: **www.urban.org**

Useem, Jerry. 2000. "Welcome to the New Company Town." *Fortune* (January 10): 62–70.

Uvin, Peter. 1998. *Aiding Violence.* West Hartford, CT: Kumarian Press.

Vail, D. Angus. 1999. "Tattoos Are Like Potato Chips . . . You Can't Have Just One: The Process of Becoming and Being a Collector." *Deviant Behavior* 20: 253–273.

Van Ausdale, Debra, and Joe R. Feagin. 1996. "The Use of Racial and Ethnic Concepts by Very Young Children." *American Sociological Review* 61 (October): 779–793.

Van Ausdale, Debra, and Joe R. Feagin. 2000. *The First R: How Children Learn Race and Racism.* Lanham, MD: Rowman and Littlefield.

Van Roosmalen, Erica, and Susan A. McDaniel. 1998. "Sexual Harassment in Academia: A Hazard to Women's Health." *Women and Health* 28: 33–54.

Vanlandingham, Mark. 1993. *Two Perspectives On Risky Sexual Practices Among Northern Thai Males: The Health Belief Model and the Theory of Reasoned Action.* Ph.D. diss. Princeton University, Princeton, NJ.

Vanneman, Reeve, and Lynn Weber Cannon. 1987. *The American Perception of Class.* Philadelphia, PA: Temple University Press.

Varelas, Nicole, and Linda A. Foley. 1998. "Blacks' and Whites' Perceptions of Interrracial and Intraracial Date Rape." *Journal of Social Psychology* 138 (June): 392–400.

Vaughan, Diane. 1996. *The Challenger Launch Decision: Risky Technology, Culture, and Deviance at NASA.* Chicago, IL: University of Chicago Press.

Vazquez, Carmen. 1992. "Appearances." Pp. 157–166 in *Homophobia: How We All Pay the Price,* edited by Warren J. Blumenfeld. Boston, MA: Beacon Press.

Veblen, Thorstein. 1899 [1953]. *The Theory of the Leisure Class: An Economic Study of Institutions.* New York: New American Library.

Vega, W. A., and H. Amero. 1994. "Latino Outlook: Good Health, Uncertain Prognosis." *Annual Review of Public Health* 15: 39–67.

Verhovek, Sam Howe. 1993. "'Messiah' Fond of Bible, Rock and Women." *The New York Times* (March 3): A1.

Vessels, Jane. 1985. "Koko's Kitten." *National Geographic* 167 (January): 110–113.

Vigorito, Anthony J., and Timothy J. Curry. 1998. "Marketing Masculinity: Gender Identity and Popular Magazines." *Sex Roles* 39 (July): 135–152.

VonNeumann, John, and Oskar Morgenstern. 1944. *The Theory of Games and Economic Behavior.* Princeton, NJ: Princeton University Press.

Voss, Laurie Scarborough. 1997. "Teasing, Disputing, and Playing: Cross-Gender Interactions and Space Utilization Among First and Third Graders." *Gender & Society* 11 (April): 238–256.

Wagner, David. 1994. "Beyond the Pathologizing of Nonwork: Alternative Activities in a Street Community." *Social Work* 29 (November): 718–727.

Waite, Linda J., and Maggie Gallagher. 2000. *The Case for Marriage.* New York: Broadway Books.

Waldfogel, Jane. 1999. "Family Leave Coverage in the 1990s." *Monthly Labor Review* 122 (October): 13–21.

Waldfogel, Jane, Wen-Jui Han, and Jeanne Brooks-Gunn. 2002. "The Effects of Early Maternal Employment on Child Cognitive Development." *Demography* 39: 369–392.

Walker-Barnes, Chanequa J., and Craig A. Mason. 2001. "Perceptions of Risk Factors for Female Gang Involvement among African American and Hispanic Women." *Youth and Society* 32 (March): 303–336.

Wallerstein, Immanuel. 1974. *The Modern World System: Capitalist Agriculture and the Origins of the European World Economy in the Sixteenth Century.* New York: Academic Press.

Wallerstein, Immanuel M. 1979. *The Capitalist World-Economy.* New York: Cambridge University Press.

Wallerstein, Immanuel M. 1980. *The Modern World-System II.* New York: Academic Press.

Wallerstein, Immanuel M. 1989. *The Modern World System III: The Second Era of Great Expansion of the Capitalist World-Economy, 1730–1840.* New York: Academic Press.

Wallman, Joel. 1992. *Aping Language.* New York: Cambridge University Press.

Walsh, Anthony. 1994. "Homosexual and Heterosexual Child Molestation: Case Characteristics and Sentencing Differentials." *International Journal of Offender Therapy and Comparative Criminology* 38: 339–353.

Walster, Elaine, V. Aronson, D. Abrahams, and L. Rottman. 1966. "The Importance of Physical Attractiveness in Dating Behavior." *Journal of Personality and Social Psychology* 4: 508–516.

Walters, Suzanna. 1995. *Material Girls: Making Sense of Feminist Cultural Theory.* Berkeley, CA: University of California Press.

Warner, Rebecca L., and Brent S. Steel. 1999. "Child Rearing as a Mechanism for Social Change: The Relationship of Child Gender to Parents' Commitment to Gender Equity." *Gender & Society* 13 (August): 503–517.

Warren, Tracey, Karen Rowlingson, and Claire Whyley. 2001. "Female Finances: Gender Wage Gaps and Gender Assets Gaps." *Work, Employment, and Society* 15 (September): 465–488.

Washington, Scott. 2004. "Racial Taxonomy." Unpublished manuscript. Department of Sociology, Princeton University, Princeton, NJ.

Wasserman, Stanley, and Katherine Faust (eds.). 1994. *Social Network Analysis: Methods and Applications.* Cambridge, MA: Cambridge University Press.

Waters, Mary C. 1990. *Ethnic Options: Choosing Identities in America.* Berkeley, CA: University of California Press.

Waters, Mary C. 1998. "Multiple Ethnic Identity Choices." Pp. 28–46 in *Beyond Pluralism: The Conception of Groups and Group Identities in America,* edited by Wendy F. Katkin, Ned Landsman, and Andrea Tyree. Urbana, IL: University of Illinois Press.

Watkins, Susan. 1987. "The Fertility Transition: Europe and the Third World Compared." *Sociological Forum* 2 (Fall): 645–673.

Watt, Toni Terling, and Susan F. Sharp. 2001. "Gender Differences in Strains Associated with Suicidal Behavior among Adolescents." *Journal of Youth and Adolescence* 30 (June): 333–348.

Watts, Duncan J. 1999. "Networks, Dynamics, and the Small-World Phenomenon." *American Journal of Sociology* 105 (September): 493–527.

Watts, Duncan J., and Stephen H. Strogatz. 1998. "Collective Dynamics of 'Small World' Networks." *Nature* 393: 440–442. Website: **www .sentencing project.org**

Weaver, Frederick Stirton, and Ron Chilcote. 2000. *Latin America in the World Economy: Mercantile Colonialism to Global Capitalism.* Boulder, CO: Westview Press.

Weber, Max. 1947 [1925]. *The Theory of Social and Economic Organization.* New York: Free Press.

Weber, Max. 1958 [1904]. *The Protestant Ethic and the Spirit of Capitalism.* New York: Scribners.

Weber, Max. 1962 [1913]. *Basic Concepts in Sociology.* New York: Greenwood.

Weber, Max. 1978 [1921]. *Economy and Society: An Outline of Interpretive Sociology,* edited by Guenther Roth and Claus Wittich. Berkeley, CA: University of California Press.

Webster, Murray, Jr., and Stuart J. Hysom. 1998. "Creating Status Characteristics." *American Sociological Review* 63 (June): 351–378.

Weeks, John R. 2004. *Population: An Introduction to Concepts and Issues,* 9th ed. Belmont, CA: Wadsworth.

Wei, William. 1993. *The Asian American Movement.* Philadelphia, PA: Temple University Press.

Weibel-Orlando, Joan. 1990. "Grandparenting Styles: Native American Perspectives." Pp. 109–124 in *The Cultural Context of Aging: Worldwide Perspectives,* edited by Jay Sokolovsky. New York: Bergin and Garvey.

Weisburd, David, Cynthia M. Lum, and Anthony Petrosino. 2001. "Does Research Design Affect Study Outcomes in Criminal Justice?" *The Annals of the American Academy of Political and Social Science* 578 (November): 50–70.

Weisburd, David, Stanton Wheeler, Elin Waring, and Nancy Bode. 1991. *Crimes of the Middle Class: White Collar Defenders in the Courts.* New Haven, CT: Yale University Press.

Weitz, Rose. 2001. *The Sociology of Health, Illness, and Health Care: A Critical Approach.* Belmont, CA: Wadsworth.

Weitzman, Lenore J. 1985. *The Divorce Revolution: The Unexpected Consequences for Women and Children in America.* New York: Free Press.

Weitzman, Lenore J. 1996. "A Re-Evaluation of the Economic Consequences of Divorce: Reply to Peterson." *American Sociological Review* 61 (June): 537–538.

Welch, Susan, and Lee Sigelman. 1993. "The Politics of Hispanic Americans: Insights from the National Surveys, 1980–1988." *Social Science Quarterly* 74 (March): 76–94.

Weller, Jack, and Enrico Quarantelli. 1973. "Neglected Characteristics of Collective Behavior." *American Journal of Sociology* 79: 665–685.

Welsh, Sandy. 1999. "Gender and Sexual Harassment." *Annual Review of Sociology* 25: 169–190.

Wenk, Deeann, and Patricia Garrett. 1992. "Having a Baby: Some Predictions of Maternal Employment Around Childbirth." *Gender & Society* 6 (March): 49–65.

West, Candace, and Don Zimmerman. 1987. "Doing Gender." *Gender & Society* 1 (June): 125–151.

West, Candace, and Sarah Fenstermaker. 1995. "Doing Difference." *Gender & Society* 9 (February): 8–37.

West, Carolyn M. 1998. "Leaving a Second Closet: Outing Partner Violence in Same-Sex Couples." Pp. 163–183 in *Partner Violence: A Comprehensive Review of 20 Years of Research,* edited by Jana L. Jasinksi and Linda M. Williams. Thousand Oaks, CA: Sage.

West, Cornel. 1993. *Race Matters.* Boston, MA: Beacon Press.

Westoff, Charles F., Noreen Goldman, and Lorenzo Moreno. 1990. *Dominican Republic Experimental Study: An Evaluation of Fertility and Child Health Information.* Princeton, NJ: Office of Population Research.

Wharton, Amy S., and Deborah K. Thorne. 1997. "When Mothers Matter: The Effects of Social Class and Family Arrangements on African American and White Women's Perceived Relations with Their Mothers." *Gender & Society* 11 (October): 656–681.

Whelan, Christine B. 2004. *Self-Help Books and the Quest for Self-Control in the United States, 1950–2000.* Ph.D. diss., University of Oxford, Worcester College, Oxford, England.

White, Deborah Gray. 1985. *Ar'n't I a Woman?: Female Slaves in the Plantation South.* New York: W. W. Norton.

White, Jonathan R. 2002. *Terrorism: An Introduction.* Belmont, CA: Wadsworth.

White, Paul. 1998. "The Settlement Patterns of Developed World Migrants in London." *Urban Studies* 35 (October): 1725–1744.

Whitley, Bernard E. 2001. "Gender-Role Variables and Attitudes toward Homosexuality." *Sex Roles* 45: 691–721.

Whittier, Nancy. 1995. *Feminist Generations: The Persistence of the Radical Women's Movement.* Philadelphia, PA: Temple University Press.

Whorf, Benjamin. 1956. *Language, Thought, and Reality: Selected Writings.* Cambridge, MA: MIT Press.

Whyte, William F. 1943. *Street Corner Society.* Chicago, IL: University of Chicago Press.

Wickelgren, Ingrid. 1999. "Discovery of 'Gay Gene' Questioned." *Science* 284 (April 23): 571.

Wilcox, Clyde, Lara Hewitt, and Dee Allsop. 1996. "The Gender Gap in Attitudes Toward the Gulf War: A Cross-National Perspective." *Journal of Peace Research* 33 (February): 67–82.

Wilder, Esther I., and Toni Terling Watt. 2002. "Risky Parental Behavior and Adolescent Sexual Activity at First Coitus." *The Milbank Quarterly* 80 (September): 481–524.

Wilkie, Jane. 1993. "Changes in U.S. Men's Attitudes Toward the Family Provider Role, 1972 to 1989." *Gender & Society* 7 (June): 261–279.

Williams, Christine L. 1989. *Gender Differences at Work: Women and Men in Nontraditional Occupations.* Berkeley, CA: University of California Press.

Williams, Christine L. 1992. "The Glass Escalator: Hidden Advantages for Men in the 'Female' Professions." *Social Problems* 39 (August): 253–267.

Williams, Christine L. 1995. *Still a Man's World: Men Who Do Women's Work.* Berkeley, CA: University of California Press.

Williams, David R., and Rith Morris Williams. 2000. "Racism and Mental Health: The African American Experience." *Ethnicity and Health* 5 (August–November): 243–268.

Williams, Lena. 1987. "Race Bias Found in Location of Toxic Dumps." *The New York Times* (April): A20.

Williams, Norma. 1990. *The Mexican American Family: Tradition and Change.* Dix Hills, NY: General Hall.

Willie, Charles Vert. 1979. *The Caste and Class Controversy.* Bayside, NY: General Hall.

Willis, Jon. 1999. "Dying in Country: Implications of Culture in the Delivery of Palliative Care in

Indigenous Australian Communities." *Anthropology and Medicine* 6 (December) 423–435.

Willis, Leigh A., David W. Coombs, and William C. Cockerham. 1999. "Ready to Die: A Postmodern Interpretation of the Increase of African American Adolescent Males Suicide." Paper presented at the annual meeting of the American Sociological Association, Chicago, IL.

Willis, Paul. 1977. *Learning to Labor: How Working Class Kids Get Working Class Jobs.* New York: Columbia University Press.

Willits, Fern K., and Donald M. Crider. 1989. "Church Attendance and Traditional Religious Beliefs in Adolescence and Young Adulthood: A Panel Study." *Review of Religious Research* 31 (September): 68–81.

Wilson, Barbara J., Stacy L. Smith, W. James Potter, Dale Kunkel, Daniel Linz, Carolyn M. Colvin, and Edward Donnerstein. 2002. "Violence in Children's Television Programming: Assessing the Risks." *Journal of Communication* 52 (March): 5–35.

Wilson, Bryan. 1982. *Religion in Sociological Perspective.* New York: Oxford University Press.

Wilson, James Q., and Richard J. Herrnstein. 1985. *Crime and Human Nature.* New York: Simon and Schuster.

Wilson, John. 1994. "Returning to the Fold." *Journal for the Scientific Study of Religion* 33 (June): 148–161.

Wilson, William Julius. 1978. *The Declining Significance of Race: Blacks and Changing American Institutions.* Chicago, IL: University of Chicago Press.

Wilson, William Julius. 1987. *The Truly Disadvantaged: The Inner City, the Underclass, and Public Policy.* Chicago, IL: University of Chicago Press.

Wilson, William Julius. 1996. *When Work Disappears: The World of the New Urban Poor.* New York: Knopf.

Winnick, Louis. 1990. "America's 'Model Minority.'" *Commentary* 90 (August): 222–229.

Winseman, Albert L. 2002. "Religion and Gender: A Congregation Divided." *The Gallup Poll.* Princeton, NJ: Gallup Organization. Website: **www.gallup.com**

Winseman, Albert L. 2004. "'Born Agains' Wield Political, Economic Influence." *The Gallup Poll.* Princeton, NJ: Gallup Organization. Website: **www.gallup.com**

Winship, Scott, and Christopher Jencks. 2002. "The Well-Being of Single Mothers after Welfare Reform, as Measured by Changes in Food Security." *Policy Brief,* Vol. 4, No. 7. Joint Center for Poverty Research, Northwestern University. Website: **www.jcpr.org**

Wirth, Louis. 1938. "Urbanism as a Way of Life." *American Journal of Sociology* 40: 1–24.

Wolf, Naomi. 1991. *The Beauty Myth: How Images of Beauty Are Used Against Women.* New York: William Morrow.

Wolfinger, Nicholas. 2003. "Family Structure Homogamy: The Effects of Parental Divorce on Partner Selection and Marital Stability." *Social Science Research* 32 (March): 80–97.

Wolmsley, Roy. 2000. *World Prison Population List,* 2nd ed. London: United Kingdom Home Office Research, Development, and Statistics Directorate.

Woo, Deborah. 1998. "The Gap Between Striving and Achieving." Pp. 247–256 in *Race, Class, and Gender: An Anthology,* 3rd ed., edited by

Margaret L. Andersen and Patricia Hill Collins. Belmont, CA: Wadsworth.

Wood, Julia T. 1994. *Gendered Lives: Communication, Gender, and Culture.* Belmont, CA: Wadsworth.

Wood, Richard L. 2002. *Faith in Action: Religion, Race, and Democratic Organizing in America.* Chicago, IL: University of Chicago Press.

Woodberry, Robert D. 1998. "Comment: When Surveys Lie and People Tell the Truth: How Surveys Oversample Church Attendance." *American Sociological Review* 63 (February): 119–121.

Woodruff, J. T. 1985. "Premarital Sexual Behavior and Religious Adolescents." *Journal for the Scientific Study of Religion* 25 (December): 343–386.

Wootton, Barbara H. 1997. "Gender Differences in Occupational Employment." *Monthly Labor Review* 170 (April): 15–24.

Worchel, Stephen, Joel Cooper, George R. Goethals, and James L. Olsen. 2000. *Social Psychology.* Belmont, CA: Wadsworth/Thomson Learning.

Workman, Jane E., and Elizabeth W. Freeburg. 1999. "An Examination of Date Rape, Victim Dress, and Perceiver Variables within the Context of Attribution Theory." *Sex Roles* 41 (August): 261–277.

World Bank. 2004. "World Development Indicators." Website: **www.worldbank.org**

World Health Organization. 2000. "Women and HIV/AIDS." Website: **www.who.int**

World Health Organization. 2002. Website: **www.who.org**

Wright, Erik Olin. 1979. *Class Structure and Income Determination.* New York: Academic Press.

Wright, Erik Olin. 1985. *Classes.* London: Verso.

Wright, Erik Olin, Karen Shire, Shu-Ling Hwang, Maureen Dolan, and Janeen Baxter. 1992. "The Non-Effects of Class on the Gender Division of Labor in the Home: A Comparative Study of Sweden and the United States." *Gender & Society* 6 (June): 252–282.

Wright, Laurence. 1994. "One Drop of Blood." *The New Yorker* 70 (July 25): 46–55.

Wright, Steve. 2002. "'A Love Born of Hate': Autonomist Rap in Italy." *Theory, Culture, and Society* 17 (June): 117–136.

Wright, Stuart A. 1991. "Reconceptualizing Cult Coercion and Withdrawal: A Comparative Analysis of Divorce and Apostasy." *Social Forces* 70 (September): 125–145.

Wrong, Dennis. 1961. "The Oversocialized Conception of Man in Modern Sociology." *American Sociological Review* 26 (April): 183–192.

Wuthnow, Robert (ed.). 1994. *I Come Away Stronger: How Small Groups Are Shaping American Religion.* Grand Rapids, MI: William B. Eerdmans.

Wuthnow, Robert. 1994. *Sharing the Journey: Support Groups and America's New Quest for Community.* New York: Free Press.

Wuthnow, Robert. 1998. *After Heaven: Spirituality in America Since the 1950s.* Berkeley, CA: University of California Press.

Wuthnow, Robert, and Marsha Witten. 1988. "New Directions in the Study of Culture." *Annual Review of Sociology* 14: 49–67.

Wyman, David. 1984. *The Abandonment of the Jews: America and the Holocaust, 1941–1945.* New York: Pantheon.

Wysocki, Diane Kholos. 1998. "Let Your Fingers Do the Talking: Sex on an Adult Chat-Line." *Sexualities* 1 (November): 425–452.

Xu, Wu, and Ann Leffler. 1992. "Gender and Race Effects on Occupational Prestige, Segregation, and Earnings." *Gender & Society* 6 (September): 376–392.

Yalom, Marilyn. 1997. *A History of the Breast.* New York: Ballantine.

Yoder, Janice. 1991. "Rethinking Tokenism: Looking Beyond Numbers." *Gender & Society* 5 (June): 178–192.

Young, J. W. 1994. "Differential Prediction of College Grades by Gender and by Ethnicity: A Replication Study." *Educational and Psychological Measurement* 54: 1022–1029.

Youniss, J., and J. Smollar. 1985. *Adolescent Relations with Mothers, Fathers, and Friends.* Chicago, IL: University of Chicago Press.

Zajonc, Robert B. 1968. "Attitudinal Effects of Mere Exposure." *Journal of Personality and Social Psychology* (Monograph Suppl., Part 2): 1–29.

Zald, Mayer, and John McCarthy. 1975. "Organizational Intellectuals and the Criticism of Society." *Social Service Research* 49: 344–362.

Zelizer, Viviana. 1985. *Pricing the Priceless Child: The Changing Social Value of Children.* Princeton, NJ: Princeton University Press.

Zicklin, Gilbert. 1995. "Deconstructing Legal Rationality: The Case of Lesbian and Gay Family Relationships." *Marriage and Family Review* 21: 55–76.

Zimbardo, Phillip G., Ebbe B. Ebbesen, and Christina Maslach. 1977. *Influencing Attitudes and Changing Behavior.* Reading, MA: Addison-Wesley.

Zisook, S., S. Shuchter, and P. Sledge. 1994. "Diagnostic and Treatment Considerations in Depression Associated with Late Life Bereavement." Pp. 419–435 in *Diagnosis and Treatment of Depression in Late Life: Results of the NIH Consensus Development Conference,* edited by L. S. Schneider, C. F. Reynolds, B. D. Lebowitz, and A. J. Friedhoff. Washington, DC: American Psychiatric Press.

Zola, Irving Kenneth. 1989. "Toward the Necessary Universalizing of a Disability Policy." *Milbank Quarterly* 67 (Suppl. 2): 401–428.

Zola, Irving Kenneth. 1993. "Self, Identity and the Naming Question: Reflections on the Language of Disability." *Social Science and Medicine* 36 (January): 167–173.

Zones, Jane Sprague. 1997. "Beauty Myths and Realities and Their Impact on Women's Health." Pp. 249–275 in *Women's Health: Complexities and Differences,* edited by Sheryl B. Ruzek et al. Columbus, OH: Ohio State University Press.

Zsembik, Barbara A., and Audrey Singer. 1990. "The Problems of Defining Retirement Among Minorities: Mexican Americans." *The Gerontologist* 30: 749–757.

Zuberi, Tukufu. 2001. *Thicker Than Blood: How Racial Statistics Lie.* Minneapolis, MN: University of Minnesota Press.

Zweigenhaft, Richard L., and G. William Domhoff. 1998. *Diversity in the Power Elite: Have Women and Minorities Reached the Top?* New Haven, CT: Yale University Press.

Zwerling, Craig, and Hilary Silver. 1992. "Race and Job Dismissals in a Federal Bureaucracy." *American Sociological Review* 57 (October): 651–660.

Credits

Chapter 1
p. 3 Lara Jo Regan/Liaison/Getty Images; p. 5 Spencer Platt/Getty Images; p. 8/left © Jessica Rinaldi/Getty Images; p. 8/right © Lindsay Hebberd/Corbis; p. 9 © Richard Lord/The Image Works; p. 13 A. Ramey/PhotoEdit Inc.; p. 15/top © Culver Pictures; p. 15/bottom © Bettmann/Corbis; p. 16/top © Bettmann/Corbis; p. 16/bottom AKG London; p. 18 © CORBIS; p. 19 © Courtesy of University of Massachusetts at Amherst

Chapter 2
p. 28 © Brian Strickland/Zuma Press. © Copyright 2004 by Brian Strickland; p. 33 © Rob Crandall/The Image Works; p. 35 © Chris Brown/Corbis ; p. 36 Jeff Greenberg/Photo Researchers, Inc.; p. 37 © Mark Richards/PhotoEdit; p. 38 AP Wide World Photo/Rodney White; p. 42 © Bettmann/Corbis; p. 45 © Jeremy Hartley/Panos Pictures

Chapter 3
p. 55 top left Francene Keery/Stock Boston; p. 55/top right © Bob Krist/Corbis; p. 55/bottom left © Annie Griffiths Belt/Corbis; p. 55/bottom right © Lawrence Migdale/Photo Researchers, Inc.; p. 56 © Tami Chappell/Reuters/Landov; p. 57 AP/Wide World Photos; p. 64 © Art Kane Estate. All rights reserved; p. 67 © Jeffrey Greenburg/PhotoEdit; p. 76/top John Phillips/Time Pix/Getty Images; p. 76/bottom © Kathy McLaughlin/The Image Works; p. 77 © Christopher Bissell/Stone/Getty Images

Chapter 4
p. 84 © Keith Meyers/NYT Pictures; p. 88/bottom ©Amy Etra/PhotoEdit; p. 88/top AP/Wide World Photos/James Woodcock/Billings Gazette; p. 90 Matt Suess; p. 94 © Paul Chesley/Stone/Getty Images; p. 98 © Caroline Penn/Corbis; p. 104/top left ©Juliet Highet/The Hutchinson Library; p. 104/top right © Judy Griesedieck/Corbis; p. 104/bottom left © E. A. Heiniger/Photo Researchers, Inc.; p. 104/bottom right © Sami Sallinen/Panos Pictures; p. 105 © Photodisc Blue/Getty Images; p. 106 © Getty Images

Chapter 5
p. 114 © Lon C. Diehl/PhotoEdit; p. 115 © Nick Robinson/Panos Pictures; p. 118 © Ryan McVay/PhotoDisc/Getty Images ; p. 124/top Dr. Fumio Hara, Science University of Tokyo; p. 124/bottom © Paul Ekman, from the Nebraska Symposium on Motivation, 1972. Courtesy of the Human Interaction Laboratory, UCSF; p. 125 © James Marshall/Corbis; p. 126 © Nina Leen/Time Life Pictures/Getty Images; p. 131 © Penny Tweedie/Corbis; p. 136 © David Young Wolff/PhotoEdit

Chapter 6
p. 141 © John Eastcott/Yva Momatiuk/The Image Works; p. 142 © Joseph Sohm, Chromo Sohm Inc./Corbis; p. 147 Source: Milgram, Stanley. 1974. Obedience to Authority: An Experimental View. New York: Harper and Row, p. 25. © Copyright 1965 by Stanley Milgram.; p. 150 © Pohle/Sutcliffe/Sipa Press; p. 154/left © The New York Times/Sipa Press; p. 154/top right AP Wide World Photos/NASA ; p. 154/bottom right AP Wide World Photo/Tyler Morning Telegraph, Dr. Scott Lieberman; p. 155 © Billy Barnes/Photo Edit; p. 157 top Rick Maimen/Corbis; p. 157/bottom © PEMCO/Corbis; p. 158 © Y. Ishii/PPS/Photo Researchers, Inc.

Chapter 7
p. 168 RAWA/WorldPicture News; p. 169 © M.Sofronski/Sipa Press; p. 171 © Christopher Brown/Stock Boston; p. 172 © Terry Eiler/Stock Boston; p. 175 © Viviane Moos/Corbis; p. 176 AP/Wide World Photos; p. 177 © Reuters News Media Inc./Corbis; p. 181 © Mark Peterson/Corbis Saba

Chapter 8
p. 194 © Robert Yager/Stone/Getty Images; p. 197/all AP/Wide World Photos; p. 203 © Fritz Hoffmann/The Image Works; p. 207 AP/Wide World Photos

Chapter 9
p. 213/left © Tony Freeman/PhotoEdit ; p. 213/right © Anne Dowie; p. 214/top © Topham/The Image Works ; p. 214/center Jon Riley/Stone/Getty Images ; p. 214/bottom © Jim Lo Scalzo/Corbis Sygma; p. 226/top AP/Wide World Photos; p. 226/bottom © Lea Suzuki/The San Francisco Chronicle; p. 227 © Most Wanted/303/Zuma Press. © Copyright 2004 by Most Wanted/303; p. 234 © Tony Freeman/PhotoEdit; p. 237 © Allen Russell/Index Stock Imagery

Chapter 10
p. 247 © Marie Dorigny/REA/Corbis Saba; p. 248/left © EPA/Wilson Wen/Landov; p. 248/right © Peter Cade/Stone/Getty Images; p. 250 © Reuters NewMedia Inc./Corbis; p. 255 © Willitz/Network/Corbis Saba; p. 256 © Rachel Epstein/PhotoEdit; p. 260 Reuters/Damir Sagolj/Landov; p. 261/left © R. Azzi/Woodfin Camp & Associates; p. 261/right AP/Wide World Photos/Denis Poroy

Chapter 11
p. 270/bottom © Cameramann/The Image Works; p. 270/top © Hazel Hankin; p. 272 © David Maxwell/EPA/Landov; p. 279 © Douglas Burrows/Liaison/Getty Images; p. 284/top Reuters/Jason Reed/Landov; p. 284/bottom AP/Wide World Photos; p. 285 Courtesy of the Schlesinger Library, Radcliffe College; p. 288 © Bob Daemmrich/The Image Works; p. 289/top © Steven Gold; p. 289/bottom © Brown Brothers; p. 292 © Bernard Boutrit/Woodfin Camp & Associates; p. 294/top © Owen Franken/Corbis; p. 294/bottom Alison Davis

Chapter 12
p. 303 © Peter Menzel/Stock Boston; p. 306 © Bob Sacha; p. 307/left © Tom Pettyman/PhotoEdit; p. 307/right © Myrleen Ferguson Cate/PhotoEdit Inc.; p. 308/left © AFP/Getty Images; p. 308/right AP/Wide World Photos; p. 311 © Jacksonville Journal Courier/The Image Works; p. 316 AP/Wide World Photos; p. 317 © Richard Hutchings/PhotoEdit; p. 319 © Spencer Grant/PhotoEdit; p. 321 © John Eastcott/Yva Momatiuk/Woodfin Camp & Associates; p. 327 © Rick Steele/UPI/Landov

Chapter 13
p. 335 © Serena Nanda; p. 336 © John Burke/SuperStock; p. 347 © Terry Schmitt/UPI/Landov; p. 349 © Sam Yeh/AFP Photo/Getty Images; p. 353 © David Young Wolfe/PhotoEdit; p. 356 AP/Wide World Photos; p. 357 © Tim Crosby/Liaison/Getty Images; p. 353 Spencer Grant/PhotoEdit

Chapter 14
p. 364 © Mark Allan/Alpha/Globe Photos; p. 365 © Walter Hodges/Stone/Getty Images; p. 369 © Lawrence Migdale; p. 371 Frank Fournier/Contact

Press Images; p. 377 © A. Ramey/Woodfin Camp & Associates; p. 378 AP/Wide World Photo; p. 380/left AP/Wide World Photos; p. 380/right © Deborah Davis/PhotoEdit; p. 382/left top Lafargue Frederic/Gamma; p. 382/left bottom AP/Wide World Photos; p. 382/right © Victor Englebert; p. 383 AP/Wide World Photos

Chapter 15

p. 392 Courtesy of Marion and Eilene Hull; p. 398 © Don Smetzer/Stone/Getty Images; p. 402 © Andrew Ramey/Stock Boston; p. 404 © David Young-Wolff/PhotoEdit; p. 405 © Shelley Gazin/Corbis; p. 409 Tom Carter/PhotoEdit; p. 412 © Ansell Horn; p. 417/top © Ellen Senisi/The Image Works; p. 417/bottom © Robert Brenner/PhotoEdit, Inc.

Chapter 16

p. 424 © Rick Scott/Index Stock Imagery; p. 430 © Kaku Kurita/Gamma-Liaison; p. 437 © Lara Jo Regan/Liaison/Getty Images ; p. 441 © Will & Deni McIntyre/Photo Researchers, Inc.

Chapter 17

p. 449 © Bill Gillette/Stock Boston; p. 460 © Michael S. Yamashita/Corbis; p. 461 © C. W. McKeen/The Image Works; p. 463 AP/Wide World Photos; p. 466 © Myrleen Cate/Stone/Getty Images; p. 469 © David Butow/Corbis Saba; p. 471 © Ethel Wolvovitz/The Image Works; p. 472 AP/Wide World Photos

Chapter 18

p. 478 © Mystic Seaport, Rosenfeld Collection, Mystic CT, neg. 1634.; p. 481 © Mark Petersen/Corbis Saba; p. 483 AP/Wide World Photos; p. 484 © AFP/Getty Images; p. 490 © Chapman/The Image Works; p. 493 AP/Wide World Photos; p. 496 © Fritz Hoffmann/The Image Works; p. 500 © Jim Pickerall/Stock Boston

Chapter 19

p. 514 AP/Wide World Photos; p. 515 © Najilah Feanny/Corbis Saba; p. 521 Reuters/Goran Tomasevic/Landov; p. 522 AP/Wide World Photos; p. 526 © United Nations; p. 527 © Bill Swersey/Liaison/Getty Images; p. 529 AP/Wide World Photos

Chapter 20

p. 536/left © Mystic Seaport, Rosenfeld Collection #98, Mystic CT; p. 536/right © Ernst Peters/The Ledger/© 2004. All rights reserved.; p. 546 © Nina Berman/Sipa Press; p. 547 EPA/Michael Reynolds/Landov; p. 551 © Peter Hvizdak/The Image Works; p. 554 © Mike Siluk/The Image Works; p. 555 © Bob Daemmrich/The Image Works; p. 556 © Flashlight/Stock Boston

Chapter 21

p. 565 Jacob Lawrence, *The Migration of the Negro Panel No. 3.* © 2004 Gwendolyn Knight Lawrence/Artists Rights Society (ARS), New York.; p. 569 Jacob Van Oost the Younger, *Saint Macavius of Ghent Succors the Plague Victims*, the Louvre, Paris. Photo ©

Scala/Art Resource; p. 572 © Dilip Mehta/Contact Press Images; p. 575 © Liba Taylor/The Hutchison Library; p. 579 © Ted Spiegel/Corbis; p. 583/left Courtesy Regional History Center, University of Southern California; p. 583/right © Richard Eller/Aerial Images; p. 584 Courtesy of American Honda Motor Co., Inc.

Chapter 22

p. 592 © Tricia Wachtendorf (Courtesy of the Disaster Research Center, University of Delaware); b AP/Wide World Photos; p. 595 © Sipa Press; p. 598 AP/Wide World Photos; p. 600 Thomas Frey/DPA/Landov; p. 602/left © Michelle Gabel/The Image Works; p. 602/right © Mystic Seaport, Rosenfeld Collection, Mystic, CT; p. 604 © Cynthia Howe/Corbis Sygma; p. 605 Reuters/Jim Ruymen/Landov; p. 606 AP/Wide World Photos; p. 609 AP/Wide World Photos

Chapter 23

p. 618 © "Disappearing World," Granada TV/The Hutchison Library; p. 619/left © Mystic Seaport, Rosenfeld Collection, Mystic CT; p. 619/right © Paul L. Sutton/Duomo; p. 620/left © Sabrina Howell; p. 620/right © Carolina Salguero/Sipa Press; p. 623 © Bettmann/Corbis; p. 626 © Terry Wild Studio; p. 629 © Figaro Magahn/Photo Researchers, Inc.; p. 629 EPA/WEDA/Landov; p. 631 © Jane Latta/Photo Researchers, Inc.; p. 632 © Tony Freeman/PhotoEdit; p. 633 © Mystic Seaport, Rosenfeld Collection, Mystic CT, neg. ANN4303

Name Index

Subject Index